EPR Spectroscopy: Fundamentals and Methods

eMagRes Books

eMagRes (formerly the *Encyclopedia of Magnetic Resonance*) publishes a wide range of online articles on all aspects of magnetic resonance in physics, chemistry, biology, and medicine. The existence of this large number of articles, written by experts in various fields, is enabling the publication of a series of *eMagRes* Books – handbooks on specific areas of NMR and MRI. The chapters of each of these handbooks will comprise a carefully chosen selection of *eMagRes* articles.

Published *eMagRes* Books

NMR Crystallography
Edited by Robin K. Harris, Roderick E. Wasylishen, Melinda J. Duer
ISBN 978-0-470-69961-4

Multidimensional NMR Methods for the Solution State
Edited by Gareth A. Morris, James W. Emsley
ISBN 978-0-470-77075-7

Solid-State NMR Studies of Biopolymers
Edited by Ann E. McDermott, Tatyana Polenova
ISBN 978-0-470-72122-3

NMR of Quadrupolar Nuclei in Solid Materials
Edited by Roderick E. Wasylishen, Sharon E. Ashbrook, Stephen Wimperis
ISBN 978-0-470-97398-1

RF Coils for MRI
Edited by John T. Vaughan, John R. Griffiths
ISBN 978-0-470-77076-4

MRI of Tissues with Short T_2s or T_2^*s
Edited by Graeme M. Bydder, Gary D. Fullerton, Ian R. Young
ISBN 978-0-470-68835-9

NMR Spectroscopy: A Versatile Tool for Environmental Research
Edited by Myrna J. Simpson, André J. Simpson
ISBN 978-1-118-61647-5

NMR in Pharmaceutical Sciences
Edited by Jeremy R. Everett, Robin K. Harris, John C. Lindon, Ian D. Wilson
ISBN 978-1-118-66025-6

Handbook of Magnetic Resonance Spectroscopy In Vivo: MRS Theory, Practice and Applications
Edited by Paul A. Bottomley, John R. Griffiths
ISBN 978-1-118-99766-6

EPR Spectroscopy: Fundamentals and Methods
Edited by Daniella Goldfarb, Stefan Stoll
ISBN 978-1-119-16299-5

Forthcoming *eMagRes* Books

Handbook of High Frequency Dynamic Nuclear Polarization NMR
Edited by Vladimir Michaelis, Robert G. Griffin, Björn Corzilius, Shimon Vega
ISBN 978-1-119-44164-9

Handbook of Safety and Biological Aspects in MRI
Edited by Devashish Shrivastava, John T. Vaughan
ISBN 978-1-118-82130-5

eMagRes

Edited by Roderick E. Wasylishen, Edwin D. Becker, Marina Carravetta, George A. Gray, John R. Griffiths, Tatyana Polenova, André J. Simpson, Myrna J. Simpson, Ian R. Young.

eMagRes (formerly the *Encyclopedia of Magnetic Resonance*) is based on the original publication of the *Encyclopedia of Nuclear Magnetic Resonance*, first published in 1996 with an updated volume added in 2000. The *Encyclopedia of Magnetic Resonance* was launched in 2007 online with all the existing published material and was later relaunched as *eMagRes* in 2013. *eMagRes* captures every aspect of the interdisciplinary nature of magnetic resonance, providing all the essential information on the science, methodologies, engineering, technologies, applications, and the history of magnetic resonance, while encompassing a whole range of techniques, including MRI, MRS, NMR, and EPR/ESR.

For more information, see http://www.wileyonlinelibrary.com/ref/eMagRes.

EPR Spectroscopy: Fundamentals and Methods

Editors

Daniella Goldfarb
Weizmann Institute of Science, Rehovot, Israel

Stefan Stoll
University of Washington, Seattle, WA, USA

WILEY

This edition first published 2018

Registered office
John Wiley & Sons Ltd, The Atrium, Southern Gate, Chichester, West Sussex, PO19 8SQ, UK

Library of Congress Cataloging-in-Publication Data is available for this title.

978-1-119-16299-5 (hardback)

A catalogue record for this book is available from the British Library.

Cover Design: Wiley
Cover Images: (Background) © kWaiGon/Gettyimages; (Top, left to right) Pulse shape image, courtesy of Stefan Stoll; HYSCORE, courtesy of Sabine Van Doorslaer; Energy Level Diagram, courtesy of Jeffrey Harmer; Tyrosyl Radical, courtesy of Franke Neese; (Bottom, left to right) DEER, courtesy of Gunnar Jeschke; Spectra, courtesy of Enrica Bordignon

Set in 9.5/11.5 pt TimesLTStd by SPi Global, Chennai, India

10 9 8 7 6 5 4 3 2 1

FSC www.fsc.org
MIX
Paper from responsible sources
FSC® C013604

This edition first published 2018

Registered office
John Wiley & Sons Ltd, The Atrium, Southern Gate, Chichester, West Sussex, PO19 8SQ, UK

Library of Congress Cataloging-in-Publication Data is available for this title.

978-1-119-16299-5 (hardback)

A catalogue record for this book is available from the British Library.

Cover Design: Wiley
Cover Images: (Background) © kWaiGon/Gettyimages; (Top, left to right) Pulse shape image, courtesy of Stefan Stoll; HYSCORE, courtesy of Sabine Van Doorslaer; Energy Level Diagram, courtesy of Jeffrey Harmer; Tyrosyl Radical, courtesy of Franke Neese; (Bottom, left to right) DEER, courtesy of Gunnar Jeschke; Spectra, courtesy of Enrica Bordignon

Set in 9.5/11.5 pt TimesLTStd by SPi Global, Chennai, India

10 9 8 7 6 5 4 3 2 1

MIX
Paper from responsible sources
FSC® C013604

EPR Spectroscopy: Fundamentals and Methods

Editors

Daniella Goldfarb
Weizmann Institute of Science, Rehovot, Israel

Stefan Stoll
University of Washington, Seattle, WA, USA

WILEY

eMagRes

MRI & MRS

John R. Griffiths
Cancer Research UK
Cambridge Research Institute
Cambridge
UK

Ian R. Young
Imperial College
London
UK

HISTORICAL PERSPECTIVES

Edwin D. Becker
National Institutes of Health
Bethesda, MD
USA

International Advisory Board

Contents

Contributors

Marina Bennati
Max Planck Institute for Biophysical Chemistry and University of Göttingen, Göttingen, Germany
Chapter 5: EPR Interactions – Hyperfine Couplings

Aharon Blank
Schulich Faculty of Chemistry, Technion, Haifa, Israel
Chapter 24: EPR Microscopy

Christoph Boehme
University of Utah, Salt Lake City, UT, USA
Chapter 26: Electrically Detected Magnetic Resonance (EDMR) Spectroscopy

Peter P. Borbat
Cornell University, Ithaca, NY, USA
Chapter 20: Dipolar Spectroscopy – Single-resonance Methods

Enrica Bordignon
Ruhr-Universität Bochum, Bochum, Germany
Chapter 14: EPR Spectroscopy of Nitroxide Spin Probes

Alice M. Bowen
Goethe University Frankfurt, Frankfurt am Main, Germany
Chapter 21: Shaped Pulses in EPR

Michael K. Bowman
The University of Alabama, Tuscaloosa, AL, USA
Chapter 15: FT-EPR

Hanjiao Chen
The University of Alabama, Tuscaloosa, AL, USA
Chapter 15: FT-EPR

David Collison
The University of Manchester, Manchester, UK
Chapter 4: EPR Interactions – Coupled Spins

Gareth R. Eaton
University of Denver, Denver, CO, USA
Chapter 9: Relaxation Mechanisms
Chapter 23: Rapid-scan EPR

Sandra S. Eaton
University of Denver, Denver, CO, USA
Chapter 9: Relaxation Mechanisms
Chapter 23: Rapid-scan EPR

Burkhard Endeward
Goethe University Frankfurt, Frankfurt am Main, Germany
Chapter 21: Shaped Pulses in EPR

Boris Epel *University of Chicago, Chicago, IL, USA*
Chapter 13: EPR Imaging

Akiva Feintuch *Weizmann Institute of Science, Rehovot, Israel*
Chapter 8: Spin Dynamics

Jack H. Freed *Cornell University, Ithaca, NY, USA*
Chapter 20: Dipolar Spectroscopy – Single-resonance Methods

Peter Gast *Leiden University, Leiden, The Netherlands*
Chapter 2: EPR Interactions – *g*-Anisotropy

Daniella Goldfarb *Weizmann Institute of Science, Rehovot, Israel*
Chapter 6: EPR Interactions – Nuclear Quadrupole Couplings
Chapter 17: Hyperfine Spectroscopy – ELDOR-detected NMR

Etienne Goovaerts *University of Antwerp, Antwerp, Belgium*
Chapter 25: Optically Detected Magnetic Resonance (ODMR)

Edgar J.J. Groenen *Leiden University, Leiden, The Netherlands*
Chapter 2: EPR Interactions – *g*-Anisotropy

Howard J. Halpern *University of Chicago, Chicago, IL, USA*
Chapter 13: EPR Imaging

Jeffrey R. Harmer *University of Queensland, St Lucia, Queensland, Australia*
Chapter 16: Hyperfine Spectroscopy – ENDOR

Gunnar Jeschke *ETH Zürich, Zürich, Switzerland*
Chapter 19: Dipolar Spectroscopy – Double-resonance Methods

Hans Malissa *University of Utah, Salt Lake City, UT, USA*
Chapter 26: Electrically Detected Magnetic Resonance (EDMR) Spectroscopy

Alexander G. Maryasov *V. V. Voevodsky Institute of Chemical Kinetics and Combustion, Novosibirsk, Russia*
Chapter 15: FT-EPR

Eric J.L. McInnes *The University of Manchester, Manchester, UK*
Chapter 4: EPR Interactions – Coupled Spins

John J.L. Morton *University College London, London, UK*
Chapter 22: Pulse Techniques for Quantum Information Processing

Frank Neese *Max Planck Institute for Chemical Energy Conversion, Mülheim an der Ruhr, Germany*
Chapter 7: Quantum Chemistry and EPR Parameters

Thomas F. Prisner *Goethe University Frankfurt, Frankfurt am Main, Germany*
Chapter 21: Shaped Pulses in EPR

Edward Reijerse
Max Planck Institute for Chemical Energy Conversion, Mülheim an der Ruhr, Germany
Chapter 12: EPR Instrumentation

Anton Savitsky
Max Planck Institute for Chemical Energy Conversion, Mülheim an der Ruhr, Germany
Chapter 12: EPR Instrumentation

Philipp Schöps
Goethe University Frankfurt, Frankfurt am Main, Germany
Chapter 21: Shaped Pulses in EPR

Alexander Schnegg
Helmholtz-Zentrum Berlin für Materialien und Energie, Berlin, Germany
Chapter 27: Very-high-frequency EPR

Philipp E. Spindler
Goethe University Frankfurt, Frankfurt am Main, Germany
Chapter 21: Shaped Pulses in EPR

Stefan Stoll
University of Washington, Seattle, WA, USA
Chapter 6: EPR Interactions – Nuclear Quadrupole Couplings
Chapter 11: Pulse EPR

Joshua Telser
Roosevelt University, Chicago, IL, USA
Chapter 3: EPR Interactions – Zero-field Splittings

Art van der Est
Brock University, Ontario, Canada
Chapter 1: Continuous-Wave EPR

Sabine Van Doorslaer
University of Antwerp, Antwerp, Belgium
Chapter 18: Hyperfine Spectroscopy – ESEEM

Shimon Vega
Weizmann Institute of Science, Rehovot, Israel
Chapter 8: Spin Dynamics

Stefan Weber
Albert-Ludwigs-Universität Freiburg, Freiburg, Germany
Chapter 10: Transient EPR

Gary Wolfowicz
University of Chicago, Chicago, IL, USA
Chapter 22: Pulse Techniques for Quantum Information Processing

Series Preface

The *Encyclopedia of Nuclear Magnetic Resonance* was originally published in eight volumes in 1996, in part to celebrate the fiftieth anniversary of the first publications describing the discovery of NMR (nuclear magnetic resonance) in January 1946. Volume 1 contained a historical overview and 200 articles by prominent NMR practitioners, whilst the remaining seven volumes consisted of 500 articles on a wide variety of topics in NMR, including MRI (magnetic resonance imaging). A ninth volume was brought out in 2000 and two 'spin-off' volumes incorporating the articles on MRI and MRS (together with some new ones) were published in 2002. In 2006, the decision was taken to publish all the articles electronically with the resulting Encyclopedia becoming available online in 2007. Since then, new articles have been published online every 3 months and many of the original articles have been updated. To recognize the fact that the *Encyclopedia of Magnetic Resonance* is a true online resource, the website was redesigned and new functionalities added, with a relaunch in January 2013 in a new Volume and Issue format, under the new name *eMagRes*. In December 2012, a new print edition of the *Encyclopedia of Nuclear Magnetic Resonance* was published in 10 volumes (6200 pages). This much needed update of the 1996 edition of the Encyclopedia encompassed the entire field of NMR.

As part of the development of *eMagRes*, a series of printed handbooks on specific areas of magnetic resonance have been introduced. The handbooks are planned in advance by specially selected editors, and new articles written to give appropriate complete coverage of the subject area. The handbooks are intended to be of value and interest to research students, postdoctoral fellows, and other researchers learning about the topic in question and undertaking relevant experiments, whether in academia or industry. All of the previous handboooks have dealt with topics that fall under the general area of NMR. Under the broader heading of 'Magnetic Resonance', we now include 'Electron Paramagnetic Resonance'. This new handbook is the first that deals with electron paramagnetic resonance (EPR) and will, of course, be of interest to scientists working in this area but also to those across the NMR community. With the exciting developments in dynamic nuclear polarization NMR, understanding the fundamentals of EPR is essential and this new handbook will be a welcome addition to the community.

Roderick E. Wasylishen

November 2017

Consult the *eMagRes* website at (www.wileyonlinelibrary.com/ref/eMagRes) for details of all our *eMagRes* handbooks.

Preface

This is a multi-author graduate-level textbook on modern EPR (electron paramagnetic resonance) spectroscopy. In the last two decades, EPR spectroscopy has witnessed tremendous progress, in terms of both new methods and new applications. These developments are of course captured in the primary scientific literature, but are scattered across many articles in very diverse journals. For a researcher entering the field, such as a beginning graduate student or postdoc, this scatter makes it quite difficult to quickly obtain a coherent overview of the field. In particular, the wide variety of methods, encompassing continuous-wave EPR, pulse EPR, high-field methods, as well as optical and electrical detection, can be bewildering to a novice. In addition, the physical principles underlying EPR are non-trivial and are not easy material for researchers without a substantial physics background. There is no textbook that one could hand a beginning non-expert researcher with the words 'Read this to learn about EPR!'

This book is intended to at least partially fill this gap. It provides an introductory, but fairly comprehensive overview of the current field of EPR. The 27 chapters of the book cover the theoretical principles, the common experimental techniques, and several important application areas of modern EPR spectroscopy. The book is not intended as a review or monograph, but rather as a graduate-level textbook. It is not a replacement for more advanced and comprehensive texts such as the 2001 monograph on pulse EPR by Schweiger and Jeschke, though it does cover new developments not covered in that book. Rather, it is intended as a preparatory text that will prepare – and hopefully entice – readers to study more advanced treatments and the original literature.

When we started to approach leaders in the field about contributing to a textbook on modern EPR, we received an overwhelmingly positive response. This demonstrated that the need for such a book is strongly perceived across the entire EPR research community. Many of the authors have been active in EPR education, particularly as lecturers at several EPR Summer Schools, organized by the European Federation of EPR Groups (EFEPR), and held every 2–3 years in different places across Europe. Since its first edition in Caorle, Italy, in 1999, this Summer School has helped train and inspire new generations of EPR spectroscopists. This book draws substantially from the experience of lecturers at the Summer Schools.

This book would not have been possible without the impressive commitment and tireless effort of all the contributing authors. They not only spent significant time and effort in putting together accessible introductions to specific topics, but also put up with us as editors. We are grateful that all contributors were extraordinarily patient with us as they diligently and graciously accommodated our multiple rounds of revision requests, which ranged from complete rearrangements to tedious notational and typographical adjustments.

Another big Thank You goes to the reviewers. All chapters have been reviewed by at least one outside expert, typically the author of another chapter, in addition to us. These reviewers added substantial value by pointing out misunderstandings, omissions, and opportunities for pedagogical improvements. Of course, any remaining issues are solely our fault.

It was one of our goals to create a set of chapters that is maximally coherent and consistent in notation and terminology, so as to minimize the potential for confusion for the intended audience. It turned out that it is an almost impossible feat to achieve complete consistency. There are remaining differences among chapters. For example, some utilize spin Hamiltonians in energy units, whereas others prefer angular-frequency units. These choices reflect the reality of different practices in various subfields of EPR.

We hope that this book will help beginning graduate students and postdocs get a grasp of the theoretical and experimental principles of modern EPR. Of course, 'modern' is a relative term and needs to be understood as of the year 2017. We have no illusion at all that many parts of this book will be rendered outdated in the near future by the continuing rapid progress in experimental techniques and applications of EPR. Until then, we hope it is useful.

Stefan Stoll
University of Washington, Seattle, WA, USA

Daniella Goldfarb
Weizmann Institute of Science, Rehovot, Israel

December 2017

Abbreviations and Acronyms

1D	one-dimensional
AA	Aharonov–Anandan
ADMR	absorption-detected magnetic resonance
ADP	adenosine diphosphate
AFC	automatic frequency control
AFM	atomic force microscopy
AFP	adiabatic fast passage
AILFT	ab initio ligand-field theory
AO	atomic orbital
AOM	angular overlap model
ARS	Advanced Research Systems
ATP	adenosine triphosphate
AWG	arbitrary waveform generator
BDPA	α,β-bisphenylene-β-phenylallyl-benzolate
BEBOP	broadband excitation by optimized pulses
BIR4	B_1 insensitive rotation
BO	Born–Oppenheimer
BP	Breit–Pauli
BPP	Bloembergen, Purcell, Pound
BW	bandwidth
BWO	backward-wave oscillator
C-NOT	controlled-NOT
C-PHASE	controlled-phase
CAP	constant adiabaticity pulse
CASPT2	complete active space second-order perturbation theory
CASSCF	complete active space self-consistent field
CC	coupled cluster
CESR	conduction electron-spin resonance
CFT	crystal field theory
CIDEP	chemically induced dynamic electron polarization
CIDME	chirp-induced dipolar modulation enhancement
CLR	cross-loop resonator
CP	combination-peak
CPMG	Carr–Purcell–Meiboom–Gill
CPT	coherence pathway transfer
CSR	coherent synchrotron radiation
CT	clock transitions
CT	coherence transfer
CW	continuous wave
DAC	digital-to-analog converter
DC	direct current
DD	dynamical decoupling
DDBSQ	2,5-dichloro-3,6-dihydroxy-1,4-benzosemiquinone
DDS	direct digital synthesizer
DEER	double electron–electron resonance
DEER ESE	double electron–electron resonance in electron spin-echo
DEFENCE	deadtime-free ESEEM by nuclear coherence-transfer echoes
DFDMR	delayed fluorescence detection of magnetic resonance
DFG	Deutsche Forschungsgemeinschaft
DFT	density functional theory
DKH	Douglas–Kroll–Hess
DNP	dynamic nuclear polarization
DONUT	double nuclear coherence transfer
DPPH	2,2-diphenyl-1-picrylhydrazyl
DQ	double-quantum
DQC	double-quantum coherence
DQF	DQ filtering
DQM	double-quantum modulation
DSV	diameter spherical volume
EC	electron coherence
ED-EPR	echo-detected EPR
EDDEER	electrically detected double electron–electron resonance
EDMR	electrically detected magnetic resonance
EFG	electric field gradient
EIK	extended interaction klystron
EIO	extended interaction oscillator
ELDOR	electron-electron double resonance
ELDOR-detected NMR	electron-electron double resonance-detected NMR
ENDOR	electon-nuclear double resonance
EP	electron polarization
EPR	electron paramagnetic resonance
EPRI	EPR imaging
EPRM	EPR microscopy
ESE	electron spin echo
ESEEM	electron spin echo envelope modulation
ET	electron transfer
EZ	electron Zeeman
f.w.h.h.	full width at half height
FAD	flavin adenine dinucleotide
FBP	filtered backprojection
FD-EPR	frequency-domain EPR
FD-FT THz-EPR	frequency-domain Fourier-transform THz-EPR
FDMR	fluorescence detection of magnetic resonance
FDMR	frequency-domain magnetic resonance
FEL	free-electron laser
FFT	fast Fourier transformation
FID	free induction decay
FJ	field-jump
FMR	ferromagnetic resonance
FPGA	field-programmable gate array
FR	Faraday rotator
FRET	fluorescence resonance energy transfer
FT	Fourier transform
FT	Fourier transformation
FWHM	full width at half maximum
GGA	generalized gradient approximation
GIAOs	gauge-including atomic orbitals
GM	Gifford–McMahon
HAS	hindered amine stabilizers
HF	Hartree–Fock
HF	high magnetic field
HF	hyperfine
HFCs	hyperfine couplings

HFEPR	high-frequency and high-field EPR
HFHF	high-field high-frequency
HFI	hyperfine interaction
HFML	High Field Magnet Laboratory
HOMO	highest occupied molecular orbital
HS	high-spin
HTA	high-turning-angle
HWHH	half width at half height
hwhm	half-width at half maximum
HYSCORE	hyperfine sublevel correlation
ID	instantaneous diffusion
IF	intermediate frequency
INS	inelastic neutron scattering
IQ	in-phase quadrature
IRESE	inversion recovery electron spin-echo
ISC	intersystem crossing
ISHE	inverse spin Hall effect
JB	Jeener–Broekaert
KS	Kohn–Sham
KSM	Kaplan, Solomon, and Mott
LAC	level anticrossing
LAN	local area network
LCP	left circular polarization
LDA	"local density" approximation
LF-DFT	ligand-field density functional theory
LFT	ligand-field theory
LGR	loop-gap resonator
LHS	left-hand side
LiPc	lithium phthalocyanine
LMO	localized molecular orbital
LNA	low-noise amplifier
LO	local oscillator
LP	lone-pair
LUMO	lowest unoccupied molecular orbital
LWHH	linewidth at half height
MCD	magnetic circular dichroism
MCDA	magnetic circular dichroism in absorption
MCP	magnetic circular polarization
MCPE	magnetic circular polarization in emission
MEH-PPV	polymer poly[2-methoxy-5-(2-ethylhexyloxy)-1,4-phenylenevinylene]
MFG	modulated field gradient
MHF-EPR	multi-high-frequency EPR
MO	molecular orbital
MR	magnetic resonance
MTSL	methanethiosulfonato spin label
MVNA	mm-wave network vector analyzer
MW	microwave
NARS	nonadiabatic rapid scan
NC	nuclear coherence
NCO	numerically controlled oscillator
NEP	noise equivalent power
NEVPT2	second-order N-electron valence perturbation theory
NHMFL	National High Magnetic Field Laboratory
NMR	nuclear magnetic resonance
NP	nuclear polarization
NPA	natural population analysis
NQ	nuclear quadrupole
NQI	nuclear quadrupole interaction
NQR	nuclear quadrupole frequencies
NV	nitrogen vacancy
OCT	optimum control theory
ODENDOR	optically detected ENDOR
ODMR	optically detected magnetic resonance
ODMRI	optically detected magnetic resonance imaging
ODNP	Overhauser dynamic nuclear polarization
OLED	organic light-emitting diode
OPO	optical parametric oscillator
P3HT	poly(3-hexylthiophene)
PAS	principal axis system
PC	paramagnetic center
PCET	proton-coupled electron transfer
PDB	protein data bank
PDMR	phosphorescence detection of magnetic resonance
PDS	pulse dipolar EPR spectroscopy
PE	primary echo
pEDMR	pulse electrically detected magnetic resonance
PEDOT:PSS	poly(3,4-ethylenedioxythiophene)
PELDOR	pulse electron–electron double resonance
pEPR	pulse EPR
PET	polyethylene terephthalate
PFU	pulse-forming unit
PID	proportional–integral–derivative
PMMA	polymethyl methacrylate
PNT	perinaphthenyl radical
PO	product operator
PP	point-to-point
PP	polaron pair
PR	projection reconstruction
QCISD	quadratic configuration interaction
QDPT	quasi-degenerate perturbation theory
QE	quantum efficiency
QIP	quantum information processing
QPT	quantum process tomography
RCP	right circular polarization
RE	refocused echo
RF	radio frequency
RHS	right-hand side
RIDME	relaxation-induced dipolar modulation enhancement
RNR	ribonucleotide reductase
ROCIS	restricted open-shell configuration interaction method with single excitations
ROHF	restricted open-shell Hartree–Fock
ROS	reactive oxygen species
RS	rapid scan
RSE	refocused stimulated echo
RVE	refocused virtual echo
RWA	rotating wave approximation
RYDMR	reaction-yield detected magnetic resonance
SC	semiconductor
SCRP	spin-correlated radical pair
SDSL	site-directed spin labeling
SE	singlet exciton
SE	stimulated echo
SEDOR	spin echo double-resonance
SH	spin Hamiltonian
SHF-EPR	single-high-frequency EPR

SIFTER	single-frequency technique for refocusing
SLE	stochastic Liouville equation
SMART	single pulse matched resonance transfer
SNR	signal-to-noise ratio
SOC	spin-orbit coupling
SOMF	spin-orbit mean-field
SOMO	singly occupied molecular orbital
SOO	spin-other-orbit
SORF	second-order rotating frame
SOS	sum over states
SPAM	state preparation and measurement
SPF	spin projection factor
SPI	single-point imaging
SPU	signal processing unit
SQ	single-quantum
SQC	single-quantum coherence
SRSL	slowly relaxing local structure
SSC	spin–spin coupling
SSO	spin-same-orbit
STE	self-trapped exciton
STH	self-trapped hole
T4L	T4 lysozyme
TCSQ	tetrachloro-1,4-benzosemiquinone
TD	Townes–Dailey
TE	triplet exciton
TEM	transmission electron microscope
TEMPO	2,2,6,6-tetra-methyl-piperidine-1-oxyl
TEs	triplet excitons
THYCOS	triple resonance hyperfine sublevel correlation spectroscopy
THz-TDS	THz time-domain spectroscopy
TMBSQ	2,3,5,6-tetramethoxy-1,4-benzosemiquinone
TMI	transition-metal-ion
trEPR	transient EPR
TTA	triplet–triplet annihilation
TTET	triplet–triplet energy transfer
TWT	traveling wave tube
TWTA	traveling wave tube amplifiers
UC	upconversion
UHF	unrestricted Hartree–Fock
UR	universal rotation
UV	ultraviolet
UWB	ultra-wideband
VHF	very high frequency
VHF-EPR	very-high-frequency EPR
VTI	variable temperature insert
VTVH	variable-temperature and variable-field
WAHUHA	Waugh–Huber–Haeberlen sequence
WURST	wideband, uniform rate, smooth truncation
XC	exchange-correlation
ZEFOZ	zero first-order Zeeman shift
ZF	zero magnetic field
ZFS	zero-field splitting
ZORA	0th-order regular approximation

PART A
Fundamental Theory

Chapter 1
Continuous-Wave EPR

Art van der Est
Brock University, Ontario, Canada

1.1 INTRODUCTION

The vast majority of EPR data in the literature are CW field-swept spectra collected using field-modulation detection. Despite the many advances in pulse techniques, it is not feasible to collect the frequency spectrum as the Fourier transform of the time-dependent response to a pulse in all but a few special cases. Hence, EPR spectra are usually measured using continuous microwave irradiation at a fixed frequency, while the magnetic field is varied over a region of interest. The difficulty in measuring the frequency spectrum is the extremely broad spectral widths encountered in EPR, which arise because electrons are not spatially localized. The delocalization means that the orbital angular momentum contributes to the total magnetic moment of the electron and that spin–orbit coupling is important. In some instances, the zero-field splitting (ZFS) arising from spin–orbit coupling is the dominant interaction in the Hamiltonian. Moreover, the dipolar coupling between unpaired electrons can be very strong if they are colocalized in close proximity on the same atom(s). A further challenge is that the energies associated with EPR transitions are typically much less than thermal energy (k_BT), and as a result the population differences are very small and the absorbance is weak. To overcome the problem of weak absorbance, the signal is usually detected using a resonator, although at very high frequency direct measurement of the absorbance is possible. The gain in sensitivity obtained using a resonator is paid for with a narrow bandwidth, which only allows the frequency spectrum to be collected over a narrow range. Hence, the spectrum is measured by keeping the frequency fixed at the resonance frequency of the resonator and by sweeping the magnetic field. Further sensitivity enhancement is gained by modulating the field and using lock-in detection. This combination of sweeping the field and using field-modulation detection gives EPR spectra some unusual properties not encountered in other forms of spectroscopy. In this chapter, a brief introduction to CW-EPR spectroscopy will be given, with emphasis on important experimental features.

EPR Spectroscopy: Fundamentals and Methods.
Edited by Daniella Goldfarb and Stefan Stoll
 ISBN: 978-1-119-16299-5
Also published in eMagRes (online edition)
DOI: 10.1002/9780470034590.emrstm1508

1.2 BASIC DESIGN AND OPERATION OF A CW-EPR SPECTROMETER

1.2.1 Field/Frequency Combinations

The microwave components needed for EPR spectrometers are available only at specific narrow frequency bands designated by letters. These letter codes were introduced during the development of radar technology and do not carry any specific meaning. X-band (9–10 GHz, 0.32–0.36 T for $g=2$) EPR spectrometers are by far the most common because the field can be easily generated with an electromagnet and microwave components are inexpensive at this frequency. Q-band (33–35 GHz, 1.2–1.3 T for $g=2$) instruments are also widespread and are the highest frequency spectrometers that use electromagnets. At W-band (95 GHz, 3.4 T for $g=2$), the need for a superconducting magnet adds substantially to the cost. Nonetheless, such spectrometers have become more common in recent years. EPR spectrometers that operate at frequencies above 95 GHz are only found in a handful of laboratories. Low-frequency spectrometers, for example, at S-band (2–4 GHz, 0.07–0.42 T for $g=2$), are also commercially available but not as widely used as X- and Q-band instruments. The choice of the appropriate field/frequency for a CW-EPR experiment is dictated primarily by the properties of the species under investigation. Solution spectra of free radicals are dominated by hyperfine interactions and thus high field/frequency does not provide better resolution, but measurements at several frequencies can be helpful for characterizing the motional dynamics of the radicals. For solids or frozen solutions, the $\boldsymbol{g}$-tensor components of spin-1/2 metal centers are often well resolved at X- or Q-band. For most organic radicals, 95 GHz or higher is required to resolve the g-anisotropy. For systems with $S>1/2$, high field/frequency can be advantageous for determining the ZFS parameters.

1.2.2 The Microwave Bridge

Figure 1.1 shows a schematic diagram of a simple CW-EPR spectrometer. The microwave source is typically a Gunn diode oscillator, or a klystron in older instruments. The emitted microwaves from the source are divided into signal and reference arms of the microwave bridge to allow phase-sensitive detection. A small amount of power is also usually diverted to a frequency counter to monitor the microwave frequency. The microwave power in the reference arm is controlled by the bias attenuator, and the relative phase between the signal and reference arms is controlled by the phase shifter. In the signal arm, the microwaves are passed through a circulator to the resonator and an attenuator allows the amount of power reaching the sample to be regulated. The circulator ensures that only reflected power returning from the resonator reaches the detector. For tuning, the main source, or a separate voltage-controlled oscillator source, is swept rapidly over a narrow frequency range around the resonance frequency of the resonator and the reflected microwave power is monitored as a function of the frequency as illustrated in Figure 1.2. The resonator is coupled to the bridge using an adjustable tuning element such as an iris, which allows the amount of power entering the resonator to be controlled. When the coupling element is adjusted so that no power is reflected at the resonance frequency of the resonator, the resonator is critically coupled (Figure 1.2b). The microwave source is then locked to the resonance frequency of the resonator using the automatic frequency control (AFC). The spectrum is measured by sweeping the field over a region of interest and when an EPR transition comes into resonance, microwave absorption by the sample occurs, the critical coupling is disturbed, and power is reflected. Thus, the reflected power reports indirectly on the power absorbed by the sample, and because the amount of reflected power is greater than would be absorbed by simply passing the microwaves through the sample, amplification of the signal is achieved.

The detector is usually a Schottky diode, which rectifies the microwave signal to give a DC voltage. In the so-called square-law region, the voltage produced by the diode is proportional to the microwave power. However, this relationship breaks down at very low microwave power. Because the resonator is critically coupled and no power reaches the detector except when EPR absorption occurs, the detector must be biased to bring it into the square-law region. Thus, the reference arm not only serves to provide phase-sensitive detection but also acts as a bias for the detector.

1.2.3 The Resonator

The most common resonators used in CW-EPR are rectangular and cylindrical cavities. Split-ring and dielectric ring resonators are also sometimes

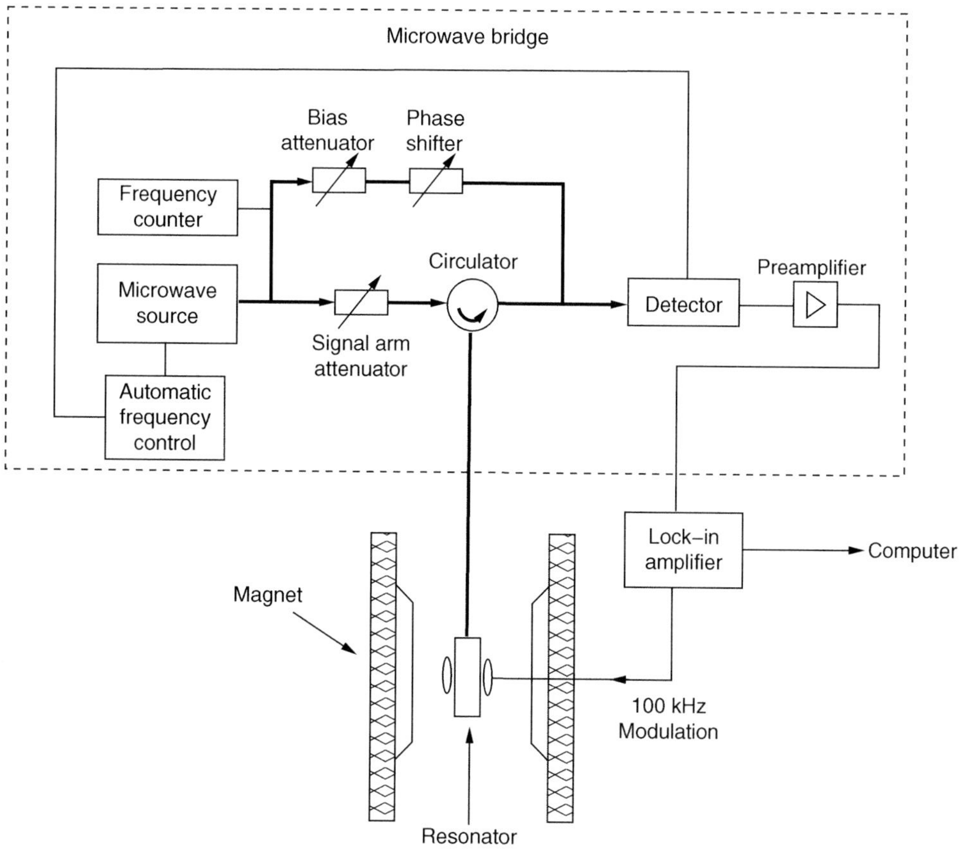

Figure 1.1. Schematic diagram of a simple CW-EPR spectrometer. (Reproduced from Molecular Biophysics for the Life Sciences, Electron Paramagnetic Resonance Spectroscopy, 2015, 175-213, J. H. Golbeck; A. van der Est © With permission of Springer)

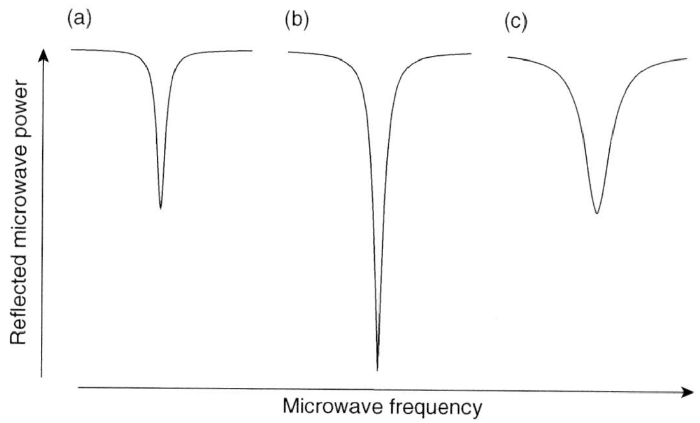

Figure 1.2. Tuning mode patterns for different coupling regimes: (a) undercoupled; (b) critically coupled; and (c) overcoupled

used, particularly for measuring aqueous samples at room temperature, but they are better suited for time-resolved and pulsed EPR experiments. Some high-field/frequency instruments (95 GHz and higher) employ Fabry–Perot resonators.[1] In all cases, the resonance mode of the resonator is chosen such that the magnetic lines of force are concentrated in the center of the cavity where the sample is placed. Figure 1.3 shows the magnetic (B_1) and electric (E_1) components of the microwave field in a rectangular TE_{102} resonator, which is widely used in X-band spectrometers. For a given mode TE_{lmn} of a rectangular resonator, the dimensions *a, b,* and *d* of the resonator are related to the wavelength, λ, and frequency, ν, of the microwaves by

$$\frac{\nu}{c} = \frac{1}{\lambda} = \frac{1}{2}\left(\frac{l}{a^2} + \frac{m}{b^2} + \frac{n}{d^2}\right)^{1/2} \tag{1.1}$$

For the TE_{102} mode shown in Figure 1.3, dimensions of approximately 2 and 4 cm for *a* and *d*, respectively, correspond to $\lambda \approx 3.3$ cm and $\nu \approx 9$ GHz. In Figure 1.3(a), the magnetic and electric lines of force at the center of the resonator and the position of the sample tube are shown. In Figure 1.3(b), the squares of the amplitudes of the electric and magnetic components of the microwave are plotted. Figure 1.3 shows that in the center of the cavity, there is a maximum in the B_1 field, while the E_1 field has a node. This distribution of the field components is particularly important when solvents with high dielectric constants such as liquid water are used. Such solvents absorb microwaves strongly through the interaction with the E_1 field. To avoid this unwanted absorption, the sample must be placed in a region of the cavity in which the E_1 field is minimal and the B_1 field is maximal. For the TE_{102} mode of a rectangular resonator, this can be achieved using a flat cell, which restricts the sample to a thin plane at the center of the resonator. Placing the sample in a thin capillary or bundle of capillaries also limits the unwanted interaction with the E_1 field, especially if a cylindrical resonator is used. As can be seen in equation (1.1), the dimensions of the resonator scale with the wavelength of the microwaves or inversely with the spectrometer frequency. Thus, as the spectrometer frequency increases, it becomes increasingly difficult to measure high dielectric samples owing to the decreasing size of the resonator.

The sensitivity of the resonator is governed by the quality factor and the filling factor. The quality factor is given by $Q = \omega/\Delta\omega$, where ω is the resonator frequency and $\Delta\omega$ the width of the resonance. The filling factor, η, is the fraction of the total magnetic energy stored by the microwave in the resonator cavity that

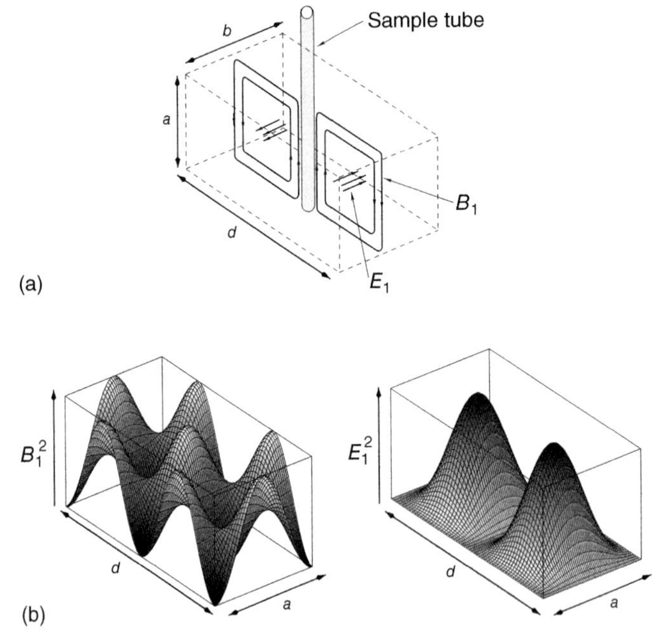

Figure 1.3. Magnetic and electric field intensity distribution in a TE_{102} rectangular resonator

interacts with the sample and is given by

$$\eta = \frac{\int_{\text{sample}} B_1^2 \mathrm{d}V}{\int_{\text{resonator}} B_1^2 \mathrm{d}V} \tag{1.2}$$

For a microwave field that is uniform throughout the resonator (a situation that is difficult to achieve in practice), η is the ratio of the sample volume to the volume of the resonator. Both the filling factor and the quality factor should be as large as possible, but optimizing one of them often leads to a decrease in the other. For example, increasing the sample size so that it interacts with a greater fraction of the B_1 field gives a larger value of η but it also leads to damping of the resonator and an increase in width of the resonance and a drop in Q. In the case of the rectangular resonator in the TE_{102} mode, the value of Q is high but this is offset by a poor filling factor, particularly if it is necessary to use a flat cell.

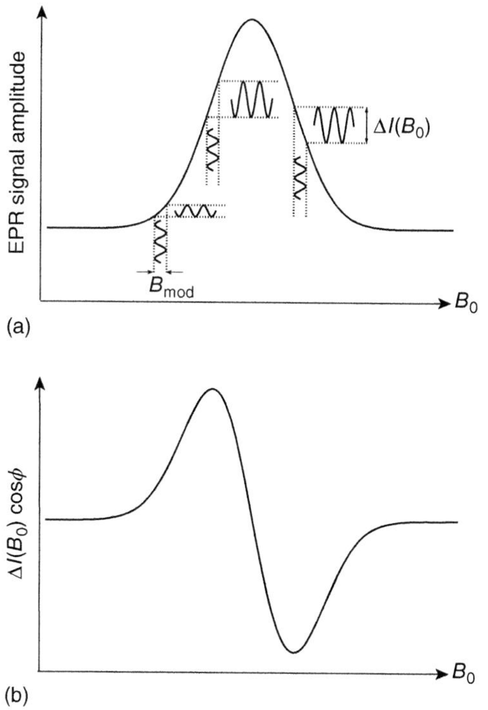

Figure 1.4. Field modulation of an EPR absorption line. (a) A Gaussian absorption line and the modulated signal at several positions in the line. (b) The resulting derivative-like line shape produced by the lock-in amplifier

1.2.4 The Magnet and Lock-in Detection

As the magnetic field must be swept, it is simplest to use an electromagnet. With a gap of a few centimeters between the pole faces, fields up to about 1.5–2.0 T can be achieved with relative ease. For higher field strengths, superconducting magnets fitted with sweep coils that allow the field to be varied over a range of 200–600 mT are used. In addition to the main field, a small modulation field is applied using a further set of coils that are usually mounted on or in the resonator. During a measurement, the main field is swept slowly across the region of interest and the modulation field is applied, typically with a frequency of 100 kHz or lower. As a result of the field modulation, the observed EPR signal oscillates at the modulation frequency as illustrated in Figure 1.4(a). The signal also contains components at zero frequency and at multiples of the modulation frequency, but these are discarded when it is amplified. In general, the signal is the sum of components that can be written[2] as

$$s(t) \propto \Delta I \cos(\omega_s t + \phi) \tag{1.3}$$

where ϕ is the phase difference between the signal from the detector and the field modulation. The amplitude ΔI of the oscillating signal is proportional to the magnitude of the change in amplitude of the absorption line, between $B_0 - B_{\text{mod}}$ and $B_0 + B_{\text{mod}}$. In the lock-in amplifier (Figure 1.1), the signal from the microwave detector is mixed with a reference signal with the same frequency and phase as the field modulation:

$$r(t) \propto \cos(\omega_r t) \tag{1.4}$$

The output of the mixer is

$$S_{\text{out}}(t) \propto \Delta I[\cos((\omega_s - \omega_r)t + \phi) + \cos((\omega_s + \omega_r)t + \phi)] \tag{1.5}$$

The DC component of the mixer output signal, resulting from the first term in equation (1.5) with $\omega_s = \omega_r$, is proportional to $\Delta I \cos\phi$. As can be seen in Figure 1.4, the phase difference, ϕ, is zero on the rising flank of the peak, whereas it is 180° on falling flank. As a result, the derivative-like line shape shown in Figure 1.4(b) is obtained. The advantage of this method of detection is that the DC signal can be separated from the noise using a low-pass filter, which gives a large increase in the signal-to-noise ratio. The first-derivative-like line shape also emphasizes shoulders in the absorption spectrum making partially resolved splitting easier to detect and quantify.

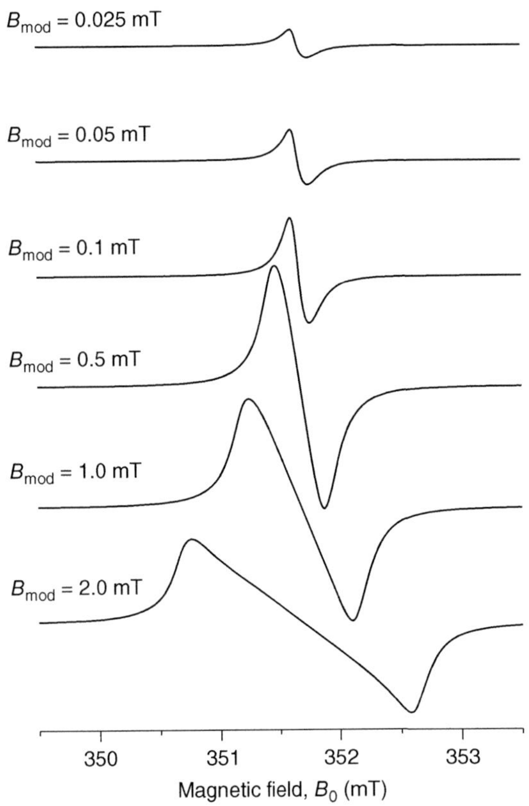

Figure 1.5. The influence of the field-modulation amplitude on the intensity and shape of the EPR spectrum. The spectra are of a sample of DPPH (2,2-diphenyl-1-picrylhydrazyl) measured with the field-modulation amplitudes given above each trace

However, a distinct drawback of field-modulation detection is that the signal amplitude and the line shape depend on the magnitude of the modulation field, B_{mod}. This effect is illustrated in Figure 1.5, which shows spectra of DPPH (2,2-diphenyl-1-picrylhydrazyl) measured with different modulation fields. When the field-modulation amplitude is small compared to the 0.14 mT width of the absorption line, the signal intensity is proportional to B_{mod}, thus in Figure 1.5 the amplitude of the spectrum doubles as B_{mod} is increased from 0.025 to 0.05 mT and from 0.05 to 0.1 mT. However, when B_{mod} becomes larger than the absorption line width, the peak is broadened and in the limit of $B_{mod} \gg \Delta B_{1/2}$, the peak-to-peak line width is equal to B_{mod}. This is seen in the lower two spectra in Figure 1.5, with $B_{mod} = 1.0$ and 2.0 mT. Because of these effects, care must be taken in selecting the modulation amplitude and interpreting peak heights when spectral features with different widths are present. If the absorption spectrum of a system has a narrow component superimposed on a broad component, the narrow component is emphasized in the field-modulated spectrum if the modulation amplitude is smaller than its line width but it is distorted and suppressed if the modulation amplitude is larger than its line width. As a result, obtaining the relative amounts of different species in a mixture from the CW-EPR spectrum is not trivial.

The modulation frequency can also affect the appearance of the EPR spectrum. If the width of the EPR line is comparable to, or smaller than, the modulation frequency expressed in field units, i.e., if $\Delta B_{1/2} \approx h\nu_{mod}/g\mu_B$, sidebands are observed in the spectrum at multiples of $h\nu_{mod}/g\mu_B$. For $\nu_{mod} = 100$ kHz, this means that sidebands occur when $\Delta B_{1/2}$ is around 0.0035 mT or less. Linewidths this narrow are not common, but are sometimes observed. In such cases, a lower modulation frequency is used to avoid the sidebands.

1.2.5 Power Broadening and Saturation

The intensity and microwave power dependence of a CW-EPR spectrum depend on the properties of the spin system being measured. The ratio of the populations of a pair of spin states, given by the Boltzmann factor $\exp(-\Delta E/k_B T)$, is important for the intensity. Spin–lattice or longitudinal relaxation, which restores the spin system to thermal equilibrium following a perturbation and is governed by the lifetime T_1, plays a role in determining the microwave power dependence of the intensity. Transverse or spin–spin relaxation, governed by the lifetime T_2, is also important for the power dependence. In a CW-EPR experiment, the signal intensity depends on the steady-state population difference between the two resonant spin states under continuous microwave irradiation. The microwave-induced transitions tend to equalize the populations taking the system away from thermal equilibrium and this is opposed by spin–lattice relaxation. To observe the CW-EPR spectrum, these two factors must be balanced. If the microwave power is too high or the spin–lattice relaxation is too slow, the spin system will saturate and the observed signal is weak. On the other hand, if the microwave power is low or the relaxation rate is high, the magnetization remains aligned along the external field and again a weak signal is obtained. Between these two limits there is an optimal value of the microwave power that gives the maximum signal. Using the steady-state solution of the Bloch equations (see Chapter 8), the power dependence of the peak-to-peak amplitude of the first-derivative-like EPR line shape can be shown to follow[3,4]

$$I \propto \frac{\sqrt{P}}{(1 + P/P_{1/2})^{b/2}} \tag{1.6}$$

where P is the microwave power and $P_{1/2}$ the power at which the saturation factor $1/(1 + (g\mu_B/\hbar)^2 B_1^2 T_1 T_2)$ is equal to 1/2. The parameter b is determined by the line shape and varies between 3 for a homogeneously broadened line and 1 for an inhomogeneously broadened line. A plot of equation (1.6) is shown in Figure 1.6(a). At low power, when the rate of the microwave-induced transitions is much less than the relaxation rate, the signal intensity varies linearly with $\sqrt{P}$. At high power, strong deviation from linear behavior is found and the intensity decreases due to saturation.

The width of the EPR lines is also influenced by the microwave power as shown in Figure 1.6(b) for the case of a homogeneously broadened line. For low power, when the intensity of the EPR signal is proportional to $\sqrt{P}$, the linewidth is determined by T_2 and is unaffected by the microwave power. However, as the signal saturates at higher power, the spectrum broadens.

The relaxation time constants, T_1 and T_2, and the population difference are both temperature dependent and they make the signal amplitude temperature dependent. If the population difference is the dominant factor, then the signal amplitude follows Currie Law behavior and is inversely proportional to the temperature. Thus, the signal intensity is higher at lower temperature. If the relaxation is the dominant factor, i.e., if T_1 and T_2 are short, the spectrum will be weak and broad. In some cases, this effect can make the EPR spectrum impossible to detect. Because the relaxation times usually become longer as the temperature decreases, the signal will again be stronger at lower temperature. However, as the relaxation times become longer, the signal will saturate at lower microwave power and the maximum intensity in the saturation curve decreases. Thus, there is also often an optimal temperature at which the highest signal intensity can be obtained.

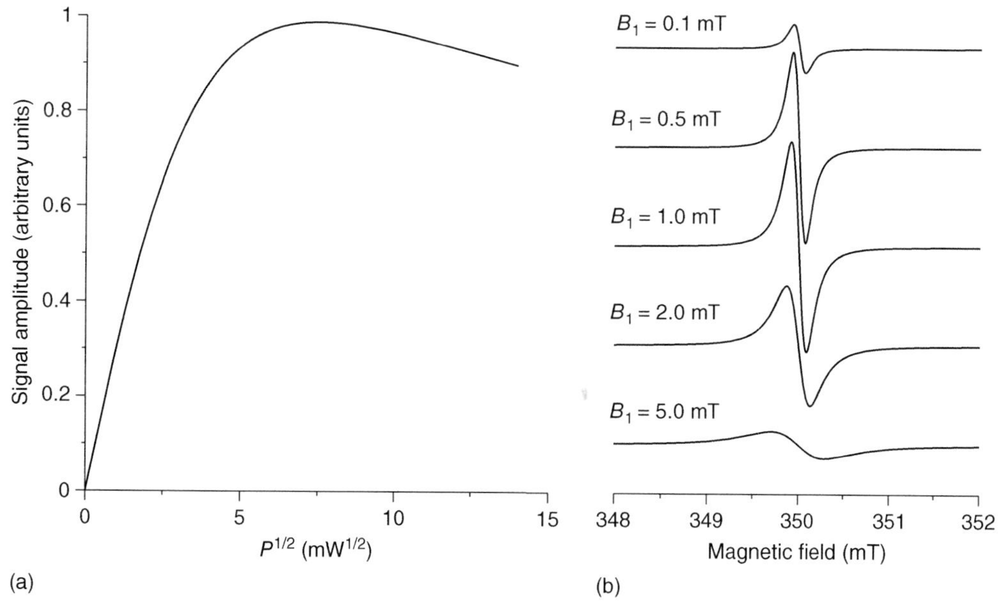

Figure 1.6. The EPR signal intensity and line shape as a function of the microwave power. (a) Saturation curve calculated using equation (1.6) with $P_{1/2}$ = 24.8 mW and b = 1.44. (b) Spectra calculated using the steady-state solution of the Bloch equations with $T_1 = T_2 = 50$ ns and $\omega_{res}/\hbar g_e \mu_B = 352$ mT

1.2.6 Internal Standards and Calibration

One of the main pieces of information that can be obtained from a CW-EPR spectrum is the isotropic g value or principal values of the $\boldsymbol{g}$-tensor (see Chapter 2) of the species being measured. The determination of g values requires that the magnetic field and microwave frequency are known. The frequency can be read using a frequency counter and the magnet current can be calibrated to the field. However, the field calibration can be difficult particularly at high fields. Thus, the use of an internal standard is necessary at high field/frequency and improves the accuracy of experimental g values from low-field instruments. The most common standards are DPPH,[5] Mn^{2+}-doped alkaline earth oxides[6,7] and Li-doped LiF.[8] Other substances such as P-doped Si[9] and K_3CrO_8-doped K_3NbO_8[10] have also been used. More recently, atomic hydrogen trapped in an octaisobutylsilsesquioxane nanocage[11] has proven to be a good standard because it has two very sharp lines with accurately known g values that do not overlap with those of organic radicals.

The concentration of radicals in a sample can also be determined from the EPR spectrum. However, this is not trivial. The concentration is related to the signal intensity through a proportionality constant that depends on the transition probability, relaxation times, Q and η of the resonator, microwave power, and modulation amplitude. The spectrum must be measured below the saturation limit and because lock-in detection is used, double integration must be carried out to obtain the intensity. Usually, baseline corrections must be applied to remove broad background signals from the resonator, sample tube, etc. To obtain the proportionality constant, a calibration standard is necessary, which must have the same properties as the species of interest and must be measured under identical conditions. The difficulty in achieving truly identical conditions for the standard and the sample limits the accuracy of concentration determination by EPR. Nonetheless, quantitative EPR studies are important, for example, for determining the stoichiometry of active sites in biological samples.

1.3 SOLUTION SPECTRA

To describe the CW-EPR spectrum of a spin system, the energies of the spin states are calculated using a spin Hamiltonian that contains terms for the interactions that are important under the conditions of the experiment. The energy differences between the spin states then give the positions of the lines in the spectrum. The motion of the spin system plays a crucial role in determining the appearance of the spectrum and the nature of the Hamiltonian. For a rapidly and isotropically tumbling paramagnetic species, all anisotropic interactions in the spin system are averaged to zero. The rate of tumbling of a molecule in solution is characterized by the rotational correlation time, τ_c, which is the average time needed for the molecule to rotate through 1 radian. To effectively average the anisotropic interactions, the inverse of the rotational correlation time, $1/\tau_c$, must be much larger than the anisotropy of the spectrum in frequency units. The inverse rotational correlation time of a molecule with a hydrodynamic radius of 10 Å in a low viscosity solvent at 300 K is ≈25 ns^{-1}. This is sufficient to average the g and hyperfine anisotropies of organic radicals at X-band, which are typically in the range of 1–100 MHz. Thus, the condition of the isotropic limit is met quite well for such species. In addition, the Zeeman energy is usually much larger than the hyperfine splitting. Together, these conditions mean that for a rapidly, isotropically tumbling radical being measured at X-band, the Hamiltonian contains only the secular parts of the isotropic Zeeman and hyperfine interactions:

$$\hat{H} = g_{\text{iso}}\mu_B B_0 \hat{S}_z + \sum_i a_{\text{iso},i}\hat{S}_z\hat{I}_{z,i} \tag{1.7}$$

The sum is over all magnetic nuclei with nuclear spin quantum number $I \neq 0$. The corresponding energies are

$$E = g_{\text{iso}}\mu_B B_0 m_S + \sum_i a_{\text{iso},i} m_S m_{I,i} \tag{1.8}$$

EPR transitions occur when only the electron spin is flipped and hence the selection rules are $\Delta m_S = 1$ and $\Delta m_I = 0$. This leads to transition energies:

$$h\nu = g_{\text{iso}}\mu_B B_0 + \sum_i a_{\text{iso},i} m_{I,i} \tag{1.9}$$

Thus, the spectrum consists of a pattern of lines due to the isotropic hyperfine couplings centered at the field corresponding to the isotropic g value. For coupling to a single nucleus with spin I, $2I + 1$ lines of equal intensity are obtained. This is illustrated in the top spectrum in Figure 1.7, which shows the pattern for a nitroxide, in which the electron spin is coupled to a single nitrogen nucleus ^{14}N with $I = 1$. In a field-swept spectrum, the separation between adjacent lines, which differ by 1 in the value of m_I, is $a_{\text{iso}}/g_{\text{iso}}\mu_B$. For a

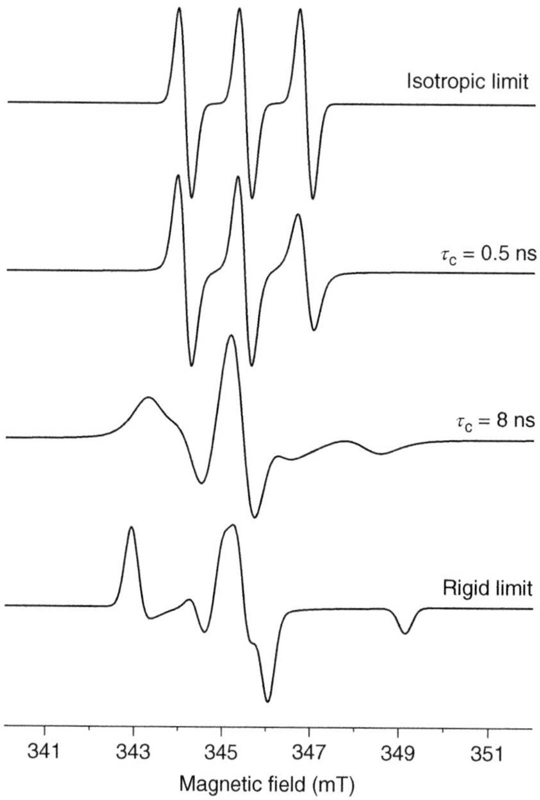

Figure 1.7. Calculated spectra of a nitroxide radical showing the effect of molecular motion as a function of the rotational correlation time τ_c. The spectra have been calculated using EasySpin[13] and the principal g values ($g_{xx} = 2.0089$, $g_{yy} = 2.0061$, $g_{zz} = 2.0027$) and ^{14}N hyperfine couplings ($A_{xx} = 14.6$ MHz, $A_{yy} = 14.6$ MHz, $A_{zz} = 86.9$ MHz) are taken from the literature[18]

group of n equivalent nuclei, a pattern of $2nI + 1$ lines is obtained. If $I = 1/2$, then the intensities of the lines follow Pascal's triangle as in proton solution NMR. The magnitude of the splitting is proportional to the electron spin density at the site of the nucleus (see Chapters 5 and 7). Because electrons are delocalized, there can be considerable spin density at many nuclei, particularly if the unpaired electron is in a delocalized π- or π^*-orbital of a conjugated system. If the electron is coupled to several groups of equivalent nuclei, the number of lines in the hyperfine pattern is

$$n_{\text{total}} = \prod_i (2n_i I_i + 1) \tag{1.10}$$

which can become very large for even a moderate number of nuclei. For example, the spectrum of the verdazyl radical, which has 2 pairs of equivalent nitrogen nuclei and two equivalent methyl groups, has 175 lines. Additional satellite peaks are also often observed from coupling to minor isotopes such as ^{13}C. Extracting the hyperfine coupling constants from such spectra is usually only feasible by numerical fitting; programs such as the commercial package XSophe[12] or the freely available EasySpin[13] toolbox for Matlab® are typically used.

Whether or not values of the hyperfine couplings can be extracted from an EPR spectrum depends crucially on the spectral resolution. Power broadening (Section 1.2.5) and overmodulation (Section 1.2.4) can both lead to loss of resolution and interfere with determination of the couplings. In addition, collisions between paramagnetic species can influence the spectrum via the Heisenberg spin-exchange mechanism. When two such species collide, the exchange interaction between them alters the positions of the EPR transitions. If the rate of collisions is high, an exchange-narrowed average spectrum is obtained, while in the limit of no collisions, the spectrum is unaffected. In the slow collision rate regime, the lines in the EPR spectrum are broadened and the change in the linewidth is proportional to the collision frequency. Because the collision frequency is directly proportional to the concentration, the line width of an EPR spectrum and hence the spectral resolution depend on the concentration of the paramagnetic species. Thus, in general, the concentration of the sample should be kept as low as possible to prevent such broadening. Typically, this means the concentration of solution samples should not be greater than $\approx 10^{-4}$ M. Molecular oxygen is paramagnetic and its concentration in solutions prepared in air under normal atmospheric pressure is high enough to cause significant exchange broadening. Thus, for liquid solution samples, it can be necessary to remove oxygen by performing freeze–pump–thaw cycles, saturating the solution with nitrogen or argon, or preparing the solution under an oxygen-free atmosphere, if high resolution is required.

For paramagnetic metal centers, solution spectra are relatively rare for several reasons. First, the relaxation times are often too short so that the spectrum cannot be detected. Second, the g-anisotropy is often too large to be completely averaged by the tumbling motion, therefore, the spectrum is broadened. Third, for centers with $S > 1/2$, the ZFS is also usually too large to be completely averaged.

1.4 THE EFFECT OF MOTION ON CW-EPR SPECTRA

The effect of motion on CW-EPR spectra is important and, in general, four motional regimes can be defined. In the isotropic limit, the anisotropy is averaged to zero. In the fast-motion regime, the system is close to the isotropic limit but some residual effects of the anisotropy are observed. In the rigid limit, no averaging of the anisotropy occurs. The slow-motion regime is close to the rigid limit but some motional averaging occurs. Motion intermediate between the fast- and slow-motion regimes is also possible. Figure 1.7 shows calculated spectra of a nitroxide radical that illustrate the four regimes. The top trace is the isotropic limit where three equal intensity peaks due to hyperfine coupling to $I = 1$ ^{14}N nucleus are obtained. The spectrum is centered at the field corresponding to the isotropic g value. When the rotational correlation time, τ_c, is greater than ≈50 ps, the fast-motion regime applies and changes in the relative intensities of the three peaks are seen. With $\tau_c = 500$ ps (Figure 1.7, second trace), these effects are quite pronounced and the intensity of the high-field peak is about half that of the other two peaks. In the rigid limit (Figure 1.7, bottom trace), the spectrum shows features associated with the three principal g values and the associated principal hyperfine couplings. With $\tau_c = 8$ ns, the system is in the slow-motion regime and the spectrum resembles the rigid limit spectrum but the features are broadened and shifted. This sensitivity of the CW-EPR spectra to the rotational correlation time has been used for determining the secondary structure of proteins.[14] In site-directed spin-labeling studies, the line shape is parameterized to allow residues to be grouped into areas of high or low mobility as a way of locating secondary structural elements and identifying structural changes. A more rigorous analysis of the spectra requires simulation based on the stochastic Liouville equation (see Chapter 14 for details).[15,16]

1.5 SPECTRA OF SOLIDS

A large percentage of the CW-EPR spectra reported in the literature are taken under conditions of the rigid limit or slow motion. This is partly because low temperatures are often needed to slow T_1 and T_2 relaxation and increase the initial magnetization. Under these conditions, the anisotropic interactions also contribute. Solid-state samples fall into two categories: single crystals and powders. The latter encompasses not only actual powders but also any sample in which there is no orientational ordering. Partially ordered samples constitute a third category but they will not be discussed here.

1.5.1 Single Crystals

In a single crystal, the positions of the EPR lines depend on the crystal orientation. If the spectrum is dominated by the Zeeman interaction, a single line is obtained for each radical in a unique orientation in the unit cell. If a series of spectra are measured as the crystal is rotated about a particular axis, the positions of the lines change and form a rotation pattern. The analysis of such patterns provides the principal values of the $\boldsymbol{g}$-tensor and the orientation of the principal axes relative to the axes of rotation. If the orientation of the rotation axes relative to the crystallographic axes is known and the structure of the molecules in the unit cell is available from X-ray scattering data, the location of the principal axes of the $\boldsymbol{g}$-tensor in the molecule can be determined. When resolved hyperfine patterns are present, a corresponding analysis of the splittings gives the principal values and axes of the hyperfine coupling. The details of the determination of the principal values and axes of the $\boldsymbol{g}$-tensor from rotation patterns are described in Chapter 2.

1.5.2 Powder Patterns

In most cases, single crystals are not available and randomly ordered solids are measured. These can be either powders or frozen solutions. In the absence of resolved hyperfine couplings, systems with a single unpaired electron give a powder pattern of the $\boldsymbol{g}$-tensor. The shape of this spectrum is determined by the symmetry of the $\boldsymbol{g}$-tensor and there are four characteristic line shapes as illustrated in Figure 1.8. If two threefold or higher rotation axes perpendicular to one another are present, e.g., tetrahedral or octahedral symmetry, second-rank tensorial quantities are isotropic. Hence, $g_{xx} = g_{yy} = g_{zz}$ and the powder pattern consists of a single peak located at $B_0 = h\nu/g_{iso}\mu_B$. In the case of a single symmetry axis with threefold or higher rotational symmetry, the $\boldsymbol{g}$-tensor is axial and $g_{xx} = g_{yy} \neq g_{zz}$, where z is the symmetry axis. Here, g_{zz} can be either larger or smaller than the other two principal g values,

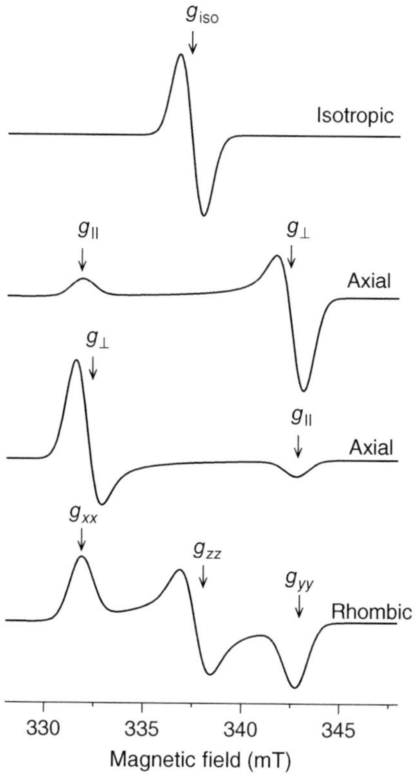

Figure 1.8. Characteristic powder patterns for $S = 1/2$ and the different possible symmetries of the **g**-tensor

and therefore, two different characteristic patterns are possible. Finally, in the case of rhombic or lower symmetry, all three principal g values are different. As can be seen in Figure 1.8, when these powder patterns are measured using field modulation, the main features are observed at positions corresponding to the principal g values because the slope of the absorption spectrum is large at these positions and small in between them. The intensity of the features also depends on the number of orientations of the molecule corresponding to g values at or near each of the principal g values with respect to the external magnetic field. For axial symmetry, for example, any orientation in which the external field is in the xy-plane will have $g = g_{\perp}$, while only those orientations in which the field is parallel or nearly parallel to the z-axis will have $g \approx g_{\|}$. Thus, the peak corresponding to $g_{\perp}$ is much stronger than that from $g_{\|}$. For organic radicals, the g-anisotropy is typically too small to be resolved at X-band and an isotropic spectrum is usually obtained, while at high field such species display axial or rhombic patterns.

For transition metal ions or clusters, the g-anisotropy is often well resolved at X-band. However, when the ion has nonzero nuclear spin, strong hyperfine splitting is superimposed on the features corresponding to the principal g values. Hyperfine splittings to ligand nuclei are also sometimes resolved, particularly if the coordination is through nitrogen. In general, the hyperfine coupling can be divided into isotropic and anisotropic components. The isotropic part is associated with the Fermi contact term, and the anisotropic part arises from the spin and orbital dipolar terms. For more details, see Chapter 5. For first-row transition metal ions, both of these components of the hyperfine coupling are typically on the order 10^2–10^3 MHz and have the same sign for one principal direction and opposite sign for the other two directions. Thus, the hyperfine splitting from the metal nucleus is typically very large in one direction and small in the other two. The effect of hyperfine coupling is illustrated in Figure 1.9 using calculated spectra of Cu(II) porphyrin. In the absence of any hyperfine coupling (Figure 1.9a), the spectrum is that of an axial **g**-tensor, owing to the D_{4h} symmetry of the porphyrin ring. The out-of-plane direction has the largest g value ($g = 2.19$) and corresponds to the peak at about 307 mT. The two in-plane principal g values ($g = 2.03$) give the positive/negative feature at about 331 mT. When hyperfine coupling to the $I = 3/2$ copper nucleus is added (Figure 1.9b), the low-field peak is split into four components, three of which are clearly visible with a splitting of 608 MHz. The in-plane features are also split, but the pattern is more complicated because of the overlap with one of the out-of-plane hyperfine pattern features and because other orientations also contribute. Finally, when the coupling to the four equivalent nitrogen nuclei is added, a large number of additional lines appear. The simulation and analysis of such spectra is not trivial because of the number of parameters involved, especially if the relative orientations of the hyperfine tensors and the **g**-tensor are unknown and the effect of slow motion must be taken into account. However, when the principal g values and hyperfine coupling can be determined, they provide information about the spin density distribution and spin–orbit coupling in the molecule as described in detail in Chapters 2, 5 and 7.

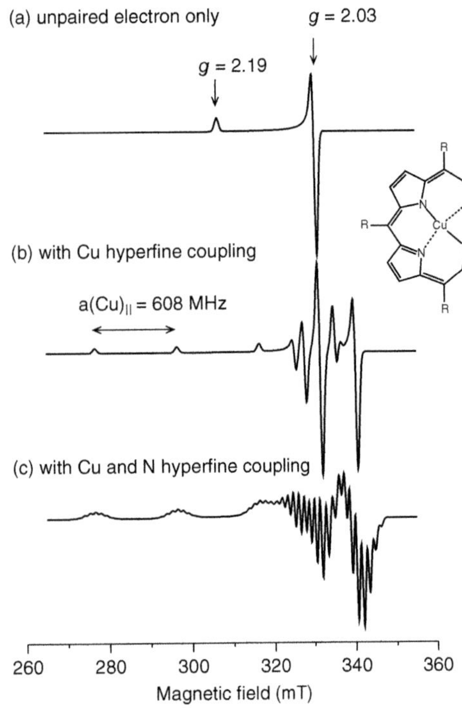

Figure 1.9. Calculated spectra of copper(II) porphyrin showing the contribution of hyperfine coupling. (a) The spectrum in the absence of hyperfine coupling. (b) The spectrum with hyperfine coupling to the $I = 3/2$ copper nucleus included. (c) The spectrum with hyperfine coupling to Cu and four equivalent coordinating N atoms from the porphyrin ring. Principal g values of Cu(II): 2.03, 2.03, 2.19; principal values of the Cu hyperfine coupling (MHz): 75, 75, 608; N hyperfine coupling (isotropic) 40 MHz

1.5.3 High-spin Systems

In systems with more than one unpaired electron (total spin $S > 1/2$), the ZFS plays an important role (see Chapter 3). As the name implies, this interaction, which is due to the dipole–dipole and spin–orbit coupling, lifts the degeneracy of the spin states and does not depend on the magnetic field strength. It is characterized by the ZFS tensor, which can be written in terms of the axial and rhombic parameters D and E. For triplet states of organic molecules and for transition metal ions for which the free ion has $L = 0$ (d^5 and f^7 electron configurations), the ZFS is typically smaller than the Zeeman energy and leads to splitting of the transitions $-m_S \leftrightarrow -m_S+1$ and $m_S - 1 \leftrightarrow m_S$, which would be degenerate in the absence of ZFS. For other transition metal ions, the strength of the ZFS is usually significantly larger than the Zeeman energy. Such ions can be broken into two groups: those with half-integer spin and those with whole-integer spin. In the half-integer spin case, the ZFS results in a series of doubly degenerate states with $m_S = \pm 1/2, \pm 3/2, \ldots$ referred to as Kramers doublets. If the ZFS is large, the applied field of the EPR spectrometer lifts the degeneracy of the Kramer's doublets but there is minimal mixing of states with different values of m_S. EPR transitions are observable between the pairs of states within each m_S manifold. In a powder spectrum of such species, the main features are observed from molecular orientations in which the field is parallel to the principal axes of ZFS tensor. The positions of these features depend on the ratio of the two ZFS parameters E/D but not on the magnitude of D provided that it is much larger than the Zeeman energy. The range of possible values for this ratio is from zero to 1/3. Thus, it is possible to generate a plot of possible positions of the features the EPR spectrum as a function of E/D. Such a plot is referred to as a *rhombogram* because the parameter E/D is a measure of the rhombicity of the ZFS tensor. With $E/D = 0$, it has axial symmetry, and with $E/D = 1/3$, it is maximally rhombic. Generally, the line positions in a rhombogram are given as effective g values $g_{\text{eff}} = \Delta E/\mu_B B_0$, where ΔE is the energy of the transition, and the positions of the lines in the field-swept spectrum can be obtained as $h\nu/g_{\text{eff}}\mu_B$. The use of effective g values has the advantage that the same rhombogram can be used for different microwave frequencies. Figure 1.10 shows the rhombogram for $S = 3/2$ and a calculated example spectrum. Transition metal ions with $S = 3/2$ have a configuration of either d^3 or high-spin d^7. In biological systems, some iron–sulfur clusters also have $S = 3/2$. The top part of the rhombogram shows the effective g values of the transitions involving the $m_S = \pm 3/2$ levels with the field along the principal x, y, and z axes of the ZFS tensor. The corresponding g values for the transitions between the $m_S = \pm 1/2$ levels are shown in the bottom part of the rhombogram (Figure 1.10b). The spectrum in Figure 1.10(c) is calculated for $E/D = 0.32$ indicated by the vertical dotted line in the rhombogram.

The spin–lattice relaxation of high-spin transition metals is usually very fast and so their spectra are normally collected at low temperature. The ZFS is typically on the order of 1–10 K in temperature units and hence there can be a significant population difference between the $m_S = \pm 3/2$ and $m_S = \pm 1/2$ level pairs at

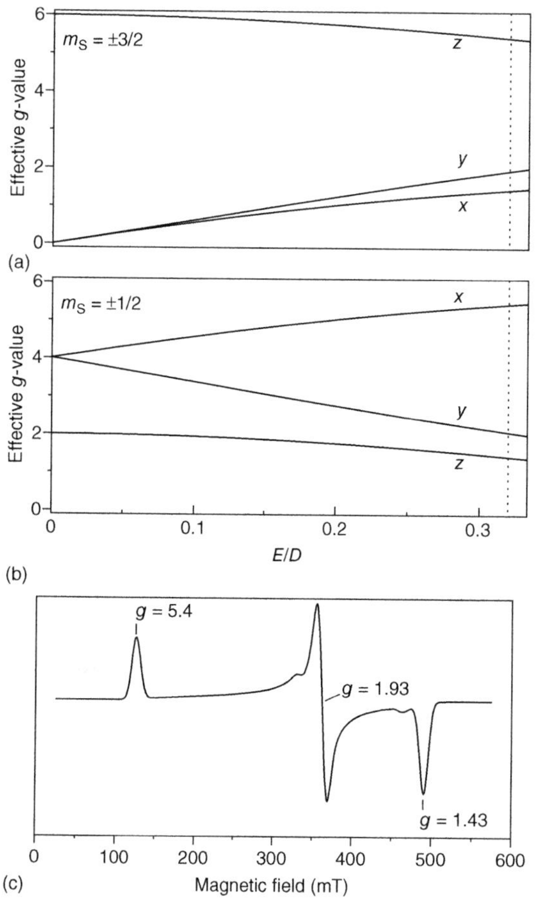

Figure 1.10. Rhombogram and example spectrum for $S = 3/2$. The spectrum is calculated for $E/D = 0.32$ and $D = -1.1 \times 10^5$ MHz. Microwave frequency = 9.75 GHz. The features at $g_{\mathrm{eff}} = 5.4$ arise from the $m_S = \pm 1/2$ transition with the field along the x-direction and the $m_S = \pm 3/2$ transition with the field along z. The peak at $g_{\mathrm{eff}} = 1.93$ is from $m_S = \pm 1/2$, y and $m_S = \pm 3/2$, y and x. The feature at $g_{\mathrm{eff}} = 1.43$ is from $m_S = \pm 1/2$, z

5 K. Which of the two level pairs has the higher population depends on the sign of D. With D positive, the $m_S = \pm 1/2$ levels are lower in energy and hence have higher population. If features from both manifolds are present in the spectrum, the sign of D can be estimated by the temperature dependence of their intensities. Transitions within the lower lying m_S level pair increase with decreasing temperature, while those from the higher lying level pair decrease.

As can be seen from the rhombogram in Figure 1.10, the positions of some of the features in the spectrum depend quite strongly on the value of E/D. This ratio is determined by the geometry of the metal center and often there is a significant spread of values of E/D as a result of a distribution of geometries. This is referred to as E-strain (or E/D strain) and it results in broadening of the lines in the spectrum. The amount of broadening caused by E-strain depends on the slope of the curves in the rhombogram and the effective g value of the feature. Because the spectrum is measured by sweeping the field and the resonant field is inversely dependent on the g value, components with high-effective g values are compressed at the low-field end of the spectrum, while those with low g values are expanded on the high-field end. As a result, E-strain has a larger effect at high field and the components at low field (high-effective g value) are most prominent in the spectrum.

For systems with whole-integer values of S, the m_S sublevels do not form Kramers doublets and are split even at zero field if the symmetry is nonaxial. If the ZFS is large, such systems often do not have transitions that are observable by EPR at conventional frequencies and fields. An unusual aspect of these non-Kramers systems is that the transition moment for EPR transitions has a significant component along the z-direction parallel to the magnetic field. Thus, they are sometimes studied using parallel-mode EPR, in which the microwave magnetic field B_1 is parallel to the static field B_0. Such experiments are most conveniently carried out using a resonator, which has two modes of similar frequency but different orientations of the B_1 field. At X-band, a rectangular resonator as shown in Figure 1.2 is often used. The dimensions a and b are chosen so that the modes TE_{102} and TE_{012} have frequencies around 100 MHz apart. The spectrometer can then be coupled to either mode and the spectrum measured with the B_1 field perpendicular to the external field using the TE_{102} mode and parallel to it with the TE_{012} mode.

1.6 CONCLUDING REMARKS

CW-EPR experiments can provide a wealth of information about paramagnetic systems and such spectra can usually be acquired quickly and easily. More detail of the interactions that can be probed in such experiments is given in the following chapters in Part A of this book. While CW-EPR remains the most widely used form of EPR spectroscopy, its main limitation is the inability to resolve the weak couplings, which leads to line broadening. Such weak interactions can contain extremely important information, particularly for biological systems. For example, weak hyperfine couplings to ligands can provide insight into the nature of protein-cofactor interaction. The weak dipolar coupling between spin labels is also of great importance in structure determination. The chapters in Part C of this book deal with the wide range of more advanced methods that are able to resolve these weak interactions.

ACKNOWLEDGMENT

This work was supported by a Discovery grant from the Natural Sciences and Engineering Research Council, Canada.

FURTHER READING

W. R. Hagen, Biomolecular EPR Spectroscopy, CRC Press: Boca Raton, FL, 2009.

M. Brustolon and E. Giamello, Electron Paramagnetic Resonance: A Practioner's Toolkit, John Wiley & Sons, Inc.: Hoboken, NJ, 2009.

G. R. Eaton, S. S. Eaton, D. R. Barr, and R. T. Weber, Quantitative EPR, Springer: Vienna, 2010.

REFERENCES

1. M. R. Fuchs, T. F. Prisner, and K. Möbius, *Rev. Sci. Instr.*, 1999, **70**, 3681.
2. M. L. Meade, Lock-in Amplifiers Principles and Applications, P. Pereginus Ltd: London, 1983, 17.
3. T. Castner Jr, *Phys. Rev.*, 1959, **115**, 1506.
4. M. Sahlin, A. Gräslund, and A. Ehrenberg, *J. Magn. Reson.*, 1986, **67**, 135.
5. N. Yordanov, *Appl. Magn. Reson.*, 1996, **10**, 339.
6. O. Burghaus, M. Rohrer, T. Götzinger, M. Plato, and K. Möbius, *Meas. Sci. Tech.*, 1992, **3**, 765.
7. J. Rosenthal and L. Yarmus, *Rev. Sci. Instr.*, 1966, **37**, 381.
8. A. Stesmans and Y. Wu, *J. Phys. D*, 1988, **21**, 1205.
9. A. Stesmans, *J. Magn. Reson.*, 1988, **76**, 14.
10. B. Cage, A. Weekley, L.-C. Brunel, and N. S. Dalal, *Anal. Chem.*, 1999, **71**, 1951.
11. S. Stoll, A. Ozarowski, R. D. Britt, and A. Angerhofer, *J. Magn. Reson.*, 2010, **207**, 158.
12. G. R. Hanson, K. E. Gates, C. J. Noble, M. Griffin, A. Mitchell, and S. Benson, *J. Inorg. Biochem.*, 2004, **98**, 903.
13. S. Stoll and A. Schweiger, *J. Magn. Reson.*, 2006, **178**, 42.
14. W. L. Hubbell, D. S. Cafiso, and C. Altenbach, *Nat. Struct. Mol. Biol.*, 2000, **7**, 735.
15. D. J. Schneider and J. H. Freed, in Biological Magnetic Resonance, Spin Labeling: Theory and Applications, eds L. J. Berliner and J. Reuben, Springer: New York, 1989, Vol. 8, Chap. 1.
16. K. A. Earle and D. E. Budil, in Advanced ESR Methods in Polymer Research, ed S. Schlick, John Wiley & Sons, Inc.: New Jersey, 2006, Chap. 3.
17. J. H. Golbeck and A. van der Est, in Molecular Biophysics for the Life Sciences, eds N. M. Allewell, L. O. Narhi and I. Rayment, Springer: New York, 2015.
18. O. H. Griffith, D. W. Cornell, and H. M. McConnell, *J. Chem. Phys.*, 1965, **43**, 2909.

Chapter 2
EPR Interactions – *g*-Anisotropy

Peter Gast and Edgar J.J. Groenen

Leiden University, Leiden, The Netherlands

2.1 INTRODUCTION

EPR spectroscopy concerns the study of magnetic-dipole transitions induced by microwave irradiation of a paramagnetic system placed in a static magnetic field. Although EPR has a fundamental basis in common with NMR, these techniques have developed along distinct lines, both experimentally and theoretically, as a consequence of the larger magnetic moment of the electron as compared to that of nuclei and the essentially delocalized nature of the electron spin.

The magnetic moment of the electron that couples to the magnetic field is proportional to its angular momentum

$$\boldsymbol{\mu} = -g\mu_B \boldsymbol{S} \qquad (2.1)$$

EPR Spectroscopy: Fundamentals and Methods.
Edited by Daniella Goldfarb and Stefan Stoll
 ISBN: 978-1-119-16299-5
Also published in eMagRes (online edition)
DOI: 10.1002/9780470034590.emrstm1500

where μ_B represents the Bohr magneton. The proportionality constant, the gyromagnetic ratio, deviates from the classically expected value by the *g*-factor. For a free electron, this *g*-factor is known to incredible accuracy[1]

$$g_e = 2.00231930436082(52) \qquad (2.2)$$

A hypothetical EPR experiment on such a system does not entail new information. For systems of interest here, the electron is bound and the *g*-factor deviates from g_e. For organic radicals, this deviation is small, commonly less than half a percent. For transition metal ions, the *g*-factor spans a wide range of values. Like the chemical shift for NMR, the *g*-factor is a fingerprint of a material. The description, determination, and interpretation of the deviation of the *g*-factor from g_e are the subject of this chapter.

2.2 THE ZEEMAN INTERACTION FOR SPIN ½

In a material environment, the angular momentum of an electron contains contributions other than from pure spin. Although commonly (for orbitally nondegenerate states) its orbital angular momentum is quenched, the electron feels the environment through coupling of the spin and orbital angular momenta. The angular momentum then acquires spatial contributions in addition to the spin. Thanks to the ingenious insight of the founding fathers of magnetic resonance, the description of EPR spectra can be based on an effective spin Hamiltonian, which contains the electron and nuclear spin operators only. Here, we restrict ourselves to the term in the spin Hamiltonian that represents the interaction of the electron spin S with the external magnetic field B. This electronic Zeeman term is of the form

$$\hat{H}_Z = -\boldsymbol{B}\hat{\boldsymbol{\mu}} = \mu_B \boldsymbol{B}\boldsymbol{g}\hat{\boldsymbol{S}} \tag{2.3}$$

All spatial information is contained in the $\boldsymbol{g}$-tensor (formally a matrix), which is in general anisotropic and only reduces to a scalar for a paramagnetic system of cubic or higher symmetry. Equation (2.3) can be looked upon as to represent the interaction of an effective momentum $-\mu_B \boldsymbol{g}\boldsymbol{S}$ with the external magnetic field or, alternatively, as the interaction of the spin momentum with the local magnetic field $\boldsymbol{g}^T\boldsymbol{B}/g_e$. While a free spin would be quantized along the external field, the bound spin is quantized along the effective local field.

The energies of the two stationary states of the electron spin in the magnetic field follow from the diagonalization of the matrix representation of the spin Hamiltonian $\hat{H}_Z$ in the basis of eigenstates $|m_S\rangle$ ($m_S = +{}^1\!/\!_2, -{}^1\!/\!_2$) of $\hat{S}_z$. In the principal axes system x, y, z of $\boldsymbol{g}$, the spin Hamiltonian becomes

$$\hat{H}_Z = \mu_B B(n_x n_y n_z)\begin{pmatrix} g_x & & \\ & g_y & \\ & & g_z \end{pmatrix}\begin{pmatrix} \hat{S}_x \\ \hat{S}_y \\ \hat{S}_z \end{pmatrix} \tag{2.4a}$$

$$= \mu_B B \sum_{i=x,y,z} n_i g_i \hat{S}_i \tag{2.4b}$$

where n_i are the three components of the unit vector $\boldsymbol{n}$ along the direction of the external magnetic field and B is the magnitude of that field. The principal values of the $\boldsymbol{g}$-tensor are indicated by g_x, g_y, and g_z. In the absence of symmetry, $g_x \neq g_y \neq g_z$, and the $\boldsymbol{g}$-tensor is referred to as rhombic. In case two principal g values are equal, the $\boldsymbol{g}$-tensor is referred to as axial, and when all principal g values are equal as cubic or isotropic. The next step is to define

$$\hat{S} \equiv {}^1\!/\!_2\hat{\sigma} \tag{2.5}$$

and insert the Pauli matrices

$$\sigma_x = \begin{pmatrix} 0 & 1 \\ 1 & 0 \end{pmatrix} \sigma_y = \begin{pmatrix} 0 & -\mathrm{i} \\ \mathrm{i} & 0 \end{pmatrix} \sigma_z = \begin{pmatrix} 1 & 0 \\ 0 & -1 \end{pmatrix} \tag{2.6}$$

in equation (2.4b). In this way, we obtain the matrix representation of the spin Hamiltonian $\hat{H}_Z$ in the basis of eigenstates $|m_s\rangle$

$$\hat{H}_Z = {}^1\!/\!_2\mu_B B \begin{pmatrix} n_z g_z & n_x g_x - \mathrm{i}n_y g_y \\ n_x g_x + \mathrm{i}n_y g_y & -n_z g_z \end{pmatrix} \tag{2.7}$$

Diagonalization of this matrix yields the energies of the two eigenstates

$$\begin{aligned} E(m_s) &= m_s \mu_B B\left(\sum_i n_i^2 g_i^2\right)^{1/2} = m_s \mu_B [B(\boldsymbol{n}^T\boldsymbol{g})(\boldsymbol{g}^T\boldsymbol{n})B]^{1/2} \\ &= m_s \mu_B B|\boldsymbol{g}^T\boldsymbol{n}| \end{aligned} \tag{2.8}$$

with $m_s = \pm\,{}^1\!/\!_2$. The expression between square brackets represents the inner product of the local-field vector with itself. The energy difference between the two states translates into the resonance condition, which forms the basis of EPR spectroscopy:

$$E(+{}^1\!/\!_2) - E(-{}^1\!/\!_2) = h\nu_{mw} = (|\boldsymbol{g}^T\boldsymbol{n}|)\mu_B B \equiv g\mu_B B \tag{2.9}$$

Here, h is the Planck constant and ν_{mw} the microwave frequency at which the EPR experiment is performed. The g introduced on the right side of equation (2.9) refers to the operational g value calculated from the experimental spectrum. The norm in equation (2.9) is invariant under orthogonal transformation, which means that this equation holds for any Cartesian coordinate system. We have used the principal axes system of the $\boldsymbol{g}$-tensor in the derivation because it simplified the calculation.

The product matrix $\boldsymbol{g}\boldsymbol{g}^T$ is symmetric, even when $\boldsymbol{g}$ is not. According to equation (2.9), a complete EPR experiment yields the six unique elements of this product matrix. The sign of g remains undetermined and is taken positive for most systems, in line with the positive value of g_e.

For g-anisotropy or different paramagnetic species, the difference in resonance field of two transitions corresponding to g_1 and g_2 becomes

$$\Delta B = \frac{h\nu_{mw}}{\mu_B}\left(\frac{1}{g_1} - \frac{1}{g_2}\right) \tag{2.10}$$

This shows that the spectral resolution increases proportional to the microwave frequency, and corresponding resonance field, as long as the linewidths do not increase with field.

An expression like equation (2.3) also describes the electron Zeeman interaction for spins other than ${}^1\!/\!_2$, but in these cases, the spin Hamiltonian contains additional terms. This will be treated in Chapter 3, and we will shortly refer to a special case in Section 2.5.

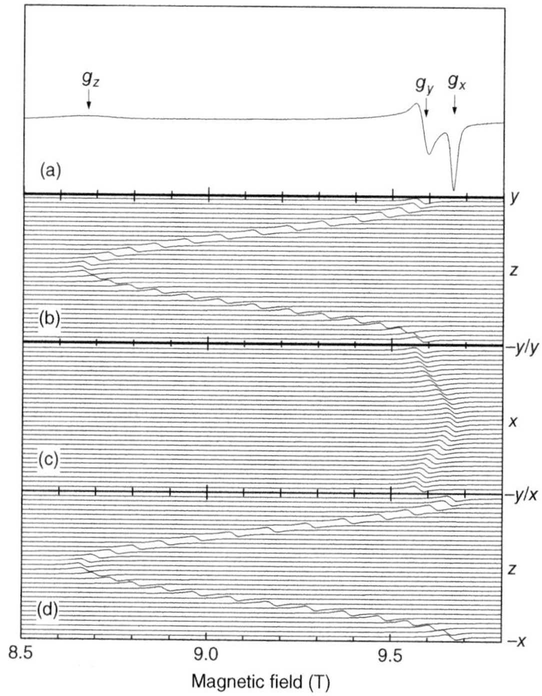

Figure 2.1. (a) Experimental continuous-wave EPR spectrum of azurin recorded at 20 K and at 275 GHz; (b) to (d). Simulated fields of resonance at 275 GHz as a function of the orientation of the magnetic field in the principal planes of the **g**-tensor, the *yz*-, the *xy*-, and the *zx*-plane with parameters: $g_x = 2.0393$, $g_y = 2.0568$, $g_z = 2.273$, intrinsic linewidth = 5 mT

2.3 *g*-ANISOTROPY IN THE SOLID STATE

Consider an ensemble of randomly oriented spins in a solid, be it a frozen solution or a powder sample. All orientations of the paramagnetic center with respect to the magnetic field are equally probable or, equivalently, all orientations of the magnetic field with respect to the paramagnetic center are equally probable, and no orientation has a special status.

Figure 2.1(a) shows the continuous-wave EPR spectrum of a frozen solution of the blue-copper protein azurin taken at the microwave frequency of 275 GHz. In the oxidized form, the metal center of this protein concerns a Cu(II) ($S = 1/2$) site. At this high microwave frequency, about 30 times the classical X-band frequency of 9 GHz, the interaction of the spins with the external magnetic field of about 10 T largely dominates the intrinsic magnetic interactions of the system. In zeroth order, the shape of the spectrum can be described taking into account only the *g*-anisotropy: $g_z = 2.273$, $g_y = 2.0568$, and $g_x = 2.0393$.[2] Neither the hyperfine interaction with the copper nucleus nor the superhyperfine interaction with the nitrogen nuclei of the histidine ligands are resolved owing to the inhomogeneous broadening, which at the high magnetic fields corresponding to 275 GHz is dominated by anisotropic *g*-strain.[2,3] The term *g*-strain refers to a distribution of *g* values, which for metalloproteins derives from the distribution of slightly different conformations of the coordination sphere of the transition metal ion.[4]

The derivative-like shape of the spectrum derives from the fact that field modulation is used in combination with phase-sensitive detection (see Chapter 1). The spectrum extends from 8.5 to 9.7 T, which covers the range of magnetic fields corresponding to all orientations of the azurin molecules with respect to the magnetic field. Equation (2.9) provides the basis for understanding the spectrum of a spin-$1/2$ system. The absorption at lowest magnetic field derives from molecules in the frozen solution for which the magnetic field is parallel to their principal *z*-axis (corresponding to the largest principal *g* value): the value of *B* corresponds to that obtained from equation (2.9) for $n_x = n_y = 0$, $n_z = 1$. The absorption at highest magnetic field derives from molecules for which the magnetic field is parallel to their principal *x*-axis (corresponding to the smallest principal *g* value): the value of *B* corresponds to that obtained from equation (2.9) for $n_x = 1$, $n_y = n_z = 0$. For orientations away from *z*(*x*), the magnetic field of resonance increases (decreases), which leads to an accumulation of resonances around the magnetic field corresponding to g_y.

Figure 2.1(b–d) illustrates the contributions to the EPR spectrum for orientations of the magnetic field in the principal planes of the **g**-tensor. For the random distribution of paramagnetic centers in a frozen solution or powder, the knowledge of the principal values of the **g**-tensor suffices to understand the spectrum. Vice versa, the EPR spectrum of a frozen solution or powder allows the determination of the principal values of the **g**-tensor, but not of the direction of the principal axes.

A second example is shown in Figure 2.2, the EPR spectrum of the nitroxide radical TEMPOL in frozen aqueous solution at different microwave frequency/magnetic field settings.[5] The spectra are plotted relative to the fixed g_z value and emphasize the increasing *g*-resolution with increasing frequency/field. The shape of the spectrum taken at 360 GHz around 12.9 T reflects the rhombicity of the

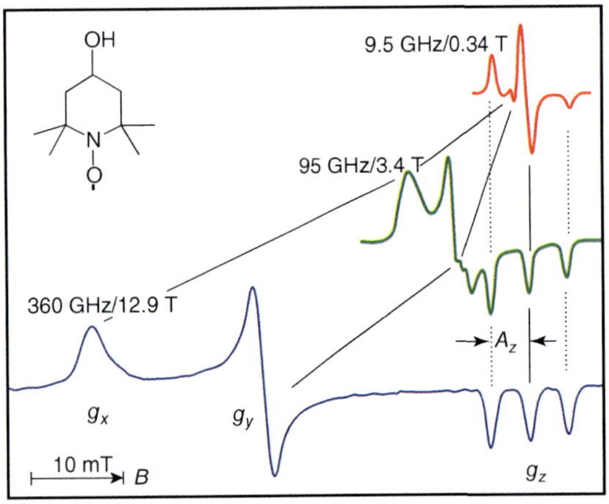

Figure 2.2. Continuous-wave EPR spectra of the nitroxide radical TEMPOL in a frozen water solution at different microwave frequency/magnetic field settings. Spectra are plotted relative to the fixed g_z value. (K. Möbius and A. Savitsky, High-Field EPR Spectroscopy on Proteins and their Model Systems, Ch. 5. The Royal Society of Chemistry: Cambridge, 2009 – Reproduced by permission of The Royal Society of Chemistry)

g-tensor of this radical. This rhombicity distinguishes the spectrum from that in Figure 2.1(a) for azurin with its nearly axial **g**-tensor. In addition, for orientations along z, the hyperfine interaction (see Chapter 5) with the nitrogen nucleus ($I = 1$) is large enough to be resolved even at the high magnetic field. This hyperfine splitting dominates the spectrum at 9.5 GHz in which the g-anisotropy is no longer resolved. The spectrum at 95 GHz is in between and together the three spectra in Figure 2.2 illustrate the power of a multifrequency approach in EPR. If, besides the electronic Zeeman interaction, internal magnetic interactions, like the hyperfine interaction, contribute to the EPR spectrum, the combination of experiments at different frequencies can help to disentangle complicated spectra. The higher the external magnetic field, the more the Zeeman interaction will dominate the EPR spectrum, i.e., the more prominent the g-anisotropy.

Determination of the complete **g**-tensor, the principal values and the direction of the principal axes, requires an oriented spin system, and a single crystal provides the optimum situation for such experiments. Consider a single crystal with one paramagnetic molecule per unit cell. The EPR spectrum will depend on the direction of the magnetic field with respect to the crystal, in contrast to the case of the randomly oriented sample discussed above. We introduce a crystal-fixed reference axes system ξ, η, ζ in which the elements of the symmetric matrix $\boldsymbol{gg}^{\mathrm{T}}$ are indicated by $(\boldsymbol{gg}^{\mathrm{T}})_{ij}$ with $i, j = \xi, \eta, \zeta$. This matrix has six unknowns, which according to equation (2.9) can be determined from EPR spectra measured for different directions of the magnetic field defined in the coordinate system ξ, η, ζ. Subsequent diagonalization of the $\boldsymbol{gg}^{\mathrm{T}}$ matrix yields the orientation of the principal axes of the $\boldsymbol{gg}^{\mathrm{T}}$ matrix in the axes system ξ, η, ζ. The square root of the diagonal matrix yields the principal values g_x, g_y, and g_z.

In practice, most spectrometers allow rotation of the crystal around one axis only, which means that the direction of the magnetic field can be varied in one plane only. For orientations in this plane, say the $\eta\zeta$-plane, equation (2.9) reduces to

$$g^2 = (\boldsymbol{gg}^{\mathrm{T}})_{\eta\eta}\sin^2\theta + 2(\boldsymbol{gg}^{\mathrm{T}})_{\eta\zeta}\sin\theta\cos\theta + (\boldsymbol{gg}^{\mathrm{T}})_{\zeta\zeta}\cos^2\theta \tag{2.11}$$

where the angle θ is the angle between the magnetic field direction and the ζ-axis. Measurements of the EPR spectrum for directions of the magnetic field in this plane result in values of three components of the $\boldsymbol{gg}^{\mathrm{T}}$ matrix. Determination of the complete matrix requires remounting the crystal in such a way that additional measurements can be performed in another plane. Figure 2.1(b–d) illustrates the variation of the resonance field as described by equation (2.11). In this simulation, the reference axes system ξ, η, ζ has been taken along the principal axes of the **g**-tensor.

If the spectrometer is equipped with a 3D-goniometer, all orientations can be investigated with one crystal mounting, which enhances the accuracy. In this case, a different approach is possible as well. The direction of the principal g_z (g_x) axes and the value of g_z (g_x) can be determined by varying the direction of the magnetic field with respect to the crystal until the lowest (highest) field of resonance is obtained. These fields are stationary fields, in other words, the field of resonance increases (decreases) for any change in the direction of the magnetic field. Once these directions have been determined accurately, the direction of the third principal axis g_y follows from its orthogonality to the first two and the spectrum for this direction of the magnetic field yields g_y. The accuracy can be optimized by additional measurements at other orientations of the magnetic field.

In order to translate the obtained orientation of the principal g-axes system from the crystal-fixed reference axes system to a molecule-fixed axes system coupled to the paramagnetic molecule, additional information is needed about the orientation of the molecule in the reference coordinate system. Commonly, this is obtained from an X-ray diffraction study on the crystal as mounted for the EPR experiment.

For more than one molecule in the unit cell, a particular direction of the magnetic field with respect to the crystal yields an EPR spectrum that is a superposition of the spectra of the individual molecules, which will have different resonance fields corresponding to their orientations. The number of distinct resonance fields depends on the space group and the number of molecules in (the asymmetric unit of) the unit cell.

A study on a single crystal of azurin serves as an illustration. This study was performed by electron spin-echo-detected EPR at 95 GHz and 1.2 K.[2] A complete orientation study was performed along the lines outlined above for a single mounting of a submillimeter crystal of the protein. It is impossible to accurately mount such a small and gel-like crystal in a predefined orientation, not to say doing this twice in a mutually well-defined way. The crystal was mounted in a capillary in an arbitrary orientation and a laboratory-fixed axes system was taken as reference. The capillary, which could be rotated around its axis, was in a horizontal plane in the probe head in the cryostat. The cryostat was in the room-temperature bore of a superconducting split-coil magnet. The magnetic field was in the horizontal plane and the magnet could be rotated around the vertical axis. By a combination of rotations of the capillary and the magnet, all mutual orientations of magnetic field and crystal could be realized. The complete orientation study yielded the principal g values and the direction in the reference axes system of the principal x- and z-axes (and thereby the y-axes) of the **g**-tensors of all 16 molecules in the unit cell. The crystal belongs to the $P2_12_12_1$ space group with four azurin molecules in the asymmetric unit. The crystallographic axes are twofold rotation axes, and their directions were obtained from the observed symmetry in the orientations of the principal g-axes. Their labeling as a, b, or c was derived from a parallel X-ray diffraction study. The result is summarized in the Wulffnet projection in Figure 2.3. Subsequently, the directions of the g-axes with respect to a, b, and c were translated into directions with respect to the copper site of the azurin molecule on the basis of the structure of the unit cell as obtained by the X-ray diffraction study.

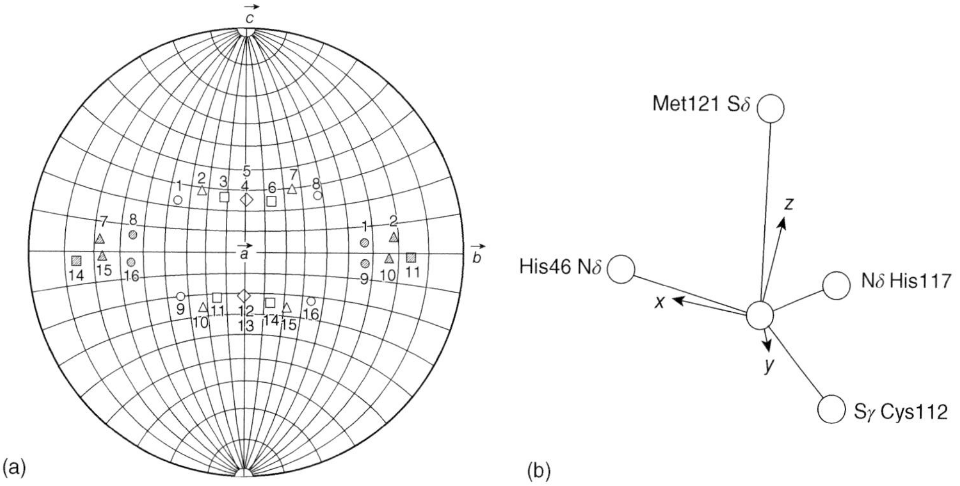

Figure 2.3. (a) Wulffnet projection of the orientation of the principal axes of the **g**-tensors of the azurin molecules in the $P2_12_12_1$ unit cell with respect to the crystallographic a, b, and c axes. The open symbols indicate the directions of the x axes, the shaded symbols those of the z axes. The different symbols refer to the four molecules in the asymmetric unit, and a total of 16 molecules in the unit cell. All points lie at the front side of the globe. The direction of the crystallographic axes has been derived from the direction of the **g**-tensor axes as obtained by EPR. (b) The direction of the **g**-tensor axes in the copper site. The copper is bound to two histidine nitrogens, a cysteine sulfur, and a methionine sulfur

2.4 *g*-ANISOTROPY IN THE LIQUID STATE

So far we considered a paramagnetic molecule in a solid, be it a frozen solution, a powder, or a single crystal. For those samples, the molecule has a fixed orientation with respect to the external magnetic field. In solution on the other hand, the molecule tumbles, and the EPR spectrum shows effects of orientation averaging of the *g*-anisotropy. The extent of averaging depends on the rate of reorientation of the molecule as compared to the time scale of the observation. If we look fast enough, i.e., at high enough microwave frequencies, the molecules seem frozen and the spectrum still looks like from a solid. If the molecules tumble fast enough, i.e., at low enough microwave frequencies, the anisotropy will be completely averaged out and the resonance will be observed at the magnetic field corresponding to $g_{\mathrm{iso}} = \frac{1}{3}\mathrm{tr}(\boldsymbol{g})$, where tr() indicates the trace (sum of the diagonal elements) of the matrix. The intermediate situation is most interesting, and what the EPR spectrum looks like at a particular microwave frequency depends on the rotational dynamics of the molecule in the solution. In terms of the rotational correlation time τ_c, slow motion corresponds to $\tau_c > 1/\Delta\nu$ and fast motion to $\tau_c < 1/\Delta\nu$, where $\Delta\nu = \Delta g \mu_B B/h$ is the total width of the spectrum in frequency units.

The effect of rotational averaging is illustrated in Figure 2.4, which shows the EPR spectrum at 275 GHz of the nitroxide radical TEMPONE in a water/glycerol mixture (50/50% by volume) as a function of temperature.[6] At −30 °C, the spectrum corresponds to that of a frozen solution. Upon temperature increase, tumbling of the nitroxide induces (partial) averaging of the *g*-anisotropy. The spectrum gradually narrows, and at 30 °C, the spectrum corresponds to that of a free rotating molecule, with a transition at g_{iso} split by the isotropic nitrogen hyperfine interaction.

In first approximation, the rate of tumbling is determined by the hydrodynamic volume of the paramagnetic molecule and the viscosity and temperature of the solution. According to the Stokes–Einstein equation, the rotational correlation time can be written as

$$\tau_c = \frac{4\pi\eta R^3}{3k_B T} \qquad (2.12)$$

where η represents the viscosity of the solution, R the hydrodynamic radius, k_B the Boltzmann constant, and T the absolute temperature.

A more sophisticated description requires the consideration of the rotational anisotropy of the molecule or molecular fragment that carries the electron spin and the local viscosity.

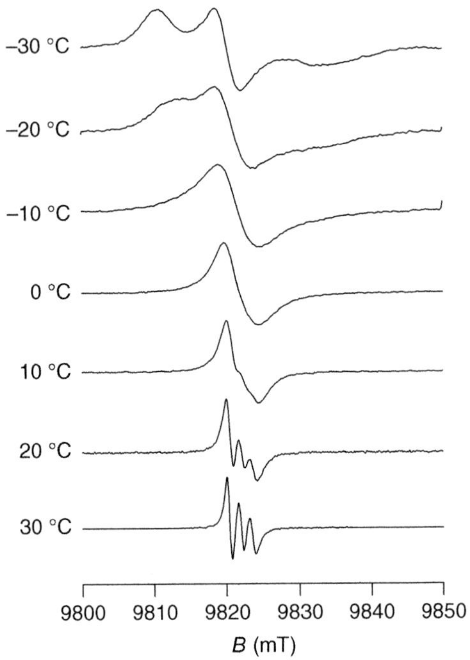

Figure 2.4. Continuous-wave EPR spectra at 275 GHz of TEMPONE in H_2O/glycerol (50/50% v/v) as a function of temperature. (Reprinted with permission from M. Azarkh and E.J.J. Groenen, J. Phys. Chem. B, 2015, 119, 13416. © 2015, American Chemical Society)

The connection between local rotational mobility and the manifestation of *g*-anisotropy implies that EPR spectroscopy provides a way to study rotational dynamics. As will be clear from the explanation in the previous paragraph, this research largely benefits from multifrequency EPR.

The simulation of the EPR spectrum of a model radical, only taking into account the Zeeman interaction in terms of *g*-anisotropy, as a function of the rotation–correlation time at 36 and 275 GHz is shown in Figure 2.5. It illustrates the sensitivity of the EPR spectrum to the microwave frequency. At 275 GHz, the range of τ_c from 0.1 to 8 ns covers more or less the slow-motion regime. At 36 GHz, significant broadening starts only for τ_c above 1 ns.

A recent application of multifrequency EPR concerned the study of the dynamics of spin-labeled T4 lysozyme.[8] A wide range of frequencies (9, 95, 170, and 240 GHz) and temperatures (2–32 °C) was applied, and the effects of faster internal modes of

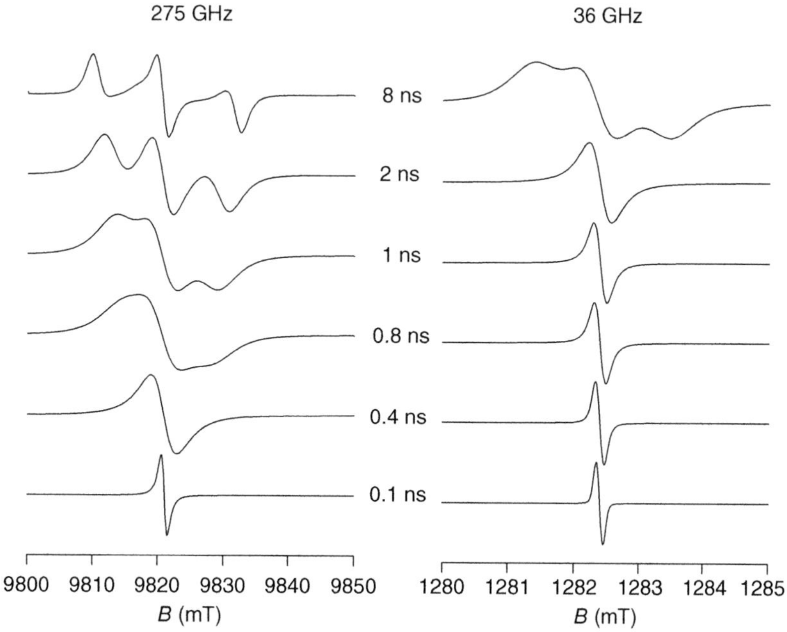

Figure 2.5. Simulations of the EPR spectrum of an $S = 1/2$ radical in the slow-motion regime at 275 and 36 GHz using Easyspin.[7] Conditions $g_x = 2.0083$, $g_y = 2.0058$, $g_z = 2.0030$, intrinsic linewidth = 0.1 mT; values of the rotation–correlation time, τ_c, are given in the figure

motion were separated from slower overall motions of the protein. Representative EPR spectra are shown in Figure 2.6.

2.5 FICTITIOUS SPIN 1/2

For systems with $S > 1/2$, spin–spin and spin–orbit interactions will lift the degeneracy of magnetic sublevels already in the absence of an external magnetic field. The consequence of this so-called zero-field splitting (ZFS) for the EPR spectrum will be described in Chapter 3. Here, we consider only briefly a special case.

For systems with an odd number of spins, time-reversal symmetry dictates the conservation of twofold degeneracy of magnetic sublevels in the absence of a magnetic field, whatever the nature and magnitude of the intrinsic interactions.[9] For $S > 1/2$, these Kramers doublets are separated by the ZFS. In case the ZFS is much larger than the electronic Zeeman interaction with the external magnetic field, each doublet can be considered an effective two-level system and modeled as a fictitious spin 1/2 system. The matrix representation of the Hamiltonian $\hat{H}_Z$ in the basis of zero-field eigenstates of this fictitious spin results in a two-dimensional Hermitian matrix. This matrix, like any two-by-two traceless Hermitian matrix, can be written as a linear combination of the Pauli matrices. The real coefficients depend on the ZFS. The result resembles equation (2.7) with effective g values, which contain the parameters D and E that represent the ZFS (cf. equations 2.14 to 2.16). Consequently, the resonance condition is described by equation (2.9) like for a normal $S = 1/2$ system, be it in terms of an effective $\boldsymbol{g}$-tensor, $\boldsymbol{g}_{\mathrm{eff}}$. The components of $\boldsymbol{g}_{\mathrm{eff}}$ can adopt values largely different from g_e.

As an example,[10] consider the high-spin cobalt(II) complex $Co[(SPPh_2)(SP^iPr_2)N]_2(Co^{iPrPh}L_2)$, in which the bidentate disulfidoimidodiphosphinato ligands make up for a pseudotetrahedral sulfur coordination of the transition metal. For the $(3d)^7$ configuration of the cobalt ion, high spin means $S = 3/2$. In zero field, the four magnetic sublevels give rise to two Kramers doublets,

$$\cos\alpha\left|\frac{3}{2},\pm\frac{3}{2}\right\rangle + \sin\alpha\left|\frac{3}{2},\mp\frac{1}{2}\right\rangle \qquad (2.13a)$$

$$\cos\alpha\left|\frac{3}{2},\pm\frac{1}{2}\right\rangle - \sin\alpha\left|\frac{3}{2},\mp\frac{3}{2}\right\rangle \qquad (2.13b)$$

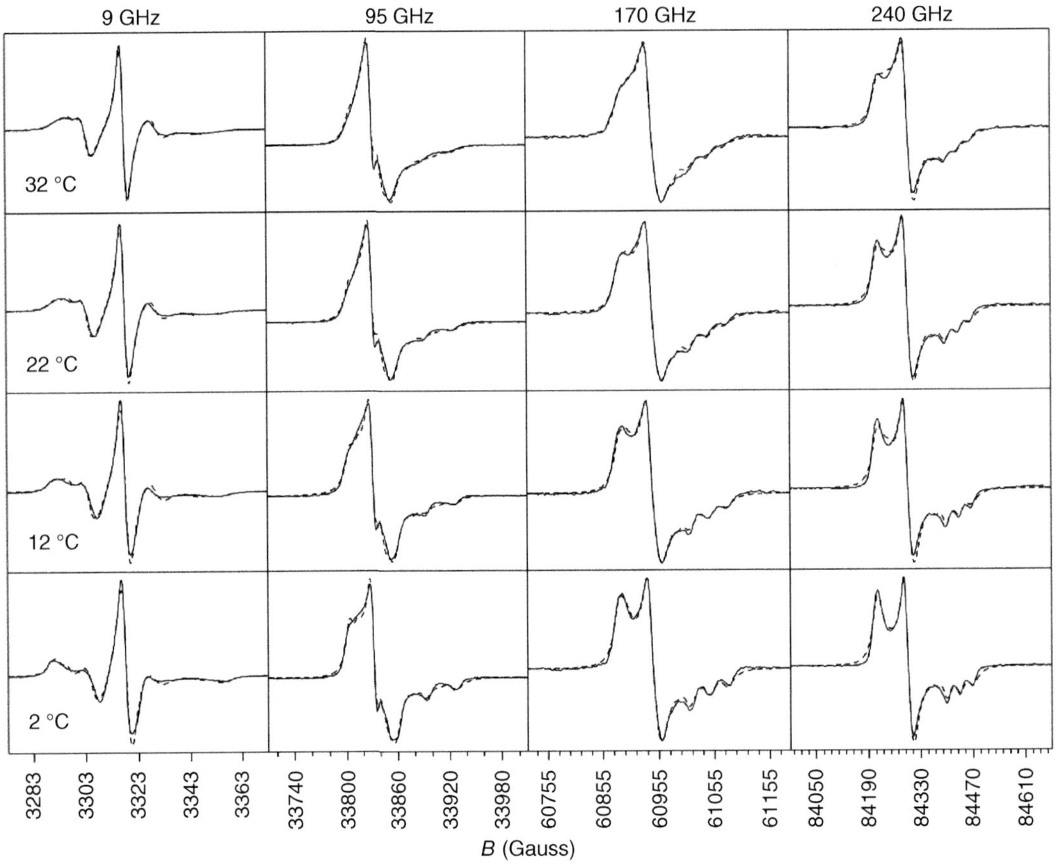

Figure 2.6. Multifrequency EPR spectra of 131R2-labeled T4 Lysozyme recorded in 25 w/v% Ficoll solution at 2, 12, 22, and 32 °C; experimental data (solid lines), simulations (dashed lines). (Reprinted with permission from Z. Zhang, M.R. Fleissner, D.S. Tipikin, Z. Liang, J.K. Moscicki, K.A. Earle, W.L. Hubbell, J.H. Freed, J. Phys. Chem. B., 2010, 114, 5503. © 2010, American Chemical Society)

where the angle α depends on the ZFS:

$$\tan 2\alpha = \sqrt{3}E/D \tag{2.14}$$

Although m_s is not a good quantum number, the doublets are commonly referred to as ±3/2 and ±1/2, respectively. The $\boldsymbol{g}_{\text{eff}}$-tensor becomes

$$\boldsymbol{g}_{\text{eff},ij} = \boldsymbol{g}_{ij}\omega_j \tag{2.15}$$

with

$$\omega_x = (2\sqrt{3}\cos\alpha\sin\alpha + 2\sin^2\alpha) \tag{2.16a}$$

$$\omega_y = (2\sqrt{3}\cos\alpha\sin\alpha - 2\sin^2\alpha) \tag{2.16b}$$

$$\omega_z = (3\cos^2\alpha - \sin^2\alpha) \tag{2.16c}$$

for the ±3/2 doublet. The corresponding expressions for the ±1/2 doublet are obtained by replacing $\cos\alpha$ by $-\sin\alpha$ and $\sin\alpha$ by $\cos\alpha$. Here, *x*, *y*, and *z* refer to the principal axes of the zero-field splitting tensor, which need not be parallel to the principal axes of the *g*-tensor.

The complete effective *g*-tensor for $Co^{iPrPh}L_2$ has been determined for the doublet of lowest energy from a study of the X-band continuous-wave EPR spectrum at 5 K of a 1% $Co^{iPrPh}L_2/Zn^{iPrPh}L_2$ single crystal as a function of the orientation of the magnetic field with respect to the crystal. The analysis led to the principal values $g_{\text{eff},x} = 1.62$, $g_{\text{eff},y} = 2.38$, and $g_{\text{eff},z} = 6.44$. The doublet of lowest energy can be either the ±1/2 or the ±3/2 doublet. In this particular case, the EPR

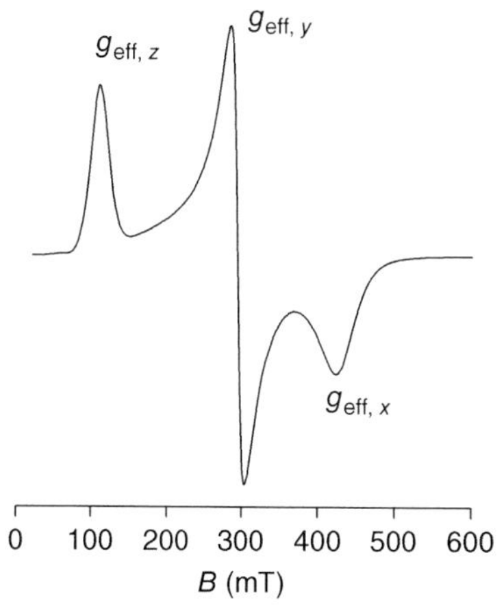

Figure 2.7. Continuous-wave EPR spectrum of a powder sample of $Co^{iPrPh}L_2$ at X-band and 8 K

experiment cannot discriminate between these assignments, because the ZFS is rhombic, i.e., the absolute value of $E/D = 1/3$. The powder spectrum in Figure 2.7 is consistent with these values and indeed shows the characteristic shape of an EPR spectrum for a spin $^1/_2$ system, be it with g values largely different from g_e. At higher magnetic fields, the electron Zeeman interaction need no longer be much smaller than the ZFS. Deviations from the $S = {}^1/_2$ picture may become noticeable. The g_{eff} values become frequency dependent, which was observed for $Co^{iPrPh}L_2$ at 94 and 275 GHz.

2.6 THE ORIGIN OF THE *g*-ANISOTROPY

The anisotropy of the $\boldsymbol{g}$-tensor derives from the fact that the electron is not free, but confined in an electrostatic spatial environment. Besides its spin angular momentum, the electron carries orbital angular momentum $\boldsymbol{L}$. The magnetic moment that couples to the magnetic field is the sum of two terms

$$\boldsymbol{\mu} = -\mu_B(\boldsymbol{L} + g_e\boldsymbol{S}) \qquad (2.17)$$

In addition, we have to take into account the coupling of the spin and orbital angular momenta

$$\hat{H}_{SO} = \lambda\hat{\boldsymbol{L}}\hat{\boldsymbol{S}} \qquad (2.18)$$

where λ represents the spin–orbit-coupling constant, which is characteristic for a particular L-state of an atom. Spin–orbit coupling is a relativistic effect. From the point of view of the electron in an atom, the positively charged nucleus moves around the electron. This results in an additional local magnetic field at the position of the electron, which interacts with its spin magnetic moment. This interaction is inversely proportional to the third power of the distance of the electron to the nucleus, which makes spin–orbit coupling a highly localized effect. Consequently, spin–orbit coupling in a molecule to very good approximation can be taken as the sum of contributions of individual nuclei.[11]

First consider an atom in an orbitally nondegenerate spin-$^1/_2$ ground state. In order to find the expression for the $\boldsymbol{g}$-tensor, we follow Pryce[12] and apply perturbation theory to find the spin operator whose eigenvalues describe the energies of the magnetic sublevels correct to second order. The zero-order states are the eigenstates of the Hamiltonian including the kinetic energy and the electrostatic potential energy. The perturbation involves the magnetic interactions

$$\hat{H}' = \mu_B\boldsymbol{B}(\hat{\boldsymbol{L}} + g_e\hat{\boldsymbol{S}}) + \lambda\hat{\boldsymbol{L}}\hat{\boldsymbol{S}} \qquad (2.19)$$

The description is of course limited to cases where the energy difference between the zero-order states is large compared to the energy shifts resulting from the perturbation. Only the second term of $\hat{H}'$,

$$\mu_B g_e\boldsymbol{B}\hat{\boldsymbol{S}} \qquad (2.20)$$

contributes to the first-order correction to the energy, because the diagonal matrix elements of $\hat{\boldsymbol{L}}$ vanish. The orbital angular momentum is quenched for the orbitally nondegenerate ground state. According to Pryce, the second-order correction to the energy contains a term linear in $\boldsymbol{B}$ and $\hat{\boldsymbol{S}}$, which derives from cross terms of the first and third terms in equation (2.19). It concerns the eigenvalue of an operator that can be expressed as

$$-2\mu_B\lambda\boldsymbol{B}\boldsymbol{\Lambda}\hat{\boldsymbol{S}} \qquad (2.21)$$

where the tensor $\boldsymbol{\Lambda}$ is defined by the elements

$$\Lambda_{ij} = \sum_{n\neq0}\frac{\langle\phi_0|\hat{L}_i|\phi_n\rangle\langle\phi_n|\hat{L}_j|\phi_0\rangle}{E_n - E_0}\quad(i,j = x,y,z) \qquad (2.22)$$

where ϕ_0 and ϕ_n are the atomic orbitals that are singly occupied in the ground and the n-th excited states. The combination of equations (2.20) and (2.21) results in the spin Hamiltonian

$$\hat{H}_Z = \mu_B \boldsymbol{B}(g_e 1 - 2\lambda \boldsymbol{\Lambda})\hat{\boldsymbol{S}} \tag{2.23}$$

and comparison with equation (2.3) yields

$$\boldsymbol{g} = g_e 1 - 2\lambda \boldsymbol{\Lambda} \tag{2.24}$$

For a molecule, where the electron spin is delocalized over several atoms that contribute to the spin–orbit coupling, the definition of the reference point for the orbital angular momentum becomes relevant. This problem of the gauge invariance of the $\boldsymbol{g}$-tensor of a molecule has been treated by Stone.[13] To a good approximation, the diagonal elements of the $\boldsymbol{g}$-tensor can be written as

$$g_{ii} = g_e - 2\sum_{n\neq 0}\sum_{\mu,\nu} \frac{\langle \psi_0|\lambda^\mu \hat{L}_i^\mu|\psi_n\rangle\langle\psi_n|\hat{L}_i^\nu|\psi_0\rangle}{\varepsilon_n - \varepsilon_0} \tag{2.25}$$

where ψ_0 and ψ_n are the molecular orbitals that are singly occupied in the ground and excited states and ε_0 and ε_n the corresponding orbital energies. In order to apply equation (2.25), the molecular orbitals have to be expanded in terms of atomic orbitals, and the superscripts μ and ν refer to the atoms. In calculating the matrix elements of the components of the orbital angular momentum operator, the overlap between atomic orbitals on different nuclei has to be set to zero, which amounts to neglecting the spin-other-orbit contribution, i.e., the angular momentum of an electron on one atom about the nucleus of another. The λ^μ in equation (2.25) represents the spin–orbit-coupling constant for the atomic orbital on atom μ. In practice, the orbital energies are commonly taken as the energies E_0 and E_n of the ground and excited states.

To illustrate the application of equation (2.25) in the calculations for molecules, we consider the examples introduced above, a nitroxide and azurin, based on a simplified view of their electronic structures. This approximate description provides insight into the origin of the deviation of their g values from g_e.

For TEMPOL, the nitroxide group (Figure 2.8) carries the unpaired electron. In the ground state, this electron occupies the antibonding π^* molecular orbital, which to good approximation is delocalized over the $2p_z$ atomic orbitals of the nitrogen and oxygen.[14] According to equation (2.25), excited states at low excitation energy will contribute mostly. Here, only the lowest excited state is taken into account, an $n\pi^*$ state, which corresponds to the promotion of an electron from a lone-pair orbital on oxygen (a linear combination of the 2s, $2p_x$, and $2p_y$ atomic orbitals) to the π^* molecular orbital. In this approximation, the elements of the $\boldsymbol{g}$-tensor become

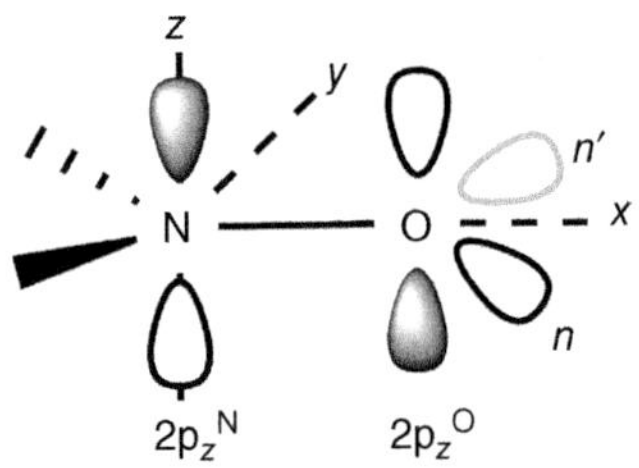

Figure 2.8. Schematic representation of the atomic orbitals involved in the singly occupied molecular orbitals for the NO bond in the ground state and the excited state: the $2p_z$ orbitals on nitrogen and oxygen and the lone-pair orbitals n and n' on oxygen

$$g_x \approx g_e + 2\lambda^O \rho_\pi^O c_{ny}^2/(E_{n\pi^*} - E_0) \tag{2.26a}$$

$$g_y \approx g_e + 2\lambda^O \rho_\pi^O c_{nx}^2/(E_{n\pi^*} - E_0) \tag{2.26b}$$

$$g_z \approx g_e \tag{2.26c}$$

where λ^O is the spin–orbit-coupling constant of 2p electrons on oxygen (151 cm^{-1}),[11] $\rho_\pi{}^O$ the spin density on oxygen in the π^* orbital, and c_{ny}, c_{nx} the coefficients of the $2p_y$, $2p_x$ orbitals in the lone-pair orbital on oxygen. The result derives from the fact that

$$\hat{L}_x(2p_y) = i(2p_z) \tag{2.27}$$

and similar relations are obtained by cyclic permutation of x, y, and z. In addition, the overlap between the $2p_z$ atomic orbitals on nitrogen and oxygen has been neglected, which means that only matrix elements of components of the orbital angular momentum operator between atomic orbitals on oxygen contribute.

With the simplest approximation of the molecular orbitals ($c_{ny} = {}^1\!/\!_2\sqrt{3}$, $c_{nx} = {}^1\!/\!_2$, $\rho_\pi^O = {}^1\!/\!_2$) and $E_{n\pi^*} - E_0 = 2.7$ eV,[14] equation (2.25) leads to $g_x = 2.0075$, $g_y = 2.0040$, and $g_z = 2.0023$. Comparison with the experimental values ($g_x = 2.0090$, $g_y = 2.0061$, and $g_z = 2.0021$)[15] shows that the simplified treatment correctly predicts that relative to the free-electron value (i) g_z hardly deviates, (ii) g_x and g_y increase, and (iii) g_x deviates most. The simple model even correctly predicts that the value of g_x is most sensitive to hydrogen bonding of the nitroxide group. The decrease of c_{ny}^2 upon hydrogen bonding, which

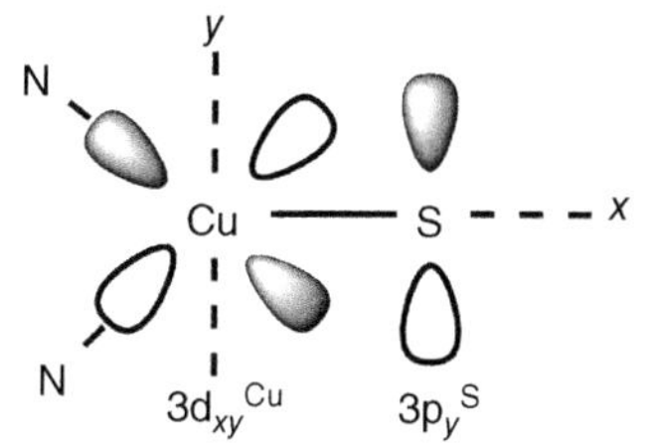

Figure 2.9. Schematic representation of the atomic orbitals involved in the singly occupied molecular orbital for the CuS bond in the ground state: the $3d_{xy}$ orbital on copper and the $3p_y$ orbital on sulfur. For azurin, the copper is less than 0.1 Å above the NNS (xy)-plane, and the drawing is a projection on this plane

arises from the delocalization of the lone-pair electrons into the H_2O orbitals, acts in the same direction as the concomitant decrease of ρ_π^O and the increase of $(E_{n\pi^*} - E_0)$.[14]

For azurin (Figure 2.9), the paramagnetic center corresponds to copper bound to five ligands in a geometry best described as a trigonal bipyramid.[16,17] The copper is approximately in the plane of the coordinating atoms of the three strongly bound ligands, the N_δ nitrogens of two histidines and the S_γ sulfur of the cysteine. In view of the difference in spin–orbit-coupling constants ($\lambda^N(2p) = 76$ cm^{-1}, $\lambda^S(3p) = 382$ cm^{-1}, and $\lambda^{Cu}(3d) = 829$ cm^{-1}),[11] it suffices to include only copper and sulfur to qualitatively explain the deviation of the **g**-tensor from the free-electron value. In the ground state, the $(3d)^9$ configuration of Cu(II) corresponds in zeroth order to an unpaired electron in the π-antibonding molecular orbital, a linear combination of the $3d_{xy}$ atomic orbital on copper and the $3p_y$ atomic orbital on sulfur.[18] The dominant spin–orbit-coupling-based mixing of states derives from the first excited state, which is much closer in energy to the ground state than the other excited states.[19] In this state, the unpaired electron occupies the molecular orbital that is dominated by the σ-bonding combination of the $3d_{x^2-y^2}$ atomic orbital on copper and the $3p_x$ atomic orbital on sulfur.

From equation (2.25), the elements of the **g**-tensor become

$$g_x \approx g_e \tag{2.28a}$$

$$g_y \approx g_e \tag{2.28b}$$

$$g_z \approx g_e - 2(\lambda^S c_x^* c_y - 2\lambda^{Cu} c_{xy} c_{x^2-y^2}^*) \times (c_x^* c_y - 2c_{xy} c_{x^2-y^2}^*)/(E_1 - E_0) \tag{2.28c}$$

where c_x and c_{xy} refer to the coefficients of the $3p_y$ and $3d_{xy}$ atomic orbitals of sulfur and copper in the molecular orbital that is singly occupied in the ground state, and c_x^* and $c_{x^2-y^2}^*$ to the coefficients of the $3p_x$ and $3d_{x^2-y^2}$ atomic orbitals of sulfur and copper in the molecular orbital that is singly occupied in the first excited state. The energies E_0 and E_1 refer to the ground state and the first excited state.

At this level of approximation, only g_z is predicted to deviate significantly from g_e, which agrees with the experimental observation (Figure 2.1) that the deviation for g_x (0.0370) and g_y (0.0545) is much less than for g_z (0.2707). In addition, comparison of the expression for g_z and its experimental value points to a significant spin density on the cysteine sulfur, which nicely illustrates the large covalence of the copper-sulfur bond. In order to explain the deviation of g_x and g_y from g_e, the $3d_{yz}$ and $3d_{zx}$ atomic orbitals on copper have to be included.[20] The small nonaxiality of the **g**-tensor largely derives from spin–orbit coupling on sulfur.[18,19]

Both for the nitroxide and for the copper site of azurin, the perturbation method in combination with the simplest description of the electronic structure elucidates the source of the anisotropy of the **g**-tensor. This approach is clearly not sufficient for a quantitative interpretation, which requires a full quantum-chemical treatment. The calculation of the **g**-tensor to high precision is a real challenge, but worth the effort because this tensor is a very sensitive probe of the electronic structure of a paramagnetic molecule or center.

Inspired by the advances in experimental EPR spectroscopy, the development of pulsed excitation besides continuous-wave excitation and both at multiple microwave frequencies, the past 25 years have shown large efforts in the quantum-chemical calculation of spin Hamiltonian parameters. Following initial work based on semiempirical methods, density functional theory-based approaches are applied with great success presently. The systematic further development and applicability of ab initio quantum chemistry will determine the future progress in the understanding of spin Hamiltonian parameters. For a description of the present status, we refer to Chapter 7.

REFERENCES

1. R. S. van Dyck, P. B. Schwinberg, and H. G. Dehmelt, *Phys. Rev. D*, 1986, **34**, 722. http://physics.nist.gov/constants

2. J. W. A. Coremans, O. G. Poluektov, E. J. J. Groenen, G. W. Canters, H. Nar, and A. Messerschmidt, *J. Am. Chem. Soc.*, 1994, **116**, 3097.
3. W. R. Hagen, *J. Magn. Reson.*, 1981, **44**, 447.
4. J. Fritz, R. Anderson, J. Fee, G. Palmer, R. H. Sands, J. C. M. Tsibris, I. C. Gunsalus, W. H. Orme-Johnson, and H. Beinert, *Biochim. Biophys. Acta*, 1971, **253**, 110.
5. K. Möbius and A. Savitsky, *High-Field EPR Spectroscopy on Proteins and their Model Systems*, The Royal Society of Chemistry: Cambridge, 2009, Ch. 5.
6. M. Azarkh and E. J. J. Groenen, *J. Phys. Chem. B*, 2015, **119**, 13416.
7. S. Stoll and A. Schweiger, *J. Magn. Res.*, 2006, **178**, 42.
8. Z. Zhang, M. R. Fleissner, D. S. Tipikin, Z. Liang, J. K. Moscicki, K. A. Earle, W. L. Hubbell, and J. H. Freed, *J. Phys. Chem. B.*, 2010, **114**, 5503.
9. A. Abragam and B. Bleaney, *Electron Paramagnetic Resonance of Transition Ions*, Clarendon Press: Oxford, 1970, Ch. 15.
10. D. Maganas, S. Milikisyants, J. M. A. Rijnbeek, S. Sottini, N. Levesanos, P. Kyritsis, and E. J. J. Groenen, *Inorg. Chem.*, 2010, **49**, 595.
11. A. Carrington and A. D. Mclachlan, *Introduction to Magnetic Resonance*, New York: Harper & Row, 1967.
12. M. H. L. Pryce, *Proc. Phys. Soc. London A.*, 1950, **63**, 25.
13. A. J. Stone, *Proc. Roy. Soc. London. A.*, 1963, **271**, 424.
14. M. Plato, H. J. Steinhoff, C. Wegener, J. T. Torring, A. Savitsky, and K. Möbius, *Mol. Phys.*, 2002, **100**, 3711.
15. O. Y. Grinberg, A. A. Dubinskii, and Y. S. Lebedev, *Uspekhi Khimii*, 1983, **52**, 1490.
16. E. N. Baker, *J. Mol. Biol.*, 1988, **203**, 1071.
17. H. Nar, A. Messerschmidt, R. Huber, M. van de Kamp, and G. W. Canters, *J. Mol. Biol.*, 1991, **221**, 765.
18. K. W. Penfield, A. A. Gewirth, and E. I. Solomon, *J. Am. Chem. Soc.*, 1985, **107**, 4519.
19. M. van Gastel, J. W. A. Coremans, H. Sommerdijk, M. C. van Hemert, and E. J. J. Groenen, *J. Am. Chem. Soc.*, 2002, **124**, 2035.
20. M. van Gastel, G. W. Canters, H. Krupka, A. Messerschmidt, E. C. de Waal, G. C. M. Warmerdam, and E. J. J. Groenen, *J. Am. Chem. Soc.*, 2000, **122**, 2322.

Chapter 3
EPR Interactions – Zero-field Splittings

Joshua Telser

Roosevelt University, Chicago, IL, USA

3.1 INTRODUCTION

One fundamental feature of magnetic resonance spectroscopy, such as in EPR, is the application of an external magnetic field, B_0. The concept, then, of zero-field splitting (ZFS) would seem out of place in a volume devoted to EPR spectroscopy. This is not the case, as the phenomenon of ZFS, when it is present, profoundly affects a given EPR spectrum. When is ZFS present? This phenomenon manifests itself when there are multiple unpaired electrons in a system of interest. A single unpaired electron gives rise to a spin doublet ground state, $S = 1/2$ (and potential investigation by EPR); two unpaired electrons give rise to spin triplet ground state, $S = 1$; three gives rise to a spin quartet, $S = 3/2$; and so forth. This chapter will consider systems with multiple unpaired electrons (i.e., $S > 1/2$) only when this situation results from the electrons being on the same paramagnetic center. Typical examples of this situation are organic triplets (whether in their ground or excited states; the term 'organic' could apply to p block radicals in general[1]) or 'high-spin' d or f block ions – the term 'high-spin' in the present context simply being $S > 1/2$. The phenomenon of ZFS in d block complexes has been reviewed in the context primarily of magnetometry[2,3] and in terms of EPR,[4,5] particularly at high frequency and high field.[6–8] Classic examples of these are complexes of half-integer spin (Kramers) ions such as Cr(III) ($3d^3$, $S = 3/2$) or Mn(II) ($3d^5$, $S = 5/2$) and integer-spin (non-Kramers) ions such as Cr(II) ($3d^4$, $S = 2$) or Fe(II) ($3d^6$, $S = 2$). Complexes of paramagnetic f block ions, with the exception of Gd(III) ($4f^7$, $S = 7/2$), are not well modeled by the simple ZFS model described in this chapter. The interested reader is directed to Chapters 5 and 6 (also Chapter 18) of Anatole Abragam and Brebis Bleaney's monumental text.[9,10]

Another scenario by which ZFS can arise from multiple unpaired electrons is when these are located on separate atoms/ions, but they are magnetically interacting.[11] The classic example is copper acetate (formally tetra-μ^2-acetatodicopper(II)), wherein two Cu(II) ($3d^9$, $S = 1/2$) ions are antiferromagnetically coupled to give a spin singlet ground state and a thermally accessible spin triplet excited state, which exhibits ZFS, and can be investigated by both conventional and high-frequency and -field EPR (HFEPR) at temperatures at which this excited state is sufficiently

EPR Spectroscopy: Fundamentals and Methods.
Edited by Daniella Goldfarb and Stefan Stoll

ISBN: 978-1-119-16299-5
Also published in eMagRes (online edition)
DOI: 10.1002/9780470034590.emrstm1501

populated.[12] Such systems are treated in Chapter 4 and will not be discussed further here; however, the effect of ZFS on an EPR spectrum of a coupled (high-)spin system is essentially the same as described in this chapter.

The statement that ZFS arises from multiple unpaired electrons gives no hint as to its physical origin. In brief, ZFS arises from two interactions: spin–orbit coupling (SOC), which is the interaction of (unpaired) electron spin angular momentum with its electron orbital angular momentum, and spin–spin coupling (SSC), which is the direct, dipolar interaction among pairs of unpaired electrons. In organic biradicals, SSC is the dominant contribution to ZFS; in d block (transition metal) ion complexes, the primary focus of this chapter, SOC is usually dominant. Clearly, there must be at least two unpaired electrons to have SSC, and thus ZFS, but SOC can (and does) occur with only a single unpaired electron. In this $S = 1/2$ case, the SOC interaction can lead only to g anisotropy, not to ZFS; see Chapter 2. All of the illustrative or pedagogical examples used in this chapter will employ an isotropic $g = 2.00$ ($\approx g_e = 2.0023$). With a single unpaired electron, there is only a single magnitude of the spin angular momentum projection: $m_S = \pm 1/2$; $|m_S| = 1/2$. However, in the case of at least two unpaired electrons, there is more than a single magnitude of the spin angular momentum projection, e.g., for the simplest case, $S = 1$: $m_S = 0, \pm 1$, $|m_S| = 0, 1$. These states with different $|m_S|$ values couple differently to the orbital angular momentum ($L > 0$) that may be present (both from ground and excited electronic states) so that the energies of the various $|m_S|$ levels become different. Thus, for $S = 1$, the energies in zero applied magnetic field (i.e., in the absence of any magnetic Zeeman interaction) of the $|m_S| = 0$ and $|m_S| = 1$ states are different, hence ZFS. When an external field is applied and EPR spectroscopy performed, the EPR spectra of systems in the presence and absence of ZFS still are drastically different.

ZFS, however, is not merely a complication or annoyance in recording and interpreting EPR spectra of high-spin systems. The ZFS parameters that can be extracted from such spectra provide valuable chemical information on structure and bonding within the complex of interest. This information is both qualitative and quantitative. Qualitatively, the magnitude of ZFS can be a reflection of the deviation from high (i.e., cubic) symmetry. Quantitatively, ZFS can be estimated using quantum chemical calculations. Indeed, the accurate and precise measurement of ZFS parameters of d block complexes, in particular by HFEPR,[6–8,13] (see Chapter 27) has gone hand in hand with the development of quantum chemical theory methods for calculation of these parameters and a detailed explanation as to their physical origin (see Chapter 7).[14–26]

3.2 EFFECTS OF ZFS ON THE EPR SPECTRUM

3.2.1 Spin Triplet, $S = 1$

3.2.1.1 Spin Triplet, S = 1, No ZFS

In the absence of any ZFS effects, whether from SSC or SOC, the EPR spectrum of a spin triplet is indistinguishable from that of a simple spin doublet. The two magnetic-dipole allowed transitions, $\Delta m_S = \pm 1$, are $|S, m_S\rangle = |1, 0\rangle \leftrightarrow |1, \pm 1\rangle$ and are degenerate. There is the possibility of a single-photon, double-quantum $\Delta m_S = \pm 2$ transition, but this is totally forbidden unless there is some ZFS. This transition will be described in detail in Section 3.2.1.3. There is also the possibility of a two-photon (“double-quantum”) transition, $|S, m_S\rangle = |1, \pm 1\rangle \leftrightarrow |1, \mp 1\rangle$, but this requires high power levels and would occur near the allowed, single-photon transitions and also requires some ZFS. That a simple EPR spectrum would occur from a spin triplet d block ion complex might seem unlikely, but this is indeed the case. A simple example is $[Ni(NH_3)_6]I_2$, an octahedral, homoleptic complex of Ni(II) ($3d^8$, $S = 1$). This complex has ideal octahedral symmetry about the Ni(II) ion. As a result, as shown in Figure 3.1, there is only a single EPR transition observed for the powdered complex, recorded using a routine EPR spectrometer, operating at X-band frequency and liquid-nitrogen temperature. The perfectly Lorentzian linewidth is quite broad, 800 MHz half-width at half maximum (hwhm), as expected given that this is not a magnetically dilute sample (see Chapter 9). That there is a perfectly homogenous lineshape suggests that ZFS contributions from both SSC and SOC are negligible in this system. Transition metal (d block) complexes in the context of ZFS will be discussed in Section 3.4.

The observed, isotropic g value is greater than g_e, as expected for a spin system that has more than half-filled d orbitals (i.e., d^m, with $m > 5$), and is the consequence of SOC. In this case, the SOC acts isotropically so there is no g anisotropy (see

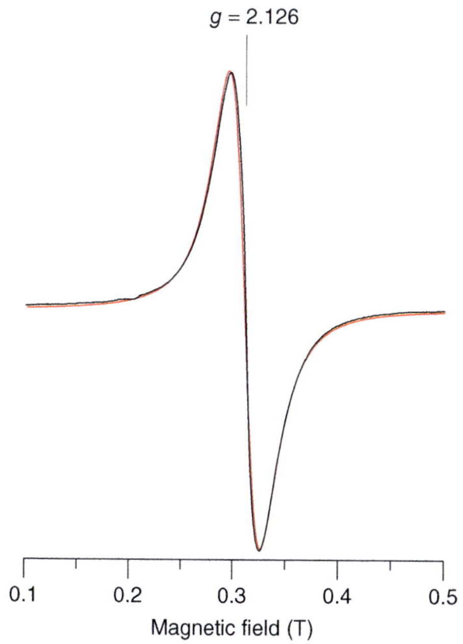

Figure 3.1. X-band EPR spectrum of a powder of $[Ni(NH_3)_6]I_2$ (black trace) recorded at 77 K, 9.235 GHz, with simulation (red trace) generated using $S' = 1/2$, $g = 2.126$, $W = 800$ MHz (Lorentzian, hwhm)

Chapter 2). More importantly, this simple spectrum challenges the widely held belief that integer spin (non-Kramers) systems are 'EPR silent'. As will be discussed in the following paragraphs, although ZFS can have a great effect on an EPR spectrum, if the magnitude of ZFS is within range of the microwave photon energy and/or magnetic fields available, then EPR spectra from integer spin systems with $S = 1, 2, 3, \ldots$ can be readily observed.

3.2.1.2 *Spin Triplet, S = 1, with ZFS, No External Magnetic Field*

A more typical situation for a spin triplet is one in which the SSC interaction is not insignificant and/or the SOC interaction is anisotropic. In this case, the spin state energy levels are nondegenerate. Before proceeding with a description of the EPR spectra of such systems, it is first necessary to define the spin Hamiltonian used to describe them. In addition to the electronic Zeeman term, a term describing the ZFS, often referred to as the fine structure, is also needed:

$$\hat{\mathcal{H}}_{\mathrm{ZFS}} = \hat{\boldsymbol{S}} \cdot \boldsymbol{D} \cdot \hat{\boldsymbol{S}} \tag{3.1}$$

This gives the following overall spin Hamiltonian with an isotropic g[27]:

$$\begin{aligned}\hat{\mathcal{H}} &= g\mu_{\mathrm{B}}\boldsymbol{B} \cdot \hat{\boldsymbol{S}} + \hat{\boldsymbol{S}} \cdot \boldsymbol{D} \cdot \hat{\boldsymbol{S}} \\ &= g\mu_{\mathrm{B}}(B_x\hat{S}_x + B_y\hat{S}_y + B_z\hat{S}_z) + (D_x\hat{S}_x^2 + D_y\hat{S}_y^2 + D_z\hat{S}_z^2)\end{aligned}$$

Remembering that $(\hat{S}_x{}^2 + \hat{S}_y{}^2 + \hat{S}_z{}^2)\psi = S(S+1)\psi$, it is easy to check that $D_x\hat{S}_x{}^2 + D_y\hat{S}_y{}^2 + D_z\hat{S}_z{}^2$ equals the following ($\hat{1}$ is the identity operator):

$$\begin{aligned}&\frac{D_x + D_y + D_z}{3}\{S(S+1)\hat{1}\} \\ &+ \frac{2D_z - D_x - D_y}{2}\left\{\hat{S}_z^2 - \frac{1}{3}S(S+1)\hat{1}\right\} \\ &+ \frac{D_x - D_y}{2}(\hat{S}_x^2 - \hat{S}_y^2)\end{aligned} \tag{3.2}$$

The first term causes no splitting of the m_S levels and may be dropped, leaving us with the commonly known ZFS spin Hamiltonian:

$$\hat{\mathcal{H}}_{\mathrm{ZFS}} = D\left\{\hat{S}_z^2 - \frac{1}{3}S(S+1)\hat{1}\right\} + E(\hat{S}_x^2 - \hat{S}_y^2) \tag{3.3}$$

where $D = \frac{2D_z-(D_x+D_y)}{2}$ and $E = \frac{D_x-D_y}{2}$. The 'real' ZFS tensor, $\boldsymbol{D}$, is not necessarily traceless, but a traceless tensor ($D_x + D_y + D_z = 0$) is always obtained from EPR. Adding a common constant of arbitrary magnitude to each of D_x, D_y, and D_z will not change D or E. A nontraceless ZFS tensor, for example, obtained from DFT calculations, can always be safely made traceless by subtracting $\frac{D_x+D_y+D_z}{3}$ from each of its diagonal components. For a traceless tensor, we have

$$D = \frac{3}{2}D_z, \quad E = \frac{1}{2}(D_x - D_y) \tag{3.4}$$

where D is the axial (sometimes called 'uniaxial') ZFS parameter and E the rhombic ZFS parameter. The commonly used limitation $|E| \le \frac{|D|}{3}$ is the consequence of a convention where x, y, and z are chosen such that $|D_z| > |D_y| > |D_x|$ can always be achieved by a choice of the coordinate system. It may thus sometimes be impossible to label the most readily distinguished g component as g_z while still obeying this convention (see Chapter 2). The overall situation is analogous to the nuclear quadrupole interaction (see Chapter 6), and indeed, the $\hat{\mathcal{H}}_{\mathrm{ZFS}}$ term is sometimes referred to as the electronic quadrupole interaction.

At zero field, the spin triplet eigenfunctions and eigenvalues (E_i) are very simple for an axial system:

$$|S, m_S\rangle = |1,0\rangle, |1,-1\rangle, |1,+1\rangle$$
$$E_i = -\frac{2}{3}D, \quad +\frac{1}{3}D, \quad +\frac{1}{3}D \qquad (3.5)$$

so that the energy separation between the nondegenerate levels is D. For a rhombic system (here using D, $E > 0$), the situation is more complicated, as the rhombic ZFS interaction, given by the operator, $\frac{1}{2}(\hat{S}_+^2 + \hat{S}_-^2)$, which is equivalent to $(\hat{S}_x^2 - \hat{S}_y^2)$, mixes the $|\pm 1\rangle$ and $|\mp 1\rangle$ states ($\Delta m_S = \pm 2$), so that the degeneracy is lifted and the three wavefunctions and energies are

$$|S, m_S\rangle = |1,0\rangle,$$
$$\frac{1}{\sqrt{2}}\{-|1,-1\rangle + |1,+1\rangle\},$$
$$\frac{1}{\sqrt{2}}\{|1,-1\rangle + |1,+1\rangle\}$$
$$E_i = -\frac{2}{3}D, \quad +\frac{1}{3}D - E, \quad +\frac{1}{3}D + E \qquad (3.6)$$

As a result, there is a separation of $2E$ within the upper doublet, the center of which is again at D from the ground singlet, i.e., giving relative energies with the ground state set to zero as 0, $D - E$, and $D + E$. This is still the case in the limit of $|E| = \frac{|D|}{3}$, except these energies then simplify to 0, $\frac{2}{3}D$, $\frac{4}{3}D$. For overall reference, a composite figure of the zero-field energies of rhombic spin triplets through sextets is shown in Figure 3.2. The higher spin systems will be discussed in turn in the following section.

3.2.1.3 *Spin Triplet, S = 1, with ZFS, with External Magnetic Field*

If an external magnetic field is applied along the principal axis (z) of the ZFS tensor, then the eigenfunctions in the limit as $B_0 \to 0$ are as follows[27,28]:

$$|T_x\rangle = \frac{1}{\sqrt{2}}\{|-1\rangle - |+1\rangle\},$$
$$|T_y\rangle = \frac{i}{\sqrt{2}}\{|-1\rangle + |+1\rangle\}, \quad |T_z\rangle = |0\rangle \qquad (3.7)$$

Here, the symbol T for triplet is traditionally used and the quantum number $S = 1$ that appears in equation (3.6) is omitted.

An energy level diagram as a function of external field for a spin triplet is shown in Figure 3.3. This example uses a magnitude of ZFS ($D/hc = 0.15\ \text{cm}^{-1}$) that is suitable for measurement by EPR at conventional frequencies and fields (i.e., X-band: $\nu \approx 9$ GHz; $B_0 \leq 0.6$ T) and presents results for both strictly axial symmetry ($E = 0$) and a modest level of rhombicity: $|E|/hc = |D|/6hc = 0.025\ \text{cm}^{-1}$.

In the axial case, the energy levels for the external magnetic field along the ZFS z-axis (i.e., parallel)

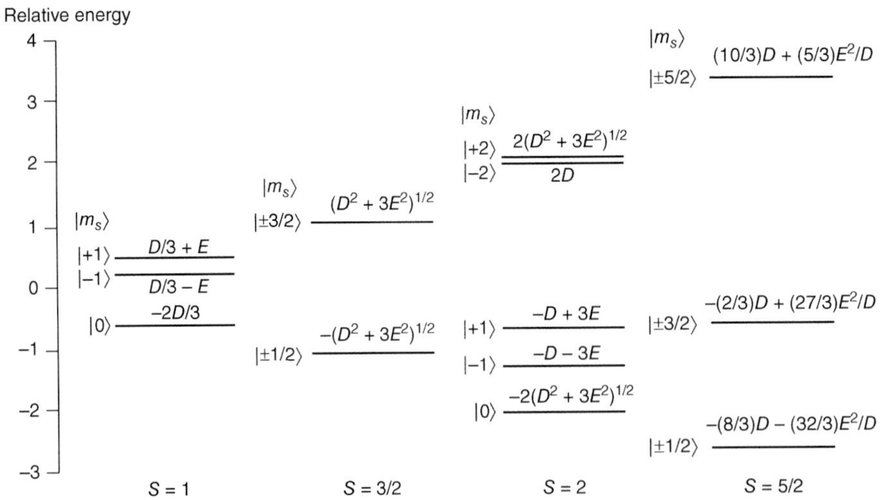

Figure 3.2. Relative energy levels in units of D of spin triplets through sextets as a result of rhombic ZFS ($|E/D| = 0.1$). The ordering is for positive D; for negative D, the order would be reversed. The energy values for triplets through quintets are exact; those for the sextet use second-order perturbation theory but are accurate to within a few percent of an exact calculation

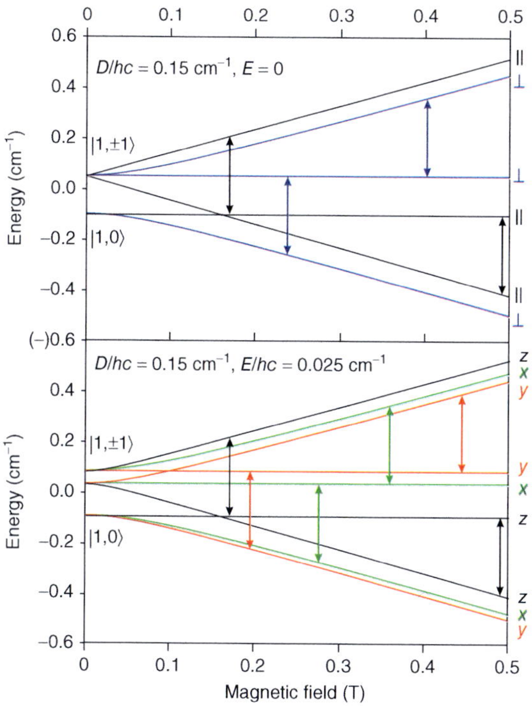

Figure 3.3. Diagrams of energy levels (eigenvalues) as a function of external magnetic field for a spin triplet with the following spin Hamiltonian parameters: $S=1$, $g_{iso}=2.00$, $D/hc=+0.15\,cm^{-1}$, upper plot: $E=0$; lower plot: $E/hc=+0.025\,cm^{-1}$. Black lines indicate energies with the magnetic field along the ZFS z-axis (parallel) in both plots; in the upper plot, blue traces indicate energies with the magnetic field along the ZFS x-axis (= y-axis; perpendicular); in the lower plot, green traces are for the field along the ZFS x-axis and red traces for the field along the ZFS y-axis. Arrows indicate allowed EPR transitions with a microwave quantum of 9.3 GHz ($0.31\,cm^{-1}$), with their colors corresponding to the transition orientation

are very simple, as the spin Hamiltonian matrix is diagonal[28]:

$$E_{+1} = +\frac{1}{3}D + g_{||}\mu_B B; E_{-1} = +\frac{1}{3}D - g_{||}\mu_B B;$$
$$E_0 = -\frac{2}{3}D \qquad (3.8)$$

Here, the energies can be directly correlated with the m_S quantum number. For the magnetic field along x (= y, i.e., perpendicular), the energy levels are the roots of a secular determinant, as follows[28]:

$$E_1 = \frac{1}{2}\left[-\frac{D}{3} + \sqrt{D^2 + (2g_\perp\mu_B B)^2}\right], \quad E_2 = +\frac{D}{3},$$
$$E_3 = \frac{1}{2}\left[-\frac{D}{3} - \sqrt{D^2 + (2g_\perp\mu_B B)^2}\right] \qquad (3.9)$$

Note that $g_{||}$ and $g_\perp$ (see Chapter 2) are used in equations (3.8) and (3.9) for generality (with the $\boldsymbol{g}$- and $\boldsymbol{D}$-tensors aligned).

With the application of an external magnetic field parallel to the $\boldsymbol{D}$-tensor principal z-axis, the energy levels for a rhombic system become[28]

$$E_1 = +\frac{1}{3}D - \sqrt{E^2 + (g_{||}\mu_B B)^2};$$
$$E_2 = +\frac{1}{3}D + \sqrt{E^2 + (g_{||}\mu_B B)^2}; \quad E_3 = -\frac{2}{3}D \qquad (3.10)$$

For the case of a perpendicular field (i.e., along the ZFS $x = y$-axis), the energies are[28]

$$E_1 = +\frac{D}{3} - E;$$
$$E_2 = \frac{1}{2}\left[-\frac{D}{3} + E + \sqrt{(D+E)^2 + (2g_\perp\mu_B B)^2}\right]$$
$$E_3 = \frac{1}{2}\left[-\frac{D}{3} + E - \sqrt{(D+E)^2 + (2g_\perp\mu_B B)^2}\right] \qquad (3.11)$$

The magnitude of ZFS used for illustration here is characteristic of what exists for organic triplet biradicals, which were extensively investigated most notably in the 1960s, using EPR at conventional fields and frequencies.[29] In addition to ground-state triplets, triplet excited states can also be generated (see Chapter 25). A specific, and classic, early example is naphthalene in its triplet excited (phosphorescent) state,[30] for which $D/hc = \pm 0.1003(6)\,cm^{-1}$ and $E/hc = \mp 0.0137(2)\,cm^{-1}$. HFEPR can be used to investigate triplets with much larger ZFS, as found in transition metal systems, of which V(III) ($3d^2$) and Ni(II) ($3d^8$) are the most common examples. An example of a less common transition metal ion that can have $S = 1$ is Fe(IV) ($3d^4$), for which $D/hc = +22$ and $+28\,cm^{-1}$, respectively, as found in an HFEPR study (see Chapter 27) of two tetragonal Fe(IV) complexes.[31]

Figure 3.3 presents the energy dependence of the m_S levels for a spin triplet as a function of magnetic field. It is also instructive to see how the resulting EPR transitions among these levels (arrows in Figure 3.3) behave as a continuous function of field orientation – at a fixed microwave frequency, for the

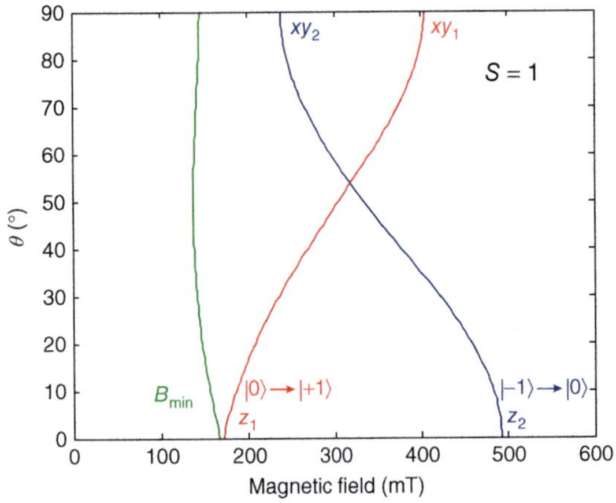

Figure 3.4. X-band (9.3 GHz) EPR transition energies calculated using EasySpin for a spin triplet with $g = 2.00$, $D/hc = 0.15\ \text{cm}^{-1}$ as a function of the angle θ between the external magnetic field and the ZFS tensor z-direction. The correspondence between transitions labels z_1 (red trace) and z_2 (blue trace) and $\Delta m_S = \pm 1$ transitions (where this is a good quantum number, i.e., at $\theta = 0°$) is indicated

major spin systems. For clarity, only axial systems are shown, so that only the angle θ (i.e., the angle between the external field, B_0, and the ZFS tensor z-axis) will be varied in the relevant range, which is $0 \le \theta \le 90°$. Figure 3.4 thus presents this behavior for a spin triplet.

Building on what was shown in Figures 3.3 and 3.4, we can finally turn to a powder pattern (i.e., random orientation) EPR spectrum. The simulated X-band (9.3 GHz; $\nu/c = 0.31\ \text{cm}^{-1}$) powder pattern spectrum for a spin triplet with parameters ($D/hc = +0.15\ \text{cm}^{-1}$, $E/hc = 0$ and $+0.01\ \text{cm}^{-1}$) roughly equivalent to those of an organic triplet such as naphthalene is shown in Figure 3.5. In this case, the spectrum of a hypothetical molecule with axial symmetry ($E = 0$) will be discussed first (red traces in Figure 3.5). What is readily apparent is the 'Pake pattern' shape of the signal, i.e., the pair of powder pattern envelopes with 'turning points' (i.e., derivative or absorption shaped features) corresponding to resonances of the molecules with their canonical $\boldsymbol{D}$-tensor axes being aligned with the external field. The transitions are labeled indicating the orientation of the transition for each of the patterns.[28] Next considering the rhombic case (green traces in Figure 3.5), it can be seen that the perpendicular turning points in the axial case split into the x- and y-directional resonances (x_1, y_1 and x_2, y_2), while the z-resonances are unaffected.

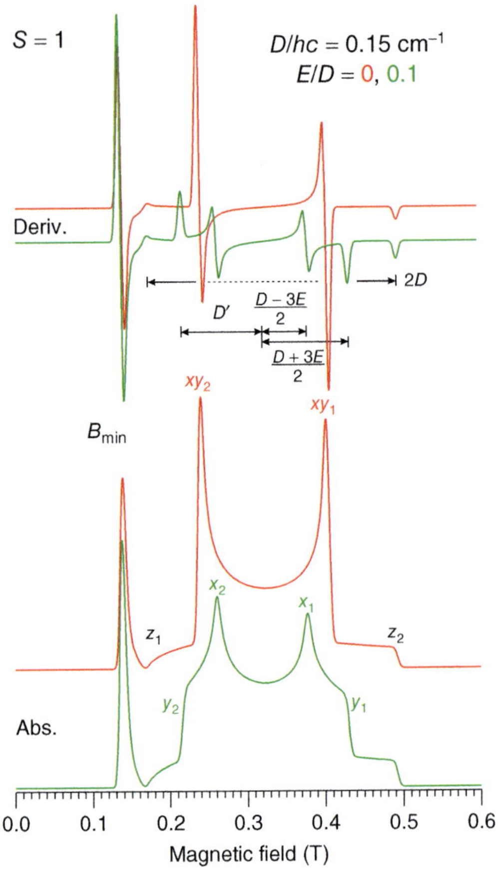

Figure 3.5. Simulated EPR spectra for a spin triplet with $g_{iso} = 2.00$, axial ZFS, $D/hc = +0.15\ \text{cm}^{-1}$; with rhombic splitting ($E/hc = +0.015\ \text{cm}^{-1}$; green traces) and without rhombic splitting (red traces). Single-crystal Gaussian linewidths (hwhm) of 100 MHz were used. A typical X-band microwave frequency of 9.3 GHz ($\approx 0.31\ \text{cm}^{-1}$) was used. Both an absorption presentation (lower traces) and a first derivative (upper traces) are shown. The turning points are labeled using standard nomenclature for triplets.[29] The relation between splittings of some of the turning points and ZFS parameters is also indicated. These splittings are shown in magnetic field quantities, not in energy, and thus need to be converted to energy quantities using $g\mu_B$ as described in the text

What is less expected is the additional, low-field resonance, traditionally known as B_{min} (or H_{min} in older, non-SI literature), although B_{dq} (dq = double-quantum) is more intuitive. This transition can in fact be dominant in triplet spectra of powders or frozen solutions, particularly when the ZFS is small relative to the microwave quantum, as is the case in Figure 3.5. Because B_{min} occurs at low field, field-dependent line broadening effects are minimal compared to the higher field single-quantum transitions. B_{min} corresponds to an off-axis turning point of $\Delta m_S = \pm 2$ (see equations (3.10) and (3.11)) and occurs at the following magnetic field[27]:

$$B_{min} = \frac{1}{2g\mu_B}\sqrt{(h\nu)^2 - 2(D_x{}^2 + D_y{}^2 + D_z{}^2)}$$
$$= \frac{1}{g\mu_B}\sqrt{\frac{(h\nu)^2}{4} - \frac{D^2 + 3E^2}{3}} \qquad (3.12)$$

The B_{min} transition can thus be used to estimate the ZFS. For example, in Figure 3.4, B_{min} appears at $\approx$0.14 T, and the use of the abovementioned equation, with $\mu_B = 0.46686\,\text{cm}^{-1}\,\text{T}^{-1}$, $h = 6.62608 \times 10^{-34}$ J · s, $\nu = 9.30 \times 10^9\,\text{s}^{-1}$ (so $\frac{\mu_B}{hc} = 0.46686\,\text{cm}^{-1}\,\text{T}^{-1}$), and $g_e = 2.0023$, gives a calculated average or effective ZFS equal to $D'/hc = \sqrt{D^2 + 3E^2} = 0.1576\,\text{cm}^{-1}$ versus $D'/hc = 0.1522\,\text{cm}^{-1}$ calculated using the parameters employed for the simulation. If other turning points can be observed, then the extraction of ZFS parameters is more accurate. As shown in Figure 3.5, the separation of the parallel features (z_1 and z_2) gives 2|*D*| and that of the perpendicular features (xy_1 and xy_2 in axial symmetry) gives |*D*|. This perpendicular separation is $\approx$0.16 T; this is converted using the factor $\frac{g_e\mu_B}{hc} = 0.93480\,\text{cm}^{-1}\,\text{T}^{-1}$ to give $|D|/hc \approx 0.15\,\text{cm}^{-1}$. In a rhombic spectrum, the splittings between the respective x_i and y_i turning points can be used to estimate 3|*E*|, and the separation between the centers of the xy_1 and xy_2 turning points allows the estimation of |*D*|, as in the axial case. This is shown in Figure 3.5 as well as elsewhere.[27,28]

The best method of analysis, of course, is to employ a simulation program that performs an exact calculation by matrix diagonalization, with these estimated ZFS parameters as a starting point. Many such programs are currently available. These include the widely used and versatile software package EasySpin,[32,33] used for many of the spectra presented, as well as programs such as SPIN,[34] and SpinCount.[35] Some of the simulated spectra presented in this chapter use software available from the author.

The abovementioned examples shown in Figures 3.3–3.5 used a magnitude of ZFS that corresponds to $D < h\nu$. However, as the ZFS increases relative to the microwave quantum, the spectral appearance changes dramatically and can become difficult to interpret. This is shown in Figure 3.6,

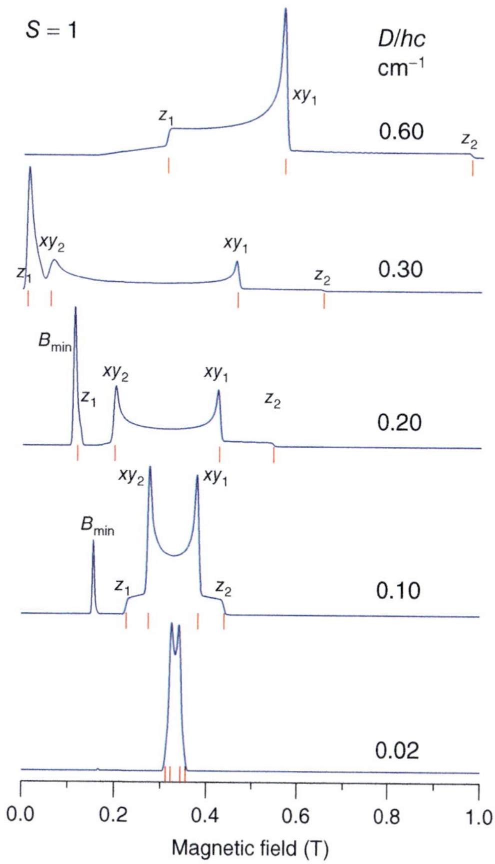

Figure 3.6. Simulated EPR spectra (absorption presentation) for spin triplets with axial ZFS. The *D*/*hc* values (in cm^{-1}) are indicated in the figure; isotropic $g = 2.00$; single-crystal Gaussian linewidths (hwhm) of 100 MHz; microwave frequency 9.3 GHz ($\approx$0.31 cm^{-1}). The spectra are calculated using a powder pattern with the single-crystal resonant fields along the canonical ZFS tensor axes indicated by red bars. The turning points are labeled using a standard nomenclature for triplets

which presents the simulated X-band spectra for a series of triplets with axial symmetry (and isotropic *g*) as a function of *D*. Both powder pattern and single-crystal spectra, calculated at the canonical ***D***-tensor directions, are shown. In the case of $D \ll h\nu$, the spectrum is nearly isotropic and B_{min} is barely perceptible. One should therefore not be misled into thinking that this is the spectrum of a spin doublet (as might occur with the spectrum in Figure 3.1). For $D/hc = 0.1\,\text{cm}^{-1} \approx h\nu/3$, the classical triplet pattern already shown in Figure 3.5 obtains. But once $D \approx h\nu$

obtains, then the spectrum is not easily recognized and the situation becomes more difficult the larger D becomes. For $D/hc = 1.0\,\text{cm}^{-1} \approx 3h\nu/c$ (not shown), there is only a single transition, observable at $\approx$0.68 T and corresponding to xy_2, which has the appearance of a single $S = 1/2$ species, but with an unusual observed g value (here, $g_{\text{obs}} \approx 0.98$). Such a signal is indeed seen for solid dioxygen (O_2(s)), present as a contaminant in many frozen solution EPR samples, provided the spectrum is recorded to high field (by X-band standards), as the signal is seen at $\approx$1.2 T, ($g_{\text{obs}} \approx 0.6$) because the ZFS is large: $D/hc = +3.572\,\text{cm}^{-1}$.[36] Other p block triplets, including isoelectronic (i.e., the same valence electron configuration) analogs to O_2 such as SO, S_2, NF, NCl, and PF, can exhibit sizable ZFS ($|D|/hc > 2\,\text{cm}^{-1}$).[28] Thus, over a conventional X-band field sweep, triplets with $D \gg h\nu$ can be 'EPR silent'. It is, however, sometimes possible to observe the forbidden $\Delta m_S = \pm 2$ transition even in these cases of large $|D|$ if $|E|$ is small so that the $|1, \pm 1\rangle$ levels are close together. For example, this transition has been seen in a V(III) ($3d^2$, $S = 1$) system.[37] Use of high frequencies and high fields (HFEPR) allows the observation of spectra and thence determination of ZFS parameters as illustrated in Section 3.4.3.

3.2.2 Spin Quartet, $S = 3/2$

3.2.2.1 *Spin Quartet, S = 3/2, No ZFS, with External Magnetic Field*

As with the spin triplet, in the absence of ZFS, there is nothing notable about the EPR spectrum of a spin quartet ($S = 3/2$) system. The only difference is that there are now three allowed ($\Delta m_S = \pm 1$) transitions: $|S, m_{S_{\text{initial}}}\rangle \rightarrow |S, m_{S_{\text{final}}}\rangle = |3/2, -3/2\rangle \rightarrow |3/2, -1/2\rangle; |3/2, -1/2\rangle \rightarrow |3/2, +1/2\rangle; |3/2, +1/2\rangle \rightarrow |3/2, +3/2\rangle$, but these are degenerate in the absence of ZFS. A spin quartet totally absent ZFS is very rare; however, truly octahedral transition metal complexes with nd^3 electronic configurations often exhibit very small ZFS. An example analogous to the case of $[Ni(NH_3)_6]I_2$ (Figure 3.1) is thus the Cr(III) congener, $[Cr(NH_3)_6]I_3$.[38] In this case as well, there is only a single EPR resonance, with an isotropic $g = 1.9857(5)$, but the slight angular variation in its linewidth (2.1–2.6 mT) observed in single-crystal measurements indicates that there is a small, but measurable, ZFS, $|D|/hc = 0.019\,\text{cm}^{-1}$ ($|D|/h = 0.57$ GHz). A powder pattern, however, such as shown in Figure 3.1, would not reveal any ZFS for this system. Transition metal (d block) complexes will be discussed further in Section 3.4.

3.2.2.2 *Spin Quartet, S = 3/2, with ZFS, No External Magnetic Field*

The spin quartet is the simplest Kramers (half-integer spin) system for which ZFS is relevant. The spin Hamiltonian is the same as for the spin triplet. In the absence of an external magnetic field, for an axial system, the situation is very simple: $|S, m_S\rangle = |3/2, \pm 1/2\rangle$ at energy equal to $-D$ and $|S, m_S\rangle = |3/2, \pm 3/2\rangle$ at $+D$, so that the energy splitting is equal to $2D$, rather than D for $S = 1$. For a rhombic system, the wavefunctions are not pure Zeeman states; the rhombic interaction operator, $\frac{1}{2}(\hat{S}_+^2 + \hat{S}_-^2)$, mixes the $|\pm 1/2\rangle$ and $|\mp 3/2\rangle$ states (i.e., $\Delta m_S = \pm 2$). Nevertheless, even for $|E| = |D|/3$, the mixing is small so that the $|3/2, \pm 1/2\rangle$ and $|3/2, \pm 3/2\rangle$ descriptions can still be used. The energy separation between these two states is given by: $\Delta = 2\sqrt{D^2 + 3E^2}$, as shown in Figure 3.2.

3.2.2.3 *Spin Quartet, S = 3/2, with ZFS, with External Magnetic Field*

As a result of being a Kramers system, there is always a spin doublet, $m_S = \pm 1/2$, within which a magnetic-dipole allowed EPR transition can occur at any frequency/field. Despite this ability to observe EPR, when $D \gg h\nu$, the resulting spectra can be rather uninformative with respect to measuring ZFS, as will be discussed in the following paragraph.

As was done earlier for the spin triplet, it is useful to describe the energy levels of the spin quartet as a function of external magnetic field. In an axial system, with the field parallel to the ZFS z-axis, the energy level situation is again very simple[28]:

$$E_{\pm 3/2} = +D \pm \frac{3}{2} g_{||} \mu_B B;\ E_{\pm 1/2} = -D \pm \frac{1}{2} g_{||} \mu_B B \tag{3.13}$$

For the field perpendicular to z, then the energy levels are[28]

$$E_{\pm 3/2} = +\sqrt{\left(D \pm \frac{1}{2} g_{\perp} \mu_B B\right)^2 + \frac{3}{4}(g_{\perp} \mu_B B)^2} \pm \frac{1}{2} g_{\perp} \mu_B B$$

$$E_{\pm 1/2} = -\sqrt{\left(D \pm \frac{1}{2} g_{\perp} \mu_{\mathrm{B}} B\right)^{2} + \frac{3}{4}(g_{\perp} \mu_{\mathrm{B}} B)^{2}} \pm \frac{1}{2} g_{\perp} \mu_{\mathrm{B}} B \tag{3.14}$$

Note that $g_{\parallel}$ and $g_{\perp}$ are used in equations (3.13) and (3.14) for generality (with the $\boldsymbol{g}$- and $\boldsymbol{D}$-tensors aligned, which is usually assumed, but requires single-crystal measurements for confirmation). As $D \to 0$, these equations become the simple case of four levels separated by $g\mu_{\mathrm{B}}B/2$, as described in Section 3.2.2.1. An energy level diagram as a function of external field for a spin quartet is shown in Figure 3.7. As in Figure 3.3, a magnitude of ZFS ($D/hc = 0.15\ \mathrm{cm}^{-1}$) suitable for X-band EPR observation is used and results for both axial and modestly rhombic symmetry are shown.

The next consideration is what a powder pattern EPR spectrum for such a quartet looks like. This is shown in Figure 3.8, again for both axial and slightly rhombic symmetry. The labeling of the transitions as xy_1, z_1, etc., is as given elsewhere.[28]

The spectral appearance is already much more complicated than in the triplet. Nevertheless, it is possible to measure splittings, such as between z_1 and xy_1, to obtain initial estimates of the D value, and in the rhombic case between x_1 and y_1 to estimate E. Note again that in a rhombic quartet, the ZFS is a combination of D and E, equal to $2\sqrt{D^2 + 3E^2}$, rather than separable as in a rhombic triplet (equation (3.6)). Note further that caution must be used because for a quartet, in contrast to a triplet, there is an 'extra line' at a field just higher than the xy_2 (and barely visible z_2) transitions.[28] This is an 'off-axis' turning point of the transitions, which occurs for an angle of the external field with the ZFS z-axis of $\theta \approx 39°$, where $\frac{\mathrm{d}B}{\mathrm{d}\theta}$ is zero, just like along a principal direction. The development of this feature can be seen in Figure 3.9, which presents the resonant field positions analogously to those shown in Figure 3.4. Unfortunately, the field position of this strong feature is not useful in extracting ZFS parameters, so it is at best irrelevant and at worst misleading if it is mistakenly assigned to, e.g., xy_2 or xy_3. The best scenario is if the parallel (z_1 and/or z_3) turning points can be used for analysis, but these can be low intensity and difficult to observe. Moreover, the observation of the low-field parallel turning point z_1 can be obscured by $|\Delta m_S| > 1$ transitions. These are analogous to B_{min} in a triplet but cannot be as easily utilized. Thus, the use of spectral simulation software for analysis of spin quartets with ZFS on the order of the microwave quantum is imperative.

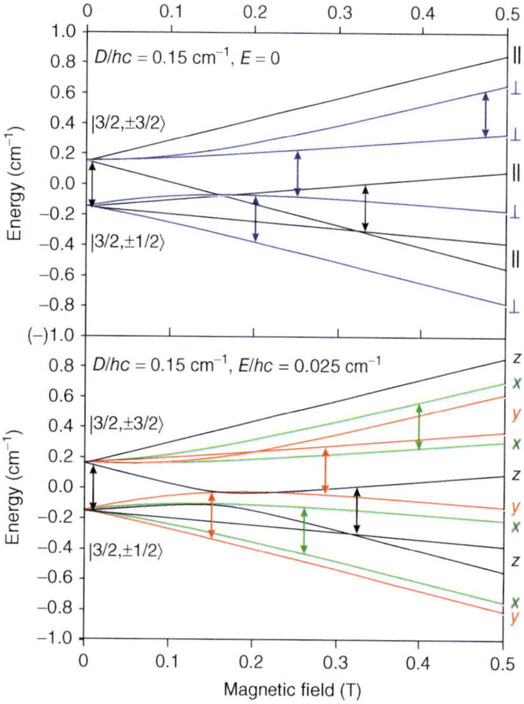

Figure 3.7. Diagrams of energy levels (eigenvalues) as a function of external magnetic field for a spin quartet with the following spin Hamiltonian parameters: $S = 3/2$, $g_{\mathrm{iso}} = 2.00$, $D/hc = +0.15\ \mathrm{cm}^{-1}$, upper plot: $E = 0$; lower plot: $E/hc = +0.025\ \mathrm{cm}^{-1}$. Black lines indicate energies with the magnetic field along the ZFS z-axis (parallel) in both plots; in the upper plot, blue traces indicate energies with the magnetic field along the ZFS x-axis (perpendicular); in the lower plot, green traces are for the field along the ZFS x-axis and red traces for the field along the ZFS y-axis. Arrows indicate allowed EPR transitions with a microwave quantum of 9.3 GHz, with their colors corresponding to the transition orientation

As shown earlier for triplets, the spectral appearance of quartets changes dramatically as a function of ZFS magnitude as shown in Figure 3.10. One point worth noting is that the xy_2 transition (which can also be seen in Figure 3.9) that corresponds to the $|-1/2\rangle \to |1/2\rangle$ transition is relatively insensitive to the magnitude of D, as long as $D \ll h\nu$, i.e., it is 'isotropic' like; however, its linewidth is proportional to the $D^2/h\nu$ in this range so that one can determine the ZFS from this line shape.

For the case of small magnitude ZFS, $D \ll h\nu$, the spectrum is relatively easily understood as described earlier. Even in the case of $D \approx h\nu$, there is potential for direct extraction of approximate ZFS parameters.

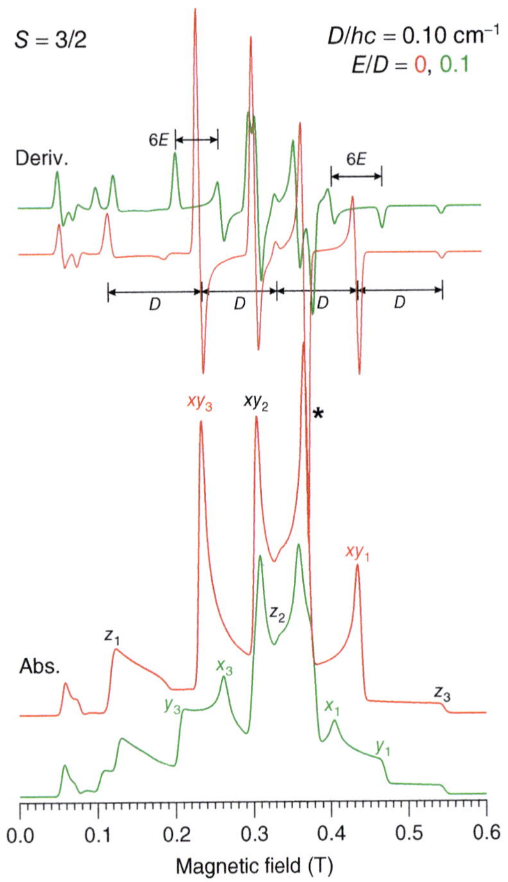

Figure 3.8. Simulated EPR spectra for a spin quartet with $g_{iso} = 2.00$, axial ZFS, $D/hc = +0.10\,cm^{-1}$; with rhombic splitting ($E/hc = +0.010\,cm^{-1}$; green traces) and without rhombic splitting (red traces). Single-crystal Gaussian linewidths (hwhm) of 100 MHz were used. A typical X-band microwave frequency of 9.3 GHz ($\approx 0.31\,cm^{-1}$) was used. Both an absorption presentation (lower traces) and a first derivative (upper traces) are shown. The turning points are labeled using standard nomenclature for quartets; green labels are unique to the rhombic case; red labels are unique to the axial case; and black labels are common to both. The asterisk indicates an off-axis turning point (Figure 3.9). The approximate relation between splittings of the turning points and ZFS parameters is also indicated. These splittings are shown in magnetic field quantities, not in energy, and thus need to be converted to energy quantities using $g\mu_B$ as described in Section 3.2.1.3

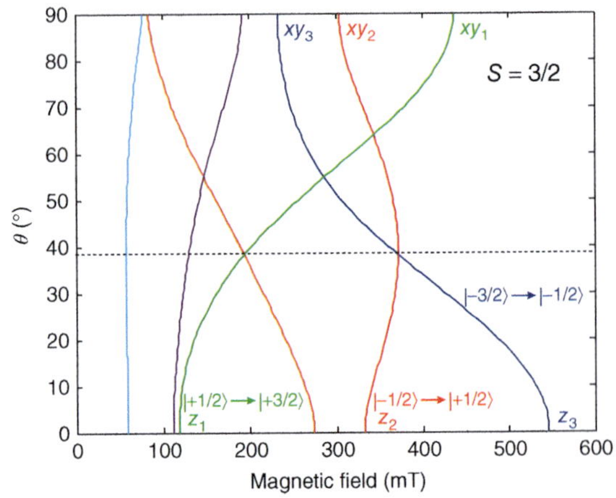

Figure 3.9. X-band (9.3 GHz) EPR transition energies calculated using EasySpin for a spin quartet with $g = 2.00$, $D/hc = 0.10\,cm^{-1}$ as a function of the angle θ between the external magnetic field and the ZFS tensor z-direction. The horizontal dashed line indicates the angle ($\theta \approx 39°$) where there is significant spectral intensity that does not correspond to either a parallel ($\theta = 0°$) or a perpendicular ($\theta = 90°$) turning point, i.e., an off-axis resonance, occurring at $dB/d\theta = 0$, as is also the case at the canonical values. The correspondence between transitions labels z_1 (green trace), z_2 (red trace), and z_3 (blue trace) and $\Delta m_S = \pm 1$ transitions (where this is a good quantum number, i.e., at $\theta = 0°$) is indicated. The unlabeled traces (cyan, violet, and orange) are less easily classified and give rise to the lower intensity transitions as seen at low field in Figure 3.8

However, when $D \gg h\nu$, an EPR spectrum is still visible, but it is essentially insensitive to the magnitude of D. In this case, for positive D ($|\pm 1/2\rangle$ doublet lowest in energy), only the xy_3 and z_2 transitions are observable (as can be seen developing in the uppermost trace in Figure 3.10) and can be described by effective (observed) g values as follows[28,39]:

$$g_{\perp}^{\text{eff}} \approx 2g_{\perp}\left[1 - \frac{3}{16}\left(\frac{g_{\perp}\mu_B B}{D}\right)^2\right] \approx 2g_{\perp}(\lim_{D\to\infty});$$
$$g_{\parallel}^{\text{eff}} \approx g_{\parallel} \tag{3.15}$$

Thus, ignoring any intrinsic g anisotropy, the parallel transition (z_2) is seen at $g_{obs} \approx 2$ and the perpendicular transition (xy_3) at $g_{obs} \approx 4$. This situation occurs for many transition metal complexes with $S = 3/2$ ground states and $D > 0$.[5,40] In many cases, the spectra are rhombic, meaning that there are two resonances in the perpendicular region. A relatively simple, but useful, analysis of such spectra is to extract the rhombicity

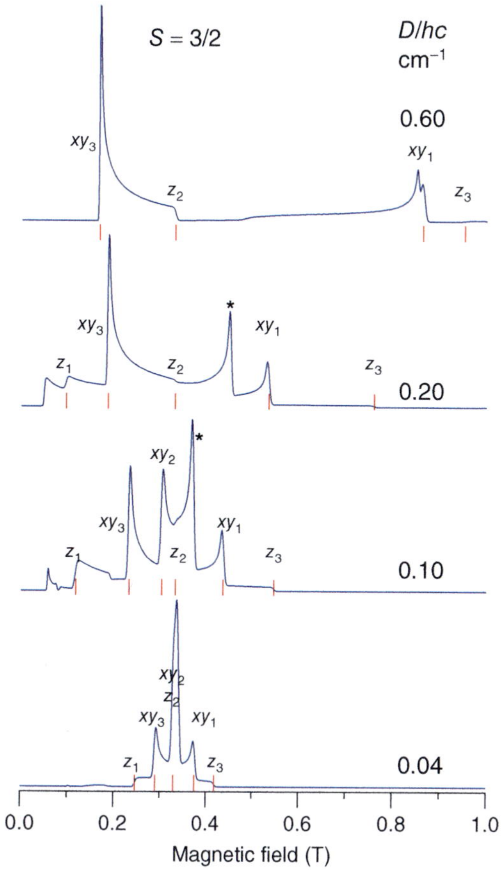

Figure 3.10. Simulated EPR spectra (absorption presentation) for spin quartets with $g_{iso} = 2.00$, axial ZFS. The D/hc values (in cm^{-1}) are indicated; single-crystal Gaussian linewidths (hwhm) of 100 MHz; microwave frequency 9.3 GHz ($\approx 0.31\ cm^{-1}$). The spectra are calculated using a powder pattern with the single-crystal resonant fields along the canonical ZFS tensor axes indicated by red bars. The turning points are labeled using standard nomenclature; an off-axis turning point (see text) is indicated by an asterisk

($r = |E/D|$; λ is often used for this ratio) and the intrinsic g values from the observed set of g^{eff} values using the following equations from Banci *et al.* (changing their equation for g_z^{eff} so that a positive value results)[41]:

$$g_x^{eff} = g_x\left[\frac{1-3r}{\sqrt{1+3r^2}}+1\right],\quad \lim_{r\to 0} = 2g_x \approx 4$$

$$g_y^{eff} = g_y\left[\frac{1+3r}{\sqrt{1+3r^2}}+1\right],\quad \lim_{r\to 0} = 2g_y \approx 4$$

$$g_z^{eff} = g_z\left[\frac{2}{\sqrt{1+3r^2}}-1\right],\quad \lim_{r\to 0} = g_z \approx 2 \quad (3.16)$$

Assuming that the intrinsic g values are similar, then these equations would assign g_{max} to g_y^{eff} and g_{mid} to g_x^{eff} (g_{min} is g_z^{eff}) so that the g values in order of increasing field would be y, x, and z.

However, if D is negative ($|\pm 3/2\rangle$ doublet lowest in energy), then the EPR transitions (at low temperatures typically employed, so that the $|\pm 1/2\rangle$ excited state doublet is unpopulated) yield the following effective g values[28]:

$$g_{\perp}^{eff} \approx 2g_{\perp}\left[\frac{3}{16}\left(\frac{g_{\perp}\mu_B B}{D}\right)^2\right] \approx 0(\lim_{D\to\infty});\ g_{\parallel}^{eff} \approx 3g_{\parallel} \quad (3.17)$$

This scenario yields atypical X-band EPR spectra as the parallel transition is seen at $g_{obs} \approx 6$ (≈ 0.1 T at X-band) and the perpendicular transition occurs at very low g_{obs} (e.g., ≈ 1) and thus observable only in scans to relatively high field (by X-band standards; $B_0 > 0.7$ T). Examples include complexes of Co(II) ($3d^7$, $S = 3/2$), both with low-coordination number[42] and hexa-coordinate ones.[43,44] In these cases, there is often well-resolved hyperfine coupling (see Chapter 5) in the parallel region, $A_{\parallel}(^{59}Co)$, owing to the low-field minimizing 'g-strain' (field-dependent broadening effects; see Chapter 2 and Section 3.4.6). Analogous to the abovementioned, equations are given by Banci *et al.* for rhombic spectra of this type (i.e., negative D; changing their equation for g_y^{eff} so that a positive value results)[41]:

$$g_x^{eff} = g_x\left[-\frac{1-3r}{\sqrt{1+3r^2}}+1\right],\quad \lim_{r\to 0} = 0$$

$$g_y^{eff} = g_y\left[+\frac{1+3r}{\sqrt{1+3r^2}}-1\right],\quad \lim_{r\to 0} = 0$$

$$g_z^{eff} = g_z\left[+\frac{2}{\sqrt{1+3r^2}}+1\right],\quad \lim_{r\to 0} = 3g_z \approx 6 \quad (3.18)$$

Assuming that the intrinsic g values are similar, then these equations would assign g_{min} to g_y^{eff} and g_{mid} to g_x^{eff} (g_{max} is g_z^{eff}) so that the g values in order of increasing field would be z, x, and y.

The optimal experimental approach for these cases of $D \gg h\nu$ at X-band is to use a larger microwave quantum (and correspondingly larger field sweep), i.e., HFEPR.[6–8] This is demonstrated in Figure 3.11, which presents the simulated EPR spectra for an axial spin quartet with $D/hc = +3.0\,\text{cm}^{-1}$ at both 35 GHz (Q-band (which is K_a-band using IEEE designations[45]); $1.17\,\text{cm}^{-1}$) and 330 GHz (a typical HFEPR frequency; $11.0\,\text{cm}^{-1}$). The increase in roughly an order of magnitude in both frequency and field sweep converts an uninformative spectrum into one that directly provides the ZFS. Note that even at 330 GHz, an off-axis turning point is observed along with nominally forbidden low-field transitions. Still higher frequencies would be needed to obtain the simple pattern as seen in Figure 3.9 for $D/hc = 0.04\,\text{cm}^{-1}$.

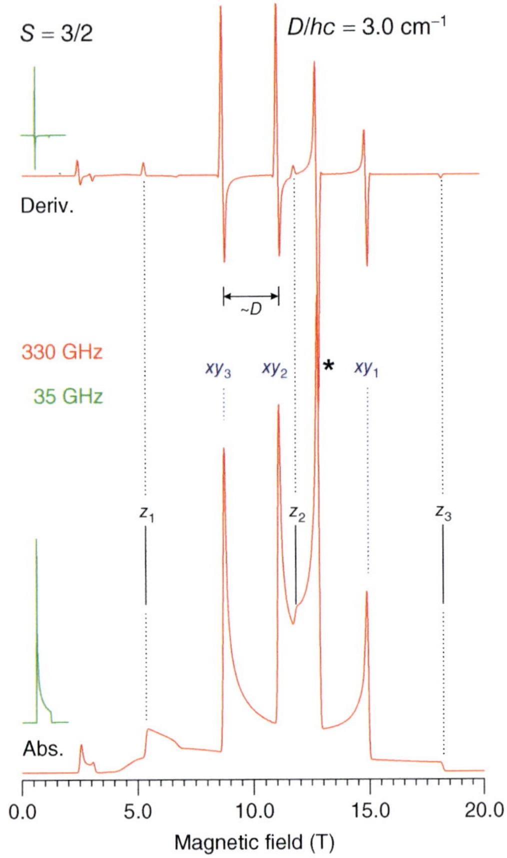

Figure 3.11. Simulated EPR spectra for a spin quartet with axial ZFS, $D/hc = +3.0\,\text{cm}^{-1}$ ($E = 0$; $g = 2.00$) using both 35 GHz (green traces; 2 T field sweep) and 330 GHz (red traces; 20 T field sweep). Single-crystal Gaussian linewidths (fwhm) of 300 MHz were used for the 35-GHz spectrum and 3000 MHz for the 330-GHz spectrum. Both an absorption presentation (lower traces) and a first derivative (upper traces) are shown. All intensities are arbitrarily scaled. The turning points are labeled using standard nomenclature for quartets. The 35- GHz spectrum is not labeled; its perpendicular (lower field) feature is xy_3 and the parallel feature (higher field) is z_1

3.2.3 Spin Quintet, $S = 2$

3.2.3.1 *Spin Quintet,* S = 2*, No ZFS*

We now return to a non-Kramers (integer-spin) system, $S = 2$. In principle, with no ZFS, this too would yield a very simple EPR spectrum, with four degenerate, magnetic-dipole allowed transitions:

$$|S, m_{S_{\text{initial}}}\rangle \rightarrow |S, m_{S_{\text{final}}}\rangle = |2,-2\rangle \rightarrow |2,-1\rangle; |2,-1\rangle \rightarrow |2,0\rangle |2,0\rangle \rightarrow |2,+1\rangle; |2,+1\rangle \rightarrow |2,+2\rangle$$

In contrast, however, to the cases of spin triplets and quartets, even high symmetry (i.e., cubic) $S = 2$ systems would exhibit ZFS owing to higher order effects (see Section 3.2.3.2), so that there are no examples of simple EPR spectra such as for $S = 1$ (a d^8 complex shown in Figure 3.1) and $S = 3/2$ (a d^3 complex described in Section 3.2.2.1).[38,46–48]

3.2.3.2 *Spin Quintet,* S = 2*, with ZFS, No External Magnetic Field*

In the case of a spin quintet, the spin Hamiltonian has the same form as employed earlier for spin triplets and quartets, except that there can be additional ZFS terms. First, we give the energies as a result of the second-order ZFS terms already described (equation (3.3)). For an axial system ($D > 0$), the ground state is $|S, m_S\rangle = |2,0\rangle$ at $-2D$, $|2,+1\rangle$ and $|2,-1\rangle$ at $-D$, and then $|2,+2\rangle$ and $|2,-2\rangle$ at $+2D$ (Figure 3.2). In terms of energies relative to the ground state, these are at 0, D, and $4D$, respectively. If there is rhombic ZFS, then the $|2,0\rangle$ and $|2,\pm 2\rangle$ states are mixed, in addition to the mixing of the $|2,\pm 1\rangle$ states as already seen for the spin triplet (equation (3.6)). This leads to the following zero-field energy levels, in ascending order for $D > 0$[28]:

$$E_1 = -2D\sqrt{1 + 3\left(\frac{E}{D}\right)^2}; E_2 = -D - 3E;$$
$$E_3 = -D + 3E; E_4 = +2D; E_5 = +2D\sqrt{1 + 3\left(\frac{E}{D}\right)^2} \quad (3.19)$$

The corresponding wavefunctions are[28]

$$\begin{aligned}
\psi_1 &= |0'\rangle = \frac{1}{2}\sqrt{1-\beta}\{|2,+2\rangle + |2,-2\rangle\} \\
&\quad - \frac{1}{\sqrt{2}}\sqrt{1+\beta}|2,0\rangle, \lim_{E\to 0} = -|2,0\rangle \\
\psi_2 &= |\pm 1^a\rangle = \frac{1}{\sqrt{2}}\{|2,+1\rangle - |2,-1\rangle\} \\
\psi_3 &= |\pm 1^s\rangle = \frac{1}{\sqrt{2}}\{|2,+1\rangle + |2,-1\rangle\} \\
\psi_4 &= |\pm 2^a\rangle = \frac{1}{\sqrt{2}}\{|2,+2\rangle - |2,-2\rangle\} \\
\psi_5 &= |\pm 2^{s'}\rangle = \frac{1}{2}\sqrt{1+\beta}\{|2,+2\rangle + |2,-2\rangle\} \\
&\quad + \frac{1}{\sqrt{2}}\sqrt{1-\beta}|2,0\rangle, \lim_{E\to 0} = \frac{1}{\sqrt{2}}\{|2,+2\rangle + |2,-2\rangle\} \\
\beta &= \frac{1}{\sqrt{1+3\left(\frac{E}{D}\right)^2}}, \lim_{E\to 0}\beta = 1
\end{aligned} \tag{3.20}$$

Equation (3.20) is a treatment analogous to what was used for $S \leq 3/2$. However, we now encounter a situation where additional, higher order ZFS terms can appear. The ZFS terms introduced for $S = 1$ in Section 3.2.1.2 (e.g., equation (3.3)) were second-order terms, i.e., they contained the operators $\hat{S}_z^2, \hat{S}_x^2$, and $\hat{S}_y^2$ (recall that $(\hat{S}_x^2 - \hat{S}_y^2) = \frac{1}{2}(\hat{S}_+^2 + \hat{S}_-^2)$). In the case of $S = 2$, fourth-order terms are now possible, i.e., ones that contain the operators $\hat{S}_z^4$ and $\frac{1}{2}(\hat{S}_+^4 + \hat{S}_-^4)$. The reason for this is that the order (or degree) of each ZFS term in the spin Hamiltonian must be $\leq 2S$,[10] because terms with order $>2S$ are zero as can be seen using, e.g., the $\hat{S}_+^4$ operator on $S = 3/2$. Thus, for $S = 1$ and 3/2, only second ($2 \leq 2S$)-order terms are allowed, but for $S = 2$, fourth ($4 = 2S$)-order terms become allowed. For reasons of time reversal symmetry,[10,49] spin operators only of even degree are allowed, so there are no third-order ZFS terms for $S > 3/2$.

Fourth-order ZFS terms are generally small in magnitude and thus are usually ignored in lower symmetry (i.e., axial or rhombic) systems where second-order ('normal', i.e., D and E) ZFS terms dominate. However, in high symmetry (i.e., cubic) systems with $S > 2$, they merit consideration. Examples include ions such as Mn(III) or Cr(II) (both $3d^4$, $S = 2$) as dopants in either octahedral oxide lattices (corundum or rutile)[50] or hydrate salts, e.g., Tutton's salts: $(M^I)_2[M^{II}(H_2O)_6](SO_4)_2$, M^I = Rb^+, Cs^+; M^{II} = Cr^{2+}[51,52]; alums: $M^I[M^{III}(H_2O)_6](SO_4)_2 \cdot 6H_2O$, M^I = Cs^+; M^{III} = Mn^{3+}[53,54]; and other hydrates: $Cs[Fe(H_2O)_6]PO_4$ (Fe(II), $3d^6$, $S = 2$).[55]

In contrast to the nearly universal use of D and E parameters, there is a great deal of variation in the definition of fourth-order ZFS parameters. The cubic ZFS parameter, namely that which can be present even in truly octahedral or tetrahedral environments, can be denoted B_4 but is frequently redefined as $a = 120B_4$. In the absence of any additional ZFS terms (i.e., a truly cubic system), a spin quintet splits into two states at $+0.6a$ and three states at $-0.4a$. The corresponding wavefunctions are as follows:

$$\begin{aligned}
|S, m_S\rangle &= -|2,-1\rangle, +|2,+1\rangle; \\
&\frac{1}{\sqrt{2}}\{-|2,-2\rangle + |2,+2\rangle\}, \\
&\frac{1}{\sqrt{2}}\{|2,-2\rangle + |2,+2\rangle\}, |2,0\rangle
\end{aligned} \tag{3.21}$$

Note that this cubic ZFS preserves the energy degeneracy of the $|2, \pm 1\rangle$ states, as is the case with axial ZFS, but the $|2, 0\rangle$ state is degenerate with the $|2, \pm 2\rangle$ states.

3.2.3.3 Spin Quintet, S = 2, with ZFS, with External Magnetic Field

Upon application of an external magnetic field, the spin Hamiltonian matrix for a spin quintet becomes relatively complicated and can be found elsewhere.[10,28,56] A diagram of the field dependence of the energy levels, using a magnitude of ZFS ($D/hc = 0.15\,\text{cm}^{-1}$) suitable for X-band EPR observation is shown in Figure 3.12.

The X-band powder pattern spectra for one such case are shown in Figure 3.13. Here, the EPR transitions are numbered from low to high field. For the case of $|D| \ll g\mu_B B$, the pattern is symmetric about the field corresponding to g_e. However, here the ZFS is sufficiently large so that the pattern is noticeably asymmetric and the indicated splittings give only estimates as to the magnitudes of D and E.

Also notable, and complicating analysis, are two 'off-axis' features that occur for angles of the external field with the ZFS z-axis of $\theta \approx 18°$ and $59°$.[56] This can be seen more easily in Figure 3.14, which presents the angular (θ) dependence of the EPR transitions, as was done previously for spin triplets (Figure 3.4) and quartets (Figure 3.9). Each of the two 'inner' transitions,

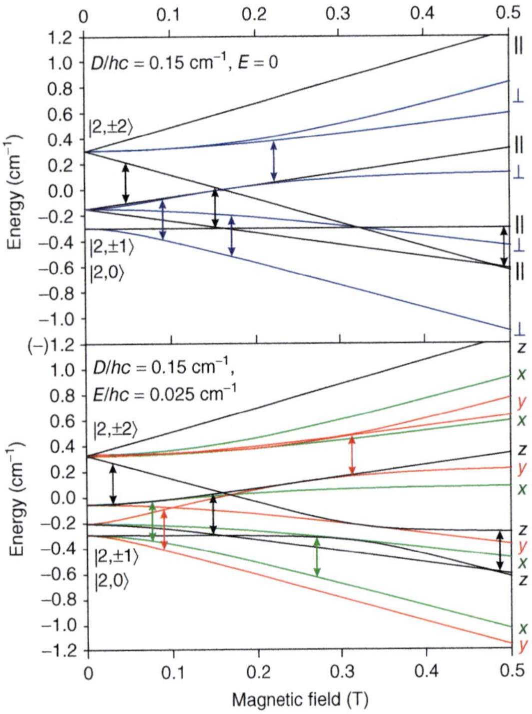

Figure 3.12. Diagrams of energy levels (eigenvalues) as a function of external magnetic field for a spin quintet with the following spin Hamiltonian parameters: $S = 2$, $g_{iso} = 2.00$, $D/hc = +0.15\,\text{cm}^{-1}$, upper plot: $E = 0$; lower plot: $E/hc = +0.025\,\text{cm}^{-1}$. Black lines indicate energies with the magnetic field along the ZFS z-axis (parallel) in both plots; in the upper plot, blue traces indicate energies with the magnetic field along the ZFS x-axis (perpendicular); in the lower plot, green traces are for the field along the ZFS x-axis and red traces for the field along the ZFS y-axis. Arrows indicate allowed EPR transitions with a microwave quantum of 9.3 GHz ($\approx$0.31 cm^{-1}), with their colors corresponding to the transition orientation

i.e., those involving $|m_S\rangle = 0$ ($|2,0\rangle \leftrightarrow |2,\pm 1\rangle$) has an off-axis feature.

Of perhaps greater importance are the features at low field. At the particular D value employed, the z_1 transition is very near zero field; however in the rhombic case, there is an additional low-field feature at a field slightly above that for z_1. Inspection of the zero-field energies (equation (3.19)) and wavefunctions (equation (3.20)) explain this phenomenon. The relevant ones of each are reproduced below:

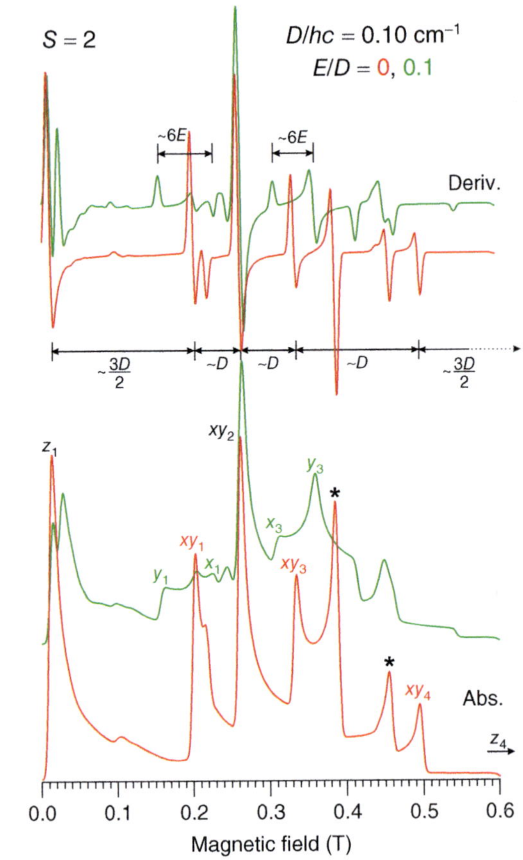

Figure 3.13. Simulated EPR spectra for a spin quintet with $g_{iso} = 2.00$, axial ZFS, $D/hc = +0.10\,\text{cm}^{-1}$; with rhombic splitting ($E/hc = +0.010\,\text{cm}^{-1}$; green traces) and without rhombic splitting (red traces). Single-crystal Gaussian linewidths (hwhm) of 100 MHz were used. A typical X-band microwave frequency of 9.3 GHz ($\approx$0.31 cm^{-1}) was used. Both an absorption presentation (lower traces) and a first derivative (upper traces) are shown. The turning points are labeled using standard nomenclature for quintets; green labels are unique to the rhombic case; red labels are unique to the axial case; and black labels are common to both. The asterisks indicate off-axis turning points (Figure 3.14). The approximate relation between splittings of the turning points and ZFS parameters is also indicated. These splittings are shown in magnetic field quantities, not in energy, and thus need to be converted to energy quantities using $g\mu_B$ as described in the text

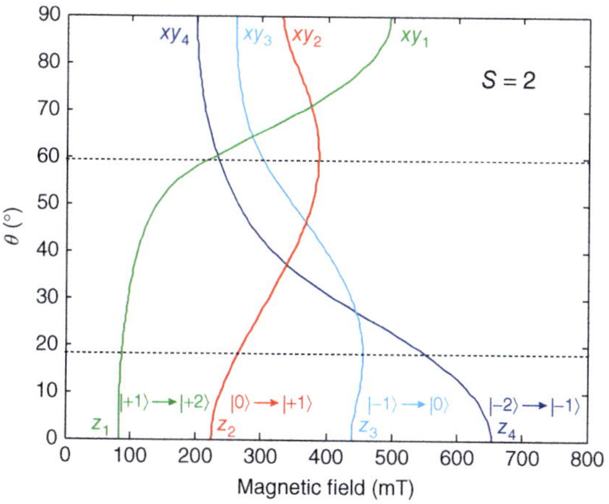

Figure 3.14. X-band (9.3 GHz) EPR transition energies calculated using EasySpin for a spin quintet with $g = 2.00$, $D/hc = 0.10\,\text{cm}^{-1}$ as a function of the angle θ between the external magnetic field and the ZFS tensor z-direction. The horizontal dashed lines indicate the angles ($\theta \approx 18°$ (transition 3, cyan trace) and 59° (transition 2, red trace)) where there is significant spectral intensity that does not correspond to either a parallel ($\theta = 0°$) or a perpendicular ($\theta = 90°$) turning point, i.e., an off-axis resonance, occurring at $dB/d\theta = 0$, as is also the case at the canonical values. The correspondence between transitions labels z_1 (green trace), z_2 (red trace), z_3 (cyan trace), and z_4 (blue trace) and $\Delta m_S = \pm 1$ transitions (where this is a good quantum number, i.e., at $\theta = 0°$) is indicated

$$E_4 = +2D, E_5 = \frac{+2D}{\beta}$$
$$\psi_4 = |\pm 2^a\rangle = \frac{1}{\sqrt{2}}\{|2,+2\rangle - |2,-2\rangle\}, \psi_5 = |\pm 2^{s\prime}\rangle$$
$$= \frac{1}{2}\sqrt{1+\beta}\{|2,+2\rangle + |2,-2\rangle\} + \frac{1}{\sqrt{2}}\sqrt{1-\beta}|2,0\rangle$$
$$\beta = \frac{1}{\sqrt{1+3\left(\frac{E}{D}\right)^2}}, \lim_{E\to 0}\beta = 1 \qquad (3.22)$$

Alternatively, the mixing coefficient can be described as $a^{\pm} = \frac{1}{\sqrt{2}}\sqrt{1 \pm \beta}$.[57] The zero-field energy separation between E_4 and E_5 is small (maximally ≈15% of D), so an X-band (or even Q-band) microwave quantum can often effect a $\Delta m_S = \pm 2$ transition between the two levels at low field. The specific resonance position can give an estimate of the ZFS as it equals $2D\left(\frac{1}{\beta} - 1\right)$ plus a small Zeeman term. The use of parallel-mode EPR detection ($B_1 \parallel B_0$), as opposed to conventional, perpendicular-mode detection ($B_1 \perp B_0$), most conveniently achieved using a dual-mode cavity (see Chapter 12), leads to an enhanced signal intensity of this transition. This procedure has been used to analyze a variety of (bio)inorganic systems, chiefly of high-spin Fe(II)[57,58] and also of high-spin d^4 systems such as Fe(IV)[59,60] and Mn(III).[61,62]

As we have seen for lower spin systems, the EPR spectral appearance changes massively as the ZFS increases. This is shown for an axial spin quintet in Figure 3.15.

The spectra speak for themselves, but one point worth noting is that when the condition $D \gtrsim g\mu_B B$ is achieved, the spectra are not readily recognizable as arising from a spin quintet, or even from any high-spin system, especially if only a conventional X-band field range is used. For example, for $D/hc = 1.0\,\text{cm}^{-1}$, the xy_2 transition occurs at 0.39 T (i.e., little shifted from the field for $D/hc = 0.4\,\text{cm}^{-1}$), and thus not worth presenting in Figure 3.15, while z_2 moves to 0.74 T ($g_{obs} = 1.7$), with negligible intensity and beyond a standard X-band field sweep. Thus, one might interpret the spectrum as due to a spin doublet with isotropic $g \approx 1.7$ – far from the truth.

For $D \gg g\mu_B B$, no conventional EPR spectra are observable, at least with axial symmetry. This is the case in square pyramidal complexes of Mn(III), e.g., Mn(TPP)Cl (TPP = tetraphenylporphyrin dianion), which effectively has C_{4v} point group symmetry and axial ZFS with $D/hc = -2.3\,\text{cm}^{-1}$ ($E = 0$) as directly measured by HFEPR.[63,64] HFEPR has been very effective in determining ZFS in $S = 2$ systems, as under these experimental conditions, $D \lesssim g\mu_B B$ so that spectral patterns resembling the lower traces shown in Figure 3.15 result, except recorded over an order of magnitude larger field range (e.g., 0–15 T).[6–8,13] Even in such HFEPR experiments, spectral simulation is crucial in the accurate determination of ZFS parameters. What is particularly useful is to plot the resonance fields (turning points) versus microwave (submillimeter wave) frequency and simultaneously fit the transitions to a single set of spin Hamiltonian parameters. Such plots have now become commonplace in papers reporting HFEPR studies, e.g., of Mn(IV) ($3d^3$, $S = 3/2$)[65] and Cr(II) ($3d^4$, $S = 2$) complexes.[66] The presence of fourth-order ZFS is an added complication, but these can at times be extracted from HFEPR spectra of doped single crystals[50,54] and even from powder pattern spectra.[67]

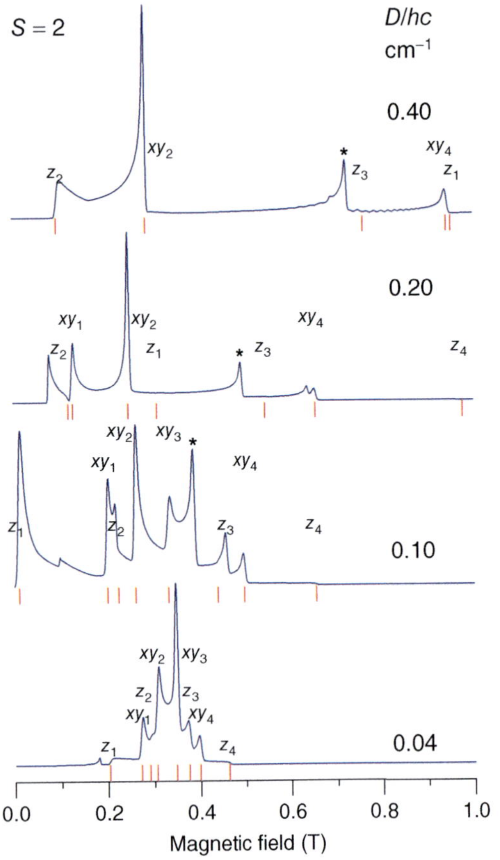

Figure 3.15. Simulated EPR spectra (absorption presentation) for spin quintets with $g_{iso} = 2.00$, axial ZFS. The D/hc values (in cm^{-1}) are indicated; single-crystal Gaussian linewidths (hwhm) of 100 MHz; microwave frequency 9.3 GHz ($\approx 0.31\ cm^{-1}$). The spectra are calculated using a powder pattern with the single-crystal resonant fields along the canonical ZFS tensor axes indicated by red bars. The turning points are labeled using standard nomenclature; an off-axis turning point (see text) is indicated by an asterisk

3.2.4 Spin Sextet, $S = 5/2$

3.2.4.1 *Spin Sextet, S = 5/2, No ZFS, with External Magnetic Field*

We now return to a Kramers ion, as was the case for the spin quartet. In the absence of ZFS, there are now five allowed ($\Delta m_S = \pm 1$) transitions:

$$|S, m_{S_{\text{initial}}}\rangle \rightarrow |S, m_{S_{\text{final}}}\rangle = |5/2, -5/2\rangle \rightarrow |5/2, -3/2\rangle;$$

$$|5/2, -3/2\rangle \rightarrow |5/2, -1/2\rangle; |5/2, -1/2\rangle \rightarrow |5/2, +1/2\rangle;$$

$$|5/2, +1/2\rangle \rightarrow |5/2, +3/2\rangle; |5/2, +3/2\rangle \rightarrow |5/2, +5/2\rangle$$

As discussed earlier for the spin quintet, systems with $S > 2$ can exhibit fourth-order ZFS effects, so that in principle there should be no cases of spin sextets absent ZFS. However, the electronic ground state of octahedral or tetrahedral HS d^5 complexes has a totally symmetric electronic configuration (one unpaired electron in each d orbital), so the ZFS is very small.[17,68] We have no experimental data on the paradigm complex, $[Mn(NH_3)_6]^{2+}$, but detailed quantum chemical calculations show that this complex would have $D_{SOC} = 0$ and $D_{SSC}/hc = +0.030\ cm^{-1}$.[68,69]

3.2.4.2 *Spin Sextet, S = 5/2, with ZFS, No External Magnetic Field*

As discussed earlier for a spin quintet, a spin sextet can experience fourth-order ZFS. In contrast to the quintet, for the sextet, such fourth-order ZFS is much more easily measured as it may be the only ZFS possible in the spherically symmetric high-spin d^5 configuration (see Section 3.2.4.1). The zero-field energy levels using the cubic ZFS parameter, a, are as follows[10]:

$$E_{1,2} = +a; \quad E_{3,4} = -2a; \quad E_{5,6} = +a \tag{3.23}$$

What spin states do these correspond to? The wavefunctions are quite different from second-order ZFS:

$$\psi_{1,2} = \left|\frac{5}{2}, \pm\frac{1}{2}\right\rangle;$$

$$\psi_{3,4} = \pm\frac{1}{\sqrt{6}}\left|\frac{5}{2}, \pm\frac{5}{2}\right\rangle \mp \sqrt{\frac{5}{6}}\left|\frac{5}{2}, \mp\frac{3}{2}\right\rangle;$$

$$\psi_{5,6} = \pm\sqrt{\frac{5}{6}}\left|\frac{5}{2}, \mp\frac{5}{2}\right\rangle \pm \frac{1}{\sqrt{6}}\left|\frac{5}{2}, \pm\frac{3}{2}\right\rangle \tag{3.24}$$

Qualitatively, these wavefunctions are the result of a fourth-order ZFS operator mixing the $m_S = \pm 5/2$ and $m_S = \mp 3/2$ states (i.e., states separated by $\Delta m_S = \pm 4$). As summarized elsewhere,[4,10] $3d^5$ ions such as Cr(I), Mn(II), and Fe(III) in octahedral oxide or fluoride lattices or in tetrahedral chalcogenide lattices do exhibit very small, but measurable ZFS, especially by use of single-crystal EPR. The typical range of cubic ZFS values is $1 \times 10^{-4} < |a|/hc < 100 \times 10^{-4}\ cm^{-1}$.

For the common situation of axial symmetry, where ZFS is dominated by second-order terms (i.e., D), the zero-field energies simplify as follows (using the same labels as in equation (3.23)):

$$E_{1,2} = -\frac{8}{3}D; E_{3,4} = -\frac{2}{3}D; E_{5,6} = +\frac{10}{3}D \quad (3.25)$$

These conveniently give relative energies of 0, $2D$, and $4D$ and correspond to the three Kramers doublets, respectively, $|5/2, \pm 1/2\rangle$, $|5/2, \pm 3/2\rangle$, $|5/2, \pm 5/2\rangle$. However, once rhombic ZFS is included, then the situation becomes complicated, although, in contrast to the non-Kramers quintet, the degeneracy of each Kramers doublet is preserved (Figure 3.2). The rhombic ZFS interaction mixes the $|5/2, \pm 3/2\rangle$ and $|5/2, \mp 1/2\rangle$ states (i.e., states separated by $\Delta m_S = \pm 2$) as described earlier for the spin quartet and also mixes the $|5/2, \pm 5/2\rangle$ and $|5/2, \pm 1/2\rangle$ states (i.e., an additional separation by $\Delta m_S = \pm 2$). Second-order perturbation theory results (Figure 3.2), which are accurate to within a few percent for modest rhombicity (e.g., $|E/D| \lesssim 0.1$), are as follows:

$$E_{1,2} = -\frac{8}{3}D - \frac{32}{3}\frac{E^2}{D}; E_{3,4} = -\frac{2}{3}D + \frac{27}{3}\frac{E^2}{D};$$
$$E_{5,6} = +\frac{10}{3}D + \frac{5}{3}\frac{E^2}{D} \quad (3.26)$$

3.2.4.3 Spin Sextet, S = 5/2, with ZFS, with External Magnetic Field

With the application of an external field, there are no analytical equations, but a plot of the energies as a function of magnetic field as determined by numerical matrix diagonalization as done for the previous, lower-spin case is helpful and is shown in Figure 3.16.

As done previously, we present in Figure 3.17 the angular dependence of the EPR transitions for the spin sextet, in this case using $D/hc = 0.10\,\text{cm}^{-1}$, the same value used in all such plots to facilitate direct comparison among these. As can thus be seen by comparison with the corresponding lower spin figures ($S = 1$, 3/2, and 2; Figures 3.4, 3.9, and 3.14, respectively), the patterns become progressively more intricate owing to the increasing number of transitions and their wide field variation. The development of two off-axis resonances can also be seen in Figure 3.17, which occur for different transitions and orientations, but at the same magnetic field, which makes it a strong transition indeed.

We lastly present in Figure 3.18 the powder pattern X-band EPR spectra for the case of $D/hc = 0.04\,\text{cm}^{-1}$,

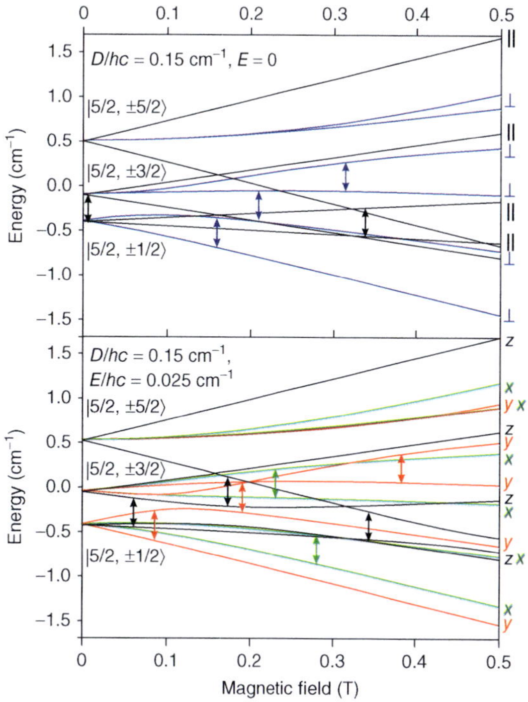

Figure 3.16. Diagrams of energy levels (eigenvalues) as a function of external magnetic field for a spin sextet with the following spin Hamiltonian parameters: $S = 5/2$, $g_{iso} = 2.00$, $D/hc = +0.15\,\text{cm}^{-1}$, upper plot: $E = 0$; lower plot: $E/hc = +0.025\,\text{cm}^{-1}$. Black lines indicate energies with the magnetic field along the ZFS z-axis (parallel) in both plots; in the upper plot, blue traces indicate energies with the magnetic field along the ZFS x-axis (perpendicular); in the lower plot, green traces are for the field along the ZFS x-axis and red traces for the field along the ZFS y-axis. Arrows indicate allowed EPR transitions with a microwave quantum of 9.3 GHz ($\approx 0.31\,\text{cm}^{-1}$), with their colors corresponding to the transition orientation

a magnitude much lower than in previous examples. This is because a given D value leads to a larger overall ZFS for a spin sextet than in lower spin cases, as can be seen in Figure 3.2. In the example shown here, a relatively interpretable spectrum is still evident, but it is not the idealized situation that obtains for $D \ll g\mu_B B$.[28] For example, an off-axis feature is already developing, in contrast to the spin quartet at this small D magnitude. Moreover, in contrast to the non-Kramers spin quintet, the sextet shows no low-field resonances of significant intensity or of any utility for analysis. Some $|\Delta m_S| > 1$ transitions are observable so that the

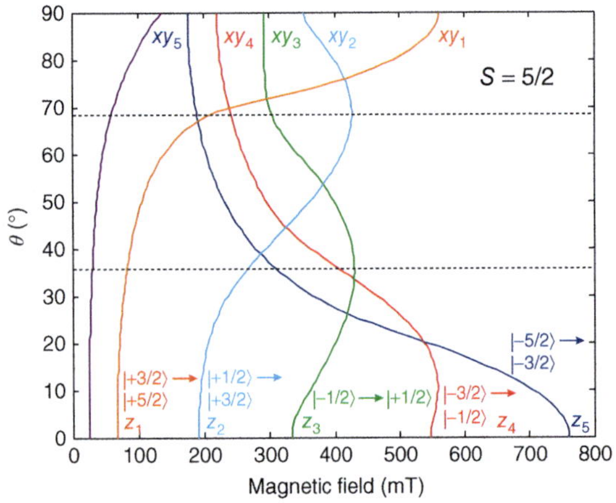

Figure 3.17. X-band (9.3 GHz) EPR transition energies calculated using EasySpin for a spin sextet with $g = 2.00$, $D/hc = 0.10\,\text{cm}^{-1}$ as a function of the angle θ between the external magnetic field and the ZFS tensor z-direction. The horizontal dashed lines indicate the angles ($\theta \approx 69°$ (transition 2, cyan trace), 36° (transition 3, green trace), and 10° (transition 4, not shown)) where there is significant spectral intensity that does not correspond to either a parallel ($\theta = 0°$) or a perpendicular ($\theta = 90°$) turning point, i.e., an off-axis resonance, occurring at $\mathrm{d}B/\mathrm{d}\theta = 0$, as is also the case at the canonical values. The correspondence between transitions labels z_1 (orange trace), z_2 (cyan trace), z_3 (green trace), z_4 (red trace), and z_5 (blue trace) and $\Delta m_S = \pm 1$ transitions (where this is a good quantum number, i.e., at $\theta = 0°$) is indicated

low-field edge (i.e., z_1 transition) appears different from the high-field edge (i.e., z_5 transition), making accurate identification of the former a challenge.

The situation continues to worsen in terms of ease of interpretation as the magnitude of D increases as shown in Figure 3.19. The large number of lines, even in this axial case, makes analysis difficult, even with spectral simulation software. What is helpful is the use of multiple frequencies, especially HFEPR, so that the simpler situation of $D \ll g\mu_B B$ can be restored. Examples of the use of HFEPR to unravel complicated sextet spectra are studies of low-symmetry complexes of Mn(II)[70] and Fe(III).[71]

There are two special cases of sextet EPR spectra that are worth mentioning. The first is the case of the so-called $g \approx 4.3$ signal.[72] This resonance (at this observed g value) is often referred to in the bioinorganic community as the 'junk iron' signal as it is widely found for Fe(III) in a wide variety of environments, both desired and adventitious. The key point

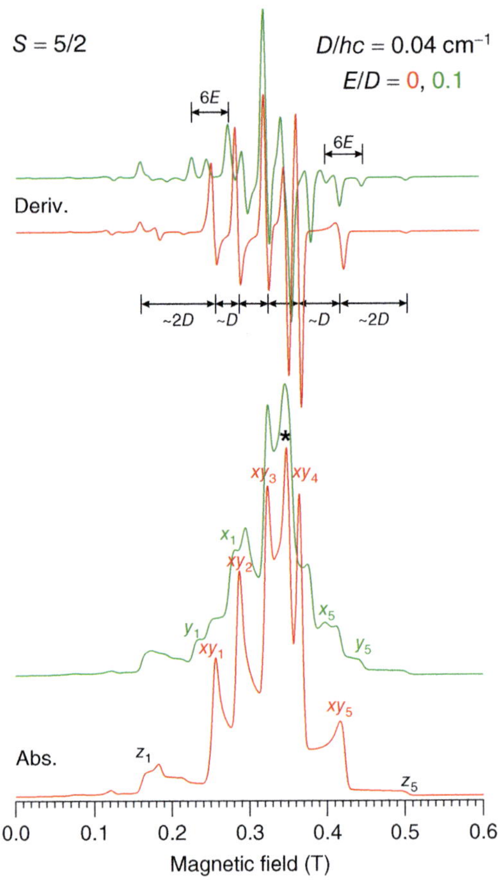

Figure 3.18. Simulated EPR spectra for a spin sextet with $g_{\text{iso}} = 2.00$, axial ZFS, $D/hc = +0.04\,\text{cm}^{-1}$, and with rhombic splitting ($E/hc = +0.004\,\text{cm}^{-1}$; green trace) and without rhombic splitting (red trace). Single-crystal Gaussian linewidths (fwhm) of 100 MHz were used. A typical X-band microwave frequency of 9.3 GHz was used. Both an absorption presentation (lower traces) and a first derivative (upper traces) are shown. The turning points are labeled using standard nomenclature; green labels are unique to the rhombic case; red labels are unique to the axial case, and black labels are common to both. The asterisk indicates an off-axis feature. The approximate relation between splittings of the turning point and ZFS parameters is also indicated. These splittings are shown in magnetic field quantities, not in energy, and thus need to be converted to energy quantities using $g\mu_B$ as described in the text

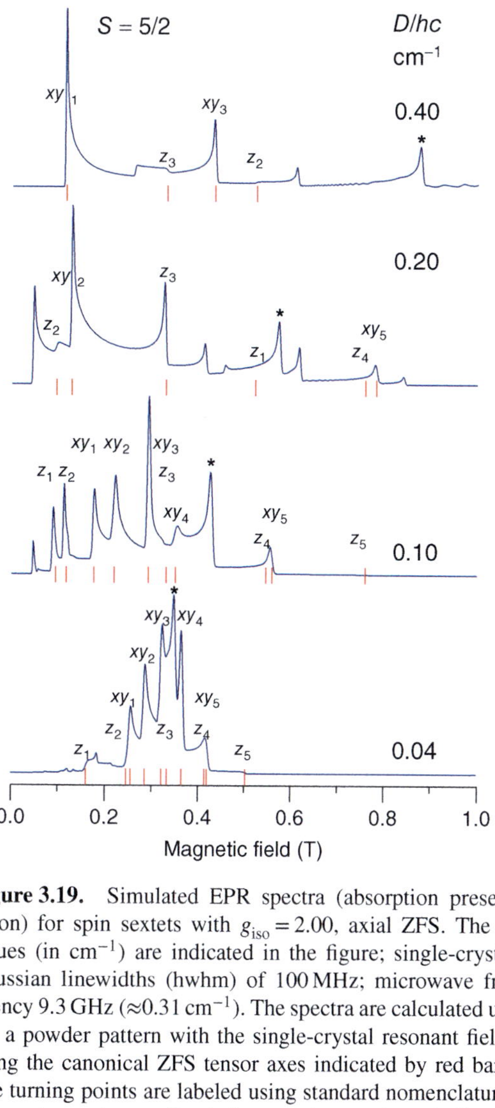

Figure 3.19. Simulated EPR spectra (absorption presentation) for spin sextets with $g_{iso} = 2.00$, axial ZFS. The D values (in cm^{-1}) are indicated in the figure; single-crystal Gaussian linewidths (hwhm) of 100 MHz; microwave frequency 9.3 GHz (≈0.31 cm^{-1}). The spectra are calculated using a powder pattern with the single-crystal resonant fields along the canonical ZFS tensor axes indicated by red bars. The turning points are labeled using standard nomenclature; the most prominent off-axis features (see text) are indicated by an asterisk

is that for Fe(III) in many, if not most, low-symmetry environments, the rhombicity is nearly maximal ($|E/D| \approx 1/3$). In this case, the transition within the middle Kramers doublet occurs at $g_{obs} \approx 4.3$, regardless of orientation of the external magnetic field, leading to an intense resonance. For example, use of parameters that are close to those for $(NH_4)[Fe(EDTA)]$ (EDTA = ethylenediamine tetraacetatetetraanion),[72,73] namely, $D/hc = +0.9\,cm^{-1}$, $E/hc = +0.3\,cm^{-1}$, $g_{iso} = 2.00$, $B_0 = 0.155\,T$, and $\nu = 9.3\,GHz$, leads to a calculated resonance between states 3 and 4 at $g_{obs} = 4.268$, 4.276, and 4.276 for B_0 along each of the x, y, and z ZFS tensor directions. The transition can be crudely described as within the $|5/2, \pm 3/2\rangle$ (i.e., middle) Kramers doublet, but the mixing-in of the other two doublets makes it magnetic-dipole allowed. Although it is not a ground-state transition, the small-magnitude ZFS typically seen for Fe(III) means that the relevant Boltzmann population under typical X-band conditions (≈80 K) is sufficient for easy observation.

The second is the case analogous to what was described earlier for spin quartets when $D \gg g\mu_B B$. This situation can be seen already developing in Figure 3.19. As D increases in magnitude (and is positive), the only transitions that remain observable are xy_1 and z_3, which are essentially invariant in field position. For example, a calculation using $g_{iso} = 2.00$ and $\nu = 9.3\,GHz$, with both $D/hc = +1.0\,cm^{-1}$ and $+10.0\,cm^{-1}$, yields the same resonance fields, xy_1 ($g_{\perp}^{eff} \approx 6.0$) at 0.111 T and z_3 at 0.332 T ($g_{||}^{eff} \approx 2.0$). However, this coincidence of resonances is not the case under all conditions; only at lower frequencies. For example, in a study of met-myoglobin (which contains ferric heme, Fe(III) $3d^5$, $S = 5/2$), careful measurement of the perpendicular (xy_1) transition over a wide range of frequencies showed slight shifts in resonant field (i.e., in $g_{\perp}^{eff}$).[74] This shift is due to varying amounts of mixing-in of the higher energy Kramers doublets $|S, m_S\rangle = |5/2, \pm 5/2\rangle, |5/2, \pm 3/2\rangle$, with the latter being much more significant due to being at only $2D$ above the $|5/2, \pm 1/2\rangle$ doublet as opposed to $6D$ for the former. The mixing is greater under HFEPR conditions because the large Zeeman splitting brings these excited states closer in energy to the ground state. This effect is shown by the following equations for the effective g values[28,39]:

$$g_{\perp}^{eff} \approx 3g_{\perp}\left[1 - \frac{1}{2}\left(\frac{g_{\perp}\mu_B B}{D}\right)^2\right]$$
$$\approx 3g_{\perp}(\lim_{D\to\infty}) \approx 6;\; g_{||}^{eff} \approx g_{||} \approx 2 \qquad (3.27)$$

The resonant field for xy_1 at X-band is ≈0.111 T, while at 400 GHz, it is ≈4.76 T, so use of $D/hc = +10.0\,cm^{-1}$ gives $g_{\perp}^{eff} = 6.00$ for the former but $g_{\perp}^{eff} = 5.41$ for the latter. Simulations using exact diagonalization will allow fine-tuning of the field/frequency dependence. Thus, HFEPR can assist in estimating ZFS even when it is too large to observe directly (i.e., the inter-Kramers doublet transitions are still larger than the microwave quantum).

3.2.5 Spin Septet, $S = 3$

3.2.5.1 *Spin Septet, S = 3, with ZFS, No External Magnetic Field*

The spin septet, $S = 3$, is not a situation that can occur for a mononuclear d block ion coordination complex. However, it can occur for special cases such as Mn(I) ($3d^5 4s^1$) as found in the matrix isolated species MnH[75] and MnF.[76] A spin septet can also result for dinuclear complexes, such as of two ferromagnetically coupled $S = 3/2$ Cr(III) ions (see Chapter 4).[77,78] The use of $S_{total} = 3/2 + 3/2 = 3$ follows the 'Giant Spin Model', which has its limitations.[78] The spin septet is worth mentioning also because it is the first case in which, in principle, sixth-order ZFS terms need to be considered in the spin Hamiltonian. As described earlier for spin quintets, even-order spin operators of degree $\leq 2S$ are possible, hence, for $S = 3$, degree six is possible.[10] Considering only axial second-order ZFS, the zero-field energy levels of a spin septet are $|S, m_S\rangle = |3, 0\rangle$ at $-4D$, $|3, \pm 1\rangle$ at $-3D$, $|3, \pm 2\rangle$ at 0, and $|3, \pm 3\rangle$ at $+5D$, so that the energy average is zero and the splitting between the levels is, respectively, D, $3D$, and $5D$. Rhombic splitting mixes the $|3, \pm 1\rangle$ and $|3, \mp 1\rangle$ states as described far earlier for the spin triplet and the $|3, 0\rangle$ and $|3, \pm 2\rangle$ states as in the quintet and also mixes the $|3, \pm 1\rangle$ and $|3, \pm 3\rangle$ states. As a result, there are no degenerate zero-field levels for a spin septet with rhombic ZFS; the $|3, \pm 1\rangle$ doublet exhibits a relatively large splitting with the other two exhibiting small splittings; and the $|3, \pm 3\rangle$ interdoublet splitting remains small even for $|E/D| = 1/3$.

3.2.5.2 *Spin Septet, S = 3, with ZFS, with External Magnetic Field*

As long as $D < g\mu_B B$, such as for $D/hc = 0.10\,\mathrm{cm}^{-1}$, a spin septet can give a rich X-band EPR spectrum. Even at $D/hc = 0.60\,\mathrm{cm}^{-1}$, an X-band signal would be observed (over a conventional magnetic field range), although its correct assignment, and even the spin state of the system, absent other information, is highly unlikely. HFEPR, such as applied to dinuclear Cr(III),[77] could well be fruitful.

3.2.6 Spin Octet, $S = 7/2$

3.2.6.1 *Spin Octet, S = 7/2, No ZFS, with External Magnetic Field*

The situation of a spin octet with very small ZFS can exist for the $4f^7$, $S = 7/2$ ions, Gd(III), Eu(II), and the much rarer Tb(IV),[79] by virtue of their spherically symmetric $4f^7$ electronic configuration (i.e., one unpaired electron in each f orbital) and the weak crystal field experienced by f block ions.[80] Nevertheless, for ZFS to be totally absent in an f^7 species is unlikely.[81–83]

3.2.6.2 *Spin Octet, S = 7/2, with ZFS, No External Magnetic Field*

As introduced earlier for the spin septet, a spin octet can require sixth-order ZFS terms for analysis.[84] For the high symmetry found in many early studies, this involves the addition of only a single, sixth-order cubic term.[10] For comparison with the septet, we mention that for only axial second-order ZFS, the zero-field situation for a spin octet is $|S, m_S\rangle = |7/2, \pm 1/2\rangle$ at $-5D$, $|7/2, \pm 3/2\rangle$ at $-3D$, $|7/2, \pm 5/2\rangle$ at $+D$, and $|7/2, \pm 7/2\rangle$ at $+7D$, so that the energy average is zero and the splittings between the levels are, respectively, $2D$, $4D$, and $6D$. Rhombic splitting mixes the $|7/2, \pm 3/2\rangle$ and $|7/2, \mp 1/2\rangle$ states as described far earlier for the spin quartet and the $|7/2, \pm 5/2\rangle$ and $|7/2, \pm 1/2\rangle$ states as in the quintet, and also mixes the $|7/2, \pm 7/2\rangle$ and $|7/2, \pm 3/2\rangle$ states, but retains the degeneracy of each Kramers doublet, in contrast to non-Kramers ions, as seen earlier for lower spin systems.

3.2.6.3 *Spin Octet, S = 7/2, with ZFS, with External Magnetic Field*

The energy levels of a spin octet as a function of applied magnetic field are not shown here but can be found elsewhere.[28] A point worth noting is that as S becomes large, the X-band spectrum for a system with modest ZFS begins to look much the same, regardless of S. Thus, the spectra for a triplet versus a quartet (Figures 3.5 and 3.8, respectively) are immediately differentiated, while the spectra for a septet versus an octet tend to blur together. Note that for the sextets in Figures 3.18 and 3.19, there is no clue as to the spin of the system from the EPR spectrum alone. In a mononuclear system, one would presumably have an idea as to what spin is reasonable (e.g., $S = 7/2$ for Gd(III)), but a coupled system (see Chapter 4), in which there may be more than one spin state present, each with its own ZFS, can present a significant challenge for analysis.

As was the case for lower spin Kramers states, as the condition $D \gg g\mu_B B$ approaches, there remain only two transitions, which are essentially invariant in field, xy_1 and z_4. Analogously to the situation for the spin sextet (equation (3.27)), these, respectively, correspond to $g_{\perp}^{\text{eff}} \approx 8$ (resonant at ≈0.083 T) and $g_{||}^{\text{eff}} \approx 2$. However, in contrast to the sextet, the axial strong field environment (such as in porphyrins), which would give rise to such simple spectra, is not readily found in $S = 7/2$ (i.e., nf^7) complexes. The spectrum reported for GdF_3 isolated in a neon matrix[85] approaches this scenario, owing to $D/hc = +0.433\ \text{cm}^{-1}$, as it is dominated by the xy_1 transition at $g_{\perp}^{\text{eff}} \approx 8$, although other off-axis resonances are seen. A more typical situation for Gd(III) is shown in Section 3.4.7.

3.3 ORIGINS OF ZFS

The reader is referred to Chapter 7 and other sources,[2–4,86,87] as well as the usual EPR-related ones[10,27,28] and other classics.[88] Moreover, useful are texts on crystal/ligand-field theory[46,47,89] as it is the variation in coordination environment about a paramagnetic center, usually a d block ion, that is manifest in the ZFS and, conversely, the ZFS provides information about coordination chemistry.

3.3.1 Spin–Spin Coupling (SSC)

The SSC interaction,[27] arising directly from magnetic dipole–magnetic dipole interactions among unpaired electrons (i.e., closed shells do not contribute, unless they are spin polarized), is crucial in giving rise to ZFS in organic triplet species where the single-electron SOC constant, ζ, is small, e.g., $30 < \zeta/hc < 40\ \text{cm}^{-1}$ for carbon.[90] As has been demonstrated recently, it can still be significant in d block complexes as well, even though ζ, and hence SOC interactions, is much larger.[91]

The SSC interaction between spins 1 and 2 is of the following form[10,27,28]:

$$\hat{\mathcal{H}}_{\text{SSC}}(\boldsymbol{r}) = \frac{\mu_0}{4\pi} g_e^{\,2} \mu_B^{\,2} \left\{ \frac{\hat{\boldsymbol{S}}_1 \cdot \hat{\boldsymbol{S}}_2}{r^3} - \frac{3\left(\hat{\boldsymbol{S}}_1 \cdot \boldsymbol{r}\right)\left(\hat{\boldsymbol{S}}_2 \cdot \boldsymbol{r}\right)}{r^5} \right\},\quad r^2 = x^2 + y^2 + z^2 \tag{3.28}$$

where $\boldsymbol{r}$ is the interelectron vector. The abovementioned equation is analogous to the electron magnetic dipole–nuclear magnetic dipole interaction that gives rise to dipolar hyperfine coupling (see Chapter 5). This equation can be expanded to give in the end[27,28]:

$$\hat{\mathcal{H}}_{\text{SSC}}(\boldsymbol{r}) = \hat{\boldsymbol{S}} \cdot \boldsymbol{D}_{\text{SSC}} \cdot \hat{\boldsymbol{S}};\ \boldsymbol{D}_{\text{SSC}} = \frac{\mu_0}{8\pi} g_e^{\,2} \mu_B^{\,2} \begin{bmatrix} \left\langle \frac{r^2-3x^2}{r^5} \right\rangle & \left\langle \frac{-3xy}{r^5} \right\rangle & \left\langle \frac{-3xz}{r^5} \right\rangle \\ \left\langle \frac{-3xy}{r^5} \right\rangle & \left\langle \frac{r^2-3y^2}{r^5} \right\rangle & \left\langle \frac{-3yz}{r^5} \right\rangle \\ \left\langle \frac{-3xz}{r^5} \right\rangle & \left\langle \frac{-3yz}{r^5} \right\rangle & \left\langle \frac{r^2-3z^2}{r^5} \right\rangle \end{bmatrix} \tag{3.29}$$

where $\hat{\boldsymbol{S}}$ is the total spin, $\hat{\boldsymbol{S}} = \hat{\boldsymbol{S}}_1 + \hat{\boldsymbol{S}}_2$, and the interelectron vector has components x, y, and z and magnitude r. The angular brackets in equation (3.29) mean that an average over the electron spatial distribution is performed. As can be seen, $\boldsymbol{D}_{\text{SSC}}$ is a symmetric and traceless matrix, as mentioned far earlier in the discussion of spin triplets. For example, if spin 2 is along the z-axis of spin 1, $(x, y, z) = (0, 0, r)$, then equation (3.29) simplifies to

$$\boldsymbol{D}_{\text{SSC}} = \frac{\mu_0}{4\pi} g_e^{\,2} \mu_B^{\,2} \begin{bmatrix} +\frac{1}{2}\left\langle \frac{1}{r^3} \right\rangle & 0 & 0 \\ 0 & +\frac{1}{2}\left\langle \frac{1}{r^3} \right\rangle & 0 \\ 0 & 0 & -\left\langle \frac{1}{r^3} \right\rangle \end{bmatrix} \tag{3.30}$$

which is the simplest form of a dipole–dipole coupling equation with the inverse-cube distance dependence.

The SSC interaction is not readily calculated using classical ligand-field methods, as it requires radial averages.[10] It is possible, however, to include SSC in a multielectron spin–orbital (i.e., L, S) Hamiltonian of the following form[10]:

$$\hat{\mathcal{H}}_{\text{SSC}} = -\rho \left\{ \left(\hat{\boldsymbol{L}} \cdot \hat{\boldsymbol{S}}\right)^2 + \frac{1}{2}(\hat{\boldsymbol{L}} \cdot \hat{\boldsymbol{S}}) - \frac{1}{3} L(L+1) S(S+1) \hat{1} \right\} \tag{3.31}$$

where ρ is the SSC constant.[10,92–94] The magnitude of ρ/hc is $\approx 0.2(1)\ \mathrm{cm}^{-1}$ for 3d ions. Modern quantum chemical software (see Chapter 7),[21] specifically ORCA,[95,96] calculates the SSC contribution to ZFS directly, using theory developed many years ago.[97] The studies by Neese and coworkers show that the SSC contribution to ZFS for d block complexes is on the order of at most a few tenths of a wavenumber (e.g., in four-coordinate Ni(II)[98]), but further generalization is unwarranted and a case-by-case analysis is needed.

3.3.2 Spin–Orbit Coupling (SOC)

For d block complexes, the dominant source of ZFS is from SOC and by analogy with D_{SSC} in Section 3.3.1, is denoted D_{SOC}. For organic triplets, D_{SOC} is usually negligible. The SOC interaction also gives rise to g anisotropy (see Chapter 2).[27] A simple physical picture of this effect arises from the spatially distinct nature of the different m_l values (projection of orbital angular momentum), $m_l = 0, \pm1, \pm2, \pm3$, for s, p, d, and f electrons, while the m_s values ($m_s = \pm 1/2$) have no spatial dependence. Thus, when there is no interaction between orbital and spin angular momentum (i.e., no SOC), then there is no basis for a single electron experiencing the electronic Zeeman interaction to behave other than as a free electron ($g_e = 2.0023$). However, when this interaction (i.e., SOC) exists, then the effects of different m_l values, with their associated directional differences, on the magnitude of spin angular momentum ($s = |m_s| = 1/2$) lead to g anisotropy (see Chapter 2). We will use both the p and the d orbitals as examples of this interaction, and in their real forms, as these are easier to visualize, which are related to their m_l values as follows:

$$\begin{aligned}
\mathrm{p}_{\pm1} &= \mp\frac{1}{\sqrt{2}}(\mathrm{p}_x \pm i\mathrm{p}_y), \mathrm{p}_0 = \mathrm{p}_z \\
\mathrm{d}_{\pm2} &= \frac{1}{\sqrt{2}}(\mathrm{d}_{x^2-y^2} \pm i\mathrm{d}_{xy}), \\
\mathrm{d}_{\pm1} &= \mp\frac{1}{\sqrt{2}}(\mathrm{d}_{xz} \pm i\mathrm{d}_{yz}), \mathrm{d}_0 = \mathrm{d}_{z^2}
\end{aligned} \tag{3.32}$$

Using the following definition of the orbital angular momentum operator, $\hat{\boldsymbol{l}}$:

$$\begin{aligned}
\hat{l}_x &= \frac{1}{2}(\hat{l}_+ + \hat{l}_-);\ \hat{l}_y = -\frac{i}{2}(\hat{l}_+ - \hat{l}_-) \\
\hat{l}_z|m_l\rangle &= m_l|m_l\rangle \\
\hat{l}_\pm|m_l\rangle &= \sqrt{l(l+1) - m_l(m_l \pm 1)}|m_l \pm 1\rangle
\end{aligned} \tag{3.33}$$

Table 3.1 results,[4] which gives an idea as to how the spatially different real-valued p and d orbitals are connected/transformed by this operator.

Thus, instead of having a pure spin angular momentum, the spatial characteristics of the different orbitals are mixed into $|m_S| = 1/2$, resulting in different g values in different directions (see Chapter 2).

If there are multiple unpaired electrons, then there are various magnitudes of spin angular momentum for a given total spin S ($S = 1$, $|m_S| = 0, 1$; $S = 3/2$, $|m_S| = 1/2, 3/2$, etc.), and each of these differing magnitudes couples differently with the various magnitudes of orbital angular momentum ($|m_L| = 0, 1, 2, 3, \ldots$; note that L not l is used as these are multiple electron states). The overall, qualitative result is that the energies of the various $|m_S|$ levels are different, as shown in Figure 3.2.

In ligand-field theory (LFT), there are several approaches to generating ZFS using the interaction described earlier. One is to use a multielectron (L, S) Hamiltonian of the following form,[10] including the orbital Zeeman interaction:

$$\hat{H}_{LS} = \lambda(\hat{\boldsymbol{L}} \cdot \hat{\boldsymbol{S}}) + \mu_B \boldsymbol{B} \cdot (\hat{\boldsymbol{L}} + g_e\hat{\boldsymbol{S}}) \tag{3.34}$$

where λ is the multielectron SOC constant.[10] For a shell that is less than half filled, λ is positive; and for a shell that is greater than half filled, λ is negative (a half-filled shell has $L = 0$, so there is no SOC). The absolute energies in zero field are given by[4]

$$E_{LS} = (\lambda/2)[J(J+1) - L(L+1) - S(S+1)] \tag{3.35}$$

As a simple illustration of SOC alone, a free d^4 ion (^{5}D, $L = 2$, $S = 2$), i.e., that with electronic configuration using $\mathrm{d}_{m_l}^{m_s}$ notation (1/2 not written out for m_s) corresponding to $\mathrm{d}_{+2}^{+}\mathrm{d}_{+1}^{+}\mathrm{d}_0^{+}\mathrm{d}_{-1}^{+}$, equation (3.35) gives the following absolute energies ($^{2S+1}L_J$ terms in parentheses): -6λ ($^5\mathrm{D}_0$); -5λ ($^5\mathrm{D}_1$); -3λ ($^5\mathrm{D}_2$); 0 ($^5\mathrm{D}_3$); $+4\lambda$ ($^5\mathrm{D}_4$), so that the average energy equals zero. These ZFS values correspond to relative energies (with their multiplicities in parentheses): 0 (1), λ (3), 3λ (5), 6λ (7), and 10λ (9). Given that for 3d ions, λ/hc values are in the tens (or even hundreds) of wavenumbers, the splitting between the ground state and the lower excited states for a system where this model is applicable (i.e., unquenched orbital angular momentum) would be much larger than expected where a spin Hamiltonian (equation (3.3)) is suitable (i.e., largely quenched orbital angular momentum). Moreover, this spin Hamiltonian predicts for d^4 a first excited state of multiplicity two (at energy λ) and a second excited state of multiplicity two (at energy 3λ). However, as seen here, a free d^4 ion has a first excited state of multiplicity three (not two) and a second excited state of multiplicity five (not two), along with two higher excited states not accounted for by the spin Hamiltonian.

This multielectron (L, S) method can actually underestimate ZFS because it treats only a single ${}^{2S+1}L$ term, usually the ground state, so that contributions to ZFS from excited ${}^{2S+1}L$ terms are not included, such as for d^4 from the triplet and singlet excited states (free-ion terms: ^{3}H, ^{1}I, etc.). It is therefore preferable to use a single-electron (l, s) Hamiltonian as follows:

$$\mathcal{H}_{ls} = \sum_{i}^{m} \zeta_i(\hat{l}_i \cdot \hat{s}_i) + \mu_B \boldsymbol{B} \cdot (\hat{l}_i + g_e \hat{s}_i) \qquad (3.36)$$

where the sum is over all of the m valence electrons in the nd^m ion and ζ is the single-electron SOC constant, which is related to the multielectron constant – for the ground ${}^{2S+1}L$ term – by $\lambda = \pm\zeta/2S$, where the positive sign is for a less than half-filled shell and a negative sign if for a greater than half-filled shell. The interelectronic repulsion terms, which provide the energy separation among the various ${}^{2S+1}L$ terms, usually given by Racah parameters (B, C) are not included here.[46] Recommended values of ζ and Racah parameters are given elsewhere.[99,100] What is also not included in either of the equations, (3.34) or (3.36), are the effects of bonding on the ion,[10] whether using a crystal field theory (CFT) treatment[47] or a covalent bonding (i.e., LFT) model, such as the angular overlap model (AOM).[46,101] These covalency effects tend to reduce (quench) the orbital angular momentum (SOC) so that the resulting ZFS can be of lower magnitude than expected. Software is available that can perform the calculation of energy levels incorporating all of these contributions (i.e., those in equation (3.34), plus interelectronic repulsion and CFT or LFT parameters), and hence, ZFS for d^m (and p^m, and even for selected f^m) systems, such as the very powerful program Ligfield,[102] as well as the local programs DDN (for d^m) and DPN (for p^m). The output of such programs is the energy levels of the various microstates and the splitting among the lowest states gives the ZFS. For example, in a spin triplet with axial symmetry, the three lowest m_S levels would be degenerate in the absence of SOC but would separate into a singlet and a doublet in the presence of SOC, with the separation between the levels corresponding to $|D|$ as in equation (3.6) (strictly speaking only $|D_{SOC}|$). Before the availability of personal computers, which can easily diagonalize the large matrices (e.g., 210×210 to correspond to the microstates for d^4 and d^6) that result from

Table 3.1. Effect of orbital angular momentum operator, $\hat{l}$, on real-valued p and d orbitals

p orbital	$\hat{l}_x$	$\hat{l}_y$	$\hat{l}_z$
p_x	0	$-i\|p_z\rangle$	$i\|p_y\rangle$
p_y	$i\|p_z\rangle$	0	$-i\|p_x\rangle$
p_z	$-i\|p_y\rangle$	$i\|p_x\rangle$	0
d orbital	$\hat{l}_x$	$\hat{l}_y$	$\hat{l}_z$
$d_{x^2-y^2}$	$-i\|d_{yz}\rangle$	$-i\|d_{xz}\rangle$	$2i\|d_{xy}\rangle$
d_{xy}	$i\|d_{xz}\rangle$	$-i\|d_{yz}\rangle$	$-2i\|d_{x^2-y^2}\rangle$
d_{xz}	$-i\|d_{xy}\rangle$	$-\sqrt{3}i\|d_{z^2}\rangle + i\|d_{x^2-y^2}\rangle$	$i\|d_{yz}\rangle$
d_{yz}	$\sqrt{3}i\|d_{z^2}\rangle + i\|d_{x^2-y^2}\rangle$	$i\|d_{xy}\rangle$	$-i\|d_{xz}\rangle$
d_{z^2}	$-\sqrt{3}i\|d_{yz}\rangle$	$\sqrt{3}i\|d_{xz}\rangle$	0

equation (3.36) plus the bonding and interelectronic terms, it was possible to derive second-order perturbation theory equations for ZFS in specific d^m configurations with, e.g., tetragonally distorted octahedral geometry. Examples of such equations are given in Sections 3.4.3–3.4.6. These methods, while quantitatively less than ideal, are very helpful in relating ZFS to simple coordination chemistry. The basis for these specific equations is the following second-order perturbation theory expression for the components of the $\boldsymbol{D}$-tensor originating from an SOC tensor, $\boldsymbol{\Lambda}$[28]:

$$\boldsymbol{D} = \varsigma^2 \Lambda,\ \Lambda_{ij} = -\sum_{n \neq 0} \frac{\langle\psi_0|\hat{L}_i|\psi_n\rangle\langle\psi_n|\hat{L}_j|\psi_0\rangle}{E_n - E_0},\ i,j = x,y,z \qquad (3.37)$$

where E_0 is the ground state energy (which could be set to zero), E_n the energy of the nth excited state, and $\hat{L}_i$ (or $\hat{l}_i$, depending on how the wavefunctions, $\psi_{0,n}$, are defined, i.e., multiple electron – equation (3.34) or single electron – equation (3.36), respectively) the orbital angular momentum operators as in equation (3.33). The varying effects of the $\hat{L}_i(\hat{l}_i)$ operators on the different orbitals (Table 3.1), together with the possible energies of the excited states, i.e., only low lying states typically contribute significantly, give an overall picture as to the origin of ZFS.

Alternatively, the contemporary method is for the molecular structure and spin ground state to be dropped into a quantum chemical theory program such as ORCA[95] (see Chapter 7) and the ZFS tensor (with its orientation in the molecular frame) directly provided. The analysis as to the origin of the components of the calculated ZFS requires a real understanding of the software, and it is difficult to provide a chemical basis for the calculated ZFS, while use of, e.g., Ligfield allows one to vary bonding or other parameters and see how the calculated ZFS is affected. Recently, what may be the 'best of both worlds' has been developed, namely a hybrid methodology, ligand-field density functional theory (LF-DFT, alternatively called ab initio ligand-field theory (AILFT)), which combines a multideterminant DFT-based method with LFT to provide directly AOM or other such bonding parameters. These methods have been used to understand the electronic structure of a variety of transition metal ion complexes.[21,24,25,65,80,103,104]

3.4 EXAMPLES OF ZFS IN ACTUAL SYSTEMS

3.4.1 Overview

As discussed earlier, ZFS is manifest in systems with multiple unpaired electrons, whether p, d, or f block molecules. The p block systems are generally triplets with relatively small ZFS that can largely, although not entirely, be explained by SSC.[90] Their current importance in areas such as photovoltaics and molecular devices warrants a brief Section 3.4.2. The f block complexes (except for f^7) represent the other extreme, where SOC is the dominant interaction so that a spin Hamiltonian model such as in equation (3.3) is not operative and is beyond the scope of this chapter. The d block complexes, therefore, represent the ideal scenario to illustrate ZFS in an EPR context – like the Baby Bear's porridge preferred by Goldilocks,[105] the d block ions have ZFS neither too small nor too large. Only two basic coordination numbers of these ions will be considered, six-coordinate and four-coordinate, and each with distortion from ideal octahedral and tetrahedral or square geometry, depending on the specific case for illustrative purposes. Selected $d^{(m,10-m)}$, $m = 2–5$, configurations will be very briefly considered in the context of its ZFS, usually revealed by HFEPR spectroscopy, to show how the ZFS is related to the electronic structure, i.e., d orbital configuration, of the particular complex.

3.4.2 p-Block Triplets

Organic triplets were of interest in the golden era of physical organic chemistry in the mid to late twentieth century. A classic example is mentioned earlier in Section 3.2.1.3, the triplet excited state of naphthalene.[30] Many other examples date from that era.[29,106] Organic triplets are also important in photosynthesis, as revealed many years ago.[107] Work actively continues in this area with the goal of developing artificial photosynthetic systems and other solar energy conversion devices.[108] An example from the Wasielewski laboratory at Northwestern University is shown in Figure 3.20, which presents the X-band EPR spectrum and simulation of the triplet state of a semisynthetic chlorophyll *a* derivative, zinc methyl 3-ethyl-pyrochlorophyllide. This triplet is formed through intersystem crossing from the excited singlet state (see Chapter 25). The data were collected in absorption mode, but owing to the polarized nature of the triplet levels, an absorption/emission spectrum is observed.[107]

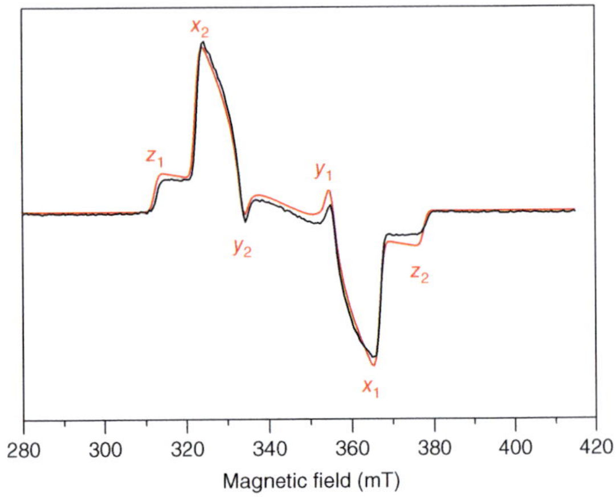

Figure 3.20. X-band ($\nu = 9.68$ GHz) EPR spectrum (black trace) of the triplet state of a semisynthetic chlorophyll *a* derivative, zinc methyl 3-ethyl-pyrochlorophyllide, with simulation (red trace). The simulation uses $S = 1$, $g = 2.00285$, $D/h = 916$ MHz, $E/h = 110$ MHz ($D/hc = 0.0306$ cm^{-1} $E/hc = 0.0037$ cm^{-1}), peak-to-peak linewidth = 1.494 mT, with the absorption/emission appearance, a consequence of the triplet's polarization, achieved using the experimental temperature function of EasySpin. (Data from Matthew Krzyaniak)

There has also been a renaissance in such systems, for an application that was only the most fantastic dream 40 years ago, namely the desire to create molecular-based quantum computing devices.[109] One such system is the nitrogen vacancy (NV) center in diamond,[110,111] which is a point defect consisting of a substitutional nitrogen neighboring a vacant carbon lattice point. There can be a negatively charged NV center (NV^-), which is a spin triplet, as well as a neutral center (NV^0), which is a spin quartet.[111]

The ZFS of the NV^- triplet is $D/h = +2.87$ GHz ($D/hc = +0.0957$ cm^{-1}), which can be measured by ODMR (see Chapter 25).[112] Spin polarization and readout can be accomplished by green (532 nm) excitation and red (637–800 nm) fluorescence detection, of relevance for the actual quantum computing use of such a center.[110]

3.4.3 d^2 and d^8 Ions

The d^2 or d^8 (i.e., two electrons or two holes) ion is one with a number of spin triplet states, one of which is the ground state. There are also singlet excited states that may interact via SOC (Section 3.3.2, e.g., equation (3.37)) with the triplet states. However, truly tetrahedral d^2 and octahedral d^8 systems exhibit little or no ZFS, as seen far earlier in Figure 3.1. This is also seen in EPR results for high-symmetry d^2 (V(III),[113] Cr(IV),[114] and Mn(V)[115]) and d^8 systems (mostly Ni(II) and also Co(I) and Cu(III)).[4] Distortion from octahedral geometry can lead to significant ZFS, as measured by HFEPR for tetragonally distorted Ni(II) complexes.[34,116] This ZFS primarily arises from SOC involving triplet excited states; perturbation theory gives the following expression for ZFS[10]:

$$D = -4\lambda^2\left(\frac{1}{\Delta_0} - \frac{1}{\Delta_1}\right)$$
$$= -\zeta^2\left(\frac{1}{\Delta E\left({}^3B_2 - {}^3B_1\right)} - \frac{1}{\Delta E({}^3E - {}^3B_1)}\right) \quad (3.38)$$

where we have provided a form that uses both more generic symbols (Δ_i) for ligand-field energies and definitions of the relevant energy splittings in modern (Mulliken) notation for C_{4v} or D_4 symmetry (also with the single-electron SOC constant). In ideal octahedral symmetry, $\Delta_0 = \Delta_1 = \Delta$, so $D = 0$, as seen for the octahedral Ni(II) complex described in Figure 3.1. In contrast, the Ni(II) complex, 2{[Ni(L-tyrosine)2(bpy)]} · $3H_2O \cdot CH_3OH$, where bpy = 2,2′-bipyridine, contains a highly distorted octahedral center (*cis*-$N_2N_2'O_2$ donor set) so that the Δ_i values in equation (3.38) are unequal (moreover, equation (3.38) would have to be modified for rhombic symmetry) and thus exhibits ZFS with $|D/hc| \approx 3$ cm^{-1}, which was measurable only by HFEPR.[116] For illustrative purposes, two spectra from this work, along with simulations, are shown in Figure 3.21. The key point is that, although the ZFS and thus the required microwave energies are roughly 30-fold the magnitudes used to generate Figure 3.5,

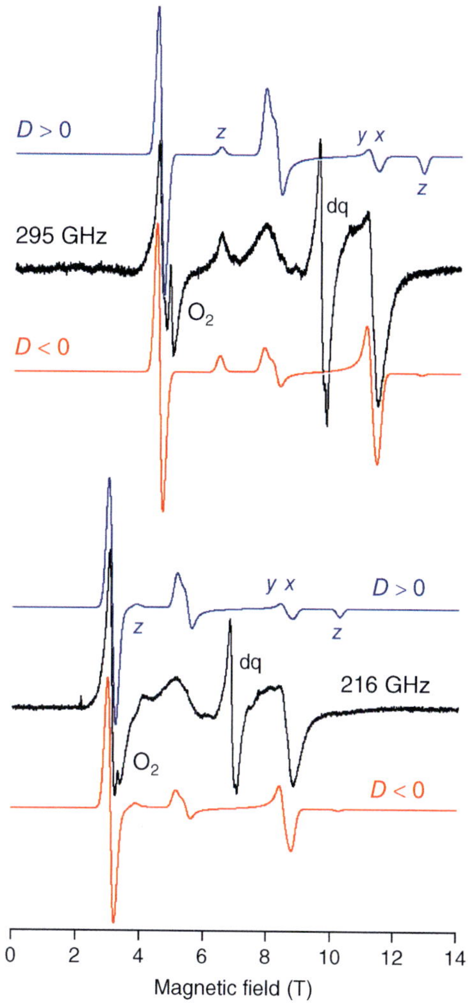

Figure 3.21. HFEPR spectra measured for a six-coordinate Ni(II) complex (see the text for description) as a powder at 10 K with the microwave frequencies 216.0 and 295.2 GHz. The simulations (colored traces) use $S = 1$, $g_x = 2.138$, $g_y = 2.156$, $g_z = 2.188$, $|D/hc| = 3.262$ cm^{-1}, and $|E/hc| = 0.1094$ cm^{-1}. Simulations with positive (blue traces) and negative (red traces) D and E values are shown, which indicate a better match of intensities using negative ZFS parameters. Resonances marked with O_2 are due to solid dioxygen adsorbed on the powder sample, which is typically observed in HFEPR below 30 K. DQ marks the 'double-quantum transition' (see Section 3.2.1.3), characteristic of Ni(II) systems.[117] (Reprinted with permission from A. Wojciechowska, A. Gągor, M. Duczmal, Z. Staszak, and A. Ozarowski, Inorg. Chem., 2013, 52, 4360. Copyright 2013 American Chemical Society)

the overall appearance of these triplet spectra is essentially the same (except for the double-quantum feature, which is commonly found in Ni(II) HFEPR spectra[117]). Another point worth noting is how comparison of the simulated intensities with experiment allowed determination of the sign of D; something which is often difficult to determine by magnetometry. In addition, relevant is the use of multiple frequencies to provide a tight determination of all of the spin Hamiltonian parameters (see Figure 3.21 caption for these values).

The case of tetrahedral d^8 with trigonal distortion (idealized C_{3v} symmetry as found, e.g., in complexes of type L_3NiX) has been treated computationally in detail,[98] using previous HFEPR data[118] and complete expressions for the relationships between components of D and ligand-field energies, wavefunction coefficients, and SOC constants provided.[98]

3.4.4 d^3 and d^7 Ions

Because of the Kramers nature of these ions, they are much more amenable to EPR investigation and have been far more widely studied than the $d^{2,8}$ ions. Extensive early EPR studies were made on a plethora of octahedral d^3 ions (V(II), especially Cr(III), Mn(IV), and even Mo(III), Tc(IV), and Re(IV)).[4] EPR of Mn(IV) has been recently reviewed in the context of the magnetic properties of its complexes.[119] The ZFS is generally quite small ($|D|/hc < 1\ \mathrm{cm}^{-1}$) in these systems. Most of the examples of high-spin d^7 are of tetrahedral Co(II) complexes.[43,44] In exact analogy with d^8 given earlier, the ground state of d^7 in ideal geometry exhibits no ZFS; this effect arises only from distortion from ideal symmetry and only by mixing-in of excited states, the most significant being the lowest lying quartet excited state. The ZFS, which is $2D$ for a spin quartet (see Section 3.2.2.2), is given by perturbation theory as[10]

$$2D = -8\lambda^2\left(\frac{1}{\Delta_0} - \frac{1}{\Delta_1}\right) = -\frac{8}{9}\zeta^2\left(\frac{1}{\Delta E\left({}^4B_2 - {}^4B_1\right)} - \frac{1}{\Delta E\left({}^4E - {}^4B_1\right)}\right) \tag{3.39}$$

where, as in equation (3.38), two forms are given.[120] A recent experimental and computational study on five-coordinate Co(II) complexes provides an example of the use of the latest techniques to understand d^7 systems, including trigonally distorted complexes of type L_3CoX.[25] Concerning d^3 ions, a recent HFEPR study on a series of Mn(IV) complexes included the latest computational analysis of their ZFS parameters.[65] As an illustration of the spectra used to extract these parameters for spin quartets, Figure 3.22 presents an HFEPR spectrum of a six-coordinate Mn(IV) complex with a *trans*-N2O4 donor set. As was the case for $S = 1$ (Figure 3.21), the sign of D could be determined by simulation and the use of numerous microwave frequencies in the range 60–170 GHz.[65] The similarity with the idealized spin quartet X-band spectrum shown in Figure 3.8 is apparent.

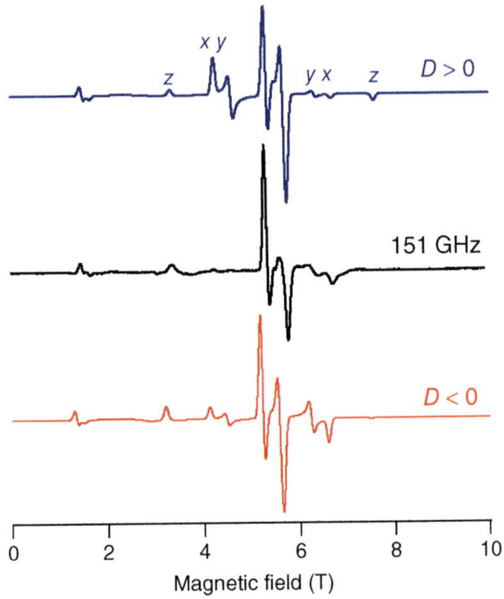

Figure 3.22. HFEPR spectrum measured for a six-coordinate Mn(IV) complex (see the text for description) as a powder at 4.5 K with the microwave frequency 151.2 GHz. Experimental (black trace) and simulated (colored traces). The simulations (colored traces) use $S = 3/2$, $g_{iso} = 1.98$, $|D/hc| = 0.98\ \mathrm{cm}^{-1}$, and $|E/hc| = 0.056\ \mathrm{cm}^{-1}$. Simulations with positive (blue traces) and negative (red traces) D and E values are shown, which indicate a better match of intensities using negative ZFS parameters. The spectra are approximately normalized to the amplitude of the two central, strongest resonances. (Reprinted with permission from M. Zlatar, M. Gruden, O. Y. Vassilyeva, E. A. Buvaylo, A. N. Ponomarev, S. A. Zvyagin, J. Wosnitza, J. Krzystek, P. Garcia-Fernandez, and C. Duboc, Inorg. Chem., 2016, 55, 1192. Copyright 2016 American Chemical Society)

3.4.5 d^4 and d^6 Ions

Tetrahedral d^4 and octahedral d^6 both have spin quintet ground states with unquenched orbital angular momentum, and hence large SOC, so the systems are difficult to study, even by HFEPR.[55,121] Octahedral d^4 and tetrahedral d^6 ions both have spin quintet ground states with little orbital contribution making such systems more amenable to study by EPR. There are early examples EPR on d^4 (Cr(II), Mn(III)), and d^6 systems (Fe(II)).[4] EPR of Mn(III) has been recently reviewed in the context of the magnetic properties of its complexes.[119] For illustrative purposes, we use both octahedral d^4 and tetrahedral d^6 with tetragonal distortions, the respective examples being $[Cr(H_2O)_6]^{2+}$[122] and $[Fe(SPh)_4]^{2-}$,[123] each studied by HFEPR. The cubic symmetry state splits into states in which the hole(d^4)/electron(d^6) is in either the d_{z^2} orbital or the $d_{x^2-y^2}$ orbital. The relevant expressions for ZFS are given in the following equation, which also include rhombic effects[6,10,63]

$$D = -3\left(\frac{\lambda^2}{\Delta} + \rho\right)\cos(2\delta),$$
$$E = -\sqrt{3}\left(\frac{\lambda^2}{\Delta} + \rho\right)\sin(2\delta) \qquad (3.40)$$

where Δ is the crystal field splitting ($\Delta = 10Dq$) and ρ is the SSC constant as in equation (3.31).[10] The parameter δ is an angle that accounts for distortions of the crystal field from simple cubic symmetry, which leads to splitting (and potentially mixing) between the d_{z^2} and $d_{x^2-y^2}$ orbitals:

$$\cos\delta\left(\frac{1}{\sqrt{2}}\left\{|m_L = +2, m_S\rangle + |m_L = -2, m_S\rangle\right\}\right) + \sin\delta|m_L = 0, m_S\rangle \qquad (3.41)$$

For $\delta = (0, \pi)$, the system is axial; and for $\delta \neq (0, \pi)$, there is rhombic distortion causing mixing between the $|L, m_L, S, m_S\rangle = |2, \pm 2, 2, (\pm 2, \pm 1, 0)\rangle$ spin quintet, which corresponds to the hole in the $d_{x^2-y^2}$ orbital, and the $|L, m_L, S, m_S\rangle = |2, 0, 2, (\pm 2, \pm 1, 0)\rangle$ spin quintet, which corresponds to the hole in the d_{z^2} orbital. For an octahedral d^4 system with tetragonal *elongation*, e.g., $[Cr(H_2O)_6]^{2+}$,[122] $\delta \approx 0$, so $D < 0$; for a tetrahedral d^6 system with tetragonal *compression*, e.g., $[Fe(SPh)_4]^{2-}$,[123] $\delta \approx \pi$, so $D > 0$. This simple correlation of the sign of D with distortion geometry is generally successful for $d^{4,6}$ except in cases involving heavy atoms, e.g., iodine.[124] In these cases, contributions to ZFS arise from states with significant spin density on the iodido ligands, which have very large free-atom SOC ($\zeta/hc = 5069\ cm^{-1}$ for $I^{\bullet}$[118]). Recent studies on Mn(III)[18,125] and Cr(II)[66] complexes using HFEPR and computational methods are recommended for giving the state-of-the art on $d^{4,6}$ systems. For illustration, Figure 3.23 presents HFEPR spectra recorded for the square planar Cr(II) complexes $[Cr(N(TMS)_2)_2(L)_2]$ (where TMS = Me_3Si^-, L = pyridine (py) or tetrahydrofuran (thf); donor sets *trans*-N2N′2 or *trans*-N2O2, respectively),[66] demonstrating the spectral appearance and extraction of ZFS parameters. The slight change in ZFS parameters as a function of this change in ligand was investigated computationally. The comparison between these spin quintet patterns at roughly 200 GHz and the simulation using X-band (Figure 3.14) is not as close as in the triplets (Section 3.4.3) because of the inherently more complicated nature of a higher spin system.

3.4.6 d^5 Ions

High-spin d^5 complexes, most commonly those of Mn(II) and Fe(III), have totally (i.e., spherically) symmetric ground states owing to the half-occupancy of all five d orbitals. EPR of Mn(II) has been recently reviewed in the context of the magnetic properties of its complexes.[119] If the symmetry is truly cubic, then there is no D_{SOC}, as calculated for $[Mn(NH_3)_6]^{2+}$, although a small D_{SSC} can still be present.[68] Experimentally, however, EPR spectra for Mn(II), in particular, can present a challenge in determining ZFS because the $|5/2, \pm 5/2\rangle \rightarrow |5/2, \pm 3/2\rangle$ and $|5/2, \pm 3/2\rangle \rightarrow |5/2, \pm 1/2\rangle$ transitions end up merely as shoulders on the dominant $|5/2, \pm 1/2\rangle \rightarrow |5/2, \mp 1/2\rangle$ transition. This is a consequence not only of the ZFS being small but also because these transitions are strongly affected by D-strain (i.e., a distribution in D values, similar to a distribution in g values; see Chapter 2)[73,126] and are conflated with unresolved hyperfine splitting from ^{55}Mn ($I = 5/2$, 100%), themselves affected by A-strain (i.e., a distribution in A values; see Chapter 5). As a result, in the conventional first-derivative mode, such broad transitions are suppressed and only a hyperfine-split sextet signal of the $|5/2, \pm 1/2\rangle \rightarrow |5/2, \mp 1/2\rangle$ transition is seen.[127] If EPR spectra in absorption mode can be recorded by pulse EPR (see Chapter 11), then observing and quantifying the modest ZFS in such Mn(II) species is facilitated.[69]

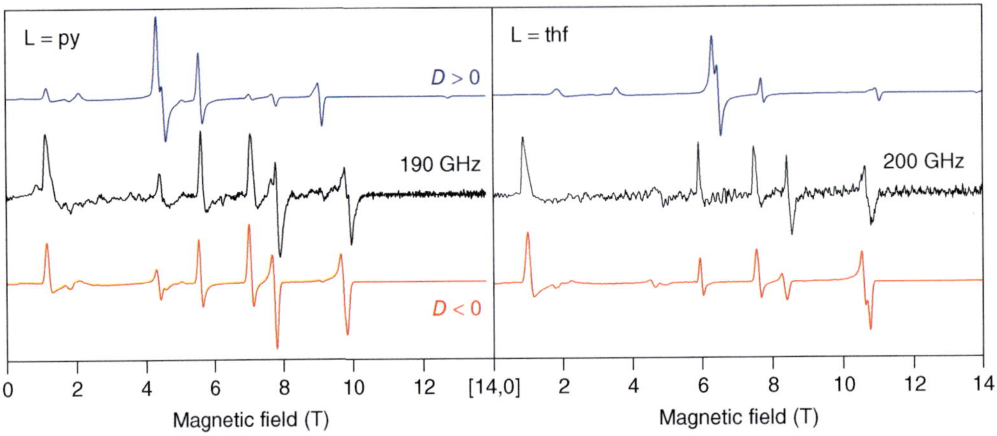

Figure 3.23. HFEPR spectra of powder $[Cr(N(TMS)_2)_2(L)_2]$, L = py (left) and L = thf (right) (black traces) with their simulations (colored traces) at 4.2 K with the frequencies indicated. Red traces are spectra simulated using positive D values, while blue traces are spectra simulated using negative D values. Simulation parameters for L = py: $S = 2$, $|D/hc| = 1.80\ cm^{-1}$, $|E/hc| = 0.020\ cm^{-1}$, $g_x = 1.97$, $g_y = 1.98$; for L = thf: $S = 2$, $|D/hc| = 2.00\ cm^{-1}$, $|E/hc| = 0.025\ cm^{-1}$, $g_x = 1.98$, $g_y = 1.99$. Precise values for g_z could not be determined but the range $1.0 < g_z < 2.0$ gave acceptable fits. (Deng, http://pubs.rsc.org/en/content/articlehtml/2015/cc/c5cc07025b. Used under CC-BY 3.0 https://creativecommons.org/licenses/by/3.0/)

If there are larger distortions from cubic symmetry,[72,73] then there is splitting within the various spin-quartet and -doublet excited states and these can lead to significant ZFS (i.e., $D/hc > 0.1\ cm^{-1}$) in the spin-sextet ground state. A perturbation theory expression for ZFS arising from coupling to the most relevant quartet excited state exists:[5,127]

$$D = \frac{1}{10}\zeta^2\left(\frac{2}{E_z} - \frac{1}{E_x} - \frac{1}{E_y}\right), E = \frac{1}{10}\zeta^2\left(\frac{1}{E_x} - \frac{1}{E_y}\right) \quad (3.42)$$

where $E_i, i = x, y, z$ corresponds to the energies for excited states that involve electron pairing of the electron in the $d_{x^2-y^2}$ orbital of the sextet ground state into the d_{yz}, d_{xz}, and d_{xy} orbitals, respectively. Thus, for cubic symmetry, $1/E_z = 1/E_x = 1/E_y$ so $D = 0$, and for axial symmetry, $1/E_x = 1/E_y$ so $E = 0$. This simple model was used to understand the ZFS, measured by 35-GHz EPR, in a highly distorted octahedral Mn(II) site in the metalloenzyme FosA.[127] For a complete understanding of ZFS in d^5 complexes, such as of Mn(II),[15,17,68] a full quantum chemical computational analysis is needed.

A classic example of ZFS in a spin sextet is the low-symmetry complex $[Fe(EDTA)]^-$,[72] mentioned in Section 3.2.4.3. $[Fe(EDTA)]^-$ was recently used as a test case for the development of a new method for analyzing D-strain in high-spin systems.[73] Briefly, in the high-field limit ($D \ll h\nu$), the ZFS parameters can be readily determined by simulation, and indeed, the sextet powder pattern seen at 275 GHz (Figure 3.24) is relatively easy to interpret by inspection, while the patterns at lower frequency are not easily explained and that at X-band is uninformative. The high-field-derived principal values of the $\boldsymbol{D}$-tensor are put into the center of a grid, and a distribution is set up around each of them. The match between experimental and simulated spectra at all frequencies is optimized following a grid-of-errors approach. The ability using this method to match exactly the linewidths of multiple EPR transitions is clearly shown in Figure 3.24, which presents experimental and simulated spectra for $[Fe(EDTA)]^-$ recorded at multiple frequencies. The parameters are given in the figure caption, but the key point is that this grid-of-errors D-strain method allows a unique set of best fit parameters, matching in particular the lower frequency spectra that are dominated by ZFS effects and hence strain as well (most notably at Q-band). The D-strain parameters themselves might be related to the structural distortions of different types within the molecule, leading to a broader distribution of say, D_z versus D_x.

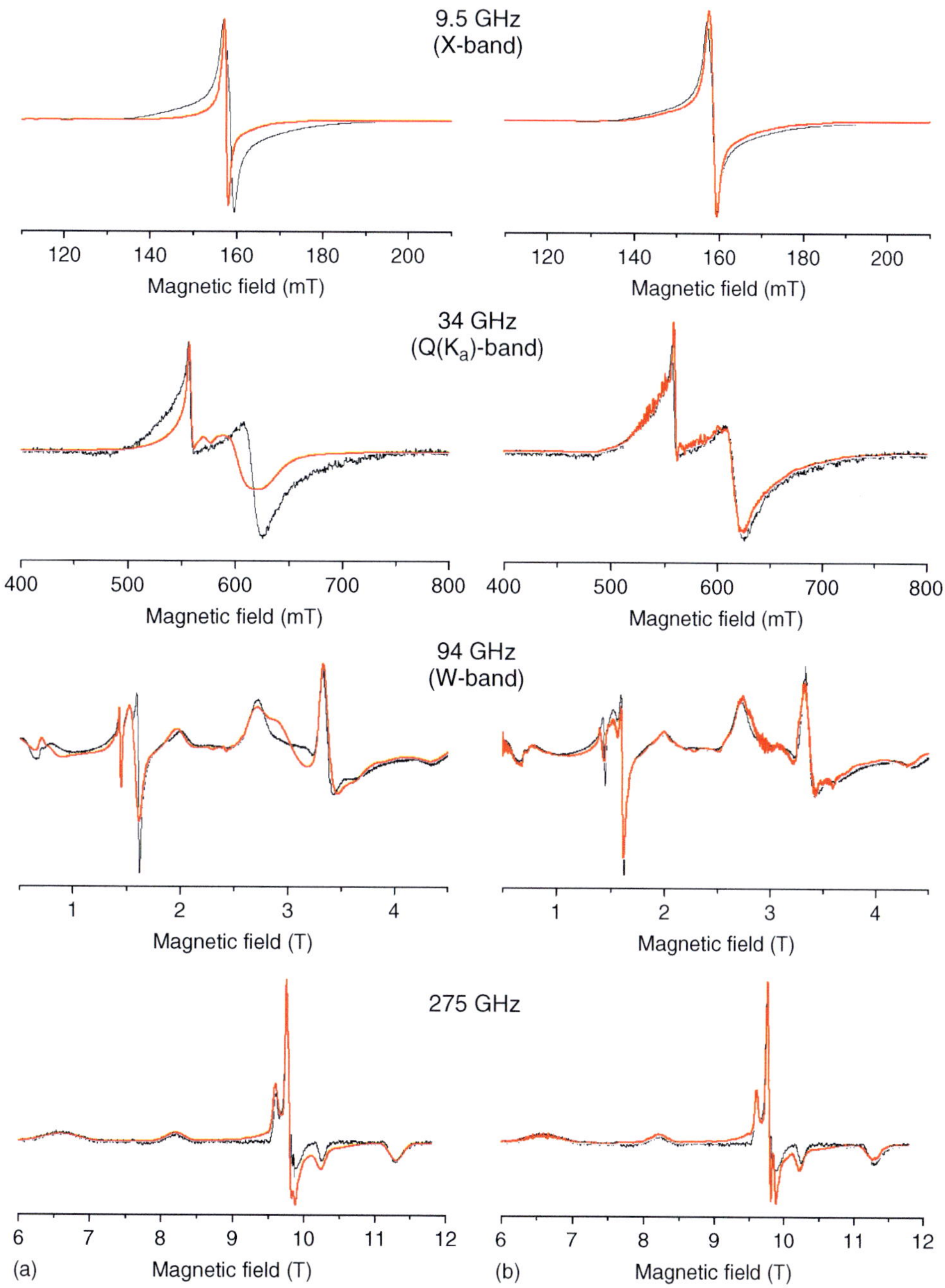

Figure 3.24. Experimental (black traces) and simulated (red traces) multifrequency EPR spectra for frozen aqueous Na[Fe(EDTA)] at pH 5, 40 K (10 K for 275 GHz spectrum), and the microwave frequencies indicated. (a) Simulations with the standard D-strain approximation built into EasySpin. (b) Simulations with the distributions of the ZFS parameters optimized from the grid-of-errors method.[73] Spin Hamiltonian parameters for the simulations: $S = 5/2$, $g_x = 2.005$, $g_y = 2.007$, $g_z = 2.004$; $D_x/h = 0.75$ GHz, $D_y/h = 14.45$ GHz, $D_z/h = -15.20$ GHz (corresponding to $D/h = -22.80$ GHz ($D/hc = 0.76$ cm^{-1}), $E/h = -6.85$ GHz ($E/hc = 0.23$ cm^{-1})). The strain in the ZFS was taken into account as Gaussian distributions in the D and E parameters with fwhm of 4.3 and 1.7 GHz, respectively. (Reproduced with permission from Ref. 73. © Elsevier, 2015)

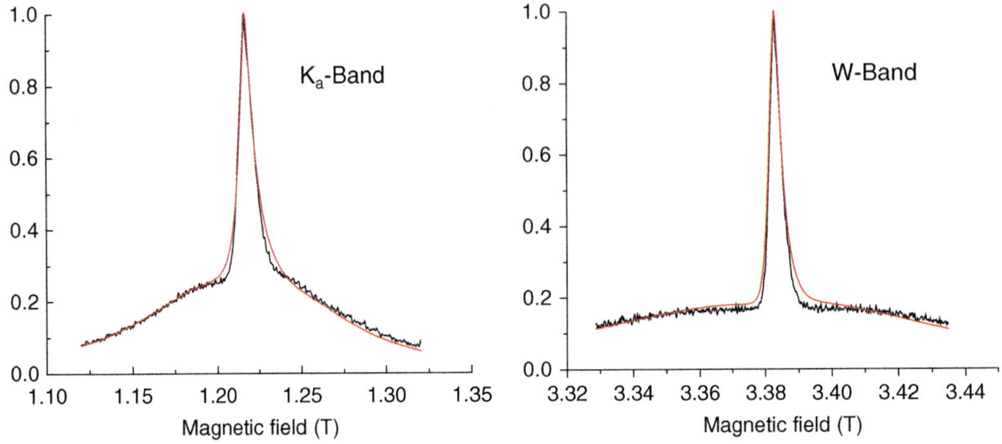

Figure 3.25. Experimental K_a-band ($\nu \approx 34$ GHz) and W-band ($\nu \approx 95$ GHz) electron-spin-echo-detected EPR spectra of a Gd(III) complex (black traces) in frozen aqueous solution and simulations (red traces) with $D/g\mu_B = 37$ and 39 mT (respectively, K_a-band and W-band; their average is $D/hc \approx 0.035$ cm^{-1}). The sharp peak is from the $|7/2, \pm 1/2\rangle \leftrightarrow |7/2, \mp 1/2\rangle$ transition; the other fine structure transitions are broadened out owing to a distribution of the ZFS parameters (D-strain). (Reprinted with permission from A. M. Raitsimring, C. Gunanathan, A. Potapov, I. Efremenko, J. M. L. Martin, D. Milstein, and D. Goldfarb, J. Am. Chem. Soc., 2007, 129, 14138. Copyright 2007 American Chemical Society)

3.4.7 f^7 Ions

The spectra shown earlier for Fe(III), an $S = 5/2$ system, clearly exhibited the effects of D-strain at all frequencies, but at W-band (and higher), multiple transitions can be identified, and even a simple D-strain protocol simulates the experimental spectrum reasonably well (Figure 3.24a). In contrast, for the most typical $S = 7/2$ system, namely Gd(III) in coordination complexes, the D-strain situation is more extreme. Such complexes are of interest as both MRI contrast agents[83] and spin labels for distance measurements (see Chapter 19).[82] As a result, the EPR spectrum, even at W-band, exhibits only a sharp signal from the $|S, m_S\rangle = |7/2, -1/2\rangle \leftrightarrow |7/2, +1/2\rangle$ transition with the others broadened out.[81,82] Despite this lack of resolution of the other $\Delta m_S = \pm 1$ transitions (e.g., $|S, m_S\rangle = |7/2, \pm 7/2\rangle \leftrightarrow |7/2, \pm 5/2\rangle$, matching the breadth of the pattern nevertheless allows determination of a D value. This is shown in Figure 3.25 for a Gd(III) complex in frozen aqueous solution.[82]

3.5 CONCLUSION

ZFS is the dominant phenomenon in EPR spectroscopy of p and d block systems with multiple unpaired electrons. The presence of ZFS greatly complicates such spectra relative to spin doublet systems, leading to multiple fine structure transitions, often with complicated field and frequency dependences. However, the ZFS parameters contain within themselves much valuable information on coordination geometry and bonding, especially for d block (transition metal) ion complexes. The advent of high-frequency and high-field EPR (HFEPR), along with other techniques such as inelastic neutron scattering (INS), along with well-established variable-temperature and variable-field (VTVH) methods such as magnetometry and Mössbauer and magnetic circular dichroism (MCD) spectroscopies has allowed the accurate measurement of ZFS parameters for a wide range of such species. Concurrent with this experimental progress, great advances have been made in theory and computational methodology so that the correlation between ZFS information, chiefly from EPR and also other techniques, and chemically useful understanding is now well established.

ACKNOWLEDGMENTS

The author thanks Drs Jurek Krzystek and Andrew Ozarowski, both of the National High Magnetic Field Laboratory (NHMFL), Tallahassee, Florida, USA, for many years of fruitful and enjoyable scientific and personal collaboration and, in particular, for many helpful comments on this chapter and providing the HFEPR data reproduced in Figures 3.21–3.23; the NHMFL for support of the experimental work that inspired this chapter and our many collaborators who have provided challenging systems from which ZFS parameters can be extracted and analyzed; Professor Stefan Stoll for spectral simulations and Professors Edgar Groenen and Daniella Goldfarb for previously published data reproduced in Figures 3.24 and 3.25, respectively, and Professor Michael R. Wasielewski and Dr Matthew D. Krzyaniak, Northwestern University, for unpublished data shown in Figure 3.20; and Professor William Weltner, Jr, University of Florida, Gainesville, Florida, USA, for revealing the wonders of ZFS to the author while in graduate school.

3.6 RECOMMENDED READING

In addition to A^2B^2,[10] the text by Weltner, which unfortunately is no longer in print, is also highly recommended for both theory and experimental data on spin triplets, as well as for high-spin systems of all types – up to spin nonets ($S = 9$).[28] Also recommended is the well-known EPR text by Weil and Bolton (see chapter 'Systems with More Than One Unpaired Electron' for high-spin systems).[27] There are many other classic, out-of-print texts that explain the ZFS phenomenon in the context of EPR (as well as every other aspect of EPR), such as by Pake and Estle,[49] Carrington and McLachlan,[128] and Atherton.[129] The Atherton text is available in an updated edition, but which is also out of print.[130] A handbook that is currently available is that edited by Misra,[131] which provides complete listing of the experimentally determined values of multifrequency transition-ion spin Hamiltonian parameters as well as a thorough review of computational modeling. For transition metal ion electronic structure, i.e., LFT, the out-of-print text by Ballhausen,[47] and that by Figgis and Hitchman,[46] which miraculously is still in print, are highly recommended. The out-of-print text by Drago covers EPR and most other spectroscopic techniques relevant to inorganic chemistry.[48]

REFERENCES

1. R. T. Boeré, in *Electron Paramagnetic Resonance: Volume 23*, The Royal Society of Chemistry: London, 2013, Vol. 23, Chap. Inorganic and organometallic radicals of main group elements, 22.
2. R. Boča, *Coord. Chem. Rev.*, 2004, **248**, 757.
3. R. Boča, *Struct. Bond.*, 2006, **117**, 1.
4. B. R. McGarvey, in *Transition Metal Chemistry*, ed R. L. Carlin, Marcel Dekker: New York, 1966, Chap. Electron spin resonance of transition metal complexes, 89.
5. J. R. Pilbrow, Transition Ion Electron Paramagnetic Resonance, Clarendon Press: Oxford, 1990, Vol. xx, 717.
6. J. Telser, J. Krzystek, and A. Ozarowski, *J. Biol. Inorg. Chem.*, 2014, **19**, 297.
7. J. Telser, A. Ozarowski, and J. Krzystek, in *Electron Paramagnetic Resonance: Volume 23*, The Royal Society of Chemistry: London, 2013, Vol. 23, Chap. High-frequency and -field electron paramagnetic resonance of transition metal ion (d block) coordination complexes, 209.
8. J. Krzystek, A. Ozarowski, and J. Telser, *Coord. Chem. Rev.*, 2006, **250**, 2308.
9. A. Abragam and B. Bleaney, Electron Paramagnetic Resonance of Transition Ions, Dover Publications, Inc.: Mineola, NY, 1986.
10. A. Abragam and B. Bleaney, Electron Paramagnetic Resonance of Transition Ions (Oxford Classic Texts in the Physical Sciences), Oxford University Press: Oxford, UK, 2012.
11. R. Boča and R. Herchel, *Coord. Chem. Rev.*, 2010, **254**, 2973.
12. A. Ozarowski, *Inorg. Chem.*, 2008, **47**, 9760.
13. J. Krzystek, S. A. Zvyagin, A. Ozarowski, S. Trofimenko, and J. Telser, *J. Magn. Reson.*, 2006, **178**, 174.
14. F. Neese, *J. Chem. Phys.*, 2007, **127**, 164112.
15. S. Zein, C. Duboc, W. Lubitz, and F. Neese, *Inorg. Chem.*, 2008, **47**, 134.
16. M. Sundararajan, D. Ganyushin, S. Ye, and F. Neese, *Dalton Trans.*, 2009, 6021.
17. C. Duboc, M.-N. Collomb, and F. Neese, *Appl. Magn. Reson.*, 2010, **37**, 229.
18. C. Duboc, D. Ganyushin, K. Sivalingam, M.-N. Collomb, and F. Neese, *J. Phys. Chem. A*, 2010, **114**, 10750.
19. D. Maganas, S. Sottini, P. Kyritsis, E. J. J. Groenen, and F. Neese, *Inorg. Chem.*, 2011, **50**, 8741.
20. R. Maurice, K. Sivalingam, D. Ganyushin, N. Guihéry, C. de Graaf, and F. Neese, *Inorg. Chem.*, 2011, **50**, 6229.

21. M. Atanasov, D. Ganyushin, K. Sivalingam, and F. Neese, in *Molecular Electronic Structures of Transition Metal Complexes II*, eds D. M. P. Mingos, P. Day and J. P. Dahl, Springer: Berlin, 2012, Chap. A modern first-principles view on ligand field theory through the eyes of correlated multireference wavefunctions, 149.
22. D. Ganyushin and F. Neese, *J. Chem. Phys.*, 2013, **138**, 104113.
23. M. Retegan, N. Cox, D. A. Pantazis, and F. Neese, *Inorg. Chem.*, 2014, **53**, 11785.
24. M. Atanasov, D. Aravena, E. Suturina, E. Bill, D. Maganas, and F. Neese, *Coord. Chem. Rev.*, 2015, **289–290**, 177.
25. D. Schweinfurth, M. G. Sommer, M. Atanasov, S. Demeshko, S. Hohloch, F. Meyer, F. Neese, and B. Sarkar, *J. Am. Chem. Soc.*, 2015, **137**, 1993.
26. D. Ganyushin, N. Gilka, P. R. Taylor, C. M. Marian, and F. Neese, *J. Chem. Phys.*, 2010, **132**, 144111.
27. J. A. Weil and J. R. Bolton, Electron Paramagnetic Resonance: Elementary Theory and Practical Applications, 2nd edn, John Wiley & Sons, Inc.: Hoboken, NJ, 2007.
28. W. Weltner Jr, Magnetic Atoms and Molecules, Dover Publications, Inc.: Mineola, NY, 1983, 156.
29. E. Wasserman, L. C. Snyder, and W. A. Yager, *J. Chem. Phys.*, 1964, **41**, 1763.
30. C. A.Hutchison Jr and B. W. Mangum, *J. Chem. Phys.*, 1961, **34**, 908.
31. J. Krzystek, J. England, K. Ray, A. Ozarowski, D. Smirnov, L.Que Jr, and J. Telser, *Inorg. Chem.*, 2008, **47**, 3483.
32. S. Stoll and A. Schweiger, *J. Magn. Reson.*, 2006, **178**, 42.
33. S. Stoll EasySpin. http://www.easyspin.org/
34. A. Pladzyk, A. Ozarowski, and Ł. Ponikiewski, *Inorg. Chim. Acta*, 2016, **440**, 84.
35. M. P. Hendrich SpinCount. http://www.chem.cmu.edu/groups/hendrich/facilities/
36. L. A. Pardi, J. Krzystek, J. Telser, and L.-C. Brunel, *J. Magn. Reson.*, 2000, **146**, 375.
37. R. W. Schwartz and R. L. Carlin, *J. Am. Chem. Soc.*, 1970, **92**, 6763.
38. B. R. McGarvey, *J. Chem. Phys.*, 1962, **37**, 3020.
39. J. R. Pilbrow, *J. Magn. Reson.*, 1978, **31**, 479.
40. J. C. Hempel, L. O. Morgan, and W. B. Lewis, *Inorg. Chem.*, 1970, **9**, 2064.
41. L. Banci, A. Bencini, C. Benelli, D. Gatteschi, and C. Zanchini, *Struct. Bond.*, 1982, **52**, 37.
42. J. A. Przyojski, H. D. Arman, and Z. J. Tonzetich, *Organometallics*, 2013, **32**, 723.
43. W. K. Myers, C. P. Scholes, and D. L. Tierney, *J. Am. Chem. Soc.*, 2009, **131**, 10421.
44. W. K. Myers, E. N. Duesler, and D. L. Tierney, *Inorg. Chem.*, 2008, **47**, 6701.
45. Wikipedia https://en.wikipedia.org/wiki/Q_band.
46. B. N. Figgis and M. A. Hitchman, Ligand Field Theory and its Applications, Wiley-VCH: New York, 2000.
47. C. J. Ballhausen, *Introduction to Ligand Field Theory*, New York: McGraw-Hill, 1962.
48. R. S. Drago, Physical Methods for Chemists, 2nd edn, Saunders College Publishing: Ft. Worth, TX, 1992.
49. G. E. Pake and T. L. Estle, in The physical principles of electron paramagnetic resonance, ed W. A. Benjamin, Advanced Book Program: London, ON, Canada, 1973, 306.
50. H. J. Gerritsen and E. S. Sabisky, *Phys. Rev.*, 1963, **132**, 1507.
51. C. Dobe, C. Noble, G. Carver, P. L. W. Tregenna-Piggott, G. J. McIntyre, A.-L. Barra, A. Neels, S. Janssen, and F. Juranyi, *J. Am. Chem. Soc.*, 2004, **126**, 16639.
52. C. Dobe, T. Strässle, F. Juranyi, and P. L. W. Tregenna-Piggott, *Inorg. Chem.*, 2006, **45**, 5066.
53. P. L. W. Tregenna-Piggott, H. Weihe, and A.-L. Barra, *Inorg. Chem.*, 2003, **42**, 8504.
54. I. Krivokapic, C. Noble, S. Klitgaard, P. L. W. Tregenna-Piggott, H. Weihe, and A.-L. Barra, *Angew. Chem. Int. Ed.*, 2005, **44**, 3613.
55. G. Carver, C. Dobe, T. B. Jensen, P. L. W. Tregenna-Piggott, S. Janssen, E. Bill, G. J. McIntyre, and A.-L. Barra, *Inorg. Chem.*, 2006, **45**, 4695.
56. J. Baranowski, T. Cukierda, B. Jeżowska-Trzebiatowska, and H. Kozłowski, *J. Magn. Reson.*, 1979, **33**, 585.
57. M. P. Hendrich and P. G. Debrunner, *Biophys. J.*, 1989, **56**, 489.
58. E. Münck, K. K. Surerus, and M. P. Hendrich, *Methods Enzymol. Part D*, 1993, **227**, 463.
59. J. England, M. Martinho, E. R. Farquhar, J. R. Frisch, E. L. Bominaar, E. Münck, and L.Que Jr, *Angew. Chem. Int. Ed.*, 2009, **48**, 3622.

60. D. C. Lacy, R. Gupta, K. L. Stone, J. Greaves, J. W. Ziller, M. P. Hendrich, and A. S. Borovik, *J. Am. Chem. Soc.*, 2010, **132**, 12188.

61. S. L. Dexheimer, J. W. Gohdes, M. K. Chan, K. S. Hagen, W. H. Armstrong, and M. P. Klein, *J. Am. Chem. Soc.*, 1989, **111**, 8923.

62. R. Gupta, T. Taguchi, A. S. Borovik, and M. P. Hendrich, *Inorg. Chem.*, 2013, **52**, 12568.

63. J. Krzystek, J. Telser, L. A. Pardi, D. P. Goldberg, B. M. Hoffman, and L.-C. Brunel, *Inorg. Chem.*, 1999, **38**, 6121.

64. J. Krzystek, L. A. Pardi, L.-C. Brunel, D. P. Goldberg, B. M. Hoffman, S. Licoccia, and J. Telser, *Spectrochim. Acta A*, 2002, **58**, 1113.

65. M. Zlatar, M. Gruden, O. Y. Vassilyeva, E. A. Buvaylo, A. N. Ponomarev, S. A. Zvyagin, J. Wosnitza, J. Krzystek, P. Garcia-Fernandez, and C. Duboc, *Inorg. Chem.*, 2016, **55**, 1192.

66. Y.-F. Deng, T. Han, Z. Wang, Z. Ouyang, B. Yin, Z. Zheng, J. Krzystek, and Y.-Z. Zheng, *Chem. Commun.*, 2015, **51**, 17688.

67. G. Aromí, J. Telser, A. Ozarowski, L.-C. Brunel, H.-M. Stoeckli-Evans, and J. Krzystek, *Inorg. Chem.*, 2005, **44**, 187.

68. C. Duboc, M.-N. Collomb, J. Pécaut, A. Deronzier, and F. Neese, *Chem. Eur. J.*, 2008, **14**, 6498.

69. A. Sharma, E. K. Gaidamakova, V. Y. Matrosova, B. Bennett, M. J. Daly, and B. M. Hoffman, *Proc. Natl. Acad. Sci. U. S. A.*, 2013, **110**, 5945.

70. C. Mantel, C. Baffert, I. Romero, A. Deronzier, J. Pécaut, M.-N. Collomb, and C. Duboc, *Inorg. Chem.*, 2004, **43**, 6455.

71. F. Biaso, C. Duboc, B. Barbara, G. Serratrice, F. Thomas, D. Charapoff, and C. Béguin, *Eur. J. Inorg. Chem.*, 2005, **2005**, 467.

72. R. Aasa, *J. Chem. Phys.*, 1970, **52**, 3919.

73. M. Azarkh and E. J. J. Groenen, *J. Magn. Reson.*, 2015, **255**, 106.

74. P. J. M. van Kan, E. van der Horst, E. J. Reijerse, P. J. M. van Bentum, and W. R. Hagen, *J. Chem. Soc. Faraday Trans.*, 1998, **94**, 2975.

75. P. S. Bagus and H. F. Schaefer III, *J. Chem. Phys.*, 1973, **58**, 1844.

76. T. C. DeVore, R. J. Van Zee, and W. Weltner Jr, *J. Chem. Phys.*, 1978, **68**, 3522.

77. V. V. Semenaka, O. V. Nesterova, V. N. Kokozay, V. V. Dyakonenko, R. I. Zubatyuk, O. V. Shishkin, R. Boča, J. Jezierska, and A. Ozarowski, *Inorg. Chem.*, 2010, **49**, 5460.

78. T. J. Morsing, H. Weihe, and J. Bendix, *Inorg. Chem.*, 2016, **55**, 1453.

79. J. M. Baker, J. R. Chadwick, G. Garton, and J. P. Hurrell, *Proc. R. Soc. Lond. A*, 1965, **286**, 352.

80. A. Borel, L. Helm, and C. A. E. Daul, *Chem. Phys. Lett.*, 2004, **383**, 584.

81. A. M. Raitsimring, A. V. Astashkin, D. Baute, D. Goldfarb, and P. Caravan, *J. Phys. Chem. A*, 2004, **108**, 7318.

82. A. M. Raitsimring, C. Gunanathan, A. Potapov, I. Efremenko, J. M. L. Martin, D. Milstein, and D. Goldfarb, *J. Am. Chem. Soc.*, 2007, **129**, 14138.

83. M. F. Tweedle and K. Kumar, in *Metallopharmaceuticals II: Diagnosis and Therapy*, eds M. J. Clarke and P. J. Sadler, Springer: Weinheim, Germany, 1999, Chap. Magnetic resonance imaging (MRI) contrast agents, 1.

84. M. Karbowiak, C. Rudowicz, and T. Ishida, *Inorg. Chem.*, 2013, **52**, 13199.

85. C. A. Baumann, R. J. Van Zee, K. J. Zeringue, and W.Weltner Jr, *J. Chem. Phys.*, 1981, **75**, 5291.

86. R. Boča, Theoretical Foundations of Molecular Magnetism, Elsevier: Amsterdam, 1999.

87. A. Bencini and D. Gatteschi, EPR of Exchange Coupled Systems, Springer Verlag: Berlin, Germany, 1990.

88. J. S. Griffith, *The Theory of Transition-Metal Ions*, Cambridge University Press: Cambridge, UK, 1964; Chapter 4.

89. D. S. McClure, Electronic spectra of molecules and ions in crystals. Part II. Spectra of ions in crystals. In *Solid State Physics*, Seitz, F.; Turnbull, D., Eds. Academic Press: New York, 1959; Vol. 9, pp 399–525.

90. S. Sinnecker and F. Neese, *J. Phys. Chem. A*, 2006, **110**, 12267.

91. F. Neese, *J. Am. Chem. Soc.*, 2006, **128**, 10213.

92. M. H. L. Pryce, *Phys. Rev.*, 1950, **80**, 1107.

93. R. E. Watson and M. Blume, *Phys. Rev.*, 1965, **139**, A1209.

94. R. E. Watson and M. Blume, *Phys. Rev.*, 1969, **184**, 606.

95. F. Neese, *Wiley Interdisciplinary Rev. Comput. Mol. Sci.*, 2012, **2**, 73.

96. F. Neese, *ORCA – an ab initio, Density Functional and Semiempirical Program Package*, 3.0.3,

Max Planck Institut für Chemische Energiekonversion: Mülheim an der Ruhr, Germany, 2014.

97. R. McWeeny and Y. Mizuno, *Proc. R. Soc. Lond. A*, 1961, **259**, 554.

98. S. Ye and F. Neese, *J. Chem. Theory Comput.*, 2012, **8**, 2344.

99. J. Bendix, M. Brorson, and C. E. Schäffer, *Inorg. Chem.*, 1993, **32**, 2838.

100. M. Brorson and C. E. Schäffer, *Inorg. Chem.*, 1988, **27**, 2522.

101. C. E. Schäffer, *Struct. Bond.*, 1968, **5**, 68.

102. J. Bendix, Ligfield. In *Comprehensive Coordination Chemistry II, Volume 2: Fundamentals: Physical Methods, Theoretical Analysis, and Case Studies*, Lever, A. B. P., Ed. Elsevier: Amsterdam, 2003; Vol. 2, pp 673–676.

103. M. Gruden-Pavlović, M. Perić, M. Zlatar, and P. Garcia-Fernández, *Chem. Sci.*, 2014, **5**, 1453.

104. M. Atanasov, C. Rauzy, P. Baettig, and C. Daul, *Int. J. Quantum. Chem.*, 2005, **102**, 119.

105. DTLK Goldilocks. http://www.dltk-teach.com/rhymes/goldilocks_story.htm

106. J. A. Weil, J. R. Bolton, and J. E. Wertz, Electron Paramagnetic Resonance: Elementary Theory and Practical Applications, John Wiley & Sons, Inc: New York, 1994, Vol. xxi, 568.

107. M. C. Thurnauer, J. J. Katz, and J. R. Norris, *Proc. Natl. Acad. Sci. U. S. A.*, 1975, **72**, 3270.

108. M. R. Wasielewski, *Acc. Chem. Res.*, 2009, **42**, 1910.

109. C. G. Yale, B. B. Buckley, D. J. Christle, G. Burkard, F. J. Heremans, L. C. Bassett, and D. D. Awschalom, *Proc. Natl. Acad. Sci. U. S. A.*, 2013, **110**, 7595.

110. K. Bayat, J. Choy, M. Farrokh Baroughi, S. Meesala, and M. Loncar, *Nano Lett.*, 2014, **14**, 1208.

111. S. Felton, A. M. Edmonds, M. E. Newton, P. M. Martineau, D. Fisher, and D. J. Twitchen, *Phys. Rev. B*, 2008, **77**, 081201.

112. K. Sasaki, Y. Monnai, S. Saijo, R. Fujita, H. Watanabe, J. Ishi-Hayase, K. M. Itoh, and E. Abe, *Rev. Sci. Instrum.*, 2016, **87**, 053904.

113. P. J. Alonso, J. Forniés, M. A. García-Monforte, A. Martín, and B. Menjón, *Chem. Commun.*, 2001, 2138.

114. P. J. Alonso, J. Forniés, M. A. García-Monforte, A. Martín, B. Menjón, and C. Rillo, *Chem. Eur. J.*, 2002, **8**, 4056.

115. P. Jiang, J. Li, A. Ozarowski, A. W. Sleight, and M. A. Subramanian, *Inorg. Chem.*, 2013, **52**, 1349.

116. A. Wojciechowska, A. Gągor, M. Duczmal, Z. Staszak, and A. Ozarowski, *Inorg. Chem.*, 2013, **52**, 4360.

117. P. J.van Dam, A. A. K. Klaassen, E. J. Reijerse, and W. R. Hagen, *J. Magn. Reson.*, 1998, **130**, 140.

118. P. J. Desrochers, J. Telser, S. A. Zvyagin, A. Ozarowski, J. Krzystek, and D. A. Vicic, *Inorg. Chem.*, 2006, **45**, 8930.

119. C. Duboc, *Chem. Soc. Rev.*, 2016, **45**, 5834.

120. J. P. Jesson, *J. Chem. Phys.*, 1968, **48**, 161.

121. G. Carver, P. L. W. Tregenna-Piggott, A.-L. Barra, A. Neels, and J. A. Stride, *Inorg. Chem.*, 2003, **42**, 5771.

122. J. Telser, L. A. Pardi, J. Krzystek, and L.-C. Brunel, *Inorg. Chem.*, 1998, **37**, 5769.

123. M. J. Knapp, J. Krzystek, L.-C. Brunel, and D. N. Hendrickson, *Inorg. Chem.*, 2000, **39**, 281.

124. S. Mossin, H. Weihe, and A.-L. Barra, *J. Am. Chem. Soc.*, 2002, **124**, 8764.

125. S. Romain, C. Duboc, F. Neese, E. Rivière, L. R. Hanton, A. G. Blackman, C. Philouze, J.-C. Leprêtre, A. Deronzier, and M.-N. Collomb, *Chem. Eur. J.*, 2009, **15**, 980.

126. W. R. Hagen, *Mol. Phys.*, 2007, **105**, 2031.

127. S. K. Smoukov, J. Telser, B. A. Bernat, C. L. Rife, R. N. Armstrong, and B. M. Hoffman, *J. Am. Chem. Soc.*, 2002, **124**, 2318.

128. A. Carrington and A. D. McLachlan, Introduction to Magnetic Resonance with Applications to Chemistry and Chemical Physics, Harper & Row: New York, 1967, Vol. xix, 266.

129. N. M. Atherton, Electron Spin Resonance. Theory and Applications, Halsted Press: New York, London, Sydney, Toronto, 1973, Vol. x, 438.

130. N. M. Atherton, Electron Spin Resonance. Theory and Applications, Ellis Horwood: Chichester, UK, 1993, Vol. ix, 585.

131. S. K. Misra Handbook of Multifrequency Electron Paramagnetic Resonance: Data and Techniques, John Wiley & Sons, Inc. Chichester. http://www.wiley.com/WileyCDA/WileyTitle/productCd-3527412220.html

Chapter 4
EPR Interactions – Coupled Spins

Eric J.L. McInnes and David Collison
The University of Manchester, Manchester, UK

EPR Spectroscopy: Fundamentals and Methods.
Edited by Daniella Goldfarb and Stefan Stoll

ISBN: 978-1-119-16299-5
Also published in eMagRes (online edition)
DOI: 10.1002/9780470034590.emrstm1502

4.1 INTRODUCTION

Compared to nuclei, electrons have a very large magnetic moment and hence electron–electron interactions can be very strong and can act over much greater distances. There are simple dipolar interactions, and there are exchange interactions. The former are covered in detail in Chapters 19 and 20 of this encyclopedia; the latter are the subject of this chapter. The underlying theory goes back to Heisenberg, Dirac, and van Vleck and is covered in any good text on quantum mechanics: we do not cover it here. Rather, we are interested in the effects of exchange coupling on the observable EPR properties of species that contain more than one paramagnetic center. We do this using the spin Hamiltonian formalism (and we assume the reader has a working knowledge of this) where the isotropic exchange can be treated phenomenologically as the dot product between electron spin operators. For the reasons stated above, exchange effects are often the dominant interaction but we will also explore situations where this is not the case. We restrict ourselves to discrete systems, i.e., containing a finite number of paramagnets, so we do not cover extended lattices. Such discrete systems are found across the scientific spectrum, from condensed matter physics to biology. This chapter is intended as an introduction, so for many examples we discuss the simplest exchange-coupled system, that of two coupled $s = 1/2$ spins. For the reader who wishes to delve deeper, we recommend the authoritative text on EPR of exchange-coupled systems by Bencini and Gatteschi.[1]

4.2 THE ISOTROPIC EXCHANGE INTERACTION

When two electron spins s_1 and s_2 interact, their individual electron spin quantum numbers s_1 and s_2 are no longer good quantum numbers for the pair. Instead we can define a *total* electron spin quantum number S. (In this chapter, we will use an uppercase S to refer to a total spin resulting from the interaction of two local spins described by lowercase s.) The total spin of the pair S can take values from $s_1+s_2, s_1+s_2-1, \ldots |s_1-s_2|$ according to the usual angular momentum (vector addition) rules. To take the simplest example, two individual electron spins $s = {}^1\!/_2$ will interact to generate a total spin $S=1$ (a spin triplet) and 0 (a spin singlet); one of these will be an excited state and the other the ground state of the pair (Figure 4.1). If the maximum spin is the ground state, this is referred to as a *ferromagnetic* interaction; an *antiferromagnetic* interaction gives the lower spin ground state.

If the two electrons are in degenerate orbitals on the same ion, then Hund's rules dictate that the ground state has maximum S, i.e., a ferromagnetic interaction results, because this minimizes electron–electron repulsions. So, for example, a Ni(II) ion in an octahedral ligand environment ($t_{2g}{}^6e_g{}^2$ configuration) or the dianion of benzene $C_6H_6{}^{2-}$ (π^{*2} configuration) have $S=1$ ground states. In this case, the interaction is very strong, i.e., excited states are at very high energy (and may be observed by optical spectroscopy) and we can consider the system as an isolated $s=1$ as far as most magnetic measurements are concerned, including magnetic resonance; chemists are possibly guilty of forgetting that this is a result of electron exchange in the first place. The EPR spectra of such isolated $s > {}^1\!/_2$ species are covered elsewhere in this encyclopedia (see Chapter 3).

If the spins are on different centers, then the interaction will be much weaker and can be ferro- or antiferromagnetic. If the interaction occurs via direct orbital overlap, then this is sometimes termed direct exchange; if it is mediated by a diamagnetic bridging pathway, then it is often referred to as superexchange. We will not discuss the actual mechanism in terms of molecular orbital overlaps here: we refer the interested

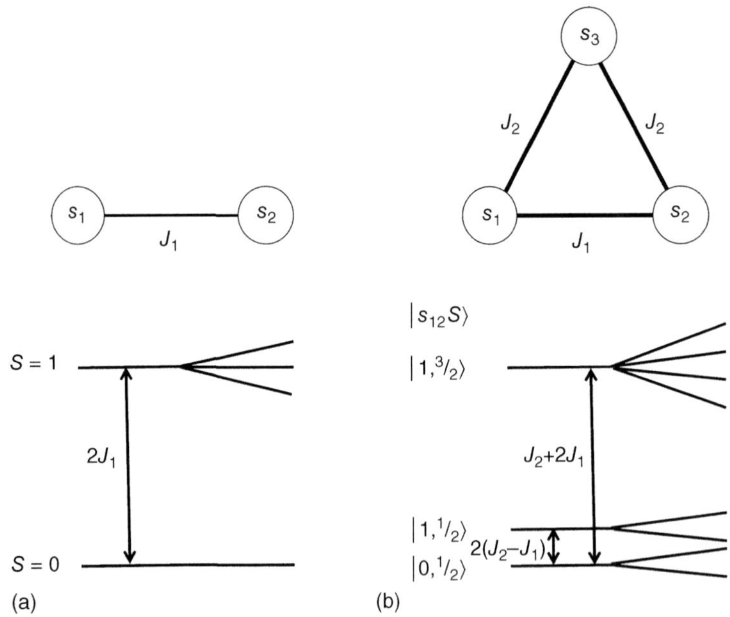

Figure 4.1. (a) Total spin states arising from isotropic exchange between two $s = {}^1\!/_2$ paramagnets according to Hamiltonian (4.1), and effect of an applied magnetic field. (b) States arising from an isosceles triangle of three interacting $s = {}^1\!/_2$ paramagnets as drawn in the scheme and in Hamiltonian (4.8). The states are labeled according to $|s_{12}S\rangle$; if $J_1 = J_2$ then the two $s = {}^1\!/_2$ states are degenerate

reader to the textbook by Bencini and Gatteschi.[1] Exchange interactions between spins on different centers are typically in the range 0–1000 cm^{-1} and in this case, for an ensemble, we will have a Boltzmann population distribution across the different total spin states and this needs to be considered explicitly in magnetic measurements.

Mathematically, in the spin Hamiltonian formalism, the isotropic exchange interaction (J) is described by an operator of the form $\hat{s}_1 \cdot \hat{s}_2$. In the form most often used by chemists, the exchange Hamiltonian is

$$\hat{H} = -2J\hat{s}_1 \cdot \hat{s}_2 \tag{4.1}$$

It strictly has energy units but, as in many areas of magnetism, SI units are clumsy and EPR spectroscopists tend to use energy equivalent units such as cm^{-1}, K, meV, or MHz, as is the case for hyperfine interactions etc. We will use cm^{-1} in this chapter. Equation (4.1) assumes that the exchange is isotropic (J is a scalar) – we will return to this assumption later. Note that different forms of equation (4.1) are used in the literature, including $-J\hat{s}_1 \cdot \hat{s}_2$ and $+J\hat{s}_1 \cdot \hat{s}_2$ (the latter being common among physicists): it is vital to define which formalism is being used. Here, where we quote literature values, we have converted the numbers to be consistent with equation (4.1), which gives the singlet–triplet energy gap as $2J$, with positive and negative signs denoting ferro- and antiferromagnetic interactions, respectively. We will now explore the basics of this formalism.

4.3 TWO COUPLED SPINS s_i: THE HAMILTONIAN MATRIX AND EPR

For $s_1 = s_2 = 1/2$, there are four possible combinations of the magnetic quantum numbers m_i. These define the 'uncoupled' basis for the pair: $|m_1 m_2\rangle = |+1/2,+1/2\rangle$, $|-1/2,-1/2\rangle$, $|+1/2,-1/2\rangle$, and $|-1/2,+1/2\rangle$ or $|\alpha\alpha\rangle$, $|\beta\beta\rangle$, $|\alpha\beta\rangle$, and $|\beta\alpha\rangle$. Hence we have states with $\Sigma m_i =$ +1, −1 and 0 (twice). [We could label the states more fully as $|s_1 s_2 m_1 m_2\rangle$ but the s_1, s_2 terms are constant for simple pairs.]

Expanding the dot product of the vector spin operators in equation (4.1) we have

$$\begin{aligned}\hat{H} &= -2J(\hat{s}_{1z}\hat{s}_{2z} + \hat{s}_{1x}\hat{s}_{2x} + \hat{s}_{1y}\hat{s}_{2y}) \\ &= -2J[\hat{s}_{1z}\hat{s}_{2z} + (\hat{s}_{1+}\hat{s}_{2-} + \hat{s}_{1-}\hat{s}_{2+})/2]\end{aligned} \tag{4.2}$$

The latter form, involving the raising and lowering operators, is the most convenient to work with. When evaluating the $\langle m_1 m_2|\hat{H}|m_1 m_2\rangle$ matrix elements, remember that $\hat{s}_i$ only operates on the spin i component of $|m_1 m_2\rangle$. This gives the energy matrix

	$\|\alpha\alpha\rangle$	$\|\beta\beta\rangle$	$\|\alpha\beta\rangle$	$\|\beta\alpha\rangle$
$\langle\alpha\alpha\|$	$-J/2$	0	0	0
$\langle\beta\beta\|$	0	$-J/2$	0	0
$\langle\alpha\beta\|$	0	0	$+J/2$	$-J$
$\langle\beta\alpha\|$	0	0	$-J$	$+J/2$

(4.3)

We find two eigenvalues of $\hat{H}$ immediately, being $-J/2$ with eigenfunctions $|\alpha\alpha\rangle$ and $|\beta\beta\rangle$. Diagonalizing the remaining 2×2 block, we get

	$\|\alpha\alpha\rangle$	$\|\beta\beta\rangle$	$\|\alpha\beta+\beta\alpha\rangle/\sqrt{2}$	$\|\alpha\beta-\beta\alpha\rangle/\sqrt{2}$
$\langle\alpha\alpha\|$	$-J/2$	0	0	0
$\langle\beta\beta\|$	0	$-J/2$	0	0
$\langle\alpha\beta+\beta\alpha\|/\sqrt{2}$	0	0	$-J/2$	0
$\langle\alpha\beta-\beta\alpha\|/\sqrt{2}$	0	0	0	$+3J/2$

(4.4)

giving the remaining two eigenvalues of $-J/2$ and $+3J/2$ with eigenfunctions $|\alpha\beta+\beta\alpha\rangle/\sqrt{2}$ and $|\alpha\beta-\beta\alpha\rangle/\sqrt{2}$, respectively.

The result is three states (i.e., a triplet) at energy $-J/2$ with $\Sigma m_i = +1, -1$, and 0, and one state (a singlet) at $+3J/2$ with $\Sigma m_i = 0$. [Note the triplet spin functions are symmetric with respect to exchange of the electrons; the singlet function is antisymmetric.] This is the result we had earlier, and we can recast in the 'coupled basis' $|S,M\rangle$, i.e., with total $S=1$ with $M=+1$, -1, 0 and total $S=0$ with $M=0$. The $|S,M\rangle$ states are linear combinations of the uncoupled $|s_1 s_2 m_1 m_2\rangle$ basis states where $M = m_1 + m_2$. The coefficients in these linear combinations are called Clebsch–Gordan coefficients: we have derived them by matrix diagonalization in this example, but they can be calculated from standard rules for coupling angular momentum states in quantum mechanics with knowledge of s_1, s_2, m_1, m_2, S, and M. The singlet is the ground state if J is negative, with the definition in equation (4.1), and the separation between singlet and triplet states is $2J$.

We now include the Zeeman interaction terms for the individual spins

$$\hat{H} = -2J\hat{s}_1 \cdot \hat{s}_2 + \sum_i \mu_B \boldsymbol{B} \cdot \boldsymbol{g}_i \cdot \hat{s}_i \tag{4.5}$$

Taking isotropic $g_1 = g_2$, and the magnetic field along the z-axis, then the Zeeman term can be simplified to $\mu_B B(g_1\hat{s}_{1z} + g_2\hat{s}_{2z})$ and we have

	$\|\alpha\alpha\rangle$	$\|\beta\beta\rangle$	$\|\alpha\beta+\beta\alpha\rangle/\sqrt{2}$	$\|\alpha\beta-\beta\alpha\rangle/\sqrt{2}$
$\langle\alpha\alpha\|$	$-J/2+(g_1+g_2)\mu_B B/2$	0	0	0
$\langle\beta\beta\|$	0	$-J/2-(g_1+g_2)\mu_B B/2$	0	0
$\langle\alpha\beta+\beta\alpha\|/\sqrt{2}$	0	0	$-J/2$	0
$\langle\alpha\beta-\beta\alpha\|/\sqrt{2}$	0	0	0	$+3J/2$

(4.6)

If we now consider a perturbing Hamiltonian $\hat{H}_1$ for the microwave magnetic field that is proportional to $(\hat{s}_{1x}+\hat{s}_{2x})$, we find the only nonzero matrix elements are of the form $\langle\alpha\alpha|\hat{H}_1|\alpha\beta+\beta\alpha\rangle/\sqrt{2}$ and $\langle\beta\beta|\hat{H}_1|\alpha\beta+\beta\alpha\rangle/\sqrt{2}$. In other words, the only allowed EPR transitions are the $\Delta M=\pm1$ transitions *within* the spin triplet. The $\Delta M=\pm1$ transitions between the $|S,M\rangle=|0,0\rangle$ and $|1,\pm1\rangle$ states are forbidden, and we have a selection rule that the total spin is conserved: $\Delta S=0$. This can be relaxed if there are significant perturbations that mix the singlet and triplet states (see later).

A famous example of a strongly coupled dimer, and a landmark in the history of EPR, is that of copper(II) acetate hydrate.[2] This complex is a dimer of formula $[Cu_2(O_2CMe)_4(H_2O)_2]$. The EPR spectrum is that of an isolated $S=1$ with zero-field splitting ($D=-0.3\,cm^{-1}$); the intensity decreases with decreasing temperature. The reason is that the two Cu(II) ($s=1/2$) ions are antiferromagnetically coupled with $2J\approx-300\,cm^{-1}$. Thermal population of the excited triplet state gives rise to the EPR spectrum; on cooling, this state is depopulated in favor of the singlet, with the material becoming diamagnetic (only the singlet state significantly populated) below about 50 K. It was EPR that led to the dimer formulation, preceding the X-ray diffraction determination of the structure. For exchange interactions in this range, the value of J is accurately determined by modeling the magnetic moment measured as a function of temperature. However, in some cases, $|J|$ can be too large to do this. For example, thiazyl radicals tend to exist as head-to-head dimers in the solid state that are so strongly antiferromagnetically coupled (weak intermolecular sulfur–sulfur bonding) that they are diamagnetic at room temperature as measured by a magnetometer. This is true of $[Cl\text{-}CNSSS](AsF_6)$ (Figure 4.2), but despite this, a well-resolved triplet EPR spectrum is observed.[3] Following the EPR intensity as a function of temperature gives $2J=-1900\,cm^{-1}$. This means that <0.1% of dimers are in the triplet state at room temperature, explaining the diamagnetism, and also illustrates that EPR is a very sensitive technique.

4.4 COUPLING TWO GENERAL SPINS

There are four eigenstates for the problem with two spins $s=1/2$ because there are four possible combinations of m_1 and m_2. A general spin s has a $2s+1$ degeneracy hence coupling two spins s_1 and s_2 gives a total degeneracy of $(2s_1+1)\times(2s_2+1)$, and this is the dimension of the Hamiltonian matrix. Hamiltonian (4.1) gives energies of the resultant total spin states as

$$E=-J[S(S+1)-s_1(s_1+1)-s_2(s_2+1)] \quad (4.7)$$

This can be derived by expressing the total spin in terms of $\hat{S}^2$ operators

$$\hat{S}^2=(\hat{s}_1+\hat{s}_2)^2=\hat{s}_1^2+\hat{s}_2^2+2\hat{s}_1\hat{s}_2$$

Hence, $\hat{H}=-2J\hat{s}_1\hat{s}_2=-J(\hat{S}^2-\hat{s}_1^2-\hat{s}_2^2)$ and the standard angular momentum relationship that $\hat{s}^2|s\rangle=s(s+1)|s\rangle$ gives equation (4.7). The energy gap between adjacent total spin multiplets S and $S-1$ for an arbitrary pair is $2JS$, which is a form of the Landé interval rule. For example, coupling two $s=1$ ions gives nine coupled states grouped as $S=0$ (singlet); $S=1$ (triplet); and $S=2$ (quintet) at energies $4J$, $2J$, and $-2J$, respectively. If these gaps are comparable to k_BT, then EPR spectra may be observable from multiple spin states. Figure 4.3 shows a well-resolved recent example of this, in an antiferromagnetically coupled Cr(III) ($s=3/2$) dimer where EPR transitions are observed within each of the total spin $S=1$, 2, and 3 excited mulitplets.[4]

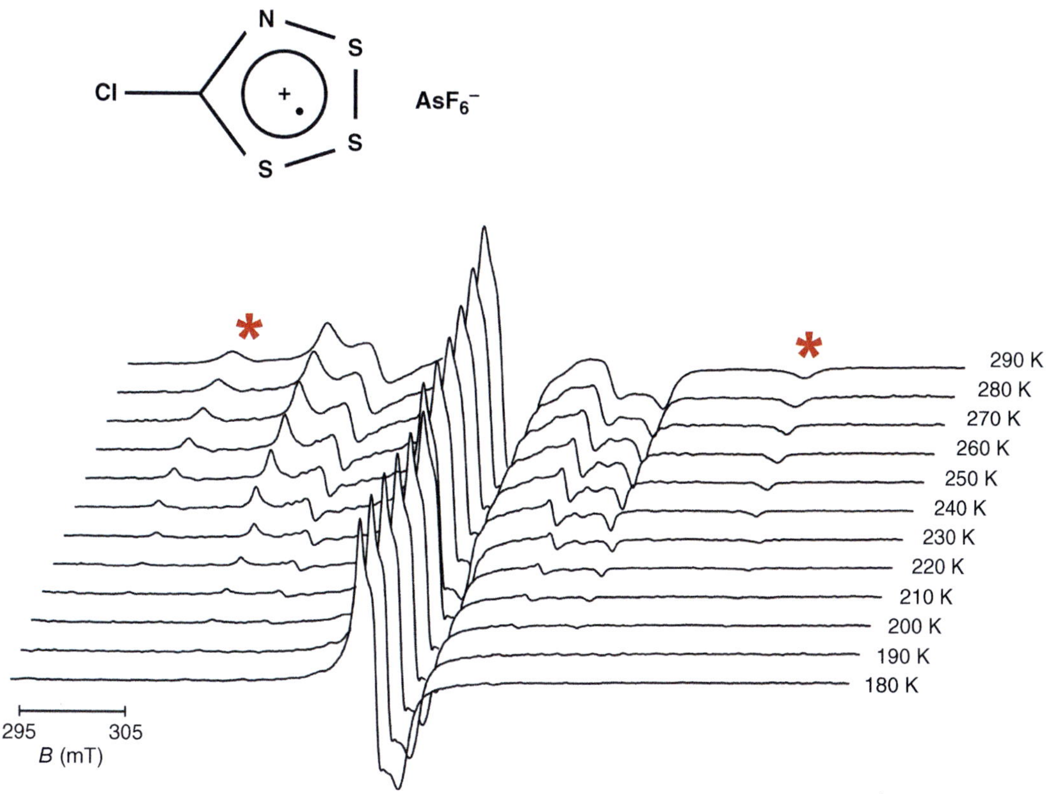

Figure 4.2. X-band EPR of a powder of [ClCNSSS](AsF_6), which exists as a head-to-head dimer in the solid state, at 290 to 180 K in 10 K increments. The spin triplet spectrum arises from the excited multiplet, and the singlet–triplet separation can be determined by modeling its intensity (e.g., the z orientation transitions marked *) as a function of temperature (giving $2J = -1900\,\text{cm}^{-1}$). The central feature is due to $s = {}^1/_2$ lattice defects (where half a dimer is missing; about 0.1%)

4.5 COUPLING MORE THAN TWO SPINS

Where more than two spins are involved, more than one multiplet can arise with the same total spin quantum number. These can sometimes be treated by a sequential coupling scheme. For example, taking a trimer of $s = {}^1/_2$ and Hamiltonian (4.8) that assumes the symmetry of an isosceles triangle with $s_1 s_2$ as the unique edge (Figure 4.1)

$$\hat{H} = -2J_1 \hat{s}_1 \cdot \hat{s}_2 - 2J_2(\hat{s}_2 \cdot \hat{s}_3 + \hat{s}_3 \cdot \hat{s}_1) \tag{4.8}$$

we can visualize coupling spins s_1 and s_2 to give an intermediate spin quantum number s_{12} (=0 and 1), and coupling this result with s_3. This gives three total spin multiplets, which are uniquely labeled in the coupled basis by $|s_{12}S\rangle = |0,{}^1/_2\rangle$, $|1,{}^1/_2\rangle$, and $|1,{}^3/_2\rangle$.

Equation (4.8) has eigenvalues

$$\begin{aligned} E = &-J_2[S(S+1) - s_{12}(s_{12}+1) - s_3(s_3+1)] \\ &-J_1[s_{12}(s_{12}+1) - s_1(s_1+1) - s_2(s_2+1)] \end{aligned}$$

Note there are two different $S = {}^1/_2$ multiplets, one at $E_{|0,{}^1/_2\rangle} = 3J_1/2$ and the other at $E_{|1,{}^1/_2\rangle} = (4J_2 - J_1)/2$. For an equilateral triangle ($J_1 = J_2$), these two doublets are degenerate.

However, in general, it is not possible to derive analytical solutions for the eigenvalues of the exchange Hamiltonian, and full diagonalization of the Hamiltonian matrix is required. For a system comprising n coupled spins s_i, we have a matrix of dimension of $\prod_n (2s_i + 1)$. Even for the simple isotropic Hamiltonian (4.1), which gives the zero-field energy spectrum, diagonalizing the matrix becomes computationally expensive for large clusters. For high-symmetry species

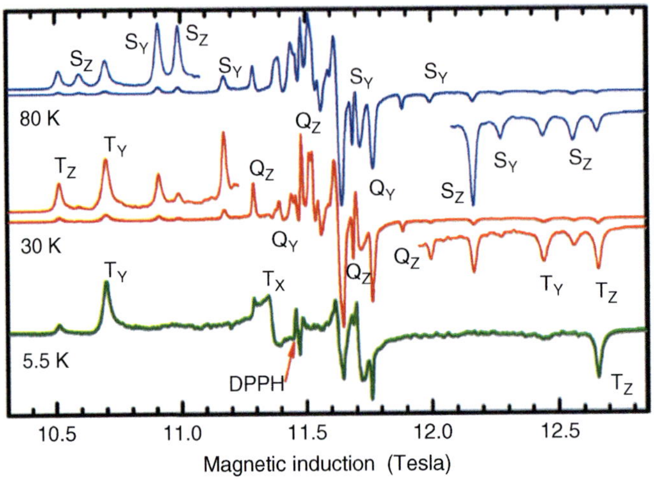

Figure 4.3. 322 GHz CW EPR spectra of $[Cr_2Zn_2(NCS)_4(dea)_2(Hdea)_2]$ (Hdea = diethanolamine) as a powder. EPR transitions are observed within the $S = 1$ (labeled T for triplet), 2 (Q for quintet), and 3 (S for septet) excited multiplets of the pair ($2J = -14\,cm^{-1}$; X, Y, Z denote the molecular orientations, with respect to the magnetic field, giving rise to the transitions). At 5.5 K, only the lowest excited multiplet ($S = 1$) is observed due to Boltzmann depopulation effects. (Reprinted with permission from V. V. Semenaka, O. V. Nesterova, V. N. Kokozay, V. V. Dyakonenko, R. I. Zubatyuk, O. V. Shishkin, R. Boča, J. Jezierska and A. Ozarowski, Inorg. Chem. 2010, 49, 5460. © 2010, American Chemical Society)

group theoretical methods can be applied to simplify the problem, and there are algorithms that exploit the fact that these matrices tend to be very sparse (most of the matrix elements are nil).[5] Some of these methods have been applied to calculate EPR spectra of large exchange-coupled systems,[6] but this is still computationally demanding given that the matrix must be diagonalized for many different magnetic fields and orientations.

However, in EPR spectroscopy, we are usually only dealing with resonances within the ground multiplet or within the lowest few multiplets that are thermally populated. In this case, the standard approach is to treat the problem as a set of fictional total spins S to describe these multiplets. Taking an extreme example, there has been a great deal of interest in large exchange-coupled molecules that exhibit magnetic memory effects. The first of these 'single-molecule magnets' was the dodecametallic cluster $[Mn^{III}{}_8Mn^{IV}{}_4O_{12}(O_2CMe)_{16}(H_2O)_4]$: six-coordinate Mn(III) and Mn(IV) have $s = 2$ and $^3/_2$, respectively, and the total matrix dimension is $5^8 \times 4^4 = 10^8$. However, EPR spectra at low temperatures[7] can be interpreted by treating the molecule as an isolated $S = 10$ spin (Figure 4.4): with $D = -0.5\,cm^{-1}$, the full spectrum is about 20 T wide in the high-field limit, hence requiring very high frequency/field methods.

4.6 RELATIONSHIP BETWEEN SPIN HAMILTONIAN PARAMETERS OF THE TOTAL SPIN STATES AND THOSE OF THE UNCOUPLED PARAMAGNETS

So far we have seen how an exchange Hamiltonian, in an uncoupled spin basis, gives rise to a set of multiplets that can be described in a total spin (coupled) basis. When we do an EPR measurement, we find that these total spin multiplets have characteristic spin Hamiltonian parameters: g values, zero-field splitting, hyperfine, etc. These must in some way be related back to the Hamiltonian parameters for the individual spins and those of the exchange Hamiltonian. We can get some insight by comparing the equivalent parts of the spin Hamiltonian matrices. We will work this through for the simple example of two coupled $s = {}^1\!/_2$ spins.

For example, Hamiltonian (4.5) (isotropic exchange + isotropic Zeeman interaction with $g_1 = g_2$ and field along the z-axis) gives matrix (4.6). Compare the upper-left 3×3 block, which describes the spin triplet state, with the matrix for an isolated $S = 1$ in the in the $|M\rangle$ basis (4.9)

$$\begin{array}{cccc} & |+1\rangle & |-1\rangle & |0\rangle \\ \langle +1| & +g\mu_B B & 0 & 0 \\ \langle -1| & 0 & -g\mu_B B & 0 \\ \langle 0| & 0 & 0 & 0 \end{array} \quad (4.9)$$

They are related by $g_{S=1} = (g_1 + g_2)/2$ (in addition to a uniform shift of all three triplet states by the exchange term).

We can do a similar analysis regarding the ZFS of the spin triplet. ZFS arises from the magnetic dipolar interaction and/or an anisotropic exchange interaction between the spins (see below). [If the spins are $s > \frac{1}{2}$, then they also have an intrinsic ZFS arising from spin–orbit coupling (SOC) to excited states.] Both contribute an *anisotropic* component to the interaction between the ions and can be treated by the second term in Hamiltonian (4.10).

$$\hat{H} = -2J\hat{s}_1 \cdot \hat{s}_2 + \hat{s}_1 \cdot \boldsymbol{D}_{12} \cdot \hat{s}_2 \quad (4.10)$$

Expressing the anisotropic $\boldsymbol{D}_{12}$ tensor in its Cartesian components (omitting the 12 subscript), and applying to the eigenfunctions of equation (4.1), gives

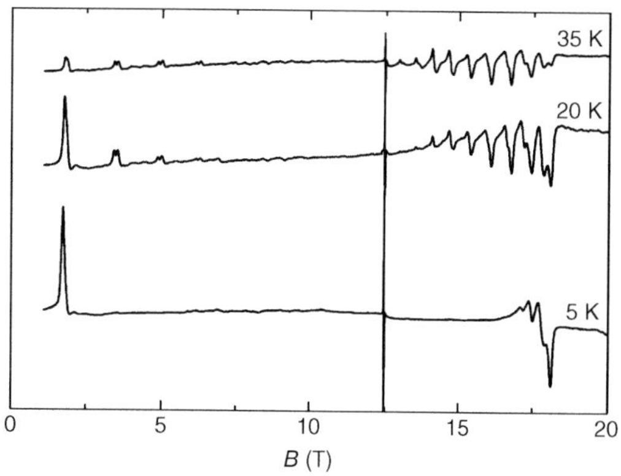

Figure 4.4. 349 GHz CW. EPR spectrum of a powder of $[Mn_{12}O_{12}(MeCO_2)_{16}(H_2O)_4]$. The molecule has an $S = 10$ ground state with $D \approx -0.5\,\mathrm{cm}^{-1}$ and at 5 K only the lowest M states ($= \pm 10$) within this multiplet are populated significantly. At 5 K, the lowest field transition corresponds to $M = -10 \rightarrow -9$ for the z orientation of the molecule (defined by the ZFS tensor). On warming, higher M states are populated and the $M = -9 \rightarrow -8$, etc., transitions become observed. In the high-field limit (which is not quite reached here), the z resonance fields would be given by $B = \frac{h\nu}{g\mu_B} - \frac{D}{2}(2M - 1)$ where M is the upper state in the transition. (Reproduced with permission from Ref. 7. © John Wiley & Sons Ltd., 2005)

$$\begin{array}{ccccc} & |\alpha\alpha\rangle & |\beta\beta\rangle & |\alpha\beta + \beta\alpha\rangle/\sqrt{2} & |\alpha\beta - \beta\alpha\rangle/\sqrt{2} \\ \langle\alpha\alpha| & -\frac{J}{2} + D_{zz}/4 & (D_{xx} - D_{yy})/4 & 0 & 0 \\ \langle\beta\beta| & (D_{xx} - D_{yy})/4 & -\frac{J}{2} + D_{zz}/4 & 0 & 0 \\ \langle\alpha\beta + \beta\alpha|/\sqrt{2} & 0 & 0 & -\frac{J}{2} - D_{zz}/2 & 0 \\ \langle\alpha\beta - \beta\alpha|/\sqrt{2} & 0 & 0 & 0 & +3J/2 \end{array} \quad (4.11)$$

Again, we can compare the 3×3 block with the matrix for an isolated $S = 1$ with a ZFS $(\hat{\boldsymbol{S}} \cdot \boldsymbol{D}_{S=1} \cdot \hat{\boldsymbol{S}})$ in the $|M\rangle$ basis

$$\begin{array}{cccc} & |+1\rangle & |-1\rangle & |0\rangle \\ \langle +1| & D_{zz}/2 & (D_{xx} - D_{yy})/2 & 0 \\ \langle -1| & (D_{xx} - D_{yy})/2 & D_{zz}/2 & 0 \\ \langle 0| & 0 & 0 & -D_{zz} \end{array} \quad (4.12)$$

These are equivalent with $\boldsymbol{D}_{S=1} = \boldsymbol{D}_{12}/2$.

Unsurprisingly, given that the total spin Hamiltonian describes a subspace of the coupled Hamiltonian, the parameters of the former are somehow inherited from those of the latter. General expressions for the projection of one Hamiltonian onto a subspace of another can be developed using the Wigner–Eckart theorem. Application of such methods to the problem of a total spin S derived from coupling spins s_1 and s_2 [with individual $\boldsymbol{g}$- and ZFS tensors $\boldsymbol{g}_i$ and $\boldsymbol{D}_i$, respectively] gives equations (4.13–4.15)[1]

$$\boldsymbol{g}_S = c_1\boldsymbol{g}_1 + c_2\boldsymbol{g}_2$$

$$\boldsymbol{D}_S = d_1\boldsymbol{D}_1 + d_2\boldsymbol{D}_2 + d_{12}\boldsymbol{D}_{12} \quad (4.13)$$

where the coefficients c_i, d_i, and d_{12} (known as vector coupling or projection coefficients) are given by

$$c_1 = \frac{1+c}{2},\ c_2 = \frac{1-c}{2}$$

$$d_1 = \frac{c_+ + c_-}{2},\ d_2 = \frac{c_+ - c_-}{2},\ d_{12} = \frac{1-c_+}{2}$$

and

$$c = \frac{s_1(s_1+1) - s_2(s_2+1)}{S(S+1)}$$

$$c_+ = \frac{3[s_1(s_1+1) - s_2(s_2+1)]^2 + S(S+1)[3S(S+1) - 3 - 2s_1(s_1+1) - 2s_2(s_2+1)]}{(2S+3)(2S-1)S(S+1)}$$

$$c_- = \frac{4S(S+1)[s_1(s_1+1) - s_2(s_2+1)] - 3[s_1(s_1+1) - s_2(s_2+1)]}{(2S+3)(2S-1)S(S+1)} \quad (4.14)$$

If hyperfine interactions $(\hat{s}_i \cdot \boldsymbol{A}_i \cdot \hat{\boldsymbol{I}}_i)$ are also being considered, then coupling of the total spin S to nucleus i (=1,2) is given by

$$\boldsymbol{A}_S^1 = c_1\boldsymbol{A}_1,\ \boldsymbol{A}_S^2 = c_2\boldsymbol{A}_2 \quad (4.15)$$

where we have neglected any hyperfine interaction of the electrons at 1 with the nucleus at 2 and vice versa. If we include the latter, then for each nucleus

$$\boldsymbol{A}_S^i = c_1\boldsymbol{A}_1^i + c_2\boldsymbol{A}_2^i$$

where, for example, $\boldsymbol{A}_1^1$ and $\boldsymbol{A}_2^1$ are the contributions to the hyperfine to nucleus 1 from the electrons at 1 and 2, respectively.

It is important to remember that $\boldsymbol{g}$, $\boldsymbol{D}$, and $\boldsymbol{A}$ are tensors and their relative orientations have to be taken into account. If the two centers are symmetry related, then their interaction tensors must also obey these relationships, which can simplify modeling. For example, if the two centers are related by an inversion center, then their various interaction tensors must be colinear.

We highlight two examples where the experimentally measured spin Hamiltonian parameters in the total spin formalism at first seem strange. The heterodimer $[LCr(PyA)_3Ni]^{2+}$ (L = 1,4,7-Me_3-1,4,7-triazacyclononane; PyA^- = monoanion of pyridyl-2-aldoxime) contains an antiferromagnetically coupled Cr(III) ($s = {}^3/_2$) and Ni(II) (s = 1) pair giving an S = ½ ground state ($2J = -18\,cm^{-1}$).[8] EPR gives the ground state g = 1.88, which is outside the range of typical g values for Ni(II) (about 2.2) and Cr(III) (about 2.0). However, equations (4.13, 4.14) give $g_{S=1/2} = +5g_{Cr}/3 - 2g_{Ni}/3$, and substitution of the typical values reproduces the experimental value. Ultimately, this is because the two spins do not project evenly on the ground state and highlights the vector nature of the spin coupling.

The Cr(III) homodimer $[LCr(\mu\text{-}OH)_3CrL]$ is strongly antiferromagnetically coupled ($2J = -130\,cm^{-1}$) giving well-isolated total spin multiplets of S = 0, 1, 2, and 3 at relative energies of 0, $2J$, $6J$, and $12J$, respectively (as in Figure 4.3).[9] The molecule has a threefold axis along the Cr ... Cr direction and so the $\boldsymbol{D}_{Cr}$ tensors of the individual ions are coparallel. Single-crystal EPR spectroscopy resolves transitions within each of the three excited multiplets, giving $D_{S=1}$, $D_{S=2}$, and $D_{S=3}$ of +2.3, +0.08, and $-0.24\,cm^{-1}$, respectively. It seems odd that the ZFS of the S = 1 multiplet is very large, but that of the S = 2 is very small. However, application of equations (4.13, 4.14) gives $D_{S=1} = 2D_{Cr}/5 + 3D_{CrCr}/10$; $D_{S=2} = D_{CrCr}/2$; $D_{S=3} = -12D_{Cr}/5 + 17D_{CrCr}/10$, which reveal that the projection of D_{Cr} on the S = 2 multiplet is nil. Solving these simultaneous equations gives $D_{Cr} = -0.84\,cm^{-1}$, and the interaction term (including dipolar and anisotropic exchange) $D_{CrCr} = 0.16\,cm^{-1}$. Such relationships have been tested by doping small amounts of the paramagnetic ion into isostructural and diamagnetic analogs.

Bigger clusters (three spins or more) are of course more complicated, but are often approximated by coupling the spins pair-wise and applying the relationships above at each stage.

4.7 BREAKDOWN OF THE STRONG EXCHANGE LIMIT

The validity of everything stated until now depends on the 'strong exchange limit' being valid. 'Strong' and 'weak' refer to the strength of the isotropic exchange J compared to other terms in the spin Hamiltonian, for example, hyperfine interactions, zero-field splittings, or even differences in the Zeeman energies of the interacting spins. If the exchange term in Hamiltonian (4.1) is much larger than all these other terms, then the latter can be considered as a perturbation on the result of the former. If this is not the case, then the total spin S is not a good quantum number. We will consider some examples below.

4.7.1 Isotropic Exchange between Dissimilar Radicals with $g_1 \neq g_2$

Let us return to Hamiltonian (4.5) but, instead of applying the Zeeman terms to the eigenfunctions of the exchange term, we evaluate the matrix elements in the $|m_1 m_2\rangle$ basis. For simplicity, we take the magnetic field along the z-axis (or, equivalently, consider that we are in fluid solution).

$$\begin{array}{c|cccc} & |\alpha\alpha\rangle & |\beta\beta\rangle & |\alpha\beta\rangle & |\beta\alpha\rangle \\ \hline \langle\alpha\alpha| & -\frac{J}{2}+(g_1+g_2)\mu_B B/2 & 0 & 0 & 0 \\ \langle\beta\beta| & 0 & -\frac{J}{2}-(g_1+g_2)\mu_B B/2 & 0 & 0 \\ \langle\alpha\beta| & 0 & 0 & +\frac{J}{2}+(g_1-g_2)\mu_B B/2 & -J \\ \langle\beta\alpha| & 0 & 0 & -J & +\frac{J}{2}+(-g_1+g_2)\mu_B B/2 \end{array} \tag{4.16}$$

If $g_1 = g_2$, then diagonalization of equation (4.16) gives the result described before with isolated triplet and singlet states, and EPR transitions only allowed within the triplet. This is also a good description for $g_1 \neq g_2$ as long as $|J| \gg (g_1 - g_2)\mu_B B$ and hence that the difference in Zeeman energy is working as a perturbation on the exchange.

If, on the other hand, the exchange is nil, then the simple $|m_1 m_2\rangle$ basis states are acceptable eigenfunctions. There are four allowed EPR transitions, but only two unique energies, giving resonance fields corresponding to the g_i values of the uncoupled radicals (Figure 4.5). If we now allow $|J| \neq 0$ but $\ll (g_1 - g_2)\mu_B B$, i.e., such that the exchange is a perturbation on the Zeeman terms, then we can to a first approximation neglect the off-diagonal terms in equation (4.16). The eigenvalues of the two middle states are modified by $+J/2$, thus breaking the degeneracy of the two sets of resonances (Figure 4.5). Our two peaks are split into doublets, centered on g_1 and g_2, with splittings of $2J$.

This is actually a very familiar picture in magnetic resonance – we have just described an AX spin system such as is common in NMR spectroscopy. AX systems are common in NMR because nuclear exchange couplings (usually formulated as $J\hat{\boldsymbol{I}}_1 \cdot \hat{\boldsymbol{I}}_2$, hence giving J as the doublet splitting) are generally weak compared to the nuclear Zeeman (chemical shift) differences for protons.

Clear-cut examples in CW EPR are relatively rare because when $|J|$ is small enough to meet the weak exchange limit, the splitting is usually lower than the intrinsic linewidth of the spectrum. For this reason, it is a treatment more commonly invoked in pulsed EPR. Note that for anisotropic systems, the difference in g could be caused by relative orientation of the spins.

Nevertheless, there are examples in CW EPR for well-separated spins of very different g, for example, nitroxides bound to Cu(II) (Figure 4.6).[10]

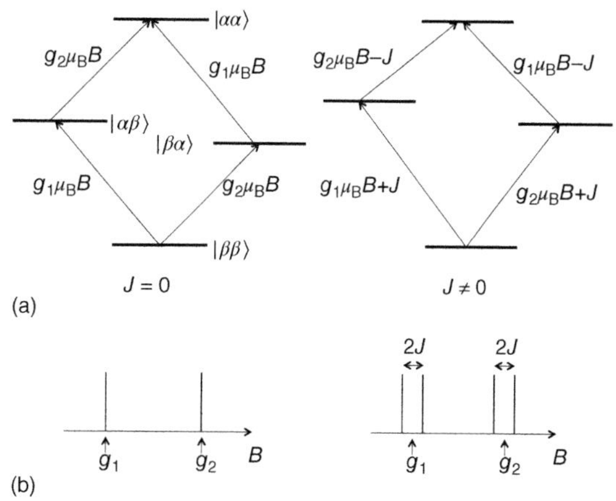

Figure 4.5. (a) Energy level diagrams for two weakly coupled $s = {}^1\!/_2$ radicals with different g values ($g_1 > g_2$), for a fixed magnetic field B, with allowed EPR transitions and energies. (b) The resulting EPR (stick) spectra. The left picture is for $|J| = 0$; the right picture is for $|J| \neq 0$ but $\ll (g_1 - g_2)\mu_B B$

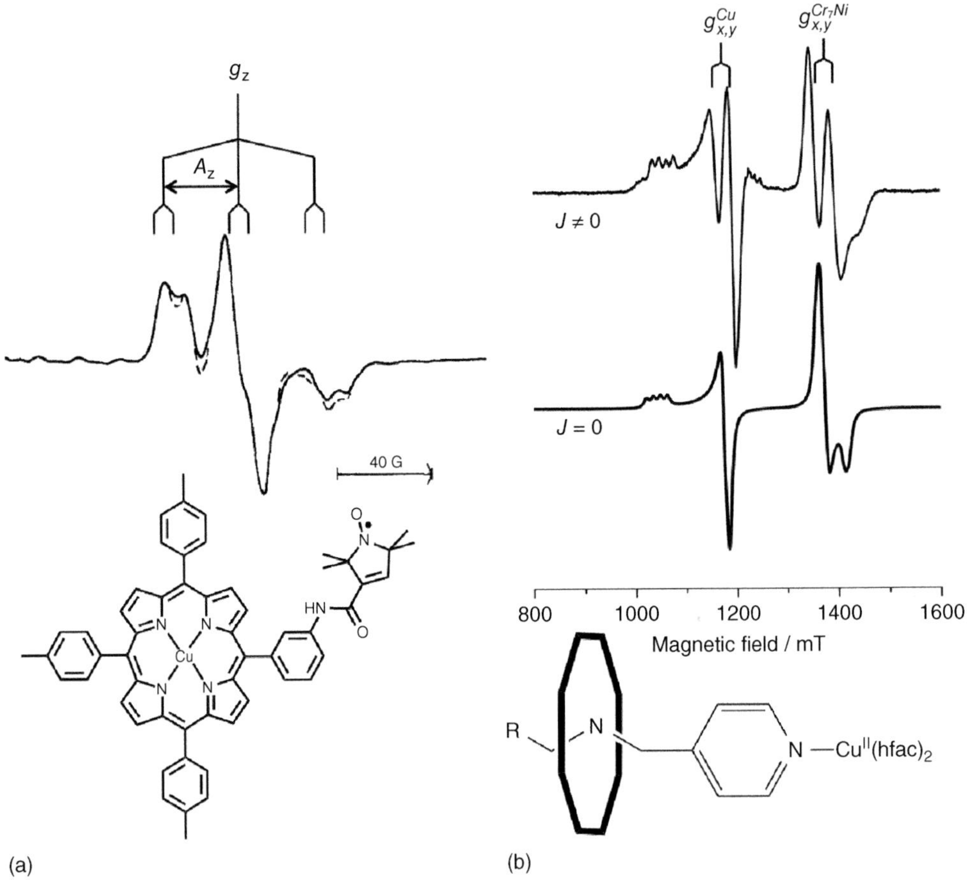

Figure 4.6. (a) X-band EPR spectrum of a weakly coupled Cu ... nitroxide complex in frozen solution with stick spectrum highlighting exchange splitting of the nitroxyl g_z features (dotted line: simulation). (Reprinted from J. Magn. Reson., 52, S. Eaton, K. M. More, B. M. Sawat, M. Sawant, P. M. Boumel and G. R. Eaton, Metal-nitroxyl interactions. 29. EPR studies of spin-labeled copper complexes in frozen solution, 435. © 1983, with permission from Elsevier) (b) Q-band spectrum of a powder of a {Cr_7Ni} ... Cu species at 5 K: experimental (top), and calculated for the uncoupled limit (bottom). The exchange splitting of the $g_{x,y}$ features of the Cu and Cr_7Ni resonances into doublets is highlighted by sticks. The cyclic {Cr_7Ni} cluster (s = ½ ground state) is represented as an octagon

In Figure 4.6, the doublet splittings on the nitroxide spectrum are about 8×10^{-4} cm^{-1}; if we approximate g = 2.1 and 2.0 for Cu(II) and nitroxide, respectively, then $\Delta g \mu_B B$ at 0.3 T is 140×10^{-4} cm^{-1}. Care needs to be taken in interpreting solid-state spectra because in general the splittings are a result of exchange *and* dipolar couplings; only the former contribute to fluid solution measurements. Such systems have been reviewed elsewhere.[11,12] An unusual but well-resolved supramolecular example is {[$Cu(hfac)_2$][$PyCH_2NH_2CH_2CH_2Ph$][$Cr_7NiF_8(O_2C^tBu)_{16}$]} (Figure 4.6; hfac = hexafluoroacetylacetonate) where a cyclic {Cr_7Ni} cluster templates about an organic thread, which binds to a Cu(II) ion.[13] The strongly coupled {Cr_7Ni} has a well-isolated s = ½ ground state with $g \approx 1.8$. Both the Cu and {Cr_7Ni} regions split as doublets at Q-band ($\Delta g \mu_B B$ ca 0.17 cm^{-1} at 1.2 T) giving $2J = -0.032$ cm^{-1}.

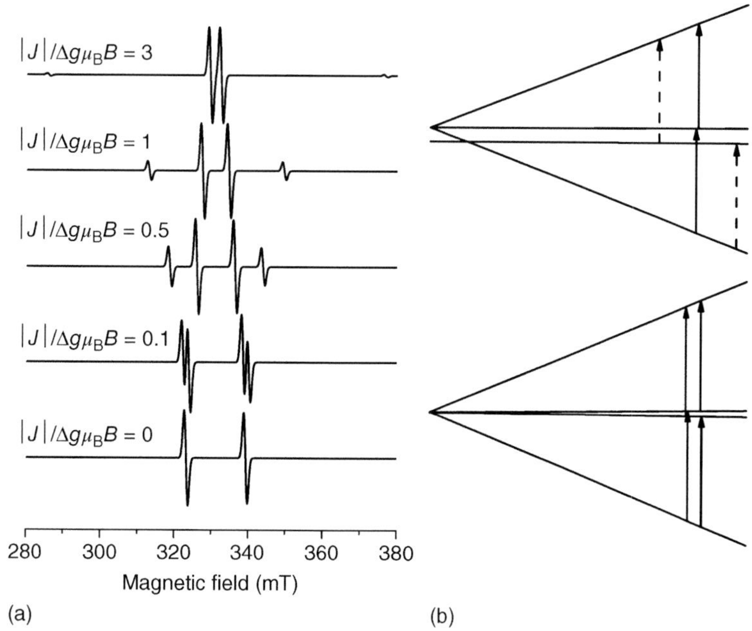

Figure 4.7. (a) Calculated 9.5 GHz EPR spectra for two dissimilar $s = ½$ spins (isotropic $g_1 = 2.1$, $g_2 = 2.0$) as a function of $|J|$. From bottom to top: $|J|/\Delta g\mu_B B = 0, 0.1, 0.5, 1$, and 3 ($B = 0.32$ T). (b) Energy level diagram (as a function of the magnetic field) for $|J|/\Delta g\mu_B B = 0.1$ (bottom) and 3 (top). The former is an AX spin system with four allowed EPR transitions (arrows) associated with g_1 and g_2. The latter behaves as a spin singlet and triplet, with mixing induced by the Zeeman terms leading to formally forbidden transitions (dashed arrows)

When the exchange energy begins to approach that of the difference in Zeeman energy, this simple AX description breaks down and complicated spectra can result (these are referred to as AB systems): calculated examples for varying $|J|/\Delta g\mu_B B$ are given in Figure 4.7. As $|J|/\Delta g\mu_B B$ increases, the two inner transitions move together, and the two outer transitions diverge and become weaker. In Figure 4.7, the outer transitions are still observed for the highest ratio considered even though the description as a singlet and triplet seems reasonable. We can now consider the Zeeman terms to be mixing the singlet and triplet, hence partially allowing transitions between the $|S,M\rangle = |1,\pm1\rangle$ and $|0,0\rangle$ states.

Finally, we note that for such systems, the description of 'weak exchange' will also depend on the magnetic field of the measurement – the same molecule could be strongly coupled in weak field and weakly coupled in strong field!

4.7.2 Isotropic Exchange between Similar Radicals with *J* Hyperfine

Other terms in the Hamiltonian can also break the $\Delta S = 0$ selection rule where the strong exchange limit seems a reasonable approximation. Taking the case of a homodimer of $s_i = ½$, each with a single nuclear spin $I_i = ½$ and assuming isotropic $g_1 = g_2$ $(= g)$ and $A_1 = A_2$ $(= A)$ for simplicity:

$$\hat{H} = -2J\hat{s}_1 \cdot \hat{s}_2 + \mu_B gB(\hat{s}_{1z} + \hat{s}_{2z}) + A(\hat{s}_{1z} \cdot \hat{I}_{1z} + \hat{s}_{2z} \cdot \hat{I}_{2z}) \quad (4.17)$$

The matrix dimension is now $\Pi(2s_i+1)(2I_i+1) = 16$. The basis states are product functions of the electron spins and nuclear spin functions: $|m_{s1}m_{s2}m_{I1}m_{I2}\rangle$. If we choose the electron spin singlet and triplet basis functions, then one general block of the matrix is shown in equation (4.18).[14]

$$
\begin{array}{lcccc}
 & |\alpha\alpha m_{I1}m_{I2}\rangle & |\beta\beta m_{I1}m_{I2}\rangle & |(\alpha\beta+\beta\alpha)m_{I1}m_{I2}\rangle/\sqrt{2} & |(\alpha\beta-\beta\alpha)m_{I1}m_{I2}\rangle/\sqrt{2} \\
\langle\alpha\alpha m_{I1}m_{I2}| & -\frac{J}{2}+g\mu_B B+A(m_{I1}+m_{I2})/2 & 0 & 0 & 0 \\
\langle\beta\beta m_{I1}m_{I2}| & 0 & -\frac{J}{2}-g\mu_B B-A(m_{I1}+m_{I2})/2 & 0 & 0 \\
\langle(\alpha\beta+\beta\alpha)m_{I1}m_{I2}|/\sqrt{2} & 0 & 0 & -J/2 & A(m_{I1}-m_{I2})/2 \\
\langle(\alpha\beta-\beta\alpha)m_{I1}m_{I2}|/\sqrt{2} & 0 & 0 & A(m_{I1}-m_{I2})/2 & +3J/2
\end{array}
\tag{4.18}
$$

There are four such blocks, one for each combination of m_{I1} and m_{I2}. For clarity, we will evaluate two of the elements arising from the hyperfine operator explicitly. Firstly, operating on the $|\alpha\alpha m_{I1}m_{I2}\rangle$ ket (the $M = +1$ component of the triplet):

$$
\begin{aligned}
&A(\hat{s}_{1z}\cdot\hat{I}_{1z}+\hat{s}_{2z}\cdot\hat{I}_{2z})|\alpha\alpha m_{I1}m_{I2}\rangle \\
&\quad=\frac{A}{2}(m_{I1}+m_{I1})|\alpha\alpha m_{I1}m_{I2}\rangle
\end{aligned}
\tag{4.19}
$$

hence giving rise to a diagonal term in equation (4.18). Secondly, operating on the $|(\alpha\beta m_{I1}m_{I2}+\beta\alpha m_{I1}m_{I2})\rangle/\sqrt{2}$ ket (the $M = 0$ component of the triplet)

$$
\begin{aligned}
&A(\hat{s}_{1z}\cdot\hat{I}_{1z}+\hat{s}_{2z}\cdot\hat{I}_{2z})|(\alpha\beta m_{I1}m_{I2}+\beta\alpha m_{I1}m_{I2})\rangle/\sqrt{2} \\
&=\frac{A}{2}(m_{I1}-m_{I2})|\alpha\beta m_{I1}m_{I2}\rangle/\sqrt{2} \\
&+\frac{A}{2}(-m_{I1}+m_{I2})|\beta\alpha m_{I1}m_{I2}\rangle/\sqrt{2} \\
&=\frac{A}{2}(m_{I1}-m_{I2})|(\alpha\beta m_{I1}m_{I2}-\beta\alpha m_{I1}m_{I2})\rangle/\sqrt{2}
\end{aligned}
\tag{4.20}
$$

hence giving rise to an off-diagonal term in equation (4.18), connecting to the $M = 0$ component of the singlet, and we can already see how the hyperfine can lead to singlet–triplet mixing.

From equation (4.18), if $|J| \gg |A|$, then the off-diagonal elements can be neglected and we have $\Delta M = \pm 1$ transition energies within the triplet state of $g\mu_B B + A(m_{I1} + m_{I2})/2$. The hyperfine splitting in the EPR spectrum is halved from that of the parent spin, but we couple to two equivalent nuclei. Note this is the result you would derive from equations (4.14, 4.15). This is an intuitive result: strong exchange results in the electron interacting equally with both nuclei, but for 'half the time'.

If $|J| \ll |A|$, then the spectrum will be identical to that of the independent radicals. Spectra in the intermediate regime are complicated, but simulation can then be used to determine $|J|$ (Figure 4.8). This has been studied most extensively for bis-nitroxides and excellent examples of these three regimes can be found, giving variously 1:1:1 multiplets with measured isotropic hyperfine splitting $\approx$ 14 G due to ^{14}N (i.e., $|J| \ll |a_{iso}|$), through to 1:2:3:2:1 multiplets with measured splitting $\approx$ 7 G (i.e., $|J| \gg |a_{iso}|$) depending on the linker.[15]

The off-diagonal elements in equation (4.18) can also lead to the singlet–triplet EPR transitions becoming allowed if $|J|$ is in the correct regime for the microwave energy $h\nu$. As with the $g_1 \neq g_2$ example above, these transitions would be weak and appear in the wings of the spectrum and resolution of such transitions allows determination of $|J|$. For example, this is the case at X-band ($h\nu \approx$ 0.3 cm^{-1}) for the dimers $[Cu_2(tren)_2(OCN)_2](BPh_4)_2$ (tren = 2,2′,2″-triamino-triethylamine) where $2J \approx$ −0.2 cm^{-1}; Cu hyperfine interactions are of the order 10^{-2} cm^{-1}.[16]

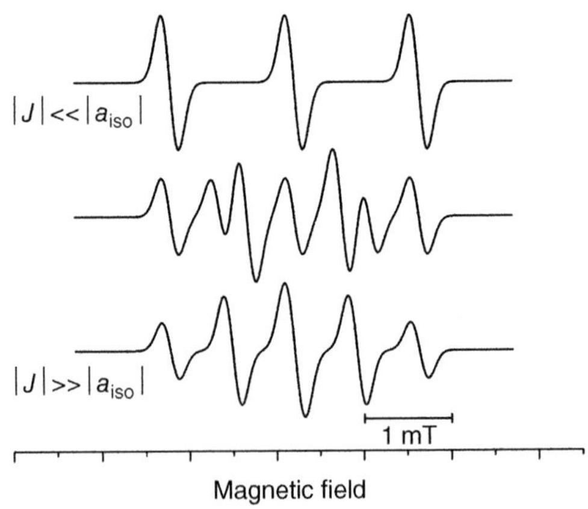

Figure 4.8. Calculated X-band solution EPR spectra (2 mT linewidth) for two $s = 1/2$ radicals with $g_1 = g_2$, a single ^{14}N nuclear spin each with isotropic hyperfine coupling constant of $|a_{iso}| = 44$ MHz (typical of a nitroxyl), and $|J|$ of (top) 1 MHz, (middle) 100 MHz, and (bottom) 1 GHz

Even where J is dominant, the interplay among exchange, hyperfine, and dipolar effects in hyperfine-resolved spectra can give an accurate method of determination of interspin geometry from CW EPR spectra. Important examples include the cofacial di(copper)porphyrins, where the 'half-field' transitions (see Chapter 3) are sensitive only to $\boldsymbol{g}_i$, $\boldsymbol{A}_i$ and the angle between the Cu ... Cu direction and the coordination planes, while the allowed transitions are also sensitive to the Cu ... Cu distance and J.[17] Hyperfine effects in dimers have been discussed in detail by Pilbrow.[18]

4.7.3 Isotropic Exchange with Dipolar Interactions ≫ J

For long-range interactions, even though the electron–electron dipolar interactions are weak they become dominant over the exchange (the latter falls off with distance quicker than the former)[19] and both interactions become undetectable by CW EPR. We do not discuss such species further here other than to note that extremely weak exchange interactions can be measured,[20,21] along with the dipolar interaction, by pulsed methods such as PELDOR/DEER, which are covered elsewhere in this encyclopedia (see Chapters 19 and 20).

4.7.4 Isotropic Exchange between Spins with ZFS

If the individual spins have $s > 1/2$, then they can have intrinsic ZFS, and these can also lead to a breakdown of the strong exchange limit. For example, a dimer of $s = 1$ will have total spin multiplets of $S = 0$, 1, and 2 as discussed above. If we include an axial ZFS term (D_i) for both spins,

$$\hat{H} = -2J\hat{s}_1 \cdot \hat{s}_2 + \sum_i D_i[\hat{s}_{iz}^2 - s_i(s_i+1)/3] \quad (4.21)$$

then the degeneracies of these total spins are broken. If the strong exchange limit is valid, then this can be described by a ZFS parameter for each total spin multiplet according to equations (4.13, 4.14). However, there is also one nonzero off-diagonal matrix element between the different total spin blocks introduced by the second term in equation (4.21), specifically $\langle 0,0|\hat{H}_{\text{zfs}}|2,0\rangle$, where $|S,M\rangle$ are the eigenfunctions of the exchange term. We will evaluate this for clarity: when expressed in the $|m_1m_2\rangle$ basis, these eigenfunctions are $|0,0\rangle \equiv (|+1,-1\rangle + |-1,+1\rangle - |0,0\rangle)/\sqrt{3}$ and $|2,0\rangle \equiv (|+1,-1\rangle + |-1,+1\rangle + 2|0,0\rangle)/\sqrt{6}$, respectively. Then, with $D_1 = D_2\ (= D)$:

$$\left\langle \frac{(\langle +1,-1| + \langle -1,+1| - \langle 00|)}{\sqrt{3}} \middle| D\left(\hat{s}_{1z}^2 + \hat{s}_{2z}^2 - \frac{4}{3}\right) \middle| \frac{(|+1,-1\rangle + |-1,+1\rangle + 2|0,0\rangle)}{\sqrt{6}} \right\rangle$$
$$= \left\langle \frac{(\langle +1,-1| + \langle -1,+1| - \langle 00|)}{\sqrt{3}} \middle| \frac{2D}{3} \frac{(|+1,-1\rangle + |-1,+1\rangle - 4|0,0\rangle)}{\sqrt{6}} \right\rangle$$
$$= \frac{2D}{3\sqrt{3}\sqrt{6}}[\langle +1,-1|+1,-1\rangle + \langle -1,+1|-1,+1\rangle + 4\langle 0,0|0,0\rangle] = \frac{2\sqrt{2}D}{3} \quad (4.22)$$

Hence, the ZFS induces a mixing between the $M = 0$ components of the $S = 0$ and 2 multiplets, which would affect the spacing of the states, and therefore EPR transitions, in the latter. In the absence of ZFS, the $S = 0$ and 2 multiplets are separated by $6J$ (equation (4.7)), hence the extent of this state mixing (and whether it can be treated as a perturbation) depends on this gap. An example where this has been observed (for coupled $s = 1$) is in the six-coordinate Ni(II) dimer $[Ni_2(en)_4Cl_2]Cl_2$ (en = ethylenediamine) where $D_{\text{Ni}} = -4.8\,\text{cm}^{-1}$ and $2J = +10\,\text{cm}^{-1}$.[22] When EPR spectra are being simulated according to a total spin model, such effects can manifest themselves as the need to include higher order ZFS terms (in principle allowed up to order $2S$) to reproduce the 'nonstandard' spacings. A nicely resolved example is the Cr(III) dimer $[LCr\{(dmg)_3Zn\}CrL](ClO_4)_2$ (dmg^{2-} = dimethylglyoximate), where $2J = -9\,\text{cm}^{-1}$, and simulation of the $S = 2$ multiplet requires introduction of fourth-order

ZFS parameters, but can also be explained in an exchange model with a D_{Cr} of about $+0.2\,cm^{-1}$.[23] Such effects have been studied deeply in molecular magnetism because they are thought to give rise to important magnetic phenomena, for example, quantum tunneling of the magnetization observed for some single-molecule magnets. State-mixing effects due to comparable ZFS and exchange are also important in the EPR of biological clusters, e.g., in manganese clusters found in several enzymes. For example, in several Mn(II)Mn(III) pairs, it is found that $|J| \approx |D_{Mn(III)}|$, with J being weakly antiferromagnetic, with the result that there is substantial mixing between the $S = 1/2$ ground and the $S = 3/2$ excited multiplets giving rise to very anomalous ^{55}Mn hyperfine patterns.[24,25]

Of course, if $|D_i| \gg |J|$, then the concept of a total spin quantum number has no meaning. Taking another example from biology, cyctochrome c oxidase has a weakly coupled Fe(III) heme ($s = 5/2$) and Cu(II) at the catalytic site. EPR spectra are only interpretable with $D_{Fe} = +5\,cm^{-1}$ and an antiferromagnetic interaction of $|J| \approx 1\,cm^{-1}$.[26] The Fe site, therefore, has a lowest Kramers doublet separated by $10\,cm^{-1}$ ($2D_{Fe}$) from the first excited doublet. Switching on the very weak exchange gives four levels well separated in energy from the other eight (Figure 4.9), and there is no even vaguely meaningful description as total spin $S = 2$ and 3 multiplets.

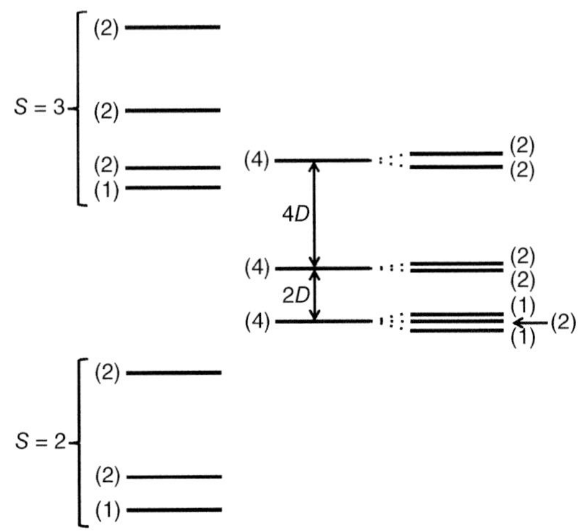

Figure 4.9. Spin states at zero field for a coupled $s = 5/2$, with axial and positive D, and $s = 1/2$ pair with $J = 0$ (uncoupled; center), $2J = -5D$ (approaching the strong exchange limit, such that states can be grouped as total spin multiplets S; left), $2J = -D/5$ (weak exchange; right). The parentheses show the degeneracies of the states. (Reprinted with permission from M. R. Cheesman, V. S. Oganesyan, N. J. Watmough, C. S. Butler and A. J. Thomson, J. Am. Chem. Soc. 2004, 126, 4157. © 2010, American Chemical Society)

4.8 OTHER TERMS IN THE EXCHANGE HAMILTONIAN

The most general form of expressing the interaction between spins s_1 and s_2 is by a 3×3 interaction tensor $\boldsymbol{J}$, as in equation (4.23). Any 3×3 tensor can be deconstructed into an isotropic part, a traceless and symmetric anisotropic part, and an antisymmetric part. In terms of the spin vector operators[1]:

$$\hat{H} = \hat{s}_1 \cdot \boldsymbol{J} \cdot \hat{s}_2 = -2J\hat{s}_1 \cdot \hat{s}_2 + \hat{s}_1 \cdot \boldsymbol{D}_{12} \cdot \hat{s}_2 + \boldsymbol{d}_{12} \cdot \hat{s}_1 \times \hat{s}_2 \quad (4.23)$$

The first term in equation (4.23) is the isotropic exchange and $-2J$, a scalar quantity, is one third of the trace of $\boldsymbol{J}$ [= $(J_{xx} + J_{yy} + J_{zz})/3$]. [Here it would be more convenient to stick to a $+J\hat{s}_1 \cdot \hat{s}_2$ formalism throughout but old habits die hard.] The second term in equation (4.23), where $\boldsymbol{D}_{12}$ is a traceless 3×3 tensor, is the anisotropic part that (as stated above) has contributions from the dipolar spin–spin and an anisotropic exchange interaction. The third term in equation (4.23) is the antisymmetric part, where $\boldsymbol{d}_{12}$ is a vector and × indicates the cross product.

In terms of orbital interactions, the isotropic exchange arises from electron exchange between the ground states of the two paramagnetic centers. Both the anisotropic and antisymmetric exchange interactions involve excited states. The mechanism for ground–excited state interactions is SOC, hence these terms are expected to be more important for heavier elements, and relatively unimportant for coupled organic radicals. In the strong exchange limit, $\boldsymbol{D}_{12}$ will contribute to the ZFS of the total spin states but, as with the other Hamiltonian terms discussed above, can also provide a mechanism for state mixing as the strong-exchange limit breaks down.

Note that the dot product between spin vectors in the isotropic and anisotropic terms expresses their tendency to align the spins (anti)parallel. The antisymmetric term involves the cross product. This favors a perpendicular alignment, and it is the antisymmetric (also known as the Dzyaloshinskii–Moriya) exchange that gives rise to phenomena such as canted antiferromagnetism in extended lattices. $\boldsymbol{d}_{12}$ can have a profound influence on EPR spectra and has been studied most thoroughly in triangular trimetallic clusters.[27] We take the simplest example, a triangle of $s = ^1\!/_2$.

There are symmetry rules associated with the $\boldsymbol{d}_{12}$ vector.[28] For a pair of spins s_1 and s_2:

1. if s_1 and s_2 are related by an inversion center, then $\boldsymbol{d}_{12} = 0$,
2. if s_1 and s_2 lie on a n-fold axis (say, y), then only $d_{12,y} \neq 0$.
3. if s_1 and s_2 lie on a mirror plane (say, xy), then only $d_{12,z} \neq 0$.
4. if s_1 and s_2 are bisected by a mirror plane (say, xz), then only $d_{12,x,z} \neq 0$.
5. if s_1 and s_2 are bisected by a C_2-axis (x), then only $d_{12,y,z} \neq 0$.

For an equilateral triangle $s_1s_2s_3$ defining the xy-plane, this means that only the z components of the $\boldsymbol{d}_{ij}$ vectors are nonzero, giving Hamiltonian (4.24). If we assume $|J| \gg |d|$, then we can apply the antisymmetric exchange term (4.24) [which assumes $d_{12,z} = d_{23,z} = d_{31,z}$ $(= d_z)$] to the eigenfunctions of the isotropic exchange (Figure 4.1). As an illustrative example, we evaluate the matrix element between the $M = +^1\!/_2$ states of the two $S = ^1\!/_2$ multiplets, i.e., $|s_{12}SM\rangle = |0,^1\!/_2,+^1\!/_2\rangle$ and $|1,^1\!/_2,+^1\!/_2\rangle$. Expressed in the uncoupled $|m_1m_2m_3\rangle$ basis, these eigenfunctions are $(|\alpha\beta\alpha\rangle - |\beta\alpha\alpha\rangle)/\sqrt{2}$ and $(2|\alpha\alpha\beta\rangle - |\alpha\beta\alpha\rangle - |\beta\alpha\alpha\rangle)/\sqrt{6}$, respectively.

$$\hat{H} = d_z(\hat{s}_{1z} \times \hat{s}_{2z} + \hat{s}_{2z} \times \hat{s}_{3z} + \hat{s}_{3z} \times \hat{s}_{1z}) \qquad (4.24)$$

Equation (4.24) is easier to evaluate by expanding each cross product as

$$\hat{s}_{iz} \times \hat{s}_{jz} = \hat{s}_{ix}\hat{s}_{jy} - \hat{s}_{iy}\hat{s}_{jx} = \frac{i}{2}(\hat{s}_{i+}\hat{s}_{j-} - \hat{s}_{i-}\hat{s}_{j+})$$

so, for example,

$$(\hat{s}_{1z} \times \hat{s}_{2z})\left|\frac{\alpha\beta\alpha - \beta\alpha\alpha}{\sqrt{2}}\right\rangle = \frac{i}{2}(\hat{s}_{1+}\hat{s}_{2-} - \hat{s}_{1-}\hat{s}_{2+})\left|\frac{\alpha\beta\alpha - \beta\alpha\alpha}{\sqrt{2}}\right\rangle = \frac{i}{2}\left|\frac{-\alpha\beta\alpha - \beta\alpha\alpha}{\sqrt{2}}\right\rangle$$

and

$$(\hat{s}_{2z} \times \hat{s}_{3z})\left|\frac{\alpha\beta\alpha - \beta\alpha\alpha}{\sqrt{2}}\right\rangle = \frac{i}{2}(\hat{s}_{2+}\hat{s}_{3-} - \hat{s}_{2-}\hat{s}_{3+})\left|\frac{\alpha\beta\alpha - \beta\alpha\alpha}{\sqrt{2}}\right\rangle = \frac{i}{2}\left|\frac{\alpha\alpha\beta}{\sqrt{2}}\right\rangle$$

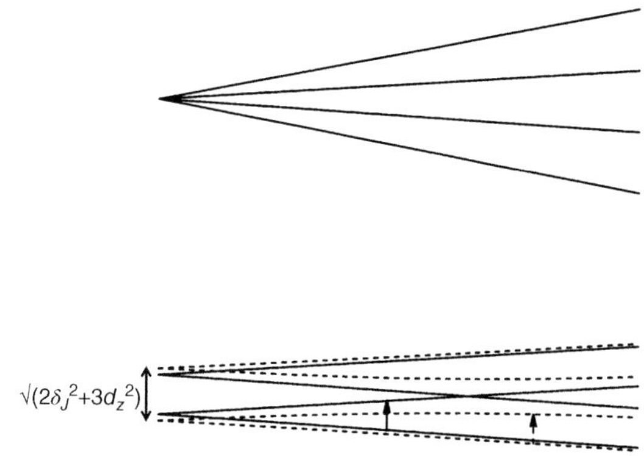

Figure 4.10. Zeeman diagram for an antiferromagnetically coupled isosceles triangle of $s = ^1\!/_2$, with the magnetic field in the xy-plane. Dashed lines show the effect of a small d_z ($\ll |J|$) antisymmetric exchange leading to mixing between the bottom two doublets. The EPR transition in the lowest Kramers doublet (arrows) gives a small effective g

with a similar term resulting from $\hat{s}_{3z} \times \hat{s}_{1z}$. Thus, we find a nonzero matrix element

$$\begin{aligned}&\left\langle\frac{2\alpha\alpha\beta - \alpha\beta\alpha - \beta\alpha\alpha}{\sqrt{6}}\right| d_z(\hat{s}_{1z} \times \hat{s}_{2z} + \hat{s}_{2z} \times \hat{s}_{3z} + \hat{s}_{3z} \times \hat{s}_{1z})\left|\frac{\alpha\beta\alpha - \beta\alpha\alpha}{\sqrt{2}}\right\rangle \\ &= \left\langle\frac{2\alpha\alpha\beta - \alpha\beta\alpha - \beta\alpha\alpha}{\sqrt{6}}\right|\frac{id_z}{2}\left|\frac{-\beta\alpha\alpha - \alpha\beta\alpha + 2\alpha\alpha\beta}{\sqrt{2}}\right\rangle \\ &= \frac{i}{4\sqrt{3}}d_z[\langle\alpha\beta\alpha|\alpha\beta\alpha\rangle + \langle\beta\alpha\alpha|\beta\alpha\alpha\rangle + 4\langle\alpha\alpha\beta|\alpha\alpha\beta\rangle] \\ &= \sqrt{3}id_z/2\end{aligned} \qquad (4.25)$$

The mixing between the two $S = ^1\!/_2$ multiplets provides a mechanism for breaking their degeneracy. EPR is very sensitive to this mixing and the relative magnitude of $h\nu$, d_z, and any isosceles distortion of the isotropic exchange ($J_1 \neq J_2$).[27] Even where the latter results in a spectroscopically isolated doublet ($|J_1 - J_2| = \Delta_J > h\nu$), this can give strange effective g values (Figure 4.10). For example, several trimetallic Cu(II) complexes show $g_{x,y} \ll 2.0$ from the ground doublet.[28] Such effects have also been observed in biological FeS clusters.[37] The antisymmetric exchange can also break the degeneracy of $S > ^1\!/_2$ multiplets,[29] and this can explain very large ZFS effects in some heterometallic clusters.[30]

Finally, there are some studies that invoke a biquadratic exchange term in the isotropic interaction

$$\hat{H} = -2J\hat{s}_1 \cdot \hat{s}_2 + j(\hat{s}_1 \cdot \hat{s}_2)^2 \qquad (4.26)$$

For a simple pair s_1s_2, the main effect is the breakdown of the Landé interval rule. However, because the relative energies of the total spin S multiplets are rarely detected directly by EPR, such a term has rarely been invoked in EPR studies.[31] However, they can be determined accurately by other measurements such as inelastic neutron scattering (INS), which can induce $\Delta S = 0, \pm 1$ transitions directly.[32]

4.9 NONCOINCIDENCE EFFECTS BETWEEN THE INTERACTION MATRICES: $g, A, J, D, J_{dip}, D_{12}$, ETC.

For the sake of clarity, all the worked examples above have assumed coincident principal axes (i.e., the reference frame in which the tensor is diagonal) for the various interaction tensors. Of course the Hamiltonian must obey the point symmetry of the paramagnet, and hence the latter may dictate some relationships between principal axes and values. If the interaction matrices do not share principal axis systems, then they must be transformed into a common global reference frame via the appropriate rotation matrices. This leads to off-diagonal matrix elements in the Hamiltonian matrix in the global reference frame, and hence further state mixings, and this must be taken into account when dealing with anisotropic quantities. The greater the anisotropy of the quantity, the more important this is. There are thorough accounts with worked examples for exchange-coupled species in the textbooks of, for example, Benicini and Gatteschi[1] and of Pilbrow.[18] As an example, Bencini and Gatteschi show that a purely dipolar coupled $s = ½$ dimer with anisotropic $\boldsymbol{g}_i$, but where $\boldsymbol{J}_{dip}$ (the dipolar interaction) has a different principal axis system to $\boldsymbol{g}_i$, in itself gives rise to isotropic, anisotropic, and antisymmetric components in the interaction matrix when expressed as $\hat{s}_1 \cdot \boldsymbol{J} \cdot \hat{s}_2$.

4.10 EXCHANGE BETWEEN ORBITALLY DEGENERATE SPECIES

Up to now we have only considered exchange interactions between paramagnetic centers where orbital angular momentum is quenched. The theory for exchange between orbitally degenerate ions is much more complicated, having to consider the spin *and* orbital functions (L, M_L, S, M_S) explicitly, hence including the crystal field and SOC interactions. The user-friendly phenomenological spin Hamiltonian is not valid.

There are some approximations that have been applied successfully to modeling magnetic data, notably the Lines model (which includes an isotropic exchange between the true spins in the spin–orbit basis functions),[33] but the applicability to EPR, which is much more sensitive to the parameters involved, is not clear. In fact, there are relatively few detailed EPR studies of such species, but there are some d-block ions and f-block ions examples and these tend to be for Kramers ions (half-integer spin).

For orbitally degenerate Kramers ions, the combined result of the crystal field and SOC is often to leave a ground Kramers doublet that is well separated (with respect to the low temperatures necessary to observe their EPR spectra) from the lowest excited states. The ion can then be treated as an effective spin $s^{eff} = ½$ characterized by its effective $\boldsymbol{g}^{eff}$ tensor. For d-block ions, the crystal field is dominant over the spin–orbit interaction, e.g., octahedral Co(II) (probably the most studied example in an EPR context) has a $^4T_{1g}$ ground term, which after SOC gives a lowest energy doublet, with quartet and sextet excited states. These are *not* spin functions, but rather spin–orbit functions depending on both M_L and M_S (a fictitious orbital angular momentum quantum number L, representing the triplet term, is often invoked). For lanthanides, SOC is dominant over the crystal field. The Russell–Saunders coupling scheme can be applied to generate the $^{2S+1}L_J$ term, where J is the total angular momentum; the crystal field breaks the degeneracy and the eigenfunctions are linear combinations of the $|J,M_J\rangle$ states. Either way, the ground Kramers doublet $\boldsymbol{g}^{eff}$ can be calculated with a good knowledge of the electronic structure, e.g., from *ab initio* calculations, or measured from EPR studies on the isolated paramagnet.

For a pair of such ions, an effective exchange interaction ($\hat{H} = \hat{s}_1^{eff} \cdot \boldsymbol{J}^{eff} \cdot \hat{s}_2^{eff}$) can then be introduced. Such a pair is then being treated as a four-level system, with the assumption that all excited states are at much higher energy. If the lowest excited states of the individual ions are at an energy comparable to $k_B T$, or to the exchange, then this approach is not valid. The effective exchange matrix $\boldsymbol{J}^{eff}$ is in general highly anisotropic (purely dipolar at its simplest), as are the $\boldsymbol{g}^{eff}$, so it is important to consider the relative orientations. In terms of EPR, this approach has been used successfully for Co(II) dimers[34] and lanthanide(III) dimers.[35] It has also been used for interpretation of INS spectra.[36] It should be stressed that $\boldsymbol{J}^{eff}$ is *not* the true exchange between the spins, but is a useful tool to parameterize and therefore to determine the low-lying level structure. The relationship between such parameters and a more physically meaningful spin–orbit Hamiltonian has been discussed.[33]

4.11 CONCLUDING COMMENTS

We conclude with some general comments on measuring exchange interactions. Very large interactions, say those associated with two electrons in degenerate orbitals on a single site, are in the domain of optical spectroscopies. The most common method to measure exchange couplings between two different paramagnetic sites is variable temperature magnetometry. This is not a direct measurement because the experimental observable is a weighted thermodynamic sum over all populated states, rather than a direct measure of a splitting between states as is the case with spectroscopic methods. The magnitude of exchange that can be determined by such thermodynamic methods (also including heat capacity methods) is then limited by the temperature of the experiments because it is necessary to have $k_B T$ in the ballpark of $|J|$ to vary the populations significantly. For a conventional magnetometer, this means down to around 2 K using standard liquid ^{4}He cooling, but much lower temperatures can be reached in specialized labs. INS is an extremely powerful method, with complementary selection rules to EPR, for splittings over the range about 10^{-1}–10^3 cm^{-1}. At the lower end, INS is limited by the intrinsic linewidths of the experiment. As for EPR, we have shown use of variable temperature methods to determine $|J|$ up to 10^3 cm^{-1} (simply exploiting the spectrometer as a sensitive magnetometer), and by direct measurement of singlet–triplet transitions if there is a suitable perturbation that mixes the states: this also requires $h\nu \approx |J|$, although observing such transitions will become less likely for large $|J|$ because the perturbation would need to be correspondingly large. We have also shown that weak exchange can be determined directly when it is comparable to differences in Zeeman energy, or comparable to *resolved* hyperfine coupling. Below or on this scale, $|J|$ may contribute to line broadening in CW spectra: it can then in principle be determined if the intrinsic linewidth of the isolated paramagnets is known. At very low-energy scales, pulsed EPR methods come into play, for example, PELDOR/DEER can give the magnitude and sign of J; the latter because J either adds or subtracts from the dipolar interaction the sign of which is known.[20] PELDOR has been used to measure isotropic $|J|$ down to about 10^{-5} cm^{-1}.

ACKNOWLEDGMENT

We are grateful to Dr N. F. Chilton (Manchester) for helpful discussions.

FURTHER READING

A. Abragam and B. Bleaney, Electron Paramagnetic Resonance of Transition Ions, OUP: London, 1970.

O. Kahn, Molecular Magnetism, VCH: New York, 1993.

F. E. Mabbs and D. Collison, Electron Paramagnetic Resonance of d Transition Ion Compounds, Elsevier: Amsterdam, 1992.

J. A. Weil, J. R. Bolton, and J. E. Wertz, Electron Paramagnetic Resonance: Elementary Theory and Applications, Wiley-Interscience: New York, 1994.

REFERENCES

1. A. Bencini and D. Gatteschi, Exchange Coupled EPR, Springer Verlag: Berlin, 1990.
2. B. Bleaney and K. D. Bowers, *Proc. Roy. Soc. (London)*, 1952, **214**, 451.
3. T. S. Cameron, A. Decken, R. M. Kowalczyk, E. J. L. McInnes, J. Passmore, J. M. Rawson, K. V. Shuvaev, and L. T. Thompson, *Chem. Commun.*, 2006, 2277.
4. V. V. Semenaka, O. V. Nesterova, V. N. Kokozay, V. V. Dyakonenko, R. I. Zubatyuk, O. V. Shishkin, R. Boča, J. Jezierska, and A. Ozarowski, *Inorg. Chem.*, 2010, **49**, 5460.
5. J. Schnack and J. Ummethum, *Polyhedron*, 2013, **66**, 28.
6. S. Piligkos, E. Bill, D. Collison, E. J. L. McInnes, G. A. Timco, H. Weihe, R. E. P. Winpenny, and F. Neese, *J. Am. Chem. Soc.*, 2007, **129**, 760.
7. A.-L. Barra, D. Gatteschi, R. Sessoli, and L. Sorace, *Magn. Reson. Chem.*, 2005, **43**, S183.
8. S. Ross, T. Weyermüller, E. Bill, K. Weighardt, and P. Chaudhuri, *Inorg. Chem.*, 2001, **40**, 6656.
9. S. Kremer, *Inorg. Chem.*, 1985, **24**, 887.
10. S. Eaton, K. M. More, B. M. Sawat, M. Sawant, P. M. Boumel, and G. R. Eaton, *J. Magn. Reson.*, 1983, **52**, 435.
11. G. R. Eaton and S. S. Eaton, *Acc. Chem. Res.*, 1988, **21**, 107.

12. R. Calvo, *Appl. Magn. Reson.*, 2007, **31**, 271.

13. A. Fernandez, E. M. Pineda, C. A. Muryn, S. Sproules, F. Moro, G. A. Timco, E. J. L. McInnes, and R. E. P. Winpenny, *Angew. Chem. Int. Ed.*, 2015, **54**, 10858.

14. N. M. Atherton, Principles of Electron Spin Resonance, Ellis Horwood: Chichester, 1993.

15. F. Gerson and W. Huber, Electron Spin Resonance Spectroscopy of Organic Radicals, Wiley-VCH: Weinheim, 2003.

16. D. M. Duggan and D. N. Hendrickson, *Inorg. Chem.*, 1974, **13**, 2929.

17. S. S. Eaton, G. R. Eaton, and C. K. Chang, *J. Amer. Chem. Soc.*, 1985, **107**, 3177.

18. J. R. Pilbrow, Transition Ion Electron Paramagnetic Resonance, OUP: Oxford, 1990.

19. R. E. Coffman and G. R. Buettner, *J. Phys. Chem.*, 1979, **18**, 2387.

20. D. Margraf, P. Cekan, T. F. Prisner, S. T. Siggurdsson, and O. Schiemann, *Phys. Chem. Chem. Phys.*, 2009, **11**, 6708.

21. A. Weber, O. Schiemann, B. Bode, and T. F. Prisner, *J. Magn. Reson.*, 2002, **157**, 277.

22. R. Herchel, R. Boča, J. Krzystek, A. Ozarowski, M. Durán, and J. van Slageren, *J. Am. Chem. Soc.*, 2007, **129**, 10306.

23. D. Burdinski, E. Bill, F. Birkelbach, K. Wieghardt, and P. Chaudhuri, *Inorg. Chem.*, 2001, **40**, 1160.

24. M. Zheng, S. V. Khangulov, G. C. Dismukes, and V. V. Barynin, *Inorg. Chem.*, 1994, **33**, 382.

25. N. Cox, W. Ames, B. Epel, L. V. Kulik, L. Rapatskiy, F. Neese, J. Messinger, K. Weighardt, and W. Lubitz, *Inorg. Chem.*, 2011, **50**, 8238.

26. M. R. Cheesman, V. S. Oganesyan, N. J. Watmough, C. S. Butler, and A. J. Thomson, *J. Am. Chem. Soc.*, 2004, **126**, 4157.

27. Y. V. Rakitin, Y. V. Yablokov, and V. V. Zelentsov, *J. Magn. Reson.*, 1981, **43**, 288.

28. R. Boča and R. Herchel, *Coord. Chem. Rev.*, 2010, **254**, 2973.

29. M. I. Belinsky, *Inorg. Chem.*, 2008, **47**, 3521.

30. S. A. Magee, S. Sproules, A.-L. Barra, G. A. Timco, N. F. Chilton, D. Collison, R. E. P. Winpenny, and E. J. L. McInnes, *Angew. Chem. Int. Ed.*, 2014, **53**, 5310.

31. E. A. Harris, *J. Phys. C*, 1972, **5**, 338.

32. R. Basler, C. Boskovic, G. Chaboussant, H. U. Güdel, M. Murrie, S. T. Ochsenbein, and A. Sieber, *ChemPhysChem*, 2003, **4**, 910.

33. A. Palii, B. Tsukerblat, J. M. Clemente-Juan, and E. Coronado, *Int. Rev. Phys. Chem.*, 2010, **29**, 135.

34. A. B. Boeer, A.-L. Barra, L. F. Chibotaru, D. Collison, E. J. L. McInnes, R. A. Mole, G. S. Simeoni, G. A. Timco, L. Ungur, T. Unruh, and R. E. P. Winpenny, *Angew. Chem. Int. Ed.*, 2011, **50**, 4007.

35. O. Guillot-Noël, P. Goldner, P. Higel, and D. Gourier, *J. Phys.: Condens. Matter*, 2004, **16**, R1.

36. H. Andres, M. Aebersold, H. U. Güdel, J. M. Clemente, E. Coronado, H. Buttner, C. Kearly, and M. Zolliker, *Chem. Phys. Lett.*, 1998, **289**, 224.

37. Y. Sannakis, A. L. Macedo, I. Moura, J. J. G. Moura, V. Papaefthymiou, and E. Münck, *J. Am. Chem. Soc.*, 2000, **122**, 11855.

Chapter 5
EPR Interactions – Hyperfine Couplings

Marina Bennati
Max Planck Institute for Biophysical Chemistry and University of Göttingen, Göttingen, Germany

5.1 INTRODUCTION

Hyperfine couplings arise from the magnetic interaction of an electron spin with nuclear spins in its close surroundings. This is one of the most important sources of information from EPR spectroscopy, enabling not only to identify paramagnetic centers but also to give insights into the more complex structure of their ligand sphere. Information from hyperfine couplings plays a key role in a large number of applications ranging from structural investigation of proteins and metallo enzymes to catalysis, materials science, and quantum information research.

Because of their physical origin, hyperfine splittings observable in EPR methods cover a wide frequency range from gigahertz down to a few kilohertz. Their size and relative isotropic and anisotropic contributions uniquely reflect the ground-state electronic wave function of a paramagnetic center. Isotropic couplings generally produce well-resolved splittings in the EPR spectra of liquid solutions; however, most paramagnetic centers are not detectable under this condition because of their reactivity, fast relaxation, or sensitivity issues. In the solid state, such as powders or frozen solutions of diluted paramagnetic centers, the large number of hyperfine lines and their anisotropy produce inhomogeneous broadening of the EPR line. Furthermore, other anisotropic electron spin interactions might overlay the hyperfine splittings and resolution is lost. Fortunately, the establishment of sophisticated pulse EPR methods in combination with low- and high-frequency EPR permits nowadays to disentangle hyperfine interactions in solid-state spectra with great details (see Chapters 16–18). The main goal of this chapter is to summarize the physical principles of the hyperfine interaction and how it contributes to EPR spectra at different frequencies. For simplicity, we will restrict the theoretical discussion to paramagnetic species with small or negligible g anisotropy.

EPR Spectroscopy: Fundamentals and Methods.
Edited by Daniella Goldfarb and Stefan Stoll
 ISBN: 978-1-119-16299-5
Also published in eMagRes (online edition)
DOI: 10.1002/9780470034590.emrstm1503

5.2 DIPOLAR INTERACTION AND HYPERFINE HAMILTONIAN

The magnetic interaction between an electron and a nuclear spin can be derived from the classical energy between two magnetic point dipoles. However, some additional consideration is required that takes into account the different particle properties of electrons and nuclei. The most important aspect is the fact that electrons and their magnetic moments are not localized in space as nuclei are. In molecules, electrons are described by wave functions that are delocalized over several atoms. Consequently, one electron spin can be magnetically coupled to a large number of nuclei over distances similar to the size of a molecule or even of molecular chains (for instance in conjugated polymers). To derive the hyperfine Hamiltonian, we first consider the idealized case of one electron spin 'localized' at a well-defined position relative to the nucleus defined by the vector $\boldsymbol{r}$ (Figure 5.1).

Their magnetic interaction corresponds to the classical point–dipole interaction between their magnetic moments. Following classical electrodynamic theory,[1] the magnetic moment $\boldsymbol{\mu}_1$ of spin 1 creates a vector potential of the form $\boldsymbol{A} = k_m(\boldsymbol{\mu}_1 \times \boldsymbol{r})/|\boldsymbol{r}|^3$, where $k_m = \mu_0/4\pi$ in SI units. The origin of the coordinate $\boldsymbol{r}$ is chosen at the nuclear spin, which is considered a point in space. The potential $\boldsymbol{A}$ generates a magnetic field $\boldsymbol{B}(\boldsymbol{r}) = \nabla \times \boldsymbol{A}\ (\boldsymbol{r})$, where ∇ is the nabla vector operator with components $(\partial/\partial x, \partial/\partial y, \partial/\partial z)$. The expression for $\boldsymbol{A}$ points to the difficulty in deriving a general hyperfine Hamiltonian, as the potential has a singularity at $\boldsymbol{r} = 0$ where it becomes infinite. The case of $\boldsymbol{r} = 0$ corresponds to a spatial overlap of the two magnetic moments, for which the dipolar interaction becomes independent on $\boldsymbol{r}$ (isotropic case). We will first consider the case $\boldsymbol{r} \neq 0$. Carrying out the vector product for $\boldsymbol{B}(\boldsymbol{r})$ one obtains the expression for the dipolar field at a distance r from the generating dipole $\boldsymbol{\mu}_1$ (Figure 5.1):[1]

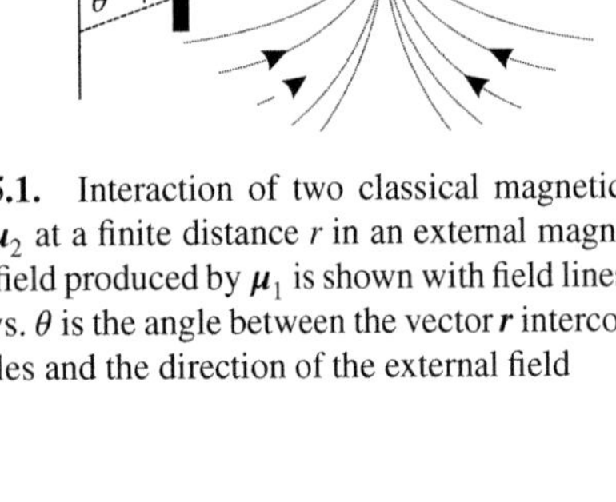

Figure 5.1. Interaction of two classical magnetic dipoles $\boldsymbol{\mu}_1$ and $\boldsymbol{\mu}_2$ at a finite distance r in an external magnetic field B_0. The field produced by $\boldsymbol{\mu}_1$ is shown with field lines marked by arrows. θ is the angle between the vector $\boldsymbol{r}$ interconnecting the dipoles and the direction of the external field

$$\boldsymbol{B}(\boldsymbol{r}) = k_m \left\{ \frac{3\boldsymbol{n}\left(\boldsymbol{n}\cdot\boldsymbol{\mu}_1\right) - \boldsymbol{\mu}_1}{r^3} \right\} = k_m \left\{ \frac{3\boldsymbol{r}\left(\boldsymbol{r}\cdot\boldsymbol{\mu}_1\right)}{r^5} - \frac{\boldsymbol{\mu}_1}{r^3} \right\} \quad (5.1)$$

where $\boldsymbol{n}$ is the unit vector in the direction of $\boldsymbol{r}$, i.e., $\boldsymbol{n} = \boldsymbol{r}/|\boldsymbol{r}|$. For simplicity, we omit the transpose superscript in all scalar products ($\boldsymbol{r}^T\cdot\boldsymbol{\mu} = \boldsymbol{r}\cdot\boldsymbol{\mu}$). For $\boldsymbol{r} > 0$, the dipolar energy experienced by the magnetic moment $\boldsymbol{\mu}_2$ is readily obtained as $E = -\boldsymbol{\mu}_2\cdot\boldsymbol{B}(\boldsymbol{r})$. The transition to the spin Hamiltonian is carried out by introducing the spin operators $\hat{\boldsymbol{I}}$ and $\hat{\boldsymbol{S}}$ for the magnetic moments, i.e., $\hat{\boldsymbol{\mu}}_1 = \hat{\boldsymbol{\mu}}_n = +g_n\mu_N\hat{\boldsymbol{I}}$ for the nuclear magnetic moment and $\hat{\boldsymbol{\mu}}_2 = \hat{\boldsymbol{\mu}}_e = -g_e\mu_B\hat{\boldsymbol{S}}$ for the electron spin moment:

$$\hat{H}_{dip} = k_m g_e \mu_B g_n \mu_N \left[\frac{3\left(\hat{\boldsymbol{S}}\cdot\boldsymbol{r}\right)(\boldsymbol{r}\cdot\hat{\boldsymbol{I}})}{r^5} - \frac{\hat{\boldsymbol{S}}\cdot\hat{\boldsymbol{I}}}{r^3} \right] \quad (5.2)$$

$\hat{H}_{dip}$ is usually written in the form:

$$\hat{H}_{dip} = \hat{\boldsymbol{S}}\cdot\boldsymbol{T}\cdot\hat{\boldsymbol{I}} \qquad \boldsymbol{T} = k_m g_e \mu_B g_n \mu_N \left[\frac{3\boldsymbol{rr}}{r^5} - \frac{\mathbf{1}}{r^3} \right] \quad (5.3)$$

where $\boldsymbol{T}$ is a 3×3 second-rank tensor (in energy units), $\boldsymbol{rr}$ ($\equiv \boldsymbol{r}\cdot\boldsymbol{r}^T$) the so-called outer product with elements $(\boldsymbol{rr})_{ij} = r_i r_j$, and $\mathbf{1}$ the 3×3 unit matrix. While the magnetic moment of the nucleus can be safely assumed as a point in space, it is now the question how the Hamiltonian (5.2) can be applied to a delocalized electron spin. This is achieved by integrating over the spatial electron probability density distribution so that generally the elements of $\boldsymbol{T}$ take the form:

$$T_{ij} = k_m g_e \mu_B g_n \mu_N \left\langle \psi_0 \left| \frac{3r_i r_j}{r^5} - \frac{\delta_{ij}}{r^3} \right| \psi_0 \right\rangle \quad (5.4)$$

where only the ground-state wave function ψ_0 is required for the spin-only contribution to the magnetic moment. A more detailed discussion of (5.4) is given in Chapter 7. For the coupling of one electron spin with the nuclear spin on the same atom, the integration of $\hat{H}_{dip}$ over the orbital coordinates is nonzero for all orbitals except for s electrons.[2] Thus, there is no through-space dipolar coupling between an unpaired electron in a pure s orbital and the nuclear spin of the same atom.

Now we turn to the important case when $\boldsymbol{r}$ approaches zero. The integral of the dipole field $\boldsymbol{B}(\boldsymbol{r})$ over a spherical electron density distribution of radius $\boldsymbol{r}$, which overlays a spherical charge distribution of radius $\boldsymbol{R} > \boldsymbol{r}$ generating the nuclear dipole $\boldsymbol{\mu}_n$, is given by[1] $\int_{r<R} \boldsymbol{B}(\boldsymbol{r})\mathrm{d}^3\boldsymbol{r} = \frac{2}{3}\mu_0\boldsymbol{\mu}_n$. Thus, the contribution of the dipolar field at $\boldsymbol{R} \rightarrow 0$ is given by a delta function $\frac{2}{3}\mu_0\boldsymbol{\mu}_n\delta(\boldsymbol{r})$. This expression was first derived by Fermi in 1930 using the relativistic Dirac Hamiltonian for the electron.[3] Fermi's expression for the magnetic coupling of one electron in an s orbital to the nuclear spin of the same atom is given by

$$E = -\frac{2\mu_0}{3}|\psi_0(\boldsymbol{r} = 0)|^2\boldsymbol{\mu}_e\boldsymbol{\mu}_n \qquad (5.5)$$

$|\psi_0(\boldsymbol{r}=0)|^2$ is the density of the s electron at the nucleus in units m^{-3} (note that $\boldsymbol{\mu}$ has units of $\mathrm{J\,T^{-1}}$ and μ_0 has units of $\mathrm{T^2{\cdot}m^3\,J^{-1}}$). For an unpaired electron in a multielectron atom or molecule, a more generalized expression of equation (5.5) is given by replacing $|\psi_0(\boldsymbol{r}=0)|^2$ with the spin density at the observed nucleus N. This quantity is represented by a function in the three-dimensional space $\rho^{\alpha-\beta}(\boldsymbol{r})$, as defined in Chapter 7, which gives the difference between the average number of electrons with spin state α versus β. Because of this, the spin density can be either positive or negative. For a mathematical definition and general treatment of the spin density operator, we refer to Chapter 7. The isotropic hyperfine coupling at a certain nucleus N results in a contribution to the Hamiltonian of

$$\hat{H} = a_{\mathrm{iso}}\hat{\boldsymbol{S}} \cdot \hat{\boldsymbol{I}} \text{ with } a_{\mathrm{iso}} = \frac{2\mu_0}{3}g_e\mu_B g_n \mu_N \rho_N^{\alpha-\beta} \qquad (5.6)$$

where a_{iso} is in energy units and $\rho_N^{\alpha-\beta}$ the spin density at nucleus N. Moreover, the Bohr magneton μ_B and the nuclear magneton μ_N are constants but the nuclear g_n factor is specific for each nucleus. In this context, it is interesting to compare a_{iso} values for most abundant magnetic nuclei, when one unpaired electron resides in the outer valence s orbital of their atoms.[4] These values together with the respective g_n values are listed in Table 5.1. We note that not only g_n but also $|\psi_0(\boldsymbol{r}=0)|^2$ varies for each atom. The results in Table 5.1 have provided qualitative insight into the dependence of a_{iso} on the atomic mass and the g_n of the nucleus. For instance, it is evident that ^{19}F is among the most sensitive nuclei to detect s spin density with an a_{iso} (^{19}F) of 52870 MHz (1886.5 mT) as compared to 1420 MHz (50.68 mT) for the H atom.[4] Other halogens such as ^{35,37}Cl or ^{79,81}Br are much less sensitive owing to their lower g_n factors. While in equation (5.6) hyperfine couplings are in energy units, they are often given in frequency units (A/h; MHz) or field units ($A/g_e\mu_B$; mT).

Table 5.1. Hyperfine coupling constants for atoms containing 100% spin population in the valence s orbital from Ref. 4

Nucleus	Spin	g_n	a_{iso}/h (MHz)
^{1}H	1/2	5.58569468	1420
^{2}H	1	0.8574382	220
^{13}C	1/2	1.4048236	3777
^{14}N	1	0.40376100	1811
^{19}F	1/2	5.257736	52870
^{31}P	1/2	2.26320	13306
^{35}Cl	3/2	0.5479162	5723

Data from J. R. Morton, K. F. Preston, J. Magn. Reson., 1978, 30, 577.

The complete hyperfine Hamiltonian should also consider the coupling of the electron orbital momentum with the nuclear spin. This contribution is called pseudo-hyperfine interaction. In the nonrelativistic approximation of the electron Hamiltonian, the field generated by the orbital motion of the electron charge also couples in first order with the nuclear magnetic moment.[5] The energy has the form $E \propto (\boldsymbol{l} \cdot \boldsymbol{\mu}_n)/r^3$, where $\boldsymbol{l} = \boldsymbol{r} \times m\boldsymbol{v}$ is the orbital momentum of the electron. However, for an unpaired electron in the ground state, the orbital momentum is quenched, and only second-order terms play a role, as previously discussed for the $\boldsymbol{g}$ matrix (Chapter 2). In the very early literature,[6] it has been formulated that the effective electron magnetic moment described by the product $-\mu_B \cdot \boldsymbol{g} \cdot \hat{\boldsymbol{S}}$ contains the contributions to second order of the spin–orbit coupling that also account to a pseudo-hyperfine interaction. However, a more rigorous treatment of the spin–orbit coupling (see Chapter 7 and also previous literature,[7]) points out that a relationship between the $\boldsymbol{g}$ matrix and the hyperfine tensor is not at all straightforward, particularly for the interaction of a metal with a ligand nucleus, where the latter does not carry spin density by its own. Taking into account also the pseudo-hyperfine interaction, the generalized form of the hyperfine tensor is written $\boldsymbol{T}'$ and the associated Hamiltonian is

$$\hat{H} = \hat{\boldsymbol{S}} \cdot \boldsymbol{A} \cdot \hat{\boldsymbol{I}} \text{ with } \boldsymbol{A} = a_{\mathrm{iso}}\boldsymbol{1} + \boldsymbol{T}' \qquad (5.7)$$

If the pseudo-hyperfine interaction does not contribute, then $\boldsymbol{T}'$ equals $\boldsymbol{T}$ from (5.3). This is the case for most ligand hyperfine couplings in transition metal complexes and for all couplings in organic radicals.

5.3 EPR TRANSITIONS IN THE GENERAL HIGH MAGNETIC FIELD CASE

The general spin Hamiltonian that describes an electron spin $S = 1/2$ coupled to one or more nuclei k is given by[2,8]

$$\hat{H} = \mu_B \boldsymbol{B} \cdot \boldsymbol{g} \cdot \hat{\boldsymbol{S}} + \sum_k (-\mu_N g_n^k \boldsymbol{B} \cdot \hat{\boldsymbol{I}} + \hat{\boldsymbol{S}} \cdot \boldsymbol{A}^k \cdot \hat{\boldsymbol{I}}) \quad (5.8)$$

The first and second terms in equation (5.8) describe the Zeeman interaction of the electron and nuclear spins, respectively, while the third term describes the hyperfine couplings. Depending on the relative size of these three terms, three cases can be distinguished, as will be discussed in the following sections. For an electron spin with isotropic electron g-factor g_e in the most common case (*case 1*), particularly at standard EPR frequencies (X-band), the electron Zeeman term dominates but the hyperfine coupling might exceed the nuclear Zeeman interaction. In this case, the electron spin can be considered as quantized along the external magnetic field direction and terms containing $\hat{S}_x$ and $\hat{S}_y$ can be neglected. However, the quantization axis of the nuclear spin depends on the direction of the hyperfine field (see Section 5.8), and $\hat{I}_x$ and $\hat{I}_y$ terms cannot be neglected in the hyperfine term. If both the Zeeman terms are much larger than the hyperfine interaction (*case 2*), the so-called high-field limit can be evoked (mostly valid for ^{1}H at W-band frequencies or higher) and only $\hat{S}_Z$ and $\hat{I}_Z$ are considered in the hyperfine term. Finally, at low magnetic fields (*case 3*), the hyperfine terms might be on the order or larger than the electron Zeeman interaction. This last case is becoming rare in modern EPR spectroscopy as increasingly high magnetic fields are employed; however, it shows some interesting effect that will be briefly illustrated in Section 5.7.

In the general *case 1*, Hamiltonian (5.8) coupled to only one nuclear spin with $g_n > 0$ (for instance ^{1}H) can be rewritten as

$$\hat{H} \approx \mu_B g_e B_0 \hat{S}_z - \mu_N g_n B_0 \hat{I}_z + A_{zz}\hat{S}_z\hat{I}_z + A_{zy}\hat{S}_z\hat{I}_y + A_{zx}\hat{S}_z\hat{I}_x \quad (5.9)$$

where x, y, and z represent the principal axes of a laboratory-fixed frame with the static field B_0 along the z-direction. The explicit matrix form of this *first-order* Hamiltonian can be obtained by expanding the spin matrices according to the total spin space with dimension $(2S+1)(2I+1)$. For $S = 1/2$ and $I = 1/2$, the dimension is 4 and the product spin wave functions are $|\alpha_e\alpha_n\rangle$, $|\alpha_e\beta_n\rangle$, $|\beta_e\alpha_n\rangle$ and $|\beta_e\beta_n\rangle$. The two last terms of (5.9) produce mixing of nuclear spin states within the individual electron manifolds α_e and β_e. The 4×4 Hamiltonian matrix contains 2×2 blocks that can be diagonalized independently:

$$\begin{array}{c} \\ \langle\alpha_e\alpha_n| \\ \langle\alpha_e\beta_n| \\ \langle\beta_e\alpha_n| \\ \langle\beta_e\beta_n| \end{array}
\begin{array}{cccc} |\alpha_e\alpha_n\rangle & |\alpha_e\beta_n\rangle & |\beta_e\alpha_n\rangle & |\beta_e\beta_n\rangle \\
\frac{1}{2}\mu_B g_e B_0 - \frac{1}{2}\mu_N g_n B_0 + \frac{A_{zz}}{4} & \frac{1}{4}\left(A_{zx} - iA_{zy}\right) & 0 & 0 \\
\frac{1}{4}(A_{zx} + iA_{zy}) & \frac{1}{2}\mu_B g_e B_0 + \frac{1}{2}\mu_N g_n B_0 - \frac{A_{zz}}{4} & 0 & 0 \\
0 & 0 & -\frac{1}{2}\mu_B g_e B_0 - \frac{1}{2}\mu_N g_n B_0 - \frac{A_{zz}}{4} & -\frac{1}{4}(A_{zx} - iA_{zy}) \\
0 & 0 & -\frac{1}{4}(A_{zx} + iA_{zy}) & -\frac{1}{2}\mu_B g_e B_0 + \frac{1}{2}\mu_N g_n B_0 + \frac{A_{zz}}{4}
\end{array}$$

After diagonalization, the energies are[9]

$$E_{1,2} = \frac{\mu_B g_e B_0}{2} \pm \frac{1}{2}K(+) \text{ (for } \alpha_e\text{, i.e., } m_S = +1/2) \quad (5.10)$$

$$E_{3,4} = -\frac{\mu_B g_e B_0}{2} \pm \frac{1}{2}K(-) \text{ (for } \beta_e\text{, i.e., } m_S = -1/2) \quad (5.11)$$

$$\text{with } K(\pm) = \sqrt{\frac{1}{4}(A_{zy}^2 + A_{zx}^2) + \left(-\mu_N g_n B_0 \pm \frac{1}{2}A_{zz}\right)^2} \quad (5.12)$$

We have labeled with $|1\rangle \cdots |4\rangle$ the corresponding states in the new diagonal frame, i.e., $|\alpha_e\alpha_n\rangle \rightarrow |1\rangle$, etc., anticipating from Section 5.8 that for small off-diagonal elements A_{zx} and A_{zy}, the m_S and m_I are still good quantum numbers. Consequently, the two allowed EPR transitions that arise from the selection

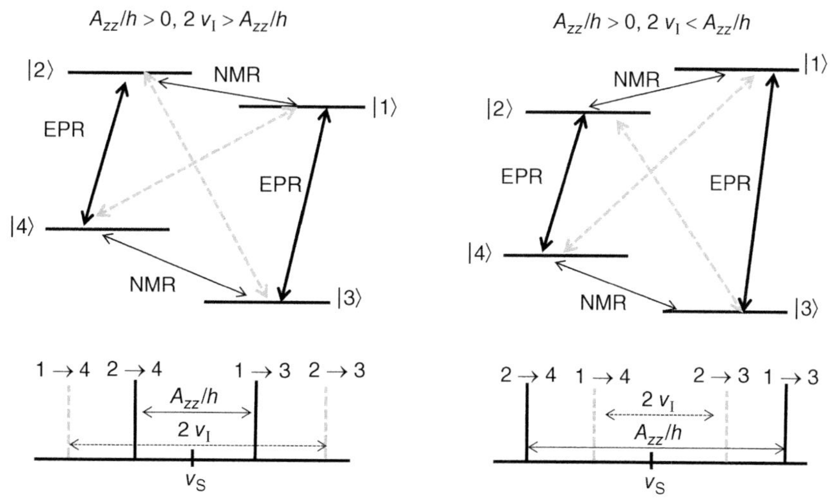

Figure 5.2. Energy levels and resonance position of allowed (solid lines) and forbidden (dashed lines) EPR transitions in the high-field limit according to equations (5.16) and (5.17) for an $S = 1/2$ electron spin coupled to a $I = 1/2$ nuclear spin with positive g_n

rules $\Delta m_S = \pm 1$, $\Delta m_I = 0$ are

$$E_2 - E_4 = \Delta E_{24} = \mu_B g_e B_0 - \frac{1}{2}K(+) + \frac{1}{2}K(-) \tag{5.13}$$

$$E_1 - E_3 = \Delta E_{13} = \mu_B g_e B_0 + \frac{1}{2}K(+) - \frac{1}{2}K(-) \tag{5.14}$$

The nuclear NMR or ENDOR transitions result from the selection rules $\Delta m_I = \pm 1$, $\Delta m_S = 0$ and correspond to

$$\begin{aligned} E_1 - E_2 &= \Delta E_{12} = K(+) \\ E_3 - E_4 &= \Delta E_{34} = K(-) \end{aligned} \tag{5.15}$$

When not only the electron but also the nuclear Zeeman term is larger than the hyperfine interaction, the so-called high-field limit (case 2), the off-diagonal elements A_{zx} and A_{zy} can be neglected as compared to the energy differences between the diagonal elements they are connecting, as is evident from (5.12). In this limit, the Hamiltonian (5.9) is diagonal, and equations (5.13–5.15) reduce to

$$\Delta E_{\text{EPR,allowed}} \approx \mu_B g_e B_0 \pm \frac{1}{2}A_{zz} \tag{5.16}$$

$$\Delta E_{\text{ENDOR}} \approx \mu_N g_n B_0 \pm \frac{1}{2}A_{zz} \tag{5.17}$$

It is evident from (5.16) that the nuclear Zeeman term is not important when considering the EPR spectrum. We note that equations (5.16, 5.17) are valid also for a large, pure isotropic hyperfine coupling (i.e., $A_{zz} = a_{iso}$) with $|a_{iso}|/h > |2\mu_N g_n B_0/h|$ as long as $|a_{iso}|/h \ll |2\mu_B g_e B_0|/h$, as we have neglected terms in $\hat{S}_x$ and $\hat{S}_y$. All transitions are represented schematically in Figure 5.2. Two important cases can be distinguished, i.e., $|A_{zz}/h| < |2\nu_I|$ and $|A_{zz}/h| > |2\nu_I|$, with the nuclear Zeeman frequency $\nu_I = {}^{\mu_N g_n B_0}/_h$. In these two cases, the resonance positions of allowed and forbidden transitions in the EPR spectrum are interchanged.

According to (5.9–5.12), the anisotropy of the hyperfine interaction is responsible for mixing nuclear states and for deviation from the high-field limit. Thus, it is important to rationalize the angular dependence of the hyperfine splitting given by the elements A_{zz}, A_{zx}, and A_{zy} in equation (5.12), which depends on the orientation of a specific molecule with respect to the external magnetic field. To explicitly obtain this dependence, the matrix elements A_{ij} have to be expressed as a function of the principal component values (A_1, A_2, and A_3) of $\boldsymbol{A}$. This is performed *via* a coordinate transformation of the $\boldsymbol{A}$ tensor from its principal axes to a laboratory-fixed frame with the z-axis along $\boldsymbol{B}_0$, $\boldsymbol{A}_{\text{Diag}} \xrightarrow{\boldsymbol{R}(\phi,\theta,0)} \boldsymbol{A}_{\text{Lab}}$:

$$\boldsymbol{A}_{\text{Lab}} = \boldsymbol{R}(\phi, \theta, 0) \cdot \boldsymbol{A}_{\text{Diag}} \cdot \boldsymbol{R}^{-1}(\phi, \theta, 0) \tag{5.18}$$

where the angles ϕ and θ are Euler angles and follow the rotation convention (z, y', z'') with the first rotation around z with the angle ϕ. We note that in equation (5.18) the rotation around the third Euler angle ψ is not required and can be set to zero for the purpose of obtaining A_{zz}, A_{zx}, and A_{zy} in equation (5.12). Analytic expressions for these elements are[10]

$$\begin{aligned} A_{zz} &= A_1 \sin^2\theta \cdot \cos^2\phi + A_2 \sin^2\theta \cdot \sin^2\phi + A_3 \cos^2\theta \\ A_{zy} &= -A_1 \sin\phi \cos\phi \sin\theta + A_2 \sin\phi \cos\phi \sin\theta \\ A_{zx} &= A_1 \cos\theta \cos^2\phi \sin\theta + A_2 \cos\theta \sin^2\phi \sin\theta \\ &\quad - A_3 \cos\theta \sin\theta \end{aligned} \tag{5.19}$$

For a tensor $\boldsymbol{A} = \mathrm{a_{iso}}\mathbf{1} + \boldsymbol{T}$ with an axially symmetric $\boldsymbol{T}$ and principal values $A_3 = A_{\|} = \mathrm{a_{iso}} + T_{\|}$ and $A_1 = A_2 = \mathrm{a_{iso}} + T_{\perp}$ and $-T_{\|}/2 = T_{\perp} = T$, we can rewrite equation (5.19) as

$$\begin{aligned} A_{zz} &= A_{\|}\cos^2\theta + A_{\perp}\sin^2\theta = a_{iso} + T(3\cos^2\theta - 1) \\ A_{zx} &= (A_{\|} - A_{\perp})\cos\theta \sin\theta = 3T\cos\theta \sin\theta \\ A_{zy} &= 0 \end{aligned} \tag{5.20}$$

For the theoretical treatment, we have considered only the simplest case of an electron spin $S = 1/2$ coupled to one nucleus $I = 1/2$. However, unpaired electrons in molecules are usually delocalized over several atoms and couple to many nuclear spins. From the Hamiltonian (5.8), it follows that the number of allowed EPR transitions increases *multiplicatively* with the number of coupled nuclei. For an electron $S = 1/2$ coupled to m nuclei with spin I_k, the number of allowed EPR transitions is given by[2]

$$N_{\mathrm{EPR}} = \prod_{k=1}^{m} (2I_k + 1) \tag{5.21}$$

The number of EPR lines reduces as compared to the number of transitions if some of the nuclei have identical hyperfine couplings (*equivalent* nuclei). For

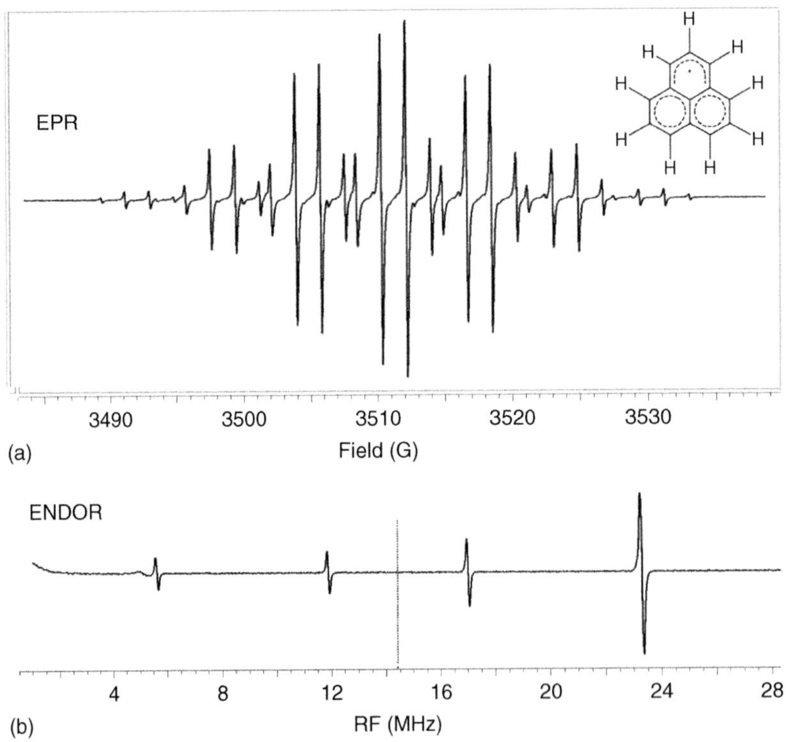

Figure 5.3. (a): CW EPR spectrum of the PNT radical in solution and (b) CW ENDOR spectrum. (Courtesy of Peter Höfer)

a set of equivalent $I = 1/2$ nuclei, the intensity ratio of the lines follows the Pascal's triangle. In contrast, the number of NMR/ENDOR transitions is given by

$$N_{\mathrm{NMR}} = 4\sum_{k=1}^{m} I_k \tag{5.22}$$

A comparative example of an EPR and ENDOR spectrum is given in Section 5.4 (Figure 5.3). The ENDOR experiment detects directly the ENDOR transitions by irradiating a radiofrequency field and is described in detail in Chapter 16. Accordingly, the ENDOR hyperfine pattern is highly simplified with respect to the EPR line pattern. As the only drawback, information about the number of equivalent nuclei is partially lost. In contrast, in the EPR spectrum, the number of contributing nuclei determines the total number of lines and the splitting pattern. Thus, a combined analysis of EPR and ENDOR spectra might be the best approach to extract both the number of coupled nuclei and their hyperfine couplings.

5.4 EXAMPLES OF EPR SPECTRA IN LIQUID AND FROZEN SOLUTION

Typical hyperfine multiplets as described by (5.21) can be best observed in CW EPR spectra of organic radicals in solution, which are characterized by narrow lines due to motional averaging of anisotropies, i.e., for $A_{zz} \rightarrow a_{\mathrm{iso}}$. The different information content of high-resolution EPR versus ENDOR spectra is illustrated by the perinaphthenyl radical, called PNT, in Figure 5.3. The EPR spectrum consists of 28 lines arising from a septet of lines owing to six equivalent ortho protons with a coupling constant of 17.6 MHz (0.63 mT in field units), which is split into seven quartets by three other equivalent protons with a smaller coupling of 5.18 MHz (0.185 mT). Note the intensity ratios in the CW EPR spectrum, which follow the Pascal's triangle, i.e., 1 : 6 : 15 : 20 : 15 : 6 : 1 for the septet and 1 : 3 : 3 : 1 for the quartets. The ENDOR spectrum instead consists only of four lines, a pair for each of the two sets of equivalent couplings.

In contrast to the liquid case, in solids or frozen solutions, the hyperfine anisotropy broadens each hyperfine line, and the resolution in EPR spectra is lost. Nevertheless, if the principal components of the hyperfine tensor (A_1, A_2, and A_3) exceed the inhomogeneous EPR line width and the anisotropy of the g-factor is resolved, hyperfine splittings become distinguishable at the canonical orientations of the EPR powder spectrum for $B_0 \| g_x$, g_y, and g_z. Resolved g anisotropy is usually encountered at X-band (≈9.5 GHz) frequency in spectra of metal ions and at higher frequencies (W-band (≈95 GHz) and higher) for organic radicals. As a first example, Figure 5.4 shows the X-band CW EPR spectrum of a Cu(II) ion, a $3d^9$ system with total $S = 1/2$, embedded in a DNA quadruplex**[11]**. The Cu(II) ion is surrounded by four pyridine ligands in a square-planar arrangement, resulting in an axially symmetric g (see Chapter 2). In frozen solution, the CW EPR spectrum dominated by g anisotropy shows two distinct hyperfine splitting patterns.

A large hyperfine splitting of four lines ($A_{\|} \approx 550$ MHz) appears at the low-field side around $g_{\|}$. This is characteristic for the coupling to the ^{63}Cu ($I = 3/2$) and ^{65}Cu ($I = 3/2$) nuclei. The two isotopes

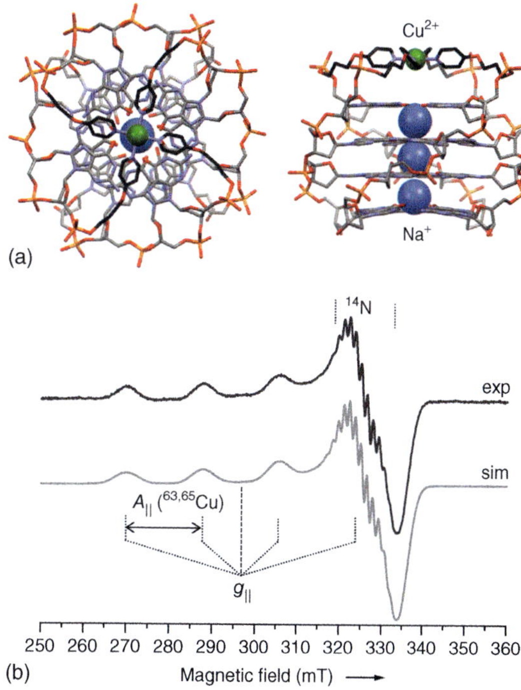

Figure 5.4. (a): Structure of an artificial, guanine-rich oligonucleotide stabilized by a terminal tetra-pyridine complexation with a Cu(II) ion. (b) The nature of the metal coordination was provided by the characteristic X-band CW EPR spectrum recorded at 70 K. Spectral regions with visible hyperfine splittings are marked. (Reproduced with permission from Ref. 11. © John Wiley and Sons, 2013)

are usually indistinguishable and together give a 100% natural abundance of a $I = 3/2$ spin. The $A_\perp$ component of the $\boldsymbol{A}$ tensor is not resolved. We point out that the quadrupole coupling associated with nuclei of $I > 1/2$ is usually not observed in EPR spectra (see Chapter 6). Instead, a so-called super-hyperfine structure of nine lines, with a splitting of ≈40 MHz, is found around the resonance of $g_\perp$. This structure arises from the coupling to four equivalent ^{14}N ($I = 1$) nuclei on the pyridine ligands. The absence of an ^{14}N splitting at $g_\parallel$ indicates that the hyperfine constant in this spectral region is smaller than the inhomogeneous line width and not resolved. It should be noted that the line shape of EPR spectra with large anisotropic g factors and hyperfine couplings might be affected by anomalies such as the occurrence of additional singularities (called *extra absorptions*) in the $g_\perp$ region arising from combinations of $\boldsymbol{g}$ and $\boldsymbol{A}$ anisotropies. These extra lines were observed and theoretically described for some Cu(II) complexes.[12]

5.5 EXAMPLE OF EPR SPECTRA AT MULTIPLE FREQUENCIES

In modern applications of EPR, spectra of diluted paramagnetic centers in proteins or materials are often recorded at low temperatures in the solid state and several paramagnetic species might contribute to the EPR spectrum. This fact taken together with the effect of $\boldsymbol{g}$ and $\boldsymbol{A}$ anisotropy might render the EPR spectrum quite difficult to interpret. Spectral resolution can be increased in a multifrequency EPR approach, eventually combined also with the use of pulse EPR methods to filter different EPR signals.[13]

As a representative example, we consider a tyrosyl radical, which is the most frequently encountered amino acid radical in biology, formed during enzymatic reactions and photosynthesis. In this radical, the unpaired electron density is delocalized over a phenol ring and couples to aromatic protons as well as to the methylene protons of the side chain. The alternating spin density distribution in the aromatic system (discussed in Figure 5.1 of Chapter 7) gives rise to hyperfine splittings that are well visible in EPR spectra at high magnetic fields. Derivatives of tyrosines with slightly perturbed redox potentials have been recently introduced as probes for proton-coupled electron transfer in biology. An example is 3-aminotyrosine that acts as radical trap (Figure 5.5, inset).[14]

Figure 5.5 illustrates multifrequency echo-detected EPR spectra of an aminotyrosyl radical formed at a specific residue (Y_{731}) in the enzyme *Escherichia coli* ribonucleotide reductase. Hyperfine structure is not discerned in the 34 GHz spectrum as g anisotropy ($g_{1,2,3}$ = 20051, 2.0041, 2.0022) is only weakly resolved. The shape of the spectrum changes completely at 94 GHz and hyperfine splitting appear at resonant fields parallel to the canonical orientations of $\boldsymbol{g}$. Nonetheless, because of the small difference in the g_x and g_y values, best resolution is obtained at 263 GHz where hyperfine multiplets are completely separated. For instance, the hyperfine structure at $B \| g_z$ (marked in the figure) shows a splitting into a triplet (intensities 1 : 1 : 1) due to the coupling with ^{14}N ($I = 1$), which is split again due to the coupling with two protons, which were assigned to $H_{\beta 1}$ and $H_{\beta 2}$. A comparison of the multifrequency pulse EPR spectra of single mutant (α-NH_2Y_{731}) and a double mutant (α-$NH_2Y_{731}/R_{411}A$, Figure 5.5, right upper inset) pointed to a large difference in the radical structures.[14] This was established through the difference in the observed splitting of $H_{\beta 1}$ and $H_{\beta 2}$, i.e., the C_β methylene protons, which provide unique information on the amino acid side chain orientation with respect to the peptide backbone (Figure 5.5, right). An explanation for the mechanisms of this hyperfine coupling is given in the following section.

5.6 MECHANISMS OF HYPERFINE COUPLING

It was already noted in Section 5.2 that the isotropic part of the hyperfine interaction relates to the spin density at the observed nucleus N and the anisotropic part to $\langle 1/r^3 \rangle$. The dominant parts of these interactions stem from the orbital the unpaired electron resides in. However, as atoms and molecules contain many electrons, the spins of which are coupled together, the unpaired electron will spin polarize other orbitals and therefore generate additional contributions to the hyperfine coupling. Therefore, quantitative interpretation of hyperfine couplings requires quantum chemical calculations (see Chapter 7). However, a number of semiempirical rules have been developed that in most cases allow rationalizing the observed couplings. In most cases, these rules relate the experimentally observed isotropic couplings a_{iso} with *unpaired electron (or spin) populations* on selected local regions or

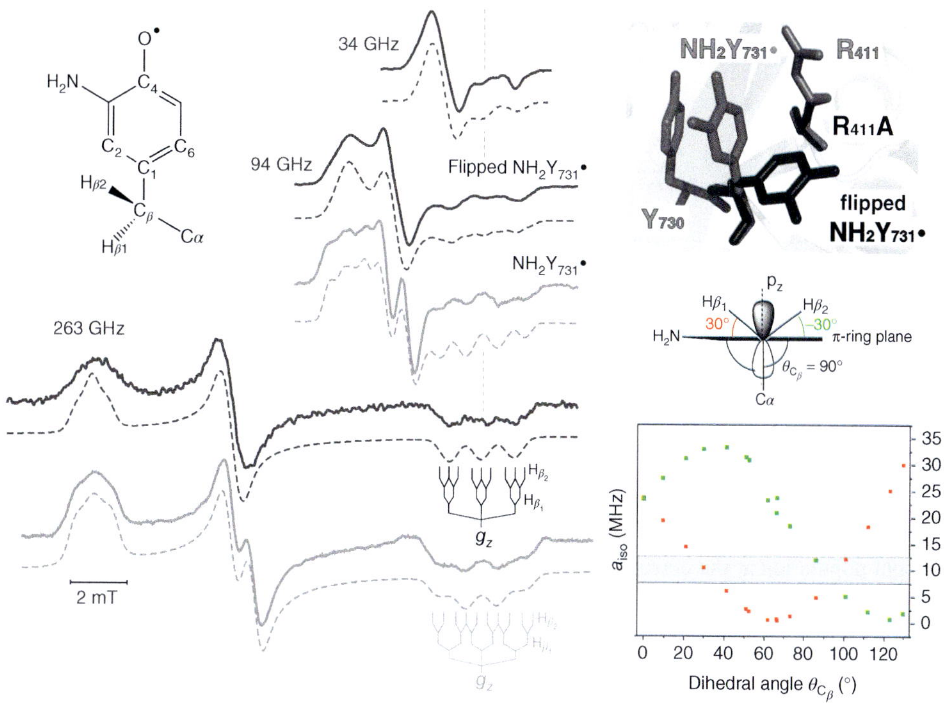

Figure 5.5. Derivatives of echo-detected EPR spectra of a 3-aminotyrosyl radical trapped during enzymatic reaction of *E. coli* ribonucleotide reductase recorded at three different EPR frequencies. Dashed lines indicate simulations with the hyperfine parameters reported in Ref. 14. Spectra are aligned such that $g \approx 2.002$ overlay. Buffer was exchanged to D_2O and amino protons do not contribute in the spectrum. Two detected radical conformations are shown in the top right inset in black and gray, respectively. Orientations of the C_β-protons with respect to phenol ring are shown schematically (right, middle). Dependence of the isotropic C_β-proton hyperfine couplings on the C_α—C_β—C_1—C_6 dihedral angle ($\theta_{C\beta}$) was calculated from a density functional theory DFT model for a 3-aminotyrosyl radical (right, bottom). Top left: Chemical structure of a 3-aminotyrosyl radical

atoms. Extensive reviews can be found in EPR text books.[2,15]

The intriguing feature of isotropic hyperfine couplings is that the unpaired electron in molecules does not usually localize in s orbitals. Classical examples are aromatic radicals, where the unpaired spin is distributed in π orbitals built from carbon, nitrogen, and oxygen 2p atomic orbitals, or transition metal ions with their valence electrons of mainly 3d character. Still, large isotropic couplings with their nuclei are observed. The mechanism responsible for generating s spin density at the nuclei is called spin polarization and arises from electron–electron exchange interactions.[16] The most illustrative example is the coupling of an aromatic proton to the π-spin density on the adjacent carbon atom (C—H fragment) due to the exchange interaction between the electrons in the C—H σ bond and the π electrons. Using perturbation theory, the McConnell relation relates the observed isotropic coupling (in field units) to the spin population at the carbon $2p_z$ orbital, $\rho_{p_z,C}$:[17]

$$a_{iso}(H) = Q^H \cdot \rho_{p_z,C} \text{ with } Q^H \approx -2.3 \text{ mT} \quad (5.23)$$

Note that the constant Q^H is negative meaning that the spin polarization at the proton is opposite to the one in the carbon $2p_z$ orbital.

The treatment has been subsequently extended by Karplus and Fraenkel to rationalize ^{13}C hyperfine couplings owing to the considerably more complicated electronic environment of a carbon nucleus.[18] They found that the σ–π interaction in π-radicals polarizes both the 1s and 2s electrons of the carbon but with different signs. In addition, contributions from neighboring atoms, on which spin density is present, have also to be considered. Thus, the ^{13}C isotropic coupling of an sp^2 hybridized carbon contains three terms with different signs:

$$a_{\mathrm{iso}}(^{13}\mathrm{C}) = \left(Q^{\mathrm{C}}_{1\mathrm{s}} + \sum_{j=1}^{3} Q_{\mathrm{CX}_j} \right) \rho_{\mathrm{C}} + \sum_{j=1}^{3} Q_{\mathrm{X}_j\mathrm{C}} \cdot \rho_{\mathrm{X}_j} \tag{5.24}$$

The index $\mathrm{CX_j}$ is defined as the contributions of the polarized bond between the atoms C and X_j owing to the spin population on atom C (and vice versa for X_jC). For an aromatic ^{13}C, the neighbor atoms X_j are usually carbons, protons, nitrogens, or oxygens. Q^{C} values were reported only for $X_j = \mathrm{C}$: $Q^{\mathrm{C}}_{1\mathrm{s}} = -1.27$ mT, $Q_{\mathrm{CC}_j} = 1.44$ mT, $Q_{\mathrm{C}_j\mathrm{C}} = -1.39$ mT, and $Q^{\mathrm{C}}_{\mathrm{CH}} = 1.95$ mT. It becomes clear that a_{iso} (^{13}C) results from counteracting effects of the spin population on the observed nucleus and from the neighboring atoms. Thus, it might result in a very small value, although the spin population at the detected ^{13}C atom is large. An illustrative recent example is provided by the semidione radical formed during the catalytic reaction of *E. coli* ribonucleotide reductase with a mutant at the active site (Figure 5.6).[19] Density functional theory (DFT) calculations on the formed radical intermediate indicated that the unpaired electron is entirely localized on a semidione moiety on the ribose ring, with $\rho_{\mathrm{p_z,C2}}$, $\rho_{\mathrm{p_z,C3}} \approx 0.2$, $\rho_{\mathrm{p_z,O2}}$, $\rho_{\mathrm{p_z,O3}} \approx 0.3$, and $\rho_{\mathrm{p_z,C1}}$ or $\rho_{\mathrm{p_z,C4}} = 0$, respectively. However, ^{13}C couplings were clearly observed only for ^{13}C1 and ^{13}C4 and not for ^{13}C2 and ^{13}C3. When inserting these $\rho_{\mathrm{p_z}}$ values in equation (5.24), one obtains $a_{\mathrm{C4}} = Q_{\mathrm{C3C4}} \cdot \rho_{\mathrm{C3}} \approx -0.27$ mT and $a_{\mathrm{C1}} \approx -0.29$ mT, which are close to the experimental values of -0.5 mT. Most interestingly, spin polarization contributions cancel each other on ^{13}C2 and ^{13}C3. For an estimate of a_{iso} at C2 and C3, the coefficients Q_{CO} and Q_{OC} are required but unknown. However, if all other values are inserted in equation (5.24), one arrives at the expression: $a_{\mathrm{C3}} = 0.01\ \mathrm{mT} + Q_{\mathrm{C3,O}} \cdot 0.2 + Q_{\mathrm{OC3}} \cdot 0.3$. As Q_{CO} and Q_{OC} might have similar values but opposite signs, just as $Q_{\mathrm{C'C}}$ and $Q_{\mathrm{CC'}}$, polarization contributions would cancel, leading to an almost vanishing ^{13}C isotropic hyperfine coupling at the O-bonded carbon atoms carrying the spin density, in agreement with the experiment.[19]

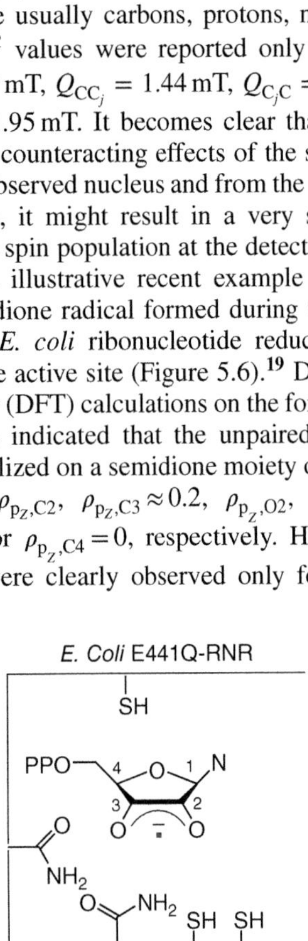

Figure 5.6. Structure of the semidione radical intermediate formed on the ms timescale during the reaction of a E441Q-α mutant of *E. coli* RNR with substrate CDP and allosteric effector TTP. The box around the radical represents the active site pocket with the essential groups involved in chemistry. Numbering of the ^{13}C atoms is shown. (Zipse, http://pubs.acs.org/doi/abs/10.1021/ja806693s. Used under CC-BY-SA 4.0 https://creativecommons.org/licenses/by/4.0/)

One more mechanism should be mentioned that is called *hyperconjugation*[15] and gives rise to isotropic hyperfine couplings. It has been reported mostly for methyl or methylene groups of organic radicals where protons are attached to a carbon (denoted C_β) located next to a carbon carrying an unpaired electron density (denoted C_α), i.e., $\mathrm{H{-}C_\beta{-}C_\alpha{}^\bullet}$. Specifically, the observed values of $a_{\mathrm{iso}}(\mathrm{H{-}C_\beta})$ depend on the dihedral angle θ defined by the direction of the $\mathrm{p_z}$ orbital on C_α, the $\mathrm{C_\alpha{-}C_\beta}$ bond and the direction of the proton bound to C_β (Figure 5.5, right middle). This results in an angle-dependent McConnell relation – compare (5.23) – which is commonly used to estimate θ from $a_{\mathrm{iso}}(\mathrm{H{-}C_\beta})$:

$$a_{\mathrm{iso}}(\mathrm{H\text{-}C_b}) \approx B_1 \cdot \rho_{\mathrm{C}_\alpha} \cdot \cos^2(\theta) \tag{5.25}$$

B_1 is a semiempirical constant and $\rho_{\mathrm{C}\alpha}$ is the $\mathrm{p_z}$ spin population at carbon C_α. It was found experimentally that $B_1 \approx 162$ MHz.[20] However, in contrast to equation (5.23), a_{iso} here has the same sign as the spin population on C_α. This indicates that the mechanism is not based on exchange interaction but rather a mixing of the molecular orbitals of the protons with the π system originating from C_α.[15] An illustrative example for the observation of this coupling and how it relates to structural information was given in the previous section (Figure 5.5).

Finally, large isotropic hyperfine couplings are also visible in spectra of transition metal ions. Many of the transition metal atoms possess magnetic nuclei with $I \geq 1/2$ in high natural abundance, resulting in characteristic hyperfine splittings of $2I+1$ equally intense lines, see equation (5.21) and Figure 5.4. Moreover, couplings to the nucleus of the metal atom usually largely exceed the couplings to ligands so that the observed hyperfine patterns are characteristic for the metal ion. The most widely encountered cases are EPR resonances of 3d metal ions, among which the magnetic nuclei of ^{55}Mn ($I = 5/2$), ^{63,65}Cu ($I = 3/2$), ^{51}V ($I = 7/2$), and ^{59}Co ($I = 7/2$) occur in nature basically with 100% abundance. A more extensive treatment of EPR spectra from metal ions with $S > 1/2$ is given in Chapter 3 in conjunction with the treatment of zero-field splitting. The interesting aspect about the mechanism of hyperfine couplings is that, similarly to π-radicals, the unpaired electrons in d orbitals do not have any density at the nucleus. Therefore, no isotropic coupling can arise from the d electrons directly. However, there is an indirect mechanism that brings spin density to the nucleus. The unpaired d-orbital electron(s) interact differently with electrons in the states α and β in the valence and core s orbitals of the ion (1s, 2s, 3s). As a result, the α and the β parts of these orbitals (e.g., 2s α and 2s β) are spatially slightly different. As s orbitals have nonzero density at the nucleus, this unbalance leads to nonzero spin density at the nucleus and thereby to nonzero isotropic hyperfine coupling. This mechanism is called spin polarization. For more details, see Chapter 7.

5.7 EPR SPECTRA IN THE LOW-FIELD CASE

It was mentioned earlier that transition metal ions display large hyperfine couplings to the nucleus of the own atom. At low magnetic fields, such as that commonly used at X-band, when this coupling is not any more negligible compared to the electron Zeeman interaction, additional effects might be observed in the EPR spectra. In this situation, the $\hat{S}_x$ and $\hat{S}_y$ components of the electron spin vector in equation (5.9) cannot be neglected. The case can be discussed analytically for isotropic hyperfine couplings:[2,8]

$$\hat{H} = \mu_B g_e B_0 \hat{S}_z - \mu_N g_n B_0 \hat{I}_z + a_{iso}(\hat{S}_z\hat{I}_z + \hat{S}_y\hat{I}_y + \hat{S}_x\hat{I}_x) \tag{5.26}$$

The last two terms mix the states $|\alpha_e\beta_n\rangle = |2\rangle$ and $|\beta_e\alpha_n\rangle = |3\rangle$, whereas states $|\alpha_e\alpha_n\rangle = |1\rangle$ and $|\beta_e\beta_n\rangle = |4\rangle$ remain pure eigenstates. This Hamiltonian was first treated by Breit and Rabi in 1931 to investigate the nuclear spin of alkali atoms.[21] After diagonalization of the block 2 × 2 matrix for states $|2\rangle$ and $|3\rangle$, the following energies were obtained:[2,8]

$$\begin{aligned} E_1 &= \frac{\mu_B g_e B_0}{2} - \frac{\mu_N g_n B_0}{2} + \frac{a_{iso}}{4} \\ E_{2,3} &= -\frac{a_{iso}}{4} \pm \frac{1}{2}\sqrt{(\mu_B g_e B_0 + \mu_N g_n B_0)^2 + a_{iso}^2} \\ E_4 &= -\frac{\mu_B g_e B_0}{2} + \frac{\mu_N g_n B_0}{2} + \frac{a_{iso}}{4} \end{aligned} \tag{5.27}$$

As an important feature, equation (5.27) indicates a splitting of the electron spin energies also in the absence of a magnetic field. This is a general property of two spins coupled by an interaction of the form $a\hat{\mathbf{S}} \cdot \hat{\mathbf{I}}$. The coefficients of the wave functions indicated that at very low fields ($|\mu_B g_e B_0| \ll |a_{iso}|$) the system is best described in a new basis set $|F, m_F\rangle$ with the quantum numbers given by the total spin angular momentum $\hat{\mathbf{F}} = \hat{\mathbf{S}} + \hat{\mathbf{I}}$ with $F = 0$, 1 and $m_F = 0$ and −1, 0, +1. At zero magnetic field, these eigenfunctions describe a singlet and triplet state.[2] As an example, the hyperfine interaction of the proton in the hydrogen atom is considered. A plot of the energies as a function of the magnetic field from equation (5.27) is illustrated in Figure 5.7.

The above energy states (5.27) lead to the so-called second-order effects in EPR spectra at low fields and frequencies, such as hyperfine line shifts and additional splittings in multiplets from equivalent nuclei. Examples were encountered already in the early days of EPR spectroscopy in CW EPR spectra of metal ions such as Mn^{2+} or Cu^{2+} at X-band, where hyperfine

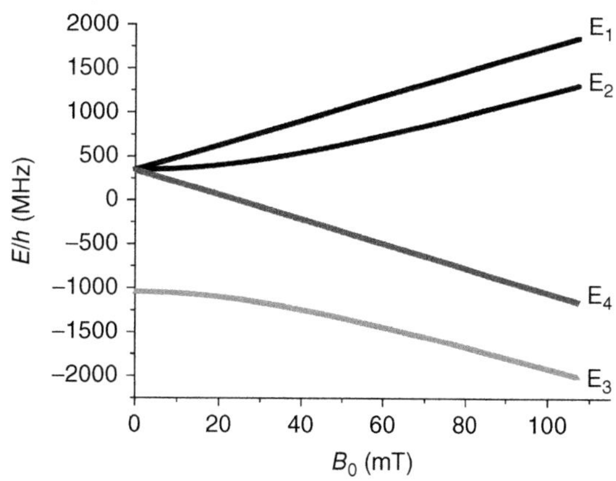

Figure 5.7. Spin energy levels for the ground-state electron in the hydrogen atom at low magnetic fields. The isotropic hyperfine constant is $a_{iso}/h \approx 1.4$ GHz[22]. In the high-field limit, level 1 will cross level 2 at a field around 16 T

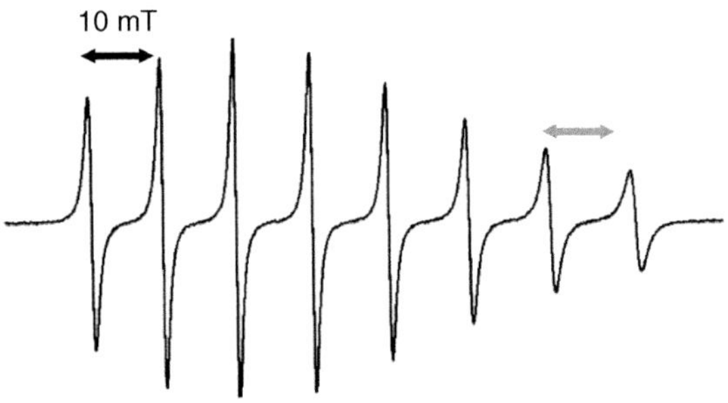

Figure 5.8. CW EPR spectrum of the vanadyl acetylacetonate complex in toluene solution at room temperature at X-band frequencies. Note also the line width and intensity variation due to incomplete motional averaging.[24] (Courtesy of Peter Höfer)

lines due to the coupling to the magnetic nucleus on the same ion were not equidistant.[23] This is also illustrated in Figure 5.8 by the X-band EPR spectrum of the vanadyl acetylacetonate complex ($S = 1/2$ and $I = 7/2$) in toluene. The large hyperfine coupling produces a nonequidistant line separation, which amounts to about 10 mT between the first and second hyperfine line at the low-field side and increases up to 12 mT at the high-field site.

5.8 TRANSITION PROBABILITIES

The nuclear state mixing in equation (5.9) due to hyperfine anisotropy affects also transition probabilities, and forbidden EPR transitions might become relevant for specific pulse EPR experiments, for instance in electron spin echo envelope modulation (ESEEM) and electron–electron double resonance (ELDOR)-detected NMR spectroscopy (see Chapters 17 and 18). Best insight into the effect of the hyperfine anisotropy, specifically how the vectorial direction of the hyperfine field determines transition probability, can be achieved by treating Hamiltonian (5.9) with the product operator formalism (see Chapter 8).[25] For this, it is convenient to rewrite equation (5.9) in angular frequency units under high-field conditions and neglecting g anisotropy:[26]

$$\hat{\mathcal{H}} = \omega_S \hat{S}_z + \omega_I \hat{I}_z + A\hat{S}_z\hat{I}_z + B\hat{S}_z\hat{I}_x \qquad (5.28)$$

We have introduced $\hat{\mathcal{H}} = \hat{H}/\hbar$, $\omega_S = \mu_B g_e B_0/\hbar$, $\omega_I = -\mu_N g_n B_0/\hbar$, $A = A_{zz}/\hbar$, and $B = \sqrt{A_{zy}^2 + A_{zx}^2}/\hbar$; A and B are called the secular and pseudo-secular hyperfine coupling. In addition, for simplification, in equation (5.28), the coordinate system has been rotated around the z-axis to eliminate the $\hat{I}_y$ term. We first introduce in equation (5.28) the spin polarization operators[25] for the system of two coupled spins $1/2$, i.e., $1 = \hat{S}^\alpha + \hat{S}^\beta$ and $\hat{S}_z = \frac{1}{2}(\hat{S}^\alpha - \hat{S}^\beta)$, where α and β refer here to the electron spin manifolds:

$$\hat{\mathcal{H}} = \omega_S \hat{S}_z + \left(\omega_I + \frac{A}{2}\right)\hat{S}^\alpha \hat{I}_z + \frac{B}{2}\hat{S}^\alpha \hat{I}_x + \left(\omega_I - \frac{A}{2}\right)\hat{S}^\beta \hat{I}_z - \frac{B}{2}\hat{S}^\beta \hat{I}_x \qquad (5.29)$$

Now one can diagonalize equation (5.29) by employing operators that act on the individual electron spin manifolds. In a general form, this is best done with fictitious $S = 1/2$ operators.[27] We illustrate here specifically these operators because they permit to treat analytically also more complex spin systems. For two coupled spin $1/2$, the operator $\hat{S}^\alpha \hat{I}_z$ is equivalent to a single transition (fictitious) spin $1/2$ operator denoted $\hat{I}_z^{12}$, which acts only on the transition between the states $|1\rangle$ and $|2\rangle$ and disregards all others. The same applies for $\hat{S}^\alpha \hat{I}_x$ and $\hat{S}^\alpha \hat{I}_y$. The second and third terms of equation (5.29) refer to a transition within the electron spin manifold α and can be treated separately from fourth and fifth terms, which refer exclusively to a transition in the electron spin manifold β. An illustrative example for the description of the hyperfine Hamiltonian with fictitious spin operators was given in a very recent paper to describe electron–nuclear polarization transfer.[28] Diagonalization of (5.29) is now carried out by transforming the Hamiltonian into a tilted frame, in which off-diagonal elements become zero. The transformation $\boldsymbol{UHU}^{-1}$ is defined by two consecutive unitary transformations $\boldsymbol{U} = \boldsymbol{U}_y^\alpha \boldsymbol{U}_y^\beta$ that act individually on the electron manifolds α and β with $\boldsymbol{U}_y^\alpha = \mathrm{e}^{-\mathrm{i}\eta_\alpha \hat{S}^\alpha \hat{I}_y}$ and $\boldsymbol{U}_y^\beta = \mathrm{e}^{-\mathrm{i}\eta_\beta \hat{S}^\beta \hat{I}_y}$. For instance, $\boldsymbol{U}_y^\alpha$ acts only on the second and third terms of equation (5.29). By substituting $\hat{S}^\alpha \hat{I}_Z = \hat{I}_Z^{12}$ and $\hat{S}^\alpha \hat{I}_X = \hat{I}_X^{12}$ in equation (5.29), the transformation is easily carried out:

$$\begin{aligned}
&\left(\omega_I + \frac{A}{2}\right)\hat{I}_z^{12} + \frac{B}{2}\hat{I}_x^{12} \xrightarrow{\eta_\alpha \hat{I}_y^{12}} \left(\omega_I + \frac{A}{2}\right) \\
&[\hat{I}_z^{12}\cos\eta_\alpha + \hat{I}_x^{12}\sin\eta_\alpha] + \frac{B}{2}[\hat{I}_x^{12}\cos\eta_\alpha - \hat{I}_z^{12}\sin\eta_\alpha] \\
&= \left[\left(\omega_I + \frac{A}{2}\right)\cos\eta_\alpha - \frac{B}{2}\sin\eta_\alpha\right]\hat{I}_z^{12} \\
&\quad + \left[\left(\omega_I + \frac{A}{2}\right)\sin\eta_\alpha + \frac{B}{2}\cos\eta_\alpha\right]\hat{I}_x^{12} \qquad (5.30)
\end{aligned}$$

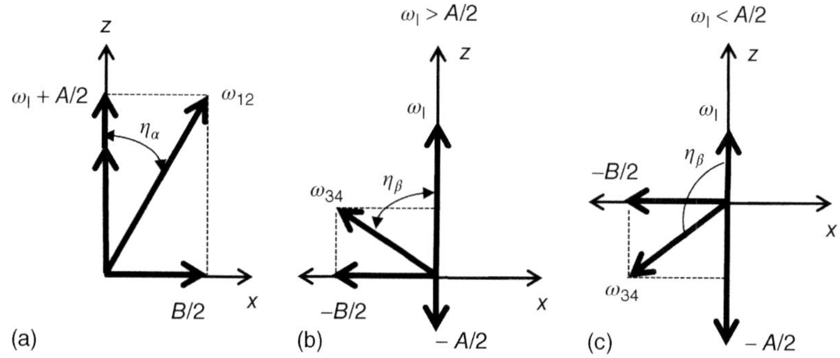

Figure 5.9. Vector representations of the nuclear effective fields ω_{12} (in the electron spin manifold α_e) and ω_{34} (in the electron spin manifold β_e) and their direction with respect to the nuclear Zeeman quantization axis z from equation (5.29). The coordinate system refers to directions defined by the spin vector components $\hat{I}_x^{12}$ and $\hat{I}_z^{12}$ for the α_e manifold (a) and $\hat{I}_x^{34}$ and $\hat{I}_z^{34}$ for manifold β_e (center and right). For β_e, ω_I and A have different signs in equation (5.31), and two cases are distinguished, i.e., $\omega_I > A/2$ or $\omega_I < A/2$ (b and c), respectively. In all cases, the effective fields are simply the vectorial sum of the field components along the z and x axes, respectively. The relevant angles, either η_α or η_β, are those between the effective field and the z-axis

After transformation, the Hamiltonian is in a so-called tilted frame. This part (5.30) of the Hamiltonian describing the α subspace is diagonal if the second term of equation (5.30) vanishes. The transformation for subspace β is analogous. Setting the second term of equation (5.30) to zero (and similarly for the β manifold), it follows for the rotation angles $\eta_{\alpha,\beta}$:

$$\tan\eta_\alpha = \frac{-B/2}{\omega_I + A/2} \quad \tan\eta_\beta = \frac{B/2}{\omega_I - A/2} \tag{5.31}$$

A positive sign of η denotes a counter clockwise rotation around the y-axis. The effect of this transformation can be visualized by a vectorial picture of the various fields acting on the nuclear spin (Figure 5.9).

The nuclear spin experiences an effective total field, which is the vectorial sum of hyperfine and nuclear Zeeman fields. This is expressed through the Hamiltonian:

$$\hat{\mathcal{H}} = \omega_S\hat{S}_z + \omega_{12}\hat{I}^{12}_{z,\text{tilted}} + \omega_{34}\hat{I}^{34}_{z,\text{tilted}} \tag{5.32}$$

The effective fields ω_{12} and ω_{34} correspond exactly to the energies given in equation (5.15), with their values given from the Pythagoras relationships for the vector components along x and z. The frequencies of the allowed and forbidden EPR transitions in Figure 5.2 can be expressed through ω_{12} and ω_{34} as

$$\omega_{13} = \omega_S + \frac{\omega_{12}}{2} - \frac{\omega_{34}}{2};\ \omega_{24} = \omega_S - \frac{\omega_{12}}{2} + \frac{\omega_{34}}{2}; \tag{5.33}$$

$$\omega_{23} = \omega_S + \frac{\omega_{12}}{2} + \frac{\omega_{34}}{2};\ \omega_{14} = \omega_S - \frac{\omega_{12}}{2} - \frac{\omega_{34}}{2};$$

These equations for ω_{13} and ω_{24} are entirely equivalent to equations (5.13) and (5.14), when using $\omega_{12}\cdot\hbar = K(+)$ and $\omega_{34}\cdot\hbar = K(-)$. The vectorial picture in Figure 5.9 provides the basis to rationalize transition probabilities. The larger the hyperfine anisotropy (described by B), the larger is the deviation of the nuclear from electron spin quantization axis and subsequently from the high-field limit. The transition probabilities are[26] proportional to $\cos^2\eta$ and $\sin^2\eta$ for allowed and forbidden transitions, respectively, where $\eta = {}^{(\eta_\alpha-\eta_\beta)}/_2$. Forbidden transitions might not be visible in the EPR spectra; however, they play a key role in a number of EPR experiments (as mentioned at the beginning of the paragraph) and also in dynamic nuclear polarization.

ACKNOWLEDGMENT

The author would like to thank Karin Halbmair, Müge Kasanmascheff, Nikolay Enkin, Tomas Orlando, Isabel Bejenke, Guoquan Liu, Thomas Nick, Monika Frömel, and Igor Tkach, for reading and discussing the manuscript and explicitly thank Daniella Goldfarb and Stefan Stoll for reading and commenting on this manuscript with great attention.

FURTHER READING

A. Carrington and A. D. McLachlan, Introduction to Magnetic Resonance, Harper & Row, Inc.: New York, 1967.

A. Schweiger and G. Jeschke, Principles of Pulse Electron Paramagnetic Resonance, Oxford University Press: Oxford, 2001.

J. A. Weil and J. R. Bolton, 2nd edn, John Wiley & Sons: New Jersey, 2007.

REFERENCES

1. J. D. Jackson, Classical Electrodynamics, 2nd edn, John Wiley & Sons, Inc.: New York, 1975.
2. J. A. Weil and J. R. Bolton, 2nd edn, John Wiley & Sons: New Jersey, 2007.
3. E. Fermi, *Z. Phys.*, 1930, **60**, 320.
4. J. R. Morton and K. F. Preston, *J. Magn. Reson.*, 1978, **30**, 577.
5. A. Abragam, Principles of Nuclear Magnetism, Oxford University Press: Oxford, 1978.
6. M. H. McConnel, *Proc. Natl. Acad. Sci. U.S.A.*, 1958, **44**, 766.
7. R. Lefebvre, *Mol. Phys*, 1967, **12**, 417.
8. C. P. Poole and H. A.Farach Jr, Theory of Magnetic Resonance, 2nd edn, John Wiley & Sons, Inc.: New York, 1987.
9. M. Bennati and D. M. Murphy, in Electron Paramagnetic Resonance, A Practitioner's Toolkit, eds M. Brustolon and E. Giamello, John Wiley & Sons: New Jersey, 2009.
10. M. Mehring and V. A. Weberruss, Object-Oriented Magnetic Resonance, Academic Press: London, 2001.
11. D. M. Engelhard, R. Pievo, and G. Clever, *Angew. Chem. Int. Ed.*, 2013, **52**, 12843.
12. R. Neiman and D. Kivelson, *J. Chem. Phys.*, 1961, **35**, 156.
13. M. Bennati and T. Prisner, *Rep. Prog. Phys.*, 2005, **68**, 411.
14. M. Kasanmascheff, W. Lee, T. U. Nick, J. Stubbe, and M. Bennati, *Chem. Sci.*, 2016, **7**, 2170.
15. A. Carrington and A. D. McLachlan, Introduction to Magnetic Resonance, New York: Harper & Row, Inc., 1967.
16. H. M. McConnell, *J. Chem. Phys.*, 1956, **24**, 764.
17. H. M. McConnell, *J. Chem. Phys.*, 1958, **28**, 1188.
18. M. Karplus and G. K. Fraenkel, *J. Chem. Phys.*, 1961, **35**, 1312.
19. H. Zipse, E. Artin, S. Wnuk, G. J. S. Lohman, D. Martino, R. G. Griffin, S. Kacprzak, M. Kaupp, B. Hoffman, M. Bennati, J. Stubbe, and L. Lees, *J. Am. Chem. Soc.*, 2009, **131**, 200.
20. R. W. Fessenden and R. H. Schüler, *J. Chem. Phys.*, 1963, **39**, 2147.
21. G. Breit and I. I. Rabi, *Phys. Rev.*, 1931, **38**, 2082.
22. B. D. Perlson and J. A. Weil, *J. Magn. Reson.*, 1974, **15**, 594.
23. B. Bleaney and K. W. H. Stevens, *Rep. Prog. Phys.*, 1953, **16**, 108.
24. P. Höfer, G. G. Maresch, D. Schmalbein, and K. Holczer, *Buker Report*, 1996, **142**, 15.
25. C. Gemperle and A. Schweiger, *Chem. Rev.*, 1991, **91**, 1481.
26. A. Schweiger and G. Jeschke, Principles of Pulse Electron Paramagnetic Resonance, Oxford University Press: Oxford, 2001.
27. S. Vega, *J. Chem. Phys.*, 1978, **68**, 5518.
28. R. Rizzato, I. Kaminker, S. Vega, and M. Bennati, *Mol. Phys.*, 2013, **111**, 2809.

Chapter 6

EPR Interactions – Nuclear Quadrupole Couplings

Stefan Stoll[1] and Daniella Goldfarb[2]

[1] *University of Washington, Seattle, WA, USA*
[2] *Weizmann Institute of Science, Rehovot, Israel*

6.1 THE NUCLEAR QUADRUPOLE INTERACTION

The nuclear quadrupole interaction (NQI) is the result of the interaction of the electric quadrupole moment of an atomic nucleus with the electric-field gradient (EFG) at its location. In this section, we will introduce the physics behind this interaction and its description in the spin Hamiltonian.

EPR Spectroscopy: Fundamentals and Methods.
Edited by Daniella Goldfarb and Stefan Stoll

ISBN: 978-1-119-16299-5
Also published in eMagRes (online edition)
DOI: 10.1002/9780470034590.emrstm1504

Atomic nuclei with spin I larger than 1/2 have a nonspherical charge distribution that is determined by the internal nuclear structure. The deviation from a spherical distribution is quantified by the nuclear quadrupole moment, denoted as Q. A positive Q indicates an elongated (prolate) charge distribution (akin to an egg or a football), whereas a negative Q indicates a flattened (oblate) distribution (burger shaped). This is illustrated in Figure 6.1. All nuclei have cylindrical symmetry, even when they possess a nonzero quadrupole moment. The larger the magnitude of the nuclear quadrupole moment, the larger the distortion from spherical symmetry. In addition, the quadrupole moment is proportional to the charge Z of the nucleus—for the same nuclear shape, a larger Z gives a larger Q. A classical analog would be an ellipsoid with uniform charge density. Its quadrupole moment is $\frac{2}{5}Z(r_z^2 - r_{xy}^2)$, where Z is the total charge, r_z the radius along the unique axis, and r_{xy} the radius perpendicular to the unique axis.

The conventional unit for Q is

$$1 \text{ barn} = 1 \text{ b} = 10^{-28} \text{ m}^2 = 100 \text{ fm}^2 \tag{6.1}$$

The atomic unit for the nuclear quadrupole moment, occasionally used in quantum chemistry software, is $ea_0^2 = 4.486551484(28) \times 10^{-40}$ C m^2. Note that this includes the electric charge, contrary to the conventional unit. The nuclear quadrupole moment is an isotope-specific property, and values for most isotopes

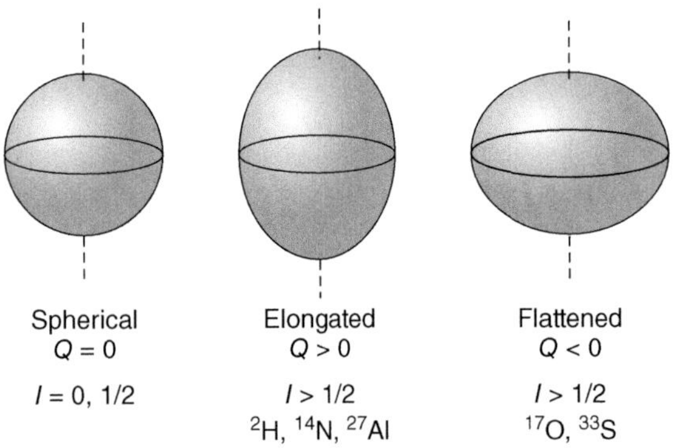

Figure 6.1. Illustration of the relation between the nuclear quadrupole moment Q, the effective shape of the nuclear charge distribution, and the nuclear spin quantum number I, with examples of some $I > 1/2$ isotopes often investigated in EPR. The vertical dashed line indicates the axis of cylindrical symmetry

Table 6.1. List of nuclear quadrupole moments of representative nuclei

Isotope	Spin I	Quadrupole moment Q (b)
^{2}H	1	+0.00286(2)
^{6}Li	1	−0.000806(6) [smallest]
^{14}N	1	+0.02044(3) [most accurate]
^{17}O	5/2	−0.0256(2)
^{27}Al	5/2	+0.1466(10)
^{33}S	3/2	−0.0678(13)
^{55}Mn	5/2	+0.330(10)
^{63}Cu	3/2	−0.220(15)
^{197}Au	3/2	+0.547(16)

are available in published tables.[1,2] Some representative values are listed in Table 6.1.

The nonspherical nature of the nuclear charge distribution leads to an orientation-dependent energy of the nucleus in the presence of an inhomogeneous external electric field. Figure 6.2 illustrates this orientation dependence. In the lowest-energy orientation, the long axis of the nucleus faces the more negative regions, and vice versa in the highest-energy orientation. For intermediate orientations, the nucleus experiences a torque toward the lowest energy orientation. The electric-field inhomogeneity at the position $\boldsymbol{R}$ of a nucleus is described by the 3×3 matrix of the second derivatives of the electrostatic potential at the nucleus, $V(\boldsymbol{R})$, due to all charges external to the nucleus

$$eq = \begin{pmatrix} V_{xx} & V_{xy} & V_{xz} \\ V_{yx} & V_{yy} & V_{yz} \\ V_{zx} & V_{zy} & V_{zz} \end{pmatrix} \text{ with } V_{ij} = \frac{\partial^2 V(\boldsymbol{R})}{\partial i \partial j} \qquad (6.2)$$

where $i, j = x, y, z$ and e is the elementary charge. Conventionally, this matrix of second derivatives of the electrostatic potential is referred to as the EFG tensor. We will adhere to this convention, even though this matrix is the negative of the actual electric-field gradient. (The electric-field vector $\mathcal{E}(\boldsymbol{R})$ is the negative derivative of $V(\boldsymbol{R})$: $\mathcal{E} = -\nabla V$.) Quantum chemistry programs can be used to calculate V_{ij} for nuclei in molecular systems. Typically, the programs return values in atomic units, $E_\mathrm{h}/ea_0^2 = 97.17362356(60)$ V Å^{-2}, where E_h is the hartree and a_0 the Bohr radius.

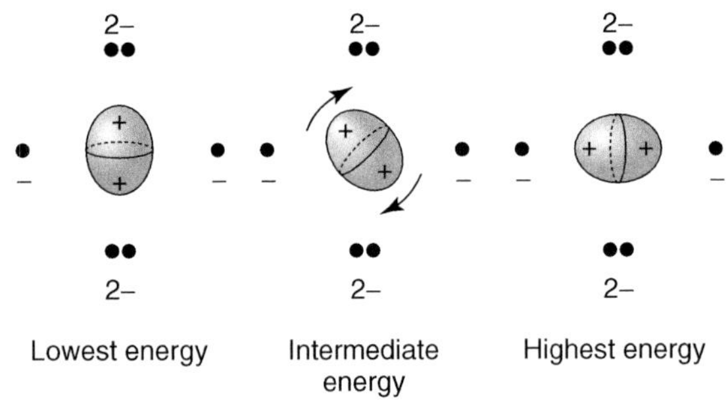

Figure 6.2. Illustration of the orientation dependence of the energy of a quadrupolar nucleus (here, $Q > 0$) in an external inhomogeneous electric field produced by a set of negatively charged particles (electrons). The arrows indicate the torque

The EFG tensor in equation (6.2) is symmetric ($V_{ij} = V_{ji}$) and traceless ($V_{xx} + V_{yy} + V_{zz} = 0$). Owing to these four relations between the nine elements of the matrix, there are only five independent parameters. There always exists a coordinate system in which the tensor is diagonal:

$$eq = \begin{pmatrix} V_{xx} & 0 & 0 \\ 0 & V_{yy} & 0 \\ 0 & 0 & V_{zz} \end{pmatrix} \qquad (6.3)$$

This axis system is called the *principal axis system* or *eigenframe*.

The charge density $\rho(\boldsymbol{r})$ at position $\boldsymbol{r} = (x, y, z)$ makes a contribution of $\frac{\rho(r)}{r^3}\left(\frac{3ij}{r^2} - \delta_{ij}\right)$ to the total EFG element V_{ij} at the origin. The inverse cubic dependence on r shows that the EFG is a local property and mostly determined by charges close to the nucleus.

The orientation-dependent electrostatic energy of the nucleus in the inhomogeneous electric field is described by the Hamiltonian[3]

$$\hat{H}_{\mathrm{NQI}} = \frac{1}{6}\sum_{i,j=x,y,z} V_{ij}\hat{Q}_{ij} \tag{6.4}$$

where $\hat{Q}_{ij}$ are the components of the nuclear-quadrupole tensor operator, which describes the nonspherical nature of the nuclear charge distribution due to the protons in the nucleus. It would be a formidable challenge to deal with this operator, as it depends on the individual positions of all the protons in the nucleus (e.g., 7 in ^{14}N). However, with the help of a profound result of quantum theory, the Wigner–Eckart theorem,[3] $\hat{Q}_{ij}$ in equation (6.4) can be replaced by an equivalent form containing only the operators of the total nuclear spin, $\hat{I}_x$, $\hat{I}_y$, and $\hat{I}_z$,

$$\hat{Q}_{ij} = \frac{eQ}{I(2I-1)}\left[\frac{3}{2}\left(\hat{I}_i\hat{I}_j + \hat{I}_j\hat{I}_i\right) - \delta_{ij}\hat{\boldsymbol{I}}^2\right] \tag{6.5}$$

where Q is the (scalar) quadrupole moment for a given nuclear isotope (see above) and $\hat{\boldsymbol{I}}$ is the vector of spin operators ($\hat{I}_x$, $\hat{I}_y$, $\hat{I}_z$) of the nucleus with spin quantum number I. This form is equivalent in the sense that the resulting effective Hamiltonian has the same matrix elements as the original Hamiltonian in equation (6.4) and therefore describes the same energy levels. This is clearly convenient, as it is now a spin-only expression that can easily be included in the spin Hamiltonian together with other terms such as the nuclear Zeeman interaction and the hyperfine interaction. One should not forget, however, that the underlying physical interaction does not involve the nuclear spin directly and is not magnetic but of purely electrostatic nature.

In the principal axes system of the EFG tensor, the NQI spin Hamiltonian simplifies to

$$\hat{H}_{\mathrm{NQI}} = \frac{e^2qQ}{4I(2I-1)}[(3\hat{I}_z^2 - \hat{\boldsymbol{I}}^2) + \eta(\hat{I}_x^2 - \hat{I}_y^2)] \tag{6.6}$$

where e is the elementary charge. e^2qQ factors into the two contributions eQ from the nuclear quadrupole moment and eq from the EFG. The conventional quantities eq (called the EFG) and η (the asymmetry) are related to the elements V_{ii} of the diagonal EFG matrix in equation (6.3) via

$$eq = V_{zz} \quad \eta = \frac{V_{xx} - V_{yy}}{V_{zz}} \tag{6.7}$$

e^2qQ/h (with the Planck constant h) is called the quadrupole coupling constant and is typically given in MHz. In the Hamiltonian in equation (6.6), it is divided by $4I(2I-1)$, so that a large value for e^2qQ/h implies a large quadrupole effect only if I is small. Alternatively, it is common to abbreviate the prefactor in equation (6.6), divided by h, as

$$K = \frac{e^2qQ/h}{4I(2I-1)} \tag{6.8}$$

which is more directly related to the size of the observed spectral shifts and splittings.

Equation (6.6) can also be written as a bilinear product of vector spin operators

$$\hat{H}_{\mathrm{NQI}} = h\,\hat{\boldsymbol{I}}\cdot\boldsymbol{P}\cdot\hat{\boldsymbol{I}} \tag{6.9}$$

with the NQI tensor $\boldsymbol{P}$, a 3×3 symmetric and traceless matrix, defined here in frequency units. In its eigenframe, $\boldsymbol{P}$ is a diagonal matrix

$$\boldsymbol{P} = \begin{pmatrix} P_x & 0 & 0 \\ 0 & P_y & 0 \\ 0 & 0 & P_z \end{pmatrix} = \begin{pmatrix} K(-1+\eta) & 0 & 0 \\ 0 & K(-1-\eta) & 0 \\ 0 & 0 & 2K \end{pmatrix} \tag{6.10}$$

and the Hamiltonian in this frame is $\hat{H}_{\mathrm{NQI}} = P_x\hat{I}_x^2 + P_y\hat{I}_y^2 + P_z\hat{I}_z^2$. The axis convention common in EPR and NMR is to order and label the principal axes such that the axis with the largest-magnitude principal value is labeled z, and the one with the smallest is labeled x

$$|P_x| \le |P_y| \le |P_z| \tag{6.11}$$

As a consequence of this labeling convention, the asymmetry parameter η can assume values only in the range $0 \le \eta \le 1$. A tensor with $\eta = 0$ is called *axial* and has principal values $P_x = P_y = -P_z/2$, and a tensor with $\eta > 0$ is called *rhombic*. In the fully rhombic limit $\eta = 1$, its principal values are $P_x = 0$ and $P_z = -P_y$.

It is noteworthy that the form of the NQI Hamiltonian in equation (6.9) is equivalent to the form $\hat{\boldsymbol{S}}\cdot\boldsymbol{D}\cdot\hat{\boldsymbol{S}}$ of the zero-field splitting (ZFS) Hamiltonian for electron spins > 1/2 (see Chapter 3). A consequence of this is that many of the expressions for energy level shifts and transition frequencies are analogous. However, it is absolutely crucial to bear in mind that the NQI is an electrostatic effect, whereas the ZFS is of purely magnetic origin (spin–spin and spin–orbit coupling).

6.2 ENERGY LEVELS AND SPECTRA

The presence of the NQIs constitutes a second torque that is acting on the nucleus in addition to the magnetic field. Consequently, the NQI affects the quantization directions of the nucleus, shifts its energy levels, and shifts energies of magnetic resonance transitions. In general, the size of these effects depends on the strength of the NQI relative to the total magnetic field experienced by the nucleus. The latter is the combination of the externally applied static field and the internal field due to the hyperfine coupling to unpaired electrons. To first order, this field is proportional to

$$\nu_{\text{eff}}(m_S) = \nu_I + m_S A \tag{6.12}$$

where ν_{eff} is an effective nuclear Zeeman frequency that depends on the electron spin state. In equation (6.12), ν_I is the nuclear Zeeman frequency $\nu_I = -g_n \mu_N B/h$ and A is the effective hyperfine coupling along the external field direction, in frequency units.

For the relative strength of the NQI, three regimes can be distinguished: (i) the high-field regime, where the magnetic field dominates ($|\nu_{\text{eff}}| \gg |K|$); (ii) the low-field regime, where the quadrupole interaction dominates ($|K| \gg |\nu_{\text{eff}}|$); and (iii) an intermediate regime, where the two are of comparable magnitude ($|\nu_{\text{eff}}| \approx |K|$). As ν_{eff} depends on m_S, the regime can be different for different electron manifolds m_S.

It is not possible to derive usable analytical expressions for the energy levels that are valid for all regimes and all nuclear spin quantum numbers I. Numerical diagonalization of the Hamiltonian is generally necessary. It is, however, possible to give concise analytical expressions for the high-field and the zero-field regime, and for spin $I = 1$ for all regimes, if there is only isotropic hyperfine coupling present. These cases are discussed in the following sections.

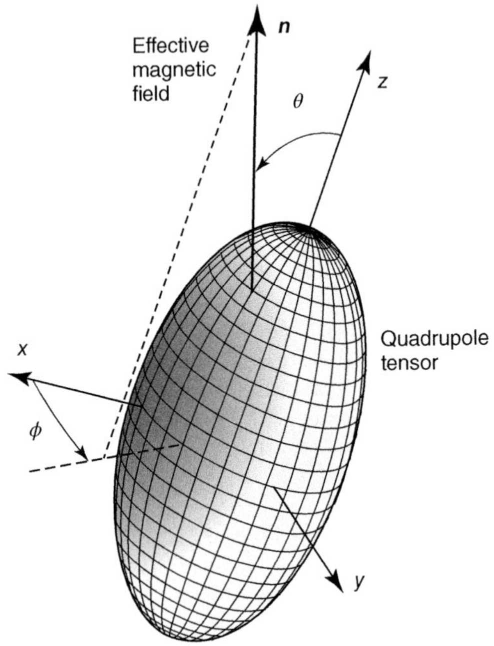

Figure 6.3. Relative orientation of the nuclear quadrupole tensor and the effective magnetic field experienced by a nucleus. The ellipsoid illustrates the quadrupole tensor for $Q > 0$, and x, y, and z indicate its principal axes. θ is the angle between the z axis and the effective magnetic field direction $\boldsymbol{n}$, ϕ is the angle between the x axis and the projection of the effective magnetic field onto the xy plane

6.2.1 High-field Regime

In the high-field regime, where the effective nuclear Zeeman frequency ν_{eff} dominates over the quadrupole coupling K, the NQI is a small perturbation and shifts each nuclear Zeeman level m_I to first order by

$$\Delta E_{\text{NQI}}(m_I) = \frac{1}{2}P[3m_I^2 - I(I+1)] \quad (|\nu_{\text{eff}}| \gg |K|) \tag{6.13}$$

where P is the effective quadrupole coupling along the direction of the effective magnetic field $\boldsymbol{n} = (\sin\theta\cos\phi, \sin\theta\sin\phi, \cos\theta)$, as illustrated in Figure 6.3. P is given by

$$P = \boldsymbol{n}^{\text{T}}\boldsymbol{P}\boldsymbol{n} = K[(3\cos^2\theta - 1) + \eta\sin^2\theta\cos 2\phi] \tag{6.14}$$

Note that P depends on m_S through $\boldsymbol{n}$ (which is affected by the hyperfine field). Unfortunately, the quadrupole tensor and the effective quadrupole coupling both have the same conventional symbol P, so they must be distinguished by their context.

The first-order energy level shifts of equation (6.13) are shown in Figure 6.4. One key property is that levels with the same $|m_I|$ have identical shifts. The EPR transitions remain unaffected to first order, but the frequencies for the nuclear transitions $(m_S, m_I) \leftrightarrow (m_S, m_I + 1)$ are shifted by $\frac{3}{2}P(2m_I + 1)$, giving first-order nuclear frequencies of

$$\nu(m_S, m_I \leftrightarrow m_I + 1) = \nu_I + m_S A + 3P(m_I + 1/2) \tag{6.15}$$

This equation shows that the $m_I = -1/2 \leftrightarrow +1/2$ transition is independent of P, unlike all other transitions. Although this high-field limit is relevant in NMR, it is much less so in EPR, where the high-field condition $|\nu_{\text{eff}}| \gg |K|$ is fulfilled only for nuclei with small quadrupole couplings at high fields, for instance for ^{2}H with $|K| < 0.1$ MHz and $|\nu_{\text{eff}}| \approx 22$ MHz at W-band (3.4 T; see example in Section 6.5.1), or

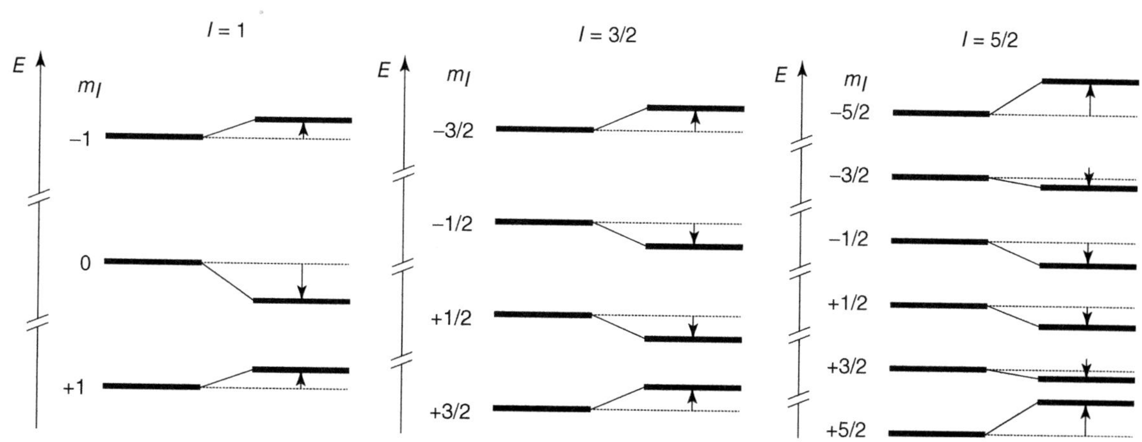

Figure 6.4. Schematic illustration of first-order energy level shifts due to the quadrupole interaction in the high-field regime. All diagrams assume $P > 0$, $\eta = 0$, and a nucleus with $g_n > 0$. The magnetic field is assumed parallel to the z axis of the NQI tensor

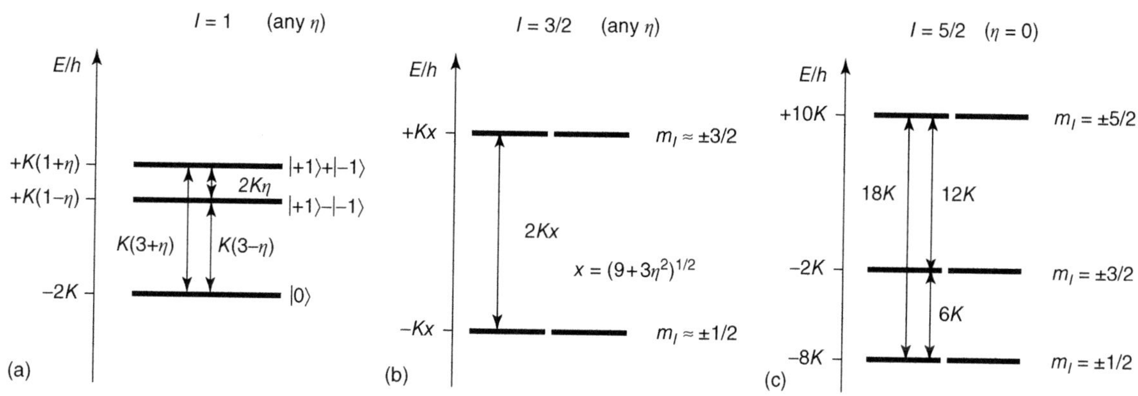

Figure 6.5. Zero-field energy levels and transitions of a spin-1 nucleus (a), a spin-3/2 nucleus (b), and a spin-5/2 nucleus (c). The plot assumes $K > 0$ and $0 \leq \eta \leq 1$. For $K < 0$, the order of the levels is inverted

for nuclei with large hyperfine couplings, such as in the ^{51}V example discussed in Section 6.5.5. For other nuclei at lower fields and with small hyperfine couplings, second-order shifts of order $|K|^2/|\nu_{\text{eff}}|$ are significant.[4] In these cases, the first-order expression can only give insight into the direction of a level or peak shift due to quadrupole coupling. In general, it is preferable to solve the general problem (for $I = 1$, see Section 6.2.3).

6.2.2 Zero-field Limit

In zero magnetic field, the nuclear transition frequencies are determined exclusively by the quadrupole coupling. Analytical expressions are available for $I = 1$ and $I = 3/2$. For an $I = 1$ nucleus, the energy levels are at $-2K$, $K(1 - \eta)$, and $K(1 + \eta)$. The three associated transition frequencies are

$$\nu_+ = K(3 + \eta) \qquad \nu_- = K(3 - \eta) \qquad \nu_0 = 2K\eta \quad (6.16)$$

where ν_+ is the largest in magnitude, for the transition from the lowest- to the highest-energy state. This is illustrated in Figure 6.5(a). The frequencies in equation (6.16) are referred to as the 'pure nuclear quadrupole frequencies (NQR)', even though the transitions are magnetic-dipole transitions.

For $I = 3/2$, the zero-field energy levels are split into two doublets with energies of $\pm K\sqrt{9 + 3\eta^2}$ (see

Figure 6.5b). Therefore, there is only one nuclear transition. This is a rare case where increasing the spin leads to a simpler spectrum. For $I > 3/2$, it is not possible to give a general expression valid for all η. In general, any half-integer spin (>1/2) will lead to a set of $I + 1/2$ pairs of doubly degenerate energy levels. Figure 6.5(c) illustrates the zero-field situation for an $I = 5/2$ spin for the limiting case $\eta = 0$.

A situation with zero effective magnetic field ($\nu_{eff}(m_S) = 0$, see equation (6.12)) can also arise in the presence of an external magnetic field, if the hyperfine field cancels the Zeeman field, $\nu_I \approx -m_S A$. This regime is called the *cancellation regime*. As the cancellation condition depends on m_S, it can be fulfilled only for one of the electron spin manifolds. For that manifold, sharp peaks at the NQR frequencies given in equation (6.16) are observed. In EPR, the zero-field (cancellation) regime is often encountered for nuclear spectra of ^{14}N-containing ligands coordinated to paramagnetic transition metals. ^{14}N nuclei with different hyperfine couplings require different spectrometer fields B_0 to reach the cancellation regime.

6.2.3 General Solution for $I = 1$

For $I = 1$ and isotropic hyperfine coupling, analytic equations can be derived that can be examined for all regimes using a graphical method.[5–7] This method provides systematic insight into the orientation dependence of the nuclear transition frequencies. Figure 6.6 illustrates its application to a case in the strong-field regime, with details of the method spelled out in the figure caption. The graph shows that the two single-quantum (sq) transitions ($\Delta m_I = \pm 1$) for a given m_S manifold are anisotropic, i.e., their position depends strongly on the orientation of the magnetic field relative to the NQI tensor, and they form broad peaks in the spectrum of a powder. In contrast, the double-quantum (dq) transition ($\Delta m_I = \pm 2$)

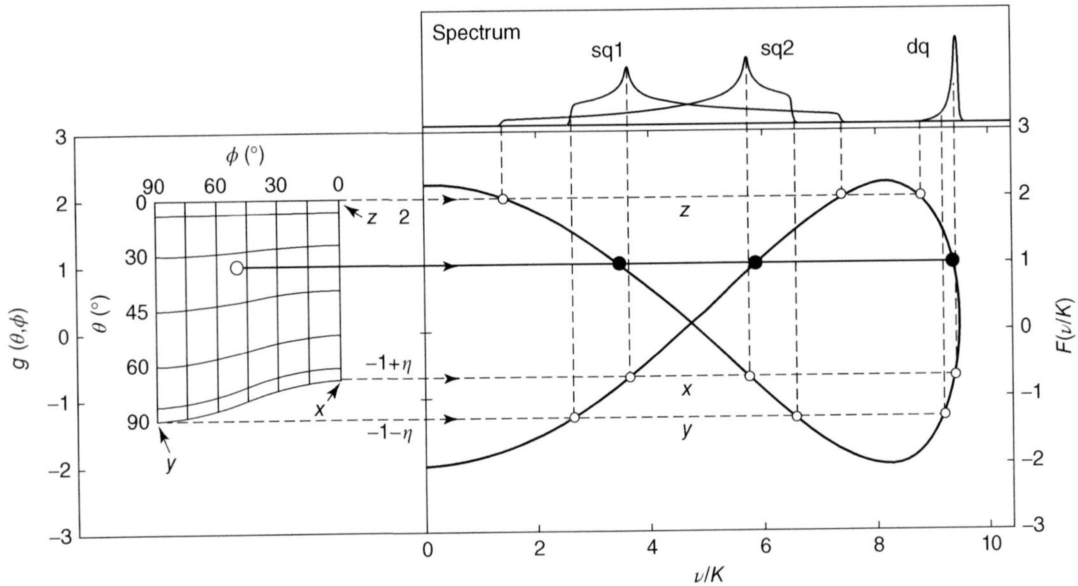

Figure 6.6. Graphical method to determine the three nuclear transition frequencies for an $I = 1$ nucleus for a given orientation (θ, ϕ) under the assumption of an isotropic hyperfine coupling. Left: function $g(\theta, \phi) = (3 - \eta \cos 2\phi)\cos^2\theta + \eta \cos 2\phi - 1$, right: function $F(\nu/K) = (2/w^2)(1 - \eta^2) \pm (3^{-3/2}/w^2)(x^2 - x_g^2)(4x_g^2 - x^2)^{1/2}$ with $x = \nu/K$, $w = \nu_{eff}/K$, $x_g^2 = w^2 + 3 + \eta^2$, and $\nu_{eff} = |\nu_I + m_S a_{iso}|$. All $g(\theta, \phi)$ values lie between $(-1 - \eta)$ and 2. $F(\nu/K)$ is defined for $0 \le \nu/K \le x_g^2$. Specific values used for the plot: $\nu_I = 1.08$ MHz (^{14}N at 350 mT), $a_{iso} = 4$ MHz, $m_S = +1/2$, $K = 0.7$ MHz, $\eta = 0.3$. Top right: powder nuclear spectrum showing the two single-quantum and the double-quantum transitions. To determine the nuclear frequencies for a given (θ, ϕ), draw a horizontal line from the corresponding grid point on the left across $F(\nu/K)$ on the right. The nuclear frequencies are at the ν/K positions where this line intersects the $F(\nu/K)$ curve. This is shown for the three principal directions x, y, and z, as well as for $(\theta, \phi) = (35°, 50°)$

is much less anisotropic and is sharp even in a powder spectrum.

To first order, the dq transition frequency does not depend on K or η (see Figure 6.4, where the $m_I = \pm 1$ levels are shifted by the NQI by identical amounts). However, to higher order, the dq transitions appear at

$$\nu_{\rm dq}(m_S) \approx 2\sqrt{\nu_{\rm eff}^2(m_S) + K^2(3+\eta^2)} \quad (|\nu_{\rm eff}| > |K|) \tag{6.17}$$

for $m_S = \pm 1/2$, with $\nu_{\rm eff}(m_S) = \nu_I + m_S a_{\rm iso}$, as given by equation (6.12) for the case $A = a_{\rm iso}$ (Figure 6.6). In this regime, it is possible to use equation (6.17) to estimate the isotropic hyperfine coupling constant from the two double-quantum transition frequencies of the two m_S manifolds: $a_{\rm iso} \approx [\nu_{\rm dq}^2(+1/2) - \nu_{\rm dq}^2(-1/2)]/8\nu_I$. If the assignment of the double-quantum peaks to the $m_S = \pm 1/2$ manifolds is not possible, the sign of A cannot be determined.

6.3 ANALYSIS OF EFG TENSORS

As with any other interaction that affects EPR spectra and is represented in the spin Hamiltonian, the NQI is determined by the underlying molecular and electronic structure of the paramagnetic center. Therefore, examination of the physical mechanism that leads to a particular set of measured quadrupole parameters, e^2qQ/h and η, provides structural insight. In this section, we discuss two approaches for the analysis of quadrupole parameters, one based on a simple model and the other based on a full quantum-chemical calculation.

6.3.1 Townes–Dailey Model

A simple and broadly applied approximate model to analyze quadrupole couplings was developed by Townes and Dailey.[8] In the Townes–Dailey (TD) model, the EFG at a given nucleus in a molecule is attributed exclusively to the valence p or d orbitals centered at that nucleus. Contributions from the core orbitals centered on that nucleus are neglected on the grounds that fully occupied core shells have spherically symmetric electron density and, therefore, no EFG. In addition, contributions from electrons in any orbitals centered on other nuclei in the molecule are neglected, as it is assumed that their contributions are cancelled by the positive nuclei. The TD model is a reasonable approximation that permits a semiquantitative rationalization of the sign, size, and orientation of the EFG tensor. Among a series of similar compounds, it allows the interpretation of trends. The model is predominantly applied to light atoms such as nitrogen, oxygen, and halogens.

In the TD model, each valence p orbital is associated with a nonspherical electron density distribution and, therefore, contributes a field gradient whose magnitude is proportional to its electron occupancy N (up to two electrons: $0 \le N \le 2$). For example, the contribution to the EFG tensor at the nucleus due to a valence p_z orbital is

$$\boldsymbol{eq} = \frac{1}{2} q_0 \begin{pmatrix} N_z & 0 & 0 \\ 0 & N_z & 0 \\ 0 & 0 & -2N_z \end{pmatrix} \tag{6.18}$$

Here, N_z is the electron population of the p_z orbital and $q_0 > 0$ is a constant determined by the shape and extent of the orbital. The tensor is axial with its unique axis parallel to the orbital axis, z. Note that the EFG along this direction, $V_{zz}(p_z)$, is negative.[4] Expressions for the contributions from the p_x and the p_y orbitals are obtained by cyclic permutation. The total field gradient at the nucleus is the sum of the contributions from the three p orbitals. For example, the zz component of the total EFG tensor is

$$V_{zz} = V_{zz}(p_x) + V_{zz}(p_y) + V_{zz}(p_z) = \frac{1}{2} q_0 [N_x + N_y - 2N_z] \tag{6.19}$$

where N_x, N_y, and N_z are the electron occupancies (populations) of the p_x, p_y, and p_z orbital, respectively. If the three orbitals are equally populated, then $N_x = N_y = N_z$, and the total field gradient at the nucleus vanishes. Only if there is an imbalance of occupancies will there be a nonzero total gradient. Therefore, within this TD model, it is possible to draw conclusions about the relative occupancies of the p orbitals at each nucleus in a molecule given the experimental values of e^2qQ/h and η for this nucleus. There are several mechanisms that can lead to an effective p-orbital population imbalance. First, different p orbitals can be involved in different types of bonds (σ or π), either directly or via their contribution to hybrid orbitals. Second, bonding p orbital populations depend on the degree of ionicity of the associated bond.

The TD model can be easily applied to sets of sp^n hybrid orbitals. One example is an imino nitrogen such as in pyridine or imidazole/histidine,[9–12] with a C–N–C bond angle of 2θ (Figure 6.7). There are three in-plane hybrid orbitals, one of them a lone-pair (LP) orbital and the other two involved in σ bonds to the neighboring C atoms. Finally, there is a π p orbital perpendicular

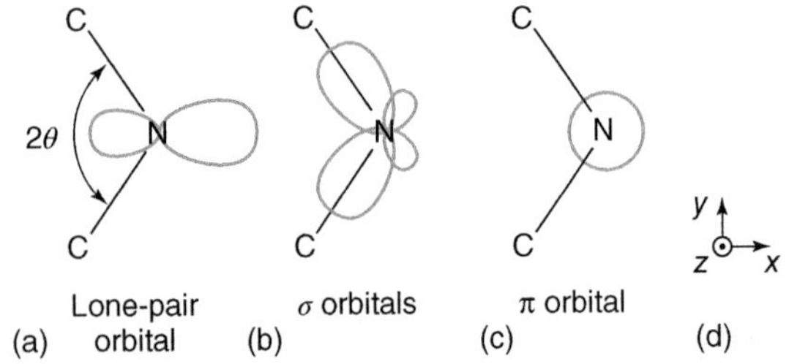

Figure 6.7. Valence orbitals used in the Townes–Dailey analysis for an uncoordinated imino nitrogen such as in imidazole or pyridine. (a) In-plane lone-pair orbital, (b) two in-plane σ-bonding orbitals, and (c) π orbital perpendicular to the paper plane. (d) Definition of local coordinate system

to the plane. The local molecular coordinate frame is defined such that x is along the LP orbital axis, and z is along the π orbital axis. The four valence orbitals can be represented as linear combinations of s and p orbitals

$$\begin{aligned} \psi_{\mathrm{LP}} &= -\alpha\psi_{\mathrm{s}} + \beta\psi_{\mathrm{p}_x} \\ \psi_{\sigma 1} &= (\beta\psi_{\mathrm{s}} + \alpha\psi_{\mathrm{p}_x} + \psi_{\mathrm{p}_y})/\sqrt{2} \\ \psi_{\sigma 2} &= (\beta\psi_{\mathrm{s}} + \alpha\psi_{\mathrm{p}_x} - \psi_{\mathrm{p}_y})/\sqrt{2} \\ \psi_{\pi} &= \psi_{\mathrm{p}_z} \end{aligned} \tag{6.20}$$

where $\alpha = \cot\theta$ and $\beta = \sqrt{1-\alpha^2}$. Each of these orbitals will contribute to the total EFG tensor an axial tensor that is aligned with its main orbital axis and proportional to its electron population N_i. Assuming equal populations in the two σ-bonding orbitals ($N_{\sigma 1} = N_{\sigma 2} = N_\sigma$), the total EFG tensor is aligned with the x, y, and z directions, and its principal values are

$$\begin{pmatrix} V_{xx} \\ V_{yy} \\ V_{zz} \end{pmatrix} = \frac{1}{2}q_0\left[N_\pi\begin{pmatrix} 1 \\ 1 \\ -2 \end{pmatrix} + N_{\mathrm{LP}}\begin{pmatrix} -2\beta^2 \\ \beta^2 \\ \beta^2 \end{pmatrix} + N_\sigma\begin{pmatrix} 1-2\alpha^2 \\ -2+\alpha^2 \\ 1+\alpha^2 \end{pmatrix}\right] \tag{6.21}$$

From an experimental EFG tensor and a known geometry (2θ), these equations can be used to infer information about the electron populations N_{LP}, N_σ, and N_π, which in turn give insight into the local bonding situation. For imidazole, $2\theta \approx 105°$, giving $\alpha^2 \approx 0.59$ and $\beta^2 \approx 0.41$. N_σ and N_π are similar and typically somewhat larger than 1, and $N_{\mathrm{LP}} \approx 2$ for an uncoordinated imino nitrogen. Consequently, the largest-magnitude value of the EFG tensor is negative, and its associated principal axis points along x, along the LP orbital. The coordination to a cation via the lone pair leads to a decrease in N_{LP} and, due to inductive effects, to increases in both N_σ and N_π. This leads to a decrease in e^2qQ/h and a concomitant increase in η, with an approximately linear relationship between $(e^2qQ/h)^{-1}$ and η, allowing the determination of the donor orbital population from the EFG tensor.[9]

The TD model can be applied to valence d shells as well. Each d orbital gives an axial contribution to the total EFG tensor with specific directionality and proportional to its electron occupancy. Any imbalance in the d orbital occupancies will lead to a nonzero total EFG tensor. For example, for a d^1 configuration with one electron in the d_{z^2} orbital, the EFG tensor will be axial ($\eta = 0$), with the unique axis along z and $V_{zz} < 0$. In a d^9 configuration with a missing electron in d_{z^2}, the tensor has the same symmetry, but an inverted sign. For a d^5 configuration, each d orbital is singly occupied, resulting in a spherical valence electron density and an expected vanishing EFG tensor. However, a small EFG is observed, indicating that the TD model is only approximate.

The TD model is also not applicable to deuterium, which only has an s orbital but no p or d orbitals in its valence shell. As experimental deuterium quadrupole couplings are nonzero, this indicates that contributions beyond the local valence shell are important in this case. These can be understood only from a full quantum-chemical analysis, which is discussed in the following section.

6.3.2 Full Quantum Chemical Analysis

Generally, the electrostatic potential, and therefore the EFG, at a nucleus is a sum of two contributions: (i) an electronic part that depends on the electron density in the molecule and (ii) a nuclear part that is due to the presence of positively charged nuclei. The electronic part can be calculated directly from the electronic wavefunction, and the nuclear part is a function of the positions and charges of the nuclei in the molecule. In order to gain physical insight into the various contributions, the EFG tensor can be decomposed in different ways, for example, using atomic orbitals (AOs) or some type of localized molecular orbitals (LMOs).[13]

In terms of AOs, the total electronic field gradient is the sum of integrals of the type $\langle\phi_{\mathrm{A},k}|\hat{V}_{ij}(\boldsymbol{R}_{\mathrm{C}})|\phi_{\mathrm{B},q}\rangle$, where $\hat{V}_{ij}(\boldsymbol{R}_{\mathrm{C}})$ is the operator for the ij component of the EFG at the position of the nucleus C and $\phi_{\mathrm{A},k}$ and $\phi_{\mathrm{B},q}$ are two occupied AOs centered on atoms A and B, respectively. Depending on the identity of atoms A and B compared to C, different types of contributions in this AO decomposition are distinguished:

(1a) local (one-center) core contributions, if both AOs are core orbitals on the atom of interest (A = B = C)
(1b) local valence contributions, if both AOs are valence orbitals on the atom of interest (again A = B = C)
(2a) two-center 'point-charge' contributions, if both AOs are centered on another atom (A = B ≠ C); this includes the nuclear contribution to the EFG
(2b) two-center 'bond' contributions, where one AO is centered on the atom of interest (A = C), and the other one is centered on another atom (B ≠ C), and
(3) three-center contributions, where A, B and C refer to three different atoms.

The total EFG is the sum of all these contributions. In this terminology, the TD model only takes the one-center valence contribution (1b) into account and neglects all others. (1b) is indeed the dominant contribution in many cases, except for deuterium.

Deuterium is special as it lacks valence p or d orbitals. Let us examine the EFG at the hydrogen positions in water, measured via the NQI of deuterons (we do not need an unpaired electron, as the NQI is purely electrostatic). The experimental quadrupole parameters for the deuterons are $e^2qQ/h = 0.31$ MHz and $\eta = 0.14$ (gas phase),[14] the z axis of the NQI tensor points along the O–H bond, and its principal value is positive. A quantum chemical calculation reveals that the point-charge contribution (2a) from the oxygen is positive and the most significant, but it is partially offset by a smaller negative value of the oxygen bond contribution (2b). Local core and valence as well as three-center contributions are negligible. Hydrogen bonding reduces the magnitude of the EFG at ^2H: in ice,[15] $e^2qQ/h = 0.214$ MHz and $\eta = 0.12$. Similarly, the ^2H NQI tensors of water coordinated to tyrosine,[16] tryptophan,[17] semiquinone,[18] and nitroxide radicals[19] are also smaller than the gas-phase value.

6.4 EXPERIMENTAL EXAMPLES – EPR SPECTRA

The NQI, being traceless, averages out to zero in liquid samples where paramagnetic centers tumble rapidly. Therefore, it is relevant for the description of the energy levels only in anisotropic samples, namely crystalline or other solid samples, frozen solutions, liquid crystals, and viscous liquids. However, in these environments, the EPR spectrum is inhomogeneously broadened and, therefore, the effect of the NQI on the EPR spectrum is expected to be small and hard to detect. It does not introduce additional splittings but causes shifts in the components of the hyperfine multiplets that can be tracked by simulations of the EPR spectrum. This is demonstrated in the example shown in Figure 6.8. The EPR spectrum of isotopically pure bis(diethyldithiocarbamato)copper(II) ($^{63}Cu(dtc)_2$, Figure 6.8a) in a frozen solution was measured at S-, X-, and Q-band and the simulations of all three spectra gave the ^{63}Cu quadrupole and hyperfine tensors.[20] Figure 6.8(b) shows the experimental and simulated X-band EPR spectra of this complex. To illustrate the effect of the ^{63}Cu ($I = 3/2$) NQI on the EPR spectrum, we simulated this X-band spectrum with and without including the quadrupole interaction in the spin Hamiltonian; the comparison is shown in Figure 6.8(c). The differences are subtle but clear and appear mainly in the g_x, g_y region. The simulation parameters are given in the figure caption. In these simulations, the slight misalignment of the $\boldsymbol{g}$, $\boldsymbol{A}$, and $\boldsymbol{P}$ tensors as well as the anisotropic linewidth used in the original simulations were neglected.[20]

6.5 EXPERIMENTAL EXAMPLES – ENDOR/ESEEM SPECTRA

In contrast to EPR spectra, nuclear spectra obtained by pulse-EPR-based hyperfine spectroscopy techniques, such as ENDOR or ESEEM (see Chapters 16 and 18), which give directly the nuclear frequencies of nuclei coupled to the unpaired electron(s), are highly sensitive to the NQI. Such spectra often reveal quadrupolar splittings, as in NMR, and can be used to determine the NQI parameters, which in turn provide geometrical and electronic structure information. Several such examples are presented next. They are ordered according to the spin of the nucleus involved, starting with $I = 1$ (^2H and ^{14}N), proceeding to $I = 5/2$ (^{17}O and ^{27}Al), and ending with $I = 7/2$ (^{51}V).

6.5.1 $I = 1$: ^2H

The NQI of the ^2H nucleus is a very attractive probe for extracting structural information because its

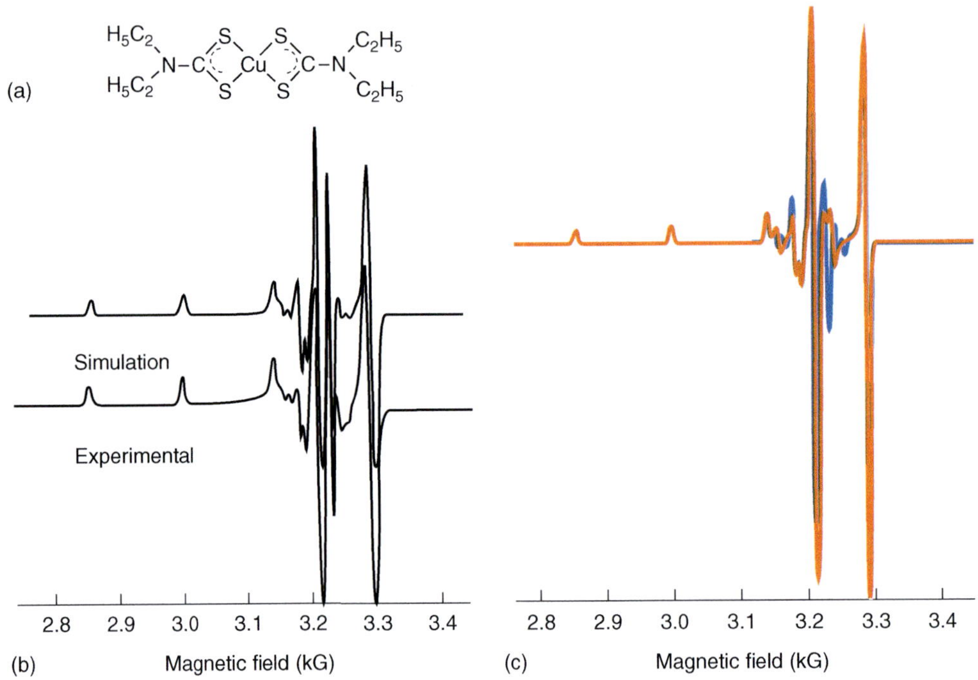

Figure 6.8. (a) The structure of $Cu(dtc)_2$, (b) experimental and simulated X-band spectrum of $Cu(dtc)_2$ in pyridine.[20] (Reprinted with permission from Liczwek, D. L.; Belford, R. L.; Pilbrow, J. R.; Hyde, J. S., Evaluation of Copper Nuclear-Quadrupole Coupling in Thio Complexes by Completion of the Coordination Sphere. J. Phys. Chem. 1983, 87, 2509–2512. Copyright 1983 American Chemical Society) (c) Simulated 9.1077 GHz spectra using EasySpin[21] with $(g_x, g_y, g_z) = (2.030\ 2.033\ 2.120)$, $(A_x, A_y, A_z) = (-43, -76, 423)$ MHz, $(P_x, P_y, P_z) = (-7.17, -7.17, +14.34)$ MHz, and all tensors were taken as collinear. The blue spectrum was calculated with $(P_x, P_y, P_z) = (0,0,0)$

quadrupole coupling constant, e^2Qq/h, depends on the bonding environment. Typical values for e^2Qq/h in rigid C–^{2}H bonds depend on the hybridization of the carbon and range from about 150 to 200 kHz, and they are somewhat larger for N–^{2}H and O–^{2}H bonds (see Section 6.3.2). The largest known e^2Qq/h for ^{2}H is 1.41 MHz in $^2H_2$.[22] Furthermore, the asymmetry parameter η is usually smaller than 0.1 and the principal axis of the interaction typically points along the ^{2}H-X direction. This makes it an excellent probe for determining bond orientations. Alternatively, if the bond orientation is known, one can use it to determine the orientation of the **g**-tensor as will be illustrated in the following example.

Figure 6.9(a) shows a binuclear, mixed-valence copper complex with a $[Cu^{+1.5}, Cu^{+1.5}]$ redox state and $S = 1/2$ stabilized with a rigid azacryptand ligand, RGT. In this system, the unpaired electron is delocalized equally over the two copper ions.[26] The echo-detected EPR spectrum of this complex in a frozen solution is shown in Figure 6.9(b). From this spectrum, one can immediately see that the **g**-tensor is axially symmetric and determine the $g_\perp$ and $g_\parallel$ values. However, the orientation of the g-principal axis system within the molecular structure remains unknown. This information was obtained through measurements of the ^{2}H ENDOR spectra of the exchangeable protons marked in red in Figure 6.9(a).

For this, the Cu_2RGT complex was prepared in D_2O, such that all the exchangeable protons were exchanged with deuterons and the ^{2}H Mims ENDOR spectrum was measured at a field position corresponding to $g_\parallel$, marked with an arrow in Figure 6.9(b). At this field position, only complexes for which the $g_\parallel$ direction is along the magnetic-field direction contribute to the ENDOR spectrum, yielding a single-crystal-like spectrum (see Chapter 11, orientation selectivity discussion). The spectrum, shown in Figure 6.9(c), is well resolved, displaying a major doublet, with a splitting of 0.67 MHz, which corresponds to the hyperfine

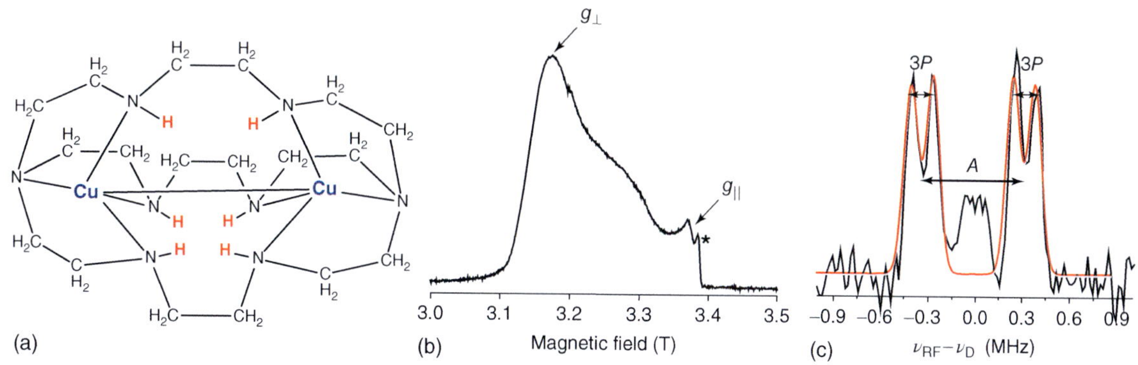

Figure 6.9. (a) Schematic structure of Cu_2RGT. The exchangeable protons are indicated in red. (b) The W-band echo-detected EPR spectrum of Cu_2RGT measured at 6.5 K, (c) ^{2}H-Mims W-band ENDOR spectrum of Cu_2RGT/D_2O recorded at $g_{||}$ ($B_0 = 3.386$ T), labeled in (b), and the corresponding simulations (red lines) obtained with the parameters (A_x, A_y, A_z) = (−0.8, −0.66, −0.2) MHz, $e^2qQ/h = 220$ kHz, $\eta = 0.14$ and the appropriate orientations of the hyperfine and quadrupole tensors. For convenience, the intensity was multiplied by −1. * marks an impurity signal. (Reprinted with permission from Kababya, S.; Nelson, J.; Calle, C.; Neese, F.; Goldfarb, D., Electronic structure of binuclear mixed valence copper azacryptates derived from integrated advanced EPR and DFT calculations. J. Am. Chem. Soc. 2006, 128 (6), 2017-29. Copyright 2006 American Chemical Society.)

coupling, A, of these deuterons. Each of the doublet components shows an additional splitting, due to the quadrupolar interaction, of 140 kHz. The frequencies of the four peaks are described by equation (6.15), which for the two transitions $m_I = 0 \leftrightarrow 1$ and $-1 \leftrightarrow 0$ give

$$\nu(m_S, \pm) = \nu_I + m_S A \pm \frac{3}{2} P \tag{6.22}$$

with the effective quadrupole coupling P given in equation (6.14). The splitting between the two inner doublets therefore is $3P$. A, similar to P, is the effective hyperfine coupling, which is orientation dependent as well.

Using equation (6.14) and the known e^2Qq/h of N–D bonds, the angle θ between the N–D bond and the $g_{||}$ direction can be determined from the experimental P value. Neglecting η as it is relatively small and taking e^2Qq/h of the N–D deuterons as 210 kHz[24,25] gives $\theta = 79°$. The angle between the N–D bond and the Cu–Cu direction in the crystal structure is 89°. Considering the uncertainty in e^2Qq/h and the assumption that $\eta = 0$, this provides experimental evidence for $g_{||}$ being along the Cu–Cu direction as expected for the trigonal bi-pyramidal coordination geometry of the copper ions.[26]

6.5.2 $I = 1$: ^{14}N

The following example considers the directly coordinated ^{14}N nuclei in a DOTA chelate with a coordinated Gd(III) ion ($S = 7/2$) (Gd-DOTA) (Figure 6.10a).[28] Here, the method of choice was also W-band Mims ENDOR. Gd(III) is an f ion and exhibits very little delocalization of the spin onto the coordinated nitrogen ligands. Consequently, the hyperfine coupling is small and the NQI dominates the spectrum with well-resolved splittings as shown in Figure 6.10(c). The ENDOR spectrum was recorded by setting the field to the maximum of the EPR spectrum (Figure 6.10b), where the Gd(III) central EPR transition, $m_S = -1/2 \leftrightarrow +1/2$, dominates (see Chapter 3). Therefore, the spectrum can be analyzed assuming an effective $S = 1/2$. Because Gd(III) has an isotropic g, at high field and for a small enough ZFS, the central transition is isotropic and consists of contributions of all orientations of the complex with respect to the magnetic field. Therefore, the ^{14}N ENDOR spectrum constitutes a complete powder pattern, namely, there is no orientation selection. The largest splitting, due to quadrupolar coupling, is indicated by $3P$, and the smaller splitting, due to the hyperfine interaction, is indicated by A. Whereas in the previous example

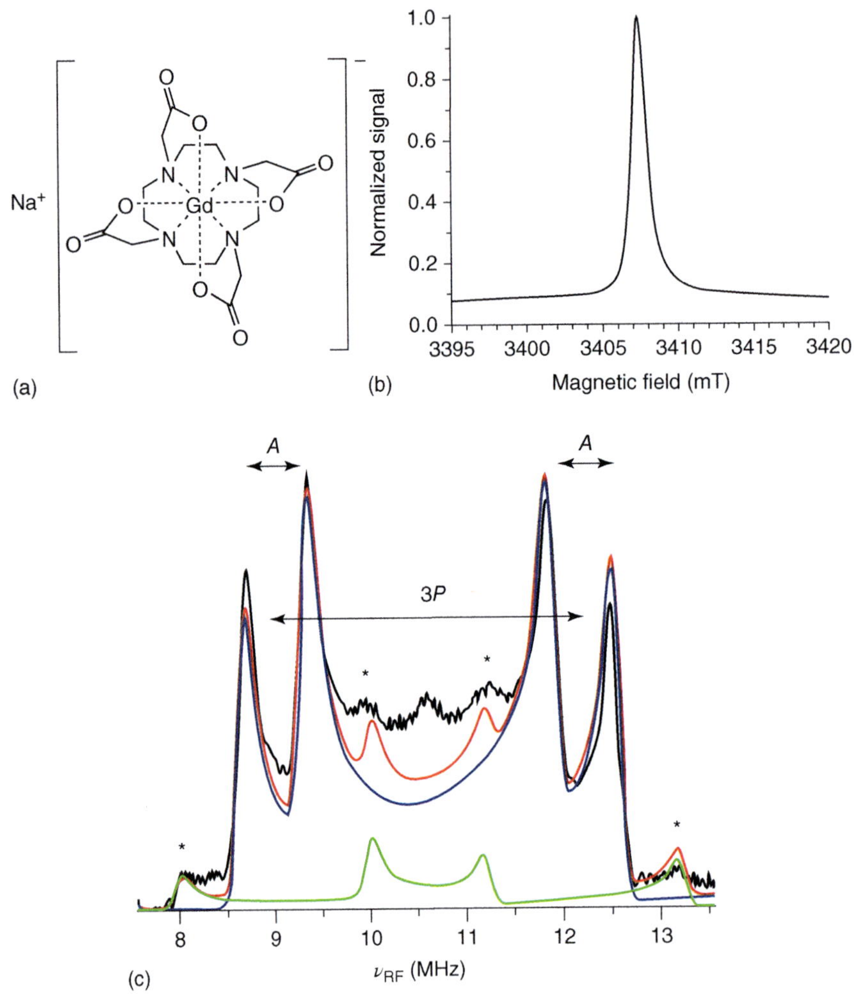

Figure 6.10. (a) The Gd-DOTA complex. (b) W-band echo-detected EPR spectrum of Gd-DOTA in the region of the central transition recorded at 10 K. (c) W-band ^{14}N Mims ENDOR spectrum of Gd-DOTA, recorded at 10 K and a magnetic field of 3407 mT, with assignments of the hyperfine, *A*, and the quadrupolar, 3*P*, splittings. The spectral features marked with * correspond to contributions attributed to the $m_S = \pm 3/2$ electron spin manifolds, while all other features are due to the $m_S = \pm 1/2$ electron spin manifolds (central transition). The blue curve represents the simulation of the ENDOR frequencies due to the central transition with the parameters presented in the text, the green curve represents the simulation of the contribution from the $m_S = \pm 3/2$ electron spin manifolds (see text), and the red trace is their sum. For the simulations, the anisotropic hyperfine coupling, $T_\perp = 0.30$ MHz, was derived from an average Gd-N distance of 2.66 Å obtained from the crystal structure of Gd-DOTA,[27] the isotropic part was $a_{iso} = -0.37$ MHz, $e^2Qq/h = 4.31$ MHz, and $\eta = 0$. At this field $\nu_I = 10.6$ MHz. (Reproduced with permission from Ref. 28 © Elsevier, 2016)

(Figure 6.9) the hyperfine coupling dominated and only a few orientations contributed to the spectrum, here the quadrupole coupling dominates and all orientations contribute to the spectrum. The best-fit simulation is shown in blue in Figure 6.10(c). The contribution of the $m_S = \pm 3/2$ electron spin manifolds was taken into account approximately by calculating the spectrum with hyperfine values multiplied by a factor of 3 as compared to the value used for effective $S = 1/2$ $m_S = \pm 1/2$ manifolds.[27]

The following example deals with a system of an $S = 1/2$ electron spin coupled to ^{14}N ($I = 1$), where the ^{14}N hyperfine coupling is within the cancellation regime at typical X-band frequencies. The specific example was taken from a study of a membrane-bound [NiFe]-hydrogenase of *Ralstonia eutropha* (*Re* MBH) compared with [NiFe]-hydrogenase from *Desulfovibrio vulgaris* (*Dv* H_2ase).[29] The paramagnetic 'ready inactive' Ni_r–B state of *Re* MBH is characterized by a total $S = 1/2$ for the nickel-iron center with principal g-values of 2.33, 2.16, and 2.01. Figure 6.11 shows the three-pulse ESEEM spectrum of the Ni_r–B state of the *Dv* H_2ase (top) measured at the $g = 2.16$ position. The spectrum is dominated by three sharp lines at low frequencies. These lines show only a minor field dependence in their positions and can therefore be attributed to the so-called zero-field transitions from equation (6.16). In addition, at around 4 MHz, two weak signals are visible, which arise from two double-quantum transitions, suggesting the presence of two slightly different species. This pattern is typical for a nitrogen in the cancellation regime. The hyperfine and quadrupole coupling values derived from these frequencies using equations (6.16) and (6.17) are consistent with an imidazole nitrogen of a histidine close to the Ni atom, which is a common feature in many [NiFe]-hydrogenases. This example demonstrates that ESEEM spectroscopy can reveal the presence of a histidine in the coordination sphere of paramagnetic transition metal centers.

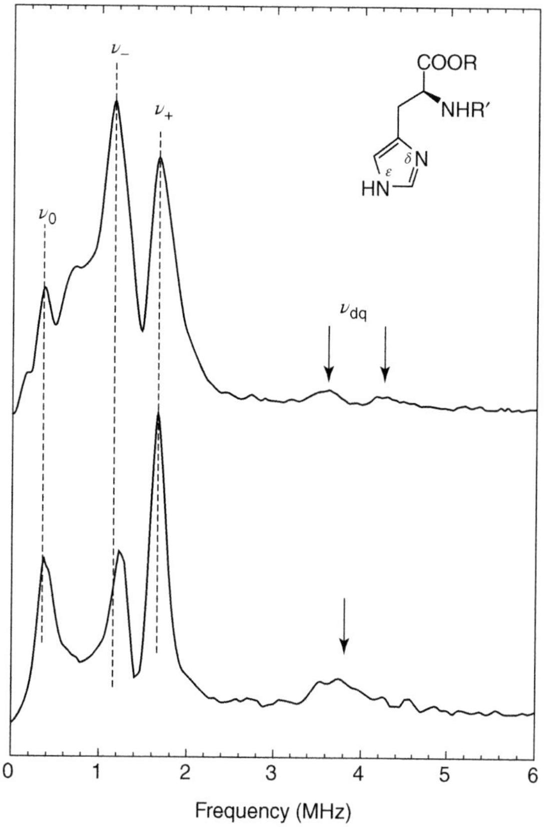

Figure 6.11. Three-pulse ESEEM spectra from *Dv* [NiFe]-hydrogenase (top) and *Re* MBH (bottom) in the Ni_r–B state, recorded at $g = 2.16$. Experimental conditions: $T = 8$ K, microwave frequency 9.7 GHz. From the frequencies of the peaks, the values $e^2Qq/h = 1.94$ MHz, $\eta = 0.38$, and $A = 1.4$–1.6 MHz were derived for *Re* MBH and $e^2Qq/h = 1.90$ MHz, $\eta = 0.37$, and $A = 1.4$–1.6 MHz for *Dv* H_2ase. The ^{14}N Larmor frequency at this field is 0.99 MHz. Inset: Structure of histidine. (Reproduced with permission from Ref. 29 © Royal Society of Chemistry, 2010)

ESEEM has been extensively used to study Cu(II) sites in proteins, which often involve histidine ligands.[30] While the directly bound nitrogen has a rather large hyperfine coupling of around 20 MHz, the hyperfine coupling of the remote nitrogen, N_δ, is within the cancellation condition at X-band frequencies. Therefore, similar to the previous example, its NQI parameters can be readily extracted. A study of Cu(II)-coordinated substituted imidazole model compounds has shown that the NQI parameters are related to the polarization of the N–H bond.[11] This study provides a very detailed analysis on how the substitution at the carbon adjacent to the remote nitrogen, alkylation at the remote nitrogen, and hydrogen bonding alter the NQI parameters.

As the ^{14}N quadrupole coupling parameters reflect the bonding characteristics of the ^{14}N atom (see Section 6.3.1), they can be used to differentiate nitrogen atoms in various types of chemical environments in paramagnetic centers. Figure 6.12 presents a compilation of experimentally determined ^{14}N quadrupolar parameters for semiquinones ($S = 1/2$) bound to different proteins that interact with nitrogen atoms from the protein backbone or its side chains. The plot of e^2Qq/h versus η reveals clear groupings of various types of nitrogens. Comparison of experimentally determined NQI parameters with such correlation plots can help identify the type of residue bearing ^{14}N in the close vicinity of the paramagnetic center. Interestingly, within some of these groups, such as B (backbone nitrogen, peptide, di- and tripeptide, and

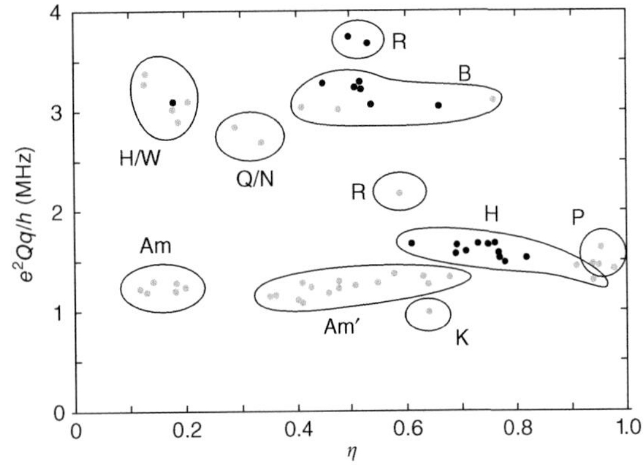

Figure 6.12. Correlation plot of ^{14}N quadrupolar parameters of ^{14}N interacting with protein-bound semiquinones as measured by ESEEM (black dots) and ^{14}N nuclear quadrupole resonance spectroscopy of amino acids and peptides (gray dots). (H/W) Indole, tryptophan, and histidine N_ε nitrogen; (Q/N) glutamine and asparagine NH_2 nitrogen; (B) backbone nitrogen, peptide, di- and tripeptide, triglycine; (Am) and (Am′) NH_3^+ amino group nitrogen; (K) NH_3^+ lysine nitrogen; (P) proline nitrogen; (H) histidine N_δ nitrogen; and (R) arginine N_ε. (Reproduced with permission from Ref. 31 © Royal Society of Chemistry, 2011)

triglycine), H (histidine N_δ nitrogen), and Am′ (NH_3 + amino group nitrogen), one can see a large variation in η while e^2Qq/h remains rather unchanged. This in turn can give detailed structural information within the group as discussed earlier for the Cu-imidazole coordination case.

6.5.3 $I = 5/2$: ^{17}O

Nuclei with $I = 5/2$ that are of interest are ^{17}O, ^{27}Al, and ^{55}Mn. A nucleus with $I = 5/2$ with a nonzero quadrupolar interaction, coupled to one unpaired electron with $S = 1/2$, features a total of 12 levels at high field, six for each electron spin manifold as shown in Figure 6.4. The quadrupolar splittings of $I = 5/2$ nuclei are usually not resolved in the EPR spectrum, but they can be resolved in ESEEM and ENDOR spectra, which report nuclear frequencies just as described earlier for the $I = 1$ case.

One example of ^{17}O is from a study of aconitase, an enzyme that interconverts citrate and isocitrate.[32] The enzyme contains a diamagnetic iron–sulfur cluster that can be readily reduced by one electron to $[4Fe-4S]^+$, which is paramagnetic with $S = 1/2$. To characterize the nature of substrate binding to the iron–sulfur cluster, ^{17}O ENDOR measurements were carried out. The ENDOR spectrum of the reduced enzyme prepared in $H_2{}^{17}O$ in the presence of citrate, shown in Figure 6.13, displays the pattern expected for an ^{17}O ENDOR signal: a doublet, each component of which splits into $2I = 5$ lines. The higher-frequency component is more intense, probably due to hyperfine enhancement (see Chapter 16) and/or due to RF conversion efficiency of the RF coils. Moreover, the low-field manifold shows only three of the expected five lines. Here, unlike the ^2H case recorded at W-band, the spectrum was recorded at X-band, where the coupling falls within the strong-coupling regime ($|A/2| > |\nu_I|$) such that the ENDOR frequencies are centered about $A/2$ (see equation (6.15))

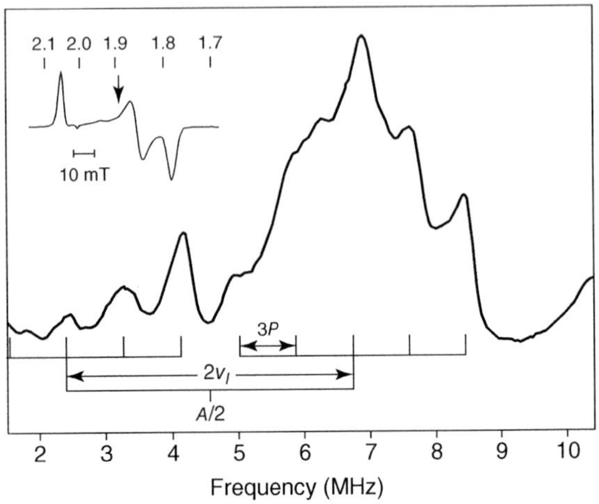

Figure 6.13. Inset, X-band (9.2325 GHz) EPR spectrum of $[4Fe-4S]^+$ aconitase in the presence of citrate. The field position and g value at which the ENDOR spectrum was recorded is indicated with an arrow. Main figure, ^{17}O ENDOR spectrum at $g = 1.88$ for this sample in $H_2{}^{17}O$ solution, recorded at 365 mT and 2 K. (Data provided by the authors of Ref. 32)

$$\nu_\pm(m_I \leftrightarrow m_I + 1) = \frac{A}{2} \pm [\nu_I + 3P(m_I + 1/2)] \quad (6.23)$$

where ± indicates $m_S = \pm 1/2$. The separation between the two main doublet components is twice the ^{17}O Larmor frequency at X-band, $2\nu_I = 4.215$ MHz. From the separation of adjacent lines within each multiplet ($=3P$), a quadrupolar splitting of $P = 0.28$ MHz was obtained. The center position of the two groups of lines is at $A/2$, yielding $A = 8.65$ MHz. As the ENDOR spectrum shown in Figure 6.13 is well resolved, with well-defined quadrupole splitting, it was concluded that the ^{17}O signal is associated with a single oxygen-containing species that coordinates to the $[4Fe–4S]^+$. However, this measurement does not identify the molecule containing ^{17}O. This assignment would require further selective ^{17}O labeling.

We have shown earlier that ESEEM measurements, which provide nuclear frequencies, can resolve the NQI for $I = 1$, particularly in the cancellation regime. It turns out that they can also resolve the quadrupolar splittings of ^{17}O, through the sum combination lines that appear in two-pulse and four-pulse ESEEM. Peaks at the fundamental nuclear frequencies, given by equation (6.15), outside the cancellation conditions are often too broad to resolve the quadrupolar splitting. The situation is different for the sum frequency peak that for $I = 1/2$ appears at $\nu_\alpha + \nu_\beta$, where ν_α and ν_β are the nuclear frequencies in the $m_S = +1/2$ and $m_S = -1/2$ electron spin manifolds, respectively. For a small anisotropic hyperfine coupling, $\nu_\alpha + \nu_\beta \approx 2\nu_I$. For an $I > 1/2$ nucleus with a nonzero quadrupole interaction, we expect $2I$ nuclear frequencies within each electron spin manifold. The sum combination line will have contributions from all combinations of frequencies in the two electron spin manifolds (see equation (6.23)). Out of these combination lines, the

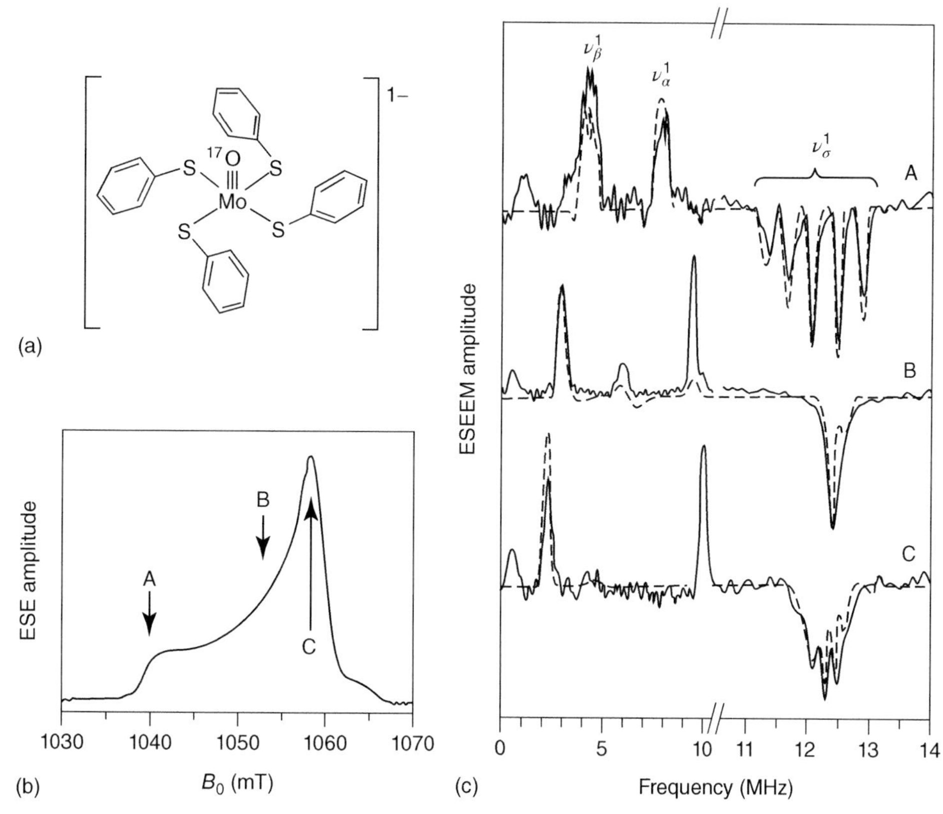

Figure 6.14. (a) Structure of $[Mo^{17}O(SPh)_4]^{1-}$. (b) Echo-detected EPR spectrum (MW frequency 29.372 GHz). (c) Cosine Fourier transforms (low-frequency part) of integrated four-pulse ESEEM obtained at magnetic fields A, B, and C, respectively (see (b)). The sum combination line is a quintet and indicated by ν_σ^1. Dashed traces are the corresponding simulated spectra of the integrated four-pulse ESEEM. ν_α^1 and ν_β^1 correspond to the nuclear frequencies in the α and β electron spin manifolds for which $\Delta m_I = 1$, and ν_σ^1 is their sum (see equation (6.24)).[33] (Reprinted with permission from Astashkin, A. V.; Neese, F.; Raitsimring, A. M.; Cooney, J. J.; Bultman, E.; Enemark, J. H., Pulsed EPR investigations of systems modeling molybdenum enzymes: hyperfine and quadrupole parameters of oxo-^{17}O in $[Mo^{17}O(SPh)_4]$. J. Am. Chem. Soc. 2005, 127 (47), 16713–22. Copyright 2005 American Chemical Society)

most interesting is the one that corresponds to the same $m_I \leftrightarrow m_I + 1$ transition within each electron spin manifold, as it can resolve the quadrupolar splitting according to

$$\nu_{\text{sum}}(m_I) = 2\nu_I + 6P(m_I + 1/2) \tag{6.24}$$

Hence, the sum combination peak splits into a quintet for $I = 5/2$, with a splitting between adjacent lines of

$$\Delta\nu_{\text{sum}} = \nu_{\text{sum}}(m_I + 1) - \nu_{\text{sum}}(m_I) = 6P \tag{6.25}$$

This approach has been applied in the study of the complex $[Mo^{17}O(SPh)_4]^-$ shown in Figure 6.14(a). In this work, to increase the resolution of the sum combination line,[34] the four-pulse ESEEM technique[35] with integration over the time interval between the first two pulses was used.[36] The integrated four-pulse ESEEM spectra of $[Mo^{17}O(SPh)_4]^{1-}$ obtained at the magnetic field position indicated in Figure 6.14(b) by A, along the $g_{\|}$ direction, is shown in Figure 6.14(c). The spectrum A shows a clear quintet for the combination line with a splitting of about 0.4 MHz (=6*P*). A quintet splitting is also visible in spectrum C, which was recorded along $g_{\perp}$, but in this case, the quintet splitting is about half as large, 0.2 MHz. In spectrum B, no splitting is observed. This behavior is consistent with an axial or nearly axial quadrupole tensor with a principal axis frame coinciding with that of the ***g***-tensor. From the combined analysis of the splittings at all fields (see equation (6.14)), $e^2Qq/h = 1.48$ MHz was obtained.

6.5.4 $I = 5/2$: ^{27}Al

As the last example, we present the manifestation of the quadrupole interaction of ^{27}Al ($I = 5/2$) in the ENDOR spectrum of Cu^{2+} exchanged into the zeolite faujasite, a microporous alumino-silicate.[37] The purpose of the ENDOR measurements was to localize the Cu^{2+} site with respect to the framework of the zeolite. The measurements were carried out on a dehydrated sample, and W-band ENDOR spectra were recorded by setting the field at various positions along the EPR spectrum. The spectrum recorded by setting the magnetic field to the $g_{\|}$ position, shown in Figure 6.15, generated the best resolved spectrum, just as in the examples shown in Figures 6.9 and 6.14, because Cu^{2+} sites with only a few orientations contribute to the ENDOR spectrum. The ^{27}Al signal is centered about its Larmor frequency (3.8 MHz) and consists of a major doublet with a splitting of 4.2 MHz due to the hyperfine interaction, with each component further split into a quintet with a separation of 1.6 MHz due to the quadrupole coupling. Simulations of the whole set of orientation-selective spectra gave $e^2Qq/h = 14.8$ MHz and $\eta = 0.7$. The Cu^{2+} location was found to be in the center of the six-membered ring (Figure 6.15b), coordinated to three framework oxygens in a symmetric manner.[37]

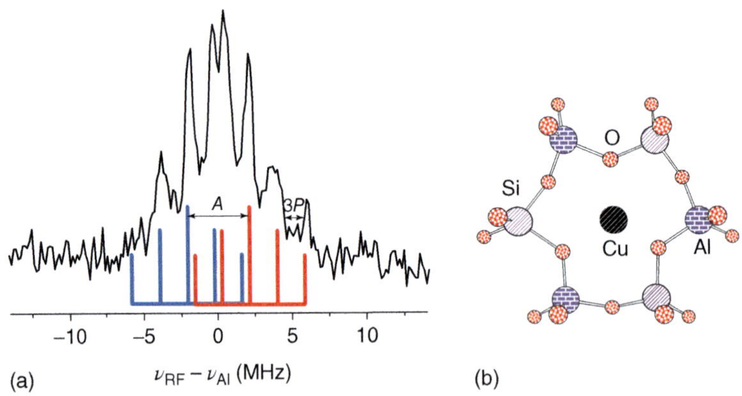

Figure 6.15. W-band Davies ^{27}Al ENDOR spectrum of dehydrated Cu^{2+} exchanged faujasite with Si/Al = 1 measured at the $g_{\|}$ field position ($g = 2.351$, 2.8799 T) and 5 K. (a) The blue and red sticks represent the two quintets of the two electron spin manifolds. (b) A schematic of the local environment of the Cu^{2+} ion. (Reprinted with permission from Carl, P. J.; Vaughan, D. E. W.; Goldfarb, D., High-field ^{27}Al ENDOR reveals the coordination mode of Cu^{2+} in low Si/Al zeolites. J. Am. Chem. Soc. 2006, 128 (22), 7160–7161. Copyright 2006 American Chemical Society)

6.5.5 $I = 7/2$: ^{51}V

The example for ^{17}O shown in Figure 6.14 made use of the sum combination frequencies intrinsic to the two- and four-pulse ESEEM to determine the quadrupole parameters. In the strong-coupling regime, if the fundamental single-quantum nuclear frequencies are resolved, the quadrupolar interaction can be determined from their difference rather than their sum. This approach was applied to determine the ^{51}V ($I = 7/2$) quadrupolar interaction in a number of vanadyl complexes.[38] Figure 6.16(a) shows the structure of the planar vanadyl acetyl acetonate complex, $VO(acac)_2$, and presents the X-band echo-detected EPR spectrum of $VO(acac)_2$ in a frozen solution. In the presence of axial ligands, the EPR spectra are virtually identical. The ^{51}V hyperfine splitting is large and well resolved, particularly for the $A_{\parallel}$ component along the V–O bond. At the low-field and high-field ends of the spectrum, corresponding to the EPR transitions with $m_I = -7/2$ and $+7/2$, respectively, only orientations with the magnetic field parallel to the V–O bond are excited, and consequently, highly resolved single-crystal-like ENDOR spectra with only two peaks are obtained.

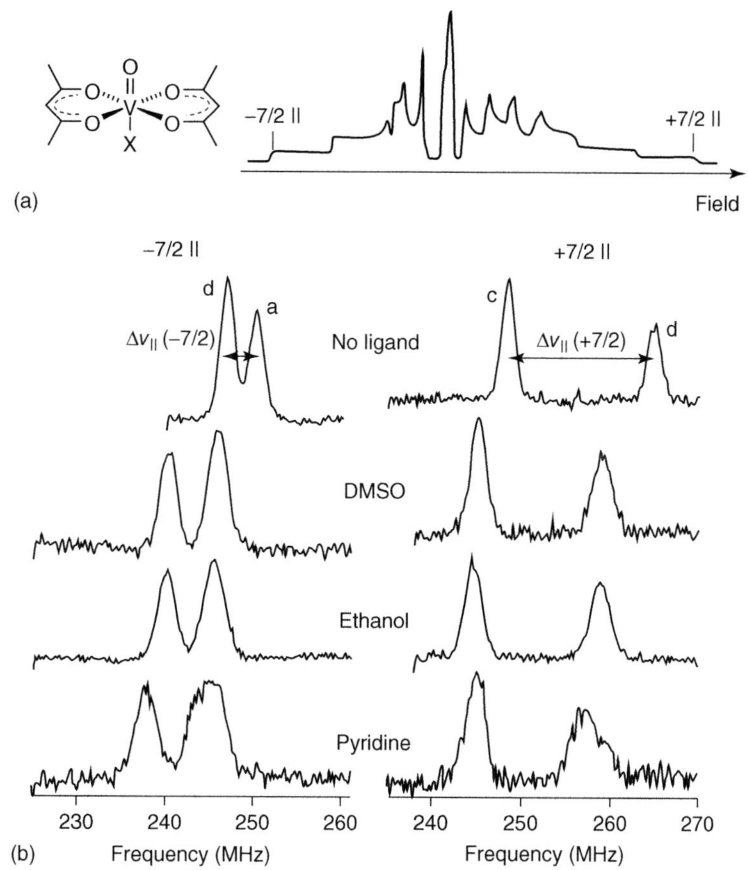

Figure 6.16. (a) Chemical structure of $VO(acac)_2$ with position X for an additional axial ligand and an X-band (10.297 GHz) echo-detected EPR spectrum (15 K) of $VO(acac)_2$ without X; (b) Davies ENDOR of $VO(acac)_2$ taken at the $m_I = -7/2$ hyperfine component (left) and the $m_I = +7/2$ hyperfine component (right), as indicated in the EPR spectrum in (a). (Reprinted with permission from Grant, C. V.; Ball, J. A.; Hamstra, B. J.; Pecoraro, V. L.; Britt, R. D., V-51 ESE-ENDOR studies of oxovanadium(IV) complexes: Investigation of the nuclear quadrupole interaction. J. Phys. Chem. B 1998, 102 (42), 8145–8150. Copyright 1998 American Chemical Society)

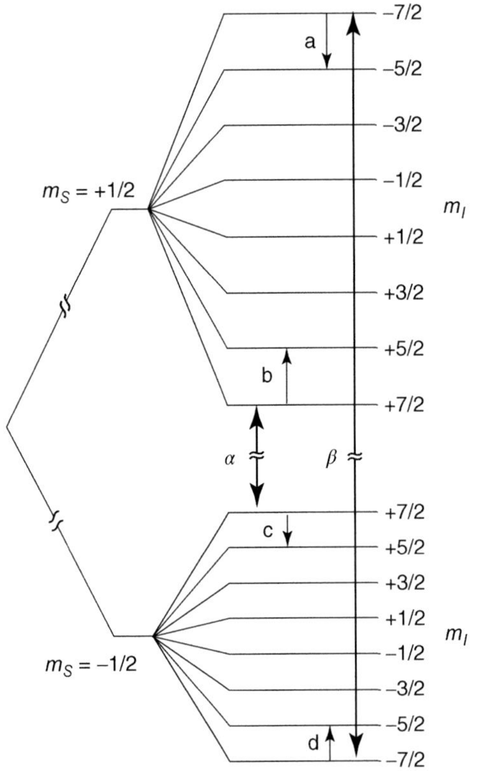

Figure 6.17. Energy level diagram for vanadyl ($S=1/2$, $I=7/2$), where β represents the EPR transition for $m_I=-7/2$ and a and d represent the associated ENDOR transitions. The EPR transition for $m_I=+7/2$ is represented by α, where b and c represent the associated ENDOR transitions. (Reprinted with permission from Grant, C. V.; Ball, J. A.; Hamstra, B. J.; Pecoraro, V. L.; Britt, R. D., V-51 ESE-ENDOR studies of oxovanadium(IV) complexes: Investigation of the nuclear quadrupole interaction. J. Phys. Chem. B 1998, 102 (42), 8145–8150. Copyright 1998 American Chemical Society)

Figure 6.17 shows a schematic energy level diagram that describes this situation. In an ENDOR experiment with the field set at the $m_I=-7/2$ low-field EPR transition, marked as β in Figure 6.17, one observes two ENDOR frequencies, $m_I=-7/2\leftrightarrow-5/2$ in the $m_S=-1/2$ manifold, marked with d, and $m_I=-5/2\leftrightarrow-7/2$ in the $m_S=+1/2$ manifold, marked with a. Similarly, with the field set at the EPR transition with $m_I=+7/2$, marked with α in Figure 6.17, the ENDOR spectrum will consist of two peaks corresponding to $m_I=+5/2\leftrightarrow+7/2$ in the $m_S=-1/2$ manifold, marked with c, and $m_I=+7/2\leftrightarrow+5/2$ in the $m_S=+1/2$ manifold, marked with b.[38] The corresponding spectra are shown in Figure 6.16(b), for cases without and with additional ligands in the axial position. For each spectrum, the splitting between the two $m_I\leftrightarrow m_I+1$ ENDOR frequencies is indicated as $\Delta\nu_{\parallel}(m_I)$ and can be calculated from equation (6.23). The difference between the two splittings is

$$\begin{aligned}\Delta\Delta\nu_{\parallel} &= \Delta\nu_{\parallel}(+7/2)-\Delta\nu_{\parallel}(-7/2)\\ &= 2\nu_I(B_{-7/2})-2\nu_I(B_{+7/2})-24P_{\parallel}\end{aligned}\tag{6.26}$$

where $P_{\parallel}=3K=3P/2$ is related to the effective quadrupole coupling P along the orientation of the field for which $\theta=0$ (see equation (6.14)). This applies for axial and collinear hyperfine and quadrupolar interactions. Using this relation, $P_{\parallel}$ values of −0.858, −0.448, −0.50, and −0.22 MHz were obtained for $VO(acac)_2$ without an additional axial ligand and with DMSO, ethanol, and pyridine, respectively. From this, it was concluded that the bound ligands contribute electron density and result in a diminished EFG at the vanadium nucleus and a reduced quadrupole coupling. The negative sign of the quadrupole coupling can be easily rationalized using the TD model and the fact that the unpaired electron resides in an in-plane d orbital. A similar approach has been applied to determine the ^{55}Mn quadrupolar interaction for the Mn^{2+} site in single crystals of the protein concanavalin A.[39]

6.6 CONCLUSIONS

The quadrupole interaction affects nuclei with $I>1/2$ and is quantified by the spin Hamiltonian parameters e^2Qq/h and η and by the orientation of the quadrupole tensor. These parameters can be determined by ENDOR or ESEEM spectroscopy, where they affect peak positions and lead to additional splittings. From the obtained values, structural insight into the local bonding environment of an atom can be derived, including the presence/absence of a ligand, the type of chemical environment, the orientation of a bond, or the occupancy of molecular orbitals.

ACKNOWLEDGMENTS

The work in the Stoll research group is currently supported by the National Science Foundation, the National Institutes of Health, the Research Corporation for Science Advancement, the American Chemical Society, and the American Heart Association. Daniella Goldfarb holds the Erich Klieger professorial chair in Physical Chemistry and the historic generosity of the Harold Perlman Family is acknowledged (D.G.).

FURTHER READING

Most EPR books contain little information about this topic. The standard monograph on pulse EPR spectroscopy, Schweiger/Jeschke,[40] briefly discusses it in 3.1.4. The book by Slichter has a dedicated insightful chapter (chapter 10 in the 3rd edition)[3] that is a worthwhile read. An article by Autschbach and Schurko contains useful information on both the TD model and on quantum-chemical calculations of the EFG.[13] The book by Dikanov and Tsvetkov[41] on ESEEM summarizes the EPR-relevant theory of nuclear quadrupole interaction and gives many examples. The old book by Lucken is dedicated to nuclear quadrupole coupling constants.[42]

REFERENCES

1. P. Pyykkö, *Mol. Phys.*, 2008, **106**, 1965.
2. N. J. Stone, *At. Data Nucl. Data Tables*, 2016, **111–112**, 1.
3. C. P. Slichter, Principles of Magnetic Resonance, Springer: Berlin, 1996.
4. C. P. Scholes, A. Lapidot, R. Mascarenhas, T. Inubushi, R. A. Isaacson, and G. Feher, *J. Am. Chem. Soc.*, 1982, **104**, 2724.
5. S. A. Dikanov, Y. D. Tsvetkov, M. K. Bowman, and A. V. Astashkin, *Chem. Phys. Lett.*, 1982, **90**, 149.
6. A. V. Astashkin, S. A. Dikanov, and Y. D. Tsvetkov, *J. Struct. Chem.*, 1984, **25**, 45.
7. H. L. Flanagan and D. J. Singel, *J. Chem. Phys.*, 1987, **87**, 5606.
8. C. H. Townes and B. P. Dailey, *J. Chem. Phys.*, 1949, **17**, 782.
9. C. I. H. Ashby, C. P. Cheng, and T. L. Brown, *J. Am. Chem. Soc.*, 1978, **100**, 6057.
10. T. G. Brown and B. M. Hoffman, *Mol. Phys.*, 1980, **39**, 1073.
11. F. Jiang, J. McCracken, and J. Peisach, *J. Am. Chem. Soc.*, 1990, **112**, 9035.
12. M. M. Dicus, A. Conlan, R. Nechushtai, P. A. Jennings, M. L. Paddock, R. D. Britt, and S. Stoll, *J. Am. Chem. Soc.*, 2010, **132**, 2037.
13. J. Autschbach, S. Zheng, and R. W. Schurko, *Concepts Magn. Reson. A*, 2010, **36A**, 84.
14. H. Huber, *Chem. Phys. Lett.*, 1984, **112**, 133.
15. D. T. Edmonds and A. L. Mackay, *J. Magn. Reson.*, 1975, **20**, 515.
16. S. Kessen, C. Teutloff, J. Kern, A. Zouni, and R. Bittl, *ChemPhysChem*, 2010, **11**, 1275.
17. S. Stoll, H. S. Shafaat, J. Krzystek, A. Ozarowski, M. J. Tauber, J. E. Kim, and R. D. Britt, *J. Am. Chem. Soc.*, 2011, **133**, 18098.
18. M. Flores, R. A. Isaacson, R. Calvo, G. Feher, and W. Lubitz, *Chem. Phys.*, 2003, **294**, 401.
19. T. I. Smirnova, A. I. Smirnov, S. V. Paschenko, and O. G. Poluektov, *J. Am. Chem. Soc.*, 2007, **129**, 3476.
20. D. L. Liczwek, R. L. Belford, J. R. Pilbrow, and J. S. Hyde, *J. Phys. Chem.*, 1983, **87**, 2509.
21. S. Stoll and A. Schweiger, *J. Magn. Reson.*, 2006, **178**, 42.
22. G. L. Hoatson and R. L. Vold, in NMR Basic Principles Progress, Springer-Verlag: Berlin, 1994, Vol. 32.
23. S. Kababya, J. Nelson, C. Calle, F. Neese, and D. Goldfarb, *J. Am. Chem. Soc.*, 2006, **128**, 2017.
24. C. A. Michal, J. C. Wehman, and L. W. Jelinski, *J. Magn. Reson. B*, 1996, **111**, 31.
25. M. G. Usha, W. L. Peticolas, and R. J. Wittebort, *Biochemistry*, 1991, **30**, 3955.
26. P. M. H. Kroneck, Binuclear Copper: CuA Copper, John Wiley & Sons, Inc.: New York, 2006, Vol. 6.
27. C. A. Chang, L. C. Francesconi, M. F. Malley, K. Kumar, J. Z. Gougoutas, M. F. Tweedle, D. W. Lee, and L. J. Wilson, *Inorg. Chem.*, 1993, **32**, 3501.
28. A. Collauto, A. Feintuch, M. Qi, A. Godt, T. Meade, and D. Goldfarb, *J. Magn. Reson.*, 2016, **263**, 156.
29. M. Saggu, C. Teutloff, M. Ludwig, M. Brecht, M. E. Pandelia, O. Lenz, B. Friedrich, W. Lubitz, P. Hildebrandt, F. Lendzian, and R. Bittl, *Phys. Chem. Chem. Phys.*, 2010, **12**, 2139.
30. Y. Deligiannakis, M. Louloudi, and N. Hadjiliadis, *Coord. Chem. Rev.*, 2000, **204**, 1.
31. F. MacMillan, S. Kacprzak, P. Hellwig, S. Grimaldi, H. Michel, and M. Kaupp, *Faraday Discuss.*, 2011, **148**, 315.

32. J. Telser, M. H. Emptage, H. Merkle, M. C. Kennedy, H. Beinert, and B. M. Hoffman, *J. Biol. Chem.*, 1986, **261**, 4840.
33. A. V. Astashkin, F. Neese, A. M. Raitsimring, J. J. Cooney, E. Bultman, and J. H. Enemark, *J. Am. Chem. Soc.*, 2005, **127**, 16713.
34. A. M. Tyryshkin, S. A. Dikanov, and D. Goldfarb, *J. Magn. Reson. A*, 1993, **105**, 271.
35. C. Gemperle, G. Aebli, A. Schweiger, and R. R. Ernst, *J. Magn. Reson.*, 1990, **88**, 241.
36. A. V. Astashkin, *J. Magn. Reson.*, 2011, **209**, 69.
37. P. J. Carl, D. E. W. Vaughan, and D. Goldfarb, *J. Am. Chem. Soc.*, 2006, **128**, 7160.
38. C. V. Grant, J. A. Ball, B. J. Hamstra, V. L. Pecoraro, and R. D. Britt, *J. Phys. Chem. B*, 1998, **102**, 8145.
39. K. V. Narasimhulu, R. Carmieli, and D. Goldfarb, *J. Am. Chem. Soc.*, 2007, **129**, 5391.
40. A. Schweiger and G. Jeschke, Principles of Pulse Electron Paramagnetic Resonance, Oxford University Press: Oxford, 2001.
41. S. A. Dikanov and Y. D. Tsvetkov, Electron Spin Echo Envelope Modulation (ESEEM) Spectroscopy, CRC Press: Boca Raton, 1992.
42. E. A. C. Lucken, Nuclear Quadrupole Coupling Constants, Academic Press: London, 1969.

Chapter 7
Quantum Chemistry and EPR Parameters

Frank Neese
Max Planck Institute for Chemical Energy Conversion, Mülheim an der Ruhr, Germany

7.1 INTRODUCTION

Modern electron paramagnetic resonance (EPR) spectroscopy has reached an impressive level of sophistication. Modern pulse techniques allow for the detection of very weak hyperfine couplings (HFCs) and multidimensional experiments allow for cross-correlations between these couplings, thus allowing for geometric structures to be deduced from spectroscopic measurements. Furthermore, electron–electron double resonance techniques are used to measure distances over the range of several nanometers. High-field/high-frequency techniques allow for high-precision measurements of EPR parameters and partially synchrotron-assisted terahertz techniques are used to measure very large zero field splittings (ZFSs).

EPR Spectroscopy: Fundamentals and Methods.
Edited by Daniella Goldfarb and Stefan Stoll
 ISBN: 978-1-119-16299-5
Also published in eMagRes (online edition)
DOI: 10.1002/9780470034590.emrstm1505

The mathematical workhorse for the EPR spectroscopist is the spin Hamiltonian (SH). An analysis of a set of experiments will lead to a set of SH parameters. Once inserted back into the SH, solution of the associated time-dependent Schrödinger equation will reproduce the experimental data with good to excellent precision. In this context, 'analysis' refers to least-squares fitting of SH parameters to experimental data. These phenomenological SH parameters then represent a concise summary of the experimental results, i.e., a common language in which the results of diverse magnetic resonance measurements are expressed and discussed.

Quantum chemistry can be useful in these studies in a number of ways:

1. It can be used to establish magneto-structural correlations that help the experimentalist to recognize patterns in the experimental data. This will allow one to maximize the chemical information content of the measurements.
2. It can be used to analyze the SH parameters in terms of electronic structure, not just geometric structure. Since electronic structure is intimately related to reactivity, this helps to understand the reactive properties of molecules.
3. If a structure is known at least to a reasonable approximation, the quantum chemical calculations can be used to generate a set of SH parameters that can be used to start simulations or least-squares fits. This is an invaluable use of quantum chemistry since for present-day high-resolution measurements, the necessary least-squares fits reach formidable complexity.

4. The quantum chemical calculations are more rigorous than many of the phenomenological methods that are used to analyze experimental data, for example point-dipole approximations or McConnell-type relationships (vide infra). Hence, they can be used to assess the range of validity of such approximations and the extent to which they are quantitatively correct.
5. Accurately measured SH parameters, in turn, serve as invaluable benchmarks against which theoretical methods can be tested. In this respect, the scarce availability of gas-phase EPR data is regrettable and renewed activity in this field would certainly be highly welcomed by theorists.

Given the attractive opportunities offered by quantum chemistry, the increased availability of sophisticated, user-friendly and efficient software packages, and the ready availability of powerful computational hardware, it is no surprise that complementing experimental EPR measurements with quantum chemical calculations has become commonplace. While these calculations generally do not require dedicated theoretical experts to be carried out successfully, there are a few pitfalls that are worthwhile being aware of if quantum chemical studies of EPR parameters are undertaken.

The present chapter is an attempt to provide a condensed summary of the necessary theoretical background that is required to understand the principles of applying quantum chemistry to EPR parameters. Some relevant practical aspects will be discussed as well. (It should be noted that there is no universal agreement on these practical aspects. Hence, it is unavoidable that some of the material will involve personal preferences. Note also that the number of allowed references is strongly limited such that an extensive coverage of the literature is, unfortunately, not possible. The selection made is not meant to be disrespectful.)

7.2 THEORY

7.2.1 Preliminary Remark: Fitting Versus Physics

In approaching the subject of the SH, one aspect might require some careful preliminary thought. As outlined above, the SH is a concise summary of the experimental data. As such one can take two extreme approaches. The first approach consists of regarding the SH parameters simply as fit parameters. In this case one could write the SH in the form of a polynomial expansion of the field and spin degrees of freedom, for example:

$$\hat{H}_{\text{Spin}} = \sum_{R,S,T=0,\ldots,\infty} \sum_{r,s,t=x,y,z} \sum_{K,L} B_r^R S_s^{(K)S} I_t^{(L)T} W_{KL,rst}^{RST}$$

where $W_{KL,rst}^{RST}$ is a high-dimensional tensor that couples the various powers of the spin- and field operators together (more general forms would obviously be possible but for the sake of clarity are not attempted here). In the case that enough experimental data is available to allow for an unambiguous least-squares fit, the values of $W_{KL,rst}^{RST}$ would summarize the information content of the experimental data. It would then be left to theoretical analysis to interpret these data in some form. In principle, this would be a perfectly valid approach to the analysis of EPR data and most certainly many empirical rules and trends could be established for the various coefficients of the polynomial fit. However, the drawback of this fitting approach is that the valuable relationship between molecular geometric and electronic structure is lost.

The second point of view is to insist on physical transparency and demand that all terms that occur in the SH must be derivable from first-principles physics. Hence, there must be an unbroken chain of logic that leads from the exact, relativistic many-particle Schrödinger equation to the SH that is used for the analysis of the experiment. In this case one would have full control over any numerical approximation that is made by simplifying the, in general, intractable problem of solving the relativistic many-body Schrödinger equation in the presence of external fields exactly. Thus, each and every approximation should be judged by how well it reproduces the next higher level of theory. In this approach, terms that do not follow to a given order in perturbation theory from first principles are not permitted.

Many intermediate approaches can be envisioned, and to some extent, are in actual use. For example, the higher-order (quartic) terms in the ZFS can be derived using group theoretical arguments for perfectly cubic systems. However, applying these same terms in the SH to arbitrary systems represents a case of polynomial curve fitting of uncertain relation to first-principle physics.

While most certainly both extreme points of view have their merits, in this chapter we take the latter point of view and insist on a well-defined and unambiguous definition of the SH terms in terms of fundamental electronic structure theory.

7.2.2 Terms in the Hamiltonian

The properties of molecules are determined by the many-particle Schrödinger equation. In its time-independent form it reads:

$$\hat{H}(\boldsymbol{x}_1, \ldots, \boldsymbol{x}_N)\Psi_I(\boldsymbol{x}_1, \ldots, \boldsymbol{x}_N) = E_I\Psi_I(\boldsymbol{x}_1, \ldots, \boldsymbol{x}_N) \tag{7.1}$$

where $I = 0, \ldots, \infty$ labels the electronic states of the system and the electronic coordinates $\boldsymbol{x}_i$ comprise the three spatial coordinates $\boldsymbol{r}_i$ and one spin coordinate σ_i of the ith electron. The Hamiltonian operator $\hat{H}$ contains all terms that contribute to the total energy of the system. It can be conveniently divided into three parts, a nonrelativistic part, a relativistic part and a part describing the interaction of the particles in the system with external fields. The nonrelativistic Hamiltonian for a system with clamped nuclei (Born–Oppenheimer (BO) approximation) is of deceptive simplicity (in atomic units).

$$\hat{H}_{\mathrm{BO}} = \underbrace{-\frac{1}{2}\sum_i \vec{\nabla}_i^2}_{\hat{T}_{\mathrm{e}}} \underbrace{-\sum_{A,i} Z_A r_{Ai}^{-1}}_{\hat{V}_{\mathrm{eN}}} \underbrace{+\frac{1}{2}\sum_{i \neq j} r_{ij}^{-1}}_{\hat{V}_{\mathrm{ee}}} \underbrace{+\frac{1}{2}\sum_{A \neq B} Z_A Z_B R_{AB}^{-1}}_{\hat{V}_{\mathrm{NN}}} \tag{7.2}$$

Here, i, j label electronic coordinates and A, B nuclear coordinates. Z_A is the charge of nucleus A, r_{Ai} is the distance of the i'th electron to the A'th nucleus, and analogous for r_{ij} and R_{AB}. Since the BO Hamiltonian is independent of the electron spin, the eigenfunctions of the BO Hamiltonian can be characterized by the additional good quantum numbers S and M, where S is the total spin of the system in a given electronic state and M is its projection onto the z-axis, thus leading to the notation Ψ_I^{SM}. Note that all $2S+1$ members of a given multiplet $\Psi_I^{SS}, \Psi_I^{SS-1}, \ldots, \Psi_I^{S,-S}$ are degenerate at the level of the BO Hamiltonian.

The associated time-independent BO Schrödinger equation can still not be solved in closed form, not even for the simplest two-electron systems. Hence, it is necessary to approximate its solution, and a plethora of different approximate approaches of varying sophistication exist, the most important of which will be briefly discussed in the following text. For the conceptual discussion, we will assume that we know all exact eigenstates of the BO Hamiltonian. Where necessary, the nuclear spin degrees of freedom will be taken into account by taking a product ansatz, where the total wavefunction will be written as a product of the electronic part and a nuclear part $\Psi_K^{SM} \otimes \Theta_K(\boldsymbol{R}_1, \ldots, \boldsymbol{R}_M, I^{(1)}, \ldots I^{(M)})$, with Θ_K being the nuclear wavefunction that depends on the nuclear spatial and spin-degrees of freedom.

At the next level of sophistication, one has to introduce additional smaller terms in the Hamiltonian as well as terms containing external electric and magnetic fields.

Smaller terms in the Hamiltonian arise from the theory of relativity.[1] The relativistic many-particle Schrödinger equation initially is in 'four-component' form and contains electronic and positronic degrees of freedom and, for both of those, two spin components. In a first step, the positronic degrees of freedom are decoupled from the electronic ones using one of the many approximate schemes that lead to 'two-component' approaches. In the second step, the resulting two-component Hamiltonian is decomposed into spin-independent ('scalar relativistic') and spin-dependent ('spin-orbit') terms.[2] The scalar relativistic terms can be conveniently treated on equal footing with the terms in the BO Hamiltonian and are usually expressed in the form of an effective one-particle Hamiltonian:

$$-\frac{1}{2}\sum_i \vec{\nabla}_i^2 - \sum_{A,i} Z_A r_{Ai}^{-1} \equiv \sum_i \hat{h}(\boldsymbol{x}_i) \xrightarrow{\text{scalar-relativity}} \sum_i \hat{h}^{\mathrm{SR}}(\boldsymbol{x}_i) \tag{7.3}$$

where $\hat{h}^{\mathrm{SR}}(\boldsymbol{x}_i)$ is the appropriately modified one-particle Hamiltonian. Popular schemes include the Breit–Pauli (BP) approach, the Douglas–Kroll–Hess (DKH[3]) method, or the 0th-order regular approximation (ZORA[4]). Details cannot be provided here. Scalar relativistic effects have some intricate consequences on EPR parameters that will be discussed later.

There are a number of terms in the Hamiltonian that depend on the electron and nuclear spins and their mutual interaction. A comprehensive treatment is not possible here (see Harriman,[5] McWeeny [6,7] and Refs 8–11). The most important spin-dependent term is the spin-orbit coupling (SOC). In the simplest (BP) version, the operator takes the form:

$$\hat{H}_{SOC} = \frac{\alpha^2}{2}\sum_i\sum_A Z_A r_{iA}^{-3}\hat{l}_{iA}\hat{s}_i - \frac{\alpha^2}{2}\sum_i\sum_{j\neq i} r_{ij}^{-3}\hat{l}_{ij}(\hat{s}_i + 2\hat{s}_j) \tag{7.4}$$

where $\hat{s}_i$ is the spin-operator for the i'th electron, $\alpha \approx \frac{1}{137}$ is the fine structure constant in atomic units, $\hat{l}_{iA}$ is the angular momentum of the i'th electron relative to the A'th nucleus, $\hat{l}_{ij}$ is the angular momentum of the i'th electron relative to the j'th electron. It is important to note that the SOC operator is a sum of a one-electron and a two-electron operator. The two contributions in the second, two-electron, term in equation (7.4) are referred to as spin-same-orbit (SSO) and spin-other-orbit (SOO). The latter arises from a relativistic correction to the electron–electron interaction. This term is not small and if it is neglected (as in certain approximate treatments based on effective potentials), significant errors (up to 30%[12]) will arise. A good approximation to the SOC operator can be obtained by a spin-orbit mean-field (SOMF) treatment.[12,13] Without going into details, it can be mentioned that this results in a more readily handled quasi-one-electron form of the SOC:

$$\hat{H}_{SOMF} = \sum_i z(\boldsymbol{r}_i)\hat{s}_i \tag{7.5}$$

where $z(\boldsymbol{r}_i)$ contains three spatial components.

One subtlety that is frequently overlooked is that the SOC operator contains the spin of the *individual* electrons while the wavefunction is cast in terms of the *total* spin of the system. In particular, it is important to understand that the form of the SOC operator that one frequently meets in the transition-metal EPR literature, namely:

$$\hat{H}_{LS} = \lambda_{SOC}\boldsymbol{lS} \tag{7.6}$$

is very misleading. This operator is cast in terms of the total spin- and orbital angular momenta. It will never lead to coupling of states of different multiplicity, while the true SOC operator clearly does. This is of utmost importance for the theory of the $\boldsymbol{g}$-matrix and the ZFS tensor that they cannot be related to a common tensor $\boldsymbol{A}$ (see Chapters 2 and 3), with:

$$A_{kl} = \sum_{n\neq 0}\frac{\langle 0|L_k|n\rangle\langle n|L_l|0\rangle}{E_n - E_0} \tag{7.7}$$

where n labels electronically excited states and L refers to the total angular momentum. The SOC operator equation (7.6) introduces an effective many-electron SOC 'constant' λ_{SOC} that is ill-defined for molecules. The operator $\hat{H}_{LS}$ has been derived for application inside a single Russell–Saunders term in atomic spectroscopy[14] and should not be used outside this realm. Hence, use of this SOC operator is generally discouraged for *all* molecular applications, no matter whether transition metal complexes or radicals, since it may lead to erroneous conclusions and ill-defined results.

Next in the development are the terms that depend on external fields. Of those, we only consider the terms that couple the electrons (spin- and orbital-Zeeman) to the magnetic field:

$$\hat{H}_{SB+LB} = \mu_B\boldsymbol{B}\sum_i(\hat{l}_i + g_e\hat{s}_i) \tag{7.8}$$

where μ_B is the Bohr magneton, $g_e = 2.002319\ldots$ is the free electron g-value and $\boldsymbol{B}$ is the magnetic field. Note that the angular momentum $\hat{\boldsymbol{l}}_i = \boldsymbol{r}_i \times \boldsymbol{p}_i$ is referred to the global origin, e.g. the origin that $\boldsymbol{r}_i$ is referred to. This is potentially problematic since it leads to a lack of gauge invariance, or, in other words, the results depend on the choice of origin, which is clearly unphysical (vide infra).

Last, we need to consider terms that couple the electronic and nuclear spin degrees of freedom. We will restrict attention to just two terms that are important for the discussion. These are the coupling of the nuclear spin to the electron orbital angular momenta:

$$\hat{H}_{LI} = \mu_B\sum_{i,A} g_N^{(A)}\mu_N r_{iA}^{-3}\hat{\boldsymbol{I}}^{(A)}\hat{\boldsymbol{l}}_{iA} \tag{7.9}$$

where μ_N is the nuclear magneton and $g_N^{(A)}$ is the nuclear g-value of the A'th nucleus. Furthermore, the coupling of the nuclear spin to the electron spins is given by:

$$\hat{H}_{SI} = g_e\mu_B\sum_{i,A} g_N^{(A)}\mu_N r_{iA}^{-3}\left\{\hat{\boldsymbol{I}}^{(A)}\hat{\boldsymbol{s}}_i - 3\left(\hat{\boldsymbol{s}}_i\boldsymbol{n}_{iA}\right)\left(\hat{\boldsymbol{I}}^{(A)}\boldsymbol{n}_{iA}\right)\right\} \tag{7.10}$$

where $\boldsymbol{n}_{iA}$ is a unit vector in the direction of $\boldsymbol{R}_A - \boldsymbol{r}_i$, i.e., $\boldsymbol{n}_{iA} = (\boldsymbol{R}_A - \boldsymbol{r}_i)r_{iA}^{-1}$.

Finally, we will also consider the direct dipole–dipole interaction between electrons, leading to the term

$$\hat{H}_{SS} = \frac{g_e^2\alpha^2}{8}\sum_{i\neq j} r_{ij}^{-3}\{\hat{\boldsymbol{s}}_i\hat{\boldsymbol{s}}_j - 3(\hat{\boldsymbol{s}}_i\boldsymbol{n}_{ij})(\hat{\boldsymbol{s}}_j\boldsymbol{n}_{ij})\} \tag{7.11}$$

Here, $\boldsymbol{n}_{ij}$ is the unit vector pointing from electron i to electron j (note that both the dipolar and Fermi contact contribution to the hyperfine interaction arise from this operator, vide infra). While there are many other small terms in the Hamiltonian,[5] we will, for the purpose of this chapter, contain ourselves with collecting all discussed terms in the perturbing Hamiltonian $\hat{H}_1$:

$$\hat{H}_1 = \hat{H}_{\text{SOMF}} + \hat{H}_{\text{SS}} + \hat{H}_{\text{LI}} + \hat{H}_{\text{SI}} + \hat{H}_{\text{SB+LB}} \quad (7.12)$$

7.2.3 Effective Hamiltonian Theory of Spin-Hamiltonian Parameters

The SH is typically written in the form:

$$\hat{H}_{\text{Spin}} = \mu_{\text{B}} \boldsymbol{B}\boldsymbol{g}\hat{\boldsymbol{S}} + \hat{\boldsymbol{S}}\boldsymbol{D}\hat{\boldsymbol{S}} + \sum_A \hat{\boldsymbol{S}}\boldsymbol{A}^{(A)}\hat{\boldsymbol{I}}^{(A)} \quad (7.13)$$

where $\boldsymbol{g}$ is the $\boldsymbol{g}$-matrix, $\boldsymbol{D}$ the ZFS tensor, $\boldsymbol{A}^{(A)}$ the HFC tensor for the A'th nucleus, $\hat{\boldsymbol{I}}^{(A)}$ is the nuclear spin and $\hat{\boldsymbol{S}}$ is a fictitious spin operator for the electron spin. This Hamiltonian acts on a set of product states:

$$|SM; I^{(1)}M_I^{(1)} \cdots I^{(M)}M_I^{(M)}\rangle = |SM\rangle \otimes |I^{(1)}M_I^{(1)} \cdots I^{(M)}M_I^{(M)}\rangle \quad (7.14)$$

The quantities $\boldsymbol{g}, \boldsymbol{D}, \boldsymbol{A}^{(A)}$ are purely numerical parameters in this treatment. Hence, unlike the 'true' Hamiltonian $\hat{H}_{\text{BO}} + \hat{H}_1$, the SH $\hat{H}_{\text{Spin}}$ is typically of low dimension (just $2S+1$ for the electronic degrees of freedom) and, hence, reasonably easily handled with paper and pencil or numerical simulation programs such as EasySpin.[15] The important task of theory is to cleanly connect the 'true' Hamiltonian $\hat{H}_{\text{BO}} + \hat{H}_1$ to the very much simpler $\hat{H}_{\text{Spin}}$. A transparent treatment has first been suggested by McWeeny[16] and further refined in Ref. 17. Let us assume that the ground-state multiplet Ψ_0^{SM} is 'sufficiently well' (discussed below) isolated from the remaining electronic states $\Psi_n^{S'M'}$ ($n = 1, \ldots, \infty$). Let us furthermore make the bold assumption that the BO problem has been solved completely with solutions Ψ_I^{SM}, e.g. $\hat{H}_{\text{BO}}\Psi_I^{SM} = E_I\Psi_I^{SM}$. Since the set of many-particle eigenfunctions forms a complete set, the fully relativistic eigenstates of $\hat{H}_{\text{BO}} + \hat{H}_1$ can be expanded in terms of the BO eigenfunctions, e.g.,

$$\Phi_X(\boldsymbol{x}_1, \ldots, \boldsymbol{x}_N) = \sum_{I,S,M} c_{ISM}\Psi_I^{SM}(\boldsymbol{x}_1, \ldots, \boldsymbol{x}_N) \quad (7.15)$$

($X = 0, \ldots, \infty$), which emphasizes that the relativistic eigenstates involve, in principle, all magnetic sublevels of all multiplets with any multiplicity. Thus, $\hat{H}_1$ is coupling different S and M. Therefore, relativistic eigenstates cannot be assigned good quantum numbers S and M although in many cases one particular S will be dominant in the expansion such that there is an approximate and identifiable spin-multiplicity associated with that state.

The unknown expansion coefficients can be determined by means of the variational principle, which immediately leads to a linear eigenvalue equation $\boldsymbol{H}\boldsymbol{c} = \bar{E}\boldsymbol{c}$, (the overbar is a reminder that this will be the fully relativistic total energy) with matrix elements:

$$H_{ISM,JS'M'} = \langle\Psi_I^{SM}|\hat{H}_{\text{BO}} + \hat{H}_1|\Psi_J^{S'M'}\rangle = \delta_{ISM,JS'M'}E_I + \langle\Psi_I^{SM}|\hat{H}_1|\Psi_J^{S'M'}\rangle \quad (7.16)$$

Under the simplifying assumptions above, the eigenvalue problem can be partitioned into an 'a-set' that only contains the $2S+1$ members of the electronic ground-state multiplet and a 'b-set' that consists of the remaining states. Thus:

$$\begin{pmatrix} \boldsymbol{H}^{aa} & \boldsymbol{H}^{ab} \\ \boldsymbol{H}^{ba} & \boldsymbol{H}^{bb} \end{pmatrix} \begin{pmatrix} \boldsymbol{c}^a \\ \boldsymbol{c}^b \end{pmatrix} = \bar{E} \begin{pmatrix} \boldsymbol{c}^a \\ \boldsymbol{c}^b \end{pmatrix} \quad (7.17)$$

Without loss of generality, this equation system can be solved for $\boldsymbol{c}^b$ which then can be inserted into the equation for $\boldsymbol{c}^a$ giving the effective nonlinear Hamiltonian:

$$[\boldsymbol{H}^{aa} - \boldsymbol{H}^{ab}(\boldsymbol{H}^{bb} - \boldsymbol{1}^b\bar{E})^{-1}\boldsymbol{H}^{ba}]\boldsymbol{c}^a = \bar{E}\boldsymbol{c}^a \quad (7.18)$$

where $\boldsymbol{1}^b$ is the unit matrix in b-space.

In order to make progress, this equation must be simplified. First, one may choose to neglect the coupling of the b-states through $\hat{H}_1$ resulting in $H_{KL}^{bb} = \delta_{KL}E_K$, and secondly one needs to replace the unknown energy $\bar{E}$ in the matrix inverse by a known energy. Since one is searching for solutions in the vicinity of the ground state multiplet energy, a reasonable choice is $\bar{E} \rightarrow E_0$ thus giving an effective Hamiltonian for the a-set of the form:

$$H_{MM'}^{\text{eff}} = \delta_{MM'}E_0 + \langle\Psi_0^{SM}|\hat{H}_1|\Psi_0^{SM'}\rangle - \sum_{KS'M''} \Delta_K^{-1}\langle\Psi_0^{SM}|\hat{H}_1|\Psi_K^{S'M''}\rangle\langle\Psi_K^{S'M''}|\hat{H}_1|\Psi_0^{SM'}\rangle \quad (7.19)$$

($\Delta_K = E_K - E_0$). The ground-state energy in the first term can be dropped since it shifts all levels equally.

Disregarding the nuclear spin degrees of freedom for the moment, this Hamiltonian is of the same dimension as the SH but is entirely cast in terms of the nonrelativistic BO eigenstates. (Note also, that in this treatment the true spin of the ground state is equal to the fictitious spin of the SH treatment. It is possible to formulate SHs when this is not the case. However, in this author's opinion, such attempts do not add to clarity. An exception arises if the SH itself is broken down into smaller pieces, such as individual Kramer's doublets inside a half-integer-spin multiplet that are treated as effective $S = 1/2$ systems. These cases arise from an effective Hamiltonian treatment of the SH formalism itself, rather than representing weakly justified derivations of an SH from first principles.) The route to explicit expressions for the SH parameters is now to compare the matrix elements of the effective Hamiltonian to the spin-Hamiltonian one-by-one. The actual process involves a fair bit of algebra and the use of the Wigner–Eckart theorem and will hence not be pursued here.[17] However, the effective-Hamiltonian equation is already sufficiently suggestive of the process. For example, for the HFC we look for terms correlating electron and nuclear spin. The direct coupling is provided by the term $\widehat{H}_{\mathrm{SI}}$ and, hence, there will be a first-order contribution. On the other hand, the cross products of $\widehat{H}_{\mathrm{LI}}$ and $\widehat{H}_{\mathrm{SOMF}}$ will also be proportional to S times I and, hence, there will also be a contribution from the final, infinite sum-over-states (SOS) term that represents a second-order contribution. The $\boldsymbol{g}$-matrix (apart from a few small first-order terms) will arise from second-order terms between $\widehat{H}_{\mathrm{SOMF}}$ and $\widehat{H}_{\mathrm{SB+LB}}$ while the ZFS will arise from the first-order term $\widehat{H}_{\mathrm{SS}}$ and the second-order term arising from twice $\widehat{H}_{\mathrm{SOMF}}$.

The final equations for the SH parameters (labels: SB = first-order free-electron $\boldsymbol{g}$-value contribution to the $\boldsymbol{g}$-matrix, RMC = relativistic mass correction, GC = gauge correction, also called diamagnetic contribution, OZ/SOC = second-order contribution from orbital Zeeman and SOC, also called paramagnetic contribution; FC = Fermi contact, SD = spin-dipole, SOC; A = spin–orbit contribution to the hyperfine coupling of nucleus A; SS = first-order spin–spin contribution to the ZFS, SOC = second-order SOC contribution to the ZFS) read:[10,17,18]

$$\boldsymbol{g} = \boldsymbol{g}^{(\mathrm{SB})} + \boldsymbol{g}^{(\mathrm{RMC})} + \boldsymbol{g}^{(\mathrm{GC})} + \boldsymbol{g}^{(\mathrm{OZ/SOC})} \tag{7.20}$$

$$\boldsymbol{A}^{(A)} = \boldsymbol{A}^{(\mathrm{FC};A)} + \boldsymbol{A}^{(\mathrm{SD};A)} + \boldsymbol{A}^{(\mathrm{SOC};A)} \tag{7.21}$$

$$\boldsymbol{D} = \boldsymbol{D}^{(\mathrm{SS})} + \boldsymbol{D}^{(\mathrm{SOC}-(0))} + \boldsymbol{D}^{(\mathrm{SOC}-(+1))} + \boldsymbol{D}^{(\mathrm{SOC}-(-1))} \tag{7.22}$$

with individual components being given by:

$$g_{KL}^{(\mathrm{SB})} = \delta_{KL} g_{\mathrm{e}} \tag{7.23}$$

$$g_{KL}^{(\mathrm{RMC})} = \delta_{KL} \frac{\alpha^2}{2} \frac{1}{S} \frac{g_{\mathrm{e}}}{2} \left\langle \Psi_0^{SS} \left| \sum_i \nabla_i^2 \widehat{s}_{z;i} \right| \Psi_0^{SS} \right\rangle \tag{7.24}$$

$$g_{KL}^{(\mathrm{GC})} = \frac{1}{S} \left\langle \Psi_0^{SS} \left| \sum_{i,A} \xi\left(r_{iA}\right) \{\boldsymbol{r}_{iA}\boldsymbol{r}_i - r_{iA;K} r_{i;L}\} s_{z;i} \right| \Psi_0^{SS} \right\rangle \tag{7.25}$$

$$g_{KL}^{(\mathrm{OZ/SOC})} = -\frac{1}{S} \sum_{b(S_b=S)} \Delta_b^{-1} \left\{ \left\langle \Psi_0^{SS} | \sum_i \widehat{l}_{i;K} | \Psi_b^{SS} \right\rangle \times \left\langle \Psi_b^{SS} | \sum_i z_{L;i} \widehat{s}_{z;i} | \Psi_0^{SS} \right\rangle + \left\langle \Psi_0^{SS} | \sum_i z_{K;i} \widehat{s}_{z;i} | \Psi_b^{SS} \right\rangle \times \left\langle \Psi_b^{SS} | \sum_i \widehat{l}_{i;L} | \Psi_0^{SS} \right\rangle \right\} \tag{7.26}$$

$$A_{KL}^{(\mathrm{FC};A)} = \delta_{KL} \frac{8\pi}{3} \frac{1}{S} \mu_{\mathrm{B}} g_{\mathrm{e}} \mu_{\mathrm{N}}^{(A)} g_{\mathrm{N}}^{(A)} \left\langle \Psi_0^{SS} | \sum_i \delta\left(r_{iA}\right) \widehat{s}_{z;i} | \Psi_0^{SS} \right\rangle \tag{7.27}$$

$$A_{KL}^{(\mathrm{SD};A)} = \frac{1}{S} \mu_{\mathrm{B}} g_{\mathrm{e}} \mu_{\mathrm{N}}^{(A)} g_{\mathrm{N}}^{(A)} \times \left\langle \Psi_0^{SS} | \sum_i \widehat{s}_{z;i} r_{iA}^{-5} \left\{ \delta_{KL} r_{iA}^2 - 3 r_{iA;K} r_{iA;L} \right\} | \Psi_0^{SS} \right\rangle \tag{7.28}$$

$$A_{KL}^{(\mathrm{SOC};A)} = -\frac{1}{S} \mu_{\mathrm{B}} g_{\mathrm{e}} \mu_{\mathrm{N}}^{(A)} g_{\mathrm{N}}^{(A)} \sum_{b(S_b=S)} \Delta_b^{-1} \times \left\{ \left\langle \Psi_0^{SS} | \sum_i \widehat{l}_{i;K}^{(A)} r_{iA}^{-3} | \Psi_b^{SS} \right\rangle \left\langle \Psi_b^{SS} | \sum_i z_{L;i} \widehat{s}_{z;i} | \Psi_0^{SS} \right\rangle + \left\langle \Psi_0^{SS} | \sum_i z_{K;i} \widehat{s}_{z;i} | \Psi_b^{SS} \right\rangle \left\langle \Psi_b^{SS} | \sum_i \widehat{l}_{i;L}^{(A)} r_{iA}^{-3} | \Psi_0^{SS} \right\rangle \right\} \tag{7.29}$$

$$D_{KL}^{(\mathrm{SS})} = \frac{1}{2} \frac{\alpha^2}{S(2S-1)} \left\langle \Psi_0^{SS} \left| \sum_i \sum_{j \neq i} \frac{r_{ij}^2 \delta_{KL} - 3(\boldsymbol{r}_{ij})_K (\boldsymbol{r}_{ij})_L}{r_{ij}^5} \times \left\{ 2\widehat{s}_{zi}\widehat{s}_{zj} - \widehat{s}_{xi}\widehat{s}_{xj} - \widehat{s}_{yi}\widehat{s}_{yj} \right\} \right| \Psi_0^{SS} \right\rangle \tag{7.30}$$

$$D_{KL}^{\mathrm{SOC}-(0)} = -\frac{1}{S^2} \sum_{b(S_b=S)} \Delta_b^{-1} \left\langle \Psi_0^{SS} \left| \sum_i z_{K;i} \widehat{s}_{i;z} \right| \Psi_b^{SS} \right\rangle \times \left\langle \Psi_b^{SS} \left| \sum_i z_{L;i} \widehat{s}_{i,z} \right| \Psi_0^{SS} \right\rangle \tag{7.31}$$

$$D_{KL}^{\mathrm{SOC}-(-1)} = -\frac{1}{S(2S-1)}$$

$$\times \sum_{b(S_b=S-1)} \Delta_b^{-1} \left\langle \Psi_0^{SS} \left| \sum_i z_{K;i} \hat{s}_{i,+1} \right| \Psi_b^{S-1,S-1} \right\rangle$$

$$\times \left\langle \Psi_b^{S-1,S-1} \left| \sum_i z_{L;i} \hat{s}_{i,-1} \right| \Psi_b^{SS} \right\rangle \quad (7.32)$$

$$D_{KL}^{\mathrm{SOC}-(+1)} = -\frac{1}{(S+1)(2S+1)}$$

$$\times \sum_{b(S_b=S+1)} \Delta_b^{-1} \left\langle \Psi_0^{SS} \left| \sum_i z_{K;i} \hat{s}_{i,-1} \right| \Psi_b^{S+1,S+1} \right\rangle$$

$$\times \left\langle \Psi_b^{S+1,S+1} \left| \sum_i z_{L;i} \hat{s}_{i,+1} \right| \Psi_0^{SS} \right\rangle \quad (7.33)$$

We note a few important features of these equations:

1. These equations are entirely cast in terms of the $S=M$ components of each state (e.g. the 'principal component' of the multiplet). All summations over intermediate M''-labels ($M''=S'$, $S'-1, \ldots, -S'$) have been carried out in the derivation.
2. The second-order terms contain an infinite sum over excited states that, in practice, cannot be performed.
3. The equations are correct to second order.
4. A contact term exists for the first-order contribution to the ZFS tensor but is usually dropped because it shifts all levels equally.
5. The second-order contributions to the **g**-matrix and the second-order contribution to the HFC tensor involve only excited states of the same spin multiplicity. These contributions show certain similarities that, for example, in ligand field theory one can employ to set them proportionally to each other. Numerically, the proportionality is only fair since the nucleus-orbit integral in the hyperfine interaction is fairly local while the angular momentum integral in the **g**-matrix is not.
6. The **g**-'tensor' is not symmetric and, hence, does not behave like a proper tensor. Hence, it is often referred to as '**g**-matrix' (see Chapter 2). However, $\boldsymbol{g}\boldsymbol{g}^{\mathrm{T}}$ behaves like a real tensor but upon taking the square-root of its eigenvalues the sign information on the g-values is lost.
7. In the chosen formulation, the **g**-matrix is not gauge invariant since the angular momentum operator is referred to a global origin. This problem has a known cure developed in the realm of NMR spectroscopy.[19,20] It consists of making the basis functions used in the calculation magnetic field dependent (so-called 'gauge-including atomic orbitals', GIAOs) by including a magnetic field dependent phase factor. A detailed discussion would become very technical and is, hence, avoided here.
8. The **D**-tensor and the **g**-matrix are not even approximately proportional to each other. The **D**-tensor contains contributions from excited states of different spin multiplicity while the **g**-matrix does not. As explained elsewhere, they also behave differently with respect to different excitation classes.[10] The **g**-matrix changes sign by going from transitions into the SOMO to transitions out of the SOMO while the **D**-tensor does not. The **D**-tensor and the Λ-tensor (equation (7.7)) are symmetric while the **g**-matrix is not. Thus, approaches that relate $\boldsymbol{D}$ and $\boldsymbol{g}$ to a common tensor $\boldsymbol{\Lambda}$ are mathematically as well as physically wrong. Hence, these approaches are discouraged (the $\boldsymbol{\Lambda}$-tensor arises from using the SOC operator $\hat{H}_{\mathrm{LS}}$ in the development, which is one example why this operator is discouraged as well).
9. It may appear that in the presence of scalar relativistic potentials, the very same equations apply. This is not the case due to the 'picture change effect' that will be briefly discussed below.[21]

7.2.4 Implementation: Linear Response

The equations for the SH parameters are general and concise, but to some extent they represent a somewhat academic exercise since they assume the knowledge of the complete eigenspectrum of the BO Hamiltonian, which is never available. It is, however, possible to cast the equations in a different form that enables a clean and satisfactory connection to approximate electronic structure methods.

The main problem with the theory described above, besides the unavailability of the exact BO eigenfunctions, is the infinite SOS in the second-order contributions to the effective Hamiltonian. In order to come to an equivalent but more practical formulation that is readily applied to approximate electronic structure methods consider a perturbing one-electron Hamiltonian of the general form:

$$\lambda\hat{H}_\lambda = \lambda\sum_i \hat{h}_\lambda(\boldsymbol{x}_i) \tag{7.34}$$

where λ is a formal perturbation parameter that will be set to unity at the end of the calculation and $\hat{H}_\lambda$ is some perturbing one-electron operator, e.g. any of the one-electron terms in $\hat{H}_1$ discussed above. We will rewrite in second quantization as:

$$\lambda\hat{H}_\lambda = \lambda\sum_{pq} h_{pq}^{(\lambda)}\hat{a}_p^+\hat{a}_q \tag{7.35}$$

where a set of orthonormal one-particle functions $\{\psi_p\}$ has been introduced (vide infra), $h_{pq}^{(\lambda)} = \langle\psi_p|\hat{h}_\lambda|\psi_q\rangle$ and $\hat{a}_p^+\hat{a}_q$ are standard Fermion creation and annihilation operators. In a nutshell, the operator $\hat{a}_p^+\hat{a}_q$ replaces orbital q with orbital p provided the state that it acts on is a state where q is occupied and p is unoccupied. Otherwise it will give zero. If we add this Hamiltonian to the BO Hamiltonian, we can expand the ground-state energy in a Taylor series:

$$E_0(\lambda) = E_0(0) + \lambda\left.\frac{\partial E_0}{\partial\lambda}\right|_{\lambda=0} + \cdots \tag{7.36}$$

By making use of the Hellmann–Feynman theorem:

$$\left.\frac{\partial E_0}{\partial\lambda}\right|_{\lambda=0} = \left\langle\Psi_0|\sum_{pq} h_{pq}^{(\lambda)}\hat{a}_p^+\hat{a}_q|\Psi_0\right\rangle \tag{7.37}$$

$$= \sum_{pq} h_{pq}^{(\lambda)}\langle\Psi_0|\hat{a}_p^+\hat{a}_q|\Psi_0\rangle \tag{7.38}$$

$$= \sum_{pq} h_{pq}^{(\lambda)} D_{pq}$$

where the last line defines the 'reduced density matrix' D_{pq} of the electronic ground state. It defines the expectation value of any one-electron operator of the ground state. This equation establishes that the first derivative of the ground-state energy with respect to the perturbation parameter λ is equivalent to first-order perturbation theory and merely involves the expectation value of the perturbation over the ground-state wavefunction. Introducing a second perturbation μ:

$$\hat{H}_\mu = \mu\sum_{pq} h_{pq}^{(\mu)}\hat{a}_p^+\hat{a}_q \tag{7.39}$$

Finally, there might be mixed terms of the form:

$$\hat{H}_{\lambda\mu} = \lambda\mu\sum_{pq} h_{pq}^{(\lambda,\mu)}\hat{a}_p^+\hat{a}_q \tag{7.40}$$

and taking the derivative, we get:

$$\frac{\partial^2 E_0}{\partial\lambda\partial\mu} = \sum_{pq} h_{pq}^{(\lambda,\mu)} D_{pq} + \sum_{pq} h^{(\lambda)}\frac{\partial D_{pq}}{\partial\mu} \tag{7.41}$$

The first term is simply the expectation value of the second derivative of the Hamiltonian with respect to both perturbations. It corresponds precisely to the first-order term in the effective Hamiltonian. The derivative of the density matrix is:

$$\frac{\partial D_{pq}}{\partial\mu} = \left\langle\frac{\partial\Psi_0}{\partial\mu}|\hat{a}_p^+\hat{a}_q|\Psi_0\right\rangle + \left\langle\Psi_0|\hat{a}_p^+\hat{a}_q|\frac{\partial\Psi_0}{\partial\mu}\right\rangle \tag{7.42}$$

Since the eigenfunctions of the BO Hamiltonian form a complete set in N-particle space, we can expand the derivatives in terms of the unperturbed wavefunctions:

$$\left|\frac{\partial\Psi_0}{\partial\mu}\right\rangle = \sum_{n>0} c_n^{(\mu)}|\Psi_n\rangle \tag{7.43}$$

and first-order perturbation theory tells us that:

$$c_n^{(\mu)} = -\frac{\langle\Psi_0|H_\mu|\Psi_n\rangle}{E_n - E_0} \tag{7.44}$$

Hence, we conclude (cc = complex conjugate):

$$\frac{\partial D_{pq}}{\partial\mu} = -\sum_{n>0}\Delta_n^{-1}\{\langle\Psi_0|H_\mu|\Psi_n\rangle\langle\Psi_n|a_p^+a_q|\Psi_0\rangle + \text{cc}\} \tag{7.45}$$

which means that:

$$\frac{\partial^2 E_0}{\partial\lambda\partial\mu} = \sum_{pq} h_{pq}^{(\lambda,\mu)} D_{pq} + \sum_{pq} h_{pq}^{(\mu)}\frac{\partial D_{pq}}{\partial\lambda}$$

$$= \langle\Psi_0|\hat{H}_{\lambda\mu}|\Psi_0\rangle - \sum_{n>0}\Delta_n^{-1}\{\langle\Psi_0|\hat{H}_\lambda|\Psi_n\rangle\langle\Psi_n|\hat{H}_\mu|\Psi_0\rangle + \langle\Psi_0|\hat{H}_\mu|\Psi_n\rangle\langle\Psi_n|\hat{H}_\lambda|\Psi_0\rangle\} \tag{7.46}$$

This is a powerful identity since it establishes that the second derivative of the ground-state energy is equivalent to the second-order effective Hamiltonian and implicitly contains an infinite sum over states (equation (7.19)). This then provides a recipe to calculate molecular properties in practice: whatever the approximate method, calculate the density matrix (equation (7.38)) and its first derivative (equation (7.42)). Thus, it is never necessary to really evaluate an infinite sum over states. All that is required is to evaluate derivatives. This is also referred to as 'linear response theory'. For example, the equivalent for the second-order contribution to the $\boldsymbol{g}$-matrix (equation (7.26)) in linear-response language is:

$$g_{KL}^{(\mathrm{OZ/SOC})} = -\frac{1}{S}\sum_{\mu\nu}\frac{\partial D_{\mu\nu}^{\alpha-\beta}}{\partial B_K}\langle\mu|z_L|\nu\rangle \qquad (7.47)$$

where $D_{\mu\nu}^{\alpha-\beta}$ is the spin-density matrix expressed in terms of the orbital basis set $\{\varphi_\mu\}$ that is used in the calculation (vide infra). How to calculate the derivative of $D_{\mu\nu}^{\alpha-\beta}$ depends on the underlying electronic structure method. The theory of how to actually take the derivatives of approximate wavefunctions is a highly developed art in quantum chemistry that we cannot go into here.[22]

7.2.5 Where Does Linear Response Fail?

It is important to understand the limitations of the treatment just outlined. It is most apparent from the SOS formula (equation (7.19)): if the ground-state multiplet is not well-isolated from the excited-state manifold, the perturbation sum will not converge, the response of the system to the perturbation will become nonlinear and either SOS methods or the linear-response approach will fail. This situation will occur if the energy denominator Δ_n is not substantially larger (say, 3–5 times) than the perturbing matrix elements $\langle\Psi_0|\hat{H}_\mu|\Psi_n\rangle\langle\Psi_n|\hat{H}_\lambda|\Psi_0\rangle$. While for many systems linear response works well, there are classes of compounds that cannot be treated this way. Foremost these are the very heavy elements, where the SOC matrix elements reach several thousand wavenumbers; secondly, transition metal complexes with orbitally degenerate or nearly degenerate ground states. These systems play a particularly important role in the field of single-molecule magnetism. A comprehensive recent review has discussed how to deal with this situation.[11] In a nutshell, if there are low-lying electronic states that couple strongly to the ground state, they will dominate the magnetic response of the system. One is then well-advised to actually calculate these excited states explicitly and treat the perturbation variationally by diagonalizing a truncated version of equation (7.16). This is referred to as 'quasi-degenerate perturbation theory' (QDPT).[23]

It is important to understand that in the case of a sufficiently well-isolated electronic ground-state multiplet, QDPT and the effective Hamiltonian truncated to the same number of contributing states provide virtually indistinguishable results. Furthermore, it is also important to understand that once the QDPT matrix has been diagonalized, the definition of the SH parameters becomes more difficult, since the solution necessarily involves all possible orders of perturbation theory and, hence, any power of field and spin-operators. For example, the bilinear ZFS term is not straightforwardly separated from higher-order contributions. Gerloch and McMeeking have discussed this situation for ligand field theory.[24] In the context of quantum chemistry, the effective Hamiltonian-based techniques described by Maurice *et al.* for the ***D***-tensor[25] and Bolvin for the ***g***-matrix can be used in this case to define appropriate ***D***-tensors and ***g***-matrices.[26]

We note in passing that the so-called two-component methods of any flavor are unsatisfactory to treat problems of this kind as discussed in detail in Ref. 11. To date, the only methods that can deal properly with this type of problem are multiconfigurational wavefunction methods. In recent years, these methods have become applicable to larger molecules, for example in the form of the complete active space self-consistent field (CASSCF) method followed by a correlation treatment such as the complete active space second-order perturbation theory (CASPT2) or the second-order N-electron valence perturbation theory (NEVPT2). These methods will not be further discussed in this review.

7.3 ELECTRONIC STRUCTURE METHODS

At this point in the development, a method is required that delivers a well-defined, approximate ground-state energy E_0. The SH parameters are then defined by means of linear-response theory through the derivatives of this energy. Of the myriad of available electronic structure methods, we will only mention Hartree–Fock (HF), coupled cluster (CC), and density functional theory (DFT).

7.3.1 Hartree–Fock Theory

The complexity of the exact ground-state N-particle wavefunction is bewildering and a good ansatz for it is difficult to achieve. Hence, it is customary to take a step-by-step approach that converges toward the exact wavefunction. A convenient first step is the HF method. Here, the N-particle wavefunction is constructed as an antisymmetrized product (a Slater determinant) of one-particle wavefunctions (orbitals). The HF method consists of applying the variational principle in order to find the shapes of the orbitals that minimize the energy under the constraint that the orbitals remain orthonormal. This energy is always higher than the true ground-state energy. In fact, HF theory might be considered to be exceptionally successful as it leads to an energy that is typically around 99.8% correct. However, the remaining 0.2% of the energy (defined as 'correlation energy') is very important and amount to hundreds of kcal mol^{-1} since the total energy of an atom or molecule is a large number.

Physically speaking, the HF model views electrons as moving in the electric field of the nuclei and the average field of the other electrons. What is missing from HF theory is hence a form of 'instantaneous' electron interaction. It turns out that this interaction depends on the spin of the two electrons. For two parallel-spin electrons, the correlation is relatively weak, since already in the HF approximation they cannot come close (the 'Fermi hole'; the Fermi hole is a direct consequence of the Pauli principle, which demands antisymmetry of the many-particle wavefunction with respect to particle interchange). However, two antiparallel-spin electrons can, with finite probability, occupy the same point in space. The exact wavefunction has a cusp at the interelectronic coalescence point (the 'Coulomb cusp'). It is this form of correlation that is difficult to describe with high accuracy. It is completely missing from the HF model. Thus, since the HF method already contains a Fermi hole for parallel-spin electrons but no Coulomb hole for antiparallel-spin electrons, it is highly biased in favor of parallel-spin electrons. This has strong to disastrous consequences for the treatment of open-shell systems. The HF method will strongly overestimate the stability of high-spin states over low-spin states. It will also frequently converge to poly-radical type solutions that are unphysical. Lastly, it will also strongly overestimate spin-polarization effects that are of crucial importance for HFCs (vide infra).

Without going into too much mathematical detail, it can be observed that the orbitals in HF theory are satisfying the nonlinear HF equations. In their canonical form they read:

$$\hat{F}[\{\psi\}]\psi_i(\boldsymbol{x}) = \varepsilon_i\psi_i(\boldsymbol{x}) \tag{7.48}$$

where $i = 1, \ldots, N$ and $\hat{F}$ is the Fock operator. The quantity ε_i is the orbital energy, which, by means of Koopmans' theorem, is approximately equal to minus the ionization energy for the i'th electron. Multiplying from the left and integrating:

$$\langle\psi_i|\hat{F}|\psi_i\rangle = \langle\psi_i|\hat{h}|\psi_i\rangle + \sum_j \langle\psi_i\psi_j|\psi_i\psi_j\rangle - \langle\psi_i\psi_j|\psi_j\psi_i\rangle \tag{7.49}$$

with the two-electron integrals:

$$\langle\psi_i\psi_j|\psi_i\psi_j\rangle = \int\int \frac{\psi_i^*(\boldsymbol{r}_1)\psi_j^*(\boldsymbol{r}_2)\psi_i(\boldsymbol{r}_1)\psi_j(\boldsymbol{r}_2)}{|\boldsymbol{r}_1 - \boldsymbol{r}_2|} \mathrm{d}\boldsymbol{r}_1\mathrm{d}\boldsymbol{r}_2 \tag{7.50}$$

Thus, the Fock operator depends on solutions through the sum over j and the equations must be solved iteratively starting from some initial guess for the orbitals. It is worth mentioning that the first term in the sum over j can be more conveniently written as:

$$\sum_j \langle\psi_i\psi_j|\psi_i\psi_j\rangle = \int |\psi_i(\boldsymbol{r}_1)|^2 \underbrace{\int \frac{\rho(\boldsymbol{r}_2)}{|\boldsymbol{r}_1 - \boldsymbol{r}_2|}\mathrm{d}\boldsymbol{r}_2}_{\hat{V}_J(\boldsymbol{r}_1)}\mathrm{d}\boldsymbol{r}_1 \tag{7.51}$$

since the spin of the electrons readily integrates out and the charge density is given by:

$$\rho(\boldsymbol{r}) = \sum_j |\psi_j(\boldsymbol{r})|^2 \tag{7.52}$$

Thus, this term is the quasi-classical repulsion of the electron with the charge distribution $|\psi_j(\boldsymbol{r})|^2$ and the local potential $V_J(\boldsymbol{r}_1)$ created by the other electrons.

The exchange term, the second term in the sum in equation (7.46), is less readily interpreted. It represents the self-interaction of an 'interference density' $\psi_i^*(\boldsymbol{x}_1)\psi_j(\boldsymbol{x}_1)$. It is readily shown that the exchange integrals are always non-negative and are zero unless the spins of the two electrons are equal. Thus, there is an energy stabilization for each pair of parallel-spin electrons. This is the energetic manifestation of the Fermi hole and the spin-bias alluded to above. It is important to point out that this term does not represent a genuine 'exchange interaction', but is a consequence of electron–electron repulsion in conjunction with the antisymmetry requirement of fermionic wavefunctions. Of special interest is the term $i = j$ since in this case the sum of Coulomb and exchange terms equals zero, which cancels an unphysical self-interaction of the electrons with itself.

As a matter of fact, the HF equations, while being very much simpler than the N-particle Schrödinger equation, can still not be solved exactly given the complicated shapes of the orbitals. Hence, the orbitals must either be discretized or expanded in a set of auxiliary functions. The latter is the most common way of approaching the problem. Hence, one introduces a basis set $\{\varphi\}$ and writes:

$$\psi_i(\boldsymbol{x}) = \sum_{\mu} c_{\mu i}\varphi_{\mu}(\boldsymbol{x}) \tag{7.53}$$

Inserting this ansatz into the HF equations and solving for the unknown expansion coefficients **c** turns the HF problem into an iterative generalized matrix-eigenvalue problem that is very computer friendly. Computer programs that allow for the efficient solutions of these equations are highly developed (for a review of some of the techniques involved see Ref. 27).

If $\{\varphi\}$ is mathematically complete, equation (7.53) is an exact identity and the solutions are genuine HF solutions (the 'HF limit'). Since in practice this is not the case, one always has to live with a finite basis set and a basis-set incompleteness error. The latter is entirely unphysical which leads to the unfortunate situation that one always needs to investigate whether the chosen basis set allows for a sufficiently accurate approximation to the HF limit. For EPR spectroscopy, this can be a fairly stringent requirement as discussed below.

For open-shell systems, two different approaches can be chosen. The most straightforward approach consists of attaching a spin-factor α or β to each orbital. Then one constructs a determinant with N_α spin-up orbitals and N_β spin-down orbitals, where $N_\alpha - N_\beta = 2S$. This resulting unrestricted Hartree–Fock (UHF) determinant is an eigenfunction of $\hat{S}_z$ but not an eigenfunction of S^2. It is, however, meant to approximate the $M = S$ component of a multiplet with spin S. The expectation value of $\langle \hat{S}^2 \rangle$ is readily calculated. It should equal $S(S+1)$ for a spin-eigenfunction. The deviations from that value are taken as a measure of 'spin-contamination' (an unwanted effect). However, such deviations are always present and also describe the (desired) effect of 'spin-polarization' (vide infra). As a rule of thumb, if $\langle \hat{S}^2 \rangle$ deviates less than, say 0.2, from the desired value the calculations are usually accepted while for large deviations of, say >0.5, the calculations most certainly converged to an electronic structure that should be more correctly described by a multiconfigurational method. While such results are frequently dismissed as useless, there almost always is a sensible physical content that is best assessed via the so-called corresponding orbital transformation as elaborated in Ref. 28.

Alternatively, one can divide the electrons in the system into a closed-shell and a singly occupied part where all singly occupied orbitals are occupied with a spin-up electron (restricted open-shell Hartree–Fock treatment, ROHF). This also results in a single determinant and defines the ROHF wavefunction. The ROHF wavefunction is more difficult to optimize than the UHF wavefunction. It is a proper spin-eigenfunction but does not include any spin-polarization. Since it has fewer variational degrees of freedom, its energy is always higher than the UHF energy (it is one manifestation of Löwdin's 'symmetry dilemma' that the more correct wavefunction does provide a higher, hence inferior, energy). The ROHF formalism is well suited to also treat more complicated open-shell cases[29] (although it is rarely used in this way), while the UHF wavefunction is bound to approximate a single high-spin determinant.

7.3.2 Coupled-cluster Theory

Since the basis set is always larger than the number of electrons in the system, the solution of the HF equations yields a set of occupied orbitals $\{\psi_i\}$ and unoccupied (virtual) orbitals $\{\psi_a\}$. The problem of calculating the correlation energy can now be approached by relaxing the requirement that the N-particle wavefunction is a single determinant and construct 'excited' determinants by replacing 1, 2, ..., N orbitals at a time in the HF determinant by virtual orbitals. If this is done to completion and the variational principle applied to solve for the many-determinant wavefunction, the resulting method is called '*full configuration interaction*' (*FCI*). It represents the exact solution of the nonrelativistic Schrödinger equation *in the given basis set*. Approaching the basis set limit of the FCI method yields the exact many-particle wavefunction.

The effort to solve the FCI equations in its naïve form scales factorially with the number of electrons and hence, is completely impractical for chemical applications. Hence, there are many approximate approaches, all of which strive for approaching the FCI limit in the same basis set. Of the available approaches, arguably, CC theory[30] is the method that combines the largest number of advantages. CC theory is based on an exponential ansatz for the wavefunction

$$|\Psi_0\rangle = \exp(\hat{T})|\Psi_{\mathrm{HF}}\rangle \tag{7.54}$$

where $|\Psi_{\mathrm{HF}}\rangle$ is the Hartree–Fock determinant of the ground state (in fact, alternative choices are possible and occasionally advantageous). The 'cluster operator' $\hat{T}$ is:

$$\hat{T} = \hat{T}_1 + \hat{T}_2 + \hat{T}_3 + \ldots \tag{7.55}$$

where $\widehat{T}_n$ represents all possible n-fold excitations, i.e.,

$$\widehat{T}_1 = \sum_{i,a} t_a^i \widehat{a}_a^+ \widehat{a}_i; \quad \widehat{T}_2 = \frac{1}{4} \sum_{i,j,a,b} t_{ab}^{ij} \widehat{a}_a^+ \widehat{a}_b^+ \widehat{a}_j \widehat{a}_i \quad (7.56)$$

where the unknown 'cluster amplitudes' t_a^i, t_{ab}^{ij} must be determined by solving the CC equations.[30] If $n = N$, the CC method converges to the FCI result. Fortunately, convergence with n is very fast, owing to the exponential ansatz, and, hence, the cluster operator can be truncated at low n while still obtaining an accurate approximation to the FCI result. In practice, truncation at $n = 2$ defines the CCSD method, which is an excellent electronic-structure method. Supplementing it with a perturbative correction for the effect of $\widehat{T}_3$ defines the 'gold-standard' method of quantum chemistry, CCSD(T). Is has proven countless times to be an excellent approximation to FCI, provided that a single-determinant wavefunction is a good starting point. The computational effort to solve the CCSD equations unfortunately scales as the sixth power of molecular size and CCSD(T) scales as the seventh power. Hence, both methods can, in rigorous form, only be applied to reasonably small molecules, with perhaps 10–20 atoms. However, great progress has been made in linear-scaling approximations to CCSD(T)[31] and in the foreseeable future, these will also be available for the calculation of EPR parameters. The formalism to calculate derivatives of the CC energy is elaborate,[32,33] but well developed and, hence, there are no fundamental obstacles in applying CC theory to EPR spectroscopy.[34,35]

7.3.3 Density Functional Theory

While CC theory is a highly systematic and accurate ab initio approach to solving the Schrödinger equation, the high computational effort, for the time being, renders its large-scale, routine application still challenging. In fact, in quantum chemistry, DFT is currently the most widely used method. Note, that unlike in solid state physics, DFT is *not* considered to be a genuine ab initio method by most quantum chemists, since it cannot converge systematically towards the FCI result. Rather, DFT is a well-founded, efficient, pragmatic and highly successful approach to electronic structure.[36] It can be formally justified by the Hohenberg–Kohn theorems and the Kohn–Sham (KS) construction (for reviews in the context of molecular spectroscopy see Refs 37, 38).

The Hohenberg–Kohn theorems prove that the exact ground-state energy of a nondegenerate electronic state is a unique (but unknown) functional of just the electron density alone, i.e., $E[\rho]$. Thus, all contributions to the energy can, in principle, be expressed in terms of the density. This is immediately obvious for the electron-nuclear attraction and the Coulomb interaction. For the kinetic energy, the clever KS construction allows for the major part of it to be expressed in terms of KS orbitals. What is then missing is the exchange energy, the correlation energy and the remaining part of the kinetic energy, all of which must be expressed in terms of the density. The resulting terms are collected in the unknown exchange-correlation (XC) functional $E_{XC}[\rho]$. Functional variation of the resulting $E[\rho]$ results in some Euler–Lagrange equations that resemble the HF equations, but with the exchange term replaced by a local potential that is the functional derivative of $E_{XC}[\rho]$ with respect to $\rho(\mathbf{r})$, i.e.,

$$\left\{ \widehat{h} + \widehat{V}_J + \frac{\delta E_{XC}[\rho]}{\delta\rho(\boldsymbol{r})} \right\} \psi_i(\boldsymbol{x}) = \varepsilon_i \psi_i(\boldsymbol{x}) \quad (7.57)$$

If $E_{XC}[\rho]$ were known exactly, the KS method would provide the exact energy. Since $E_{XC}[\rho]$ is not known, one has to approximate it. There have been hundreds if not thousands of proposals on how to accomplish this. The principles behind the construction of approximate density functionals range from highly formal reasoning to highly parameterized semi-empirical approaches. We cannot hope to review this large field here, but rather provide a crude overview:

One of the earliest DFT methods that was applied to chemistry was invented by Slater. It did not have any correlation term, but was introduced as a simplification of the HF method, even prior to the formulation of the HK theorems, using an exchange potential proportional to $\rho^{1/3}$ ('Slater exchange'). The resulting method is called X_α and is only of historical value. Most, but not all, functionals in existence make some reference to the homogeneous electron gas as a starting point. Combining Slater exchange with local density correlation defines the 'local density' approximation (LDA), still a major workhorse of solid state physics but rarely used in chemistry. A major step forward was the introduction of terms depending on the gradient of the density ('generalized gradient approximation', GGA). These are still widely used functionals with typical members being the BP86, PBE, or BLYP. Adding the kinetic energy density defines meta-GGA functionals, with TPPS being the most popular choice. These functionals are strictly local. In 1993, Becke showed that a substantial improvement resulted from mixing in some fraction of the HF exchange, thus defining hybrid functionals.[39] The major workhorse for chemistry, the B3LYP functional, is of that kind, as is PBE0 or TPSSh. The functionals of the Minnesota group are based on extensive parameterization and have recently gained popularity in chemistry, most noticeable M06L and M06-2X. Grimme has proposed to also mix in a fraction of wavefunction-like perturbative correlation, this defining 'double hybrid' functionals,[40] with the first and prototypical member being B2PLYP.

We close this section with a few comments:

1. The density functionals in existence are essentially all constructed toward aiming for good total energies that most closely resemble experiment or highly accurate wavefunction energies. However, even if they deliver excellent energies, it is not guaranteed that they also deliver excellent molecular properties. While for wavefunction methods such as CC all properties, including the total energy, co-converge toward the FCI result as the cluster operator is taken to higher orders and the one-particle basis set is reaching completion, the same is not true for DFT. Hence, there is an enormous literature that is concerned with statistical evaluations of which functionals perform well for a given property.
2. Secondly, it should be noted that by replacing the HF exchange partially or completely by an approximate exchange, the self-interaction terms, alluded to above, will no longer cancel. Hence, all present-day DFT methods (except those based on the optimized effective potential[41,42] or Görling's construction[43]) are inflicted with a non-negligible self-interaction error. This error must be partially cancelled by the correlation functional. To which extent this is physical and systematic is open to debate.
3. The effective Kohn–Sham potentials of present-day functionals show wrong asymptotic behavior close to the nucleus as well as at long range. The former affects EPR properties while the latter adversely affects results for weak intermolecular interactions. The latter can be partially cured by the hugely successful Grimme correction for the dispersion energy.[44,45] The latter does not change the KS wavefunction and, hence, has no effect on computed EPR properties.
4. The consistent formulation of DFT in the presence of a magnetic field is an unsolved problem, which is still under investigation.[46]
5. No consistent and generally accepted multiconfigurational DFT theory has yet emerged. Hence, only states that are conceptually well described by a single determinant are also expected to be well treated by DFT.
6. The majority of functionals are constructed by using insights obtained from modeling the inhomogeneous electron gas. It is not clear whether a functional that models the inhomogeneous electron gas perfectly would be successful in chemistry. It would certainly *not* coincide with FCI. Given their common origin, most functionals tend to have the same flaws and, hence, the confidence gained by applying a wide variety of functionals to a given problem and obtaining similar results may not be fully justified.

7.4 ILLUSTRATION: HYPERFINE COUPLINGS

7.4.1 Introduction: Spin Density Versus Spin Population

In this section, we will discuss some aspects that concern the calculation of HFCs as a representative SH parameter and will also focus on the first-order contributions. While each and every other term in the SH would deserve a careful and detailed treatment in its own right, the available space allows only for the discussion of one particular subject.

We start from the general expression for the spin-dipolar part of the HFC:

$$A_{KL}^{(\mathrm{SD};A)} = \frac{P^{(A)}}{S}\left\langle \Psi_0^{SS} \left| \sum_i \hat{s}_{z;i} r_{iA}^{-5} \left\{ \delta_{KL} r_{iA}^2 - 3 r_{iA;K} r_{iA;L} \right\} \right| \Psi_0^{SS} \right\rangle \tag{7.58}$$

$$P^{(A)} \equiv g_e g_N^{(A)} \mu_B \mu_N$$

Note that the integral contains the operator $\hat{s}_{z;i}$. For all integrals of the form $\langle \Psi_0^{SS} | \sum_i \hat{s}_{z;i} \hat{o}(\boldsymbol{r}_i) | \Psi_0^{SS} \rangle$ with a spin-independent one-electron operator $\hat{o}(\boldsymbol{r}_i)$, these simply give:

$$\left\langle \Psi_0^{SS} \left| \sum_i \hat{s}_{z;i} \hat{o}\left(\boldsymbol{r}_i\right) \right| \Psi_0^{SS} \right\rangle = \frac{1}{2} \sum_{pq} D_{pq}^{\alpha-\beta} \langle \psi_p | \hat{o} | \psi_q \rangle \tag{7.59}$$

with the spin-density matrix:

$$D_{pq}^{\alpha-\beta} = \left\langle \Psi_0^{SS} | \sum_{pq} \left(\hat{a}_{p\alpha}^{+} \hat{a}_{q\alpha} - \hat{a}_{p\beta}^{+} \hat{a}_{q\beta} \right) | \Psi_0^{SS} \right\rangle \quad (7.60)$$

Thus, the trace of the spin-density matrix equals $2S$.

Associated with the spin-density matrix is the spin-density distribution function $\rho^{\alpha-\beta}(\boldsymbol{r})$. This is simply a function of three-dimensional space that can be positive or negative and integrates to $2S$. For a single-determinant KS or HF determinant, the spin-density distribution function is:

$$\rho_{\mathrm{HF/KS}}^{\alpha-\beta}(\boldsymbol{r}) = \sum_{\mu\nu} D_{\mu\nu}^{\alpha-\beta} \varphi_{\mu}(\boldsymbol{r}) \varphi_{\nu}(\boldsymbol{r}) \quad (7.61)$$

where the spin-density matrix has been expressed in terms of basis functions. For a spin-unrestricted treatment:

$$\begin{aligned} \rho_{\mathrm{HF/KS}}^{\alpha-\beta}(\boldsymbol{r}) &= \sum_{i_\alpha} |\psi_i^{\alpha}(\boldsymbol{r})|^2 - \sum_{i_\beta} |\psi_i^{\beta}(\boldsymbol{r})|^2 \\ &= \sum_{\mu\nu} D_{\mu\nu}^{\alpha-\beta} \varphi_{\mu}(\boldsymbol{r}) \varphi_{\nu}(\boldsymbol{r}) \quad (7.62) \end{aligned}$$

The integral over the dipole operator involves the inverse third power of the electron–nuclear distance (r^{-3}), while the volume element in spherical coordinates involves r^2. Hence, the integrals are potentially divergent. For *s*-functions, there is no additional power of *r* entering the integral from the wavefunction and, hence, for such orbitals the integral is potentially divergent. However, it is readily shown using integration by parts that there arises a boundary term at $r=0$ that is $\frac{4\pi}{3}\delta(r)$ and hence, the Fermi-contact term arises naturally from the spin–nucleus dipole–dipole interaction. It does not represent a genuinely new physical interaction.

In the early history of EPR spectroscopy, relationships have been established between the molecular orbital coefficients of singly occupied molecular orbitals (SOMOs) and values of HFCs.[47] Most famous among the many empirical relationships is the very successful McConnell relation that relates isotropic HFCs on hydrogen atoms at planar aromatic rings to squares of MO coefficients from a Hückel treatment ($A_{\mathrm{iso}}^{(\mathrm{H})} = \rho_{\pi,\mathrm{C}} Q$, where $\rho_{\pi,\mathrm{C}} = C_{\mu,\mathrm{C}}^2$ is the Hückel spin population in the SOMO and $Q = -22.5$ G is an empirical constant). Ever since the widespread use of such relationships, it has been common practice to use the terms spin population and spin density almost interchangeably.

However, it is important to conceptually differentiate the spin density (a three-dimensional function of space and a physical observable) from the spin population (a more or less arbitrary division and integration of spin density onto atoms or fragments and not a physical observable).

In order to appreciate the difference, please consider the spin-density distribution of a phenoxyl radical as shown in Figure 7.1. In Table 7.1, we provide spin populations predicted by a few popular analysis schemes (Mulliken,[48] Löwdin, natural population analysis,[49] and Hirshfeld[50]). It is evident that there are noticeable

Table 7.1. Spin populations (percent) of the derivatized phenoxyl radical as predicted by various analysis schemes (NPA = natural population analysis)

	Mulliken	Löwdin	Hirshfeld	NPA
C1	−13.2	−7.3	−3.5	−11.5
C2	28.2	22.2	18.7	26.2
C3	−06.7	−1.4	2.4	−6.2
C4	28.2	22.3	18.7	26.2
C5	−13.2	−7.3	−3.5	−11.5
C6	36.4	28.8	23.9	34.0
C7	−1.8	0.6	1.6	−1.5
C8	2.1	2.2	2.0	2.1
C9	40.5	39.1	36.9	41.7

The calculations were done with the B3LYP functional and the EPRII basis set at the BBE-D3(BJ)/def2-TZVP geometry.

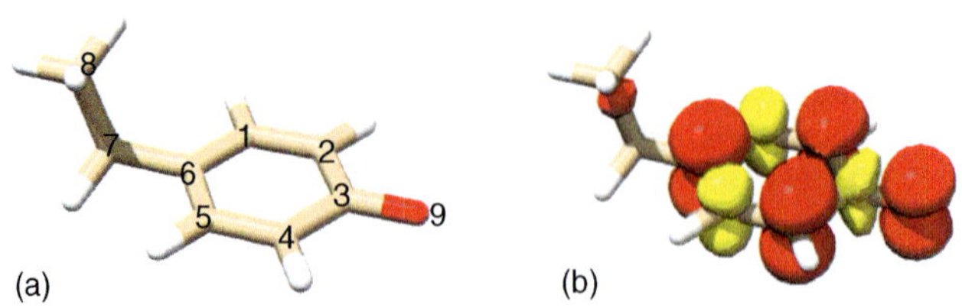

Figure 7.1. (a) Numbering scheme for a derivatized phenoxyl radical. (b) Spin-density distribution contoured at ±0.003 electrons/a_0 (red = positive spin-density, yellow = negative spin-density)

differences (10% and more) in the spin populations delivered by the various analysis schemes despite the fact that all take the same physical spin-density function as input. For the small spin populations not even the sign is unanimous amongst the various schemes. Hence, taking these values as guides for the signs of HFCs would produce inconsistent results while the direct calculation of the HFCs is unambiguous. Hence, trends and variations in these numbers across chemistry are certainly helpful (provided the same functional, basis set, and analysis scheme is used) but it should be realized that there is no deep fundamental physical significance in the spin populations.

A more subtle case has been analyzed a few years ago.[51] Many Cu(II) complexes display prominent HFC to nitrogen-based ligands. Depending on the chemical nature of the nitrogen ligand, the donor orbital maybe a formally sp^2 or sp^3 hybridized lone pair. The former has a higher s-contribution than the latter and, hence, qualitatively speaking, the ratio of isotropic to dipolar HFC will be higher for sp^2 nitrogens than for sp^3 nitrogens. It has often been attempted to deduce the hybridization state from the EPR data. However, the analysis demonstrated that these attempts are misleading since the actual HFCs are: (i) inflicted by spin polarization effects that change the formal hybridization ratio and (ii) the isotropic and dipolar HFCs show a poor correlation with the spin populations in the nitrogen s- and p-orbitals (Figure 7.2). The reason for the latter observation is that the populations make no reference to the actual radial distribution function of the orbital that contributes to the spin density. However, the latter changes with the chemical environment. Hence, the simple picture breaks down.

All of these results show that in EPR spectroscopy one should not rely on spin populations if the aim is to calculate HFCs but there is no substitute for considering the actual spin-density distribution. Correlations involving spin populations have a qualitative value and can be fruitfully used to interpret chemical trends. However, care should be taken not to over-interpret these numbers.

7.4.2 The Isotropic Hyperfine Coupling

7.4.2.1 Spin Polarization

If the electronic structure of a radical were perfectly well described by a single ROHF Slater determinant, the spin-density distribution would be precisely defined by the square of the SOMO. In this case, the spin density would be positive everywhere in space and no negative isotropic HFCs could arise (for isotopes that have a positive value of $P^{(A)}$). However, negative isotropic HFCs are commonplace in EPR spectroscopy, with the textbook example being given by the methyl radial. The radical is planar and the SOMO consists of the p_z orbital of the central carbon. Hence, there is

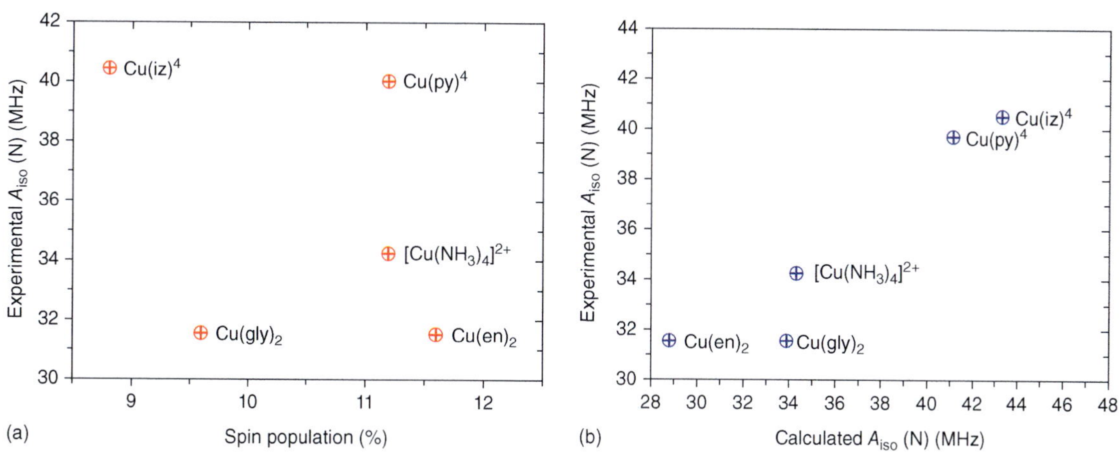

Figure 7.2. (a) Lack of correlation between total s-orbital spin population and hyperfine coupling for nitrogen in a series of square-planar Cu(II)N_4 complexes. (b) Good correlation between calculated and experimental HFCs for the same series of complexes, thus demonstrating that the HFC is not determined by the spin population (iz = imidazole, py = pyridine, en = ethylene-diamine, gly = glycine)

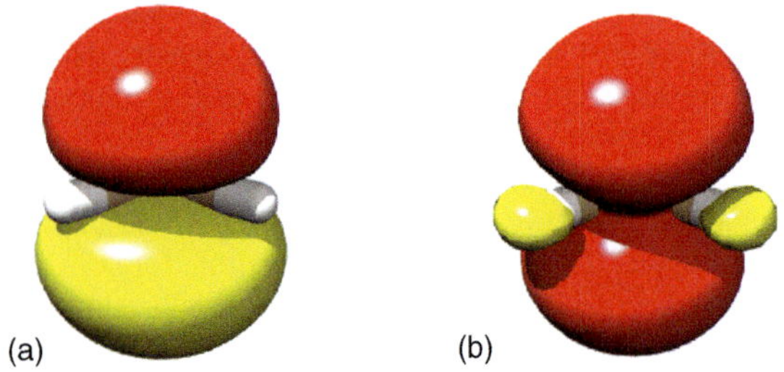

Figure 7.3. (a) Singly occupied molecular orbital of the CH_3 radical contoured at ± 0.05 (electrons/a_0^3)$^{1/2}$. (b) Spin density of the CH_3 radical contoured at ± 0.003 electrons/a_0^3. Red = positive, yellow = negative. It is evident that there is negative spin density at the hydrogen positions although the SOMO has no amplitude there. (B3LYP/cc-pCVTZ)

no amplitude of the SOMO at the hydrogen positions. Nevertheless, experiment shows that there is negative spin-density at the hydrogens, despite the fact that the hydrogens sit in a node of the SOMO (Figure 7.3).

The same holds for hydrogens in aromatic radicals. The phenomenon was early on attributed by McConnell and Chesnut to 'spin polarization'.[52] The latter arises quite naturally in the spin-unrestricted formalism through the separate potentials felt by the spin-up and spin-down electrons. However, as pointed out above, spin polarization in the UHF formalism necessarily also comes with spin contamination, and to separate the desired from the undesired effect is not straightforward. We will see below that the UHF method itself drastically overestimates the spin polarization. One might then ask the question how the spin polarization arises in a more rigorous treatment than the UHF method.

In order to approach the problem, consider a simple model system that has an unpaired electron in a π-orbital, a lower-lying, nearby C—H σ-bonding orbital and an empty σ^*_{CH} orbital. The ground-state configuration is then simply $(\sigma_{CH})^2(\pi)^1(\sigma^*_{CH})^0$. One can perform three types of single excitations on this model system, namely $\sigma_{CH}\to\pi$, $\pi\to\pi^*$ and $\sigma_{CH}\to\sigma^*_{CH}$. The first two excitations still lead to single-determinantal wavefunctions. Their mixing with the ground state is very small due to Brillouin's theorem. However, the third excitation leads to three unpaired electrons and two spin-doublet configuration state functions:

$$|\Psi^{(a)}_{\sigma\to\sigma^*}\rangle = \frac{1}{\sqrt{2}}\{|\sigma_{CH}\pi\bar{\sigma}^*_{CH}| - |\bar{\sigma}_{CH}\pi\sigma^*_{CH}|\} \qquad (7.63)$$

$$|\Psi^{(b)}_{\sigma\to\sigma^*}\rangle = \frac{1}{\sqrt{6}}\{|\sigma_{CH}\pi\bar{\sigma}^*_{CH}| + |\bar{\sigma}_{CH}\pi\sigma^*_{CH}| + 2|\sigma_{CH}\bar{\pi}\sigma^*_{CH}|\} \qquad (7.64)$$

(the | … | denotes a Slater determinant and the overbar indicated occupation with a spin-down electron). Of these, $|\Psi^{(a)}_{\sigma\to\sigma^*}\rangle$ will also not interact directly with the ground-state determinant due to Brillouin's theorem. However, $|\Psi^{(b)}_{\sigma\to\sigma^*}\rangle$ corresponds to triplet coupling of the excited electron congruent with a spin-flip in the SOMO (a so-called 'trip-doublet' function). This function contains a doubly excited determinant relative to the ground state and it will interact with it via a matrix element $\frac{1}{2}\sqrt{6}\langle\sigma_{CH}\pi|\pi\sigma^*_{CH}\rangle$. Evaluation of the spin-density operator of the admixed ground state then provides the actual spin-density function, which is readily shown to feature negative spin density at the hydrogen positions. Thus, a successful ab initio treatment needs to incorporate the trip-doublet excitations in a balanced manner, which will then bring in the spin polarization without spoiling the nature of the wavefunction as a spin-eigenfunction. However, it will not be sufficient to just consider the trip-doublets in a configuration expansion. The reason is that the orbitals are optimized for the ground state and, hence, the trip-doublets with an electron in a σ^* orbital will be much too high in energy and consequently mix too little with the ground state. Hence, electronic relaxation needs to be present in form of more highly excited determinants that provide a balanced treatment of the differential correlation between the ground state and the trip-doublets. This is not easy to achieve by means of linear-configuration expansions such as the restricted open-shell configuration interaction method with single excitations (ROCIS[53,54]) that explicitly includes such excitations. However, CC theory appears to be fairly successful in this respect. This is apparently not just due to the presence of the $\exp(\hat{T}_1)$ (which provides the main part of the electronic relaxation), since the simpler 'quadratic configuration interaction' (QCISD) method also provides excellent HFCs and it contains only up to quadratic terms in the cluster amplitudes.[55]

7.4.2.2 *Some Numerical Illustrations*

It is instructive to look at the simple example of the H_2O^+ cation radical. Its electronic structure is shown in Figure 7.4.

From the molecular orbital scheme it is evident that the SOMO is a nonbonding oxygen lone pair of

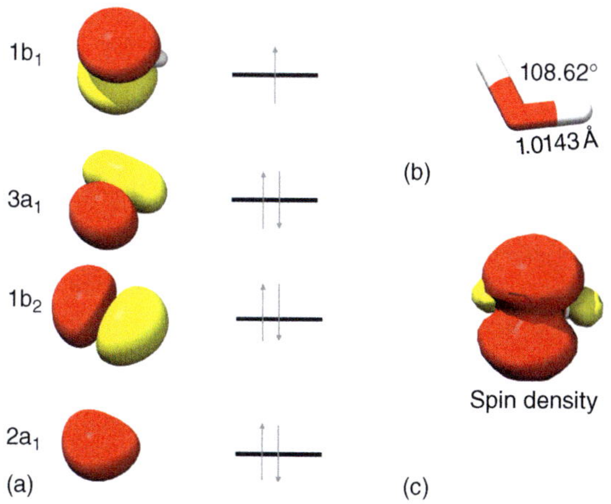

Figure 7.4. Electronic structure of the H_2O^+ cation radical in its $^2B_1$ ground state. (a) Molecular orbital diagram of valence orbitals. (b) Optimized structure, (c) spin-density plot (contour levels for orbitals ±0.05 (electrons/a_0^3)$^{1/2}$ and for spin densities ±0.003 electrons/a_0^3)

Table 7.2. CCSD results for the isotropic hyperfine couplings of the H_2O^+ radical in the $^2B_1$ ground state (in MHz) (uc = s-functions have been uncontracted; all other basis sets are standard basis sets)

	Frozen core	A_{iso}(H)	A_{iso}(O)
cc-pVDZ	No	−75.25	−158.66
cc-pVTZ	No	−77.46	−25.77
cc-pVQZ	No	−74.36	−39.76
cc-pCVDZ	Yes	−75.25	−49.55
cc-pCVTZ	Yes	−77.18	−75.89
cc-pCVQZ	Yes	−74.27	−77.22
cc-pCVDZ	No	−75.24	−50.53
cc-pCVTZ	No	−77.29	−74.79
cc-pCVQZ	No	−74.43	−77.82
uc-cc-pCVDZ	No	−71.28	−78.59
uc-cc-pCVTZ	No	−73.11	−77.31
uc-cc-pCVQZ	No	−74.37	−77.31
Exp.[56]	—	−73	−83

p-character (MO 1b$_1$) that has a node in the H—O—H plane. Hence, there is no direct spin density on the hydrogen atoms and no s-spin density at the oxygen atom. Hence, all isotropic HFC in this radical must arise from spin polarization. For the hydrogen nuclei, this is the valence-shell spin polarization. For the oxygen nucleus, it is an interplay of valence-shell spin polarization (providing spin density in the 2s orbital) and core-level spin polarization (providing spin density in the 1s orbital). It is clearly seen in the spin-density plot that negative spin density (shown in yellow) is induced at the hydrogen positions. The spin density at the oxygen nucleus is somewhat covered, but from the negative sign of the calculated HFCs, it is clear that it is positive ($P^{(O)} = -72.3588$ MHz/a_0^3).

The first thing that we investigate is the dependence of the results on the basis set using as a reference result the CCSD method (Table 7.2). In order to be systematic, we use the series of correlation-consistent basis sets of double- through quadruple-ζ quality.[57] The first thing that is observed is that the hydrogen HFCs are reasonably insensitive to the basis set and all results scatter around −74 ± 3 MHz. For the oxygen nucleus, the results with the standard cc-pVnZ basis sets scatter erratically between −25 and −150 MHz and, hence, these basis sets are completely useless to calculate this property. What is necessary is to take proper account of the core-level spin polarization. For correlated ab initio methods, this requires to add some tight polarization functions and, hence, perhaps also to supply added flexibility in the s-basis functions. Indeed, if one adds such functions, as in the cc-pCV*n*Z series, the results immediately improve. For the double-ζ basis sets, the results remain poor but as soon as one supplies at least a triple-ζ basis set with core-polarization functions, the results have been observed to stabilize to a final value of around −75 to −77 MHz. For the double-ζ basis sets, the results become acceptable after decontracting the s-functions (in the uc-cc-pCV*n*Z series). Perhaps somewhat surprisingly, neither the explicit inclusion of the core electrons in the calculation nor the decontraction of the s-functions has a strong effect for the triple- and quadruple-ζ calculations that all provide nearly converged results. The final best numbers of −74.4 MHz for the hydrogen HFC and −77.3 MHz for the oxygen HFC are in reasonably good agreement with the experimental numbers of −73 MHz for hydrogen and −83 MHz for oxygen.[56] Further improvement would require careful vibrational averaging.

Having established the suitability of the CCSD results, we can now test how well a range of selected density functionals reproduce the ab-initio results (Table 7.3). The first result that is immediately evident is the extremely poor performance of the UHF method. This is testimony to the strong overestimation of spin-polarization effects in the UHF method and shows that electron correlation goes a long way in

Table 7.3. Comparison of CCSD results for H_2O^+ radical with UHF and a range of DFT methods (EPR-II basis set; all results in MHz)

	$A_{iso}(H)$	$A_{iso}(O)$
CCSD	−84.3	−78.5
UHF	−120.8	−140.5
LDA	−75.5	7.0
BP86	−67.5	−22.0
PBE	−65.3	−37.1
TPSS	−66.7	−35.5
TPSSh	−70.6	−44.4
B3LYP	−72.4	−57.1
M06L	−95.6	−194.9
M06-2X	−82.2	−134.1
B2PLYP	−71.7	−76.1
CAM-B3LYP	−71.1	−49.7
Exp.[56]	−73	−83

repairing this deficiency of HF theory. For the hydrogen HFC, most DFT methods are reasonable but underestimate the valence-shell spin polarization by about 10%, which appears to be acceptable. Exceptions are the M06L functional that overestimates the hydrogen HFC (in absolute value) and the M06-2X functional that is spot on with the CCSD results. The results for the more problematic oxygen HFC scatter wildly. The LDA method provides a completely useless result of the wrong sign and the Minnesota functionals show a strong overestimation by a factor of 2–3. All 'pure' functionals underestimate the oxygen HFC, which is quite typical of DFT functionals. Adding HF exchange improves the results. However, this is probably mostly due to the fact that the HF method itself overestimates spin polarization strongly. Hence, a DFT error cancels partially with a HF error in this case. Among the tested methods, the B2PLYP results appear to be the best and the B3LYP results appear to be acceptable. None of the other functionals can really be recommended for such studies. It is clear that these are isolated results on a single molecule that should not be over-interpreted. Nevertheless, it is quite typical of the difficulties one meets in applying DFT to EPR spectroscopy. Unconditional trust in calculated HFCs is definitely not warranted.

In order to assess the importance of relativistic effects, Table 7.4 collects some B3LYP calculations on alkali atoms using a large, uncontracted basis set. These systems have one unpaired electron in their valence s-orbital. Hence, their HFCs arise mostly directly from the SOMO and less directly from core-level spin polarization. In these calculations, the DKH method is used to incorporate scalar relativistic effects into the calculations. However, it is immediately obvious from the results that it is not enough to merely include the DKH scalar relativistic potential. The latter is known to diverge at the nucleus with a point nucleus.[1] Hence the relativistic orbitals will also show a weak divergence and the result is a numerically unstable result that will strongly overestimate the HFCs. In order to come to a proper scalar relativistic result, one needs to either back-transform the scalar relativistic wavefunction into the four-component picture or transform the property operator to the scalar relativistic framework.[21,59–62] Secondly, the calculations need to incorporate a nucleus of finite size in order to prevent the divergence inherent in a relativistic treatment with a point nucleus. The results collected in Table 7.4, show that the relativistic effects are below 2% for H and Li but already lead to sizeable changes for Na and K. For Cs it is absolutely necessary to incorporate relativistic effects. Table 7.4 also demonstrates that most of the improvement comes from incorporating picture-change effects while the finite-nucleus model only leads to modest changes in the results, except for Cs where apparently it worsens the agreement with experiment. The agreement with experiment is reasonable but not spectacular which is mostly blamed on the shortcomings of the B3LYP method rather than a shortcoming in the relativistic treatment.

Since neither the incorporation of the DKH correction nor the calculation of the picture-change effect is computationally expensive, it can be generally recommended for HFC calculations on molecules that involve elements beyond Ne.

It should be noted that the importance of picture-change effects is quite different for different relativistic methods since, in general, the orbital shapes delivered by different scalar relativistic approaches can be fairly different, in particular close to the nucleus. To illustrate this effect, we have collected the same data for the series of alkali atoms in conjunction with the ZORA relativistic correction in Table 7.5. The numbers indicate that the picture-change effects are somewhat smaller with the ZORA method compared to DKH but neither method appears to be uniformly more accurate than the other.

Table 7.4. Scalar relativistic effects on the hyperfine couplings of the alkali atoms using the DKH2 relativistic correction, the B3LYP functional and the decontracted def2-QZVPP basis set

Atom	Nonrelativistic	+DKH2	+Picture Change	+Finite Nucleus	Experiment
H	1414.9	1416.4	1413.1	1413.1	1416.8
Li	448.3	458.0	440.4	440.4	401.8
Na	966.7	1173.4	914.8	914.7	885.8
K	234.3	370.0	220.6	220.4	230.9
Rb	861.6	1993.8	874.9	873.7	1011.9
Cs	1617.7	14087.7	2082.6	1900.7	2183.3

All numbers in MHz.
Experimental numbers quoted from Ref. 58.

Table 7.5. Scalar relativistic effects on the hyperfine couplings of the alkali atoms using the ZORA relativistic correction, the B3LYP functional and the decontracted def2-QZVPP basis set

Atom	Nonrelativistic	+ZORA	+Picture change	Experiment
H	1414.9	1415.2	1415.0	1416.8
Li	448.3	449.4	448.5	401.8
Na	966.7	1002.9	977.3	885.8
K	234.3	263.5	243.3	230.9
Rb	861.6	1228.6	988.1	1011.9
Cs	1617.6	4954.8	2389.1	2183.3

All numbers in MHz.

7.4.3 The Dipolar Hyperfine Coupling

Having established the principles that determine the isotropic HFCs, we now turn our attention to the calculation of the dipolar part of the hyperfine interaction. We make use of the fact that the spin density can be expressed in terms of atom-centered basis functions. For the HFC of nucleus C:

$$A_{KL}^{(C)} = \frac{P^{(C)}}{S} \sum_{\mu_A \nu_B} D_{\mu_A \nu_B}^{\alpha-\beta} \langle \mu_A | r_C^{-5} \{ \delta_{KL} r_C^2 - 3 r_{C;K} r_{C;L} \} | \nu_B \rangle \tag{7.65}$$

where μ_A is short for basis function φ_μ attached to atom A. Since A, B, C can be all different centers, there are contributions involving 1-, 2-, and 3-centers (PC = 'point charge', BO = 'bond').

$$A_{KL}^{(C-1c)} = \frac{P^{(C)}}{S} \sum_{\mu_C \nu_C} D_{\mu_C \nu_C}^{\alpha-\beta} \times \langle \mu_C | r_C^{-5} \{ \delta_{KL} r_C^2 - 3 r_{C;K} r_{C;L} \} | \nu_C \rangle \tag{7.66}$$

$$A_{KL}^{(C-2cPC)} = \frac{P^{(C)}}{S} \sum_{\mu_{A\neq C} \nu_{A\neq C}} D_{\mu_A \nu_A}^{\alpha-\beta} \times \langle \mu_A | r_C^{-5} \{ \delta_{KL} r_C^2 - 3 r_{C;K} r_{C;L} \} | \nu_A \rangle \tag{7.67}$$

$$A_{KL}^{(C-2cBO)} = \frac{2P^{(C)}}{S} \sum_{\mu_C \nu_{A\neq C}} D_{\mu_C \nu_A}^{\alpha-\beta} \times \langle \mu_C | r_C^{-5} \{ \delta_{KL} r_C^2 - 3 r_{C;K} r_{C;L} \} | \nu_A \rangle \tag{7.68}$$

$$A_{KL}^{(C-3c)} = \frac{2P^{(C)}}{S} \sum_{\mu_{A\neq B,C} \nu_{B\neq A,C}} D_{\mu_A \nu_B}^{\alpha-\beta} \times \langle \mu_A | r_C^{-5} \{ \delta_{KL} r_C^2 - 3 r_{C;K} r_{C;L} \} | \nu_B \rangle \tag{7.69}$$

These contributions to the dipolar HFC are physically quite distinct.

The most prominent terms are of one-center nature, i.e., $A_{KL}^{(C-1c)}$. This is to be expected as the HFC operator depends on the inverse third power of the distance of the electron from nucleus C. Thus, unpaired electrons located in basis functions on atoms that are far away from atom C cannot contribute to the HFC. That is the reason why the 3-center terms are generally considered to be negligible.

For illustrative purposes these terms have been collected for the simple case of the ^{14}N-HFCs in the $[Cu(NH_3)_4]^{2+}$ transition metal complex in Table 7.6. The data in the Table support the general notion that the one-center terms are the largest contributions to the HFC and the three-center terms are negligible. However, it is probably not commonly appreciated that two-center bond terms even exist. These illustrative calculations show that they can be of the same order of magnitude as the more commonly anticipated point-charge terms that will be discussed later. In this particular example, the bond and point-charge

Table 7.6. One, two, and three center contributions to the calculated ^{14}N-HFCs (in MHz) in $[Cu(NH_3)_4]^{2+}$ using the B3LYP functional

	A^N_{iso}	A^N_{max}	A^N_{mid}	A^N_{min}
1-Center	34.96	−4.93	−4.95	9.87
2-Center point charge	−0.02	−0.68	−0.53	1.21
2-Center bond	2.76	+0.47	+0.45	−0.91
3-Center	0.06	<0.01	<0.01	<0.01
Total	37.76	−5.09	−5.03	10.12

From Ref. 51.

contributions almost cancel each other. However, this is not a general conclusion and the relative signs of the point-charge and bond contributions certainly depend on the geometry of the particular system.

Physically speaking, the one-center terms mostly reflect the spin-density distribution around the center in question. For example, for spin-density in a p-orbital the integral evaluates to $+\frac{4}{5}\langle r^{-3}\rangle_{\mu\nu}\frac{P^{(C)}}{S}$ along the lobe of the p-orbital and $-\frac{2}{5}\langle r^{-3}\rangle_{\mu\nu}\frac{P^{(C)}}{S}$ perpendicular to it. With reasonable estimates of $\langle r^{-3}\rangle$ and tabulated values for $P^{(C)}$ one can deduce the spin population in the p-orbitals of a given atom. However, there are variations in $\langle r^{-3}\rangle$ with the specific bonding interaction and, hence, tabulated atomic values for $\langle r^{-3}\rangle$ have a limited range of validity. Nevertheless, assuming that all multicenter terms are zero (with reference to the fast decay of the r^{-3} operator) has been a cornerstone in EPR theory for a long time.

The two-center bond contribution arises from the interaction of the unpaired spin-density in the bonds around the nucleus in question. They will fall off exponentially with the distance of atom B relative to C because this is the decay rate of the differential overlap of basis functions μ_C and ν_B. Hence, contributions are essentially limited to the nearest neighbors of atom C. Conversely, if one is discussing the anisotropic HFC of an atom C that is in the vicinity of another atom that carries significant spin, the two-center bond terms cannot be neglected. In transition metal complexes this certainly concerns the HFCs of the directly coordinated atoms.

The two-center point-charge contribution is perhaps more interesting since it maps onto the widely used point-dipole approximation. For atoms B that are sufficiently far from atom C, the detailed shape of the orbitals that carry the spin becomes rather irrelevant and one may condense the basis functions to simple Dirac-Delta functions, thus obtaining:

$$A_{KL}^{(A-2cPD)} = \frac{P^{(A)}}{S}\sum_B Q_B R_{AB}^{-5}\left\{\delta_{KL}R_{AB}^2 - 2R_K^{AB}R_L^{AB}\right\} \tag{7.70}$$

where $\boldsymbol{R}^{AB} = \boldsymbol{R}_A - \boldsymbol{R}_B$ is the distance vector between atoms A and B with magnitude R_{AB}. Q_B is the one-center spin-population on atom B ($Q_B = \sum_{\mu_B\nu_B} D^{\alpha-\beta}_{\mu_B\nu_B}$). It is not commonly appreciated that this value is different from the spin population delivered by any population analysis scheme. The reason is that any population analysis scheme contains contributions from two-center terms in the spin-density matrix $\boldsymbol{D}^{\alpha-\beta}$ while the summation that leads to Q_B is strictly confined to one-center terms. The differences may not be overwhelmingly large but this fact should be considered upon applying point-dipole approximations to the analysis of HFCs in conjunction with quantum chemical population analysis.

The second important question to consider is how far is far enough for a point-dipole approximation to work well? This question is answered by a model calculation involving a hydrogen 1s orbital and a metal d-orbital in Figure 7.5. It is observed that already at about 2 Å the agreement between the point-dipole formula and the actual quantum chemical calculation is pretty good. Thus, if nucleus C is more than, say, 3 Å

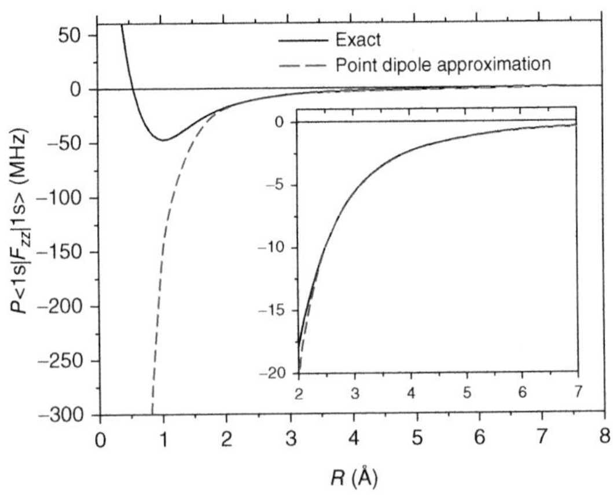

Figure 7.5. Validity of the point-dipole approximation for a model system consisting of a metal d-orbital and a hydrogen 1s orbital. The black line is the correct quantum mechanical calculation and the broken red line represents the point-dipole prediction

away from any center with significant spin population, then the point-dipole approximation can be safely used. However, it should not be forgotten that a transition metal complex is *not* well approximated by a picture where *all* spin population is located at the metal. There will be non-negligible spin population on the directly coordinating atoms. For example, for a water ligand, the oxygen will carry some spin population. It will be much smaller than that of the metal center. However, its contributions to the hydrogen HFC will be strongly amplified by the R^{-3} operator that strongly enhances the contributions of nearby atoms. In such a situation the uncritical use of the point-dipole approximation cannot be recommended, at least if more than a quite rough estimate is the purpose of the investigation.

A final but important remark concerns the common practice of incorporating the $\boldsymbol{g}$-matrix in using equation (7.70). For a transition metal complex, this equation might be expressed as:[63] $\boldsymbol{A}^A = A_{\text{iso}}\mathbf{1} + \boldsymbol{T} + \boldsymbol{T}\Delta\boldsymbol{g}/g_e$ where $\boldsymbol{T}$ is a point-dipole tensor and it is assumed that the ligand carries no spin-density on its own and all spin resides on the metal. The plausible physical argument is that this represents the full magnetic dipole moment of the unpaired electrons that is interacting with the nuclear spin. However, from the point of view of first-principles theory, this treatment is not correct. The $\boldsymbol{g}$-matrix provides the linear response to a magnetic field, which should not be intermingled with the HFC term that represents a dipole–dipole interaction. At least to second order, the point-dipole formula should only contain the isotropic, free-electron g-value, and not the full $\boldsymbol{g}$-matrix. If one writes this in terms of linear response theory, this would correspond to a fourth derivative property $\frac{\partial^4 E}{\partial B \partial S^2 \partial I^{(A)}}$. It is evident that such contributions must exist in the SH, but they would not show up until a fourth-order treatment is pursued and would then be proportional to $(\boldsymbol{BS})(\boldsymbol{I}^A\boldsymbol{S})$. Thus, it appears to be impossible to derive a term of the form $\boldsymbol{S}\boldsymbol{A}^A\boldsymbol{I}^{(A)}$ with $\boldsymbol{A}^A$ containing a contribution of the form $A_{\text{iso}}\mathbf{1} + \boldsymbol{T} + \boldsymbol{T}\Delta\boldsymbol{g}/g_e$ by second-order perturbation or linear response theory. It also does not represent the limiting form of any electronic structure approximation that this author is aware of. Hence, the use of this equation is considered one instance of breaking the logical connection between the SH and first-principles physics as discussed in Section 7.2.1. This should not be confused with the SOC contribution to the HFC, which, under certain approximations,[10] takes the form $\boldsymbol{A}^{(\text{SOC};A)} \approx \mu_B g_e \mu_N^{(A)} g_N^{(A)} \Delta\boldsymbol{g}$. However, as discussed in the literature[10,51] this only holds for the metal nucleus (where it is also not a particularly good approximation[18]), but not the ligand nuclei in a transition metal complex. Note also, that the $\boldsymbol{T}$ operator does not occur in this equation.

It is noted in passing that a closely related comment applies to the point-dipole approximation to the ZFS that also should not involve any 'site $\boldsymbol{g}$-matrix' but only the free-electron g-value.[64]

The basis set dependence of the dipolar HFC is not overly pronounced. Most of it arises from the one-center contributions, where the matrix elements over the r^{-3} operator react sensitively to small changes in the radial functions of the orbitals, which, in turn change slightly with the basis set. As a numerical illustration, Table 7.7 collects calculated dipolar HFCs for the carbon center in the methyl-radical as a function of the basis set. Again, adding tight functions greatly stabilizes the results and one reaches the basis set limit pretty much with a core-polarized triple-ζ basis set. Uncontracting the basis set appears to be unnecessary once tight functions have been added.

Table 7.7. Basis set dependence of the largest dipolar hyperfine component of the carbon center in the CH_3 radical (B3LYP density functional; 'uc' = uncontracted)

Basis set	$A_{\text{dip}}^{\|\|\|}$
cc-pVDZ	147.7
cc-pVTZ	157.6
cc-pVQZ	161.2
cc-pV5Z	164.0
cc-pCVDZ	153.0
cc-pCVTZ	166.7
cc-pCVQZ	166.0
cc-pCV5Z	166.6
uc-cc-pCVQZ	166.5

All numbers in MHz.

7.4.4 Large Systems and Environment Effects

Keeping in mind all the technicalities described above, actual applications are typically performed on larger systems in the condensed phase. Hence, some treatment of the environment must be included in the calculations. While space is insufficient to cover any of the relevant subjects in detail, some issues must be mentioned.

7.4.4.1 *Vibrational Averaging*

Radicals and transition metal complexes are not static but undergo molecular motions on the time scale of the EPR experiment. Since SH parameters are sensitive functions of molecular geometry, these motions have significant effects on the calculated SH parameters.[55] For small-amplitude motions, one can include vibrational averaging effects. This is particularly important for isotropic HFCs that arise dominantly from spin polarization. For example, for the methyl radical, the out-of-plane motion has a big effect on the calculated HFCs. For the planar configuration the hydrogens are situated in a node of the SOMO and the HFC arises solely from spin polarization. For an out-of-plane arrangement, the hydrogens start to overlap with a lobe of the p_z orbital, which contributes positively to the spin population at the hydrogens. This reasoning nicely explains the property surface in Figure 7.6 that shows a nearly perfectly quadratic behavior of the hydrogen HFC along the umbrella mode of CH_3 with the lowest value being obtained at the planar geometry.

In general, one can expand a given molecular property P in a Taylor series in terms of the normal coordinates q_i of the system:

$$P(\boldsymbol{q}) = P(0) + \sum_i \frac{\partial P}{\partial q_i} q_i + \frac{1}{2} \sum_{i,j} \frac{\partial^2 P}{\partial q_i \, \partial q_j} q_i \; q_j + \cdots \tag{7.71}$$

where $P(0)$ is the property at equilibrium geometry. The lowest vibrational eigenfunction along the ith mode is a simple Gaussian function:

$$\Theta_0(q_i) = \left(\frac{m_i \omega_i}{\pi}\right)^{1/4} \exp\left(-\frac{1}{2} m_i \omega_i q_i^2\right) \tag{7.72}$$

where m_i is the reduced mass and ω_i the harmonic vibrational frequency of the mode. Hence, the expectation value of the property along a given mode becomes:

$$\langle \Theta_0 | P | \Theta_0 \rangle = P(0) + \frac{1}{4} \sum_i \frac{\partial^2 P}{\partial q_i^2} \frac{1}{m_i \omega_i} + \cdots \tag{7.73}$$

The linear term vanishes since the harmonic-oscillator wavefunctions are symmetric around the origin and a linear function is antisymmetric. Thus, the first surviving term features the second derivative of the property with respect to normal modes. Hence, either a nonlinear geometry dependence of the given property or anharmonic motion is necessary in order for the property to change upon vibrational averaging. Furthermore, for very low-energy motions, excited vibrational states will be populated that have their peak amplitudes at larger values of the normal mode coordinates and consequently, the vibrational averaging effects will become larger and temperature dependent. It is perhaps fair to say that for larger molecules no standardized software exists that would allow one to perform such calculations in a routine fashion.

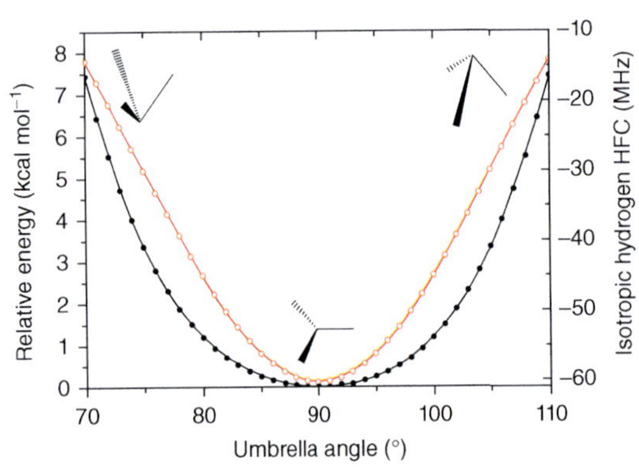

Figure 7.6. Energy and property surface for the methyl radical along the out-of-plane 'umbrella' mode. The calculations were done with the B3LYP functional and the cc-pCVTZ basis set (black, full dots = relative total energy, red, open dots = hyperfine coupling)

7.4.4.2 *Solvation*

For solvated molecules, it is customary to include solvation effects via an electrostatic continuum model. These models are highly useful to obtain rough estimates of solvation energies. Since the models will also change the electronic structure of the solute, a certain effect on molecular properties arises as well. However, these models only capture electrostatic effects in an average fashion. Whenever there is a specific solvent–solute interaction, for example a hydrogen bond, there is no substitute for incorporating solvent molecules into the quantum system. The question then arises how many solvent molecules should be included and in which orientation? Clearly, the associated geometry optimizations are tricky, since the solvent–solute interaction is weak, the intermolecular modes are floppy and may contain many local minima. The optimized geometries represent a zero-Kelvin situation that does not properly represent the situation in fluid solution. For example, optimizations on water clusters will converge to a solution representing one of the many different forms of ice, which is significantly more compact than water clusters in fluid solution. Thus, such calculations, while being undeniably useful and preferable to pure gas-phase calculations, always contain a certain amount of arbitrariness and will ultimately require significant manual labor in order to reach convincing and stable conclusions.

7.4.4.3 Molecular Dynamics

For larger-scale and significantly anharmonic molecular motions, it is necessary to perform large-scale molecular dynamics simulations that provide the molecular structure as a function of simulation time.[65,66] Using established molecular-dynamics techniques for sampling and averaging, the average value and statistical distribution function of the given molecular property can be obtained. In this way, one also gains access to relaxation times.[67–70] The details, however, are outside the scope of this chapter.

7.4.4.4 Macromolecules

For enzymes and other macromolecules, one can incorporate the environment via combined force-field and quantum chemical calculations (QM/MM) approach in the EPR parameter studies.[71–76] QM/MM modeling is highly useful but also very labor intensive and contains many pitfalls of technical nature. Alternatively, present-day hard- and software allows for QM-only cluster models of significant size. Perhaps, 200–300 atoms can be treated in a routine fashion, which allows for the active site and the first and second coordination sphere to be explicitly included. In many cases, this is enough to properly represent the electronic structure of the radical or metal site in question. Cluster calculations are much easier to perform than QM/MM calculations. In a few cases it has been shown that the results of QM/MM and cluster calculations converge to the same answers once the cluster models are made large enough.[77]

As an example of such a cluster calculation on an enzyme-based radical, we briefly discuss the case of ribonucleotide reductase (RNR).[78,79] The so-called class I ribonucleotide reductases catalyze the conversion of nucleotides to deoxynucleotides. They are composed of two subunits: α2 and β2. The β2 subunit contains a stable di-iron tyrosyl radical ($Y_{122}^{\bullet}$) cofactor required to generate a thiyl radical ($C_{439}^{\bullet}$) in α2 over a distance of 35 Å, which in turn initiates the chemistry of the reduction process (Figure 7.7). It has been established that the radical transfer process occurs by means of proton-coupled electron transfer (PCET). It involves the pathway $Y_{122} \rightarrow W_{48} \rightarrow Y_{356}$ in β2, then crosses the subunit border and proceeds

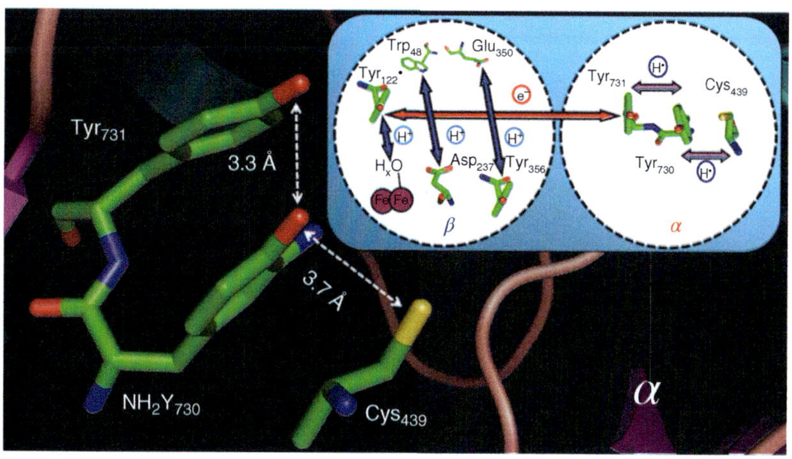

Figure 7.7. Crystal structure of the $Y_{730}NH_2Y$-$\alpha 2$ mutant of *Escherichia coli* RNR (molecule A) showing the residues Y_{731}, NH_2Y_{730}, and C_{439} in the PCET pathway. The O—O distance and the O—S distance between the essential amino acid residues are marked. The inset shows the current model for the long-range (≈35 Å), reversible PCET starting at Y_{122} in β2 and reaching C_{439} in α2. Note that the position of Y_{356} is unknown as the C-terminus is disordered in all X-ray structures of β2. (Reprinted with permission from T. Argirević, C. Riplinger, J. Stubbe, F. Neese, and M. Bennati, J. Am. Chem. Soc. 2012, 134, 17661. © 2012, American Chemical Society)

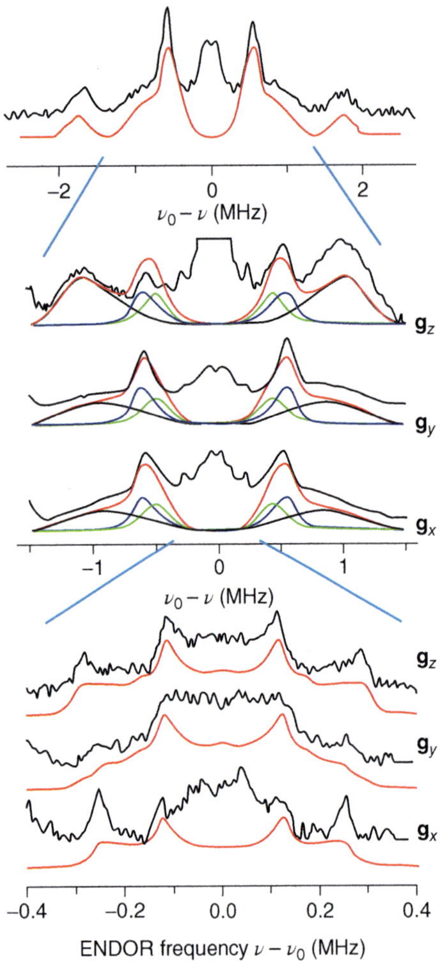

Figure 7.8. ^{2}H-Mims ENDOR spectrum of $NH_2Y_{730}^{\bullet}$ in D_2O assay buffer of the weakly coupled regions (black) and simulation (red). (Reprinted with permission from T. Argirević, C. Riplinger, J. Stubbe, F. Neese, and M. Bennati, J. Am. Chem. Soc. 2012, 134, 17661. © 2012, American Chemical Society)

via $Y_{731} \rightarrow Y_{730}$ finally to C_{439} in α2. In order to experimentally investigate this intriguing sequence, a 3-amino tyrosine (NH_2Y) modification of Y_{730} in α2 was synthesized and subjected to detailed spectroscopic analysis. Specifically, a ($NH_2Y_{730}^{\bullet}$) radical was generated in ^2H_2O. [^{2}H]-Electron-nuclear double resonance (ENDOR) spectra were obtained at 94 GHz.[79]

The resulting spectra are rich and complicated and their interpretation is anything but straightforward (Figure 7.8). It is difficult to document the process that finally led to the simulations in Figure 7.8. However, these simulations were not achieved without the input from quantum chemical calculations. In general, the number of HFC parameters that need to be fitted to experiment is very large. For each nucleus, there are six parameters, three principal values of the HFC tensor, and three Euler angles that relate the tensor to the global frame of reference. In particular, the latter are difficult to determine by least-squares fits. Furthermore, it needs to be determined how many nuclei contribute to the spectral envelope and to which precise nucleus each tensor belongs.

DFT calculations provide approximate answers to all of these questions. One can approach the problem by constructing cluster models for the active site and consider different possible conformations or the presence of additional water molecules. The calculations will then provide complete HFC tensors for all nuclei.

In the present example, it was possible to establish that there must be a water hydrogen bonded to the oxygen of Tyr_{730} (Figure 7.9). This is a significant finding since it impacts the energetics of the electron transfer and, hence, also on the kinetics of the process. The differences are energetically subtle, but are large enough such that the presence or absence of the hydrogen bond can turn electron transfer and, hence, enzymatic activity on or off (Figure 7.10). It would have been difficult to reach this significant conclusion without the careful combination of high-resolution EPR experiments and quantum chemical calculations.

In general, DFT calculated HFCs will typically not be highly accurate, but good enough to differentiate which nuclei contribute to the spectral envelope and which nuclei provide large, medium, and small HFCs. Quite typically, if the structure is correct, the principal values of the HFC tensor components of the larger HFCs can be expected to be within 10–20% of experiment, for the small couplings a factor of two is not uncommon. More importantly, the Euler angles predicted by the calculations tend to be good. Thus, a successful strategy involves fixing the Euler angles to their calculated values and allowing for 10–20% variation of the principle values in the simulation. This has led to the successful simulation of a variety of complex EPR and ENDOR spectra that could not be achieved 'by hand'.

A second example pertains to a low-molecular weight model complex, namely a delocalized mixed-valence $Cu^{1.5} \cdots Cu^{1.5}$ compound analyzed in Ref. 80. The structure of the complex is shown in Figure 7.11.

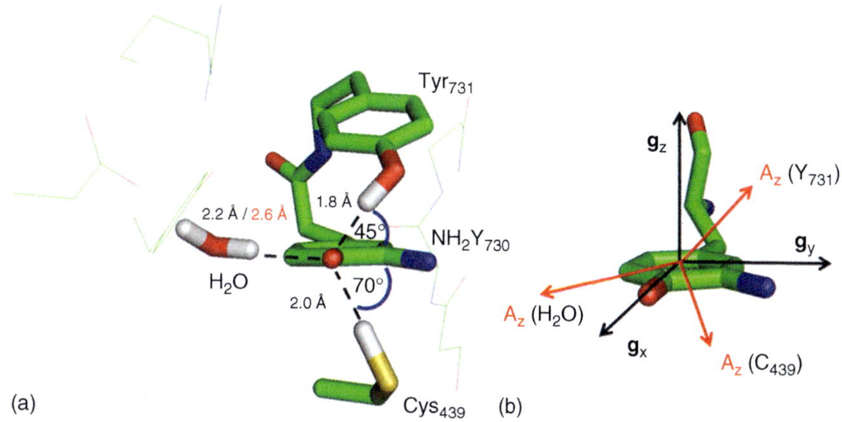

Figure 7.9. Combined EPR and DFT model for the hydrogen bond network around $NH_2Y_{730}^{\bullet}$ in α2. (a) (Side view): distances and out-of-plane angles of the hydrogen bonds and to a water molecule within the optimized DFT model 4. (b) Top view. Inset: location of the ***g*** tensor within the $NH_2Y^{\bullet}$ structure. The direction of the hyperfine tensor component A_z of the three hydrogen/deuteron couplings is also schematically drawn

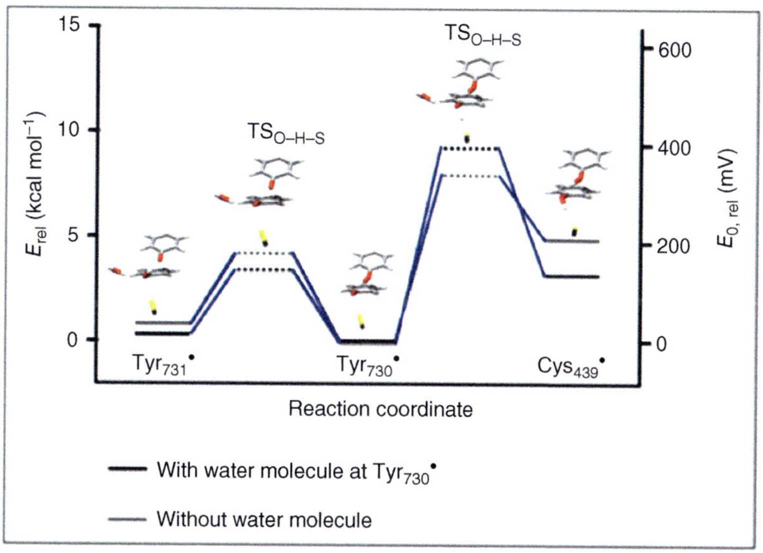

Figure 7.10. Energy level diagram including the transition states for the PCET pathway residues $Y_{731}-Y_{730}-C_{439}$. The reaction coordinate was investigated using large models with and without the water molecule

The large number of protons in this complex and the high degree of electron and spin delocalization renders the interpretation of the HYSCORE and ENDOR spectra rather challenging. Experimentally, four groups of protons could be identified based on the HFC coupling strength. However, the assignment of these spectrally identifiable protons to actual protons in the structure was not achievable without the input from quantum chemical calculations. In particular, attempts to simulate the ^{1}H-Davies W-band ENDOR spectra were hampered by the uncertainty in the appropriate Euler angles for each group of protons. Hence, simulations were started from the predicted HFC and Euler-angle values from quantum chemistry and limited variations were allowed in the least-square fitting process. Eventually, the simulations illustrated

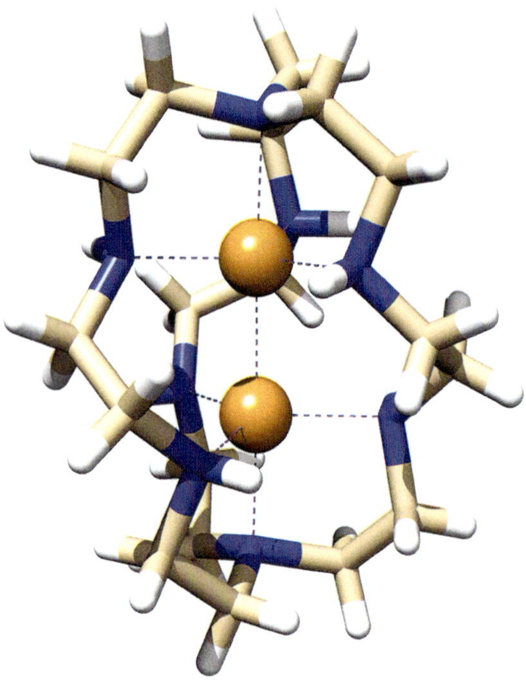

Figure 7.11. Structure of the $Cu^{1.5} \cdots Cu^{1.5}$ mixed valence model complex analyzed in Ref. 80. (Reprinted with permission from S. Kababya, J. Nelson, C. Calle, F. Neese, and D. Goldfarb, J. Am. Chem. Soc. 2006, 128, 2017. © 2006, American Chemical Society)

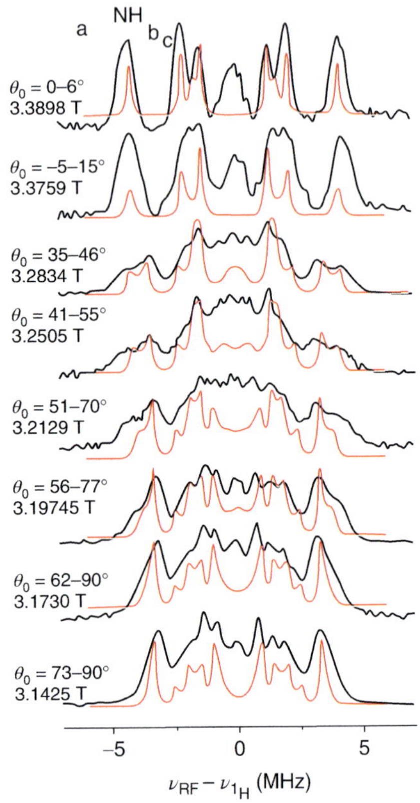

Figure 7.12. Comparison of experimental and simulated ^{1}H-Davies W-band ENDOR spectra of $[Cu_2(RGT)]^{3+}$ from Ref. 80. (Reprinted with permission from S. Kababya, J. Nelson, C. Calle, F. Neese, and D. Goldfarb, J. Am. Chem. Soc. 2006, 128, 2017. © 2006, American Chemical Society)

in Figure 7.12 were achieved. This combination of theory and experiment hence: (i) allowed for a successful interpretation of the experimental results, (ii) clarified the assignment of HFCs to the actual geometric structure of the compound, and (iii) provided an experimentally validated picture of the electronic structure of the compound, which shows an interesting metal–metal bond of bond-order 0.5. Further details are discussed in Ref. 80.

7.5 CONCLUDING REMARKS

In this short chapter, it was attempted to provide an overview over the intellectual framework that underlies the SH parameters in EPR spectroscopy, to hint at practical electronic structure methods and to show how they are connected to EPR spectroscopy. The subtleties that arise in the calculation of these properties were exemplified for isotropic and dipolar HFCs. Finally, it was hinted at how to incorporate environment, dynamic, and temperature effects into the calculations. While none of the subjects touched in this chapter could be treated in great depth, it is hoped that the selected material is helpful as a 'bird's eye view' for the theoretician entering the field or for an experimentalist who wants to supplement his or her experiments with theoretical calculations.

ACKNOWLEDGMENTS

I gratefully appreciate the financial support of this work and the work cited herein by the Max Planck society and the Deutsche Forschungsgemeinschaft. I also gratefully acknowledge the careful reading of the manuscript and many helpful suggestions by Profs. Stefan Stoll, Daniella Goldfarb, and Gunnar Jeschke.

REFERENCES

1. P. Strange, *Relativistic Quantum Mechanics*, Cambridge University Press: Cambridge, 1998.
2. L. Visscher and E. van Lenthe, *Chem. Phys. Lett.*, 1999, **306**, 357.
3. B. A. Hess, *Phys. Rev. A*, 1986, **333**, 3742.
4. E. van Lenthe, J. G. Snijders, and E. J. Baerends, *J. Chem. Phys.*, 1996, **105**, 6505.
5. J. E. Harriman, *Theoretical Foundations of Electron Spin Resonance*, Academic press: New York, 1978.
6. R. McWeeny, *Methods of Molecular Quantum Mechanics*, Academic Press: London, 1992.
7. R. McWeeny, *Spins in Chemistry*, Academic Press: New York, 1970.
8. F. Neese, in *Multifrequency Electron Paramagnetic Resonance. Theory and Applications*, ed. S. Misra, Wiley-VCH: Weinheim, 2011, 297.
9. F. Neese, in *Biol. Mag. Res*, eds. G. Hanson and L. Berliner, Springer: Heidelberg, 2009, Vol. **28**, 175.
10. F. Neese and E. I. Solomon, in *Magnetoscience – From Molecules to Materials. Volume IV*, eds J. S. Miller and M. Drillon, Wiley VCH: Weinheim, 2003, 345.
11. M. Atanasov, D. Aravena, E. Suturina, E. Bill, D. Maganas, and F. Neese, *Coord. Chem. Rev.*, 2015, **289**, 177.
12. F. Neese, *J. Chem. Phys.*, 2005, **122**, 034107/1.
13. B. A. Hess, C. M. Marian, U. Wahlgren, and O. Gropen, *Chem. Phys. Lett.*, 1996, **251**, 365.
14. E. U. Condon and G. H. Shortly, *The Theory of Atomic Spectra*, Camebridge University Press: Camebridge, 1959.
15. S. Stoll and A. Schweiger, *J. Magn. Reson.*, 2006, **178**, 42.
16. R. McWeeny, *J. Chem. Phys.*, 1965, **42**, 1717.
17. F. Neese and E. I. Solomon, *Inorg. Chem.*, 1998, **37**, 6568.
18. F. Neese, *J. Chem. Phys.*, 2003, **118**, 3939.
19. R. Ditchfield, *Molec. Phys.*, 1974, **27**, 789.
20. K. Wolinski, J. F. Hinton, and P. Pulay, *J. Am. Chem. Soc.*, 1990, **112**, 8251.
21. V. Kellö and A. J. Sadlej, *Int. J. Quant. Chem.*, 1998, **68**, 159.
22. Y. Yamaguchi, Y. Osamura, J. Goddard, and H. F. Schaefer III, *A New Dimension to Quantum Chemistry Analytic Derivative Methods in Ab Initio Molecular Electronic Structure Theory*, Oxford University Press: Oxford, 1994.
23. D. Ganyushin and F. Neese, *J. Chem. Phys.*, 2006, **125**, 024103.
24. M. Gerloch and R. F. McMeeking, *J. Chem. Soc. Dalton Trans.*, 1975, 2443.
25. R. Maurice, R. Bastardis, C. de Graaf, N. Suaud, T. Mallah, and N. Guihery, *J. Chem. Theor. Comp.*, 2009, **5**, 2977.
26. H. Bolvin, *Chem. Phys. Chem.*, 2006, **7**, 1575.
27. F. Neese, in *Linear Scaling Methods in Quantum Chemistry*, eds. R. Zalesny, M.G. Papadopoulos, P.G. Mezey, and J. Leszczynski, Springer: Heidelberg, 2010, Vol, **13**, pp 227–261.
28. F. Neese, *J. Phys. Chem. Solids*, 2004, **65**, 781.
29. W. D. Edwards and M. C. Zerner, *Theor. Chim. Acta*, 1987, **72**, 347.
30. R. J. Bartlett, *Rev. Mod. Phys.*, 2007, **79**, 291.
31. C. Riplinger, P. Pinski, U. Becker, E. F. Valeev, and F. Neese, *J. Chem. Phys.*, 2016, **144**, 024109.
32. M. Kallay and J. Gauss, *J. Chem. Phys.*, 2004, **120**, 6841.
33. M. Kallay, J. Gauss, and P. G. Szalay, *J. Chem. Phys.*, 2003, **119**, 2991.
34. J. Gauss, M. Kallay, and F. Neese, *J. Phys. Chem. A*, 2009, **113**, 11541.
35. M. L. Munzarova and M. Kaupp, *J. Phys. Chem. A*, 1999, **103**, 9966.
36. R. G. Parr and W. Yang, *Density Functional Theory of Atoms and Molecules*, Oxford University Press: Oxford, 1989.
37. F. Neese, *Coord. Chem. Rev.*, 2009, **253**, 526.
38. J. Autschbach and T. Ziegler, *Coord. Chem. Rev.*, 2003, **238/239**, 83.
39. A. D. Becke, *J. Chem. Phys.*, 1993, **98**, 1372.
40. S. Grimme, *J. Chem. Phys.*, 2006, **124**, 34108.
41. J. Garza, J. A. Nichols, and D. A. Dixon, *J. Chem. Phys.*, 2000, **112**, 7880.
42. O. Gritsenko, R. VanLeeuwen, and E. J. Baerends, *Int. J. Quant. Chem.*, 1996, **57**, 17.

43. F. Della Sala and A. Görling, *J. Chem. Phys.*, 2001, **115**, 5718.

44. S. Grimme, J. Antony, S. Ehrlich, and H. Krieg, *J. Chem. Phys.*, 2010, **132**, 19.

45. J. Antony and S. Grimme, *Phys. Chem. Chem. Phys.*, 2006, **8**, 5287.

46. S. Reimann, U. Ekstrom, S. Stopkowicz, A. Teale, A. Borgoo, and T. Helgaker, *Phys. Chem. Chem. Phys.*, 2015, **17**, 18834.

47. H. M. McConnell, *J. Chem. Phys.*, 1956, **24**, 764.

48. R. S. Mulliken, *J. Chem. Phys.*, 1955, **23**, 1833.

49. A. E. Reed, R. B. Weinstock, and F. Weinhold, *J. Chem. Phys.*, 1985, **83**, 1736.

50. F. L. Hirshfeld, *Theoret. Chim. Acta*, 1977, **44**, 129.

51. F. Neese, *J. Phys. Chem. A.*, 2001, **105**, 4290.

52. H. M. McConnell and D. B. Chesnut, *J. Chem. Phys.*, 1958, **28**, 107.

53. M. Roemelt and F. Neese, *J. Phys. Chem. A*, 2013, **117**, 3069.

54. M. Roemelt, D. Maganas, S. DeBeer, and F. Neese, *J. Chem. Phys.*, 2013, **138**, 204101.

55. R. Improta and V. Barone, *Chem. Rev.*, 2004, **104**, 1231.

56. B. Engels, L. A. Eriksson, and S. Lunell, *Adv. Quant. Chem.*, 1996, **27**, 83.

57. T. H. Dunning, *J. Chem. Phys.*, 1989, **90**, 1007.

58. A. W. Hauser, T. Gruber, M. Filatov, and W. E. Ernst, *Chem. Phys. Chem.*, 2013, **14**, 716.

59. F. Neese, A. Wolf, T. Fleig, M. Reiher, and B. A. Hess, *J. Chem. Phys.*, 2005, **122**.

60. B. Sandhöfer, S. Kossmann, and F. Neese, *J. Chem. Phys.*, 2013, **138**, 104102.

61. E. Malkin, I. Malkin, O. L. Malkina, V. G. Malkin, and M. Kaupp, *Phys. Chem. Chem. Phys.*, 2006, **8**, 4079.

62. I. Malkin, O. L. Malkina, V. G. Malkin, and M. Kaupp, *Chem. Phys. Lett.*, 2004, **396**, 268.

63. A. Schweiger and G. Jeschke, *Principles of Pulse Electron Paramagnetic Resonance*, Oxford University Press: Oxford, 2001.

64. C. Riplinger, J. P. Y. Kao, G. M. Rosen, V. Kathirvelu, G. R. Eaton, S. S. Eaton, A. Kutateladze, and F. Neese, *J. Am. Chem. Soc.*, 2009, **131**, 10092.

65. J. R. Asher and M. Kaupp, *Chemphyschem*, 2007, **8**, 69.

66. M. Pavone, C. Benzi, F. De Angelis, and V. Barone, *Chem. Phys. Lett.*, 2004, **395**, 120.

67. P. Cimino, F. Neese, and V. Barone, in *Computational Spectroscopy: Methods, Experiments and Applications*, ed. J. Grunenberg, Wiley-VCH Verlag GmbH & Co. KGaA: Weinheim, Germany, 2011, 63.

68. M. Zerbetto, S. Carlotto, A. Polimeno, C. Corvaja, L. Franco, C. Toniolo, F. Formaggio, V. Barone, and P. Cimino, *J. Phys. Chem. B*, 2007, **111**, 2668.

69. V. Barone and P. Polimeno, *Phys. Chem. Chem. Phys.*, 2006, **8**, 4609.

70. V. Barone and A. Polimeno, *Phys. Chem. Chem. Phys.*, 2006, **8**, 4609.

71. C. Riplinger, E. Bill, A. Daiber, V. Ullrich, H. Shoun, and F. Neese, *Chem. Eur. J.*, 2014, **20**, 1602.

72. M. Sundararajan and F. Neese, *J. Chem. Theory Comput.*, 2012, **8**, 563.

73. C. Riplinger and F. Neese, *ChemPhysChem*, 2011, **12**, 3192.

74. A. Altun, D. Kumar, F. Neese, and W. Thiel, *J. Phys. Chem. A*, 2008, **112**, 12904.

75. S. Sinnecker and F. Neese, *J. Comput. Chem.*, 2006, **27**, 1463.

76. J. C. Schöneboom, F. Neese, and W. Thiel, *J. Am. Chem. Soc.*, 2005, **127**, 5840.

77. M. Retegan, F. Neese, and D. A. Pantazis, *J. Chem. Theory Comput.*, 2013, **9**, 3832.

78. T. U. Nick, W. Lee, S. Kossmann, F. Neese, J. Stubbe, and M. Bennati, *J. Am. Chem. Soc.*, 2015, **137**, 289.

79. T. Argirević, C. Riplinger, J. Stubbe, F. Neese, and M. Bennati, *J. Am. Chem. Soc.*, 2012, **134**, 17661.

80. S. Kababya, J. Nelson, C. Calle, F. Neese, and D. Goldfarb, *J. Am. Chem. Soc.*, 2006, **128**, 2017.

Chapter 8
Spin Dynamics

Akiva Feintuch and Shimon Vega

Weizmann Institute of Science, Rehovot, Israel

8.1 INTRODUCTION

In this chapter, we provide a comprehensive summary of the theoretical tools that are presently used to design and analyze microwave (MW) and radiofrequency (RF) pulse experiments in pulse EPR spectroscopy. Many of the modern experimental techniques used in EPR and NMR rely on the propagation of the electron and nuclear spin systems between and after a set of short MW and RF irradiation periods. The time evolution of these spin systems during these experiments can be derived from first principles using the basic quantum mechanics equations of motion. General introductions to quantum mechanics,[1–3] as well as its application to EPR and NMR,[4–8] can be found in many basic textbooks. This chapter is strongly based on these important works; however, for simplicity, we only refer to them here, and we will not repeat these references anymore along the way.

For the convenience of the readers, we begin with an overview of the quantum theory applied to magnetic resonance (MR). To describe the evolution of the electron and nuclear spin systems in MR, we start with presenting the time dependence of the quantum spin states as determined by the solution of the Schrödinger equation and governed by the spin Hamiltonian defining the interactions between the spins and with their environment. However, as in a typical MR experiment we deal with a statistical mixture (ensemble) of spin systems, we must introduce the spin density operator, which describes an average over the ensemble. The evolution of this operator can be evaluated by solving the Liouville–von Neumann equation.

The quantum angular momentum operators form a basis in terms of which both the spin Hamiltonian and the spin density operators can be expressed. Consequently, these operators form the 'language of the theory' of MR. Thus, in order to understand MR experiments, we have to get accustomed to this language so that we can communicate with each other and evaluate our results. Insertion of the operator expansions of the spin Hamiltonian and spin density operator into the Liouville–von Neumann equation enables us to follow the quantum evolution by solving this equation. To do so, we must establish transformation rules that enable analytic solutions for small spin systems and their estimation for larger ones. Thus, in the following sections, we introduce the necessary transformation rules and apply them to a set of common pulse EPR experiments.

EPR Spectroscopy: Fundamentals and Methods.
Edited by Daniella Goldfarb and Stefan Stoll

ISBN: 978-1-119-16299-5
Also published in eMagRes (online edition)
DOI: 10.1002/9780470034590.emrstm1506

To include spin relaxation in the description of the spin evolution, it is common to expand the Hilbert space, having a dimension equal to the number of spin states of the system, to Liouville space having a dimension equal to the square of the number of spin states. This extension allows the introduction of longitudinal and transverse relaxation rates that must be derived from relaxation theory.[9,10] In this chapter, we will only discuss the basic principles for defining Liouville space and the general form of the master equation replacing the Liouville–von Neumann equation. However, actual derivations of relaxation rates and their implementation into the master equation for solving the spin evolution with relaxation are outside the scope of this chapter.

This chapter is structured as follows. In Section 8.2, we introduce the quantum description of the electron spin and derive its evolution in an external magnetic field, introducing the necessary angular momentum operators and their transformation rules. Because of space restrictions, we consider only systems with spins with a quantum number equal to ½. We then describe two very basic pulse sequences, the Hahn echo sequence and the stimulated echo sequence. Section 8.3 deals with a two-electron spin system and its description in terms of angular momentum product operators and fictitious spin-1/2 operators. Here, transformation rules between these operators are used to discuss the basic spin evolution during solid-echo experiments and the three-pulse double electron–electron resonance (DEER, also called PELDOR) pulse sequence. In Section 8.4, we discuss the spin evolution of an electron-nucleus two-spin system. Here again angular momentum operators are utilized together with their transformation rules to present the evolution during the two-pulse electron spin echo envelope modulation (ESEEM) pulse sequence and the basics of electron-nuclear double-resonance (ENDOR) experiments. The sequences discussed in this chapter are part of the many pulse EPR techniques that are currently commonly used.

8.2 THE SINGLE-ELECTRON SPIN SYSTEM

8.2.1 The Electron Eigenstates and their Energies

An electron has a spin angular moment quantum number $S = 1/2$ and becomes quantized in its eigenstates, in Dirac notation, $|\alpha\rangle$ 'up' or $|\beta\rangle$ 'down' spin state, when positioned in an external magnetic field B_0. When this field is along the z-direction of a Cartesian frame of reference, these states are the eigenstates of the z-component of the linear electron spin angular momentum vector operator $\hat{\boldsymbol{S}} = (\hat{S}_x, \hat{S}_y, \hat{S}_z)$, with

$$\hat{S}_z|\alpha\rangle = \frac{\hbar}{2}|\alpha\rangle \text{ and } \hat{S}_z|\beta\rangle = -\frac{\hbar}{2}|\beta\rangle \tag{8.1}$$

The spin Hamiltonian of the electron spin in a magnetic field, B_0, is defined by the Zeeman interaction and has the form

$$\hat{H}_Z = -\hat{\mu}_z B_0 = g_e \mu_B \hbar^{-1} B_0 \hat{S}_z = \omega_0 \hat{S}_z \tag{8.2}$$

where $\hat{\mu}_z$ is the z-component of the electron magnetic moment vector operator. This operator is proportional to $\hat{S}_z$ and for the electron with a charge $-e$ is given by $\hat{\mu}_z = -g_e\mu_B\hbar^{-1}\hat{S}_z$, where g_e is the free-electron g-factor ≈ 2.002, $\hbar$ is the reduced Planck constant, and the Bohr magneton is $\mu_B = e\hbar/2m_e$. The values of these two constants in the expression for $\hat{\mu}_z$ are $\hbar = 1.054 \times 10^{-34}$ J s^{-1} and $\mu_B = 9.27 \times 10^{-24}$ J T^{-1}. When the B_0 field is measured in units of tesla, the energies of the electron in the eigenstates of $\hat{H}_Z$ in units of joule are

$$E_\alpha = \langle\alpha|\hat{H}_Z|\alpha\rangle = \frac{\hbar}{2}\omega_0; \quad E_\beta = \langle\beta|\hat{H}_Z|\beta\rangle = -\frac{\hbar}{2}\omega_0 \tag{8.3}$$

When we express these two energies in units of s^{-1}, i.e. angular frequency in terms of radians per second, using $\omega_{\alpha/\beta} = E_{\alpha/\beta}/\hbar$, they become $\omega_\alpha = -\omega_\beta = 1/2 \cdot \omega_0 = 0.88 \times 10^{11} \cdot B_0 \text{s}^{-1}$. Transferring these values into units of hertz, we get that $\nu_\alpha = -\nu_\beta = (1/2 \cdot \omega_0)/2\pi = 14.01 \times 10^9 \cdot B_0$ Hz and the $(|\alpha\rangle \leftrightarrow |\beta\rangle)$ transition frequency $(\nu_\alpha - \nu_\beta)$ per unit of tesla of the external field becomes 28.02 GHz T^{-1}.

An arbitrary spin state of the electron at time t can be written as a linear combination of the two eigenstates, with complex coefficients $c_\alpha(t)$ and $c_\beta(t)$

$$|\psi(t)\rangle = c_\alpha(t)|\alpha\rangle + c_\beta(t)|\beta\rangle \tag{8.4}$$

where normalization demands that $|c_\alpha(t)|^2 + |c_\beta(t)|^2 = 1$. As these coefficients are complex, they can be expressed in the form of a real amplitude and a phase

$$c_\alpha(t) \equiv a_\alpha(t)\mathrm{e}^{-\mathrm{i}\varphi_\alpha(t)} \text{ and } c_\beta(t) = a_\beta(t)\mathrm{e}^{-\mathrm{i}\varphi_\beta(t)} \tag{8.5}$$

and the normalization requires that $a_\alpha^2(t) + a_\beta^2(t) = 1$.

8.2.2 The Angular Momentum Operators

The three components of the spin-½ angular momentum vector operator $(\hat{S}_x, \hat{S}_y, \hat{S}_z)$ satisfy the important commutation relation

$$[\hat{S}_x, \hat{S}_y] = \mathrm{i}\hbar\hat{S}_z \tag{8.6}$$

and its cyclic permutations. Applying these operator components on $|\alpha\rangle$ and $|\beta\rangle$, and taking into account that these two states are orthonormal, i.e., $\langle\alpha|\beta\rangle = 0$ and $\langle\alpha|\alpha\rangle = \langle\beta|\beta\rangle = 1$, each of the three $\hat{S}_p$ operators $(p = x, y, z)$ can be represented in matrix form with elements $\langle\pi|\hat{S}_p|\pi'\rangle$ for $\pi, \pi' = \alpha, \beta$.

We define the angular momentum operators from now on in units of $\hbar$, and, as a result, the eigenvalues of the Hamiltonian $\hat{H}_Z = \omega_0\hat{S}_z$ are expressed in units of s^{-1}

$$E_a = \frac{\omega_0}{2} \text{ and } E_\beta = \frac{\omega_0}{2} \tag{8.7}$$

and all energies are expressed in these angular frequency units. The angular momentum components now satisfy the commutation relation $[\hat{S}_x, \hat{S}_y] = \mathrm{i}\hat{S}_z$ and its cyclic permutations, and their matrix representations in the complete set of states $\{|\alpha\rangle, |\beta\rangle\}$ are

$$\hat{S}_x = 1/2\begin{pmatrix} 0 & 1 \\ 1 & 0 \end{pmatrix} \quad \hat{S}_y = 1/2\begin{pmatrix} 0 & -\mathrm{i} \\ \mathrm{i} & 0 \end{pmatrix}$$
$$\hat{S}_z = 1/2\begin{pmatrix} 1 & 0 \\ 0 & -1 \end{pmatrix} \tag{8.8}$$

which have the form of the well-known Pauli matrices.

The signals measured during EPR measurements are proportional to the expectation values $m_p(t)$ of the electron magnetic moment operator components $\hat{\mu}_p$ with $p = x, y, z$, which in turn are proportional to the expectation values of $\hat{S}_p$,

$$\begin{pmatrix} m_x(t) \\ m_y(t) \\ m_z(t) \end{pmatrix} = -g_e\mu_B\begin{pmatrix} \langle\psi(t)|\hat{S}_x|\psi(t)\rangle \\ \langle\psi(t)|\hat{S}_y|\psi(t)\rangle \\ \langle\psi(t)|\hat{S}_z|\psi(t)\rangle \end{pmatrix} \tag{8.9}$$

where for now we assume that g_e is isotropic. For an electron in state $|\psi(t)\rangle$ positioned in an external magnetic field B_0, we get from equations (8.4) and (8.5) and the Pauli matrices, that

$$\begin{pmatrix} m_x(t) \\ m_y(t) \\ m_z(t) \end{pmatrix} = -g_e\mu_B\begin{pmatrix} a_\alpha(t)\,a_\beta(t)\cos(\varphi_\alpha(t) - \varphi_\beta(t)) \\ a_\alpha(t)a_\beta(t)\sin(\varphi_\alpha(t) - \varphi_\beta(t)) \\ 1/2\{a_\alpha^2(t) - a_\beta^2(t)\} \end{pmatrix} \tag{8.10}$$

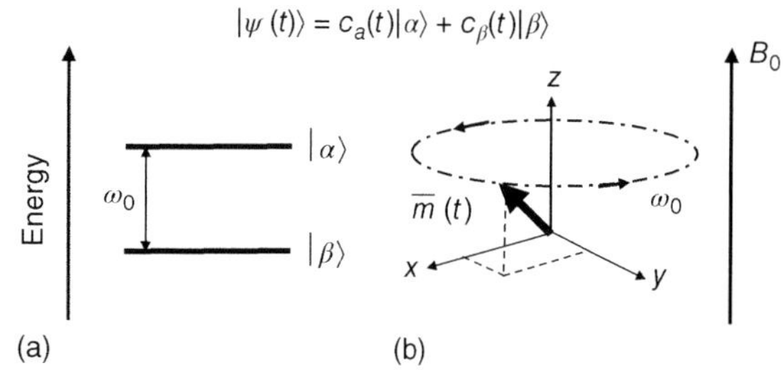

Figure 8.1. The Cartesian representation of the magnetic moment vector of an electron described by a spin state $|\psi(t)\rangle$ expressed in terms of the eigenstates $|\alpha\rangle$ and $|\beta\rangle$ of its Hamiltonian of the form $\hat{H} = \omega_0\hat{S}_z$ (a) together with its energy level diagram. (b) The evolution of this magnetization vector in 3D space is described by equation (8.14)

This result demonstrates the correlation between the spin state coefficients and the direction of the magnetic moment in the coordinate system of the B_0 field (Figure 8.1).

8.2.3 The Schrödinger Equation

The time evolution of the magnetic moment vector composed of the values $m_p(t)$ of an electron with some spin Hamiltonian $\hat{H}(t)$ depends on the time evolution of its spin state $|\psi(t)\rangle$. This state develops according to the solution of the Schrödinger equation,

$$\frac{\mathrm{d}}{\mathrm{d}t}|\psi(t)\rangle = -\mathrm{i}\hat{H}(t)|\psi(t)\rangle \tag{8.11}$$

Choosing a coordinate system and a basis set of states equal to the eigenstates of $\hat{S}_z$, the spin state of the electron can be expanded as in equation (8.4), and equation (8.11) can be replaced by the vector equation

$$\frac{\mathrm{d}}{\mathrm{d}t}\begin{pmatrix} c_\alpha(t) \\ c_\beta(t) \end{pmatrix} = -\mathrm{i}\begin{pmatrix} H_{\alpha\alpha}(t) & H_{\alpha\beta}(t) \\ H_{\beta\alpha}(t) & H_{\beta\beta}(t) \end{pmatrix}\begin{pmatrix} c_\alpha(t) \\ c_\beta(t) \end{pmatrix} \tag{8.12}$$

where the Hamiltonian is now represented by a matrix with elements $H_{\pi\pi'}(t) = \langle\pi|\hat{H}(t)|\pi'\rangle$.

For a single electron spin in a fixed magnetic field B_0, the matrix representation of its Zeeman interaction Hamiltonian $\hat{H}(t) = \omega_0\hat{S}_z$ is time independent and has only diagonal elements $H_{\alpha\alpha} = \frac{1}{2}\omega_0$ and $H_{\beta\beta} = -\frac{1}{2}\omega_0$. Then the solution of equation (8.12) for an initial state $|\psi(0)\rangle = c_\alpha(0)|\alpha\rangle + c_\beta(0)|\beta\rangle$ becomes simply

$$\begin{pmatrix} c_\alpha(t) \\ c_\beta(t) \end{pmatrix} = \begin{pmatrix} \mathrm{e}^{-\mathrm{i}/2\omega_0 t}c_\alpha(0) \\ \mathrm{e}^{\mathrm{i}/2\omega_0 t}c_\beta(0) \end{pmatrix} = \begin{pmatrix} a_\alpha(0)\,\mathrm{e}^{-\mathrm{i}\varphi_\alpha(t)} \\ a_\beta(0)\mathrm{e}^{-\mathrm{i}\varphi_\beta(t)} \end{pmatrix} \tag{8.13}$$

where the phase angles of $c_{\alpha/\beta}(t) = a_{\alpha/\beta}(0)e^{-i\varphi_{\alpha/\beta}(t)}$ become time dependent as $\varphi_\alpha(t) = \varphi_\alpha(0) + \frac{1}{2}\omega_0 t$ and $\varphi_\beta(t) = \varphi_\beta(0) - \frac{1}{2}\omega_0 t$, and the amplitudes stay constant $a_{\alpha/\beta}(t) = a_{\alpha/\beta}(0)$.

Insertion of these amplitudes and phase dependences into equation (8.10) results in the expressions for the time dependence of the magnetization vector $(m_x(t), m_y(t), m_z(t))$

$$\begin{pmatrix} m_x(t) \\ m_y(t) \\ m_z(t) \end{pmatrix} = -g_e\mu_B \begin{pmatrix} a_\alpha(0)\,a_\beta(0)\cos(\Delta\varphi(0) + \omega_0 t) \\ a_\alpha(0)a_\beta(0)\sin(\Delta\varphi(0) + \omega_0 t) \\ 1/2(a_\alpha^2(0) - a_\beta^2(0)) \end{pmatrix} \tag{8.14}$$

with $\Delta\varphi(0) = \varphi_\alpha(0) - \varphi_\beta(0)$. Thus, the time evolution of an electron spin in an external magnetic field can be described by a magnetic moment of constant length rotating around the direction of the magnetic field with a frequency ω_0. Of course, this result corresponds to the Cartesian vector representation of Feynmann–Vernon–Hellwarth[11] of a two-level system (Figure 8.1). An evolution time τ with $\omega_0\tau = 2\pi$ causes a full 2π rotation of the magnetization vector and only π rotations of $\varphi_{\alpha/\beta}(t+\tau) = \varphi_{\alpha/\beta}(t) + \pi$, i.e., in order to get $c_{\alpha,\beta}(t+\tau) = c_{\alpha,\beta}(t)$, the value of τ must satisfy $\omega_0\tau = 4\pi$. Such an object that returns to its original state after a 4π rotation is called a *spinor* and has been utilized in MR.[12]

8.2.4 Ensemble of Spin Systems and the Spin Density Operator

In order to treat a macroscopic system, it is common to decompose it in the form of an ensemble of K non-interacting equivalent small spin systems $k = 1,..., K$. In this case, each individual small system has its own spin Hamiltonian $\hat{H}_k(t)$ and evolves according to its own Schrödinger equation. We then assume that the physical properties of the macroscopic system are determined by statistical averaging of the ensemble of individual ones. For simplicity, we will first consider an ensemble of single spin-1/2 systems and then generalize the ensemble averaging approach to larger spin systems.

We begin by representing the spin state $|\psi_k(t)\rangle$ of each individual system k by an operator $\hat{C}_k(t)$ with a matrix representation with elements $\langle\pi_k|\psi_k(t)\rangle\langle\psi_k(t)|\pi'_k\rangle$ where $\pi_k, \pi'_k = \alpha_k, \beta_k$ and $\{|\alpha_k\rangle, |\beta_k\rangle\}$ is an orthonormal set of spin states. Expanding the spin state as $|\psi_k(t)\rangle = c_\alpha^k(t)|\alpha_k\rangle + c_\beta^k(t)|\beta_k\rangle$, this matrix gets the form

$$\hat{C}_k(t) = \begin{pmatrix} c_\alpha^k(t)\,c_\alpha^{k*}(t) & c_\alpha^k(t)c_\beta^{k*}(t) \\ c_\beta^k(t)c_\alpha^{k*}(t) & c_\beta^k(t)c_\beta^{k*}(t) \end{pmatrix} \tag{8.15}$$

and contains all available information about $|\psi_k(t)\rangle$. Evaluating the time derivative of the elements of $\hat{C}_k(t)$ using equation (8.12) results straightforwardly in the rate equation

$$\frac{d}{dt}\hat{C}_k(t) = -i[\hat{H}_k(t), \hat{C}_k(t)] \tag{8.16}$$

where $\hat{H}_k(t)$ is represented in its matrix form with elements $\langle\pi_k|\hat{H}_k(t)|\pi'_k\rangle$ as in equation (8.12). At this point, we assume that the Hamiltonians of all systems in the ensemble have the same form and that each $\hat{C}_k(t)$ matrix is represented in equivalent basis sets $\{|\alpha\rangle, |\beta\rangle\}$. Then we can define an ensemble averaged spin density matrix, which in the case of single spin-½ systems is defined by

$$\hat{\sigma}(t) = \begin{pmatrix} \overline{c_\alpha(t)\,c_\alpha^*(t)} & \overline{c_\alpha(t)c_\beta^*(t)} \\ \overline{c_\beta(t)c_\alpha^*(t)} & \overline{c_\beta(t)c_\beta^*(t)} \end{pmatrix} \tag{8.17}$$

where the elements represented by $\overline{c_\pi(t)c_{\pi'}^*(t)}$ are averages of all $c_\pi^k(t)c_{\pi'}^k(t)$ elements of the individual $\hat{C}_k(t)$ matrices. The rate equation describing the time evolution of the elements of $\hat{\sigma}(t)$ known as the *Liouville–von Neumann (L-vN) equation* has the same form as equation (8.16)

$$\frac{d}{dt}\hat{\sigma}(t) = -i[\hat{H}(t), \hat{\sigma}(t)] \tag{8.18}$$

The actual values of the spin density matrix elements, as well as the elements of the matrix representation of $\hat{H}(t)$, depend on the choice of the basis set $\{|\alpha\rangle, |\beta\rangle\}$. Equation (8.18) is of course valid for any orthonormal basis set spanning the individual spin systems; the spin density matrix defined by a chosen basis set is one of the representations of the spin density operator. In this chapter, any matrix representation of the spin density operator will also be represented by the symbol $\hat{\sigma}$.

This derivation can be generalized for dealing with an ensemble of spin systems $k = 1,.., K$ that can be described by an equivalent set of N orthonormal spin states $\{|\varphi_i\rangle\}_{i=1}^N$ and that have Hamiltonians $\hat{H}_k(t)$ with matrix elements $\langle\varphi_i|\hat{H}_k(t)|\varphi_j\rangle$ $(i,j = 1, \ldots, N)$. We can then define for each individual system in a state $|\psi_k(t)\rangle = \sum_{i=1}^{N} c_i^k|\varphi_i\rangle$ a $\hat{C}_k(t)$ with elements $\langle\varphi_i||\psi_k(t)\rangle\langle\psi_k(t)||\varphi_j\rangle = c_i^k(t)c_j^{k*}(t)$ and introduce an ensemble averaged spin density operator $\hat{\sigma}(t)$, whose matrix representation in the manifold of states $\{|\varphi_i\rangle\}_{i=1}^{N}$ has elements $\overline{c_i(t)c_j^*(t)}$. These matrix representations satisfy the L-vN equation in equation (8.18).

Continuing for the moment with the N-level spin systems, we can now assume that all experiments start at a time $t=0$ when the individual spin systems with time-independent Hamiltonians $\hat{H}_k = \hat{H}_{\text{eq}}$ interact with their surrounding lattice and are at thermal equilibrium defined by $\hat{C}_k^{\text{eq}}$ matrices that satisfy Boltzmann statistics. When the Hamiltonians $\hat{H}_{\text{eq}}$ are represented in their own eigenstates $\{|\varphi_i\rangle\}_{i=1}^{N}$, their matrix representation is diagonal, and the diagonal elements $c_{i,\text{eq}}^k c_{i,\text{eq}}^{k*}$ of the matrices $\hat{C}_k^{\text{eq}}$ in the same representation are constant. Consequently, the diagonal elements $\sigma_{ii}^{\text{eq}} = \overline{|c_{i,\text{eq}}|^2}$ of the matrix representation of the spin density operator $\hat{\sigma}_{\text{eq}}$ in the eigenstate manifold are constants as well, and they present the probability of each system to be found in its eigenstate $|\varphi_i\rangle$, i.e., the 'population' of the state. The ensemble averages of the off-diagonal elements of $\hat{C}_k^{\text{eq}}$ are $\sigma_{ij}^{\text{eq}} = \overline{c_{i,\text{eq}}c_{j,\text{eq}}^*}$, which have the form $\overline{a_{i,\text{eq}}a_{j,\text{eq}}\text{e}^{-\text{i}(\varphi_{i,\text{eq}}-\varphi_{j,\text{eq}})}}$. These are the probabilities of being in a superposition state of $|\varphi_i\rangle$ and $|\varphi_j\rangle$ and are called *coherences*.

At thermal equilibrium, due to entropy maximization one can assume that the constant phase angles $\varphi_{i,\text{eq}}$ at equilibrium are uniformly distributed in the interval [0, 2π) and, consequently, that the ensemble average of all $\text{e}^{-\text{i}(\varphi_{i,\text{eq}}-\varphi_{j,\text{eq}})}$ is zero and therefore $\sigma_{ij}^{\text{eq}}(0) = 0$ for $i \neq j$. Thus, the spin density matrices in the eigenstate representation of the Hamiltonian at thermal equilibrium are diagonal and their average values $\sigma_{ii}^{\text{eq}}(0)$ are, as mentioned earlier, equal to the occupation probability of $|\varphi_i\rangle$. These populations follow Boltzmann statistics and for a given temperature T satisfy

$$\hat{\sigma}_{ii}^{\text{eq}} = \frac{1}{Z}\text{e}^{-\langle\varphi_i|\hat{H}_{\text{eq}}|\varphi_i\rangle/k_{\text{B}}T} \tag{8.19}$$

where Z is the partition function $\sum_i \text{e}^{-\langle\varphi_i|\hat{H}_{\text{eq}}|\varphi_i\rangle/k_{\text{B}}T}$. These can now be considered as the initial condition for $\hat{\sigma}(0)$ in the solution of equation (8.18) of any experiment described by a Hamiltonian $\hat{H}(t)$ with $\hat{H}(0) = \hat{H}_{\text{eq}}$. Of course, this spin density operator can also be represented in a basis set different from the eigenstates of $\hat{H}_{\text{eq}}$ by applying a simple complete basis set transformation.

In the forthcoming paragraphs, we will continue to consider small spin systems, while using the L-vN equation and its solution with a dimension governed by the dimension of these systems, while realizing that we are dealing with ensemble averages of the systems.

The solution of the L-vN equation has the 'sandwich' form

$$\hat{\sigma}(t) = \hat{U}(t)\hat{\sigma}(0)\hat{U}^{-1}(t) \tag{8.20}$$

where the evolution operator $\hat{U}(t)$ for a time-dependent Hamiltonian can be calculated via a stepwise integration defined by the Dyson time ordering operator $\hat{T}$

$$\hat{U}(t) = \hat{T}\text{e}^{-\text{i}\int_0^t \hat{H}(t')\text{d}t'} \tag{8.21}$$

and

$$\hat{T}\text{e}^{\int_0^t \hat{F}(t)\text{d}t} \equiv \lim_{\substack{N\to\infty;\,\Delta t\to 0\\ N\Delta t=t}} \prod_{k=1,N} \text{e}^{\hat{F}(k\Delta t)\Delta t} \tag{8.22}$$

For a time-independent Hamiltonian

$$\hat{U}(t) = \text{e}^{-\text{i}\hat{H}t} \tag{8.23}$$

The introduction of the $\hat{\sigma}(t)$ operator is very convenient when we are interested in evaluating the ensemble average expectation value of some operator $\hat{O}$. Choosing any complete set of basis states $\{|\varphi_i\rangle\}_{i=1,N}$ of the individual systems k and again writing their state as $|\psi_k(t)\rangle = \sum_i c_i^k(t)|\varphi_i\rangle$, we can present $\hat{O}$ and $\hat{\sigma}(t)$ in their matrix form in this basis and calculate the ensemble average $\langle\hat{O}\rangle(t)$ value

$$\begin{aligned}\langle\hat{O}\rangle(t) &= \sum_{i,j} \overline{c_j(t)c_i^*(t)}\langle\varphi_i|\hat{O}|\varphi_j\rangle\\ &= \sum_{i,j} \sigma_{ji}(t)O_{ij} = \text{Tr}(\hat{\sigma}(t)\hat{O})\end{aligned} \tag{8.24}$$

where the trace, Tr, of a matrix is the sum of all diagonal elements of that matrix. Choosing a different basis set of states will result in different matrix representations of $\hat{\sigma}(t)$, $\hat{O}$, and $\hat{\sigma}(t)\hat{O}$, but as the trace of a matrix is invariant under any linear transformation of its orthonormal basis set, the value of $\langle\hat{O}\rangle(t)$ is also invariant.

8.2.5 The Liouville Space Representation and Spin Relaxation

At this point, we return to the single-electron spin system. Considering again the matrix representation of $\hat{\sigma}(t)$ in the manifold $\{|\alpha\rangle, |\beta\rangle\}$ the L-vN equation, in equation (8.18), can be transformed to a vector differential equation by representing the spin density matrix elements in vector form $\vec{\sigma}(t)$ with elements $\{\sigma_{\alpha\alpha}, \sigma_{\beta\beta}, \sigma_{\alpha\beta}, \sigma_{\beta\alpha}\}$. In this Liouville representation, the rate equation for the matrix elements of $\hat{\sigma}(t)$ gets the form

$$\frac{\mathrm{d}}{\mathrm{d}t}\vec{\sigma}(t) = -i\hat{\hat{L}}(t)\vec{\sigma}(t) \tag{8.25}$$

where equation (8.18) dictates that the 4×4 Liouville superoperator $\hat{\hat{L}}(t)$ has the form

$$\hat{\hat{L}}(t) = \begin{pmatrix} 0 & 0 & -H_{\beta\alpha} & H_{\alpha\beta} \\ 0 & 0 & H_{\beta\alpha} & -H_{\alpha\beta} \\ -H_{\alpha\beta} & H_{\alpha\beta} & H_{\alpha\alpha} - H_{\beta\beta} & 0 \\ H_{\beta\alpha} & -H_{\beta\alpha} & 0 & H_{\beta\beta} - H_{\alpha\alpha} \end{pmatrix}(t) \tag{8.26}$$

The solution for the $\vec{\sigma}(t)$ vector requires an evaluation of the propagator $\hat{\hat{U}}(t)$ in this Liouville space representation

$$\vec{\sigma}(t) = \hat{\hat{U}}_L(t)\vec{\sigma}(0) \ : \ \hat{\hat{U}}_L(t) = \hat{T}\mathrm{e}^{-\mathrm{i}\int_0^t \hat{\hat{L}}(t')\mathrm{d}t'} \tag{8.27}$$

The reason for transforming the L-vN equation in equation (8.18) to equation (8.26) is that we are now able to introduce relaxation.

Above we have discussed the evolution of the spin systems as if they are isolated from the dynamic properties of their atomic and molecular surroundings that form the lattice of the system. We now want to discuss very briefly the influence of the fluctuations of the lattice on $\hat{\sigma}(t)$. These fluctuations can be visualized by a Hamiltonian composed of interaction terms with coefficients that are fluctuating in time. These time dependences are mostly random in nature and require a nontrivial evaluation of the time propagation of $\hat{\sigma}(t)$. In fact, the lattice has its own quantum energy levels and the fluctuating spin–lattice interactions should have been defined in the product basis of the states of the spins and the lattice. The projection of the influence of these interactions back onto the spin system can be presented as relaxation processes that drive $\hat{\sigma}(t)$ back to its thermal equilibrium $\hat{\sigma}_{\mathrm{eq}}$ defined in equation (8.19) with the average time-independent Hamiltonian $\hat{H}_{\mathrm{eq}}$. Expressions for the relaxation rates determining the time course of these processes are described in Chapter 9.

The relaxation of the spin-½ system to thermal equilibrium can in the Liouville space representation be represented by two rate constants when choosing $\{|\alpha\rangle, |\beta\rangle\}$ as the eigenstates of $\hat{H}_{\mathrm{eq}}$: The longitudinal relaxation rate constant R_1 that determines the timescale it takes for the populations $\sigma_{\alpha\alpha}(t)$ and $\sigma_{\beta\beta}(t)$ to reach their thermal equilibrium values $\sigma_{\alpha\alpha}^{\mathrm{eq}}$ and $\sigma_{\beta\beta}^{\mathrm{eq}}$ and the transverse relaxation rate constant R_2 that describes the timescale it takes to eliminate the coherences $\sigma_{\alpha\beta}(t)$ and $\sigma_{\beta\alpha}(t)$.[6] The equilibrium values are determined by the two conditions

$$\begin{aligned} &\sigma_{\alpha\alpha}^{\mathrm{eq}} + \sigma_{\beta\beta}^{\mathrm{eq}} = \sigma_{\alpha\alpha}(t) + \sigma_{\beta\beta}(t) = 1 \ \text{ and} \\ &\sigma_{\alpha\alpha}^{\mathrm{eq}}/\sigma_{\beta\beta}^{\mathrm{eq}} = \mathrm{e}^{-\hbar\omega_0/k_{\mathrm{B}}T} \equiv \varepsilon_{\mathrm{B}} \end{aligned} \tag{8.28}$$

which yield the rate equations for the populations

$$\begin{aligned} \frac{\mathrm{d}}{\mathrm{d}t}\sigma_{\alpha\alpha}(t) &= -R_1(\sigma_{\alpha\alpha}(t) - \sigma_{\alpha\alpha}^{\mathrm{eq}}) \\ &= \frac{-R_1}{(1+\varepsilon_{\mathrm{B}})}(\sigma_{\alpha\alpha}(t) - \varepsilon_{\mathrm{B}}\sigma_{\beta\beta}(t)) \\ \frac{\mathrm{d}}{\mathrm{d}t}\sigma_{\beta\beta}(t) &= -R_1(\sigma_{\beta\beta}(t) - \sigma_{\beta\beta}^{\mathrm{eq}}) \\ &= \frac{-R_1}{(1+\varepsilon_{\mathrm{B}})}(\varepsilon_{\mathrm{B}}\sigma_{\beta\beta}(t) - \sigma_{\alpha\alpha}(t)) \end{aligned} \tag{8.29}$$

This process is called *spin–lattice relaxation* and is typically described by the time constant $T_1 = \frac{1+\varepsilon_{\mathrm{B}}}{R_1}$. The relaxation rate equations for the coherences have the form

$$\begin{aligned} \frac{\mathrm{d}}{\mathrm{d}t}\sigma_{\alpha\beta}(t) &= -R_2\sigma_{\alpha\beta}(t) \\ \frac{\mathrm{d}}{\mathrm{d}t}\sigma_{\beta\alpha}(t) &= -R_2\sigma_{\beta\alpha}(t) \end{aligned} \tag{8.30}$$

and is typically described by introducing the spin–spin relaxation time constant $T_2 = \frac{1}{R_2}$.

Now equation (8.26) can be extended by taking relaxation into account

$$\frac{\mathrm{d}}{\mathrm{d}t}\vec{\sigma}(t) = (-\mathrm{i}\hat{\hat{L}} + \hat{\hat{R}})\vec{\sigma}(t) \tag{8.31}$$

with the relaxation superoperator

$$\hat{\hat{R}} = \begin{pmatrix} \dfrac{-R_1}{(1+\varepsilon_{\mathrm{B}})} & \dfrac{\varepsilon_{\mathrm{B}}R_1}{(1+\varepsilon_{\mathrm{B}})} & & \\ \dfrac{R_1}{(1+\varepsilon_{\mathrm{B}})} & \dfrac{-\varepsilon_{\mathrm{B}}R_1}{(1+\varepsilon_{\mathrm{B}})} & & \\ & & -R_2 & \\ & & & -R_2 \end{pmatrix} \tag{8.32}$$

8.2.6 The Orthogonal Angular Momentum Operators

Any time-dependent Hamiltonian affecting the physical state of the electron spin can be expressed in terms of a sum of the spin angular momentum operators $\hat{S}_p$ $(p=x,y,z)$ with real coefficients $h_p(t)$

$$\hat{H}(t) = h_x(t)\hat{S}_x + h_y(t)\hat{S}_y + h_z(t)\hat{S}_z \qquad (8.33)$$

The matrix representations of $\hat{S}_p$ and $\hat{S}_0$, defined by $\hat{S}_0|\psi(t)\rangle = |\psi(t)\rangle$, in some manifold of states $\{|\varphi_1\rangle, |\varphi_2\rangle\}$ form a set of orthogonal matrices, which fulfill $\mathrm{Tr}(\hat{S}_p\hat{S}_q)=0$ for $p\neq q$ and $\mathrm{Tr}(\hat{S}_p\hat{S}_p)=1/2$ for $p=x,y,z$ and which can span any 2×2 Hermitian matrix. This set of matrices can thus also be used to express the spin density matrix representation by writing

$$\hat{\sigma}(t) = 2s_x(t)\hat{S}_x + 2s_y(t)\hat{S}_y + 2s_z(t)\hat{S}_z + s_0\hat{S}_0 \qquad (8.34)$$

with real coefficients $s_p(t)$ and where we introduced the factors of 2 to simplify the expression for the expectation values realizing that $\mathrm{Tr}(\hat{S}_p\hat{S}_p)=1/2$. This enables us to express $\hat{\sigma}(t)$ in an operator form without choosing a specific basis set. Following equation (8.19), an example of a spin density operator is the Taylor expansion of the equilibrium spin density matrix elements of an electron with a Hamiltonian $\hat{H}_{\mathrm{eq}} = \omega_0\hat{S}_z$ in the form

$$\hat{\sigma}_{\mathrm{eq}} = \frac{1}{Z}\left[\left\{(1 + \frac{1}{4\times 2!}\left(\frac{\omega_0}{k_B T}\right)^2 + \ldots\right\}\hat{S}_0 - \left\{\left(\frac{\omega_0}{k_B T}\right) + \frac{1}{4\times 3!}\left(\frac{\omega_0}{k_B T}\right)^3 + \ldots\right\}\hat{S}_z\right] \qquad (8.35)$$

with

$$Z = \mathrm{e}^{\omega_0/2k_B T} + \mathrm{e}^{-\omega_0/2k_B T};\quad (\hat{S}_z)^{2n} = 4^{-n}\hat{S}_0 \qquad (8.36)$$

Thus, the equilibrium spin density operator has the general form

$$\hat{\sigma}_{\mathrm{eq}} = s_0\hat{S}_0 + 2s_z^{\mathrm{eq}}\hat{S}_z \qquad (8.37)$$

The operator form of $\hat{\sigma}(t)$ is very convenient when we are interested in evaluating expectation values as with the abovementioned orthogonality relation, we get

$$\langle\hat{S}_p\rangle(t) = \mathrm{Tr}(\hat{\sigma}(t)\hat{S}_p) = s_p(t) \qquad (8.38)$$

In summary, we thus realize that the Hamiltonian $\hat{H}(t)$ and the spin density operator $\hat{\sigma}(t)$ are both expressed in terms of the same set of independent operators $\hat{S}_p$ whose expectation values are given by equation (8.38) and can be evaluated by solving $\hat{U}(t)\hat{\sigma}(0)\hat{U}^{-1}(t)$. Thus, the task of any calculation for obtaining expectation values is to solve expressions of the form

$$s_p(\kappa\Delta t) = 2\mathrm{Tr} \times\left(\sum_{r=x,y,z} s_r((\kappa-1)\Delta t)\left\{\mathrm{e}^{-\mathrm{i}\Delta t\hat{H}(\kappa\Delta t)}\hat{S}_r\mathrm{e}^{\mathrm{i}\Delta t\hat{H}(\kappa\Delta t)}\right\}\hat{S}_p\right) \qquad (8.39)$$

or for time-independent Hamiltonians

$$s_p(t) = 2\mathrm{Tr}\left(\sum_{r=x,y,z}\hat{s}_r(0)\left\{\mathrm{e}^{-\mathrm{i}\hat{H}t}\hat{S}_r\mathrm{e}^{\mathrm{i}\hat{H}t}\right\}\hat{S}_p\right) \qquad (8.40)$$

Therefore, a basic requirement for calculating the time evolution of a spin system is to know how to calculate expressions of the form

$$\mathrm{e}^{-\mathrm{i}\,\theta\,\hat{S}_p}\hat{S}_q\mathrm{e}^{\mathrm{i}\,\theta\,\hat{S}_p} \qquad (8.41)$$

These types of calculations can be performed using the Taylor expansion of the exponential function as can be found in basic textbooks for quantum mechanics.[3] For two operators $\hat{A}$ and $\hat{B}$ defined in the same Hilbert space, we can expand the following transformation:

$$\mathrm{e}^{-\mathrm{i}\theta\hat{A}}\hat{B}\mathrm{e}^{\mathrm{i}\theta\hat{A}} = \hat{B} + \mathrm{i}\theta[\hat{B},\hat{A}] + \left(\frac{1}{2!}\right)(\mathrm{i}\theta)^2[\hat{A},[\hat{B},\hat{A}]] + \left(\frac{1}{3!}\right)(\mathrm{i}\theta)^3[\hat{A},[\hat{A},[\hat{B},\hat{A}]]] + \cdots \qquad (8.42)$$

When we now introduce a third operator $\hat{C}$ satisfying $[\hat{A},\hat{B}] = \mathrm{i}\hat{C}$ and $[\hat{A},\hat{C}] = -\mathrm{i}\hat{B}$, we can simplify this expansion and get

$$\mathrm{e}^{-\mathrm{i}\theta\hat{A}}\hat{B}\mathrm{e}^{\mathrm{i}\theta\hat{A}} = \hat{B}\left[1 - \frac{\theta^2}{2!}. + \cdots\right] + \hat{C}\left[\theta - \frac{\theta^3}{3!} + \cdots\right] = \hat{B}\cos\theta + \hat{C}\sin\theta \qquad (8.43)$$

We therefore obtain the following rule:

$$\text{“When}\left\{\begin{array}{c}[\hat{A},\hat{B}] = \mathrm{i}\hat{C}\\ [\hat{A},\hat{C}] = -\mathrm{i}\hat{B}\end{array}\right\}\text{then}\quad \mathrm{e}^{-\mathrm{i}\theta\hat{A}}\hat{B}\mathrm{e}^{\mathrm{i}\theta\hat{A}} = \hat{B}\cos\theta + \hat{C}\sin\theta\text{”} \qquad (8.44)$$

As mentioned previously, the spin-$^1/_2$ angular momentum operators $(\hat{S}_x, \hat{S}_y, \hat{S}_z)$ satisfy the cyclic commutation relations described in equation (8.6), and therefore, any operator transformation of the type in equation (8.41) can be evaluated using this rule. It also forms an essential part of spin evolution calculations based on the product operator formalism that has been used extensively to describe the evolution of a spin system under complex pulse sequences and interactions in NMR and EPR.[6–8,13]

In fact, in this chapter, we will try to provide the reader with the necessary tools to implement the product operator and fictitious spin-1/2 operator formalisms. To do so, we will discuss several basic pulse EPR sequences in terms of these sets of operators that also satisfy the abovementioned transformation rule. To simplify the text and assist the reader in utilizing the well-established product operator formalism, we introduce the notion of three *t*ransformation-*r*elated *i*ndependent *o*perators (trios) satisfying equation (8.44). To write this rule in short, we introduce the notation

$$\{\hat{A} : \hat{B} \rightarrow \hat{C}\}_1 \tag{8.45}$$

which means a transformation defined by some scalar θ times an operator $\hat{A}$ applied [:] on an operator $\hat{B}$ evolves [→] into an operator $\hat{C}$ when the commutation relations in equation (8.44) are valid. The subscript '1' indicates that the time evolution of the coefficients of $\hat{B}$ and $\hat{C}$ is the *sine* and *cosine* of one times θ.

8.2.7 The Rotating Frame and the Experimental Observables

When a MW irradiation field B_1 is applied along the x-axis, perpendicular to the external magnetic field B_0 determining the z-axis, with a frequency ω_{MW}, a phase φ_{MW}, and an amplitude $2\omega_1 = g_e\mu_B\hbar^{-1}B_1$ (the factor 2 is added to express the possible decomposition of the linear *cosine* oscillating B_1 vector into two counter rotating vectors of equal length ω_1), the single-electron spin Hamiltonian in equation (8.2) becomes time dependent

$$\hat{H}(t) = \omega_0\hat{S}_z + \hat{H}_{\mathrm{MW}}(t) = \omega_0\hat{S}_z + 2\omega_1\hat{S}_x\cos(\omega_{\mathrm{MW}}t + \varphi_{\mathrm{MW}}) \tag{8.46}$$

As a result, the solution of the L-vN equation requires a stepwise integration to obtain $\hat{U}(t)$ as in equations (8.21) and (8.22). A common way to overcome the stepwise integration is to transfer the spin density operator to a 'rotating frame' in which the Hamiltonian can be made time independent. In this frame, we then can solve the L-vN equation using equation (8.23).

Introducing the MW rotating frame transformation operator,

$$\hat{U}^{\mathrm{RoF}} = \mathrm{e}^{-\mathrm{i}\omega_{\mathrm{MW}}t\hat{S}_z} \tag{8.47}$$

we define the rotating-frame spin density operator by the transformation of $\sigma(t)$,

$$\hat{\sigma}^{\mathrm{RoF}}(t) = \hat{U}^{\mathrm{RoF}^{-1}}(t)\hat{\sigma}(t)\hat{U}^{\mathrm{RoF}}(t) \tag{8.48}$$

and derive the L-vN equation for $\hat{\sigma}^{\mathrm{RoF}}(t)$ by straightforwardly calculating the time derivative of $\hat{\sigma}^{\mathrm{RoF}}(t)$ and using the laboratory-frame L-vN equation in equation (8.18). This equation gets the form

$$\frac{\mathrm{d}}{\mathrm{d}t}\hat{\sigma}^{\mathrm{RoF}}(t) = -\mathrm{i}[\hat{H}^{\mathrm{RoF}}(t), \hat{\sigma}^{\mathrm{RoF}}(t)] \tag{8.49}$$

where

$$\hat{H}^{\mathrm{RoF}}(t) = \hat{U}^{\mathrm{RoF}^{-1}}(t)\hat{H}(t)\hat{U}^{\mathrm{RoF}}(t) - \omega_{\mathrm{MW}}\hat{S}_z \tag{8.50}$$

To derive the explicit expression for the rotating-frame Hamiltonian $\hat{H}^{\mathrm{RoF}}(t)$, we must transform the Hamiltonian in equation (8.46) using the trio $\{\hat{S}_z : \hat{S}_x \rightarrow \hat{S}_y\}_1$

$$\begin{aligned}\hat{U}^{\mathrm{RoF}^{-1}}(t)\hat{H}(t)\hat{U}^{\mathrm{RoF}}(t) &= \omega_0\hat{S}_z + 2\omega_1 (\hat{S}_x\cos\omega_{\mathrm{MW}}t - \hat{S}_y\sin\omega_{\mathrm{MW}}t)\cos(\omega_{\mathrm{MW}}t + \varphi_{\mathrm{MW}})\\ &= \omega_0\hat{S}_z + \omega_1(\hat{S}_x\cos\varphi_{\mathrm{MW}} + \hat{S}_y\sin\varphi_{\mathrm{MW}})\\ &\quad + \omega_1\{\hat{S}_x\cos(2\omega_{\mathrm{MW}}t + \varphi_{\mathrm{MW}})\\ &\quad - \hat{S}_y\sin(2\omega_{\mathrm{MW}}t + \varphi_{\mathrm{MW}})\}\end{aligned} \tag{8.51}$$

The time-dependent terms oscillating at a frequency $2\omega_{\mathrm{MW}}$ are in general ignored based on the rotating wave approximation (RWA), assuming that these terms with $2\omega_{\mathrm{MW}} \gg \omega_1$ are not actively influencing $\hat{\sigma}^{\mathrm{RoF}}(t)$. In fact, they cause a small shift of the transition frequency ω_0, called the *Bloch–Siegert shift.*[14] Disregarding this small shift, the rotating-frame Hamiltonian in equation (8.49) gets the form

$$\hat{H}^{\mathrm{RoF}}(t) = \Delta\omega_0\hat{S}_z + \omega_1(\hat{S}_x\cos\varphi_{MW} + \hat{S}_y\sin\varphi_{\mathrm{MW}}) \tag{8.52}$$

with the off-resonance frequency $\Delta\omega_0 = \omega_0 - \omega_{\mathrm{MW}}$.

The rotating-frame spin density operator can of course again be expressed as a linear combination of the orthogonal angular momentum operators as in equation (8.34)

$$\hat{\sigma}^{\mathrm{RoF}}(t) = 2s_x(t)\hat{S}_x + 2s_y(t)\hat{S}_y + 2s_z(t)\hat{S}_z + s_0\hat{S}_0 \tag{8.53}$$

where the coefficients are now determining the spin evolution in the rotating frame. During pulse EPR experiments, one typically measures the transverse magnetization of the ensemble of electrons. The heterodyne detection scheme of the spectrometers results in signals that are proportional to these magnetization components as if they are monitored in the MW rotating frame. Calling these components themselves x- and

y-signals, during quadrature detection, we can observe

$$m_x = -g_e\mu_B \mathrm{Tr}(\hat{\sigma}^{\mathrm{RoF}}(t)\hat{S}_x);\; m_y = -g_e\mu_B \mathrm{Tr}(\hat{\sigma}^{\mathrm{RoF}}(t)\hat{S}_y) \tag{8.54}$$

Combining these two signals in the form of a single 'complex signal', we define

$$m(t) = m_x(t) + \mathrm{i}m_y(t) \tag{8.55}$$

8.2.8 Expectation Values in the Liouville Space Representation

To simplify the addition of relaxation to the evolution, it is possible to replace the L-vN equation for a single electron with a Hamiltonian $\hat{H}^{\mathrm{RoF}} = \sum_{p=x,y,z} h_p(t)\hat{S}_p$ by a rate equation of its expectation values $s_p(t) = \mathrm{Tr}(\hat{\sigma}^{\mathrm{RoF}}(t)\hat{S}_p)$

$$\begin{aligned}\frac{\mathrm{d}}{\mathrm{d}t}s_p(t) &= -2\mathrm{i}\sum_{q,r=x,y,z} h_q(t)s_r(t)\mathrm{Tr}\{[\hat{S}_q,\hat{S}_r]\hat{S}_p\}\\ &= \sum_{\substack{r=x,y,z\\ r\neq p}}\left\{\sum_{\substack{q=x,y,z\\ q\neq p,r}} \varepsilon_{p,q,r}h_q(t)\right\}s_r(t)\\ &\times\begin{cases}\varepsilon_{x,y,z} = -\varepsilon_{x,z,y} = 1\\ \varepsilon_{y,z,x} = -\varepsilon_{y,x,z} = 1\\ \varepsilon_{z,x,y} = -\varepsilon_{z,y,x} = 1\end{cases}\end{aligned} \tag{8.56}$$

Presenting this rate equation as

$$\frac{\mathrm{d}}{\mathrm{d}t}\vec{s}(t) = \hat{\hat{L}}(t)\vec{s}(t) \tag{8.57}$$

where $\vec{s}(t)$ is the vector $(s_x(t), s_y(t), s_z(t))$, the matrix representing the Liouville superoperator $\hat{\hat{L}}(t)$ gets the form

$$\hat{\hat{L}}(t) = \begin{pmatrix} 0 & -h_z(t) & h_y(t)\\ h_z(t) & 0 & -h_x(t)\\ -h_y(t) & h_x(t) & 0\end{pmatrix} \tag{8.58}$$

When adding the spin relaxation processes, which drive the $\vec{s}(t)$ to its equilibrium vector $\vec{s}^{\mathrm{eq}} = (0,0,s_z^{\mathrm{eq}})$, we must again distinguish between a transverse relaxation rate constant R_2, causing the $s_x(t)$ and $s_y(t)$ to approach zero, and a longitudinal relaxation rate constant R_1, causing the value of $s_z(t)$ to approach the equilibrium s_z^{eq}. Equation (8.57) becomes

$$\frac{\mathrm{d}}{\mathrm{d}t}\vec{s}(t) = -\mathrm{i}\hat{\hat{L}}(t)\vec{s}(t) - \hat{\hat{R}}(\vec{s}(t) - \vec{s}^{\mathrm{eq}}) \tag{8.59}$$

with the relaxation superoperator

$$\hat{\hat{R}} = \begin{pmatrix} R_2 & & \\ & R_2 & \\ & & R_1\end{pmatrix} \tag{8.60}$$

Again realizing that the magnetization components $m_p(t)$ are proportional to $s_p(t)$, and using $\omega_{1x} = \omega_1\cos\varphi_{\mathrm{MW}}$ and $\omega_{1y} = \omega_1\sin\varphi_{\mathrm{MW}}$ in equation (8.52), the rate equation for $\vec{m}(t) = -g_e\mu_B\vec{s}(t)$ gets the form of the famous Bloch equation, with $\hat{H}^{\mathrm{RoF}} = \Delta\omega_0\hat{S}_z + \omega_{1x}\hat{S}_x + \omega_{1y}\hat{S}_y$

$$\begin{aligned}\frac{\mathrm{d}}{\mathrm{d}t}\begin{pmatrix} m_x(t)\\ m_y(t)\\ m_z(t)\end{pmatrix} &= \begin{pmatrix} 0 & -\Delta\omega_0 & \omega_{1y}(t)\\ \Delta\omega_0 & 0 & -\omega_{1x}(t)\\ -\omega_{1y}(t) & \omega_{1x}(t) & 0\end{pmatrix}\begin{pmatrix} m_x(t)\\ m_y(t)\\ m_z(t)\end{pmatrix}\\ &+\begin{pmatrix} -R_2 & & \\ & -R_2 & \\ & & -R_1\end{pmatrix}\begin{pmatrix} m_x(t)\\ m_y(t)\\ m_z(t) - m_z^{\mathrm{eq}}\end{pmatrix}\end{aligned} \tag{8.61}$$

The solution of this equation in the presence and absence of MW irradiation is explicitly discussed elsewhere (see Chapters 1 and 11).

8.2.9 The g-tensor Anisotropy

Before we discuss examples of pulse sequences, we will be more precise about the Hamiltonian of a single electron in an external magnetic field B_0. The general form for the Hamiltonian in the laboratory frame with its z-axis in the direction of the field, while taking the g anisotropy into account, is

$$\hat{H}_{\mathrm{Lab}} = \mu_B\hbar^{-1}\sum_{p=x,y,z} B_0 g_{zp}(\Omega_g)\hat{S}_p \tag{8.62}$$

with $g_{zp}(\Omega_g)$ the **g**-tensor matrix elements in the laboratory frame. These elements are a function of the relative orientation Ω_g defined by the Euler angles between the **g**-tensor principal axis system and the laboratory frame. Truncating this Hamiltonian for anisotropic **g**-tensor matrix elements that are much smaller than the isotropic g value, $\Delta g_{zp} \ll g_{\mathrm{iso}}$, it simplifies to

$$\hat{H}_{\mathrm{Lab}} = (\omega_0^{\mathrm{iso}} + \Delta\omega_z(\Omega_g))\hat{S}_z \tag{8.63}$$

with $\omega_0^{\mathrm{iso}} = g_{\mathrm{iso}}\mu_B\hbar^{-1}B_0$ and $\Delta\omega_z(\Omega_g) = \Delta g_{zz}(\Omega_g)\mu_B\hbar^{-1}B_0$. This approximation applies to organic radicals and many crystal defects but not to many paramagnetic transition metal and rare-earth metal ions (see Chapter 2).

8.2.10 Ideal MW Pulses

We will now use the basic tools we have introduced to describe some simple MW pulse experiments. While various MW pulse techniques are explicitly described in different chapters of the book, including off-resonance effects and relaxation effects during MW irradiation, we will here assume that the MW pulses are ideal, meaning that the amplitudes of the pulses are $\omega_1 \to \infty$ and their durations $t_{MW} \to 0$ and that they are solely characterized by a 'flip angle' defined by $\omega_1 t_{MW} = \theta$ and a phase φ_{MW} defining the direction of the MW field in the rotating frame. For a pulse in the $\pm x$ directions, the phase equals $\varphi_{MW} = 0, \pi$ and in the $\pm y$ directions $\varphi_{MW} = \pm \pi/2$. Thus, an MW pulse transfers $\hat{\sigma}^{RoF}(t)$ according to

$$\hat{\sigma}^{RoF}(t^+) = \hat{P}_\theta^{\varphi_{MW}} \hat{\sigma}^{RoF}(t^-) \hat{P}_\theta^{\varphi_{MW}^{-1}} \tag{8.64}$$

where t^- is the time just before the pulse and t^+ just after the pulse with $(t^+ - t^-) \to 0$. The pulse propagators $\hat{P}_\theta^{\varphi_{MW}}$ have the form

$$\hat{P}_\theta^{\varphi_{MW}} = e^{-i\theta(\cos\varphi_{MW}\hat{S}_x + \sin\varphi_{MW} S_y)} \tag{8.65}$$

and we will use the definitions $\hat{P}_\theta^{x/-x} \equiv \hat{P}_\theta^{0/\pi}$ and $\hat{P}_\theta^{\pm y} \equiv \hat{P}_\theta^{\pm\pi/2}$. For example, a $\theta = \pi/2$ pulse in the x- or y-direction applied on the equilibrium spin density operator as in equation (8.37) $\hat{\sigma}_{eq}(0^-) = 2s_z^{eq}\hat{S}_z + s_0\hat{S}_0$ transforms it to $\hat{\sigma}^{RoF}(0^+) = -2s_z^{eq}\hat{S}_y + s_0\hat{S}_0$ or $\hat{\sigma}^{RoF}(0^+) = 2s_z^{eq}\hat{S}_x + s_0\hat{S}_0$, respectively, where we used the trios of the form $\{\hat{S}_x : \hat{S}_z \to -\hat{S}_y\}_1$ and $\{\hat{S}_y : \hat{S}_z \to \hat{S}_x\}_1$. The $\hat{\sigma}^{RoF}(0^+)$ operator after a y-pulse of a system with a Hamiltonian $\hat{H}^{RoF} = \Delta\omega_0\hat{S}_z$ evolves according to

$$\begin{aligned}\hat{\sigma}^{RoF}(t) &= e^{-i\Delta\omega_0\hat{S}_z t}(2s_z^{eq}\hat{S}_x + s_0\hat{S}_0)e^{i\Delta\omega_0\hat{S}_z t}e^{-R_2 t}\\ &= s_0\hat{S}_0 + 2s_z^{eq}(\cos\Delta\omega_0 t\,\hat{S}_x + \sin\Delta\omega_0 t\,\hat{S}_y)e^{-R_2 t}\end{aligned} \tag{8.66}$$

where R_2 relaxation was introduced phenomenologically. Applying the definition of the complex signal in equations (8.54) and (8.55), this signal evolves then like

$$m(t) = -g_e\mu_B s_z^{eq}(\cos\Delta\omega_0 t + i\sin\Delta\omega_0 t)e^{-R_2 t} \tag{8.67}$$

which is also called the *free induction decay* (*FID*) *signal* after a single pulse.

As the identity matrix is a constant of the motion under any transformation of the type $e^{-i\theta\hat{S}_p t}\hat{S}_0 e^{i\theta\hat{S}_p t} = \hat{S}_0$, we simplify the equations from now on by following the time evolution of the reduced spin density operator

$$\hat{\sigma}(t) \equiv \hat{\sigma}^{RoF}(t) - s_0\hat{S}_0 \tag{8.68}$$

8.2.11 Echo Pulse Sequences

8.2.11.1 The Hahn Echo Sequence

For our first example, we chose the basic Hahn echo pulse sequence that consists of two MW pulses, the first a $\pi/2$-pulse applied at time $t = 0$ and the second a π-pulse at $t = \tau$, and two evolution delays with equal time durations τ, as shown in Figure 8.2(a). Operating on $\hat{\sigma}_{eq}$ of the single-electron system with a rotating-frame Hamiltonian $H^{RoF} = \Delta\omega_0\hat{S}_z$ and a propagator $U^{RoF}(t) = e^{-i\Delta\omega_0\hat{S}_z t}$, this sequence with a π-pulse in the x-direction can be represented as

$$\{\underset{\downarrow 0^+}{\hat{P}_{\pi/2}^{x/y}} - \tau - \underset{\downarrow \tau^- \ \downarrow \tau^+}{\hat{P}_\pi^x} - \tau - \underset{\downarrow 2\tau}{}\}$$

The purpose of this sequence is to retrieve at time $t = 2\tau$ the signal that is obtained after the first pulse at $t = 0^+$ (Figure 8.2a). This sequence is used as the standard signal detection block in pulse EPR experiments as the *FID* is difficult to detect due to receiver dead time. Assuming that after the first pulse $\hat{\sigma}(0^+) = 2s_p\hat{S}_p$ with p equal to x or to y, depending on the phase of this pulse, the complex signal (see equation (8.55)) becomes equal to $m_x(0^+)$ or $im_y(0^+)$ and an ideal Hahn echo sequence must result in $m_p(2\tau) = \pm m_p(0^+)$ with $p = x$ or $p = y$ and thus

$$\mathrm{Tr}(\hat{\sigma}(0^+)\hat{S}_p) = \pm\mathrm{Tr}(\hat{\sigma}(2\tau)\hat{S}_p) \tag{8.69}$$

where for simplicity, we ignore all relaxation effects during the 2τ time interval. The evolution of the spin

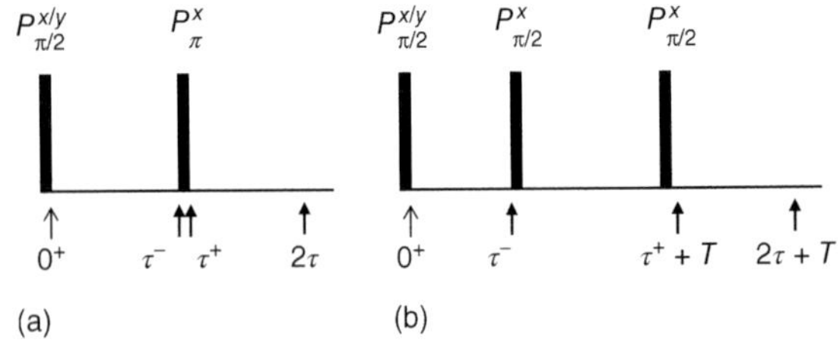

Figure 8.2. The two basic pulse-EPR sequences that are applied on a single electron spin as discussed in Section 8.2: (a) the Hahn echo sequence and (b) the stimulated echo sequence

density operator during the pulse sequence until $t = 2\tau$ can straightforwardly be calculated using the operator transformations as described earlier. However, when we will consider more complex echo sequences applied on larger spin systems later on, these calculations can become cumbersome as each pulse and free evolution period can increase the number of operators involved. To simplify calculations we use here a theoretical approach for determining the efficiency of the echo sequence.

When we again assume that $\hat{\sigma}(0^+) = 2s_p\hat{S}_p$ and calculate the expression for $\hat{\sigma}(2\tau)$, we can introduce an echo (efficiency) factor $E^p_{\text{Hahn-echo}}$ defined by

$$m_p(2\tau) = E^p_{\text{Hahn-echo}} m_p(0^+) \tag{8.70}$$

In the ideal case, one should get that $E^p_{\text{Hahn-echo}} = \pm 1$. To calculate this factor, we can write that

$$E^p_{\text{Hahn-echo}} = \frac{m_p(2\tau)}{m_p(0^+)} = \frac{2s_p \text{Tr}(\hat{\sigma}(2\tau)\hat{S}_p)}{2s_p \text{Tr}(\hat{\sigma}(0^+)\hat{S}_p)} = \frac{\text{Tr}(\hat{\sigma}(2\tau)\hat{\sigma}(0^+))}{\text{Tr}(\hat{\sigma}(0^+)^2)} \tag{8.71}$$

The trace in the numerator can be written explicitly as

$$\begin{aligned}\text{Tr}\left\{\hat{\sigma}(2\tau)\hat{\sigma}(0^+)\right\} &= \text{Tr}([\hat{U}(\tau)\hat{P}^x_\pi\hat{U}(\tau)\hat{\sigma}(0^+)\hat{U}^{-1}(\tau)\hat{P}^{x-1}_\pi\hat{U}^{-1}(\tau)]\hat{\sigma}(0^+)) \\ &= \text{Tr}([\hat{P}^x_\pi\hat{U}(\tau)\hat{\sigma}(0^+)\hat{U}^{-1}(\tau)\hat{P}^{x-1}_\pi][\hat{U}^{-1}(\tau)\hat{\sigma}(0^+)\hat{U}(\tau)])\end{aligned} \tag{8.72}$$

with $\hat{U}(t) = \hat{U}^{\text{RoF}}(t)$ and where we used that $\text{Tr}(AB) = \text{Tr}(BA)$. Inspection of the argument of this trace shows that we can make a distinction between operator $\hat{P}^x_\pi\hat{U}(\tau)\hat{\sigma}(0^+)\hat{U}^{-1}(\tau)\hat{P}^{x-1}_\pi$ and $\hat{U}^{-1}(\tau)\hat{\sigma}(0^+)\hat{U}(\tau)$. The first one represents the spin density operator $\hat{\sigma}^\tau \equiv \hat{U}(\tau)\hat{\sigma}(0^+)\hat{U}^{-1}(\tau)$ at time τ transformed by the echo pulse $\hat{P}^x_\pi\hat{\sigma}^\tau\hat{P}^{x-1}_\pi$. The second one looks like a spin density operator $\hat{\sigma}^{-\tau}$ that starts at $\hat{\sigma}(0^+)$ and evolves backward in time $\hat{\sigma}^{-\tau} \equiv \hat{U}^{-1}(\tau)\hat{\sigma}(0^+)\hat{U}(\tau)$ during a period $-\tau$. Thus, a calculation of the trace in equation (8.69) requires first a calculation of $\hat{\sigma}^\tau$ and $\hat{\sigma}^{-\tau}$ and then a transformation $\hat{P}^x_\pi\hat{\sigma}^\tau\hat{P}^{x-1}_\pi$. In the special case that $\hat{P}^x_\pi\hat{\sigma}^\tau\hat{P}^{x-1}_\pi = \pm\hat{\sigma}^{-\tau}$ the trace becomes equal to $\pm\text{Tr}((\hat{P}^x_\pi\hat{\sigma}^\tau\hat{P}^{x-1}_\pi)^2)$. Using again that $\text{Tr}(\hat{A}\hat{B}) = \text{Tr}(\hat{B}\hat{A})$ and the definition of $\hat{\sigma}^\tau$, this last trace is equal to $\text{Tr}(\hat{\sigma}(0^+)^2)$. Going back to the expression for the echo factor $E^p_{\text{Hahn-echo}}$ in equation (8.71), we can now write it as

$$E^p_{\text{Hahn-echo}} = \frac{\text{Tr}([\hat{P}^x_\pi\hat{\sigma}^\tau\hat{P}^{x-1}_\pi]\,\hat{\sigma}^{-\tau})}{\text{Tr}(\hat{\sigma}(0^+)^2)} \tag{8.73}$$

It is now apparent that to reach a value ± 1 the echo sequence must result in $\hat{P}^x_\pi\hat{\sigma}^\tau\hat{P}^{x-1}_\pi = \pm\sigma^{-\tau}$. If not, the echo factor becomes $|E_{\text{Hahn-echo}}| < 1$. We should mention here again that we ignored any spin–spin relaxation effects on the spin system.

We can now apply this method of calculating the echo factor on the basic Hahn echo sequence. Beginning with the first ideal π/2-pulse, applied in the x- or y-direction and using the trios mentioned earlier, we get that $\hat{\sigma}(0^+)$ equals to $-2s^{\text{eq}}_z\hat{S}_y$ or $2s^{\text{eq}}_z\hat{S}_x$, respectively. Now we can calculate $\hat{\sigma}^\tau$ and $\hat{\sigma}^{-\tau}$ and the transformation $\hat{P}^x_\pi\hat{\sigma}^\tau\hat{P}^{x-1}_\pi$. When the Hamiltonian of the electron has the form $\hat{H}^{\text{RoF}} = \Delta\omega_0\hat{S}_z$, the results of these calculations are

$$\hat{\sigma}(0^+) = 2s^{\text{eq}}_z\hat{P}^x_{\pi/2}\hat{S}_z\hat{P}^{x-1}_{\pi/2} = -2s^{\text{eq}}_z\hat{S}_y \tag{8.74}$$

$$\hat{\sigma}^\tau = 2s^{\text{eq}}_z(-\hat{S}_y\cos\Delta\omega_0\tau + \hat{S}_x\sin\Delta\omega_0\tau) \tag{8.75}$$

$$\hat{\sigma}^{-\tau} = 2s^{\text{eq}}_z(-\hat{S}_y\cos\Delta\omega_0\tau - \hat{S}_x\sin\Delta\omega_0\tau) \tag{8.76}$$

$$\hat{P}^x_\pi\hat{\sigma}^\tau\hat{P}^{x-1}_\pi = 2s^{\text{eq}}_z(\hat{S}_y\cos\Delta\omega_0\tau + \hat{S}_x\sin\Delta\omega_0\tau) \tag{8.77}$$

$$E^y_{\text{echo}} = -1 \tag{8.78}$$

or

$$\hat{\sigma}(0^+) = 2s^{\text{eq}}_z\hat{P}^y_{\pi/2}\hat{S}_z\hat{P}^{y-1}_{\pi/2} = 2s^{\text{eq}}_z\hat{S}_x \tag{8.79}$$

$$\hat{\sigma}^\tau = 2s^{\text{eq}}_z(\hat{S}_x\cos\Delta\omega_0\tau + \hat{S}_y\sin\Delta\omega_0\tau) \tag{8.80}$$

$$\hat{\sigma}^{-\tau} = 2s^{\text{eq}}_z(\hat{S}_x\cos\Delta\omega_0\tau - \hat{S}_y\sin\Delta\omega_0\tau) \tag{8.81}$$

$$\hat{P}^x_\pi\hat{\sigma}^\tau\hat{P}^{x-1}_\pi = 2s^{\text{eq}}_z(\hat{S}_x\cos\Delta\omega_0\tau - \hat{S}_y\sin\Delta\omega_0\tau) \tag{8.82}$$

$$E^x_{\text{echo}} = +1 \tag{8.83}$$

These calculations show that when the first π/2 pulse is applied in the same x-direction the echo signal is inverted $m_y(2\tau) = -m_y(0^+)$ with respect to the initial signal and when this pulse is applied in the y-direction the echo signal is in phase $m_x(2\tau) = m_x(0^+)$. In a typical echo experiment, the spins have many different off-resonance frequencies, all evolving at 2τ to the same value, resulting in the formation of an echo.

The repetitive application of π-pulses in the x-direction with delays 2τ, called the *Carr–Purcell sequence*, $\left\{\hat{P}^x_{\pi/2} - \left(\tau - \hat{P}^x_\pi - \tau\right)_{\times n}\right\}$, results for $p = x$ in alternating echo signals $m_y(2n\tau) = (-1)^{(2n-1)}m_y(0^+)$ and for $p = y$, called the *Carr–Purcell–Meiboom–Gill* sequence, $\left\{\hat{P}^y_{\pi/2} - \left(\tau - \hat{P}^x_\pi - \tau\right)_{\times n}\right\}$, in equal-sign signals $m_x(2n\tau) = m_x(0^+)$, again ignoring relaxation effects.

8.2.11.2 The Stimulated Echo Sequence

During the stimulated echo sequence, the π-pulse of the Hahn echo sequence is split into two $\pi/2$-pulses with a delay T between them (Figure 8.2b)

$$\{\overset{0^+}{\downarrow}\hat{P}^{x/y}_{\pi/2} - \tau - \overset{\tau^-}{\downarrow}\hat{P}^{x}_{\pi/2} - T - \hat{P}^{x}_{\pi/2}\overset{\tau^++T}{\downarrow} - \tau - \overset{2\tau+T}{\downarrow}\}$$

In this case, we can again start with $\hat{\sigma}(0^+) = s_z^{\text{eq}}\hat{S}_p$ and use the same methodology as above except that we must replace the π pulse transformation in equation (8.73) with the transformation $\hat{P}^x_T\hat{\sigma}^\tau\hat{P}^{x-1}_T$ where $\hat{P}^x_T \equiv \hat{P}^x_{\pi/2}\hat{U}(T)\hat{P}^x_{\pi/2}$. In that case we can again define an echo factor according to

$$m_p(2\tau + T) = E^p_{\text{st-echo}} m_p(0^+) \qquad (8.84)$$

with

$$E^p_{\text{st-echo}} = \frac{\text{Tr}([\hat{P}^x_T\hat{\sigma}^\tau\hat{P}^{x-1}_T]\,\hat{\sigma}^{-\tau})}{\text{Tr}(\hat{\sigma}(0^+)^2)} \qquad (8.85)$$

Introducing in this case a phase memory relaxation mechanism with a time constant T_{m} that is short enough to eliminate all coherences during time T, we can ignore the term in the spin density operator of the form $2s_x(t)\hat{S}_x$ or $2s_y(t)\hat{S}_y$ after the second $\pi/2$-pulse of the sequence. Consequently, the $\hat{S}_x$ term of $\hat{\sigma}^\tau$ is eliminated because it remains the same during the time T, but the $\hat{S}_y$ term remains because it turns into $\hat{S}_z$ during T. To obtain the echo factor for the stimulated echo sequence, we obtain, taking the T_{m} relaxation into account, and assuming $T_{\text{m}} < T$

$$\hat{\sigma}(0^+) = 2s_z^{\text{eq}}\hat{P}^x_{\pi/2}\hat{S}_z\hat{P}^{x-1}_{\pi/2} = -2s_z^{\text{eq}}\hat{S}_y \qquad (8.86)$$

$$\hat{\sigma}^\tau = 2s_z^{\text{eq}}(-\hat{S}_y\cos\Delta\omega_0\tau + \hat{S}_x\sin\Delta\omega_0\tau) \qquad (8.87)$$

$$\hat{\sigma}^{-\tau} = 2s_z^{\text{eq}}(-\hat{S}_y\cos\Delta\omega_0\tau - \hat{S}_x\sin\Delta\omega_0\tau) \qquad (8.88)$$

$$\hat{P}^x_T\hat{\sigma}^\tau\hat{P}^{x-1}_T = 2s_z^{\text{eq}}\hat{S}_y\cos\Delta\omega_0\tau \qquad (8.89)$$

$$E^y_{\text{st-echo}} = -\cos^2\Delta\omega_0\tau \qquad (8.90)$$

or

$$\hat{\sigma}(0^+) = 2s_z^{\text{eq}}\hat{P}^y_{\pi/2}\hat{S}_z\hat{P}^{y-1}_{\pi/2} = 2s_z^{\text{eq}}\hat{S}_x \qquad (8.91)$$

$$\hat{\sigma}^\tau = 2s_z^{\text{eq}}(\hat{S}_x\cos\Delta\omega_0\tau + \hat{S}_y\sin\Delta\omega_0\tau) \qquad (8.92)$$

$$\hat{\sigma}^{-\tau} = 2s_z^{\text{eq}}(\hat{S}_x\cos\Delta\omega_0\tau - \hat{S}_y\sin\Delta\omega_0\tau) \qquad (8.93)$$

$$\hat{P}^x_T\hat{\sigma}^\tau\hat{P}^{x-1}_T = -2s_z^{\text{eq}}\hat{S}_y\sin\Delta\omega_0\tau \qquad (8.94)$$

$$E^x_{\text{st-echo}} = +\sin^2\Delta\omega_0\tau \qquad (8.95)$$

Here, we used that the numerator $2s_z^{\text{eq}}\text{Tr}(-\cos^2\Delta\omega_0\tau(\hat{S}_y)^2)$ of $E^y_{\text{st-echo}}$ equals to $-\cos^2\Delta\omega_0\tau$ times its denominator $2s_z^{\text{eq}}\text{Tr}((\hat{S}_y)^2)$ and similarly for $E^x_{\text{st-echo}}$.

8.3 TWO COUPLED ELECTRONS

8.3.1 The Product Operators and Their Commutation Relations

The matrix representation of the spin density operator $\hat{\sigma}(t)$ of a single electron in a magnetic field can be easily extended to a system of two coupled electrons $\{e_A - e_B\}$ with spin quantum numbers $S_A = S_B = 1/2$ by choosing an appropriate basis set of spin states. A convenient basis set of states can be composed of the four product-eigenstates of $\omega_0(\hat{S}_{A,z} + \hat{S}_{B,z})$

$$|1\rangle = |\alpha_A\alpha_B\rangle;\ |2\rangle = |\alpha_A\beta_B\rangle;\ |3\rangle = |\beta_A\alpha_B\rangle;\ |4\rangle = |\beta_A\beta_B\rangle \qquad (8.96)$$

(with decreasing eigenvalues) as shown in Figure 8.3(a). In analogy to the discussion in Section 8.2.4 the elements of the 4×4 matrix representation of $\hat{\sigma}(t)$ are the values of the averages $\sigma_{ij}(t) = \overline{c_i(t)c_j^*(t)}$ $(i,j = 1,..,4)$ of an ensemble of K two-electron systems in states $|\psi_k(t)\rangle = \sum_{i=1,4} c_i^k(i)|i\rangle$ where $k = 1, \ldots, K$. The

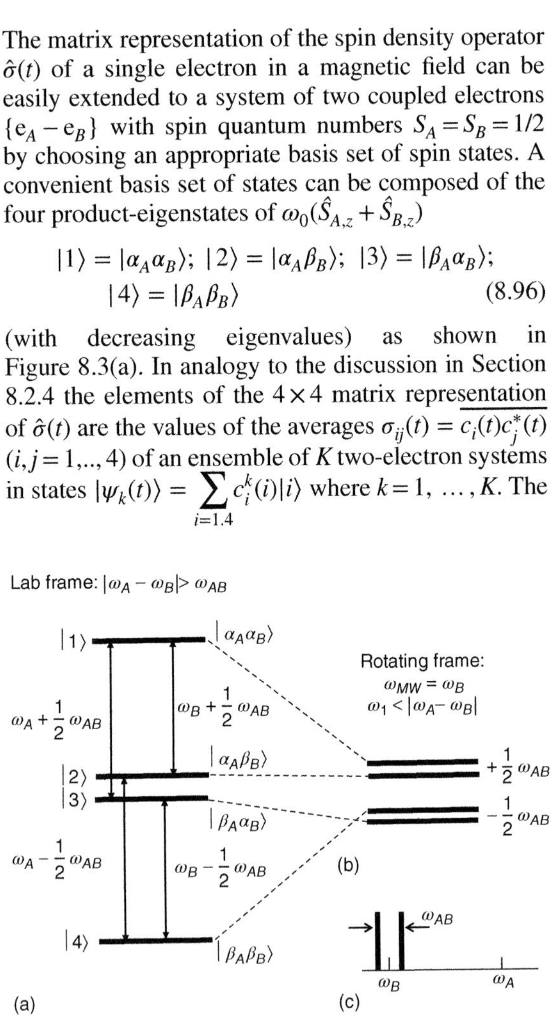

Figure 8.3. The energy level diagram of a two-electron spin system under 'weak coupling' conditions in (a) the laboratory frame and (b) the MW rotating frame in the case of a semiselective irradiation at the frequency of spin e_B. (c) A stick diagram of the spectrum obtained after a semiselective $\pi/2$ pulse at $\omega_{\text{MW}} = \omega_B$

diagonal elements $\sigma_{ii}(t)$ represent the probabilities of the systems to be in state $|i\rangle$, i.e., populations, and the off-diagonal elements $\sigma_{ij}(t)$ the probabilities of being in a superposition state of $|i\rangle$ and $|j\rangle$, i.e., coherences. The elements $\sigma_{12}(t)$ and $\sigma_{34}(t)$ are the coherences of the two $(\alpha_B \leftrightarrow \beta_B)$ single-quantum transitions of electron e_B and $\sigma_{13}(t)$ and $\sigma_{24}(t)$ the two $(\alpha_A \leftrightarrow \beta_A)$ single-quantum transitions of electron e_A. $\sigma_{23}(t)$ is the coherence of the zero-quantum (zq) transition and $\sigma_{14}(t)$ the coherence of the double-quantum (dq) transition. The fact that $\sigma_{ji}(t) = \sigma_{ij}^*(t)$ implies that the $\hat{\sigma}(t)$ matrix representation is Hermitian. The values of $\sigma_{ij}(t)$ can be determined by solving the 4×4 L-vN equation of the form as in equation (8.18), this time with a 4×4 Hamiltonian matrix with elements $H_{ij}(t) = \langle i|\hat{H}(t)|j\rangle$.

Similar to the single-electron case, the 4×4 $\hat{\sigma}(t)$ matrix representation can now be expressed in terms of a sum of 16 orthogonal Hermitian matrices. These matrices can again be derived from the electron spin angular momentum operators represented in the basis of equation (8.96).

Different sets of orthogonal operators can be considered, and here we choose the common angular momentum product operators. In addition to the 4×4 unit operator $\hat{S}_0^{(AB)}$ with nonzero matrix elements $\langle i|\hat{S}_0^{(AB)}|i\rangle = 1$, we have the 15 operators (their matrix representations are shown in Figure 8.4), grouped as follows:

- the four linear single-quantum operators

$$\hat{S}_{x0}^{(AB)} = \hat{S}_{A,x};\ \hat{S}_{y0}^{(AB)} = \hat{S}_{A,y};\ \hat{S}_{0x}^{(AB)} = \hat{S}_{B,x};\ \hat{S}_{0y}^{(AB)} = \hat{S}_{B,y} \tag{8.97}$$

- the four bilinear single-quantum operators

$$\hat{S}_{xz}^{(AB)} = 2\hat{S}_{A,x}\hat{S}_{B,z};\ \hat{S}_{yz}^{(AB)} = 2\hat{S}_{A,y}\hat{S}_{B,z}$$
$$\hat{S}_{zx}^{(AB)} = 2\hat{S}_{A,z}\hat{S}_{B,x};\ \hat{S}_{zy}^{(AB)} = 2\hat{S}_{A,z}\hat{S}_{B,y} \tag{8.98}$$

- the four bilinear {zero+double}-quantum operators

$$\begin{aligned}
\hat{S}_{xx}^{(AB)} &= 2\hat{S}_{A,x}\hat{S}_{B,x} \\
&= 1/2(\hat{S}_A^+\hat{S}_B^+ + \hat{S}_A^-\hat{S}_B^- + \hat{S}_A^+\hat{S}_B^- + \hat{S}_A^-\hat{S}_B^+) \\
\hat{S}_{yy}^{(AB)} &= 2\hat{S}_{A,y}\hat{S}_{B,y} \\
&= -1/2(\hat{S}_A^+\hat{S}_B^+ + \hat{S}_A^-\hat{S}_B^- - \hat{S}_A^+\hat{S}_B^- - \hat{S}_A^-\hat{S}_B^+) \\
\hat{S}_{xy}^{(AB)} &= 2\hat{S}_{A,x}\hat{S}_{B,y} \\
&= -i/2(\hat{S}_A^+\hat{S}_B^+ - \hat{S}_A^-\hat{S}_B^- - \hat{S}_A^+\hat{S}_B^- + \hat{S}_A^-\hat{S}_B^+) \\
\hat{S}_{yx}^{(AB)} &= 2\hat{S}_{A,y}\hat{S}_{B,x} \\
&= -i/2(\hat{S}_A^+\hat{S}_B^+ - \hat{S}_A^-\hat{S}_B^- + \hat{S}_A^+\hat{S}_B^- - \hat{S}_A^-\hat{S}_B^+)
\end{aligned} \tag{8.99}$$

Combining these {zero+double}-quantum operators can result in the zq and dq operators

$$\begin{aligned}
\hat{S}_{zq,x}^{(AB)} &\equiv 1/2(\hat{S}_{xx}^{(AB)} + \hat{S}_{yy}^{(AB)}); \\
\hat{S}_{dq,x}^{(AB)} &\equiv 1/2(\hat{S}_{xx}^{(AB)} - \hat{S}_{yy}^{(AB)}) \\
\hat{S}_{zq,y}^{(AB)} &\equiv 1/2(\hat{S}_{yx}^{(AB)} - \hat{S}_{xy}^{(AB)}); \\
\hat{S}_{dq,y}^{(AB)} &\equiv 1/2(\hat{S}_{yx}^{(AB)} + \hat{S}_{xy}^{(AB)})
\end{aligned} \tag{8.100}$$

- and in addition three independent z-component operators

$$\hat{S}_{z0}^{(AB)} = \hat{S}_{A,z};\ \hat{S}_{0z}^{(ab)} = \hat{S}_{B,z};\ \hat{S}_{zz}^{(AB)} = 2\hat{S}_{A,z}\hat{S}_{B,z} \tag{8.101}$$

which generate the dq and zq z-operators

$$\hat{S}_{dq,z}^{(AB)} = 1/2(\hat{S}_{z0}^{(AB)} + \hat{S}_{0z}^{(AB)});\ \hat{S}_{zq.z}^{(AB)} = 1/2(\hat{S}_{z0}^{(AB)} - \hat{S}_{0z}^{(AB)}) \tag{8.102}$$

The nomenclature presented here for the operators was chosen to simplify the equations expressing the expansions and transformations in the forthcoming discussion without losing the information about their composition. The spin density operator of the two-electron spin system can now be expanded as a linear combination of the 16 orthogonal operators with time-dependent coefficients of the form $s_{pq}^{(AB)}(t)$ $(p, q = 0, x, y, z)$

$$\begin{aligned}
\hat{\sigma}(t) = &\sum_{p=x,y,z} \left(s_{p0}^{(AB)}\hat{S}_{p0}^{(AB)} + s_{0p}^{(AB)}\hat{S}_{0p}^{(AB)}\right) \\
&+ \sum_{p=x,y} \left(s_{pz}^{(AB)}\hat{S}_{pz}^{(AB)} + s_{zp}^{(AB)}\hat{S}_{zp}^{(AB)}\right) \\
&+ \sum_{p,q=x,y} s_{pq}^{(AB)}\hat{S}_{pq}^{(AB)} + s_{zz}^{(AB)}\hat{S}_{zz}^{(AB)} + s_0\hat{S}_0^{(AB)}
\end{aligned} \tag{8.103}$$

where for simplicity we used $s_{pq}^{(AB)} \equiv s_{pq}^{(AB)}(t)$. The expectation values of the operators $\hat{S}_{pq}^{(AB)}$ can be obtained by calculating

$$\langle \hat{S}_{pq}^{(AB)} \rangle = \mathrm{Tr}(\hat{\sigma}(t)\hat{S}_{pq}^{(AB)}) = s_{pq}^{(AB)}(t) \tag{8.104}$$

realizing that in the two-electron spin system $\mathrm{Tr}\left(\left(\hat{S}_{pq}^{(AB)}\right)^2\right) = 1$. To evaluate the time evolution of these expectation values, we must solve the L-vN equation, where now the Hamiltonian has the

$$S_{0x}^{(AB)}=\frac{1}{2}\begin{pmatrix}0&1&&\\1&0&&\\&&0&1\\&&1&0\end{pmatrix};\quad S_{0y}^{(AB)}=\frac{1}{2}\begin{pmatrix}0&-i&&\\i&0&&\\&&0&-i\\&&i&0\end{pmatrix};\quad S_{x0}^{(AB)}=\frac{1}{2}\begin{pmatrix}0&&1&\\&0&&1\\1&&0&\\&1&&0\end{pmatrix};\quad S_{y0}^{(AB)}=\frac{1}{2}\begin{pmatrix}0&&-i&\\&0&&-i\\i&&0&\\&i&&0\end{pmatrix}$$

$$S_{zx}^{(AB)}=\frac{1}{2}\begin{pmatrix}0&1&&\\1&0&&\\&&0&-1\\&&-1&0\end{pmatrix};\quad S_{zy}^{(AB)}=\frac{1}{2}\begin{pmatrix}0&-i&&\\i&0&&\\&&0&i\\&&-i&0\end{pmatrix};\quad S_{xz}^{(AB)}=\frac{1}{2}\begin{pmatrix}0&&1&\\&0&&-1\\1&&0&\\&-1&&0\end{pmatrix};\quad S_{yz}^{(AB)}=\frac{1}{2}\begin{pmatrix}0&&-i&\\&0&&i\\i&&0&\\&-i&&0\end{pmatrix}$$

$$S_{xx}^{(AB)}=\frac{1}{2}\begin{pmatrix}&&&1\\&&1&\\&1&&\\1&&&\end{pmatrix};\quad S_{yy}^{(AB)}=\frac{1}{2}\begin{pmatrix}&&&-1\\&&1&\\&1&&\\-1&&&\end{pmatrix};\quad S_{xy}^{(AB)}=\frac{1}{2}\begin{pmatrix}&&&-i\\&&i&\\&-i&&\\i&&&\end{pmatrix};\quad S_{yx}^{(AB)}=\frac{1}{2}\begin{pmatrix}&&&-i\\&&-i&\\&i&&\\i&&&\end{pmatrix}$$

$$S_{0z}^{(AB)}=\frac{1}{2}\begin{pmatrix}1&&&\\&-1&&\\&&1&\\&&&-1\end{pmatrix};\quad S_{z0}^{(AB)}=\frac{1}{2}\begin{pmatrix}1&&&\\&1&&\\&&-1&\\&&&-1\end{pmatrix};\quad S_{zz}^{(AB)}=\frac{1}{2}\begin{pmatrix}1&&&\\&-1&&\\&&-1&\\&&&1\end{pmatrix};\quad S_{0}^{(AB)}=\begin{pmatrix}1&&&\\&1&&\\&&1&\\&&&1\end{pmatrix}$$

Figure 8.4. The matrix representations of the 16 product operators presented in equations (8.97)–(8.102) in the basis set of states defined by equation (8.96) and shown in Figure 8.3. These operators are used to express the Hamiltonian and the spin density operator of two-electron spin systems

general form

$$\begin{aligned}\hat{H}(t) = &\sum_{p=x,y,z}\left(h_{p0}^{(AB)}\hat{S}_{p0}^{(AB)} + h_{0p}^{(ab)}\hat{S}_{0p}^{(AB)}\right)\\ &+\sum_{p=x,y}\left(h_{pz}^{(AB)}\hat{S}_{pz}^{(AB)} + h_{zp}^{(AB)}\hat{S}_{zp}^{(AB)}\right)\\ &+\sum_{p=x,y} h_{pq}^{(AB)}\hat{S}_{pq}^{(AB)} + h_{zz}^{(AB)}\hat{S}_{zz}^{(AB)}\end{aligned} \tag{8.105}$$

The solution of the L-vN equation is again of the form of equation (8.20), and the type of transformations we must perform are

$$\left\{ e^{-ih_{pq}^{(AB)}\hat{S}_{pq}^{(AB)}}\hat{S}_{p'q'}^{(AB)}e^{ih_{pq}^{(AB)}\hat{S}_{pq}^{(AB)}} \right\} \tag{8.106}$$

The actual calculation of the solution of the L-vN equation cannot, in general, be performed analytically, and numerical procedures are required to evaluate $\hat{\sigma}(t)$. However, these numerical calculations do not provide the physical insight necessary to design new experiments or evaluate experimental results. Thus, we should at least be able to estimate the characteristic features of the time evolution of $\hat{\sigma}(t)$, and a way of doing so requires the knowledge of performing transformations of the type of equation (8.106).

Using the Taylor expansion as we did for the derivations of the transformation rule for the trios in equation (8.44), we can generalize this rule by claiming that for any scalar θ and integer κ

$$\text{“When } \left\{\begin{matrix}[\hat{A},\hat{B}] = i\kappa\hat{C}\\ [\hat{A},\hat{C}] = -i\kappa\hat{B}\end{matrix}\right\} \text{ then}\quad e^{-i\theta\hat{A}}\hat{B}e^{i\theta\hat{A}} = \hat{B}\cos\kappa\theta + \hat{C}\sin\kappa\theta\text{”} \tag{8.107}$$

Although this rule is a direct consequence of equation (8.44), we do introduce it here because it becomes convenient during the forthcoming derivations.

In the following sections, we will demonstrate the use of the operators in equations (8.97)–(8.102) to describe the evolution of some of the classic sequences used in pulse EPR. For the two-electron spin case, the rule in equation (8.107) will thus be useful during the calculations and again to simplify the text we will represent the three transformation-related independent operators (trios) of equation (8.107) as

$$\{\hat{A} : \hat{B} \rightarrow \hat{C}\}_{\kappa} \tag{8.108}$$

We now present a set of trios that will be used later in this chapter. In the expressions for the following trios, we use that $p, q, r = x, y, z$ or its cyclic permutations:

- The trivial trios of the linear operators

$$\{\hat{S}_{p0}^{(AB)} : \hat{S}_{q0}^{(AB)} \rightarrow \hat{S}_{r0}^{(AB)}\}_1; \{\hat{S}_{0p}^{(AB)} : \hat{S}_{0q}^{(AB)} \rightarrow \hat{S}_{0r}^{(AB)}\}_1 \quad (8.109)$$

- The trios between linear and bilinear single quantum operators and zz-operators

$$\begin{aligned}
&\{\hat{S}_{p0}^{(AB)} : \hat{S}_{qz}^{(AB)} \rightarrow \hat{S}_{rz}^{(AB)}\}_1;\\
&\{\hat{S}_{pz}^{(AB)} : \hat{S}_{q0}^{(AB)} \rightarrow \hat{S}_{rz}^{(AB)}\}_1;\\
&\{\hat{S}_{pz}^{(AB)} : \hat{S}_{qz}^{(AB)} \rightarrow \hat{S}_{r0}^{(AB)}\}_1\\
&\{\hat{S}_{0p}^{(AB)} : \hat{S}_{zq}^{(AB)} \rightarrow \hat{S}_{zr}^{(AB)}\}_1;\\
&\{\hat{S}_{zp}^{(AB)} : \hat{S}_{0q}^{(AB)} \rightarrow \hat{S}_{zr}^{(AB)}\}_1;\\
&\{\hat{S}_{zp}^{(AB)} : \hat{S}_{zq}^{(AB)} \rightarrow \hat{S}_{0r}^{(AB)}\}_1
\end{aligned} \quad (8.110)$$

- The trios including the {zero+double} quantum operators

$$\begin{aligned}
&\{\hat{S}_{p0}^{(AB)} : \hat{S}_{qq}^{(AB)} \rightarrow \hat{S}_{rq}^{(AB)}\}_1;\\
&\{\hat{S}_{pp}^{(AB)} : \hat{S}_{q0}^{(AB)} \rightarrow \hat{S}_{rp}^{(AB)}\}_1;\\
&\{\hat{S}_{pp}^{(AB)} : \hat{S}_{qp}^{(AB)} \rightarrow \hat{S}_{r0}^{(AB)}\}_1\\
&\{\hat{S}_{0p}^{(AB)} : \hat{S}_{qq}^{(AB)} \rightarrow \hat{S}_{qr}^{(AB)}\}_1;\\
&\{\hat{S}_{pp}^{(AB)} : \hat{S}_{0q}^{(AB)} \rightarrow \hat{S}_{pr}^{(AB)}\}_1;\\
&\{\hat{S}_{pp}^{(AB)} : \hat{S}_{pq}^{(AB)} \rightarrow \hat{S}_{0r}^{(AB)}\}_1
\end{aligned} \quad (8.111)$$

Two examples of trios that can be derived straightforwardly from the trios in equations (8.109–8.111) are

$$\begin{aligned}
&\{\hat{S}_{zz}^{(AB)} : (\hat{S}_{0x}^{(AB)} + \hat{S}_{x0}^{(AB)}) \rightarrow (\hat{S}_{yz}^{(AB)} + \hat{S}_{zy}^{(AB)})\}_1\\
&e^{-i\theta\hat{S}_{zz}^{(AB)}}(\hat{S}_{0x}^{(AB)} + \hat{S}_{x0}^{(AB)})e^{i\theta\hat{S}_{zz}}\\
&= (\hat{S}_{0x}^{(AB)} + \hat{S}_{x0}^{(AB)})\cos\theta + (\hat{S}_{yz}^{(AB)} + \hat{S}_{zy}^{(AB)})\sin\theta
\end{aligned} \quad (8.112)$$

where a Hamiltonian term proportional to $S_{zz}^{(AB)}$ transfers single-quantum linear to bilinear operators and

$$\begin{aligned}
&\{(\hat{S}_{0x}^{(AB)} + \hat{S}_{x0}^{(AB)}) : (\hat{S}_{yz}^{(AB)} + \hat{S}_{zy}^{(AB)}) \rightarrow (\hat{S}_{zz}^{(AB)} - \hat{S}_{yy}^{(AB)})\}_2\\
&e^{-i\theta(\hat{S}_{0x}^{(AB)}+\hat{S}_{x0}^{(AB)})}(\hat{S}_{yz}^{(AB)} + \hat{S}_{zy}^{(AB)})e^{i\theta(\hat{S}_{0x}^{(AB)}+\hat{S}_{x0}^{(AB)})}\\
&= (\hat{S}_{yz}^{(AB)} + \hat{S}_{zy}^{(AB)})\cos 2\theta + (\hat{S}_{zz}^{(AB)} - \hat{S}_{yy}^{(AB)})\sin 2\theta
\end{aligned} \quad (8.113)$$

where a Hamiltonian term $\hat{S}_{x0}^{(AB)} + \hat{S}_{0x}^{(AB)}$ (as one finds, e.g., in the presence of MW irradiation) transfers single-quantum bilinear operators to {zero+double) quantum operators. Here, the coefficient $\kappa = 2$ implies that a transformation with $\theta = \pi/2$ inverts the single-quantum bilinear operators.

These and similar transformations are relevant during the analysis of pulse EPR experiments, in particular when one term $h_{pq}^{(AB)}\hat{S}_{pq}^{(AB)}$ of the Hamiltonian dominates the spin evolution. This term then governs the time evolution of the coefficients of $s_{p'q'}^{(AB)}(t)\hat{S}_{p'q'}^{(AB)}$ in the expansion of $\hat{\sigma}(t)$. Before discussing these pulse EPR sequences, we should mention that all the above-mentioned trios are independent of their possible matrix representations.

8.3.2 The Fictitious Spin-half Operators and Their Commutation Relations

The actual calculations involving the product operators can become cumbersome, and it is sometimes worth the effort to use operators that are linear combinations of the ones in equations (8.97)–(8.102). Thus, in the following paragraphs, we will also make use of an additional set of independent operators that provide some additional insight in the spin evolution of our four-level spin system.

The additional set we chose here are the fictitious spin-1/2, or single-transition, operators.[4,15,16] We introduce these operators by defining their matrix representation with respect to the complete $\{|i\rangle\}_{i=1,4}$ set of basis spin states in equation (8.96). In this representation, we define 18 operators $\hat{S}_p^{i-j}$ with $p = x, y, z$, $i, j = 1,..,4$ and $i < j$ that have each only two nonzero matrix elements

$$\begin{aligned}
\langle i|\hat{S}_x^{i-j}|j\rangle &= \langle j|\hat{S}_x^{i-j}|i\rangle = 1/2\\
\langle i|\hat{S}_y^{i-j}|j\rangle &= -\langle j|\hat{S}_y^{i-j}|i\rangle = -i/2\\
\langle i|\hat{S}_z^{i-j}|i\rangle &= -\langle j|\hat{S}_z^{i-j}|j\rangle = 1/2
\end{aligned} \quad (8.114)$$

A simple inspection reveals that the six $\hat{S}_z^{i-j}$ operators are dependent on one another, via equalities such as $\hat{S}_z^{i-j} + \hat{S}_z^{j-k} = \hat{S}_z^{i-k}$, and thus to generate a set of 15 independent orthogonal operators we have to choose only three of those. A simple analysis of the matrix representations of the product operators in Figure 8.4 yields the following relations between the fictitious spin-1/2 operators and the set of Cartesian product operators defined in equations (8.97)–(8.102):

$$\hat{S}_{x0}^{(AB)} = (\hat{S}_x^{1-3} + \hat{S}_x^{2-4});\ \hat{S}_{y0}^{(AB)} = (\hat{S}_y^{1-3} + \hat{S}_y^{2-4});$$
$$\hat{S}_{0x}^{(AB)} = (\hat{S}_x^{1-2} + \hat{S}_x^{3-4});\ \hat{S}_{0y}^{(AB)} = (\hat{S}_y^{1-2} + \hat{S}_y^{3-4});$$
$$\hat{S}_{xz}^{(AB)} = (\hat{S}_x^{1-3} - \hat{S}_x^{2-4});\ \hat{S}_{yz}^{(AB)} = (\hat{S}_y^{1-3} - \hat{S}_y^{2-4});$$
$$\hat{S}_{zx}^{(AB)} = (\hat{S}_x^{1-2} - \hat{S}_x^{3-4});\ \hat{S}_{zy}^{(AB)} = (\hat{S}_y^{1-2} - \hat{S}_y^{3-4});$$
$$\hat{S}_{xy}^{(AB)} = (\hat{S}_y^{1-4} - \hat{S}_y^{2-3});\ \hat{S}_{yx}^{(AB)} = (\hat{S}_y^{1-4} + \hat{S}_y^{2-3});$$
$$\hat{S}_{xx}^{(AB)} = (\hat{S}_x^{1-4} + \hat{S}_x^{2-3});\ \hat{S}_{yy}^{(AB)} = -(\hat{S}_x^{1-4} - \hat{S}_x^{2-3});$$
$$\hat{S}_{z0}^{(AB)} = (\hat{S}_z^{1-3} + \hat{S}_z^{2-4});\ \hat{S}_{0z}^{(AB)} = (\hat{S}_z^{1-2} + \hat{S}_z^{3-4});$$
$$\hat{S}_{zz}^{(AB)} = (\hat{S}_z^{1-2} - \hat{S}_z^{3-4}) \tag{8.115}$$

For completeness, we also notice that

$$\hat{S}_x^{i-j} = \hat{S}_x^{j-i},\quad \hat{S}_y^{j-i} = -\hat{S}_y^{i-j} \text{ and } \hat{S}_z^{j-i} = -\hat{S}_z^{i-j} \tag{8.116}$$

These fictitious spin-1/2 operators can thus also be used to expand the Hamiltonians $\hat{H}(t)$ and the spin density operator $\hat{\sigma}(t)$ of the spin system.

The $\hat{S}_p^{i-j}$ operators also fulfill commutation relations. For each single transition $\{|i\rangle \leftrightarrow |j\rangle\}$, they are

$$[\hat{S}_p^{i-j}, \hat{S}_q^{i-j}] = \mathrm{i}\hat{S}_r^{i-j} \tag{8.117}$$

and for two transitions with one common state $\{|i\rangle \leftrightarrow |j\rangle\}$; $\{|j\rangle \leftrightarrow |k\rangle\}$ they are

$$[\hat{S}_x^{i-j}, \hat{S}_x^{j-k}] = \frac{\mathrm{i}}{2}\hat{S}_y^{i-k} \quad ; \quad [\hat{S}_y^{i-j}, \hat{S}_y^{j-k}] = -\frac{\mathrm{i}}{2}\hat{S}_y^{i-k}$$
$$[\hat{S}_x^{i-j}, \hat{S}_y^{j-k}] = -\frac{\mathrm{i}}{2}\hat{S}_x^{i-k} ; \quad [\hat{S}_y^{i-j}, \hat{S}_x^{j-k}] = -\frac{\mathrm{i}}{2}\hat{S}_x^{i-k} \tag{8.118}$$

where $p, q, r = x, y, z$ or its cyclic permutations. Using the definition of the trios in equation (8.107) and the commutation relation in equations (8.117) and (8.118), we can derive

$$\{\hat{S}_x^{i-j} : \hat{S}_y^{i-j} \to \hat{S}_z^{i-j}\}_1$$
$$\{\hat{S}_x^{i-j} : \hat{S}_x^{j-k} \to \hat{S}_y^{i-k}\}_{1/2} ; \ \{\hat{S}_y^{i-j} : \hat{S}_y^{j-k} \to -\hat{S}_y^{i-k}\}_{1/2}$$
$$\{\hat{S}_x^{i-j} : \hat{S}_y^{j-k} \to -\hat{S}_x^{i-k}\}_{1/2} ; \ \{\hat{S}_y^{i-j} : \hat{S}_x^{j-k} \to -\hat{S}_x^{i-k}\}_{1/2} \tag{8.119}$$

These results show that, for example, a major term in $\hat{H}(t)$ in equation (8.105) of the form $h_x^{i-j}\hat{S}_x^{i-j}$ can transfer part of a $s_x^{j-k}(t)\hat{S}_x^{j-k}$ term of $\hat{\sigma}(t)$ to a term of the form $s_y^{i-k}(t)\hat{S}_y^{i-k}$.

8.3.3 The Spin Hamiltonian of Two Coupled Electrons

The electron–electron dipolar interaction term $\hat{H}_D$ in the Hamiltonian of the two-electron spin system $\{e_A - e_B\}$ in the laboratory frame with its z-axis in the direction of the external magnetic field is in general expressed as a linear combination of the components of the irreducible second-rank tensor operator $T_m^{(2)}$ $(m = -2, .., 2)$.[6]

$$\hat{H}_D = \sum_{m=0,\pm1,\pm2} (-1)^m \omega_m^{AB}(\theta_{AB}, \varphi_{AB})\hat{T}_m^{(2)} \tag{8.120}$$

This expansion is in particular convenient when the dependence of the ω_m^{AB} coefficients in $\hat{H}_D$ on the polar angles $(\theta_{AB}, \varphi_{AB})$ of the dipolar vector $\vec{r}_{AB}$ connecting e_A and e_B in the laboratory frame must be considered. These coefficients are

$$\omega_0^{AB}(\theta_{AB}, \varphi_{AB}) = \sqrt{\frac{3}{2}}\omega_D^{AB}(1 - 3\cos^2\theta_{AB})$$
$$\omega_{\pm1}^{AB}(\theta_{AB}, \varphi_{AB}) = \frac{3}{4}\omega_D^{AB} \sin\theta_{AB} \cos\theta_{AB} e^{-\mathrm{i}\varphi_{AB}}$$
$$\omega_{\pm2}^{AB}(\theta_{AB}, \varphi_{AB}) = -\frac{3}{4}\omega_D^{AB} \sin^2\theta_{AB} e^{-2\mathrm{i}\varphi_{AB}} \tag{8.121}$$

where the dipolar interaction strength equals $\omega_D^{AB} = g_A g_B \mu_B^2 / r_{AB}^3$, and the $T_m^{(2)}$ operators for the two-electron system expressed in terms of the 15 orthogonal product operators in equations (8.97)–(8.102) have the form

$$\hat{T}_0^{(2)} = \frac{1}{\sqrt{6}}(\hat{S}_{zz}^{(AB)} - 1/2(\hat{S}_{xx}^{(AB)} + \hat{S}_{yy}^{(AB)}))$$
$$T_{\pm1}^{(2)} = (\hat{S}_{xz}^{(AB)} + \hat{S}_{zx}^{(AB)}) \pm i(\hat{S}_{yz}^{(AB)} + \hat{S}_{zy}^{(AB)})$$
$$\hat{T}_{\pm2}^{(2)} = (\hat{S}_{xx}^{(AB)} - \hat{S}_{yy}^{(AB)}) \pm i(\hat{S}_{xy}^{(AB)} + \hat{S}_{yx}^{(AB)}) \tag{8.122}$$

In the rotating frame defined by the MW irradiation frequency ω_{MW}, the spin density operator becomes $\hat{\sigma}^{RoF}(t) = \hat{U}_{MW}^{-1}(t)\hat{\sigma}(t)\hat{U}_{MW}(t)$ with $U_{MW}(t) = e^{-\mathrm{i}\omega_{MW}(\hat{S}_{z0}^{(AB)} + \hat{S}_{0z}^{(AB)})t}$. The dipolar Hamiltonian terms proportional to $\hat{T}_m^{(2)}$ with $m = \pm 1, \pm 2$ remain time dependent and can be neglected because $\omega_{MW} \gg |\omega_0^{AB}|$. The dipolar Hamiltonian then becomes

$$\hat{H}_D^{RoF} = \frac{1}{2}\omega_{AB}\left\{\hat{S}_{zz}^{(AB)} - \frac{1}{2}(\hat{S}_{xx}^{(AB)} + \hat{S}_{yy}^{(AB)})\right\} \tag{8.123}$$

with the dipolar frequency coefficient

$$\omega_{AB} = \omega_D^{AB}(1 - 3\cos^2\theta_{AB}) \tag{8.124}$$

The two electrons can also be coupled via an isotropic exchange coupling. The strength of this coupling can be defined by the exchange frequency coefficient J in the exchange interaction Hamiltonian term

$$\hat{H}_J = J(\hat{S}_{xx}^{(AB)} + \hat{S}_{yy}^{(AB)} + \hat{S}_{zz}^{(AB)}) \tag{8.125}$$

The term is invariant under the rotating-frame transformation and thus in the rotating frame we can define a combined Hamiltonian term of the form

$$\hat{H}^{\text{RoF}}_{\text{D+J}} = \hat{H}^{\text{RoF}}_{\text{D}} + \hat{H}_{\text{J}} = \left(\frac{1}{2}\omega_{AB} + J\right)\hat{S}^{(AB)}_{zz} - 1/2\left(\frac{1}{2}\omega_{AB} - 2J\right)(\hat{S}^{(AB)}_{xx} + \hat{S}^{(AB)}_{yy}) \quad (8.126)$$

8.3.4 Liouville Space

When relaxation effects must be taken into account during spin density matrix calculations, we can transfer the L-vN equation to Liouville space and generate the Liouville superoperator either in (i) the state vector $\vec{\sigma}(t)$ representation, where $\vec{\sigma}(t)$ is the vector composed of all $\sigma_{ij}(t)$ elements of the matrix representation of $\hat{\sigma}(t)$ in some basis set $\{|\varphi_i\rangle\}_{i=1,4}$; or in (ii) the expectation value $\vec{s}(t)$ representation, where $\vec{s}(t)$ is the vector composed of all $s^{(AB)}_{pq}$.

(i) In the first case, the 16×16 $\hat{\hat{L}}(t)$ Liouville superoperator matrix elements are determined by the matrix elements of $\hat{H}(t)$, while $\vec{\sigma}(t)$ comprises 4 populations $\sigma_{ii}(t)$ and 12 coherences $\sigma_{ij}(t)$

$$\frac{\mathrm{d}}{\mathrm{d}t}\vec{\sigma}(t) = (-i\hat{\hat{L}}(t) + \hat{\hat{R}})\vec{\sigma}(t) \quad (8.127)$$

where the matrix elements of $\hat{\hat{L}}$ for $H_{ij}(t) = \langle\varphi_i|\hat{H}(t)|\varphi_j\rangle$ derived from the L-vN equation for $\hat{\sigma}(t)$ are given by

$$\begin{aligned} L_{ij,ij}(t) &= H_{ii}(t) - H_{jj}(t) \\ L_{ij,ik}(t) &= -H_{kj}(t);\quad L_{ij,kj}(t) = H_{ik}(t) \\ L_{ii,ii}(t) &= 0;\quad L_{ij,kl}(t) = 0 \end{aligned} \quad (8.128)$$

and the relaxation superoperator $\hat{\hat{R}}$ has matrix elements

$$R_{ii,ii} = \sum_{k\neq i}\frac{-R_{1,ik}}{1+\varepsilon_{ik}};\quad R_{ii,jj} = \frac{\varepsilon_{ij}R_{1,ij}}{1+\varepsilon_{ij}}\quad R_{ij,ij} = -R_{2,ij} \quad (8.129)$$

Here, we chose the matrix representation according to the equilibrium Hamiltonian with eigenstates $\{|\varphi_i\rangle\}^N_{i=1}$ and assigned to every single transition $|\varphi_i\rangle \leftrightarrow |\varphi_j\rangle$ with a transition frequency ω_{ij} a separate longitudinal rate constant $R_{1,ij}$, driving $\sigma_{ii}(t)$ and $\sigma_{jj}(t)$ to their Boltzmann ratio $\varepsilon_{ij} = \sigma^{\text{eq}}_{ii}(t)/\sigma^{\text{eq}}_{jj}(t) = \mathrm{e}^{-\hbar\omega_{ij}/k_{\text{B}}T}$ with $\sigma^{\text{eq}}_{ii}(t) + \sigma^{\text{eq}}_{jj}(t) = \sigma_{ii}(t) + \sigma_{jj}(t)$, and a transverse relaxation rate constant $R_{2,ij}$, decaying $\sigma_{ij}(t)$ to zero.

(ii) In the second case, we can construct the $\hat{\hat{L}}(t)$ superoperator using $\mathrm{Tr}(\hat{S}^{(AB)}_{pq}\hat{S}^{(AB)}_{p'q'}) = 0$ and inserting the operator expansion of $\hat{\sigma}(t)$ and $\hat{H}(t)$ in

$$\frac{\mathrm{d}}{\mathrm{d}t}s^{(AB)}_{pq}(t) = -\mathrm{i}\sum_{\substack{p',q',r',p'',q''\\=0,x,y,z}}\mathrm{Tr}(\hat{S}^{(AB)}_{pq}[\hat{S}^{(AB)}_{p''q''},\hat{S}^{(AB)}_{p'q'}])\, h^{(AB)}_{p''q''}(t)s^{(AB)}_{p'q'}(t) \quad (8.130)$$

Adding relaxation effects to this rate equation is not straightforward[17] and is therefore not discussed any further in this chapter.

8.3.5 Selective and Non-selective MW Pulses and Signal Detection

The impact of a short MW pulse, during which all relaxation effects are ignored, on the two-electron spin system can be calculated by solving the L-vN equation in the rotating frame. Depending on the frequency ω_{MW}, intensity ω_1, phase φ_{MW}, and duration t_{MW} of the MW irradiation, we can calculate the changes in the $s^{(AB)}_{pq}(t)$ coefficients from before to after a pulse. In this chapter, we will not elaborate on the changes resulting from on- and off-resonance during the MW pulse. Here, we will restrict ourselves to very short MW pulses that modify the populations and coherences of the two-electron system in an infinitesimal short time. In the rotating frame defined by

$$\hat{\sigma}^{\text{RoF}}(t) = \hat{U}^{\text{RoF}^{-1}}(t)\hat{\sigma}(t)\hat{U}^{\text{RoF}};\quad \hat{U}^{\text{RoF}}(t) = \mathrm{e}^{-\mathrm{i}\omega_{\text{MW}}(\hat{S}^{(AB)}_{z0}+\hat{S}^{(AB)}_{0z})t} \quad (8.131)$$

The MW term in the Hamiltonian of a pulse in the x-direction including its off-resonance terms, while ignoring the dipolar and spin exchange terms by assuming that $\omega_1 \gg \omega_{AB}, J$, are

$$\hat{H}^{\text{RoF}}_{\text{MW}} = \Delta\omega_A\hat{S}^{(AB)}_{z0} + \omega_1\hat{S}^{(AB)}_{x0} + \Delta\omega_B\hat{S}^{(AB)}_{0z} + \omega_1\hat{S}^{(AB)}_{0x} \quad (8.132)$$

where $\Delta\omega_A = \omega_A - \omega_{\text{MW}}$ and $\Delta\omega_B = \omega_B - \omega_{\text{MW}}$ and ω_A and ω_B are the laboratory transition frequencies of e_A and e_B. Depending on the values of $\Delta\omega_A$ and $\Delta\omega_B$, this Hamiltonian will affect all linear operators of both electrons or only those of e_A or e_B. When $\omega_1 \gg \Delta\omega_A, \Delta\omega_B$, we will talk about a nonselective pulse, and when $|\Delta\omega_A| < \omega_1$ and $|\Delta\omega_B| \gg \omega_1$ or $|\Delta\omega_B| < \omega_1$ and $|\Delta\omega_A| \gg \omega_1$ about a semiselective pulse because we ignored in equation (8.132) all interaction terms causing possible frequency shifts.

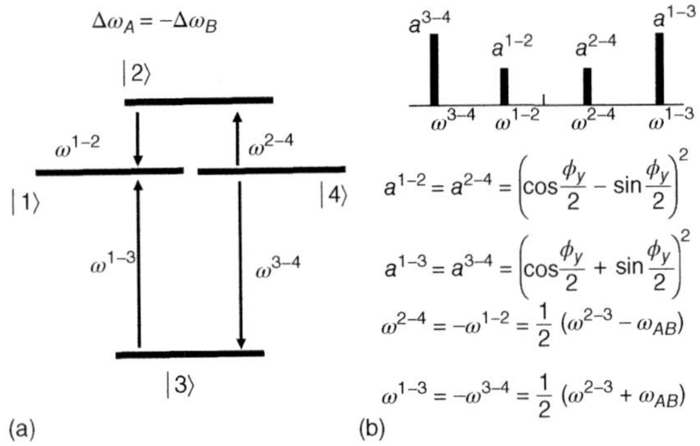

Figure 8.5. (a) The energy level diagram of a two-electron spin system in the MW rotating frame defined by the irradiation frequency $\omega_{\mathrm{MW}} = \frac{1}{2}(\omega_A + \omega_B)$. The rotating frame Hamiltonian of this system is shown in equation (8.150) with $\Delta\omega_A = -\Delta\omega_B$. The eigenstates $|1\rangle$ and $|4\rangle$ are degenerate such that the transition frequencies form a symmetric spectrum with $\omega^{1-3} = -\omega^{3-4}$ and $\omega^{1-2} = -\omega^{2-4}$. (b) The stick diagram of the spectrum obtained after a nonselective $\pi/2$ pulse composed of four lines with frequencies and amplitudes given in the figure and defined in equations (8.152), (8.153), and (8.154)

Furthermore, we will assume that the pulses are much shorter than any relevant time evolution affecting $\hat{\sigma}^{\mathrm{RoF}}(t)$. Similar to the single-spin case, we will present the pulses as instantaneous transformations of the form

$$\hat{\sigma}^{\mathrm{RoF}}(t^+) = \hat{P}_\theta^{s,p}\hat{\sigma}^{\mathrm{RoF}}(t^-)\hat{P}_\theta^{s,p^{-1}} \tag{8.133}$$

where the pulse propagator is characterized by its flip angle $\theta = \omega_1 t_{\mathrm{MW}}$, its direction $p = \pm x, \pm y$ in the rotating frame, and its selectivity $s = A, B, (A+B)$ (Figure 8.5).

In order to calculate the response of the application of a sequence of these ideal MW pulses on the equilibrium $\hat{\sigma}^{\mathrm{eq}} = s_z^{\mathrm{eq}}(\hat{S}_{z0}^{(AB)} + \hat{S}_{0z}^{(AB)}) + s_0\hat{S}_0^{(AB)}$ (in analogy with equations (8.35)–(8.37)), we must choose the appropriate rotating frame and calculate the complex signal

$$m_x(t) + \mathrm{i}m_y(t) = -g_e\mu_B\mathrm{Tr}\{\hat{\sigma}^{\mathrm{RoF}}(t)[(\hat{S}_{x0}^{(AB)} + \hat{S}_{0x}^{(AB)}) + \mathrm{i}(\hat{S}_{y0}^{(AB)} + \hat{S}_{0y}^{(AB)})] \tag{8.134}$$

As in the single-spin case, from now on we only refer to the reduced spin density operator in the rotating frame

$$\hat{\sigma}(t) \equiv \hat{\sigma}^{\mathrm{RoF}}(t) - s_0\hat{S}_0^{(AB)} \tag{8.135}$$

8.3.5.1 *Three Special Selective and Nonselective Pulses*

We will now demonstrate the evolution of a two-spin system with a dipolar interaction frequency ω_{AB} and an exchange frequency J for three cases of ideal pulses: (A) a semiselective pulse on e_A with $\Delta\omega_A = 0$, (B) a semiselective pulse on e_B with $\Delta\omega_B = 0$, and $(A+B)$ a nonselective pulse with $\Delta\omega_A = \Delta\omega_B \approx 0$. In addition, we will assume for cases (A) and (B) that $|\Delta\omega_B - \Delta\omega_A| \gg |\omega_{AB}|$ (the weak coupling limit), while in case $(A+B)$ $|\Delta\omega_B - \Delta\omega_A| \ll |\omega_{AB}|$ (strong coupling limit). Using the trios in equations (8.109)–(8.111) transferring *sq* linear operators to *sq* bilinear operators, we can derive the following independent trios:

$$\begin{aligned}&\{\hat{S}_{zz}^{(AB)} : (\hat{S}_{x0}^{(AB)} + \hat{S}_{0x}^{(AB)}) \to (\hat{S}_{yz}^{(AB)} + \hat{S}_{zy}^{(AB)})\}_1\\ &\{-(\hat{S}_{xx}^{(AB)} + \hat{S}_{yy}^{(AB)}) : (\hat{S}_{x0}^{(AB)} + \hat{S}_{0x}^{(AB)}) \to (\hat{S}_{yz}^{(AB)} + \hat{S}_{zy}^{(AB)})\}_1\end{aligned} \tag{8.136}$$

thus showing that $\hat{S}_{zz}^{(AB)}$ and $-(\hat{S}_{xx}^{(AB)} + \hat{S}_{yy}^{(AB)})$ independently transfer $(\hat{S}_{x0}^{(AB)} + \hat{S}_{0x}^{(AB)})$ to $(\hat{S}_{yz}^{(AB)} + \hat{S}_{zy}^{(AB)})$. The time evolution $\hat{\sigma}^{(AB)}(t) = U^{\mathrm{RoF}}(t)\hat{\sigma}(0^+)U^{\mathrm{RoF}-1}(t)$ in the three cases can now be derived by (i) choosing the appropriate rotating frame, (ii) truncating the Hamiltonian $\hat{H}^{\mathrm{RoF}}$, (iii) applying the $\pi/2$ pulse (here in the y-direction), and (iv) propagating $\hat{\sigma}(0^+)$ using $\hat{U}^{\mathrm{RoF}}(t) = \mathrm{e}^{-\mathrm{i}H^{\mathrm{RoF}}t}$. Using the trios in equation (8.111) for (A) and (B) and in equation (8.136) for $(A+B)$, we get

$$\text{(A)} : |\Delta\omega_A - \Delta\omega_B| \gg |\omega_{AB}| \text{ and } \Delta\omega_B \gg \omega_1 \tag{8.137}$$

$$\text{(i) } \hat{H}^{\mathrm{RoF}} \approx \Delta\omega_A\hat{S}_{z0}^{(AB)} + \Delta\omega_B\hat{S}_{0z}^{(AB)} + \left(\frac{1}{2}\omega_{AB} + J\right)\hat{S}_{zz}^{(AB)} \tag{8.138}$$

$$\text{(ii) } \hat{\sigma}(0^+) = \hat{P}_{\pi/2}^{A,y}\hat{\sigma}^{\mathrm{eq}}\hat{P}_{\pi/2}^{A,y^{-1}} = s_z^{\mathrm{eq}}(\hat{S}_{x0}^{(AB)} + \hat{S}_{0z}^{(AB)}) \tag{8.139}$$

$$\begin{aligned}\text{(iii) } \hat{\sigma}(t) = s_z^{\mathrm{eq}}(&\cos\Delta\omega_A t\cos\left(\frac{\omega_{AB}}{2} + J\right)t\,\hat{S}_{x0}^{(AB)}\\ &+ \sin\Delta\omega_A t\cos\left(\frac{\omega_{AB}}{2} + J\right)t\,\hat{S}_{y0}^{(AB)}\\ &+ \cos\Delta\omega_A t\sin\left(\frac{\omega_{AB}}{2} + J\right)t\,\hat{S}_{yz}^{(AB)}\\ &- \sin\Delta\omega_A t\sin\left(\frac{\omega_{AB}}{2} + J\right)t\,\hat{S}_{xz}^{(AB)} + \hat{S}_{0z}^{(AB)})\end{aligned} \tag{8.140}$$

(B) : $|\Delta\omega_A - \Delta\omega_B| \gg |\omega_{AB}|$ and $|\Delta\omega_A| \gg \omega_1$ (8.141)

$$\text{(i)}\ \hat{H}_D^{\text{RoF}} \approx \Delta\omega_A \hat{S}_{z0}^{(AB)} + \Delta\omega_B \hat{S}_{0z}^{(AB)} + \left(\frac{1}{2}\omega_{AB} + J\right)\hat{S}_{zz}^{(AB)} \tag{8.142}$$

$$\text{(ii)}\ \hat{\sigma}(0^+) = \hat{P}_{\pi/2}^{B,y}\hat{\sigma}^{\text{eq}}\hat{P}_{\pi/2}^{B,y^{-1}} = s_z^{\text{eq}}(\hat{S}_{z0}^{(AB)} + \hat{S}_{0x}^{(AB)}) \tag{8.143}$$

$$\begin{aligned}\text{(iii)}\ \hat{\sigma}(t) = s_z^{\text{eq}}&(\cos\Delta\omega_B t\cos\left(\frac{\omega_{AB}}{2}+J\right)t\,\hat{S}_{0x}^{(AB)}\\ &+\sin\Delta\omega_B t\cos\left(\frac{\omega_{AB}}{2}+J\right)t\,\hat{S}_{0y}^{(AB)}\\ &+\cos\Delta\omega_B t\sin\left(\frac{\omega_{AB}}{2}+J\right)t\,\hat{S}_{zy}^{(AB)}\\ &-\sin\Delta\omega_B t\sin\left(\frac{\omega_{AB}}{2}+J\right)t\,\hat{S}_{zx}^{(AB)} + \hat{S}_{z0}^{(AB)})\end{aligned} \tag{8.144}$$

(A + B) : $\Delta\omega_A = \Delta\omega_B = 0$ and $\omega_1 \gg |\omega_{AB}|$ (8.145)

$$\begin{aligned}\text{(i)}\ \hat{H}^{\text{RoF}} &= \left(\frac{1}{2}\omega_{ab} + J\right)\hat{S}_{zz}^{(AB)}\\ &-\frac{1}{2}\left(\frac{1}{2}\omega_{ab} - 2J\right)(\hat{S}_{xx}^{(AB)} + \hat{S}_{yy}^{(AB)})\}\end{aligned} \tag{8.146}$$

$$\text{(ii)}\ \hat{\sigma}(0^+) = \hat{P}_{\pi/2}^{A+B,y}\hat{\sigma}^{\text{eq}}\hat{P}_{\pi/2}^{A+B,y^{-1}} = s_z^{\text{eq}}(\hat{S}_{x0}^{(AB)} + \hat{S}_{0x}^{(AB)}) \tag{8.147}$$

$$\begin{aligned}\text{(iii)}\ \hat{\sigma}(t) = s_z^{\text{eq}}&\left(\cos\frac{3\omega_{AB}}{4}t\left(\hat{S}_{x0}^{(AB)} + \hat{S}_{0x}^{(AB)}\right)\right.\\ &\left.+\sin\frac{3\omega_{AB}}{4}t\left(\hat{S}_{zy}^{(AB)} + \hat{S}_{zy}^{(AB)}\right)\right)\end{aligned} \tag{8.148}$$

The signals in the three cases become

$$\begin{aligned}&(\text{A}) : m_x(t) + im_y(t)\\ &= -g_e\mu_B s_z^{\text{eq}}\cos\left(\frac{\omega_{AB}}{2}+J\right)t\times e^{i\Delta\omega_A t}e^{-R_2 t}\\ &(\text{B}) : m_x(t) + im_y(t)\\ &= -g_e\mu_B s_z^{\text{eq}}\cos\left(\frac{\omega_{AB}}{2}+J\right)t\times e^{i\Delta\omega_B t}e^{-R_2 t}\\ &(\text{A}+\text{B}) : m_x(t) + im_y(t)\\ &= -2g_e\mu_B s_z^{\text{eq}}\left\{\cos\frac{3\omega_{AB}}{4}t\right\}e^{-R_2 t}\end{aligned} \tag{8.149}$$

Fourier transformation of the signals in the three cases after the ideal pulses and adding R_2 relaxation results for (A) in two spectral lines at $\Delta\omega_A \pm 1/2\omega_{AB}$, for (B) in $\Delta\omega_B \pm 1/2\omega_{AB}$, and for (A+B) in $\pm 3/2\omega_{AB}$. In the strong-coupling case, where all trios in equations (8.137)–(8.140) have to be taken into account, the dipolar frequency splitting is increased by 50% with respect to the dipolar splitting in the weak coupling cases.

8.3.5.2 *A Nonselective Pulse*

At this point, we should also be able to derive the signal when $\Delta\omega_A \neq \Delta\omega_B \neq 0$. This can be achieved straightforwardly when we use the fictitious spin-1/2 operators. Adding to the dipolar Hamiltonian in equation (8.146), the off-resonance terms with $\Delta\omega_A = -\Delta\omega_B$ and $\Delta\omega_{AB} = \Delta\omega_A - \Delta\omega_B$, we can express the sum using the operators presented in equations (8.115), (8.116) in the form (for $J = 0$)

$$H^{\text{RoF}} = \frac{1}{2}\omega_{AB}(\hat{S}_z^{1-2} - S_z^{3-4}) - \frac{1}{2}\omega_{AB}\hat{S}_x^{2-3} + \Delta\omega_{AB}\hat{S}_z^{2-3} \tag{8.150}$$

To evaluate $\hat{\sigma}(t)$ after the pulse, we must first express $\hat{\sigma}(0^+)$ just after a nonselective $\pi/2$ pulse (here in the y-direction) in terms of the $\hat{S}_p^{i-j}$ operators

$$\hat{\sigma}(0^+) = s_z^{\text{eq}}(\hat{S}_x^{1-3} + \hat{S}_x^{2-4} + \hat{S}_x^{1-2} + \hat{S}_x^{3-4}) \tag{8.151}$$

At this point, we must diagonalize the Hamiltonian using $\{\hat{S}_y^{2-3} : \hat{S}_z^{2-3} \to \hat{S}_x^{2-3})_1$ and $\hat{U}_y^{2-3} = e^{-i\varphi_y\hat{S}_z^{2-3}}$

$$\begin{aligned}\hat{\Lambda}^{\text{RoF}} &= \hat{U}_y^{2-3}\hat{H}^{\text{RoF}}\hat{U}_y^{2-3^{-1}}\\ &= \frac{1}{2}\omega_{AB}(\hat{S}_z^{1-2} - \hat{S}_z^{3-4}) + \omega^{2-3}\hat{S}_z^{2-3}\end{aligned} \tag{8.152}$$

with

$$\omega^{2-3} = \sqrt{\omega_{AB}^2/4 + \Delta\omega_{AB}^2};\ \sin\varphi_y = \frac{\omega_{AB}/2}{\omega^{2-3}};\quad \cos\varphi_y = \frac{\Delta\omega_{AB}}{\omega^{2-3}} \tag{8.153}$$

The transition frequencies of Λ^{RoF} are then equal to

$$\begin{aligned}\omega^{1-2} &= -\omega^{2-4} = -\frac{1}{2}\omega^{2-3} + \frac{1}{2}\omega_{AB}\\ \omega^{3-4} &= -\omega^{1-3} = -\frac{1}{2}\omega^{2-3} - \frac{1}{2}\omega_{AB}\end{aligned} \tag{8.154}$$

as schematically shown in Figure 8.6. Inserting the value of φ_y into $\hat{U}_y^{2-3}$, it is now possible to evaluate $\hat{\sigma}(t)$ via

$$\hat{\sigma}(t) = \hat{U}_y^{2-3^{-1}}e^{-i\hat{\Lambda}^{\text{RoF}}t}\hat{U}_y^{2-3}\hat{\sigma}(0^+)\hat{U}_y^{2-3^{-1}}e^{i\hat{\Lambda}^{\text{RoF}}t}\hat{U}_y^{2-3} \tag{8.155}$$

To do so, we can rely on the trios $\{\hat{S}_y^{2-3} : \hat{S}_{x/y}^{1-2} \to \hat{S}_{x/y}^{1-3}\}_{1/2}$, $\{\hat{S}_y^{2-3} : \hat{S}_{x/y}^{1-3} \to -\hat{S}_{x/y}^{1-2}\}_{1/2}$, $\{\hat{S}_y^{2-3} : \hat{S}_{x/y}^{2-4} \to \hat{S}_{x/y}^{3-4}\}_{1/2}$, and $\{\hat{S}_y^{2-3} : \hat{S}_{x/y}^{3-4} \to -\hat{S}_{x/y}^{2-4}\}_{1/2}$, and after a straightforward calculation, this results in

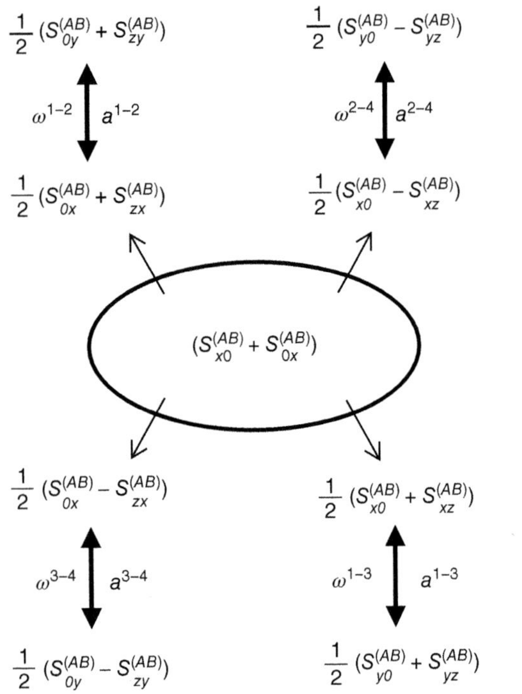

Figure 8.6. The expression for the time evolution of the spin density operator $\hat{\sigma}^{\mathrm{RoF}}(t)$ of a two-electron spin system after a single nonselective $\pi/2$ pulse in terms of its fictitious spin-1/2 operators is given in equation (8.156). Using the expansions in equations (8.115), (8.116), these operators can be expressed in terms of the product operators $\hat{S}_{pq}^{(AB)}$. The resulting expression for $\hat{\sigma}^{\mathrm{RoF}}(t)$ is shown here schematically, where the time evolutions of the product operator terms are represented by arrows indicating oscillations at frequencies ω^{i-j} between pairs of terms the form $a^{i-j}(\hat{S}_{pq}^{(AB)} \pm \hat{S}_{p'q'}^{(AB)})$

$$\begin{aligned}\hat{\sigma}(t) = {} & \left(\cos\frac{\varphi_y}{2} - \sin\frac{\varphi_y}{2}\right)^2 \\ & \times (\cos\omega^{1-2}t\,\hat{S}_x^{1-2} + \sin\omega^{1-2}t\,\hat{S}_y^{1-2}) \\ & + \left(\cos\frac{\varphi_y}{2} + \sin\frac{\varphi_y}{2}\right)^2 \\ & \times (\cos\omega^{1-3}t\,\hat{S}_x^{1-3} + \sin\omega^{1-3}t\,\hat{S}_y^{1-3}) \\ & \left(\cos\frac{\varphi_y}{2} - \sin\frac{\varphi_y}{2}\right)^2 \\ & \times (\cos\omega^{2-4}t\,\hat{S}_x^{2-4} + \sin\omega^{2-4}t\,\hat{S}_y^{2-4}) \\ & + \left(\cos\frac{\varphi_y}{2} + \sin\frac{\varphi_y}{2}\right)^2 \\ & \times (\cos\omega^{3-4}t\,\hat{S}_x^{3-4} + \sin\omega^{3-4}t\,\hat{S}_x^{3-4}) \end{aligned} \tag{8.156}$$

where the angle $\varphi_y/2$ is a consequence of the fact that the abovementioned trios have a value $\kappa = 1/2$. This result can of course also be expressed in terms of the product operators when we express the $\hat{S}_p^{i-j}$ operators in terms of the $\hat{S}_{pq}^{(AB)}$ operators using equations (8.115), (8.116). The result is shown schematically in Figure 8.6. Simplifying the expression in equation (8.156) using equation (8.154) and $\hat{\sigma}(t) = \sum_{p=x,y,z;i<j} s_p^{i-j}(t) S_p^{i-j}$ and $s_p^{i-j} \equiv s_p^{i-j}(t)$ the signal $m(t) = m_A(t) + m_B(t)$,

$$\begin{aligned} m(t) = {} & -g_e\mu_B(s_x^{1-2} + s_x^{1-3} + s_x^{2-4} + s_x^{3-4}) \\ & - g_e\mu_B(s_x^{1-2} + s_x^{1-3} + s_x^{2-4} + s_x^{3-4}) \end{aligned} \tag{8.157}$$

after the pulse becomes

$$\begin{aligned} m(t) = {} & -2\mu_B g_e\left(\cos\frac{\varphi_y}{2} - \sin\frac{\varphi_y}{2}\right)^2 \cos(\omega^{1-2}t) \\ & - 2\mu_B g_e\left(\cos\frac{\varphi_y}{2} + \sin\frac{\varphi_y}{2}\right)^2 \cos(\omega^{3-4}t) \end{aligned} \tag{8.158}$$

This $m(t)$ corresponds to the signals in equations (8.149) when $\Delta\omega_{AB} \gg \omega_{AB}$ and $\varphi_y \approx 0$ resulting in four frequency terms with equal amplitudes of 1 oscillating at $\pm\frac{1}{2}(\Delta\omega_{AB} \pm \omega_{AB})$. For the case of equation (8.149) with $\Delta\omega_{AB} = 0$ and $\varphi_y = \pi/2$, we get $\left(\cos\frac{\varphi_y}{2} - \sin\frac{\varphi_y}{2}\right)^2 = 0$ and $\left(\cos\frac{\varphi_y}{2} + \sin\frac{\varphi_y}{2}\right)^2 = 2$ resulting in two frequency terms with amplitudes 2 oscillating at $\pm\frac{3}{4}\omega_{AB}$.

8.3.6 Echo Pulse Sequences

In this section, we will present short descriptions of two basic pulse sequences, utilizing the product operator formalism. All these sequences are elaborately discussed in other chapters; thus, here we will not go into any experimental details or applications but restrict ourselves to the use of the $\hat{S}_{pq}^{(AB)}$ operators.

8.3.6.1 The Solid-Echo Sequence

The solid-echo pulse sequence of the form

$$\{\overset{0^+}{\downarrow}\hspace{-0.8em}\hat{P}_{\pi/2}^{A+B,y} - \tau - \overset{\tau^-}{\downarrow}\ \overset{\tau^+}{\downarrow}\hspace{-0.8em}\hat{P}_{\pi/2}^{A+B,x} - \tau - \overset{2\tau}{\downarrow}\}$$

with nonselective pulses as shown in Figure 8.7(a), can be applied to the two-electron spin system for the purpose of eliminating the effect of the dipolar and spin exchange interactions on the spin evolution at a time 2τ. For a pair of electron spins irradiated at a frequency $\omega_{MW} = 1/2(\omega_A + \omega_B)$, the Hamiltonian in the rotating frame $\hat{H}^{\mathrm{RoF}}$ has the form

$$\hat{H}^{\mathrm{RoF}} = \left(\frac{\omega_{AB}}{2} + J\right)\hat{S}_{zz}^{(AB)} - \frac{1}{2}\left(\frac{\omega_{AB}}{2} - 2J\right) \times (\hat{S}_{xx}^{(AB)} + \hat{S}_{yy}^{(AB)}) + \frac{1}{2}\Delta\omega_{AB}(\hat{S}_{z0}^{(AB)} - \hat{S}_{0z}^{(AB)}) \tag{8.159}$$

where the last term is the off-resonance term of the two electrons. After applying the first ideal nonselective y-pulse on the spin density operator at thermal equilibrium, we get

$$\hat{\sigma}(0^+) = s_z^{\mathrm{eq}}(\hat{S}_{x0}^{(AB)} + \hat{S}_{0x}^{(AB)}) \tag{8.160}$$

At first, let us consider equivalent electrons with $\omega_A = \omega_B$. Then, following the calculation procedure of the echo sequences in Section 8.2.10, we can obtain an echo factor $E^x_{\text{dip-echo}}$ correlating the signals at 0^+ and 2τ

$$m(2\tau) = E^x_{\text{solid-echo}} m(0^+) \tag{8.161}$$

$$E^p_{\text{solid-echo}} = \frac{\mathrm{Tr}\{[\hat{P}_{\pi/2}^{A+B,x}\hat{\sigma}^\tau \hat{P}_{\pi/2}^{A+B,x^{-1}}]\,\hat{\sigma}^{-\tau}\}}{\mathrm{Tr}\{\hat{\sigma}(0^+)^2\}} \tag{8.162}$$

where the evolution of the spin density operator $\hat{\sigma}^\tau = \hat{\sigma}(\tau)$ after the first pulse equals to the expression in equation (8.148) and $\hat{\sigma}^{-\tau}$ is obtained by replacing τ by $-\tau$ in the same equation. Thus

$$\begin{aligned}\hat{\sigma}^\tau &= s_z^{\mathrm{eq}}\left\{\left(\hat{S}_{x0}^{(AB)} + \hat{S}_{0x}^{(AB)}\right)\cos\left(\frac{3}{4}\omega_{AB}\tau\right) + \left(\hat{S}_{yz}^{(AB)} + \hat{S}_{zy}^{(AB)}\right)\sin\left(\frac{3}{4}\omega_{AB}\tau\right)\right\}\\ \hat{\sigma}^{-\tau} &= s_z^{\mathrm{eq}}\left\{\left(\hat{S}_{x0}^{(AB)} + \hat{S}_{0x}^{(AB)}\right)\cos\left(\frac{3}{4}\omega_{AB}\tau\right) - \left(\hat{S}_{yz}^{(AB)} + \hat{S}_{zy}^{(AB)}\right)\sin\left(\frac{3}{4}\omega_{AB}\tau\right)\right\}\end{aligned} \tag{8.163}$$

The effect of the $\pi/2$ x-pulse on $\hat{\sigma}^\tau$ can be derived from the trio

$$\{(\hat{S}_{x0}^{(AB)} + \hat{S}_{0x}^{(AB)}) : (\hat{S}_{yz}^{(AB)} + \hat{S}_{zy}^{(AB)}) \rightarrow (\hat{S}_{zz}^{(AB)} - \hat{S}_{yy}^{(AB)}\}_2 \tag{8.164}$$

indicating that this pulse inverts $(\hat{S}_{yz}^{(AB)} + \hat{S}_{zy}^{(AB)})$ because the value of κ of the trio is 2

$$\hat{P}_{\pi/2}^{A+B,x}\hat{\sigma}^\tau \hat{P}_{\pi/2}^{A+B,x^{-1}} = (\hat{S}_{x0}^{(AB)} + \hat{S}_{0x}^{(AB)})\cos\left(\frac{3}{4}\omega_{AB}\tau\right) - (\hat{S}_{yz}^{(AB)} + \hat{S}_{zy}^{(AB)})\sin\left(\frac{3}{4}\omega_{AB}\tau\right) \tag{8.165}$$

From the fact that $\hat{\sigma}^{-\tau} = \hat{P}_{\pi/2}^{A+B,x}\hat{\sigma}^\tau \hat{P}_{\pi/2}^{A+B,x^{-1}}$ and following the discussion in Section 8.2.11.1, we can conclude that $\mathrm{Tr}\{[\hat{P}_{\pi/2}^{A+B,x}\hat{\sigma}^\tau \hat{P}_{\pi/2}^{A+B,x^{-1}}]^2\} = \mathrm{Tr}\{\hat{\sigma}(0^+)^2\}$. This results in $E^x_{\text{solid-echo}} = 1$ and shows that the echo sequence refocuses the influence of both the dipolar and J exchange interactions.

This result will not be valid in the presence of a nonzero $\frac{1}{2}\Delta\omega_{AB}(\hat{S}_{z0}^{(AB)} - \hat{S}_{0z}^{(AB)})$ term in $\hat{H}^{(AB)\mathrm{RoF}}$. In the presence of this term, it is still possible to calculate the evolution of nonselective pulses; however, it requires diagonalization of the Hamiltonian, and the calculation of $E^x_{\text{solid-echo}}$ becomes cumbersome. To avoid showing even more complicated expressions, we will not show this calculation here.

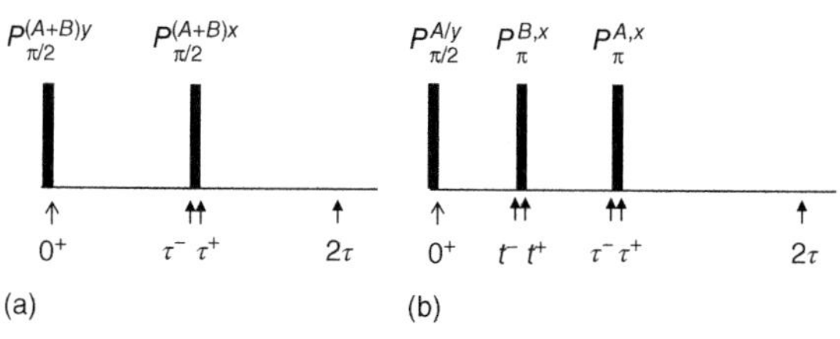

Figure 8.7. The two basic pulse EPR sequences that are applied on a two-electron spin system as discussed in Section 8.3: (a) the solid-echo pulse sequence and (b) the DEER/PELDOR pulse sequence

8.3.6.2 *The Three-pulse DEER/PELDOR Sequence*

A pulse sequence, which is in broad use for measuring distances in biological systems, is the double resonance DEER sequence also known as *PELDOR* (see Chapter 19). We will discuss here the three-pulse version given by[18]

$$\{\overset{0^+}{\downarrow}\hat{P}_{\pi/2}^{A,y} - t - \overset{t^-}{\downarrow}\hat{P}_{\pi}^{B,x}\overset{t^+}{\downarrow} - (\tau - t) - \overset{\tau^-}{\downarrow}\hat{P}_{\pi}^{A,y}\overset{\tau^+}{\downarrow} - - \tau - - \overset{2\tau}{\downarrow}\}$$

and use it here as an example demonstrating the use of the product operators to calculate the evolution of a two-electron spin system. Let us consider a two-electron spin system $\{e_A - e_B\}$ where the difference $\Delta\omega_{AB} = |\omega_A - \omega_B|$ is much larger than the dipolar frequency ω_{AB}. Furthermore, we assume that the MW pulses are semiselective with amplitudes satisfying the condition $\omega_{AB} \ll \omega_1 \ll \Delta\omega_{AB}$. In this case, we can excite the pairs of transitions corresponding to e_A or e_B selectively. We start our calculations with choosing the rotating frame of the MW irradiation at $\omega_{MW} = \omega_A$. The Hamiltonian in this frame takes the form

$$\hat{H}^{\text{RoF}} = \Delta\omega_{AB}\hat{S}_{0z}^{(AB)} + \frac{1}{2}\omega_{AB}\hat{S}_{zz}^{(AB)} \tag{8.166}$$

where for simplicity we assume that $J = 0$. Ignoring the π-pulse on e_B, the sequence looks like a simple echo experiment with a semiselective $\pi/2$-pulse on e_A followed after a time τ by a π-pulse again on e_A and after an additional τ detection of the signal $m_x^A(2\tau)$. Because the signal of e_B is not monitored and the term $s_z^{\text{eq}}\hat{S}_{0z}^{(AB)}$ in the initial equilibrium spin density operator $\hat{\sigma}^{\text{eq}} = s_z^{\text{eq}}(\hat{S}_{z0}^{(AB)} + \hat{S}_{0z}^{(AB)})$ is not affected by the pulse sequence without the π-pulse on e_B, we can restrict our calculation solely to the evolution of $\hat{\sigma}_A(t) = s_z^{\text{eq}}\hat{S}_{z0}^{(AB)}$ and derive an echo factor $E_{\text{echo}}^{A,x}$ by calculating $\hat{\sigma}_A^{\tau}$, $\hat{\sigma}_A^{-\tau}$ and $\hat{P}_\pi^{A,y}\hat{\sigma}_A^{\tau}\hat{P}_\pi^{A,y^{-1}}$

$$\hat{\sigma}_A(0^+) = s_z^{\text{eq}}\hat{S}_{x0}^{(AB)} \tag{8.167}$$

$$\hat{\sigma}_A^{\tau} = s_z^{\text{eq}}\left(\hat{S}_{x0}^{(AB)}\cos\frac{\omega_{AB}}{2}\tau + \hat{S}_{yz}^{(AB)}\sin\frac{\omega_{AB}}{2}\tau\right) \tag{8.168}$$

$$\hat{\sigma}_A^{-\tau} = s_z^{\text{eq}}\left(\hat{S}_{x0}^{(AB)}\cos\frac{\omega_{AB}}{2}\tau - \hat{S}_{yz}^{(AB)}\sin\frac{\omega_{AB}}{2}\tau\right) \tag{8.169}$$

$$\hat{P}_\pi^{A,y}\hat{\sigma}_A^{\tau}\hat{P}_\pi^{A,y^{-1}} = s_z^{\text{eq}}(\left(-\hat{S}_{x0}^{(AB)}\cos\frac{\omega_{AB}}{2}\tau + \hat{S}_{yz}^{(AB)}\sin\frac{\omega_{AB}}{2}\tau\right) \tag{8.170}$$

where we again used $\{\hat{S}_{zz}^{(AB)} : \hat{S}_{x0}^{(AB)} \rightarrow \hat{S}_{yz}^{(AB)}\}_1$ for the evolution during the τ delays and $[\hat{S}_{y0}^{(AB)}, \hat{S}_{yz}^{(AB)}] = 0$ for the π-pulse transformation. Defining the echo factor for the e_A signal as $m_x^A(2\tau) = E_{\text{echo}}^{A,x} m_x^A(0^+)$ and with $\hat{\sigma}_A^{-\tau} = -\hat{P}_\pi^{A,y}\hat{\sigma}_A^{\tau}\hat{P}_\pi^{A,y^{-1}}$ from equations (8.169), (8.170)

$$E_{\text{echo}}^{A,y} = \frac{\text{Tr}\{\hat{P}_\pi^{A,y}\hat{\sigma}_A^{\tau}\hat{P}_\pi^{A,y^{-1}}\hat{\sigma}_A^{-\tau}\}}{\text{Tr}\{\hat{\sigma}_A(0^+)^2\}} = -1 \tag{8.171}$$

Again, we must mention that we ignored here any relaxation effects.

At this point, it is rather easy to consider the full three-pulse DEER sequence, which contains in addition to the simple echo sequence a π-pulse on e_B at a variable time t during the first τ time interval. To take this pulse into account, we first realize that this additional pulse inverts the $s_z^{\text{eq}}\hat{S}_{0z}^{(AB)}$ part of $\hat{\sigma}^{\text{eq}}$, which stays further invariant during the rest of the DEER sequence and does not contribute to signal $m_{A,x}(2\tau)$. Evaluating the effect of the additional π-pulse at time t, we must recalculate the expression for $\hat{\sigma}_A^{\tau}$ in equation (8.168), resulting in $E_{\text{DEER}}^{A,x}$. The MW term of this pulse when added to the Hamiltonian in the $\omega_{MW} = \omega_A$ rotating frame stays time dependent and looks like

$$\begin{aligned} H_{\text{MW}}^{\text{RoF}} &= \omega_1^x(\hat{S}_{x0}^{(AB)} + \hat{S}_{0x}^{(AB)})\cos\Delta\omega_{AB}t \\ &+ (\hat{S}_{y0}^{(AB)} + \hat{S}_{0y}^{(AB)})\sin\Delta\omega_{AB}t \end{aligned} \tag{8.172}$$

These irradiation terms together with the $\Delta\omega_{AB}S_{zz}^{(AB)}$ term in $\hat{H}^{\text{RoF}}$ can cause an inversion of the $\hat{S}_{pz}^{(AB)}$ operators of $\hat{\sigma}(t^-)$ during the transformation $\hat{\sigma}_A(t^+) = \hat{P}_\pi^{B,x}\hat{\sigma}_A(t^-)\hat{P}_\pi^{B,x^{-1}}$:

$$\hat{\sigma}_A(0^+) = s_z^{\text{eq}}(\hat{S}_{x0}^{(AB)} + \hat{S}_{0z}^{(AB)}) \tag{8.173}$$

$$\hat{\sigma}_A(t^-) = s_z^{\text{eq}}\left(\hat{S}_{x0}^{(AB)}\cos\frac{\omega_{AB}}{2}t + \hat{S}_{yz}^{(AB)}\sin\frac{\omega_{AB}}{2}t + \hat{S}_{0z}^{(AB)}\right) \tag{8.174}$$

$$\hat{\sigma}_A(t^+) = s_z^{\text{eq}}\left(\hat{S}_{x0}^{(AB)}\cos\frac{\omega_{AB}}{2}t - \hat{S}_{yz}^{(AB)}\sin\frac{\omega_{AB}}{2}t - \hat{S}_{0z}^{(AB)}\right) \tag{8.175}$$

$$\begin{aligned} \hat{\sigma}_A^{\tau} = \hat{\sigma}_A(\tau) = s_z^{\text{eq}}\Big(&\hat{S}_{x0}^{(AB)}\cos\frac{\omega_{AB}}{2}(\tau - 2t) \\ &+\hat{S}_{yz}^{(AB)}\sin\frac{\omega_{AB}}{2}(\tau - 2t) - \hat{S}_{0z}^{(AB)}\Big) \end{aligned} \tag{8.176}$$

$$\begin{aligned} \hat{P}_\pi^{A,y}\sigma_A^{\tau}\hat{P}_\pi^{A,y^{-1}} = s_z^{\text{eq}}\Big(&-\hat{S}_{x0}^{(AB)}\cos\frac{\omega_{AB}}{2}(\tau - 2t) \\ &+\hat{S}_{yz}^{(AB)}\sin\frac{\omega_{AB}}{2}(\tau - 2t) - \hat{S}_{0z}^{(AB)}\Big) \end{aligned} \tag{8.177}$$

where we again used the trios $\{\hat{S}_{zz}^{(AB)} : \hat{S}_{x0}^{(AB)} \rightarrow \hat{S}_{yz}^{(AB)}\}_1$ and $\{\hat{S}_{0x}^{(AB)} : \hat{S}_{yz}^{(AB)} \rightarrow -\hat{S}_{yy}^{(AB)}\}_1$.

Insertion of the results of equations (8.173)–(8.177) into equation (8.171) results in

$$\begin{aligned} E_{\text{DEER}}^{A,x}(t) &= -\cos\frac{\omega_{AB}}{2}(\tau - 2t)\cos\frac{\omega_{AB}}{2}\tau \\ &- \sin\frac{\omega_{AB}}{2}(\tau - 2t)\sin\frac{\omega_{AB}}{2}\tau \end{aligned} \tag{8.178}$$

and the echo signal $m_x^A(2\tau)$ becomes a function of the position in time of the π-pulse on e_B

$$m_x^A(2\tau) = -\cos\omega_{AB}t\, m_x^A(0^+) \tag{8.179}$$

This is the result of an ideal DEER experiment for a single orientation. In a typical DEER experiment, one measures a powder average over many dipolar orientations.

8.4 AN ELECTRON COUPLED TO A NUCLEUS WITH $I = 1/2$

8.4.1 The Spin Density Operator and Its Basis Representations

In a manner similar to the two-electron system, we can now present the spin density operator and its matrix representations for a system containing one electron ($S = 1/2$) and one nucleus with spin quantum number $I = 1/2$. The dimension of the spin density matrix representations of this system is 4×4 and it can again be expanded in terms of 16 independent operators. When choosing again the product operator expansion, the independent operators are the same as those in equations (8.97)–(8.102) after changing the superscript from (AB) to (en) and the operators to $\hat{S}_{pq}^{(en)} = 2\hat{S}_p\hat{I}_q$, $\hat{S}_{q0}^{(en)} = \hat{S}_q$ and $\hat{S}_{0q}^{(en)} = \hat{I}_q$ with $p, q = x, y, z$. In this way, we get that

- the eight independent single-quantum operators for $p = x, y$ are

$$\hat{S}_{p0}^{(en)} = \hat{S}_p;\ \hat{S}_{0p}^{(en)} = \hat{I}_p;\ \hat{S}_{pz}^{(en)} = 2\hat{S}_p\hat{I}_z;\ \hat{S}_{zp}^{(en)} = 2\hat{S}_z\hat{I}_p \quad (8.180)$$

- the four {zero+double}-quantum operators get the form again for $p, q = x, y$

$$\hat{S}_{pq}^{(en)} = 2\hat{S}_p\hat{I}_q \quad (8.181)$$

- and the three z-component operators

$$\hat{S}_{z0}^{(en)} = \hat{S}_z;\ \hat{S}_{0z}^{(en)} = \hat{I}_z;\ \hat{S}_{zz}^{(en)} = 2\hat{S}_z\hat{I}_z \quad (8.182)$$

These operators form again a complete set, including the unit operator $\hat{S}_0^{(en)}$, and can be used to generate the spin density operator of the form of equation (8.103), by replacing (AB) by (en). Therefore, the product operator formalism can be applied in the same way as shown before also to the {e-n} spin system. Matrix representations of these operators in the set of eigenstates

$$|1\rangle = |\alpha_e\beta_n\rangle;\ |2\rangle = |\alpha_e\alpha_n\rangle;\ |3\rangle = |\beta_e\beta_n\rangle;\ |4\rangle = |\beta_e\alpha_n\rangle \quad (8.183)$$

of the Zeeman Hamiltonian $(\omega_e\hat{S}_z - \omega_n\hat{I}_z)$, with decreasing eigenvalues for $\omega_e > \omega_n > 0$, are shown in Figure 8.8. We should mention here that the $|1\rangle \leftrightarrow |2\rangle$ and $|3\rangle \leftrightarrow |4\rangle$ transitions are the nuclear transitions and the $|1\rangle \leftrightarrow |3\rangle$ and $|2\rangle \leftrightarrow |4\rangle$ the electron spin transitions. The actual calculations involving these operators can again become cumbersome, and it is sometimes worth the effort to use operators that are linear combinations of the ones in equations (8.180)–(8.182). These new operators can be the fictitious spin-1/2 operators with matrix representations each with two elements defined in the set of eigenstates of equation (8.183) (Figure 8.8). The formal definitions of these operators, their commutation relations, and trios are presented in equations (8.114)–(8.119). Simple inspection of the matrix representations in Figure 8.8 reveals that we have the relations, similar to equation (8.103)

$$\begin{aligned}
&\hat{S}_{0x}^{(en)} = (\hat{S}_x^{1-2} + \hat{S}_x^{3-4});\ \hat{S}_{0y}^{(en)} = -(\hat{S}_y^{1-2} + \hat{S}_y^{3-4});\\
&\hat{S}_{x0}^{(en)} = (\hat{S}_x^{1-3} + \hat{S}_x^{2-4});\ \hat{S}_{y0}^{(en)} = (\hat{S}_y^{1-3} + \hat{S}_y^{2-4});\\
&\hat{S}_{zx}^{(en)} = (\hat{S}_x^{1-2} - \hat{S}_x^{3-4});\ \hat{S}_{zy}^{(en)} = (-\hat{S}_y^{1-2} + \hat{S}_y^{3-4});\\
&\hat{S}_{xz}^{(en)} = (-\hat{S}_x^{1-3} + \hat{S}_x^{2-4});\ \hat{S}_{yz}^{(en)} = (-\hat{S}_y^{1-3} + \hat{S}_y^{3-4});\\
&\hat{S}_{xx}^{(en)} = (\hat{S}_x^{1-4} + \hat{S}_x^{2-3});\ \hat{S}_{yy}^{(en)} = (\hat{S}_x^{1-4} - \hat{S}_x^{2-3});\\
&\hat{S}_{xy}^{(en)} = (-\hat{S}_y^{1-4} + \hat{S}_y^{2-3});\ \hat{S}_{yx}^{(en)} = (\hat{S}_y^{1-4} + \hat{S}_y^{2-3});\\
&\hat{S}_{0z}^{(en)} = (-\hat{S}_z^{1-4} + \hat{S}_z^{2-3});\ \hat{S}_{z0}^{(en)} = (\hat{S}_z^{1-4} + \hat{S}_z^{2-3});\\
&\hat{S}_{zz}^{(en)} = (-\hat{S}_z^{1-2} + \hat{S}_z^{3-4})
\end{aligned} \quad (8.184)$$

8.4.2 The Hyperfine Interaction Between the Electron and the Nucleus

The hyperfine interaction Hamiltonian representing the interaction between the electron and the nucleus (see Chapter 5) is a sum of an isotropic interaction (Fermi contact) and the electron–nuclear dipolar coupling that, similar to the two electron case (equation (8.114)), can be expressed as a linear combination of the components of the irreducible second-rank tensor operator T_m^2 $(m = -2, .., 2)$

$$\begin{aligned}
\hat{H}_{\mathrm{HF}} = &\ \omega_{\mathrm{iso}}^{en}(\hat{S}_x\hat{I}_x + \hat{S}_y\hat{I}_y + \hat{S}_z\hat{I}_z)\\
&+ \sum_{m=-2,2} (-1)^m \omega_m^{(en)}(\theta_{en}, \varphi_{en})\hat{T}_m^{(2)}
\end{aligned} \quad (8.185)$$

The Fermi contact coefficient is given by $\omega_{\mathrm{iso}}^{en} = \frac{2}{3} g_e g_n \mu_B \mu_n \rho^{\alpha-\beta}(0)$ ($\omega_{\mathrm{iso}}^{en} \equiv a_{\mathrm{iso}}$ in Chapter 5), where $\rho^{\alpha-\beta}(0)$ is the value of the spin density function at the nucleus. The coefficients $\omega_m^{(en)}$ and the operators $\hat{T}_m^{(2)}$ are given by expressions similar to those in equations (8.115), (8.116) and (8.117), (8.118), where the interaction strength ω_{D}^{AB} should be replaced with $\omega_{\mathrm{D}}^{en} = g_e g_n \mu_B \mu_n / r_{AB}^3$ ($\omega_{\mathrm{D}}^{en} \equiv T$ in Chapter 5) and the operators $\hat{S}_{pq}^{(AB)}$ by $\hat{S}_{pq}^{(en)}$. At high fields ($\omega_0 \gg \omega_{\mathrm{D}}^{en}$), all terms in the Hamiltonian containing operators of the form $\hat{S}_{xp}^{(en)}$ and $\hat{S}_{yp}^{(en)}$

$$S^{(en)}_{0x}=\frac{1}{2}\begin{pmatrix}0&1&&\\1&0&&\\&&0&1\\&&1&0\end{pmatrix};\quad S^{(en)}_{0y}=\frac{1}{2}\begin{pmatrix}0&i&&\\-i&0&&\\&&0&i\\&&-i&0\end{pmatrix};\quad S^{(en)}_{x0}=\frac{1}{2}\begin{pmatrix}0&&&\\&0&&1\\1&&0&\\&1&&0\end{pmatrix};\quad S^{(en)}_{y0}=\frac{1}{2}\begin{pmatrix}0&&-i&\\&0&&-i\\i&&0&\\&i&&0\end{pmatrix}$$

$$S^{(en)}_{zx}=\frac{1}{2}\begin{pmatrix}0&1&&\\1&0&&\\&&0&-1\\&&-1&0\end{pmatrix};\quad S^{(en)}_{zy}=\frac{1}{2}\begin{pmatrix}0&i&&\\-i&0&&\\&&0&-i\\&&i&0\end{pmatrix};\quad S^{(en)}_{xz}=\frac{1}{2}\begin{pmatrix}0&&-1&\\&0&&1\\-1&&0&\\&1&&0\end{pmatrix};\quad S^{(en)}_{yz}=\frac{1}{2}\begin{pmatrix}0&&i&\\&0&&-i\\-i&&0&\\&i&&0\end{pmatrix}$$

$$S^{(en)}_{xx}=\frac{1}{2}\begin{pmatrix}&&&1\\&&1&\\&1&&\\1&&&\end{pmatrix};\quad S^{(en)}_{yy}=\frac{1}{2}\begin{pmatrix}&&&1\\&&-1&\\&-1&&\\1&&&\end{pmatrix};\quad S^{(en)}_{xy}=\frac{1}{2}\begin{pmatrix}&&&i\\&&-i&\\&i&&\\-i&&&\end{pmatrix};\quad S^{(en)}_{yx}=\frac{1}{2}\begin{pmatrix}&&&-i\\&&-i&\\&i&&\\i&&&\end{pmatrix}$$

$$S^{(en)}_{0z}=\frac{1}{2}\begin{pmatrix}-1&&&\\&1&&\\&&-1&\\&&&1\end{pmatrix};\quad S^{(en)}_{z0}=\frac{1}{2}\begin{pmatrix}1&&&\\&1&&\\&&-1&\\&&&-1\end{pmatrix};\quad S^{(en)}_{zz}=\frac{1}{2}\begin{pmatrix}-1&&&\\&1&&\\&&1&\\&&&-1\end{pmatrix};\quad S^{(AB)}_{0}=\begin{pmatrix}1&&&\\&1&&\\&&1&\\&&&1\end{pmatrix}$$

Figure 8.8. The matrix representations of the 16 product operators presented in equations (8.180)–(8.182) in the basis set of states defined by equation (8.183). These operators are used to express the Hamiltonian and the spin density operator of electron–nucleus spin systems

with $p=x,y,z,0$ (the *nonsecular* terms) are truncated, while terms with $\hat{S}^{(en)}_{zx}$ and $\hat{S}^{(en)}_{zy}$ are kept (the *pseudo-secular* terms) resulting in the axial hyperfine Hamiltonian

$$\hat{H}_{\text{HF}}=\frac{A}{2}\hat{S}^{(en)}_{zz}+\frac{B}{2}\hat{S}^{(en)}_{zx} \tag{8.186}$$

where the coefficients are given by

$$A=\omega^{en}_{\text{iso}}+\omega^{en}_{\text{D}}(3\cos^2\theta_{en}-1);\ B=3\omega^{en}_{\text{D}}\sin\theta_{en}\cos\theta_{en} \tag{8.187}$$

Thus, the hyperfine interaction term together with the nuclear Zeeman term in the MW (at ω_{MW}) rotating frame Hamiltonian of a $\{e–n\}$ spin system can now be expressed, in terms of both independent sets of operators, as

$$\begin{aligned}\hat{H}^{\text{RoF}}&=\Delta\omega_0\hat{S}_{z0}+\frac{A}{2}\hat{S}^{(en)}_{zz}+\frac{B}{2}\hat{S}^{(en)}_{zx}-\omega_n\hat{S}^{(en)}_{0z}\\&=\Delta\omega_0(\hat{S}^{1-3}_z+\hat{S}^{2-4}_z)+\left(\omega_{\text{n}}-\frac{A}{2}\right)\hat{S}^{1-2}_z\\&\quad+\frac{B}{2}\hat{S}^{1-2}_x+\left(\omega_{\text{n}}+\frac{A}{2}\right)\hat{S}^{3-4}_z-\frac{B}{2}\hat{S}^{3-4}_x\end{aligned} \tag{8.188}$$

where $\Delta\omega_0$ is the electron off-resonance value $(\omega_0-\omega_{\text{MW}})$. In the $\hat{S}^{i-j}_p$ expansion, the $(1-2)$ operators (which keep the electron state at $|\alpha_e\rangle$) have matrix elements between the states $|\alpha_e\beta_n\rangle\leftrightarrow|\alpha_e\alpha_n\rangle$ and the $(3-4)$ operators ($|\beta_e\rangle$ invariant) between the states $|\beta_e\beta_n\rangle\leftrightarrow|\beta_e\alpha_n\rangle$.

8.4.3 The Free Induction Decay Signal

At this point, we should be able to evaluate the signal of the electron after a short MW $\pi/2$-pulse excitation. In this section, we present a full derivation of the time dependence of this signal, emphasizing the complexity of the time evolution of the coefficients of the spin density operator during this rather simple experiment. The formal solution must be derived by inserting $\hat{H}^{\text{RoF}}$ into the L-vN equation and solving the time evolution of the spin density operator $\hat{\sigma}(t)$ (in its reduced form in the rotating frame as in equation (8.135)) of the $\{e–n\}$ system. After an ideal MW $\pi/2$ pulse in the y-direction, the initial spin density can be expressed as $\hat{\sigma}(0^+)=$

$s_z^{eq}\hat{S}_{x0}^{(en)} = s_{e,z}^{eq}(\hat{S}_x^{1-3} + \hat{S}_x^{2-4})$, where we ignore the nuclear z-operator term in the equilibrium $\hat{\sigma}^{(e-n)eq} = s_{e,z}^{eq}\hat{S}_{z0}^{(en)} + s_{n,z}^{eq}\hat{S}_{0z}^{(en)}$ with its relatively small Boltzmann prefactor. It should now be possible to calculate the evolution of the spin density operator as a function of time and derive the electron spin FID signal. To do so, it is necessary to propagate $\hat{\sigma}(0^+)$ by applying the operator $U^{\text{RoF}}(t) = \text{e}^{-\text{i}\hat{H}^{\text{RoF}}t}$, which in the following will be accomplished using the fictitious spin-1/2 operators. Expanding the spin density operator as

$$\hat{\sigma}(t) = \sum_{\substack{r=x,y,z\\ i<j=1,4}} s_r^{i-j}(t)\hat{S}_r^{i-j} \tag{8.189}$$

we must determine the time evolution of its coefficients $s_p^{i-j}(t)$.

The matrix representation of the Hamiltonian in the manifold of states in equation (8.183) is not diagonal. This can be seen easily when we consider its expression in terms of the $\hat{S}_p^{i-j}$ operators. Realizing that $\text{e}^{-\text{i}(\hat{S}_x^{i-j}+\hat{S}_y^{i-j})} \neq \text{e}^{-\text{i}\hat{S}_x^{i-j}}\text{e}^{-\text{i}\hat{S}_y^{i-j}}$, it is not straightforward to use the trios derived in Section 8.3 for obtaining $\hat{\sigma}(t) = \hat{U}^{\text{RoF}}(t)\hat{\sigma}(0^+)\hat{U}^{\text{RoF}^{-1}}(t)$. Thus, to solve this time evolution, we must diagonalize $\hat{H}^{\text{RoF}}$, i.e., apply a transformation that yield a form for $\hat{H}^{\text{RoF}}$ that contains only $\hat{S}_z^{i-j}$ terms.

Inspection of the Hamiltonian expansion in equation (8.188) reveals that we can diagonalize the part belonging to the (1−2) transition and the part belonging to the (3−4) transition separately. Introducing transformation operators of the form

$$\hat{U}_{\theta,p}^{i-j} \equiv \text{e}^{-\text{i}\theta\hat{S}_p^{i-j}} \tag{8.190}$$

the diagonalization is achieved by transforming the operators of these two transitions by $\hat{U}_{\varphi_y^{1-2},y}^{1-2}$ and $\hat{U}_{\varphi_y^{3-4},y}^{3-4}$, respectively

$$\begin{aligned}\hat{A}^{\text{RoF}} &= \Delta\omega_0(\hat{S}_z^{1-3} + \hat{S}_z^{2-4})\\ &\quad + \hat{U}_{\varphi_y^{1-2},y}^{1-2}\left\{\left(\omega_\text{n} - \frac{A}{2}\right)\hat{S}_z^{1-2} + \frac{B}{2}\hat{S}_x^{1-2}\right\}\hat{U}_{\varphi_y^{1-2},y}^{1-2^{-1}}\\ &\quad + \hat{U}_{\varphi_y^{3-4},y}^{3-4}\left\{\left(\omega_\text{n} + \frac{A}{2}\right)\hat{S}_z^{3-4} - \frac{B}{2}\hat{S}_x^{3-4}\right\}\hat{U}_{\varphi_y^{3-4},y}^{3-4^{-1}}\\ &\equiv \Delta\omega_0(\hat{S}_z^{1-3} + \hat{S}_z^{2-4}) + \omega_\alpha\hat{S}_z^{1-2} + \omega_\beta\hat{S}_z^{3-4}\end{aligned} \tag{8.191}$$

Note that the transformation operators commute with the off-resonance term. In equation (8.191), we choose the $\omega_{\alpha/\beta}$ frequencies to be positive

$$\begin{aligned}\omega_\alpha &= \sqrt{(\omega_\text{n} - A/2)^2 + (B/2)^2}\\ \omega_\beta &= \sqrt{(\omega_\text{n} + A/2)^2 + (B/2)^2}\end{aligned} \tag{8.192}$$

and

$$\begin{aligned}\sin\varphi_y^{1-2} &= \frac{-1/2B}{\omega_\alpha};\ \cos\varphi_y^{1-2} = \frac{(\omega_\text{n} - A/2)}{\omega_\alpha}\\ \sin\varphi_y^{3-4} &= \frac{-1/2B}{\omega_\beta};\ \cos\varphi_y^{3-4} = \frac{(\omega_\text{n} + A/2)}{\omega_\beta}\end{aligned} \tag{8.193}$$

The diagonalized Hamiltonian has four eigenstates with frequency differences

$$\omega^{i-j} = \langle i|\hat{A}^{\text{RoF}}|i\rangle - \langle j|\hat{A}^{\text{RoF}}|j\rangle \tag{8.194}$$

$$\begin{aligned}\omega^{1-3} &= \Delta\omega_0 + \frac{1}{2}(\omega_\alpha - \omega_\beta) \equiv \Delta\omega_0 - \Omega_e\\ \omega^{2-4} &= \Delta\omega_0 - \frac{1}{2}(\omega_\alpha - \omega_\beta) \equiv \Delta\omega_0 + \Omega_e\\ \omega^{1-4} &= \Delta\omega_0 + \frac{1}{2}(\omega_\alpha + \omega_\beta) \equiv \Delta\omega_0 + \Omega_{en}\\ \omega^{2-3} &= \Delta\omega_0 - \frac{1}{2}(\omega_\alpha + \omega_\beta) \equiv \Delta\omega_0 - \Omega_{en}\end{aligned} \tag{8.195}$$

The frequency shifts with respect to the off-resonance frequency $\Delta\omega_0$ are $\pm\Omega_e$ and $\pm\Omega_{en}$, which become for $A, B < \omega_n$ about equal to $|\Omega_e| = \left|-\frac{1}{2}(\omega_\alpha - \omega_\beta)\right| \approx |A/2|$ and $|\Omega_{en}| = \left|-\frac{1}{2}(\omega_\alpha + \omega_\beta)\right| \approx \omega_\text{n}$. In the case that $A, B > \omega_\text{n}$ $|\Omega_e| \to 0$ $|\Omega_{en}| \approx |A/2|$. The energy level diagram of $A_{(e-n)}^{\text{RoF}}$ is shown in Figure 8.9. We must be aware of the fact that in this derivation the electron frequency Ω_e and the dq frequency Ω_{en} can be positive or negative, depending on the relative magnitudes of ω_α and ω_β. The four frequencies in equation (8.195) already determine the positions of the lines in the EPR spectrum of $(e\text{–}n)$. A discussion about their actual values for positive and negative values of A and B or for $|A| > \omega_\text{n}$ can be found in Chapter 18.

It is now possible to continue the derivation in terms of the $\hat{S}_{pq}^{(en)}$ product operator expansion. To do so, we rewrite the $\hat{U}_{\text{tr}}$ transformation operator in the form

$$\hat{U}_{\text{tr}} = \text{e}^{-\text{i}(\varphi^{1-2}\hat{S}_y^{1-2}+\varphi^{3-4}\hat{S}_y^{3-4})} = \text{e}^{-\text{i}(\Sigma\varphi\,\hat{S}_{0y}^{(en)}+\Delta\varphi\,\hat{S}_{zy}^{(en)})} \tag{8.196}$$

Here, we used equation (8.184) with $\hat{S}_y^{1-2} = -\frac{1}{2}(\hat{S}_{0y}^{(en)} + \hat{S}_{zy}^{(en)})$ and $S_y^{3-4} = -\frac{1}{2}(S_{0y}^{(en)} - S_{zy}^{(en)})$ and,

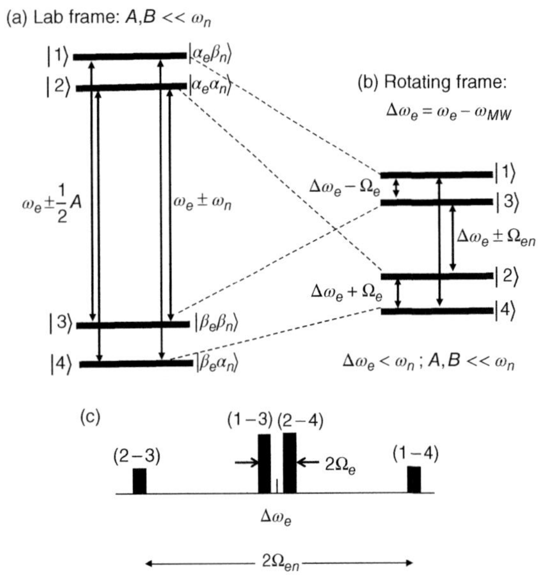

Figure 8.9. The energy level diagram of an electron–nucleus spin system with a hyperfine coupling frequency smaller than the nuclear Larmor frequency in (a) the laboratory frame and (b) in the MW rotating frame with $\omega_{\mathrm{MW}} = \omega_0 - \Delta\omega_0$. The definitions of Ω_e and Ω_{en} are given in equations (8.194), (8.195). (c) A stick diagram of the electron spectrum obtained after a single non-selective $\pi/2$ pulse showing the line splittings of $2\Omega_e$ and $2\Omega_{en}$ around $\Delta\omega_e$. The amplitudes of these lines can be found in equations (8.203), (8.204)

as a result

$$\Sigma\varphi = -\frac{1}{2}(\varphi_y^{1-2} + \varphi_y^{3-4}); \quad \Delta\varphi = -\frac{1}{2}(\varphi_y^{1-2} - \varphi_y^{3-4}) \tag{8.197}$$

and with $\hat{S}_{0z}^{(en)} = -\frac{1}{2}(\hat{S}_z^{1-2} + \hat{S}_z^{3-4})$ and $\hat{S}_{zz}^{(en)} = -\frac{1}{2}(\hat{S}_z^{1-2} - \hat{S}_z^{3-4})$, the diagonal Hamiltonian $\hat{\Lambda}^{\mathrm{RoF}}$ becomes

$$\hat{\Lambda}^{\mathrm{RoF}} = \Delta\omega_0 \hat{S}_{z0}^{(en)} - \Omega_{en}\hat{S}_{0z}^{(en)} + \Omega_e \hat{S}_{zz}^{(en)} \tag{8.198}$$

At this point, we must (i) transfer $\hat{\sigma}(0^+) = s_z^{\mathrm{eq}}\hat{S}_{x0}^{(en)}$ to the representation in which $\hat{H}^{\mathrm{RoF}}$ becomes diagonal, (ii) let it evolve under the influence of $\hat{\Lambda}^{\mathrm{RoF}}$, and (iii) transfer it back to the original representation. As the transformation operator commutes with the off-resonance terms of $\hat{\Lambda}^{\mathrm{RoF}}$, we first perform the three steps of the calculation setting $\Delta\omega_0 = 0$ and after reaching the result for $\hat{\sigma}(t)$ (iv) apply the off-resonance propagator $\mathrm{e}^{-\mathrm{i}\Delta\omega_0 \hat{S}_{z0}^{(en)} t}$.

(i)

$$\hat{\sigma}_{\mathrm{tr}}(0^+) = s_z^{\mathrm{eq}}\hat{U}_{\mathrm{tr}}\hat{S}_{x0}^{(en)}\hat{U}_{\mathrm{tr}}^{-1} = \cos\Delta\varphi\,\hat{S}_{x0}^{(en)} + \sin\Delta\varphi\,\hat{S}_{yy}^{(en)} \tag{8.199}$$

where we used $\{\hat{S}_{zy}^{(en)} : \hat{S}_{x0}^{(en)} \to \hat{S}_{yy}^{(en)}\}_1$ and $[\hat{S}_{0y}^{(en)}, \hat{S}_{x0}^{(en)}] = 0$

(ii)

$$\hat{\sigma}_{\mathrm{tr}}(t) = \cos\Delta\varphi\,\cos\Omega_e t\,\hat{S}_{x0}^{(en)} + \cos\Delta\varphi\,\sin\Omega_e t\,\hat{S}_{yz}^{(en)} + \sin\Delta\varphi\,\cos\Omega_{en} t\,\hat{S}_{yy}^{(en)} + \sin\Delta\varphi\,\sin\Omega_{en} t\,\hat{S}_{yx}^{(en)} \tag{8.200}$$

where we used $\{\hat{S}_{zz}^{(en)} : \hat{S}_{x0}^{(en)} \to \hat{S}_{yz}^{(en)}\}_1$, $\{\hat{S}_{0z}^{(en)} : \hat{S}_{yy}^{(en)} \to -\hat{S}_{yx}^{(en)}\}_1$ and $[\hat{S}_{0z}^{(en)}, \hat{S}_{x0}^{(en)}] = [\hat{S}_{zz}^{(en)}, \hat{S}_{yy}^{(en)}] = 0$

(iii)

$$\begin{aligned}
\hat{\sigma}(t) &= (\cos\Delta\varphi\,\cos\Omega_e t\,\cos\Delta\varphi \\
&\quad + \sin\Delta\varphi\cos\Omega_{en} t\sin\Delta\varphi)\,\hat{S}_{x0}^{(en)} \\
&\quad + (\cos\Delta\varphi\sin\Omega_e t\,\cos\Sigma\varphi + \sin\Delta\varphi \\
&\quad \sin\Omega_{en} t\sin\Sigma\varphi)\hat{S}_{yz}^{(en)} \\
&\quad + (\sin\Delta\varphi\cos\Omega_{en} t\cos\Delta\varphi \\
&\quad - \cos\Delta\varphi\,\cos\Omega_e t\,\sin\Delta\varphi)\,\hat{S}_{yy}^{(en)} \\
&\quad + (\sin\Delta\varphi\sin\Omega_{en} t\,\cos\Sigma\varphi \\
&\quad - \cos\Delta\varphi\sin\Omega_e t\,\sin\Sigma\varphi)\,\hat{S}_{yx}^{(en)} \\
&\equiv s_{x0}^{(en)}(t)\hat{S}_{x0}^{(en)} + s_{yz}^{(en)}(t)\hat{S}_{yz}^{(en)} + s_{yy}^{(en)}(t)\hat{S}_{yy}^{(en)} \\
&\quad + s_{yx}^{(en)}(t)\hat{S}_{yx}^{(en)}
\end{aligned} \tag{8.201}$$

where we used $(-\hat{S}_{0y}^{(en)} : \hat{S}_{yz}^{(en)} \to -\hat{S}_{yx}^{(en)}\}_1$, $\{-\hat{S}_{0y}^{(en)} : \hat{S}_{yx}^{(en)} \to \hat{S}_{yz}^{(en)}\}_1(-\hat{S}_{zy}^{(en)} : \hat{S}_{x0}^{(en)} \to -\hat{S}_{yy}^{(en)}\}_1$ and $(-\hat{S}_{zy}^{(en)} : \hat{S}_{yy}^{(en)} \to \hat{S}_{x0}^{(en)}\}_1$.

The final evolution in equation (8.201) shows the complex evolution of a $\{e\text{–}n\}$ spin system even for $\Delta\omega_e = 0$. It is composed of single quantum operators and two (double+zero) quantum operators. The expansion for $\hat{\sigma}^{(en)}(t)$ after applying the off-resonance propagator contains twice as much operators. As a result, the full expression can be derived from equation (8.201) using the coefficients $s_{pq}^{(en)}(t)$ of the four operators

$$\begin{aligned}
\sigma(t) &= s_{x0}^{(en)}(t)\,(\hat{S}_{x0}^{(en)}\cos\Delta\omega_0 t + \hat{S}_{y0}^{(en)}\sin\Delta\omega_0 t) \\
&\quad + s_{yz}^{(en)}(t)(\hat{S}_{yz}^{(en)}\cos\Delta\omega_0 t - \hat{S}_{xz}^{(en)}\sin\Delta\omega_0 t) \\
&\quad + s_{yy}^{(en)}(t)(\hat{S}_{yy}^{(en)}\cos\Delta\omega_0 t - \hat{S}_{xy}^{(en)}\sin\Delta\omega_0 t) \\
&\quad + s_{yx}^{(en)}(t)\,(\hat{S}_{yx}^{(en)}\cos\Delta\omega_0 t - \hat{S}_{xx}^{(en)}\sin\Delta\omega_0 t)
\end{aligned} \tag{8.202}$$

Thus, to describe the evolution of the density operator after the single π/2 excitation pulse, we require 8 of the 15 independent product operators. The electron spin complex signal after this pulse depends only on the expectation values of the two in-phase linear electron operators

$$m(t) = -g_e\mu_B(\langle\hat{S}_{x0}^{(en)}\rangle(t) + i\langle\hat{S}_{y0}^{(en)}\rangle(t) \tag{8.203}$$

Taking the effect of $\Delta\omega_0$ into account, the signal becomes

$$m(t) = -g_e\mu_B s_z^{eq}(\cos^2\Delta\varphi\cos\Omega_e t + \sin^2\Delta\varphi\cos\Omega_{en}t)$$
$$(\cos\Delta\omega_0 t + i\sin\Delta\omega_0 t) \tag{8.204}$$

Multiplying $m(t)$ with a decaying $e^{-R_2 t}$ and applying a Fourier transformation, the resulting spectrum shown in Figure 8.9 consists of two lines at $\Delta\omega_0 \pm \Omega_e$ with intensities proportional to $\cos^2\Delta\varphi$, the 'allowed' spectral resonances close to $\Delta\omega_0 \pm A/2$ when $B \ll \omega_n$, and two lines at $\Delta\omega_0 \pm \Omega_{en}$ with intensities proportional to $\sin^2\Delta\varphi$, the 'forbidden' resonances again close to $\Delta\omega_0 \pm \omega_n$ when $B \ll \omega_n$. For further discussions of the meaning of this result, we refer to Chapters 16, 17 and 18.

8.4.4 The Liouville Representation and Relaxation

As in Section 8.2.5, the spin density operator expanded in terms of the independent operators can be represented as a vector $\overline{\sigma}(t)$ composed now of all $s_p^{i-j}(t)$. The Liouville superoperator in the rate equation for $\overline{\sigma}(t)$ can again be derived from the matrix elements of the Hamiltonian as in equation (8.128), and a set of longitudinal and transverse relaxation rate constants can be added as in equation (8.129)

$$\frac{d}{dt}\overline{\sigma}(t) = (-i\hat{\hat{L}}(t) + \hat{\hat{R}})\overline{\sigma}(t) \tag{8.205}$$

In particular, we can distinguish between R_{1e} electronic relaxation rate constants of the electron transitions ($|1\rangle - |3\rangle$) and ($|2\rangle - |4\rangle$) and R_{1n} nuclear relaxation rate constants of the nuclear transition ($|1\rangle - |2\rangle$) and ($|3\rangle - |4\rangle$). In the same way, R_{2e} and R_{2n} relaxation rate constants can be added to $\hat{\hat{R}}$ for the coherence decay of the respective transitions. In addition, double- and zero-quantum relaxation rates can be introduced for the ($|1\rangle - |4\rangle$) and ($|2\rangle - |3\rangle$) transitions, respectively.

8.4.5 EPR Pulse Sequences

At this point, we use the methodology of calculating echo factors in Sections 8.2.10 and 8.2.11 to describe two basic echo pulse sequences that are commonly used in pulse EPR spectroscopy, ESEEM and ENDOR. This description will not deal with practical issues or the characterization of samples in terms of homogenous or inhomogeneous EPR lineshapes. The purpose of our discussion is solely to show the derivations of $E^x_{\text{ESEEM}}(t)$ and $E^x_{\text{ENDOR}}(t)$ factors. A detailed description of the ESEEM (electron-spin echo envelope modulation) and ENDOR (electron–nuclear double resonance) sequences can be found in Chapters 16 and 18.

8.4.5.1 *The Two-pulse ESEEM Sequence*

In this example, we describe the ESEEM pulse sequence applied to the single two-spin $\{e–n\}$ system experiencing a Hamiltonian of the form as in equation (8.188)[19]

$$\{\underset{\downarrow}{\overset{0^+}{}}\hat{P}^{e,y}_{\pi/2} - \tau - \underset{\downarrow}{\overset{\tau^-}{}}\hat{P}^{e,y}_{\pi}\underset{\downarrow}{\overset{\tau^+}{}} - \tau - \underset{\downarrow}{\overset{2\tau}{}}\}$$

The pulse sequence is shown in Figure 8.10(a) and is in fact a standard echo pulse sequence applied to the electron of the $\{e–n\}$ system. Following the procedure used in Sections 8.2.10 and 8.2.11 to evaluate the effect of the echo sequences on the echo signal at 2τ. We start with introducing an echo factor $E^x_{\text{ESEEM}}(t)$

$$m^e_x(2\tau) = E^x_{\text{ESEEM}}(\tau)m^e_x(0^+) \tag{8.206}$$

where the x-axis signal after the initial pulse in the y-direction at $t = 0^+$ and at the maximum echo position at $t = 2\tau$ are $m_{e,x}(t) = -g_e\mu_B s_x^{(en)}(t)$. The expression for $E^x_{\text{ESEEM}}(t)$ is

$$E^x_{\text{ESEEM}}(\tau) = \frac{\text{Tr}(\hat{P}^{e,y}_{\pi}\hat{\sigma}^{\tau}\hat{P}^{e,y}_{\pi}\hat{\sigma}^{-\tau})}{\text{Tr}(\hat{\sigma}(0^+))^2} \tag{8.207}$$

where $\hat{\sigma}^{\tau}$ is the spin density operator at $t = \tau^-$ just before the second pulse, starting with $\hat{\sigma}(0^+) = s_z^{eq}\hat{S}_{0x}^{(en)}$, and $\hat{\sigma}^{-\tau}$ after an evolution during a time $t = -\tau$, also starting from $\hat{\sigma}(0^+)$. To evaluate these expressions, we can use the derivation in Section 8.4.3 and in particular the expression for $\hat{\sigma}(t)$ in equation (8.202) and evaluate $\hat{\sigma}^{\pm\tau} = \hat{\sigma}(\pm\tau)$. The effect of the π-pulse on $\hat{\sigma}^{\tau}$ can easily be evaluated when we assume that an ideal y-pulse inverts only the operators $\hat{S}_{x0}^{(en)}$, $\hat{S}_{xz}^{(en)}$, and $\hat{S}_{xx}^{(en)}$. When

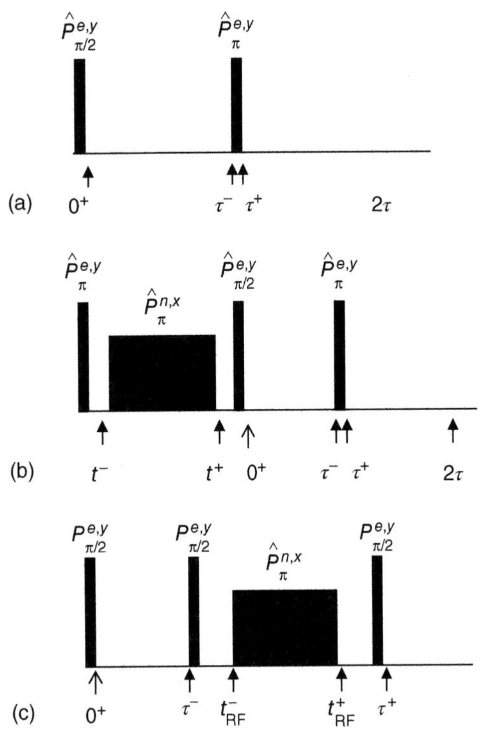

Figure 8.10. The three basic pulse EPR sequences that are applied on an electron–nucleus spin system as discussed in Section 8.4: (a) the Hahn echo sequence resulting in the ESEEM modulations, (b) the Davies ENDOR pulse sequence, and (c) the Mims ENDOR pulse sequence

we rewrite equation (8.202) in the form

$$\begin{aligned}\hat{\sigma}^{\tau} = &\{s_{x0}^{(en)}(\tau)\hat{S}_{x0}^{(en)} + s_{yy}^{(en)}(\tau)\hat{S}_{yy}^{(en)}\\ &+ s_{yz}^{(en)}(\tau)\hat{S}_{yz}^{(en)} + s_{yx}^{(en)}(\tau)\hat{S}_{yx}^{(en)}\}\cos\Delta\omega_0\tau\\ &+ \{s_{x0}^{(en)}(\tau)\hat{S}_{y0}^{(en)} - s_{yy}^{(en)}(\tau)\hat{S}_{xy}^{(en)}\\ &- s_{yz}^{(en)}(\tau)\hat{S}_{xz}^{(en)} - s_{yx}^{(en)}(\tau)\hat{S}_{xx}^{(en)}\}\sin\Delta\omega_0\tau \quad (8.208)\end{aligned}$$

then

$$\begin{aligned}&\hat{P}_{\pi}^{e,y}\hat{\sigma}^{\tau}\hat{P}_{\pi}^{e,y-1}\\ &= \{-s_{x0}^{(en)}(\tau)\hat{S}_{x0}^{(en)} + s_{yy}^{(en)}(\tau)\hat{S}_{yy}^{(en)}\\ &\quad + s_{yz}^{(en)}(\tau)\hat{S}_{yz}^{(en)} + s_{yx}^{(en)}(\tau)\hat{S}_{yx}^{(en)}\}\cos\Delta\omega_0\tau\\ &\quad + \{s_{x0}^{(en)}(\tau)\hat{S}_{y0}^{(en)} + s_{yy}^{(en)}(\tau)\hat{S}_{xy}^{(en)}\\ &\quad + s_{yz}^{(en)}(\tau)\hat{S}_{xz}^{(en)} + s_{yx}^{(en)}(\tau)\hat{S}_{xx}^{(en)}\}\sin\Delta\omega_0\tau\end{aligned} \quad (8.209)$$

To evaluate $\hat{\sigma}^{-\tau}$, all we need to do is to use the fact that $\sin(-\alpha) = -\sin\alpha$ and $\cos(-\alpha) = \cos\alpha$ and starting from equations (8.208) and (8.209) we obtain

$$\begin{aligned}\hat{\sigma}^{-\tau} = &\{s_{x0}^{(en)}(\tau)\hat{S}_{x}^{(en)} + s_{yy}^{(en)}(\tau)\hat{S}_{yy}^{(en)}\\ &- s_{yz}^{(en)}(\tau)\hat{S}_{yz}^{(en)} - s_{yx}^{(en)}(\tau)\hat{S}_{yx}^{(en)}\}\cos\Delta\omega_0 t\\ &- \{s_{x0}^{(en)}(\tau)\hat{S}_{y0}^{(en)} - s_{yy}^{(en)}(\tau)\hat{S}_{xy}^{(en)}\\ &+ s_{yz}^{(en)}(\tau)\hat{S}_{xz}^{(en)} + s_{yx}^{(en)}(\tau)\hat{S}_{xx}^{(en)}\}\sin\Delta\omega_0 t\end{aligned} \quad (8.210)$$

Expanding the echo factor $E_{\text{ESEEM}}^{x}(\tau)$ in terms of the sum of the products of the coefficients of (8.209) and (8.210) and using $\text{Tr}((\hat{S}_{pq}^{(en)})^2) = 1$ and $\text{Tr}(\hat{S}_{pq}^{(en)}\hat{S}_{p'q'}^{(en)}) = 0$ with $p', q' \neq p, q$ yields

$$\begin{aligned}E_{\text{ESEEM}}^{x}(\tau) = &-\{s_{x0}^{(en)}(\tau)\}^2 + \{s_{yy}^{(en)}(\tau)\}^2\\ &- \{s_{yz}^{(en)}(\tau)\}^2 - \{s_{yx}^{(en)}(\tau)\}^2\end{aligned} \quad (8.211)$$

which is independent of the off-resonance frequency $\Delta\omega_0\tau$. This means that the off-resonance frequency is refocused, and the hyperfine and nuclear Zeeman terms modulate $E_{\text{ESEEM}}^{x}(\tau)$. Insertion of the expressions for the $s_{pq}^{(en)}(\tau)$ coefficients given in equation (8.202) into equation (8.211) results in

$$\begin{aligned}E_{\text{ESEEM}}^{x}(\tau) = &-(\cos^2\Delta\varphi - \sin^2\Delta\varphi)\\ &(\cos^2\Delta\varphi\cos^2\Omega_e\tau - \sin^2\Delta\varphi\cos^2\Omega_{en}\tau)\\ &- 4\cos^2\Delta\varphi\sin^2\Delta\varphi\cos\Omega_e\tau\cos\Omega_{en}\tau\\ &- \cos^2\Delta\varphi\sin^2\Omega_e\tau - \sin^2\Delta\varphi\sin^2\Omega_{en}\tau\end{aligned} \quad (8.212)$$

Following some trigonometric manipulations to isolate single frequencies, one obtains the well-known ESEEM formula

$$\begin{aligned}E_{\text{ESEEM}}^{x}(\tau) = 1 - \frac{k}{4}[&2 - 2\cos(\omega_\alpha\tau) - 2\cos(\omega_\beta\tau)\\ &+ \cos(2\Omega_e\tau) + \cos(2\Omega_{en}\tau)]\end{aligned} \quad (8.213)$$

where $k = \sin^2 2\Delta\varphi$. In cases where the pseudo-secular hyperfine coefficient B is small with respect to the nuclear Zeeman frequency, we get the following frequencies:

$$\begin{aligned}|\omega_\alpha| \approx \omega_n - A/2; \quad |\omega_\beta| \approx \omega_n + A/2\\ |2\Omega_e| \approx A; \quad |2\Omega_{en}| \approx 2\omega_n\end{aligned} \quad (8.214)$$

8.4.5.2 *Pulse ENDOR*

In the last examples, we consider again an electron–nucleus $\{e-n\}$ system and give a short description of the common Davies[20] and Mims[21] ENDOR pulse sequences involving RF irradiation at the frequencies of the nuclear transitions $\omega_\alpha, \omega_\beta$. We restrict our discussion to the weak-coupling case with $|A, B| < \omega_n$.

Davies ENDOR

The Davies ENDOR sequence, illustrated in Figure 8.10(b), starts with a selective on-resonance π pulse at a frequency ω_{MW} equal to one of the two electron frequencies $\omega_0 + \Omega_e$ or $\omega_0 - \Omega_e$. This is immediately followed by an RF irradiation at a variable frequency ω_{RF},

$$\{\hat{P}_{\pi}^{e,y} - \hat{P}^{n,x}_{\omega_{1,RF}t_{RF}} \overset{\overset{t}{\downarrow}}{} - \overset{\overset{0^+}{\downarrow}}{\hat{P}_{\pi/2}^{e,y}} - \tau - \overset{\overset{\tau^-}{\downarrow}}{} \overset{\overset{\tau^+}{\downarrow}}{\hat{P}_{\pi}^{e,y}} - \tau - \overset{\overset{2\tau}{\downarrow}}{} \}$$

Here we only consider values for ω_{RF} equal to the nuclear frequencies ω_α or ω_β. The sequence ends with a frequency-selective echo sequence again with ω_{MW} equal to $\omega_0 + \Omega_e$ or $\omega_0 - \Omega_e$. In the following discussion, we divide the sequence into two parts: (i) The preparation sequence that first inverts an electron transition at $\omega_0 \pm \Omega_e$ and is followed by an RF irradiation at $\omega_{\alpha/\beta}$ of a duration t and (ii) a standard selective echo sequence again at $\omega_0 \pm \Omega_e$. Thus, we first calculate the spin density operator at t and use this as the initial operator of the echo sequence $\hat{\sigma}^{(i)}(t) = \hat{\sigma}^{(ii)}(0^-)$.

Because all pulses are selective, and the frequencies of the $\{e\text{–}n\}$ system are determined by the diagonal Hamiltonian as in equation (8.198), we can describe the spin evolution in the diagonal representation of the Hamiltonian after the application of the transformation $\hat{U}_{tr}$ on the Hamiltonian and the spin density operator.

At this point, we choose the conditions that the MW irradiation is applied at the frequency $\omega^{1-3} = \Delta\omega_0 - \Omega_e$ and the RF irradiation at the frequency $\omega^{1-2} = \omega_\alpha$ (see equation (8.195) and Figure 8.9). To follow the spin density operator, we transfer the Hamiltonian to the $\Delta\omega_0 - \Omega_e$ rotating frame and represent it in terms of the S_r^{i-j} operators to fully benefit from the fact that we are dealing with selective pulses. The diagonal form of the Hamiltonian is, according to equation (8.191),

$$\begin{aligned}\hat{\Lambda}^{\mathrm{RoF}}_{(e-n)} &= \Omega_e(\hat{S}_z^{1-3} + \hat{S}_z^{2-4}) + \omega_\alpha \hat{S}_z^{1-2} + \omega_\beta \hat{S}_z^{3-4} \\ &= 2\Omega_e S_z^{2-4} + \Omega_{en}(S_z^{1-2} - S_z^{3-4}) \qquad (8.215)\end{aligned}$$

where we used the definitions of Ω_e and Ω_{en} in equation (8.195). The transition frequencies in this rotating frame are as expected

$$\begin{aligned}&\omega^{1-3} = 0;\ \omega^{2-4} = 2\Omega_e;\ \omega^{1-4} = \Omega_{en} + \Omega_e; \\ &\omega^{2-3} = -(\Omega_{en} - \Omega_e) \qquad (8.216)\end{aligned}$$

The MW Hamiltonian in this diagonal frame representing the selective excitations at ω^{1-3} has the form

$$\hat{\Lambda}^{\mathrm{RoF}}_{\mathrm{MW}} = \omega_{1,\mathrm{MW}} \cos\Delta\varphi\, \hat{S}_y^{1-3} \qquad (8.217)$$

and the selective time-dependent RF Hamiltonian representing the irradiation at $\omega^{1-2} = \omega_\alpha$ is

$$\hat{\Lambda}^{\mathrm{RoF}}_{\mathrm{RF}} = 2\omega_1^{\mathrm{RF}}(\cos\varphi_y^{1-2}\hat{S}_x^{1-2} - \sin\varphi_y^{1-2}\hat{S}_z^{1-2})\cos\omega_\alpha t \qquad (8.218)$$

In both irradiation Hamiltonians, we ignored all off-resonance terms because $\omega_{1,\mathrm{MW}} \ll \Omega_e$ and $\omega_{1,\mathrm{RF}} \ll |\omega_\alpha - \omega_\beta|$. The initial spin density operator is $\hat{\sigma}_{tr}^{(i)}(0) = s_z^{eq}(\hat{S}_z^{1-3} + \hat{S}_z^{2-4})$, and the first selective pulse inverts $\hat{S}_z^{1-3}$ with $\omega_{1,\mathrm{MW}} \cos\Delta\varphi \times t_{\mathrm{MW}} = \pi$

$$\begin{aligned}\hat{P}_\pi^{(1-3),y}\hat{\sigma}_{tr}^{(i)}(0)\hat{P}_\pi^{(1-3)y} &= s_z^{eq}(-\hat{S}_z^{1-3} + \hat{S}_z^{2-4}) \\ &= s_z^{eq}(-\hat{S}_z^{1-2} + \hat{S}_z^{3-4}) \qquad (8.219)\end{aligned}$$

Next, we must follow the effect of the RF irradiation Hamiltonian $\hat{\Lambda}^{\mathrm{RoF}}_{\mathrm{RF}}$. This requires an additional transformation to the RF rotating frame, a propagation during the RF pulse of duration t_{RF} by a time-independent RF Hamiltonian $\hat{H}_{\mathrm{RF}}^{\mathrm{RoF_{MW}+RoF_{RF}}} = \omega_{1,\mathrm{RF}}(\cos\varphi_y^{1-2}\hat{S}_x^{1-2} - \sin\varphi_y^{1-2}\hat{S}_z^{1-2})$, and then a transformation back to the MW rotating frame. When we are only interested in the time evolution of the coefficient of S_z^{1-2} and ignore all coefficients of operators of the form $\hat{S}_x^{1-2}$ and $\hat{S}_y^{1-2}$ that oscillate at a frequency ω_α around the z-direction during t_{RF} in the MW rotating frame, the effect of the RF irradiation is

$$\begin{aligned}\hat{\sigma}_{tr}^{(i)}(t_{\mathrm{RF}}) &= s_z^{eq}(-f^{1-2}(t_{\mathrm{RF}})\hat{S}_z^{1-2} + \hat{S}_z^{3-4}) \\ &= s_z^{eq}(\frac{1}{2}(-f^{1-2}(t_{\mathrm{RF}}) + 1)(\hat{S}_z^{1-2} + \hat{S}_z^{3-4}) \\ &\quad + \left(\frac{1}{2}\left(-f^{1-2}\left(t_{\mathrm{RF}}\right) - 1\right)\left(\hat{S}_z^{1-3} - \hat{S}_z^{2-4}\right)\right) \qquad (8.220)\end{aligned}$$

with

$$f^{1-2}(t_{\mathrm{RF}}) = \cos^2\varphi_y^{1-2}\cos\omega_{1,\mathrm{RF}}t_{\mathrm{RF}} + \sin^2\varphi_y^{1-2} \qquad (8.221)$$

derived from a rotation of $\hat{S}_z^{1-2}$ by $\hat{H}_{\mathrm{RF}}^{\mathrm{RoF_{MW}+RoF_{RF}}}$. The spin density operator in equation (8.220) must now be considered as the initial operator $\hat{\sigma}_{tr}^{(ii)}(0^-)$ of the selective echo pulse sequence. The resulting echo signal originates only from the coefficient of the $\hat{S}_z^{1-3}$ term in $\hat{\sigma}_{tr}^{(ii)}(0^-)$ and must be compared with an echo signal, without the preparation sequence, originating from the coefficient s_z^{eq} of $\hat{S}_z^{1-3}$ in $\hat{\sigma}^{eq}(0^-)$. As a result, we get that the ENDOR echo factor has the form

$$E^{x}_{\mathrm{ENDOR}} = \frac{\frac{1}{2}s_z^{\mathrm{eq}}(-f^{1-2}(t_{\mathrm{RF}}) - 1)}{s_z^{\mathrm{eq}}} = -\frac{1}{2}(1 + \cos^2\varphi_y^{1-2}\cos\omega_{1,\mathrm{RF}}t_{\mathrm{RF}} + \sin^2\varphi_y^{1-2}) \tag{8.222}$$

This echo factor oscillates with the frequency $\omega_{1,\mathrm{RF}}$ between -1 when $t_{\mathrm{RF}} = 0$ and $-\sin^2\varphi_y^{1-2}$ when $\omega_{1,\mathrm{RF}}t_{\mathrm{RF}} = \pi$, showing the effect of the pseudo-secular terms in the original Hamiltonian via a nonzero value of φ_y^{1-2}. In the case of an isotropic hyperfine interaction, $\varphi_y^{1-2} = 0$ we get

$$E^{x}_{\mathrm{ENDOR}} = -\frac{1}{2}(1 + \cos\omega_{1,\mathrm{RF}}t_{\mathrm{RF}}) \tag{8.223}$$

Mims ENDOR

For completeness, we also briefly describe the Mims ENDOR sequence. This pulse sequence, illustrated in Figure 8.10(c),

$$\begin{array}{ccccccccc} & 0^{+} & & \tau^{-} & & t_{RF} & t_{RF}+\tau^{+} & t_{RF}+2\tau & \\ & \downarrow & & \downarrow & & \downarrow & \downarrow & \downarrow & \\ \{ & \hat{P}^{e,y}_{\pi/2} & -\tau- & \hat{P}^{e,y}_{\pi/2} & - \hat{P}^{n,x}_{\omega_{1,RF}t_{RF}} & - \hat{P}^{e,y}_{\pi/2} & -\tau- & & \} \end{array}$$

is composed of a nonselective stimulated electron echo sequence, where in the time between the second and third $\pi/2$ pulse, an RF irradiation excites selectively nuclear transitions. To evaluate the echo factor of this sequence, we can go back to the product operator representation and use the derivation in Section 8.4.5.1 concerning the ESEEM sequence. As for this sequence, we can use here again the $\hat{\sigma}^{\tau}$ and $\hat{\sigma}^{-\tau}$ spin density operators presented in equations (8.208), (8.210). Thus, what is left is to calculate the transformation $\hat{P}_{\mathrm{RF}}\hat{\sigma}^{\tau}\hat{P}^{-1}_{\mathrm{RF}}$ with $\hat{P}_{\mathrm{RF}} \equiv \{\hat{P}^{e,y}_{\pi/2} - \hat{P}^{(1-2),x}_{\omega_{1,\mathrm{RF}}t_{\mathrm{RF}}} - \hat{P}^{e,y}_{\pi/2}\}$.

After the first pulse of $\hat{P}_{\mathrm{RF}}$, we can assume that the coefficients of those product operators of $\hat{P}^{e,y}_{\pi/2}\hat{\sigma}^{\tau}\hat{P}^{e,y^{-1}}_{\pi/2}$ that do not commute with the electronic Hamiltonian oscillate during t_{RF} and decay to zero. Implementing this and performing the pulse transformations is not straightforward. To simplify the expressions for the echo factor E^{x}_{ENDOR}, we restrict our calculations to spin systems with isotropic hyperfine interactions. In this case, the angles appearing in the expression for $\hat{\sigma}^{\tau}$ are $\Delta\varphi = \Sigma\varphi = 0$, and using equation (8.201), we get

$$\hat{\sigma}^{\tau} = \cos\Omega_e\tau\,(\hat{S}^{(en)}_{x0}\cos\Delta\omega_0\tau + \hat{S}^{(en)}_{y0}\sin\Delta\omega_0\tau) + \sin\Omega_e\tau\,(\hat{S}^{(en)}_{yz}\cos\Delta\omega_0\tau - \hat{S}^{(en)}_{xz}\sin\Delta\omega_0\tau) \tag{8.224}$$

The $P^{e,y}_{\pi/2}\sigma^{\tau}P^{e,y^{-1}}_{\pi/2}$ transformation turns $\hat{S}^{(en)}_{x0}$ and $\hat{S}^{(en)}_{xz}$ into $-\hat{S}^{(en)}_{z0}$ and $-\hat{S}^{(en)}_{zz}$, respectively, and leaves $\hat{S}^{(en)}_{y0}$ and $\hat{S}^{(en)}_{yz}$ unchanged. During the RF irradiation time t_{RF}, the coefficients of $\hat{S}^{(en)}_{y0}$ and $\hat{S}^{(en)}_{yz}$ decay due to relaxation and the selective irradiation on the $(|1\rangle - |2\rangle)$ transition results in

$$\hat{P}^{(1-2),x}_{\omega_{\mathrm{RF}}t_{RF}}\hat{P}^{e,y}_{\pi/2}\sigma^{\tau}\hat{P}^{e,y-1}_{\pi/2}\hat{P}^{(1-2),x^{-1}}_{\omega_{\mathrm{RF}}t_{\mathrm{RF}}} \approx -\cos\Omega_e\tau\,\hat{S}^{(en)}_{z0}\cos\Delta\omega_0\tau + \sin\Omega_e\tau\,\hat{S}^{(en)}_{zz}\sin\Delta\omega_0\tau\left\{\frac{1}{2}\left(1 + \cos\omega_{\mathrm{RF}}t_{\mathrm{RF}}\right)\right\} \tag{8.225}$$

To derive this expression, a transformation to the RF rotating frame, a propagation with the selective RF Hamiltonian $\frac{1}{2}\omega_{1,\mathrm{RF}}(\hat{S}^{(en)}_{0x} - \hat{S}^{(en)}_{zx})$, and a transformation back to the original MW rotating frame must be performed. In addition, terms not contributing to the echo factor are ignored to simplify the expression (8.225).

After applying the last $\pi/2$ pulse of $\hat{P}_{\mathrm{RF}}$, the operators in equation (8.225) transform back to $\hat{S}^{(en)}_{x0}$ and $\hat{S}^{(en)}_{xx}$

$$\hat{P}_{\mathrm{RF}}\hat{P}^{e,y}_{\pi/2}\sigma^{\tau}\hat{P}_{\mathrm{RF}} \approx -\cos\Omega_e\tau\,\hat{S}^{(en)}_{x0}\cos\Delta\omega_0\tau + \sin\Omega_e\tau\,\hat{S}^{(en)}_{xz}\sin\Delta\omega_0\tau\left\{\frac{1}{2}\left(1 + \cos\omega_{\mathrm{RF}}t_{\mathrm{RF}}\right)\right\} \tag{8.226}$$

and with the expression for $\hat{\sigma}^{-\tau}$ derived from $\hat{\sigma}^{\tau}$ in equation (8.224),

$$\hat{\sigma}^{-\tau} = \cos\Omega_e\tau\,(\hat{S}^{(en)}_{x0}\cos\Delta\omega_0\tau - \hat{S}^{(en)}_{y0}\sin\Delta\omega_0\tau) + \sin\Omega_e\tau\,(-\hat{S}^{(en)}_{yz}\cos\Delta\omega_0\tau - \hat{S}^{(en)}_{xz}\sin\Delta\omega_0\tau) \tag{8.227}$$

the Mims-ENDOR echo factor for systems with isotropic hyperfine becomes

$$E^{x}_{\mathrm{ENDOR}} = -\left\{\cos^2\Omega_e\tau\cos^2\Delta\omega_0\tau + \frac{1}{2}\sin^2\Omega_e\tau\sin^2\Delta\omega_0\tau\left(1 + \cos\omega_{\mathrm{RF}}t_{\mathrm{RF}}\right)\right\} \tag{8.228}$$

Explicit expressions of this factor, after integration over a broad range of $\Delta\omega_0$ values, can be found in the literature and can account for the 'blind spots' of Mims ENDOR when the sum of coefficients in equation (8.226) become zero. These effects and more elaborate analysis of the Mims ENDOR sequence are outside the scope of this chapter.

We should stress here that we did not take any off-resonance integration or relaxation effects into account or discuss the differences between the Davies and Mims ENDOR sequences. The purpose of our discussions was solely to show the basic operator manipulations necessary to calculate ENDOR echo intensities.

ACKNOWLEDGMENTS

The authors would like to thank Daniella Goldfarb for her helpful comments and insights regarding both the form and the content of this chapter. This work was made possible in part by the historic generosity of the Harold Perlman Family.

FURTHER READING

All the pulse sequences described in this chapter are explained and described in detail in the monumental book of G. Jeschke and A. Schweiger.[7] A comprehensive description of product operator formalism can be found in Sorenson *et al.*[13]

REFERENCES

1. C. Cohen Tannoudji, B. Diu, and F. Laloe, Quantum Mechanics, John Wiley & Sons, Inc.: New York, 1977.
2. A. Messiah, Quantum Mechanics, John Wiley & Sons, Inc.: North Holland, 1966.
3. J. J. Sakurai, Modern Quantum Mechanics, Addison Wesley: Reading, 1994.
4. A. Abragam, Principles of Nuclear Magnetism, Clarendon: Oxford, 1961.
5. A. Abragam and B. Bleaney, Electron Paramagnetic Resonance of Transition Ions, Clarendon: Oxford, 1970.
6. C. P. Slichter, Principles of magnetic Resonance, 3rd edn, Springer-Verlag: Berlin, 1990.
7. A. Schweiger and G. Jeschke, Principles of Pulse Electron Paramagnetic Resonance, Oxford University Press: New York, 2001.
8. M. H. Levitt, Spin Dynamics, Basics of Nuclear Magnetic Resonance, John Wiley & Sons, Ltd: Chichester, 2001.
9. N. Bloembergen, E. M. Purcell, and R. V. Pound, *Phys. Rev.*, 1948, **73**, 679.
10. A. G. Redfield, in Advances in Magnetic Resonance, ed J. S. Waugh, Academic Press, 1965, Vol. **1**, 1.
11. R. P. Feynman, F. L. Vernon, and R. W. Hellwarth, *J. Appl. Phys.*, 1957, **28**, 49.
12. D. Suter, A. Pines, and M. Mehring, *Phys. Rev. Lett.*, 1986, **57**, 242.
13. O. W. Sørensen, G. w. Eich, M. H. Levitt, G. Bodenhausen, and R. R. Ernst, *Prog. Nucl. Magn. Reson. Spectrosc.*, 1983, **16**, 163.
14. F. Bloch and A. Siegert, *Phys. Rev.*, 1940, **57**, 522.
15. A. Wokaun and R. R. Ernst, *J. Chem. Phys.*, 1977, **67**, 1752.
16. S. Vega, *J. Chem. Phys.*, 1978, **68**, 5518.
17. M. H. Levitt and L. D. Bari, *Bull. Magn. Reson.*, 1994, **16**, 94.
18. A. D. Milov, K. M. Salikhov, and M. D. Shirov, *Sov. Phys. Solid State*, 1981, **24**, 565.
19. L. G. Rowan, E. L. Hahn, and W. B. Mims, *Phys. Rev. A*, 1965, **137**, 61.
20. E. R. Davies, *Phys. Lett. A*, 1974, **47**, 1.
21. W. B. Mims, *Proc. R. Soc. London*, 1965, **283**, 452.

Chapter 9
Relaxation Mechanisms

Sandra S. Eaton and Gareth R. Eaton

University of Denver, Denver, CO, USA

9.1 INTRODUCTION

Relaxation refers to the return to equilibrium by an unpaired electron spin system that is not in thermal equilibrium. Readers who come to the field of EPR with a background in optical spectroscopy may be surprised by the need to consider relaxation in designing and interpreting EPR experiments. The situation for optical spectroscopy is as follows. (i) The energy separation between the ground and first excited states is so large that the Boltzmann population of the excited state at ambient temperature is very low. A significant number of molecules can be excited from the ground state to the first excited state, while still maintaining a large population difference between the two states. (ii) The time constants for the processes by which the molecules return from the electronic excited state to the ground state are on the order of nanoseconds to picoseconds, so equilibrium is rapidly restored. The situation is very different for EPR. (i) The most commonly utilized EPR frequency of about 9 GHz (denoted as X-band) corresponds to a resonance magnetic field of 0.35 T for $g = 2$. At 298 K, the energy separation ΔE between the ground and excited states of an electron spin-1/2 ($m_S = \pm 1/2$ spin states) is only 6.3×10^{-24} J and the Boltzmann ratio, $\Delta E/k_B T$, is about 1.5×10^{-3}. As a result of this very small energy separation, the Boltzmann populations of the two spin states differ by only about 0.1%. The excitation of even a small fraction of the spins from the ground state to the excited state approximately equalizes the populations with the result that absorption of microwaves equals emission and there is no net EPR signal. (ii) For organic radicals at temperatures near 298 K, spin–lattice relaxation time constants are commonly of the order of a microsecond when rapidly tumbling and may be tens of microseconds when immobilized.[1] These relaxation times are long enough that use of high microwave B_1 can put energy into the system faster than it is transferred to the lattice by relaxation, which results in 'saturation' of the spin system (see Chapter 1). The saturation factor, s, can be expressed

EPR Spectroscopy: Fundamentals and Methods.
Edited by Daniella Goldfarb and Stefan Stoll

ISBN: 978-1-119-16299-5
Also published in eMagRes (online edition)
DOI: 10.1002/9780470034590.emrstm1507

as in equation (9.1).

$$s = \frac{1}{1 + \gamma^2 B_1^2 T_1 T_2} \tag{9.1}$$

where γ is the electron magnetogyric ratio, B_1 the microwave magnetic field at the sample, T_1 the electron spin–lattice relaxation time, and T_2 the electron spin–spin relaxation time, as discussed in the following section. When $s < 1$, the amplitude of the CW EPR signal decreases and broadening of the signal may be observed.[2] As s depends on T_1 and T_2 for the sample, the B_1 that can be used without causing saturation is sample specific.

The goal of this chapter is to provide the reader with an overview of the most common relaxation mechanisms and an appreciation of the changes in relaxation time constants that result from changes in temperature, motion, and molecular structure. After an overview of processes that contribute to the temperature and motion dependence of relaxation, the remainder of the chapter is devoted to illustrative examples. The focus is on $S = 1/2$ molecules in magnetically dilute samples.

Electron spin relaxation is described in terms of two time constants, T_1 and T_2, that are also discussed in the chapter on spin dynamics (see Chapter 8). Both time constants contribute to the saturation factor, s, equation (9.1). T_1 is the spin–lattice relaxation time constant, which is the time constant for return of magnetization along the z-axis (magnetic field B_0-axis) to its equilibrium value. T_2 is the spin–spin relaxation time constant, which is also referred to as the coherence time or decoherence time in the physics literature. It is the time constant for return to equilibrium of magnetization in the x,y-plane. For a sample in which there is no unresolved nuclear hyperfine splitting, the lineshape of the unsaturated CW signal is Lorentzian and the peak-to-peak linewidth of the first-derivative signal, ΔB_{pp} is related to T_2 as shown in equation (9.2).

$$\Delta B_{pp} = \frac{2}{\sqrt{3}\gamma T_2} \tag{9.2}$$

When there is unresolved hyperfine splitting, the lineshape is described as inhomogeneously broadened and the lineshape approaches Gaussian when the inhomogeneous contribution dominates. For a sample with no unresolved hyperfine splitting, the time constant for a free induction decay (FID) is T_2. For an inhomogeneously broadened line, the time constant for the FID is T_2^*. T_2^* is shorter than T_2. In some experiments, it is also important to consider spectral diffusion, which encompasses all processes that stochastically change the resonance frequency of spins (see Chapter 11).

There are several methods that are commonly used to measure T_1 (Figure 9.1).[1] In a saturation recovery experiment, a long saturating pulse is followed by low-power nonsaturating CW observation (Figure 9.1a). The length of the saturating pulse is chosen to be long enough to equalize the populations of all EPR transitions that are accessible from the pumped transition by spectral diffusion processes. This is achieved by increasing the length of the pulse until there is no further increase in the recovery time constant that is calculated by fitting the data to an exponential. The advantage of the saturation recovery experiment is that the recovery time constant does not contain contributions from spectral diffusion processes. However, the disadvantage is that the sensitivity is low because of the need to use very low CW observe power. In an inversion recovery experiment (Figure 9.1b), a π pulse takes the net magnetization from the $+z$-axis to the $-z$-axis. After a variable delay time T, a $\pi/2-\tau-\pi-\tau-$echo spin-echo sequence (see Chapter 11) is used to monitor the net magnetization along the z-axis. The echo amplitude recovers toward its thermal equilibrium value as a function of the time delay T. Analysis of the recovery curve gives the recovery time constant, which is the approximate value of T_1. A major advantage of the inversion recovery experiment is that all spins that are encompassed within the bandwidths of the pulses are flipped by the pulses and contribute to the detected signal, so sensitivity is higher than for saturation recovery. However, the disadvantage is that the recovery time constant may have contributions from spectral diffusion. A variation of the inversion recovery experiment is to use a long low-power saturating pulse (Figure 9.1c) or a picket fence of inverting pulses to saturate the spectral diffusion processes. When the spacing between the pulses in the picket fence is selected to match the time constant for spectral diffusion, the recovery time constant should be T_1, with no contribution from spectral diffusion. For the methods illustrated in Figure 9.1, a single microwave frequency is used. A two-frequency electron–electron double-resonance (ELDOR) experiment can be used to characterize the processes that contribute to spectral diffusion. In an ELDOR experiment, the pump pulse is at one microwave frequency and the observe pulses or CW observation is at a second frequency.[3,4] Comparison of time constants obtained by several methods is important for selection

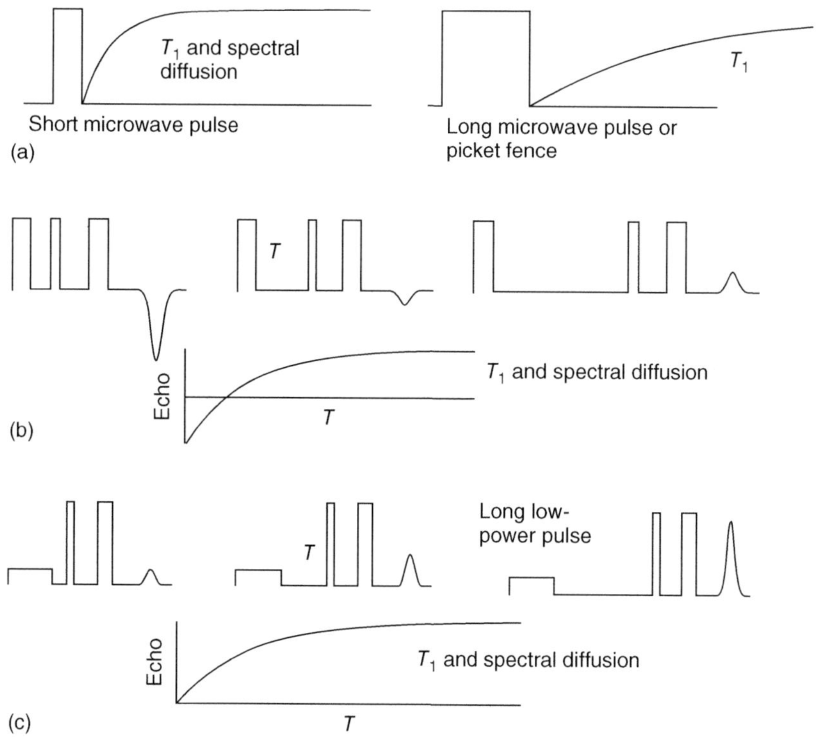

Figure 9.1. Methods to measure T_1. (a) *Saturation recovery*. The length of the saturating pulse is increased until there is no further increase in the recovery time constant. A picket fence of pulses can be used in place of the single saturating pulse. (b) *Inversion recovery*. A π pulse is used to invert the spins. A $\pi/2-\tau-\pi-\tau-$echo spin-echo sequence is used to monitor the magnetization along the z-axis after delay time T. (c) *Echo-detected saturation recovery*. A long low-power pulse is used to saturate the spin system. A $\pi/2-\tau-\pi-\tau-$echo spin-echo sequence is used to monitor the magnetization along the magnetic z-axis after delay time T

of the method that is appropriate for a particular species and motional regime.

There are several ways of measuring T_2. If the EPR lineshape is precisely Lorentzian with no inhomogeneous broadening, the value of T_2 can be calculated from the peak-to-peak linewidth of the CW EPR spectrum by rearranging equation (9.2) to obtain the relationship T_2 (s) = 6.56×10^{-8} mT s/(ΔB_{pp}/mT). This is a special case that is relatively rare. Examples in fluid solution are peroxylamine disulfonate (Fremy's salt) and 1,4-benzosemiquinone in aqueous solution. Inhomogeneous broadening includes the contributions to linewidths from unresolved nuclear hyperfine splittings in solids or solutions and from anisotropy, g-strain, and dipole–dipole interactions in solids (see Section 9.2.2). For the more general case of inhomogeneously broadened spectra, pulse methods are required to measure T_2. In a two-pulse $\pi/2-\tau-\pi-\tau-$echo spin-echo sequence, the time between pulses is varied. The echo decay time constant that is calculated from a plot of echo intensity as a function of the time between pulses is designated as the phase memory decay time, T_m. This time constant includes contributions from many processes, as well as T_2.[1] T_m includes the effects of all processes that move spins off resonance on the timescale of the spin-echo experiment. One contribution is instantaneous diffusion, in which the second pulse flips a neighboring spin to which the observed spin is dipole coupled. Flipping the neighboring spin changes the resonance of the observed spin and moves it off resonance. The contribution from instantaneous diffusion can be decreased by decreasing the excitation bandwidth of the $\pi/2$ pulse.[5] T_m also includes contributions from dynamic processes such as rotation of methyl groups that average inequivalent nuclei to which the unpaired electron is coupled and averaging of g and A anisotropy by librational motion when these occur on the timescale of the spin-echo experiment. Each of these motions modulates the environment of the unpaired electron spin and/or averages interactions with nuclear spins that impact the resonance energy. Examples are given in the following paragraphs. In fluid solution, it is customary to equate T_2 and T_m, and for rapidly tumbling molecules, $T_1 \approx T_2$. The contributions to T_2 in fluid solution that arise from collisions and from incomplete motional averaging of g and A anisotropy are equal to $1/T_2 - 1/T_1$.

9.2 RELAXATION PROCESSES

9.2.1 Spin–Lattice Relaxation

The underlying principle is that electron spin–lattice relaxation involves an exchange of energy between the spin system and the 'lattice' that establishes thermal equilibrium. These processes require net conservation of energy and can occur by a single-phonon process in which the energy released matches the energy of a lattice mode. However, in many cases, the energy released to the lattice is equal to the difference between real or virtual energy states of the lattice, which is denoted as a two-phonon process. Early studies of relaxation focused primarily on paramagnetic centers in crystalline solids and the 'lattice' literally meant the crystal lattice.[6,7] The concept has evolved to define the 'lattice' as whatever environment surrounds the unpaired electron. As the size of the molecule increases, the molecule itself makes an increasingly large contribution to the 'lattice'. The classic papers upon which subsequent work on spin–lattice relaxation relies are described by Abragam and Bleaney,[8] Kevan and Schwartz,[9] Poole and Farach,[10] Standley and Vaughn,[6] and Manenkov and Orbach.[11]

The descriptions for several relaxation processes (the direct, the Raman, and the Orbach processes) rely on the Debye model of phonon vibrational modes in solids, in which there is a continuum of modes extending up to a cutoff at the Debye temperature, θ_D. The cutoff frequency is $\nu_{max} = k_B\Theta_D/h$, where k_B is Boltzmann's constant and h Planck's constant. For glassy matrices and large molecular solutes, the distribution of energies deviates from the theoretically predicted frequency squared dependence of the Debye model,[6,12] but the concept of a Debye temperature remains useful.

9.2.1.1 *Direct Process*

In this process, there is an exact match of the unpaired electron spin flip energy (the Zeeman energy) with a lattice phonon energy and direct transfer of energy occurs. For Kramers systems, including $S = 1/2$, the matrix elements for the transition probability that govern this spin–lattice relaxation become nonzero when there is mixing with excited states. The relaxation rate is then given by equation (9.3)

$$\frac{1}{T_1} = A_{dir}B^4T \tag{9.3}$$

where A_{dir} is an experimentally determined adjustable parameter, B the magnetic field, and T the temperature. The B^4 field dependence and the linear dependence on temperature are characteristic of the direct process. As shown in the case studies in the following sections, the direct process typically dominates relaxation at relatively low temperature. However, there are many cases in the literature in which the relaxation rates for samples with relatively high-spin concentrations vary linearly with temperature at much higher temperatures than would be expected for the direct process. For these samples, it is likely that cross relaxation is occurring with mutual electron–electron spin flips of neighboring spins.[13,14]

9.2.1.2 *Raman Process*

The two-phonon Raman process involves a virtual excited state with an energy that is less than the Debye temperature. The energy that is transferred from the spin system to the lattice is the difference between the energies for absorption to and emission from this virtual excited state.[15] For $S = 1/2$, the contribution to T_1 from the Raman process is given by equation (9.4).[16,17]

$$\frac{1}{T_1} = A_{Ram}\left(\frac{T}{\theta_D}\right)^9 J_8\left(\frac{\theta_D}{T}\right) \tag{9.4}$$

where A_{Ram} is an experimentally determined adjustable parameter and J_8 the transport integral equation (9.5) that was developed to model heat capacity.

$$J_8\left(\frac{\theta_D}{T}\right) = \int_0^{\theta_D/T} x^8 \frac{e^x}{(e^x - 1)^2}dx \tag{9.5}$$

Well below the Debye temperature, the contribution from the Raman process is proportional to T^9, but above the Debye temperature the limiting temperature dependence is T^2. This temperature dependence of $1/T_1$ is characteristic of the Raman process and has been observed for a wide range of species.[1]

9.2.1.3 Local-mode Process

The term 'local' distinguishes this mechanism from those that involve long-range (phonon) modes. It involves a discrete vibrational frequency, usually for a molecular species. This contribution to relaxation is described by equation (9.6).[17,18]

$$\frac{1}{T_1} = A_{\mathrm{loc}} \frac{\mathrm{e}^{\Delta_{\mathrm{loc}}/k_{\mathrm{B}}T}}{(\mathrm{e}^{\Delta_{\mathrm{loc}}/k_{\mathrm{B}}T} - 1)^2} \tag{9.6}$$

where A_{loc} is an experimentally determined adjustable parameter and Δ_{loc} the energy of the local mode.

In some cases, the energy Δ_{loc} that is found by fitting the temperature dependence of $1/T_1$ can be matched to a specific vibrational frequency of the molecule studied.

9.2.1.4 Orbach Process

This is a two-phonon process, but unlike the Raman process, it involves an actual excited state of a molecule, with an energy that is less than the Debye temperature of the system. This process is described by equation (9.7).[18]

$$\frac{1}{T_1} = A_{\mathrm{orb}} \frac{\Delta_{\mathrm{orb}}^3}{\mathrm{e}^{\Delta_{\mathrm{orb}}/k_{\mathrm{B}}T} - 1} \approx A_{\mathrm{orb}} \Delta_{\mathrm{orb}} \mathrm{e}^{-\Delta_{\mathrm{orb}}/k_{\mathrm{B}}T} \tag{9.7}$$

where A_{orb} is an experimentally determined adjustable parameter and Δ_{orb} the energy of the low-lying excited state, and the approximation is valid for $\Delta_{\mathrm{orb}} \gg k_{\mathrm{B}}T$. For systems in which the Orbach process dominates relaxation, modeling the temperature dependence of $1/T_1$ provides a way to measure the energy for the low-lying excited state.

9.2.1.5 Thermally Activated Process

Thermally activated processes such as rotation of methyl groups, rotation of amino groups, or hopping of a hydrogen-bonded proton can absorb energy from the spin system when the rate of the process is comparable to the resonance frequency of the spin system. The temperature dependence of this contribution is described by equation (9.8).[19]

$$\frac{1}{T_1} = A_{\mathrm{therm}} \left[\frac{2\tau_{\mathrm{c}}}{1 + \omega^2 {\tau_{\mathrm{c}}}^2} \right] \tag{9.8}$$

where A_{therm} is an experimentally determined adjustable parameter, τ_{c} the correlation time for the thermally activated process, $\tau_{\mathrm{c}} = \tau_{\mathrm{c}}^{\mathrm{o}} e^{E_{\mathrm{a}}/k_{\mathrm{B}}T}$, E_{a} the activation energy for the thermally activated process, $\tau_{\mathrm{c}}^{\mathrm{o}}$ the preexponential factor, and ω the electron spin Larmor frequency in angular frequency units. The function in equation (9.8) reflects the fact that when the correlation time is stated to be τ_{c}, there actually is a distribution of correlation times, with τ_{c} as the average value. Unlike the Raman, local-mode, or Orbach processes, this contribution to relaxation depends on the Larmor frequency. It has maximum impact when $\omega\tau_{\mathrm{c}} \approx 1$, so when this process dominates, the relaxation rate exhibits a maximum as a function of temperature that is dependent on the Larmor frequency.

9.2.1.6 Tumbling-dependent Processes

When rotational tumbling correlation times, τ_{R}, are of the order of nanoseconds or shorter, tumbling can be a very effective relaxation mechanism. Tumbling contributes to relaxation both by spin rotation (equation (9.9)) and by modulation of anisotropic interactions (equations (9.10), (9.12), and (9.13)). The importance of spin rotation, equation (9.9), was recognized in early studies of radicals in fluid solution.[20,21] If τ_{R} is known, there are no adjustable parameters in equation (9.9). This contribution to T_1 is independent of resonance frequency.

$$\frac{1}{T_1^{\mathrm{SR}}} = \frac{\sum_{i=1}^{3} (g_i - g_{\mathrm{e}})^2}{9\tau_{\mathrm{R}}} \tag{9.9}$$

where g_i is a principle g value.

Modulation by molecular tumbling of anisotropic interaction such as g anisotropy (equation (9.10)),[22,23] hyperfine (A_i) anisotropy (equation (9.12)),[22–24] or dipolar coupling to solvent nuclei (equations (9.13))[25] is an important relaxation process in fluid solution. These processes are all frequency dependent because of the dependence of the spectral density function, $J(\omega)$ (equation (9.11)) on ω. The contribution due to g anisotropy is

$$\frac{1}{T_1^g} = \frac{2}{5} \left(\frac{\mu_B \omega}{g\beta} \right)^2 \left\{ \frac{(\Delta g)^2}{3} + (\delta g)^2 \right\} J(\omega) \tag{9.10}$$

where $\Delta g = g_{zz} - 0.5(g_{xx} + g_{yy})$, $\delta g = 0.5(g_{xx} - g_{yy})$, μ_{B} is the Bohr magneton, and $J(\omega)$ the Bloembergen,

Purcell, Pound (BPP) spectral density function (equation (9.11)), which assumes isotropic motion. The spectral density function represents the distribution of tumbling correlation times around the average value.

$$J(\omega) = \frac{\tau_R}{1 + (\omega\tau_R)^2} \quad (9.11)$$

where τ_R is the rotational tumbling correlation time.

The contribution due to modulation of hyperfine anisotropy is

$$\frac{1}{T_1^A} = \frac{2}{9}I(I+1)\sum_i (A_i - a_{iso})^2 J(\omega) \quad (9.12)$$

where A_i is a principal component of the nuclear hyperfine coupling in angular frequency units, a_{iso} the isotropic nuclear hyperfine, and I the nuclear spin. This expression was derived for nitroxide radicals, but can be used for any system with dominant hyperfine coupling to a single nuclear spin.

When other contributions to relaxation are small, it is possible to detect contributions to relaxation from modulation of dipolar interactions with nuclear spins in the solvent.[25]

$$\frac{1}{T_1^{solvent}} = C_{solvent}\frac{\tau_{solvent}}{1 + (\omega\tau_{solvent})^2} \quad (9.13)$$

where $\tau_{solvent}$ is the correlation time for motion of the solvent relative to the radical and $C_{solvent}$ a function of the dipolar interaction with solvent nuclei.

Although theoretical models for the temperature dependence of various relaxation processes fit well with experimental data, theories are not yet capable of predicting the coefficients that reflect the effectiveness of the processes for a particular system in rigid lattices. Thus, equations (9.3), (9.4), (9.6), (9.7), and (9.8) include adjustable coefficients A_{dir}, A_{Ram}, A_{loc}, A_{orb}, and A_{therm}, respectively, that are determined experimentally. Processes in fluid solutions described by equations (9.9), (9.10), and (9.12) do not have adjustable parameters.

9.2.2 Spin–Spin Relaxation

In a rigid lattice, if there are no nuclear spins interacting with the electron spin, T_2 (T_m) is dominated by dipolar interaction between the unpaired electrons. Bloembergen *et al.*[26] obtained an approximate expression that T_2 is proportional to $r_{avg}{}^3$, where r_{avg} is the average distance between unpaired spins. If the spins are uniformly distributed, $r_{avg}^{-1} = \sqrt[3]{C \times N_A}$, where C is the concentration in moles per cubic meter and N_A Avogadro's number. This dependence of T_2 on the concentration of unpaired electron spins and therefore on the average distance between spins has been observed for irradiated fused SiO_2[27,28] and for defects in diamonds.[29] For most dilute solutions of paramagnetic species immobilized in matrices that contain nuclear spins, spin-echo dephasing is dominated by nuclear spin diffusion (nuclear spin flip flops).[30,31] The nuclear spin flips modulate the electron-nuclear dipolar interaction, thereby providing a mechanism for spin-echo dephasing and loss of spin coherence in the *x,y*-plane. Even though the nuclear spin flips are relatively slow, the very large number of nuclei that are dipole coupled to the electron spin means that the net effect of many interactions makes the nuclear flip flops a major contributor to T_2. The concentration of nuclear spins is similar for a wide range of solvents and for proteins so the limiting value of T_m in a wide range of samples at low temperatures is about 4 μs.[1,32] The important role of nuclear spins has been demonstrated by replacing solvent protons by deuterons. The smaller magnetic moment of D than of H reduces the dipole–dipole interactions and dramatically increases T_m.[1,32] These nuclear spin diffusion processes exhibit little or no temperature dependence.

Spin coherence in the *x,y*-plane cannot persist longer than coherence along the z-axis, T_1 processes contributes to T_2. For most species at low temperature, T_1 is much longer than T_2, so its contribution to T_2 is negligible. T_1 typically is more strongly temperature dependent than T_2, so as temperature is increased T_1 eventually becomes the dominant contribution to T_2 and T_2 becomes equal to T_1. This happens at lower temperature for rapidly relaxing transition metals than for more slowly relaxing organic radicals. For most species, the CW lineshape is inhomogeneously broadened and the relaxation-determined contribution from T_2 is relatively small. However, if the spin–spin relaxation becomes fast enough, it may dominate the linewidths in rigid lattice CW spectra.[33] In this regime, T_2 can be calculated from the temperature-dependent contribution to the linewidths. A temperature-dependent linewidth in rigid lattice typically occurs because the strongly temperature-dependent T_1 is dominating T_2 so it is assumed that $T_1 = T_2$, which permits determination of T_1 and T_2 at temperatures where the time constants are too short to be measured directly by pulsed methods.[34]

Linewidths may also be temperature dependent because of motional processes. As motional rates increase and become comparable to the energy differences, in frequency units, between inequivalent environments for an unpaired electron, T_2 decreases and linewidths broaden. When the rate of the process is equal to the energy difference, the impact of the process on T_2 is maximum (T_2 is shortest) and the linewidth is the largest. Further increase in the rate of the process leads to more effective averaging of the inequivalent environments, T_2 increases, and the lines narrow. The motional processes may be tumbling of the molecule that averages g and A anisotropy or internal motions such as rotation of methyl groups or ring inversion that interchanges axial and equatorial substituents.

In fluid solution, the modulation of anisotropic interactions frequently dominates T_2. Kivelson[35] and Wilson and Kivelson[20] summarized contributions to EPR linewidths near the rapid tumbling limit as in equation (9.14).

$$\frac{1}{T_2}(m_I) = A + Bm_I + Cm_I^2 + \cdots \qquad (9.14)$$

where m_I is the nuclear spin quantum number and A, B, and C the functions of the tumbling correlation time, g anisotropy, and hyperfine anisotropy.[36] Simulations of the spectra provide values of A, B, and C, from which the tumbling correlation time can be calculated.

Freed and coworkers have developed models and computational methods to simulate the lineshapes for nitroxide radicals over a wide range of motional regimes. Extensive studies of nitroxide lineshapes in fluid solution have provided detailed understanding of molecular motion (see Chapter 14).[37–39]

9.3 RELAXATION OF TRIARYLMETHYL (TRITYL) RADICALS

Because the peak-to-peak linewidths of their CW spectra are about 10 μT in fluid solution and electron spin relaxation times are relatively long, trityl radicals are powerful probes for in vivo oximetry.[40] Understanding the relaxation mechanisms is key to designing and interpreting the oximetry experiments. The isotropic g values for trityls are between 2.002 and 2.003,[41] which are close to the free electron g value and indicate that spin–orbit coupling is weak. Trityls for imaging, including trityl-CD_3 (**I**), trityl-CH_3 (**II**), OX63 (**III**), and OX31 (**IV**), are designed to minimize electron-nuclear spin–spin interaction and proton/deuteron hyperfine couplings are small. Replacement of the protons by deuterons decreases the linewidths because the smaller magnetic moment of D compared to H reduces the hyperfine splitting proportional to $\gamma_D/\gamma_H \approx 0.15$. The very small g anisotropy and very small hyperfine couplings limit the effectiveness of tumbling-dependent relaxation processes that are major contributors to relaxation for many other organic radicals.

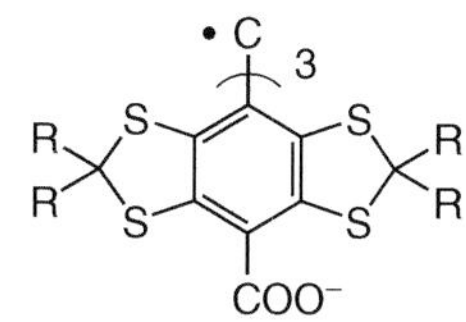

R = CD_3	trityl-CD_3	I
R = CH_3	trityl-CH_3	II
R = CH_2CH_2OH	OX63	III
R = $CH_2OCH_2CH_2OH$	OX31	IV

9.3.1 T_1 and T_m in Rigid Lattice

The temperature dependence of $1/T_1$ and $1/T_m$ at X-band for trityl-CD_3 (**I**), trityl-CH_3 (**II**), OX63 (**III**), and OX31 (**IV**) in 1 : 1 water : glycerol glasses between 80 and 300 K is shown in Figure 9.2.[41] Values of T_m for the four trityls at 100 K are between 3 and 5 μs and are the same at X-band and W-band. The dephasing is assigned to nuclear spin diffusion (see Section 9.2.2). The values of T_m for the trityls are longer than for nitroxides or semiquinones in the same solvent at this temperature. Because of the large size of the trityls, and the deliberate design of these trityls, the nearest protons are further away from the unpaired electron than for smaller radicals, which decreases the electron-nuclear dipolar coupling and thus decreases the effectiveness of nuclear spin diffusion in spin-echo dephasing. For OX63 (**III**) and OX31 (**IV**), T_m is essentially temperature independent from 80 to almost 200 K, which also is indicative of a dominant nuclear spin diffusion mechanism. For OX63 (**III**) in glycerol-d_8 : D_2O, T_m below about 130 K is substantially longer than in solvents with a natural isotope distribution, which supports the assignment of spin diffusion as the dominant contribution to T_m in this temperature range. For trityl-CH_3 (**II**), spin-echo dephasing is enhanced (T_m is decreased) between about 100 and 200 K relative to that for OX63 (**III**) or OX31 (**IV**) (Figure 9.2). The enhancement is assigned to rotation

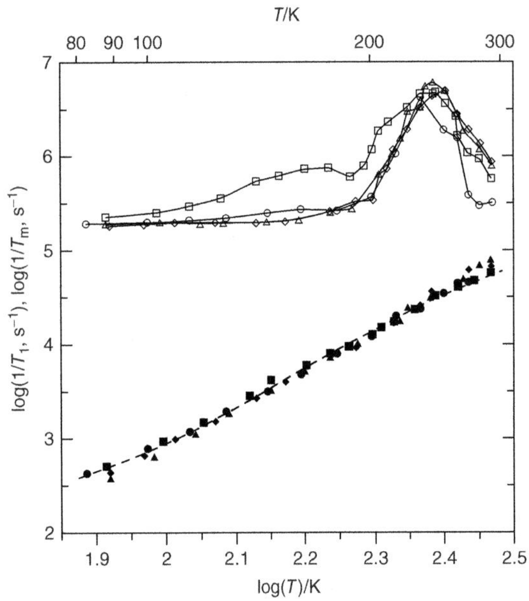

Figure 9.2. Temperature dependence of T_1 at X-band for 0.2 mM solutions in 1 : 1 water : glycerol for trityl-CD_3 (**I**) (•), trityl-CH_3 (**II**) (■), OX31 (**III**) (◆), and OX63 (**IV**) (▲). The dashed line is a fit to the experimental data including contributions from the Raman process and local-mode process. T_m is shown for trityl-CD_3 (O), trityl-CH_3 (□), OX31 (◇), and OX63 (Δ). The solid line connects the points for T_m.[41] (Reproduced from Appl. Magn. Reson., Multifrequency EPR of four triarylmethyl radicals, 28, 2005, 231, Fielding, A. J.; Carl, P. J.; Eaton, G. R.; Eaton, S. S. © With permission of Springer)

of the methyl substituents at rates that are comparable to the anisotropy in electron–proton couplings, which is an effective spin-echo dephasing mechanism.[41] The impact of methyl rotation on T_m for trityl-CD_3 (**I**) is smaller than for trityl-CH_3 (**II**) because the electron-nuclear couplings are smaller. Between about 200 and 300 K, spin-echo dephasing is enhanced for all four trityls (Figure 9.1). This is the temperature range in which the water : glycerol glass softens and becomes a liquid. The increased motion of the radicals averages *g* anisotropy. Even for the rather small *g* anisotropy of the trityls, the averaging of *g* anisotropy is an effective spin-echo dephasing mechanism.

The temperature dependence of $1/T_1$ at X-band (Figure 9.2) is indistinguishable for the four trityls. In the temperature interval between 200 and 300 K where increasing molecular motion enhances spin-echo dephasing, the temperature dependence of $1/T_1$ does not exhibit a change in slope, which indicates that $1/T_1$ relaxation processes that depend on τ_R are not effective relaxation mechanisms for the trityls at X-band. At 293 K, $1/T_1$ at X-band is independent of solvent viscosity, which supports the conclusion that tumbling-dependent processes are not major contributors to relaxation.[25] The dashed line in Figure 9.2 is the fit to the experimental $1/T_1$ data for the sum of contributions from the Raman process (equation (9.4)) and a local mode (equation (9.6)) with an energy of 590 cm^{-1}. The fit line was calculated including data at lower temperatures that are not shown in Figure 9.2.[42,43] The Raman process dominates below about 100 K, and the local mode makes significant contributions above about 100 K. The infrared spectrum of trityl-CD_3 (**I**) has bands in the range 518–586 cm^{-1}, which have substantial C-S stretch character.[43] It has been proposed that these vibrational modes are the local modes that contribute to the spin–lattice relaxation. The absence of a change in slope in the plot of log($1/T_1$) vs log(*T*) as the glass softens and melts and the good fit of the data as the sum of contributions from the Raman and local-mode processes at temperatures up to 300 K indicate that the local mode continues to contribute to the relaxation in fluid solution.

9.3.2 T_1 in Fluid Solution

The data in Figure 9.2 show that at X-band, T_1 at 300 K is the same for the four trityls.[25] The frequency dependence of $1/T_1$ for the trityls in solution at 293 K is shown in Figure 9.3.[25] The observation of frequency dependence is inconsistent with domination by spin rotation (equation (9.9)) or a local mode (equation (9.6)), which are frequency independent. In addition, the very small *g* anisotropy for the trityls decreases the potential effectiveness of the spin-rotation process (equation (9.9)). Examination of the dependence of relaxation on solvent viscosity as well as frequency demonstrated that the enhanced relaxation at lower frequencies is due to modulation of electron–proton dipolar interaction by molecular tumbling (equation (9.13)). The impact of this mechanism is maximized when $\omega\tau_R \approx 1$. For these relatively large molecules in solvents such as water or 1 : 1 water : glycerol, this condition is satisfied at ambient temperature when the Larmor frequency, ω,

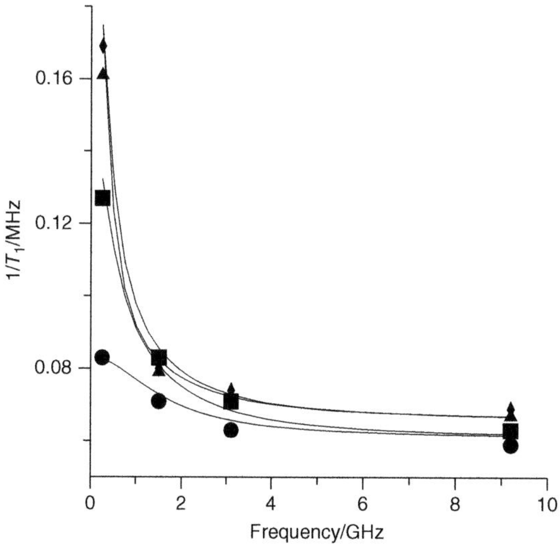

Figure 9.3. Frequency dependence of $1/T_1$ for trityl-CD_3 (**I**) (●), trityl-CH_3 (**II**) (■), OX63 (**III**) (▲), and OX31 (**IV**) (♦) in water at 293 K. The fit lines (____) include contributions from the local mode plus modulation of intermolecular and intramolecular electron–proton interactions by molecular tumbling and solvent motion.[25] (Reprinted from J. Magn. Reson., 172, Owenius, R.; Eaton, G. R.; Eaton, S. S., Frequency (250 MHz to 9.2 GHz) and viscosity dependence of electron spin relaxation of triarymethyl radicals at room temperature, 168. © 2005, with permission from Elsevier)

is ≈250 MHz.[25] This low frequency is important for in vivo application of EPR (see Chapter 13).

9.3.3 T_2 in Fluid Solution

Incomplete motional averaging of g anisotropy makes significant contributions to T_2 in water and water : glycerol solutions at frequencies between 250 MHz and 9 GHz.[25,43] These changes in T_2 are large enough to impact the CW EPR linewidths. To be useful for in vivo imaging, it is advantageous to have the narrowest possible linewidths.

9.4 RELAXATION OF SEMIQUINONES

Semiquinones are ubiquitous in biology, from photosynthesis to environmental toxicology,[44,45] and are involved in many electron-transfer reactions.[46] The isotropic g values are in the range 2.0023–2.0065,[47] which are significantly larger than that for trityls, indicating larger spin–orbit coupling. Semiquinones with a range of structures and solubilities in a variety of solvents permit study of the effects of solvent and molecular structure on relaxation.

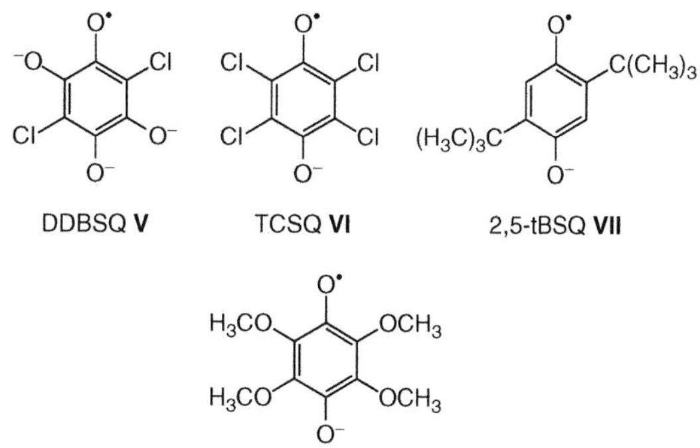

9.4.1 T_1 in Rigid Lattice

Molecular structure and solvent dependence of T_1 can be seen when data for 2,5-dichloro-3,6-dihydroxy-1,4-benzosemiquinone (DDBSQ, **V**) and tetrachloro-1,4-benzosemiquinone (TCSQ, **VI**) are compared with 2,5-tBSQ, **VII** (Figure 9.4).[42] For these three radicals, the temperature dependence of relaxation below the glass transition temperature could be modeled with contributions from the direct (equation (9.3)), Raman (equation (9.4)), and local-mode (equation (9.6)) processes. Above the glass transition temperatures, tumbling-dependent processes contribute to the relaxation, but were not included in the modeling of the relaxation that is shown in Figure 9.4. Relaxation rates for DDBSQ (**V**) in 1 : 1 water : glycerol are consistently about 50% slower than for 2,5-tBSQ (**VII**) in 1 : 1 ethanol : glycerol, which is attributed to greater rigidity of the water : glycerol glass.[42] The isotropic g value for DDBSQ (2.0044) is less than that for 2,5-tBSQ (2.0047), indicating smaller spin–orbit coupling for DDBSQ.[42] Smaller spin–orbit coupling contributes to slower relaxation for DDBSQ than for 2,5-tBSQ and impacts both the effectiveness of the Raman process (equation (9.4)) and the local-mode process (equation (9.6)).[42] The faster relaxation for TCSQ is attributed to the higher g values, indicating larger spin–orbit coupling.

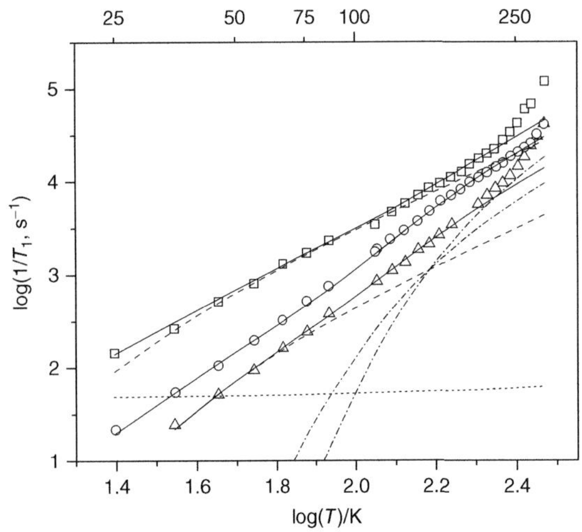

Figure 9.4. Temperature dependence of $1/T_1$ at X-band for 2,5tBSQ (**VII**) in 1 : 1 ethanol : glycerol (O), DDBSQ (**V**) in 1 : 1 water : glycerol (Δ), and TCSQ (**VI**) in 1 : 1 triacetin : hexamethyl-phosphoramide (□). The solid lines are the fits, which are the sum of contributions from the direct, Raman, and local-mode processes. The contributions to the fit line from the Raman process (----) and a local-mode process (-·-·-) for TCSQ and DDBSQ and direct process (· · ·) for TCSQ are also given.[42] (Reproduced from Appl. Magn. Reson., Impact of Chlorine Substitution on Spin-Lattice Relaxation of Triarylmethyl and 1,4-Benzosemiquinone Radicals in Glass-Forming Solvents Between 25 and 295 K, 37, 2010, 649, Kathirvelu, V.; Eaton, G. R.; Eaton, S. S. © With permission of Springer)

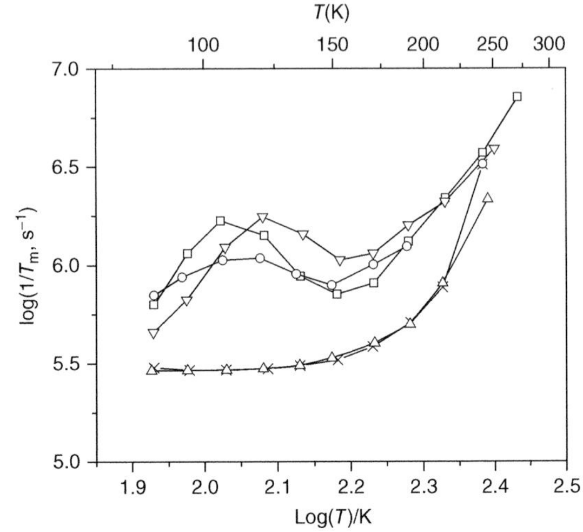

Figure 9.5. Temperature dependence of spin-echo dephasing rates in triethanolamine at X-band for 2,5-PSQ (**X**) (Δ), THSQ (**IX**) (×), 2,5-tASQ (**XII**) (O), 2,6-tBSQ (**XI**) (□), and 2,5-tBSQ (**VII**) (∇).[42] (Reprinted from Appl. Magn. Reson., 37, Kathirvelu, V.; Eaton, G. R.; Eaton, S. S., Electron spin relaxation rates for semiquinones between 25 and 295 K in glass-forming solvents, 649. © 2010, with permission from Elsevier)

9.4.2 T_m in Rigid Lattice

The temperature dependence of spin-echo dephasing rates between 80 and 250 K for five semiquinones is shown in Figure 9.5.[48] The limiting value of T_m of about 3 μs at about 80 K is similar to that observed at low temperature for nitroxide and trityl radicals and is attributed to spin dephasing of solvent nuclei[32] (see Section 9.2.2). Enhanced spin-echo dephasing rates between about 85 and 150 K for methyl-containing 2,5-tASQ (**XII**), 2,6-tBSQ (**XI**), and 2,5-tBSQ (**VII**) are due to rotation of the methyl substituents at rates comparable to the anisotropy of the electron–proton couplings, analogous to the contributions of methyl groups to spin-echo dephasing for trityl-CH_3 (**II**)[41] and for methyl-containing nitroxides and Cr(V) complexes[49] in this temperature range.

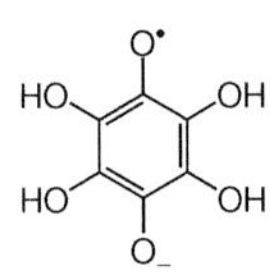

THSQ **IX**

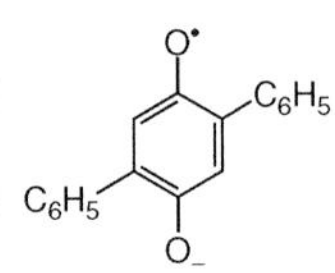

2,5-PSQ **X**

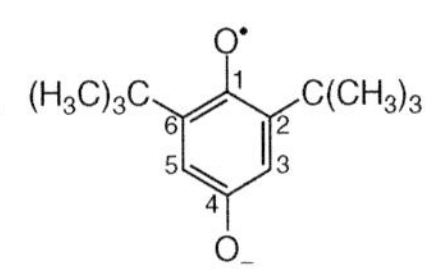

2,6-tBSQ **XI**

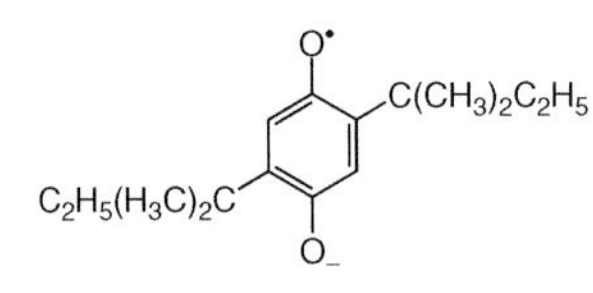

2,5-tASQ **XII**

9.4.3 T_1 in Fluid Solution

Extensive studies by Rengan and coworkers examined the temperature and solvent (viscosity) dependence of T_1 for semiquinones and demonstrated two contributions, one of which was spin rotation (equation (9.9)).[47] The larger g values for semiquinones makes spin rotation more effective for semiquinones than for trityls.[50] To distinguish between potential contributions to the spin–lattice relaxation rates at 293 K, three anionic semiquinones (2,5-tBSQ (**VII**), 2,6-tBSQ (**XI**), and 2,3,5,6-tetramethoxy-1,4-benzosemiquinone (TMBSQ) (**VIII**)) were studied at up to eight frequencies between 250 MHz and 34 GHz in ethanol or methanol solution at 20 °C.[50] Solvent deuteration was used to determine the contribution of protons/deuterons in the solvent to relaxation. The relaxation rates were modeled as the sum of two frequency-independent contributions {spin rotation (equation (9.9)) and a local mode (equation (9.6))} and two frequency-dependent contributions {modulation of dipolar interaction with solvent nuclei (equation (9.13)) and a much smaller contribution from modulation of g anisotropy (equation (9.10))}. The modulation of g anisotropy by molecular tumbling is significant for the semiquinones only at Q-band. The correlation time for modulation of the interaction with solvent nuclei is longer than the tumbling correlation time of the semiquinone. Modulation of hydrogen bonding to the semiquinone oxygens is the most probable source of the second frequency-dependent contribution to relaxation.[50]

9.5 RELAXATION OF NITROXIDES

Because of the extensive use of nitroxides as spin labels and spin probes for biological systems (see Chapter 14), the relaxation times have been studied more extensively than for other radicals. The large anisotropic nitrogen hyperfine coupling plays a major role in relaxation in fluid solution and the substantial g anisotropy impacts relaxation both in rigid lattice and in fluid solution.

9.5.1 T_1 in Rigid Lattice

The spin–lattice relaxation rates for nitroxides such as tempone (**XIII**) in rigid lattice are substantially faster than for trityl-CD_3 (**I**) (Figure 9.6), which is attributed to larger spin–orbit coupling.[51] The fit to the temperature dependence of $1/T_1$ for both species modeled as the sum of contributions from the direct (equation (9.3)), Raman (equation (9.4)), and local-mode (equation (9.6)) processes is shown in Figure 9.6.[51] The direct process impacts relaxation only at the lower temperatures. The deviation from the fit line calculated based only on the direct and Raman processes (dashed lines in Figure 9.6) demonstrates the need to include the local mode. The coefficients for the Raman and local-mode processes are correlated, which indicates that both are impacted by spin–orbit coupling and interaction with the surroundings.[52] Sucrose octaacetate is a convenient solvent for these studies because the glass transition temperature is about 345 K. This glass is softer than the hydrogen-bonded 1 : 1 water : glycerol glass, which results in substantially faster relaxation in sucrose octaacetate than in 1 : 1 water : glycerol (Figure 9.6).[53] Analogously,

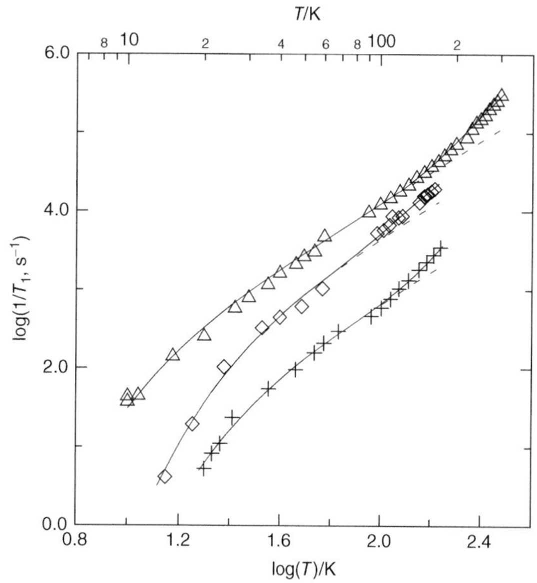

Figure 9.6. Temperature dependence of X-band spin–lattice relaxation rates for 0.2 mM trityl-CD_3 (**I**) in 1 : 1 water : glycerol (+), 1 mM tempone (**XIII**) in 1 : 1 water : glycerol (◊), and 1 mM tempone (**XIII**) in glassy sucrose octaacetate (Δ). The data for tempone were obtained in the center of the spectrum. The solid lines are the fits to the data including contributions from the direct, Raman, and local-mode processes. The dashed lines are the values calculated based only on the direct and Raman processes.[51] (Reprinted from J. Magn. Reson., 139, Zhou, Y.; Bowler, B. E.; Eaton, G. R.; Eaton, S. S., Electron Spin Lattice Relaxation Rates for S = 1/2 Molecular Species in Glassy Matrices or Magnetically Dilute Solids at Temperatures between 10 and 300 K, 165. © 2010, with permission from Elsevier)

nitroxide relaxation is slower in a hydrogen-bonded sorbitol matrix than in sucrose octaacetate.[54] (Sorbitol is an isomer of the sugar mannitol). In a wide range of glassy solvents, T_1 for tempone at 100 K decreases as the nitrogen nuclear hyperfine component A_{zz} decreases.[55] Decreases in A_{zz} can occur due to increased librational mobility, decreased solvent polarity, and decreased hydrogen bonding. These factors, as well as changes in the rigidity (hardness) of the glassy matrix, impact relaxation. Increasing the molecular mass of the nitroxide also decreases relaxation rates.[54] Attachment of a nitroxide to a protein has little impact on T_1 in a rigid lattice, so slower relaxation for higher molar mass nitroxides is an effect of local motion and spin delocalization and not of global motion of the molecule to which it is attached.

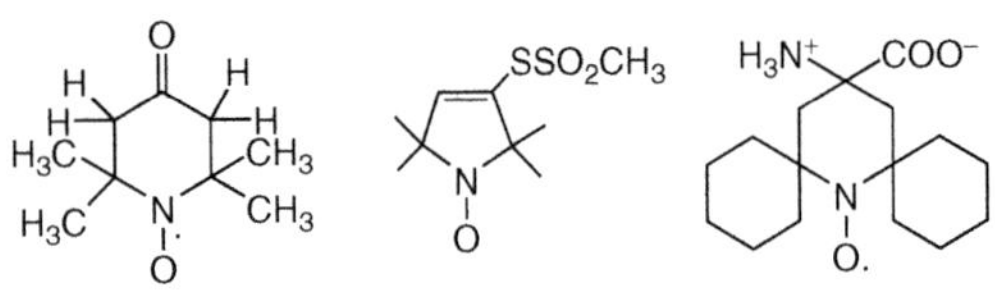

In glassy solvents and doped solids, T_1 for tempone (**XIII**) is orientation dependent with longer values along the g_{zz}-axis and shorter values in the perpendicular plane.[55] This orientation dependence of T_1 is consistent with the orientation dependence of spin–orbit coupling and with symmetry arguments concerning the orientation dependence of vibrations.

9.5.2 T_m in Rigid Lattice

At temperatures below about 80 K, T_m for nitroxides are independent of temperature and the values are solvent dependent.[32] In solvents that do not contain methyl groups, nuclear spin diffusion dominates relaxation. In solvents that contain methyl groups, rotation of the methyl groups enhances relaxation. The effects of solvent methyls on T_m are weakly dependent on temperature because there is a cumulative effect of interactions for a wide distribution of distances between dipoles. Most stable nitroxides, including tempone and common spin labels, have gem-dimethyl groups to sterically protect the N–O moiety. The intramolecular electron-nuclear couplings to these protons are relatively large and the couplings to the three protons are inequivalent. When the rate of rotation of the methyl

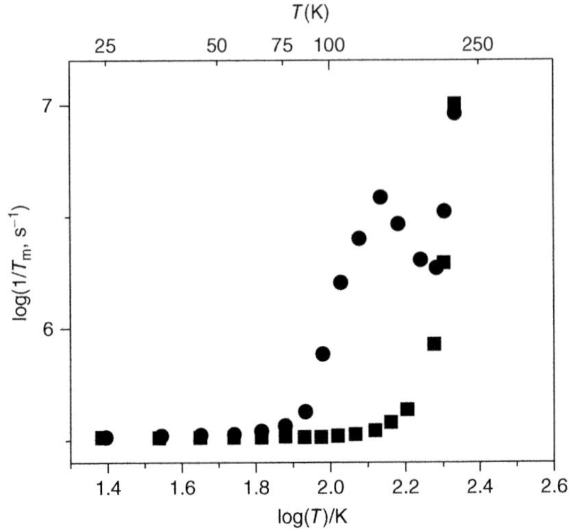

Figure 9.7. Temperature dependence of T_m in 1 : 1 water : glycerol at X-band for the spin label MTSSL (**XIV**) (•) and for a nitroxide in which the methyl groups were replaced by spirocyclohexyl groups (**XV**) (■). (Reproduced with permission from Ref. 56. © John Wiley & Sons, 2010)

groups is comparable to the inequivalence between the hyperfine couplings in hertz, methyl rotation is a very effective spin-echo dephasing mechanism. The temperature dependence of T_m for the spin label MTSSL (**XIV**) in 1 : 1 water : glycerol is shown in Figure 9.7 and is typical of many nitroxides.[56] T_m becomes dramatically shorter and $1/T_m$ is enhanced between about 80 and 200 K. These enhanced dephasing rates at temperatures above about 80 K restricted double electron–electron resonance (DEER) measurements of interspin distances in proteins to temperatures below about 80 K (see Chapter 19).[57,58] When the gem-dimethyl groups are replaced by spirocyclohexyl groups, as in spirocyclohexyl nitroxide **XV**, T_m in 1 : 1 water : glycerol becomes very weakly dependent on temperature up to about 180 K.[56,59] The rapid increase in $1/T_m$ above about 180 K is due to the onset of molecular motion as the glassy matrix softens and becomes a liquid. These motions modulate g and A anisotropy and cause substantial shortening of T_m. In regimes where molecular motion dominates T_m, T_2 commonly is defined as equal to T_m. If a matrix such as sucrose octaacetate or trehalose that has an above-ambient glass transition temperature is used to immobilize the nitroxide, T_m at ambient temperature remains sufficiently long that DEER experiments can be performed at 293 K.[60]

9.5.3 T_1 in Fluid Solution

In fluid solution, tumbling-dependent relaxation processes dominate T_1 for nitroxides. Characterization of the frequency dependence of relaxation is important to distinguish between potential contributions outlined in Section 9.2.1.6. Experiments have been performed at 250 MHz to 34 GHz for nitroxides in solvents that give τ_R between 4 and 26 ps. In this fast-tumbling regime, $1/T_2 \approx 1/T_1$ at frequencies up to about 9 GHz. The contribution to $1/T_1$ from spin rotation (equation (9.9)) is independent of frequency and decreases as τ_R increases. Spin rotation dominates $1/T_1$ at 34 GHz for τ_R in the range studied, and at all frequencies studied for $\tau_R = 4$ ps. The contribution to $1/T_1$ from modulation of nitrogen hyperfine anisotropy (equation (9.12)) increases as frequency decreases and as τ_R increases; it dominates at low frequencies for $\tau_R \gtrsim 15$ ps. The contribution from modulation of g anisotropy (equation (9.10)) is significant only at 34 GHz and above. Inclusion of a thermally activated process was required to account for the observation that for most of the nitroxides studied and in this tumbling range, $1/T_1$ was smaller at 250 MHz than at 1 to 2 GHz. The significant $^{15}N/^{14}N$ isotope effect, the small H/D isotope effect, and the viscosity dependence of the magnitude of the contribution from the thermally activated process suggest that it arises from intramolecular motions of the nitroxide ring that modulate the isotropic A values. Studies of nitroxide T_1 as a function of temperature in fluid solution have confirmed the contribution to $1/T_1$ from a thermally activated process.[61] (see Chapter 14).

9.5.4 T_2 in Fluid Solution

Nitroxide lineshapes and T_2 in fluid solution are exquisitely sensitive to motion (see Chapter 14), which has been extensively exploited to study motion of small molecules and biomolecules.[37] These experiments have been pioneered by Jack Freed and coworkers and the reader is referred to reviews by that group.[62–64] Lineshapes can be analyzed to distinguish motional models and restricted motion.

9.6 RELAXATION OF CU(II) COMPLEXES

As d^9 Cu(II) has an $S = 1/2$ spin, the relaxation times are substantially longer than for metal ions with larger S, which makes it possible to measure T_1 in rigid lattices directly by inversion recovery and saturation recovery at temperature up to about 120 K.[1,51] Relaxation rates for Cu(II) in fluid solution are too fast to be measured directly by currently available pulse methods that are described in Section 9.1.

9.6.1 T_1 in Rigid Lattice

The temperature dependence, between 10 and 120 K, of electron spin–lattice relaxation at X-band was analyzed for a series of copper(II) complexes with varying ligands and geometry.[65] The relaxation rates are strongly dependent on coordination geometry. The fit to the saturation recovery curves was better for a distribution of values of T_1 than for a single time constant, which suggests that distributions of relaxation times may be characteristic of Cu(II) in glassy solution. The presence of a distribution is analogous to the widely observed phenomenon known as g-strain and A-strain.[66] The temperature dependence of spin–lattice relaxation rates was analyzed in terms of contributions from the direct process (equation (9.3)), the Raman process (equation (9.4)), and local-mode process (equation (9.6)). There was a small contribution from the direct process at low temperature. The Raman process was the dominant contribution to relaxation between about 20 and 60 K. Debye temperatures were between 80 and 120 K. For samples with similar Debye temperatures, the coefficient of the Raman process tended to increase as g_z increased, as expected if modulation of spin–orbit coupling is a major factor in relaxation rates. Above about 60 K, local modes with energies in the range 260–360 K (180–250 cm^{-1}) dominated the relaxation. For molecules with similar geometry, relaxation rates were faster for more flexible molecules than for more rigid ones with comparable coordination spheres. At each temperature studied, the range of relaxation rates for the Cu(II) complexes was less than an order of magnitude. The spread was smaller between 20 and 60 K where the Raman process dominates, than at higher temperatures where local modes dominate the relaxation.

9.6.2 T_m in Rigid Lattice

It is difficult to obtain accurate values of T_m for Cu(II) in many complexes with nitrogen donors or distant nitrogens in the ligands because of deep nitrogen modulation on the 2-pulse echoes at X-band. At low temperature, the limiting values for T_m for Cu(II) complexes are similar to those for organic radicals and are dominated by nuclear spin diffusion (see Section 9.2.2).[67] Values commonly are orientation dependent because even low-amplitude librational motion takes spins off resonance on the time scale of the pulse experiments. Values of T_m are longer along the principal axes and shorter for intermediate orientations because the same amplitude of motion has less impact on the resonance field/frequency along a principal axis than at intermediate orientations.[67] This occurs because of the sin/cos variation of the EPR interactions.

9.7 RELAXATION OF IRON–SULFUR CLUSTERS

In the examples discussed in the previous sections, there has been no mention of a contribution from an Orbach process (equation (9.7)) because there are no accessible low-lying excited states for those species. Iron–sulfur clusters play key roles in many electron transport chains and have low-lying excited states.[68,69] The dominant contributions to the relaxation for the EPR signal are the Raman and Orbach processes for both $[2Fe\text{-}2S]^+$[68–70] and $[4Fe\text{-}4S]^+$ clusters.[71–73] The spin–lattice relaxation rates that result from both processes are determined by the spin–orbit coupling,[69,74] and stronger coupling enhances relaxation. The Orbach energy, Δ_{orb}, is the energy separation between the ground state and the lowest excited state, which is determined by the spin–spin interaction within the iron–sulfur cluster. An example of the temperature dependence of relaxation for an $[4Fe\text{-}4S]^+$ cluster is shown in Figure 9.8.[75] Values of T_1 below about 15 K were measured directly by inversion recovery. Values between about 25 and 50 K were calculated from the temperature-dependent contribution to the CW linewidths and the assumption that $T_1 \approx T_2$ (see Section 9.2.2). The determination of the Orbach energies provides crucial insight into the electronic structures of the clusters.

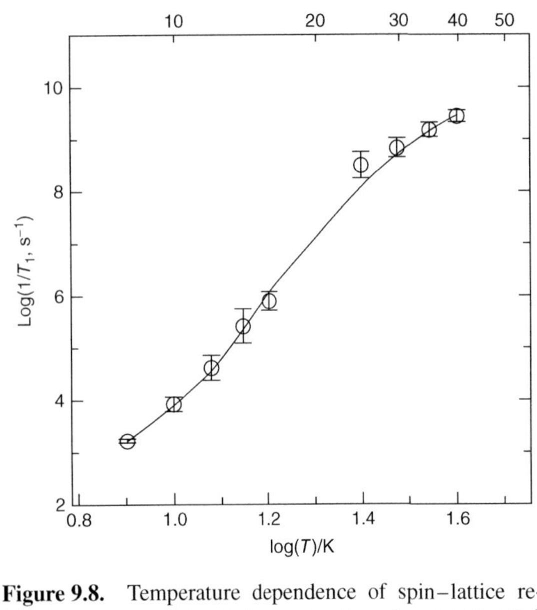

Figure 9.8. Temperature dependence of spin–lattice relaxation rates at X-band for dithionite reduced $[4Fe\text{-}4S]^+$ in *Rhodobacter sphaeroides* electron-transfer flavoprotein ubiquinone oxidoreductase. The fit line is the sum of contributions from the Raman and Orbach processes. The energy of the low-lying excited state was $\Delta E = 210$ K (146 cm^{-1}).[72] (Reprinted with permission from Usselman, R. J.; Fielding, A. J.; Frerman, F. E.; Eaton, G. R.; Eaton, S. S. Biochemistry 2008, 47, 92. © 2008, American Chemical Society)

9.8 RELAXATION IN IRRADIATED ORGANIC SOLIDS

The relaxation processes for radicals formed by irradiation of organic solids are the same as for radicals in glassy matrices that have been emphasized in the prior sections. Examples of the impact of a thermally activated process can be found in the effect of methyl group rotation. A striking example is the 4-Me-2,6-*t*-Bu-phenoxy radical formed by γ-irradiation of 4-Me-2,6-*t*-Bu-phenol **XVI** (Figure 9.9).[4] For this sample, rates of spectral diffusion are comparable to spin–lattice relaxation, so the effective value of the recovery time constant is strongly dependent on the experimental method. ELDOR was used to separate the contributions and obtain values of T_1 shown in Figure 9.9. For this sample, there is a local maximum in relaxation rates as a function of temperature at about 25 K at S-band (3.0 GHz) that shifts to about 35 K at X-band. The frequency dependence of the relaxation rates and the maxima in their temperature dependence are characteristic of the thermally activated process (equation (9.8)). The hyperfine coupling to the 4-methyl group in this radical is about 1.0 mT, so rotation of the methyl on the timescale of the Larmor frequency is a very effective relaxation mechanism.

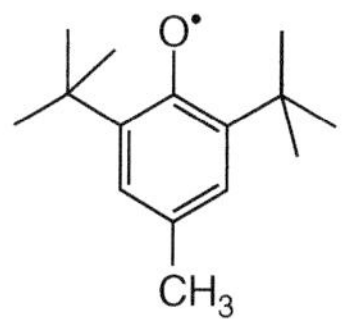

4-Me-2,6-*t*-Bu-phenoxy **XVI**

9.9 METALS WITH $S > {}^1/_2$

Examples of relaxation in species with $S > {}^1/_2$ are in Refs 6–10. A few examples are discussed here. $S > {}^1/_2$ produces additional relaxation mechanisms. The electron spins couple and energy level splittings can occur even without an applied magnetic field, which is called zero-field splitting (ZFS) (see Chapter 3). The magnitude of the ZFS depends strongly on the coordination environment of the metal, and hence is sensitive to vibrations of the metal complex and the lattice, and to collisions with solvent. These are very effective relaxation mechanisms, so T_1 of metals with $S > {}^1/_2$ usually is very short.

The main exceptions are metals with S ground states. The S state ions are special cases, as a pure S state does not have orbital angular momentum. As mentioned for $S = {}^1/_2$ spin systems, coupling of electron spin and orbital angular momentum is an important relaxation mechanism, so to a first approximation, the absence of orbital angular momentum except in a vibrationally distorted ion means that a major relaxation mechanism is absent. The most common examples of electronic S state ions are Mn(II) ($3d^5$, $S = 5/2$) and Gd(III) ($4f^7$, $S = 7/2$). For both of these ions, the lowest electronic excited states are high enough in energy that they are not significantly populated at the temperatures at which EPR spectra usually are recorded. EPR of these ions can be observed at room temperature.

$3d^3$ Cr(III) complexes have fairly fast relaxation, but can be studied over a wide range of temperatures. The relaxation time of Cr(III) can be long enough that CW spectra can be observed of complexes in fluid solution at room temperature. For discussions of CW lineshapes and relaxation for Cr(III), it is important that there are three transitions, $m_S = 3/2 \rightarrow 1/2$, $1/2 \rightarrow -1/2$, and $-1/2 \rightarrow -3/2$, which have the same g values, but may have different relaxation times. Early relaxation measurements for Cr(III) were comprehensively reviewed in Refs 6, 76, 77.

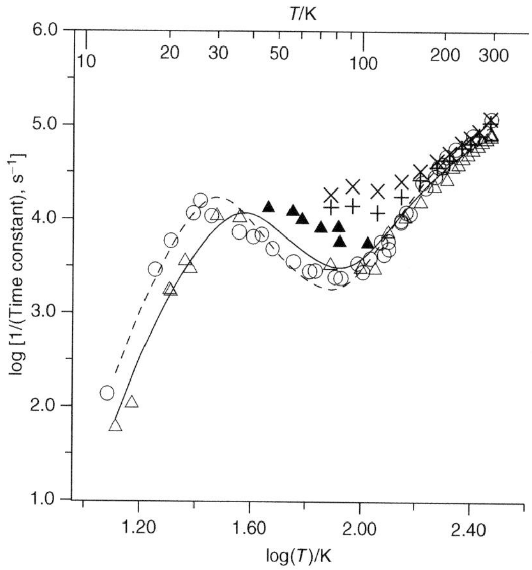

Figure 9.9. Temperature dependence of the recovery rate constant obtained by (×) X-band inversion recovery, (+) X-band echo-detected saturation recovery, (△) X-band saturation recovery or ELDOR, (O) S-band saturation recovery, and (▲) X-band saturation recovery values for 4-Me-2,6-t-Bu-phenoxy radical (**XVI**). The significance of contributions from spectral diffusion varies with temperature and detection method. Fits to the X-band (__) or S-band (_ _ _) data are based on a thermally activated process. Rates by various methods that were shorter than that from ELDOR were not included in the fitting.[4] (Reprinted with permission from Harbridge, J. R.; Eaton, S. S.; Eaton, G. R. J. Phys. Chem. A 2003, 107, 598. © 2003, American Chemical Society)

A detailed description of iron EPR is in Konig.[78] Many early relaxation measurements of Fe doped in various solids are summarized in Ref. 6. T_1 is in the millisecond range at 4.2 K and below, and the relaxation rate increases as T^9 at higher temperatures in most cases tested, but T^5, T^7, and T^8 dependence have been reported. One special case that spectroscopists should always be alert to is that the iron impurity in laboratory glassware is easily observed at $g \approx 4.3$ at room temperature. Low-spin Fe(III) ($S = 1/2$) usually has fast relaxation times because of low-lying excited states. EPR of Fe(II) is rarely observed because the low-spin state is diamagnetic in most symmetries, and the high-spin state is the even spin $S = 2$.

$3d^7$ Co(II) can be low spin, with one unpaired electron ($S = 1/2$), or high-spin with three unpaired electrons ($S = 3/2$). In high-spin Co(II), the ground state is ^4F, where an O_h crystal field splits into a singlet and two triplets, with one of the triplets being lowest. With spin–orbit coupling, a Kramers doublet is the lowest. The EPR spectrum has $g \approx 4.3$. In many cases, the ZFS is large enough that high magnetic fields and frequencies are needed to record all of the transitions. Relaxation is usually fast enough that T_1 influences linewidth above a few K, and $1/T_1$ increases with a high power of temperature via the Raman process.

9.10 SUMMARY

The relative importance of various relaxation processes for particular paramagnetic species is strongly dependent on temperature, electronic, and molecular structure as shown in the selected examples. The Raman and local-mode processes make significant contributions to T_1 relaxation in rigid and semirigid lattices for a wide range of species. The Orbach process requires a low-lying excited state. The thermally activated process is significant when a stochastic process modulates inequivalences on a timescale comparable to the Larmor frequency. Spin-echo dephasing at low temperatures is dominated by nuclear spin diffusion. It is enhanced by dynamic processes that average interactions with inequivalently coupled nuclei on the time scale of the hyperfine interaction and by motions that average g and A anisotropy. In fluid solution, molecular tumbling provides additional relaxation mechanisms. Experiments as a function of both temperature and resonance frequency are key to distinguishing between mechanisms in solution. Additional examples of contributions to relaxation for organic radicals are summarized in Ref. 79.

ACKNOWLEDGMENTS

Funding from the National Institutes of Health and the National Science Foundation that supported our work on electron spin relaxation is gratefully acknowledged. The insights summarized here are based on the excellent work by many former and present coworkers whose names are on the papers cited.

REFERENCES

1. S. S. Eaton and G. R. Eaton, *Biol. Magn. Reson.*, 2000, **19**, 29.
2. G. R. Eaton, S. S. Eaton, D. P. Barr, and R. T. Weber, Quantitative EPR, Springer-Verlag/Wein: New York, 2010.
3. J. S. Hyde, W. Froncisz, and C. Mottley, *Chem. Phys. Lett.*, 1984, **110**, 621.
4. J. R. Harbridge, S. S. Eaton, and G. R. Eaton, *J. Phys. Chem. A*, 2003, **107**, 598.
5. S. S. Eaton and G. R. Eaton, *J. Magn. Reson.*, 1993, **102**, 354.
6. K. J. Standley and R. A. Vaughan, Electron Spin Relaxation Phenomena in Solids, Plenum Press: New York, 1969.
7. M. K. Bowman and L. Kevan, in Time Domain Electron Spin Resonance, eds L. Kevan and R. N. Schwartz, John Wiley & Sons, Inc: New York, 1979, 68.
8. A. Abragam and B. Bleaney, Electron Paramagnetic Resonance of Transition Ions, Oxford University Press: Oxford, 1970.
9. L. Kevan and R. N. Schwartz (eds), Time Domain Electron Spin Resonance, John Wiley & Sons, Inc: New York, 1979.
10. C. P. Poole and H. Farach, Relaxation in Magnetic Resonance, Academic Press: New York, 1971.
11. A. A. Manenkov and R. Orbach (eds), Spin-Lattice Relaxation in Ionic Solids, Harper & Row: New York, 1966.
12. G. P. Srivastava, The Physics of Phonons, Adam Hilger: Bristol, 1990.
13. V. M. Meyer, S. S. Eaton, and G. R. Eaton, *Appl. Magn. Reson.*, 2013, **44**, 509.
14. H. Sato, V. Kathirvelu, G. Spagnol, S. Rajca, S. S. Eaton, and G. R. Eaton, *J. Phys. Chem. B.*, 2008, **112**, 2818.
15. J. W. Orton, Electron Paramagnetic Resonance: An Introduction to Transition Group Ions in Crystals, Gordon and Breach: New York, 1968.
16. A. Abragam, The Principles of Nuclear Magnetism, London: Oxford University Press, 1961, 405.
17. J. Murphy, *Phys. Rev.*, 1966, **145**, 241.
18. J. G. Castle Jr, and D. W. Feldman, *Phys. Rev. A*, 1965, **137**, 671.
19. V. A. Atsarkin, V. V. Demikov, and G. A. Vasneva, *Phys. Rev. B*, 1997, **56**, 9448.

20. R. Wilson and D. Kivelson, *J. Chem. Phys.*, 1966, **44**, 4445.
21. P. W. Atkins, in Electron Spin Relaxation in Liquids, eds L. T. Muus and P. W. Atkins, Plenum Press: New York, 1972, 279.
22. B. H. Robinson, D. A. Haas, and C. Mailer, *Science*, 1994, **263**, 490.
23. B. H. Robinson, C. Mailer, and A. W. Reese, *J. Magn. Reson.*, 1999, **138**, 199.
24. C. Mailer, R. D. Nielsen, and B. H. Robinson, *J. Phys. Chem. A*, 2005, **109**, 4049.
25. R. Owenius, G. R. Eaton, and S. S. Eaton, *J. Magn. Reson.*, 2005, **172**, 168.
26. N. Bloembergen, E. M. Purcell, and R. V. Pound, *Phys. Rev.*, 1948, **73**, 679.
27. Y. Mizuta, M. Kohno, and K. Fujii, *Jpn. J. Appl. Phys.*, 1994, **33**, 1885.
28. B. T. Ghim, S. S. Eaton, G. R. Eaton, R. W. Quine, G. A. Rinard, and S. Pfenninger, *J. Magn. Reson. A*, 1995, **115**, 230.
29. M. E. Newton, *Electron Paramagn. Reson.*, 2007, **20**, 131.
30. I. M. Brown, in Time Domain Electron Spin Resonance, eds L. Kevan and R. N. Schwartz, John Wiley & Sons, Inc: New York, 1979, 195.
31. K. M. Salikhov and Y. D. Tsvetkov, in Time Domain Electron Spin Resonance, eds L. Kevan and R. N. Schwartz, John Wiley & Sons, Inc: New York, 1979, 232.
32. A. Zecevic, G. R. Eaton, S. S. Eaton, and M. Lindgren, *Mol. Phys.*, 1998, **95**, 1255.
33. Y. Zhou, B. E. Bowler, G. R. Eaton, and S. S. Eaton, *J. Magn. Reson.*, 2000, **144**, 115.
34. A. J. Fielding, R. J. Usselman, N. Watmough, M. Simkovic, F. E. Frerman, G. R. Eaton, and S. S. Eaton, *J. Magn. Reson.*, 2008, **190**, 222.
35. D. Kivelson, *J. Chem. Phys.*, 1960, **33**, 1094.
36. N. D. Chasteen and M. W. Hanna, *J. Phys. Chem.*, 1972, **76**, 3951.
37. J. H. Freed, in Spin Labeling: Theory and Applications, ed L. J. Berliner, Academic Press: New York, 1976, 53.
38. S. A. Zager and J. H. Freed, *J. Chem. Phys.*, 1982, **77**, 3360.
39. G. L. Millhauser and J. H. Freed, *J. Chem. Phys.*, 1984, **81**, 37.
40. B. Epel, S. V. Sundramoorthy, E. D. Barth, C. Mailer, and H. Halpern, *J. Med. Phys.*, 2011, **38**, 2045.
41. A. J. Fielding, P. J. Carl, G. R. Eaton, and S. S. Eaton, *Appl. Magn. Reson.*, 2005, **28**, 231.
42. V. Kathirvelu, G. R. Eaton, and S. S. Eaton, *Appl. Magn. Reson.*, 2010, **37**, 649.
43. L. Yong, J. Harbridge, R. W. Quine, G. A. Rinard, S. S. Eaton, G. R. Eaton, C. Mailer, E. Barth, and H. J. Halpern, *J. Magn. Reson.*, 2001, **152**, 156.
44. Y. Song and G. R. Buettner, *Free Rad. Biol. Med.*, 2010, **49**, 919.
45. S. Mishra, P. Malhotra, A. K. Gupta, P. K. Singh, S. Javed, and R. Kumar, *Envir. Toxicol. Pharm.*, 2014, **37**, 553.
46. G. Armendariz-Vidales, L. S. Hermandez-Munoz, F. J. Gonzalez, A. A. de Souza, F. C. de Abreu, G. A. M. Jardim, E. N. da Silva, M. O. Gulart, and C. Frontana, *J. Org. Chem.*, 2014, **79**, 5201.
47. S. K. Rengan, M. P. Khakhar, B. S. Prabhananda, and B. Venkataraman, *Pramana*, 1974, **3**, 95.
48. V. Kathirvelu, H. Sato, S. S. Eaton, and G. R. Eaton, *J. Magn. Reson.*, 2009, **198**, 111.
49. K. Nakagawa, M. B. Candelaria, W. W. C. Chik, S. S. Eaton, and G. R. Eaton, *J. Magn. Reson.*, 1992, **98**, 81.
50. H. B. Elajaili, J. R. Biller, S. S. Eaton, and G. R. Eaton, *J. Magn. Reson.*, 2014, **247**, 81.
51. Y. Zhou, B. E. Bowler, G. R. Eaton, and S. S. Eaton, *J. Magn. Reson.*, 1999, **139**, 165.
52. H. Sato, S. E. Bottle, J. P. Blinco, A. S. Micallef, G. R. Eaton, and S. S. Eaton, *J. Magn. Reson.*, 2008, **191**, 66.
53. Y. Zhou, R. Mitri, G. R. Eaton, and S. S. Eaton, *Curr. Top. Biophys.*, 1999, **23**, 63.
54. H. Sato, V. Kathirvelu, A. J. Fielding, S. E. Bottle, J. P. Blinco, A. S. Micallef, S. S. Eaton, and G. R. Eaton, *Mol. Phys.*, 2007, **105**, 2137.
55. J.-L. Du, G. R. Eaton, and S. S. Eaton, *J. Magn. Reson. A*, 1995, **115**, 213.
56. A. Rajca, V. Kathirvelu, S. K. Roy, M. Pink, S. Rajca, S. Sarkar, S. S. Eaton, and G. R. Eaton, *Chem. Eur. J.*, 2010, **16**, 5778.
57. P. G. Fajer, L. Brown, and L. Song, *Biol. Magn. Reson.*, 2007, **27**, 95.
58. G. Jeschke, V. Chechik, P. Ionita, A. Godt, H. Zimmermann, J. Banham, C. R. Timmel, D. Hilger, and H. Jung, *Appl. Magn. Reson.*, 2006, **30**, 473.

59. V. Kathirvelu, C. Smith, C. Parks, M. A. Mannan, Y. Miura, K. Takeshita, S. S. Eaton, and G. R. Eaton, *Chem. Commun.*, 2009, 454.

60. V. Meyer, M. A. Swanson, L. Clouston, P. J. Boratyński, R. Stein, H. Mchaourab, A. Rajca, S. S. Eaton, and G. R. Eaton, *Biophysical J.*, 2015, **108**, 1213.

61. K. Kundu, D. R. Kattnig, B. Y. Mladenova, G. Grampp, and R. Das, *J. Phys. Chem. B*, 2015, **119**, 4501.

62. S. Lee, B. R. Patyal, and J. H. Freed, *J. Chem. Phys.*, 1993, **63**, 165.

63. D. E. Budil, K. A. Earle, and J. H. Freed, *J. Phys. Chem.*, 1993, **97**, 1294.

64. J. H. Freed, *Biol. Magn. Reson.*, 2005, **24**, 239.

65. A. J. Fielding, S. Fox, G. L. Millhauser, M. Chattopadhyay, P. M. H. Kroneck, G. Fritz, G. R. Eaton, and S. S. Eaton, *J. Magn. Reson.*, 2006, **179**, 92.

66. W. Froncisz and J. S. Hyde, *J. Chem. Phys.*, 1980, **73**, 3123.

67. J. L. Du, K. M. More, S. S. Eaton, and G. R. Eaton, *Israel J. Chem.*, 1992, **32**, 351.

68. P. Bertrand, J. P. Gayda, J. A. Fee, D. Kuila, and R. Cammack, *Biochim. Biophys. Acta*, 1987, **916**, 24.

69. M. Sarewicz, M. Dutka, R. Pietras, A. Borek, and A. Osyczka, *Phys. Chem. Chem. Phys.*, 2015, **17**, 25297.

70. J. P. Gayda, P. Bertrand, A. Deville, C. More, G. Roger, J. F. Gibson, and R. Cammack, *Biochim. Biophys. Acta*, 1979, **581**, 15.

71. P. Bertrand, J. P. Gayda, and K. K. Rao, *J. Chem. Phys.*, 1982, **76**, 4715.

72. R. J. Usselman, A. J. Fielding, F. E. Frerman, G. R. Eaton, and S. S. Eaton, *Biochemistry*, 2008, **47**, 92.

73. H. Blum, M. A. Cusanovich, W. V. Sweeney, and T. Ohnishi, *J. Biol. Chem.*, 1981, **256**, 2199.

74. H. Rupp, A. de la Torre, and D. O. Hall, *Biochim. Biophys. Acta*, 1979, **548**, 552.

75. M. A. Swanson, R. J. Usselman, F. E. Frerman, G. R. Eaton, and S. S. Eaton, *Biochemistry*, 2008, **47**, 8894.

76. I. Bertini, G. Martini, and C. Luchinat, in Handbook of Electron Spin Resonance: Data Sources, Computer Technology, Relaxation, and ENDOR, eds C. P. Poole Jr, and H. A. Farach, American Institute of Physics: New York, 1994, 79.

77. I. Bertini, G. Martini, and C. Luchinat, in Handbook of Electron Spin Resonance, eds C. P. Poole Jr, and H. A. Farach, American Institute of Physics: New York, 1994, 51.

78. E. Konig (ed), The Organic Chemistry of Iron, Elsevier: Amsterdam, 1978, Vol. **1**, 257.

79. V. Meyer, S. S. Eaton, and G. R. Eaton, *Appl. Magn. Reson.*, 2014, **45**, 993.

PART B

Basic Techniques and Instrumentation

Chapter 10
Transient EPR

Stefan Weber
Albert-Ludwigs-Universität Freiburg, Freiburg, Germany

10.1 INTRODUCTION

Transient EPR (trEPR) spectroscopy is a time-resolved detection technique that has been pioneered by Soon Sam Kim and Samuel I. Weissman in the late 1970s.[1,2] It has proven to be a powerful method to study light-generated, short-lived paramagnetic species, such as triplet states and radical pairs, in the time range of a few nanoseconds to hundreds of microseconds. The method derives from conventional CW EPR (see Chapter 1) insofar as the signal is observed in the presence of a weak microwave magnetic field, unlike in pulse EPR (see Chapter 11), where the EPR response to a manipulation of a spin system is detected with the microwave being turned off. Following a short light pulse that generates the paramagnetic species and that simultaneously serves as a trigger to start data acquisition, the time-dependent EPR signal is recorded at a fixed magnetic field value B_0 without application of any modulation, such as periodic variations of the magnetic field strength B_0, the microwave frequency ω, or the light intensity, in order not to constrict the optimal time resolution. After the accumulation of sufficient trEPR time profiles (also called transients) to obtain a desired signal-to-noise ratio (SNR), the magnetic field is increased by ΔB_0, and the next transient is recorded. This is repeated until the full spectral width of all resonances is covered. The primary data set obtained is then a two-dimensional variation of the trEPR signal amplitude with respect to the time t and the magnetic field B_0: Int(t, B_0). trEPR spectra at selected times t after the laser pulse may be obtained as slices parallel to the magnetic field axis (Figure 10.1). As signal detection at high temporal resolution (i.e., at broad detection bandwidth) is inherently rather insensitive, the method is confined to spin systems that exhibit large electron spin polarization, i.e., the respective energy level populations strongly deviate from those at thermal equilibrium at the given temperature and magnetic field B_0. In such cases, the method even allows for detection of transverse magnetization of paramagnetic species in proteins under physiological conditions, as the magnetization vector is driven and probed under the influence of a constant microwave magnetic-field strength B_1. This makes observation of trEPR less dependent on relaxation effects as compared to pulse EPR, whose variants are typically not applicable to proteins in aqueous solution at ambient temperatures owing to the inherent short T_2 relaxation times. Hence, trEPR may be the only method available to study such systems, provided their paramagnetic states can be

EPR Spectroscopy: Fundamentals and Methods.
Edited by Daniella Goldfarb and Stefan Stoll

ISBN: 978-1-119-16299-5
Also published in eMagRes (online edition)
DOI: 10.1002/9780470034590.emrstm1509

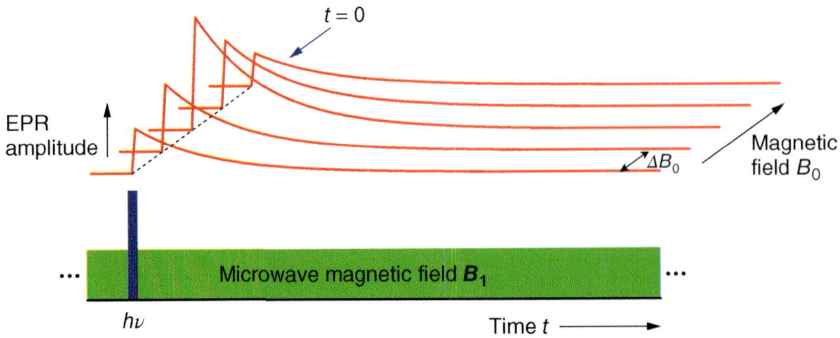

Figure 10.1. Schematic drawing on how trEPR data sets are collected. EPR time profiles are detected as a function of time t after pulsed photoexcitation of a sample. The instant of the laser pulse defines $t = 0$. Time-dependent EPR signals are recorded in the presence of a weak microwave magnetic field $\boldsymbol{B}_1$. The typical time resolution of a spectrometer operating at X-band microwave frequencies is in the range 1–10 ns and does not suffer from any dead-time problems

generated by photoexcitation. trEPR is nowadays well established and has been comprehensively reviewed previously by several authors, most frequently in the context of its application to studies of the primary electron transfer processes in photosynthetic reaction centers.[3–8]

10.2 EXPERIMENTAL CONSIDERATIONS

Microwave bridges dedicated for trEPR detection at high temporal resolution are nowadays commercially available. Alternatively, the bridge of a conventional CW-EPR spectrometer can be modified to record trEPR data with relatively little technical efforts and at moderate costs. In both cases, the required pulse laser system is the costliest acquisition. Figure 10.2 shows a schematic drawing of typical CW-EPR instrumentation with the additional components required for time-dependent detection of trEPR signals.

10.2.1 Laser

A pulse laser is of course essential. It is required to optically excite a given photoactive system, such as a protein, a low-molecular-weight organic molecule, or a polymer carrying chromophores, thereby producing transient paramagnetic species to be examined in a time-resolved manner. Ideally, a laser system should be chosen that provides a broad tunable range of excitation wavelengths from the near-ultraviolet (UV) to the near-infrared of the electromagnetic spectrum. Laser pulses need to be short enough so as not to constrict the desired time resolution of the experiment (see the following discussion). In most cases, excitation pulses with lengths of a few nanoseconds are sufficient even for studies of coherence phenomena, such as quantum beats,[5,9–14] transient nutations,[15,16] or coherences induced by magnetic nuclei.[12,17,18] Longer laser pulses, of course, smear out coherent photoexcitation of the sample and should therefore be avoided, especially when kinetics on very short time scales need to be measured. Pulse repetition frequencies should be chosen such that the sample can relax back to the ground state (the state before pulse laser excitation) within the time between two consecutive laser pulses, in particular when a number of signals need to be accumulated to obtain a favorable SNR. But even in cases where deactivation of photochemical states is efficient, care should be taken when applying pulses at a very high rate because of the dielectric heating of the microwave resonator, thereby spoiling critical coupling (see also the following discussion). Typical laser pulse repetition frequencies are in a range from 1 to 20 Hz. Similar considerations apply to the ideal energy of the laser pulses. With the laser beam it should be possible to optically excite the entire sample in the active part of the microwave resonator (where B_1 is maximal). This critically depends on the sample tube diameter (the path length) as well as the sample's extinction coefficient at the given wavelength. To avoid photobleaching caused by excessive excitation energies, it might therefore be advantageous to perform a trEPR experiment at higher microwave frequency bands, where sample tubes of smaller diameters are used owing to the smaller dimensions of the respective microwave

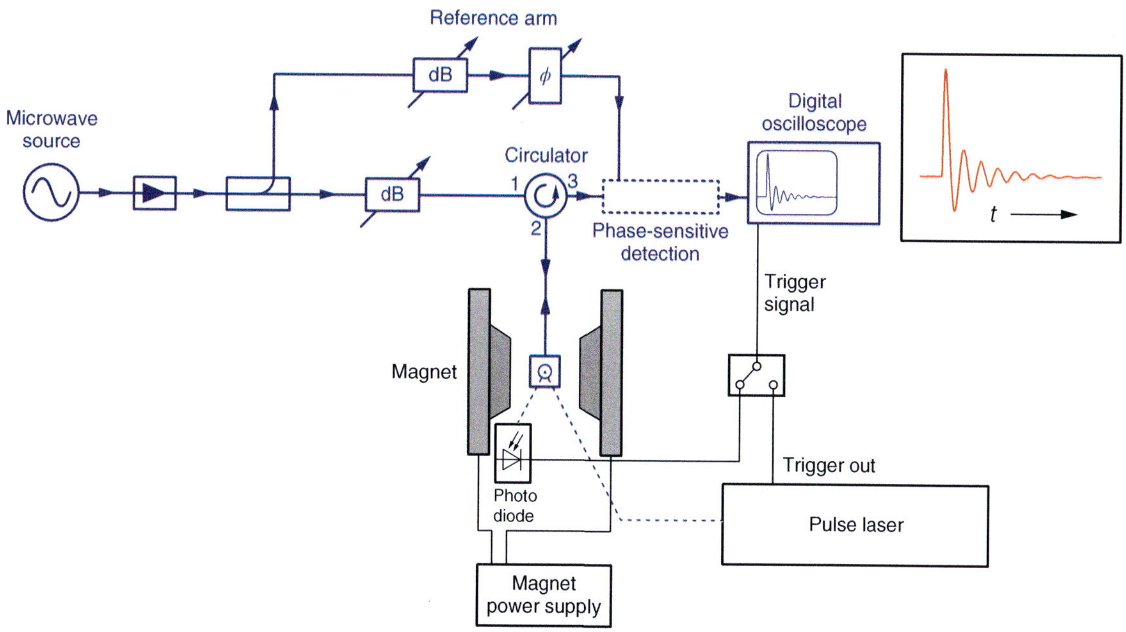

Figure 10.2. Schematic diagram of a conventional CW-EPR spectrometer with the additional components required to examine short-lived photoexcited paramagnetic species with high time resolution. A laser and a fast transient recorder or a digital oscilloscope are required to record the trEPR response as a function of time *t* following a laser pulse. The digital oscilloscope is triggered either by a TTL pulse provided by the laser electronics or by a photodiode that detects reflections of the laser beam and converts them into a fast trigger signal. To record trEPR in a phase-sensitive mode, a detection system with suitably fast response time is required (dashed rectangle). Its components are described in more detail in Figure 10.3

cavities. Laser pulses should be highly reproducible in terms of intensity and shape in order to have comparable relative signal amplitudes while accumulating EPR transients over a range of the magnetic fields. The laser's electronics typically supplies a trigger pulse that ideally has little to no jitter with respect to the timing of the laser light. This trigger signal may be used to start signal acquisition by a digital oscilloscope or a transient recorder. Alternatively, a photodiode equipped with a suitably fast driver can be placed near the laser beam to register beam reflections and convert them into a trigger signal. As light source, most modern trEPR instrumentations have pulsed Nd:YAG lasers that typically emit high-intensity laser pulses with a wavelength of 1064 nm in the infrared. They may be efficiently frequency doubled, tripled, or even quadrupled to generate laser light at wavelengths of 532, 355, and 266 nm, respectively. These higher harmonics of the fundamental wavelength can be used to drive an optical parametric oscillator (OPO) that converts an input laser (pump) wave into two output waves with longer wavelengths by means of second-order nonlinear optical excitation. With such a system, laser pulses covering virtually the entire spectral range of visible light at pulse energies of around 10–20 mJ at repetition rates of 1–20 Hz can be generated.

10.2.2 Microwave Resonator

The resonator used in trEPR should ideally provide one or several slits to allow for direct sample irradiation by the pulse laser. Alternatively, the sample can be illuminated by light guided through a glass or quartz fiber that is immersed into the sample tube in the resonator. The requirements for resonators used in trEPR experiments in terms of their quality factor Q are similar to those of pulse EPR (see Chapter 12).[19,20] Higher Q promises better sensitivity and hence, a higher SNR. However, as the resonator needs to be critically coupled because of trEPR signal detection in the presence of the microwave magnetic field $\boldsymbol{B}_1$, a high Q can cause problems on detuning and dematching the resonator by the subtle heating caused by the incident laser beam. Such effects typically induce strong background signals that need to be corrected for. Resonators with lower Q are less susceptible to transient detuning/dematching by resonator heating effects of the laser pulse. High-Q resonators furthermore restrict the response time (τ_R) of the resonator according to $\tau_R = Q/(\pi\nu_{res})$, where ν_{res} is the resonance frequency of the resonator. Therefore, a low Q is also preferred for experiments at the highest possible time resolution. The concomitant decrease in sensitivity, and hence SNR, can be compensated or in most cases even overcompensated by the inherent electron-spin-polarized state of the spin system that is generated as a consequence of optical sample excitation (see the following discussion). Frequently used resonators for trEPR are in the order of increasing Q: (bridged) loop-gap resonators,[21–24] split-ring resonators (commercially available), or dielectric resonators (commercially available), all of them allowing rather high filling factors.[25,26] Rectangular or circular cavity resonators can accommodate larger samples for which, however, complete optical sample excitation may be difficult to achieve. The calculation of the filling factor η of a cavity in a trEPR experiment is obtained by calculating the ratio of B_1 squared integrated over the optically excited sample volume to B_1 squared integrated over the resonator volume:

$$\eta = \frac{\int_{\text{optically excited sample}} B_1^2 \mathrm{d}V}{\int_{\text{cavity}} B_1^2 \mathrm{d}V} \tag{10.1}$$

This value is typically smaller than that of a CW- or pulsed EPR experiment on a 'dark-stable' paramagnetic sample, for which a similar expression is used (see Chapter 12), but the integration in the numerator is carried out over the entire sample volume.[25,26]

10.2.3 Microwave Bridge

A microwave bridge suitable for trEPR detection with high time resolution can be built by modifying a conventional CW microwave bridge with moderate efforts. In most cases, the rather narrow bandwidth diode detection system of a commercially available EPR spectrometer needs to be replaced or extended such that dual operation becomes possible. Figure 10.3 shows that a set of three electromechanical switches is needed to run the microwave bridge in the standard CW mode with diode detection (①) or in the time-resolved mode with a frequency mixer detection (②). In the latter mode, the signal coming back from the microwave resonator is first amplified by a low-noise broadband microwave preamplifier. Thereafter, a directional coupler diverges a small fraction of the microwaves that are subsequently rectified by a diode. The resulting low-frequency output of the diode is then fed into the standard automatic frequency control (AC- or DC-AFC) that ensures that the microwave source (typically a klystron or a Gunn oscillator) adjusts its output frequency in case of slow drifts of the resonance frequency of the resonator. The amplified microwaves of frequency f_{sig} from the cavity are fed as input signal into a frequency mixer with a bandwidth ideally in the range 100–250 MHz to provide the necessary time resolution, which is at X-band frequency at best about 1–10 ns. Together with the microwaves coming from the reference arm, which are of the same frequency, $f_{ref} = f_{sig}$, the mixer produces sum ($f_{sig} + f_{ref}$) and difference ($f_{sig} - f_{ref}$) frequencies, the latter virtually devoid of the carrier frequency but containing the time-resolved signal. If a quadrature mixer is used, two output signals are generated, which are 90° phase shifted with respect to each other. If necessary, this allows adjusting the phase of the trEPR signal at a later stage. The signal at the output of the mixer is then amplified with a broad bandwidth video amplifier (100–500 MHz, depending on the required time resolution). Thereafter, it is fed into a digital oscilloscope or a transient recorder, in which data accumulation at best possible digital resolution with respect to both amplitude and time can take place following a trigger signal from either the laser or a photodiode (see earlier).

10.2.4 trEPR Recording Scheme

A typical trEPR experiment is conducted without (amplitude) modulation of the magnetic field, amplitude or frequency modulation of the microwave frequency (Of course with the exception of a small modulation of the microwaves, typically of around 75 kHz, which is used by the AFC to stabilize the microwave frequency output of the microwave source to the resonance frequency ν_{res} of the resonator.) or the light amplitude in order not to constrict the desired time resolution. This renders signal detection rather insensitive as compared to CW-EPR that uses a lock-in detection scheme with respect to magnetic-field modulation for improving

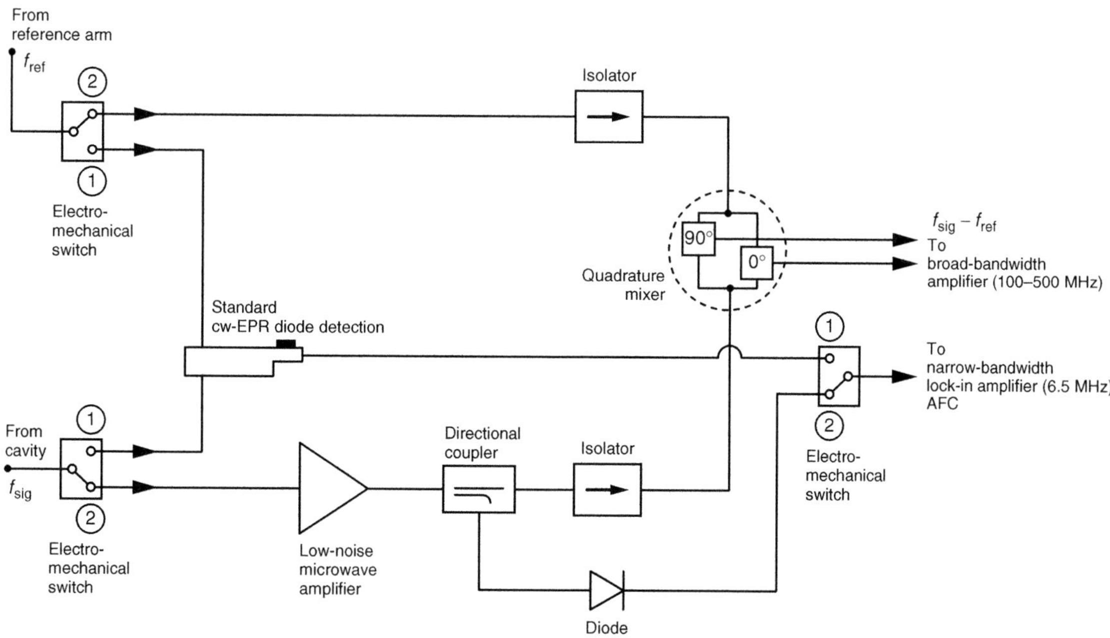

Figure 10.3. Typical phase-sensitive detection scheme used for trEPR at X-band microwave frequencies. A commercially available microwave bridge can be modified such that recording of trEPR data at high time resolution (all three electromechanical switches in position ②) is possible, or, alternatively, the accumulation of standard CW-EPR using a diode (all three electromechanical switches in position ①). For a description of the components, see text

the noise figure. However, the strong electron-spin polarization, i.e., the non-Boltzmann population of spin states encountered in photochemically generated spin systems (see the following discussion) in many cases compensates for the lack of signal-to-noise gain a lock-in detection provides.[8] Nevertheless, the accumulation of a number of signals to increase the SNR of trEPR spectra and transients is routinely applied. Unlike CW-EPR, by which transitions are detected as first derivatives of the detector output with respect to the magnetic field, by trEPR resonances are collected in the direct detection mode in which the sign of a signal directly reflects the polarization of the resonant transition: positive and negative trEPR signals represent (enhanced) absorptive (A) and emissive (E) EPR transitions, respectively.

Data processing typically comprises subtraction of a background signal that may be superimposed on the desired trEPR data owing to dielectric heating of the microwave resonator or the sample from optical excitation. For this purpose, a time profile recorded at an off-resonance magnetic field position is collected and subtracted from all transients of the two-dimensional data set. trEPR spectra may be obtained by extracting the signal amplitude at a certain time t after the laser pulse or by integrating over a time window Δt and plotting the result as a function of B_0.

Recording trEPR data from a photo-induced spin system has some inherent advantages as compared to data collection by pulse EPR methods, such as electron-spin echo detection as a function of the magnetic field B_0 and time t after pulsed photoexcitation of the sample. Although both methods in principle reveal similar information, pulse EPR can only be applied to spin systems that have sufficiently long electron-spin relaxation times T_1 and T_2, so that coherences have not already decayed within the sequence of microwave pulses. To render relaxation suitable for pulse EPR, the spin system typically needs to be cooled down to a frozen state, which in the case of proteins may be far from their physiological working conditions. In trEPR, by contrast, the transient magnetization is probed in the presence of the weak microwave magnetic field $\boldsymbol{B}_1$, similar to CW-EPR, and hence, fast relaxation is much less problematic than in the pulsed methods. Therefore, trEPR data of light-induced paramagnetic

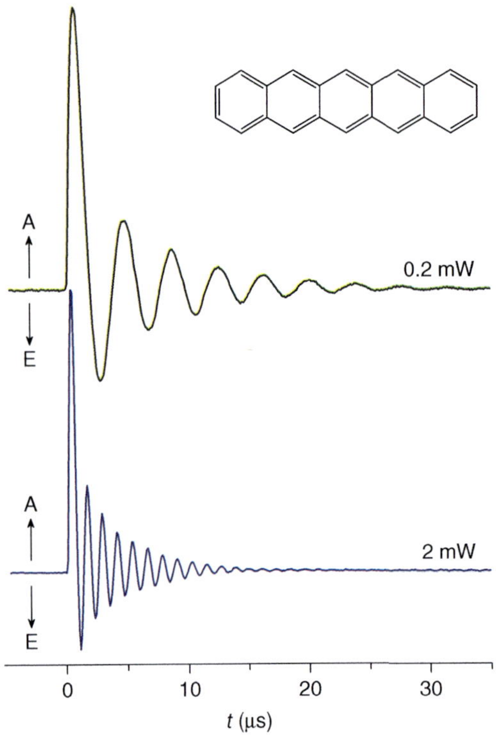

Figure 10.4. trEPR transients of the triplet state ($S = 1$) of a single crystal of 0.1% pentacene in *para*-terphenyl. A: enhanced absorption and E: emission. The oscillations in the time domain are the so-called Torrey or B_1 oscillations of the magnetization vector $\boldsymbol{M}$ about $\boldsymbol{B}_1$, whose frequency ω_1 is proportional to the square root of the microwave power

states can often be recorded at elevated temperatures, in proteins even in liquid aqueous solution near or at room temperature.

Furthermore, trEPR has a certain (multiplex) advantage over the echo-detected EPR methods because the EPR response to a laser pulse at a given B_0 is probed directly as a function of time t. Such a time profile, collected with one single laser flash, does contain information on spin relaxation as well as chemical kinetics (e.g., the lifetime of the spin system). By contrast, to probe relaxation times and time constants of kinetics, two-pulse electron-spin echo detection with a pulse sequence $(h\nu)_{\text{laser}} - t_{\text{DAF}} - (\pi/2)_x - \tau - (\pi)_x - \tau -$ echo would need to be performed at various delay times t_{DAF} and τ, thus rendering data collection much more time consuming than in trEPR and, therefore, more straining to the investigated sample so that photobleaching effects need to be considered.

Figure 10.4 shows typical trEPR transients obtained from the photo-generated triplet state of pentacene distributed at low concentration (0.1%) in a host crystal of *para*-terphenyl. Transient nutations are observed, which are a signature of the precessing motion of the magnetization vector $\boldsymbol{M}$ about the microwave magnetic field $\boldsymbol{B}_1$.[3] The frequency of these so-called Torrey oscillations, ω_1, depends on the strength of $\boldsymbol{B}_1$. The oscillations are damped by relaxation: Transient nutations can be observed only in cases with sufficiently slow spin–spin relaxation, $1/T_2 \ll \omega_1$, as, for example, is the case when the paramagnetic molecule is tightly held in place in a crystalline surroundings (i.e., under solid-state conditions). At high microwave power P, the dominant relaxation process is transverse relaxation T_2, and at lower power, it is T_1. The effective decay time T_{eff} in an intermediate range of $\omega_1 = \gamma_e B_1$ is given by $1/T_{\text{eff}} = 1/T_1 + \omega_1^2 T_2$.[27] Figure 10.4 shows transient nutations for two different values of the microwave power. The frequency in the time profiles can be used to directly measure the magnetic field strength B_1 in a resonator at a given microwave power, and by doing so, one can determine the resonator's conversion factor C, which depends on the quality factor Q and the incident microwave power P according to $B_1 = C \times (QP)^{1/2}$ (see Chapter 12).[26]

As in many other magnetic-resonance techniques, the trend of performing the experiment at higher magnetic fields and correspondingly higher microwave frequencies also applies to trEPR.[28–38] The main motivation is the higher spectral resolution obtained at stronger magnetic fields and also the intrinsically higher temporal resolution available at higher microwave frequencies (Figure 10.5). At X-band (9–10 GHz), a time resolution of 1–10 ns can be reached. For higher frequencies, time resolutions of less than 1 ns are in principle feasible, provided that the bandwidth of the detection system (resonator, amplifiers, and detection mixer) is broad enough.[39] As photoexcitation of samples typically leads to electron-spin-polarized spin states,[6,8,40,41] the higher magnetic fields do, however, not contribute to obtaining higher population differences of transitions because generation of electron-spin polarization is in most cases independent of the strength of the external magnetic field B_0.

It should be mentioned, however, that trEPR signal acquisition in the presence of the microwave magnetic field requires additional efforts in spectral simulations as the interaction of the spin(s) with the microwave

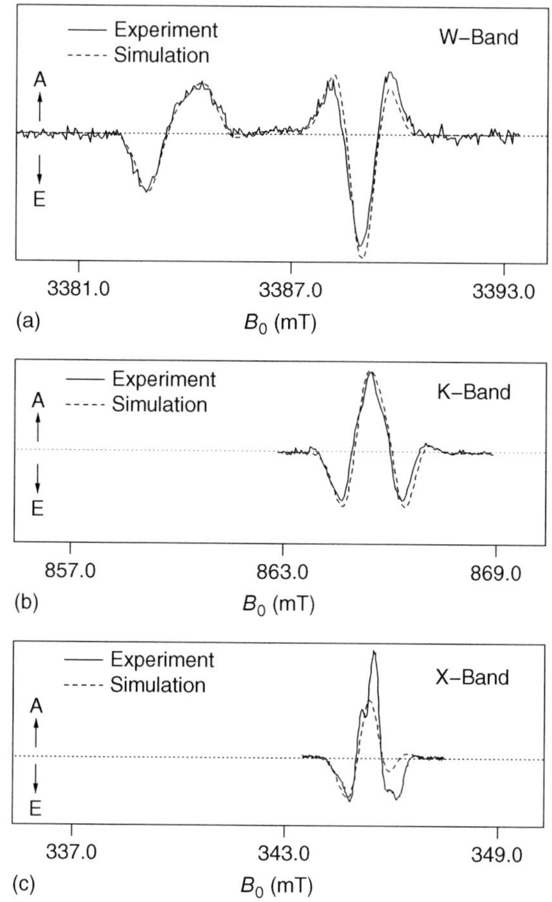

Figure 10.5. trEPR spectra from the secondary radical pair of photosynthetic charge separation in plant photosystem I recorded at different spectrometer frequencies: X-band (9.706 GHz, c), K-band (24.269 GHz, b), and W-band (95.462 GHz, a). The dashed lines show spectral simulations. Higher frequency trEPR provides better spectral resolution for systems with high $\boldsymbol{g}$-anisotropy. Hence, the resonances of the two radicals, $P_{700}^{\bullet+}$ and $A_1^{\bullet-}$, of the coupled radical pair are best separated at W-band frequency. (Reprinted with permission from A. van der Est, T. Prisner, R. Bittl, P. Fromme, W. Lubitz, K. Möbius and D. Stehlik, J. Phys. Chem. B, 1997, 101, 1437. Copyright 1997 American Chemical Society)

magnetic field $\boldsymbol{B}_1$ needs to be explicitly taken into account if the time evolution of the signal has to be analyzed with respect to coherence phenomena and relaxation rates. Such treatments are nowadays routinely performed.[42]

10.3 APPLICATIONS OF TRANSIENT EPR

Most recent applications of trEPR deal with the detection of radical pairs in the course of electron transfer in proteins or at the surface of proteins,[37,43–48] and in donor–acceptor electron-transfer models,[49–54] or higher multiplet states, such as triplets or even quartets, in π-conjugated organic systems or in materials relevant for light-to-energy conversion and spintronics.[35,36,55–65] To demonstrate the power of trEPR spectroscopy for the examination of short-lived paramagnetic species, some representative examples of its application will be presented in more detail in this section.

10.3.1 Triplet States

trEPR is ideally suited to study photo-induced triplet states, usually generated by intersystem crossing from an excited singlet-state precursor. In the triplet, the dipole–dipole coupling between the two individual electron spins, S_1 and S_2, removes the energetic degeneracy of the triplet sublevels. Hence, the eigenstates denoted T_X, T_Y, and T_Z of the Hamiltonian

$$\hat{H}_{ZFS} = \hat{\boldsymbol{S}}\boldsymbol{D}\hat{\boldsymbol{S}} \tag{10.2}$$

are energetically different even in the absence of an external magnetic field. In equation (10.2), $\hat{\boldsymbol{S}}$ is the total spin angular momentum operator, $\hat{\boldsymbol{S}} = \hat{\boldsymbol{S}}_1 + \hat{\boldsymbol{S}}_2$, and $\boldsymbol{D}$ the symmetric and traceless dipolar interaction tensor. The spin alignment in the excited singlet state is projected onto the triplet state's three zero-field sublevels depending on the symmetries of the singlet and triplet wave functions.[66] These initial spin state populations typically deviate strongly from those at thermal equilibrium (Boltzmann population) (Figure 10.6). They are converted into electron-spin polarization in the presence of an external magnetic field, i.e., the eigenstates T_+, T_0, and T_- that diagonalize the spin Hamiltonian

$$\hat{H} = \hat{\boldsymbol{S}}\boldsymbol{D}\hat{\boldsymbol{S}} + \mu_B \boldsymbol{B}\boldsymbol{g}\hat{\boldsymbol{S}} \tag{10.3}$$

at a given strength of the external magnetic field $\boldsymbol{B}$ are also populated in a manner strongly deviating from thermal population. This gives rise to enhanced absorptive and emissive EPR transitions, depending on the orientation of the principal axes X, Y, and Z of $\boldsymbol{D}$ with respect to $\boldsymbol{B}$ (see Chapter 3). In a nonoriented rigid sample, the spectra of all orientations of

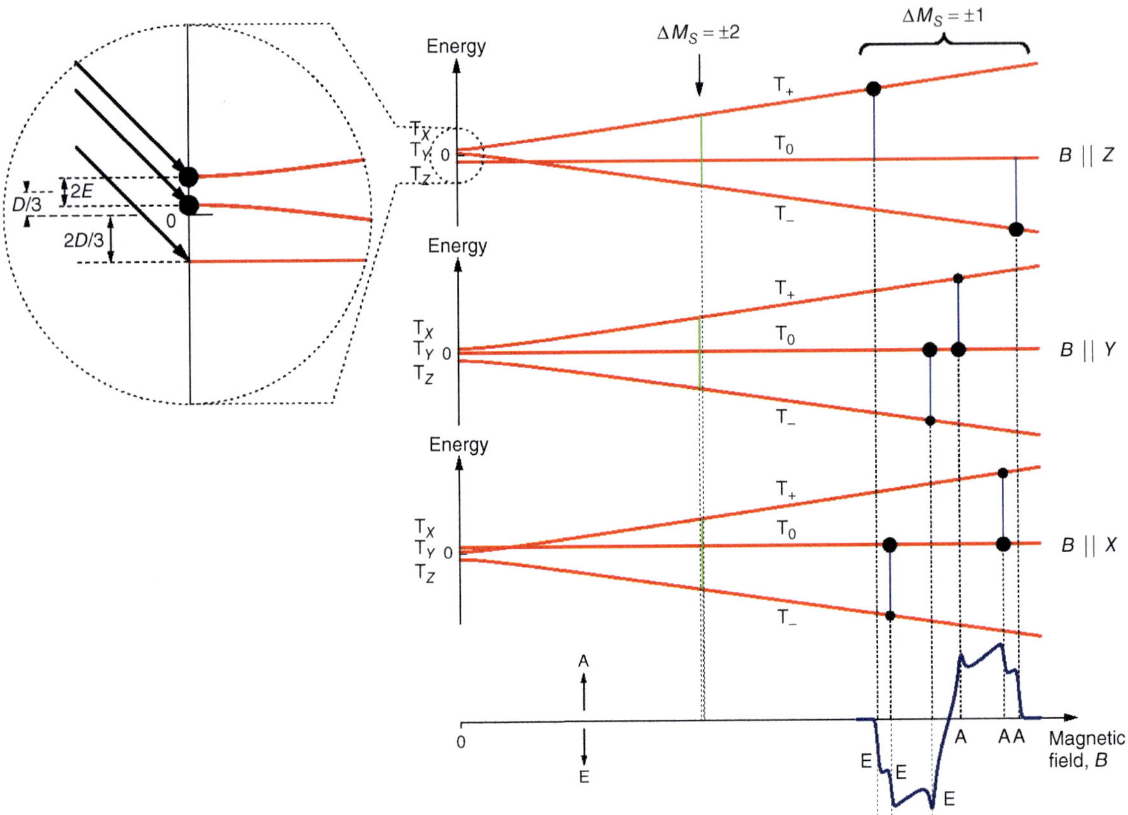

Figure 10.6. Energy level diagram of a triplet state with dipolar coupling between the individual electron spins $\boldsymbol{S}_1$ and $\boldsymbol{S}_2$. The zero-field eigenstates are denoted T_X, T_Y, and T_Z, where X, Y, and Z are the principal axes of the dipolar coupling tensor $\boldsymbol{D}$. T_+, T_- and T_0 are the triplet eigenstates in the presence of an external magnetic field $\boldsymbol{B}$. The population of the zero-field eigenstates is typically sublevel specific and depends on the symmetries of the excited singlet and the triplet wavefunctions. In the present example, the zero-field populations, indicated by the filled black circles, are $p_Z = 0$ and $p_X = p_Y = 0.5$. The nonthermal population of T_X, T_Y, and T_Z is converted into electron-spin polarization at high magnetic field, i.e., the high-field eigenstates T_+, T_-, and T_0 are also populated in a nonthermal manner (see filled black circles; their diameters indicate the extend of population of a certain state), depending on the orientation of the principal axes of $\boldsymbol{D}$ with respect to $\boldsymbol{B}$. This results in (enhanced) absorptive (A) EPR transitions (selection rule $\Delta M_S = \pm 1$), see, e.g., the high-field transition $T_- \leftrightarrow T_0$ in the $\boldsymbol{B}$||Z case ($p_- > p_0$), or emissive (E) EPR transitions, see, e.g., the low-field transition $T_+ \leftrightarrow T_0$ in the $\boldsymbol{B}$||Z case ($p_+ > p_0$). In a rigid and nonoriented sample, spectra from all orientations of $\boldsymbol{D}$ with respect to $\boldsymbol{B}$ sum up to a powder pattern with characteristic (enhanced) absorptive and emissive spectral features (bottom). In cases, where the dipolar interaction $\boldsymbol{D}$ is very strong, the $\Delta M_S = \pm 2$ transition may also be observed

$\boldsymbol{D}$ relative to $\boldsymbol{B}$ add up to a characteristic powder pattern (see the bottom of Figure 10.6). Spin–lattice relaxation promotes the decay of the nonthermal triplet spin state populations into that at thermal equilibrium. Sublevel-specific triplet decay to the singlet ground state may alter the initial polarization pattern.

The triplet electron-spin polarization can be observed for a relatively short time after the laser pulse provided that the detection system is fast enough. trEPR spectra of triplet states are dominated by the zero-field splitting interaction parameters D and E that parameterize the traceless dipolar coupling tensor $\boldsymbol{D}$ (see Chapter 3) (Figure 10.6). The zero-field splitting parameter D is a measure of the delocalization of the unpaired electron spin of the triplet over the molecule, and E describes the nonaxial asymmetry (rhombicity) of delocalization. Smaller D values are, for example, observed for molecular triplet states that are spread over large planar π-ring systems, whereas larger values are expected for spatially more confined triplets. In addition to the zero-field splitting parameters, the relative populations p_X, p_Y, and p_Z (normalized to $p_X + p_Y + p_Z = 1$) of the zero-field eigenstates T_X, T_Y, and T_Z, respectively, can be obtained from simulations of the spectral trEPR lineshape.

Studies of triplet states using trEPR become increasingly popular in the thriving field of materials research for organic photovoltaic devices.[59,67] In energy conversion, one interesting topic that needs to be addressed is whether the triplet state represents a deactivation pathway, thus inhibiting charge separation from the singlet state, and thereby decreasing the quantum efficiency of a solar cell, or a reactive state from which charge separation might also take place[68,69] (Figure 10.7). In the latter case, owing to the conservation of the spin angular momentum, a triplet-born radical pair is generated, from which charge recombination might even be slowed down, because it is spin forbidden, to regenerate the diamagnetic ground state with two electrons in a triplet configuration, thus leading to an enhanced conversion efficiency. Hence, studies of the triplet state, the delocalization of its frontier orbital, and its initial spin-state populations generated by intersystem crossing are essential to optimize materials for organic solar cells.

Figure 10.8 depicts trEPR spectra of poly[*N*-9′-heptadecanyl-2,7-carbazole-*alt*-5,5-(4′,7′-di-2-thienyl-2′,1′,3′-benzothiadiazole)] (PCDTBT, see Figure 10.8b), which were extracted at 500 ns after pulsed laser excitation at a wavelength of 566 nm generated with an OPO pumped with an Nd:YAG laser.[59] PCDTBT is increasingly replacing poly(3-hexylthiophene) (P3HT) as a benchmark material and is a candidate for large-scale processing and device manufacturing for real-world applications.[70] Rather broad spectra, typical of a spin-polarized triplet state, with emissive (E) and enhanced absorptive (A) features covering a magnetic-field range of nearly 100 mT are observed. The transitions stem from the repeat unit of the polymer depicted in Figure 10.8(b). The zero-field splitting parameters D and E can be extracted from spectral simulations: $|D|/h = 1268$ MHz and $|E|/h = 101$ MHz. Good starting points for the least-squares spectral fittings could be obtained from the outer inflection points at around 300 and 390 mT, which are related to $2|D|$, whereas inner maxima and inflection points in a powder-averaged triplet spectrum

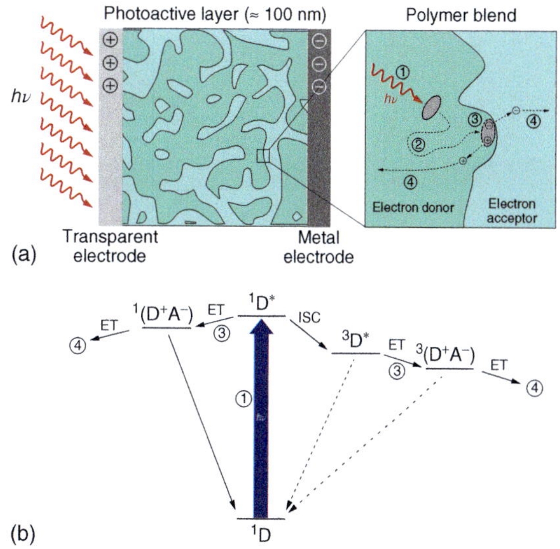

Figure 10.7. (a) Schematic drawing of a dispersed-junction organic photovoltaic cell with a photoactive layer consisting of a nanoscale polymer blend of donor and acceptor materials between the electrodes (light-green and dark-green areas, respectively). Photoexcitation of the donor (①) leads to the formation of a singlet exciton that may convert by intersystem crossing (ISC) to a triplet exciton. Singlet or triplet excitons diffuse to the heterojunction (②), where charge separation takes place (③). The charges migrate to the electrodes (④). (b) Energy level diagram of the photophysical processes that take place in a photovoltaic device. A donor molecule D in its (singlet) ground state, ^{1}D, is promoted by light into its excited (singlet) state, ^{1}D*. ^{1}D* can either undergo ISC to form a triplet state, ^{3}D*, or perform electron transfer (ET) to an acceptor molecule A to form a singlet radical-pair state $^1(D^{\bullet+}A^{\bullet-})$, the charge-separated state. ^{3}D* may or may not undergo charge separation to form a triplet radical pair state $^3(D^{\bullet+}A^{\bullet-})$. In the latter case, it is a deactivation pathway because the return of ^{3}D* to the ground state (dashed arrow) is spin forbidden, hence, slow. In the singlet channel, subsequent charge migration steps lead to more separated radical pair states; however, (spin-allowed) backward ET to regenerate the ground state ^{1}D competes with charge separation

are defined by $|D| - 3|E|$ and $|D| + 3|E|$. The anisotropy of the $\boldsymbol{g}$-tensor is small as compared to that of $\boldsymbol{D}$, and hence, at X-band microwave frequencies, it is not resolved: $g_{\mathrm{iso}} = 2.002$. In addition to the zero-field splitting parameters, the zero-field populations were also obtained from the spectral simulations (Because the signs of the zero-field splitting parameters D and E were not determined in this study, it is not possible to unambiguously assign the zero-field populations p_1,

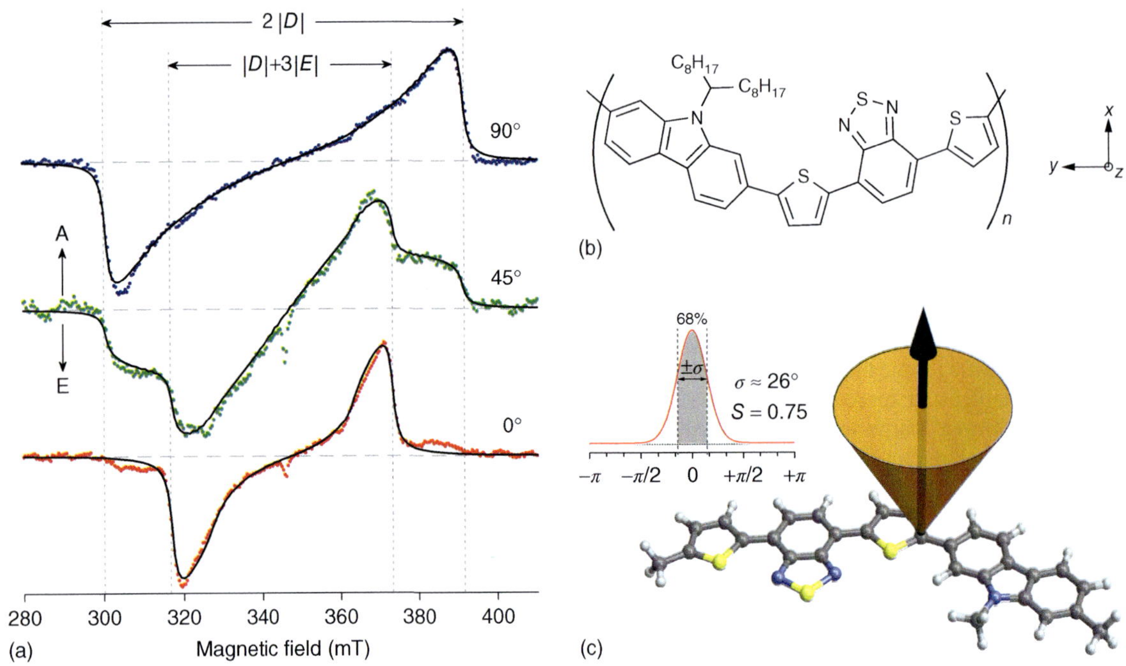

Figure 10.8. (a) The triplet state of poly[*N*-9′-heptadecanyl-2,7-carbazole-*alt*-5,5-(4′,7′-di-2-thienyl-2′,1′,3′-benzothiadiazole)] (PCDTBT, for the molecular structure of the chromophoric unit, see (b)) drop cast onto a PET film and detected by trEPR at three different orientations of the film with respect to the direction of the external magnetic field (dots, experimental data points recorded 500 ns after pulsed laser excitation; drawn black lines, spectral simulations performed under consideration of ordering of the chromophore). (c) The orientation of the chromophoric unit with respect to the PET film normal: 68% of all molecules assume an angle θ between their π-plane and the PET film normal that is within the 26°-limit depicted in orange. This corresponds to an order parameter of $S = 0.75$. (Reproduced with permission from Ref. 59. © John Wiley and Sons, 2015)

p_2, and p_3 to p_X, p_Y, and p_Z.): $p_1 = 0.304$, $p_2 = 0.331$, and $p_3 = 0.365$.

In this chapter, it was interesting to demonstrate that the experimentally determined trEPR spectra could not be simulated with the assumption that the repeat units of the polymer have fully random orientation with respect to the direction of the external magnetic field, even though the sample was drop cast onto a polyethylene terephthalate (PET) film as a substrate that was subsequently placed into the EPR resonator without extra measures to orient the sample. Rather, ordering effects were observed when rotating the PET film with the drop-cast PCDTBT in the resonator relative to the direction of the external magnetic field $\boldsymbol{B}$. The trEPR spectrum collapsed to about half the width at an angle of 0° as compared to 90° of the plane normal with respect to $\boldsymbol{B}$. Satisfactory spectral simulations could only be obtained when ordering effects of the chromophoric unit of the polymer were taken into account. Careful analysis revealed that the aromatic rings of PCDTBT orient parallel to the film surface, whereas the backbone shows no preferred orientation. When considering the plane normal (axis z in Figure 10.8b), about 68% of all molecules were found to be within ±26° from the PET film normal (see Figure 10.8c). This corresponds to a very high order parameter of $S = \langle (3\cos^2\theta - 1)/2 \rangle = 0.75$ because of the narrow spread of the angle θ between the z-axis of PCDTBT and the PET film normal, thus demonstrating that quite substantial ordering took place when drop casting the sample onto a surface. These results show that trEPR is an ideal tool to resolve local morphologies of polymers.[71] Given the important role of morphology for device efficiency, the method is becoming increasingly popular for device characterization, because charge-separated states can also be examined by this technique.[56,62,72] The influence on light-induced paramagnetic states on the photocurrent of photovoltaic devices can nowadays also be studied using trEPR in combination with transient electrically detected magnetic resonance (see Chapter 26).[73]

10.3.2 Radical-pair States

Radical pairs generated by photo-induced electron transfer (which may subsequently be followed by one or several proton transfer steps)

$$A + B \xrightarrow{h\nu} A^{\bullet +} + B^{\bullet -}$$

exhibit, similar to triplet states, pronounced electron-spin polarization. The situation is best described by evaluating the spin Hamiltonian of a weakly coupled radical pair:

$$\hat{H} = \mu_B \boldsymbol{B} \boldsymbol{g}_A \hat{\boldsymbol{S}}_A + \mu_B \boldsymbol{B} \boldsymbol{g}_B \hat{\boldsymbol{S}}_B - 2J\hat{\boldsymbol{S}}_A\hat{\boldsymbol{S}}_B + 2\hat{\boldsymbol{S}}_A \boldsymbol{D} \hat{\boldsymbol{S}}_B \quad (10.4)$$

Here, $\boldsymbol{g}_A$ and $\boldsymbol{g}_B$ are the $\boldsymbol{g}$-tensors of radicals A and B, respectively, J the (isotropic) exchange interaction parameter (see Chapter 4), and $\boldsymbol{D}$ the dipolar coupling tensor. For simplicity, hyperfine interactions of magnetic nuclei to either radical A or radical B are neglected here. The principal axes of the three tensors, $\boldsymbol{g}_A$, $\boldsymbol{g}_B$, and $\boldsymbol{D}$, are typically not collinear. By choosing the principal axes of one of the tensors as a reference coordinate system, two sets of Euler angles are required to unambiguously describe the orientations of all three tensors with respect to each other. One further set of angles sets the radical pair in relation to the laboratory frame defined by the direction of the external magnetic field $\boldsymbol{B}$.

Diagonalization of the radical pair Hamiltonian for a specific orientation of interaction tensors in equation (10.4) with respect to the magnetic field $\boldsymbol{B}$ yields four eigenstates with the corresponding energies[74]

$$|\psi_1\rangle = |\alpha_A\alpha_B\rangle \equiv |T_+\rangle \qquad E_1 = \hbar\omega_0 - J + d \quad (10.5a)$$

$$|\psi_2\rangle = \cos\delta\,|S\rangle + \sin\delta|T_0\rangle \qquad E_2 = -d + \hbar\Omega \quad (10.5b)$$

$$|\psi_3\rangle = -\sin\delta\,|S\rangle + \cos\delta\,|T_0\rangle \qquad E_3 = -d - \hbar\Omega \quad (10.5c)$$

$$|\psi_4\rangle = |\beta_A\beta_B\rangle \equiv |T_-\rangle \qquad E_4 = -\hbar\omega_0 - J + d \quad (10.5d)$$

The eigenfunctions $|\psi_2\rangle$ and $|\psi_3\rangle$ are linear combinations of the singlet ($|S\rangle$; $S=0$) and the triplet ($|T_0\rangle$; $S=1$) wavefunctions, both with $M_S=0$:

$$|S\rangle = \frac{1}{\sqrt{2}}(|\alpha_A\beta_B\rangle - |\beta_A\alpha_B\rangle) \quad (10.6a)$$

$$|T_0\rangle = \frac{1}{\sqrt{2}}(|\alpha_A\beta_B\rangle + |\beta_A\alpha_B\rangle) \quad (10.6b)$$

The coefficients that weigh the singlet and triplet contributions in the eigenfunctions (10.5b) and (10.5c) depend on the strength of the dipolar and exchange couplings between the two radicals,

$$\sin(2\delta) = \frac{\Delta\omega}{\Omega} \qquad \cos(2\delta) = \frac{J+d}{\hbar\Omega} \quad (10.7)$$

$$\Omega = \sqrt{\frac{1}{\hbar^2}(J+d)^2 + (\Delta\omega)^2} \quad (10.8)$$

with $d = D_{zz}/3$, as well as on the difference in the Larmor precession frequencies of the two spins:

$$\omega_0 = \frac{1}{2}(\omega_A + \omega_B) \qquad \Delta\omega = \frac{1}{2}(\omega_A - \omega_B) \quad (10.9)$$

$$\omega_i = \frac{\mu_B}{\hbar} g_{zz,i} B; \quad i \in \{A, B\} \quad (10.10)$$

The energy eigenvalues are depicted graphically in Figure 10.9. For a singlet-born radical pair, only the eigenstates $|\psi_2\rangle$ and $|\psi_3\rangle$ are populated according to their weighting factors of the singlet wavefunction $|S\rangle$. The eigenstates $|\psi_1\rangle$ and $|\psi_4\rangle$ remain unpopulated because they are pure triplet states. EPR transitions are allowed between eigenstates with different triplet characters. This gives rise to four resonances, two of which are emissive and the other two absorptive. In a nonoriented sample, all individual orientations of the radical pair, i.e., its $\boldsymbol{g}$- and $\boldsymbol{D}$-tensors, with respect to $\boldsymbol{B}$ need to be taken into account, and the resulting single-orientation trEPR spectra summed up to a spectral powder pattern that may look complicated in shape, see, e.g., the trEPR spectra of Figure 10.5, but that can be analyzed by spectral simulations. Integration of the trEPR signal amplitude over the entire range of resonances yields net-zero electron-spin polarization for a singlet-born radical pair.[75] For a triplet-born radical pair, the situation is more complex. In this case, in principle, all four eigenstates may be populated, depending on the populations p_X, p_Y, and p_Z of the eigenfunctions of the triplet precursor. Typically, either net-absorptive or net-emissive electron-spin polarization is then observed on integrating the trEPR

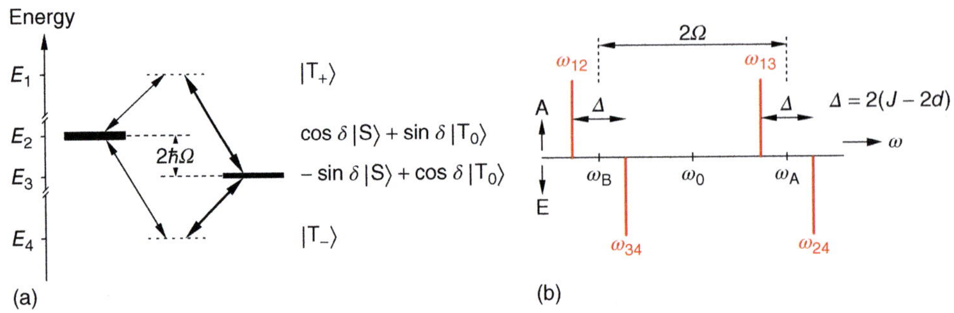

Figure 10.9. (a) Energy eigenvalues and corresponding eigenfunctions of a spin-correlated coupled radical pair, $A^{\bullet+}B^{\bullet-}$. The populations of the eigenstates are shown for a singlet precursor of radical pair formation. The thickness of the horizontal lines indicates their relative population. (b) Schematic trEPR spectrum of a single radical pair with one specific orientation with respect to the direction of the external magnetic field $\boldsymbol{B}$. Positive and negative signal amplitudes signify (enhanced) absorptive and emissive electron spin polarization, respectively

signal amplitude over the entire magnetic-field range (Figure 10.10).[77,78]

trEPR fully proves its benefits in the characterization of short-lived pairs of correlated spins in proteins. The wealth of information drawn from trEPR studies on the radical pair states formed in the course of photosynthetic charge separation has been summarized in a number of excellent reviews.[3–7] In this chapter, the focus is set on paramagnetic species generated in proteins of the photolyase/cryptochrome family, which are blue-light active by virtue of their flavin adenine dinucleotide (FAD) cofactor.[79,80] Similar to the quinone electron acceptor in photosynthetic charge separation, the 7,8-dimethyl isoalloxazine moiety of FAD can assume three different redox and various protonation states that are biologically relevant in the diverse functions of this protein family: fully oxidized (FAD^{ox}), one-electron reduced ($FAD^{\bullet-}$/$FADH^{\bullet}$), and two-electron (fully) reduced ($FADH^-$/$FADH_2$).[80]

Cryptochromes are involved in a number of biological processes, some of which are light dependent.[81–83] Magnetoreception is one of the most intensely discussed light-dependent processes, in which cryptochromes are thought to play a prominent role in some animals that use light to perceive direction derived from the Earth's magnetic field.[84–86] The FAD cofactor is believed to be essential for magnetic sensing. Upon photoexcitation with blue light, cryptochromes readily form radical pairs,[76,87] via an electronically excited state of FAD^{ox}, by electron abstraction from a nearby tryptophan that is part of a cascade of three (and in some cases even four) conserved tryptophan residues, often called Trp_A, Trp_B, and Trp_C (and in the case of a tetrad, Trp_D), in the order of increasing distance from the FAD (see Figure 10.11a). Trp_C (or alternatively Trp_D) is in most cryptochrome proteins close to the protein surface and, once one-electron oxidized as cationic $Trp_{C/D}{}^{\bullet+}$ or deprotonated neutral $Trp_{C/D}{}^{\bullet}$, mediates electron transfer from exogenous electron donors. Together with the FAD radical (neutral, $FADH^{\bullet}$, or anionic, $FAD^{\bullet-}$), the $Trp_{C/D}$ radical forms a pair of coupled radicals. This radical pair is generated in a spin-correlated manner because the initial electron-transfer steps $Trp_A \rightarrow FAD^{ox}$, $Trp_B \rightarrow Trp_A{}^{\bullet+}$, and $Trp_C \rightarrow Trp_B{}^{\bullet+}$ (and possibly also $Trp_D \rightarrow Trp_C{}^{\bullet+}$) are completed within tens of picoseconds,[89] which is much faster than decoherence to become efficient; thus, the precursor multiplicity of FAD^{ox}, excited singlet or triplet, from which radical pair formation started out, is preserved in the terminal radical pair state. Correlated spin pairs of that type were predicted by Klaus Schulten to be essential in a magnet compass that migratory birds may use to navigate along the magnetic-field lines of the Earth (Figure 10.12).[91–93] In 2000, cryptochrome was put forward by Schulten and Ritz as the specific protein responsible for the birds' sense for magnetic fields.[94] In brief, this hypothesis is based on the fact that singlet- and triplet-born radical pairs, $^1[A^{\bullet+}\cdots B^{\bullet-}]$ and $^3[A^{\bullet+}\cdots B^{\bullet-}]$, respectively, have different fates: recombination to form the diamagnetic ground-state molecules (AB) can only take place from the singlet state but not from the triplet manifold because this process is spin forbidden in the latter.[90] The coherent mixing, i.e., the interconversion between the different multiplicities of the radical pair state is governed by an external magnetic field $\boldsymbol{B}$. The strength of this field therefore has a decisive influence on the yield of radicals.

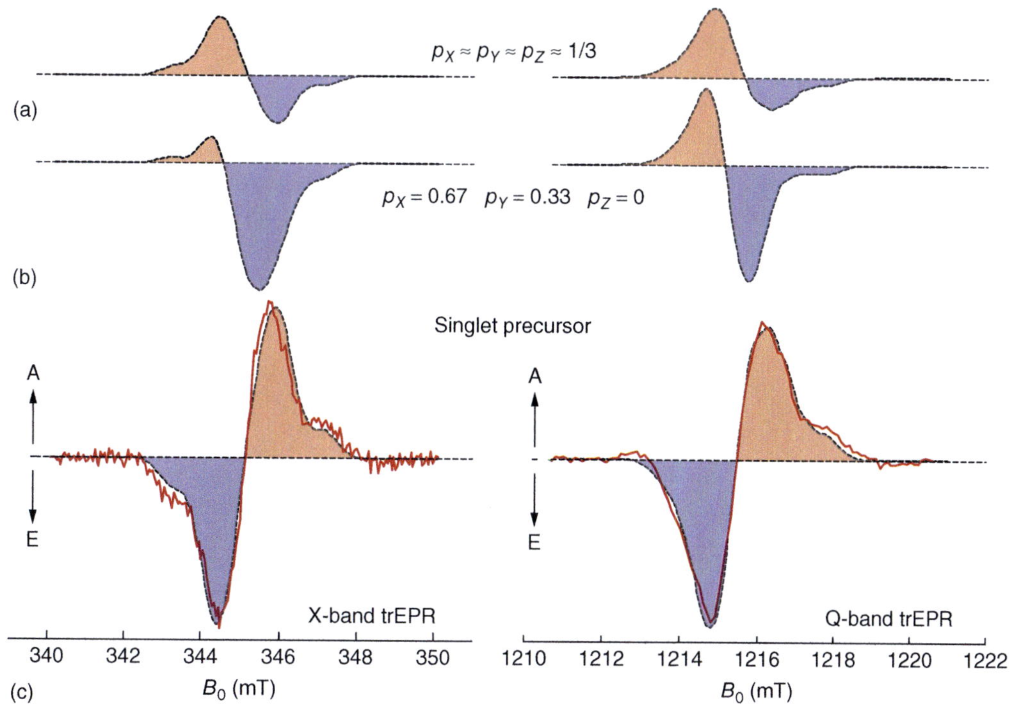

Figure 10.10. trEPR spectra of the $FAD^{•}\cdots Trp_C^{•}$ radical pair of *Xenopus laevis* cryptochrome recorded at 274 K at 9.68 GHz (X-band, left)[76] and 272 K at 34.08 GHz (Q-band, right). (c) Experimental spectra (red curves) recorded at 1 μs (X-band) and 130 ns (Q-band) after pulsed laser excitation (460 nm). Spectral simulations based on a pure singlet-precursor state are superimposed as dashed black curves (c). Simulations of radical pair spectra generated from an FAD triplet precursor with zero-field spin-state populations at thermal equilibrium ($p_X \approx p_Y \approx p_Z \approx 1/3$) and from a spin-polarized triplet state ($p_X = 0.67, p_Y = 0.33, p_Z = 0$) (FAD triplet parameters $D/h = 1.919$ GHz and $E/h = -524.9$ MHz) are shown in panels (a) and (b), respectively. For a singlet-born radical pair, emissive (E, blue-shaded area) and absorptive (A, red-shaded area) polarizations of transitions cancel to zero upon integration over the entire magnetic-field range at X-band and at Q-band microwave frequencies (c). For radical pairs generated from triplet precursor states, net emissive or absorptive spin polarization is expected in at least one microwave-frequency band. (Reprinted with permission from S. Weber, T. Biskup, A. Okafuji, A. R. Marino, T. Berthold, G. Link, K. Hitomi, E. D. Getzoff, E. Schleicher and J. R. Norris, J. Phys. Chem. B, 2010, 114, 14745–14754. 2010 American Chemical Society)

Several conditions need to be met in order to make cryptochrome an efficient photo-magnetoreceptor:[90] Radical pairs (i) are required to be born in a spin-correlated manner, (ii) they should not be too short lived, (iii) they must react in a spin-selective way, and (iv) should exhibit reasonably slow electron-spin relaxation. Furthermore, (v) the mutual coupling of the two unpaired electron spins, dipolar and exchange, should be small as compared to the strength of the external magnetic field that needs to be sensed and (vi) at least one of the electron spins needs to be anisotropically hyperfine coupled to one or several nearby magnetic nuclei. Most of these properties may be favorably probed by trEPR spectroscopy, even under physiologically relevant conditions, but not by pulse EPR, the application of which is mostly restricted to the frozen protein state.

The precursor state and hence, the multiplicity of the radical pair, singlet versus triplet, can be nicely probed by trEPR at two or more different microwave frequency bands, e.g., at X- and Q-band, and correspondingly different magnetic field ranges.[77] A singlet-born spin-correlated radical pair always exhibits net-zero electron-spin polarization, i.e., integrated over the entire magnetic-field range covering all resonances, emissive, and enhanced

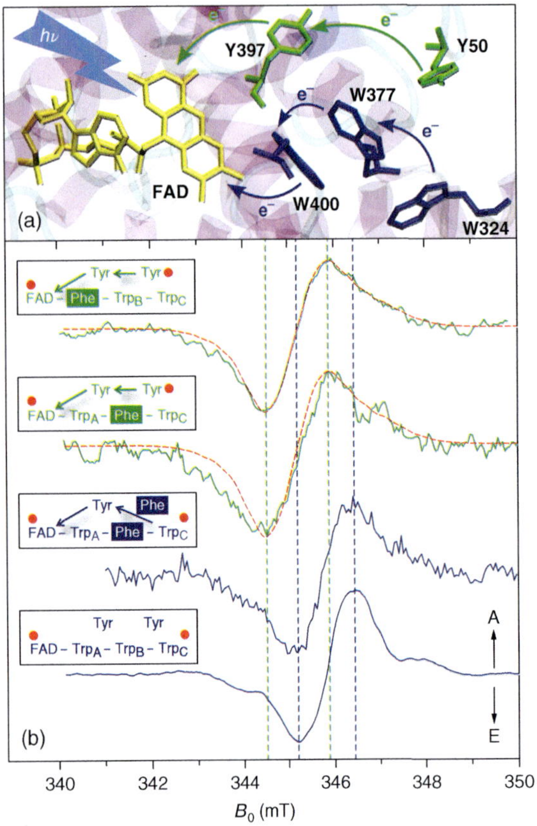

Figure 10.11. trEPR spectra of radical pairs in wild-type and mutant *Xenopus laevis* cryptochromes recorded at 274 K. From (a) to (b): W400F (=Trp_A mutant), W377F (=Trp_B mutant), W377F/Y50F (=Trp_B+Tyr double mutant), and wild type. Blue curves: $FAD^{\bullet}\cdots Trp_C^{\bullet}$ radical pair spectra; green curves: $FAD^{\bullet}\cdots Tyr50^{\bullet}$ radical pair spectra. The latter are slightly shifted toward lower magnetic fields because of the higher ***g*** principal values of tyrosine radicals as compared to tryptophan radicals. The dashed lines show spectral simulations of the $FAD^{\bullet}\cdots Tyr50^{\bullet}$ radical pair state, which were performed using the ***g***-tensor components of flavin and tyrosine radicals, and spin–spin couplings (dipolar and exchange) for the radical pair state. (For further details, see Ref. 88)

absorptive transitions of the radical pair state cancel to zero. A triplet-born radical pair, on the other hand, shows nonzero (net) electron spin polarization predominantly in lower magnetic field ranges, and the amount of net polarization changes with the frequency band. An unequivocal assignment may be obtained by spectral simulations of trEPR data. An example is shown in Figure 10.10. trEPR spectra

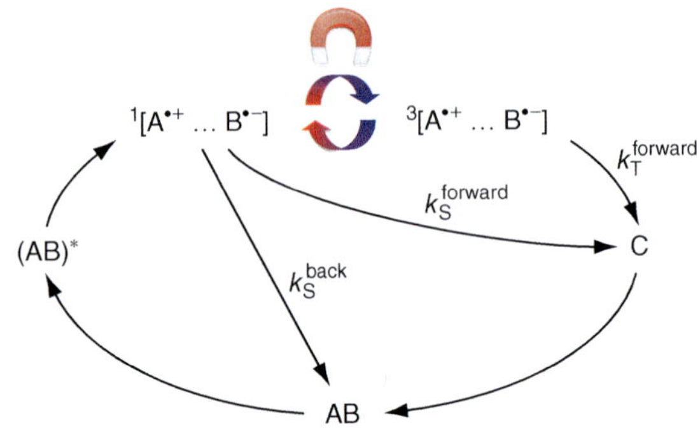

Figure 10.12. Schematic diagram of the functioning of a radical-pair-based magnetic compass. A spin-correlated radical pair is formed from an excited singlet state of a precursor species (AB)*. Interconversion between singlet- and triplet-configured radical pairs is governed by an external magnetic field. By backward electron transfer (k_S^{back}), only the singlet-configured radical pair can return to the ground state, AB. The respective return to the ground state is spin-forbidden for the radical pair in the triplet configuration. Therefore, further reaction steps, such as subsequent electron or proton transfer, to stabilize the radical pair state and generate a signaling state (here denoted C) are more likely to take place from the triplet radical pair ($k_T^{forward}$) than from the singlet radical pair ($k_S^{forward}$). In a magnetic compass, an applied magnetic field can alter the yield of the signaling state by regulating the competition between its formation and the return to the ground state, AB. (Adapted with permission from Ref. 90. © National Academy of Sciences of the United States of America, 2009)

of light-generated spin-correlated radical pairs in *Xenopus laevis* cryptochrome recorded at X- and Q-band microwave frequencies are shown, together with calculated spectra.[77] The agreement between experiment and simulation is best with a singlet-born radical pair. Hence, by trEPR, it can be demonstrated that electron transfer in cryptochromes starts out from an excited singlet state of FAD^{ox}, and this spin configuration is preserved in subsequent sequential electron transfer. This shows that each individual electron transfer step must be fast so that singlet-to-triplet interconversion is inefficient in the early radical pair states.

The radical pair trEPR spectra are determined by the ***g***-tensors of both radicals, their mutual coupling (exchange and dipolar), and also hyperfine couplings arising from magnetic nuclei in direct vicinity to the paramagnetic moieties. trEPR performed at high magnetic fields and correspondingly high microwave frequencies may help to spread out resonances along the magnetic field domain, thereby making determination of magnetic-resonance parameters more reliable.[32,95,96] But even with X-band trEPR, different types of radical pairs can be distinguished, which is shown in panel B of Figure 10.11,[88] provided that most of the abovementioned parameters (**g**-tensor components, spin–spin couplings, and inhomogeneous linewidths) are known from independent experiments. In such cases, FAD–tryptophan and FAD–tyrosine radical pairs can be discerned, and, by examination of mutant proteins, alternative electron transfer paths unraveled. The latter may serve as backup channels in cryptochrome mutants, in which the conserved tryptophan triad is interrupted. In *Xenopus laevis* cryptochrome, for example, tyrosines act as stepping stones to maintain FAD cofactor photoreduction in cases when electron transfer along the conventional tryptophan cascade is blocked.

In very weakly coupled radical pairs, where contributions of dipolar and exchange interactions to inhomogeneous line broadening are weak, also hyperfine couplings of the individual radicals to magnetic nuclei in their vicinity may be resolved that allow gaining insight into the electronic structure of the protein's functional relevant working states (Figure 10.13). In *Drosophila melanogaster* cryptochrome, for example, a protein with proven magnetoreceptor capacity,[85] the conserved electron-transfer cascade of three tryptophans is extended by a fourth one, Trp_D, that prolongs the electron transfer channel and forms, together with the FAD, a weakly coupled radical pair state $FAD^{\bullet -}\cdots Trp_D^{\bullet(+)}$. A mutant protein, in which Trp_D is replaced with phenylalanine, is still

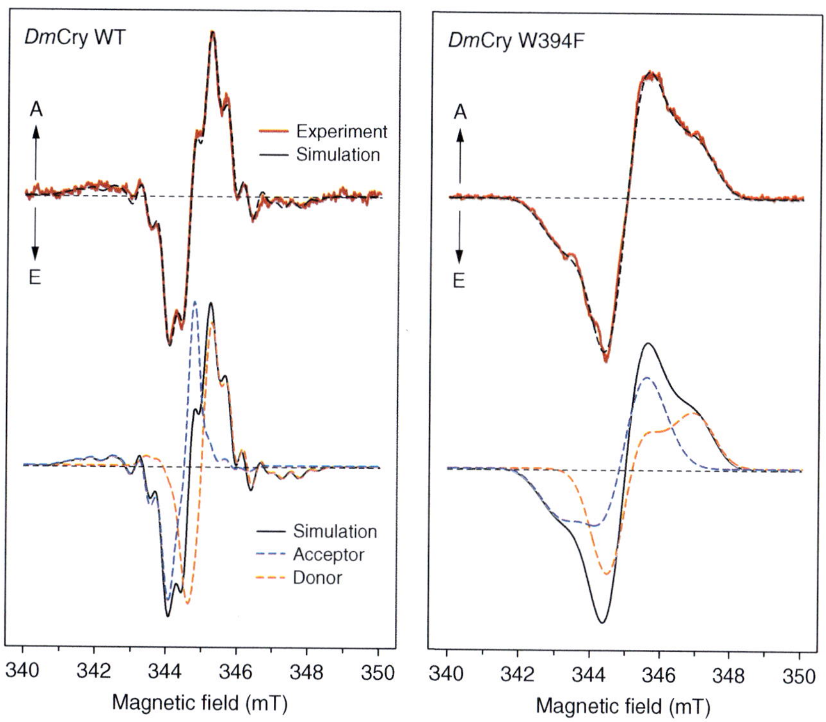

Figure 10.13. trEPR spectra of wild-type (left) and W394F (=Trp_D mutant) (right) *Drosophila melanogaster* cryptochromes recorded at 274 K. A and E arrows indicate enhanced absorptive and emissive signals, respectively. Experimental spectra (red curves, top) are superimposed with spectral simulations (dashed black curves) based on an FAD singlet-state precursor: the wild-type and the mutant spectra have been calculated using parameters (**g**-tensors of FAD and tryptophan radicals, dipolar and exchange couplings, and selected proton hyperfine couplings of the Trp radical) of an $FAD^{\bullet -}\cdots Trp_D^{\bullet}$ (left) and an $FAD^{\bullet -}\cdots Trp_C^{\bullet}$ (right) radical pair, respectively. The lower panels show the contributions of the donor (Trp) and acceptor (FAD) radicals to the overall radical pair spectra. (Reproduced with permission from Ref. 43. © Elsevier, 2016)

photoactive but forms an $FAD^{\bullet -}\cdots{}^{\bullet}Trp_C^{\bullet(+)}$ radical pair, whose trEPR spectrum does not exhibit any hyperfine structure owing to increased broadening by dipolar and exchange interactions.[43] The fact that animal cryptochromes utilize four instead of three stepping stones for the formation of spin-correlated radical pairs with a microsecond lifetime may shed some light on the still unsolved puzzle of the animal magnet compass. Charge separation over longer distances is certainly useful to slow down backward electron transfer that would regenerate the diamagnetic ground state and to decrease mutual spin–spin coupling,[97,98] which may render a radical-pair-based magnet sensor more responsive to the weak magnetic fields of the Earth. Clearly, trEPR with its high temporal and spectral resolution can contribute to unraveling the working states of this fascinating class of flavoproteins under physiologically relevant conditions.

10.4 CONCLUSIONS

trEPR can be applied to study photoinduced paramagnetic species, such as radical pairs and molecular triplet states, in real time in a time-resolved manner with nanosecond to microsecond temporal resolution. Following pulsed photoexcitation with a laser, the detection of resonant transitions is performed in the presence of a weak microwave magnetic field, thus rendering the method less sensitive to fast relaxation as compared to pulse EPR methods. Hence, short-lived radical pairs generated by light-induced electron transfer in proteins may be studied under physiological conditions, i.e., in aqueous solution at ambient temperatures. In addition, more strongly coupled spin states with higher spin quantum number, $S \geq 1$ (e.g., molecular triplet and quartet states), can be characterized at more elevated temperatures as compared to pulse EPR methods. trEPR has proven to provide useful information on the identity of radical pair partners, their mutual spin–spin couplings, hyperfine interactions of the paramagnetic moieties to nearby magnetic nuclei, and the spin multiplicity of radical pairs generated, e.g., in charge separation of photosynthesis and the blue-light receptor cryptochrome as well as in blends used in organic photovoltaic devices, in which morphology effects and their influence on cell efficiency may also be examined.

ACKNOWLEDGMENTS

Over the past years, a number of coworkers, BSc, MSc, Diploma, and PhD students, postdocs as well as colleagues from all over the world, whose names appear in the respective references, have contributed to the work presented here. To all of them, I want to express my gratitude. I am furthermore grateful to Till Biskup for critically reading this manuscript. This work was supported by the Deutsche Forschungsgemeinschaft (DFG, FOR 1976).

REFERENCES

1. S. S. Kim and S. I. Weissman, *J. Magn. Reson.*, 1976, **24**, 167.
2. S. I. Weissman, *Annu. Rev. Phys. Chem.*, 1982, **33**, 301.
3. D. Stehlik, C. H. Bock, and M. C. Thurnauer, in Advanced EPR. Applications in Biology and Biochemistry, ed. A. J. Hoff, Elsevier: Amsterdam, Oxford, New York, Tokyo, 1989, ch. 11, 371.
4. H. Levanon, in Biophysical Techniques in Photosynthesis, eds J. Amesz and A. J. Hoff, Kluwer Academic Publishers: Dordrecht, 1996, Vol. **3**, 211.
5. G. Kothe, J. R. Norris, O. G. Poluektov, and M. C. Thurnauer, in Biophysical Techniques in Photosynthesis, eds T. J. Aartsma and J. Matysik, Springer: Dordrecht, 2008, Vol. **26**, ch. 15, 305.
6. A. van der Est, *Photosynth. Res.*, 2009, **102**, 335.
7. G. Kothe and M. C. Thurnauer, *Photosynth. Res.*, 2009, **102**, 349.
8. M. D. E. Forbes, L. E. Jarocha, S. Sim, and V. F. Tarasov, *Adv. Phys. Org. Chem.*, 2013, **47**, 1.
9. G. Kothe, S. Weber, R. Bittl, E. Ohmes, M. C. Thurnauer, and J. R. Norris, *Chem. Phys. Lett.*, 1991, **186**, 474.
10. G. Kothe, S. Weber, E. Ohmes, M. C. Thurnauer, and J. R. Norris, *J. Phys. Chem.*, 1994, **98**, 2706.
11. K. Laukenmann, S. Weber, G. Kothe, C. Oesterle, A. Angerhofer, M. R. Wasielewski, W. A. Svec, and J. R. Norris, *J. Phys. Chem.*, 1995, **99**, 4324.
12. R. Bittl, A. van der Est, A. Kamlowski, W. Lubitz, and D. Stehlik, *Chem. Phys. Lett.*, 1994, **226**, 349.
13. G. Link, T. Berthold, M. Bechtold, J.-U. Weidner, E. Ohmes, J. Tang, O. Poluektov, L. Utschig, S. L. Schlesselman, M. C. Thurnauer, and G. Kothe, *J. Am. Chem. Soc.*, 2001, **123**, 4211.

14. U. Heinen, L. M. Utschig, O. G. Poluektov, G. Link, E. Ohmes, and G. Kothe, *J. Am. Chem. Soc.*, 2007, **129**, 15935.

15. R. Furrer, F. Fujara, C. Lange, D. Stehlik, H. M. Vieth, and W. Vollmann, *Chem. Phys. Lett.*, 1980, **75**, 332.

16. R. Furrer and M. C. Thurnauer, *Chem. Phys. Lett.*, 1981, **79**, 28.

17. S. Weber, E. Ohmes, M. C. Thurnauer, J. R. Norris, and G. Kothe, *Proc. Natl. Acad. Sci. U. S. A.*, 1995, **92**, 7789.

18. S. Weber, T. Berthold, E. Ohmes, M. C. Thurnauer, J. R. Norris, and G. Kothe, *Appl. Magn. Reson.*, 1996, **11**, 461.

19. S. K. Misra, Multifrequency Electron Paramagnetic Resonance. Theory and Applications, Wiley-VCH Verlag & Co. KGaA: Weinheim, 2011.

20. A. Schweiger and G. Jeschke, Principles of Pulse Electron Paramagnetic Resonance, Oxford University Press: Oxford, 2001.

21. W. Froncisz and J. S. Hyde, *J. Magn. Reson.*, 1982, **47**, 515.

22. A. Webb, *Prog. Nucl. Magn. Reson. Spectrosc.*, 2014, **83**, 1.

23. J. S. Hyde and W. Froncisz, in Advanced EPR. Applications in Biology and Biochemistry, ed. A. J. Hoff, Elsevier: Amsterdam, Oxford, New York, Tokyo, 1989, ch. 7, 277.

24. M. Dutka, T. Oleś, M. Mossakowski, and W. Froncisz, *J. Magn. Reson.*, 2011, **210**, 44.

25. G. R. Eaton, S. S. Eaton, D. P. Barr, and R. T. Weber, Quantitative EPR, Springer-Verlag: Wien, New York, 2010.

26. C. P. Poole Jr, Electron Spin Resonance. A Comprehensive Treatise on Experimental Techniques, Dover Publications, Inc.: Mineola, New York, 1983.

27. P. W. Atkins, K. A. McLauchlan, and P. W. Percival, *Mol. Phys.*, 1973, **25**, 281.

28. K. Möbius, A. Savitsky, A. Schnegg, M. Plato, and M. Fuchs, *Phys. Chem. Chem. Phys.*, 2005, **7**, 19.

29. M. Bortolus, M. Prato, J. van Tol, and A. L. Maniero, *Chem. Phys. Lett.*, 2004, **398**, 228.

30. J. van Tol, L.-C. Brunel, and A. Angerhofer, *Appl. Magn. Reson.*, 2001, **21**, 335.

31. R. Zeng, J. van Tol, A. Deal, H. A. Frank, and D. E. Budil, *J. Phys. Chem. B*, 2003, **107**, 4624.

32. A. van der Est, T. Prisner, R. Bittl, P. Fromme, W. Lubitz, K. Möbius, and D. Stehlik, *J. Phys. Chem. B*, 1997, **101**, 1437.

33. A. Savitsky, M. Galander, and K. Möbius, *Chem. Phys. Lett.*, 2001, **340**, 458.

34. A. Savitsky and K. Möbius, *Helv. Chim. Acta*, 2006, **89**, 2544.

35. S. Yamauchi, M. Tanabe, K. Takahashi, S. Islam, H. Matsuoka, and Y. Ohba, *Appl. Magn. Reson.*, 2010, **37**, 317.

36. H. Moons, E. Goovaerts, V. P. Gubskaya, I. A. Nuretdinov, C. Corvaja, and L. Franco, *Phys. Chem. Chem. Phys.*, 2011, **13**, 3942.

37. A. Marchanka, A. Savitsky, W. Lubitz, K. Möbius, and M. van Gastel, *J. Phys. Chem. B*, 2010, **114**, 14364.

38. T. Berthold, E. D. von Gromoff, S. Santabarbara, P. Stehle, G. Link, O. Poluektov, P. Heathcote, C. F. Beck, M. C. Thurnauer, and G. Kothe, *J. Am. Chem. Soc.*, 2012, **134**, 5563.

39. J. van Tol, L.-C. Brunel, and R. J. Wylde, *Rev. Sci. Instrum.*, 2005, **76**, Art.-No. 074101.

40. N. J. Turro, M. H. Kleinman, and E. Karatekin, *Angew. Chem. Int. Ed.*, 2000, **39**, 4436.

41. A. van der Est, *Biochim. Biophys. Acta*, 2001, **1507**, 212.

42. I. I. Rabi, N. F. Ramsey, and J. Schwinger, *Rev. Mod. Phys.*, 1954, **26**, 167.

43. D. Nohr, S. Franz, R. Rodriguez, B. Paulus, L.-O. Essen, S. Weber, and E. Schleicher, *Biophys. J.*, 2016, **111**, 301.

44. M. Fuki, H. Murai, H. Tachikawa, and Y. Kobori, *J. Phys. Chem. B*, 2016, **120**, 4365.

45. B. Ferlez, W. Dong, R. Siavashi, K. Redding, H. J. M. Hou, J. H. Golbeck, and A. van der Est, *J. Phys. Chem. B*, 2015, **119**, 13714.

46. R. Fudim, J. Mehlhorn, T. Berthold, S. Weber, E. Schleicher, J. T. M. Kennis, and T. Mathes, *FEBS J.*, 2015, **282**, 3161.

47. C. Engelhard, X. Wang, D. Robles, J. Moldt, L.-O. Essen, A. Batschauer, R. Bittl, and M. Ahmad, *Plant Cell*, 2014, **26**, 4519.

48. F. Conti, L. Panzella, A. Napolitano, M. d'Ischia, and A. Toffoletti, *Spectroscopy*, 2010, **24**, 289.

49. G. Zhu, Y. Wang, H. Fu, X. Xu, Z. Cui, X. Ji, and G. Wu, *Spectrochim. Acta Part A Mol. Biomol. Spectrosc.*, 2015, **137**, 148.

50. A. van der Est and P. K. Poddutoori, *Appl. Magn. Reson.*, 2013, **44**, 301.
51. Y. E. Kandrashkin, P. K. Poddutoori, and A. van der Est, *Appl. Magn. Reson.*, 2016, **47**, 511.
52. M. T. Colvin, R. Carmieli, T. Miura, S. Richert, D. M. Gardner, A. L. Smeigh, S. M. Dyar, S. M. Conron, M. A. Ratner, and M. R. Wasielewski, *J. Phys. Chem. A*, 2013, **117**, 5314.
53. M. T. Colvin, A. B. Ricks, A. M. Scott, D. T. Co, and M. R. Wasielewski, *J. Phys. Chem. A*, 2012, **116**, 1923.
54. M. Fathalla, J. C. Barnes, R. M. Young, K. J. Hartlieb, S. M. Dyar, S. W. Eaton, A. A. Sarjeant, D. T. Co, M. R. Wasielewski, and J. F. Stoddart, *Chem. Eur. J.*, 2014, **20**, 14690.
55. S. M. Dyar, E. A. Margulies, B. A. Horwitz, K. E. Brown, M. D. Krzyaniak, and M. R. Wasielewski, *J. Phys. Chem. B*, 2015, **119**, 13560.
56. F. Kraffert, R. Steyrleuthner, S. Albrecht, D. Neher, M. C. Scharber, R. Bittl, and J. Behrends, *J. Phys. Chem. C*, 2014, **118**, 28482.
57. C. E. Tait, P. Neuhaus, H. L. Anderson, and C. R. Timmel, *J. Am. Chem. Soc.*, 2015, **137**, 6670.
58. C. E. Tait, P. Neuhaus, M. D. Peeks, H. L. Anderson, and C. R. Timmel, *J. Am. Chem. Soc.*, 2015, **137**, 8284.
59. T. Biskup, M. Sommer, S. Rein, D. L. Meyer, M. Kohlstädt, U. Würfel, and S. Weber, *Angew. Chem. Int. Ed.*, 2015, **54**, 7707.
60. N. E. Horwitz, B. T. Phelan, J. N. Nelson, M. D. Krzyaniak, and M. R. Wasielewski, *J. Phys. Chem. A*, 2016, **120**, 2841.
61. C. E. Tait, P. Neuhaus, M. D. Peeks, H. L. Anderson, and C. R. Timmel, *Phys. Chem. Chem. Phys.*, 2016, **18**, 5275.
62. T. Miura, M. Aikawa, and Y. Kobori, *J. Phys. Chem. Lett.*, 2014, **5**, 30.
63. F. Conti, E. Negro, V. Di Noto, G. Elger, T. Berthold, and S. Weber, *Int. J. Hydrogen Energy*, 2012, **37**, 6317.
64. S. Richert, M. D. Peeks, C. E. Tait, H. L. Anderson, and C. R. Timmel, *Phys. Chem. Chem. Phys.*, 2016, **18**, 24171.
65. T. Miura, A. Kageyama, S. Torii, and H. Murai, *J. Phys. Chem. B*, 2010, **114**, 14550.
66. J. H. van der Waals and M. S. de Groot, in The Triplet State, ed A. B. Zahlan, Cambridge University Press: Cambridge, 1967, 101.
67. O. G. Poluektov, J. Niklas, K. L. Mardis, S. Beaupré, M. Leclerc, C. Villegas, S. Erten-Ela, J. L. Delgado, N. Martín, A. Sperlich, and V. Dyakonov, *Adv. Energy Mater.*, 2014, **4**, Art.-No. 1301517.
68. A. Köhler and H. Bässler, *Mater. Sci. Eng. R-Rep.*, 2009, **66**, 71.
69. A. Rao, P. C. Y. Chow, S. Gélinas, C. W. Schlenker, C.-Z. Li, H.-L. Yip, A. K.-Y. Jen, D. S. Ginger, and R. H. Friend, *Nature*, 2013, **500**, 435.
70. J. W. Kingsley, P. P. Marchisio, H. Yi, A. Iraqi, C. J. Kinane, S. Langridge, R. L. Thompson, A. J. Cadby, A. J. Pearson, D. G. Lidzey, R. A. L. Jones, and A. J. Parnell, *Sci. Rep.*, 2014, **4**, Art.-No. 5286.
71. N. E. Jackson, B. M. Savoie, T. J. Marks, L. X. Chen, and M. A. Ratner, *J. Phys. Chem. Lett.*, 2015, **6**, 77.
72. J. Behrends, A. Sperlich, A. Schnegg, T. Biskup, C. Teutloff, K. Lips, V. Dyakonov, and R. Bittl, *Phys. Rev. B*, 2012, **85**, Art.-No. 125206.
73. F. Kraffert, R. Steyrleuthner, C. Meier, R. Bittl, and J. Behrends, *Appl. Phys. Lett.*, 2015, **107**, Art.-No. 043302.
74. D. Stehlik, C. H. Bock, and J. Petersen, *J. Phys. Chem.*, 1989, **93**, 1612.
75. P. J. Hore, in Advanced EPR in Biology and Biochemistry, ed A. J. Hoff, Elsevier: Amsterdam, 1989, 405.
76. T. Biskup, E. Schleicher, A. Okafuji, G. Link, K. Hitomi, E. D. Getzoff, and S. Weber, *Angew. Chem. Int. Ed.*, 2009, **48**, 404.
77. S. Weber, T. Biskup, A. Okafuji, A. R. Marino, T. Berthold, G. Link, K. Hitomi, E. D. Getzoff, E. Schleicher, and J. R. Norris, *J. Phys. Chem. B*, 2010, **114**, 14745.
78. Y. Kobori, M. Fuki, and H. Murai, *J. Phys. Chem. B*, 2010, **114**, 14621.
79. A. Sancar, *Chem. Rev.*, 2003, **103**, 2203.
80. S. Weber, *Biochim. Biophys. Acta*, 2005, **1707**, 1.
81. I. Chaves, R. Pokorny, M. Byrdin, N. Hoang, T. Ritz, K. Brettel, L.-O. Essen, G. T. J. van der Horst, A. Batschauer, and M. Ahmad, *Annu. Rev. Plant Biol.*, 2011, **62**, 335.
82. C. Lin and D. Shalitin, *Annu. Rev. Plant Biol.*, 2003, **54**, 469.
83. C. Lin and T. Todo, *Genome Biol.*, 2005, **6**, Art.-No. 220.
84. M. Liedvogel and H. Mouritsen, *J. R. Soc. Interface*, 2010, **7**, S147.

85. R. J. Gegear, A. Casselman, S. Waddell, and S. M. Reppert, *Nature (London)*, 2008, **454**, 1014.
86. R. J. Gegear, L. E. Foley, A. Casselman, and S. M. Reppert, *Nature (London)*, 2010, **463**, 804.
87. T. Biskup, *Mol. Phys.*, 2013, **111**, 3698.
88. T. Biskup, B. Paulus, A. Okafuji, K. Hitomi, E. D. Getzoff, S. Weber, and E. Schleicher, *J. Biol. Chem.*, 2013, **288**, 9249.
89. K. Brettel and M. Byrdin, *Proc. Natl. Acad. Sci. U. S. A.*, 2012, **109**, E1462.
90. C. T. Rodgers and P. J. Hore, *Proc. Natl. Acad. Sci. U. S. A.*, 2009, **106**, 353.
91. K. Schulten, C. E. Swenberg, and A. Weller, *Z. Phys. Chem.*, 1978, **111**, 1.
92. K. Schulten, in Festkörperprobleme, ed J. Treusch, Vieweg: Braunschweig, 1982, Vol. **22**, 61.
93. K. Schulten and A. Windemuth, in Proceedings in Physics, Vol. 11: Biophyscial Effects of Steady Magnetic Fields, eds G. Maret, N. Boccara and J. Kiepenheuer, Springer: Berlin, 1986, 99.
94. T. Ritz, S. Adem, and K. Schulten, *Biophys. J.*, 2000, **78**, 707.
95. J. Tang, L. M. Utschig, O. Poluektov, and M. C. Thurnauer, *J. Phys. Chem. B*, 1999, **103**, 5145.
96. A. Savitsky, O. Gopta, M. Mamedov, J. H. Golbeck, A. Tikhonov, K. Möbius, and A. Semenov, *Appl. Magn. Reson.*, 2010, **37**, 85.
97. O. Efimova and P. J. Hore, *Biophys. J.*, 2008, **94**, 1565.
98. C. R. Timmel, U. Till, B. Brocklehurst, K. A. McLauchlan, and P. J. Hore, *Mol. Phys.*, 1998, **95**, 71.

Chapter 11
Pulse EPR

Stefan Stoll
University of Washington, Seattle, WA, USA

11.1 INTRODUCTION

EPR spectroscopy comes in two main experimental realizations: continuous-wave (CW) EPR and pulse EPR. In pulse EPR experiments, the spins of the paramagnetic centers (PCs) in the sample are typically excited via short high-power microwave (MW) pulses of tens to hundreds of nanoseconds in length, and the spins' weak transient emissive response following these pulses is recorded. In addition to allowing the recording of field-swept EPR spectra analogous to CW EPR, pulse EPR methods can resolve small interactions whose splittings are unresolved in field-swept EPR spectra, such as dipolar couplings between nearby electron spins (inter-spin distance < approx. 8 nm) or hyperfine couplings between an electron spin and neighboring magnetic nuclei (distance < approx. 0.8 nm). Pulse EPR also allows the disentanglement of overlapping signals by providing multiple ways of spreading out spectra into a second dimension. Pulse EPR can also be used to measure relaxation times. Due to its multitude of pulse sequences, pulse EPR is very versatile and has a wide range of different applications. This chapter introduces the basics of pulse EPR. It is intended for novices to pulse EPR, but assumes familiarity with CW EPR (see Chapter 1).

Experimentally, pulse EPR differs from the more common CW EPR in several major respects. Table 11.1 summarizes these differences. In CW EPR, MW irradiation of the sample is continuous and low-power, whereas in pulse EPR it is pulsed and high-power. Similarly to CW EPR, pulse EPR employs resonators to amplify the MW fields at the sample, but the resonators differ in their frequency bandwidth: For CW EPR, narrow-band resonators are used, whereas for pulse EPR, resonators with wider bandwidths are required. In CW EPR, absorption by the spins in the sample is detected (indirectly via a change in the power reflection from the resonator – see Chapter 1), whereas in pulse EPR, emission from the spins in the sample is detected. The detected signal in pulse EPR is directly caused by precessing magnetic moments of excited PCs, which induce an oscillating electric field in the resonator. This then propagates to the receiver, where a quadrature mixer is used to detect signals both in-phase and out-of-phase with respect to a reference frequency. In CW EPR, the magnetic field

EPR Spectroscopy: Fundamentals and Methods.
Edited by Daniella Goldfarb and Stefan Stoll

ISBN: 978-1-119-16299-5
Also published in eMagRes (online edition)
DOI: 10.1002/9780470034590.emrstm1510

Table 11.1. Comparison between CW EPR and pulse EPR

	CW EPR	Pulse EPR
MW irradiation	Continuous (100% on)	Pulses (<0.1% on)
MW power	Low (0.2 W at 9–10 GHz)	High (1 kW at 9–10 GHz)
Measured signal	Continuous absorption (indirectly)	Transient emission (free induction decay, echo)
Resonator bandwidth	Very narrow (a few MHz)	Wider (≈100 MHz)
Resonator coupling	Critically coupled	Overcoupled
MW detection	Diode, in-phase only	Quadrature mixer, in-phase and out-of-phase
Varied quantity	Magnetic field	Magnetic field, time delays, pulse lengths, and/or frequencies

is swept, whereas in pulse EPR, any parameter of the pulse sequence (such as a timing interval, the carrier frequency, or the length of a pulse) can be varied in addition to the magnetic field.

Most commonly, EPR samples contain PCs with one or more electron spins, where different centers are far enough apart from each other so that interactions between them can be neglected and the PCs can be treated as independent. These samples are called magnetically dilute. The vast majority of pulse EPR experiments are performed on solid-state samples such as powders, crystals, frozen solutions, or molecules immobilized in polymer matrices. This is because in magnetically dilute solids spin relaxation times are long enough to allow application of pulses, evolution of the spins, and subsequent detection on a technically feasible time scale of hundreds of nanoseconds to tens of microseconds. Lowering the temperature also prolongs relaxation times. In liquids, except for organic radicals in organic solvents, relaxation times can be too fast for pulse EPR.

11.2 PULSES AND SPINS

In this section, we outline the principles of how a MW pulse interacts with an ensemble of spins in a sample, using a simple vector picture. Chemically identical PCs in a sample usually differ in their resonance frequencies for a variety of reasons, e.g., different molecular orientations relative to the static magnetic field, slightly different molecular structures as a result of small distortions (strains) induced by the local environment, or coupling to magnetic nuclei or electron spins in different quantum states. Each subensemble of noninteracting PCs with close-to-identical resonance frequency is called a spin packet, monochromat, or isochromat, and is associated with a magnetic-moment vector. The total magnetic moment of the sample, $\boldsymbol{M} = (M_x, M_y, M_z)$, is the vector sum of the individual magnetic moments of all spin packets. In pulse EPR, the component of $\boldsymbol{M}$ that precesses in the plane (xy plane) perpendicular to the static magnetic field (along the z axis) is measured. Often, the magnetization (magnetic moment per unit volume) is used instead of the total magnetic moment. Here, we will use the latter.

11.2.1 Classical Dynamics

For a spin system that consists of a single electron spin-1/2 with an isotropic g value, and neglecting relaxation, the dynamics of the magnetic moment (both the magnetic moments of individual spin packets and the total magnetic moment) is described by the classical torque equations

$$\frac{\mathrm{d}}{\mathrm{d}t}\boldsymbol{M} = -\gamma \boldsymbol{B}(t) \times \boldsymbol{M} = -\gamma[\boldsymbol{B}_0 + \boldsymbol{B}_{1\mathrm{lin}}(t)] \times \boldsymbol{M} \quad (11.1)$$

Here, γ is the gyromagnetic ratio of the spin, which is negative for an electron: $\gamma = -g_e\mu_B/\hbar$. $\boldsymbol{B}$ is the total magnetic field, i.e., the sum of the external static magnetic field along the z axis, $\boldsymbol{B}_0$, and the linearly oscillating magnetic field component of the MW irradiation along x, $\boldsymbol{B}_{1\mathrm{lin}}(t) = 2B_1 \cos(\omega_{\mathrm{MW}} t)\boldsymbol{x}$, which is perpendicular to $\boldsymbol{B}_0$ in pulse EPR resonators. $\omega_{\mathrm{MW}} = 2\pi\nu_{\mathrm{MW}}$ is the MW angular frequency. The Bloch equations are extensions of these equations with an additional phenomenological term that models spin relaxation (see Section 11.5).

Since the only aspect of the magnetic field we are interested in is its effect on the spins, γ and $\boldsymbol{B}$ are commonly combined into the angular-frequency vector $\boldsymbol{\omega} = -\gamma\boldsymbol{B}$. With this, the torque equations read

$$\frac{\mathrm{d}}{\mathrm{d}t}\boldsymbol{M} = \boldsymbol{\omega}(t) \times \boldsymbol{M} = [\boldsymbol{\omega}_0 + \boldsymbol{\omega}_{1\mathrm{lin}}(t)] \times \boldsymbol{M} \quad (11.2)$$

with $\boldsymbol{\omega}_{1\mathrm{lin}}(t) = 2\omega_1 \cos(\omega_{\mathrm{MW}} t)\boldsymbol{x}$. At thermal equilibrium, $\boldsymbol{M}$ is aligned parallel to the external magnetic

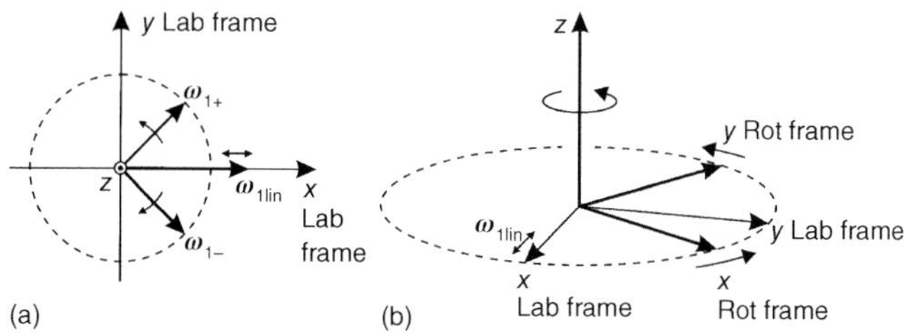

Figure 11.1. (a) Decomposition of the linearly oscillating MW field into two counter-rotating components (lab frame with z along the static magnetic field, x along the linearly oscillating MW field). (b) The relation between the laboratory frame and the rotating frame (rotating counterclockwise around the z axis for electron spins). The angular rotation frequency is $\omega_{\text{rot}} = \omega_{\text{MW}}$

field and does not change in time. The effect of a MW field applied perpendicular to $\boldsymbol{M}$ is to tilt $\boldsymbol{M}$ away from the z axis, as described by equation (11.2). As a consequence, $\boldsymbol{M}$ starts immediately to precess counterclockwise around z with the precession frequency (Larmor frequency) of the electron spins in the spin packet. In order to effectively interact with the precessing spins, the oscillating magnetic field has to follow their precession. This is not the case for the linearly oscillating MW field $\boldsymbol{\omega}_{1\text{lin}}(t)$, as it is fixed in space (along x and $-x$). However, we can decompose $\boldsymbol{\omega}_{1\text{lin}}(t)$ into two counter-rotating components $\boldsymbol{\omega}_{1\text{lin}}(t) = \boldsymbol{\omega}_{1+}(t) + \boldsymbol{\omega}_{1-}(t)$, one rotating counterclockwise, $\boldsymbol{\omega}_{1+}(t)$, and one rotating clockwise, $\boldsymbol{\omega}_{1-}(t)$. This is illustrated in Figure 11.1. Both rotate at the MW frequency, ω_{MW}. The electron spins, which are precessing counterclockwise, are affected only by the counterclockwise rotating component. The clockwise rotating component is ineffective, since it oscillates rapidly relative to the precessing magnetic moment, and its effect averages out. In other words, half the applied MW power is unproductive.

11.2.2 Rotating Frame

To visualize the spin dynamics under MW irradiation more concisely, and to simplify the theoretical description, it is advantageous to use a frame of reference that rotates around the z axis with angular frequency $\omega_{\text{rot}} = \omega_{\text{MW}}$ in the same sense as the spins and $\boldsymbol{\omega}_{1+}(t)$ (counterclockwise). This is called the rotating frame and is shown in Figure 11.1. In this frame, $\boldsymbol{\omega}_{1+}$ is time-independent. From now on, we will use ω_1 to denote ω_{1+}. The two axes of the rotating frame in the transverse plane, perpendicular to the external field, are called x and y (or sometimes u and v). Since the frame rotates, they do not correspond to fixed directions in space, even though they are sometimes referred to as 'x direction' etc. Rather, they correspond to specific phases relative to the carrier frequency ω_{MW}, so they are better called 'x phase' etc.

In the rotating frame, the evolution of the magnetic moment is described by

$$\frac{\mathrm{d}}{\mathrm{d}t}\boldsymbol{M} = (\boldsymbol{\omega}_0 - \boldsymbol{\omega}_{\text{rot}} + \boldsymbol{\omega}_1) \times \boldsymbol{M} = (\Delta\boldsymbol{\omega} + \boldsymbol{\omega}_1) \times \boldsymbol{M} \tag{11.3}$$

with the angular rotation frequency vector $\boldsymbol{\omega}_{\text{rot}} = \omega_{\text{MW}} z$ pointing along $\boldsymbol{B}_0$ and the MW amplitude vector $\boldsymbol{\omega}_1$. This equation has the same form as equation (11.2), except that the time dependence of $\boldsymbol{\omega}_1$ has been eliminated and $\boldsymbol{\omega}_0$ has been replaced by $\Delta\boldsymbol{\omega}$, which represents the offset between the resonance frequency of the spin and the rotating-frame frequency, i.e., the MW frequency. It is expressed in angular-frequency units, and its magnitude is

$$\Delta\omega = \omega_0 - \omega_{\text{rot}} = 2\pi\left(\frac{g_e \mu_B B_0}{h} - \nu_{\text{MW}}\right) \tag{11.4}$$

If the spin packet is resonant with the MW frequency, this offset and the associated effective field is zero ($\Delta\omega = 0$), and its magnetic moment $\boldsymbol{M}$ appears stationary in the rotating frame. If the resonance frequency is larger than ν_{MW}, $\Delta\omega > 0$, $\boldsymbol{M}$ precesses counterclockwise in the rotating frame; otherwise $\Delta\omega < 0$, and $\boldsymbol{M}$ precesses clockwise. Of course, in the laboratory frame, electron spins precess counterclockwise no matter what their resonance frequency is.

In the rotating frame, the effect of a MW pulse with a certain phase on resonant spins ($\Delta\omega = 0$) is simple: The pulse rotates (nutates) the magnetic-moment vector around the axis defined by its phase with a frequency proportional to the strength of the MW magnetic field. For example, pulses with x and y phase rotate magnetic moments of resonant spins around the x and y axis of the rotating frame, respectively. For a spin-1/2, the nutation frequency ω_{nut} (in angular-frequency units) is equal to the rotating MW field strength: $\omega_{nut} = \omega_1$. More generally, a quantum theory treatment shows that ω_{nut} also depends on the spin quantum number S and the projection quantum numbers of the two states involved in the transition, $m_S \leftrightarrow m'_S$, resonant with the MW, via

$$\omega_{nut} = \omega_1 \cdot \sqrt{S(S+1) - m_S m'_S} \qquad (11.5)$$

The nutation frequency is also called the Rabi frequency. The overall angle that a pulse of duration t_p rotates a resonant spin packet is called the flip angle of the pulse, θ, and is given by

$$\theta = \omega_{nut} t_p \qquad (11.6)$$

By adjusting the pulse length and/or the strength of the MW field, specific pulse flip angles can be achieved. The most common ones are $\pi/2$ (90°) and π (180°). When applied to spin packets with $\boldsymbol{M}$ initially oriented along $+z$, a $\pi/2$ pulse rotates $\boldsymbol{M}$ to the xy plane, and a π pulse takes $\boldsymbol{M}$ to $-z$. The flip angle in equation (11.6) only applies to spin packets that are at resonance ($\Delta\omega = 0$). For off-resonance spin packets, the rotation is around the tilted direction $\boldsymbol{\Delta\omega} + \boldsymbol{\omega}_1$, and the effective nutation frequency is $\omega_{eff} = \sqrt{\Delta\omega^2 + \omega_1^2}$.

In many cases, the MW field during pulses is so strong that the effects of internal magnetic interactions within the spin system (hyperfine coupling, etc.) as well as the frequency offset $\Delta\omega$ during the pulse can be neglected, i.e., $\omega_1 \gg \Delta\omega$. In this approximation, the pulses are called ideal or hard pulses, and in a perfect world would be infinitely short and of appropriate power to achieve the desired flip angle. They then also have infinite excitation bandwidth (see Section 11.5) and are nonselective with regard to transitions or molecular orientations. Ideal pulses are useful approximations for simple theoretical modeling of experiments that employ short high-power pulses such as electron spin echo envelope modulation (ESEEM) (see Chapter 18). For experiments where longer and more selective pulses are used, the excitation bandwidth can no longer be treated as infinite, and frequency offsets and internal interactions cannot be neglected. In these cases, full quantum-dynamical modeling is required.

11.3 BASIC PULSE SEQUENCES

In this section, we examine the three most fundamental pulse sequences and the resulting spin responses: a single pulse followed by the free induction decay (FID), a two-pulse sequence followed by the primary echo, and a three-pulse sequence followed by the stimulated echo. These sequences serve as building blocks for many more complicated pulse sequences. It is therefore important to understand them thoroughly. For simple spin-1/2 systems, these sequences can be visualized using a vector model, which we will use here. We will also introduce the important concept of coherence order. We will limit ourselves to ideal pulses. More details about the FID and the two-pulse echo are discussed in Chapter 15. Details about the general quantum theory using the density matrix formalism and product operators can be found in Chapter 8.

11.3.1 The Free Induction Decay

The simplest of all pulse sequences involves a single pulse, applied to an ensemble of spin packets at thermal equilibrium. It generates an emissive transient signal immediately afterward, which is called the FID. The FID amplitude is proportional to the number of excited spins, and it is maximal if the pulse flip angle equals $\pi/2$. As shown in Figure 11.2, the $\pi/2$ pulse rotates the magnetic moments of a set of near-resonance spin packets from its thermal-equilibrium orientation along the z axis into the xy plane of the rotating frame. If the pulse has $+y$ phase, the magnetic moments end up with $+x$ phase.

Immediately after the end of the hard pulse, the magnetic moments of the individual spin packets start precessing at their respective resonance frequency offsets $\Delta\omega$ around the z axis. In the rotating frame, spin packets with $\Delta\omega < 0$ (labeled as 'slow' in Figure 11.2) appear to rotate clockwise, while spin packets with $\Delta\omega > 0$ ('fast') rotate counterclockwise. The precessing total magnetic moment induces a signal – this is the FID. Since all excited spin packets start with the same phase, but have different precession frequencies, their magnetic moments start to get out of phase alignment and dephase. As a consequence of the dephasing, the total magnetic moment in the xy plane decreases over time, and the FID loses intensity until the spin packets are completely dephased and relaxed.

Quantitatively, the evolution of the magnetic moment $\boldsymbol{M} = (M_x, M_y, M_z)$ of a particular spin-1/2 packet

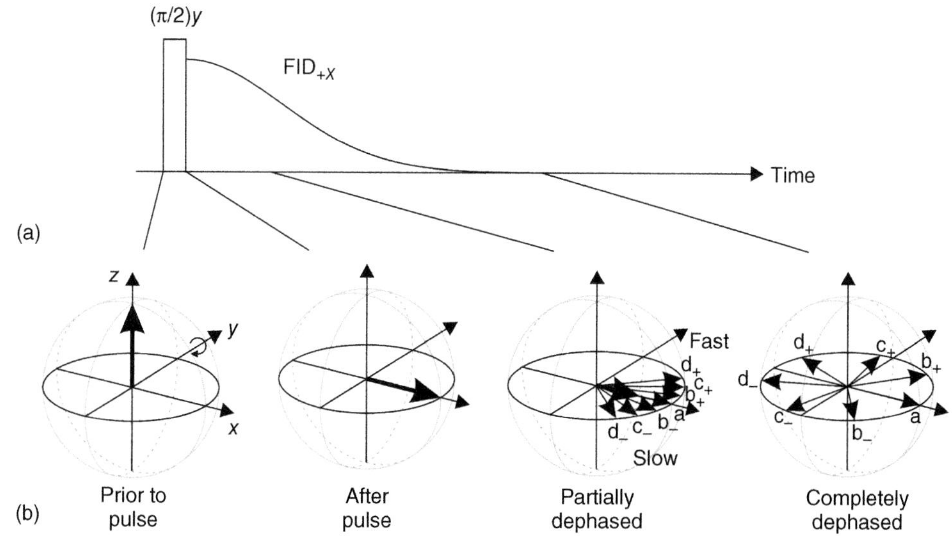

Figure 11.2. (a) Pulse sequence and FID response of a broad EPR spectrum. (b) The origin of the FID after a $(\pi/2)_y$ pulse, as seen in the rotating frame. The large arrow presents the total magnetic moment, and the small arrows indicate the magnetic moments of individual spin packets with different resonance frequencies: a is on resonance, b_+, c_+, and d_+ precess faster than the rotating frame, and b_-, c_-, and d_- precess slower

with resonance offset $\Delta\omega$ in this single-pulse sequence starts from its equilibrium alignment along z, $(0, 0, M_0)$ and gives

$$\begin{pmatrix} 0 \\ 0 \\ M_0 \end{pmatrix} \xrightarrow{(\pi/2)_y} \begin{pmatrix} M_0 \\ 0 \\ 0 \end{pmatrix} \xrightarrow{t} \begin{pmatrix} M_0 \cos \Delta\omega t \\ M_0 \sin \Delta\omega t \\ 0 \end{pmatrix} \tag{11.7}$$

where t indicates the time after the pulse. The accumulated phase in the xy plane is $\Delta\omega t$. For systems other than isolated spins-1/2, the dynamics must be described using quantum theory. A key concept in this quantum description is the electron coherence order.[1,2] The spin packet starts out at thermal equilibrium. This is coherence order $p = 0$, which corresponds to populations of the eigenstates of the spin system and is represented by the time-independent diagonal elements of the density matrix. After the pulse, the spin packet is in electron coherence order $p = \pm 1$, corresponding to precessing electron spins and represented by non-zero time-dependent off-diagonal elements of the density matrix. Therefore, the coherence transfer pathway for the one-pulse sequence is designated as $0 \rightarrow \pm 1$.

The simplest pulse EPR experiment is the field sweep, where the amplitude of the FID is measured as a function of the magnetic field. This gives an absorption spectrum instead of the derivative-like spectrum obtained by CW EPR. The pulse field sweep also has the advantage that it captures broad lines that are difficult to detect in CW EPR. With short pulses (less than about 50–100 ns) and the typical broad solid-state EPR spectra, FIDs are short. Often, they do not extend beyond the deadtime of the spectrometer and are therefore undetectable, as the deadtime prevents detection near $t \approx 0$ (see Section 11.6). However, longer pulses lead to longer FIDs that can be detected (as longer pulses have narrower excitation profiles, see Section 11.4). For broad lines, the FID is usually monotonically decaying and featureless, as illustrated in Figure 11.2. However, for organic radicals in liquid solution, where the EPR spectrum consists of a sum of narrow lines, the FID shows a lot of structure and is periodic (see Chapter 15). In general, the FID gives the Fourier transform of the excited part of the spectrum.

11.3.2 The Primary (Two-Pulse) Echo

In this pulse sequence, two pulses are applied. A first pulse with flip angle $\theta_1 = \pi/2$ is followed by a dephasing delay of duration τ and then a second pulse with flip angle $\theta_2 = \pi$. The effect of this sequence on excited spin packets is illustrated in Figure 11.3. After the first pulse, the spin packets give an FID and dephase. Some lose phase, and some gain phase relative to the rotating frame. The second pulse rotates all spin packets

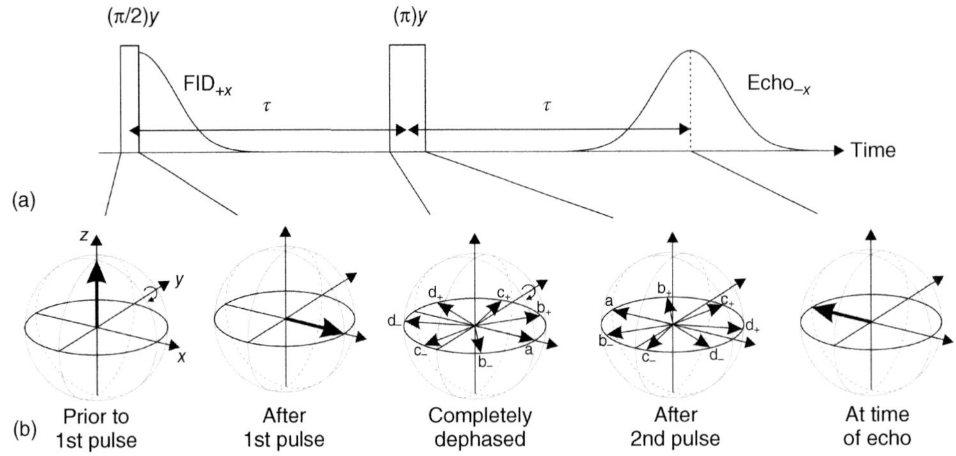

Figure 11.3. (a) The two-pulse sequence leading to the two-pulse echo. (b) The mechanism of formation of the two-pulse echo, as seen in the rotating frame. Individual spin packets are labeled: a is on resonance, b_+, c_+, and d_+ precess faster than the rotating frame ($\Delta\omega > 0$), and b_-, c_-, and d_- precess slower ($\Delta\omega < 0$)

around the y axis by π, so that they end up in the transverse plane again. Now, however, the spin packets with higher precession frequency lag in phase behind the ones with lower precession frequency. As a consequence, all spin packets catch up with each other and refocus with $-x$ phase at time τ after the second pulse. This produces an emission signal from the resonator and is called a Hahn echo, primary echo, or two-pulse echo.[3] As with the FID, the amplitude of the echo is proportional to the number of excited spins.

The primary echo is closely related to the FID. It consists of two FIDs back-to-back: a reverse FID where the spins rephase, immediately followed by a forward FID where the spins dephase again. For a specific spin packet with offset frequency $\Delta\omega$, the magnetic moment evolves in time according to

$$\begin{pmatrix} 0 \\ 0 \\ M_0 \end{pmatrix} \xrightarrow{(\pi/2)_y} \begin{pmatrix} M_0 \\ 0 \\ 0 \end{pmatrix} \xrightarrow{\tau} \begin{pmatrix} M_0 \cos \Delta\omega\tau \\ M_0 \sin \Delta\omega\tau \\ 0 \end{pmatrix}$$
$$\xrightarrow{(\pi)_y} \begin{pmatrix} M_0 \cos(\pi - \Delta\omega\tau) \\ M_0 \sin(\pi - \Delta\omega\tau) \\ 0 \end{pmatrix} \xrightarrow{\tau} \begin{pmatrix} M_0 \cos \pi \\ M_0 \sin \pi \\ 0 \end{pmatrix} = \begin{pmatrix} -M_0 \\ 0 \\ 0 \end{pmatrix} \tag{11.8}$$

From this, one can see that all spin packets have $-x$ phase at the time of the echo, independent of their offset $\Delta\omega$. In terms of coherence transfers, the first pulse converts electron coherence order $p=0$ to $p=\pm 1$, and the second pulse inverts the coherence order, $p=\pm 1 \rightarrow p=\mp 1$. The overall coherence transfer pathway leading to the primary echo is $0 \rightarrow \pm 1 \rightarrow \mp 1$. The echo is formed at the point in time when the spin packets have spent an equal amount of time in coherence orders $+1$ and -1.

A two-pulse echo appears for any combination of pulse flip angles θ_1 and θ_2, but its intensity is maximal for $\theta_1 = \pi/2$ and $\theta_2 = \pi$. In practice, there is also an FID after the second pulse. This is due to spins that relaxed back to equilibrium after the first pulse, and to spins that have not been excited by the first pulse, but are excited by the second pulse. The two-pulse sequence forms the basis of many pulse EPR experiments, where it is used as the final detection sequence. It can be used to detect a field-swept spectrum by recording the echo amplitude as a function of the magnetic field. However, unlike with FID detection, this echo detection can lead to spectral distortions due to nuclear modulation effects on the echo amplitude. The two-pulse echo is used in inversion recovery (see Chapter 9), two-pulse ESEEM (see Chapter 18), echo-detected Davies electron-nuclear double resonance (ENDOR) (see Chapter 16), electron-electron double resonance-detected NMR (ELDOR-detected NMR) (see Chapter 17), and relaxation time measurements (see below).

After the primary echo forms, the spin packets proceed to dephase again. If another π pulse is now applied after some delay τ_2, the magnetic moments refocus again at time τ_2 after the second π pulse. The resulting refocused primary echo is used in pulse experiments such as DEER (double electron–electron resonance, see Chapter 19). Refocusing pulses can be applied repeatedly to increase the signal-to-noise ratio (SNR) by recording and adding all the refocused echoes, or to suppress certain relaxation mechanisms as in the Carr–Purcell–Meiboom–Gill (CPMG) sequence.[4,5]

11.3.3 The Stimulated (Three-Pulse) Echo

After a sequence of three pulses, there appears a new echo that requires the presence of all three pulses. It appears at time τ after the third pulse, where τ indicates the separation between pulses 1 and 2. Its intensity is maximal if all three pulses have a flip angle of $\pi/2$. This is the so-called three-pulse or stimulated echo.[3] The delay between pulses 2 and 3 is conventionally denoted T.

The origin of the stimulated echo is visualized with a vector picture in Figure 11.4. The first $\pi/2$ pulse turns the magnetic moments of excited spin packets into the transverse plane. After dephasing relative to each other for the time τ, the spin packets have accumulated different phases. Some are within $\pi/2$ of the starting phase (a, b_+, b_- in the figure), and others lag or lead by more than a quarter turn (c_+, c_-, d_+, d_-). The second $\pi/2$ pulse rotates all magnetic moments from the xy plane into the yz plane. From now on, it is best to consider the magnetic moment vectors decomposed into their components along y and along z. The components along y (corresponding to electron coherence order ± 1) resume their precession. They do not contribute to the formation of the stimulated echo. In the figure, we have disregarded them after the delay T. On the other hand, the longitudinal components along the z axis (electron coherence order 0) stay put and do not precess during time T. Spin packets whose magnetic moments were within $\pi/2$ of $+x$ just before the second pulse (a, b_+, b_-) have negative z components, and those with accumulated phase lag/leads larger than $\pm\pi/2$ (c_+, c_-, d_+, d_-) have positive z components, with amplitudes depending on the offset frequency. The third $\pi/2$ pulse rotates these z components back into the transverse plane. Positive z components have $+x$ phase and dephase by more than a quarter turn during τ to end up with a component along $-x$. Negative z components start along $-x$, and they dephase by less than a quarter turn, thus maintaining a component along $-x$. Therefore, an echo forms with $-x$ phase, although the refocusing is only partial.

Quantitatively, a particular spin packet's magnetic moment evolves during the three-pulse sequence according to

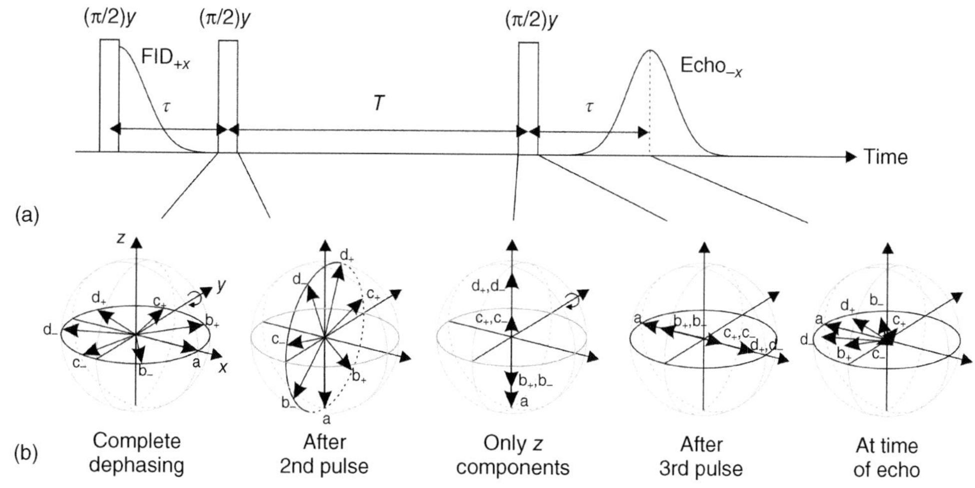

Figure 11.4. (a) The three-pulse sequence leading to the stimulated echo. (b) The mechanism of formation of the stimulated echo. Magnetic moments at various points during the sequence, as seen in the rotating frame. Individual spin packets are labeled: a is on resonance, b_+, c_+, and d_+ precess faster than the rotating frame ($\Delta\omega > 0$), and b_-, c_-, and d_- precess slower ($\Delta\omega < 0$)

$$\begin{pmatrix} 0 \\ 0 \\ M_0 \end{pmatrix} \xrightarrow{(\pi/2)_y} \begin{pmatrix} M_0 \\ 0 \\ 0 \end{pmatrix} \xrightarrow{\tau} \begin{pmatrix} M_0 \cos \Delta\omega\tau \\ M_0 \sin \Delta\omega\tau \\ 0 \end{pmatrix}$$
$$\xrightarrow{(\pi/2)_y} \begin{pmatrix} 0 \\ M_0 \sin \Delta\omega\tau \\ -M_0 \cos \Delta\omega\tau \end{pmatrix} \xrightarrow{T} \begin{pmatrix} 0 \\ \dots \\ -M_0 \cos \Delta\omega\tau \end{pmatrix}$$
$$\xrightarrow{(\pi/2)_y} \begin{pmatrix} -M_0 \cos \Delta\omega\tau \\ 0 \\ 0 \end{pmatrix} \xrightarrow{\tau} \begin{pmatrix} -M_0 \cos^2 \Delta\omega\tau \\ -M_0 \cos \Delta\omega\tau \sin \Delta\omega\tau \\ 0 \end{pmatrix} \quad (11.9)$$

As in Figure 11.4, we have dropped the y component after the delay T from consideration (indicated by …), since it does not contribute to the stimulated echo. From the final vector, it is apparent that there is a net magnetic moment with $-x$ phase, as $\cos^2\Delta\omega\tau$ is non-negative independent of the value of $\Delta\omega$. The y component averages to zero when integrated over $\Delta\omega$. The coherence transfer pathway that leads to the stimulated echo is $0 \rightarrow \pm 1 \rightarrow 0 \rightarrow \mp 1$.

Compared to the two-pulse sequence, the three-pulse sequence has the π pulse split into two $\pi/2$ pulses separated by a delay of length T. From another perspective, the first two $\pi/2$ pulses generate an offset-dependent pattern of nonequilibrium z magnetic moment $M_z/M_0 = -\cos(\Delta\omega\tau)$. Such a nonequilibrium polarization pattern is called a spectral hole. This is then converted to transverse magnetic moments by the third $\pi/2$ pulse. Therefore, the stimulated echo can be understood as an FID generated by the third pulse that reveals the spectral hole pattern 'burned' into the EPR spectrum by the first two pulses.

In general, a sequence of three finite-length pulses elicits a number of transient signals from the sample: In addition to the stimulated echo, there is an FID after each of the three pulses, there is a primary echo generated by each pair of pulses (1–2, 1–3, and 2–3), and there is an echo due to the refocusing effect of the third pulse on the spin packets that generated the 1–2 primary echo (see discussion of phase cycling in Section 11.6). The relative intensities and shapes of these FIDs and echoes depend on the flip angles and excitation bandwidths of the three pulses as well as on relaxation times.

Although the stimulated echo results from partial refocusing only and is therefore weaker than the primary echo, it is very useful since the pulse sequence includes a period T between pulses 2 and 3 with coherence order 0, where electron spins relax more slowly (see Section 11.5). This period can therefore be used for additional spin manipulation or nuclear coherence evolution. Thus, the stimulated echo forms the basis for pulse experiments such as Mims ENDOR (see Chapter 16), three-pulse ESEEM and HYSCORE (see Chapter 18), and RIDME (see Chapter 20).

11.4 BANDWIDTHS

For pulse EPR, it is very important to consider the following bandwidths, i.e., frequency ranges: (i) the excitation bandwidth of the pulses employed in the pulse EPR experiment, (ii) the width of the EPR spectrum of the sample at the given fixed magnetic field, and (iii) the bandwidth of the resonator. The excitation bandwidth of rectangular pulses is approximately the inverse of their length. For example, a 10-ns pulse has an excitation bandwidth on the order of 100 MHz. Since pulses cannot be shorter than a few nanoseconds due to technical limitations, their excitation bandwidths are limited. EPR spectra of orientationally disordered solid-state samples (frozen solutions, powders, glasses; as opposed to single crystals) tend to be very broad – the spectral width is most often much wider than the pulse bandwidth. As a consequence, only a subset of spins in the sample is excited. The PCs in this excited subset differ from the others in their orientation (in a disordered sample), or in the particular EPR transition that is resonant (in a high-spin or multi-spin system). This leads to orientation and transition selection effects, which affect the measured FIDs and echoes and what can be learned from them. Additionally, resonators admit power only over a limited bandwidth. In this section, we look at spectral bandwidths, pulse bandwidths, and their consequences in terms of orientation and transition selection. Resonator bandwidth effects are discussed in Section 11.6.

11.4.1 Pulse Excitation Bandwidth

Ideal, infinitely short pulses have infinite excitation bandwidth. Real-world pulses have finite length, and their excitation bandwidth is therefore limited. This can be rationalized in an approximate way via the 'frequency content' of the pulse as obtained by Fourier transformation of the pulse shape: the Fourier transform of a rectangular function is a sinc function, $\text{sinc}\phi = \sin\phi/\phi$, and a squared sinc function

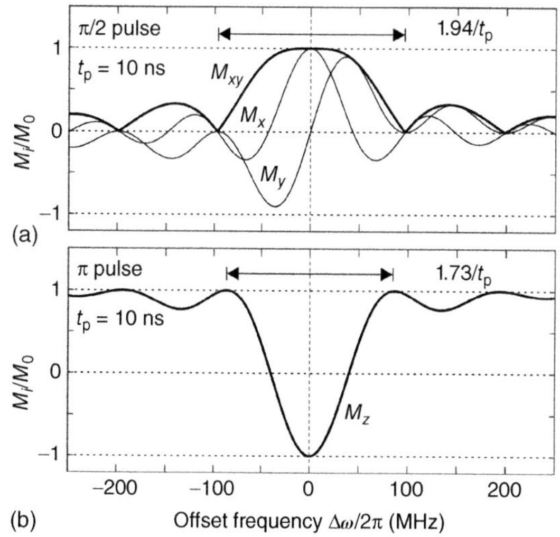

Figure 11.5. Excitation profiles of pulses with flip angle $\theta = \pi/2$ (a) and π (b) and length $t_p = 10$ ns, as a function of the offset frequency, $\Delta\omega/2\pi$. The vertical axis shows components of the magnetic moment (M_x, M_y, M_z, $M_{xy} = \sqrt{M_x^2 + M_y^2}$) at the end of the pulse applied at thermal equilibrium. The equations are[6] $M_x/M_0 = \sin(\omega_{\text{eff}} t_p)\ \omega_1/\omega_{\text{eff}}$, $M_y/M_0 = 2\sin^2(\omega_{\text{eff}} t_p/2)\Delta\omega\ \omega_1/\omega_{\text{eff}}^2$, $M_z/M_0 = 1 - 2\sin^2(\omega_{\text{eff}} t_p/2)\omega_1^2/\omega_{\text{eff}}^2$, with $\omega_{\text{eff}}^2 = \omega_1^2 + \Delta\omega^2$ and $\omega_1 = \theta/t_p$

approximates the pulse excitation profile of a π pulse reasonably well.

It might appear counterintuitive that a rectangular pulse at a single carrier frequency ν_{MW} can excite spins with resonance frequencies larger or smaller than ν_{MW}, which are off-resonant. The nature of the excitation profile of a pulse can be fully understood only by looking at the actual spin dynamics induced by the pulse as described by the torque equations or the quantum master equations.[6] Figure 11.5 illustrates this for rectangular pulses with flip angles π/2 and π. The excitation profile shows the components of the magnetic moment immediately after the pulse. It shows which spin packets are driven out of equilibrium by the pulse, as a function of $\Delta\omega$, the offset between the resonance frequency of the spin and the MW frequency of the pulse. The excitation bandwidth is inversely proportional to the pulse length, t_p. For the π pulse, the profile shows the z component. It features a central excitation band of roughly Gaussian shape with a full width at half maximum (FWHM) of approximately $0.8/t_p$. The first zeroes are separated by about $1.73/t_p$. For the π/2 pulse, the profile shows the norm of the magnetic-moment component in the xy plane. It has a broad central excitation band with a FWHM of about $1.37/t_p$ and a separation between the first zeroes of about $1.94/t_p$. In addition, shallow side bands appear at lower and higher frequency offsets for both flip angles. The depth of the sidebands depends on the rise and fall time of the rectangular pulse shape. In the figure, infinitely fast rise/fall times are assumed. In practice, rise times are finite and thus the pulses are not rectangular, and the side bands are then shallower.

It is possible to utilize pulse shapes with nonrectangular profiles combined with frequency sweeps during the pulse to achieve wider excitation bandwidths than with fixed-frequency monochromatic rectangular pulses (see Chapter 21).

11.4.2 Spectral Width

Pulse EPR sequences are performed at fixed magnetic field values. Therefore, in order to understand the spin excitation and precession dynamics in a pulse EPR sequence, it is more useful to consider the frequency-domain EPR spectrum at a fixed field, as opposed to the field-swept EPR spectrum at a fixed frequency, as obtained by CW EPR or echo or FID detection. Figure 11.6 illustrates the two types of spectra

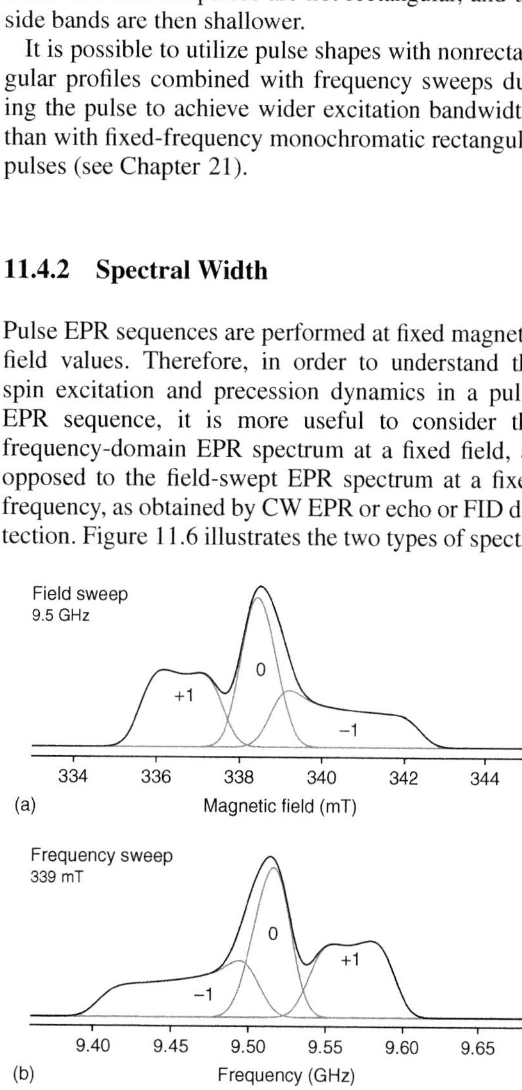

Figure 11.6. Field-swept EPR spectrum at fixed frequency (a) vs frequency-swept EPR spectrum at fixed magnetic field (b) of a frozen solution of nitroxide radical (simulated). The three subspectra corresponding to the three substates of the ^{14}N nucleus ($m_I = -1, 0, +1$) are shown separately

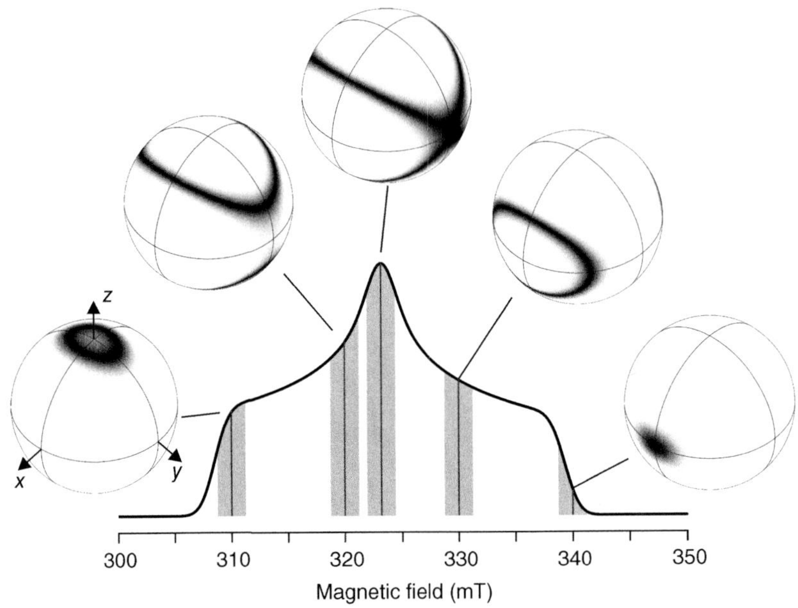

Figure 11.7. Orientation selectivity of a 9.5 GHz pulse with 100 MHz excitation bandwidth on a powder sample of spin-1/2 centers with a rhombic $\boldsymbol{g}$ tensor with principal values $g_x = 2.0$, $g_y = 2.1$, and $g_z = 2.2$, at various magnetic field values (310, 320, 324, 330, and 340 mT). x, y, and z indicate the orientations along the principal directions of the $\boldsymbol{g}$ tensor which is fixed in some chosen molecular frame. The shaded areas on the sphere plots show the set of orientations (in terms of $\boldsymbol{B}_0$ relative to the xyz frame of the $\boldsymbol{g}$ tensor) for which spins are excited. The shaded areas in the spectrum illustrate the excitation bandwidth of the pulse

for a nitroxide radical. For systems with relatively narrow spectra around $g = 2$, such as nitroxide radicals, the shapes of the field-swept and the frequency-swept spectra are essentially mirror images of each other.

EPR spectra can be very broad. For instance, a nitroxide powder spectrum at 340 mT is about 0.2 GHz wide (Figure 11.6), whereas a Cu(II) spectrum can be up to 2 GHz wide, and high-spin Fe(III) complexes have spectral bandwidths significantly beyond 10 GHz. For samples with very broad spectra, only a small subset of all spins are resonant at a given frequency and field. Therefore, sensitivity is significantly reduced compared to narrow-spectra samples with the same concentration of PCs.

11.4.3 Orientation and Transition Selection

Due to the limited excitation bandwidth of pulses and the often enormous spectral width of EPR spectra, not all spins in a sample are excited by a pulse. Pulses are therefore selective. Spins differ in resonance frequency due to different orientations of the PCs and due to different states of coupled spins (nuclei or electrons). These lead to two types of selectivity: orientation selectivity and transition selectivity.

Figure 11.7 illustrates the orientation selection effect of a narrow-band pulse on a disordered sample of a spin system with a rhombic $\boldsymbol{g}$ tensor. Depending on the field and frequency settings, the pulse excites only a subset of spins in the sample that are oriented such that their resonance frequency falls within the pulse excitation window. Therefore, in this case, the majority of spins in the sample do not contribute to FIDs and echoes, which are consequently weak. On the other hand, the ability to select a narrow subset of orientations via orientation selection can be advantageous, since these can correspond to a single-crystal-like situation. In Figure 11.7, this happens at the low- and at the high-field end, near 308 and 340 mT. Experiments such as ENDOR and ESEEM performed at these fields yield high-resolution spectra, which are more resolved than at intermediate fields. Orientation selection can be a disadvantage in methods like DEER where the data analysis typically relies on full powder excitation.

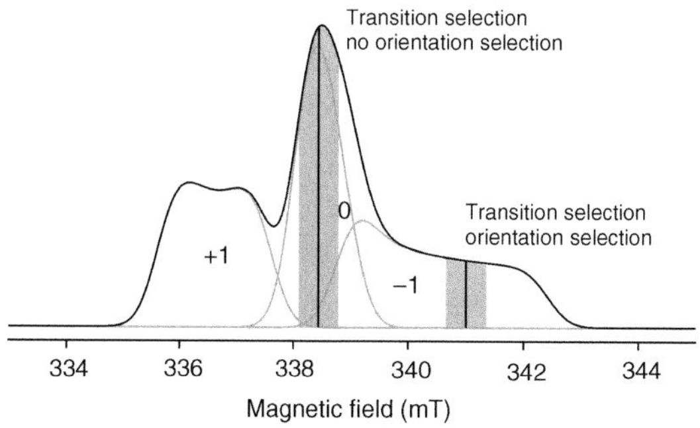

Figure 11.8. Transition and orientation selectivity of 9.5 GHz pulses with 20 MHz excitation bandwidth on a powder sample of a nitroxide radical, applied at 338.5 and at 341 mT

The second type of selectivity imposed by narrow-band pulses on broad-band spectra is transition selectivity. This occurs when the EPR spectrum is a sum of several resolved transitions, due to either hyperfine splittings or zero-field splittings. This is illustrated in Figure 11.8 for the powder spectrum of a nitroxide radical (see Chapter 14). The spectrum decomposes into three subspectra corresponding to different spin states of the spin-1 ^{14}N nucleus ($m_I = -1, 0, +1$). A 9.5 GHz pulse at 338.5 mT excites essentially only the $m_I = 0$ transition, and is only weakly orientation selective. The same pulse at 341 mT excites only the $m_I = -1$ transition, and is strongly orientation selective. As this example shows, orientation and transition selectivity often occur simultaneously. Both orientation and transition selectivity depend on the magnetic field.

11.5 NON-COHERENT EFFECTS

In the above discussion of FIDs and echoes, we have neglected relaxation and assumed that the spins stay in the nonequilibrium excited states the pulses put them in, at least for the time scale of the experiment. This is not the case in the real world – excited spins eventually relax back toward thermal equilibrium. The time scales of these relaxation processes are characterized by the longitudinal and transverse relaxation time constants, T_1 and T_2. On the simplest level, this is described by the Bloch equations[7]

$$\frac{\mathrm{d}}{\mathrm{d}t}\boldsymbol{M} = -\gamma \boldsymbol{B} \times \boldsymbol{M} - \begin{pmatrix} M_x/T_2 \\ M_y/T_2 \\ (M_z - M_0)/T_1 \end{pmatrix} \qquad (11.10)$$

which extend the classical equations in equation (11.1) by a phenomenological term designed to capture the effects of relaxation.

So far, we also have assumed that the resonance frequency of any particular spin is unchanged throughout the pulse sequence. Again, this is not always the case in practice. The resonance frequency of spins can change stochastically for a variety of reasons both between and during pulses. This results in a diffusion-like behavior where the spin's resonance frequency randomly 'diffuses' along the frequency axis to higher or lower frequencies. Both spin relaxation and diffusion are non-coherent effects, and they have practical consequences for pulse EPR experiments. Next, we examine these effects.

11.5.1 Longitudinal Relaxation

The longitudinal relaxation time constant, T_1, describes the time scale on which nonequilibrium magnetization relaxes back toward thermal equilibrium along z. T_1 can be estimated by using an inversion or saturation recovery experiment. This and the various physical processes leading to longitudinal (spin–lattice) relaxation are explained in Chapter 9. The longitudinal relaxation time is important for two practical reasons: First, T_1 limits the maximum feasible length of pulse sequences. In samples with short T_1, pulse sequences have to be correspondingly short to give detectable FIDs or echoes. Second, T_1 determines the shortest possible repetition time in pulse EPR experiments. In order to get sufficient signals, pulse sequences and the associated echo measurements are repeated many times, with repetition times t_R typically between 1 and 100 ms. Before a new pulse sequence can start, the spins in the sample have to relax back to thermal equilibrium to avoid saturation effects. This requires that the repetition time is larger than the slowest relaxation time constant in the spin system, T_1. Typically, $t_R > 5T_1$ is sought. Shorter t_R values lead to saturation. However, since faster repetition times lead to more acquired echoes per time, in many experiments there is a practical benefit of operating under partial saturation at about $t_R \approx 1.5\, T_1$.

11.5.2 Transverse Relaxation, Phase Memory

The transverse relaxation time constant, T_2, describes the time scale over which magnetic moments in the *xy* plane (precessing spins) within a spin packet lose their phase coherence and consequently cannot be refocused to an echo anymore. T_2 is always shorter than T_1. The dynamic processes underlying transverse relaxation are discussed in Chapter 9. The most common experiment to measure transverse relaxation is the two-pulse echo, acquired as a function of τ. The echo decays as a function of τ, typically in an exponential or stretched-exponential manner, $\exp(-(2\tau/T_m)^x)$, with values for x between about 1 and 3. This expression contains 2τ in the exponent, since relaxation is effective during both periods of τ. The decay constant of this echo decay, T_m, is called the phase memory time constant. It not only includes T_2 processes, but also additionally the effects of spectral and instantaneous diffusion (see below). It is therefore always shorter than the actual T_2, which is tedious to measure using pulse EPR. T_2^*, which is used in NMR spectroscopy to describe the apparent time scale of transverse relaxation including dephasing due to static-field inhomogeneity, is rarely used in EPR, since static-field inhomogeneity is usually irrelevant (small compared to other factors).

For echo experiments, it is important to know the dephasing timescale, and T_m is the relevant number (and is often sloppily denoted T_2). T_m limits the maximum length of inter-pulse delays with precessing spins in a pulse sequence. If the inter-pulse delays are too long relative to T_m, significant signal loss occurs. One example where this limit is painful is DEER spectroscopy (see Chapter 19), where very long inter-pulse delays are needed to detect long spin-spin distances. Often, the two-pulse echo is amplitude modulated as a function of τ due to nuclear modulation effects (ESEEM, see Chapter 18), which can render the decay curve sufficiently oscillating that it becomes impossible to extract a precise value for T_m.

Experimentally, there are several ways of modifying the relaxation time scales. Lowering or increasing the temperature in almost all cases prolongs or shortens relaxation times, respectively (for details, see Chapter 9). Similarly, increasing/decreasing spin concentrations shortens/prolongs relaxation times. For each new sample, it is therefore important to measure T_1 and T_m for a variety of temperatures or concentrations. This allows one to find conditions that result in the best signal-to-noise ratio for a given amount of time. In dilute solid-state samples where transverse relaxation is determined by nuclear spin diffusion, changing the matrix or solvent to contain fewer methyl groups and fewer protons (e.g., by deuteration or by using an aprotic solvent), will slow down relaxation. On the other hand, relaxation rates can be enhanced by adding fast relaxing paramagnetic species (e.g., lanthanide ions) to a sample. All these options should be explored when optimizing pulse EPR measurement conditions for a new set of samples to achieve maximum sensitivity.

In experiments involving nuclear spins, such as ENDOR, there are additional time scales that determine the experimental echo intensities. If nuclear spins are excited, then nuclear relaxation times are important. However, nuclear relaxation cannot be directly observed via electron FIDs or echoes and can thus be an unknown property. Nuclear longitudinal relaxation times, described by T_{1n}, can be very long compared to T_1 of the electron spins and can severely reduce (or even invert) signals in ENDOR spectra. In these cases, it is important to reduce the repetition rate of the pulse sequence, or to use random acquisition.[8] See Chapter 16 for more details.

11.5.3 Spectral Diffusion

In Section 11.3, we assumed that the resonance frequency of a spin is constant and does not change during the pulse sequence. This is not always fulfilled in practice. During free evolution, a spin's resonance frequency can change randomly due to relaxation-driven flips of nearby coupled nuclear or electron spins. This stochastic process is called spectral diffusion. Its effects are illustrated in Figure 11.9. For a spectral hole (frequency region of inverted spins) burned, e.g., by a π pulse, it leads to a loss in depth and increase in width of the spectral hole over time. In the case of precessing spins, it reduces the intensity of an echo since it prevents complete rephasing of all the spin packets. Therefore, spectral diffusion shortens measured inversion recovery and phase memory times.

In certain samples such as frozen aqueous solutions of spin labels, nuclear spin diffusion can be a limiting factor for relaxation times, in particular T_m. To mitigate this effect, the concentration or the magnetic moment of the diffusion-inducing spins must be reduced: if protons are the culprits, replace them with deuterons which have a magnetic moment that is 6.5 times smaller, or use CPMG-type sequences; if electrons are the culprits, reduce the sample concentration. There are cases where this effect can be utilized to determine the distance between spins (RIDME, see Chapter 20).

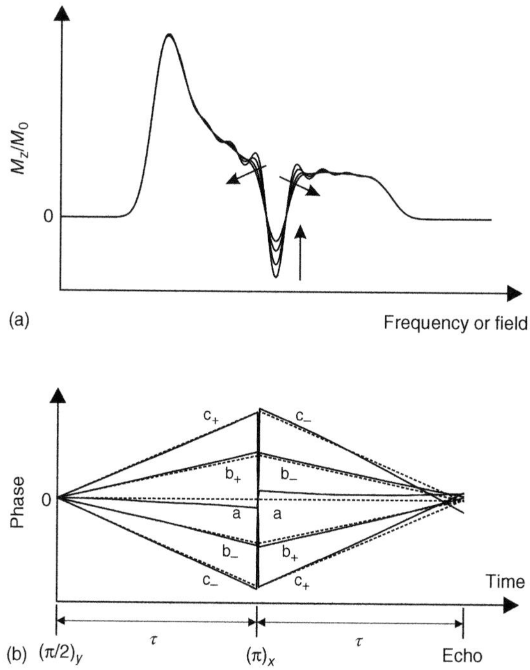

Figure 11.9. (a) Effect of spectral diffusion on a spectral hole of spin packets inverted by a rectangular pulse; over time, the hole washes out (broadens), as indicated by the arrows. T_1 relaxation would only reduce the depth of the hole. (b) Effect of spectral diffusion on the formation of a primary echo (see Figure 11.3), demonstrated with spin packets with different frequency offsets (a is on resonance, b_+ and c_+ precess faster than the rotating frame, b_- and c_- precess slower). The dashed lines indicate phase accumulation by the spin packets in the absence of spectral diffusion; they fully rephase at the time of the echo. The solid lines show the effect of spectral diffusion via stochastic frequency jumps during evolution; the rephasing is incomplete

11.5.4 Instantaneous Diffusion

The resonance frequency of a spin (say spin A) can also change during a pulse, if a nearby spin B is affected by the same pulse. Due to the through-space magnetic dipolar coupling between spins, any change in orientation of spin B changes the resonance frequency of spin A. Therefore, after the pulse, spin A can have a resonance frequency that is different from the one it had just prior to the pulse. This effect of a sudden change in resonance frequency is called instantaneous diffusion[9] and is undesired. For example, it distorts the measurement of relaxation times. To mitigate it, one can reduce the spin concentration in the sample; however, this reduces signal intensity. Another way to eliminate the instantaneous diffusion contribution to the phase memory time is to successively reduce the flip angle (shorten the length) of the second pulse in a two-pulse experiment: the smaller the flip angle, the smaller the effect of instantaneous diffusion. Extrapolation of the resulting echo decay constants to zero pulse length gives a decay time constant free of instantaneous diffusion (see Chapter 22, Figure 22.4).

11.6 SPECTROMETER

In this section, we discuss the basic elements of a pulse EPR spectrometer, to the degree that they are relevant for a fundamental understanding of the technique. More extensive technical details are given in Chapter 12.

11.6.1 Block Diagram

The block diagram of a typical pulse EPR spectrometer is shown in Figure 11.10. The MW part consists of a transmitter and a receiver. The transmitter generates the high-power pulses. A CW MW source emits low-power radiation, and a pulse-forming unit (PFU) such as a fast switch chops it so as to generate low-power pulses which are then amplified by a power amplifier. The high-power pulses are routed to the resonator via a circulator. For double-resonance experiments involving pulses at two different MW frequencies, a second MW source and a second PFU are added. For ENDOR experiments, an additional RF transmitter with RF source, PFU, and RF power amplifier is added, and the RF pulses are routed to RF coils mounted on or inside the resonator.

After the spins in the sample are excited by the pulse(s) from the transmitter, the typically very weak response (FID or echo) emitted from the sample is routed via the circulator to the receiver. To prevent the energy from the high-power pulses reflected off the resonator from destroying the sensitive components in the receiver, a protection switch is used as the first component of the receiver. It is open during pulsing and isolates the receiver from the resonator and transmitter. It is closed during detection and passes the weak signals emitted from the sample in the resonator. A sensitive low-noise amplifier (LNA) amplifies the signal before it reaches the detector. The detector is

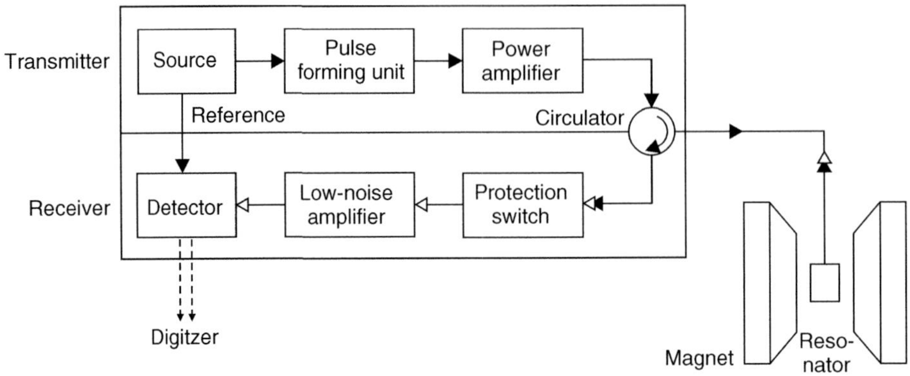

Figure 11.10. Schematic block diagram of a typical pulse EPR spectrometer consisting of transmitter, circulator, resonator in a magnet, receiver, and digitizer. Solid arrows with black arrowheads indicate low- and high-power transmitted MW, white arrowheads indicate the weak emitted MW signal, and dashed arrows indicate low-frequency (video) signals

a quadrature mixer that combines the signal with a reference MW signal routed to the detector from the CW source to give an in-phase (I) and out-of-phase (quadrature, Q) component. The resulting two signals are not MW signals anymore, but have low frequencies up to about 200 MHz. They are called video signals. After another amplification stage (not shown), they are routed to a digitizer and digitally recorded. In the computer, the I and Q signals are combined into a complex signal.

Traditionally, the transmitter/circulator/receiver system is called the MW bridge. Its most expensive component is the power amplifier. Generally, more power is desirable, as it allows one to shorten the pulses for a desired flip angle, which widens the excitation bandwidth and excites more spins, leading to higher sensitivity. Typical amplifier powers are 1 kW at 9–10 GHz (X-band), 10–300 W at 33–35 GHz (Q-band), and a few watts at 94 GHz (W-band).

11.6.2 Resonators

Like CW EPR, pulse EPR employs resonators to amplify the MW field at the sample. In order for a pulse to excite the spins in the sample, the MW power has to reach the spins in the sample within the resonator. However, the resonator has a limited bandwidth, i.e., it only admits frequencies at or near its center frequency, ν_0. MW power outside that frequency bandwidth is reflected from the resonator. The bandwidth characteristics of a resonator are quantified by its quality factor Q. (This is the 'loaded' Q, also indicated by Q_L, where the subscript L indicates that this is the quality factor for a 'loaded' resonator, i.e., a resonator attached to a waveguide or coaxial cable.) The quality factor characterizes the bandwidth $\Delta\nu$ of the resonator relative to its center frequency via $Q = \nu_0/\Delta\nu$. Here, $\Delta\nu$ is the half-power bandwidth, i.e., the difference between the two frequencies where transmitted power is 50% of its maximum at resonance (the −3 dB points). This is illustrated in Figure 11.11, which shows the power transmitted into the resonator as a function of frequency. This profile of transmitted power is the complement to the reflected-power profile (the 'tune picture' or 'dip') that is typically observed and used for tuning.

Q, and therefore the bandwidth, of a resonator depends on its intrinsic quality factor Q_U (the so-called unloaded Q) and on its coupling to the waveguide or coaxial cable. Typically, this coupling can be varied by moving a small coaxial antenna loop near the resonator, or a small metallic element near the iris hole between the resonator and the waveguide. There are three coupling regimes: undercoupled ($\beta < 1$), critically coupled or matched ($\beta = 1$), and overcoupled ($\beta > 1$), where β is the coupling parameter. The coupling co-determines Q via $Q = Q_U/(1+\beta)$. Increasing the coupling reduces Q and therefore increases the bandwidth according to

$$\Delta\nu = \frac{(1+\beta)\nu_0}{Q_U} \tag{11.11}$$

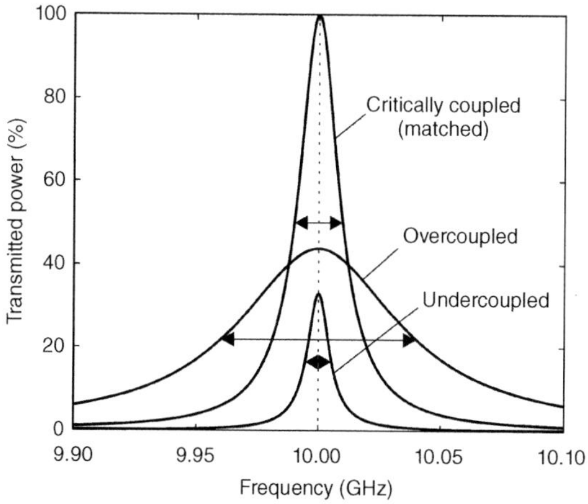

Figure 11.11. Power transmitted into a resonator as a function of MW frequency, for several couplings (overcoupled, matched/critically coupled, undercoupled). Resonance frequency $\nu_0 = 10\,\text{GHz}$, unloaded quality factor $Q_U = 1000$, coupling constants $\beta = 7$ (overcoupled), 1 (critically coupled), 0.1 (undercoupled). For pulse EPR, wide bandwidths are required, so that resonators are usually overcoupled. The expression for the transmitted power is $4\beta/[(1+\beta)^2 + Q_U^2(\nu/\nu_0 - \nu_0/\nu)^2]$

(see Figure 11.11). Since wide resonator bandwidths are desired for pulse EPR, resonators with moderate Q_U and overcoupling are commonly employed. However, the detection efficiency decreases with decreasing Q (decreasing Q_U and increasing β), and Q should be as high as possible. Therefore, it is necessary to make a trade-off between the excitation and detection requirements.[10]

11.6.3 Ringdown and Deadtime

After the end of a pulse, MW power built up in the resonator dissipates exponentially as $\exp(-t/\tau_{rd})$, with a time constant $\tau_{rd} = Q/\pi\nu_0 = 1/\pi\Delta\nu$. This is illustrated in Figure 11.12. Most of the power dissipating during ringdown propagates from the resonator toward the receiver. For several time constants, the power levels at the receiver from the ringdown far exceed the damage limit of the sensitive components such as the low-noise amplifier. Therefore, the receiver needs to be protected not only during the pulse, but also during ringdown after the pulse. The protection switch cannot be closed, and signal cannot be detected, until the high-power ringdown has subsided. The minimum time after a

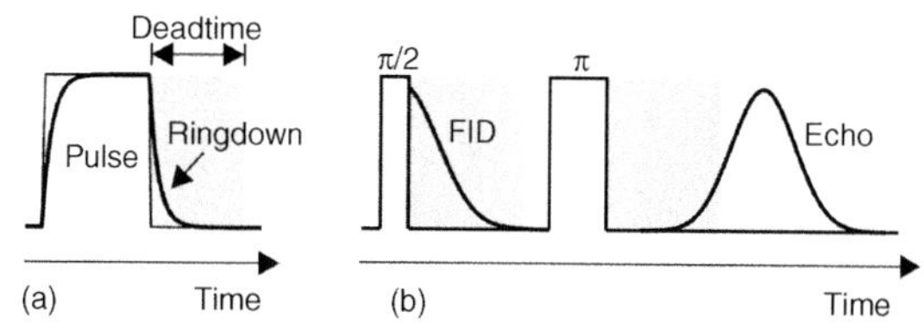

Figure 11.12. (a) Power ringdown after a pulse. The decay is $\exp(-t/\tau_{rd})$ with the decay constant $\tau_{rd} = Q/\pi\nu_0 = 1/\pi\Delta\nu$. 40 ns pulse, $Q = 100$, $\nu_0 = 9.5\,\text{GHz}$, $\Delta\nu = 95\,\text{MHz}$, $t_{rd} = 3.4\,\text{ns}$. The deadtime is a multiple of τ_{rd}. (b) The deadtime can result in loss or clipping of FIDs and echoes

high-power pulse that needs to elapse before the ringing has decayed to a level safe for the detector is called the deadtime of the spectrometer. The deadtime has critical consequences for pulse EPR measurements: It is not possible to detect FIDs and echoes that occur during the deadtime. For example, a short FID can be entirely within the deadtime, or a two-pulse echo for short τ value could be clipped. This is illustrated in Figure 11.12. If the spins relax on the timescale of the deadtime, then no echo or FID is recordable at all. Typically, for short pulses, the FID is short relative to the length of deadtime. This is the primary reason why most pulse EPR experiments are carried out using echoes. In addition to ringdown, the deadtime is also affected by power reflections within the spectrometer.

Typical deadtime durations are 50–100 ns. They tend to decrease with increasing MW frequency, since $\Delta\nu$ increases. As the resonator bandwidth determines the ringing time, increasing the resonator bandwidth has the crucial benefit of decreasing the ringdown time and shortening the deadtime. A decreased ringdown time means that the high power from a pulse is dissipated faster, and that weak signals can be detected earlier.

11.6.4 Pulse Phases and Phase Cycling

A two-pulse sequence generates an FID after each pulse and a single echo. In contrast, a three-pulse sequence generates, in addition to the three FIDs, a total of five echoes: the stimulated echo discussed above, three primary echoes (due to spins excited by the pulse combinations 1 + 2, 1 + 3, and 2 + 3), and the refocused primary echo (see Figure 11.13). These echoes are unwanted and usually occur at times different from the stimulated echo. However, they can overlap in time with the stimulated echo for particular

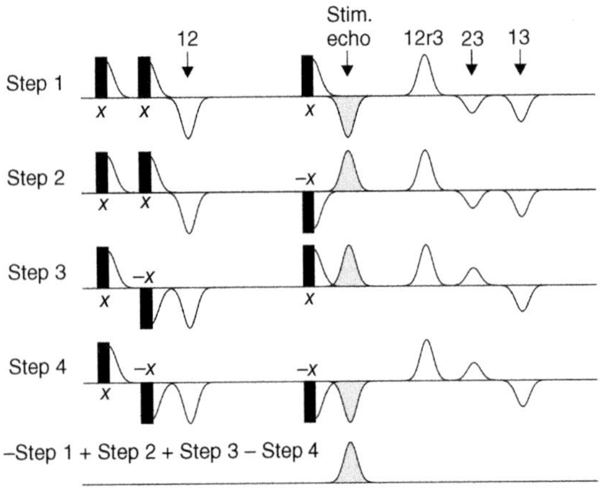

Figure 11.13. Example of a four-step phase cycling protocol for a three-pulse sequence. The second and third pulses are cycled through phases $+x$ (phase 0) and $-x$ (phase π). Linear combination of the transients from the four steps eliminates all echoes except the stimulated echo. 12, 23, 13 indicate the two-pulse echoes due to spins excited by the respective pulses. 12r3 indicates the two-pulse echo 12 refocused by the third pulse

combinations of delays ($T = 0, \tau, 2\tau$). This leads to artifacts when acquiring the stimulated echo as a function of pulse delays τ or T. The situation gets worse in sequences with more than three pulses.

To remove these unwanted signals, one can make use of the fact that the phases of different echoes depend differently on the phases of the pulses. The general rule is that the phase change of an echo, $\Delta\phi_{\text{echo}}$, due to a phase change of a pulse in the sequence, $\Delta\phi_{\text{pulse}}$, is given by

$$\Delta\phi_{\text{echo}} = -\Delta p \cdot \Delta\phi_{\text{pulse}} \tag{11.12}$$

where Δp indicates the electron coherence order change effected by the pulse for the echo pathway under consideration. For example, the π pulse in a two-pulse sequence changes the coherence order by 2 for the spins forming the echo, from ± 1 to ∓ 1. Therefore, changing the pulse phase $\Delta\phi_{\text{pulse}}$ by $\pi/2$ (for example from $+y$ to $-x$, see Figure 11.3) changes the phase of the resulting echo $\Delta\phi_{\text{echo}}$ by π (from $-x$ to $+x$ in Figure 11.3). In contrast, the first pulse of the sequence (flip angle $\pi/2$) causes a coherence order change of 1 (from 0 to ± 1), so a $\pi/2$ phase change of this pulse results only in a $\pi/2$ phase change of the echo.

The pulse phase dependence of the echoes is used to remove the unwanted echoes in a procedure called phase cycling.[1] For this, several transients measured with different pulse phase settings are linearly combined. This is illustrated in Figure 11.13 for the detection of a stimulated echo. In the four-step phase cycle shown, the phases of the second and the third pulse are systematically cycled through phases x (0) and $-x$ (π), and the resulting transients are linearly combined so as to eliminate all but the wanted echo. For pulse sequences with more pulses, phase cycles with more steps are required.

The phase cycles for the elimination of unwanted echoes can be extended with additional phase cycling steps that eliminate receiver offsets or imbalances in the I and Q channels of the quadrature detector. In general, it is important to deploy an appropriate phase cycling scheme when running a pulse experiment. Since the intensities of the unwanted echoes strongly depend on pulse flip angles and relaxation times, some unwanted echoes can have negligible intensity for specific experiments and samples, and the phase cycles can be trimmed accordingly.

11.7 PRACTICAL ASPECTS

The success of a pulse EPR experiment depends both on a good sample and on the proper choice of the many experimental parameters. In the following, we briefly discuss some important practical aspects.

11.7.1 Sample Concentration

Typical concentrations are 1 μM–1 mM (6×10^{14}–6×10^{17} PCs cm^{-3}). For pulse experiments that resolve nuclear frequencies, slightly higher concentrations can be tolerated. The upper limit of the concentration of PCs in the sample is dictated by the requirement of minimizing magnetic coupling between adjacent PCs to prevent spectral broadening and relaxation enhancement. The lower limit is not only determined by the inherent sensitivity of the spectrometer (how many excited spins are needed to achieve a certain voltage at the detector), but also the width of the spectrum: The broader the spectrum, the smaller the fraction of spins that lie within the excitation bandwidth of a pulse, and the larger the overall concentration needs to be to achieve adequate sensitivity.

11.7.2 Sample Matrix

In addition to suitable PC concentration, the composition of the matrix (solvent or crystal) can make a strong difference for sensitivity and resolution. Solvents should be chosen to give good glasses upon freezing to prevent aggregation of PCs, which enhances relaxation and broadens spectra. For aqueous solutions, glycerol is added as a glassing agent in concentrations between 10% and 60%. For organic solvents, many mixtures of solvents form good glasses. In solutions, dioxygen (O_2, a molecule with a ground-state spin triplet) can have similar effects and should therefore be removed by purging with argon or by freeze/pump/thaw cycles. O_2 can also create background contributions in a solid-state sample, particularly at very low temperatures. Magnetic nuclei in the matrix can shorten T_m and should therefore be minimized. For this reason, D_2O is preferred over H_2O in DEER experiments. In completely nonmagnetic matrices (such as a ^{28}Si crystal), phenomenally long T_m times can be achieved (see Chapter 22).

11.7.3 Frequency

Often, it is valuable to consider all available frequency bands from 2–4 GHz (S-band) to 94 GHz (W-band) and beyond and examine which band is optimal for the intended experiment and sample. In general, ENDOR experiments are best performed at higher fields, since that leads to increased spectral separation of different isotopes as well as to better first-order behavior of the peak positions (see Chapter 16). ESEEM experiments should be performed at the band where the echo envelope modulations are expected to be deepest, to maximize the sensitivity (see Chapter 18). DEER experiments with nitroxide samples can typically be done both at 9–10 GHz (X-band) and 33–35 GHz (Q-band), where the latter tends to be more sensitive. On a given spectrometer, the choice of frequency is limited by the bandwidth of the available resonators.

11.7.4 Resonator Coupling

As described in Section 11.6, pulse EPR resonators are typically overcoupled to achieve resonator bandwidths that are wide enough to admit the full excitation profile of the pulses and to minimize the length of the ringdown after the pulses. However, overcoupling also reduces the detected amplitude of the emitted FIDs and echoes. Therefore, a resonator should be overcoupled just enough to achieve the bandwidth necessary to admit the pulses and eliminate the ringdown, but not more in order not to degrade the detection sensitivity.

11.7.5 Temperature

The sample temperature has a profound effect on the longitudinal and transverse relaxation rates of the spins (see Chapter 9). To find an optimal temperature, T_1 and T_m (T_2) should be measured over a range of temperatures. Ideally, the temperature is such that both T_1 and T_m are not too short relative to the pulse sequence length. As discussed above, T_1 should not be too long, as it then limits the shortest possible repetition time. Typical temperatures are 10 K or lower for high-spin systems and metal ion clusters, 10–30 K for transition metal ions, and up to 100 K for organic radicals.

11.7.6 Repetition Time

The various relaxation rates determine how fast the spin systems return back to thermal equilibrium after a pulse sequence. This rate dictates the fastest possible repetition rate for the pulse sequence. Typically, one waits for about five times the slowest relaxation time constant before starting a new pulse sequence. At that time, $1 - e^{-5} \approx 99\%$ of all excited spins have relaxed back to thermal equilibrium. Faster repetition times are allowable if partial saturation does not affect the desired spectral shapes. Typical repetition times are in the 1–20 ms range, but can be shorter or longer in more unusual circumstances. For ENDOR experiments, nuclear relaxation times are also relevant and can be substantially longer than the electron relaxation times. If the repetition time is too fast to allow for the spins to relax back to thermal equilibrium, saturation effects ensue. In ENDOR experiments, their effect can be averaged out using random acquisition schemes, but they can also be used to determine relative signs of hyperfine couplings (see Chapter 16).

11.7.7 Pulse Lengths

The lengths of rectangular pulses are chosen based on the desired excitation bandwidth and on the available power. Experiments that rely on narrow-band selective pulses (Davies ENDOR, FID-detected field sweeps) and on exciting forbidden transitions (ELDOR-detected NMR) require long pulses. For other experiments (ESEEM), the shortest possible pulses are desired. If two MW frequencies are used, as in DEER experiments, then the bandwidth of the pulses has to be considered in order to control (or avoid) the overlap of the pulse excitation bandwidths. Frequency-swept pulses can be used to achieve broadband excitation (see Chapter 21).

11.7.8 Detection

Pulse EPR signals are detected by transient recorders and digitizers. Before transfer to the computer, the FID or echo is often integrated over a certain time window (boxcar integration). The width of this integration window affects the SNR and the spectral resolution. For integrating an echo, the best SNR is obtained when the integration window is centered around the echo maximum and of a width somewhat larger than the full width at half maximum of the echo. A wider window leads to the deterioration of the SNR. However, in addition to SNR, bandwidth considerations are also important when choosing a detection window. Narrow detection windows, or single-point detection in the extreme limit, have wide bandwidth, whereas wide detection windows have narrower bandwidth and might lead to loss of detail if echo modulations are recorded. For echo-detected field sweeps, it is best to integrate most of the echo to avoid broadening.

11.8 FURTHER READING

The standard monograph on pulse EPR spectroscopy is by Schweiger and Jeschke.[2] It contains a comprehensive overview of pulse EPR theory and extensive references. Any textbook about NMR is also a worthwhile read for pulse EPR, as many of the principles are the same. A particularly useful one at the introductory level is by Levitt.[11] The classic NMR textbook is by Ernst *et al.*[12] The EPR book edited by Brustolon and Giamello[13] contains a chapter on pulse EPR by Bowman. The EPR textbook by Weil and Bolton has a chapter on pulse EPR ('non-continuous excitation').[14] The manuals of Bruker pulse EPR spectrometers contain excellent introductory material. Several other multi-contributor books,[15–17] two book series[18,19] and a recent volume of Methods in Enzymology[20] contain chapters about specific pulse EPR experiments or specific applications of pulse EPR.

ACKNOWLEDGMENTS

The work in the Stoll research group is currently supported by the National Science Foundation, the National Institutes of Health, the Research Corporation for Science Advancement, the American Chemical Society, and the American Heart Association. The author thanks his current research group, William K. Myers (University of Oxford), Troy A. Stich (University of California Davis), and Jeffrey R. Harmer (University of Queensland) for valuable feedback on the manuscript.

REFERENCES

1. C. Gemperle, G. Aebli, A. Schweiger, and R. R. Ernst, *J. Magn. Reson.*, 1990, **88**, 241.
2. A. Schweiger and G. Jeschke, Principles of Pulse Electron Paramagnetic Resonance, Oxford University Press: Oxford, 2001.
3. E. L. Hahn, *Phys. Rev.*, 1950, **80**, 580.
4. F. Mentink-Vigier, A. Collauto, A. Feintuch, I. Kaminker, V. Tarle, and D. Goldfarb, *J. Magn. Reson.*, 2013, **236**, 117.
5. S. Meiboom and D. Gill, *Rev. Sci. Instrum.*, 1958, **29**, 688.
6. A. L. Bloom, *Phys. Rev.*, 1955, **98**, 1105.
7. F. Bloch, *Phys. Rev.*, 1946, **70**, 460.
8. B. Epel, D. Arieli, D. Baute, and D. Goldfarb, *J. Magn. Reson.*, 2003, **164**, 78.
9. J. R. Klauder and P. W. Anderson, *Phys. Rev.*, 1962, **125**, 912.
10. G. A. Rinard, R. W. Quine, S. S. Eaton, G. R. Eaton, and W. Froncisz, *J. Magn. Reson. A*, 1994, **108**, 71.
11. M. H. Levitt, Spin Dynamics: Basics of Nuclear Magnetic Resonance, 2nd edn, Wiley: Hoboken, NJ, 2008.
12. R. R. Ernst, G. Bodenhausen, and A. Wokaun, Principles of Nuclear Magnetic Resonance in One and Two Dimensions, Clarendon Press: Oxford, 1989.

13. M. Brustolon and E. Giamello, Electron Paramagnetic Resonance. A Practitioner's Toolkit, Wiley: Hoboken, NJ, 2009.
14. J. A. Weil and J. R. Bolton, Electron Paramagnetic Resonance, 2nd edn, Wiley: New York, 2007.
15. L. Kevan and R. N. Schwartz, Time Domain Electron Spin Resonance, Wiley: Weinheim, 1979.
16. A. J. Hoff, Advanced EPR. Applications in Biology and Biochemistry, Elsevier: Amsterdam, 1989.
17. L. Kevan and M. K. Bowman, Modern Pulsed and Continous-Wave Electron Spin Resonance, Wiley: Chichester, 1990.
18. B. C. Gilbert, V. Chechik, and D. M. Murphy, ed. Electron Paramagnetic Resonance, Royal Society of Chemistry: Cambridge, 2015, Vol. 24.
19. M. A. Hemmings and L. J. Berliner, ESR Spectroscopy in Membrane Biophysics, Springer: New York, 2007.
20. P. Z. Qin and K. Warncke, ed. Electron Paramagnetic Resonance Investigations of Biological Systems by Using Spin Labels, Spin Probes, and Intrinsic Metal Ions: 2015, Vol. 563.

Chapter 12
EPR Instrumentation

Edward Reijerse and Anton Savitsky

Max Planck Institute for Chemical Energy Conversion, Mülheim an der Ruhr, Germany

12.1 INTRODUCTION

The great leap forward in microwave (MW) and electronic technology, triggered by the development of radar techniques during World War II, allowed the construction of a new generation of spectrometers with the sensitivity and resolution required to detect magnetic resonance effects. In 1944, the first experiments of E.K. Zavoisky at Kazan University revealed electron spin resonance effects in solids. Since its discovery, EPR spectroscopy has undergone continuous improvements owing to new developments in spectroscopic instrumentation and methodology. During the first decade after its discovery, a large number of spectrometers were constructed at various universities and research laboratories, some of which are still in operation to date. Commercial EPR spectrometers were first available in 1956 when Varian Associates shipped six X-band CW EPR instruments. Pulse EPR methods became accessible with the availability of high-power MW sources (magnetrons and klystrons) and amplifiers (traveling wave tube). Again, numerous groups at university laboratories and research institutes took the lead in developing powerful and versatile instruments with unsurpassed performance (even to the present date). The first commercial pulse X-band EPR spectrometer became available from Bruker Biospin in 1987. A commercial high-field EPR spectrometer operating at 94 GHz (W-band) was introduced by Bruker Biospin in 1996, more than 30 years after Varian Associates had presented a 35 GHz (Q-band) instrument in 1962. The first commercial millimeter-wave 263 GHz EPR spectrometer (Bruker Biospin), which is based on a closed-cycle cryogenic superconducting magnet, was installed in late 2009. The field and frequency range of EPR instrumentation is continuously extended and currently is reaching the THz domain in specialized high magnetic field laboratories (large-scale facilities) making use of resistive and hybrid magnet technology (see Section 12.2) and multiplied solid-state millimeter-wave sources (see Section 12.3.2). In the meantime, EPR spectroscopy is approaching the level of technical sophistication necessary to become a powerful analytical method, ready to be applied to characterize a wide range of paramagnetic systems with hitherto unprecedented detail.

EPR Spectroscopy: Fundamentals and Methods.
Edited by Daniella Goldfarb and Stefan Stoll
 ISBN: 978-1-119-16299-5
Also published in eMagRes (online edition)
DOI: 10.1002/9780470034590.emrstm1511

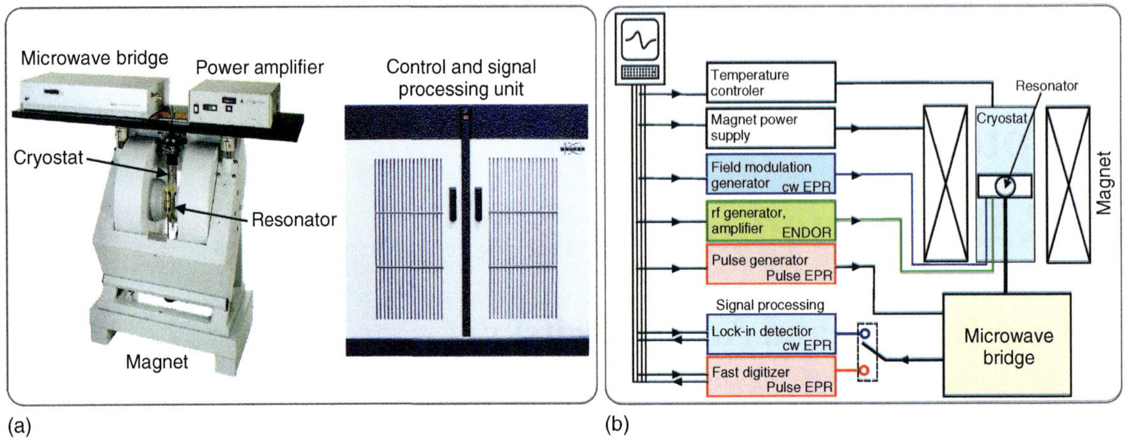

Figure 12.1. (a) X-band CW/pulsed EPR spectrometer Elexsys E580 (Bruker Biospin). (b) General block diagram of a CW/pulse EPR spectrometer. Blue, red, and green colors mark the specific components required for CW, pulse EPR, and electron nuclear double resonance (ENDOR) experiments, respectively

A typical multipurpose CW/pulse EPR spectrometer setup is depicted in Figure 12.1(a). The general block diagram of such a spectrometer is shown in Figure 12.1(b). It consists of several key components: (i) a magnet that produces a variable homogeneous magnetic field covering the field range of interest; (ii) an MW bridge containing the MW source and detector along with the necessary control electronics; (iii) a resonator that concentrates the incident MW radiation onto the sample and allows detecting the small amount of energy absorbed or emitted when the EPR resonance condition is met; (iv) a cryostat or thermostat that allows varying the sample temperature; and (v) control and signal processing electronics. In the following, we will briefly describe these components.

As of 2018, Bruker Biospin (www.bruker.com) is the major supplier of EPR CW/pulse multifrequency EPR systems. CW EPR instruments (X- and Q-band) can also be obtained from Jeol Ltd. (www.jeolusa.com). Compact benchtop X-band CW EPR instruments are supplied by Bruker Biospin (EMXnano), Magnettech (www.magnettech.de), Active Spectrum (www.activespectrum.com), Adani (www.adanisystems.com), and Keycom (www.keycom.co.jp).

12.2 MAGNET SYSTEMS

12.2.1 Iron Magnets

The standard magnet system for EPR spectrometers up to Q-band (34 GHz) consists of an iron magnet with pole diameter up to 10 inch (25 cm), air gap of typically 72 mm, and a current control system based on a Hall sensor. Using a 12 kW power supply, a standard 10″ magnet can reach up to 1.45 T. With additional pole pieces, the field range can be extended up to 1.8 T, but this configuration has a smaller air gap (56 mm). For special applications such as single-crystal studies, these magnets can be delivered on a rotating base. The homogeneity of these magnets is typically about 10^{-5} in a spherical volume of 1 cm^3, and the absolute field accuracy of the Hall controller is approximately 0.2 mT. Larger magnets (13 inch, 33 cm) with 1 Ohm resistance (fields up to 2 T) are no longer in the production program, but second-hand units (Bruker Biospin, Varian) may still be available. When iron magnets are ramped using a Hall field controller, the magnetic field axis of the experiment is well defined. The Hall sensor supplies an accurate field value as reference for the current supply. The absolute field accuracy can be substantially improved by integrating an NMR teslameter (5 ppm, 2 μT at 0.34 T) in the spectrometer system and acquisition software. Bruker Biospin also provides a field-frequency lock as accessory to reach even higher accuracy for applications involving radicals in solution. In the meantime, modern current supplies provide a 10-ppm accuracy (4 μT at 0.34 T), which can be reached without a Hall sensor as offered by Danfysik (Table 12.1). However, some inaccuracy will remain depending on the field hysteresis of the magnet's iron core.

Table 12.1. Vendors of magnets, power supplies, and field sensors

Vendor	Web site
Bruker Biospin	www.bruker.com
Jeol	www.jeolusa.com
Cryogenic Ltd	www.cryogenic.co.uk
Danfysik	www.danfysik.com
Cryomagnetics	www.cryomagnetics.com
Lakeshore	www.lakeshore.com
Ice Oxford	www.iceoxford.com
Oxford Instruments	www.oxford-instruments.com

12.2.2 Cryomagnets

For EPR applications above Q-band (34 GHz), iron magnets are no longer practical and superconducting magnets are generally employed. For EPR applications requiring a small-field range such as the study of organic radicals and triplet states, a regular NMR magnet extended with a room temperature solenoid sweep coil (0.1 T) could be used.[1,2] The advantages here are that field homogeneity is good (1 ppm) and the operating costs (helium losses) are quite low. The disadvantage, however, is that the solenoid configuration in which the magnet field is aligned along the open bore of the magnet is not practical for sample top loading in a cylindrical TE_{011} resonator (see Section 12.4). Specially designed inserts in which the resonator can be exchanged together with the sample allow operation also in a solenoid configuration.[3–5] A transverse magnetic field that is generated by a split-pair configuration is much more versatile as it allows for top loading in any resonator (both TE_{011} and Fabry–Pérot type, see Section 12.4) as well as sample rotation. The standard magnet configuration for W-band EPR, therefore, typically consists of a superconducting split-pair magnet with persistent switch (optional shims), a superconducting sweep coil, and an 80-mm warm bore accepting a helium flow cryostat in which a top-loading probehead is accommodated.[6,7] For applications requiring large field sweeps, the current controller can be switched to the main coil. To limit helium blow off, the sweep rate is usually reduced to 0.1 T per minute. A fundamental problem with superconducting magnets equipped with a thermoswitch (allowing persistent mode operation) is that the magnetic field inside the magnet is not linearly proportional to the current applied by the supply as part of the current is flowing through the shunt formed by the thermoswitch and protection diodes. By applying a current jump equal to the shunt current before starting the current ramp, the magnetic field ramp can be linearized.

The field strength can be further calibrated using standard samples generating EPR signals with known g-values.[6,8,9] For higher fields (>6 T), a split-pair configuration is no longer feasible. Therefore, EPR spectrometers with millimeter-wave frequencies above 150 GHz are exclusively based on superconducting solenoid type magnets[5,10] (see Chapter 27).

12.2.3 Closed-cycle Cryomagnets

An upcoming trend in magnetic resonance is the implementation of closed-cycle cryogenic systems for cooling the superconducting magnet and the sample space. Apart from cutting out liquid helium, which is an increasingly costly resource, a closed-cycle magnet system also provides the possibility to perform fast ($0.1-0.3\,\mathrm{T\,min^{-1}}$) and continuous field ramps without helium loss. In addition, these systems feature a graceful recovery after accidental magnet quench. Figure 12.2 shows a closed-cycle 12 T magnet system from Cryogenic Ltd.[11] A similar magnet is delivered with the Bruker 263 GHz pulsed EPR instrument (see Chapter 27). The cooling power of the cold head can be used to cool the sample space (see Section 12.5). These integrated systems in which one coldhead cools both the magnet and a Helium flow cryostat are now offered by Cryogenic Ltd (see Table 12.1).

12.2.4 Resistive Magnets

For EPR applications with fields of 2 T up to 14 T, superconducting magnets are the primary choice. For very high fields, in particular higher than ≈22 T, superconducting magnet designs are not (yet) feasible. Instead users have to turn to large-scale facilities operating resistive magnets ('Bitter' type) connected to a very large power supply (20 MW).[12,13] The magnet is built up from flat segments forming a vertical solenoid. The cooling water ($\approx 1\,\mathrm{m^3\,min^{-1}}$) is flowing vertically through a hole pattern in the flat windings. Typical Bitter magnets can reach magnetic fields up to 37.5 T in a 32 mm warm bore (see Table 12.2). Using a hybrid design, i.e., a Bitter magnet inside a superconductive magnet, fields up to 45 T can be

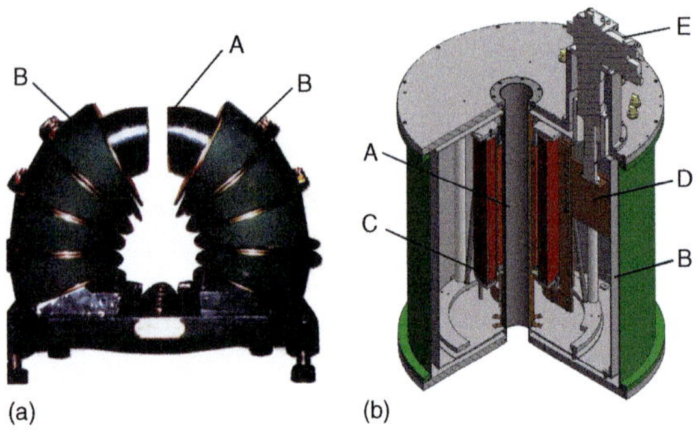

Figure 12.2. A century's evolution of EPR Magnet technology. (a): Iron magnet type used by Zavoiky in 1944 (Hartmann & Braun, Frankfurt am Main, 1928). (A) Iron core and (B) magnet coils. (b): Schematic of a cryogen-free superconducting magnet (Cryogenic Ltd, London, 2004):[11] (A) 89-mm room temperature bore, (B) radiation shield, (C) 12.1 T superconducting solenoid, (D) copper thermal links, and (E) cryocooler. (Courtesy of Cryogenic Ltd.)

reached in a bore of 32 mm. Large-scale high magnetic field facilities are available in the United States (nationalmaglab.org), Europe (www.emfl.eu), and China (http://english.hmfl.cas.cn). A major problem with Bitter magnets is the low homogeneity (10^{-2} in 1 cm diameter spherical volume (DSV).), which is suboptimal for EPR applications. Several solutions for this problem have been reported. At the magnet laboratory in Tallahassee (NHFML), a specially designed Bitter magnet reaches field homogeneity of 5×10^{-5} at 35 T. The high-field magnet laboratory in Nijmegen (HFML) has developed room temperature ferroshim inserts into regular Bitter magnets improving the homogeneity to allow solid-state NMR experiments.[14] In general, the magnet facilities at these large-scale facilities are used for EPR studies up to 25 T on molecular magnets and other high-spin materials.[13]

12.2.5 Pulse Magnets

For very special applications (e.g., ferromagnetic resonance, quantum Hall effect, and cyclotron resonance), even higher fields are desirable. These experiments are performed in pulse magnets (fields up to 100 T).[15,16] The charge providing the required peak currents is stored in large capacitor banks. Typically, only a few milliseconds of resonant field are available for each shot. Owing to the substantial time required to recharge the capacitor banks, the delay between these experiments is of the order of 1 h. The major high magnetic field facilities in the United States (nationalmaglab.org) and Europe (www.emfl.eu) also provide pulsed field magnets.

12.3 MICROWAVE BRIDGES

12.3.1 Detection Schemes

Figure 12.3(a) shows the operation principle of a reflection resonator homodyne EPR spectrometer. For

Table 12.2. Example of magnet systems used in EPR

Magnet	Field (T)	Homogeneity (1 cm DSV)[a]	Bore/gap (mm)	Power (kW)	Accuracy Abs. Gauss	Accuracy Rel. Gauss
10″ (25 cm)[b]	1.45	10^{-5}	72	12	2	<0.1
10″ (25 cm)[b]	1.7	10^{-5}	56	12	2	<0.1
Cryo split[b]	6	10^{-5}	80	—	5	<1
Cryo solenoid[c]	12	10^{-4}	80	—	1	<1
Bitter[d]	30	10^{-3}	50	17 000	—	<1
Bitter[d]	37.5	10^{-3}	32	20 700	—	<1
Hybrid[e]	45	10^{-3}	32	30 000	—	<1
Bitter[e]	30	5×10^{-5}	32	18 000	—	<1

[a]DSV, diameter spherical volume.
[b]www.bruker.com
[c]www.cryogenic.co.uk
[d]HFML Nijmegen (www.emfl.eu)
[e]NHFML Tallahassee (https://nationalmaglab.org)

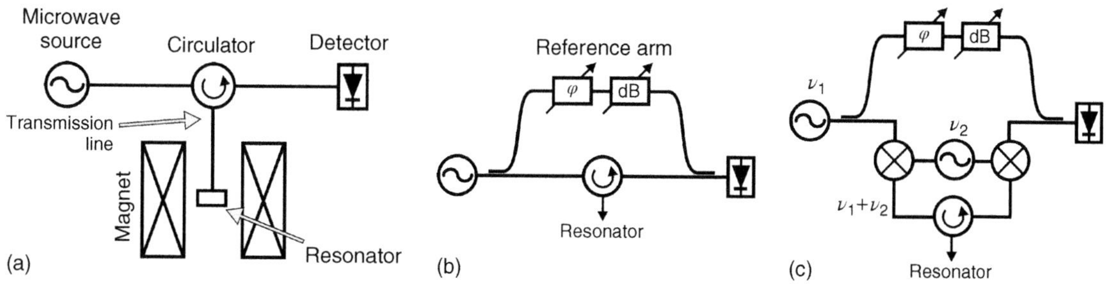

Figure 12.3. Schematic block diagrams of EPR spectrometers with (a) homodyne detection and reflection mode resonator; (b) homodyne detection with reference arm and reflection mode resonator; and (c) heterodyne detection with reflection mode resonator

optimum sensitivity, the resonator is exactly matched to the transmission line waveguide and absorbs all incident MW power. In case of EPR absorption, the matching between the resonator and the waveguide is slightly perturbed and the same amount of MW power as absorbed by the sample is reflected toward the detector. The excitation and reflected MWs are decoupled from each other by the circulator, which ensures that a signal fed into one of its ports leaves through one of adjacent ports only. If it is required that a reflection resonator spectrometer provides information about EPR absorption and dispersion separately, a reference arm should be included in the system as depicted in Figure 12.3(b). Using directional couplers, part of the MW power of the MW source can be split off, bypassing the circulator and resonator, and fed directly to the detector where it combines with the MW signal reflected from the resonator carrying the EPR information. The phase and magnitude of this reference MW signal is adjusted by a phase shifter and attenuator to obtain the required degree of suppression of either the absorption or dispersion signal. The principles, limitations, and practical realizations of the different homodyne EPR detection schemes in Figure 12.3(a-b) are described in the monographs of Wilmshurst,[17] Poole,[18] and Ingram,[19] to which we refer for further details. The homodyne detection scheme is generally used in CW and pulse EPR spectrometers up to X-band frequencies. For higher frequencies (e.g., Q- and W-bands), the homodyne scheme is employed mostly in CW EPR spectrometers, while in pulsed EPR instruments, the heterodyne detection scheme is typically used, for instance, in Bruker EPR spectrometers.[5]

The heterodyne detection scheme with reflection mode resonator is depicted in Figure 12.3(c). In this scheme, an additional MW source (local oscillator – LO) with frequency ν_2 is added. Its signal is upconverted using a Schottky diode mixer that generates the MW signal at the desired EPR operation frequency, $\nu_1 + \nu_2$, using the MW signal ν_1 of the intermediate frequency (IF) source. The subsequent EPR response from the resonator is downconverted to the frequency of the main (IF) source at which the detection takes place. An advantage of this scheme is that the all manipulations of excitation MW and amplification and phase manipulation of the EPR signal can be performed at the IF frequency using high-performance low-cost MW components. Moreover, it allows using fixed-frequency low-phase noise LO sources, which results in improved signal-to-noise ratio (SNR) of the EPR signal. The heterodyne principle allows to extent any X-band EPR spectrometer to operation at higher or lower MW frequencies. For example, the Bruker Biospin E600/680 W-band spectrometer is based on the standard X-band MW pulse bridge with the addition of an up–down converter accessory to obtain an operational frequency around 94 GHz. Spectrometers at even higher frequencies (millimeter-waves) are usually based on the same heterodyne principle but may operate with different probeheads (e.g., nonresonant sample holders) and different MW propagation techniques. These instruments are discussed elsewhere in this book (see Chapter 27).

12.3.2 Microwave Sources and Amplifiers

Selection criteria for the main MW source for EPR purposes include noise performance, tuning bandwidth, durability, convenience, and costs. Most MW sources for EPR spectroscopy can be divided into two groups: vacuum-tube oscillators and solid-state oscillators. Vacuum-tube oscillators such as klystrons, backward-wave oscillators (BWOs), gyrotrons, extended interaction oscillators (EIOs), and magnetrons have been, and still are, used as coherent sources especially for high-power high-frequency MW. Until the 1970s, the reflex klystron was the main EPR source owing to its excellent phase-noise characteristics. Reflex klystrons are capable to produce up to 1 W CW radiation at frequencies up to 220 GHz. Breakthrough discoveries in solid-state semiconductor technology in the mid-1960s, such as the 'Gunn effect' in gallium-arsenide in 1964, allowed the development of reliable solid-state MW sources. In the following years, the vacuum-tube sources were increasingly replaced by solid-state semiconductor oscillators, which are smaller, easier to operate and reach excellent performance with increased reliability. Solid-state sources are preferred because of relatively low cost and compact encasement. This simplifies the integration into the EPR MW bridge with its compact array of components. Moreover, solid-state sources do not require high control voltages, additional cooling and are generally less sensitive to stray fields from the EPR magnet. They deliver, however, relatively low output power as compared to vacuum-tube devices. EPR solid-state sources generally employ negative resistance devices, based on IMPATT (GaAs, Si) or Gunn diodes (GaAs, InP). Gunn diode sources are used in the frequency range from 1 to 150 GHz providing up to 300 mW (25 dBm) of CW radiation at 9.5 GHz, 100 mW (20 dBm) at 95 GHz, and 20 mW (13 dBm) at 140 GHz. dBm (decibel-milliwatts) is an abbreviation for the power ratio in decibels (dB) of the measured power referenced to 1 milliwatt (mW): $P(\mathrm{dBm}) = 10 \times \log_{10} \frac{P(\mathrm{mW})}{1\,\mathrm{mW}}$. A power level of 0 dBm corresponds to a power of 1 mW.

A disadvantage of free-running solid-state sources is the rather poor noise and frequency jitter characteristics, which require phase-lock loops to a reference oscillator or external cavity stabilization. For example, a free-running GaAs Gunn diode at 95 GHz has a phase noise of about $-40\,\mathrm{dBc\,Hz^{-1}}$ at 10-kHz frequency offset ($-70\,\mathrm{dBc\,Hz^{-1}}$ at 100 kHz), which is the typical range of CW EPR with field modulation. Phase noise is typically expressed in units of $\mathrm{dBc\,Hz^{-1}}$, and it represents the noise power relative to the carrier contained in a 1-Hz bandwidth centered at a certain offsets from the carrier. Phase locking to a 100 MHz low-noise quartz oscillator allows improving the noise performance. In practice, the noise performance of a 100 MHz oscillator upscaled to 95 GHz can be reached, i.e., $-100\,\mathrm{dBc\,Hz^{-1}}$ at 10 kHz and $-130\,\mathrm{dBc\,Hz^{-1}}$ at 100 kHz. This phase-noise level fulfills all requirements for modern CW EPR. Phase locking or cavity stabilization, however, limits the 'sweepability' of the MW sources, which is necessary, for example, to adjust the matching of the EPR resonator. Diode devices can be used as MW sources, amplifiers, and active multipliers. This property allows realizing different designs to obtain the desired EPR operation frequency, either by multiplying the low-frequency MWs or by combining two MW frequencies by a balanced mixer with further MW amplification. The diode devices can be expected to operate continuously for over 100 000 h, thus offering a longer life time than vacuum-tube oscillators. Most vacuum tubes have a limited lifetime, due to the filament or heater burning out or other failure modes, for instance, the typical life time of klystron operating at W-band is approximately 5000 h.

The limited power level of MW oscillators is usually not sufficient for pulse EPR applications (especially at X-band). For the generation of a 10-ns $\pi/2$-pulse in an X-band dielectric resonator (see Section 12.4.2), approximately 500 W MW power is required. Therefore, MW power amplifiers are generally employed. Until recently, the traveling wave tube (TWT) amplifier was the main type of power amplifier. The TWTs are capable of producing 1 kW output power at X-band and 200 W at Q-band. For high-power pulse electron–electron double resonance (PELDOR) experiments at W-band, a $\approx$1 kW extended interaction klystron (see www.cpii.com) has been used in combination with a quasi-optical bridge.[20,21] At X-band, however, the tube-based amplifiers can already be replaced by solid-state amplifies. The advantages of solid-state devices are the much higher duty cycle, improved noise characteristics, lower phase droop over a pulse, and a favorable gating behavior as compared to vacuum-tube devices such as the TWT. Solid-state amplifiers with 300 W power at X-band, 20 W at Q-band, and up to 2 W at W-band are available for prices comparable to those of corresponding TWT devices.

Currently, no solid-state power amplifiers are available for frequencies above 150 GHz. Therefore, the millimeter-wave signal can only be generated either using active multiplier chains (see, e.g., www.vadiodes.com) or vacuum-tube oscillators. For pulse power (100 W) at frequencies in the range 0.2–3 THz, large-scale free electron laser (FEL) facilities offer interesting possibilities.[22]

12.3.3 Hardware for Pulse Double Resonance EPR Experiments

In the past 20 years, pulse EPR methods employing multiple MW frequencies have gained importance. In particular, PELDOR (also called DEER) and electron–electron double resonance (ELDOR)-detected NMR has become widely used for distance measurements between paramagnetic centers and for the determination of nuclear hyperfine couplings, respectively (see Chapters 17 and 20). For these experiments, at least two different MW frequencies are required. There are two approaches for the generation of MWs with different frequencies. The first one, which was the standard approach until recently, is to include an additional MW sources in the pulse bridge configuration, see Figure 12.4(a). The MW pulses generated by individual pulse forming units are then combined, amplified, and delivered to the EPR resonator. The second approach is to combine the MWs of the main source with radiofrequency wave generated by an arbitrary waveform generator (AWG), analogous to the principle used in heterodyne detection scheme. An AWG can generate any arbitrarily defined wave shape (e.g., 14 bit resolution 1.6 GSa s^{-1}) as output. Mixing this AWG signal with the MW from the EPR source (see Figure 12.4b) allows generating an MW frequency with an instantaneous offset ±800 MHz with respect to the detection MW frequency. Each individual MW pulse can be arbitrarily shaped in terms of amplitude and frequency (phase). This allows for composite pulses and frequency chirps to be realized in an EPR pulse experiment (see Chapter 21). This new technology will greatly stimulate further development of pulsed EPR methodology in the near future. Bruker Biospin offers an AWG extension to the existing EPR MW bridges since 2015.

12.3.4 Microwave Transmission Lines

The MW signal in the EPR bridge propagates between the various components and the resonator/probehead through transmission lines that are either fundamental-mode rectangular waveguides or coaxial cables. The power losses in these transmission lines, especially the one carrying the EPR signal from the probehead to the detector, are a critical determinant for the sensitivity of the spectrometer. In an X-band EPR spectrometer, the typical length of the transfer line to the probehead is about 1 m. At 9.5 GHz, the WR90 rectangular TE_{10} mode waveguide attenuates the MW power by about 0.1 dB m^{-1}. Typical losses of modern semirigid or flexible low-loss coaxial cables are higher (close to 1 dB m^{-1}) but still do not seriously degrade the spectrometer performance. At 34 GHz (Q-band), however, the loss in a coaxial cable increases to 2.5 dB m^{-1}, i.e., 45% of the MW power is dissipated in a 1-m cable and reduces the EPR signal by 25%. Therefore, WR28 rectangular waveguides with losses of 0.5 dB m^{-1} are generally used as transmission lines in Q-band EPR spectrometers. Fundamental-mode rectangular waveguides become less attractive when going to higher MW frequencies. With increasing frequency, the MW power losses increase, reaching unacceptable levels (>3 dB m^{-1}) above 100 GHz. An obvious solution is to use oversized waveguides with cross sections considerably larger than normally used for the specific frequency band. Increasing the waveguide cross section leads to decreasing wall currents and, as a result, a decrease in resistive (ohmic) losses. Thus, using WR28

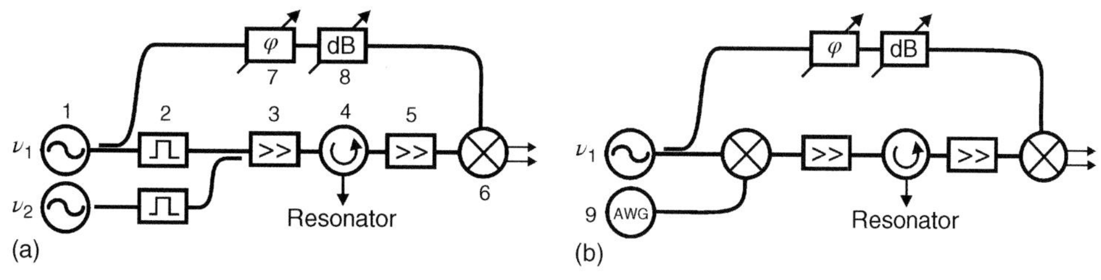

Figure 12.4. Basic schematic diagrams of pulse double/multiple resonance EPR bridges with (a) two independent microwave sources and (b) radiofrequency arbitrary wave generator. The main microwave components: 1, MW source; 2, pulse forming unit; 3, MW power amplifier; 4, circulator; 5, low-noise amplifier; 6, quadrature mixer; 7, phase shifter; 8, attenuator; and 9, arbitrary waveform generator

waveguide at 95 GHz (W-band) allows reducing the transmission losses to about 1.5 dB m^{-1} as compared to 2.5 dB m^{-1} in single-mode WR10. This solution is used, for example, in the commercial Bruker W-band probehead having a length of approximately 1 m. An additional loss of spectrometer performance comes from poor connections between the different transmission lines especially those connecting the probe head. At these interfaces, reflections will occur that lead to standing waves in the system, which interfere with the resonance mode of the probehead.

For frequencies above 140 GHz, propagation using rectangular waveguides becomes impractical altogether. At these frequencies, the MW wavelength is reduced to a few millimeters and free-space propagation using quasi-optics is the method of choice. In these millimeter-wave bridges, the millimeter-waves are launched from corrugated feed horns and propagated as Gaussian beams that can be manipulated using off-axis elliptical mirrors, wire grid polarizers, and Faraday rotators.[5,23–26] The probehead can be integrated into this system making use of a cylindrical corrugated waveguide propagating the HE mode, which is compatible with the free-space Gaussian beam. The typical insertion loss of these guides is below 1 dB m^{-1}. The Bruker 263 GHz millimeter-wave EPR spectrometer is based on such a quasi-optical bridge. An extensive overview of high-frequency EPR instrumentation is provided in Chapter 27. Table 12.3 provides an overview of the various MW frequency bands employed in EPR together with selected key parameters on the waveguide and the EPR resonance condition.

12.4 EPR RESONATORS

In this section, we will briefly discuss various MW resonators that are commonly used for EPR spectroscopy and several structures that are designed for special purposes. We will limit ourselves to a mostly qualitative and practical comparison between different resonators. An extensive overview of EPR resonator cavities, including the corresponding MW theory, can be found in the classical text by Charles Poole Jr.[18]

In general, all resonator configurations follow three basic principles: (i) The magnetic component of the MW field, B_1, in the sample should be perpendicular to the external magnetic field; (ii) the sample is located at a position with maximum B_1; and (iii) the sample is located at a position with minimum electric field, E_1. The sensitivity of an EPR experiment is critically dependent on the (loaded) quality factor of the resonator that can be expressed as

$$Q_L = 2\pi \cdot \frac{\text{stored energy}}{\text{dissipated energy in one cycle}} \quad (12.1)$$

Table 12.3. Microwave frequency bands at which EPR spectrometers have been constructed. The WR designation indicates the largest inner cross-sectional dimension of the waveguide in inch/100

Band	Frequency (GHz)	Waveguide WR	Selected ν (GHz)	Free-space λ (mm)	Energy (cm^{-1})	Res. field for $g = 2$ (T)
L	1–2	WR650	1	300	0.033	0.036
S	2–4	WR284	3	100	0.1	0.11
C	4–8	WR137	6	50	0.2	0.21
X	8–12	WR90	9.5	31.6	0.32	0.34
Ku	12–18	WR62	15	20	0.5	0.54
K	18–27	WR42	24	12.5	0.8	0.86
Ka	27–40	WR28	35	8.6	1.2	1.25
V	40–75	WR15	50	6.0	1.7	1.78
W	75–110	WR10	95	3.16	3.16	3.4
F	90–140	WR8	122	2.46	4.06	4.36
D	110–170	WR6.5	130	2.3	4.3	4.64
G	140–220	WR5.1	180	1.67	6.0	6.43
—	170–260	WR4.3	244	1.23	8.13	8.72
J	220–325	WR3.4	263	1.14	8.77	9.4
J	220–325	WR3.4	275	1.09	9.17	9.82
—	265–400	WR2.8	360	0.83	12	12.86

The quality factor is equivalently defined as the ratio of the resonance frequency ν_0 over the resonator bandwidth $\Delta\nu$: $Q_L = \nu_0/\Delta\nu$. The resonator is always connected to a transmission line (coax and waveguide) through a coupling structure (e.g., iris, antenna, and loop). Therefore, the Q_L-factor depends on both the 'unloaded' Q (Q_U), which is related to the ohmic losses in the cavity walls and dielectric losses in the sample, and the 'radiation' Q (Q_R) of the coupling structure, which is related to the losses through the coupling structure:

$$1/Q_L = 1/Q_U + 1/Q_R \tag{12.2}$$

Alternatively, the Q_L-factor can be expressed using the dimensionless coupling parameter β:

$$Q_L = Q_U/(1+\beta) \tag{12.3}$$

The optimal matching condition for energy transfer to the resonator is obtained for $Q_R = Q_U$, i.e., $\beta = 1$. The exact Q_U and coupling parameter of a resonator can be best determined using a vector network analyzer.[27] An alternative method is to analyze the signal responses after application of rectangular MW pulse (Figure 12.5).[28] From the reflected monoexponential voltage decay profile after switching off the MW pulse, the voltage decay time or 'ringing time', τ, is determined. From this, the Q_L-value can be calculated as $Q_L = \pi\nu\tau$. In addition, phase-sensitive detection of the pulse response (i.e., signal voltage V_p during the MW pulse after ring-down process) allows three coupling regimes to be distinguished: ($V_p > 0$): if the reflected MWs are *in phase* with the incident MWs, the resonator is *undercoupled* ($\beta < 1$); ($V_p < 0$): if the reflected MWs are *out of phase* with the incident MWs, the resonator is *overcoupled* ($\beta > 1$); ($V_p = 0$): if no MWs are reflected, the resonator is *critically coupled* ($\beta = 1$), see Figure 12.5(b). The reflection coefficient, the voltage ratio of the reflected to incident MW powers, $\Gamma = (P_{refl}/P_{inc})^{1/2}$, allows a direct measurement of the coupling parameter:

$$\beta = (1-\Gamma)/(1+\Gamma) \text{ for } \beta < 1$$
$$\beta = (1+\Gamma)/(1-\Gamma) \text{ for } \beta > 1 \tag{12.4}$$

In addition to the Q_L-value, the filling factor, η, which quantifies the ratio of the energy of the MW magnetic field at the sample to that integrated over the resonator volume, is of particular importance for EPR sensitivity. In the case of a critically coupled resonator ($Q_L = Q_U/2$, $\beta = 1$), the concentration sensitivity is proportional to $Q_U\,\eta\,\omega_0$ where ω_0 is the spectrometer frequency in angular frequency units, $\omega_0 = 2\pi\nu_0$. Maximizing this parameter will lead to improved CW performance. The conversion factor, K, which relates the B_1 field amplitude to the square root of the incident power P, $B_1 = K\sqrt{P}$, is proportional to $(Q_L\omega_0^{-1}V_C^{-1})^{1/2}$, where V_C is the effective volume of the resonator. Optimizing this parameter is important for pulse EPR, as a high B_1 field is required at the

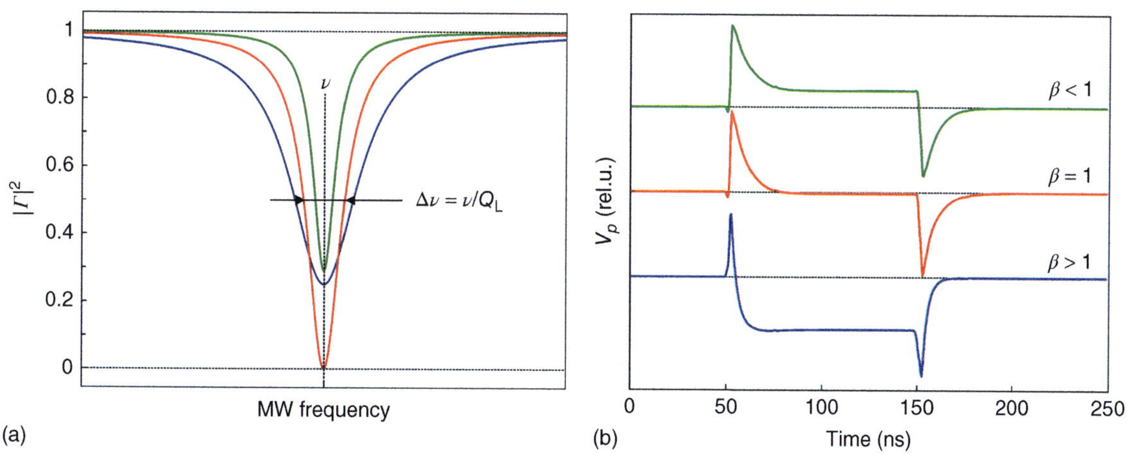

Figure 12.5. (a) The frequency dependence of MW power fraction, $|\Gamma|^2$, reflected from undercoupled ($\beta = 0.3$, green trace), critically coupled ($\beta = 1$, red trace), and overcoupled ($\beta = 3$, blue trace) resonators. The resonator bandwidth for critically coupled case is indicated. (b) Recorded profiles of 95-GHz mw pulse of 100 ns applied to TE_{011} cylindrical cavity for undercoupled ($\beta < 1$), critically coupled ($\beta = 1$) and overcoupled ($\beta > 1$) conditions. For details, see Ref. 28

sample to obtain short MW pulses. It is clear that the requirements on the resonator are different for pulse and CW EPR experiments: For CW experiments, ideally, the resonator is critically coupled providing a high Q_L-value and a narrow bandwidth, $\Delta\nu$, allowing for high sensitivity and accurately locking the EPR source to the resonance frequency of the resonator using an automatic frequency control (AFC) circuit. Pulse EPR experiments instead require a large bandwidth to accommodate both the short excitation MW pulses and the induction response of the spin system. Moreover, a large bandwidth guarantees a short dead time of the spectrometer, which is determined by the time required for the ringing MW power to decay approximately to the level of the expected EPR response. The reduction in detection sensitivity due to the low Q_L of the overcoupled ($\beta \gg 1$) resonator has to be compensated by improving the filling factor η (i.e., the volume of the resonator). As will be discussed in the following sections, resonators for pulse EPR applications are of special design.

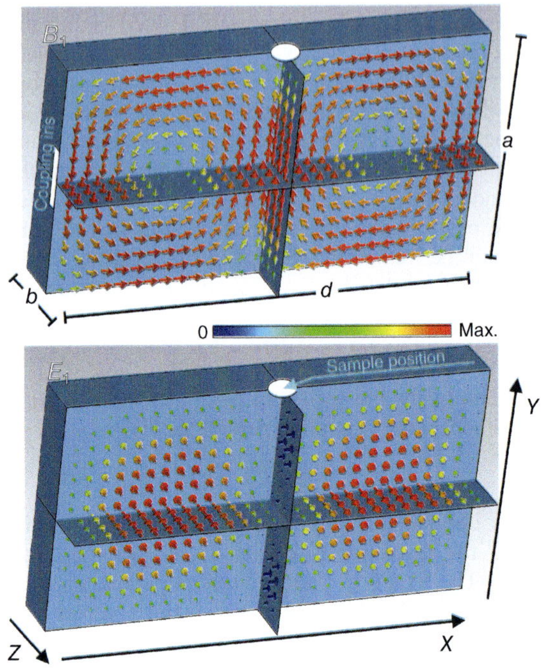

Figure 12.6. Distribution of microwave magnetic (B_1) and electric (E_1) field intensities in a TE_{102} resonator. Directions Y and Z correspond to the sample tube axis and the static magnetic field (B_0) direction, respectively

12.4.1 Single-mode Cavities

A MW cavity resonator consists of reflective and conductive surfaces that confine the MWs into a limited space forming a standing wave between the walls of the cavity. The cavity can be shaped such that positioning the sample at an optimal position with maximum magnetic component and minimum electric component of the MW field is easily possible. EPR cavities can be shaped as a rectangular or cylindrical box or be formed by two mirrors in a 'Fabry–Perot'-type resonator.

12.4.1.1 *TE_{102} Cavity*

The most commonly used general-purpose X-band EPR resonator is the TE_{102} rectangular cavity. The designation 'TE_{102}' originates from the rectangular waveguide mode from which this resonator is derived. While the electric field is perpendicular (transverse, TE) to the propagation direction, it contains *one* electric field maximum along the '*a*' direction (long side), *zero* along the '*b*' direction (short side), and *two* along the propagation direction '*d*'. The *ab* plane at *d*/2 has a magnetic field maximum (B_1) where the EPR sample is positioned, see Figure 12.6. The resonance frequency for a TE_{102} resonator is given as[18]

$$\nu = \frac{\sqrt{(1/a^2 + 4/d^2)}}{2\sqrt{\varepsilon\mu}} \approx \sqrt{(1/a^2 + 4/d^2)} \cdot 0.15\ \text{GHz} \quad \text{(for empty resonator)} \tag{12.5}$$

Here, ε and μ are the permittivity and permeability of the medium in the resonator (typically air). Thus, approximate dimensions for a TE_{102} cavity with a resonance frequency of 9.5 GHz are (a, b, d) = (24, 10, 42) mm. A sample access of 10 mm diameter allows the use of a quartz dewar insert for sample cooling. The MWs are usually coupled from a waveguide through a variable iris. The side walls of the cavity are built thin enough such as to allow for the penetration of 100 kHz sinusoidal field modulation with up to 2 mT peak-to-peak amplitude (see Chapter 23).

12.4.1.2 *TE_{104} Cavity*

This variant of the standard rectangular cavity contains two positions with a B_1 maximum allowing the insertion of two samples simultaneously. This feature is ideal for quantitative spin measurements in which one of the samples (the reference) contains a known number of electron spins. Using two sets of modulation coils (see Chapter 1) the sample signals can be recorded separately, but as both samples reside in the same resonator, the cavity parameters (Q_L-factor and coupling) affect both signals in the same way thus facilitating a reliable calibration of the spin counting measurement. In practice, however, additional parameters such as the B_1 field distribution inside the individual samples affect the signal amplitudes. Therefore, the reference sample has to be 'designed' to match the unknown sample as close as possible. Spin counting experiments have improved considerably using dedicated spectrometer software making it possible to perform spin counting in any resonator configuration once the spectrometer parameters have been calibrated and stored in a configuration file.[27]

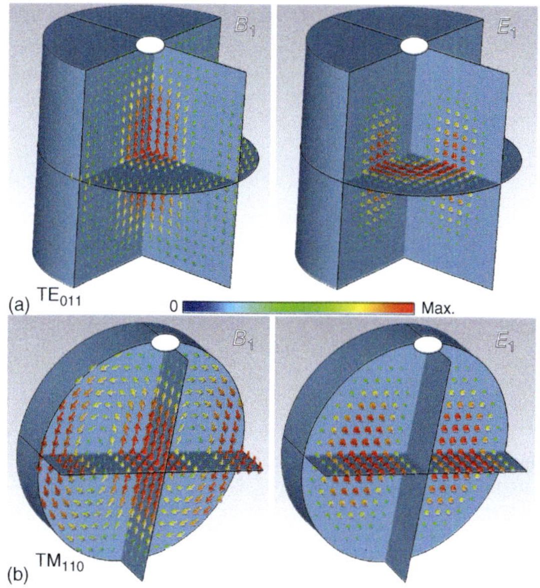

Figure 12.7. Distribution of microwave magnetic (B_1) and electric (E_1) field intensities in TE_{011} (a) and TM_{110} cavities (b)

12.4.1.3 TM_{110} Cavity

The cylindrical TM_{110} cavity is mostly used for aqueous liquid samples that are accommodated in flat cells (Figure 12.7). In this cavity, the plane with the maximum B_1 field is better separated from the electrical fields making it well suited to study aqueous samples that would lead to smaller losses as compared to the TE_{102} cavity. It has been reported that an aqueous sample in a 10-mm flat cell measured in the Bruker TM_{110} resonator provides a fourfold increase in sensitivity as compared to the same sample measured in a 1-mm i.d. capillary inside the Bruker TE_{102} resonator.[27] As alternative to flat cells, aqueous samples can also be studied using multibore capillaries. These can be used in combination with the standard TE_{102} resonators and also with a high-Q TE_{011} cavity.

12.4.1.4 TE_{011} Cavity

The cylindrical TE_{011} cavity is very versatile and can be used up to very high frequencies (e.g., 275 GHz[3]). The resonant magnetic field B_1 is concentrated along the center axis of the cylinder, while the electric field lines are following circular paths parallel to the walls of the cylinder (Figure 12.7). At X-band frequencies, the resonator is mostly used as high-sensitivity probehead: The already high intrinsic Q-value (theoretical Q_U-value is 27 000 at 9.5 GHz) is further enhanced by additional modifications (ER 4119HS and ER 4122SHQE of Bruker) making it exquisitely suited for low-loss samples. However, very high sensitivity can also be obtained for aqueous samples in the Bruker multibore Aqua-X cells. For higher MW frequencies, the main advantage of the TE_{011} cavity is that the frequency is easily adjustable by moving the plungers that do not need electrical contact with the cylinder wall. Thus, TE_{011} cavities are routinely employed for both CW and pulse EPR experiments (requiring increased bandwidth) as well as pulse ENDOR experiments at frequencies above 34 GHz. For Q-band, the electric field distribution in the resonator allows for placement of vertical conduction posts inside the resonator, thus forming an RF coil (see Section 12.4.7).[29] The large sample access (3 mm at Q-band and 1 mm at W-band) is beneficial for the concentration sensitivity of the resonator. For very wide bandwidth applications such as DEER- and ELDOR-detected NMR (see Chapters 17 and 19), Bruker provides a Q-band TE_{012} resonator (QT-II) (without ENDOR capability) that accepts large samples (3 mm o.d.) and offers bandwidth of over 100 MHz.

The TE_{011} resonator was also used in homebuilt high-field EPR spectrometers operating from 95 to 275 GHz.[5,10] Bruker also provides a single-mode

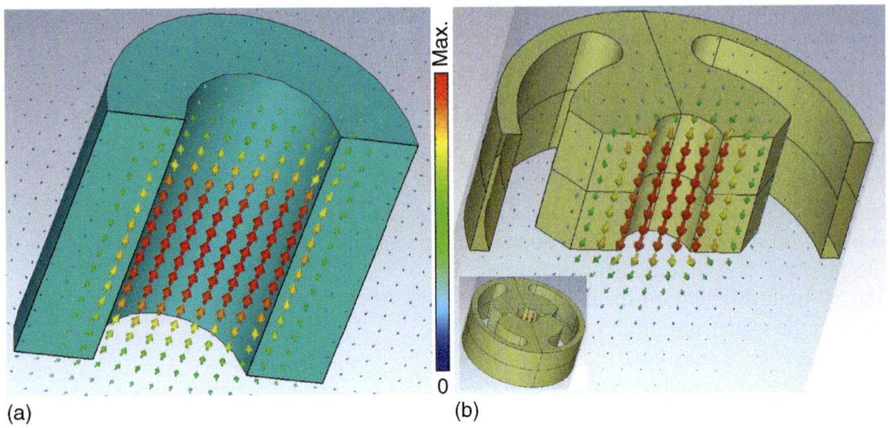

Figure 12.8. Distribution of microwave magnetic (B_1) field intensity in a dielectric ring resonator (a) and a split-ring resonator (b) (shield is omitted for clarity)

TE_{011} resonator at 263 GHz accepting 0.35 mm o.d. sample capillaries (more information in Chapter 27).

12.4.1.5 Dual-mode Cavity

For the studies of triplets, biradicals, and transition metal ions exhibiting 'forbidden' fine structures or hyperfine structure transitions, a dual-mode cavity accommodating both TE_{102} and TE_{012} modes can be used. The two modes differ slightly in frequency and have their B_1 field either perpendicular or parallel to the external magnetic field, B_0. This allows conducting EPR experiments with B_1 field either parallel or perpendicular to the external magnetic field, B_0 in one cavity. Bruker Biospin offers such a cavity for X-band EPR (ER 4116DM).

12.4.2 Dielectric Resonators

Insulators with a high dielectric permittivity (ε) and low-loss tangent tan δ (a measure of the energy loss in the medium) are well suited to construct resonance structures in the MW range. It turns out that a single crystal of sapphire ($\varepsilon \approx 10$, tan $\delta \approx 10^{-5}$ at 10 GHz) has the right parameters to construct a resonator compatible with 4.8-mm X-band tubes. A sapphire cylinder with i.d. = 5 mm, o.d. = 10 mm, and height = 13 mm positioned in a shield with diameter of 16 mm resonates around 9.5 GHz (see Figure 12.8a). The mode structure is very similar to the cylindrical TE_{011} mode (see Figure 12.7). Owing to the fact that the magnetic field lines are mostly concentrated inside the bore of the dielectric ring, this resonator has a high filling factor. The absence of wall currents also guaranties a very high Q-value. The most interesting property of this resonator is, however, that by adjusting the coupling antenna position a very high overcoupling can be achieved allowing for bandwidth of several 100 MHz (Bruker ER4118X-MD5). The high filling factor in combination with the wide coupling range makes this resonator the prime choice for pulsed EPR applications. In addition, the geometry of the electric field lines allows the insertion of an RF saddle coil inside the dielectric ring enabling ENDOR experiments (Bruker ER4118X-MD4) reducing, however, the sample access to 4 mm (o.d.)

This dielectric resonator design (including ENDOR coil) has also been scaled to Q-band frequencies (34 GHz) (Bruker EN 5102D2). The waveguide iris coupling allows for overcoupling that, in combination with intrinsically lower Q, facilitates pulse EPR experiments with bandwidths up to 100 MHz.

Employing dielectric materials with even higher permittivity will reduce the overall size of the resonator thus increasing the filling factor and, thus, the absolute spin sensitivity of the resonator. Recent studies described the performance of a dielectric ring (i.d. = 1 mm, o.d. = 2 mm, height = 3 mm) of a composite material with $\varepsilon \approx 80$, tan $\delta \approx 10^{-6}$ at X-band frequencies.[30] The resonator can be conveniently accommodated in the center of a regular TE_{102} or TE_{011} cavity.

12.4.3 Loop-gap Resonators

MW resonators based on discrete (lumped) elements (capacitors, resistors, and inductors) have become very popular during the development of EPR spectroscopy.[31–34] Loop-gap resonators (LGRs) can be realized in a large variety of designs. These structures contain one (or more) 'gaps' that act as capacitors and one (or more) 'loop' elements that act as inductors. The surface resistivity of resonator material acts as resistor.[31–34] As compared to cavity resonators, the mechanical construction of LGRs is more involved especially concerning cryogenic cooling and the implementation of magnetic field modulation. On the other hand, the filling factor of LGR's is very favorable and the larger bandwidth makes them very suitable for pulse EPR applications. The Bruker Flexline probehead series also contain an LGR design referred to as 'split ring' (Figure 12.8b) in which the B_1 field is concentrated in the central loop and the resonance mode is similar to that of the TE_{011} resonator. The X-band split-ring structures (ER4118X-MS3, -MS5) have sample access of 3 mm or 5 mm and are coupled through an antenna in a similar way as the dielectric resonator (ER4118X-MD5). Similar probeheads with 5 mm access are also available at S-band (4 GHz) and L-band (1 GHz). Split-ring structures can also be scaled down to W-band frequencies (95 GHz) where increasing the number of gaps (i.e., decreasing the effective capacitance) allows for sample access of 0.5–0.8 mm.[35,36]

Numerous variants of this basic LGR scheme have been developed for various applications. One example is the so-called bridged LGR probehead[37,38] whose physical construction is closely compatible with that of the dielectric ring resonator. The resonator is formed by conducting layers (baked colloidal silver) on a quartz cylinder. The frequency of the structure can be simply adjusted by scratching the bridging capacitor layers. Using the same technology, resonators that are used in NMR imaging (Alderman–Grant, crossed coil, and birdcage) can be miniaturized to resonate at MW frequencies.[39]

12.4.4 Microresonator Structures

In several cases, the number of spins that are available for the EPR experiment can be limited, e.g., tiny microcrystals of proteins or other materials (typical dimension 50 μm), thin layers obtained by a lithographic process, and reaction products in a microfluidic reactor. In all these cases, the EPR resonator has to be adapted to the size and shape of the particular microsample. In principle, LGRs at higher frequencies (e.g., Q-band) can be reduced in size to accommodate sample of about 0.5 mm.[40,41] Microcoils that are used in NMR experiment on microsamples often have self-resonances in the MW range, which makes them potential candidates for micro-EPR resonators.[42,43] The most recent and promising microresonator designs are based on microstrip technology. Here, again the lumped-element design principle is applied and the resonators are based on either a $\lambda/2$ resonating strip[44–46] or a surface LGR. A series of surface LGRs were developed with the main aim to maximize the absolute sensitivity with the ultimate target of single-spin detection and micro/nano imaging.[47] (see Chapter 24). A general trend is observed that smaller loop diameters lead to a decrease in the Q-value, thus counteracting, in part, the increased absolute spin sensitivity due to the higher filling factor (and power conversion factor). It turns out that the low surface quality of the substrate material (Rogers TMM10i with $\varepsilon \approx 9.9$, tan $\delta \approx 2 \times 10^{-3}$) leads to increased surface resistance that also lowers the Q-factor. By substitution of the substrate with sapphire, a Q-factor enhancement of a factor 2–3 is expected.

12.4.5 Probeheads for Very High Frequencies

For very high-frequency EPR applications (>95 GHz), the construction and operation of single-mode resonators becomes increasingly challenging, especially when also radiofrequency (RF) irradiation and light access are required. For this reason, multimode Fabry–Pérot cavities are often used. These resonators usually consist of two concave mirrors in between which a standing wave is set up with several (typically 4–8) magnetic field maxima. Although the filling factor (and power conversion) of Fabry–Pérot and other oversized resonators is significantly lower than that of a single-mode resonator, the concentration sensitivity can be, nevertheless, superior because of (i) the higher Q-value, (ii) the fact that the samples of larger volume can be inserted, and (iii) the fact that the mode is less sensitive to dielectric losses in the sample, which is critical for samples containing liquid water solutions. Owing to the intrinsically higher EPR sensitivity at high fields and frequencies, nonresonant probeheads can be a convenient alternative to resonator probeheads especially in a quasi-optical arrangement. In this case, the nonresonant sample holder is mounted at the end of the corrugated guide. The cross-polar EPR signal component in the reflected MWs is then extracted using a polarizer in the quasi-optical bridge.[20,23] Fabry–Pérot resonators and other probeheads that are used in high-frequency EPR applications are extensively discussed in Chapter 27.

12.4.6 A Practical Comparison Between Various X-band CW EPR Resonators

It is difficult to directly compare EPR resonator performances. This is because each type of resonator is best suited for a specific type of EPR experiment (CW or pulse) and sample characteristics (volume and dielectric losses). Therefore, we limited ourselves to a comparison of commercial resonators for X-band CW EPR (Figure 12.9) and include recently developed microstrip resonators. The tests were performed at room temperature using a 2 mM aqueous solution of perdeuterated TEMPOL nitroxide radical filled in quartz capillaries of different inner diameters in the range 0.1–1 mm. First, the quality factor of each critically coupled resonator containing water samples of different diameter was determined. The measurements were performed for critically coupled resonators by recording and analyzing the power ring-down time after an MW pulse as described in Section 12.4.[28] The results are shown in Figure 12.10(a). For an empty resonator, the highest quality factor $Q_L = Q_U/2$ was obtained for the dielectric (MD) and TE_{011} (SHQ) resonators. The same resonators show the strongest dependence on sample diameter as compared to moderate-Q resonators (TM and ST). It was observed that samples up to 0.2 mm diameter do not affect the quality of the resonators. Therefore, a 0.2 mm capillary was selected for quantification of the resonator performances.

For comparison, CW EPR spectra of the TEMPOL radical were recorded in each resonator as a function of the incident MW power attenuation using identical magnetic field modulation amplitude. The spectra were evaluated in terms of SNR. The results are represented in Figure 12.10(b). The SNR power dependence shows typical saturation behavior. The analysis of the saturation curves yields (i) relative conversion factor, (ii) relative sensitivity at fixed B_1 value, and (iii) relative sensitivity at the maximum reachable B_1. As expected, the conversion factors in resonators designed for pulse applications, i.e., dielectric (MD), split ring (MS), and microstrip (Ω05), are higher as compared to the CW resonators, SHQ, TM, and ST, as depicted in Figure 12.11(a). The sensitivities at fixed B_1, which can be read from the maxima of the saturation curves in Figure 12.10(b), are shown in Figure 12.11(b). Thus, the dielectric resonator shows the highest absolute sensitivity, which is about a factor 1.5 higher as compared to the SHQ cavity. Therefore, CW cavity resonators are not favorable for the detection of small samples (Figure 12.11c). For small nonlossy samples, the pulse resonators provide the best CW performance in terms of conversion factor and sensitivity.

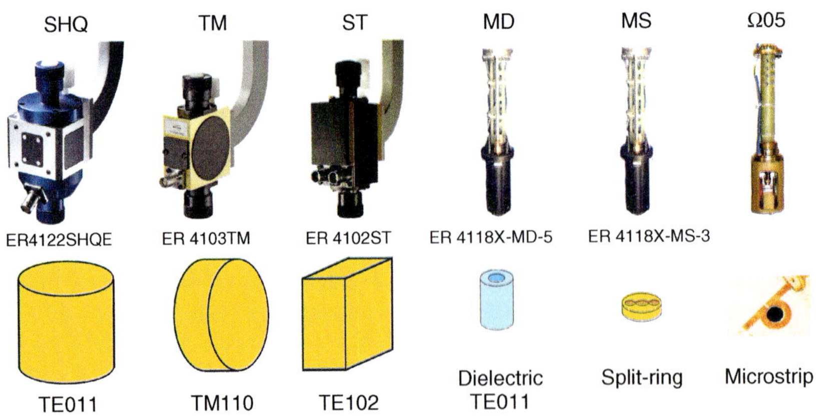

Figure 12.9. Commercial (Bruker Biospin) X-band resonators and Ω-type microstrip resonator with 0.5-mm loop diameter chosen for comparison. The exact models are indicated

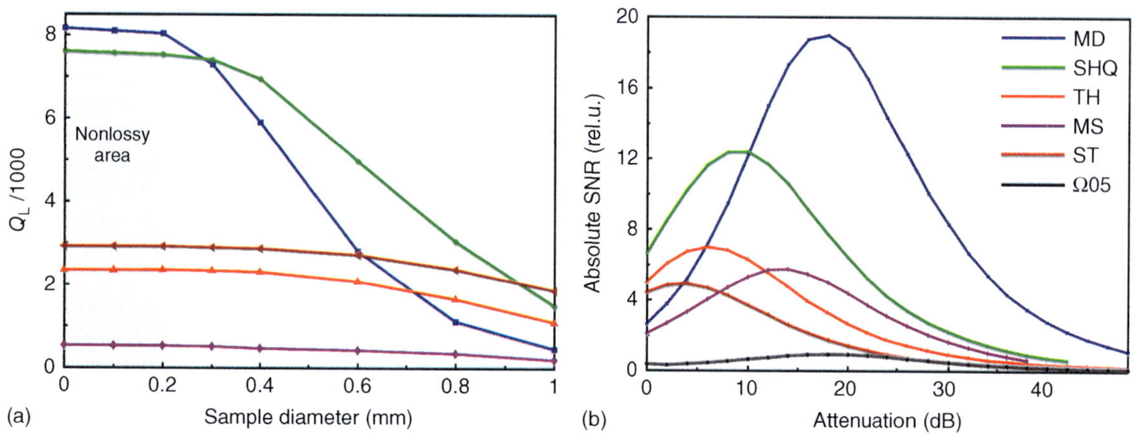

Figure 12.10. (a) Loaded quality factor, Q_L, of the different critically coupled ($\beta = 1$) resonators containing water sample in capillaries of different inner diameters. (b) Power saturation curves of TEMPOL/water sample in 0.2 mm i.d. capillary at room temperature

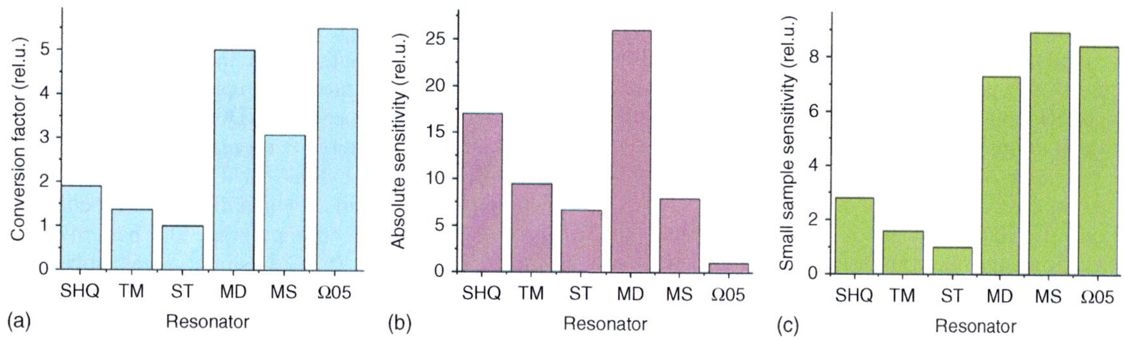

Figure 12.11. (a) The relative conversion factors, (b) relative absolute sensitivities at fixed B_1, and (c) relative small-sample sensitivities for different resonators obtained from evaluation of power saturation curves of water solution of TEMPOL radical in 0.2-mm round capillary. The small-sample sensitivity is calculated by scaling the absolute sensitivity on operation length (including B_1 and field modulation distributions) of the corresponding resonator

As was pointed out earlier, the results in Figure 12.11 show comparisons for a specific type of EPR experiment. They do not provide a general recipe for a resonator choice. For instance, the standard TM and ST cavities perform best for CW EPR experiments on lossy large volume samples as shown in Figure 12.10(a). Resonators optimized for pulse experiments are less suited for CW EPR measurements especially with large magnetic field sweeps and low temperature because of possible background signals. These signals arise, for instance, from paramagnetic impurities in the sapphire ring of the dielectric resonator. Additional distortions can also arise in split-ring resonators owing to the heating of the resonator body by the eddy currents induced by the magnetic field modulation. In addition, the pulse resonators do not allow for separate cooling of the sample owing to the small size (see Section 12.5) in contrast to CW cavity resonators. This leads to increase in background CW signals at low temperature.

12.4.7 ENDOR Resonators

Apart from the commercially available ENDOR probeheads from Bruker, numerous homebuilt resonators for particular applications have been described. In this section, we will briefly discuss the basic design principles for the resonators and the inserted RF coils.

As the introduction of conductors inside a resonator may strongly reduce the resonator quality factor, care should be taken that all conducting wires are oriented perpendicular to the electric field components of the MW mode. In addition, the magnetic component of the generated RF field should be perpendicular to the external static magnetic field. These requirements can be met in most EPR resonators, but to combine a high MW field (B_1) with a high RF field (B_2) has proven to be more challenging. For CW-ENDOR, the 'classic' resonator is based on a design by Biehl.[48] The TM_{110} resonator mode is ideal for the insertion of an RF solenoid that can cover the whole length of the probehead (see Figure 12.12a and 12.12b). The Bruker ER801 CW-ENDOR (X-band) resonator (TM_{110} mode) is compatible with various variable temperature inserts (VTIs) and can be used for both liquid and solid/frozen samples.

For single-crystal (X-band) CW-ENDOR work, a rectangular TE_{102} resonator that has a better filling factor is often preferred. Here, however, only specially shaped RF coils can be used (e.g., a saddle coil or flat RF coil) in order not to disturb the MW properties of the resonator. Several dedicated homebuilt resonators based on this principle have been described.[49]

For pulse EPR applications at X- and Q-bands, the dielectric resonators from Bruker can also be equipped with an ENDOR coil, which is inserted into two vertical groves inside the dielectric ring and positioned by a thin quartz cylinder. For the X-band MD4, this reduces the sample access to 4 mm and at Q-band (Bruker EN5102D2) to 2 mm. Other designs have been described in the literature, for instance, a homebuilt

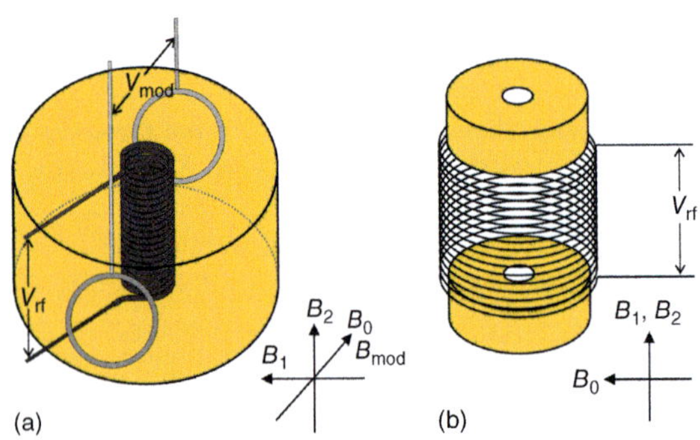

Figure 12.12. ENDOR resonator configuration: (a) CW X-band version based on TM_{110} rectangular cavity and (b) pulse W-band version based on TE_{011} cylindrical cavity

TE_{011} resonator was reported that accommodates four 0.6-mm ENDOR wires parallel to the sample tube forming an RF coil and allows for 3 mm o.d. samples. Owing to the very robust coil design, this resonator is also suitable for CW-ENDOR.[29]

Owing to their compact design, LGRs are less suitable to incorporate an ENDOR coil. Nevertheless, some successful attempts have been made for bridged LGR that allows for the insertion of an external ENDOR coil.[38]

Similar to LGRs, high-frequency (>94 GHz) single-mode cavities are too small to conveniently accommodate an ENDOR coil inside the resonator. Therefore, the cylindrical TE_{011} W-band probehead of Bruker (EN680-1021H) is built up from a solenoid coil that also functions as the wall of the resonator,[50] see Figure 12.12(b). The number of turns making up the RF coil can be defined by contacting the coax leads to the selected windings. In this way, the inductance of the RF coil can be adjusted to optimize RF performance at the RF frequency of interest. This design has the additional advantage that the coil-shaped walls also suppress unwanted MW resonant modes. Other reported high-frequency ENDOR probeheads are based on either a Fabry–Pérot resonator with internal ENDOR coil or a cylindrical TE_{011} resonator with external ENDOR coil of which the field can penetrate the resonator due to an appropriate slot pattern in its walls.[3] The Bruker 263 GHz single-mode probehead is also based on a cylindrical TE_{011} resonator with external RF coil (see Chapter 27).

In most ENDOR applications, especially at X- and Q-bands, a large RF bandwidth is required. Therefore, the RF coil is designed as part of a 50-Ω terminated transmission line. The input impedance of such a line is dependent on its length and is given by[49]

$$Z = Z_0 \frac{Z_n \cos(kl) + iZ_0 \sin(kl)}{Z_0 \cos(kl) + iZ_n \sin(kl)} \tag{12.6}$$

where Z_n represents the impedance of the coil system, Z_0 is the characteristic impedance of the line (50 Ω), l the length of the input line, and k the wavenumber. If the coil is terminated by a load of impedance Z_0, its impedance is given by $Z_n = i\omega L + Z_0$, where ω is the radial RF frequency. If the coil inductance (L) is very low, the effective impedance (Z_n) will approach the characteristic impedance of the line. Owing to the impedance mismatch, part of the power will be reflected back to the amplifier. Therefore, it is important to select an RF amplifier that can handle a large mismatch. The Bruker ENDOR resonators are delivered with a matching circuit to improve the smoothness of the ENDOR frequency response over the bandwidth.

12.5 SAMPLE CRYOSTATS

Depending on the application, it is often necessary to vary the sample temperature. A large variety of temperature control systems are available for EPR spectroscopy. In general, the sample or the whole resonator is contained in a VTI in which the temperature can be varied either directly through a gas flow or indirectly by a static exchange gas. The temperature of the gas flow or the exchange gas is adjusted by a heating element that is connected to the output of a proportional–integral–derivative (PID) temperature controller that takes as input the signal of a temperature sensor close to the heating element. The PID controller parameters are to be optimized for each specific VTI configuration. Some controllers, such as the ITC models of Oxford Instruments, also allow adjusting the gas flow in the VTI. In this section, we can only cover the basic temperature systems commonly used in EPR spectroscopy. Detailed information on cryostats, temperature sensors, and controllers can be obtained by contacting the vendors that are listed in Table 12.4.

12.5.1 'Finger' Cryostat

Most X-band CW EPR cavities have a free access bore through the resonator wall of 10 mm diameter.

Table 12.4. Vendors offering cryogenic components and services

Vendor	Web site
Bruker Biospin	www.bruker.com
Advanced Research Systems (ARS)	www.arscryo.com
Cryogenic Ltd	www.cryogenic.co.uk
SHI Cryogenics	www.shicryogenics.com
Cryomagnetics	www.cryomagnetics.com
Cold Edge Technologies	www.coldedgetech.com
Cryo Industries of America	www.cryoindustries.com
Ice Oxford	www.iceoxford.com
Janis Research	www.janis.com
Oxford Instruments	www.oxford-instruments.com
Cryomech	www.cryomech.com
Cryofab	www.cryofab.com

Thus, the temperature variation of the sample can be conveniently accomplished using a 10 mm o.d. double-wall quartz Dewar that enters into the resonator (Figure 12.13(a)). It should be noted that the resonance frequency of the resonator is strongly affected by the dielectric properties of the quartz material with a typical downshift up to 500 MHz. At the same time, the filling and conversion factors are improved owing to the focusing effect of the quartz insert, i.e., the principle used in dielectric resonators (see Section 12.4.2). EPR sample tubes up to 5 mm o.d. can be accommodated in such a finger system using simple fixtures. To reach temperatures between 100 and 500 K, this quartz Dewar is connected to a nitrogen gas flow system with heater and thermostat. To reach temperatures down to 100 K, a cold nitrogen gas flow is obtained directly from a pressurized liquid nitrogen vessel. A relatively low-cost N_2 system can be obtained from Bruker Biospin (ER 4131VT). For applications requiring temperatures in the range 400–1300 K, a special ceramic TE_{011} resonator is available with an integrated three-wall quartz finger system (ER 4114HT-1000). To reach cryogenic temperatures down to 4.2 K and lower, a helium flow cryostat is required. These systems use the same double-wall quartz insert configuration. However, in this case, the Dewar is connected to a liquid helium vessel via a liquid helium transfer line. The Bruker Biospin ER 4112HV system is specified for the temperature range 3.8–300 K and is based on the Oxford Instruments ESR900 cryostat (Figure 12.13(b)). A membrane pump is evacuating the VTI through a flexible transfer line. The cold helium gas that is extracted from the VTI cools a heat shield around the input helium channel of the transfer line which is refilling the VTI. This configuration strongly reduces the helium consumption of the system. A variant of the Oxford ESR900 system is available that contains a small helium reservoir integrated in the cryostat. This Oxford ESR910 cryostat is specified with a base temperature of 1.9 K. In practice however, such a low temperature is difficult to reach. Systems equivalent to the Oxford ESR900 cryostat are offered by Janis (ST-170, www.janis.com) and Advanced Research Systems (ARS) (www.arscryo.com/LTR.html).

12.5.2 Immersion Helium Flow Cryostats

For Q-band and pulse X-band EPR resonators, the sample access is limited by resonator design, e.g.,

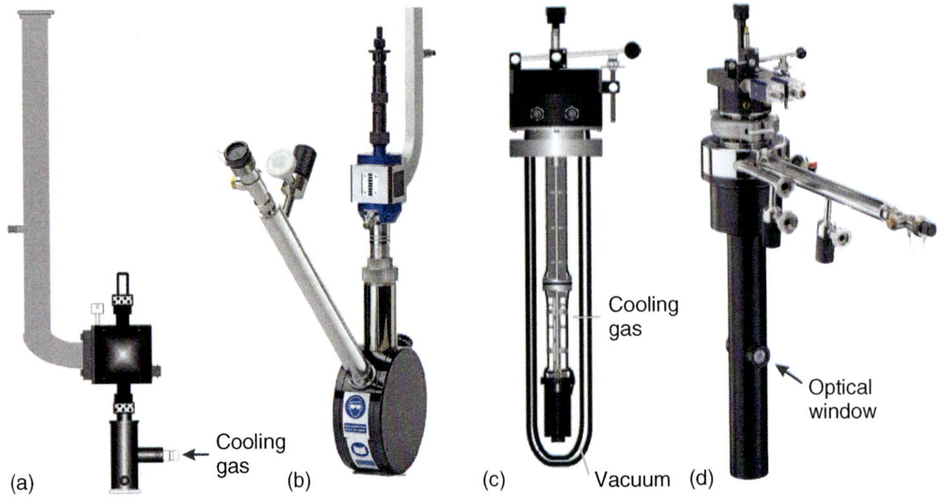

Figure 12.13. (a) Nitrogen gas controlled variable temperature setup with dewar in cavity. (b) Helium gas controlled cryostat (ESR900, Oxford Instruments) with TE_{011} cavity (ER4122SHQE, Bruker). (c) Nitrogen gas controlled variable temperature setup with cavity in dewar. (d) Helium flow cryostat (CF935, Oxford Instruments) with optical window and Flexline probehead (Bruker) inserted

1–5 mm o.d., and a dewar insert cannot be used. Therefore, these resonators are operated in an immersion helium flow system where the whole resonator including the sample is cooled. The most commonly used cryostat of this type is the Oxford CF935, see Figure 12.13(d). This system has a cold sample space of 42 mm diameter and is designed to accommodate the Bruker Q-band and Flexline series of resonators (see Section 12.4). The operation is identical to that of the ESR900 system (see 12.5.1), i.e., a cold Helium flow is extracted from a liquid Helium vessel via a flexible transfer line by the membrane pump. The resonator and sample are in direct contact with the cold Helium gas, which flows in through an entrance at the bottom of the cryostat. A heater and temperature sensor connected to a temperature PID controller adjust the temperature of the helium flow. The actual sample temperature may be higher due to the temperature gradient in the cryostat induced by thermal losses in the probehead. For a precise temperature setting, a second temperature sensor on the probehead should be installed and monitored. A second temperature PID control loop which would require a heater on the resonator itself also can be included. In practice, the CF935 cryostat operates over a temperature range of 4–300 K. Many variations of the CF935 cryostat are available. One option for the cryostat features an optical window for sample light excitation as shown in Figure 12.13(d). For the Bruker W-band probehead, an elongated version of the cryostat is supplied. For some applications, it may be desirable to avoid contact of the probehead with the cold helium flow. In these 'static' cryostats, only the wall surrounding the probehead (lower part) is cooled and the temperature at the resonator and sample is reached by convection using an exchange gas. Immersion cryostats compatible with the CF935 and corresponding transfer lines can also be supplied by Ice Oxford (Table 12.4).

12.5.3 Closed-cycle Systems

With the continuously increasing cost of helium and the advancement of cryogenic technology, closed-cycle cryogenic systems are gaining popularity in many spectroscopies, in particular NMR and EPR. Instead of consuming liquid helium, a cryogen-free cooling system is used based on a cryocooler using compressed high-purity helium gas that is recycled. The most common cryocoolers are of the Gifford–McMahon (GM) type and use a mechanical heat exchanger. A more recent development is the introduction of pulse tube cryocoolers that make use of a thermoacoustic effect and thus can be made without any mechanically moving parts. This property is of great advantage for experiments that are sensitive to mechanical vibrations, e.g., EPR, NMR, Mössbauer, and FTIR. In addition, the maintenance costs of these coolers are lower. Pulse tube coolers are, however, thermodynamically less efficient than GM coolers and require a larger (more expensive) compressor consuming more electric power. Several cryogenics vendors offer systems that are compatible with regular EPR equipment (Table 12.4).

ARS offers solutions for both CW and pulse EPR setups. The CS202_I-DMX-15 (cold finger) system is a replacement for the Oxford ESR900 cryostat. Similarly, a replacement of the Oxford CR935 cryostat is available. These systems are relatively simple (and affordable) as they are based on a GM tube that is directly cooling the cryostat walls. In this 'static' cooling configuration, the EPR probehead is indirectly cooled by Helium exchange gas. The advantage is that sample change is straightforward and the temperature is very stable. On the other hand, the base temperature is limited by the relatively large temperature gradient in the static gas column surrounding the sample. For the CF935 replacement system employing a 0.5 W@4 K GM cooler, in our laboratory (MPI CEC, Mülheim), a base temperature of 8 K can be reached using an elongated version (≈500 mm) of a homebuilt Q-band insert.[29] The mechanical vibrations induced by the GM cooler do not affect the sensitivity of pulse EPR experiments. It is, however, to be expected that for CW experiments in high-Q resonators the mechanical vibrations of the cold head will add to the detection noise.

Bruker Biospin has teamed up with Cold Edge Technologies (Table 12.4) to develop closed-cycle sample cryostats for X-band CW EPR (dewar insert type) and pulse EPR (immersion flow type). In both the systems, gaseous helium is cooled by a closed-cycle refrigerator (4 or 8 K) and fed through the sample space after which it is recompressed. An interesting spin-off of this design is the so-called Stinger system in which the VTI is replaced by a flexible transfer line that can be inserted into any Oxford helium flow cryostat including models for optical spectroscopy and microscopy.

Cryogenic Ltd (Table 12.4) also offers a closed-cycle immersion cryostat. However, instead of recirculating the helium through a compressor, it is pumped out of the VTI using a large capacity (21 $m^3 h^{-1}$) scroll pump and transferred into a dump (0.2 m^3) from where the gas is recirculated into the refrigerator (Figure 12.14). The end vacuum of the scroll pump (0.03 mbar) is sufficient to condense helium into the VTI and reach a sample temperature of less than 2 K with a Bruker Flexline resonator installed.

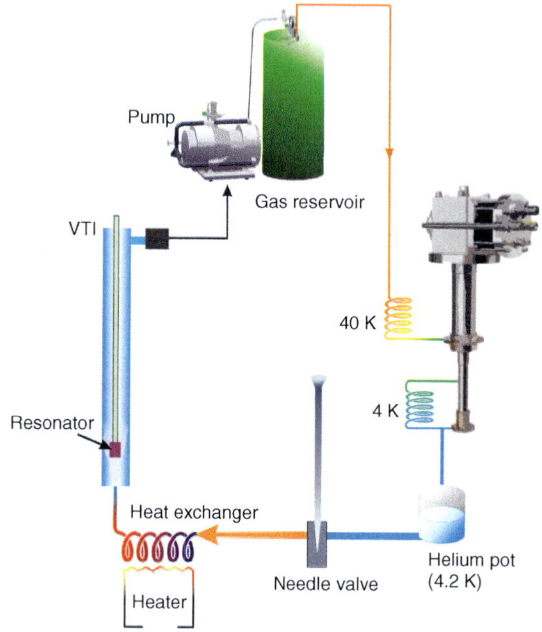

Figure 12.14. Block diagram of a closed-cycle cryostat (Cryogenic Ltd) with circulating helium. (Courtesy of Cryogenic Ltd.)

12.5.4 Bath Cryostats

For special applications, e.g., the investigation of molecular magnets, subhelium temperatures are required. In a bath cryostat, the resonator insert is immersed in liquid helium and the temperature is regulated by pumping on the helium bath. The actual sample temperature can be determined relatively accurately from the helium vapor pressure. The bath cryostats allow for a base temperature of 1.2 K. For even lower temperatures, a ^{3}He (300 mK) or ^{3}He–^{4}He (5 mK) dilution refrigerator is used.[51] These highly specialized systems are offered by several cryogenics manufacturers (Table 12.4). Commercial EPR resonator inserts are not compatible with these systems, and dedicated homebuilt probeheads need to be designed and constructed.

12.6 SPECTROMETER CONTROL AND SIGNAL DETECTION

The spectrometer control components as depicted in Figure 12.1(b) can be realized in several ways. Home-built EPR consoles are usually built up from separate components (digitizer, pulse programmer, RF generator, lock-in amplifier, and field controller) connected to a PC, e.g., through a GPIB (or USB, LAN) interface (popular vendors are listed in Table 12.5). Alternatively, an industrial PC containing various interface boards (digitizer, pulse programmer, GPIB, and digital I/O) can be configured as spectrometer console. Most groups have developed in-house control software, some of which is offered as a (non)commercial product: FSC2 (http://users.physik.fu-berlin.de/~jtt/) and Specman[52] (http://www.specman4epr.com/). These programs contain drivers for a multitude of external devices and allow configuring many types of EPR experiments including 2D pulsed EPR/ENDOR.

For reference, we provide here the specifications of the latest Bruker CW/pulse EPR spectrometers. The capabilities of the individual components are mostly on par with those of external devices used in home-built setups. The devices are, however, integrated in one console and controlled by an integrated user interface. The Bruker Elexsys I EPR spectrometers were originally delivered with a Motorola 68000 Computer system running under OS9 operation system that was imported from the older ESP300 spectrometer series. The user interface Xepr runs on a Linux PC computer that is connected via local area network (LAN) to the OS9 computer. In the most recent Elexsys II series EPR spectrometers, the modules in the console have embedded processors that communicate directly via LAN to the Linux computer. Depending on the configuration, the Elexsys console can contain the following modules:

1. Signal Channel or signal processing unit (SPU) containing a digital lock-in detection system for CW EPR. This includes a field modulation amplifier, a digitizer for transient signals, and a rapid field scan unit (20 mT) (see Chapter 23). Signal Channel specification: 0.16 ms to 5 s conversion time; 24 or 32 bit constant amplitude resolution. Fast Digitizer: 8 ns, 14 bit resolution.
2. Hall Field Controller: 2-256000 points. 32 bit Field resolution (15 μT).
3. DICE: Direct digital synthesizer (DDS) RF generator for ENDOR experiments (0.5–400 MHz)
4. PatternJet II: Pulse Programmer for Pulse EPR experiments (1 ns time resolution, 2 Mbit internal memory)
5. SpecJet II: Fast digitizer (8 bit, 1 ns-10 ms time resolution, 65535 on-board accumulations, and 128 k points)

Table 12.5. Vendors offering control and detection components for homebuilt EPR setups

Vendor	Web site
Stanford Research Systems (lock-in amplifiers)	www.thinksrs.com
Ametek (lock-in amplifiers)	www.ameteksi.com
SpinCore (pulse generators and RF generators)	www.spincore.com
Chase Scientific (Pulse generators)	www.chase-scientific.com
Tektronix (digitizers and AWG)	www.tek.com
Keysight (digitizers, AWG, and RF generators)	www.keysight.com
SP Devices (digitizers and AWG)	www.spdevices.com
Rohde & Schwarz (RF generators and RF amplifiers)	www.rohde-schwarz.com
Amplifier Research (RF amplifiers)	www.arworld.us
Active Technologies (AWG)	www.activetechnologies.it

12.6.1 Signal Detection and Digitization

The standard detection mode scheme in CW EPR is based on external magnetic field modulation in combination with phase-sensitive lock-in detection (see Chapter 1). This scheme has not changed over the past 50 years. The EPR sensitivity enhancement was only obtained owing to the improvement of the MW sources (see Section 12.3.2) and resonator designs (see Section 12.4). In the classical lock-in detection scheme, the recorded signal is affected by the time constant and the number of data points. To prevent signal distortion, the field modulation amplitude should be less than half the line width of the narrowest EPR line in the spectrum. In addition, the time constant (RC filter time) should be just small enough to follow the signal variations of the narrowest line during signal recording. In practice, the following rule of thumb is given:[27]

$$\frac{\text{spectrum width in T}}{\text{line width in T}} \times \frac{\text{time constant in s}}{\text{sweep time in s}} < 0.1 \quad (12.7)$$

The new generation of Bruker CW EPR spectrometers is equipped with a digital signal channel (SPU). The modulation frequency in this unit has a continuous range from 80 Hz to 100 kHz. The digital lock-in amplifier has a single-shot 24 or 32 bit dynamic range, and the time constant is replaced by a smoothing factor. The large dynamic range makes it possible to detect very small signals in a spectrum dominated by large signals. Furthermore, the digital lock-in scheme enables the simultaneous detection of up to five modulation harmonics and phase correction and smoothing as postprocessing step. In CW EPR experiments on slowly relaxing paramagnetic species, the use of rapid-scan techniques and corresponding data processing offer significant enhancement in sensitivity. These methods are discussed in Chapter 23. CW transient detection is also supported in the new SPU unit. The user can choose between direct detection (no field modulation) and transient detection in CW mode without low-pass filter. These detection modes are used in the study of photo-excited triplets (direct) and kinetic measurements (for instance, stopped-flow technique) (see Chapter 10).

For pulse EPR applications, an entirely different detection scheme is used (see Chapter 11). Early pulse EPR setups used a boxcar integrator to capture the echo or FID signal. With the development of fast digitizers, it is now possible to record the whole echo/FID signal profile with subnanosecond time resolution. This also opens the way for 'on the fly' data processing (FT display in Bruker software) and background correction. Currently, mostly 1 GSa s^{-1} 8 bit fast digitizers are implemented in modern EPR spectrometer acquisition systems. Owing to the limited dynamic range of these instruments, a video amplifier is required to scale the output signal from the mixer unit. Therefore, pulsed EPR spectra from different samples can only be compared when the video gain is taken into account. Moreover, excessive video gain may lead to data processing artifacts. It is expected that the next-generation detection systems will be based on 1 GSa s^{-1} 14 bit digitizers that will not need amplification (see, e.g., www.spdevices.com).

12.6.2 Pulse ENDOR Control Hardware

In a pulse ENDOR experiment, the RF pulses are generated by an RF source that can be either an analog synthesizer or a DDS. A DDS has the advantage that the RF frequency can be switched very fast (microsecond time scale). The DICE units as employed in the Bruker Elexsys Spectrometers are based on DDS modules. In the standard ENDOR mode, the RF frequency in linearly swept from the lowest to the highest value and the spin echo or FID signal is recorded synchronously. The RF pulses are usually amplified to powers in the range 100–1000 W. Under certain conditions, these high powers may cause experimental problems.

During an ENDOR experiment, the RF current in the ENDOR coil will produce eddy currents in the wall or shield of the resonator. Depending on the design of the resonator, these currents can cause heating of the probehead and, correspondingly, changes in the resonator Q-value and resonance frequency. As these effects are dependent on the RF frequency, the baseline of the ENDOR spectrum will be affected. This heating effect is particularly strong for small-size resonators such as in W-band ENDOR experiments. A very effective strategy to improve the pulse ENDOR baseline is to record the spectrum using 'random acquisition'.[53] Here, the RF frequency in the pulse ENDOR experiment is not linearly swept but randomly changed for each individual pulse sequence, thus distributing the heating effect equally over the whole spectrum. The price to pay is that the coherent heating effect is converted into random noise, thus deteriorating the SNR of the experiment. The random acquisition option is implemented in the Bruker Xepr software.

12.6.3 Pulse Programmers

For the control of pulse EPR experiments, a fast pulse programmer is essential. The first homebuilt pulse EPR setups worked with programmable delay lines (Berkeley Nucleonics) or homebuilt word generators based on ECL logic in various configurations.[54,55] Later instruments were equipped with modified single PC-board NMR pulse programmers (www.spincore.com) and dedicated very fast word generators (www.chase-scientific.com). The Bruker Elexsys II series of pulse EPR spectrometers employ dedicated independent single-channel pulse programmers with a time resolution (and shortest pulse) of 1 ns. A factor that affects the overhead of the pulse experiments (i.e., the time needed for the pulse experiment apart from the measuring time) is the reprogramming time of the pulse programmer in between two pulse sequences. The latest model pulse programmer that is included in the Bruker Elexsys II series (i.e., Patternjet II) is able to store several traces of a 2D pulse experiment in memory thus reducing overhead substantially. It is to be expected that the performance of pulse programmers will be further enhanced by the use of high-speed field-programmable gate arrays (FPGAs).[56]

12.7 INSTRUMENTATION FOR SPECIAL APPLICATIONS

12.7.1 Reaction Kinetics

For medium-speed chemical reactions (seconds-minutes), flow cell accessories that can be inserted in lieu of the sample tube are available from, e.g., Bruker. The AquaX accessory consists of multibore capillaries that can be used for an aqueous solutions and a continuous-flow system. Chemical reactions can be initiated using a fluid mixer just outside the resonator, and kinetics can be studied either in continuous-flow mode or in a stopped-flow configuration in combination with the time-resolved option of the EPR spectrometer. For fast kinetics, special resonators have been built consisting of stacked dielectric rings inside a shield. Owing to the small size of these resonators, the mixer can be very close (a few mm) to the EPR sensitive area. In our laboratory (MPI CEC, Mülheim), a setup has been built featuring 0.33 ms dead time.[57]

12.7.2 Electrochemistry

To follow electrochemical reactions occurring on the working electrode or in solution, several electrode configurations have been developed that fit into a regular 5-mm X-band EPR tube[58] or even in a 0.8 mm W-band capillary.[59] For electrochemistry in microliter volumes, an LGR (see Section 12.4) configuration has been described.[60,61] When transition metal complexes take part in the electrochemical process, the cell including the electrodes needs to be cooled down to cryogenic temperatures in order to slow down the relaxation times of these species and record their EPR spectrum. For every potential point in the electrochemical series, a large temperature cycle would be needed. This would require an extremely robust electrode system.

12.7.3 In Situ EPR

Apart from just controlling the sample temperature, the quartz finger normally inserted into the 10 mm sample space can also be used to accommodate a chemical reactor including a catalyst bed and multiple gas flow connections. In this way, e.g., gas initiated reactions on the surface of a heterogeneous catalyst can be studied as function of various reaction parameters (temperature, pressure, and gas composition). The aim of these studies would be to identify the active (paramagnetic) species in the catalyst. To get also access to the diamagnetic intermediates of the catalyst, additional spectroscopies (UV–Vis, FTIR, Raman, etc.) could be implemented in the same system. True 'Operando' EPR is achieved when also the reaction products are analyzed in parallel, e.g., through in-line mass spectrometry or gas chromatography connected to an accessory.[62,63]

12.7.4 EPR Imaging

Standard EPR instruments can be adapted to imaging applications through gradient accessories that are attached to the electromagnet. Bruker is offering imaging systems at L-band (0.8–1.4 GHz) and X-band. Biomedical imaging is covered in detail in Chapter 13 in this book, whereas EPR microscopy is treated in Chapter 24.

FURTHER READING

G. M. Smith, P. A. S. Cruickshank, D. R. Bolton, and D. A. Robertson, *Electron Paramagn. Reson.*, 2008, **21**, 216.

E. J. Reijerse, *Appl. Magn. Reson.*, 2010, **37**, 795.

A. Webb, *Prog. Nucl. Magn. Reson. Spectrosc.*, 2014, **83**, 1.

O. Y. Grinberg and L. J. Berliner (eds), in Biological Magnetic Resonance, Kluwer/Plenum Publishers: , 2004, Vol. **22**, Chap. Very high frequency (VHF) ESR/EPR.

K. Mobius and A. Savitsky, High Field EPR Spectroscopy on Proteins and their Model Systems, Royal Society of Chemistry: Cambridge, 2009.

G. R. Eaton, S. S. Eaton, D. P. Barr, and R. T. Weber, Quantitative EPR, Springer: 2010.

C. P. Poole, Electron Spin Resonance, John Wiley & Sons: , 1983.

W. R. Hagen, Biomolecular EPR Spectroscopy, Taylor & Francis: 2009.

REFERENCES

1. O. Burghaus, E. Haindl, M. Plato, and K. Mobius, *J. Phys. E-Sci. Instrum.*, 1985, **18**, 294. DOI: 10.1088/0022-3735/18/4/011.
2. E. Haindl, K. Möbius, and H. Oloff, *Z. Naturforsch. A*, 1985, **40**, 169.
3. H. Blok, J. A. J. M. Disselhorst, H. van der Meer, S. B. Orlinskii, and J. Schmidt, *J. Magn. Reson.*, 2005, **173**, 49. DOI: 10.1016/j.jmr.2004.11.019.
4. D. Goldfarb, Y. Lipkin, A. Potapov, Y. Gorodetsky, B. Epel, A. M. Raitsimring, M. Radoul, and I. Kaminker, *J. Magn. Reson.*, 2008, **194**, 8.
5. K. Mobius and A. Savitsky, High Field EPR Spectroscopy on Proteins and their Model Systems, Royal Society of Chemistry: Cambridge, 2009.
6. D. Schmalbein, G. G. Maresch, A. Kamlowski, and P. Hofer, *Appl. Magn. Reson.*, 1999, **16**, 185. DOI: 10.1007/BF03161933.
7. P. Hofer, A. Kamlowski, G. G. Maresch, D. Schmalbein, and R. T. Weber, in Biological Magnetic Resonance, eds O. Y. Grinberg and L. J. Berliner, Kluwer/Plenum Publishers: New York, 2004, Vol. **22**, Chap. *Very high frequency (VHF) ESR/EPR*, 401.
8. O. Burghaus, M. Rohrer, T. Gotzinger, M. Plato, and K. Mobius, *Meas. Sci. Technol.*, 1992, **3**, 765. DOI: 10.1088/0957-0233/3/8/013.
9. S. Stoll, A. Ozarowski, R. D. Britt, and A. Angerhofer, *J. Magn. Reson.*, 2010, **207**, 158. DOI: 10.1016/j.jmr.2010.08.006.
10. O. Y. Grinberg and L. J. Berliner, Very high Frequency (VHF) ESR/EPR, Kluwer/Plenum Publishers: New York, 2004, Vol. **22**.
11. A. I. Smirnov, T. I. Smirnova, R. L. MacArthur, J. A. Good, and R. Hall, *Rev. Sci. Instrum.*, 2006, **77**. DOI: 10.1063/1.2182571.
12. A. K. Hassan, A. L. Maniero, H. van Tol, C. Saylor, and L. C. Brunel, *Appl. Magn. Reson.*, 1999, **16**, 299. DOI: 10.1007/BF03161940.
13. L. C. Brunel, J. van Tol, A. Angerhofer, S. Hill, J. Krzystek, and A. L. Maniero, in Biological Magnetic Resonance, eds O. Y. Grinberg and L. J. Berliner, Kluwer/Plenum Publishers: New York, 2004, Vol. **22**, Chap. *Very high frequency (VHF) ESR/EPR*, 465.
14. P. J. M. van Bentum, J. C. Maan, J. W. M. Van Os, and A. P. M. Kentgens, *Chem. Phys. Lett.*, 2003, **376**, 338. DOI: 10.1016/S0009-2614(03)01014-5.
15. N. Nakagawa, T. Yamada, K. Akioka, S. Okubo, S. Kimura, and H. Ohta, *Int. J. Infrared Milli.*, 1998, **19**, 167. DOI: 10.1023/A:1022511405901.
16. T. Sakurai, A. Taketani, T. Tomita, S. Okubo, H. Ohta, and Y. Uwatoko, *Rev. Sci. Instrum.*, 2007, **78**. DOI: 10.1063/1.2746818.
17. T. H. Wilmshurst, Electron Spin Resonance Spectrometers, Adam Hilger: London, 1967.
18. C. P. Poole, Electron Spin Resonance, John Wiley & Sons: New York, 1983.
19. D. J. E. Ingram, Biological abd Biochemical Applications of Electron Spin Resonance, Adam Hilger LDT: London, 1969.
20. P. A. S. Cruickshank, D. R. Bolton, D. A. Robertson, R. I. Hunter, R. J. Wylde, and G. M. Smith, *Rev. Sci. Instrum.*, 2009, **80**. DOI: 10.1063/1.3239402.
21. W. Hofbauer, K. A. Earle, C. R. Dunnam, J. K. Moscicki, and J. H. Freed, *Rev. Sci. Instrum.*, 2004, **75**, 1194. DOI: 10.1063/1.1710700.
22. V. Zhaunerchyk, R. T. Jongma, and W. J. van der Zande Spontaneous and induced interpulse coherence in the Nijmegen THz-FEL. *Proceedings of FEL2009*, 2009. http://epaper.kek.jp/FEL2009/papers/tupc85.pdf
23. E. J. Reijerse, P. J. van Dam, A. A. K. Klaassen, W. R. Hagen, P. J. M. van Bentum, and G. M. Smith, *Appl. Magn. Reson.*, 1998, **14**, 153. DOI: 10.1007/BF03161887.
24. A. F. Gulla, P. Feenan, S. Burgess, and D. E. Budil, *Concepts Magn. Reson.*, 2002, **15**, 201. DOI: 10.1002/cmr.10040.
25. A. F. Gulla and D. E. Budil, *Concepts Magn. Reson. Part B Magn. Reson. Eng.*, 2004, **22B**, 15. DOI: 10.1002/cmr.b.20014.
26. E. J. Reijerse and E. P. R. High-Frequency, *Appl. Magn. Reson.*, 2010, **37**, 795. DOI: 10.1007/s00723-009-0070-y.
27. G. R. Eaton, S. S. Eaton, D. P. Barr, and R. T. Weber, Quantitative EPR, Springer: New York, 2010.

28. A. Savitsky, Y. Grishin, R. Rakhmatullin, E. Reijerse, and W. Lubitz, *Rev. Sci. Instrum.*, 2013, **84**. DOI: 10.1063/1.4788735.
29. E. Reijerse, F. Lendzian, R. Isaacson, and W. Lubitz, *J. Magn. Reson.*, 2012, **214**, 237. DOI: 10.1016/j.jmr.2011.11.011.
30. S. Friedlaender, O. Ovchar, H. Voigt, R. Boettcher, A. Belous, and A. Poeppl, *Appl. Magn. Reson.*, 2015, **46**, 33. DOI: 10.1007/s00723-014-0611-x.
31. W. Froncisz and J. S. Hyde, *J. Magn. Reson.*, 1982, **47**, 515. DOI: 10.1016/0022-2364(82)90221-9.
32. R. L. Wood, W. Froncisz, and J. S. Hyde, *J. Magn. Reson.*, 1984, **58**, 243. DOI: 10.1016/0022-2364(84)90214-2.
33. W. Froncisz, T. Oles, and J. S. Hyde, *Rev. Sci. Instrum.*, 1986, **57**, 1095. DOI: 10.1063/1.1138663.
34. W. L. Hubbell, W. Froncisz, and J. S. Hyde, *Rev. Sci. Instrum.*, 1987, **58**, 1879. DOI: 10.1063/1.1139536.
35. J. W. Sidabras, R. R. Mett, W. Froncisz, T. G. Camenisch, J. R. Anderson, and J. S. Hyde, *Rev. Sci. Instrum.*, 2007, **78**. DOI: 10.1063/1.2709746.
36. L. Mainali, J. W. Sidabras, T. G. Camenisch, J. J. Ratke, M. Raguz, J. S. Hyde, and W. K. Subczynski, *Appl. Magn. Reson.*, 2014, **45**, 1343. DOI: 10.1007/s00723-014-0578-7.
37. S. Pfenninger, J. Forrer, A. Schweiger, and T. Weiland, *Rev. Sci. Instrum.*, 1988, **59**, 752. DOI: 10.1063/1.1139822.
38. J. Forrer, S. Pfenninger, J. Eisenegger, and A. Schweiger, *Rev. Sci. Instrum.*, 1990, **61**, 3360. DOI: 10.1063/1.1141584.
39. A. V. Koptioug, E. J. Reijerse, and A. K. Klaassen, *J. Magn. Reson.*, 1997, **125**, 369. DOI: 10.1006/jmre.1997.1133.
40. J. Forrer, I. Garcia-Rubio, R. Schuhmam, R. Tschaggelar, and J. Harmer, *J. Magn. Reson.*, 2008, **190**, 280. DOI: 10.1016/j.jmr.2007.11.009.
41. B. Simovic, P. Studerus, S. Gustavsson, R. Leturcq, K. Ensslin, R. Schuhmann, J. Forrer, and A. Schweiger, *Rev. Sci. Instrum.*, 2006, **77**. DOI: 10.1063/1.2206776.
42. Y. Morita and K. Ohno, *J. Magn. Reson. A*, 1993, **102**, 344. DOI: 10.1006/jmra.1993.1113.
43. A. P. M. Kentgens, J. Bart, P. J. M. van Bentum, A. Brinkmann, E. R. H. Van Eck, J. G. E. Gardeniers, J. W. G. Janssen, P. Knijn, S. Vasa, and M. H. W. Verkuijlen, *J. Chem. Phys.*, 2008, **128**. DOI: 10.1063/1.2833560.
44. R. Narkowicz, D. Suter, and I. Niemeyer, *Rev. Sci. Instrum.*, 2008, **79**. DOI: 10.1063/1.2964926.
45. R. Narkowicz, D. Suter, and R. Stonies, *J. Magn. Reson.*, 2005, **175**, 275. DOI: 10.1016/j.jmr.2005.04.014.
46. R. Narkowicz, H. Ogata, E. Reijerse, and D. Suter, *J. Magn. Reson.*, 2013, **237**, 79. DOI: 10.1016/j.jmr.2013.09.017.
47. Y. Artzi, Y. Twig, and A. Blank, *Appl. Phys. Lett.*, 2015, **106**. DOI: 10.1063/1.4913806.
48. K. Mobius and R. Biehl, in Multiple Electron Resonance Spectroscopy, eds M. M. Dorio and J. H. Freed Ch. 14, Plenum: New York, 1979.
49. J. P. Willems, A. A. K. Klaassen, E. J. Reijerse, G. E. Janssen, and E. Deboer, *Rev. Sci. Instrum.*, 1992, **63**, 5362. DOI: 10.1063/1.1143403.
50. V. Weis, M. Bennati, M. Rosay, J. A. Bryant, and R. G. Griffin, *J. Magn. Reson.*, 1999, **140**, 293. DOI: 10.1006/jmre.1999.1841.
51. F. Pobell, Matter and Methos at low Temperatures, Springer: Berlin, 2007.
52. B. Epel, I. Gromov, S. Stoll, A. Schweiger, and D. Goldfarb, *Concepts Magn. Reson. Part B Magn. Reson. Eng.*, 2005, **26B**, 36. DOI: 10.1002/cmr.b.20037.
53. B. Epel, D. Arieli, D. Baute, and D. Goldfarb, *J. Magn. Reson.*, 2003, **164**, 78. DOI: 10.1016/s1090-7807(03)00191-5.
54. R. W. Quine, J. R. Harbridge, S. S. Eaton, and G. R. Eaton, *Rev. Sci. Instrum.*, 1999, **70**, 4422. DOI: 10.1063/1.1150088.
55. N. Devasahayam, S. Subramanian, and M. C. Krishna, *Rev. Sci. Instrum.*, 2008, **79**. DOI: 10.1063/1.2839398.
56. L. Sun, J. J. Savory, and K. Warncke, *Concepts Magn. Reson. Part B Magn. Reson. Eng.*, 2013, **43**, 100. DOI: 10.1002/cmrb.21240.
57. G. Lassmann, P. P. Schmidt, and W. Lubitz, *J. Magn. Reson.*, 2005, **172**, 312. DOI: 10.1016/j.jmr.2004.10.015.
58. R. D. Allendoerfer, G. A. Martinchek, and S. Bruckenstein, *Anal. Chem.*, 1975, **47**, 890. DOI: 10.1021/ac60356a023.
59. P. R. Murray, D. Collison, S. Daff, N. Austin, R. Edge, B. W. Flynn, L. Jack, F. Leroux, E. J. L. McInnes, A. F. Murray, D. Sells, T. Stevenson, J. Wolowska, and L. J. Yellowlees, *J. Magn. Reson.*, 2011, **213**, 206. DOI: 10.1016/j.jmr.2011.09.041.

60. R. D. Allendoerfer, W. Froncisz, C. C. Felix, and J. S. Hyde, *J. Magn. Reson.*, 1988, **76**, 100. DOI: 10.1016/0022-2364(88)90204-1.

61. M. A. Tamski, J. V. Macpherson, P. R. Unwin, and M. E. Newton, *Phys. Chem. Chem. Phys.*, 2015, **17**, 23438. DOI: 10.1039/c5cp04259c.

62. A. Brueckner, *Chem. Soc. Rev.*, 2010, **39**, 4673. DOI: 10.1039/b919541f.

63. T. Risse, D. Hollmann, and A. Brueckner, in Catalysis, eds J. Spivey, Y. Han and K. Dooley Ch. 1, Royal Society of Chemistry: Cambridge, 2015, Vol. **27**, 1.

Chapter 13
EPR Imaging

Boris Epel and Howard J. Halpern
University of Chicago, Chicago, IL, USA

13.1 INTRODUCTION

EPR imaging (EPRI) is a well-established method for the study of the spatial distribution of electron paramagnetic centers (PC) of molecules bearing unpaired spins, spin labels, or spin probes and their local characteristics such as kinetics, spectral shape, or relaxation. The imaging methodologies that examine the distribution of PCs and spin probes yield spatial images. Spatial images have different dimensionality: medical images are typically three-dimensional, while for material science two-dimensional or even one-dimensional images suffice. A second type of image can exhibit an additional time dimension and is used to monitor spatially resolved kinetics. These images are essentially a time-resolved version of spatial images; however, their acquisition methodology may be very different from the purely spatial ones. Spectral–spatial images that determine spatially resolved spectral shapes of PCs carry information about the PC environment including viscosity, temperature, pH, oxygen concentration, redox potential, etc. A number of pulse methodologies have been developed that focus on particular PC parameters, such as relaxation times. These methodologies are typically spatial with an additional dimension provided by the dependence of the EPR signal on a parameter of interest. We will call those images parametric images.

The development of EPR imaging went through many stages. Certain techniques that looked promising earlier on were abandoned later. This chapter covers present EPR imaging technology including instrumentation, methodology, image reconstruction, and spin probes.

13.2 BRIEF HISTORY

Early work on EPR imaging is covered in comprehensive reviews.[1–3] The work of Hoch and Day[4] on diamonds was the first in which a multidimensional EPR image was obtained using a method similar to the one used in the seminal work by Lauterbur.[5] It did not take long for EPR imaging to involve biologic specimens, like celery stems soaked in nitroxide in a study reported by Berliner and Fujii.[6] The most common

EPR Spectroscopy: Fundamentals and Methods.
Edited by Daniella Goldfarb and Stefan Stoll
 ISBN: 978-1-119-16299-5
Also published in eMagRes (online edition)
DOI: 10.1002/9780470034590.emrstm1512

imaging methodology used before 2000 was continuous wave (CW) EPR. A number of studies, however, prepared the advent of pulse imaging technology. Milov *et al.*[7] and Eaton and Eaton[8] demonstrated the application of spin echoes for imaging. Initial attempts to obtain images using FID detection showed the potential of the technique; however, the images had significant artifacts due to the dead time of the EPR imagers.[9] Emid *et al.*[10] and Balcom *et al.*[11] introduced single-point imaging (SPI) to solid-state MRI. The group of Krishna applied this technique to EPR.[12] The spin-echo imaging methodology was developed in parallel and resulted in spin-echo phase relaxation imaging[13] and later inversion recovery spin–lattice relaxation imaging (IRESE) introduced by Halpern and coworkers.[14]

Many early EPR imaging studies targeted oxygen measurements, especially for in vivo applications.[15] This field saw a consistent development over the last 35 years. EPR detection of oxygen using a nitroxide spin probe was first reported by Backer *et al.*[16] and Popp and Hyde[17] and later extensively investigated by Swartz *et al.*[18,19] In the nineties, a few groups pioneered multidimensional CW imaging on rodents in vivo[15,20] and ex vivo,[21] enabling repeated measurements of oxygen concentrations in tissues. With the development of narrow-line trityl spin probes,[22,23] the focus of preclinical oximetric imaging shifted to pulse methodologies.[13,14,24]

Separate efforts on development of EPR microscopy were carried out by Freed and coworkers (see Chapter 24).

13.3 BASICS OF IMAGE ACQUISITION

The general theory of magnetic resonance is covered in a number of books.[25,26] Here, we present the minimum required for understanding imaging. Although the theory can be extended to higher spins, we will illustrate the imaging concepts using $S = 1/2$. The resonant magnetic field can be derived from the expression

$$h\nu = g\mu_B B_0 \tag{13.1}$$

The right side of equation (13.1) represents the Zeeman energy splitting between the states of the electron spin with the magnetic field B_0. ν is the microwave (MW) or RF used. μ_B is the Bohr magneton, and h is the Planck constant. The g-factor, g, describes the ratio of the electron magnetic moment to its spin angular momentum.

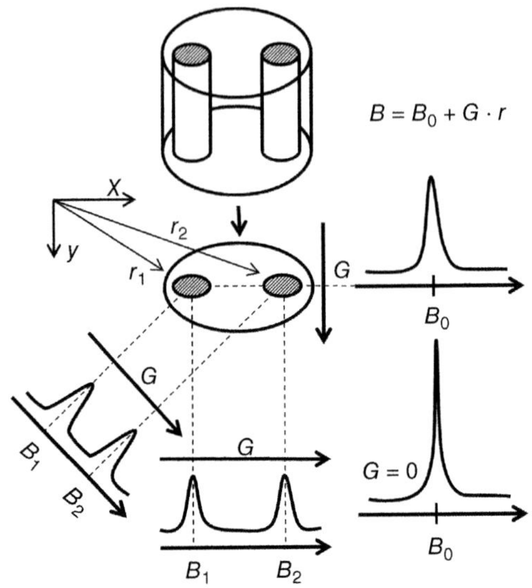

Figure 13.1. Influence of the magnetic field gradient $\boldsymbol{G}$ on the EPR spectrum of a simple object with two cylindrical EPR-active regions. Three orientations of $\boldsymbol{G}$ in the axial plane of the object are shown

For imaging, the spatial position of paramagnetic species is encoded with a linear magnetic field gradient, $\boldsymbol{G}$, that alters the amplitude of the constant magnetic field B_0. The additional magnetic field experienced by a species at position $\boldsymbol{r}$ in the sample is then $\Delta B = \boldsymbol{G} \cdot \boldsymbol{r}$ (Figure 13.1). Here, $\boldsymbol{r}$ is a three-dimensional set of spatial coordinates in vector form. The spatial direction of the gradient can be varied. Most imagers use three gradient coils along three principal directions, from which arbitrary gradients can be generated by superposition. In contrast to nuclear MRI, the six orders of magnitude faster relaxation rates of electrons compared to nuclei prohibit the use of pulsed gradients in EPRI of macroscopic samples.

13.3.1 Field-domain Image Acquisition and Reconstruction

CW EPR typically involves acquisition of spectra as a function of a changing magnetic field. During image acquisition a fixed gradient $\boldsymbol{G}$ is applied; the main magnetic field is swept through the resonance field where the signal is recorded. The EPR spectrum $s(B_{sw}, \boldsymbol{G})$ recorded in the presence of the static gradient $\boldsymbol{G}$ will

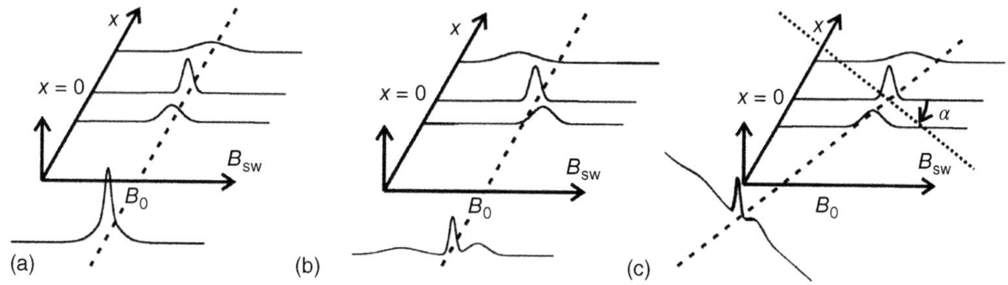

Figure 13.2. The effect of a linear magnetic field gradient on spectra recorded for a two-dimensional spectral–spatial object and the interpretation of these spectra as the Radon transformation of the object.[27] Three spatially resolved objects with different spectra centered at the same magnetic field are shown. The directions orthogonal to the projection integration directions are marked by dashed lines. (a) Magnetic field gradient is not applied. The recorded spectrum at the figure bottom is the sum of the individual spectra. (b) Positive gradient is applied; the individual spatial components pass through resonance at different magnetic fields that depend on their spatial positions. (c) With proper scaling, the spectrum shown in (b) can be interpreted as a projection of the spectral–spatial object observed with spectral projection angle α that is visualized as an angle between B_{SW} axis and projection integration dotted line

be referred to as a projection[27]:

$$s(B_{sw}, \boldsymbol{G}) = \int_{\Omega_{r,B}} f_r(B, \boldsymbol{r})\delta(B - B_0 - \boldsymbol{G}\cdot\boldsymbol{r})\mathrm{d}\boldsymbol{r}\,\mathrm{d}B \tag{13.2}$$

Here, the integral is taken over all positions in a sample and magnetic field. B_{sw} is the magnetic field swept by the imager, $B = B_{sw} - B_0 - \boldsymbol{G}\cdot\boldsymbol{r}$. $f_r(B,\boldsymbol{r})$ is the function describing the EPR line shape at every point $\boldsymbol{r}$ of a sample space $\Omega_{r,B}$, and δ is the Dirac delta function. If we assume that all the line intensity is located at one field $f_r(B,\boldsymbol{r}) = f(\boldsymbol{r})\delta(B - B_0)$ or $f_B(B) = \delta(B - B_0)$, equation (13.2) becomes identical to three-dimensional Radon transformation.[28] In the case of a sample with a spatially uniform single line, f_B, $f_r(B,\boldsymbol{r}) = f(\boldsymbol{r})\, f_B(B)$, the resulting projections are essentially a convolution of line shape independent projections $s(B,\boldsymbol{G})|_{f_B(B)=\delta(B-B_0)}$ with a line shape $f_B(B)$, $s(B,\boldsymbol{G}) = s(B,\boldsymbol{G})|_{f_B(B)=\delta(B-B_0)} * f_B(B)$. Thus, for a known line shape, $f_B(B)$, $s(B,\boldsymbol{G})|_{f_B(B)=\delta(B-B_0)}$ can be obtained using deconvolution or similar mathematical procedures.[29,30]

In the continuous data approximation, inversion of the Radon transformation recovers $f(\boldsymbol{r})$ from $s(B_{sw}, \boldsymbol{G})$ exactly. For a three-dimensional experiment, projections with a single gradient strength $|\boldsymbol{G}|$ and multiple orientations are required. The most efficient choice for the set of projection orientations is an equal solid-angle sampling.[31] It should be noted that the Radon transformation is defined in continuous space, and therefore reconstruction from a limited number of projections will inevitably result in imaging artifacts. The method most commonly used for the inversion of Radon transformation is a filtered backprojection (FBP). In our experience, this method provides the best image quality for a medium-to-large number of projections. FBP reconstruction is a direct implementation of the inverse Radon transform formula. In the first step, projection filtering is required to correct the radial frequency representation of $f_r(B,\boldsymbol{r})$. This is achieved by convolution of the projections with Ram–Lak filter, which has a linear frequency response function with a maximum frequency value at the Nyquist frequency. The term backprojection comes from the fact that $s(B,\boldsymbol{G})$ is directly mapped to $f_r(B,\boldsymbol{r})$ along the integration paths used in equation (13.2). A number of FBP optimizations were proposed including nonuniform projection acquisition[32] and projection interpolation.[33] Projections acquired using pulse spin-echo methodology are treated similarly by transforming projections from the time to the field/frequency domain using a Fourier transformation.[34] Image reconstruction is covered in more detail elsewhere.[28,35]

The treatment becomes more complex for a general case of $f_r(B,\boldsymbol{r})$. Introduction of a four-dimensional vector $\left(B, \frac{\Delta B}{\Delta L}\boldsymbol{r}\right)$, where ΔB and ΔL are spectral and spatial extents of an object, can bring equation (13.2) to a four-dimensional Radon transformation form.[27] Four-dimensional FBP may then be used to reconstruct the image. The most complex aspect of spectral–spatial FBP reconstruction is associated with the requirements it imposes on the applied gradient and field sweep. The spectral dimension is typically introduced using a spectral angle, $\alpha \in \left[-\frac{\pi}{2}; \frac{\pi}{2}\right]$ (Figure 13.2). The gradient strength and field sweep range B_{range} are spectral angle dependent[27]:

$$|\boldsymbol{G}| = \frac{\Delta B}{\Delta L}\tan(\alpha) \tag{13.3}$$

$$B_{\text{range}} = \sqrt{2}\,\Delta B\,\text{acos}(\alpha) \tag{13.4}$$

Projections with a full range of α are required for exact image reconstruction. Acquisition of a projection with spectral angles approaching $\pi/2$ is practically impossible because it requires infinitely large $|\boldsymbol{G}|$ and B_{range}. The image reconstruction methodologies that avoid these 'missing angles' are based on algebraic reconstruction.[36] However, they still need to be proven to be robust for the case of three spatial dimensions and one spectral dimension. A promising approach to the reduction of the problem dimensionality is described by Tseitlin *et al.*[37] for the two-dimensional case.

13.3.2 Time-domain Image Acquisition and Reconstruction

In pulse or time-domain EPR, the signal is acquired in a response to a series of high-power RF pulses. In time-domain imaging, the expression (13.2) for spatial projections becomes

$$s(t, \boldsymbol{k}) = \int_{\Omega_r} f(\boldsymbol{r})R(t, r)\exp[-\text{i}2\pi\boldsymbol{k}(t)\boldsymbol{r}]\text{d}\boldsymbol{r} \tag{13.5}$$

Here, $f(\boldsymbol{r})$ is the spatial distribution of the magnetization and $R(t, r)$ describes spatially nonuniform effects of phase relaxation, T_2, and inhomogeneous line broadening, $T_2^{\,*}$, on the magnetization evolution. The term

$$\boldsymbol{k}(t) = \frac{\gamma_e}{2\pi}\int_0^t \boldsymbol{G}(t')\text{d}t' \tag{13.6}$$

can be seen as a trajectory in a Fourier space of the sample (k-space).[38,39] Equation (13.5) has the same form as the expression for Fourier transformation. If k values are arranged on a rectangular grid in k-space, the object image can be recovered by application of a multidimensional Fourier transformation. Other arrangements of k-space values, such as radial distribution, require more complex reconstruction procedures involving either resampling of data to a rectangular grid or FBP. This can be artifact prone. For static gradients used for macroscopic imaging, equation (13.6) is simplified to

$$\boldsymbol{k}(t) = \frac{\gamma_e}{2\pi}\boldsymbol{G} \tag{13.7}$$

Two general ways of sampling k-space with static gradients have been developed (Figure 13.3). The first method is referred to as radial imaging. Consider a spin echo with the refocusing time (maximum) $t = 0$. At this time, $\boldsymbol{k}$ is equal to zero for any applied gradient. As time increases, $\boldsymbol{k}$ increases in magnitude and traces out a radial line starting from the origin (Figure 13.3). A change in the gradient direction changes the slope of the line and allows k-space sampling. The gradient protocol of this method is identical to the field-domain method described above. In fact, $s(t, \boldsymbol{k})$ and $\boldsymbol{s}(B, \boldsymbol{G})$

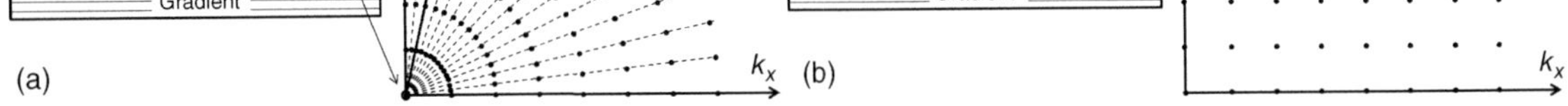

Figure 13.3. Filling of two-dimensional k-space by radial (a) and SPI (b) methodologies. Only one quadrant of k-space is shown. (a) Single echo envelope can be mapped to a k-space radial line starting from the coordinate origin that corresponds to the echo maximum. The direction of the radial line is defined by the direction of the applied gradient. Commonly used inversion recovery electron spin-echo (IRESE) and FID pulse sequences are shown. (b) Single-point imaging generates one image k-space point per experiment. The single-pulse sequence that generates an FID is shown. A single point of the FID envelope is projected to k-space depending on the applied gradient. (See the text for details)

are related through the Fourier transformation, and the same algorithms can be used for image reconstruction.

The second method is the SPI method. This method, suggested by Maresch *et al.*, was intended to overcome the influence of the EPR linewidth (or T_2*) on the image resolution.[10] Use of a single data point from the magnetization decay at delay $t = t_p$ in equation (13.7) allows for sampling at arbitrary points in k-space by choosing an appropriate $\boldsymbol{G}$. By sampling $\boldsymbol{k}$ values on the rectangular grid, image reconstruction can be reduced to a multidimensional Fourier transformation.[12,24,29]

13.3.3 Number of Projections, Discrete Image Reconstruction, and Sparse Sampling

One question that arises during selection of an imaging protocol is how many projections one should use for particular image acquisition. As practice shows, there is no unique answer to this question. Most researchers use the criteria of 'sufficient representation' of features of interest in the image rather than strict numerical criteria.

For reconstruction of a 'true' continuous object, an infinite number of projections or k-space samples are required. In case of discrete objects, the number of k values required for proper object representation can be estimated.

For SPI, discussed in the previous section, acquisition is performed in Fourier space complementary to the object space, and thus the maximum number of k-space point acquisitions is equal to the number of points in the image matrix. For radial imaging there are two interdependent requirements: the number of projections and the maximum gradient magnitude (see next section). The number of voxels in the image is decided according to the expected resolution based on the applied gradient. For a three-dimensional image with the number of image voxels in one dimension N_{1D}, the number of projections (or spokes starting from the origin of k-space) N_S[40] is

$$N_S = 4\pi N_{1D}^2 \qquad (13.8)$$

For a three-dimensional matrix with $N_{1D} = 64$ voxels, a very large number of measurements from 51 472 (radial) to 262 144 (SPI) is required. The agility of the gradients restricts acquisition rates of modern large-object imagers to only few projections per second. Thus, EPR images are typically under-sampled and acquisition methodologies are sparse.

There is a large body of MRI literature for the methods of sparse image acquisition.[41] These methods are referred to as compressed sensing since they try to reconstruct signals and images from significantly fewer measurements than were traditionally thought necessary. It is important to note that the resulting images are not a 'true' depiction of the reality in the Nyquist theorem sense but they preserve the information of interest and thus are useful.

For radial EPR imaging, optimal sampling schemes include uniform sampling schemes.[42] Nonuniform acquisition schemes calculate the most informative projection angles from the projections already acquired. The image acquisition becomes then an iterative process until image quality related criteria is satisfied.[32,43] The field range of the projections that has to be acquired for the given projection scheme for a given object shape was explored as well as acceleration of SPI acquisition.[44]

13.3.4 Image Resolution and Voxel Resolution

Estimation of image resolution is a rather complex task. Many factors contribute to this problem – image acquisition methodology, imaged sample spectral shape that causes blurring, noise, data processing artifacts, etc. Although many reconstruction algorithms allow reconstructing images with an arbitrary number of voxels, a larger number of voxels in the image do not necessarily indicate better image resolution.

A crude estimation can be obtained from the EPR linewidth at half height, LWHH, and the maximum gradient applied during imaging: resolution = LWHH/$|\boldsymbol{G}_{max}|$. Modern image processing techniques including deconvolution of the line shape allow improving resolution beyond this estimation.

Image resolution can be quantified by observing the response of an image amplitude to an abrupt change in sample density, namely, no sample at one side and constant sample density on the other side of a boundary of a homogeneous object.[33] The change of the normalized image signal intensity obtained along a line perpendicular to the boundary can be fitted with a Gaussian error function $(1 + \mathrm{erf}(x/\sigma\sqrt{2}))/2$. The width of this error function (σ) will then give an estimate of the spatial resolution.

Resolution of single-point images is approximately equal to their voxel resolution as the EPR line shape does not affect the image.

Typical small animal EPR image resolution is 0.5–2 mm with image acquisition times varying from seconds to an hour. The resolution is limited by the sensitivity of the imagers, which for animal-size resonators and typical spin probes is 10^{12}–10^{14} spins per voxel. This sensitivity allows successful imaging of small animals with average spin probe concentrations of 100 μM–1 mM. As will be discussed below, imagers operating above 1 GHz have higher sensitivity and resolution but could not be applied for animal imaging due to limitation of the MW penetration into the body. Material imaging, which is not limited by object size and MW penetration into object, can achieve better resolution (see Chapter 24).

13.4 SPIN PROBES

Highly concentrated PCs are not frequent in nature and even less frequent in living bodies. Endogenous paramagnetic species found in mammalian bodies include hemoglobin, metalloenzymes, various diffusible radicals, and molecular oxygen. Metal centers and oxygen at animal body temperature have very short relaxation times, broad lines, and, thus, can hardly be imaged using present day EPR techniques. Diffusible paramagnetic species have concentrations below tens of picomolar. At present, exogenous spin probes are the only practical reporters, and appropriate spin probes are the key to successful in vivo imaging. The line shapes, hyperfine structures, and relaxation times of the probe and their sensitivity to the parameters of interest largely define the sensitivity and precision of a methodology. Thus, development of appropriate spin probes is of utmost importance for EPR imaging.

There are two large classes of spin probes: soluble free radicals (nitroxides and trityl radicals) and insoluble paramagnetic particles (particulates).[45]

13.4.1 Nitroxides

A major aspect of the development of physiologic EPR imaging has been the synthesis of free radical reporter molecules that are specific to particular aspects of the physiological environment. Soluble probes are administered intraperitoneally,[46] intravenously,[47,48] and, more rarely, intra-arterially.[49] Historically, the first probes applied for oximetry were nitroxides.[6] Nitroxides (see Chapter 14, Figure 13.4a) are free radicals that have an unpaired electron delocalized between bonded nitrogen and oxygen. A wide variety of nitroxides are available: there are compounds that can target intracellular and extracellular compartments[50] or penetrate the blood–brain barrier.[51] Multiple classes of spin probes sensitive to pH, cell redox potential, and pO_2 are available.[52,53] Disadvantages of nitroxide reporters include metabolic conversion to diamagnetic hydroxylamines with timescale from seconds to minutes, multiple EPR lines reducing the sensitivity, and short relaxation times (broad linewidths), making pulse detection at low frequencies difficult. Optimal structure, isotopic spin labeling with ^{15}N (two hyperfine lines in the ^{15}N probe instead of three in the ^{14}N probe), and deuteration can improve the sensitivity of nitroxide imaging considerably.[54,55]

13.4.2 Trityl Radicals

The success of oximetry in the last decade is strongly linked to triarylmethyl radicals (trityls) developed by Nycomed Innovations (later acquired by GE Healthcare). Trityls possess a narrow single EPR line (Figure 13.4b).[22] The syntheses of various trityl radicals are described elsewhere.[56] Trityls that are commonly used for in vivo imaging are methyl-tris[8-carboxy-2,2,6,6-tetrakis[2-hydroxyethyl]benzo[1,2-d:4,5-d′]bis[1,3]dithiol-4-yl]-trisodium salt, OX063 (16 μT peak-to-peak, p-p), and its partially deuterated form (8 μT p-p). The relaxation rates of trityls are linearly dependent on the oxygen partial pressure, pO_2.[22] These spin probes are distributed in the extracellular fluid compartment.[57] In the blood stream of a mouse, the clearance halftime of these probes is approximately 9–10 min, whereas in tumors they remain and provide strong signals for about 40–50 min.[58] No bioreduction of trityls during this time scale was observed. The lethal dose (LD50) of OX063 is large (8 mmol kg^{-1}) that allows high dose injections.[59] Highly concentrated solutions of 80–100 mM are injected. Trityls with modified structure to simultaneously determine pH and pO_2 are available.[60]

13.4.3 Particulates

Various particulates have been used, primarily for EPR oximetry: activated charcoal, lithium phthalocyanine,[61] and a number of its derivatives with

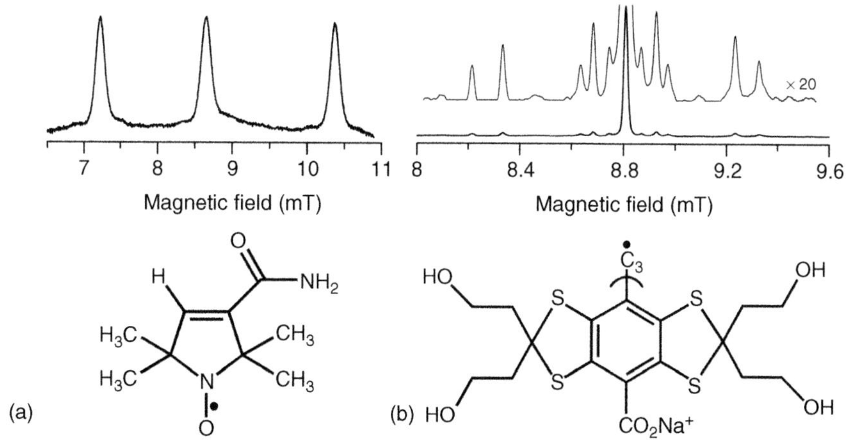

Figure 13.4. 250 MHz EPR spectra and chemical structures of selected spin probes. (a) CTPO nitroxide. Rapid scan EPR spectrum is shown. (b) OX063 spin probe. An integrated CW EPR spectrum of the deoxygenated partially deuterated spin probe is shown. The hyperfine structure due to natural abundance ^{13}C isotopes is magnified for clarity

improved oxygen sensitivity such as octa-*n*-butoxy-naphthalocyanine.[62]

The insoluble spin probes can be inserted surgically, injected in the form of a slurry, fed to an animal,[63] or implanted with tumor cells during inoculation in mice.[64,65] The limited mobility of particulates enables repeated measurements of the oxygen concentration for an extended period. A two orders of magnitude higher concentration of unpaired electrons (compared to a soluble probe) results in higher sensitivity. Ultimately, slow migration and degradation of the spin probe leads to loss of EPR signal intensity. Also, there are potential biocompatibility concerns due to direct and prolonged exposure of particulates to tissue. Encapsulation or coating of the crystalline probes in biocompatible polymer matrices is viewed as a potential strategy for overcoming some of these concerns and to enhance clinical applicability of these spin probes.[66]

13.5 IMAGING INSTRUMENTATION AND METHODOLOGY

13.5.1 Operating Frequency

The choice for the EPR imager operating frequency is governed by two factors. The sensitivity of the instrument grows with increasing frequency, $\nu^{0.8}$, which includes the increase in signal as ν^2 and an increase in the sample noise from dielectric and conductive losses as $\nu^{1.2}$, whereas the penetration depth of the RF radiation falls with increasing frequency.[67–69] The approx. 6–7 cm RF penetration depth in muscle/saline (and 20 cm in fat) at 250 MHz falls to 1–2 cm at 1 GHz and to a few millimeters at 9 GHz. Modern in vivo imagers operate in one of the frequency ranges: 250–350 MHz,[34,70,71] 550–750 MHz,[29,72] and 1–1.2 GHz or L-band (Bruker). Imagers operating at 250–350 MHz are typically used for larger animals like pigs or rabbits or full-body rodent imaging, whereas L-band imagers primarily target smaller animals or animal peripheral anatomy.

13.5.2 Gradient and Magnetic Field Systems

In addition to a regular CW or pulse EPR instrument, imagers use gradient coils around a sample to generate X, Y, and Z gradients, power supplies for these coils, gradient controlling hardware, and specialized software. Currently, the only commercial vendor of EPR imagers is Bruker BioSpin Corporation (http://www.bruker.com) with their line of 3D L-band and X-band CW and pulse products. A number of laboratories have developed their own systems with operational system ranging from 250 MHz to Ku-band. The lower-frequency units are typically used for in vivo application,[29,34,72–74] while higher-frequency units are designed for EPR microscopy (see Chapter 24).

Conventional imagers require relatively large volumes with homogeneous magnetic field strength to accommodate the specimens. Magnets capable of accommodating parts of the human anatomy were designed at the Dartmouth School of Medicine[19] and the University of Chicago. One challenging aspect of imaging magnet design is their field stabilization system, which cannot rely on magnetic field reading due to the presence of changing gradients. EPR gradient coils are similar to the ones used for MRI.[73–75] They typically reach 100–300 mT m^{-1} strength and operate in continuous mode. To allow rapid acquisition of images with multiple thousands of projections, low inductance gradient coils and high compliance power supplies are used.

13.5.3 Continuous Wave and Rapid Scan

Signal-to-noise ratio (SNR) is the major factor that limits EPR imaging applications. At small RF power, P, the EPR signal is proportional to $P^{0.5}$. However, this dependence saturates, delivering maximum signal at a power P_{max} that is determined by the relaxation properties of the sample. Most imagers have sufficient power to operate close to saturation. Conventional CW EPR uses magnetic field modulation and lock-in detection. The amplitude of the resulting first EPR signal harmonic is proportional to the modulation amplitude. The dependence of the signal on the modulation amplitude is linear until the modulation amplitude becomes comparable to the linewidth of spin probe.[76,77] Modulation amplitudes above this value distort the EPR line (see Chapter 1).

Rapid-scan (RS) EPR is an enhancement of CW methodology and a promising imaging methodology that is free of many of the drawbacks listed above (see Chapter 23).[78] The overall effect of SNR enhancement by RS EPR ranges from 3–5 to 100, depending on the sample relaxation and instrumentation used. In addition, RS instruments feature lower resonator quality factor that improves the stability of in vivo measurements. RS directly acquires the EPR line shape and not the line shape derivative like CW EPR and thus suffers less SNR losses due to the line broadening induced by the application of the imaging gradients. Finally, RS allows the acquisition of more projections per unit time.

13.5.4 Pulse Imaging

Although pulse EPR imaging can be applied only to limited classes of spin probes with slow relaxation, whenever applicable it has demonstrated a remarkable advantage over CW imaging.[79]

Pulse imaging offers a variety of pulse sequences that enable parametric imaging.[80] Figure 13.5 shows examples of pulse methodologies for spin–lattice and T_2^* imaging. When acquiring spin–lattice relaxation images, separate three-dimensional images with different delay values T in the inversion recovery electron spin-echo sequence are obtained (Figure 13.5a). For SPI, images are reconstructed at different t_p (Figure 13.5b). Then, for each voxel location of these images, the signal evolution is fitted. Logarithmically spaced delays yield a more precise determination of the relaxation time. Parametric acquisition avoids four-dimensional spectral–spatial acquisition used in CW images that reduces acquisition time and improves precision of the parameter's determination.

For radial k-space imaging (see Section 13.3.2 and Figure 13.3), the choice of readout techniques is of utmost importance. Pulse imagers do not allow signal detection immediately after the excitation pulse due to detector overload. The delay between excitation and the time of earliest access to the detector, called the 'dead time' of the instrument (t_{dead}), leads to a missing part of k-space and image distortion. To avoid this, dead-time-free sequences such as spin echo or stimulated spin echo are used.[13,34]

An important parameter of any pulse sequence is its excitation bandwidth, $\Delta\omega$. This bandwidth should be sufficient to excite simultaneously the gradient-broadened EPR line throughout the sample:

$$\Delta\omega = \gamma_e G_{max} \Delta L \qquad (13.9)$$

In this equation, G_{max} is the gradient magnitude and ΔL is the maximum spatial extent of the object. The pulse sequence is one of the key bandwidth determinants. In a resonator with infinite bandwidth, a single $\pi/2$ detection pulse with length t_{pulse} has a bandwidth of about $1/t_{pulse}$. The bandwidth of a two-pulse echo sequences with identical pulse lengths is nearly four times narrower.[81] For example, a 35–35 ns two-pulse sequence in a 19 mm loop-gap resonator with $Q = 14$ at 250 MHz achieved the bandwidth of 8.7 MHz, which is sufficient for imaging objects of ≈2 cm with 15 mT m^{-1} gradient.[80] The stimulated echo sequence

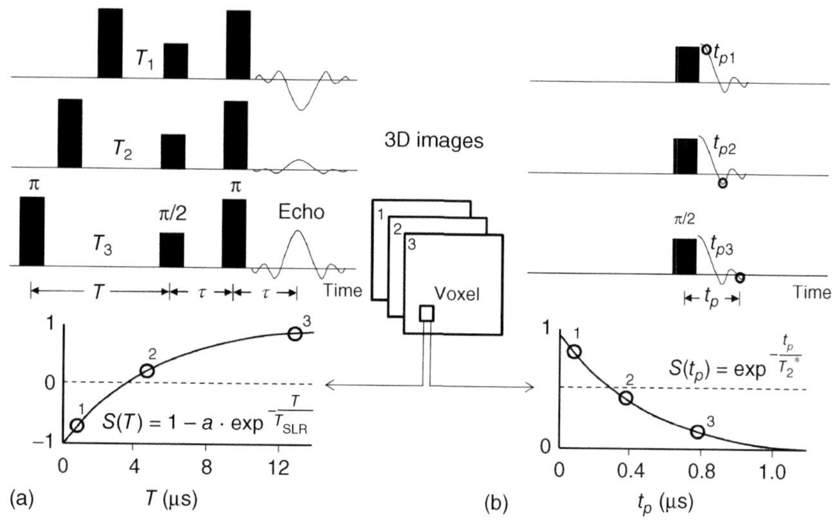

Figure 13.5. Acquisition of parametric pulse EPR images. (a) Inversion recovery spin–lattice relaxation imaging (IRESE). (b) T_2^* single-point imaging

has nearly twice the bandwidth of the two-pulse echo sequence and half the bandwidth of a single π/2 pulse.[80] In resonators whose bandwidth is comparable to the bandwidth of the pulse sequence, differences in the pulse sequence bandwidths is reduced but are still considerable. Application of tens of nanosecond long RF pulses in resonators with useful volumes from cm^3 to hundreds of cm^3 requires RF amplifiers delivering power of hundreds to thousands of watts. The available power imposes limitations on sample size and image resolution. Excitation bandwidth can be extended using multiple-window acquisition[82] and/or pulse shaping[83] (see Chapter 21). Using spin probes with a narrower linewidth allows for a reduction in the gradient magnitude and thus a reduction in the excitation bandwidth.

13.5.5 Software

13.5.5.1 Image Reconstruction Packages

A two-dimensional inverse Radon transformation (image reconstruction) routine is included in the MATLAB (Mathworks, Natick, MA) computational language imaging toolbox. A stand-alone MATLAB toolbox developed for spatial and spectral–spatial image reconstruction written at the University of Chicago is available online (http://epri.uchicago.edu).

13.5.5.2 Image Registration Packages

Image-based research requires numerous tools for data processing and analysis. In vivo EPR images carry limited anatomic information. Thus, they have to be paired with anatomic images like CT or MRI for localization of the areas of interest. In most cases, EPR and CT/MRI images are acquired using different instruments. The pairing can be performed through the procedure called image registration. This procedure establishes a transformation between image spaces, which allows transformation of the coordinate system of one image into the coordinate system of another image. To introduce spatial localization into EPR images, fiducial tubes filled with EPR material are used.[84] The animals are immobilized in dental mold material that cradles the tumor with fiducial locators. This ensures a fixed relative position between animals and fiducials during animal relocation from one imager to another. A number of elaborated free software tools for image processing including ImageJ (http://imagej.nih.gov/ij/) and Slicer (http://www.slicer.org) are available. Our laboratory uses a locally written ArbuzGUI MATLAB toolbox (http://epri.uchicago.edu).

13.5.6 Imaging Rapid Processes

One important question that always arises during live system studies is whether the observed results are time

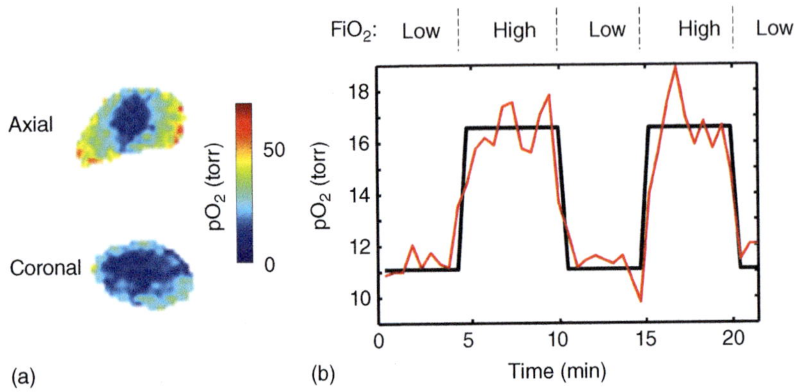

Figure 13.6. Time-resolved pO_2 imaging of a mouse leg bearing a tumor enabled by the low-rank tensor method. (a): axial and coronal slices of pO_2 image. (b): time-resolved pO_2 fluctuations overlapped with breathing gas FiO_2 time sequence in one of the pO_2 image voxels located at the tumor periphery. The animal was subjected to breathing atmosphere change between the normal air (FiO_2 = 21%) and the high-oxygen supply (FiO_2 = 95% and $FiCO_2$ = 5%). Image was obtained using deuterated OX063 spin probe and 250 MHz pulse imager

averaged over lengthy image acquisition or steady values. There is compelling evidence that in vivo parameters like oxygen partial pressure (pO_2) show relatively fast dynamics in some tissues. At normal conditions, pO_2 pressure in mammals varies from $\approx$100 torr in arterial blood to 40–50 torr in veins. Tissue pO_2 below 10 torr is considered abnormally oxygenated or hypoxic. Studies suggest that acute hypoxia (sometimes referred to as cycling or transient hypoxia) may be an important determinant of cancer progression and patient prognosis as chronic hypoxia.[85] Different approaches were undertaken to achieve rapid imaging – resolution reduction,[86] rotating gradients,[87] or principal component analysis intelligent de-noising technique.[88] The last approach allowed capturing oxygen fluctuations with a temporal resolution of better than 1 min.

Elimination of redundant data acquisition is another way of accelerating image acquisition. The low-rank tensor imaging model shows promise.[89] This model allows achieving partial separability of image time sequences into independent space and time subsets. Both subsets are acquired simultaneously: time-resolved low-resolution images – navigators and a single spatially resolved image. The low-rank tensor imaging method synthesizes these two images into a time series of spatially resolved images. Figure 13.6 shows in vivo results from a tumor-bearing mouse leg, where pO_2 fluctuations were induced by cycling the fraction of inspired oxygen (FiO_2), toggling the FiO_2 approximately every 5 min. The figure shows slices from 3D pO_2 maps at one time point as well as a graph of the pO_2 variation over time for one voxel at the center of the tumor.

13.6 APPLICATIONS

13.6.1 Oximetry

One of the most developed applications of modern EPR imaging is in vivo oximetry. This is no accident. Tissue oxygen (O_2) levels are the most important and most quantifiable stimuli to which cells and tissues respond through inducible signaling pathways. Tumor O_2 levels are major determinants of the response to cancer therapy. Developing more accurate measurement and images of tissue O_2, partial pressure (pO_2) assumes enormous practical, biological, and medical importance. The oxygen molecule has a triplet ground state with two unpaired electrons that define its paramagnetic behavior. This triplet state exhibits unmeasurably fast relaxation, and upon interaction with a spin probe, oxygen enhances spin–spin and spin–lattice relaxation rates of the probe, mainly via Heisenberg exchange.[90,91] The Smoluchowski diffusion equation predicts a linear relationship between pO_2 and homogeneous EPR line broadening ΔB.

The concentration proportionality coefficients can be determined experimentally by fitting the linewdth or relaxation rate dependence in samples with a known pO_2.[76] This allows a direct absolute measure of pO_2

in a full range of physiologically relevant oxygen pressures (0–150 torr) and is a unique feature of EPR oximetry in comparison to other noninvasive methods.[92,93] The absolute accuracy and precision of the EPR measurement exceeds 1 torr.

CW methodologies determine the spin probe line shapes. For the Lorentzian line shape, the half width at half height (HWHH) is equal to $1/\gamma_e T_2^*$. For EPR lines with multiple broadening mechanisms, $1/\gamma_e T_2^*$ describes the homogeneous broadening of the EPR line (spin-packet linewidth). The extraction of an accurate spin-packet linewidth from the total line shape requires precise knowledge of the hyperfine structure as well as instrumental parameters such as modulation field amplitude.[76,94,95] In time-domain imaging, the FID of an inhomogeneously broadened line has a non-exponential decay. However, it is common to approximate this signal with an exponential decay having a relaxation time T_2^*.[29] Spin-echo methodologies can measure T_2 and T_1 directly.[13]

Several physical parameters besides oxygen pressure affect transverse relaxation. Some of these parameters, such as temperature, viscosity, and salinity, are tightly controlled in the body of a living animal. This allows a quantitative account of their effects on relaxation and, thereby, the calibration of oxygen measurements. Other parameters such as spin probe concentration may vary and affect the accuracy of oximetry.

Heisenberg spin exchange between a spin probe and oxygen acts not only on the EPR linewidth but also on the spin probe's longitudinal relaxation.[90,91] Importantly, other relaxation processes affect the transverse and the longitudinal relaxation differently. Electron spin exchange between two trityl spin probe molecules increases the transverse relaxation rate.[90] This effect is dependent on the number of collisions between spin probes per unit time and therefore on the concentration of the spin probe. Exchange interaction between spin probes does not alter the total energy of the interacting spin pair and does not affect longitudinal relaxation rate, R_1, which shows a nearly an order-of-magnitude smaller concentration dependence in free tumbling radicals.[14,91] This makes R_1 more suitable for oximetry. Figure 13.7 shows the advantage of R_1 imaging over R_2 imaging.

13.6.2 Measurement of pH and Redox Potential

Acid–base balance is altered in a variety of common pathologies, including ischemia, renal failure, and cancer. While intracellular pH is tightly maintained, the extracellular pH may vary depending on cell metabolism. There are data indicating that altered pH

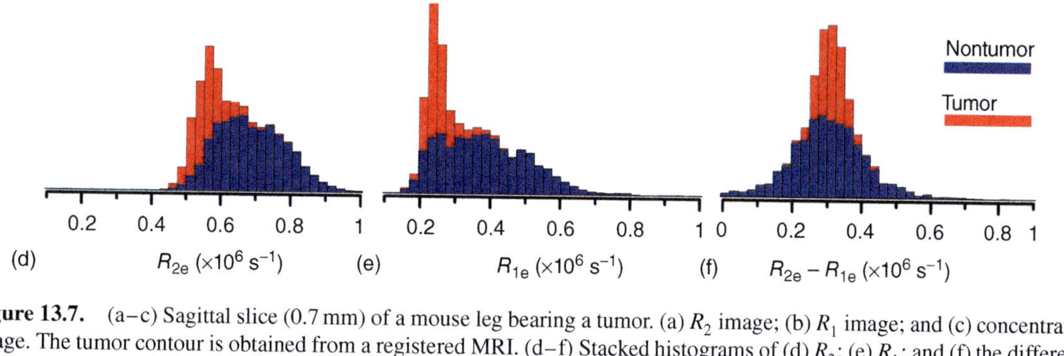

Figure 13.7. (a–c) Sagittal slice (0.7 mm) of a mouse leg bearing a tumor. (a) R_2 image; (b) R_1 image; and (c) concentration image. The tumor contour is obtained from a registered MRI. (d–f) Stacked histograms of (d) R_2; (e) R_1; and (f) the difference between R_2 and R_1. The difference between R_2 and R_1 is due to spin probe self-broadening. (Reproduced with permission from Ref. 14. © John Wiley and Sons, 2014.)

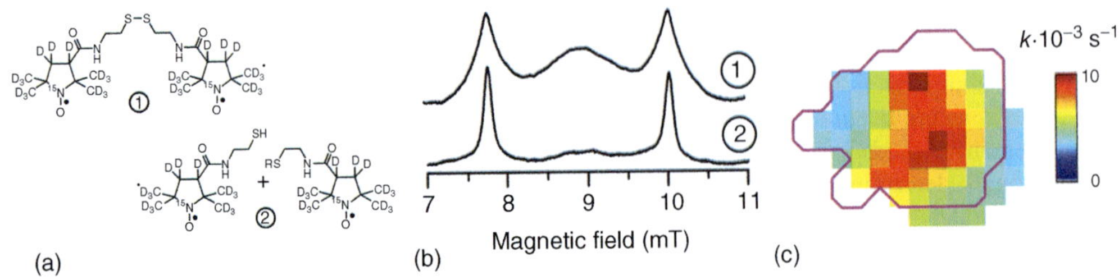

Figure 13.8. 250 MHz rapid scan image of spin probe dissociation kinetics related to cell redox status. (a) ^{2}H,^{15}N-disulfide dinitroxide spin probe.[100] Upon interaction with intracellular glutathionine, the diradical molecule (1) dissociates into two radicals (2). (b) Diradicals (1) and radicals (2) have different EPR spectra. (c) Image of dissociation kinetics observed in a mouse leg bearing a tumor. Kinetic time image was obtained by fitting image intensity of (2) to an exponential recovery function. The tumor outline is shown. Spin probe was delivered using intratumoral injection. Unpublished data

can have important health-related consequences. MRI measurements indicate that in tumors, pH gradients of up to 1.0 pH unit can exist within a 1-cm distance.[96] The EPR approach to pH imaging is based on imaging of nitroxides that have pH-sensitive hyperfine splitting and have sufficient stability to survive the living body reducing environment without penetrating the cell membrane. Imidazoline nitroxide has been proven to be useful for this application.[53] The pH is obtained from changes in hyperfine coupling of nitrogen caused by protonation/deprotonation of the N-3 atom in the radical. The changes of the nitroxide spectra are observed using spectral–spatial imaging. Recently there were successful efforts to functionalize and image pH-sensitive trityl radicals in vitro.[97]

In cells, thiols coexist with their oxidized (disulfide) counterparts. The cellular redox microenvironment is determined by the thiol–disulfide balance or thiol redox status.[98] Reactive oxygen species (ROS) are modulated by the availability of reducing equivalents provided predominantly by reduced glutathione. The change in redox status is the tissue or tumor cell's response to excessive, toxic ROS. However, in vivo, it is extremely difficult to quantitatively measure, much less image ROS.[99] A promising approach for EPR redox status imaging is based on using sulfide-bridged nitroxide compounds to monitor the dissociation kinetics using EPR imaging (Figure 13.8).[100]

13.6.3 Material Science

An interesting example of EPR imaging application to material science can be found in publications by Schlick and Kruczala[101] In these studies, the formation of stable nitroxide radicals derived from hindered amine stabilizers (HAS) was observed. EPR imaging allowed nondestructive profiling of these radicals, which gave an insight into oxidative degradation of polymers. The intensity profile in the sample depth was deduced by 1D EPRI, and the spatial variation of the EPR line shapes ('spectral profiling') was determined by 2D spectral–spatial ESRI. The method was proposed to detect spatial heterogeneities in the degradation process and to identify morphological domains that are selectively degraded.

13.7 CONCLUSION

Modern applications of EPR imaging are increasing in numbers and in complexity. In some areas, like preclinical imaging, they are becoming the method of choice, especially for oxygen and pH measurements. This demonstrates the maturity of the technology. Until recently, most of the studies were performed on home-built or custom-made instruments. We expect that increased availability of commercial EPR imaging instruments will stimulate further development of the EPR imaging field and we will see more applications of this exciting technology.

ACKNOWLEDGMENTS

This work was supported by NIH grants P41 EB 002034, R01 CA 098575 and R50 CA211408.

REFERENCES

1. K. Ohno, *Appl. Spectrosc. Rev.*, 1986, **22**, 1.
2. G. R. Eaton, S. S. Eaton, and K. Ohno, EPR Imaging and In Vivo EPR, CRC Press: Boca Raton FL, 1991.

3. S. K. Misra (ed.), Multifrequency Electron Paramagnetic Resonance: Theory and Applications, Wiley-VCH Verlag & Co.KGaA: Weinheim, Germany, 2011.
4. M. J. R. Hoch and A. R. Day, *Solid State Commun.*, 1979, **30**, 211.
5. P. C. Lauterbur, *Nature*, 1973, **242**, 190.
6. L. J. Berliner and H. Fujii, *Science*, 1985, **227**, 517.
7. A. D. Milov, A. Y. Pusep, S. A. Dzuba, and Y. D. Tsvetkov, *Chem. Phys. Lett.*, 1985, **119**, 421.
8. G. R. Eaton and S. S. Eaton, *J. Magn. Reson.*, 1987, **71**, 271.
9. M. Afeworki, G. M. van Dam, N. Devasahayam, R. Murugesan, J. Cook, D. Coffin, J. H. A. Larsen, J. B. Mitchell, S. Subramanian, and M. C. Krishna, *Magnet Reson. Med.*, 2000, **43**, 375.
10. G. G. Maresch, M. Mehring, and S. Emid, *Physica B & C*, 1986, **138**, 261.
11. B. J. Balcom, R. P. MacGregor, S. D. Beyea, D. P. Green, R. L. Armstrong, and T. W. Bremner, *J. Magn. Reson. Ser. A*, 1996, **123**, 131.
12. S. Subramanian, N. Devasahayam, R. Murugesan, K. Yamada, J. Cook, A. Taube, J. B. Mitchell, J. A. Lohman, and M. C. Krishna, *Magnet Reson. Med.*, 2002, **48**, 370.
13. C. Mailer, S. V. Sundramoorthy, C. A. Pelizzari, and H. J. Halpern, *Magnet Reson. Med.*, 2006, **55**, 904.
14. B. Epel, M. K. Bowman, C. Mailer, and H. J. Halpern, *Magnet Reson. Med.*, 2014, **72**, 362.
15. H. J. Halpern, C. Yu, M. Peric, E. Barth, D. J. Grdina, and B. A. Teicher, *Proc. Natl. Acad. Sci. U S A*, 1994, **91**, 13047.
16. J. M. Backer, V. G. Budker, S. I. Eremenko, and Y. N. Molin, *Biochim. Biophys. Acta*, 1977, **460**, 152.
17. C. A. Popp and J. S. Hyde, *J. Magn. Reson.*, 1981, **43**, 249.
18. H. M. Swartz and R. B. Clarkson, *Phys. Med. Biol.*, 1998, **43**, 1957.
19. N. Khan, B. B. Williams, H. Hou, H. Li, and H. M. Swartz, *Antioxid. Redox Signal.*, 2007, **9**, 1169.
20. P. Kuppusamy, M. Chzhan, K. Vij, M. Shteynbuk, D. J. Lefer, E. Giannella, and J. L. Zweier, *Proc. Natl. Acad. Sci. U S A*, 1994, **91**, 3388.
21. J. L. Zweier, S. Thompsongorman, and P. Kuppusamy, *J. Bioenerg. Biomembr.*, 1991, **23**, 855.
22. J. H. Ardenkjaer-Larsen, I. Laursen, I. Leunbach, G. Ehnholm, L. G. Wistrand, J. S. Petersson, and K. Golman, *J. Magn. Reson.*, 1998, **133**, 1.
23. R. Murugesan, J. A. Cook, N. Devasahayam, M. Afeworki, S. Subramanian, R. Tschudin, J. A. Larsen, J. B. Mitchell, A. Russo, and M. C. Krishna, *Magnet Reson. Med.*, 1997, **38**, 409.
24. K. Matsumoto, S. Subramanian, N. Devasahayam, T. Aravalluvan, R. Murugesan, J. A. Cook, J. B. Mitchell, and M. C. Krishna, *Magnet Reson. Med.*, 2006, **55**, 1157.
25. A. Schweiger and G. Jeschke, Principles of Pulse Electron Paramagnetic Resonance, Oxford University Press: New York, 2001.
26. N. M. Atherton, Principles of Electron Spin Resonance, Ellis Horwood Ltd.: Chichester, 1993.
27. B. B. Williams, X. C. Pan, and H. J. Halpern, *J. Magn. Reson.*, 2005, **174**, 88.
28. S. R. Deans, The Radon Transformation and Some of Its Applications, John Wiley & Sons, Inc: New York, 1983.
29. N. Devasahayam, S. Subramanian, R. Murugesan, F. Hyodo, K. Matsumoto, J. B. Mitchell, and M. C. Krishna, *Magnet Reson. Med.*, 2007, **57**, 776.
30. H. Hirata, M. Wakana, and H. Susaki, *Appl. Phys. Lett.*, 2006, **88**, 254103.
31. G. H. Glover and J. M. Pauly, *Magnet Reson. Med.*, 1992, **28**, 275.
32. R. Ahmad, B. Clymer, Y. Deng, G. He, D. Vikram, P. Kuppusamy, and J. L. Zweier, *J. Magn. Reson.*, 2006, **179**, 263.
33. K. H. Ahn and H. J. Halpern, *Med. Phys.*, 2007, **34**, 1047.
34. B. Epel, S. V. Sundramoorthy, C. Mailer, and H. J. Halpern, *Conc. Magn. Reson. B*, 2008, **33B**, 163.
35. Z.-P. Liang and P. C. Lauterbur, Principles of Magnetic Resonance Imaging: A Signal Processing Perspective, IEEE Press Marketing: New York, 2000.
36. S. Som, L. C. Potter, R. Ahmad, D. S. Vikram, and P. Kuppusamy, *J. Magn. Reson.*, 2008, **193**, 210.
37. M. Tseitlin, J. R. Biller, H. Elajaili, V. V. Khramtsov, I. Dhimitruka, G. R. Eaton, and S. S. Eaton, *J. Magn. Reson.*, 2014, **245**, 150.
38. S. Ljunggren, *J. Magn. Reson.*, 1983, **54**, 338.
39. D. B. Twieg, *Med. Phys.*, 1983, **10**, 610.

40. M. A. Bernstein, K. F. King, and X. J. Zhou, Handbook of MRI Pulse Sequences, Elsevier Inc.: London, UK, 2004.
41. M. Lustig, D. L. Donoho, J. M. Santos, and J. M. Pauly, *IEEE Signal Proc. Mag.*, 2008, **25**, 72.
42. K. H. Ahn and H. J. Halpern, *J. Magn. Reson.*, 2007, **185**, 152.
43. G. Placidi, M. Alecci, and A. Sotgiu, *J. Magn. Reson. Ser. B*, 1995, **108**, 50.
44. H. Jang, S. Matsumoto, N. Devasahayam, S. Subramanian, J. Zhuo, M. C. Krishna, and A. B. McMillan, *Magnet Reson. Med.*, 2015, **73**, 1692.
45. L. J. Berliner (ed), In Vivo EPR (ESR): Theory & Applications, Kluwer Academic: New York, 2003.
46. H. J. Halpern, M. Peric, C. Yu, E. D. Barth, G. V. Chandramouli, M. W. Makinen, and G. M. Rosen, *Biophys. J.*, 1996, **71**, 403.
47. K. Golman, J. S. Petersson, J. H. Ardenkjaer-Larsen, I. Leunbach, L. G. Wistrand, G. Ehnholm, and K. Liu, *J. Magn. Reson. Imag.*, 2000, **12**, 929.
48. M. Elas, B. B. Williams, A. Parasca, C. Mailer, C. A. Pelizzari, M. A. Lewis, J. N. River, G. S. Karczmar, E. D. Barth, and H. J. Halpern, *Magnet Reson. Med.*, 2003, **49**, 682.
49. B. Epel, C. R. Haney, D. Hleihel, C. Wardrip, E. D. Barth, and H. J. Halpern, *Med. Phys.*, 2010, **37**, 2553.
50. N. Kocherginsky and H. M. Swartz, in Nitroxide Spin Labels: Reactions in Biology and Chemistry, eds N. Kocherginsky and H. M. Swartz, CRC Press: Boca Raton, FL, 1995, Chap. Terminology, classification and distribution of the nitroxides in cells, 15.
51. M. Miyake, S. R. Burks, J. Weaver, P. Tsai, W. Liu, D. Bigio, K. S. Bauer, K. J. Liu, G. M. Rosen, and J. P. Kao, *J. Pharm. Sci.*, 2010, **99**, 3594.
52. V. V. Khramtsov, in Nitroxides – Theory, Experiment and Applications, ed A. I. Kokorin, InTech: Rijeka, 2012, Chap. In vivo spectroscopy and imaging of nitroxide probes, InTech, Online open access, http://www.intechopen.com/books/nitroxides-theory-experiment-and-applications, 317.
53. V. V. Khramtsov, *Curr. Org. Chem.*, 2005, **9**, 909.
54. S. R. Burks, J. Bakhshai, M. A. Makowsky, S. Muralidharan, P. Tsai, G. M. Rosen, and J. P. Y. Kao, *J. Org. Chem.*, 2010, **75**, 6463.
55. S. R. Burks, M. A. Makowsky, Z. A. Yaffe, C. Hoggle, P. Tsai, S. Muralidharan, M. K. Bowman, J. P. Y. Kao, and G. M. Rosen, *J. Org. Chem.*, 2010, **75**, 4737.
56. T. J. Reddy, T. Iwama, H. J. Halpern, and V. H. Rawal, *J. Org. Chem.*, 2002, **67**, 4635.
57. B. B. Williams, H. al Hallaq, G. V. Chandramouli, E. D. Barth, J. N. Rivers, M. Lewis, V. E. Galtsev, G. S. Karczmar, and H. J. Halpern, *Magn. Reson. Med.*, 2002, **47**, 634.
58. K. Matsumoto, S. English, J. Yoo, K. Yamada, N. Devasahayam, J. A. Cook, J. B. Mitchell, S. Subramanian, and M. C. Krishna, *Magnet Reson. Med.*, 2004, **52**, 885.
59. M. C. Krishna, S. English, K. Yamada, J. Yoo, R. Murugesan, N. Devasahayam, J. A. Cook, K. Golman, J. H. Ardenkjaer-Larsen, S. Subramanian, and J. B. Mitchell, *Proc. Natl. Acad. Sci. U S A*, 2002, **99**, 2216.
60. A. A. Bobko, I. Dhimitruka, D. A. Komarov, and V. V. Khramtsov, *Anal. Chem.*, 2012, **84**, 6054.
61. K. J. Liu, P. Gast, M. Moussavi, S. W. Norby, N. Vahidi, T. Walczak, M. Wu, and H. M. Swartz, *Proc. Natl. Acad. Sci. U S A*, 1993, **90**, 5438.
62. R. P. Pandian, N. L. Parinandi, G. Ilangovan, J. L. Zweier, and P. Kuppusamy, *Free Radical Biol. Med.*, 2003, **35**, 1138.
63. G. L. He, R. A. Shankar, M. Chzhan, A. Samouilov, P. Kuppusamy, and J. L. Zweier, *Proc. Natl. Acad. Sci. U S A*, 1999, **96**, 4586.
64. G. Ilangovan, A. Bratasz, and P. Kuppusamy, *Adv. Exp. Med. Biol.*, 2005, **566**, 67.
65. G. Ilangovan, A. Bratasz, H. Li, P. Schmalbrock, J. L. Zweier, and P. Kuppusamy, *Magn. Reson. Med.*, 2004, **52**, 650.
66. G. Meenakshisundaram, R. P. Pandian, E. Eteshola, S. C. Lee, and P. Kuppusamy, *J. Magn. Reson.*, 2010, **203**, 185.
67. P. Roschmann, *Med. Phys.*, 1987, **14**, 922.
68. H. J. Halpern and M. K. Bowman, in EPR Imaging and In Vivo EPR, eds G. R. Eaton, S. S. Eaton and K. Ohno, CRC Press: Boca Raton FL, 1991, Chap. Low frequency EPR spectrometers: MHz range, 45.
69. J. L. Schepps and K. R. Foster, *Phys. Med. Biol.*, 1980, **25**, 1149.
70. R. W. Quine, G. A. Rinard, S. S. Eaton, and G. R. Eaton, *J. Magn. Reson.*, 2010, **205**, 23.
71. J. Koscielniak, N. Devasahayam, M. S. Moni, P. Kuppusamy, K. Yamada, J. B. Mitchell, M. C. Krishna, and S. Subramanian, *Rev. Sci. Instrum.*, 2000, **71**, 4273.

72. H. Sato-Akaba, H. Fujii, and H. Hirata, *J. Magn. Reson.*, 2008, **193**, 191.

73. H. J. Halpern, D. P. Spencer, J. Vanpolen, M. K. Bowman, R. J. Massoth, A. C. Nelson, E. M. Dowey, and B. A. Teicher, *Rev. Sci. Instrum.*, 1989, **60**, 1040.

74. G. A. Rinard, R. W. Quine, G. R. Eaton, S. S. Eaton, E. D. Barth, C. A. Pellizari, and H. J. Halpern, *Concept Magn. Res.*, 2002, **15**, 51.

75. W. A. Anderson, *Rev. Sci. Instrum.*, 1961, **32**, 241.

76. C. Mailer, B. H. Robinson, B. B. Williams, and H. J. Halpern, *Magnet Reson. Med.*, 2003, **49**, 1175.

77. C. P. Poole, Electron Spin Resonance: A Comprehensive Treatise on Experimental Techniques, Dover Publications, Inc.: Mineola, New York, 1997.

78. S. S. Eaton, R. W. Quine, M. Tseitlin, D. G. Mitchell, G. A. Rinard, and G. R. Eaton, Rapid-Scan Electron Paramagnetic Resonance. Handbook of Multifrequency Electron Paramagnetic Resonance: Data and Techniques, Wiley-VCH Verlag GmbH & Co.: KGaA, Boschstr. 12, 69469 Weinheim, Germany, 2014, 3.

79. B. Epel, S. V. Sundramoorthy, E. D. Barth, C. Mailer, and H. J. Halpern, *Med. Phys.*, 2011, **38**, 2045.

80. B. Epel and H. J. Halpern, *J. Magn. Reson.*, 2015, **254**, 56.

81. M. K. Bowman, Fourier Transform Electron Spin Resonance. Modern Pulsed and Continuous Electron Spin Resonance, John Wiley & Sons: New York, 1990, 1.

82. P. Seifi, B. Epel, S. V. Sundramoorthy, C. Mailer, and H. J. Halpern, *Med. Phys.*, 2011, **38**, 3062.

83. N. Devasahayam, R. Murugesan, K. Matsumoto, J. B. Mitchell, J. A. Cook, S. Subramanian, and M. C. Krishna, *J. Magn. Reson.*, 2004, **168**, 110.

84. C. R. Haney, X. Fan, A. D. Parasca, G. S. Karczmar, H. J. Halpern, and C. A. Pelizzari, *Conc. Magn. Reson. B*, 2008, **33B**, 138.

85. M. W. Dewhirst, Y. Cao, and B. Moeller, *Nat. Rev. Cancer*, 2008, **8**, 425.

86. H. Yasui, S. Matsumoto, N. Devasahayam, J. P. Munasinghe, R. Choudhuri, K. Saito, S. Subramanian, J. B. Mitchell, and M. C. Krishna, *Cancer Res.*, 2010, **70**, 6427.

87. S. Subramanian, J. W. Koscielniak, N. Devasahayam, R. H. Pursley, T. J. Pohida, and M. C. Krishna, *J. Magn. Reson.*, 2007, **186**, 212.

88. G. Redler, B. Epel, and H. J. Halpern, *Magnet Reson. Med.*, 2014, **71**, 440.

89. B. Zhao, W. M. Lu, T. K. Hitchens, F. Lam, C. Ho, and Z. P. Liang, *Magnet Reson. Med.*, 2015, **74**, 489.

90. Y. N. Molin, K. M. Salikhov, and K. I. Zamaraev, Spin Exchange: Principles and Applications in Chemistry and Biology, Springer-Verlag: Berlin, 1980.

91. M. P. Eastman, R. G. Kooser, M. R. Das, and J. H. Freed, *J. Chem. Phys.*, 1969, **51**, 2690.

92. J. L. Tatum, G. J. Kelloff, R. J. Gillies, J. M. Arbeit, J. M. Brown, K. S. Chao, J. D. Chapman, W. C. Eckelman, A. W. Fyles, A. J. Giaccia, R. P. Hill, C. J. Koch, M. C. Krishna, K. A. Krohn, J. S. Lewis, R. P. Mason, G. Melillo, A. R. Padhani, G. Powis, J. G. Rajendran, R. Reba, S. P. Robinson, G. L. Semenza, H. M. Swartz, P. Vaupel, D. Yang, B. Croft, J. Hoffman, G. Liu, H. Stone, and D. Sullivan, *Int. J. Radiat. Biol.*, 2006, **82**, 699.

93. B. Gallez, C. Baudelet, and B. F. Jordan, *NMR Biomed.*, 2004, **17**, 240.

94. B. H. Robinson, C. Mailer, and A. W. Reese, *J. Magn. Reson.*, 1999, **138**, 199.

95. B. H. Robinson, C. Mailer, and A. W. Reese, *J. Magn. Reson.*, 1999, **138**, 210.

96. R. J. Gillies, N. Raghunand, M. L. Garcia-Martin, and R. A. Gatenby, *IEEE Eng. Med. Biol. Mag.*, 2004, **23**, 57.

97. H. B. Elajaili, J. R. Biller, M. Tseitlin, I. Dhimitruka, V. V. Khramtsov, S. S. Eaton, and G. R. Eaton, *Magn. Reson. Chem.*, 2015, **53**, 280.

98. V. V. Khramtsov and R. J. Gillies, *Antioxid. Redox Signal.*, 2014, **21**, 723.

99. G. B. Waypa, J. D. Marks, R. D. Guzy, P. T. Mungai, J. M. Schriewer, D. Dokic, M. K. Ball, and P. T. Schumacker, *Am. J. Respir. Crit. Care Med.*, 2013, **187**, 424.

100. H. Elajaili, J. R. Biller, G. M. Rosen, J. P. Kao, M. Tseytlin, L. A. Buchanan, G. A. Rinard, R. W. Quine, J. McPeak, Y. Shi, S. S. Eaton, and G. R. Eaton, *J. Magn. Reson.*, 2015, **260**, 77.

101. S. Schlick and K. Kruczala, *Comp. Anal. C*, 2008, **53**, 499.

Chapter 14
EPR Spectroscopy of Nitroxide Spin Probes

Enrica Bordignon

Ruhr-Universität Bochum, Bochum, Germany

14.1 INTRODUCTION

Nitroxide probes or, correctly by IUPAC nomenclature, aminoxyl radicals are molecules carrying a stable unpaired electron localized in an N–O group, with the nitrogen attached to two alkyl substituents (often in a ring structure). These probes can be covalently or noncovalently introduced into a variety of different materials to monitor by EPR the specific characteristics of their microenvironment. Such local properties can then be used to understand at the molecular level relevant characteristics of the material under investigation, such as viscosity, local dynamics, pH, polarity, H-bond networks, transition temperatures, and distances toward other nitroxide probes. Nitroxides are stable in nonreducing environments and can be used in almost any experimental condition, and in most cases, being the unique paramagnetic centers in an otherwise diamagnetic sample, they provide unequivocal information with high precision. The broad applicability of nitroxide probes makes them valuable molecular tools in many fields, from materials science to structural biology. Here we will introduce the following aspects: i) how nitroxide probes were tailored for specific purposes; ii) which relevant terms in the Hamiltonian are necessary to describe a nitroxide spin system; iii) how the spectral lineshape of nitroxides encodes molecular motions within defined 'EPR timescales' as well as dipolar and exchange interactions between neighboring spins; iv) how to monitor polarity and formation of H-bonds in their microenvironment. Some examples of applications of nitroxide probes from our research will be given.

14.2 TYPES OF NITROXIDE PROBES

A series of nitroxide molecules have been synthesized and functionalized, with special, tailored properties to probe specific local characteristics of materials and macromolecules. In the following sections, we will describe the chemical properties of the nitroxide molecules and the subsequent modifications introduced to optimize their use as probes of different

EPR Spectroscopy: Fundamentals and Methods.
Edited by Daniella Goldfarb and Stefan Stoll

ISBN: 978-1-119-16299-5
Also published in eMagRes (online edition)
DOI: 10.1002/9780470034590.emrstm1513

physicochemical properties in distinct research areas.

14.2.1 The Basic Structure of a Nitroxide Probe

The disodium nitrosodisulfonate first synthesized in 1845 and now known as Frémy's salt (Figure 14.1a) can be considered as the precursor of the stable nitroxide probes available nowadays. The di-*tert*-butyl nitroxide followed in 1961[1] (Figure 14.1b), and in 1964, the 2,2,6,6-tetramethyl-1-piperidinyloxyl was synthesized,[2] which paved the way to the development of the well-known nitroxide molecules with a five (pyrrolidine or pyrroline type)-membered or six (piperidine type)-membered ring (Figure 14.1c).

Radical stability was early recognized to be the key feature of those molecules,[3] which is achieved mainly by steric blocking of the NO group by bulky groups (usually methyl) on the ring carbon atoms adjacent to the nitrogen (Figure 14.1).

The only real chemical drawback of nitroxides is their susceptibility to reduction to the corresponding diamagnetic hydroxylamine in the presence of reducing agents. However, the radical can be easily regenerated in mild oxidizing conditions or exposure to O_2; therefore, the reduction of the NO group does not interfere with the synthetic strategies but only with its applications, especially in cellular context, as described in the following sections.

The spin population is mainly distributed on the nitrogen and oxygen atoms of the aminoxyl group of the nitroxide probe. Figure 14.1(d) shows the schematic representation of the relevant orbitals on the $NO^{\bullet}$ group based on the Hückel molecular orbital (MO) theory, namely the *n* orbitals, with the lone pair electrons on the oxygen, and the π and π* orbitals. The two $2p_z$ orbitals of the nitrogen and oxygen are coupled to form a fully occupied bonding π orbital and a half-occupied antibonding π* orbital, which is closer in energy to the nitrogen $2p_z$ orbital.[4] The energy of the nitrogen $2p_z$ orbital versus that of the oxygen determines the magnitude of the nitrogen versus oxygen character of the π* orbital. If the polarity of the nitroxide microenvironment increases, the energy of the oxygen $2p_z$ orbital will decrease relative to that of the nitrogen. Therefore, the half-occupied antibonding orbital π* will increase its nitrogen character, which results in a larger spin population on the nitrogen (ρ_π^N). This will lead to larger values of the nitrogen hyperfine coupling constants and lower values of the *x*-component of the ***g*** tensor (see Section 14.3.1). The redistribution of the spin population can also be described with the two mesomeric structures of the nitroxide molecules, sketched in Figure 14.1(e). One structure is neutral and carries the unpaired electron on the oxygen atom, and the second, with the unpaired electron on the nitrogen atom, is zwitterionic, therefore energetically favored in polar solvents and accounts for the increase in electron spin population on the nitrogen nucleus (ρ_π^N).

14.2.2 Nitroxide Probes for Biological Membranes and Membrane-mimetic Systems

Functionalized fatty acids, with a nitroxide probe attached at specific positions in the hydrocarbon chain, can be very good molecular probes for biological membrane bilayers and membrane-mimetic systems, such as micelles or nanodiscs (for a recent review, see Ref. 5). Spin-labeled stearic acids (*n*-SASL; *n* indicates the position of the nitroxide probe in the hydrocarbon chain and SASL stands for stearic acid spin label) or spin-labeled lipids can provide a detailed profile of different membrane properties, including hydrophobicity, oxygen solubility and diffusion rates, bilayer penetration of water molecules, and membrane order and fluidity. Examples of 16-SASL and 5-SASL can be found in Figure 14.1(f). Spin-labeled fatty acids are usually inserted at a 1 : 100 ratio with respect to the other lipids in the membrane bilayer, to avoid spin–spin interactions, which are detrimental for the CW EPR spectral analysis as they cause line broadening. This low level of doping with the spin-labeled probe leads to minimal perturbation of the membrane bilayers and provides high sensitivity in CW EPR. Some spin-labeled molecules have a preferential orientation of the NO group with respect to the lipid axis (see, e.g., the 5-doxyl stearic acid in Figure 14.1f), which provides fingerprint spectral features due to the anisotropy of the motion.[6] The fatty acid inserted in a membrane bilayer shows a high microscopic ordering when probed by the nitroxide at position 5 in the hydrocarbon chain, while the probe at position 16 in the middle of the bilayer shows the highest degree of conformational dynamics. At low EPR frequencies, both the overall restricted motion of

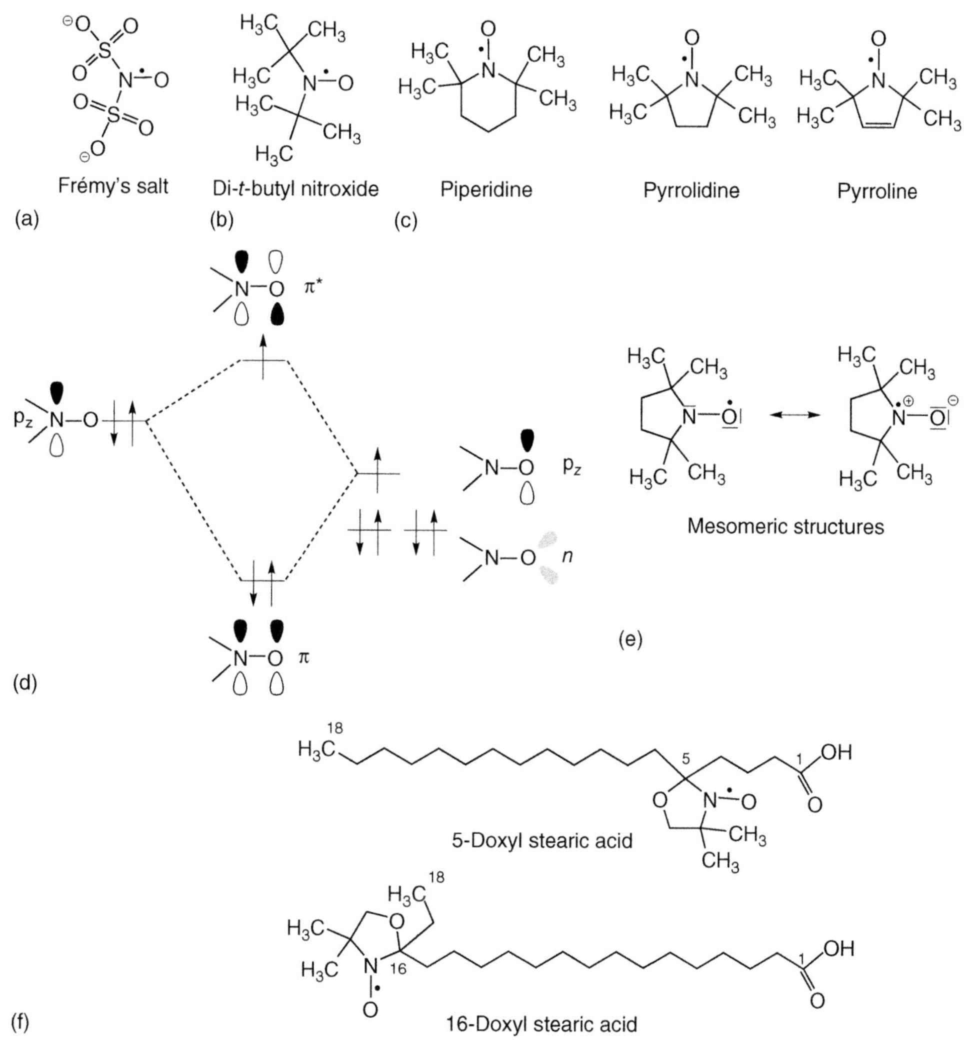

Figure 14.1. Nitroxide probes. (a) Frémy's salt. (b) Di-*tert*-butyl nitroxide. (c) Five- and six-membered ring nitroxides. (d) Schematic representation of the frontiers orbital of the NO group, based on the MO theory. On the left, the $2p_z$ orbital on the nitrogen; on the right, the $2p_z$ orbital on the oxygen and the nonbonding *n* orbitals on oxygen. (e) Two mesomeric structures of a nitroxide molecule. (f) Examples of nitroxide-based stearic acids to investigate membrane bilayers

the lipid in the membrane and the internal dynamics of the lipid chain are visible in the spectra. However, going to higher frequencies helps disentangling the two types of motion, as the higher frequencies are more sensitive to the fast motions (different 'EPR timescales', see the following sections). The hydrophobicity of the membrane interior is largely determined by the extent of water molecules being able to penetrate from the bulk solution, which defines the polarity gradient across the bilayer. The environment becomes increasingly nonpolar as one moves from the water-membrane boundary to the terminus of the lipid alkyl chains. In contrast, the membrane interior with respect to the aqueous environment is enriched in apolar oxygen molecules (see, e.g., Ref. 7). Profiles of water penetration and oxygen solubility in bilayers at physiological temperatures can be obtained, for example, via CW power saturation techniques (see

Section 14.6) based on a nitroxide scan performed with different spin-labeled lipids carrying the nitroxide probe at different depths with respect to the water-membrane boundaries. Depth-dependent water penetration is also clearly visible in spin-labeled membranes suspended in deuterated water as different modulation depths of the deuterium signal in ESEEM (electron spin echo envelope modulation) experiments on frozen samples (see Section 14.6). Depth-dependent effects on the water permeability due to membrane composition, degree of saturation of the hydrocarbon chain, presence of cholesterol, membrane-bound peptides, or integral membrane proteins, lipid oxidation, and temperature can be monitored at a molecular level via EPR.

14.2.3 Nitroxide Spin Labels in Structural Biology

Nitroxides are extensively used to label proteins and nucleic acids to extract dynamic–structural–functional properties. For the application in nucleic acid research, spin labels were designed to bind nucleic acids both in a noncovalent and site-directed fashion. A major effort was made to create rigid labels able to directly report on the frozen dynamics of the DNA or RNA by dipolar spectroscopy (see, e.g., Ref. 8). A detailed description of such spin labels and their applications can be found in a recent book chapter.[9] In this chapter, we limit our description to the use of site-directed spin labeling (SDSL) with nitroxide probes in protein science. Nitroxide spin labels were first introduced into proteins in 1965 by McConnell[10] and, thanks to the work of Hubbell,[11–14] nitroxide radicals have become the spin probes of choice for most researchers in the field.[9,15] Their characteristic stability and small molecular weight, the availability of different functional reactive groups targeting cysteines, and a vast library of isotopic, chemical, and diamagnetic variants make nitroxides the most appropriate choice for most EPR studies. Among the various types of nitroxides available, the most cited spin label in the literature is MTSL (also abbreviated as MTSSL), (1-oxyl-2,2,5,5-tetramethyl-Δ3-pyrroline-3-methyl) methanethiosulfonate (Figure 14.2a). The highly reactive methanethiosulfonate group is specific toward cysteines, and complete spin labeling with essentially 100% efficiency can be achieved for surface-exposed cysteines with a 1 : 1 Cys:MTSL stoichiometric ratio only after few seconds of incubation at room temperature. The spin labeling efficiency is the ratio of the spin concentration and the cysteine concentration in the protein. The spin concentration is calculated by comparing the double integral of the room temperature derivative CW EPR spectrum with that of a standard solution of TEMPOL in water with known concentration. The two spectra need to be detected under the same experimental conditions (especially sample volume, modulation amplitude, and microwave power). The protein concentration is usually determined by UV–vis spectroscopy. Cysteines with higher steric constraints from the neighboring amino acids are less accessible from the bulk water; therefore, a 1 : 10 Cys:MTSL stoichiometric ratio is normally used, with prolonged incubation to achieve high spin labeling efficiency. Owing to the propensity of MTSL to form disulfide bridges, its concentration in water should not exceed 200 μM to avoid creation of biradicals over time, resulting in the loss of reactive label and subsequent decreased labeling efficiency. Biradical formation during the labeling reaction can be easily recognized by the appearance of two additional lines in between the characteristic components of the three-line spectrum of a freely tumbling MTSL monoradical.[6] It is important to note that low spin labeling efficiencies are tolerable if the aim is to extract mobility or accessibility information because unlabeled proteins are EPR silent. The only spectroscopic problem is the decreased sensitivity. However, because it is mandatory to verify biochemically that labeling does not affect protein folding and functionality, a labeling efficiency below 50% impairs the reliability of biochemical assays, which probes the total molecular ensemble. Sites buried in the protein core are in general not easily labeled, and they are not preferential targets of SDSL EPR, owing to the high probability that spin labeling may interfere with the protein's conformational dynamics or folding. Thus, surface-exposed sites are preferred and it is crucial to carefully decide on the labeling sites before planning an SDSL EPR study.

MTSL bound to a protein has a high degree of flexibility owing to the five rotatable bonds linking it to the backbone of an engineered cysteine. This makes the label very adaptable to most protein sites. Another advantage of MTSL is that free software packages such as MMM[16] or MtsslWizard[17] are available, which provide rotamer libraries for MTSL bound to macromolecules. Mainly, these programs are used to predict the NO—NO distances between two labeled

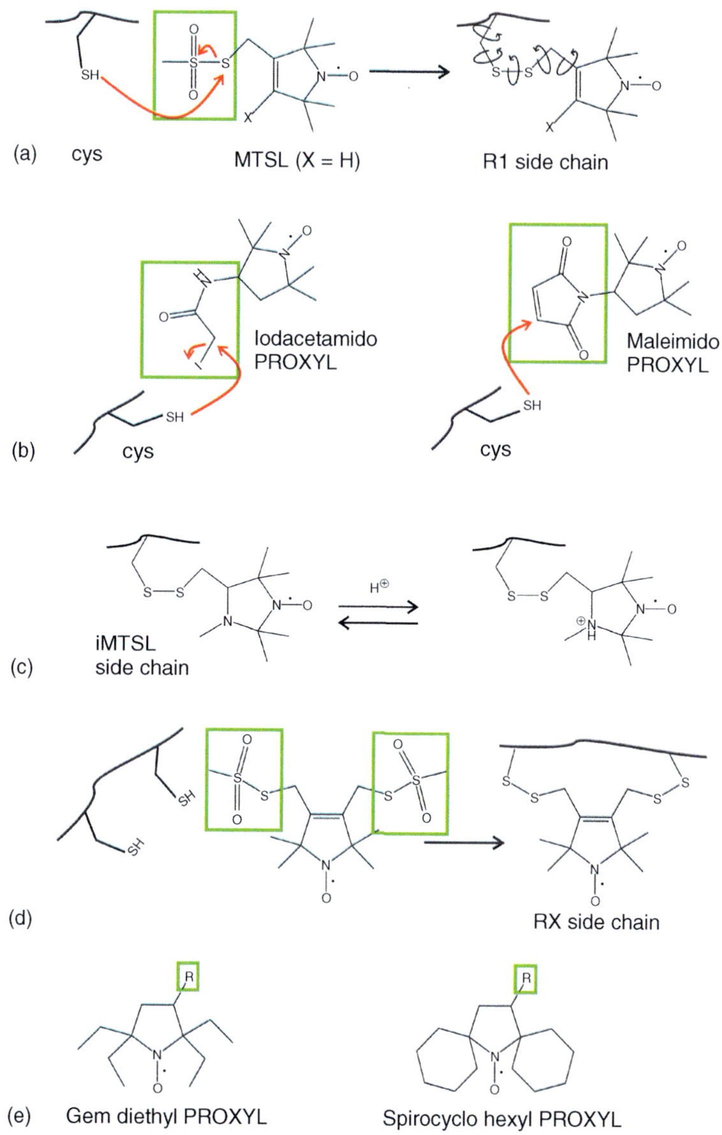

Figure 14.2. Examples of spin labels for protein studies by SDSL EPR. (a) MTSL. The reaction with the sulfhydryl group of a cysteine is schematically shown, yielding the R1 side chain with five rotatable bonds. Red arrows indicate the reaction mechanism and the creation of the leaving group. Green squares highlight the reactive functional groups. (b) Two spin labels targeting cysteines. The C—S bond linking the probe to the protein backbone is inert toward chemical reduction. (c) iMTSL, a spin label sensitive to changes in pH of the microenvironment. (d) The bifunctional MTSL, a very rigid label targeting specific neighboring cysteines in proteins. (e) Sterically hindered spin labels, resistant to chemical reduction. The spirocyclohexyl groups provide long relaxation times up to 200 K. Functional groups targeting cysteines are available (R). (f) Unnatural amino acids for specifically functionalized nitroxide probes. (g) Examples of unnatural amino acids containing the nitroxide moiety

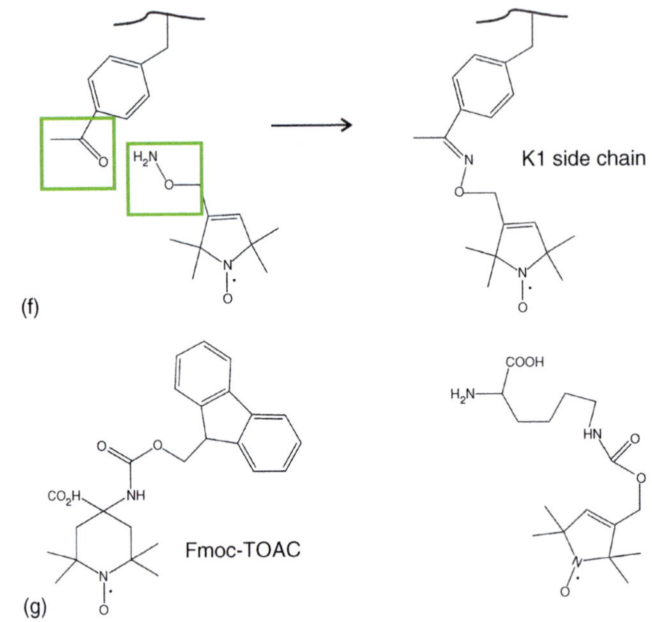

Figure 14.2. (Continued)

sites for dipolar spectroscopy applications but may serve as guidelines for choosing labeling positions in a protein for which a protein data bank (PDB) file exists. A high partition function for a site exposed to the bulk water, or in other words a high number of populated rotamers, suggests high accessibility and labeling efficiency.

The biggest disadvantage of MTSL is the reversibility of the disulfide bond linking the nitroxide moiety to the protein backbone. The appearance of the characteristic three-line spectrum of the free label is the spectroscopic fingerprint for the release of the spin label in the solution. An incompatible condition is the presence of reducing agents in the physiological milieu. This may lead not only to the breakage of the S—S bond but also to the reduction of the NO group to the diamagnetic hydroxylamine (NOH). The latter manifests itself as a decreased signal intensity, but either possibility causes a decreased labeling efficiency over time. It has been demonstrated that five-membered (pyrrolidine) ring structures,[18,19] as well as steric shielding of the nitroxide moiety,[20,21] slow down the reduction rate of nitroxides.

Chemical variants of MTSL are available with diamagnetic groups replacing the NO group, which are used for spin dilution studies. One example is the (1-acetyl-2,2,5,5-tetramethyl-Δ3-pyrroline-3-methyl) label, in which the oxygen on the NO group is replaced with an acetyl group. Other MTSL variants have different groups replacing the hydrogen at position X (Figure 14.2a) in the ring. The most common variant contains a bromine atom, which partially hinders the rotation of the nitroxide ring around the last bond of the linker. Isotopic variants are also available, with ^{15}N ($I = 1/2$) replacing the most abundant isotope ^{14}N ($I = 1$) and/or with ^{2}H replacing all protons. Changing the isotope of the nitrogen atom modifies the hyperfine pattern of the nitroxide spectrum, which can be useful to simplify the low-temperature high-field spectra in order to distinguish populations of nitroxides with different g_x. The deuteration of the molecule prolongs the phase memory time of the nitroxide for pulse EPR studies, as well as it decreases the linewidth of the CW spectra, simplifying data analysis.[22,23]

An option to eliminate the reversible reduction of the S—S bond linking MTSL to the protein backbone is the use of iodoacetamide or maleimide reactive groups (Figure 14.2b). Both groups yield an irreversible binding preferentially to cysteines with a C—S bond resistant to reducing agents. Examples of five-membered ring structures are the 3-(2-iodoacetamido)-PROXYL (3-(2-iodoacetamido)-2,2,5,5-tetramethyl-1-pyrrolidinyloxy)) radical and the 3-maleimido-PROXYL (3-maleimido-2,2,5,5-tetramethyl-1-pyrrolidinyloxy) radicals (Figure 14.2b). The main disadvantage of these labels is their lower selectivity for cysteines compared to MTSL. In fact, especially at high pH (>8), the maleimide group can react with amines, such as lysine and arginine, and even with the N-terminus of the polypeptide. The iodoacetamide group can also interact with lysine, histidine, methionine, and carboxylates (aspartate, glutamate, and C-terminus). In addition, the lower flexibility of the maleimide variants may destabilize the spin-labeled protein or hinder its conformational transitions.

An interesting variant of MTSL, which can be of relevance for the study of changes in pH during chemical reactions or proton translocation through membranes, is the imidazolidine MTSL[24] (Figure 14.2c). The reversible protonation of the tertiary amine nitrogen in the heterocycle is reflected in a change of the isotropic hyperfine coupling constant to the ^{14}N. Therefore, this spin label can be used as a local probe of changes in pH at physiological temperatures.

As stated earlier, the flexibility of MTSL allows labeling at most sites without affecting the functionality of the protein. However, this advantage comes with a price; in fact, one measures the properties of the EPR probe (i.e., the NO group, where the radical is localized) but would rather like to measure the properties of the protein backbone. Therefore, attempts were made to create more rigid labels, which can encode largely the dynamics of the backbone. A bifunctional methanethiosulfonate reagent, which creates the conformationally constrained spin-labeled side-chain RX, fulfills this requirement (Figure 14.2d).[25] Owing to its geometry, the bifunctional reagent interacts specifically with pairs of cysteine residues at positions i and $i + 3$ (or $i + 4$) in an α-helix, at i and $i + 2$ in a β-strand, or with cysteine residues in adjacent strands in a β-sheet. Being very rigid, it is a good probe for backbone motion at physiological temperatures and for distance measurements via dipolar spectroscopy (see Chapters 19 and 20).

One possibility to further increase sensitivity under physiologically relevant conditions (e.g., for studies in cellular context) is to decrease the propensity of the nitroxide group to be chemically reduced. For this aim, promising spin labels are those with the NO group sterically shielded with gem-diethyl or spirocyclohexyl groups (Figure 14.2e).[20,21] Interestingly, for labels with spirocyclohexyl groups, the phase memory time is longer than 1 μs up to 250 K, making those probes potentially useful tools for pulse EPR studies at temperatures above 200 K.[26]

If a protein contains several cysteines, whose removal may affect stability and functionality of the protein, SDSL techniques cannot be applied. To overcome this problem, unnatural amino acids could be employed, which introduce spin labels at specific positions, leaving the rest of the protein unaffected. One example is the reaction of a *p*-acetylphenylalanine, which is postsynthetically converted into a spin-labeled amino acid (K1 side chain) by reaction with a nitroxide functionalized with a ketone-specific reactive group (Figure 14.2f).[27] However, spin labeling efficiency and labeling conditions are not yet optimal.[28]

Other possibilities rely on unnatural amino acids carrying the nitroxide radical for peptide/protein synthesis or ribosomal synthesis of proteins. In Figure 14.2(g), the Fmoc-protected TOAC,[29] an unnatural amino acid used in solid-state synthesis, is shown. This unnatural amino acid, once incorporated into a peptide, provides a very rigid side chain, with characteristic conformations of the piperidine ring, which can better report the dynamics of the backbone with respect to MTSL. Genetic encoding of unnatural amino acids carrying nitroxide radicals can be used for intracellular biosynthesis of proteins, which could be purified, for in vitro studies, or directly measured in cellular environment. In fact, in the past years, new labeling strategies and the increased sensitivity, especially for detection of interspin distances in macromolecules, are paving the way toward *in-cell* EPR. Genetic encoding of a spin-labeled amino acid in T4-lysozyme in *Escherichia coli* was already shown in 1994 by Hubbell.[30] However, despite recent advances, intracellular biosynthesis of spin-labeled proteins with unnatural amino acids (one example[31] in Figure 14.2g) and direct detection of the nitroxide signal in cells remain challenging. Besides genetic encoding of amino acids carrying stable spin centers, several other strategies are also emerging: introduction of ex-situ site-directed spin-labeled proteins (with MTSL and gadolinium-based labels) directly into cells,[32–34] oocytes,[35] and mitochondrial membranes[36,37]; introduction of a nickel-tris-NTA-PROXYL targeting specifically His tags of selected proteins in cellular context[38]; and direct spin labeling of proteins (with MTSL) in the outer membrane of *E. coli* cells.[39,40]

14.3 THE HAMILTONIAN DESCRIBING THE NITROXIDE SPECTRUM

An isolated nitroxide molecule can be described as a spin system with an electron spin $S = 1/2$ coupled to a nuclear spin $I = 1$ (^{14}N), due to the fact that ≈95% of the spin population is distributed on the nitrogen and oxygen atoms (see energy level diagram in Figure 14.3a). The hyperfine interactions with nearby protons are usually very small and unresolved in the spectral linewidth. The hyperfine interaction with the naturally abundant ^{13}C in the nearby carbon atoms can produce small-intensity doublets around each of the three spectral lines, which are visible when the nitroxide is in the fast isotropic motion limit. The relevant terms to be considered in the spin Hamiltonian are therefore the electron Zeeman, the hyperfine, nuclear Zeeman, and nuclear quadrupole interactions correlated with the $I = 1$ (^{14}N) (see Chapters 2, 5, and 6):

$$\hat{\mathcal{H}} = \mu_B \boldsymbol{B}\boldsymbol{g}\hat{\boldsymbol{S}} + \hat{\boldsymbol{S}}\boldsymbol{A}\hat{\boldsymbol{I}} + \mu_N g_n \boldsymbol{B}\hat{\boldsymbol{I}} + \hat{\boldsymbol{I}}\boldsymbol{P}\hat{\boldsymbol{I}} \tag{14.1}$$

The spin Hamiltonian further simplifies for nitroxide variants containing ^{15}N ($I = 1/2$), owing to the absence of nuclear quadrupolar interactions.

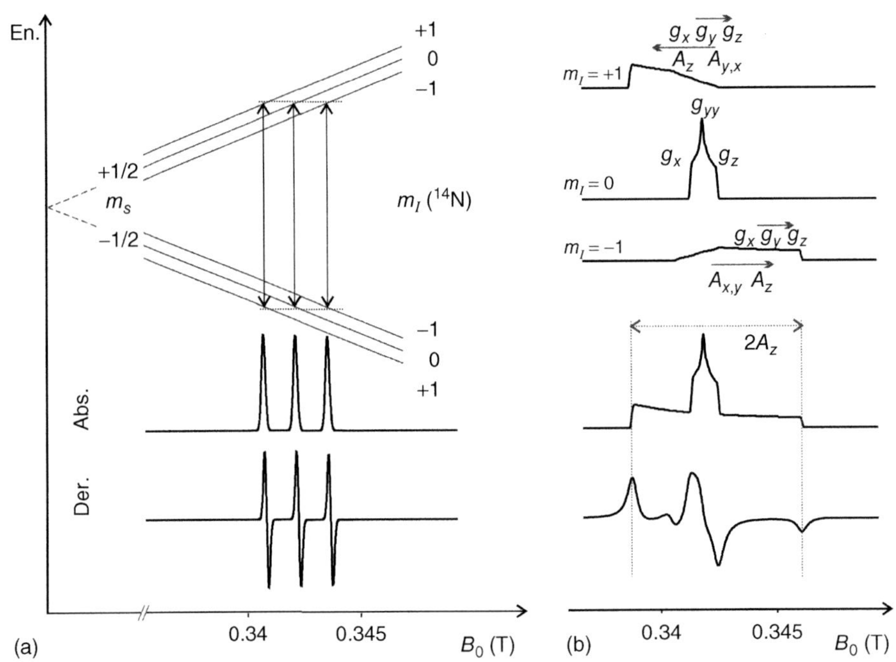

Figure 14.3. Spectral features of the nitroxide ($S = 1/2$, $I = 1$ system). (a) Energy level diagram for a nitroxide in fast motional regime. Note that the same type of diagram could also represent a single orientation of a nitroxide in the magnetic field (with the appropriate hyperfine splittings). At the bottom, the absorption and derivative X-band EPR spectrum characteristic of a free nitroxide fast tumbling in solution. (b) Effects on the three transitions associated with $m_I = 1$, 0–1 of the $\boldsymbol{g}$ and $\boldsymbol{A}$ anisotropies at X-band in powder samples. The arrows show the direction of the change in the resonance field of the corresponding transitions going from orientations of the external magnetic field parallel to x, y to z. At the bottom, the absorption and derivative X-band EPR spectrum characteristic of nitroxide in frozen conditions. The derivative spectrum shows the characteristic positive peak at low field and negative peak at high field. (Reproduced with permission from Ref. 41. © Springer, 2013). (c) Principal axis system of the nitroxide molecule. The z-axis is parallel to the direction of the $2p_z$ orbital of the nitrogen atom, and the x-axis is parallel to the N—O bond. On the right, three simulated powder spectra show the increased Zeeman resolution going from 9.5 to 95 GHz. The g and A values, which are completely resolved at W-band, are indicated in the figure

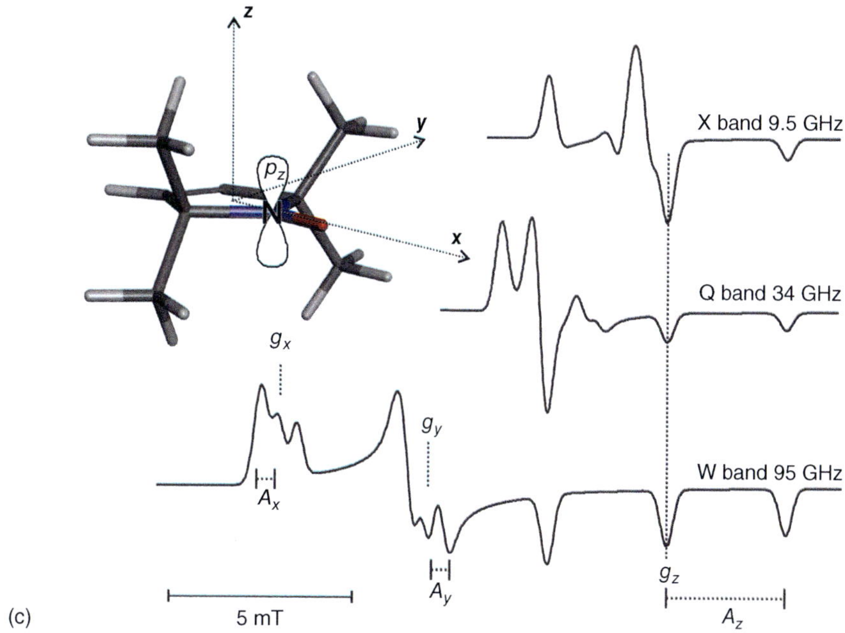

Figure 14.3. (Continued)

14.3.1 g and A Tensors

The hyperfine interaction between the unpaired electron and the $I=1$ ^{14}N nucleus accounts for the characteristic three-line spectrum of the nitroxide probes (Figure 14.3a).

The effects of the anisotropies of the $\boldsymbol{g}$ and $\boldsymbol{A}$ tensors on X-band powder spectra are shown in Figure 14.3(b), in which each transition has been plotted separately to highlight the different effects due to the anisotropies of the two tensors. The central line ($m_I=0$) is solely broadened by the anisotropy of the $\boldsymbol{g}$ tensor (Figure 14.3b). The high-field line ($m_I=-1$) has the largest spectral breadth owing to the combined effects of the two tensors. In fact, when the external B field is parallel to the z-axis of the molecule (Figure 14.3c), the largest, principal component of the $\boldsymbol{A}$ tensor, A_z, produces the largest hyperfine splitting, therefore the high-field transition moves toward higher fields; simultaneously, the small g_z principal component of the $\boldsymbol{g}$ tensor moves the central resonance of the spectrum toward higher fields. In contrast, in the low-field transition, the effects induced by the g_z and A_z are directed in opposite directions, thereby creating a narrower spectral breadth (Figure 14.3b, see arrows).

The $\boldsymbol{g}$ tensor is rhombic for nitroxide probes (reflecting the Zeeman and spin–orbit interactions of the unpaired electron spin). As an example the values for MTSL[42,43] are

$$g_x \approx 2.0083-2.0091;\ g_y \approx 2.0061;\ g_z \approx 2.0022 \qquad (14.2)$$

The principal axis system (PAS) in the nitroxide molecule is shown in Figure 14.3(c), with the z-axis being parallel to the p_z orbital of the nitrogen and the x-axis parallel to the NO bond. Examples of nitroxide powder spectra at different microwave frequencies (9.5, 34, and 95 GHz) show the achievable Zeeman resolution (Figure 14.3c). The major contribution to the polarity dependence of the nitroxide $\boldsymbol{g}$ tensor comes from the g_x principal value; hence, its value reports on the microenvironment of the nitroxide, as addressed in Section 14.5. The lower the g_x, the more protic is the environment (H-bond donor toward the NO group).[44]

The expression for g_x from Stone's perturbation treatment is given in equation (14.3)

$$g_x = g_e + \frac{2\zeta_O (C_{O,y}^{(n)})^2 \rho_\pi^O}{\Delta E_{n\pi^*}} \qquad (14.3)$$

where g_e is the isotropic free electron g value, ζ_O the spin–orbit coupling constant of oxygen, $C^{(n)}_{O,y}$ the coefficient of the oxygen $2p_y$ orbital in the lone pair orbital, ρ^O_π the spin population on the oxygen atom, and $\Delta E_{n\pi^*}$ the $n \rightarrow \pi^*$ excitation energy.[44]

In the fast motion regime, where all anisotropies are averaged out, the value of the isotropic g value, g_{iso}, also varies, depending on the environment. Such environment-dependent changes can be detected with CW EPR at W-band owing to the higher Zeeman resolution.

The $\boldsymbol{A}$ tensor of nitroxide probes has almost axial symmetry ($A_x \approx A_y \neq A_z$) with A_z having the largest value. As an example, typical values of the principal components of the hyperfine tensor $\boldsymbol{A}$ of MTSL are

$$A_x/h \approx 12\text{–}13\ \text{MHz};\ A_y/h \approx 12\text{–}13\ \text{MHz};$$
$$A_z/h \approx 92\text{–}103\ \text{MHz} \tag{14.4}$$

In conventional low-temperature X-band CW EPR spectra (Figure 14.3b), the ^{14}N hyperfine anisotropy (and especially the largest A_z value) is the dominant factor in determining the spectral breadth of the nitroxide probes. In contrast, the g-anisotropy is unresolved in the spectral linewidth. To resolve the g-anisotropy, higher field/frequency is needed. W-band (3.4 T/95 GHz) can be considered as 'high field' for nitroxide probes owing to the achieved spectral resolution in both the $\boldsymbol{g}$ and $\boldsymbol{A}$ tensors (Figure 14.3c).

The A_z value is a good indicator of the polarity of the microenvironment of the nitroxide, with lower values correlating with apolar and higher values with polar microenvironments.

The polarity dependence of A_z is determined by both the isotropic and anisotropic (i.e., dipolar) terms:

$$A_z = a_{iso} + 2T^d_\perp \tag{14.5}$$

with $-T^d_\perp$ being the perpendicular element of the traceless hyperfine tensor that arises from the electron–nucleus dipolar interaction. The latter depends directly on ρ^N_π, the unpaired spin population on the nitrogen atom.

The expression for a_{iso} based on a McConnell-type relation is given in equation (14.6).

$$a_{iso} = Q_N \rho^N_\pi + Q_{NO} \rho^O_\pi \tag{14.6}$$

where ρ^N_π and ρ^O_π are the unpaired electron spin densities on the nitrogen and oxygen atoms, respectively, and $Q_N \gg Q_{NO}$. To first order, the increase in ρ^N_π is directly proportional to the electric field along the x-axis in the PAS of the nitroxide. The electric field arises from polarization of the solvent molecules in aprotic solvent, and in the presence of water, the main effect is due to the formation of H-bonds toward the N–O group. Therefore, the larger the polarity and proticity (propensity to form H-bonds) of the environment, the higher is the value of the isotropic hyperfine constant (see, e.g., Ref. 45). The value of a_{iso} can vary by up to 3 MHz depending on the microenvironment, and it can be easily extracted from X-band CW EPR spectra of nitroxides in different liquid solvents.

The values of the principal components of the nitrogen quadrupole tensor $\boldsymbol{P}$ (reflecting the interaction of the nuclear electric quadrupole moment with the electric field gradient) can be obtained by W-band ESEEM and ELDOR-detected NMR. The P_z value is also a sensitive probe of the polarity of the microenvironment of nitroxides, as shown for five- and six-membered spin probes in different solvents. The $|P_z|$ value decreases with increasing polarity.[46–48]

14.3.2 Exchange and Dipolar Interactions

When two nitroxide probes interact, exchange and dipolar coupling terms are added to the terms from equation (14.1) (see Chapter 4):

$$\hat{\mathcal{H}} = \mu_B \boldsymbol{B g}_1 \hat{\boldsymbol{S}}_1 + \mu_B \boldsymbol{B g}_2 \hat{\boldsymbol{S}}_2 + \hat{\boldsymbol{S}}_1 \boldsymbol{A}_1 \hat{\boldsymbol{I}}_1 + \hat{\boldsymbol{S}}_2 \boldsymbol{A}_2 \hat{\boldsymbol{I}}_2$$
$$+ J \hat{\boldsymbol{S}}_1 \hat{\boldsymbol{S}}_2 + \hat{\boldsymbol{S}}_1 \boldsymbol{D} \hat{\boldsymbol{S}}_2 \tag{14.7}$$

The first two terms describe the Zeeman interactions for each electron spin, the second two terms represent the hyperfine interactions of each electron spin with its nitrogen atom (in general, the hyperfine tensor is site specific due to differences in the nitroxide microenvironment), and the last two terms describe the exchange (J) and dipolar ($\boldsymbol{D}$) interactions between the two electron spins. Such interactions can be present in biradicals, i.e., molecules carrying two nitroxide probes covalently linked through a molecular spacer, or when two mono-nitroxide probes are in close proximity. For biradicals, depending on the nature of the linker and on the degree of spin distribution, both through-space and through-bond interactions can occur. For interacting monoradicals, only a through-space interaction exists.

In contrast to the exchange coupling, which is predominantly isotropic, the dipolar coupling, described by the traceless $\boldsymbol{D}$ tensor, is anisotropic. It averages to zero when molecules are freely tumbling in solution.

A quantification of precise mean distances and distance distributions from spectral analysis in the presence of exchange coupling is a challenging task.[49] Biradicals with short and flexible linkers usually have solution spectra with two additional peaks in between the three characteristic nitroxide lines, yielding a quintet spectrum. This five-line spectrum arises when the average exchange coupling is 10 times stronger than the hyperfine interaction.[50] The spin system can therefore be seen as an $S = 1$ electron spin interacting with two equivalent ^{14}N. In contrast, the solution EPR spectra of biradicals with short rigid linkers, such as the polarizing agent TOTAPOL, have a three-line spectrum with a different degree of line broadening.[51]

For interspin distances longer than 1.5 nm, the exchange interaction is negligible, if the system is not conjugated[49]; therefore, the through-space dipolar part is the leading term in the coupling Hamiltonian. The distance information can then be precisely determined by lineshape analysis of CW EPR spectra in the static limit or by pulsed EPR methods using the point-dipole approximation, with the assumption that the unpaired electron is solely localized in the midpoint of the N—O bond of a nitroxide.[49] Experimental studies on model compounds proved that the errors introduced by this approximation are negligible compared with the uncertainty in the conformation of commonly-used nonaromatic nitroxide rings connected by a saturated linker.[52]

The dipolar interaction can be written as a sum of several components ($\hat{A}$ through $\hat{F}$), which are referred to as the 'dipolar alphabet'.[53] The $\hat{A}$ term contains only the secular part, and the $\hat{B} - \hat{F}$ terms contain the pseudo ($\hat{B}$) and nonsecular parts of the interaction (see Chapters 4 and 19). In the strong-coupling regime, where the dipolar interaction is comparable or larger than the difference between the resonance frequencies of the two investigated spins, both $\hat{A}$ and $\hat{B}$ terms are relevant. In the weak-coupling regime (inter-nitroxide distances above 1.5 nm), the matrix elements arising from the $\hat{B}$ term can be neglected.

In the static limit (powder or frozen samples), dipolar broadening for internitroxide distances in the range 1.5–2 nm affects the CW EPR lineshape. Extraction of the mean distances is done via convolution techniques based on the weak- or strong- coupling regimes. In the weak-coupling regime, the frequency separation between the perpendicular orientations in the Pake pattern is ν_{dd}. For a 1-nm interspin distance, as calculated with the free electron g_e factor, the frequency separation is 52.0 MHz and it decreases rapidly with r^{-3} dependence to 6.5 MHz for an interspin distance of 2 nm. In the strong-coupling regime, the frequency separation between the perpendicular orientations in the Pake pattern is $\frac{3}{2}\nu_{dd}$ (78.0 MHz or 9.75 MHz for distances of 1 and 2 nm, respectively).

To extract the mean distance between two nitroxide probes from CW EPR spectral analysis, one first needs to measure the low-temperature CW EPR spectrum of each single probe and the labeling efficiency (for proteins) or radical content (for biradicals). The sum of the two single-probe spectra yields the spectral shape of the noninteracting spins, which is used for the convolution with the Pake function in the strong- or weak-coupling regime. If the labeling efficiency is below 100%, one needs to consider that the experimental spectrum also contains a variable fraction of nondipolar-broadened spectrum arising from the noninteracting spins. The labeling efficiency provides the fraction of noninteracting spins to be taken into account for the spectra calculation. Notably, extraction of interspin distances from dipolar broadening of room temperature CW EPR spectra is not recommended owing to the partial averaging of the dipolar interactions caused by molecular motion.

Figure 14.4(a) shows the simulation of the dipolar broadening on low-temperature CW EPR spectra performed with the program DIPFIT[54] using a convolution method based on the strong-coupling regime. As clearly demonstrated by the simulation, the dipolar broadening is less evident if non-interacting nitroxide probes are present (see bottom spectra of Figure 14.4a). In this case, the extraction of the distance distribution function becomes more cumbersome and requires prior knowledge of the spin labeling efficiency of the sample under investigation.

The extent of broadening in the spectrum can be estimated also from the peak height ratio (d_1/d) of the low-temperature spectrum (Figure 14.4a, right panel). For singly- or doubly-labeled proteins with an interspin distance above 2 nm, a value of <0.4 is expected. Higher values indicate distances below 2 nm. However, this method is highly sensitive to the fraction of noninteracting spins in the sample, owing to their sharper lines that tend to dominate the peak height ratios.[55]

Figure 14.4(c) shows the appearance and disappearance of dipolar broadening in the experimental CW EPR spectra of spin-labeled sites attached at key positions in the periplasmic and cytoplasmic gates of a membrane protein (depicted in Figure 14.4b). Upon binding of adenosine triphosphate (ATP), the periplasmic gate (positions 168) closes and concomitantly

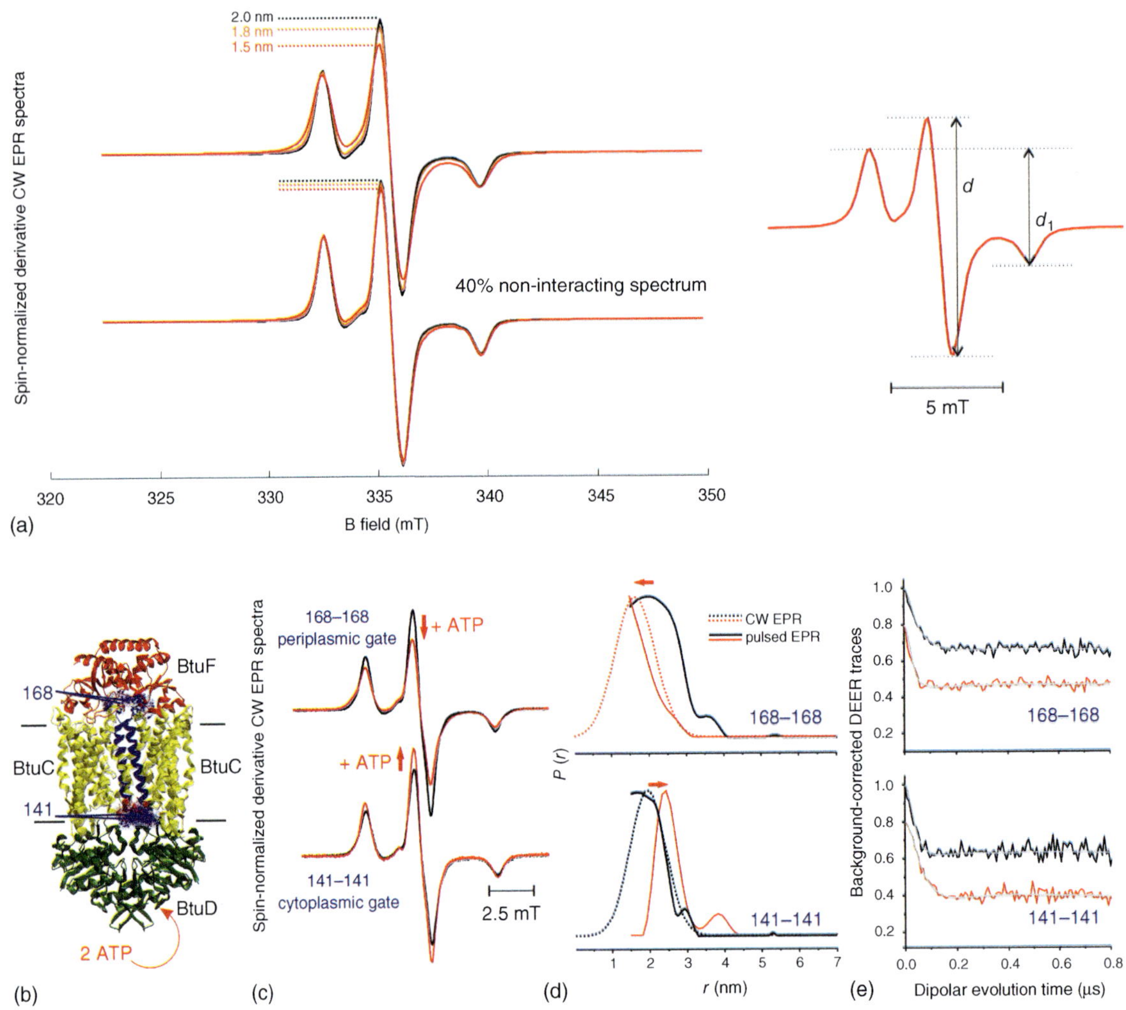

Figure 14.4. Detection of internitroxide distances. (a) Simulation of dipolar broadened powder CW EPR spectra at 9.45 GHz for mean internitroxide distances of 2, 1.8, and 1.5 nm performed with DIPFIT. The parameters used are as follows: $g_x = 2.0088$, $g_y = 2.0061$, $g_z = 2.0021$; $A_x = A_y = 14$ MHz, $A_z = 99.5$ MHz. A Voigtian lineshape is used and a 0.5-nm width of the Gaussian distance distribution. Upper spectra: 100% interacting spins, lower spectra: 40% of the total spectral area is due to the noninteracting spins in the sample. The right panel shows the peak heights to be considered for an estimation of the dipolar broadening in the low-temperature spectra. (b) Spin-labeled positions 168 (periplasmic gate) and 141 (cytoplasmic gate) in the homodimeric BtuC subunit in the transporter. (c) Nucleotide-dependent movement of periplasmic and cytoplasmic gates of BtuCD-F in proteoliposomes. Spin-normalized CW EPR spectra detected at 160 K (black, in the absence of nucleotide; red, in the presence of ATP). (d) Gaussian distance distributions obtained from low-temperature CW EPR spectra via convolution methods (dotted) superimposed to the distance distribution (solid) extracted from the X-band background-corrected DEER traces $F(t)/F(0)$ presented in panel (e) (black, in the absence of nucleotide; red, in the presence of ATP)

the cytoplasmic gate (positions 141) opens. In both cases, the distance distribution in one state is characterized by a mean distance below 2 nm; therefore, the conformational change of the protein is detectable by appearance or disappearance of dipolar broadening in the corresponding spin-normalized CW spectra at 160 K (decrease or increase in the spectral amplitude in Figure 14.4c, see arrows). The Gaussian distance distributions (dotted lines in Figure 14.4d) were obtained by fitting the experimental spectra with convolution methods using the program DIPFIT.[54]

In recent years, pulsed dipolar spectroscopy has become a popular tool in structural biology, since interspin distances up to 16 nm can be accurately extracted.[56] To this end, the main technique is double electron–electron resonance (four-pulse DEER,[57,58] also known as PELDOR). Examples of DEER traces obtained on the same spin-labeled transporter are shown in Figure 14.4(e).[59] The distributions characterized by short mean distances extracted via model-free Tikhonov regularization (see Chapter 19) agree in both cases with the Gaussian distributions obtained from the low-temperature CW EPR spectra via convolution methods (compare dotted and solid lines in Figure 14.4d). Figure 14.4 shows how the two techniques are complementary, especially in the borderline region of applicability of both. The upper distance limit for measuring dipolar coupling by CW techniques (using strong- or weak-coupling regimes) and the lower distance limit for X-band DEER are analyzed in details in.[60]

14.4 DYNAMICS OF THE NITROXIDE PROBE ENCODED IN THE SPECTRAL LINEWIDTH

One of the main advantages of nitroxide probes lies in the effective modulation of their spectral features owing to partial averaging of the ***g*** and ***A*** anisotropies by molecular motion, in particular by rotational diffusion (tumbling) of the nitroxide molecule with respect to the applied external magnetic field. The possibility to sensitively monitor at different temperatures the isotropic or anisotropic rotational diffusion of nitroxide probes dissolved in materials characterized by different viscosities or attached to macromolecules with distinct dynamic properties makes the nitroxide probes excellent tools to study dynamic processes via CW EPR.

Depending on the EPR frequency, motion on timescales between 10^{-3} and 10^{-10} s produces characteristic spectral effects, which can be investigated. The molecular motion is classified into four regimes with distinct relative magnitudes of the characteristic rotational correlation time (τ_c) of the motion and the frequency width of the spectrum (Figure 14.5a). Depending on the ratio between the spectral width $\Delta\omega$ and the characteristic rotational correlation time τ_c, different approaches exist to calculate spectral effects due to molecular motion, which leads to partial averaging of the anisotropies of the ***g*** and ***A*** tensors.

In the isotropic limit ($\tau_c^{-1}/\Delta\omega \gg 1$), the anisotropic interactions are averaged out, and one observes only the isotropic average of the magnetic interactions. In the fast motion regime ($\tau_c^{-1}/\Delta\omega > 1$), the spectrum can be considered as a sum of lines with different Lorentzian linewidths. The high-field line of the nitroxide spectrum is the most sensitive reporter of changes in the rotational correlation times from picoseconds to nanoseconds. In fact, for isotropic motions, its amplitude is the first to decrease with respect to the low-field and central lines owing to the largest spectral anisotropy arising from the combined effects of the ***g*** and ***A*** tensors (see Figure 14.3b). Notably, in this regime, the rotational correlation time can also be inferred from the comparison of the full width at half maximum of each spectral line (proportional to the inverse of the transverse relaxation time T_2) based on the Kivelson theory.[6,61]

For very long correlation times, a static distribution of all orientations of the nitroxide is observed (rigid limit and powder spectra).[62] Between the fast motion and the rigid limit, when $\tau_c^{-1}/\Delta\omega \approx 1$, the probe is in the slow motion regime for the given EPR frequency. In this complex motional regime, the SRSL (slowly relaxing local structure) approach developed by Freed and coworkers[63] provides the possibility to disentangle via multifrequency analysis the different types of coupled rotational motions (e.g., the motion of the nitroxide ring, the chemical tether linking the nitroxide to a macromolecule, and the macromolecule itself).

The effect of different τ_c values of isotropic motions on X-band CW EPR spectra (simulated with EasySpin[64]) is shown in Figure 14.5(a). At this frequency/field, the hyperfine anisotropy is the dominant contribution to the spectral width. For nitroxide radicals at 0.35 T, the transition from fast to slow motion starts at $\tau_c > 1-2$ ns.

The notion of a field/frequency-dependent 'EPR timescale' is sketched in Figure 14.5(b) with spectra simulated at three different frequencies (X, Q, and W bands) for two isotropic motions characterized by τ_c of 500 and 50 ps. The degree of averaging of the spectral anisotropies for a given τ_c depends on the spectral width. At Q- and W-bands, the effect of the ***g***-anisotropy sets in, therefore, the spectral width becomes larger, and faster motions are required to average the combined anisotropies of ***A*** and ***g*** with respect

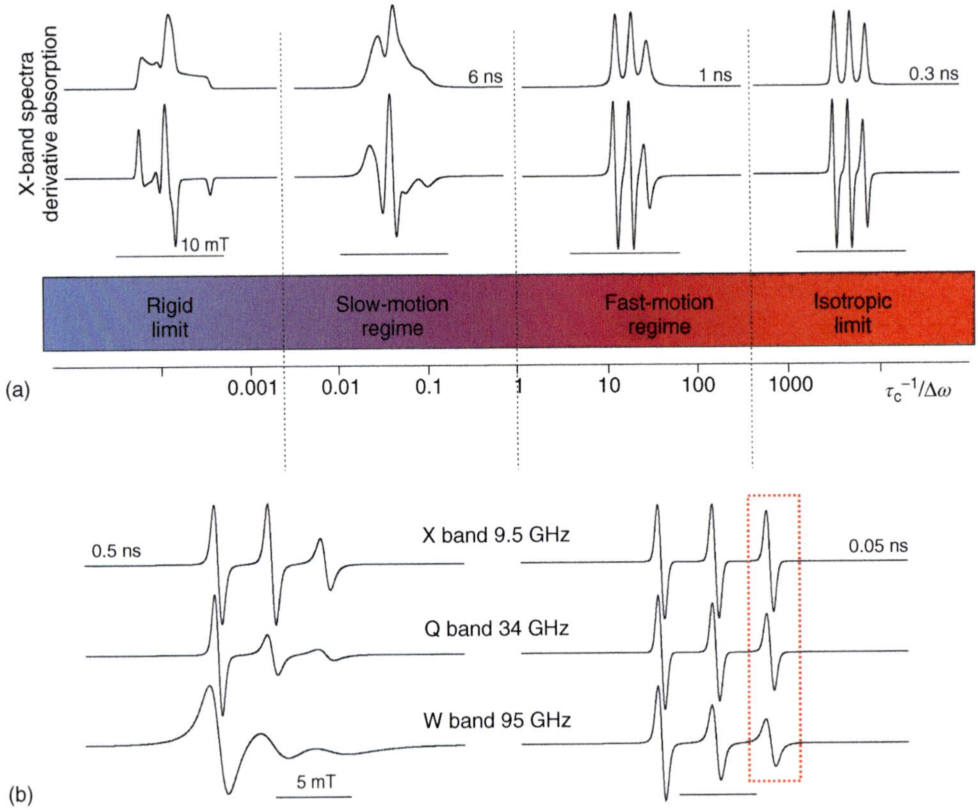

Figure 14.5. The effect of the rotational correlation times on the X-band spectrum of a nitroxide probe. (a) X-band derivative and absorption spectra of nitroxide probes in the rigid limit and in different motional regimes characterized by different isotropic rotational correlation times ($\tau_c = 6$, 1 and 0.3 ns). Principal components of the $\boldsymbol{g}$ tensor are 2.0085, 2.0061, and 2.0022 and of the $\boldsymbol{A}$ tensor 13, 13, and 92 MHz. The borders between different regimes are only indicative. (b) At high frequency/field, the spectral width $\Delta\omega$ increases due to the resolved g-anisotropy; therefore, the spectra become more sensitive to faster motions than at conventional X-band (different 'EPR timescales'). Spectra simulated with EasySpin

to X-band. As a result, the spectra simulated at higher frequencies show for the same type of motion a less complete averaging of their anisotropies with respect to those at low frequencies. This can be best appreciated by comparison of spectra with a τ_c of 50 ps. This rotational correlation time at X-band yields a spectrum close to the isotropic limit, in which the intensities of the three lines are almost equal. In contrast, at Q- and W-bands, the high-field line shows the lowest amplitude (larger linewidth) compared to the other two lines of the spectrum. Therefore, subtle changes in this very fast rotational regime can still produce measurable spectral effects, and consequently, spectra detected at higher frequencies are more sensitive to faster motions.

A quantitative description of the motion of the nitroxide probes in nonisotropic media, such as liquid crystals, membrane bilayers, or proteins, requires a multifrequency approach[63,65,66] and appropriate modeling of the motional anisotropies. The theory that describes the motional effects on the CW spectra of nitroxide probes is based on the stochastic Liouville equation.[67] Slow anisotropic motion of nitroxide probes can also be described by other approaches, based on stochastic trajectories of the nitroxide molecule or molecular dynamics simulations. For proteins carrying spin-labeled side chains, molecular dynamic simulations based on X-ray or NMR structures of the protein of interest can be used to extract the trajectories of the spin-labeled side chains and therefore their spectral fingerprint.[68–71] A recent review describes the available techniques to calculate spectral effects in this regime.[68]

To further complicate the spectral shape, several populations of nitroxides in slow exchange, which are characterized by different degrees of mobility, can be present. The case of multicomponent spectra, with distinct spectral features, which can be labeled as 'immobile' and 'mobile' components, is very often encountered. Disentangling the origin of such effects is in general very challenging.

Despite the difficulties in unraveling the physical models encoded in the nitroxide lineshape, semiquantitative information on the characteristic type of motion can be extracted from spectral inspection. A common qualitative approach to systematically compare different spectral shapes mainly at X-band frequencies consists in measuring the linewidth of the central nitroxide spectral line and using its inverse value as 'mobility' parameter. The smaller the linewidth, the higher is the mobility parameter. In composite EPR spectra, this 'mobility' parameter is strongly biased by the fastest motional component, which has the largest amplitude in the central line. The absolute value of the first or the second moment of the spectra[72] can also be used to quantitatively extract the spectral breadth, related to the line broadening and therefore to the 'mobility'. An advantage of spectral moments as 'mobility' parameters is that contributions from composite spectra are additive. In protein studies, these mobility parameters can aid distinguishing sites buried in the protein interior, in helical regions, loops, or β-barrels, based on the large amount of mobility data correlated with secondary–tertiary structures available in the literature.[41,73]

In the following discussion, we will give some examples taken from SDSL studies of proteins, for the anisotropies encoded in the spectral lines, the effects of the type of spin label linker on the mobility of the nitroxides, and the use of changes in the mobility of the spin-labeled side chain to monitor protein's conformational changes. Figure 14.6(a) shows four room temperature spectra of single cysteine mutants of a water-soluble protein labeled with MTSL. To analyze the EPR spectral shape, it is important to note that any residual unbound MTSL needs to be removed by biochemical methods. An example of the spectral features typical of residual free label fast tumbling in solution is shown in Figure 14.6(a), for position 126. The black spectrum contains the characteristic three-line spectrum of a free label (marked with asterisks) in addition to the spectrum arising from the label bound to the protein. After removal of the free label, the resulting spectrum (green) encodes solely the motional properties of the bound spin-labeled side chain at position 126.

The EPR spectrum of MTSL bound to position 126 or 72 is characteristic of a mobile nitroxide side chain located at the end of a helical secondary structure. The EPR spectrum of position 62 located in the middle of an alpha helical segment in the protein shows two distinguishable spectral fractions characterized by different degrees of mobility (labeled as 'immobile' (*i*) and 'mobile' (*m*) component). This is a commonly encountered spectral feature indicative of the anisotropic and complex motion of spin-labeled side chains in proteins. The spin label bound to the C-terminal residue (position 193) in the protein shows a narrow three-line spectrum, indicative of very fast motion of the nitroxide probe. This is due to the intrinsic dynamics of the nitroxide side chain as well as to the flexibility of the partially unfolded C-terminal region of the protein. For each spin-labeled side chain, the MTSL rotamers calculated on the available NMR structure of the protein with the software MMM2015[16] are presented.

Figure 14.6(b) shows one example of the detection of a conformational change in the secondary and tertiary structures of a protein using one spin-labeled side chain. In this specific case, we monitor the rearrangement of the 'tongue' region of the Cph2 phytochrome from a β-sheet (present in the the so-called Pr-red absorbing-state) to an α-helical structure (in the Pfr-far-red absorbing-state). The conformational change of the protein, induced by excitation of the covalently bound chromophore with far-red or red light, respectively, is captured by the spin-labeled side chain. By switching the protein to the Pfr state, the EPR spectrum becomes broader, indicating the decrease in the overall mobility of the nitroxide side chain located at a key position close to the 'tongue' region.

The sensitivity of nitroxide spectra toward different molecular motions is exemplified in Figure 14.6(c) in the study of a small spin-labeled peptide freely tumbling in solution in the presence of a membrane protein, which can bind the peptide with a certain affinity (in the micromolar range). The spectrum of the free peptide shows three narrow lines, indicative of

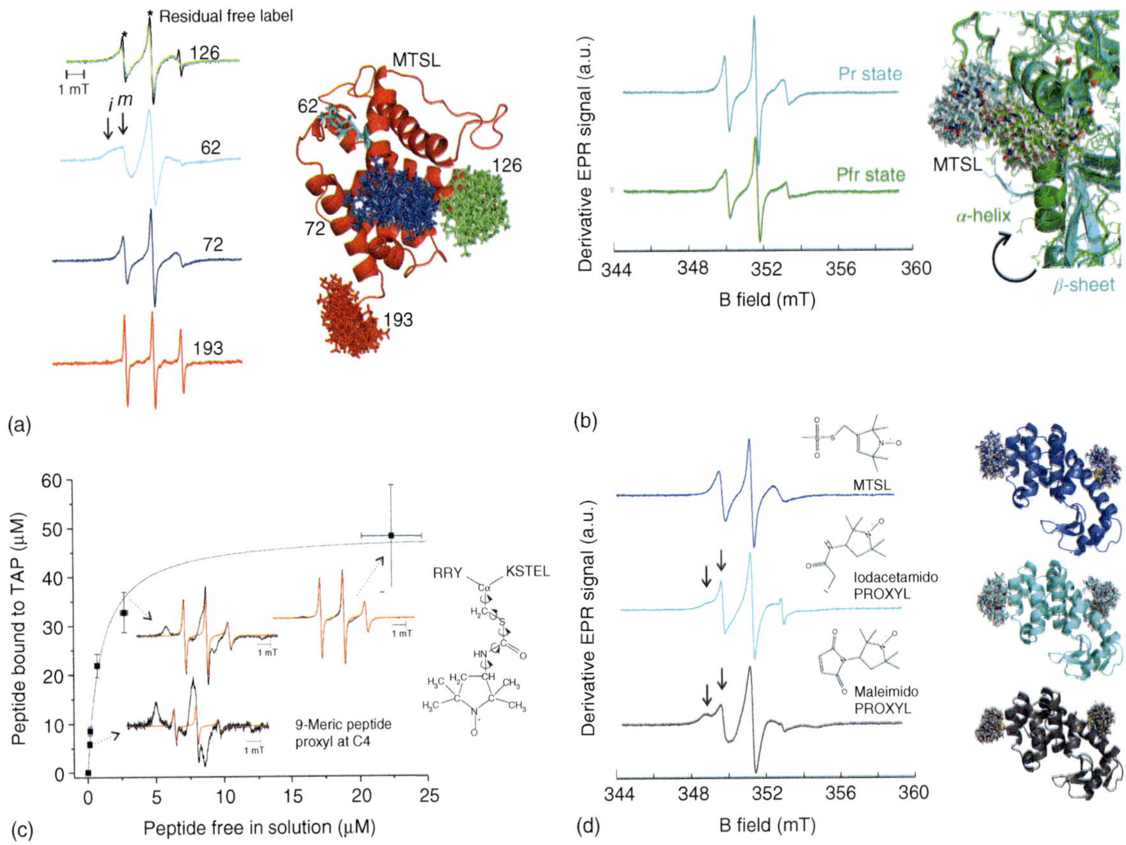

Figure 14.6. Dynamics of spin-labeled side chains encoded in room temperature continuous wave X-band EPR spectra. (a) CW EPR spectra of singly-labeled water-soluble protein Bax. Different degrees of mobility and anisotropic motions are encoded in the spectral lineshape. Residual free label is highlighted in the black spectrum of position 126. Distinct spectral components with different mobilities are highlighted in the spectrum of position 62 (*m*: mobile, *i*: immobile). On the right, ribbon representation of the protein labeled with MTSL (PDB 1F16) with multiple rotamers of the spin-labeled sites shown in sticks (calculated with MMM2015 using the ambient temperature library). (b) Spin-labeled sensory domain of the phytochrome Cph2 from *Synechocystis* sp. PCC 6803 (position 381, close to the 'tongue' region), illuminated in the resonator with light at 720 and 625 nm to create the Pr (cyan) or Pfr (green) states, respectively (unpublished results). On the right, the homologous site in the X-ray structures of DrBPhP (PDB 4O01, cyan and 4O0P, green) is spin labeled in silico with MMM2015. (c) The peptide RRYCKSTEL carrying a PROXYL label at position C4 is incubated with detergent-solubilized TAP1/2 at increasing concentrations. The spectrum of the free peptide in solution (red lines) can be recognized in the spectra due to its narrow lines, in contrast to the fraction of peptide bound to the membrane transporter TAP1/2, which shows a very slow rotational correlation time. The dissociation constant can be extracted from the fractions obtained.[74] (Reproduced with permission from M. Herget, C. Baldauf, C. Scholz, D. Parcej, K. H. Wiesmuller, R. Tampe, R. Abele, and E. Bordignon, Proc. Natl. Acad. Sci. U. S. A., 2011, 108, 1349. © National Academy of Sciences of the United States of America, 2011) (d) CW EPR spectra of bacteriophage T4 lysozyme doubly-labeled at positions 72 and 131 with MTSL (upper spectrum, blue), iodoacetamido PROXYL (central, cyan), and maleimido PROXYL (bottom, black) (unpublished results). On the right, ribbon representation of the protein (PDB 2LZM) showing the calculated rotamers (MMM2015) for each label at the two chosen positions

a rotational correlation time close to 1 ns (see simulated spectra in Figure 14.5). Once bound to the protein, the peptide shows a distinct spectrum with a prominent positive peak in the low-field region and a negative peak in the high-field region, typical of very slow motion (see rigid-limit spectrum in Figure 14.5). The defined change in rotational correlation time from free to bound peptides allows to extract with high precision the fractions of free and bound peptides at increasing concentration of peptide in the sample, which can be used to obtain the dissociation constant. Figure 14.6(d) shows how different motions of spin labels with distinct tethers bound to the same protein sites are encoded in the EPR spectral shape. Three spin labels among those presented in Section 14.2 were bound to the same double Cys mutant (^{72}C–^{131}C) of bacteriophage T4-lysozyme with high efficiency (MTSL: 95%; iodoacetamido PROXYL: 69%; and maleimido PROXYL: 100%). The spectrum of MTSL bound to the two sites is characterized by a unique and rather isotropic component; however, the spectra of the iodoacetamido and maleimido PROXYL show anisotropic features with distinguishable nitroxide populations characterized by different mobilities. Having two spin-labeled side chains, it is not possible from spectral inspection to associate the 'immobile' or 'mobile' component to a specific site. The two labels could have distinct 'mobilities,' or they could both contribute to each spectral component. Analysis of one or better both singly-labeled proteins would clarify the origin of the spectral heterogeneity observed.

14.5 POLARITY AND PROTICITY OF THE NITROXIDE MICROENVIRONMENT EXTRACTED FROM *g* AND *A* TENSOR PARAMETERS

The local properties of the nitroxide microenvironment in terms of dielectric constants of the solvent and the propensity to form H-bonds toward the nitroxide moiety can be determined by analysis of the ***g*** and ***A*** tensor principal components obtained from CW EPR spectra. In particular, the values of a_{iso} and A_z increase when the spin population on the nitrogen atom increases in polar microenvironments (see Section 14.3.1), and the value of g_x decreases (owing to its linear dependence on the spin population on the oxygen atom). In addition, polar environments and H-bonds toward the nitroxide oxygen modify the $n \rightarrow \pi^*$ excitation energy, which induces a further decrease in g_x (see Figure 14.1d and Section 14.3.1). Therefore, going from apolar–aprotic to polar–protic environments, the values of a_{iso} and A_z increase and concomitantly g_x decreases.[44]

The change in a_{iso} can provide the polarity of the microenvironment of the nitroxides at room temperature already at X-band, and it was used to characterize solvents with different polarities.[75] The A_z parameter, also reporting the local polarity, can be easily extracted from low-temperature X-band CW EPR spectra (Figure 14.3). However, to extract simultaneously A_z and g_x parameters, one needs to analyze low-temperature high-field EPR spectra, as those shown in Figure 14.7(a). The A_z and g_x parameters allow investigating the local polarity and H-bond network in a frozen snapshot of the nitroxide microenvironment.

As already shown in Figure 14.3, the resolution of low-temperature W-band CW EPR spectra is sufficient to disentangle the ***g*** and ***A*** principal components. Examples of simulated W-band spectra of nitroxide probes carrying ^{14}N or ^{15}N isotopes with different g_x and A_z values are shown in Figure 14.7(a). Best resolution is achieved with deuterated nitroxides in deuterated solvents, where the spectral linewidth is minimized and unresolved hyperfine interactions with protons are removed.[23] It is important to highlight that slightly different coordination of the NO moiety with water molecules and heterogeneity in the local polarity can result in a distribution of g_x values (g_x strain), which induces a further broadening in the g_x spectral region.

Distinct g_x values characterizing the nitroxide populations with a specific microenvironment are coupled to distinct values of A_z. With these parameters it was possible to probe the water permeabilization of the channel of bacteriorhodopsin,[23] as reviewed in.[76] Plots of g_x versus A_z show the linear dependency of the two parameters (see Section 14.3.1).[44] When a series of W-band spectra of nitroxide probes in different environments was analyzed, it was recognized that in most cases coexisting populations with up to three different g_x values could be detected.[43] The three g_x values were assigned to three populations characterized by 0, 1, and 2 H-bonds toward the nitroxide oxygen, based on the fit of the W-band spectra from a series of probes dissolved in solvents and spin-labeled proteins (Figure 14.7b and c). In the range of polarity investigated, the relative fractions

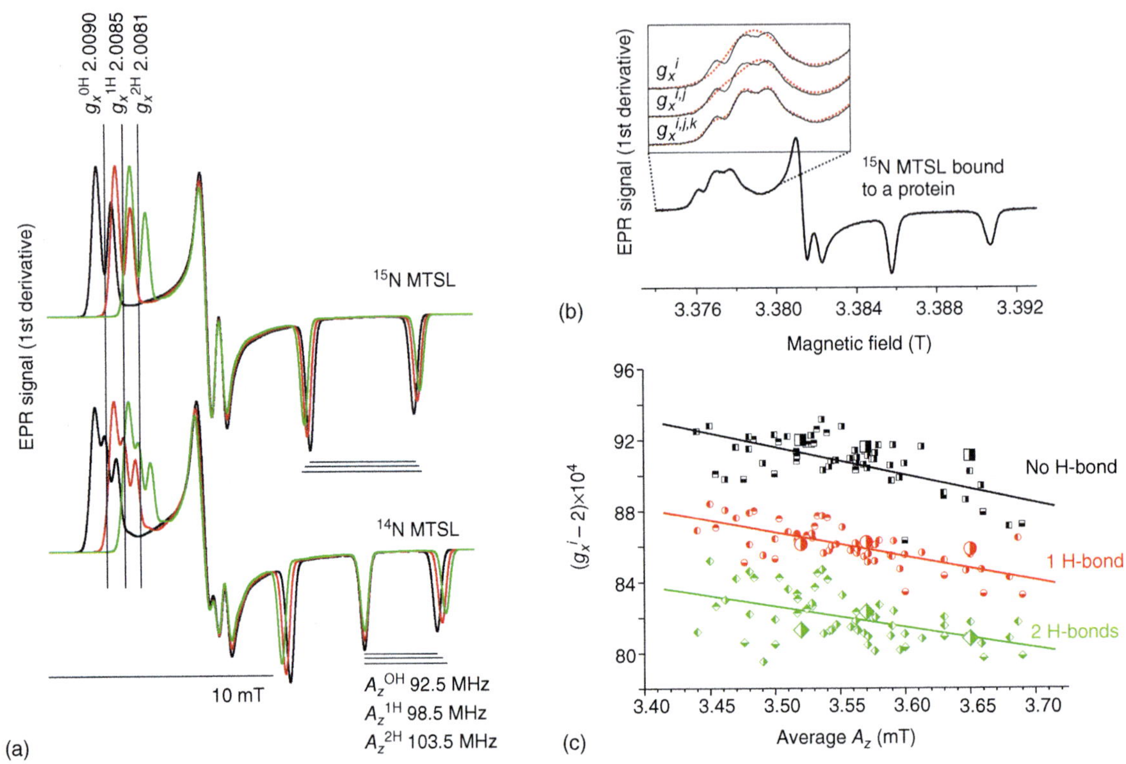

Figure 14.7. High-field CW EPR spectra probing polarity and proticity of the microenvironment of the nitroxide probe. (a) Simulated W-band powder CW spectra of MTSL carrying the two nitrogen isotopes. Each simulation is performed with one set of **A** and **g** parameters with the Easyspin. The values in the figure are those of the ^{14}N MTSL variant. The hyperfine parameters for the ^{15}N isotope are larger than those for ^{14}N by a factor of 1.4. The ^{15}N spectra facilitate the analysis of the g_x heterogeneity owing to a better resolution of the hyperfine lines in the low-field region. The three pairs of g_x and A_z corresponds to 0-, 1-, or 2-H-bonds toward the NO moiety (b) Example of three g_x fractions coexisting in the 160 K W-band spectrum when MTSL (^{15}N) is located at position 17 in sensory rhodopsin II from *N. pharaonis*. The shoulders visible in the low-field region of the spectrum can be fitted only with three components with distinct values of g_x (see inset). (c) A series of spin-labeled proteins were fitted with three g_x components, and the respective values plotted versus the average value of A_z. Different A_z components could not be extracted from the fit of the spectra owing to poor resolution in the high-field region. The lower values of the average A_z are associated with the largest fractions of 0- and of 1-H-bonded nitroxides; the higher values of A_z are associated with a decreasing fraction of 0- and an increasing fraction of the 2-H-bonded populations. (Reproduced with permission from Ref. 43. © Springer, 2010). (d) W-band CW EPR spectra of 1 mM of the perdeuterated nitroxide spin probe 3-hydroxymethyl-2,2,5,5-tetramethylpyrrolin-1-oxyl in a frozen solution of D_2O/glycerol-d_8 (50/50% v/v) recorded at 80 K before and after annealing of the sample for 1 h at 200 K (gray lines). The spectral positions corresponding to different principal g_x components (for one and two H-bonds to the nitroxide) and average A_z values are indicated. The black dashed lines are the simulated spectra. The two simulated spectral contributions corresponding to single (red) and double (green) H-bonded fractions after annealing are plotted below. The fraction of singly H-bonded fraction increases upon annealing with respect to the 2H-bonded one owing to the rearrangement of the H-bond network around spin labels close to the glass transition temperature. (Reprinted with permission from E. Bordignon, A. I. Nalepa, A. Savitsky, L. Braun, and G. Jeschke, J. Phys. Chem. B, 2015, 119, 13797. Copyright 2015 American Chemical Society). (e) The average g_x value obtained from the analysis presented in panel C plotted versus average A_z gives a correlation that can be useful for comparison of the polarity and proticity character of spin-labeled protein sites. (Reproduced with permission from Ref. 43. © Springer, 2010)

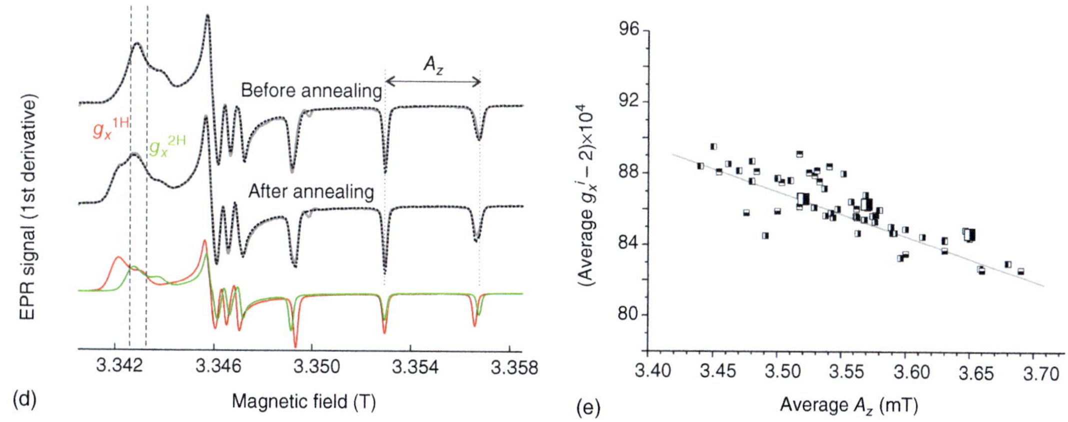

Figure 14.7. (Continued)

of the three populations varies strongly, with the 2-H-bond fraction being the largest for the most polar environment (highest A_z). This assignment was confirmed by CW spectra detected at 275 GHz, and it was found that the g_x value for each H-bond population showed minor variations in the range of investigated polarities, indicating only marginal dependency on A_z.[42] However, the fractions of the populations were confirmed to vary strongly with the polarity of the environment, which provides a consistent description of the variation of the average g_x and A_z values from site to site (Figure 14.7e).

It was recently recognized[77] that depending on the thermal history of frozen aqueous samples, different fractions of H-bonded populations could be present in the high-field spectra, raising questions on the best experimental conditions to obtain reproducible data on spin-labeled sites in proteins. In particular, when samples were measured at $T < 160$ K, directly after snap freezing them in liquid nitrogen, a larger fraction of 2H-bonded nitroxides were observed, while after annealing above 190 K, a reproducible ratio of the different H-bonded nitroxide populations could be detected, with a lower fraction of 2H-bonded nitroxides.[77] The first method provides a snapshot of the hydrogen network around nitroxides at room temperature, the second method allows rearrangements of the water molecules at the glass transition temperature of the water–glycerol mixture (175 K for a 50% v/v), yielding a stable water coordination, which can be used to characterize the water accessibility at the labeled sites. An example is shown in Figure 14.7(d) for a nitroxide probe dissolved in a water–glycerol mixture.

14.6 TECHNIQUES TO MONITOR WATER ACCESSIBILITY TOWARD THE NITROXIDES

Besides the high-field methods described above, which present some disadvantages in terms of reduced sensitivity and limited availability of high-field spectrometers, there are additional EPR techniques available to extract information on water accessibility toward nitroxide probes. In this section, the available methods with their main advantages and disadvantages are presented, with examples taken from studies of spin-labeled proteins and nitroxide probes dissolved in solvents.

The most common technique to obtain the accessibility profile of spin-labeled proteins or membrane bilayers doped with spin-labeled lipids at physiological temperature is power saturation CW EPR.[73,76,78–81] This method relies on the extraction of the relative collision frequency of each spin label with (i) freely diffusing paramagnetic O_2 (triplet ground state), which preferentially partitions in the apolar lipid region of the membrane; (ii) membrane-impermeant paramagnetic reagents, such as NiEDDA, soluble in the bulk water only (Figure 14.8a). The measurements are performed with the sample inserted into gas-permeable capillaries. Heisenberg exchange between a nitroxide and an 'exchange reagent' requires a direct contact interaction in an encounter complex, and the exchange rate (W_{ex}) is taken to be a measure of the nitroxide's exposure to the solvent containing the reagent. Owing to the Heisenberg exchange mechanism, equal changes (Δ) in the transverse and longitudinal relaxation rates of the nitroxide spin labels are expected (under the conditions that the longitudinal relaxation time of the paramagnetic reagent is smaller than the encounter complex lifetime):

$$W_{ex} = \Delta\left(\frac{1}{T_1}\right) = \Delta\left(\frac{1}{T_2}\right) \quad (14.8)$$

To extract the changes in the relaxation rates by power saturation CW EPR (see Chapter 1), the peak-to-peak intensity of the central line of the EPR derivative spectrum (I_{pp} in equation (14.9)) is plotted versus the square root of the incident microwave power, and the $P_{1/2}$ value is extracted (Figure 14.8b). The $P_{1/2}$ value is the power at which the intensity is half of the value in the absence of saturation. It is inversely correlated with the relaxation times T_1 and T_2. The fit is performed with the following function:

$$I_{pp} = a\frac{\sqrt{P}}{\left(1 + \left(2^{1/\varepsilon} - 1\right)\frac{P}{P_{1/2}}\right)^{\varepsilon}} \quad (14.9)$$

with

$$P_{1/2} = \frac{\left(2^{1/\varepsilon} - 1\right)}{\gamma^2 T_1 T_2} \quad (14.10)$$

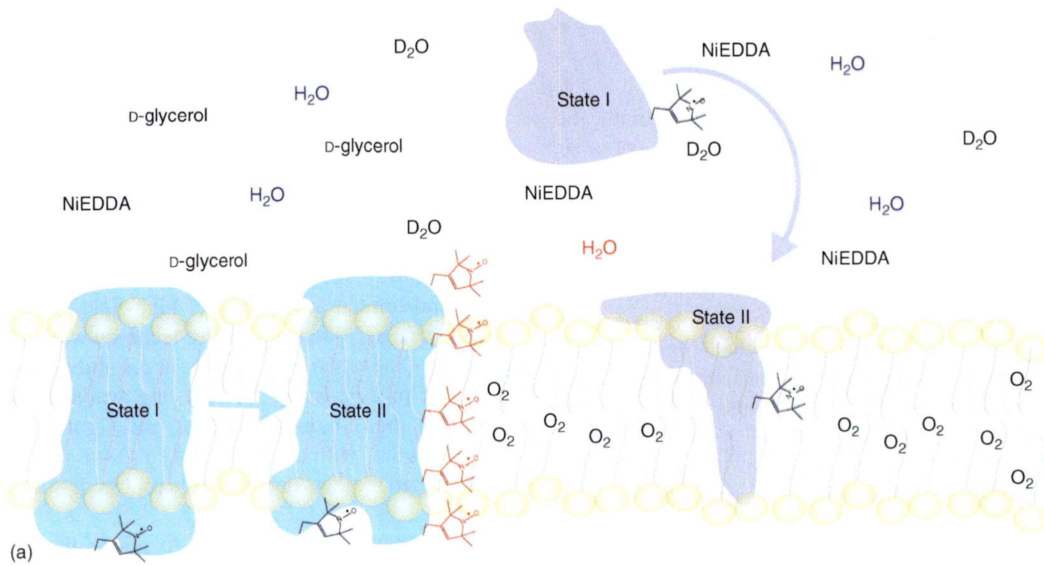

Figure 14.8. Methods to extract water accessibility toward the nitroxide probe. (a) Sketch of protein undergoing conformational transitions from state I to state II that can be monitored by change in the water accessibility at specific sites (e.g. black labels). The red labels represent sequential singly-labeled sites to be scanned to characterize, for example, the depth of a certain protein region within the membrane bilayer. The different molecules added to the sample to indirectly probe the water accessibility are drawn. (b) Examples of three power saturation curves obtained with NiEDDA, air (21% oxygen), and nitrogen (reference). The increase in the power saturation with NiEDDA is indicative of a site, which is accessible to the bulk water. The fit of those curves yields the $P_{1/2}$ values, which can be used to get a quantitative parameter related to water accessibility. (c) Example of an X-band 3P-ESEEM trace obtained on a water-exposed site on a membrane-embedded protein. The Fourier transformation of the background-corrected trace yields a peak at the Larmor frequency of ^{2}H (2.5 MHz), whose intensity is proportional to the concentration of deuterium atoms in the vicinity of the spin label. The interpulse delay is usually chosen to maximize the deuterium signal and minimize that of the protons (14 MHz). (d) Example of room temperature X-band ODNP measurements on the spin-labeled cytoplasmic gate of an ABC transporter solubilized in detergent before (open triangles) and after (closed triangles) addition of a substrate-binding protein interacting with the periplasmic region of the transporter on the other side of the protein. The plot shows the normalized intensity of the amplitude of the FID of the NMR proton signal from the water molecules versus applied microwave power. The stronger ODNP effect after complex formation is indicative of an increased accessibility of the spin label toward the water molecules. (Reproduced with permission from Ref. 82. © Elsevier, 2012)

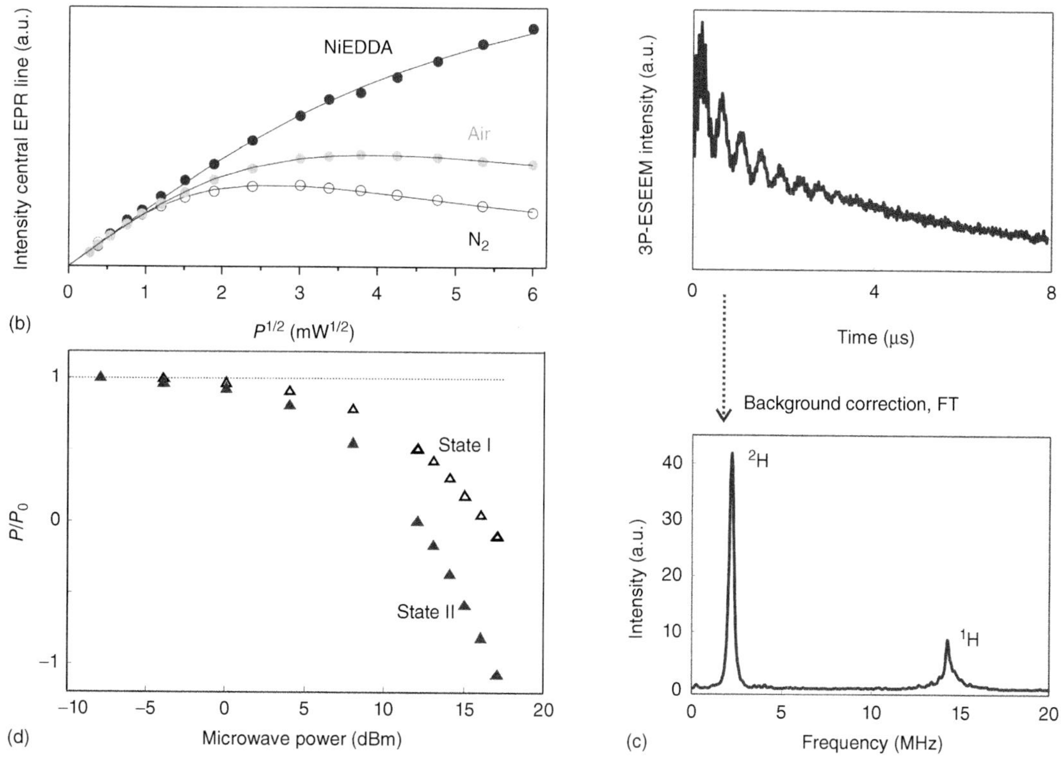

Figure 14.8. (Continued)

The parameter ε is restricted in the fitting to the range $^1/_2 - ^3/_2$ to account for the homogeneous and inhomogeneous line broadening, respectively; a is a free scaling factor. The difference between the $P_{1/2}$ values obtained in the presence and absence of paramagnetic reagents divided by the spectral linewidth is used as accessibility parameter toward the particular paramagnetic reagent (proportional to W_{ex}). A detailed description of the procedure can be found in Ref. 83 (note that in equations (14.11) and (14.12) from Ref. 83 2^{ε} should be $2^{1/\varepsilon}$). Additional information about this method to extract the membrane and water accessibility parameters can be found in.[79,80,83] Power saturation techniques address indirectly the water accessibility toward nitroxides at physiological temperatures by detecting the accessibility toward paramagnetic reagents partitioning preferentially either in aqueous or lipid environments. Comparing the effects of the water-soluble and membrane-partitioning (oxygen) reagents give the overall water accessibility of the nitroxide side chain. The concentration gradient of the oxygen along the membrane normal allows also extracting the immersion depth of the spin-labeled sites in the membrane bilayer (Figure 14.8a, red spin labels). In addition, periodical patterns appearing while scanning a region of the protein with spin labels can directly provide the secondary structure of the region under investigation. This information can be extracted for amphipathic α-helices (periodicity 3.6) of β-sheets (periodicity 2) with one side being exposed to water and one to the protein interior or the membrane bilayer. The sensitivity of the method is high and can be applied to proteins at concentrations of a few micromolar. However, the output is only an indirect measure of the water accessibility and the samples are perturbed with reagents. It was recently shown that via power saturation CW EPR performed at different temperatures, it is also possible to probe the glass transition temperature of spin-labeled polymers,[84,85] as well as of solvents in which nitroxide probes are dissolved,[77] owing to the sudden change in the relaxation rates.

Saturation recovery EPR is another related method, which requires paramagnetic reagents and provides information on protein topology at the water-membrane interfaces, also at room temperature.[80,86,87] In this case, the relaxation rates are directly measured and not extracted from the saturation curves, which gives an advantage in terms of reproducibility. In addition, different spectral components present in the CW spectra could in principle be monitored independently, another advantage with respect to the power saturation methods.

It is also possible to obtain information on the accessibility toward the aqueous milieu from the analysis of the intensity of the deuterium signal in ESEEM traces at cryogenic temperatures.[88] This experiment is based on the application of a series of microwave pulses that generate an echo, and the echo decay is followed as a function of one of the time intervals between the pulses (see Chapter 18). When an anisotropic hyperfine interaction is present, the echo decay is modulated (Figure 14.8c). For weak anisotropic hyperfine interactions, the modulation frequency is the nuclear Larmor frequency, and the modulation depth is a function of the electron–nuclear distance, the number of interacting nuclei, and their nuclear spin. In this case, the water accessibility is indirectly probed by measuring the concentrations of deuterium atoms introduced in the sample by deuterated water, glycerol, or lipids. Three-pulse (3P)-ESEEM has been used to study the topology of membrane-embedded proteins in the presence of D_2O and D-glycerol[89]; the water penetration profile in membranes suspended in D_2O by means of spin-labeled lipids[90,91]; and the topology of peptides inserted into model membranes using D_2O or specifically deuterated lipids.[92] The method is sensitive, there is no perturbation of the sample (with D_2O and D-glycerol), and the relation between the modulation pattern and the distance toward the deuterium atoms is relatively simple.

The main disadvantage of this method is the necessity to freeze the sample. In addition, if deuterated water or glycerol is used, deuteration of exchangeable protons in the protein is unavoidable. In Figure 14.8(c), an example of a 3P-ESEEM trace obtained at 50 K on a 20 μM spin-labeled protein embedded in membrane is shown. Extraction of the modulation depth of the deuterium signal via background correction and Fourier transformation yields a peak at the Larmor frequency of ^{2}H, whose intensity correlates with the concentration of deuterium atoms in the vicinity of the label.

Liquid-state Overhauser dynamic nuclear polarization (ODNP) is an emerging method to directly measure the accessibility of the spin label toward water molecules surrounding it at physiological temperatures. An in-depth description of ODNP theory can be found elsewhere.[93,94] ODNP measures the dynamic exchange of water molecules with the spin-labeled site in the protein and provides an equilibrium picture with no need of deuteration or chemical perturbation of the sample. In addition, ODNP can also be time resolved to follow slow reactions in the molecular ensemble over minutes or hours.

X-band ODNP is performed by continuously irradiating the sample at X-band frequencies, thus saturating the nitroxide central EPR transition, while simultaneously recording proton NMR signal at about 14 MHz. Upon saturation of the EPR transition, the high spin polarization of the unpaired electron in the spin label is transferred to the protons on the nearby water molecules (Overhauser effect). A negative NMR signal enhancement is observed because the translational diffusion of water is rapid enough to induce dipolar relaxation with the electron spin. Equation (14.11) describes the enhancement defined as the ratio between the ODNP-enhanced proton signal (P) and the equilibrium signal in the absence of MW irradiation (P_0):

$$\frac{P}{P_0} = 1 - \varepsilon s f \frac{\gamma_e}{\gamma_H} \tag{14.11}$$

where $\frac{\gamma_e}{\gamma_H} = 658$, ε is the coupling factor, which is used to determine dynamics and subsequently the diffusion coefficient of waters; s is the saturation factor of the electron spin and approaches 1 when nitroxide spin labels are bound to slow tumbling molecules. The leakage factor ($f = 1 - \frac{T_1}{T_{10}}$) describes the efficiency with which the electron spin facilitates the proton nuclear spin relaxation (T_1 and T_{10} are the nuclear spin–lattice relaxation times in the presence and absence of the spin label, respectively). Figure 14.8(d) shows an example of a change in water accessibility detected via ODNP on a spin-labeled membrane protein. The ratio P/P_0 is plotted versus increasing MW power. This DNP parameter contains ''water accessibility'' information in terms of local dynamics of water-label collisions (ε) and in terms of exchange between local and bulk water molecules factor (f). To compare different states of a protein, as in the example shown in Figure 14.8(d), it is mandatory that the nitroxide concentration is constant, and only minor variation in the EPR lineshape occurs, to minimize changes in the leakage factor and in the saturation of the EPR lines. ODNP has been already successfully applied to study conformational transitions of membrane proteins, and it is becoming one of the most appealing EPR tools for measuring water accessibility.[95,96]

14.7 CONCLUDING REMARKS

Since their discovery, nitroxide probes have been widely used to investigate local physicochemical properties of a vast range of materials. New labels such as metal-based or carbon-based probes were recently added to the library of available EPR tools, mostly for dipolar studies. However, the small and stable nitroxide probes, which can provide dynamics and polarity–proticity characteristics of their microenvironment at any temperature with high sensitivity, in addition to dipolar interactions, will surely stay among the most useful EPR probes in the years to come.

ACKNOWLEDGMENTS

Thanks to Tufa Assafa for his help in the preparation of Figure 14.5; to Kathrin Anders and Lars-Oliver Essen for the spin-labeled Cph2 samples shown in Figure 14.5(d); to Johann Klare for providing the DIP-FIT software; to Markus Teucher, M. Hadi Timachi, Tufa Assafa, and Stephanie Bleicken for proof reading the chapter; and to Stefan Stoll and Daniella Goldfarb for careful editing of the chapter.

REFERENCES

1. A. K. Hoffmann and A. T. Henderson, *J. Am. Chem. Soc.*, 1961, **83**, 4671.
2. O. L. Lebedev, M. L. Khidekel, and V. A. Razuvaev, *Dokl. Acad. Nauk*, 1961, **140**, 1327.
3. A. R. Forrester and R. H. Thomson, *Nature*, 1964, **203**, 74.
4. M. Huber, *Photosynth. Res.*, 2009, **102**, 305.
5. W. K. Subczynski, J. Widomska, and J. B. Feix, *Free Radical Biol. Med.*, 2009, **46**, 707.
6. L. J. Berliner, Spin Labeling: Theory and Applications, Academic Press: New York, 1976.
7. D. Marsh, B. G. Dzikovski, and V. A. Livshits, *Biophys. J.*, 2006, **90**, L49.
8. B. Endeward, A. Marko, V. P. Denysenkov, S. T. Sigurdsson, and T. F. Prisner, in Methods Enzymol, eds Z. Q. Peter and W. Kurt, Academic Press: New York, 2015, 403.
9. S. A. Shelke and S. T. Sigurdsson, in Structural Information from Spin-Labels and Intrinsic Paramagnetic Centres in the Biosciences, eds R. C. Timmel and R. J. Harmer, Springer Berlin Heidelberg: Berlin, Heidelberg, 2013, 121.
10. T. J. Stone, T. Buckman, P. L. Nordio, and H. M. McConnell, *Proc. Natl. Acad. Sci. U. S. A.*, 1965, **54**, 1010.
11. C. Altenbach, S. L. Flitsch, H. G. Khorana, and W. L. Hubbell, *Biochemistry*, 1989, **28**, 7806.
12. C. Altenbach, T. Marti, H. G. Khorana, and W. L. Hubbell, *Science*, 1990, **248**, 1088.
13. W. L. Hubbell and C. Altenbach, *Curr. Opin. Struct. Biol.*, 1994, **4**, 566.
14. A. P. Todd, J. P. Cong, F. Levinthal, C. Levinthal, and W. L. Hubbell, *Protein Struct. Funct. Genet.*, 1989, **6**, 294.
15. D. Bardelang, M. Hardy, O. Ouari, and P. Tordo, Encyclopedia of Radicals in Chemistry, Biology and Materials, John Wiley & Sons, Ltd: Chichester, 2012, 1.
16. Y. Polyhach, E. Bordignon, and G. Jeschke, *Phys. Chem. Chem. Phys.*, 2011, **13**, 2356.
17. G. Hagelueken, R. Ward, J. H. Naismith, and O. Schiemann, *Appl. Magn. Reson.*, 2012, **42**, 377.
18. F. Hyodo, K.-i. Matsumoto, A. Matsumoto, J. B. Mitchell, and M. C. Krishna, *Cancer Res.*, 2006, **66**, 9921.
19. S. Belkin, R. J. Mehlhorn, K. Hideg, O. Hankovsky, and L. Packer, *Arch. Biochem. Biophys.*, 1987, **256**, 232.
20. J. T. Paletta, M. Pink, B. Foley, S. Rajca, and A. Rajca, *Org. Lett.*, 2012, **14**, 5322.
21. A. P. Jagtap, I. Krstic, N. C. Kunjir, R. Hansel, T. F. Prisner, and S. T. Sigurdsson, *Free Radic. Res.*, 2015, **49**, 78.
22. R. Dastvan, B. E. Bode, M. P. R. Karuppiah, A. Marko, S. Lyubenova, H. Schwalbe, and T. F. Prisner, *J. Phys. Chem. B*, 2010, **114**, 13507.
23. K. Möbius, A. Savitsky, C. Wegener, M. Plato, M. Fuchs, A. Schnegg, A. A. Dubinskii, Y. A. Grishin, I. A. Grigor'ev, M. Kühn, D. Duché, H. Zimmermann, and H. J. Steinhoff, *Magn. Reson. Chem.*, 2005, **43**, S4.
24. A. I. Smirnov, A. Ruuge, V. A. Reznikov, M. A. Voinov, and I. A. Grigor'ev, *J. Am. Chem. Soc.*, 2004, **126**, 8872.

25. M. R. Fleissner, M. D. Bridges, E. K. Brooks, D. Cascio, T. Kalai, K. Hideg, and W. L. Hubbell, *Proc. Natl. Acad. Sci. U. S. A.*, 2011, **108**, 16241.

26. V. Meyer, M. A. Swanson, L. J. Clouston, P. J. Boratynski, R. A. Stein, H. S. Mchaourab, A. Rajca, S. S. Eaton, and G. R. Eaton, *Biophys. J.*, 2015, **108**, 1213.

27. M. R. Fleissner, E. M. Brustad, T. Kalai, C. Altenbach, D. Cascio, F. B. Peters, K. Hideg, S. Peuker, P. G. Schultz, and W. L. Hubbell, *Proc. Natl. Acad. Sci. U. S. A.*, 2009, **106**, 21637.

28. E. G. Evans and G. L. Millhauser, *Methods Enzymol.*, 2015, **563**, 503.

29. J. L. Flippen-Anderson, C. George, G. Valle, E. Valente, A. Bianco, F. Formaggio, M. Crisma, and C. Toniolo, *Int. J. Pept. Protein Res.*, 1996, **47**, 231.

30. V. W. Cornish, D. R. Benson, C. A. Altenbach, K. Hideg, W. L. Hubbell, and P. G. Schultz, *Proc. Natl. Acad. Sci. U. S. A.*, 1994, **91**, 2910.

31. M. J. Schmidt, J. Borbas, M. Drescher, and D. Summerer, *J. Am. Chem. Soc.*, 2014, **136**, 1238.

32. A. Martorana, G. Bellapadrona, A. Feintuch, E. Di Gregorio, S. Aime, and D. Goldfarb, *J. Am. Chem. Soc.*, 2014, **136**, 13458.

33. F. X. Theillet, A. Binolfi, B. Bekei, A. Martorana, H. M. Rose, M. Stuiver, S. Verzini, D. Lorenz, M. van Rossum, D. Goldfarb, and P. Selenko, *Nature*, 2016, **530**, 45.

34. S. Dunkel, L. P. Pulagam, H. J. Steinhoff, and J. P. Klare, *Phys. Chem. Chem. Phys.*, 2015, **17**, 4875.

35. I. Krstic, R. Hansel, O. Romainczyk, J. W. Engels, V. Dotsch, and T. F. Prisner, *Angew. Chem. Int. Ed. Engl.*, 2011, **50**, 5070.

36. S. Bleicken, G. Jeschke, C. Stegmueller, R. Salvador-Gallego, A. J. García-Sáez, and E. Bordignon, *Mol. Cell*, 2014, **56**, 496.

37. M. Robotta, H. R. Gerding, A. Vogel, K. Hauser, S. Schildknecht, C. Karreman, M. Leist, V. Subramaniam, and M. Drescher, *ChemBioChem*, 2014, **15**, 2499.

38. C. Baldauf, K. Schulze, P. Lueders, E. Bordignon, and R. Tampé, *Chem. A Eur. J.*, 2013, **19**, 13714.

39. B. Joseph, A. Sikora, E. Bordignon, G. Jeschke, D. S. Cafiso, and T. F. Prisner, *Angew. Chem. Int. Ed. Engl.*, 2015, **54**, 6196.

40. B. Joseph, A. Sikora, and D. S. Cafiso, *J. Am. Chem. Soc.*, 2016, **138**, 1844.

41. E. Bordignon, in Encyclopedia of Biophysics, ed. G. C. K. Roberts, Springer: Berlin Heidelberg, 2013, 2439.

42. P. Gast, R. T. L. Herbonnet, J. Klare, A. Nalepa, C. Rickert, D. Stellinga, L. Urban, K. Mobius, A. Savitsky, H. J. Steinhoff, and E. J. J. Groenen, *Phys. Chem. Chem. Phys.*, 2014, **16**, 15910.

43. E. Bordignon, H. Brutlach, L. Urban, K. Hideg, A. Savitsky, A. Schnegg, P. Gast, M. Engelhard, E. J. J. Groenen, K. Möbius, and H.-J. Steinhoff, *Appl. Magn. Reson.*, 2010, **37**, 391.

44. M. Plato, H. J. Steinhoff, C. Wegener, J. T. Törring, A. Savitsky, and K. Möbius, *Mol. Phys.*, 2002, **100**, 3711.

45. D. Marsh and C. Toniolo, *J. Magn. Reson.*, 2008, **190**, 211.

46. A. Savitsky, A. A. Dubinskii, M. Plato, Y. A. Grishin, H. Zimmermann, and K. Möbius, *J. Phys. Chem. B*, 2008, **112**, 9079.

47. M. Florent, I. Kaminker, V. Nagarajan, and D. Goldfarb, *J. Magn. Reson.*, 2011, **210**, 192.

48. A. Nalepa, K. Möbius, W. Lubitz, and A. Savitsky, *J. Magn. Reson.*, 2014, **242**, 203.

49. C. Riplinger, J. P. Kao, G. M. Rosen, V. Kathirvelu, G. R. Eaton, S. S. Eaton, A. Kutateladze, and F. Neese, *J. Am. Chem. Soc.*, 2009, **131**, 10092.

50. G. R. Luckhurst, in Spin Labeling Theory and Applications, ed. L. J. Berliner, Academic Press: New York, 1976, 133.

51. C. Song, K. N. Hu, C. G. Joo, T. M. Swager, and R. G. Griffin, *J. Am. Chem. Soc.*, 2006, **128**, 11385.

52. G. Jeschke, *Annu. Rev. Phys. Chem.*, 2012, **63**, 419.

53. A. Abragam, Principles of Nuclear Magnetism, Oxford University Press: Oxford, 1961.

54. H. J. Steinhoff, N. Radzwill, W. Thevis, V. Lenz, D. Brandenburg, A. Antson, G. Dodson, and A. Wollmer, *Biophys. J.*, 1997, **73**, 3287.

55. J. Z. Sun, J. Voss, W. L. Hubbell, and H. R. Kaback, *Biochemistry*, 1999, **38**, 3100.

56. G. Jeschke, *J. Chem. Theory Comput.*, 2012, **8**, 3854.

57. R. E. Martin, M. Pannier, F. Diederich, V. Gramlich, M. Hubrich, and H. W. Spiess, *Angew. Chem. Int. Ed. Engl.*, 1998, **37**, 2833.

58. M. Pannier, S. Veit, A. Godt, G. Jeschke, and H. W. Spiess, *J. Magn. Reson.*, 2000, **142**, 331.

59. B. Joseph, G. Jeschke, B. A. Goetz, K. P. Locher, and E. Bordignon, *J. Biol. Chem.*, 2011, **286**, 41008.

60. J. E. Banham, C. M. Baker, S. Ceola, I. J. Day, G. H. Grant, E. J. J. Groenen, C. T. Rodgers, G. Jeschke, and C. R. Timmel, *J. Magn. Reson.*, 2008, **191**, 202.

61. D. Kivelson, *J. Chem. Phys.*, 1960, **33**, 1094.
62. S. Stoll, in Methods Enzymol, eds Z. Qin Peter and W. Kurt, Academic Press: New York, 2015, 121.
63. A. Polimeno and J. H. Freed, *J. Phys. Chem.*, 1995, **99**, 10995.
64. S. Stoll and A. Schweiger, *J. Magn. Reson.*, 2006, **178**, 42.
65. J. P. Barnes, Z. C. Liang, H. S. Mchaourab, J. H. Freed, and W. L. Hubbell, *Biophys. J.*, 1999, **76**, 3298.
66. S. K. Misra and J. H. Freed, Multifrequency Electron Paramagnetic Resonance, Wiley-VCH Verlag GmbH & Co. KGaA: Weinheim, 2011, 497.
67. J. H. Freed, in Spin Labeling: Theory and Applications, ed. L. J. Berliner, Academic Press: New York, 1976, 53.
68. D. E. Budil, in Methods Enzymol, eds Z. Qin Peter and W. Kurt, Academic Press: New York, 2015, 143.
69. C. Beier and H.-J. Steinhoff, *Biophys. J.*, 2006, **91**, 2647.
70. D. E. Budil, K. L. Sale, K. A. Khairy, and P. G. Fajer, *J. Phys. Chem. A*, 2006, **110**, 3703.
71. V. S. Oganesyan, *Phys. Chem. Chem. Phys.*, 2011, **13**, 4724.
72. D. Marsh, *J. Magn. Reson.*, 2014, **248**, 66.
73. E. Bordignon, in EPR Spectroscopy, eds M. Drescher and G. Jeschke, Springer Berlin Heidelberg: Heidelberg, 2012, 121.
74. M. Herget, C. Baldauf, C. Scholz, D. Parcej, K. H. Wiesmuller, R. Tampe, R. Abele, and E. Bordignon, *Proc. Natl. Acad. Sci. U. S. A.*, 2011, **108**, 1349.
75. R. Owenius, M. Engstrom, M. Lindgren, and M. Huber, *J. Phys. Chem. A*, 2001, **105**, 10967.
76. E. Bordignon and Y. Polyhach, in Lipid-Protein Interactions, ed. J. H. Kleinschmidt, Humana Press: New York, 2013, 329.
77. E. Bordignon, A. I. Nalepa, A. Savitsky, L. Braun, and G. Jeschke, *J. Phys. Chem. B*, 2015, **119**, 13797.
78. C. Altenbach, W. Froncisz, and W. L. Hubbell, *Biophys. J.*, 2002, **82**, 479a.
79. C. Altenbach, D. A. Greenhalgh, H. G. Khorana, and W. L. Hubbell, *Proc. Natl. Acad. Sci. U. S. A.*, 1994, **91**, 1667.
80. C. Altenbach, W. Froncisz, R. Hemker, H. Mchaourab, and W. L. Hubbell, *Biophys. J.*, 2005, **89**, 2103.
81. C. Altenbach, C. J. Lopez, K. Hideg, and W. L. Hubbell, *Methods Enzymol.*, 2015, **564**, 59.
82. A. Doll, E. Bordignon, B. Joseph, R. Tschaggelar, and G. Jeschke, *J. Magn. Reson.*, 2012, **222**, 34.
83. E. Bordignon and H. J. Steinhoff, in ESR Spectroscopy in Membrane Biophysics, eds M. A. Hemminga and L. J. Berliner, Springer: New York, 2007, 129.
84. Y. Miwa, *Macromolecules*, 2009, **42**, 6141.
85. Y. Miwa and K. Yamamoto, *J. Phys. Chem. B*, 2012, **116**, 9277.
86. J. Pyka, J. Ilnicki, C. Altenbach, W. L. Hubbell, and W. Froncisz, *Biophys. J.*, 2005, **89**, 2059.
87. Z. Yang, M. Bridges, M. T. Lerch, C. Altenbach, and W. L. Hubbell, *Methods Enzymol.*, 2015, **564**, 3.
88. L. Liu, D. J. Mayo, I. D. Sahu, A. Zhou, R. Zhang, R. M. McCarrick, and G. A. Lorigan, *Methods Enzymol.*, 2015, **564**, 289.
89. A. Volkov, C. Dockter, T. Bund, H. Paulsen, and G. Jeschke, *Biophys. J.*, 2009, **96**, 1124.
90. D. A. Erilov, R. Bartucci, R. Guzzi, A. A. Shubin, A. G. Maryasov, D. Marsh, S. A. Dzuba, and L. Sportelli, *J. Phys. Chem. B*, 2005, **109**, 12003.
91. R. Bartucci, R. Guzzi, D. Marsh, and L. Sportelli, *Biophys. J.*, 2003, **84**, 1025.
92. R. Carmieli, N. Papo, H. Zimmermann, A. Potapov, Y. Shai, and D. Goldfarb, *Biophys. J.*, 2006, **90**, 492.
93. I. Kaminker, R. Barnes, and S. Han, *Methods Enzymol.*, 2015, **564**, 457.
94. B. D. Armstrong and S. Han, *J. Am. Chem. Soc.*, 2009, **131**, 4641.
95. C. Y. Cheng, J. Varkey, M. R. Ambroso, R. Langen, and S. Han, *Proc. Natl. Acad. Sci. U. S. A.*, 2013, **110**, 16838.
96. T. F. Segawa, M. Doppelbauer, L. Garbuio, A. Doll, Y. O. Polyhach, and G. Jeschke, *J. Chem. Phys.*, 2016, **144**, 194201.

PART C
High-Resolution Pulse Techniques

Chapter 15
FT-EPR

Michael K. Bowman[1], Hanjiao Chen[1], and Alexander G. Maryasov[2]

[1]*The University of Alabama, Tuscaloosa, AL, USA*
[2]*V. V. Voevodsky Institute of Chemical Kinetics and Combustion, Novosibirsk, Russia*

15.1 INTRODUCTION

The word 'resonance' in EPR or ESR spectroscopy implies some correspondence between features in a spectrum and energy differences among states in free radicals or paramagnetic centers (generically referred to as PCs). This correspondence is quite clear in CW spectroscopy, which is often described in terms of sweeping the magnetic field in order to match the energy difference between two states to the energy of the MW (microwave) quantum. When a match occurs, energy from the MW field can be transferred to the PCs, and the EPR signal appears as an absorption or phase shift of the MW field.

In pulse EPR spectroscopy, the signal is derived from MW radiation emitted by the sample (see Chapter 11). Usually, the spectrometer's receiver is designed to detect signals in a certain bandwidth around its operating frequency, ω_{MW}. The signal contains MW frequencies that correspond to energy differences in the system, and the signal can be analyzed, e.g., by FT in FT-EPR spectroscopy, to reveal those frequencies. Signal components outside the receiver bandwidth are missed. Often, frequency analysis in pulse EPR is implicit, e.g., in discussions of contributions from specific PCs or from a limited range of orientations (orientation selection). This chapter focuses on recovering an 'EPR spectrum' from the frequency content of pulse EPR data, and we will call such a spectrum the FT-EPR spectrum. The pulse EPR data may include information about, e.g., frequencies of nuclear transitions, dipolar couplings, correlations between spectral features, or relaxation rates. We will consider a few cases where such data can be converted into two or multidimensional spectra where at least one axis is the EPR frequency. Analogous measurements are common in NMR where enough data to construct an NMR spectrum is captured from each shot of the pulse sequence. This data is known as the 'direct' dimension of the spectrum, while the 'indirect' dimensions are constructed from changes in the signal between different steps in the pulse sequence. Because of technical limitations, many pulse

EPR Spectroscopy: Fundamentals and Methods.
Edited by Daniella Goldfarb and Stefan Stoll
 ISBN: 978-1-119-16299-5
Also published in eMagRes (online edition)
DOI: 10.1002/9780470034590.emrstm1514

EPR experiments measure only a single data point in each shot of the experiment and have no direct dimension. However, the experiment, the data, and the recovery of spectra are the same for equivalent data collected with or without a direct dimension. A further limitation is due to ringing of the resonator and recovery of circuits from overload after each MW pulse that prevents measurements during the so-called dead time t_d.

The EPR spectrum is really a 2D function of frequency, ω, and magnetic field, $\boldsymbol{B}$. In most cases, a measured EPR spectrum is simply a slice through this 2D spectrum with either ω or $\boldsymbol{B}$ constant. The typical CW-EPR spectrum is a 1D slice in which $\boldsymbol{B}$ varies while ω remains constant. The prototypical FT-EPR spectrum has one dimension in which ω varies while $\boldsymbol{B}$ remains constant. In the simplest CW- and FT-EPR spectroscopies, the EPR spectrum is a slice from the much larger 2D spectrum.

Pulse EPR experiments have the potential, in multiple-pulse experiments, to go beyond simply mapping out the energy levels in a sample. The time delays between MW pulses or an external pulse, e.g., laser or RF pulses, or even the MW-pulse durations can be independently varied to produce additional dimensions to the data and generate multidimensional spectra. The other dimensions can reveal interactions with other paramagnetic centers or nuclei, or dynamics and relaxation times.

The best known 2D pulse EPR experiment is hyperfine sublevel correlation spectroscopy (HYSCORE) (see Chapter 18) in which both dimensions correspond to electron nuclear double resonance (ENDOR) frequencies while information about EPR frequencies is usually characterized as orientation selection, referring to the orientations of the PCs that contribute to the spectrum. Conventional HYSCORE produces spectra with nuclear, but not EPR, frequency axes and would not be FT-EPR. However, spectra with high-resolution along an EPR axis are obtained in several classes of 2D experiments. Time-resolved FT-EPR spectra are used to study chemical kinetics or chemically induced dynamic electron polarization (CIDEP) following a laser pulse or other excitation.[1,2] Analogs of several NMR correlation spectroscopies, e.g., COSY, SECSY, and FOCSY, have been implemented, yielding high-resolution, 2D FT-EPR spectral information and relaxation or chemical exchange information.[3,4]

Full analysis of pulse EPR signals reveals considerable information about PCs: their high-resolution EPR spectrum and spin Hamiltonian parameters; the populations and spin polarizations of each energy level; the relaxation rates between each pair of levels; the chemical reactions of the PCs; and the interactions and distances between PCs. However, many measurements, such as relaxation rates or ENDOR, give the same at all points in the EPR spectrum of a PC and only the identity of the PC is important. In such cases, it is better to trade the EPR frequency dimension for increased sensitivity. This chapter examines FT-EPR spectroscopy as an integral part of pulse EPR measurements. It will not consider the related area of pulse EPR imaging. We start with the simple FID and examine the information it contains for isotropic and anisotropic $\boldsymbol{g}$ and $\boldsymbol{A}$. This treatment is generalized to multipulse EPR measurements, including spin echo signals, with experimental examples. The electron Zeeman interaction is implicitly assumed to be the dominant term in the spin Hamiltonian so that ω and $\boldsymbol{B}$ are linearly related at resonance. We briefly mention PCs with $S > {}^1\!/_2$.

15.2 SINGLE PULSE, $S = 1/2$

The signal, known as the FID (see Chapter 11), following one short MW pulse is the basis of the prototypical FT-EPR experiment. The FID is the signal induced in a probe or resonator by the evolution of a magnetic dipole in the absence of applied fields other than $\boldsymbol{B}_0$. In EPR, the observable is the magnetic moment, $\boldsymbol{\mu}$. The MW signal that becomes the EPR spectrum is generated through the interaction of $\boldsymbol{\mu}$ with the metal and dielectric in the macroscopic resonator. The $\boldsymbol{\mu}$ in paramagnetic samples is commonly explained by a quantity known as spin, $\boldsymbol{S}$, which is not directly observable in an EPR experiment although both are related through the density matrix, σ (see Chapter 8),[5–7]

$$\boldsymbol{\mu} = \mathrm{Tr}(\sigma\, \hat{\boldsymbol{\mu}});\ \boldsymbol{S} = \mathrm{Tr}(\sigma\, \hat{\boldsymbol{S}}) \qquad (15.1)$$

The signal measured by an EPR spectrometer comes from the oscillating projection of $\boldsymbol{\mu}$ on the direction of the MW magnetic field in the MW resonator. The $\boldsymbol{\mu}$ oscillates in a plane perpendicular to $\boldsymbol{B}_0$ and is elliptically polarized (this means that $\boldsymbol{\mu}$ draws an ellipse during free precession) in the general case. With a conventional single-mode resonator, the induced signal is proportional to the $\boldsymbol{\mu}$ component in the direction of $\boldsymbol{B}_1$ (taken here as $\boldsymbol{x}$) in the laboratory system of coordinates where z is the direction of $\boldsymbol{B}_0$. Only the component of $\boldsymbol{\mu}$ along $\boldsymbol{x}$ induces a signal into the same resonator mode that produced the $\boldsymbol{B}_1$ and can be measured. For convenience, we express $\boldsymbol{\mu}$ in terms of its component vectors along the laboratory axes as $\boldsymbol{\mu} = \boldsymbol{\mu}_x + \boldsymbol{\mu}_y + \boldsymbol{\mu}_z = (\boldsymbol{x} \cdot \boldsymbol{\mu})\boldsymbol{x} + (\boldsymbol{y} \cdot \boldsymbol{\mu})\boldsymbol{y} + (\boldsymbol{z} \cdot \boldsymbol{\mu})\boldsymbol{z}$. Because $\boldsymbol{\mu}$ and its evolution are simply the sum of these components and their evolutions, it can be very informative to consider that portion of $\boldsymbol{\mu}$ which starts, for example, along the $\boldsymbol{x}$-direction as $\boldsymbol{\mu}_x$ and evolves or nutates to a different direction and possibly different amplitude. Use of the component vectors rather than their scalar amplitudes also avoids the awkwardness of discussing rotations of a scalar quantity.

The component of $\boldsymbol{\mu}$ along $\boldsymbol{y}$, $\boldsymbol{\mu}_y$, induces no signal directly but does evolve into $\boldsymbol{\mu}_x$ and then contributes to the signal. However, with a crossed bimodal (induction) resonator, both $\boldsymbol{\mu}_x$ and $\boldsymbol{\mu}_y$ can be measured.[8,9]

The spin Hamiltonian of the system during a measurement usually has the form

$$\hat{H} = \hat{H}_0 + \hat{H}_1(t) + \hat{H}_2(t) \tag{15.2}$$

where $\hat{H}_0$, $\hat{H}_1(t)$, and $\hat{H}_2(t)$ represent the static (e.g., Zeeman and hyperfine) interactions, the MW pulses, and the other time-dependent interactions, respectively. $\hat{H}_2(t)$ causes relaxation and will be assumed to be weak and rapid, i.e., in the Redfield limit,[5,10] and $\hat{H}_1$ will be treated as a δ-function or 'hard' pulse with a width t_p that causes $\boldsymbol{\mu}$ to nutate through a turning angle, $\theta = \omega_1 t_p$, with ω_1 being the MW field strength in the angular frequency units. In such a pulse EPR measurement, the $\boldsymbol{\mu}$ of an ensemble of PCs evolves in a combination of nutations, precessions, and relaxations (with spin–lattice, T_{1e}, and spin–spin or phase, T_{2e}, relaxation times). Each pair of the energy levels of a PC potentially has an associated $\boldsymbol{\mu}$, but only a few contribute to the FID or the CW-EPR spectrum. The $\boldsymbol{\mu}$ vanishes if that transition is forbidden, e.g., by selection rules; or, if the energy difference lies outside the operating band of the EPR spectrometer, the $\boldsymbol{\mu}$ does not contribute to the signal measured by the spectrometer.

The time-domain FID is related to the frequency-domain spectrum (as in NMR) by FT, although other spectral analysis techniques can be used to recover the spectrum, see Section 15.5.2. If there are multiple PC orientations or more than one type of PC, their individual spectra or individual signals are additive provided their mutual interactions are negligible. The FT-EPR spectrum has a characteristic difference from the conventional CW-EPR spectrum. An increase in $\boldsymbol{B}$ (or ω) shifts both spectra to higher ω (or $\boldsymbol{B}$), but transitions at low $\boldsymbol{B}$ in the CW-EPR spectrum appear at high ω in the FT-EPR spectrum and vice versa (see Chapter 11).

15.2.1 Isotropic PC

An important class of FT-EPR measurements involves noninteracting $S = 1/2$ PCs where $\hat{H}_0$ is diagonal for an electron spin quantization axis parallel to $\boldsymbol{z}$. Examples include free radicals in solution, PCs in solids with isotropic $\boldsymbol{g}$ and $\boldsymbol{A}$ or with $\boldsymbol{B}_0$ parallel to a principal axis of both $\boldsymbol{g}$ and $\boldsymbol{A}$. We refer to this limiting class as 'isotropic' and write $\hat{H}_0$ as $-\boldsymbol{B}_z\hat{\boldsymbol{\mu}} = \mu_B \boldsymbol{B}_z g_{iso}\hat{\boldsymbol{S}}$ for brevity, with the hyperfine interaction (HFI) added to the applied field $\boldsymbol{B}_0$ to give a total effective field $\boldsymbol{B}_z$. When dealing only with isotropic PCs, it is common to consider $\hat{\boldsymbol{S}}$ while $\hat{\boldsymbol{\mu}}$ is more natural for anisotropic PCs. We will use the observable in EPR measurements, $\hat{\boldsymbol{\mu}}$, even for isotropic PCs in order to emphasize the relation between FT-EPR of isotropic and anisotropic PCs. Any of the equations for isotropic PCs can readily be written in terms of $\boldsymbol{S}$ by replacing $\boldsymbol{\mu}$ by $-\mu_B g_{iso}\boldsymbol{S}$.

This isotropic PC limit has strong EPR selection rules: $\Delta m_S = \pm 1$ and $\Delta m_I = 0$ with I and m_I the nuclear spin counterparts of S and m_S. Each PC energy level is involved in only one EPR transition. The FID is simply the sum of these few transitions. In this limit, the evolution of $\boldsymbol{\mu}$ for each transition is described by the well-known Bloch equation (see Chapter 8)

$$\frac{\partial}{\partial t}\boldsymbol{\mu}(t) = \frac{\mu_B g_{iso}}{\hbar}\boldsymbol{\mu}(t) \times (\boldsymbol{B}_z + \boldsymbol{B}_1) - \frac{1}{T_{2e}}(\boldsymbol{\mu}_x + \boldsymbol{\mu}_y) - \frac{1}{T_{1e}}(\boldsymbol{\mu}_z - \boldsymbol{\mu}_{eq}) \tag{15.3}$$

where $\boldsymbol{\mu}_{eq}$ is the equilibrium value. The FID signal is the sum of contributions from the $\boldsymbol{\mu}$ of all allowed EPR transitions. If there is no significant interaction between PCs, the evolution of an ensemble of PCs is just the sum of the independent evolutions of each individual PC. The FID following a single MW pulse taking time $t = 0$ as the end of that pulse is

$$V_{FID}(t) \propto \mu_x(t) = \sum_j (\mu_{x,j}(0)\cos(\omega_j t) - \text{sign}(g_{iso})\,\mu_{y,j}(0)\sin(\omega_j t))\,e^{-t/T_{2e,j}} \tag{15.4}$$

in the limit of $T_{1e} \gg T_{2e}$, where j indexes the allowed EPR transitions and ω_j is the frequency of an EPR transition. For isotropic PCs, $\boldsymbol{\mu}_{eq,x} = \boldsymbol{\mu}_{eq,y} = 0$ and there is no FID unless some of $\boldsymbol{\mu}_{eq}$ is converted into $\boldsymbol{\mu}_x$ and/or

$\boldsymbol{\mu}_y$ by a MW pulse. The sign of $g_{\rm iso}$ in equation (15.4) determines the direction of precession of $\boldsymbol{\mu}$ around $\boldsymbol{B}_0$. The experimentally measured FID arises only from $\boldsymbol{\mu}_x$ but the FID at $t > 0$ also contains information about the $\boldsymbol{\mu}_y(0)$ that has evolved into $\boldsymbol{\mu}_x$, which always occurs for isotropic PCs except in the rare case that $\omega_j = 0$. For isotropic PCs, the inability to measure $\boldsymbol{\mu}_y(t)$ results only in an inability to determine whether $g_{\rm iso}$ is positive or negative.

The FID, $\mu_x(t)$, and $\boldsymbol{\mu}$ are real-valued quantities, but the FID of each transition, j, has a phase reflecting the phase of the oscillating $\mu_{x,j}$ relative to $\omega_{\rm MW}$. The FID is typically mixed in a quadrature mixer with $\omega_{\rm MW}$, generating a signal often represented as a complex value indicating the amplitude of $\mu_x(t)$ and its phase relative to $\omega_{\rm MW}$.[9] This complex-valued signal contains frequencies at $|\omega_{\rm MW} - \omega_j|$ that fall within the bandwidth of the spectrometer. For isotropic PCs, this signal is equivalent to the 'rotating-frame' signal in an $(\boldsymbol{x}', \boldsymbol{y}', z)$ coordinate system rotating at $\omega_{\rm MW}$ in the direction that $\boldsymbol{\mu}$ precesses around $\boldsymbol{B}_0$ (see Chapter 8).[9] By adjusting the phase of the complex-valued signal, it is possible to make the real and imaginary parts of that signal coincide with the components of $\boldsymbol{\mu}$ (or $\boldsymbol{S}$) along the $\boldsymbol{x}'$ and $\boldsymbol{y}'$ axes of the rotating frame and show the subsequent evolution of $\boldsymbol{\mu}_x$ and $\boldsymbol{\mu}_y$. Sadly, the induced signals from anisotropic PCs correspond to $\boldsymbol{\mu}_x$, but not to $\boldsymbol{\mu}_y$, nor to the $\boldsymbol{\mu}_{x'}$ and $\boldsymbol{\mu}_{y'}$, as discussed in Section 15.2.3.1.

15.2.1.1 The FID

The simplicity of the FID and FT-EPR spectrum in this isotropic limit results from the structure of σ and $\widehat{H}_0$. Both can be rearranged into block diagonal form with only 2×2 blocks. This characteristic produces the strong selection rules, makes all transitions independent of each other, and gives the Bloch equations their simple form. The complex-valued signal produced in the EPR spectrometer by mixing the laboratory frame signal with $\omega_{\rm MW}$ can be expressed in a variety of ways. A form that emphasizes the shifts and splittings caused by HFI and $g_{\rm iso}$ is

$$V_{\rm FID}(t) \propto \frac{\hbar\omega_0}{2k_{\rm B}T}\sin(\theta)\,{\rm e}^{+{\rm i}\phi}\sum_j {\rm e}^{{\rm i}\,(\Delta\omega+\omega_{\alpha,j}-\omega_{\beta,j})t}\,{\rm e}^{-t/T_{2{\rm e},j}} \tag{15.5}$$

where $\Delta\omega = \omega_0 - \omega_{\rm MW}$ is the offset, positive or negative, of the spectrometer frequency from the electron Zeeman frequency ω_0, and ϕ is the phase of the MW pulse relative to $\omega_{\rm MW}$. The center frequency of the EPR spectrum can also be used to define $\Delta\omega$. In any case, $(\Delta\omega + \omega_{\alpha,j} - \omega_{\beta,j})$ are the frequencies in the rotating frame. The $\hbar\omega_{\alpha,j}$ (or $\hbar\omega_{\beta,j}$) is the difference between the energy of the level with α (or β) electron spin that participates in the jth EPR transition and its nominal Zeeman energy. The MW pulse with a phase shift ϕ relative to $\omega_{\rm MW}$ rotates $\boldsymbol{\mu}_{\rm eq}$ through a turning angle θ about $\boldsymbol{B}_1$. Phase shifting the MW pulses helps to isolate signals as mentioned later.

Formation of the FID for PCs with isotropic $\boldsymbol{g}$ is illustrated in Figure 15.1. Here, an MW pulse with $\theta = \pi/4$ and $\phi = 0$ rotates $\boldsymbol{\mu}_{\rm eq}$ of a set of PCs (all having identical EPR properties) from the magnetic field direction $\boldsymbol{B}_0$, toward $\boldsymbol{y}$. When $T_{1\rm e}$ and $T_{2\rm e}$ relaxation is negligible (Figure 15.1a), the precessing $\boldsymbol{\mu}$ vector traces a circle, shown in light blue. Its projection onto the $\boldsymbol{x},\boldsymbol{y}$-plane at the bottom also is a circle, while the projection onto the $\boldsymbol{x},z$-plane at the right is the magenta line oscillating at ω_j. $T_{2\rm e}$ relaxation causes $\boldsymbol{\mu}_x$ and $\boldsymbol{\mu}_y$ to decay slowly (Figure 15.1b), showing that $\boldsymbol{\mu}_x(0) = 0$ but that $\boldsymbol{\mu}_y(0)$ quickly evolves into $\boldsymbol{\mu}_x$. In these calculations, relaxation rates (relative to ω_j) are much faster than in typical pulse EPR experiments in order to keep the figure legible. $T_{2\rm e}$ causes the tip of $\boldsymbol{\mu}$ to follow a spiral perpendicular to $\boldsymbol{B}_0$, instead of precessing in a circle around $\boldsymbol{B}_0$. The projection of $\boldsymbol{\mu}$ onto the $\boldsymbol{x},z$-plane at the right shows μ_x oscillating at ω_j. These FID oscillations gradually decrease with a time constant $T_{2\rm e}$.

$T_{1\rm e}$ relaxation causes μ_z to recover to its original value, Figure 15.1(c). Instead of precessing in a plane, the tip of $\boldsymbol{\mu}$ spirals upward toward $\boldsymbol{\mu}_{\rm eq}$. Notably, the projection of $\boldsymbol{\mu}$ onto the $\boldsymbol{x},\boldsymbol{y}$-plane at the bottom is unaffected by $T_{1\rm e}$. The projection of $\boldsymbol{\mu}$ onto the $\boldsymbol{x},z$-plane at the right still shows that μ_x oscillates at ω_j, and the FID is unchanged. The precession is no longer confined in a plane; this projection shows μ_z increasing in the z-direction with a time constant $T_{1\rm e}$.

The $\boldsymbol{\mu}$ vector traces a circle in Figure 15.1(a) with coaxial $\boldsymbol{\mu}_{\rm eq}$ and $\boldsymbol{B}_0$ because $\boldsymbol{g}$ is isotropic. This makes it productive to go to the rotating frame, where the axes in Figure 15.1 rotate around z at the MW frequency $\omega_{\rm MW}$. This has the effect of slowing down the apparent precession frequency of $\boldsymbol{\mu}$ to $\Delta\omega$ in that frame without introducing any new time-dependent motion. It leaves the effects of $T_{1\rm e}$ and $T_{2\rm e}$ relaxation intact. The rotating frame makes $\boldsymbol{\mu}_x$ (the FID observable) appear as $\boldsymbol{\mu}_{x'}$ and $\boldsymbol{\mu}_{y'}$ in effect, measuring $\boldsymbol{\mu}_x$ and $\boldsymbol{\mu}_y$ as they are rapidly interconverting for isotropic PCs [equations (15.3) and (15.4)].

The FID that is measured by a pulse EPR spectrometer consists of a series of points, usually equally spaced in time representing the real and imaginary or in-phase

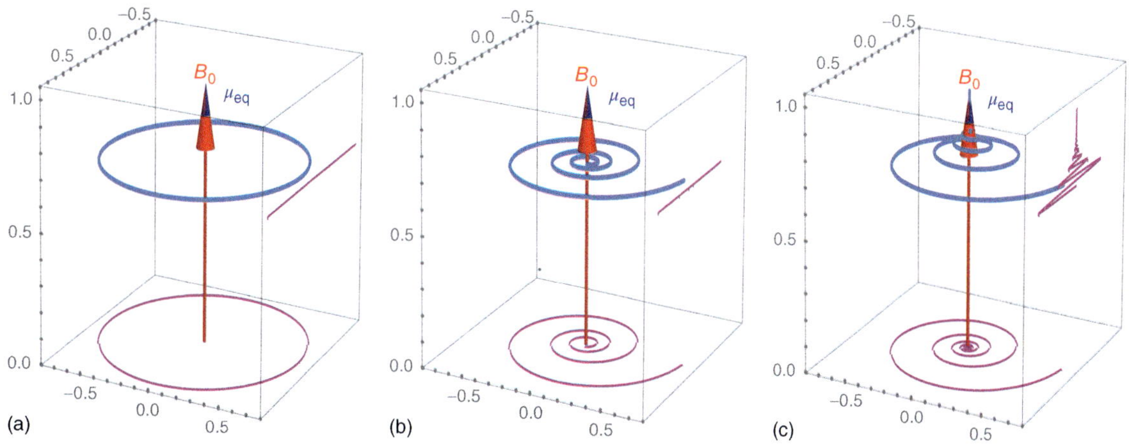

Figure 15.1. The calculated path of precession of $\boldsymbol{\mu}$ for $S = ½$ PCs with isotropic $g_{iso} = 2.00$ after $\boldsymbol{\mu}_{eq}$ is rotated by MW pulse with $\theta = \pi/4$ and $\phi = 0$. The light blue curve shows the path of the tip of $\boldsymbol{\mu}$ as it precesses around $\boldsymbol{\mu}_{eq}$. The precession is projected, as magenta curves, onto the $\boldsymbol{x},\boldsymbol{y}$-plane at the bottom and onto the $\boldsymbol{x},\boldsymbol{z}$-plane on the right-hand side. The precession paths and their projections are shown for (a) $1/T_{1e} = 1/T_{2e} = 0$; (b) $1/T_{1e} = 0$ and $1/T_{2e} \neq 0$; (c) $T_{1e}/T_{2e} = 5$. See discussion in Section 15.2.1. The laboratory $\boldsymbol{y}$-axis runs from left to right and is numbered along the bottom of these plots, and the laboratory z-axis is the vertical axis

and out-of-phase outputs from the detected signal. Plots of the FID usually connect the points, giving the impression of a continuous line. Figure 15.2 shows the calculated FID for a PC with isotropic interactions. HFI with spin ½ nucleus produces two allowed EPR transitions centered at −17.5 and −12.5 MHz from ω_{MW}; their FIDs are shown in Figure 15.2(a) and (b), respectively. Each FID consists of real and imaginary components indicated by red and blue, respectively. Each FID consists of oscillations decaying with a time constant $T_{2e} = 0.4$ μs, at the difference $\Delta\omega$ between its EPR frequency and the spectrometer frequency. The complete FID of the PC, Figure 15.2(c), is simply the sum of the FIDs from all transitions. The entire FID has beats from the interference between the frequencies of the two FIDs.

15.2.1.2 Interpreting the FID

The FID consists of a sum of damped oscillations from each allowed EPR transition, j, at its own resonant frequency in the rotating frame $(\Delta\omega + \omega_{\alpha,j} - \omega_{\beta,j})$, decaying with its own $T_{2e,j}$. The amplitude and phase of each oscillation depend on θ and ϕ, respectively. If the MW pulse is sufficiently short that negligible relaxation of $\boldsymbol{\mu}$ occurs during the pulse, then the FID provides an accurate determination of the intensity, T_{2e}, and EPR frequency of each allowed EPR transition. Alternatively, the FID can be transformed, e.g., by a FT, into a frequency-domain EPR spectrum. Each frequency component becomes a spectral line centered at that frequency with a width $1/T_{2e}$ and an integrated intensity proportional to μ_z at the instant of the MW pulse. The frequency-domain spectra are usually much easier to interpret in studies of rapid kinetics or CIDEP than are the CW-EPR transients (see Chapter 10). However, extracting T_{2e} from overlapping transitions can be difficult just as it is in CW-EPR. In one regard, the frequency-domain spectrum is simpler than the time-domain FID. The FIDs of all the transitions overlap each other, but their corresponding Lorentzian lines in the frequency domain are often well resolved when the T_{2e} relaxation is slow. However, fast relaxation or significant inhomogeneous broadening may cause significant signal loss during the spectrometer dead time, and is often the case in EPR.

Equation (15.5) describes the FID from a set of PCs with the same spin Hamiltonian. Often, there is inhomogeneous broadening caused by a distribution of PCs with slightly different spin Hamiltonian parameters or several different types of PCs in the sample. The FID from such samples is obtained simply by summing the FIDs from the entire distribution weighted appropriately for the distribution of spin

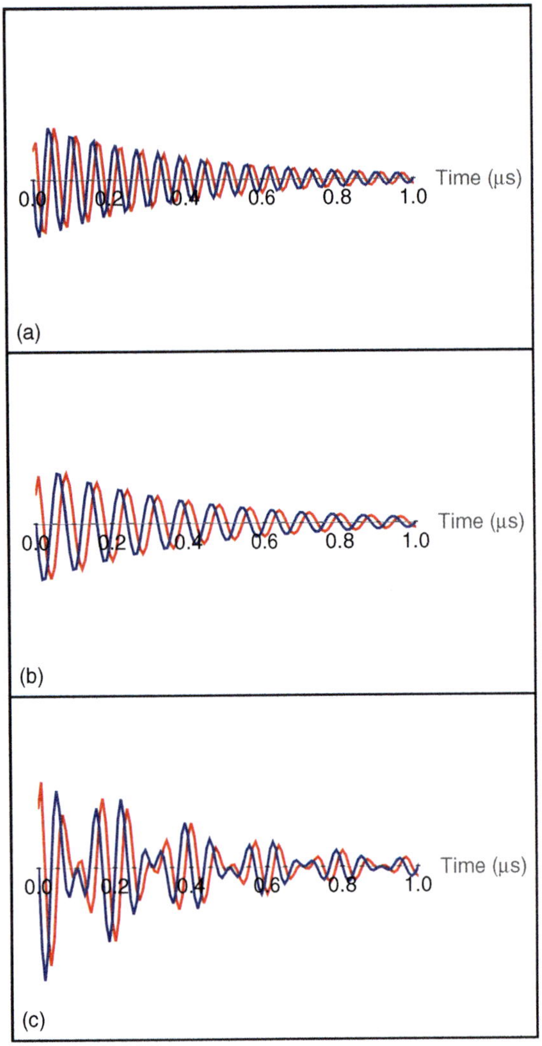

Figure 15.2. The initial portion of calculated FIDs for a PC that has isotropic g-factor and HFI with a single spin ½ nucleus with a microwave pulse phase of $\phi = 0$. The FIDs of the two EPR transitions in this example occur at (a) −17.5 MHz and (b) −12.5 MHz. Their sum is the full FID (c). The two transitions have equal intensities and identical $T_{2e} = 0.4\,\mu s$. The real and imaginary or in-phase and out-of-phase components of the FID are shown in red and blue, respectively. Each FID consists of 512 points separated by 8 ns

Hamiltonian parameters and PCs in the sample. The process is analogous to the treatment of multiple species or inhomogeneous broadening in CW-EPR. Inhomogeneous broadening typically causes the FID to disappear faster than would be expected from the T_{2e} of the transitions. The FT of the FID from noninteracting, isotropic PCs coincides with the integral of the first-derivative, frequency-domain CW-EPR spectrum if the entire spectrum is uniformly excited by the MW pulse and all the frequencies in the FID fall within the spectrometer bandwidth.

As mentioned earlier, the FID is often converted into a frequency domain spectrum by FT, which converts the FID oscillations of Figure 15.2 into the amplitudes of each frequency present. FT is a well-known tool in mathematics, signal processing and NMR. It has been described in detail from many points of view, including that of NMR[11] and EPR.[12] FT of the FIDs in Figure 15.2 produces the frequency-domain spectra in Figure 15.3. The real part of the transformations, shown in red, is the absorption-like EPR spectrum, and the imaginary part, in blue, gives a dispersion-like spectrum. The absorption-like peaks occur at the EPR difference frequencies of −17.5 MHz and −12.5 MHz. The spectrum of the entire FID of the PC (Figure 15.3c) is just the sum of the spectra from the individual EPR transitions. The spectra from these idealized FIDs are quite clear, but a number of practical and experimental difficulties can distort FT-EPR spectra, as is well known in NMR and ESEEM (electron-spin echo envelope modulation).[11,12] Some of these challenges and common responses are discussed in Section 15.5.1.

15.2.2 Anisotropic HFI

A common, but slightly more complex, situation arises when the HFI tensor (see Chapter 5), $\boldsymbol{A}$, is anisotropic and comparable to the nuclear Zeeman energy $\hbar\omega_I$. Pulse EPR experiments in this situation exhibit modulation of the spin echo intensity, known as ESEEM (see Chapter 18). This is a common occurrence in solid samples for PCs surrounded by nuclear spins. The ESEEM is accompanied by additional transitions in the CW-EPR spectrum, i.e., spin-flip satellite lines, which are seldom resolved for nuclei other than ^{1}H and ^{19}F. The new transitions result from the breakdown of the $\Delta m_I = 0$ selection rule (see Section 15.2.1). The breakdown occurs when the nuclear spin part of the Hamiltonian contains electron-spin-dependent

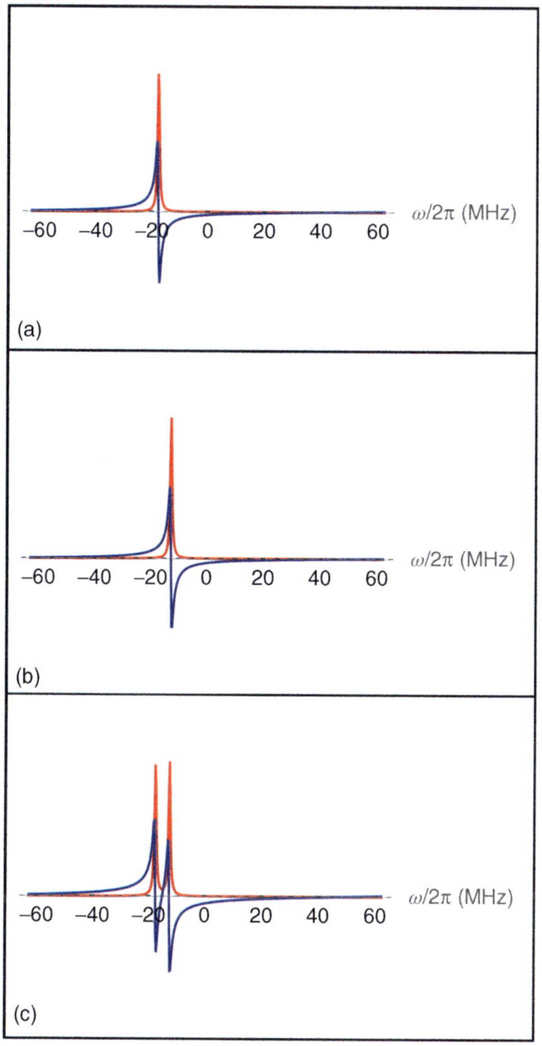

Figure 15.3. (a–c) The FT of the FIDs shown in Figure 15.2, see its caption for details

and independent terms of comparable magnitude along different spatial directions. This can occur, for example, with combinations of isotropic or anisotropic HFI with nuclear quadrupole interactions; anisotropic HFI with nuclear Zeeman interactions; or isotropic HFI with large $\boldsymbol{g}$ anisotropy.

In this event, the nuclear spin states for one electron spin state do not map perfectly onto the nuclear spin states of the other electron spin state. As a result, each energy level can have EPR transitions to several levels, resulting in a breakdown of the $\Delta m_I = 0$ selection rule that simplified the spectroscopy of isotropic PCs (see Section 15.2.1). The Mims matrix $\boldsymbol{M}$ is the overlap matrix between the nuclear spin wavefunctions of different electron spin states and governs selection rules and intensities in the EPR spectrum.[13] The magnetic moment of a PC produces a local hyperfine field at each nucleus in its vicinity. The local field strength and direction depend on the electron spin state. This means that the nuclear spin states differ for different electron spin states. There are just two states for the electron spin of 1/2, with $m_S = \pm 1/2$. Local fields from these two states have the same strength but opposite directions, i.e., they are collinear. When the HFI is anisotropic, the sums of the local field vector and the external magnetic field generally have different strengths in the two electron spin states and point in different directions. Consequently, the nuclear spin states for one electron spin state are not identical to the nuclear states of the other electron spin state. The overlap is summarized in the Mims matrix, whose elements, $M_{k,m}$, may be defined as

$$M_{k,m} = \langle \chi_k(m_S = -1/2) | \chi_m(m_S = +1/2) \rangle \quad (15.6)$$

where $|\chi_k(m_S)\rangle$ is the kth nuclear wavefunction for the m_S state of the electron spin of the PC. The Mims matrix for isotropic PCs can be written as the unit matrix with 1 along the diagonal and 0 elsewhere, corresponding to allowed and totally forbidden EPR transitions. With anisotropic PCs, each row and column of the Mims matrix can have several nonzero elements corresponding to multiple EPR transitions involving the same energy level and causing a breakdown of the standard selection rules.

Equation (15.5) for the FID after mixing with ω_{MW} can be written in terms of the elements of the Mims matrix as

$$V_{\mathrm{FID}}(t) = \frac{\hbar\omega_0}{2k_{\mathrm{B}}T} \sin(\theta) \mathrm{e}^{+\mathrm{i}\phi} \times \sum_{m,k} |M_{k,m}|^2 \mathrm{e}^{\mathrm{i}(\Delta\omega + \omega_{\alpha,k} - \omega_{\beta,m})t} \mathrm{e}^{-t/T_{2\mathrm{e},m,k}} \quad (15.7)$$

This equation is similar to that of the isotropic case in equation (15.5): a set of exponentially damped sinusoids oscillating at the frequencies of the EPR transitions in the rotating frame. The number of nonzero terms equals the number of nuclear spin states, i.e., the dimension of the Mims matrix, for isotropic PCs. However, the number of nonzero elements for anisotropic PCs can be as large as the square of the number of nuclear states. Equation (15.5) is easily converted into equation (15.7) using the property that the elements of $\boldsymbol{M}$ are either 0 or 1 for isotropic PCs and the index j is the combined m and k of the nonzero elements.

15.2.2.1 Example for I = ½

The FID signal from a PC coupled to a single $I = 1/2$ nucleus contains four frequencies corresponding to the two allowed and two forbidden lines in the CW-EPR spectrum arranged symmetrically around their center. The indices m and k in equation (15.7) correspond to the values of m_I, i.e., $\pm 1/2$, and the frequencies can be expressed in terms of the ENDOR frequencies ω_α or ω_β in the α or β electron spin manifolds, respectively (see Chapter 16). Making use of the symmetry of the Mims matrix and assuming all the T_{2e} are equal, the FID signal becomes

$$V_{\text{FID}}(t) = \frac{\hbar\omega_0}{2k_BT}\sin(\theta)\,e^{+i\phi}\,e^{-t/T_{2e}} \left(\left|M_{+\frac{1}{2},+\frac{1}{2}}\right|^2\left(e^{i\left(\Delta\omega+\left(\frac{\omega_\alpha-\omega_\beta}{2}\right)\right)t}+e^{i\left(\Delta\omega-\left(\frac{\omega_\alpha-\omega_\beta}{2}\right)\right)t}\right) +\left|M_{+\frac{1}{2},-\frac{1}{2}}\right|^2\left(e^{i\left(\Delta\omega+\left(\frac{\omega_\alpha+\omega_\beta}{2}\right)\right)t}+e^{i\left(\Delta\omega-\left(\frac{\omega_\alpha+\omega_\beta}{2}\right)\right)t}\right)\right) \tag{15.8}$$

The FID can also be written more compactly in terms of the ENDOR frequencies as[14]

$$V_{\text{FID}}(t) = \frac{\hbar\omega_0}{2k_BT}\sin(\theta)\,e^{+i\phi}\,e^{-t/T_{2e}}e^{i\Delta\omega t} \left(\cos\frac{\omega_\alpha t}{2}\cos\frac{\omega_\beta t}{2} + \cos(2\eta)\sin\frac{\omega_\alpha t}{2}\sin\frac{\omega_\beta t}{2}\right) \tag{15.9}$$

where η is half of the angle between effective fields seen by the nucleus (sum of the nuclear Zeeman, ω_I, and hyperfine fields) in the two electron manifolds and plays a role similar to the Mims matrix,

$$\sin(2\eta)^2 = (\omega_I B)^2/(\omega_\alpha\omega_\beta)^2 \tag{15.10}$$

with B the pseudo-secular (off-diagonal) part of the HFI. FT of equation (15.8) or (15.9) gives two pairs of lines: one pair of 'allowed' transitions and one pair of 'forbidden' transitions. This FT spectrum corresponds to the CW-EPR spectrum that would be measured by sweeping ω at constant $\boldsymbol{B}_0$.

The FID with anisotropic HFI usually looks more complicated because it contains additional frequencies corresponding to the formally forbidden EPR transitions. This is illustrated in Figure 15.4(a) where

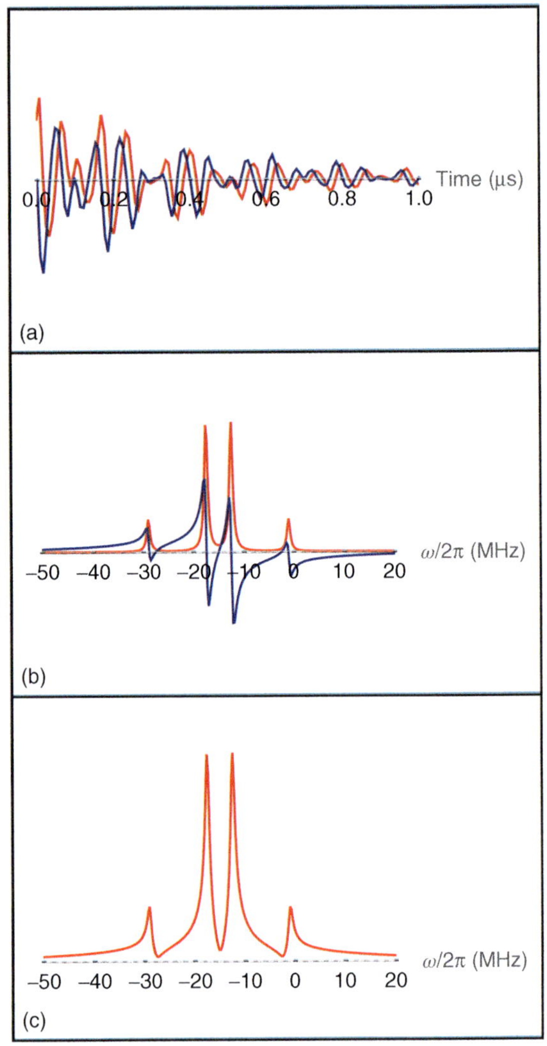

Figure 15.4. The FID and FT-EPR spectra from a PC in a single crystal with anisotropic HFI. The conditions are the same as for Figure 15.2 except that the HFI is anisotropic and the nuclear Zeeman interaction is 14 MHz. The formally allowed and forbidden EPR transitions have a 0.8 : 0.2 ratio of intensities in the CW EPR spectrum. (a) The initial portion of the FID. (b) An expanded portion of the FT of the FID, showing the real and imaginary parts as in Figure 15.3. (c) The magnitude or absolute value of the complex spectrum in (b)

the nucleus is a proton with a nuclear Zeeman frequency of 14 MHz and produces an intensity ratio of 0.8 : 0.2 between the formally allowed and forbidden transitions. This situation corresponds to a PC in an oriented single crystal. After FT, the spectrum (Figure 15.4b) contains four lines, the original pair seen in the isotropic case (Figure 15.3c) and the weaker pair split by twice the nuclear Zeeman interaction. The absorption-like spectrum in red has excellent resolution of all four lines in the spectrum. However, the dispersion-like spectrum has broad tails around each line with noticeable overlap. If the magnitude of the Fourier transformed spectrum is taken to avoid the inevitable phase corrections needed for experimental data (as is often done in ESEEM[12]), the broad tails of the imaginary spectrum distort peak positions and intensities (Figure 15.4c).

15.2.3 Anisotropic *g*

15.2.3.1 Zeeman Interaction

The Zeeman interaction of a PC may be more complex than the isotropic case described earlier. When the system ground state is a Kramers' doublet, its spin levels are split by a magnetic field so that the system can be described using a real, effective, or fictitious spin of $S = 1/2$, but the Zeeman interaction is generally anisotropic. In other words, the Zeeman splitting of the two levels depends on the orientation of the PC with respect to $\boldsymbol{B}_0$, and the transition probability depends on its orientation relative to the resonant $\boldsymbol{B}_1$. Anisotropy of the Zeeman interaction is attributed to a coupling of the orbital motion of the unpaired electrons with their spin degrees of freedom, e.g., through spin–orbit coupling (see Chapter 2).[5,6]

The anisotropy provides indirect information about PC structure, but it complicates the dynamics of $\boldsymbol{\mu}$. The PC spin vector is related to $\boldsymbol{\mu}$, but it is not always parallel to it, e.g., μ_x can include contributions from S_x, S_y, and S_z. The $\boldsymbol{g}$-tensor relates $\boldsymbol{S}$ to $\boldsymbol{\mu}$:

$$\widehat{\boldsymbol{\mu}} = -\mu_B \boldsymbol{g}\widehat{\boldsymbol{S}} \quad (15.11)$$

The Zeeman interaction Hamiltonian may be written in terms of $\widehat{\boldsymbol{S}}$, formally similar to the isotropic case, as

$$\widehat{H}_Z = \mu_B g_{\text{eff}} B_0 \widehat{S}_{z'} \quad (15.12)$$

where g_{eff} is the effective value of the $\boldsymbol{g}$-tensor, $g_{\text{eff}} = \sqrt{G_{zz}}$, with $\boldsymbol{G} = \boldsymbol{g}\boldsymbol{g}^{\text{T}}$. The z -axis here is directed along $\boldsymbol{B}_0$, while the spin quantization axis z' is parallel to $\boldsymbol{g}^{\text{T}}\boldsymbol{B}_0$ and lies along $\boldsymbol{B}_0$ only in the isotropic case or when the PC is oriented so that a principal axis of the $\boldsymbol{g}$-tensor is parallel to $\boldsymbol{B}_0$.[5] The g_{eff} also depends on the orientation of the PC in the laboratory frame. The CW-EPR signal intensity is proportional to p, the square of the transition moment, which depends on the orientation of the PC in the EPR spectrometer as[5]

$$p \propto g_1{}^2 B_1{}^2 \quad (15.13)$$

and the effective MW field strength is

$$\hbar\omega_1 = \mu_B g_1 B_1 \quad (15.14)$$

where ω_1 is the Rabi nutation frequency of the anisotropic PC. The value of g_1 relates the effective MW field strength $\boldsymbol{g}^{\text{T}}\boldsymbol{B}_1$ perpendicular to $\boldsymbol{g}^{\text{T}}\boldsymbol{B}_0$ and is given by

$$g_1 = \sqrt{G_{xx} - G_{xz}^2/G_{zz}} \quad (15.15)$$

The g_1 also scales μ_x which is the observable in most pulse EPR measurements. The $\boldsymbol{g}$ anisotropy of organic radicals is small enough that the departure of the quantization axis from $\boldsymbol{B}_0$ and the orientation dependence of p can be neglected; only the orientation dependence of g_{eff} must be considered because it determines where the EPR resonance lines appear. However, when $\boldsymbol{g}$ anisotropy is not small so that the electron spin quantization axis is significantly different from the magnetic field direction, several complications arise in the EPR spectroscopy of the PC. CW-EPR measurements on these PCs face the same complications and numerous CW-EPR studies have been successfully made on such systems. Some of the more significant differences from isotropic PCs will be mentioned.

It is often overlooked that the FID and other pulse EPR signals from very anisotropic PCs depend not only on the orientation of the PC relative to $\boldsymbol{B}_0$, but the signals also depend on the orientation of the PC relative to $\boldsymbol{B}_1$. This follows from equations (15.13–15.15), which show that the response of a PC to MW pulses and the signal amplitude it then produces depend on the orientation of its $\boldsymbol{g}$-tensor relative to $\boldsymbol{B}_1$, which defines the laboratory $\boldsymbol{x}$-axis. This effect is seen in the evolution of $\boldsymbol{S}$, which takes a form similar to equation (15.3),[15]

$$\frac{\partial}{\partial t}\boldsymbol{S}(t) = \frac{\mu_B}{\hbar}\boldsymbol{S}(t) \times \{\boldsymbol{g}^{\text{T}}(\boldsymbol{B}_0 + \boldsymbol{B}_1)\} - \frac{1}{T_{2e}}(\boldsymbol{S}_{x'} + \boldsymbol{S}_{y'}) - \frac{1}{T_{1e}}(\boldsymbol{S}_{z'} - \boldsymbol{S}_{\text{eq}}) \quad (15.16)$$

The $\boldsymbol{S}$ of each PC precesses around the instantaneous direction parallel to $\boldsymbol{g}^{\text{T}}(\boldsymbol{B}_0 + \boldsymbol{B}_1)$ so that the trajectory during the pulse depends on orientation of the PC in the spectrometer. Even during periods when the MW field is absent, the $\boldsymbol{S}$ from each PC precesses around an axis that depends on the PC's orientation in the laboratory frame. This seriously complicates the response of an anisotropic PC to MW pulses.

As discussed in Sections 15.1 and 15.2, the observable in magnetic resonance is $\boldsymbol{\mu}$ rather than $\boldsymbol{S}$. This means that equation (15.16) is not immediately applicable to the FT-EPR of anisotropic PCs. However, equations (15.1) and (15.11) can provide a basis for interpreting spectra.[16] Bloch-like equations have been obtained for the evolution of $\boldsymbol{\mu}$, but they lack the simplicity of equation (15.16).[15] However, those equations do show that each $\boldsymbol{\mu}$ precesses around its own particular direction of $\boldsymbol{\mu}_{eq}$, which is not along $\boldsymbol{B}_0$, but is along $\boldsymbol{GB}_0$. When relaxation can be neglected, the projection of $\boldsymbol{\mu}$ onto the $\boldsymbol{x,y}$-plane traces an ellipse, so that the $\boldsymbol{\mu}$ responsible for the FID has elliptical polarization rather than the circular polarization found for isotropic PCs. Because only μ_x can be measured with the usual single-mode resonator, the elliptical polarization of the oscillating magnetic field induced by $\boldsymbol{\mu}$ cannot be measured. The rotating frame has several peculiarities for anisotropic PCs that limit its usefullness.[16] One difficulty is that the rotating frame is defined for $\boldsymbol{S}$ and not for $\boldsymbol{\mu}$, but they generally point in different directions and precess about different axes.[5,15,16] So the evolution of $\boldsymbol{\mu}$, which produces the observed signal, is quite complicated in the rotating frame. In addition, the signal in a single-mode resonator comes only from μ_x, not from the rotating frame components of $\boldsymbol{S}$ or $\boldsymbol{\mu}$. The signal corresponds to the projection of $\boldsymbol{\mu}$ onto a precessing vector in the rotating frame. Another peculiarity is that the observed EPR signal comes from the two circularly polarized components that make up the elliptically polarized $\boldsymbol{\mu}$, requiring two rotating frames going in opposite directions with a fixed phase relation between them.

The equilibrium $\boldsymbol{\mu}$ generally has a nonzero component perpendicular to $\boldsymbol{B}_0$, so the center of the ellipse is offset from the origin (Figure 15.5). The $\boldsymbol{S}$ can have an oscillating component in the z-direction following an MW pulse, $\boldsymbol{\mu}$ does not. The $\boldsymbol{\mu}$ precesses in a plane that is perpendicular to $\boldsymbol{B}_0$ (for conservation of

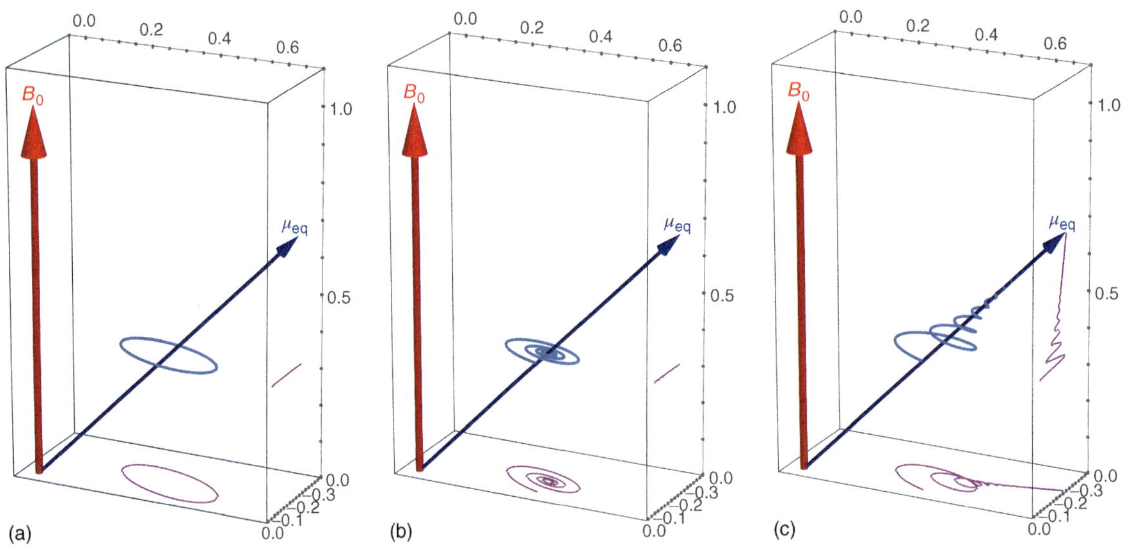

Figure 15.5. The calculated path of precession of $\boldsymbol{\mu}$ for $S = \frac{1}{2}$ PCs having their anisotropic $\boldsymbol{g} = (1, 2, 4)$ after rotation of the molecular frame by $\pi/3$ around the laboratory $\boldsymbol{y}$-axis and $\pi/4$ around the laboratory z-axis. The $\boldsymbol{\mu}_{eq}$ is rotated by MW pulse with $\theta = \pi/4$ and $\phi = 0$. The light blue curve shows the path of the tip of $\boldsymbol{\mu}$ as it precesses around $\boldsymbol{\mu}_{eq}$. The precession is projected, as magenta curves, onto the $\boldsymbol{x,y}$-plane at the bottom and onto the $\boldsymbol{x,z}$-plane on the right-hand side. The precession paths and their projections are shown for (a) $1/T_{1e} = 1/T_{2e} = 0$; (b) $1/T_{1e} = 0$ and $1/T_{2e} \neq 0$; (c) $T_{1e}/T_{2e} = 5$. See discussion in Section 15.2.3. The laboratory $\boldsymbol{y}$-axis runs from left to right and is numbered along the top in these plots, and the laboratory z-axis is the vertical axis

energy) but is generally tilted with respect to its quantization axis.

In practice, the most significant feature of $\boldsymbol{g}$ anisotropy is that the Zeeman frequency is proportional to g_{eff}, but the Rabi frequency (and also μ_x that produces the measured signal) is proportional to g_1. These two values can be quite different. Moreover, in an unoriented sample, PCs with the same resonant frequency may have different Rabi frequencies, making it impossible for them to have the same turning angle. Identical PCs with $\boldsymbol{B}_0$ in the same direction in the molecular frame, but $\boldsymbol{B}_1$ in different directions in the molecular frame, have identical EPR frequencies but respond differently to $\boldsymbol{B}_1$. This property produces a second type of orientation selection in pulse EPR experiments with PCs having several orientations in the sample because the signal intensity at each point in the spectrum does not necessarily represent the resonant PCs in proportion to their numbers.

The FID for PCs with anisotropic $\boldsymbol{g}$ is substantially different from those of isotropic PCs. These differences are shown by the contrast between the calculated FIDs for anisotropic $\boldsymbol{g}$ shown in Figure 15.5 and those for isotropic g in Figure 15.1. The two PCs are identical aside except for their $\boldsymbol{g}$. For generality, $\boldsymbol{B}_0$ and $\boldsymbol{B}_1$ are not parallel to any of the principal axes of $\boldsymbol{g}$. As a result, $\boldsymbol{\mu}_{\text{eq}}$ is not parallel to $\boldsymbol{B}_0$, but is tilted away from it. When $T_{1\text{e}}$ and $T_{2\text{e}}$ relaxation is negligible, for the same MW pulse (Figure 15.5a), the precessing $\boldsymbol{\mu}$ vector traces an ellipse in the laboratory frame, shown in light blue. $\boldsymbol{\mu}$ has a tilted precession around $\boldsymbol{\mu}_{\text{eq}}$. The projection of $\boldsymbol{\mu}$ onto the $\boldsymbol{x,y}$-plane at the bottom also is an ellipse that is generally displaced from the origin. This means that a portion of $\boldsymbol{\mu}_x$ and/or $\boldsymbol{\mu}_y$ does not precess. The projection onto the $\boldsymbol{x,z}$-plane at the right of Figure 15.5(a) shows μ_x, which induces the observable signal, oscillating at ω_j. In this particular example, the oscillation of μ_y is considerably larger and would produce a FID with a larger amplitude than the FID from μ_x. However, if $\boldsymbol{B}_1$ had been applied along the $\boldsymbol{y}$-direction, the starting $\boldsymbol{\mu}(0)$ would be quite different.

$T_{2\text{e}}$ relaxation causes μ_x and μ_y to decay slowly (Figure 15.5b). The tip of $\boldsymbol{\mu}$ follows an elliptical spiral centered on $\boldsymbol{\mu}_{\text{eq}}$ but perpendicular to $\boldsymbol{B}_0$. The projection of $\boldsymbol{\mu}$ onto the $\boldsymbol{x,z}$-plane at the right shows μ_x, which produces the observable signal, oscillating at ω_j. These FID oscillations gradually decrease with a time constant $T_{2\text{e}}$. Note that Figure 15.5(b) shows $\mu_x(0) \neq 0$ with a considerable component in the $\boldsymbol{x}$-direction.

$T_{1\text{e}}$ relaxation causes μ_z to recover to its original value (Figure 15.5c). Instead of precessing in a plane, the tip of $\boldsymbol{\mu}$ spirals toward $\boldsymbol{\mu}_{\text{eq}}$. Notably, the projection of $\boldsymbol{\mu}$ onto the $\boldsymbol{x,y}$-plane at the bottom is quite different. The projection of $\boldsymbol{\mu}$ onto the $\boldsymbol{x,z}$-plane at the right still shows that μ_x has a portion that oscillates at ω_j, and the FID is unchanged. However, μ_z and a portion of μ_x (and μ_y) change with a time constant $T_{1\text{e}}$.

The transformation into a rotating frame has several conceptual difficulties for large $\boldsymbol{g}$ anisotropy. Simply rotating the axes in Figure 15.5 around z at the MW frequency ω_{MW} makes $\boldsymbol{\mu}_{\text{eq}}$ oscillate even though it induces no signal in the EPR spectrometer. Rotating the axes around $\boldsymbol{\mu}_{\text{eq}}$ produces a large, oscillating field from $\boldsymbol{B}_0$. More complicated approaches make it very difficult to apply physical intuition to the motion of $\boldsymbol{\mu}$ which is a major attraction of the rotating frame for isotropic PCs. Moreover, $\boldsymbol{\mu}_{x'}$ and $\boldsymbol{\mu}_{y'}$ for any rotating frame, normally are not experimentally observable because μ_y is not an observable in a single-mode resonator. Fortunately, despite all the complications with anisotropic $\boldsymbol{g}$, the frequencies in the FID and those of absorption lines in the CW-EPR spectrum still correspond to energy differences between the energy levels of the PC, just as they do for isotropic PCs. The point at which it becomes necessary to explicitly consider the effects of anisotropy depends on the quality of the data and the significance attached to the EPR measurements.

15.2.3.2 Spin–Spin Interactions

A second aspect of anisotropic PCs lies in their interactions with other spins. This impacts the interpretation of spectra and interactions more than it does their measurement. Section 15.4 briefly considers how spin–spin interactions between PCs affect FT-EPR spectra, but how $\boldsymbol{g}$ anisotropy affects their interpretation is considered briefly in this chapter. The quantization axis of $\boldsymbol{S}$ deviates from $\boldsymbol{B}_0$, shifting the energy levels of the interacting spins and altering the orientation dependence of the interaction from those of isotropic PCs. In most cases, the interaction is not actually between $\boldsymbol{S}$ (and $\boldsymbol{I}$) but between their $\boldsymbol{\mu}$.[5,6] This impacts dipole–dipole interactions between PCs (in double electron-electron resonance (DEER), pulsed electron electron double resonance (PELDOR), and relaxation-induced dipolar modulation enhancement (RIDME) spectroscopies, (see Chapters 19 and 20) when there is strong $\boldsymbol{g}$ anisotropy. Because both types of spin–spin interactions are used to determine structural details in many contexts, how to interpret data is an important consideration.

The dipole–dipole interaction (with $\boldsymbol{r}_{ij}$ the vector connecting two distant PCs) is (see Chapter 9 of Ref. 5)

$$\hat{H}_{\mathrm{dd}} = \frac{1}{r_{ij}^3}\{\hat{\boldsymbol{\mu}}_i\hat{\boldsymbol{\mu}}_j - 3(\hat{\boldsymbol{\mu}}_i\boldsymbol{r}_{ij})(\hat{\boldsymbol{\mu}}_j\boldsymbol{r}_{ij})/r_{ij}^2\} \tag{15.17}$$

The secular part commutes with $\hat{H}_{\mathrm{Z}}$ of equation (15.12) for each PC. For 'unlike' PCs, i.e., when Zeeman frequencies of the PCs differ by more than their dipolar interaction, the secular part of $\hat{H}_{\mathrm{dd}}$ is[17]

$$\hat{H}_{\mathrm{dd}}^{\mathrm{sec}} = D_{ij}\hat{S}_{iz'}\hat{S}_{jz'} \tag{15.18}$$

where

$$D_{ij} = \frac{\mu_{\mathrm{B}}^2}{r_{ij}^3 g_{i,\mathrm{eff}}\, g_{j,\mathrm{eff}}}\{(\boldsymbol{G}_i\boldsymbol{k}\cdot\boldsymbol{G}_j\boldsymbol{k}) - 3(\boldsymbol{r}_{ij}\cdot\boldsymbol{G}_i\boldsymbol{k})(\boldsymbol{r}_{ij}\cdot\boldsymbol{G}_j\boldsymbol{k})/r_{ij}^2\} \tag{15.19}$$

and $\boldsymbol{k}$ is the unit vector along $\boldsymbol{B}_0$. For isotropic PCs, D_{ij} reduces to the well-known relation, $D_{ij} = \mu_{\mathrm{B}}^2 g_i g_j\{1 - 3\cos(\theta)^2\}/r_{ij}^3$ with θ the angle between $\boldsymbol{B}_0$ and $\boldsymbol{r}_{ij}$. For organic radicals, this simple relation is nearly exact and produces the classic Pake-like doublet. Figure 15.6 demonstrates how the Pake doublet of one isotropic PC coupled by a dipole interaction to a highly anisotropic PC changes with $\boldsymbol{g}$ anisotropy and mutual orientation. The dipolar interaction appears to become nonaxial and have a nonzero averaged frequency for each of Pake doublet component. Although weak dipolar interactions affect the spectral shape, they do not appreciably affect the response of either PC to a strong MW pulse.

Equations (15.17–15.19) describe the spectral broadening caused by dipole–dipole interactions between PCs observed in many experiments. Two notably and highly relevant examples are the line broadening of Er^{3+} ions doped into single crystal of $CaWO_4$, conclusively showing that dipolar broadening varies with $\boldsymbol{\mu}$ and not $\boldsymbol{S}$,[18] and measurement of the dipole–dipole interaction in a pair of PCs with one anisotropic member studied by RIDME, showing powder patterns of the type in Figure 15.6.[19]

The HFI is traditionally described by a tensor, $\boldsymbol{A}$, between spins, i.e., $\hat{\boldsymbol{S}}^{\mathrm{T}}\boldsymbol{A}\hat{\boldsymbol{I}}$. However, with very anisotropic PCs, $\hat{\boldsymbol{S}}$ is usually not the real spin but an effective or fictitious spin. Nuclei around a PC are often classified as a 'central atom' bearing most of the unpaired spin density in an atomic orbital; or as a ligand, or 'matrix' nucleus bearing little or no unpaired spin density. HFI

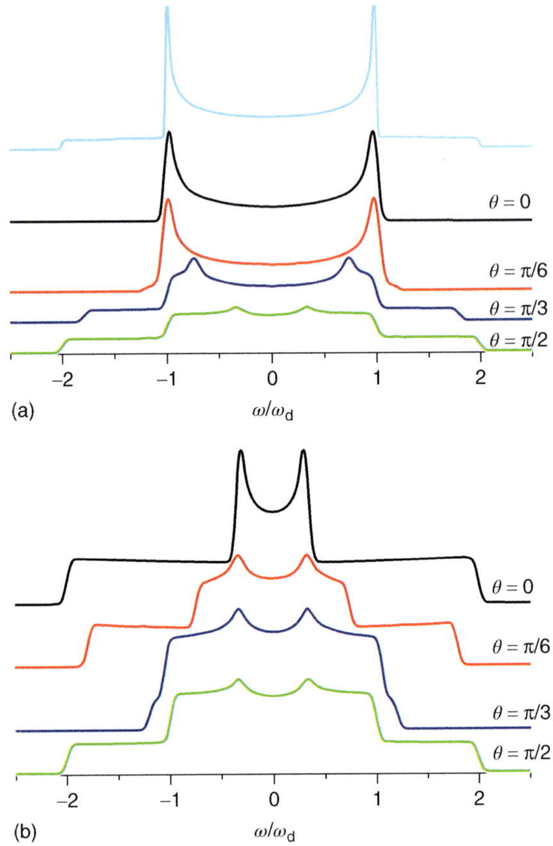

Figure 15.6. Pake-like patterns caused by interaction of isotropic PC with a partner having anisotropic axially symmetric $\boldsymbol{g}$-tensor are shown in integral form. The patterns strongly depend on the pair geometry. The line colors specify the angle between the vector r connecting the two PCs and direction of the axial symmetry of the $\boldsymbol{g}$-tensor, 0 (black), π/6 (red), π/3 (blue), and π/2 (green). (a) $g_{\mathrm{parallel}} = 1.0$, $g_{\mathrm{perp}} = 3.0$, and the standard Pake pattern is shown as the cyan line for comparison; (b) $g_{\mathrm{parallel}} = 3.0$, and $g_{\mathrm{perp}} = 1.0$. Here, $\omega_d = g_e^2\mu_B^2/2\hbar r_{ij}^3$

with ligand or matrix nuclei includes magnetic dipole interactions between the PC and the nuclei.[5,6] However, their HFI can still be written in the form

$$\hat{H}_{\mathrm{HFI}} = \hat{\boldsymbol{S}}^{\mathrm{T}}\boldsymbol{A}\hat{\boldsymbol{I}} \tag{15.20}$$

The dipolar part of the HFI with distant or matrix nuclei can be considered a local field acting on the nuclear spin, directed along the vector $\boldsymbol{A}^{\mathrm{T}}\boldsymbol{g}^{\mathrm{T}}\boldsymbol{k}$ (where $\boldsymbol{g}^{\mathrm{T}}\boldsymbol{k}$ is the quantization axis of the electron spin), with magnitude $m_S\sqrt{(\boldsymbol{g}\boldsymbol{A}\boldsymbol{A}^{\mathrm{T}}\boldsymbol{g}^{\mathrm{T}})_{zz}}/g_{\mathrm{eff}}$. When $\boldsymbol{g}$ anisotropy is significant, it influences the dipolar part of the HFI indirectly owing to the inclination of the electron spin quantization axis from the external magnetic field direction.

The HFI local field direction generally differs from that of $\boldsymbol{B}_0$, affecting the ESEEM.

15.3 MULTIPULSE FT-EPR

Most pulse EPR experiments use multiple MW pulses to measure relaxation times, such as T_{1e}, T_{2e}, or T_m; to measure HFI by ESEEM, HYSCORE, or ENDOR; or to measure dipole–dipole interactions between PCs by DEER/PELDOR or RIDME. Each of these experiments has a 'hidden' dimension – the EPR frequency. This dimension is rarely considered but can have a major impact on measurements through spectral editing and even lead to loss of information.

The signals from a sequence of pulses contain overlapping FIDs and spin echoes and rapidly become very complicated. However, echoes, such as FIDs, are measured while no MW pulses are being applied. Thus, they are in many senses 'free' induction signals that are evolving in the absence of an MW field. In particular, echoes are composed of frequencies corresponding to the energy level differences of the PC. The multiple pulses select what combination of PC properties appear in the FID or echo evolution.

Each pulse in a pulse sequence produces a FID; each pair of pulses generates an echo; and each pulse following an echo produces three additional echoes: the stimulated, virtual, and refocused echoes. Each response contains different information. The responses generally appear at different times in the sequence, with different rotation angles θ and MW phase ϕ dependences for each pulse, and these characteristics can be used to isolate a single, desired response. A pair of pulses produces only three responses and provides a manageable example of multiple pulse FT-EPR experiments although the sequences of more than six pulses are currently in use.

15.3.1 Two-pulse Sequence

Consider the signal from a pair of pulses: $p_1 \cdots \tau \cdots p_2$ where p_1 and p_2 denote MW pulses with turning angles and phases of θ_1, ϕ_1 and θ_2, ϕ_2, respectively. They produce three signals from PCs: two FIDs and one echo. The signals for isotropic PCs are fairly clear and will be the focus here, but the echo signal with anisotropic HFI will also be discussed because it is central to 2D FT-EPR spectroscopy. The signal at time t from the start of the two-pulse sequence can be represented as

$$V(\tau,t) = V_{\mathrm{FID1}}(t) + V_{\mathrm{FID2}}(t>\tau) + V_{\mathrm{Echo}}(\tau, t>\tau) \tag{15.21}$$

where τ is the time between the two MW pulses.

15.3.1.1 *The FIDs from Isotropic PCs*

The $V_{\mathrm{FID1}}(t)$ signal is just the FID from the first pulse. It is identical to the signal in equation (15.5) (or (15.7)) for $t<\tau$ with $\theta=\theta_1$ and $\phi=\phi_1$, but after the second pulse ($t>\tau$), the V_{FID1} signal is attenuated by a factor of $\cos(\theta_2/2)^2$, which is the probability for a spin not to flip from the MW pulse but to continue precession.

V_{FID2} is the FID from the second MW pulse and only exists after the second pulse when $t>\tau$. The V_{FID2} signal is the basis for saturation recovery or inversion recovery experiments using the FID to measure spin–lattice relaxation.[20] The signal, from isotropic PCs at a time $t=\tau+\delta t$, i.e., $\delta t>0$ after the second pulse, is

$$\begin{aligned} V_{\mathrm{FID2}}(\tau+\delta t) &= -\frac{\hbar\omega_{\mathrm{o}}}{2k_{\mathrm{B}}T}\sin(\theta_2)\mathrm{e}^{+\mathrm{i}\phi_2} \\ &\times \sum_j \left(\mathrm{e}^{\mathrm{i}(\Delta\omega+\omega_{\alpha,j}-\omega_{\beta,j})\,\delta t}\,\mathrm{e}^{-\delta t/T_{2e,j}}\right. \\ &\times \left.\left(1+\left(\cos\left(\theta_1\right)-1\right)\mathrm{e}^{-\tau/T_{1e,j}}\right)\right) \end{aligned} \tag{15.22}$$

Except for the final factor in parentheses, V_{FID2} has the form of equation (15.5) starting from the end of the second pulse. That final factor in the summation in equation (15.22) accounts for the decreased μ_z caused by the first pulse and for its recovery by spin–lattice relaxation during τ. Complete inversion of μ_z by the first pulse (with $\theta_1=\pi$) is the classic inversion recovery experiment, and the initially inverted FID recovers during τ at a rate $1/T_{1e}$.

Equation (15.22) shows a separation and correlation of information. Each EPR transition j appears, with its transition frequency and T_{2e}, in the FID as a function of δt after the second pulse. FT along the δt dimension, or similar frequency analysis, results in a series of FT-EPR spectra, similar to the FT-EPR spectrum from the single-pulse FID of equation (15.5). However, the intensity of each EPR transition j varies with τ according to the T_{1e} of that transition. Thus, the measurement of T_{1e} is in many ways orthogonal to (independent of) T_{2e} and EPR frequency, but each T_{1e} is 'spectrally resolved' and correlated with a transition and the corresponding T_{2e}. Analogous NMR measurements of spectrally resolved T_{1n} are common. Note that the phase of V_{FID1} varies with ϕ_1, while that of V_{FID2} depends only on ϕ_2, so that systematic cycling of pulse phases can separate these signals from each other and from the echo.[21]

15.3.1.2 The Echo for Isotropic PCs

The third response is the two-pulse spin echo centered at $t = 2\tau$, which can be written for isotropic PCs with $t = 2\tau + \delta t$ as

$$V_{\text{Echo}}(\tau, 2\tau + \delta t) = -\frac{\hbar\omega_0}{2k_B T}\sin(\theta_1)\sin\left(\frac{\theta_2}{2}\right)^2 e^{-i(\phi_1 - 2\phi_2)} \times \sum_j \left(e^{i(\Delta\omega + \omega_{\alpha,j} - \omega_{\beta,j})\delta t}\, e^{-\frac{\delta t}{T_{2e,j}}}\, e^{-\frac{2\tau}{T_{2e,j}}}\right) \quad (15.23)$$

The factors preceding the summation give the phase and turning angle dependence of the echo. The first two factors in the summation are the same damped oscillations as a function of $\delta t > 0$ that appear in equation (15.5) for the FID. The echo shape has the functional form of the FID. However, before the center of the echo at $\delta t = 0$, the shape corresponds to what the FID might have been at negative values of time. Nevertheless, the echo shape can be Fourier transformed in the δt dimension to produce the FT-EPR spectrum.

The final factor inside the summation is a simple exponential decay along τ at a rate of $2/T_{2e}$. A series of FT-EPR spectra, obtained by FT of the echo shape, shows the T_{2e} decay for each EPR transition, yielding spectrally resolved measurement of T_{2e} analogous to T_{1e} measurements discussed in Section 15.3.1.1. In practice, the decay of the echo for EPR transitions in a solid is often nonexponential, and the time constant characterizing that decay is frequently called the phase memory time or T_m. However, the 2D FT-EPR spectra just described do provide spectrally resolved measurements of the echo decays and time constants for isotropic PCs.

In many cases, the shape of the echo is not recorded in spin echo measurements, just the amplitude near the center of the echo or the integral of some portion of the echo in an effort to decrease the experimental measurement times. When the point at the center of the echo corresponding to $\delta t = 0$ is measured, then the first two factors in the summation in equation (15.23) are 1 and the signal intensity with τ is simply the sum of the decays from all EPR transitions. However, if $\delta t \neq 0$, then the first two terms in the summation lie between 1 and -1, depending on the EPR resonant frequency, and can be complex valued. This results in a subtle type of spectral editing where components of the idealized spectrum are altered or eliminated. The signal intensity as a function of τ is a weighted sum of decays from all EPR transitions for $\delta t \neq 0$ with positive, negative, and even complex weights.

On the other hand, integration of the entire echo, which is equivalent to the point in the Fourier transform with a frequency of zero, results in a signal that contains the decay of the transition resonant at ω_{MW}. If the echo is integrated over a narrower window, then the signal is the sum of the individual decays, weighted by an off-center, complex-valued sinc function corresponding to the FT of the integration window.

15.3.1.3 The Echo for Anisotropic HFI

When there is anisotropic HFI or anytime that ESEEM occurs (see Chapter 18), each EPR transition in the spectrum has different modulation.[22] The individual transitions can be recovered from the echo shape, see Section 15.3.1.2, and the intensity of EPR transition i, k, varies with τ as[23]

$$V_{i,k}(\tau) = \sin(\theta_1)\sin\left(\frac{\theta_2}{2}\right)^2 e^{-i(\phi_1 - 2\phi_2)} * \left(|M_{i,k}|^4 + \sum_{j(\neq i)} |M_{i,k}|^2 |M_{j,k}|^2 \cos(\omega_{\beta,i,j}\tau) + \sum_{n(\neq k)} |M_{i,k}|^2 |M_{i,n}|^2 \cos(\omega_{\alpha,n,k}\tau) + \sum_{j(\neq i),\, n(\neq k)} \text{Re}(M^*_{j,k} M_{j,n} M^*_{i,n} M_{i,k}) \times \cos(\omega_{\beta,i,j}\tau)\cos(\omega_{\alpha,n,k}\tau)\right) \quad (15.24)$$

where Re(…) takes the real part of a complex quantity and $\omega_{\gamma,u,v} = \omega_{\gamma,u} - \omega_{\gamma,v}$ are ENDOR frequencies with $m_S = \gamma$ and can be positive or negative. The amplitude of each EPR transition is modulated only by ENDOR frequencies (or sums and differences of ENDOR frequencies) that involve levels in the EPR transition. This characteristic can be used to determine the sign of the HFI relative the nuclear quadrupole interaction (NQI) for quadrupolar nuclei (see Chapter 6).[23] The echo signal retains the MW pulse phase and turning angle dependence of isotropic PCs.

The 2D spectrum obtained by a 2D FT along the δt and τ dimensions give a spectrum with EPR frequencies along one axis and ENDOR frequencies along the other. Interestingly, a slice along one of the ENDOR frequencies shows a spectrum of the EPR transitions involving that ENDOR frequency with comparable amplitudes for the allowed and forbidden EPR transitions. On the other hand, when the HFI and NQI are weak compared to ω_I, a slice along the 'sum' frequency at roughly $2\omega_I$ contains only the forbidden EPR transitions, while the slice at the 'difference' at the hyperfine splitting contains only the allowed EPR transitions.[23]

When relaxation rates are negligible, the portion of equation (15.24) in large parentheses can be written compactly as a function of δt and τ as[14,23]

$$e^{i\Delta\omega\delta t}\sum_{j,k,l,m} M^*_{l,j}M_{l,m}M^*_{k,m}M_{k,j}e^{-i\omega_{\beta,m,j}\tau}e^{-i\omega_{\alpha,l,k}\tau}e^{i(\omega_{\alpha,k}-\omega_{\beta,j})\delta t} \tag{15.25}$$

This gives the well-known ESEEM result of Mims at $\delta t = 0$.[24] However, when $\delta t \neq 0$, the amplitude in each factor in the summation is altered, and ENDOR frequencies can even entirely disappear from the modulation along τ, a significant, if unintended, form of spectral editing in 1D ESEEM measurements. Each EPR transition oscillates smoothly into the negative δt region, extending itself as δt changes sign. However, T_{2e} relaxation causes the signal to increase as δt becomes more negative. In most echo experiments, the growing amplitude is not obvious because the echo is symmetrically damped by strong inhomogeneous broadening. However, for negligible inhomogeneous broadening, the echo shape, at first glance, can resemble the FID from the final pulse. Yet, close examination will reveal that the signal retains the phase and turning angle dependence of the echo and that slight variations in $\boldsymbol{B}_0$ have no effect on the signal at $\delta t = 0$. Both relaxation and ESEEM alter the echo signal shape from that of two back-to-back FID signals as it is commonly approximated. Yet, the echo shape does contain information equivalent, but not identical, to that in the FID.[14]

When a PC has several nuclei with anisotropic HFI, the Mims matrix is a tensor product of the Mims matrices each nucleus would have if it were the only coupled nucleus. This makes it possible to describe the FID, ESEEM, and their FTs, when all the EPR transitions of the PC are uniformly excited, in terms of products and convolutions of the contributions from each nucleus. This 'product rule' was first noted by Mims[25] and has been elaborated for many types of echo signals.[12,14,26]

15.4 $S > 1$, TRIPLETS AND INTERACTING RADICALS

Many PCs with $S > {}^1\!/_2$ have been studied by FT-EPR. These high-spin PCs behave in some ways like those with $S = {}^1\!/_2$. Their FT-EPR signals can be calculated using the approaches discussed earlier. However, there are important differences arising from two characteristics of high-spin systems. (i) The EPR transitions of a PC with $S > {}^1\!/_2$ typically cover a wide frequency range, so that it is difficult, if not impossible, to uniformly excite them with MW pulses. Consequently, the finite MW pulse widths and amplitudes need to be explicitly considered because they have a major impact on signals. (ii) Systems with $S > {}^1\!/_2$ have levels that participate in more than one EPR transition, a characteristic shared with $S = {}^1\!/_2$ PCs having anisotropic HFI. These branching transitions create modulation in echo measurements that is the basis of DEER and PELDOR and is also manifested in nuclear ESEEM. Nevertheless, the FID signals and the echo shapes consist of damped oscillations at frequencies that correspond to lines in the CW-EPR spectrum, i.e., differences between PC energy levels. However, the intensities of transitions in the FT-EPR spectrum differ from those in the CW-EPR spectrum, just as they do in NMR,[27–29] particularly at low temperature or in nonequilibrium states, e.g., with CIDEP.[30–32]

EPR experiments require samples with many PCs. Even in dilute samples, there are interactions, primarily dipolar, between all pairs of PCs in solids. The ensemble of PCs can be viewed as a large spin system with $S \gg {}^1\!/_2$. However, it is much easier to treat them as a collection of $S = {}^1\!/_2$ states. As long as ω_1 and the distribution of ω_0 are much larger than the dipolar interactions $D_{ij}/\hbar$, only the secular part needs to be retained in calculations [see equations (15.17) and (15.18)]. The effect of such weak, dipolar interactions appears as inhomogeneous broadening or relaxation, e.g., instantaneous diffusion.[7,10] Weak spin–spin interactions can be measured in some pulse experiments, e.g., instantaneous diffusion or the Jeener–Broekaert dipolar echo. However, the variation of the pulse EPR signals with turning angle θ and pulse phase ϕ is independent of S for hard MW pulses for PCs initially at thermal equilibrium at $T \gg \hbar\omega_0/k_B$. The value of S must be determined from the spectral information or by the use of 'selective' pulses as discussed in Section 15.4.1.1.

15.4.1 Triplet States and Radical Pairs with $S = 1$

Pulse EPR is used to study triplet PCs owing to their importance in photochemical processes and molecular magnetism. Studies tend to focus on the identity and structure of the triplet PC and, in photoexcited triplets, on the dynamics of nonequilibrium spin states. The typical molecular triplet state has a large singlet–triplet separation, i.e., $|J| \gg \hbar\omega_0$, which also holds for many biradicals.

The main source of EPR spectral broadening in triplet PCs is the zero-field splitting (ZFS) (see Chapter 3). The spectral width often exceeds the MW pulse strength, i.e., $D \gtrsim \hbar\omega_1$. In such a sample, nearly every PC has two EPR transitions with frequencies that differ by more than ω_1. Consequently, the MW pulses are selective, not 'hard', and excite one transition in each triplet more than the other. The standard approach in such cases uses the effective-spin approximation to treat the excited transition as that of a fictitious $S = 1/2$ PC. However, some fraction of the triplet PCs are uniformly excited, so both the hard-pulse and selective-pulse limits will be considered.

The spin evolution between, or after, MW pulses is easy to calculate from an exact solution for the $S = 1$ spin Hamiltonian.[33,34] When the electron Zeeman interaction is much larger than the ZFS, it is convenient to write the relative energies of the three m_S (= +1, 0, and −1) levels as

$$\begin{aligned} E(+1) &= \hbar(+\omega_0 + a + d) \\ E(0) &= 0 \\ E(-1) &= \hbar(-\omega_0 - a + d) \end{aligned} \tag{15.26}$$

where a includes inhomogeneous broadening such as $\boldsymbol{g}$ anisotropy and dipolar interactions with other nonresonant electron and nuclear spins, and $2d$ is the splitting, caused by ZFS, between the two triplet transitions in the frequency domain. The two EPR transitions (and the DQ transition with $\Delta m_S = \pm 2$) have their own Mims matrices that are overlap matrices of the nuclear eigenfunctions for the two m_S levels involved in the transition. When the ZFS is comparable to the electron Zeeman interaction, the Mims matrices and nuclear eigenfrequencies are more complicated.[35]

15.4.1.1 *Selective Pulses*

The selective pulse limit applies when the MW pulse is resonant with one transition, i.e., $\omega_{MW} \approx |\omega_0 + a \pm d|$, but the other transition is far from resonance and is unaffected by the pulse. Calculation of the effect of selective pulses when relaxation is negligible during the MW pulse was developed for ESEEM but is applicable to any selective pulses.[22] The rotating-frame signal is very similar to equation (15.7) for $S = 1/2$. The similarity is because the two states of the triplet are an 'effective' $S = 1/2$ PC with a turning angle $\sqrt{2}$ times stronger. The rotating-frame signals for the $m_S = +1 \leftrightarrow m_S = 0$ transition and the $m_S = 0 \leftrightarrow m_S = -1$ transition are, respectively:

$$V_{+1,0}(t) \propto \Delta P_{+1,0} \sin(\sqrt{2}\theta)\, e^{+i\phi} \sum_{j,k} |M_{k,j}|^2 e^{i(\Delta\omega + \omega_{+1,k} - \omega_{0,j})t} \tag{15.27}$$

and

$$V_{0,-1}(t) \propto \Delta P_{0,-1} \sin(\sqrt{2}\theta) e^{+i\phi} \sum_{n,p} |M_{p,n}|^2 e^{i(\Delta\omega + \omega_{0,p} - \omega_{-1,n})t} \tag{15.28}$$

with ΔP the population difference between the two levels participating in the transition, $\Delta\omega$ is still a resonance offset, and the summations range only over the participating levels, e.g., to account for orientation selection. A narrow range of transitions are excited and detected, approximately $\omega_{MW} \pm \omega_1$, so the FID is often smooth and featureless. Nevertheless, the FID and the echo are composed only of frequencies corresponding to energy differences in the PC.

One highly significant feature of equations (15.27) and (15.28) is the dependence on the pulse turning angle. The turning angle here is defined as $\theta = \omega_1 t_p$, the same as for $S = 1/2$ PCs.[30] Sometimes, the turning angle is defined so that it depends on S, which is inconvenient when the S is unknown. With selective excitation, the triplet PC's FID is most intense at a lower MW pulse power than that of an $S = 1/2$ PC. In fact, θ can be set to suppress the FID, echo, etc. from triplets while observing those of doublet PCs and vice versa. This dependence on S occurs only for selective excitation and is the basis for some nutational spectroscopies.[36]

The echo signals from triplet PCs are very much similar to equations (15.23) and (15.24) for selective pulses but with scaled turning angles. This is clearest in the absence of anisotropic HFI where the echo signals for the single quantum transitions are similar to equation (15.23) and proportional to

$$\Delta P_{+1,0}\, e^{i(-\phi_1+2\phi_2)} \sin(\sqrt{2}\theta_1) \sin^2\left(\frac{\sqrt{2}\theta_2}{2}\right) \times \sum_j \left(e^{i(\Delta\omega+\omega_{+1,j}-\omega_{0,j})\delta t}\, e^{-\frac{\delta t}{T_{2e,j}}}\, e^{-\frac{2\tau}{T_{2e,j}}} \right) \quad (15.29)$$

for the $m_S = +1 \leftrightarrow m_S = 0$ transition, and

$$\Delta P_{0,-1}\, e^{i(-\phi_1+2\phi_2)} \sin(\sqrt{2}\theta_1) \sin^2\left(\frac{\sqrt{2}\theta_2}{2}\right) \times \sum_j \left(e^{i(\Delta\omega+\omega_{0,j}-\omega_{-1,j})\delta t}\, e^{-\frac{\delta t}{T_{2e,j}}}\, e^{-\frac{2\tau}{T_{2e,j}}} \right) \quad (15.30)$$

for the $m_S = 0 \leftrightarrow m_S = -1$ transition, where j is restricted to transitions excited by the pulses. A portion of the EPR spectrum still appears in the time domain in the δt dimension, while the T_{2e} of each of those transitions appears in the τ dimension. When anisotropic HFI results in ESEEM, the signals are similar to equation (15.24) if $\omega_1 \gg |A|/\hbar$. The ESEEM from selective excitation of triplets can be expressed using the same formalisms used for $S = 1/2$ PCs.[12,37]

15.4.1.2 *Hard or Nonselective Pulses*

The other limiting case, equal excitation of both electron spin transitions of the triplet, occurs for small ZFS when $|D| \ll \hbar\omega_1$ or at the center of the triplet EPR spectrum where $d \ll \omega_1$. In that limit, ignoring relaxation and HFI, each MW pulse excites both EPR transitions and a DQ transition, $m_S = +1 \leftrightarrow m_S = -1$. Normally, the DQ transitions are weak or strongly forbidden for triplets at thermal equilibrium at $T \gg \hbar\omega_0/k_B$ and $|J| \gg \hbar\omega_0$. However, photoexcited triplet states with strong electron spin polarization often violate the first of these requirements. Their strong spin polarization can be characterized by populations of spin levels that are linearly related to the energy of the spin levels (Zeeman polarization, a_Z) and a center level population that differs from the average of the other two (dipolar polarization, a_D). These two polarizations produce distinctly different pulse EPR signals.[30–32]

The FID signals from the Zeeman and dipolar polarizations of a triplet PC when there is no ESEEM are

$$V_{FIDZ}(t) \propto i\, a_Z\, e^{i\phi} \sin(\theta) \sum_j \cos(d_j t)\, e^{-i(-\Delta\omega-a_j)t} \quad (15.31)$$

and

$$V_{FIDD}(t) \propto a_D\, e^{i\phi} \sin(2\theta) \sum_j \sin(d_j t)\, e^{-i(-\Delta\omega-a_j)t} \quad (15.32)$$

respectively, with j summed over all EPR transitions. The two FIDs have the same frequency content: a pair of frequencies at $-a \pm d$ relative to the center EPR frequency ω_0. However, there are also important differences. The V_{FIDZ} contains a factor of $i = \sqrt{-1}$, while V_{FIDD} does not. This is a $\pi/2$ phase shift between the time-dependent parts of the two FIDs and means that after FT, the V_{FIDZ} of each PC contributes two absorptive lines or two emissive lines to the total EPR spectrum, while the V_{FIDD} contributes one absorptive and one emissive line. This characteristic of the Zeeman and dipolar polarization is also implicit in equations (15.27) and (15.28) in the selective-pulse limit. We note that in dipole-coupled radical pairs with $|J| \ll \hbar\omega_0$, the Zeeman polarization produces a signal having two phase-shifted components at temperatures below $T \approx \hbar\omega_0/k_B$.[38]

Another very significant difference between the FIDs from the two spin polarizations is their dependence on pulse turning angle. V_{FIDZ} from hard-pulse excitation of the Zeeman polarization is proportional to $\sin(\theta)$, exactly as for $S = 1/2$. Again, showing that it is impossible, in the hard-pulse limit, to use the θ dependence of V_{FIDZ} to distinguish between a triplet (or quartet, etc.) and an $S = 1/2$ PC because the FIDs for both are maximized at $\theta = \pi/2$. In contrast, V_{FIDD}, from dipolar polarization, vanishes for $\theta = \pi/2$ in the hard-pulse limit but is maximized for $\theta = \pi/4$. However, both signals have the same MW pulse phase dependence that is characteristic of all two-pulse echoes. The phase dependence of the signal provides no criteria for experimentally determining either S or polarization, although the turning angle dependence may discriminate between them, but can be complicated by the spatial variation of $\boldsymbol{B}_1$ and θ across the sample in the spectrometer.

Two- and three-pulse echoes from triplet PCs were considered in the hard-pulse limit without ESEEM.[30] Both echoes show distinct signals from the Zeeman and dipolar polarization, all of which have the MW pulse phase dependence characteristic of two-

or three-pulse echoes. However, the turning angle dependence is complicated and different for each polarization, echo type, and S. The echo shape does contain frequencies of the EPR spectrum in the δt dimension for both echoes. The T_{2e} and frequencies of the dipolar splitting, d, appear in the τ dimension of the two-pulse echo, making it possible to measure small ZFS that would not be resolved in the EPR spectrum, as shown for some biradicals.[39] Thus, the echoes provide an opportunity for 2D FT-EPR spectroscopy correlating relaxation or dipolar interactions with EPR spectral features as discussed for PCs with $S = 1/2$.

For triplets with large ZFS, i.e., $|D| \gg \hbar\omega_1$, that are not oriented in single crystals, the outer parts of the spectrum will fall into the selective pulse limit and the center of the spectrum into the hard-pulse limit. Consequently, the shape of the FT-EPR spectrum, whether derived from FIDs or echoes or echo-detected EPR (ED-EPR), will vary with θ and $\boldsymbol{B}_0$. A region appears near the center of each triplet or radical pair in ED-EPR where the hard-pulse limit applies. The FID from such a completely excited $S = 1$ PC is indistinguishable from the signal from an $S = 1/2$ PC unless there is significant dipolar spin polarization.

15.5 DIGITAL FT

A frequency-domain spectrum may be obtained by FT of an experimental, time-domain signal. The fast FT algorithm by Cooley and Tukey allowed data processing within a reasonable period of time using slow computers. A brief glance at the form of the FID signal, e.g., equation (15.7), makes the analysis of FT-EPR signals by Fourier methods seem very simple and straightforward. The formal simplicity of the fast Fourier transform masks a number of issues will be briefly mentioned in this chapter.[11] Rapid advances in computers and digital signal processing make it seem that major changes in signal processing and data collection are likely.

15.5.1 Fourier Transform Spectra

One set of problems affecting Fourier-transform spectra arises from tacit assumptions in equation (15.7): a complex-valued signal starting at time $t = 0$ and extending until relaxation reduces it to zero. In reality, there are phase shifts, time delays, gain variations, and voltage offsets between the two signal channels. The last MW pulse overloads the signal channels so that a portion of the signal is lost in the dead time, t_d, where measurements are not possible. The exact point of $t = 0$ is uncertain, and finite MW pulse widths cause second-order phase shifts in the spectrum. These same problems occur in FT-NMR and ESEEM spectroscopy, and there are fairly effective remedies, e.g., manual or automated phase correction, phase cycling, and signal reconstruction.

The problem associated with a finite t_d is illustrated in Figure 15.7. The first six points of the FID were set to zero for the anisotropic PC from Section 15.2.2.1 but with a shorter $T_{2e} = 0.1$ μs, to simulate the effect of $t_d = 48$ ns. The missing data produces strong oscillations throughout the entire spectral window (Figure 15.7b), thus making it quite difficult to recognize the four EPR lines of the spectrum. The absolute value spectrum (Figure 15.7c, in red) is more recognizable as an EPR spectrum. The spectrum is still severely distorted with the four apparent peaks shifted from their true values and with spurious intensity between the peaks and across the entire spectrum. The absolute-value spectrum for the PC with $t_d = 0$ ns is shown in green for comparison. These effects of spectrometer dead time on the spectrum are readily understood as an undesirable form of windowing.

15.5.1.1 Windowing

The signal is measured during some finite time window, e.g., for $t_d < t < T$. This means that the 'full' signal $S(t)$, for $0 \le t \le \infty$, is multiplied by a windowing or apodization function $W(t, t_d, T)$, which is equal to 1 for $t_d < t < T$ and is 0 otherwise. FT of the product of two time-domain functions gives the convolution of their individual frequency-domain spectra. The spectrum of $W(t, t_d, T)$ is a sinc function ($\text{sinc}(x) = \sin(x)/x$), which has a large maximum and oscillating wings in the frequency domain. As T increases, the sinc function approaches a delta-function, with wings that become narrower and less prominent (Figure 15.7b). In EPR experiments, the prototypical signal is a set of damped oscillations that become Lorentzian lines in the frequency domain. If the 'windowed' signal is too short in time, the oscillating sinc wings are evident around each line in the frequency-domain spectrum

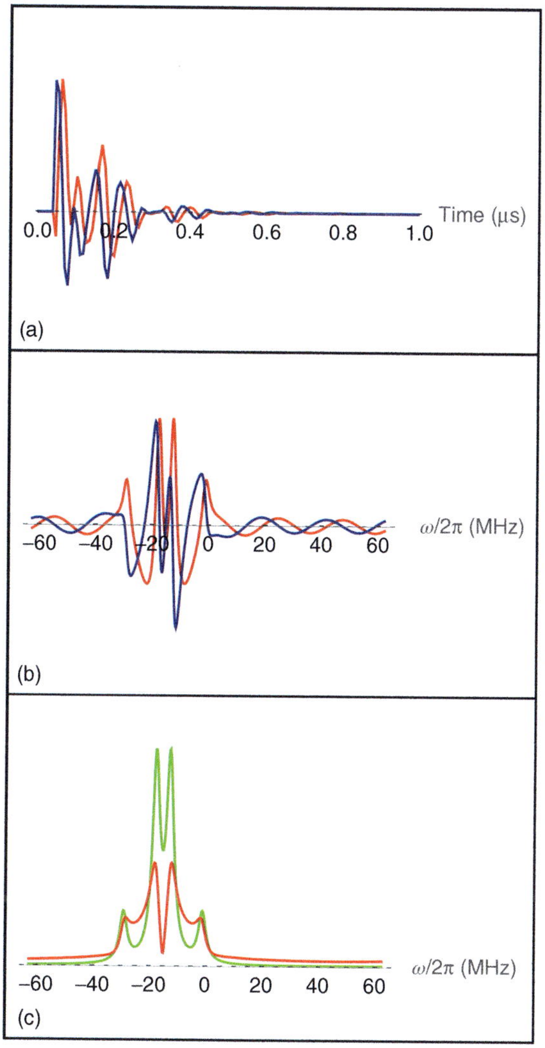

Figure 15.7. The effect of dead time on the FT-EPR spectrum using a simulated FID for the PC in Figure 15.4, except $T_{2e} = 0.1$ μs and the first six points have been set to zero, simulating a spectrometer dead time $t_d = 48$ ns. (a) The simulated FID with t_d. (b) The FT-EPR spectrum from that FID. (c) The magnitude of the complex FT-EPR spectrum for the FID with $t_d = 48$ ns in red and with $t_d = 0$ ns in green

and interferences between different transitions can obscure important spectral features. The sinc wings can be suppressed using smoother windowing functions that smoothly approach zero at $t = t_d$ and $t = T$. Suppressing these wings comes at the price of lower resolution due to additional spectral broadening. Windowing allows even more complex signal manipulation, e.g., if the window function increases linearly from 0 at $t = 0$, the frequency-domain spectrum approaches the first derivative of the EPR spectrum.

Too wide a time window is undesirable for a different reason. At large T, there is very little, if any, FID signal, only noise. So with larger T, there may be no increase in the measured signal, only additional noise, giving a decrease in the overall ratio of signal to noise.

15.5.1.2 Zero Filling

Digital FT of a signal is based on a Fourier series with the signal repeated an infinite number of times in both directions from $t = 0$. This assumes FID data at $t < 0$, which is absent unless an echo experiment captures the entire echo. Half the spectral information in the FID appears only in the dispersion-like, out-of-phase signal that is generally ignored. Zero filling, extending the FID with zeros, moves this information into the in-phase, absorption-like spectrum and smoothly interpolates between points to give smoother spectra.

15.5.1.3 Fold-over

An important parameter in FT-EPR is the sampling rate. This must be high enough to capture the highest frequency component in the signal with at least two measurements per cycle. With a time interval Δt between sampled points (dwell time), the highest frequency in the signal, ω_{max}, should satisfy the condition $\omega_{max} \le \pi/\Delta t$, also known as the Nyquist frequency. If this Nyquist criterion is not met, frequency-aliasing or fold-over occurs. This means that spectral frequencies larger than the Nyquist frequency will appear, after FT, folded over at the opposite end of the spectrum. Too slow a sampling rate therefore leads to serious distortions in the frequency-domain spectrum, while too fast a rate may drastically increase measurement time in multidimensional experiments. Setting the sampling rate needs some care because the wings or tails of a peak can extend to rather high frequencies and fold over to the opposite end of the spectrum.

15.5.2 The Future for Spectral Processing

The brief discussion of FT does not exhaust the subject but points out some important considerations. The use of homemade FT programs or commercial data processing packages, e.g., Mathematica, Matlab, Fortran, or software from EPR vendors, does not free the user from the need to consider windowing, zero filling, frequency aliasing, fold over, and a host of other effects. Computational methods that produce spectra without FT are rapidly developing and promise to radically change FT-EPR spectroscopy, particularly multidimensional spectroscopy.

FT and signal processing for FT-EPR is identical, in principle, to that in FT-NMR, but there are three practical differences that present challenges for traditional high-resolution FT-NMR approaches to signal processing. One major difference is the large dead time t_d in EPR relative to the T_{2e} or inhomogeneous broadening T_{2e}*. The t_d is typically ≈0.1 μs while T_{2e} or T_{2e}* is rarely greater than 2 μs and can even be orders of magnitude less than t_d. This results in the loss of much or all of the FID and frequently prevents reconstruction of the FT-EPR spectrum from the FID. In this regard, the FID of PCs resembles those of quadrupolar nuclei in solid-state NMR much more than those of high-resolution NMR.

A second difference is that MW pulses and pulse sequences are much less ideal than those in NMR. The pulse turning angle often varies significantly across the sample due to $\boldsymbol{B}_1$ inhomogeneity; there is significant spin relaxation during the pulse or pulse sequence; and significant spin evolution can occur during each pulse.

A third difference is that ω_1 is often much less than the width of the EPR spectrum. It is not uncommon for EPR spectra to have a width of 100 000 ppm (10% of ω_{MW}) or more. It is not practical to excite the full spectrum, nor is that much bandwidth desirable because of the loss of sensitivity as a result of the larger bandwidths and lower resonator Q-values required for full excitation (see Chapter 12).[10] FT-EPR requires methods that can make use of the limited excitation width and still produce useful spectra or spectrally resolved measurements.

Obtaining FT-EPR spectral information by spin echoes relieves much of the dead time problem but exacerbates effects of spin relaxation and evolution during the pulse sequence. Echoes are relatively sensitive to variations in pulse turning angle across a sample. Nevertheless, FT-EPR using echoes does measure EPR frequencies of the PC, which are sufficient to determine the spin Hamiltonian, as in CW-EPR. Echoes do allow resolution and measurement of spin relaxation rates of the observable transitions, providing valuable information that is difficult to obtain by CW-EPR.

15.6 FT-EPR EXAMPLES

FT-EPR spectroscopy often uses spin echoes in ways quite different from FT-NMR spectroscopy. Consequently, a few echo-based experiments and their resulting spectra are illustrated here. All examples were measured at X-band on a Bruker Biospin E680 pulse EPR spectrometer under Xepr.

15.6.1 Echo-detected EPR

The oldest way to retrieve the EPR spectrum from pulse EPR experiments is ED-EPR. Part, or all, of the echo is averaged with a square window function, also known as boxcar averaging, while $\boldsymbol{B}_0$ is swept through the EPR spectrum. This integrated signal at each $\boldsymbol{B}_0$ is the analog equivalent of the point at zero frequency in the Fourier transform of the echo at each $\boldsymbol{B}_0$ with that windowing function. As $\boldsymbol{B}_0$ is swept, the FT-EPR spectrum is traced out point by point. If the windowing function is much wider than the duration of the echo, a high-resolution spectrum can be recovered, while narrow windows broaden and distort spectra. A series of ED-EPR spectra can provide spectrally resolved ESEEM or relaxation information from the dependence of the signal on delay between pulses.[23,40]

ED-EPR is demonstrated with the substitutional nitrogen center, P1, in 10 synthetic diamond cubes ≈0.4 mm on each side.[41] This PC has isotropic $\boldsymbol{g}$ and anisotropic $\boldsymbol{A}$, but the small number of crystals only approaches the powder-average spectrum. The P1 center has virtually no ESEEM, only a miniscule amount from natural abundance ^{13}C. The full echo shape was digitized at evenly spaced values of $\boldsymbol{B}_0$. ED-EPR spectra were calculated from this single data set by digital integration with different window widths (Figure 15.8). The ED-EPR spectra consist of three hyperfine lines from nitrogen. With a narrow window, all three hyperfine lines have similar shape and intensity. However, with broader integration windows, the center line becomes sharper and more intense because it is not broadened by $\boldsymbol{g}$ or $\boldsymbol{A}$ anisotropy. The outer hyperfine lines show axial $\boldsymbol{A}$ anisotropy with broader windows, but the small number of crystals in the

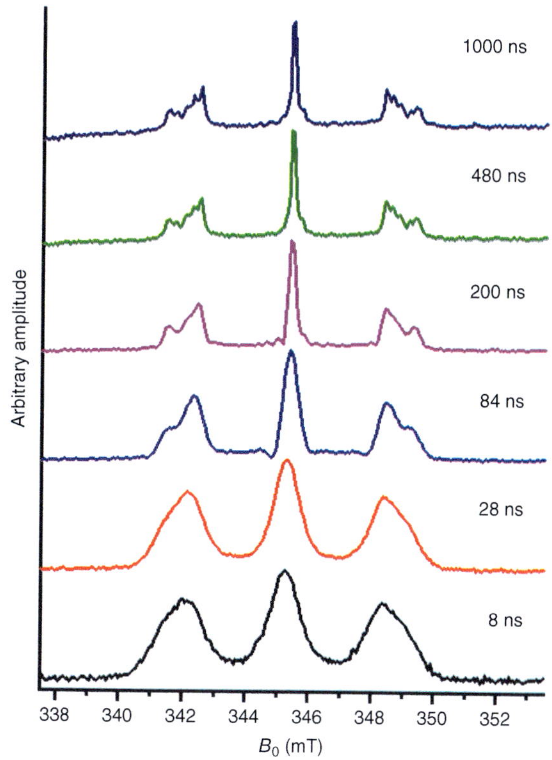

Figure 15.8. Room-temperature ED-EPR spectra of the substitutional nitrogen center, P1, in 10 synthetic diamond cubes ≈0.4 mm on each side for a total sample volume of 6×10^{-4} cm^3. The two-pulse echo was measured with $\tau = 500$ ns, and 16 ns MW pulses with turning angles of $\pi/2$ to decrease relaxation from instantaneous diffusion. Each ED-EPR spectrum was generated by integration of the echo over a rectangular window function centered on the echo. The three nitrogen hyperfine lines appear, but the small number of crystals in the sample is not sufficient to reach the full powder-average spectrum and sharp lines from individual crystals can be seen in some spectra. From bottom to the top, the widths of the integration window are 8, 28, 84, 200, 480, and 1000 ns. Spectra are normalized to the same peak intensity. Note that the relative noise decreases initially as the window becomes wider and more of the peak echo signal is integrated, but for wider windows, the noise again increases as more noise but little additional echo signal contributes to the spectrum

sample produces sharp features seen with the broadest integration windows. Note that the signal-to-noise ratio, after initially increasing as the integration window broadens from 8 to 84 ns, decreases as the windows broaden further because the broader windows include more noise but little additional signal intensity in the spectra.

ED-EPR spectral measurements generate echoes at many values of $\boldsymbol{B}_0$, and each echo contains FT-EPR information over a range $\approx B_0 \pm B_1$. However, all the spectral information is discarded except for one point. As a result, significant sensitivity or resolution is lost in the ED-EPR approach but may be recovered in analogous experiments detecting the FID from a single, very selective pulse.

15.6.2 Skew EPR

Information wasted in ED-EPR can be recovered by an approach called skew EPR.[41] The echo shapes recorded to illustrate ED-EPR in the previous section can be Fourier transformed with respect to δt into a series of FT-EPR spectra based on equation (15.23). Each spectrum is measured at a different $\boldsymbol{B}_0$ and its center fulfills a slightly different resonance condition; this is the basis for ED-EPR. The individual spectra can be shifted to bring them all into registry at an arbitrary resonance condition through a shear transform of the 2D data set, resulting in Figure 15.9. The original rectangular data set becomes a stripe across a larger data set produced by shifting the frequencies of each horizontal slice by the equivalent of the $\boldsymbol{B}_0$ offset of that slice. A sum projection onto the EPR frequency axis recovers the frequency-domain FT-EPR spectrum shown at the top. This strategy is effective even if the entire EPR spectrum is not uniformly excited. One adage in NMR holds that a series of narrow-band, optimized spectral measurements, as in this skew EPR example, should deliver better sensitivity and efficiency than the equivalent broadband, optimized measurement with lower Q and higher bandwidth.

Field-domain, skew-EPR spectra can be obtained by shearing vertical slices of the data set by its magnetic field offset. The sum projection onto the vertical axis would be equivalent to the ED-EPR spectrum but with much less noise. Each point in the skew-spectrum is the sum of signals from several slices, giving a considerable amount of signal averaging. Liberal zero filling before the FT increases the number of points or

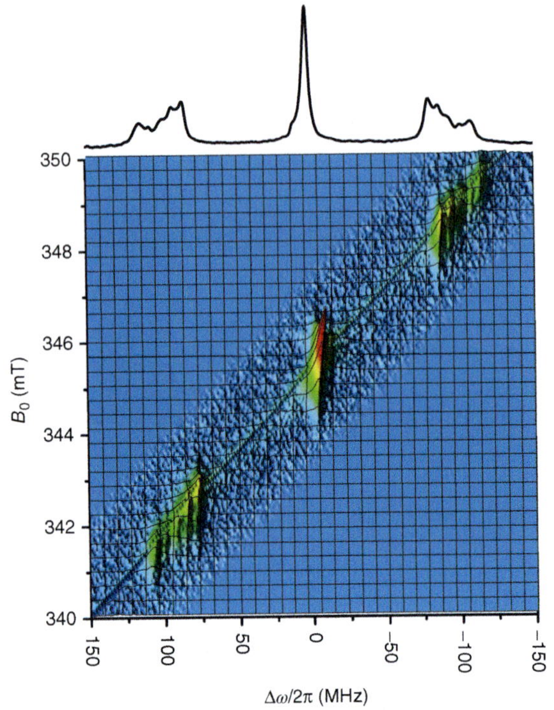

Figure 15.9. Skew EPR spectrum from the same data set used in Figure 15.8. The 2D surface shows the FT-EPR spectrum from the echo shape at a series of $\boldsymbol{B}_0$ as horizontal stripes. $\Delta\omega_{EPR}$ is the shift in EPR frequency from the nominal ω_{MW}. The individual FT-EPR spectra are horizontally shifted by a simple shear transform so that transitions in different horizontal slices line up vertically. The sum projection onto the frequency axis provides the high-resolution FT-EPR spectrum. The diagonal band of noise provides an indication of the bandwidth of the single FT-EPR measurements and the degree of shear applied to the data

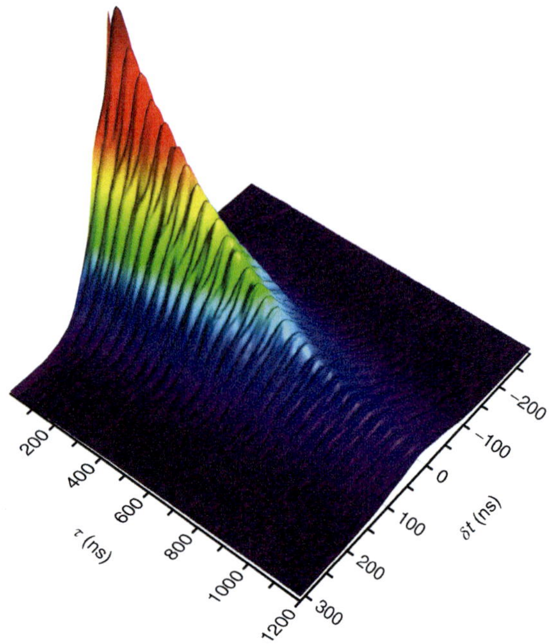

Figure 15.10. A 2D surface plot of the two-pulse spin echo using 8 and 16 ns pulses of 2 mM Finland trityl in methanol at 9.772 GHz. Note the ridges, gullies, and other texture on the side of the echo that result from variations in echo amplitude and shape as τ changes

'digital resolution' in either domain as compared to the ED-EPR spectrum from the same dataset.[41]

15.6.3 High-resolution 2D FT-EPR

Almost any property of the PC that can be reported by an echo or FID can be visualized as the second dimension of a 2D FT-EPR measurement, as demonstrated here by ESEEM from the two-pulse echo of a dilute, frozen solution of Finland trityl in methanol.[42] The EPR spectrum has one strong line flanked by spin-flip satellite lines that produce weak ^{1}H ESEEM on the echo decay.[42] The ESEEM is from the radical and the solvent and is described by equations (15.23) and (15.25). The HFI from ^{1}H is rather weak, so that ω_α and ω_β are approximately ω_I.

The echo intensity is modulated as a function of τ, (Figure 15.10) as expected for ESEEM and described by the terms containing τ in equations (15.24) and (15.25). However, close examination shows that the echo shape (in the δt dimension) varies with τ, (Figure 15.11a). There are periodic changes in echo shape, in its width at half height, and even in its peak position in the δt dimension. These are all indications that the modulation varies across the echo as predicted in equations (15.23) and (15.25). The signal intensity measured as a function of τ for values of $\delta t = -8$, 40, or 92 ns shows strikingly different modulation patterns (Figure 15.11b). It is simple to see how the ESEEM varies with δt across the echo, by subtracting the unmodulated baseline (or decay) from each slice in the τ-direction (Figure 15.12). Both the phase and

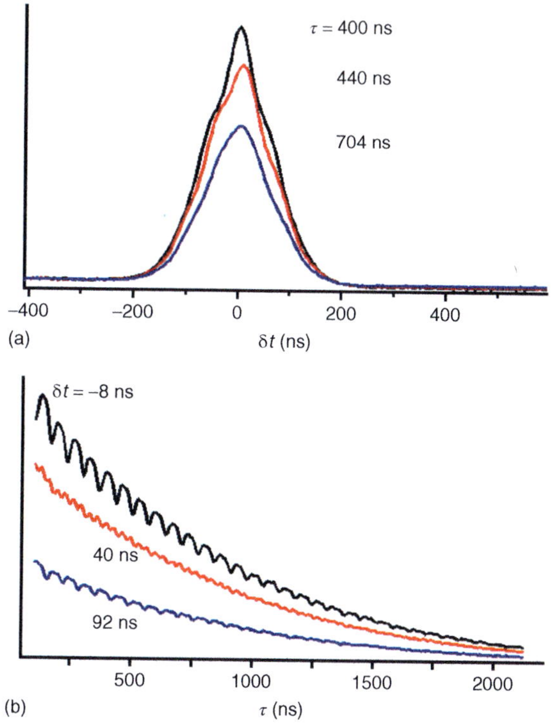

Figure 15.11. Slices through the data set in Figure 15.10. (a) Echo shape for $\tau = 400$ (black), 440 (red), and 704 ns (blue). Note how the echo shape and peak position changes with τ. (b) The echo intensity at $\delta t = -8$ (black), 40 (red), and 92 ns (blue). Note that the ESEEM frequency content of the three traces varies considerably. At $\delta t = -8$ ns, the fundamental ESEEM frequencies predominate, while only the sum combination is evident in the 40-ns trace, and all ESEEM frequencies are present at 92 ns. Slight differences in measurement conditions can have a strong influence on ESEEM spectra and the sensitivity of ESEEM spectroscopy

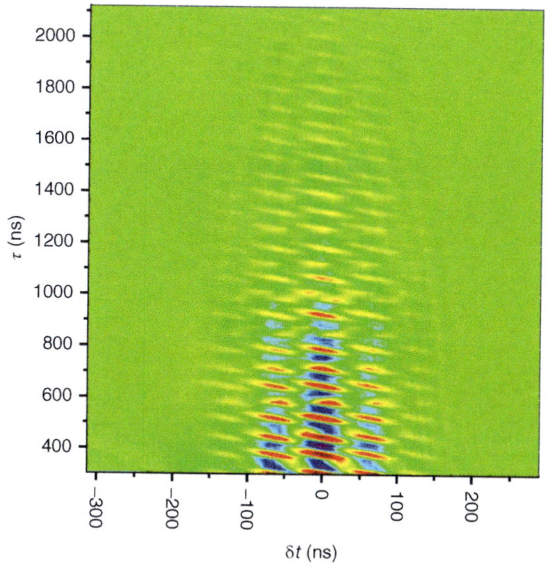

Figure 15.12. The data in Figure 15.10 with the unmodulated decay of the echo signal at each value of δt subtracted. Note that both the phase and amplitude of the ESEEM are a function of δt

the amplitude of each frequency component vary, as indicated in Figures 15.11 and 15.12.

FT in the δt dimension produces a series of FT-EPR spectra showing the sharp trityl line flanked by its two proton spin-flip satellite lines at $\approx \pm 15$ MHz[22,42] (Figure 15.13a). Amplitudes of all three lines oscillate with τ. The satellite lines are positive going at $\tau = 408$ ns, but negative at $\tau = 504$ ns. The center line also has an oscillation with τ on top of its large positive intensity. The oscillations are readily seen in slices taken along τ for different values of $\Delta\omega_{EPR}$ (Figure 15.13b). The formally forbidden EPR transitions (spin-flip satellite lines) are modulated by the ENDOR frequencies of the ^{1}H and their sums ($\sim \pm 2\omega_I$). The center line, which has the formally allowed EPR transitions, is modulated by the fundamental ESEEM frequencies and the difference combination frequency ($\sim \pm A$) that survived baseline subtraction. This is a general feature for weak HFI.[23]

The ESEEM frequencies are revealed by 2D FT of the 2D echo shape in Figure 15.10(a). A color density plot of the spectrum (Figure 15.14a) shows the FT-EPR spectrum with red positive peaks in the slice at $\Delta\omega = 0$; the formally-allowed and -forbidden EPR transitions at the fundamental frequencies $\omega_\alpha \approx \omega_\beta$; and the formally forbidden transition at $\omega_\alpha + \omega_\beta \approx \pm 2\omega_I$ as blue negative peaks. It is not apparent in Figure 15.14(a) that the formally allowed transitions in the center of the spectrum are flanked by pairs of negative lines at the $\omega_\alpha - \omega_\beta \approx 0$ difference combination frequency. Slices along $\Delta\omega_{ESEEM}$ at different values of EPR frequency, $\Delta\omega_{EPR}$ (Figure 15.14b), show ESEEM spectra at the formally allowed transitions and the spin-flip satellite lines. When there is appreciable HFI, all the ESEEM frequencies can be resolved and used to interpret spectra and assign the relative signs of spin Hamiltonian terms.[23] 2D FT-EPR spectra can be obtained for three- or four-pulse echo ESEEM by measuring

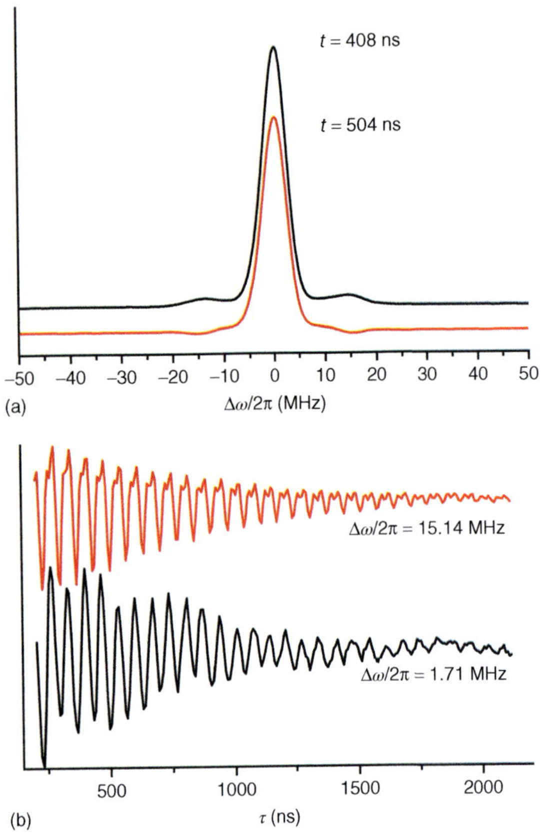

Figure 15.13. (a) FT-EPR spectra from Fourier transforms of slices from data in Figure 15.10 at $t = 408$ (black) and 504 ns (red). The spin-flip satellite lines at $\approx\pm15$ MHz have opposite intensities in these two spectra. (b) Intensity at $\Delta\omega_{EPR}/2\pi = 1.71$ (black) and 15.14 MHz (red) in the FT-EPR spectra showing the ESEEM after subtraction of the unmodulated signal

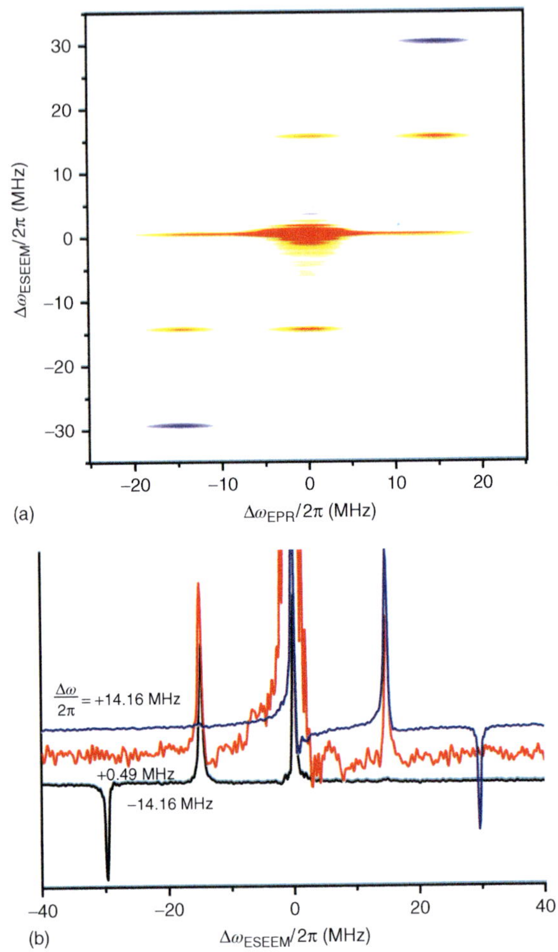

Figure 15.14. (a) The 2D spectrum of ESEEM frequency versus EPR frequency from a 2D Fourier transform of the data in Figure 15.10. Red and yellow indicate positive peaks, while blue indicates negative peaks. (b) Slices through the upper spectrum showing the ESEEM spectra for $\Delta\omega_{EPR}/2\pi = +14.16$ (blue), 0.49 (red), and -14.16 MHz (black)

the echo shape, but their patterns of peaks differ. Three-dimensional and higher spectra, e.g., from HYSCORE measurements, are possible.

ACKNOWLEDGMENTS

We thank John Weil and Kev Salikhov for helpful discussions over the years, Molly Lockart for useful suggestions, and the RFBR NO. 14-03-93180 (AGM) and NSF 1416238 (MKB) for supporting the trityl measurements.

FURTHER READING

A. Abragam, The Principles of Nuclear Magnetism, Clarendon Press: Oxford, 1961.

A. Abragam and B. Bleaney, Electron Paramagnetic Resonance of Transition Ions, Dover: New York, 1986.

S. A. Dikanov and Y. D. Tsvetkov, Electron Spin Echo Envelope Modulation (ESEEM) Spectroscopy, CRC Press: Boca Raton, 1992.

R. R. Ernst, G. Bodenhausen, and A. Wokaun, Principles of Nuclear Magnetic Resonance in One and Two Dimensions, Clarendon Press: Oxford, 1987.

L. Kevan and M. K. Bowman, Modern Pulsed and Continuous-Wave Electron Spin Resonance, Wiley: New York, 1990.

M. H. Levitt, Spin Dynamics: Basics of Nuclear Magnetic Resonance, 2nd edn, Wiley: Chichester, UK, 2008.

W. B. Mims, in Electron Paramagnetic Resonance, ed S. Geschwind, Plenum: New York, 1972, p 263.

K. M. Salikhov, A. G. Semenov, and Y. D. Tsvetkov, Electron Spin Echoes and their Applications, Nauka: Novosibirsk, 1976.

A. Schweiger and G. Jeschke, Principles of Pulse Electron Paramagnetic Resonance, Oxford University Press: Oxford, UK; New York, 2001.

C. P. Slichter, Principles of Magnetic Resonance, Springer: Berlin, 1996.

J. A. Weil and J. R. Bolton, Electron Paramagnetic Resonance Elementary Theory and Practical Applications, 2nd edn, Wiley-Interscience: New York, 2007.

REFERENCES

1. M. K. Bowman, in Modern Pulsed and Continuous Electron Spin Resonance, eds L. Kevan and M. K. Bowman, Wiley: New York, 1990, Vol. 1, p 1.
2. J. M. Lu, J. Geimer, S. Naumov, and D. Beckert, *Phys. Chem. Chem. Phys.*, 2001, **3**, 952.
3. J. Gorcester, G. L. Millhauser, and J. H. Freed, in Modern Pulsed and Continuous-Wave Electron Spin Resonance, eds L. Kevan and M. K. Bowman, Wiley: New York, 1990, Vol. 1, p 119.
4. K. A. Earle, B. Dzikovski, W. Hofbauer, J. K. Moscicki, and J. H. Freed, *Magn. Reson. Chem.*, 2005, **43**, S256.
5. A. Abragam and B. Bleaney, Electron Paramagnetic Resonance of Transition Ions, Dover: New York, 1986.
6. J. A. Weil and J. R. Bolton, Electron Paramagnetic Resonance: Elementary Theory and Practical Applications, 2nd edn, Wiley-Interscience: New York, 2007.
7. A. Schweiger and G. Jeschke, Principles of Pulse Electron Paramagnetic Resonance, Oxford University Press: Oxford, UK; New York, 2001.
8. D. T. Teaney, M. P. Klein, and A. M. Portis, *Rev. Sci. Instrum.*, 1961, **32**, 721.
9. M. H. Levitt, Spin Dynamics: Basics of Nuclear Magnetic Resonance, 2nd edn, Wiley: Chichester, UK, 2008.
10. K. M. Salikhov, A. G. Semenov, and Y. D. Tsvetkov, Electron Spin Echoes and their Applications, Nauka: Novosibirsk, 1976.
11. R. R. Ernst, G. Bodenhausen, and A. Wokaun, Principles of Nuclear Magnetic Resonance in One and Two Dimensions, Clarendon Press: Oxford, 1987.
12. S. A. Dikanov and Y. D. Tsvetkov, Electron Spin Echo Envelope Modulation (ESEEM) Spectroscopy, CRC Press: Boca Raton, 1992.
13. M. K. Bowman and R. J. Massoth, in Electronic Magnetic Resonance of the Solid State, eds J. A. Weil, M. K. Bowman, J. R. Morton and K. F. Preston, Canadian Society for Chemistry: Ottawa, 1987, p 99.
14. A. G. Maryasov, M. K. Bowman, and Y. D. Tsvetkov, *Appl. Magn. Reson.*, 2002, **23**, 211.
15. A. G. Maryasov and M. K. Bowman, *J. Magn. Reson.*, 2013, **233**, 80.
16. A. G. Maryasov and M. K. Bowman, *J. Magn. Reson.*, 2012, **221**, 69.
17. A. F. Bedilo and A. G. Maryasov, *J. Magn. Reson. A*, 1995, **116**, 87.
18. V. V. Kurshev, A. M. Raitsimring, and K. M. Salikhov, *Fiz. Tverd. Tela*, 1988, **30**, 420.
19. A. V. Astashkin, B. O. Elmore, W. H. Fan, J. G. Guillemette, and C. J. Feng, *J. Am. Chem. Soc.*, 2010, **132**, 12059.
20. M. K. Bowman, in Electron Paramagnetic Resonance – A Practitioners Toolkit, eds M. Brustalon and E. Giamello, John Wiley & Sons: Hoboken, 2009, p 159.
21. C. Gemperle, G. Aebli, A. Schweiger, and R. R. Ernst, *J. Magn. Reson.*, 1990, **88**, 241.
22. G. M. Zhidomirov and K. M. Salikhov, *Theor. Exp. Chem.*, 1968, **4**, 332.
23. J. Isoya, M. K. Bowman, J. R. Norris, and J. A. Weil, *J. Chem. Phys.*, 1983, **78**, 1735.
24. W. B. Mims, *Phys. Rev. B Solid State*, 1972, **6**, 3543.
25. W. B. Mims, *Phys. Rev. B Solid State*, 1972, **5**, 2409.
26. S. Stoll and R. D. Britt, *Phys. Chem. Chem. Phys.*, 2009, **11**, 6614.
27. S. Schäublin, A. Hohener, and R. R. Ernst, *J. Magn. Reson.*, 1974, **13**, 196.
28. R. R. Ernst, W. P. Aue, E. Barthold, A. Hohener, and S. Schäublin, *Pure Appl. Chem.*, 1974, **37**, 47.

29. S. Schäublin, A. Wokaun, and R. R. Ernst, *J. Magn. Reson.*, 1977, **27**, 273.
30. M. K. Bowman, H. Chen, and A. G. Maryasov, *Z. Phys. Chem.*, 2017, **231**, 637.
31. M. N. Uvarov, J. Behrends, A. G. Maryasov, and L. V. Kulik, *Appl. Magn. Reson.*, 2016, **47**, 781.
32. K. Hasharoni, H. Levanon, J. Tang, M. K. Bowman, J. R. Norris, D. Gust, T. A. Moore, and A. L. Moore, *J. Am. Chem. Soc.*, 1990, **112**, 6477.
33. G. M. Muha, *J. Magn. Reson.*, 1982, **49**, 431.
34. G. M. Muha, *J. Chem. Phys.*, 1980, **73**, 4139.
35. A. G. Maryasov, M. K. Bowman, and Y. D. Tsvetkov, *Appl. Magn. Reson.*, 2006, **30**, 683.
36. J. Isoya, H. Kanda, J. R. Norris, J. Tang, and M. K. Bowman, *Phys. Rev. B Condens. Matter*, 1990, **41**, 3905.
37. H. L. Yu, D. J. Sloop, S. I. Weissman, T. S. Lin, J. R. Norris, and M. K. Bowman, *J. Phys. Chem.*, 1982, **86**, 4287.
38. A. G. Maryasov, S. A. Dzuba, and K. M. Salikhov, *J. Magn. Reson.*, 1982, **50**, 432.
39. V. F. Yudanov, K. M. Salikhov, G. M. Zhidomirov, and Y. D. Tsvetkov, *Theor. Exp. Chem.*, 1969, **5**, 451.
40. S. A. Dzuba, Y. D. Tsvetkov, and A. G. Maryasov, *Chem. Phys. Lett.*, 1992, **188**, 217.
41. M. K. Bowman, M. D. Krzyaniak, A. A. Cruce, and R. T. Weber, *J. Magn. Reson.*, 2013, **231**, 117.
42. S. N. Trukhan, V. F. Yudanov, V. M. Tormyshev, O. Y. Rogozhnikova, D. V. Trukhin, M. K. Bowman, M. D. Krzyaniak, H. Chen, and O. N. Martyanov, *J. Magn. Reson.*, 2013, **233**, 29.

Chapter 16
Hyperfine Spectroscopy – ENDOR

Jeffrey R. Harmer

University of Queensland, St Lucia, Queensland, Australia

16.1 INTRODUCTION

To characterize the electronic and geometric structure of a paramagnetic center requires determining as completely as possible the couplings between the paramagnetic center and the nearby magnetic nuclei. Often the g-matrix, zero-field splitting (ZFS) (for $S > 1/2$), and largest hyperfine (HF) interactions can be determined by field-sweep EPR methods, for example, from a CW or echo-detected EPR spectrum. However, the resolution of the field-sweep approach is limited and many smaller important couplings are typically not resolved. In order to access this information, high-resolution techniques for measuring magnetic interactions between the paramagnetic center and nearby magnetic nuclei can be employed – this approach is typically referred to as HF spectroscopy and delivers information on the HF (see Chapter 5) and nuclear quadrupole (NQ) (see Chapter 6) interactions. The HF interaction is determined by the electron spin density distribution and provides information on the position of the nuclei relative to the paramagnetic center and on the spatial characteristics of the single occupied molecular orbital (SOMO) (see Chapter 7 for more detail); the NQ interaction is determined by the electric field gradient at the site of the nucleus and is thus able to characterize the bonding situation.

Electron–nuclear double resonance (ENDOR) is one of the main techniques in HF spectroscopy for measuring nuclear transition frequencies of a paramagnetic center. ENDOR involves excitation of the nuclear transitions with radio frequency (RF) irradiation and detection of these NMR transitions via a change in the EPR signal. The addition in ENDOR of an RF pulse greatly increases the resolution as compared to a field-sweep EPR spectrum enabling the measurement of very small to large HF and NQ couplings. The ENDOR technique was first introduced to study electron–nuclear couplings in radicals in the solid state by Feher in 1956,[1] and then to radicals in solution in 1964 by Hyde and Maki.[2,3]

Today, ENDOR techniques are used extensively to determine the coupling of magnetic nuclei to a paramagnetic center in a great variety of applications; a small selection of review chapters follows: chemistry review[4]; bioinorganic[5]; materials science[6]; quantum dots[7]; microporous materials[8]; metalloenzymes and transition metals[9–11]; photosynthesis and high-field EPR (HF EPR)[12–14]; HFEPR/ENDOR,[15] organics radicals[16]; and proton ENDOR in proteins.[17] Spectrometers have been developed for ENDOR measurements at an array of frequencies, see, for example,

EPR Spectroscopy: Fundamentals and Methods.
Edited by Daniella Goldfarb and Stefan Stoll

ISBN: 978-1-119-16299-5
Also published in eMagRes (online edition)
DOI: 10.1002/9780470034590.emrstm1515

at Q-band,[18] W-band,[19] 140 GHz,[20] 180 GHz,[21] 360 GHz,[22] and reviews of HF EPR with ENDOR.[23,24]

There are two main branches of ENDOR; in CW ENDOR, continuous microwave (MW) and RF excitation of the sample is employed and in pulse ENDOR, MW and RF pulses are used. Pulse ENDOR relies on detection of transverse magnetization, usually via an electron spin echo, and is thus limited to systems with phase memory times T_m that are greater than the spectrometer dead time (80 ns is a typical pulse spectrometer dead time) and spin–lattice relaxation times T_1 long compared to the ENDOR pulse sequence. CW ENDOR does not have this constraint but does require that the spin system can be partially saturated by the CW irradiation and this requires optimizing the relaxation times of the unpaired electron and nuclear spins by e.g. adjusting the measurement temperature and sample viscosity.

This chapter starts with a description of the spin Hamiltonian and the features of ENDOR spectra for simply $S = 1/2$ spin systems and in the more complex high-spin systems where $S > 1/2$, before the concepts behind CW ENDOR and pulsed ENDOR are explained. A selection of ENDOR sequences and acquisition schemes is presented throughout with the aim of covering a representative range of approaches that ENDOR offers to investigate paramagnetic centers.

16.2 STATIC SPIN HAMILTONIAN AND THE NUCLEAR FREQUENCIES

The general static spin Hamiltonian in the laboratory frame needed to describe the energy levels and hence the spectra from an ENDOR experiment can be written as[25]

$$\hat{H}_0 = \hat{H}_{EZ} + \hat{H}_{ZFS} + \hat{H}_{HF} + \hat{H}_{NZ} + \hat{H}_{NQ} \quad (16.1)$$

$$= \mu_B \boldsymbol{B}_0^T \boldsymbol{g} \hat{\boldsymbol{S}}/\hbar + \hat{\boldsymbol{S}}^T \boldsymbol{D} \hat{\boldsymbol{S}} + \sum_{k=1}^{m} \hat{\boldsymbol{S}}^T \boldsymbol{A}_k \hat{\boldsymbol{I}}_k$$

$$- \mu_N \sum_{k=1}^{m} g_{n,k} \boldsymbol{B}_0^T \hat{\boldsymbol{I}}_k/\hbar + \sum_{k=1\ I>=1}^{m} \hat{\boldsymbol{I}}_k^T \boldsymbol{P}_k \hat{\boldsymbol{I}}_k \quad (16.2)$$

The electron Zeeman (EZ), ZFS, and largest HF interaction (e.g., ^{65,67}Cu in a Cu(II) complex) typically determine the appearance of the field-sweep EPR spectrum and its simulation consequently delivers the principal values and relative orientations of these interactions. All other couplings from magnetic nuclei around the paramagnetic center are not resolved and contribute to the inhomogeneously broaden EPR linewidth. These unresolved couplings are usually the target of ENDOR experiments that can detect, for example, protons (with the largest magnetic moment) up to a distance of about 10 Å from the paramagnetic center; for instance, Ref. 26 describes the accessible distance range with an emphasis on ^{31}P nuclei.

In the simplest case, for an isotropic g-matrix and HF interaction, and in the high-field approximation where $\hat{H}_{EZ}$ dominates (and the ZFS is negligible) and thus the electron and nuclear spins are quantized along the B_0 field (along the z-axis), the spin Hamiltonian of equation (16.2) simplifies to

$$\hat{H}_0 = \left(\frac{\mu_B g_{iso} B_0}{\hbar}\right)\hat{S}_z - \left(\frac{\mu_N g_n B_0}{\hbar}\right)\hat{I}_z + A_{iso}\hat{S}_z\hat{I}_z \quad (16.3)$$

$$= \omega_S \hat{S}_z - \omega_I \hat{I}_z + A_{iso}\hat{S}_z\hat{I}_z \quad (16.4)$$

with the first-order eigenvalues (energies in radians, ω)

$$\omega(m_S, m_I) = \omega_S m_S - \omega_I m_I + A_{iso} m_S m_I \quad (16.5)$$

In an EPR experiment, the selection rules are $\Delta m_S = \pm 1$ and $\Delta m_I = 0$, and therefore, there are $2S$ allowed EPR transitions. In an ENDOR experiment, the allowed NMR transitions are driven by an RF field with the selection rules $\Delta m_S = 0$ and $\Delta m_I = \pm 1$. According to equation (16.5), the allowed nuclear frequencies that are measured in an ENDOR experiment are given by

$$\omega_N(m_S) = |\omega(m_S, m_I+1) - \omega(m_S, m_I)| = |m_S A_{iso} - \omega_I| \quad (16.6)$$

This equation shows that the nuclear frequencies $\omega_N(m_S)$ depend on the electron spin manifold via m_S.

For an $S = 1/2$ spin system, the nuclear (NMR) frequencies are, to first order, symmetrically centered around the Larmor frequency of the nucleus $|\omega_I|$ in the weak coupling case defined as $2|\omega_I| > |A|$, or $|A|/2$ in the strong coupling case, $2|\omega_I| < |A|$. Energy-level diagrams for an $S = 1/2$, $I = 1/2$ spin system for $A > 0$ and $A < 0$ (and the nuclear g-factor $g_n > 0$) in the strong and weak coupling cases are shown in Figure 16.1 (computed using EasySpin).[27] Note that the sign of the HF coupling dictates which electron spin manifold has the largest nuclear frequency.

The spin Hamiltonian in equation (16.3) assumes an isotropic HF interaction, which is almost never met when ENDOR is carried out on samples in the solid state. The anisotropic contribution to the HF interaction is a very important source of structural

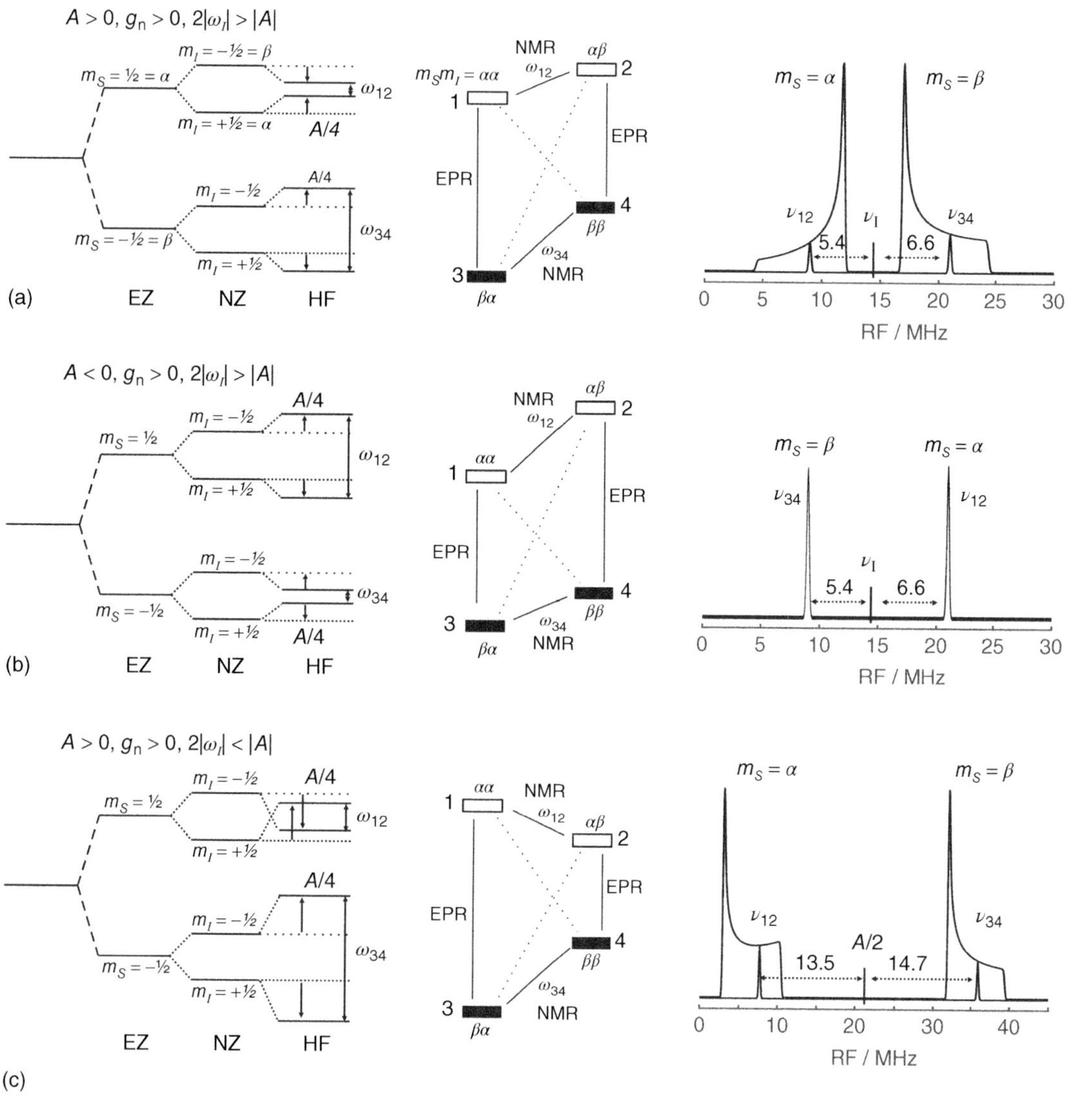

Figure 16.1. Energy-level diagrams for an $S = 1/2$, $I = 1/2$ spin system and the corresponding ENDOR (electron–nuclear double resonance) spectra. (a) Weak coupling case, $2|\omega_I| > |A|$, with $A > 0$, $g_n > 0$; the largest nuclear frequency ω_{34} is associated with the $m_S = -1/2 = \beta$ electron spin manifold. The ENDOR spectrum (right) shows the two peaks for a single orientation at $\theta = 45°$, and the powder average (note the unit conversion $\nu = \omega/2\pi$). Because the hyperfine coupling is anisotropic, the peaks are shifted to higher frequency (equations (16.8) and (16.9)) and are not centered around ω_I and split by $|A|$ (equation (16.6)) except along the principal axes where $\theta = 0°$ and $90°$. In this example, the principal values are $A(^1H) = 10 + [-5, -5, 10]$ MHz with $\nu_I = \omega_I/2\pi = 14.3$ MHz, and the frequency shifts relative to ν_I are 5.4 and 6.6 MHz as shown in the diagram. (b) Same parameters as in (a) but with $A < 0$, $g_n > 0$, and now the largest nuclear frequency ω_{12} is associated with the $m_S = +1/2 = \alpha$ electron spin manifold. (c) Strong coupling case $2|\omega_I| < |A|$, anisotropy in the hyperfine coupling shifts peaks to higher frequency and the first-order expression where peaks are centered at $|A|/2$ and split by $2|\omega_I|$ is not valid (equation (16.6)). Shown are the two peaks for a single orientation at $\theta = 45°$, and the powder average, for $A(^1H) = 40 + [-5, -5, 10]$ MHz with $\omega_I/2\pi = 14.3$ MHz. The frequency shifts at $\theta = 45°$ relative to $|A|/2$ are 13.5 MHz and 14.7 MHz as shown in the diagram

information. With anisotropy, the ENDOR peaks are shifted to higher frequencies (relative to ω_I or $A/2$) when the $\boldsymbol{B}_0$ field vector is not collinear with one of the HF principal value axes as shown in Figure 16.1 (right) for the strong and weak coupling cases. To describe this case analytically, an appropriately simple spin Hamiltonian comprising an $S = 1/2$ spin with an isotropic g-matrix is sufficient to capture the spin physics,

$$\hat{H}_0 = \omega_S \hat{S}_z + \omega_I \hat{I}_z + A\hat{S}_z\hat{I}_z + B_x\hat{S}_z\hat{I}_x + B_y\hat{S}_z\hat{I}_y \quad (16.7)$$

where A and B_x/B_y are the secular and pseudo-secular terms of the anisotropic HF interaction and ω_S and ω_I are defined in equations (16.3) and (16.4). The nuclear frequencies for the $S = 1/2$ case are given by[25]

$$\omega_N(m_S = 1/2) = \omega_\alpha = \sqrt{(\omega_I + A/2)^2 + B^2/4} \quad (16.8)$$

$$\omega_N(m_S = -1/2) = \omega_\beta = \sqrt{(\omega_I - A/2)^2 + B^2/4} \quad (16.9)$$

with $B = (B_x^2 + B_y^2)^{1/2}$. The pseudo-secular term B tilts the quantization axis of the nuclear spin away from the magnetic field vector $\boldsymbol{B}_0$ direction and differently for the α and β electron spin manifolds, and consequently, the two nuclear frequencies are no longer centered symmetrically around $|\omega_I|$ (weak coupling) or $|A|/2$ (strong coupling), as shown in Figure 16.1 (right). Note that equations (16.8) and (16.9) become approximately equal to equation (16.6) when the term $(\omega_I \pm A/2)^2 \gg B^2/4$, that is, when ω_I is large as in high-field ENDOR experiments.

In the case of an axial HF interaction,

$$A = A_{||}\cos^2\theta + A_\perp \sin^2\theta = A_{iso} + T(3\cos^2\theta - 1) \quad (16.10)$$

and

$$B = (A_{||} - A_\perp)\sin\theta\cos\theta = 3T\sin\theta\cos\theta \quad (16.11)$$

Polar angle θ is between the z-axis ($\boldsymbol{B}_0$ direction) and the electron–nuclear vector. The dipolar coupling constant T (a scalar, defined in equation (16.13)) describes the dipolar part of the HF interaction and provides very important structural information as it depends on the distance between the nucleus and the electron spin according to

$$\boldsymbol{T} = \frac{\mu_0}{4\pi\hbar} g_e \mu_B g_n \mu_N \sum_i \rho_i \frac{3\boldsymbol{n}_i\boldsymbol{n}_i^T - 1}{r_i^3} = \sum_i T_i(3\boldsymbol{n}_i\boldsymbol{n}_i^T - 1) \quad (16.12)$$

where $\boldsymbol{T}$ is the dipolar coupling tensor (of size 3×3), r_i the distance and $\boldsymbol{n}_i$ the corresponding unit vector denoting the electron–nuclear direction in a molecular frame where vector $\boldsymbol{B}_0$ is along the z-axis and T_i is given by

$$T_i = \frac{\mu_0}{4\pi\hbar} g_e \mu_B g_n \mu_N \frac{\rho_i}{r_i^3} \quad (16.13)$$

Equation (16.12) is valid for unpaired electron–nuclear separations of about >2.5 Å and allows a model to be developed where the spin density ρ_i is delocalized. This is typically done in terms of atomic spin populations (the integrated spin density of an orbital) that provide an estimate of the contribution of atomic orbitals to the SOMO.

This expression for the HF interaction needs to be corrected if the orbital magnetic moment of the paramagnetic center is large (i.e., very anisotropic g values, see Chapter 2) for the pseudo-anisotropic contribution.[28] This is achieved to a reasonably accuracy using

$$\boldsymbol{A} = A_{iso}\boldsymbol{1} + \boldsymbol{gT}/g_e \quad (16.14)$$

The isotropic component of the HF interaction, A_{iso}, enables the spin density at the nucleus to be calculated (**1** here is the identity matrix). This aids in the characterization of the SOMO, and the presence of an A_{iso} is useful to establish, for example, that a substrate is electronically connected (thus chemically bound) to a paramagnetic center of an enzyme or catalyst and to characterize spin polarization mechanisms if the sign of A_{iso} can also be determined.

$$A_{iso} = \frac{2}{3}\frac{\mu_0}{\hbar} g_e \mu_B g_n \mu_N \rho(0) \quad (16.15)$$

In equation (16.15), $\rho(0)$ is the electron spin density at the nucleus.

The inclusion of the NQ interaction for nuclear spins with $I > 1/2$ leads to a further shift of the energy levels, and the nuclear frequencies to first order are given by[25]

$$\omega_N(m_S, m_I) = \left| m_S A_{iso} - \omega_I + (2m_I - 1) \times \left(\frac{P_{zz}}{2}(3\cos^2\psi - 1) + \eta \sin^2\psi \cos 2\xi \right) \right| \quad (16.16)$$

where ξ and ψ define the principal axis orientation of the tensor **P** with respect to the magnetic field vector $\boldsymbol{B}_0$, and the quadrupole coupling constants are given by

$$P_{zz} = \frac{3e^2Qq}{4I(2I-1)\hbar} \text{ and } \eta = \frac{P_{xx} - P_{yy}}{P_{zz}} \quad (16.17)$$

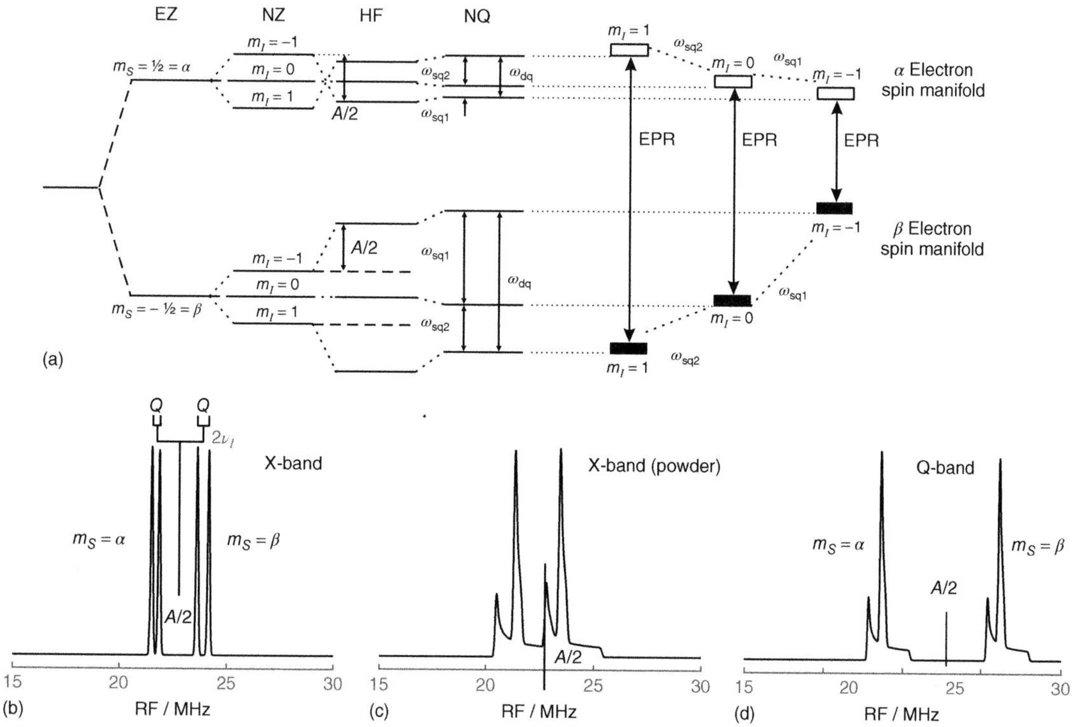

Figure 16.2. (a) Energy-level diagrams for an $S = ½$, $I = 1$ spin system in the strong coupling case $2|\omega_I| < |A|$. There are two single-quantum $\Delta m_I = \pm 1$ and one double-quantum $\Delta m_I = \pm 2$ NMR transition in each electron spin manifold. (b) ENDOR spectrum for a single orientation; because of anisotropy in A and P, the peaks are slightly shifted relative to the first-order expression (equation (16.16)) where they would be centered around $|A|/2$, split by $2|\omega_I|$, and further split by the nuclear quadrupole interaction. (c) Same parameters as in (b) but for the powder average. (d) Same parameters in (b) but at a higher magnetic field such that the nuclear Zeeman interaction is sufficient to separate the nuclear frequencies from the two electron spin manifolds. Simulation parameters: $A(^{14}N) = [44, 44, 47]$ MHz, $P_{zz} = 0.5$ MHz, $\eta = 0$ with $\omega_I/2\pi = 1.0$ MHz (X-band), 3.7 MHz (Q-band)

with $|P_{xx}| < |P_{yy}| < |P_{zz}|$ and $0 < \eta < 1$. Figure 16.2 shows the energy-level diagrams for an $S = ½$, $I = 1$ spin system. The quadrupole interaction gives information on the bonding, for example, measurement of the ^{51}V NQ coupling constant was used to characterize axial anionic ligands of oxovanadium compounds in Ref. 29.

In high-spin systems, there are $2S+1$ electron spin manifolds as equation (16.5) shows ($m_S = -S, -S+1, \ldots, S-1, S$) and each has associated nuclear transitions for coupled magnetic nuclei. The influence of the ZFS on the nuclear transitions frequencies must also be carefully considered, and the energy levels of the spin system depend critically on the ratio of the EZ to the ZFS interaction. There are two limiting situations that make high-spin systems much simpler to analyze. If the ZFS is much larger than the EZ, then the magnetic moment of the unpaired electron spin is quantized in the ZFS frame and hence related to the local structure of the paramagnetic center. If the EZ is much larger than the ZFS interaction, the so-called high-field case, the electron spin is quantized along the $\boldsymbol{B_0}$ field vector. In both the cases, ENDOR spectra are then greatly simplified.

We will look at this latter case first, and use as an example, the $S = 5/2$, $I = ½$ spin system with small ZFS and equation (16.5) to examine the energy levels. Here, each m_S electron spin manifold is split into a doublet due to the nuclear spin, and each EPR transition connecting two m_S manifolds is associated with two NMR transitions, which, to first order, are given by

$$\text{EPR}: |-5/2\rangle \leftrightarrow |-3/2\rangle, \text{NMR}: \omega(-5/2) = -5/2A - \omega_I \text{ and } \omega(-3/2) = -3/2A - \omega_I$$

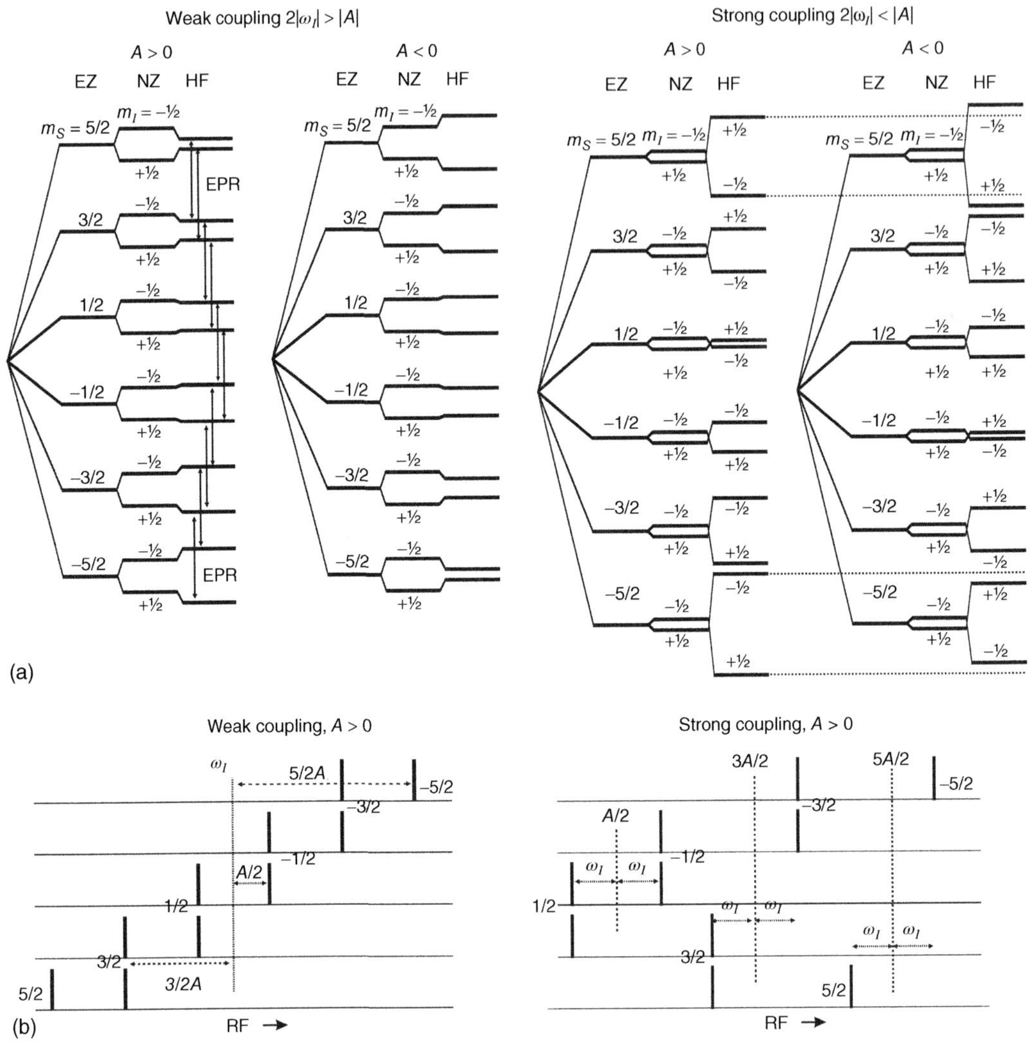

Figure 16.3. (a) Energy-level diagram for an $S = 5/2$, $I = ½$ spin system, showing the energy levels for a spin Hamiltonian comprising an electron Zeeman, then split by the nuclear Zeeman, and then shifted by the hyperfine interaction, for the weak and strong coupling cases for both $A > 0$ and $A < 0$ ($g_n > 0$ in all). The ZFS was set to zero, so equation (16.5) is valid. There are five allowed electron paramagnetic resonance (EPR) transition frequencies ($\Delta m_S = \pm 1$) and six allowed NMR transition frequencies ($\Delta m_I = \pm 1$). (b) ENDOR lines associated with the five allowed EPR transitions; numbers indicate the m_S spin state to which each NMR peak belongs. Spectra for the weak (left) and strong (right) coupling cases for $A > 0$ are shown

$$\text{EPR}: |-3/2\rangle \leftrightarrow |-1/2\rangle, \text{NMR}: \omega(-3/2) = -3/2A - \omega_I \text{ and } \omega(-1/2) = -1/2A - \omega_I$$
$$\text{EPR}: |-1/2\rangle \leftrightarrow |+1/2\rangle, \text{NMR}: \omega(-1/2) = -1/2A - \omega_I \text{ and } \omega(+1/2) = 1/2A - \omega_I$$
$$\text{EPR}: |+1/2\rangle \leftrightarrow |+3/2\rangle, \text{NMR}: \omega(+1/2) = 1/2A - \omega_I \text{ and } \omega(+3/2) = 3/2A - \omega_I$$
$$\text{EPR}: |+3/2\rangle \leftrightarrow |+5/2\rangle, \text{NMR}: \omega(+3/2) = 3/2A - \omega_I \text{ and } \omega(+5/2) = 5/2A - \omega_I$$

The associated energy-level diagrams in this case are given in Figure 16.3(a). Among the EPR transitions, only the ENDOR lines from the EPR transition $|-1/2\rangle \leftrightarrow |+1/2\rangle$ are symmetric with respect to the nuclear Zeeman frequency ω_I in the weak coupling case (Figure 16.3b left) or $|A|/2$ in the strong coupling case (Figure 16.3b right). If all the EPR transitions contribute equally to the ENDOR spectrum, then the ENDOR spectrum is symmetric about ω_I in the weak coupling case. However, if only a part of the EPR spectrum contributes to the ENDOR signal, either because of orientation selection resulting from the limited bandwidth of the irradiation compared to the EPR spectrum, or at low temperature and high-field where the EPR levels to higher energy are partially depopulated, then the ENDOR spectrum becomes distinctly asymmetric. This latter effect enables the sign of the HF interaction to be determined by an ENDOR experiment under suitable conditions as it reveals if the nuclear frequency from a known electron spin manifold lies above or below $\omega_I/|m_S A|$ in the weak/strong coupling cases, respectively. This is demonstrated in Figure 16.4 and discussed further in the following section.

When the HF coupling is not very small as compared to the EZ, as it can be for ^{55}Mn in Mn^{2+} complexes or ^{51}V in VO^{2+} complexes, then the energy levels depart considerably from the first-order expression of equation (16.16) (and equation (16.6)). In this case, equation (16.16) can be extended with the second-order contribution from the HF interaction,[30,31]

$$+\frac{A^2}{\omega_S}(-S(S+1)+m_S^2-m_S(2m_I-1)) \qquad (16.18)$$

As can be appreciated, the factor A^2/ω_S dictates the size of the second-order frequency shifts.

If the ZFS is substantial with respect to the EZ interaction, then shifts in the ENDOR frequencies occur in relation to[32,33]

$$\frac{AD^2}{\omega_S^2} = AD^2\left(\frac{\hbar}{\mu_B g B_0}\right)^2 \qquad (16.19)$$

An X-band ENDOR study analyzing ^{57}Fe signals and the effect of the ZFS is given in Ref. 32. Equation (16.18) and in particular equation (16.19) emphasize the need for high-spin systems to be studied at high field/frequency,[12] in order to simplify the analysis of the ENDOR spectrum. If experimental conditions cannot be found where the influence of the ZFS on the ENDOR frequencies cannot be suppressed, then it is relatively straightforward to compute an ENDOR spectrum using the full Hamiltonian of equation (16.2),[34] but it may be very difficult if not impossible to extract a unique set of HF parameters from the experimental data under these conditions.

In the second limiting case, where ZFS ≫ EZ, then the magnetic moment of the unpaired electron spins is quantized in the ZFS frame and hence related to the local structure of the paramagnetic center. In this case,

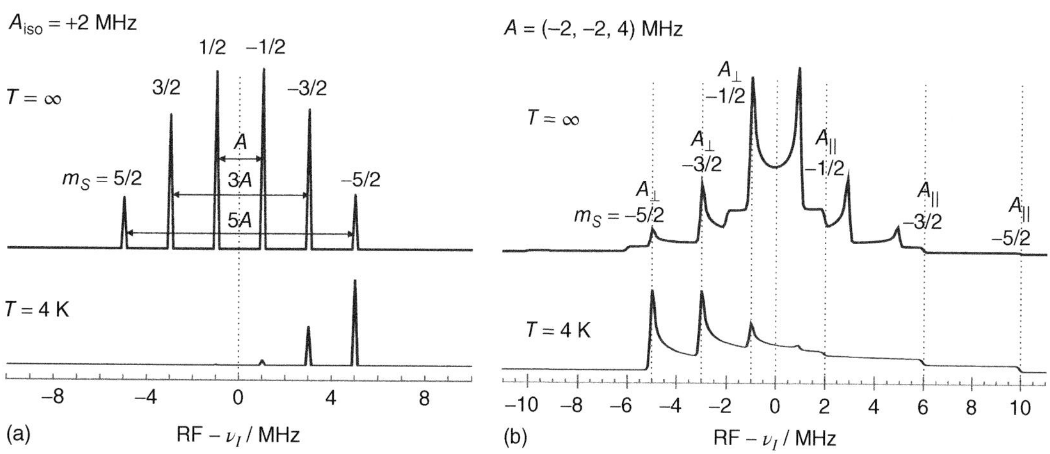

Figure 16.4. Simulated ENDOR spectra for an $S = 5/2$, $I = ^1\!/_2$ spin system at W-band in the weak coupling regime and where $g_n > 0$. (a) ENDOR spectrum for a positive isotropic hyperfine coupling, $A = +2$ MHz. The nuclear frequencies are labeled with the m_S manifold to which they belong. The spectrum at the top shows equal populations of the m_S manifolds; the bottom spectrum is at 4 K where the upper electron spin manifolds, m_S = 5/2, 3/2, 1/2, are essentially depopulated and thus give no ENDOR peaks. For a positive hyperfine coupling, the ENDOR peaks from $m_S = -5/2, -3/2, -1/2$ are above the nuclear Zeeman frequency ω_I. (b) ENDOR spectra for an anisotropic hyperfine coupling; the $A_\perp = -2$ MHz and $A_\parallel = 4$ MHz features from the selected electron spin manifolds $m_S = -5/2, -3/2, -1/2$ are marked and note the nuclear frequencies from these manifolds are below and above ω_I as $A_\perp < 0$ and $A_\parallel > 0$, respectively

the observed (effective) EPR and ENDOR spectra arise from transitions of the lowest Kramers doublet.[35] For example, the CW EPR spectrum of high-spin ferric heme proteins ($S = 5/2$) in tetragonal symmetry can be described using an $S^{\mathrm{eff}} = 1/2$ spin system with $g_{\perp}{}^{\mathrm{eff}} \approx 6$ and $g_{\parallel}{}^{\mathrm{eff}} \approx 2$. A good description of high-spin EPR spectra with a biological inorganic emphasis is given in Ref. 36. Within the lowest Kramers doublet, the spin Hamiltonian of equation (16.2) can be recast as

$$\hat{H}_0^{\mathrm{eff}} = \mu_{\mathrm{B}} \boldsymbol{B}_0^{\mathrm{T}} \boldsymbol{g}^{\mathrm{eff}} \hat{\boldsymbol{S}}^{\mathrm{eff}} / \hbar + \sum_{k=1}^{m} \hat{\boldsymbol{S}}^{\mathrm{eff}^{\mathrm{T}}} \boldsymbol{A}_k^{\mathrm{eff}} \hat{\boldsymbol{I}}_k - \mu_{\mathrm{N}} \sum_{k=1}^{m} \boldsymbol{B}_0^{\mathrm{T}} \boldsymbol{g}_{\mathrm{n},k}{}^{\mathrm{eff}} \hat{\boldsymbol{I}}_k / \hbar + \sum_{k=1,\, I>=1}^{m} \hat{\boldsymbol{I}}_k^{\mathrm{eff}^{\mathrm{T}}} \boldsymbol{P}_k{}^{\mathrm{eff}} \hat{\boldsymbol{I}}_k \tag{16.20}$$

where $\boldsymbol{g}^{\mathrm{eff}}$, $\boldsymbol{A}^{\mathrm{eff}}$, $\boldsymbol{g}_{\mathrm{n}}{}^{\mathrm{eff}}$, and $\boldsymbol{P}^{\mathrm{eff}}$ are effective coupling tensors in the lowest Kramers doublet. In the case of the $S = 5/2$ high-spin heme example, the HF tensor in the effective spin representation is given by $\boldsymbol{A}^{\mathrm{eff}} = \boldsymbol{A}\boldsymbol{g}^{\mathrm{eff}}/g_{\mathrm{e}}$, and the mixing of the $m_S = \pm 3/2$ doublet into the ground state by the EZ and electron–nuclear HF interaction requires a tensor representation for the nuclear Zeeman interaction, $\boldsymbol{g}_{\mathrm{n}}$, which is coaxial with the ZFS tensor.[35] In these spin systems, ENDOR is used to measure the nuclear transitions as for any other spin system, and the appropriate effective spin Hamiltonian (or the larger full spin Hamiltonian) is used to analyze the data and extract structural and electronic information.

Coupled spin systems, such as the Mn cluster in photosystem II,[37] or iron-sulfur clusters can also be studied effectively by EPR and ENDOR spectroscopies and the spin system described with the use of spin projection factors (SPFs, see Chapter 4).[38,39] For example, a reduced $[2\mathrm{Fe}\text{-}2\mathrm{S}]^+$ cluster contains one Fe^{3+} ion ($S_1 = 5/2$) and one Fe^{2+} ion ($S_2 = 2$), which couple antiferromagnetically to give an $S^{\mathrm{total}} = 1/2$ ground state. Using the strong-exchange limit assumption between the two spins, the SPFs K_i giving the projection of the total spin S^{total} onto the local spin sites S_1 and S_2 have the values $K_1 = +7/3$ and $K_2 = -4/3$. The HF coupling measured in an ENDOR experiment are then scaled by the SPFs. ENDOR has been used, for example, in $[2\mathrm{Fe}\text{-}2\mathrm{S}]^+$ clusters to determine the distance and orientation of the closest protons to orientate the g-matrix.[40–43] The dipolar part of the proton HF coupling is calculated by summing over contributions from the paramagnetic ions of the cluster using the SPFs with the dipolar expression

$$\boldsymbol{T} = \frac{\mu_0}{4\pi\hbar} g_{\mathrm{e}} \mu_{\mathrm{B}} g_{\mathrm{n}} \mu_{\mathrm{N}} \sum_i K_i \frac{3\boldsymbol{n}_i \boldsymbol{n}_i^{\mathrm{T}} - 1}{r_i^3} \tag{16.21}$$

where $\boldsymbol{n}_i$ is the unit vector (direction cosines of $\boldsymbol{r}_i$ in a molecular frame where $\boldsymbol{B}_0$ is along the z-axis) between the local spin site with SPF K_i and the nuclear spin whose dipolar HF coupling is being calculated.

16.3 CW ENDOR

CW ENDOR[44] is a technique where the ENDOR spectrum (which is essentially an NMR spectrum) can be recorded at a chosen B_0 position within the field-sweep EPR spectrum through application of CW MW and CW RF irradiation to the sample. The basic concepts behind the measurement are depicted in Figure 16.5 for the simple two-spin system $S = 1/2$, $I = 1/2$. The various electron (w_{e1}, w_{e2}), nuclear (w_{n1}, w_{n2}), and cross-relaxation (w_{x1}, w_{x2}) pathways are depicted by the dotted lines (middle diagram), and the MW and RF irradiation as solid thick lines for the particular case where the MW irradiation is resonant with EPR transition (2,4) and RF irradiation with NMR transition (1,2). In CW ENDOR, the NMR transitions are not observed directly but rather indirectly as a change in the MW absorption of an irradiated EPR transition. Measurement of the NMR frequencies by monitoring an EPR transition yields a sensitivity enhancement of several orders of magnitude as compared to direct detection of the NMR frequencies with NMR spectroscopy (called quantum transformation). The MW power should be sufficient to partially saturate the EPR transition (when $(\gamma_{\mathrm{e}} B_1)^2 T_{1\mathrm{e}} T_{2\mathrm{e}} > 1$, where γ_{e} is the free-electron gyromagnetic ratio and B_1 the transverse MW field), so that the polarization departs from the Boltzmann equilibrium value at the particular sample temperature. This is shown in Figure 16.5 for EPR transition (2,4) by the half-filled rectangles that indicate the relative population of the levels. The most effective relaxation path for EPR transition (2,4) is w_{e2}, but this does not lead to an ENDOR signal. Another possible relaxation pathway, referred to as a bypass route, is via $w_{1\mathrm{n}}$ $(2,1) \rightarrow w_{\mathrm{e1}}$ $(1,3) \rightarrow w_{\mathrm{n2}}$ (3,4). This bypass route is normally not very effective as the nuclear relaxation rates are slow, $w_{\mathrm{n}} \ll w_{\mathrm{e}}$, causing an increase in the population of levels 2 and 3 under MW irradiation. The bottleneck caused by the small $w_{1\mathrm{n}}$ rate may be partially removed by saturating RF irradiation (when $(\gamma_{\mathrm{n}} B_2)^2 T_{1\mathrm{n}} T_{2\mathrm{n}} > 1$, where γ_{n}

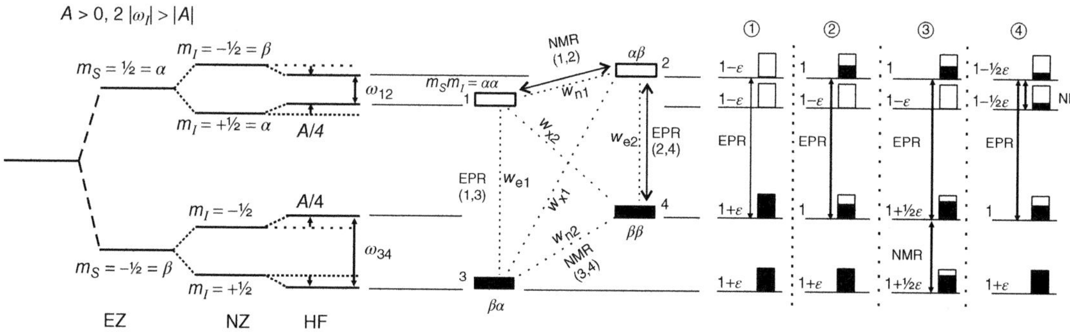

Figure 16.5. (Left) Energy-level diagram for a simple $S = 1/2$, $I = 1/2$ spin system. The two EPR and NMR transitions are indicated. The dashed lines show the electron (w_{e1}, w_{e2}), nuclear (w_{n1}, w_{n2}), and cross-relaxation (w_{x1}, w_{x2}) rates. The energy levels are labeled according to the four basic functions $|m_S m_I\rangle$; $|\beta\alpha\rangle$ and $|\beta\beta\rangle$ for the β-electron spin manifold and $|\alpha\alpha\rangle$ and $|\alpha\beta\rangle$ for the α-electron spin manifold. (①–④) The solid double-arrow lines represent CW MW and CW RF irradiation resonant with a transition and driving transitions at a rate w_{MW} and w_{RF}, respectively. $\varepsilon = (N_\beta - N_\alpha)$ defines the thermal population difference across the EPR levels, $N_\alpha/N_\beta = \exp(-g\mu_B B_0/k_B T)$. Thermal population differences between the NMR levels within an electron spin manifold are small and are neglected. ① weak MW irradiation $w_{mw} < w_e$, as in a conventional CW EPR experiment, ② saturating MW irradiation, $w_{mw} > w_e$, which equalizes the populations (2,4), this is the state for RF irradiation off resonance during the CW ENDOR experiment (in practice the (2,4) transition is only partially saturated), ③/④ saturating MW and RF irradiation and an ENDOR effect is observed as EPR transition (2,4) is partially desaturated by the RF irradiation

is a nuclear gyromagnetic ratio and B_2 the transverse RF field) resonant with NMR transition (1,2). This increases the effective electron spin relaxation rate and leads to desaturation of the EPR transition (2,4) in response to the CW RF irradiation, and thus a CW ENDOR signal. The bottleneck may also be removed by saturating RF irradiation resonant with NMR transition (3,4), which again gives rise to a CW ENDOR signal.

The strength of the CW ENDOR signal depends critically on the relaxation rates w_e and w_n, and thus unlike pulse ENDOR experiments described in the next section, the ENDOR signal is not proportional to the number of contributing nuclear spins. In the limit of strong saturation of the NMR transitions and assuming negligible cross-relaxation ($w_{x1} = w_{x2} = 0$), the signal strength depends on the enhancement factor E^{cw},

$$E^{cw} = \frac{1}{2(2 + (w_n/w_e) + (w_e/w_n))} \tag{16.22}$$

and is maximum when $w_e = w_n$. Optimization of E^{cw} and the ENDOR signal intensity is achieved by adjusting the temperature, and additionally if solution ENDOR is being carried out by adjusting the solvent viscosity as $w_e \propto 1/\tau_R$ and $w_n \propto \tau_R$, where τ_R is the rotational correlation time for Brownian diffusion. The two relaxation rates can be fine tuned by temperature and viscosity because their dependence on τ_R is different; for protons $w_n \ll w_e$ so optimal conditions occur in solvents of low viscosity near the freezing point; for solvents with a high viscosity room-temperature ENDOR may be possible.[44]

A typical solution ENDOR spectrum of an organic radical is shown in Figure 16.6.[44] Because of the phase-sensitive detection method both the CW EPR and the CW ENDOR spectrum are plotted in a first harmonic (derivative) representation. In CW EPR, the amplitude of the static field B_0 is modulated at typically 100 kHz (see Chapter 1). In CW ENDOR, either the amplitude of the RF field is modulated or the frequency of the RF field is modulated, with the component of the EPR signal at the modulation frequency of the RF irradiation being detected using a lock-in amplifier – thus a first-harmonic ENDOR spectrum is recorded.

CW ENDOR can also be used on powders and frozen solution samples where the ENDOR lines are typically broad and consequently the signal sensitivity and baseline can be a problem because the first harmonic is recorded. Nevertheless, the technique provides good data in many instances, see, for example, Ref. 45 where a metalloenzyme study at about 2 K was carried out.

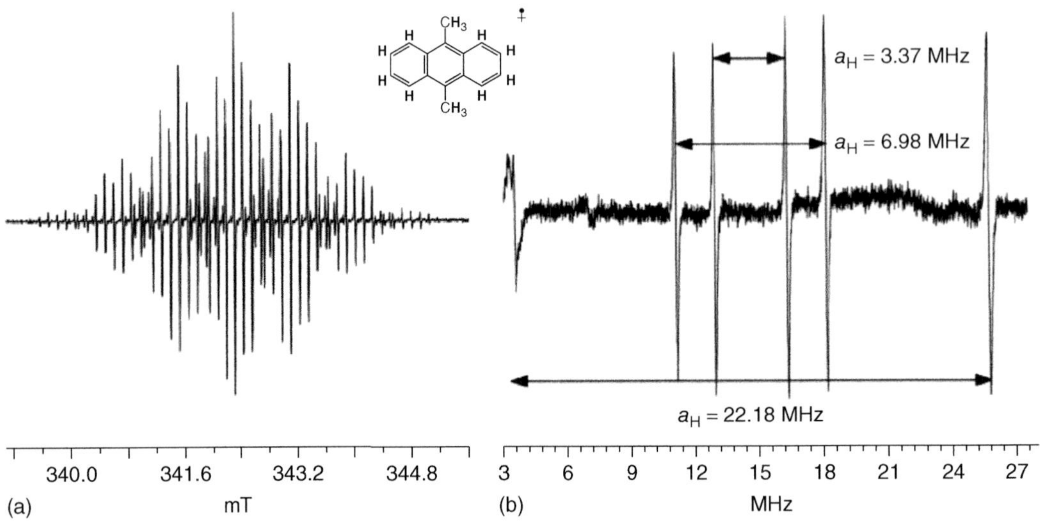

Figure 16.6. CW EPR spectrum (a) and a ^{1}H CW ENDOR spectrum (b) of the radical cation 9,10-dimethylanthracene in solution. Note that like the CW EPR spectrum, the CW ENDOR spectrum is collected as the first harmonic of the absorption spectrum. (Adapted from Ref. 44 with permission of The Royal Society of Chemistry)

16.3.1 Special Triple CW ENDOR

CW ENDOR requires desaturation of the EPR transition being irradiated as a result of saturating CW RF irradiation, and this can be achieved by optimizing the two relaxation rates w_{n1} and w_{n2} in the bypass route. If these two relaxation rates are very slow, they act as effective bottlenecks that reduce the sensitivity of the ENDOR experiment. If these relaxation pathways can be enhanced, a significant ENDOR signal enhancement should result. This can be achieved if both NMR transitions are simultaneously irradiated, which is used in a special triple CW ENDOR experiment.[46,47] An important advantage of the technique is that now the signal intensity of the ENDOR peaks reflects the ratio of the number of equivalent nuclei, provided the RF transition rates obey the condition $w_{RF} \gg w_e \gg w_n$.

Special triple CW ENDOR is carried out on weakly coupled nuclei by starting the two RF irradiation frequencies at the corresponding nuclear Zeeman and sweeping one RF irradiation to high frequency and the other to low frequency, that is, $\omega_{RF\pm} = \omega_I \pm \omega_{scan}$, where ω_{scan} is scanned. This sweeping procedure requires that the experiment is carried out in solution, or for disordered solids or single crystals under the high-field condition, where equation (16.6) or (16.16) applies and the peaks are symmetrically placed around ω_I.

16.3.2 General CW Triple ENDOR

General CW Triple ENDOR[48] enables the relative sign of HF couplings to be determined, and as such is applied to systems containing several nuclei coupled to an electron spin with resolved ENDOR lines. In the simplest case, $S = 1/2$ coupled to $I_1 = I_2 = 1/2$, there are four nuclear spin transitions and each of them is doubly degenerate. CW general triple ENDOR employs RF irradiation at two frequencies, ω_{RF1}, which is swept as normal, and ω_{RF2}, which is fixed and set to one of the nuclear transitions so that it is saturated. This saturation affects the intensity of all other connected transitions, and by comparing the intensity changes from this scheme to a standard ENDOR experiment, the relative sign the two HF couplings can be determined. A more detailed description is given for the pulse version of triple ENDOR in the following section.

16.4 PULSE ENDOR TECHNIQUES

Figure 16.7 lists a summary of the pulse ENDOR sequences discussed in this chapter. The two standard and most used sequences for measuring nuclear transition frequencies are Davies[49] and Mims ENDOR[50,51] (Figure 16.7a, a′), which are described next for a simple $S = 1/2$, $I = 1/2$ spin system. Both are based on the

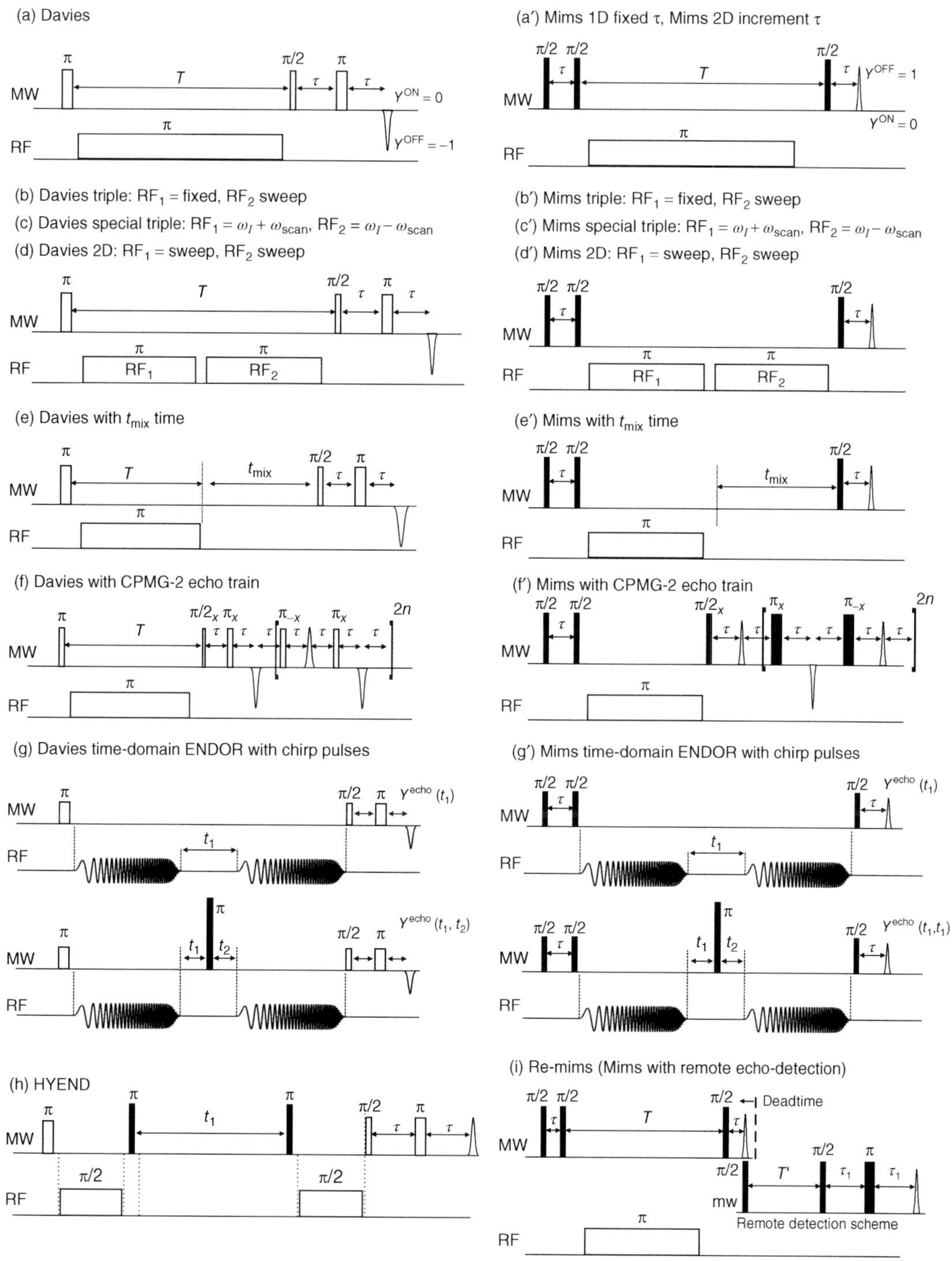

Figure 16.7. Summary of pulse sequences for ENDOR spectroscopy described in this chapter. Note that typical MW pulse lengths t_p are in the range 10–100 ns, whereas an RF π-pulse is microseconds long, often >10 μs. Nonselective pulses are filled and selective pulses are unfilled

transfer of polarization between electron and nuclear energy levels. Pulse ENDOR requires spin relaxation rates to be slow compared to the ENDOR pulse sequence, which often necessitates measurement at low temperature and thus limits the applicability of the technique, particularly for liquid samples.

16.4.1 Davies ENDOR

Figure 16.8 shows the Davies ENDOR pulse sequence and the basic spin physics demonstrating how the sequence works. The sequence starts with a selective MW π-pulse that inverts the polarization of a particular EPR transition, transition (1,3) in Figure 16.8. During the time T, a selective RF pulse is applied, which optimally has a π flip angle. If the RF pulse is resonant with a nuclear transition, either (1,2) or (3,4), the polarization of the NMR transition is inverted, which also alters the polarization of the EPR transition (1,3) that is being observed with a two-pulse Hahn echo sequence, π/2–τ–π–τ–echo, but an FID detection is also possible. An FID detection scheme is advantageous particularly when long, weak-selective MW pulses are employed to measure small couplings. The ENDOR spectrum is recorded by sweeping the RF stepwise over the desired range of frequencies and monitoring the Hahn echo intensity, $Y_{\text{echo}}(\omega_{\text{RF}})$. The ENDOR efficiency is defined as

$$F^{\text{ENDOR}} = \left| \frac{Y^{\text{ON}}_{\text{echo}} - Y^{\text{OFF}}_{\text{echo}}}{2Y^{\text{OFF}}_{\text{echo}}} \right| \tag{16.23}$$

and lies in the range $0 \leq F^{\text{ENDOR}} \leq 1$. The maximum ENDOR efficiency is obtained if the RF irradiation results in inversion of the electron spin echo being monitored. For Davies ENDOR, theoretically the maximum efficiency is $F^{\text{ENDOR}} = 0.5$ when the echo goes from $Y^{\text{OFF}}_{\text{echo}} = -1$ (RF pulse OFF resonance) to $Y^{\text{ON}}_{\text{echo}} = 0$ (RF pulse ON resonance), assuming ideal MW and RF pulses and no losses from relaxation.

Typically, a selective MW pulse around 100 ns in length is used, whereas the RF pulse is often considerably more than 10 μs. During this long RF pulse, considerable electron relaxation may occur, and consequently, the optimal signal-to-noise ratio (SNR) depends on relaxation and the length of the RF pulse and magnitude of the B_2 RF field, so the best SNR may occur with a RF pulse length shorter than that required for a π-pulse. Experimentally the optimal pulse length can be determined by measuring the rate of decay of the echo due to electron spin relaxation as time T is increased along with a nutation experiment where the echo is measured as a function of RF pulse length.

The first MW pulse in a Davies ENDOR experiment is required to be selective as described above. If this MW pulse is not selective, then both the (1,3) and (2,4) EPR transitions are inverted, and as can be readily appreciated, the RF pulse then has no effect on the Hahn echo used for detection regardless of whether or not the RF pulse is resonant with an NMR transition (Figure 16.8 ①′). Under these conditions, the inversion pulse acts as a filter for small HF couplings, and this filter can be adjusted by varying the length and thus bandwidth of the MW inversion π-pulse. The bandwidth, full width at half height (f.w.h.h.) in hertz, of a rectangular MW pulse of length t_p is given by

$$\Delta\nu \approx \frac{1}{1.6578 \times t_p} \tag{16.24}$$

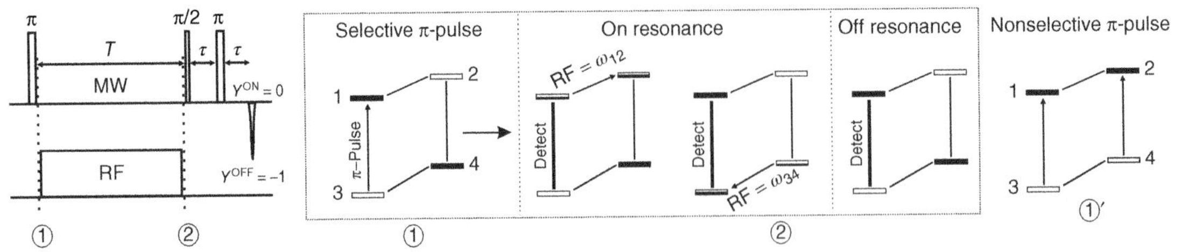

Figure 16.8. Concept behind Davies ENDOR demonstrated for an $S = 1/2$, $I = 1/2$ spin system. After the initial selective inversion MW π-pulse on the EPR transition (1,3) ①, a RF pulse resonant with NMR transition (1,2) or (3,4) ② results in a change in spin polarization across the (1,3) EPR transition (depicted as slightly less than an RF π-pulse in the diagram), which is detected via a change in the Hahn echo intensity. ①′ A nonselective MW π-pulse inverts the population difference across both EPR transitions, (1,3) and (2,4), and there is no ENDOR effect

The absolute ENDOR intensity as a function of the selectivity of the inversion π-pulse is given by[52]

$$V(\eta_s) = V_{max}\frac{\sqrt{2}\eta_s}{(\eta_s^2 + 1/2)} \quad (16.25)$$

with

$$\eta_s = \frac{t_\pi A_{iso}}{2\pi} \quad (16.26)$$

Here, V_{max} is the maximum ENDOR intensity obtained when $\eta_s = \sqrt{2}/2$ (V_{max} can be determined experimentally but clearly decreases as the pulse length increases), t_π the length of the preparation MW π-pulse, and A_{iso} the observed HF coupling (in radians). Equations (16.25) and (16.26) (when $\eta_s = \sqrt{2}/2$) are plotted in Figure 16.9(a) and (b), respectively, and indicate the optimal inversion pulse length for a give HF coupling.

This filtering technique is often very valuable at X-band where strongly and weakly coupled nuclei overlap, as, for instance, strongly coupled ^{14}N nuclei with weakly coupled ^{1}H nuclei. Figure 16.10 shows an example for a Cu(II) complex[53] coordinated by nitrogen nuclei with HF values around 29–36 MHz, so broad ENDOR peaks are approximately centered at $|A|/2 = 14.5–17$ MHz and split by the nuclear Zeeman (≈1 MHz) and the NQ interaction (not resolved). The protons are all weakly coupled and centered around their nuclear Zeeman frequency of 12.5 MHz ($g_{\|}$) and 14 MHz ($g_\perp$) and split by the HF couplings, which are $|A| < 8$ MHz, and thus the ^{14}N and ^{1}H ENDOR signals overlap. The proton signals are, however, well suppressed by a relatively hard π preparation pulse as shown by the data in Figure 16.10.

According to the above HF selectivity, there should always be a hole in a Davies ENDOR spectrum at the nuclear Zeeman of the nuclei present as any very weakly coupled nuclei should be suppressed. However, there is almost always a strong ^{1}H matrix line (unless for instance the solvent is deuterated) that results from the simultaneous inversion of a large number of ^{1}H nuclear spins very weakly coupled to the paramagnetic center by the RF pulse as described in Refs 54, 55 for both Davies and Mims ENDOR.

The inversion pulse filtering in Davies ENDOR makes the sequence insensitive for weak HF couplings because the inversion MW π-pulse is required to be very long and selective and thus the number of electron spins contributing to the experiment decreases. In addition, the τ value of the Hahn echo detection subsequence needs to be increased to accommodate the longer pulses and thus the increased time for relaxation (phase memory relaxation time T_m) further decreases the echo intensity. For weak couplings, a Davies scheme with FID detection may be advantageous, but it is often preferable to use Mims ENDOR.[26]

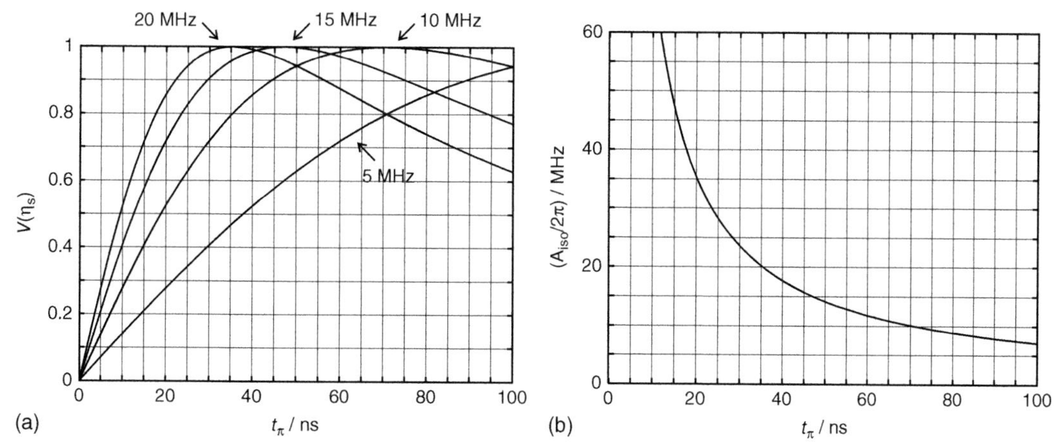

Figure 16.9. Hyperfine contrast selectivity in Davies ENDOR. (a) The absolute ENDOR intensity $V(\eta_s)$ as a function of the length and thus selectivity of the inversion π-pulse t_π (equation (16.25)) for $A_{iso} = 20$, 15, 10, 5 MHz as indicated by the arrows. V_{max} was set to 1. (b) A_{iso} versus t_π for the maximum absolute ENDOR intensity when $V(\eta_s) = 1$, i.e., when $\eta_s = \sqrt{2}/2 = t_\pi A_{iso}/2\pi$ (equation (16.26)).

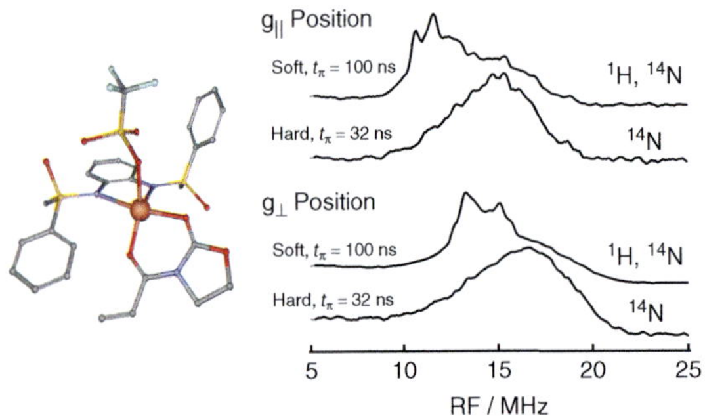

Figure 16.10. Davies ENDOR with two different hyperfine selective MW preparation π-pulses for a Cu(II) bis(sulfoximine) complex, which has overlapping ^{1}H and ^{14}N couplings. The weak ^{1}H couplings ($A < 8$ MHz) are well suppressed by adjusting the selectivity of the inversion pulse from $t_\pi = 100$ ns (soft) to $t_\pi = 32$ ns (hard), to reveal signal from strongly coupled ^{14}N nuclei.[53] (Adapted with permission from C. Bolm; M. Martin; G. Gescheidt; C. Palivan; D. Neshchadin; H. Bertagnolli; M. Feth; A. Schweiger; G. Mitrikas; J. Harmer J. Am. Chem. Soc. 2003, 125, 6222. © 2003, American Chemical Society)

16.4.2 Mims ENDOR

Mims ENDOR is based on the three-pulse stimulated echo sequence shown in Figure 16.7(a′) and employs hard nonselective MW pulses and thus has good sensitivity as far as echo intensity is concerned. The preparation pulses (π/2−τ−π/2) create a τ-dependent polarization grating (in the B_0 direction) that results in blind spots in the ENDOR spectrum according to the HF value, see equations (16.27) and (16.28). During the mixing time T, a selective RF pulse is applied that changes the polarization of the NMR transitions and consequently also the EPR transitions. The third π/2 pulse of the stimulated echo sequence creates electron coherence that refocuses at time τ into the stimulated echo (the refocusing is altered if there is an ESEEM (electron spin-echo envelope modulation) effect, i.e., if forbidden EPR transitions are excited, see Chapter 18). However, if the RF pulse is on resonance with a nuclear transition, the EPR polarization is shifted in a τ-dependent manner to a different transition with a different frequency and this magnetization does not refocus causing the stimulated echo intensity to decrease to zero for a resonant RF π-pulse. The theoretical ENDOR efficiency is dependent on the value of $A\tau$

$$F^{\text{ENDOR}} = 1/4(1 - \cos(A\tau)) \text{ and } F^{\text{ENDOR}}_{\text{max}} = 0.5 \quad (16.27)$$

The maximum efficiency value is the same as in Davies ENDOR. The Mims ENDOR spectrum is recorded by stepwise sweeping the selective RF π pulse and recording the stimulated echo intensity.

Because the echo decreases when the RF pulse is resonant with an NMR transition, the spectrum is typically represented with 'negative' peaks, as compared to Davies ENDOR. As mentioned above, the Mims ENDOR signal contains blind spots according to

$$Y^{\text{stimulated echo}} = 1/4(1 + \cos(A\tau)) \quad (16.28)$$

and thus there are maxima and blind spots (minima) in the Mims ENDOR spectrum at the HF values according to

$$A^{\text{blind-spot}} = \frac{n}{\tau},\ A^{\text{maximum}} = \frac{n + \frac{1}{2}}{\tau} \quad (16.29)$$

where n is an integer 0, 1, 2, 3,…. The blind-spot behavior is generally a disadvantage of the sequence and one may have to record spectra at several different τ values to ensure that peaks are not missing from the ENDOR spectrum. It does, however, offer a way to make the spectrum sensitive to a particular HF frequency, for example, to measure peaks with a 5 MHz HF coupling, an optimal $\tau = 100$ ns would be used. If it is advantageous to use a τ less than the spectrometer dead time, then a remote detection scheme can be used by storing the electron coherence (simulated echo) as longitudinal magnetization with an additional nonselective π/2-pulse at time $\tau + T + \tau$, and then reading out the stimulated echo intensity with a Hahn echo π/2−τ_1−π−τ_1−echo, see Figure 16.7(i).[56] Figure 16.11(a) shows a nutation experiment to determine an optimal RF pulse length and Figure 16.11(b) compares Mims ENDOR spectra recorded with $\tau = 100$, 200, 500, and 1000 ns and Davies ENDOR spectra recorded with inversion MW pulses of length $t_\pi = 32$ and 100 ns. This data highlights the large changes in the ENDOR signal intensity that can occur depending on the parameters of the sequences and emphasizes the need to carefully consider these parameters when interpreting ENDOR data.

Some points to consider when using Davies and Mims ENDOR experiments:

- Mims ENDOR is generally much more sensitive than Davies ENDOR because hard nonselective

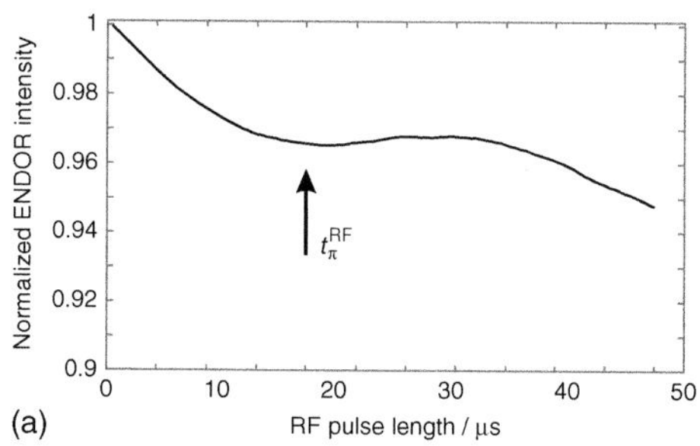

Figure 16.11. Q-band ENDOR spectra recorded at 80 K on a Bruker coal sample. (a) Nutation of the stimulated echo around the RF excitation field B_2^{eff} allowing the Rabi frequency and t_π^{RF} to be determined. The trace was recorded using Mims ENDOR with $t_{\pi/2} = 12$ ns, $T = 60$ μs, $\tau = 200$ ns, RF pulse length swept from 0.5 to 50 μs, RF frequency = 52 MHz (^{1}H Larmor frequency), and RF input power = 150 W. The ENDOR efficiency for the RF pulse $t_\pi^{\text{RF}} = 20$ μs is $F^{\text{ENDOR}} \cong (1 - 0.96)/2 = 0.02$ (equation (16.23)). (b) A comparison between Davies (top two traces) and Mims (bottom four traces) ENDOR spectra. The Davies ENDOR spectra were recorded with different hyperfine contrast selectivity as determined by the inversion pulse length with the sequence employing $t_\pi = 32$ ns enhancing the largest hyperfine couplings, $A_{\max} \cong 20$ MHz (Figure 16.9). The Mims ENDOR spectra have blind spots at the hyperfine values expected from equation (16.29): $\tau = 100$ ns, $A^{\text{blind spot}} = 0, 10, 20, \ldots$ MHz; $\tau = 200$ ns, $A^{\text{blind spot}} = 0, 5, 10, \ldots$ MHz, $\tau = 500$ ns, $A^{\text{blind spot}} = 0, 2, 4, \ldots$ MHz; $\tau = 1000$ ns, $A^{\text{blind spot}} = 0, 1, 2, \ldots$ MHz. In all the spectra, the RF pulse $t^{\text{RF}} = 20$ μs, and for Mims ENDOR, $t_{\pi/2} = 12$ ns

pulses with a large excitation bandwidth can be used, which excite many spin packets all of which contribute to the experiment.

- For small couplings, Mims ENDOR is generally preferable, as the MW pulses can be hard and τ can generally be made long enough to be sensitive to small couplings.
- For large, anisotropic HF couplings generally Davies ENDOR is the best choice. In this case, the selective MW pulse in Davies ENDOR can be made relatively broadband (equations (16.24–16.26), Figure 16.9), and there are no blind spots to complicate the analysis.

16.4.3 Effective Flip Angle of the RF Pulse

The efficiency of an ENDOR sequence is very dependent on the RF pulse flip angle, θ_{RF}, that can be achieved. For a RF pulse of length t^{RF} with the effective transverse RF field strength B_2^{eff}, the flip angle is

$$\theta_{\text{RF}} = (\gamma B_2^{\text{eff}}) t^{\text{RF}} \tag{16.30}$$

The flip angle θ_{RF} can be determined experimentally with Davies or Mims ENDOR by measuring the echo intensity as the length of an RF pulse of fixed field strength B_2 and frequency is increased, that is, by measuring the oscillation frequency of the RF polarization around the effective transverse RF excitation field B_2^{eff} (see equation (16.33) and Figure 16.11a). For this measurement, usually Mims ENDOR is preferable because of its good sensitivity. The mixing time T is made quite long (several RF π pulse lengths if possible) and the RF frequency is set typically to the proton Larmor frequency where the ENDOR signal is generally very strong and the ENDOR enhancement factor E is essentially one. The length of the RF pulse is then swept from zero to time T, and the echo intensity monitored. The result is a cosine oscillation (damped by relaxation) where the first minimum corresponds to the length of the RF π pulse, t_π^{RF}. This measurement establishes the length and field strength B_2 required for a RF π-pulse for uncoupled ($A = 0$) nuclei(usually determined for ^{1}H). To determine the required RF π-pulse length for other nuclei with weak couplings, this e.g. ^{1}H π-pulse length can be scaled by the gyromagnetic ratios of the nuclei involved, for example, for a proton and nitrogen nucleus

$$t_{\text{RF},\pi}(^{14}N) \approx \frac{\gamma(^1H)}{\gamma(^{14}N)} t_{\text{RF}}(^1H) = 13.8\, t_{\text{RF},\pi}(^1H) \tag{16.31}$$

In practice, the NMR coil impedance changes significantly with the RF. Therefore, this procedure is only a guide that should be corrected for the impedance characteristics of the coil if known, or measured again for the second nuclear spin if possible (often the RF pulse would get too long, i.e., for ^{14}N nuclei).

If the HF coupling to the nuclear spin is strong compared to ω_I, then the HF enhancement needs to be accounted for to determine the flip angle of the RF pulse. For an isotropic interaction with $B_2 \ll B_0$, the component of the HF field of the unpaired electron B_{hf} that is perpendicular to B_0 can be written as[57]

$$B_{hf} = \frac{A_{iso}}{\omega_I} B_2 \tag{16.32}$$

with the total oscillating RF field amplitude at the nucleus

$$B_2^{eff} = EB_2 \tag{16.33}$$

and the enhancement factor

$$E(m_S, A_{iso}) = \left| 1 + \frac{m_S A_{iso}}{\omega_I} \right| \tag{16.34}$$

The HF enhancement E results in a cosine modulation of the ENDOR signal versus A_{iso}, for example, in Davies ENDOR, the signal varies as

$$\begin{aligned} V^{echo}(\theta_{RF}) &\propto 1 + \cos(\theta_{RF}) = 1 + \cos(\gamma B_2^{eff} t_{RF}) \\ &= 1 + \cos(\gamma E B_2 t_{RF}) \end{aligned} \tag{16.35}$$

The enhancement factor E can be very large at the commonly used X-band fields. Consider a heme copper complex with an $S = 1/2$ electron spin and a nitrogen HF coupling of $A(^{14}N) \approx 30$ MHz. At 350 mT (X-band), the nuclear Zeeman frequency is $\omega_I/2\pi = \nu_I = 1.1$ MHz and $E = 14.6$ ($m_S = +1/2$) and 12.6 ($m_S = -1/2$). Here, an RF π-pulse for protons of length $t_\pi^{RF} = 15$ μs also has close to a π flip angle for this strongly coupled ^{14}N nucleus as the increase in the RF pulse length due to the smaller $\gamma(^{14}N)$ by a factor of $\gamma(^1H)/\gamma(^{14}N) = 13.8$ is offset by the large enhancement factors E. At higher field, this enhancement is reduced, for example, at 3.5 T (W-band) the enhancements are $E = 2.4$ ($m_S = +1/2$) and 0.4 ($m_S = -1/2$). The dependence of B_2^{eff} on m_S will obviously effect the ENDOR intensities, all other factors being equal. The enhancement factor can often mean that low field (MW frequency) is a good option for low-γ nuclei in the strong coupling regime with respect to the optimal excitation of NMR transitions with relatively short RF pulses.

One further consequence of the enhancement factor E is that as the HF coupling of a nucleus approaches the exact cancelation condition, $2|\omega_I| = |A|$, the enhancement factor in one of the electron spin manifolds goes to zero, and it is thus impossible to measure these nuclear frequencies via ENDOR. Fortunately this is exactly where ESEEM techniques are most sensitive and the missing NMR frequencies in the electron spin manifold with nuclear frequencies close to zero can be determined using these techniques (see Chapter 18).

16.4.4 Orientation Selection

In the vast majority of pulse experiments, the MW pulses only excite a small fraction of the full powder EPR spectrum (a short pulse with $t_p = 10$ ns has a bandwidth of about 60 MHz). The pulse is then said to be orientation selective as it excites spin packets from a set of molecules with particular orientations relative to the magnetic field vector $\boldsymbol{B}_0$ (see Chapter 11). Figure 16.12 shows an example of orientation-selective Davies ENDOR spectra measured at W-band on an [2Fe-2S]$^+$ cluster ($S^{total} = 1/2$) from Palustrisredoxin B from *Rhodopseudomonas palustris* (Figure 16.12a). The field-sweep EPR spectrum in Figure 16.12(b) shows the seven field positions of the Davies ENDOR measurements, along with a schematic representation of the bandwidth of the MW pulses. Figure 16.12(c) shows the corresponding ^{1}H Davies ENDOR spectra, and on the right for each spectrum, the calculated orientation selection in the g-matrix frame; blue indicates orientations off resonance. The experimental ENDOR spectra (black) were simulated with a model including eight protons around the paramagnetic center (blue line), and simulations for the two largest proton HF couplings are given in cyan and red. The HF matrices are represented as ellipses in Figure 16.12(a). The dipolar part of the HF coupling for the simulation was calculated using equation (16.21), the known SPFs, and the known proton distances and orientations from the X-ray structure. With the HF interactions computed in the molecular frame, the orientation-selective ENDOR spectra then depend only on the unknown g-matrix orientation in the molecular frame, which enables its orientation to be determined.[42]

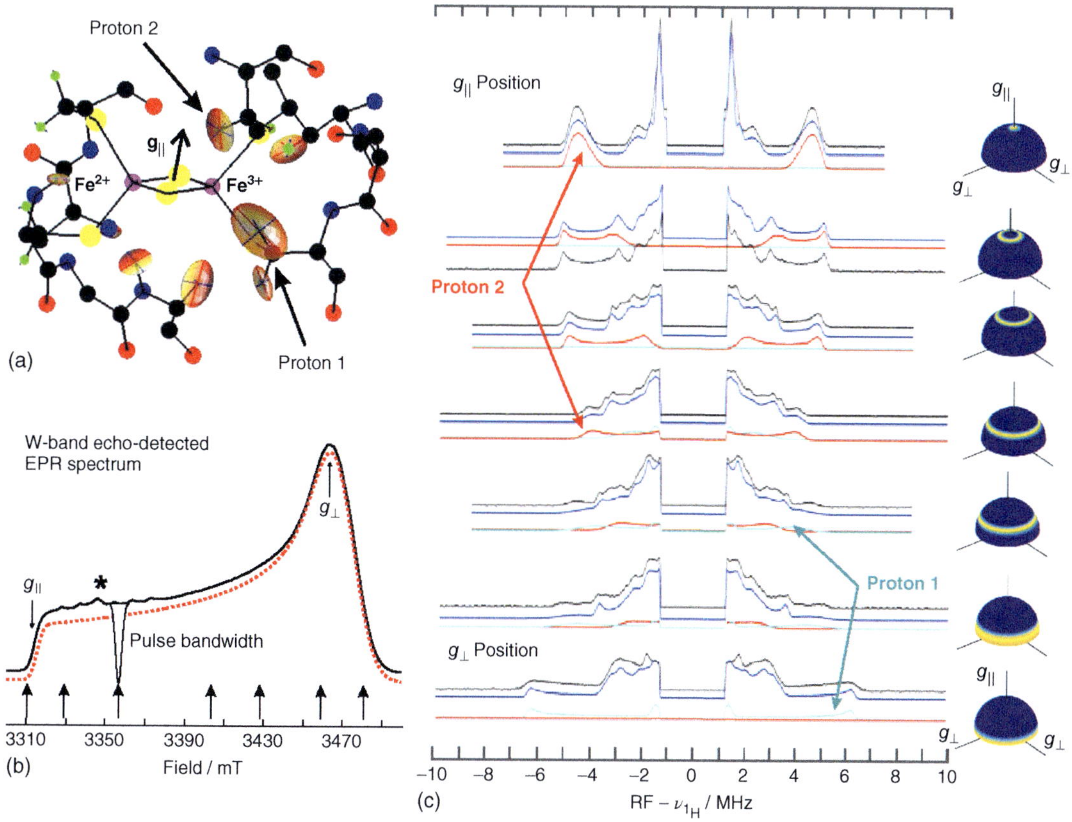

Figure 16.12. (a) X-ray structure of the reduced $[2Fe\text{-}2S]^{\bullet +}$ cluster and surrounding amino acids showing the protons that contribute to the ENDOR data. The relevant hyperfine matrices are represented as ellipses, and the $\mathbf{g}_{\|}$ vector determined from the data is shown. (b) W-band EPR spectrum; arrows show the position of ENDOR measurements. The MW pulse bandwidth is depicted for one of the positions; '*' marks resonator background signal from Mn^{2+}. (c) W-band Davies ENDOR spectra recorded at the field positions indicated in (b). Black – experiment, blue – simulation (sum of eight proton contributions), cyan – proton 1, and red – proton 2. The orientations selected by the MW pulses in the g-matrix frame are shown at the far right, and blue indicates that the ENDOR pulse sequence is not resonant with any EPR transitions at this orientation. (Adapted from Ref. 42 with permission from the PCCP Owner Societies)

16.4.5 Pulse Triple ENDOR

The concept behind this technique is similar to its CW triple ENDOR counterpart and provides information on the relative signs of the HF couplings. The triple ENDOR pulse sequence based on Davies ENDOR is shown in Figure 16.13(a).[58] It comprises two RF pulses, RF1 is the pump pulse with a fixed frequency set to one of the nuclear transitions and RF2 is swept as in normal Davies ENDOR. Consider the simplest case of an $S = 1/2$ electron spin coupled to two $I_1 = I_2 = 1/2$ nuclear spins. First, the spin system is prepared with a selective MW inversion π-pulse ②. Next, the pump pulse RF1 inverts the polarization of a chosen NMR transition in one of the electron spin manifolds, the ω_α^1 transition in Figure 16.13(b ③), which changes the polarization of all other NMR transitions that have an energy level in common with the EPR transition excited by the preparation MW π-pulse. The second RF pulse, RF2, is then swept and the echo monitored. Figure 16.13(c) shows the standard ENDOR spectrum (top), the triple ENDOR spectrum (middle), and the difference triple ENDOR spectrum (bottom). In Figure 16.13(b ④), the polarization patterns for RF1 exciting ω_α^1 and then RF2 exciting ω_α^1, ω_α^2, ω_β^1, and ω_β^2 are schematically represented

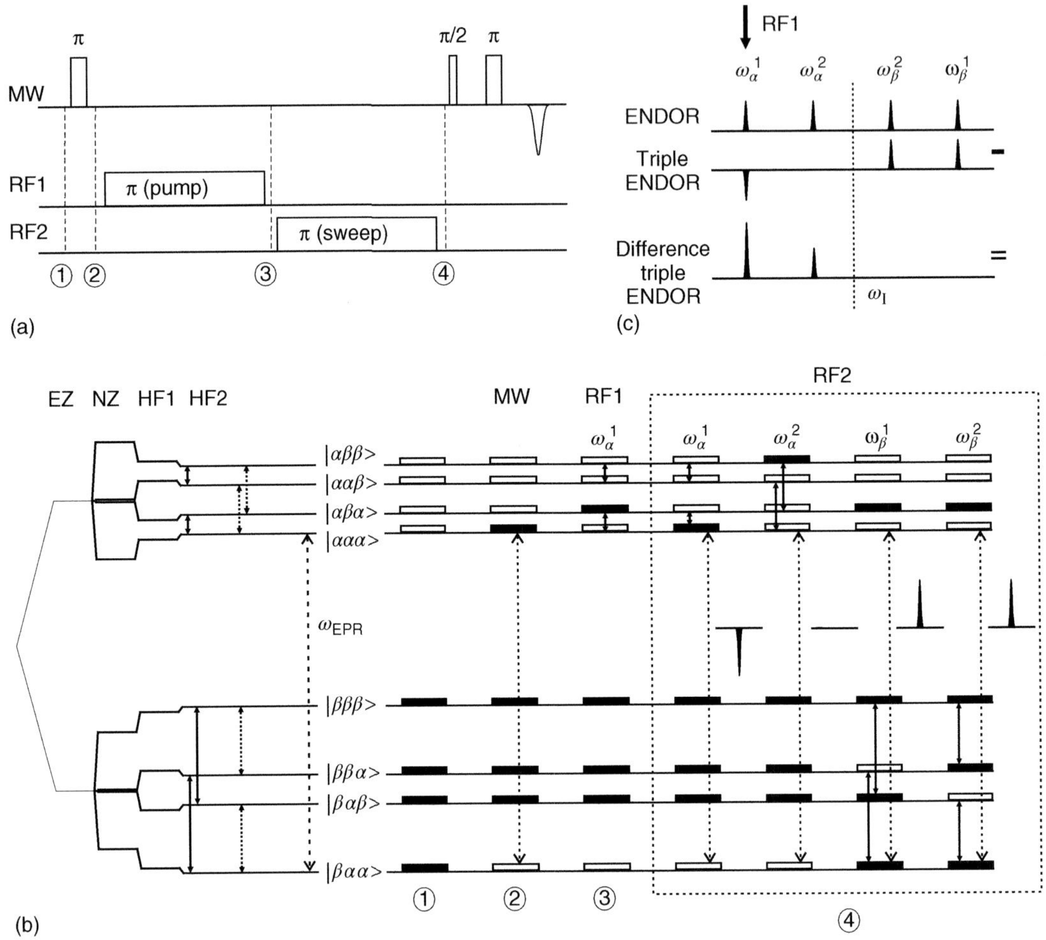

Figure 16.13. (a) Triple ENDOR sequence based on Davies ENDOR. (b) Energy-level diagram for an $S = ½$ electron spin coupled to two $I_1 = I_2 = ½$ nuclei with $A_1 > A_2 > 0$, and $g_n > 0$, from left to right are the levels for the EZ interaction, the two NZ interactions, then the HF interaction 1 and then HF interaction 2. Right depicts the level populations during a TRIPLE experiment; ① thermal equilibrium (NMR levels within each electron spin manifold are assumed to be equally populated). ② the first selective MW pulse invests the population of the $|\beta\alpha\alpha\rangle \leftrightarrow |\alpha\alpha\alpha\rangle$ EPR transition, ③ the RF1 pump pulse, at frequency $\omega_\alpha^{\,1}$ in this example, inverts the population of the two corresponding NMR transitions, which removes spin polarization from the EPR transition that will be monitored. ④ the second pulse RF2 is swept, and if it is resonant with transition $\omega_\alpha^{\,1}$, it reestablishes the polarization after the MW π-pulse and if it is resonant with transition $\omega_\alpha^{\,2}$, it has no effect on the triple spectrum (no polarization across the monitored EPR transitions), whereas if it is resonant with transition $\omega_\beta^{\,1}$ or $\omega_\beta^{\,2}$, it reestablishes the initial polarization ① of the EPR transition being monitored

along with the monitored EPR transition (dotted line) for easy of interpretation. In this example, the difference triple ENDOR peaks are both on the same side of ω_I, establishing that the two nuclei have the same relative sign (belong to the same m_S manifold). In the case of one weakly and one strongly coupled nuclei, the presence, for example, in the difference triple ENDOR spectrum of the two low-frequency ENDOR lines with respect to ω_I and $|A|/2$ would again indicate the same relative sign for the two HF couplings (consult Figure 16.1).

A related technique, a triple resonance HF sublevel correlation experiment (THYCOS), combines ELDOR (electron double resonance) (two MW frequencies) and ENDOR schemes to build a sequence for assigning nuclear frequencies to their respective paramagnetic centers and electron spin manifolds and to thus determine the relative signs of the HF couplings (see Chapter 17).[59]

16.4.6 Pulse Special Triple ENDOR

The concept behind pulse special triple ENDOR is similar to the special triple CW ENDOR experiment described previously in that two RF frequencies are employed that are swept symmetrically around the ω_I of the nucleus of interest, one RF pulse to higher frequency and the other to lower frequency, that is, $\omega_{RF\pm} = \omega_I \pm \omega_{scan}$, where ω_{scan} is scanned. The sequence based on Davies ENDOR is shown in Figure 16.14 along with the polarization at various stages during the sequence.[60] After the inversion MW and two RF π-pulses, position ③, the polarization of the EPR transition being monitored is fully inverted, and thus the ENDOR effect, $Y^{ENDOR} = 1$. The experiment requires equation (16.6) or (16.16) to be valid to a very good approximation, that is, the high-field condition where the nuclear frequencies from weakly coupled nuclei are symmetric around ω_I. This implies that knowing the nuclear frequency in one electron spin manifold means the nuclear frequency in the other manifold is known. This is not the case when the HF coupling has anisotropy, see Figure 16.1 and equations (16.8) and (16.9). For solid-state samples measured at low fields (e.g., X-band), the high-field approximation will routinely not be valid, but an optimal ENDOR effect can still be achieved theoretically using the pulse sequence in Figure 16.14(b) where the second RF π-pulse is replaced by an equivalent (besides a sign change) composite pulse, which comprises a nonselective MW π-pulse with a selective RF π-pulse with the same frequency as the first RF pulse; that is, $\xrightarrow{\pi I_x^{34}} \equiv \xrightarrow{\pi S_x} \xrightarrow{\pi I_x^{12}}$, see Figure 16.14(b).[61] As in sequence (a), the polarization of the EPR transition being monitored is now completely inverted by the RF and composite pulse and thus $F^{ENDOR} = 1$. In practice, the sensitivity advantage that the second RF pulse provides will be reduced by increased signal loss from relaxation because of the increased length of the sequence.

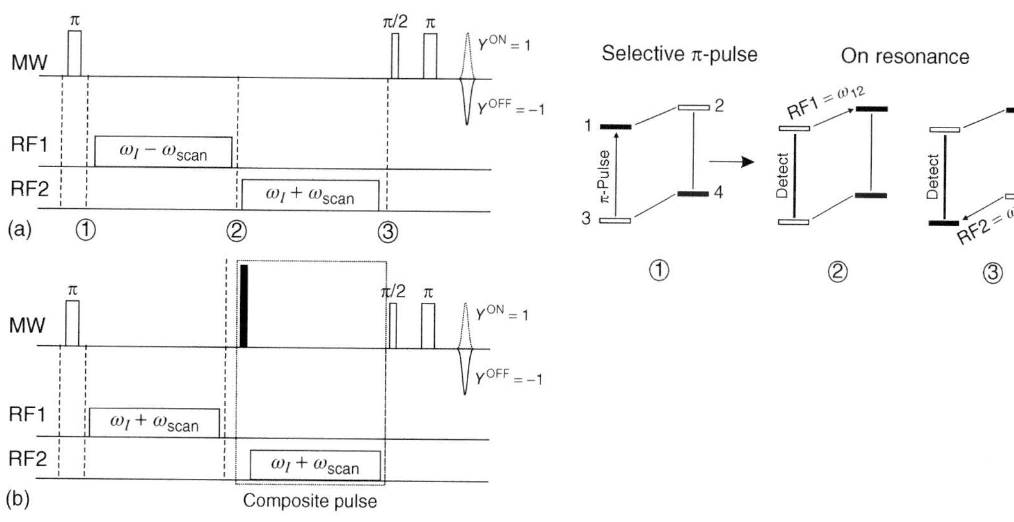

Figure 16.14. Pulse special triple ENDOR for improving sensitivity. (a) Sequence based on Davies ENDOR, where the first MW π-pulse inverts the polarization of EPR transitions (1,3) ①, and the first RF π-pulse, if resonant with NMR transition (1,2), inverts the polarization of this transition with the result of equalizing the spin polarization of EPR transitions (1,3) ②. The maximum theoretical ENDOR efficiency is 0.5 at this stage. The application of a second RF pulse resonant with NMR transition (3,4) inverts the polarization of this transition ③, and now the polarization of EPR transition (1,3) has been inverted and the ENDOR efficiency is 1. (b) If the hyperfine coupling is anisotropic, knowing the NMR frequency ω_{12} does not mean that the NMR frequency ω_{34} is also known. To overcome this, the second selective RF pulse in (a) is replaced in (b) by an equivalent (besides a sign change) composite pulse, a nonselective MW π-pulse – selective RF pulse with the frequency of the first pulse RF1

16.4.7 Time-domain ENDOR

In a time-domain ENDOR experiment,[62] the free evolution of nuclear coherence is recorded – this requires a minimum of three building blocks in the sequence – a nuclear coherence generator, a free evolution period, and a nuclear coherence detector. A Davies-type sequence is shown in Figure 16.15(a), where the nuclear coherence generator consists of a selective MW and nonselective RF $\pi/2$-pulse, the evolution period t_1 that is incremented, and detection via a nonselective RF $\pi/2$-pulse and Hahn echo. To determine the sign of the nuclear coherences relative to the RF frequency, the experiment needs to be measured with the RF pulses along the x-axis and then y-axis in the doubly rotating frame. The nuclear frequency spectrum is then obtained by a complex Fourier transformation of the ESEEM signal. The sensitivity of this type of experiment is limited by the bandwidth of the RF pulse, which also hampers the development of correlation experiments because of the narrow bandwidth. This limitation can be overcome with a chirp pulse, one in which the pulse frequency is swept over a desired frequency range over the pulse length. A Davies-type sequence using chirp RF pulses[63] is shown in Figure 16.15(b) and has the advantage that the entire ENDOR spectrum is measured without having to increment the RF frequency and repeat the experiment as in Figure 16.15(a). This sequence can be readily extended to a 2D version by adding the hard nonselective MW π-pulse (mixing pulse), which transfers nuclear coherences between the electron spin manifolds (Figure 16.15c). The echo is now a function of times t_1 and t_2 and its Fourier transformation yields a 2D spectrum where cross-peaks correlate nuclear frequency from the difference electron spin manifolds – the experiment is thus an ENDOR equivalent of the HYSCORE experiment.

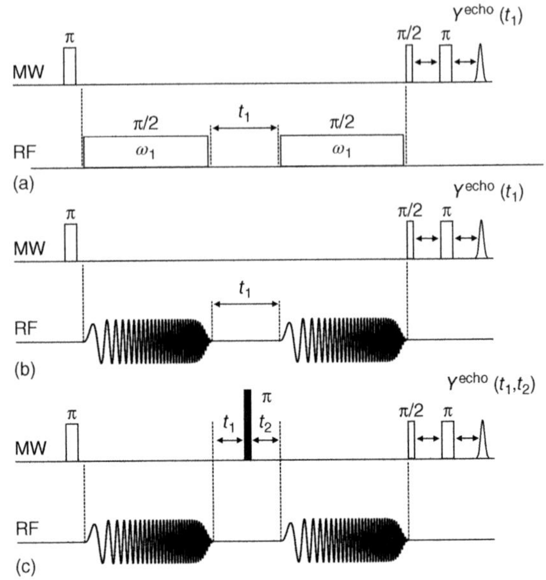

Figure 16.15. Time-domain ENDOR sequences based on the generation of nuclear coherence created by the subsequence (MW π-pulse–RF $\pi/2$-pulse). (a) Time t_1 is incremented and the Fourier transform of the modulation of the echo intensity $Y^{echo}(t_1)$ gives a 1D nuclear frequency spectrum. This sequence is comparable to the spectrum from a MW only ESEEM experiment and the ENDOR linewidth is determined by the t_1 dimension and not by the RF pulse bandwidth. (b) Same as is (a), but with a broadband chirp pulse (one in which the RF frequency is swept). (c) A HYSCORE-ENDOR sequence that is based on the nuclear coherence generating (MW π–RF chirp pulse) in (b) but with a nonselective hard mixing MW π-pulse that transfers nuclear coherences between the electron spin manifolds. A 2D Fourier transformation of the modulation of the echo intensity $Y^{echo}(t_1, t_2)$ yields a 2D frequency spectrum correlating the nuclear frequencies from different electron spin manifolds of the same paramagnetic center. The resulting 2D ENDOR spectrum is comparable to that obtained from HYSCORE

16.4.8 Two-dimensional ENDOR Techniques

The basic 1D Mims and Davies ENDOR experiments can suffer from poor separation of signal from different nuclei due to overlapping signals. To overcome these limitations, two options may be available, either a second dimension can be added to ENDOR experiments at moderate to low fields or a high-field ENDOR approach used where much better separation of the nuclear Zeeman frequencies of the various nuclei is achieved. Here, we described options that yield a second dimension to disentangle overlapping peaks primarily from different nuclei.

16.4.8.1 2D Mims ENDOR

One approach to disentangle a 1D spectrum is to correlate the ENDOR frequencies with their corresponding HF frequencies in a technique referred to as HF-correlated ENDOR spectroscopy. A simple extension of the 1D Mims ENDOR sequence (Figure 16.7a′) allows this to be achieved; in addition to the normal stepwise RF frequency sweep, the τ value is also incremented stepwise. As shown in equation (16.28), the simulated echo intensity for a fixed RF that is resonant with an NMR transition oscillates with the τ value according to $\cos(A\tau)$. Thus, a Fourier transformation of the time-domain signal collected as a function of τ results in a peak at the HF coupling. The complete experiment collects a 2D matrix of data with RF (or nuclear frequency) along one axis and τ along the second, $Y^{echo}(\omega_{RF}, \tau)$, which after Fourier transformation along the τ dimension yields a HF-correlated ENDOR spectrum. An example is shown in Figure 16.16.[64] One drawback of the technique is the experimental resolution along the HF axis as it is limited by the decay of the simulated echo with the phase memory time T_m, which is often only microseconds long in samples even measured at cryogenic temperatures.

16.4.8.2 HYEND

The HYEND experiment also achieves a correlation between the RFs and their corresponding HF couplings. The HYEND sequence based on Davies ENDOR is shown in Figure 16.17(a) and achieves the HF dimension by transferring nuclear coherence, generated by two selective coherent RF pulses, between the electron spin manifolds. The ENDOR or nuclear frequency dimension is obtained as in a 1D ENDOR experiment by stepwise incrementing the frequency of two RF $\pi/2$-pulses of the same frequency over the desired range, and the HF dimension is obtained by varying time t_1. A 2D matrix of data is thus obtained, $Y^{echo}(\omega_{RF}, t_1)$. The states during the experiment are demonstrated in Figure 16.17(b). The first selective MW pulse inverts the population on level (1,3), and the selective RF $\pi/2$-pulse on resonance with NMR transition (1,2) creates nuclear coherence at frequency ω_{12}. This coherence in the α electron spin manifold is immediately transferred to the β electron spin manifold by a hard nonselective MW π-pulse where it evolves during time t_1 with the nuclear frequency ω_{34} and gains phase by $(\omega_{12} - \omega_{34})t_1$ (weak coupling) or $(\omega_{12} + \omega_{34})t_1$ (strong coupling), relative to the phase of the RF pulses that must remain coherent during the time t_1. The second nonselective MW π-pulse then transfers this nuclear coherence back to the α electron spin manifold and the second RF $\pi/2$-pulse transfers the nuclear coherence back to electron spin polarization, which is then detected by the two-pulse Hahn

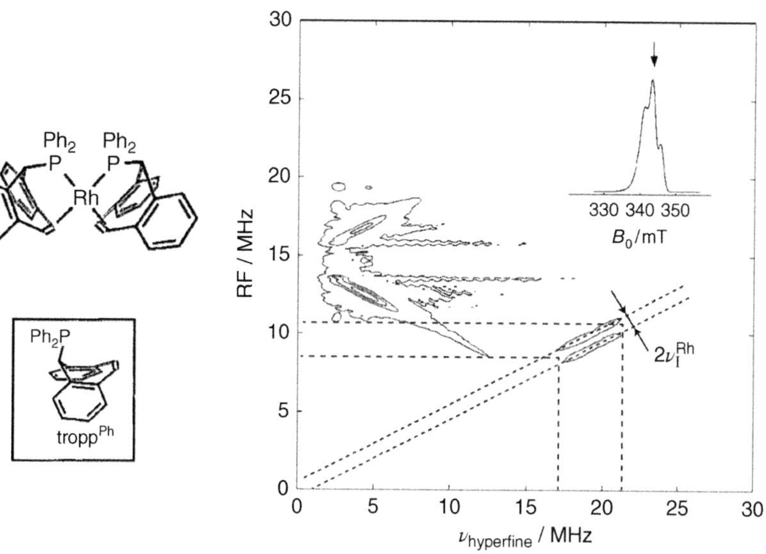

Figure 16.16. X-band 2D Mims ENDOR spectrum of [Rh(tropp^{Ph})$_2$] (left). The dashed lines are separated by $2\nu_I(^{103}Rh)$, and the Rh ENDOR signals (ν_{RF} axis) from 8 to 11 MHz indicate strong coupling ($\nu_I(^{103}Rh) = 0.47$ MHz, $I = 1/2$) and appear along the hyperfine dimension ($\nu_{hyperfine}$ axis) with the corresponding hyperfine couplings of about 16–22 MHz. The Rh ENDOR signals overlap with the weak coupled proton ENDOR signals above approximately 8 MHz but are well separated by the hyperfine dimension. (Reproduced with permission from Ref. 64. © John Wiley & Sons, Ltd., 2002)

echo. As the two RF π/2-pulses are coherent, the electron polarization oscillates with the phase accumulated by the nuclear coherence during time t_1 in the β electron spin manifold with respect to the phase of the RF excitation, this corresponds to the HF frequency $|A| = |\omega_\alpha - \omega_\beta|$. The HYEND signal for HF coupling A with the RF pulse resonant on a nuclear transition is given by

$$Y^{\text{echo}}(t_1) \cong \pm\text{sign}(2\omega_I + A)\cos((\omega_\alpha - \omega_\beta)t_1) = \pm\text{sign}(2\omega_I + A)\cos(At_1) \quad (16.36)$$

An example is shown in Figure 16.17(c),[65] where HYEND was used to separate overlapping ^1H, ^{14}N, and ^{13}C signals to identify the methyl ligand, $H_3{}^{13}C$-Ni(III), in the active site of the enzyme methyl-coenzyme reductase.

16.4.8.3 2D Triple ENDOR

The triple ENDOR sequence (Figure 16.7d,d′) can be extended into a 2D experiment by sweeping simultaneously and independently the frequencies of both RF pulses.[66,67] The 2D triple experiment provides correlations (which appear as cross-peaks) between ENDOR frequencies that belong to the sample paramagnetic center and the same m_S manifold in the same way as a 1D triple ENDOR does and the relative signs of the HF couplings are thus determined. In addition, the shape and orientation of the cross-peaks, which result when the two RF pulses excite two NMR transitions from two nuclei from the same paramagnetic center, enable the principal values of the two HF interactions and their relative orientation to be deduced. 2D triple ENDOR can resolve overlapping ENDOR powder patterns in orientationally disordered systems, particularly if one nucleus already has resolved ENDOR signals as this helps to separate out cross-peaks from different pairs of nuclei. The disadvantage of the experiment is sensitivity as each points in the 2D spectrum has to be collected stepwise.

16.4.9 Determining the Sign of the Hyperfine Coupling

Asymmetric or polarized ENDOR lines can appear at low temperature (<10 K) and high field (W-band, 96 GHz).[8,68,69] This is due to high thermal electron spin polarization in combination with an electron spin–lattice relaxation time that is short compared to the mixing time t_{mix} defined in Figure 16.7(e).

For an $S = 1/2$, $I = 1/2$ spin system, there are three types of relaxation processes with times T_{1e}, T_{1n}, and T_{1x}, (or rates $W = 1/T$) as defined in Figure 16.5. A general relaxation model was described in Ref. 70 where the ENDOR effect is extended to include the dependence on time t_{mix} and the pulse sequence repetition

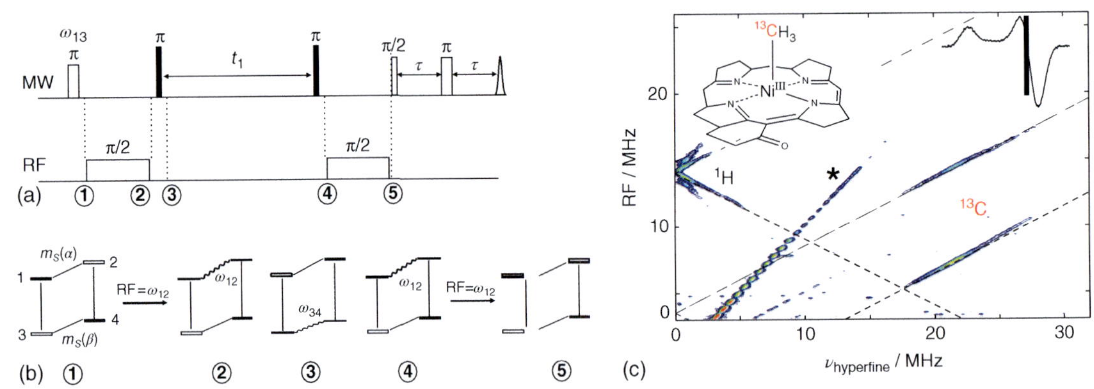

Figure 16.17. (a) HYEND sequence that correlates the nuclear frequencies ν_{RF} against the corresponding hyperfine couplings, $\nu_{\text{hyperfine}}$. (b) Basic principal of the experiment.[25] The wavy lines represents a coherent spin state across this transition. (c) Example of an X-band HYEND spectrum of the nickel-methyl species in methyl-coenzyme M reductase (inset left)[65] (Adapted with permission from N. Yang; M. Reiher; M. Wang; J. Harmer; E. C. Duin J. Am. Chem. Soc. 2007, 129, 11028. © 2007, American Chemical Society). In a 1D experiment, the ^1H and pyrrole ^{14}N signals (not seen in HYEND) overlap with the strongly coupled ^{13}C signals of the methyl group, which are cleanly separated in the HYEND experiment. The ridge extending along the diagonal marked with '*' is an artifact. (Inset right) CW EPR spectrum and HYEND measurement position

time t_R,

$$F^{\text{ENDOR}} = \left| \frac{Y^{\text{ON}}_{\text{echo}}(t_{\text{mix}}, t_R) - Y^{\text{OFF}}_{\text{echo}}(t_{\text{mix}}, t_R)}{2Y^0_{\text{echo}}} \right| \quad (16.37)$$

Here, $Y^{\text{ON}}_{\text{echo}}(t_{\text{mix}}, t_R)$ and $Y^{\text{OFF}}_{\text{echo}}(t_{\text{mix}}, t_R)$ are the echo amplitudes obtained when the RF irradiation is on and off resonance with an NMR transition in one of the electron spin manifolds, Y^0_{echo} the echo amplitude from a single two-pulse echo sequence with the same experimental settings (τ value, pulse lengths, etc.) as in the ENDOR sequence, and $t_R \gg (T_{1e}, T_{1n}, T_{1x})$. Under normal measurement conditions, $t_{\text{mix}} \approx 0$ and $F^{\text{ENDOR}} = 0.5$. Under suitable t_{mix} and t_R conditions, the amplitudes of the ENDOR lines in the different electron spin manifolds from a nuclear spin can become asymmetric, and two regimes are most useful. For long mixing times t_{mix} and negligible mixing rates, so that $t_{\text{mix}} \gg T_{1e} \gg T_{1n}, T_{1x}$, the effect on the ENDOR line intensity from the $m_S = +{}^1\!/_2$ manifold is negative and from the $m_S = -{}^1\!/_2$ manifold is positive. This allows the sign of the HF interaction to be determined as the ENDOR line can be assigned to its m_S manifold (consult Figure 16.1; in panel (a), for example, with $A > 0$, $g_n > 0$, the largest nuclear frequency belongs to the $m_S = -{}^1\!/_2$ manifold). In the second case, asymmetric lines are obtained when the cross and/or nuclear

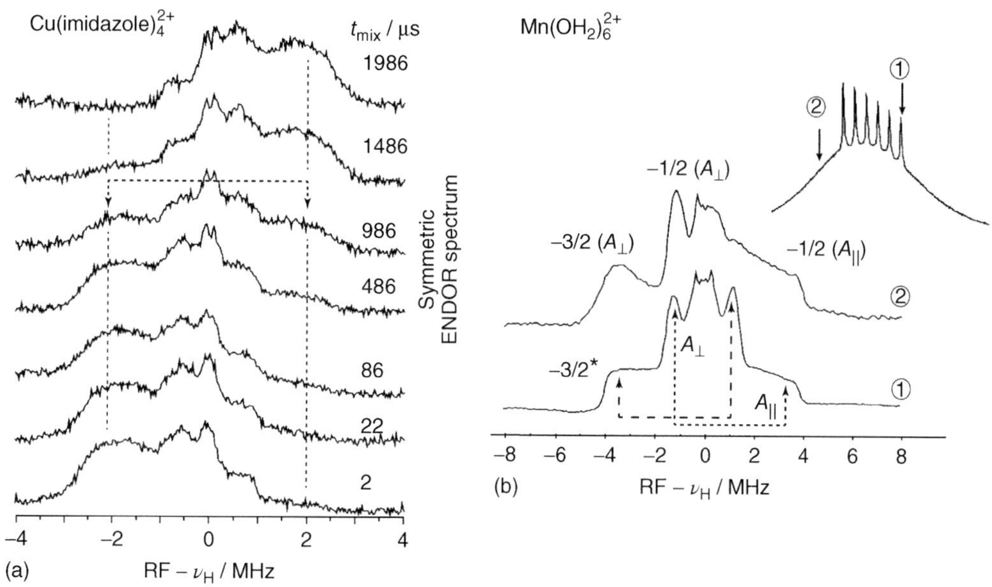

Figure 16.18. (a) Variable mixing time t_{mix} of the Davies ENDOR spectra recorded at W-band at 4.5 K from a Cu(imidazole)$_4^{2+}$ complex. For a $t_{\text{mix}} = 986$ µs, the spectrum is symmetric and shows two broad proton doublets, one at ±0.5 MHz and the other at ±2 MHz that we concentrate on. For small t_{mix}, the intensity of the +2 MHz peak is zero, identifying it with the $m_S = -1/2$ (β) manifold, and thus $A > 0$ ($g_n > 0$ for ^{1}H). As t_{mix} increases, the +2 MHz peak increases, and the −2 MHz peak decreases and becomes negative at the longest t_{mix} time, thus reaching the measurement conditions where the t_{mix} ENDOR effect is negative in the $m_S = +1/2$ (α) manifold, and thus again $A > 0$. (b) ^{1}H ENDOR spectra of Mn(OH$_2$)$_6^{2+}$ measured at 4.5 K at W-band. At observer position ①, ENDOR signals arise mainly from the $|-1/2\rangle \leftrightarrow |1/2\rangle$ EPR transition and show the expected symmetric doublet signal from protons of the coordinated water with $A_\perp = -2.6$ MHz and $A_\parallel = 7.6$ MHz ($A_{\text{iso}} = 0.8$ MHz). There is a small signal from the $m_S = |-3/2\rangle$ manifold at approximately −3.5 MHz as indicated with the '*', which becomes obvious in the ENDOR spectrum recorded at field position ②, which is dominated by the nuclear frequencies from the $m_S = -1/2$ and the $m_S = -3/2$ electron spin manifolds associate with the $|-3/2\rangle \leftrightarrow |-1/2\rangle$ EPR transition. The position of the ENDOR lines shows that $A_\parallel > 0$ and $A_\perp < 0$ (as expected for dipolar coupled protons) as the NMR frequencies for the $m_S = -3/2$ manifold are below ω_I for $A < 0$, and above ω_I for $A > 0$ in the weak coupling case, as can be appreciated by inspection of Figures 16.3 and 16.4.[69] (Adapted from Appl. Magn. Reson., High-field ENDOR and the sign of the hyperfine coupling, 21, 2001, 287, B. Epel; P. Manikandan; P. M. H. Kroneck; D. Goldfarb © With permission of Springer)

relaxation times are longer than the pulse sequence repetition time t_R; $t_{mix} \ll T_{1e} \ll t_R$ and $t_R < T_{1n}$, T_{1x}. Under these conditions, where the nuclear transitions are saturated, ENDOR lines from the $m_S = +\frac{1}{2}$ (α) manifold are positive, while those of the $m_S = -\frac{1}{2}$ (β) manifold are negative. An instructive example for the case of a Cu(imidazole)$_4^{2+}$ complex is shown in Figure 16.18(a).[69]

For high-spin systems, the sign of the HF interaction can be determined by selective excitation of specific EPR transitions. For example, in an $S = 5/2$ spin system, the assignment of the ENDOR lines to an m_S manifold can be achieved by selective excitation of the $|-3/2\rangle \rightarrow |-1/2\rangle$ EPR transition, which is highly populated relative to its symmetric counterpart $|3/2\rangle \rightarrow |1/2\rangle$ at high field and low temperature (consult Figure 16.4). Figure 16.18(b) shows an experimental example for the high-spin system $Mn(OH_2)_6^{2+}$; note that the six sharp lines in the EPR spectrum (inset) are from the $|-1/2\rangle \leftrightarrow |-1/2\rangle$ EPR transition with resolved Mn^{2+} HF couplings. The broad component is from all other EPR transitions, with integrated EPR transition intensities in the order $|-5/2\rangle \leftrightarrow |-3/2\rangle$, $|-3/2\rangle \leftrightarrow |-1/2\rangle$, $|-1/2\rangle \leftrightarrow |+1/2\rangle$, $|+1/2\rangle \leftrightarrow |+3/2\rangle$, $|+3/2\rangle \leftrightarrow |+5/2\rangle$, as a result of depopulation of the higher energy levels at the measurement temperature of 4.5 K and at the frequency of 95 GHz.

16.4.10 Stochastic RF Excitation – Avoiding Heating and Baseline Artifacts

A significant technical problem with the application of strong RF pulses in ENDOR is sample heating, which can detune the resonator and lead to large baseline drifts. This problem is exacerbated with a linear sweep of the RF frequency, which can be largely suppressed using a stochastic (i.e., random order) acquisition of the RF frequencies (ENDOR spectrum) over the RF sweep range. The improvements that this acquisition scheme produces are convincingly shown in Ref. 60 at W-band frequencies, where the baseline distortions are often very large. In additional to the heating improvement, the stochastic accumulation approach allows more time (up to a maximum of $t_R \times$ the number of points in the ENDOR spectrum) for the nuclear spin system to return to equilibrium before the next accumulation that is beneficial as generally $T_{1n} \gg T_{1e}$.

16.4.11 Echo Train Detection Schemes – Increasing Sensitivity

Typically pulse ENDOR experiments are based on the detection (integration of signal around the echo maximum usually) of a single spin echo. The total ENDOR pulse sequence is typically some tens of microseconds long, and then there is a long delay of around $5T_{1e}$ while the electron spin system recovers to thermal equilibrium, which is typically several milliseconds or more. As was shown in Ref. 71, a large signal-to-noise improvement, without increasing the experiment time, can be achieved if the single echo detection scheme is replaced by a train of echoes generated by Carr–Purcell-type sequences, that is

$\pi/2_x - (\tau - \pi_y - \tau - \text{echo})_n$ (CPMG)
$\pi/2_x - (\tau - \pi_x - \tau - \text{echo} - \tau - \pi_{-x} - \tau - \text{echo})_{n/2}$ (CPMG-2)

This alteration to the standard Davies ENDOR sequence is shown in Figure 16.19 for the CPMG sequence. Phase cycles appropriate for these sequences and the Mims ENDOR alternative are given in Ref. 71. It was found in Ref. 71 that the best signal-to-noise (SNR) results from the integration of around $n \approx 20$ echoes per sequences, for example, at W-band for a

- nitroxide biradical, ^{1}H Mims ENDOR gave a SNR(20)/SNR(1) = 3.5
- Mn^{2+}ATP sample, ^{31}P Davies ENDOR gave a SNR(19)/SNR(1) = 5.1

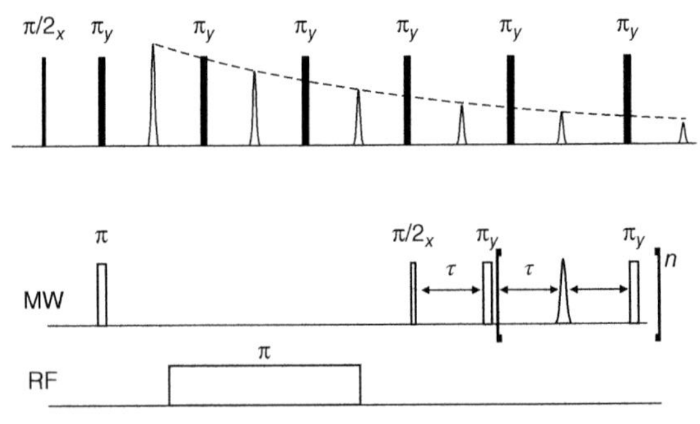

Figure 16.19. (a) The CPMG pulse sequence where the echo decays with the transverse magnetization phase memory time T_m. (b) Application of the CPMG pulse sequence to the standard Davies ENDOR sequence. For each accumulation, the summed intensity of the CPMG echoes is recorded to improve the signal-to-noise ratio. In Ref. 71, best results were obtained for $n \approx 20$

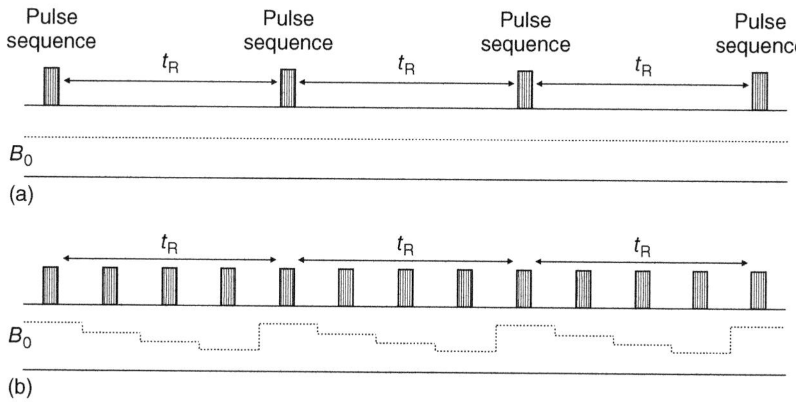

Figure 16.20. (a) A linear acquisition scheme, where data is accumulated at a set B_0 magnetic field value. (b) A parallel acquisition scheme, where the field B_0 is jumped after each acquisition to utilize the repetition time t_R to accumulate data. In (b), data at four field positions are collected per repetition time, whereas in (a), only data from one field position is collected per repetition time

- Mn^{2+}ATP sample, ^1H Mims ENDOR gave a SNR(22)/SNR(1) = 2.8.

16.4.12 Parallel Acquisition Schemes – Increasing Sensitivity

The vast majority of ENDOR measured are carried out on systems where there is orientation selection, and in this case, a set of ENDOR experiments are recorded at selected field positions B_0 across the field-sweep EPR spectrum. At each field position, the ENDOR pulse sequence must often be repeated many hundreds or thousands of times and the signal accumulated until the desired SNR is achieved. In a linear acquisition scheme, where the ENDOR spectrum at each B_0 field position is acquired sequentially, the repetition time t_R of around $5T_{1e}$ is not utilized after each ENDOR pulse sequence, while the spin system returns to thermal equilibrium; m accumulations take a time of (pulse sequence + t_R)^m. In a parallel acquisition scheme, the repetition time t_R after each ENDOR pulse sequence is utilized by rapidly jumping the magnetic field B_0 to another field position and immediately accumulating ENDOR data at this new field position from a single application of the ENDOR sequence. The field can then be jumped to another new B_0 position, or back to the first B_0 position, for the next accumulation if $\approx 5T_{1e}$ has elapsed since the last accumulation at this field position was carried out. This procedure is schematically depicted in Figure 16.20. To be able to apply the parallel acquisition scheme effectively, the spectrometer must be capable of jumping the B_0 field in a time short as compared to t_R, and by a meaningful amount, enough to change the ENDOR sequence orientation selection. The technique was demonstrated at W-band on the ENDOR, DEER, ESEEM, and ELDOR-detected NMR sequences.[72]

REFERENCES

1. G. Feher, *Phys. Rev.*, 1956, **103**, 834.
2. J. S. Hyde and A. H. Maki, *J. Chem. Phys.*, 1964, **40**, 3117.
3. J. S. Hyde, *J. Chem. Phys.*, 1965, **43**, 1806.
4. S. Van Doorslaer and D. M. Murphy, in EPR Spectroscopy: Applications in Chemistry and Biology, eds M. Drescher and G. Jeschke, Springer: Berlin, 2012, Vol. 321, 1.
5. B. M. Hoffman, *Proc. Natl. Acad. Sci. U. S. A.*, 2003, **100**, 3575.
6. D. Goldfarb, *Phys. Chem. Chem. Phys.*, 2006, **8**, 2325.
7. P. G. Baranov, S. B. Orlinskii, C. d. M. Donega, and J. Schmidt, *Appl. Magn. Reson.*, 2010, **39**, 151.
8. D. Goldfarb, K. G. Strohmaier, D. E. W. Vaughan, H. Thomann, O. G. Poluektov, and J. Schmidt, *J. Am. Chem. Soc.*, 1996, **118**, 4665.
9. S. Van Doorslaer and E. Vinck, *Phys. Chem. Chem. Phys.*, 2007, **9**, 4620.

10. B. M. Hoffman, *Acc. Chem. Res.*, 2003, **36**, 522.

11. C. Calle, A. Sreekanth, M. V. Fedin, J. Forrer, I. Garcia-Rubio, I. A. Gromov, D. Hinderberger, B. Kasumaj, P. Leger, B. Mancosu, G. Mitrikas, M. G. Santangelo, S. Stoll, A. Schweiger, R. Tschaggelar, and J. Harmer, *Helvetica Chimica Acta*, 2006, **89**, 2495.

12. A. Savitsky and K. Moebius, *Photosynth. Res.*, 2009, **102**, 311.

13. L. Kulik and W. Lubitz, *Photosynth. Res.*, 2009, **102**, 391.

14. T. Lohmiller, V. Krewald, M. P. Navarro, M. Retegan, L. Rapatskiy, M. M. Nowaczyk, A. Boussac, F. Neese, W. Lubitz, D. A. Pantazis, and N. Cox, *Phys. Chem. Chem. Phys.*, 2014, **16**, 11877.

15. M. Bennati and T. F. Prisner, *Rep. Prog. Phys.*, 2005, **68**, 411.

16. E. Schleicher and S. Weber, in EPR Spectroscopy: Applications in Chemistry and Biology, eds M. Drescher and G. Jeschke, Springer: Berlin, 2012, Vol. 321, 41.

17. D. Goldfarb and D. Arieli, *Annu. Rev. Biophys. Biomol. Struct.*, 2004, **33**, 441.

18. I. Gromov, J. Shane, J. Forrer, R. Rakhmatoullin, Y. Rozentzwaig, and A. Schweiger, *J. Magn. Reson.*, 2001, **149**, 196.

19. I. Gromov, V. Krymov, P. Manikandan, D. Arieli, and D. Goldfarb, *J. Magn. Reson.*, 1999, **139**, 8.

20. M. Bennati, C. T. Farrar, J. A. Bryant, S. J. Inati, V. Weis, G. J. Gerfen, P. Riggs-Gelasco, J. Stubbe, and R. G. Griffin, *J. Magn. Reson.*, 1999, **138**, 232.

21. M. M. Hertel, V. P. Denysenkov, M. Bennati, and T. F. Prisner, *Magn. Reson. Chem.*, 2005, **43**, S248.

22. A. Schnegg, A. A. Dubinskii, M. R. Fuchs, Y. A. Grishin, E. P. Kirilina, W. Lubitz, M. Plato, A. Savitsky, and K. Mobius, *Appl. Magn. Reson.*, 2007, **31**, 59.

23. M. Fuhs and K. Mobius, in High Magnetic Fields: Applications in Condensed Matter Physics and Spectroscopy, eds C. Berthier, L. P. Levy and G. Martinez, Springer: Berlin, 2001, Vol. 595, 476.

24. O. Grinberg and L. J. Berliner, Very High Frequency (VHF) ESR/EPR, Springer Verlag GMBH: New York, 2004, Vol. 22.

25. A. Schweiger and G. Jeschke, Principles of Pulse Electron Paramagnetic Resonance, Oxford University Press, Inc.: New York, 2001.

26. P. P. Zanker, G. Jeschke, and D. Goldfarb, *J. Chem. Phys.*, 2005, **122**, 024515.

27. S. Stoll and A. Schweiger, *J. Magn. Reson.*, 2006, **178**, 42.

28. N. M. Atherton and A. J. Horsewill, *J. Chem. Soc., Faraday Trans. 2*, 1980, **76**, 660.

29. C. P. Aznar, Y. Deligiannakis, E. J. Tolis, T. Kabanos, M. Brynda, and D. Britt, *J. Phys. Chem. A*, 2004, **108**, 4310.

30. C. V. Grant, J. A. Ball, B. J. Hamstra, V. L. Pecoraro, and R. D. Britt, *J. Phys. Chem. B*, 1998, **102**, 8145.

31. B. E. Sturgeon, J. A. Ball, D. W. Randall, and R. D. Britt, *J. Phys. Chem.*, 1994, **98**, 12871.

32. R. Vardi, M. Bernardo, H. Thomann, K. G. Strohmaier, D. E. W. Vaughan, and D. Goldfarb, *J. Magn. Reson.*, 1997, **126**, 229.

33. A. V. Astashkin and A. M. Raitsimring, *J. Chem. Phys.*, 2002, **117**, 6121.

34. S. Stoll and R. D. Britt, *Phys. Chem. Chem. Phys.*, 2009, **11**, 6614.

35. A. Abragam and B. Bleaney, Electron Paramagnetic Resonance of Transition Ions, OUP: Oxford, 1970.

36. J. Telser, J. Krzystek, and A. Ozarowski, *J. Biol. Inorg. Chem.*, 2014, **19**, 297.

37. J. M. Peloquin, K. A. Campbell, D. W. Randall, M. A. Evanchik, V. L. Pecoraro, W. H. Armstrong, and R. D. Britt, *J. Am. Chem. Soc.*, 2000, **122**, 10926.

38. A. Bencini and D. Gatteschi, EPR of Exchange Coupled Systems, Springer-Verlag: Berlin, 1990.

39. J. M. Mouesca, L. Noodleman, D. A. Case, and B. Lamotte, *Inorg. Chem.*, 1995, **34**, 4347.

40. R. Kappl, M. Ebelshauser, F. Hannemann, R. Bernhardt, and J. Huttermann, *Appl. Magn. Reson.*, 2006, **30**, 427.

41. R. Kappl, G. Bracic, and J. Huettermann, in High Resolution EPR: Applications to Metalloenzymes and Metals in Medicine, eds G. Hanson and L. Berliner, Springer: New York, 2009, Vol. 28, 63.

42. J. A. B. Abdalla, A. M. Bowen, S. G. Bell, L. L. Wong, C. R. Timmel, and J. Harmer, *Phys. Chem. Chem. Phys.*, 2012, **14**, 6526.

43. M. M. Dicus, A. Conlan, R. Nechushtai, P. A. Jennings, M. L. Paddock, R. D. Britt, and S. Stoll, *J. Am. Chem. Soc.*, 2010, **132**, 2037.

44. D. M. Murphy and R. D. Farley, *Chem. Soc. Rev.*, 2006, **35**, 249.
45. H. I. Lee, R. Y. Igarashi, M. Laryukhin, P. E. Doan, P. C. Dos Santos, D. R. Dean, L. C. Seefeldt, and B. M. Hoffman, *J. Am. Chem. Soc.*, 2004, **126**, 9563.
46. J. H. Freed, *J. Chem. Phys.*, 1969, **50**, 2271.
47. K. P. Dinse, R. Biehl, and K. Mobius, *J. Chem. Phys.*, 1974, **61**, 4335.
48. R. Biehl, M. Plato, and K. Mobius, *J. Chem. Phys.*, 1975, **63**, 3515.
49. E. R. Davies, *Phys. Lett. A*, 1974, **47**, 1.
50. W. B. Mims, in Electron Paramagnetic Resonance, ed S. Geschwind, Plenum: New York, 1972, 263.
51. W. B. Mims, *Proc. R. Soc. Lond. A Math. Phys. Sci.*, 1965, **283**, 452.
52. C. L. Fan, P. E. Doan, C. E. Davoust, and B. M. Hoffman, *J. Magn. Reson.*, 1992, **98**, 62.
53. C. Bolm, M. Martin, G. Gescheidt, C. Palivan, D. Neshchadin, H. Bertagnolli, M. Feth, A. Schweiger, G. Mitrikas, and J. Harmer, *J. Am. Chem. Soc.*, 2003, **125**, 6222.
54. P. E. Doan, N. S. Lees, M. Shanmugam, and B. M. Hoffman, *Appl. Magn. Reson.*, 2010, **37**, 763.
55. A. V. Astashkin and A. Kawamori, *J. Magn. Reson.*, 1998, **135**, 406.
56. P. E. Doan and B. M. Hoffman, *Chem. Phys. Lett.*, 1997, **269**, 208.
57. A. Schweiger, *Struct. Bond.*, 1982, **51**, R1.
58. M. Mehring, P. Hofer, and A. Grupp, *Phys. Chem. Chem. Phys.*, 1987, **91**, 1132.
59. A. Potapov, B. Epel, and D. Goldfarb, *J. Chem. Phys.*, 2008, **128**, 10.
60. B. Epel, D. Arieli, D. Baute, and D. Goldfarb, *J. Magn. Reson.*, 2003, **164**, 78.
61. C. Gemperle, O. W. Sorensen, A. Schweiger, and R. R. Ernst, *J. Magn. Reson.*, 1990, **87**, 502.
62. P. Hofer, A. Grupp, and M. Mehring, *Phys. Rev. A*, 1986, **33**, 3519.
63. G. Jeschke and A. Schweiger, *J. Chem. Phys.*, 1995, **103**, 8329.
64. S. Deblon, L. Liesum, J. Harmer, H. Schönberg, A. Schweiger, and H. Grützmacher, *Chem. Eur. J.*, 2002, **8**, 601.
65. N. Yang, M. Reiher, M. Wang, J. Harmer, and E. C. Duin, *J. Am. Chem. Soc.*, 2007, **129**, 11028.
66. D. Goldfarb, B. Epel, H. Zimmermann, and G. Jeschke, *J. Magn. Reson.*, 2004, **168**, 75.
67. B. Epel and D. Goldfarb, *J. Magn. Reson.*, 2000, **146**, 196.
68. M. T. Bennebroek and J. Schmidt, *J. Magn. Reson.*, 1997, **128**, 199.
69. B. Epel, P. Manikandan, P. M. H. Kroneck, and D. Goldfarb, *Appl. Magn. Reson.*, 2001, **21**, 287.
70. B. Epel, A. Poppl, P. Manikandan, S. Vega, and D. Goldfarb, *J. Magn. Reson.*, 2001, **148**, 388.
71. F. Mentink-Vigier, A. Collauto, A. Feintuch, I. Kaminker, V. T. Le, and D. Goldfarb, *J. Magn. Reson.*, 2013, **236**, 117.
72. I. Kaminker, M. Florent, B. Epel, and D. Goldfarb, *J. Magn. Reson.*, 2011, **208**, 95.

Chapter 17

Hyperfine Spectroscopy – ELDOR-detected NMR

Daniella Goldfarb

Weizmann Institute of Science, Rehovot, Israel

17.1 INTRODUCTION

Currently, there are three general families of experiments that are used for measuring magnetic resonance frequencies of nuclei coupled to a paramagnetic center: (i) the ENDOR family, discussed in Chapter 16; (ii) the electron–spin-echo envelope modulation (ESEEM) family, presented in Chapter 18; and (iii) the electron–electron double resonance (ELDOR)-detected NMR (EDNMR) family, which is the focus of the current chapter. Once the nuclear frequencies are determined, they can be used to extract hyperfine interaction tensors and, for $I > 1/2$ nuclei, also nuclear quadrupole tensors. These, in turn, provide important information about the geometrical and electronic structure of the paramagnetic center. The EDNMR family is the youngest of the three and holds great potential, which has not yet been fully realized. This is because it is most effective at high frequencies, such as W-band (95 GHz, ≈3 T), particularly for low γ-nuclei, because of the larger spread of nuclear frequencies, as compared with the commonly used X band (≈9.5 GHz) frequencies. Some focus on EDNMR methodology can be found in a recent review.[1] In this chapter we first present the standard EDNMR pulse sequences and qualitatively explain the inner workings of the experiment, along with a few example applications. We then present experimental practicalities and guidelines on how to optimize the experiment, followed by a comparison of ENDOR and EDNMR. Next, approaches for simulating EDNMR spectra are briefly described, and we end with a short description of two-dimensional EDNMR experiments.

17.2 THE EDNMR EXPERIMENT

The EDNMR sequence, as introduced by Schweiger and coworkers in 1994, is shown in Figure 17.1.[2] The mechanism underlying the EDNMR pulse sequence can be conceptually understood by considering the energy level diagram of one unpaired electron with $S = 1/2$ coupled via a hyperfine coupling to one nuclear spin with $I = 1/2$, shown in Figure 17.1(b). This spin

EPR Spectroscopy: Fundamentals and Methods.
Edited by Daniella Goldfarb and Stefan Stoll

ISBN: 978-1-119-16299-5
Also published in eMagRes (online edition)
DOI: 10.1002/9780470034590.emrstm1516

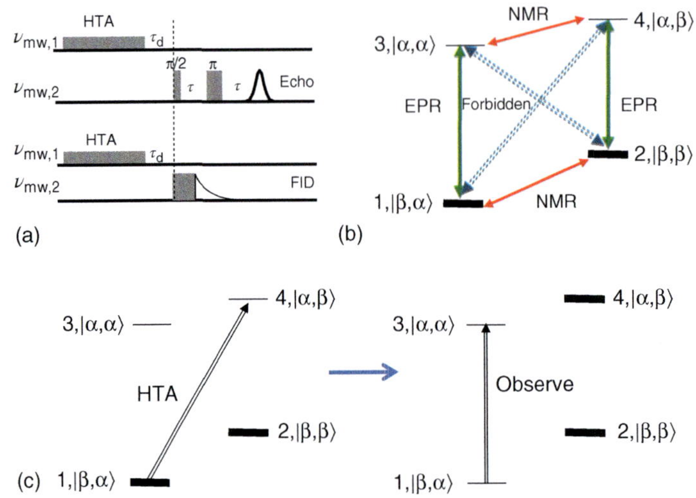

Figure 17.1. (a) The EDNMR pulse sequence with echo detection (top) or FID detection (bottom). (b) Energy level diagram of $S = 1/2$, $I = 1/2$ for the weak-coupling case. The thicknesses of the lines marking the energy level indicate their populations. The allowed EPR transitions are denoted in green, the forbidden ones are in dashed green, and the NMR transitions are in red. (c) The mechanism underlying the EDNMR experiment in terms of population transfer

system is characterized by four energy levels corresponding to the states $1 = |\beta, \alpha\rangle$, $2 = |\beta, \beta\rangle$, $3 = |\alpha, \alpha\rangle$, and $4 = |\alpha, \beta\rangle$, where the first label (α or β) represents the electron spin state and the second label the nuclear state. In such a system there are three types of transitions: (i) The so-called allowed EPR transitions 1→3 and 2→4 with frequencies ν_{13} and $\nu_{24,}$ respectively, for which $\Delta m_S = \pm 1$ and $\Delta m_I = 0$ and the transition probability are usually high. (ii) The so-called forbidden EPR transitions 1→4 and 2→3, with frequencies ν_{14} and $\nu_{23,}$ respectively, for which $\Delta m_S = \pm 1$ and $\Delta m_I = \pm 1$ and the transition probabilities are usually low. There are special cases when the hyperfine coupling is close to the nuclear Larmor frequency. Then, the transition probability of the forbidden transitions can be as high as those of the allowed EPR transitions. (iii) Allowed NMR transitions 1→2 and 3→4 with frequencies ν_{12} and $\nu_{34,}$ respectively, where $\Delta m_S = 0$ and $\Delta m_I = \pm 1$ and the transition probabilities are high under the appropriate radio frequency (RF) irradiation. The EDNMR experiment is a double-resonance experiment where the large population difference between the energy levels (i.e., polarization) across an allowed EPR transition is transferred to nuclear transitions, thus affecting the EPR signals, as illustrated in Figure 17.1(c). The experiment uses a so-called high-turning-angle (HTA) pulse of varying microwave (MW) frequency in reference to an allowed transition at a fixed MW frequency, but when applied to a forbidden transition, its turning angle is usually around π. The difference in turning angle between the allowed and forbidden EPR transitions arises from the low transition probability of the forbidden transitions (see below).

The HTA pulse, which is applied at frequency $\nu_{mw,1}$ ($=\omega_{mw,1}/2\pi$), is on resonance with a forbidden transition, 1→4 (or 2→3), and inverts the population across the 1→4 (or 2→3) transition. This pulse equalizes the population across both allowed EPR transitions 1→3 and 2→4. Then, one of the allowed EPR transitions is observed by either an echo detection sequence or a highly selective FID generated by a long soft pulse at $\nu_{mw,2}$ ($=\omega_{mw,2}/2\pi$) resonant with one of the allowed transitions. The intensity of the observed signal is considerably reduced compared with the situation where no HTA pulse was applied. The final state of the populations in the EDNMR experiment is the same as in the case of Davies ENDOR (see Chapter 16, Figure 17.1). The HTA pulse combines the action of a selective MW π pulse on one of the allowed EPR transitions and a selective RF π pulse on one of the NMR transitions.

The experiment is carried out by fixing $\nu_{mw,2}$ on the frequency of one of the allowed transitions ($\nu_{mw,2} = \nu_{13}$ or ν_{24}) and measuring the observed signal intensity as a function of $\nu_{mw,1}$. Whenever the HTA pulse frequency matches that of a forbidden

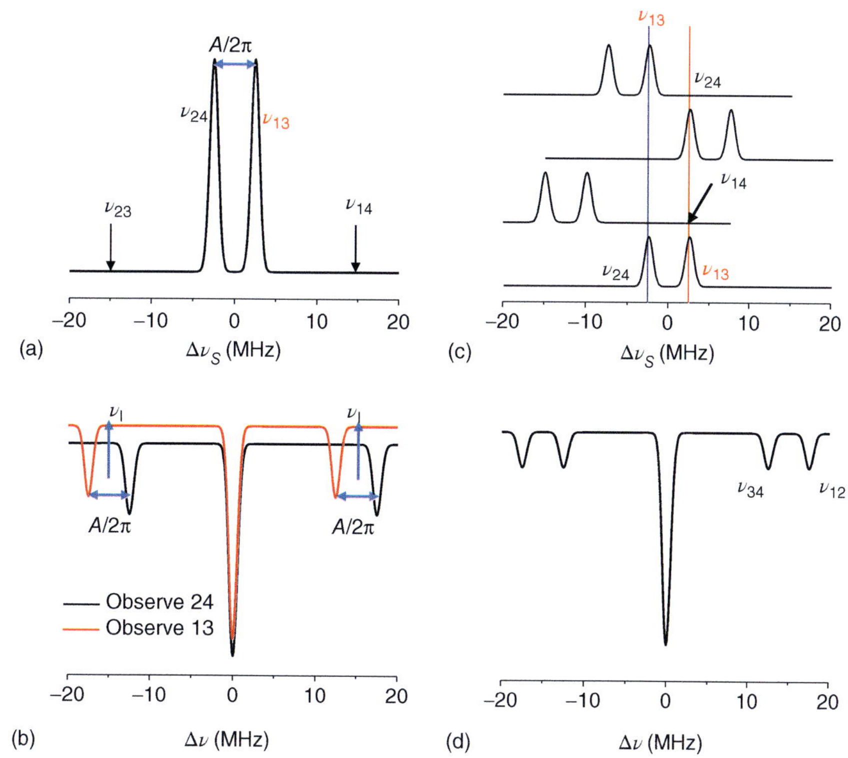

Figure 17.2. Schematics of the EPR and EDNMR spectra for $S = 1/2$, $I = 1/2$ (^{15}N) at W-band, where the Larmor frequency of ^{15}N, ν_I, is 15 MHz. (a) EPR spectrum for $A/2\pi = 5$ MHz; the positions of the weak forbidden transitions are denoted by arrows. The *x*-axis is given relative to $\omega_S/2\pi$ [see equation (17.1)]. (b) EDNMR spectra for the hyperfine-selective case. The red trace describes the spectrum when ν_{13} is probed and the black trace when ν_{24} is probed. (c) The EPR spectra of individual spin packets under conditions of inhomogeneous broadening where the spectra are shifted relative to each other due to, for example, *g*-strain. The shift can be on the order of the hyperfine coupling or larger. The final EPR spectrum is a superposition of many such individual spectra, each shifted a little relative to the other and the hyperfine coupling resolution is lost. This illustration shows that allowed and forbidden EPR transitions of different spin packets can overlap (follow the red line where the arrows point to the position of the forbidden ν_{14}) and that the two allowed transitions of different spin packets can also overlap (follow the blue line). Under such inhomogeneous broadening conditions, the EDNMR experiment is not selective with respect to the hyperfine splitting and at any selected position both EPR transitions contribute to the detected echo, in contrast to the selective case showed in (b). The resulting EDNMR spectrum is shown in (d)

transition (ν_{14} and ν_{23}), a reduction in the signal will be observed. Under these conditions, the difference $\Delta\nu = \nu_{\mathrm{mw},1} - \nu_{\mathrm{mw},2}$ matches the nuclear frequencies (ν_{12} or ν_{34}). Therefore, when the spectrum is displayed as a function of $\Delta\nu$, it directly provides the nuclear frequencies. When $\Delta\nu = 0$, the frequency of the HTA pulse matches the frequency of the observed allowed EPR transition and nearly saturates it. A schematic example of an EDNMR spectrum is shown in Figure 17.2(b); it corresponds to the EPR spectrum shown in Figure 17.2(a). It reflects the reduction in the EPR intensity (FID or echo) and therefore it is negative. When ν_{13} is observed, the red spectrum is obtained with negative peaks at $\nu_{23} - \nu_{13} = -\nu_{12}$ and $\nu_{14} - \nu_{13} = \nu_{34}$. In contrast, when ν_{24} is observed, the black spectrum is obtained with negative peaks at $\nu_{23} - \nu_{24} = -\nu_{34}$ and $\nu_{14} - \nu_{24} = \nu_{12}$.

With this conceptual understanding of the EDNMR experiment, we can proceed to derive the expressions for the various frequencies observed in the EDNMR spectrum and the associated intensities for $S = 1/2$ and $I = 1/2$. The spin Hamiltonian, in angular frequency units, describing this system is[2]

$$\hat{\mathcal{H}} = \omega_S \hat{S}_z + \omega_I \hat{I}_z + \hat{\mathbf{S}}\mathbf{A}\hat{\mathbf{I}} \tag{17.1}$$

where ω_S is the electron spin frequency with $\omega_S = g_e \mu_B B_0/\hbar$, $\omega_I = -g_n \mu_N B_0/\hbar$ is the nuclear Larmor frequency, B_0 is the external magnetic field, and $\boldsymbol{A}$ denotes the hyperfine tensor with its principal values A_x, A_y, A_z, all given in angular frequency units. Using the high-field approximation, which allows neglecting terms in $\hat{S}_{x,y}$, this reduces to

$$\hat{\mathcal{H}} = \omega_S \hat{S}_z + \omega_I \hat{I}_z + A\hat{S}_z\hat{I}_z + B\hat{S}_z\hat{I}_x \tag{17.2}$$

When $\boldsymbol{A}$ is axially symmetric, A and B are related to the principal values $A_{||} = A_z, A_\perp = A_x = A_y$ according to

$$A = A_{||}\cos^2\theta + A_\perp \sin^2\theta,\ B = (A_{||} - A_\perp)\sin\theta\cos\theta \tag{17.3}$$

where θ is the angle between the principal axis corresponding to $A_{||}$ and the external magnetic field.

The frequencies of the allowed transitions ($\omega_a = \omega_{13}, \omega_{24}$) are

$$2\pi\nu_{13} = \omega_{13} = \omega_s + \frac{1}{2}(\omega_\alpha - \omega_\beta),$$
$$2\pi\nu_{24} = \omega_{24} = \omega_s - \frac{1}{2}(\omega_\alpha - \omega_\beta) \tag{17.4}$$

The forbidden transition frequencies ($\omega_f = \omega_{23}, \omega_{14}$) are

$$2\pi\nu_{14} = \omega_{14} = \omega_s + \frac{1}{2}(\omega_\alpha + \omega_\beta),$$
$$2\pi\nu_{23} = \omega_{23} = \omega_s - \frac{1}{2}(\omega_\alpha + \omega_\beta) \tag{17.5}$$

The nuclear frequencies $\omega_\alpha = \omega_{34}$ and $\omega_\beta = \omega_{12}$ are given by

$$2\pi\nu_{\alpha,\beta} = \omega_{\alpha,\beta} = \left[\left(\pm\frac{1}{2}A + \omega_I\right)^2 + \left(\frac{1}{2}B\right)^2\right]^{1/2} \tag{17.6}$$

The transition probabilities of the forbidden transitions are:[3]

$$I_f = \frac{\left|\omega_I^2 - \frac{1}{4}(\omega_\alpha + \omega_\beta)^2\right|}{\omega_\alpha\omega_\beta} = \sin^2\eta \tag{17.7}$$

and those of the allowed transitions are

$$I_a = \frac{\left|\omega_I^2 - \frac{1}{4}(\omega_\alpha - \omega_\beta)^2\right|}{\omega_\alpha\omega_\beta} = \cos^2\eta \tag{17.8}$$

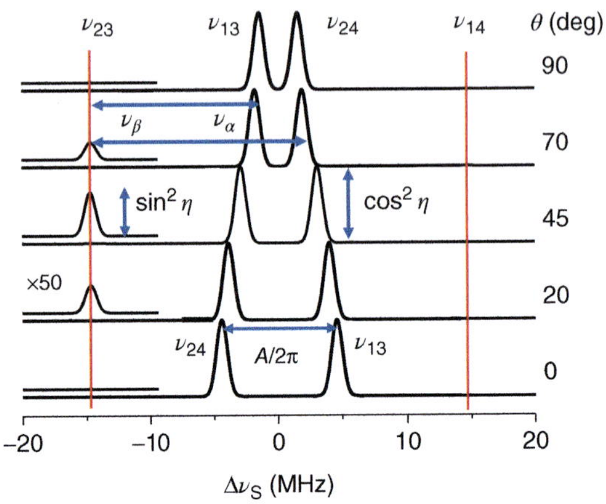

Figure 17.3. Calculated EPR spectra for $S = 1/2$ coupled with one ^{15}N nucleus at 95 GHz, with $A_{||}/2\pi = 9$ MHz and $A_\perp/2\pi = 3$ MHz, as a function of θ. All transitions are identified on the figure. The red lines denote the positions of the forbidden transitions, and on the low-frequency side, they are scaled by a factor of 50×. $\Delta\nu_S$ is given relative to $\omega_S/2\pi$

Here $\eta = (1/2)(\eta_\alpha - \eta_\beta)$, $\eta_\alpha = \arctan[-B/(A + 2\omega_I)]$, $\eta_\beta = \arctan[-B/(A - 2\omega_I)]$. I_f depends on the extent to which $(\omega_\alpha + \omega_\beta)$ is different from $2\omega_I$. The larger the B, the more intense the forbidden transition is. Equation (17.7) also shows that I_f is strongly orientation dependent and for the canonical orientations $\theta = 0°$ and $90°$, $\eta = 0$ and therefore also $I_f = 0$. For full derivation of how the frequencies and transition probabilities were obtained, see the chapter on ESEEM (Chapter 18). In ESEEM the modulation depth, k, for $S = 1/2$, $I = 1/2$ is given by $k = \sin^2(2\eta)$, and because $\eta = 0$ for the canonical orientation, the modulation depth is zero for the canonical orientations, just as for EDNMR. Figure 17.3 shows a number of calculated EPR spectra with different values of θ.

Because EPR spectra in solids or frozen solutions are usually much broader than the excitation bandwidth of the applied pulses, the EDNMR experiment creates holes in the broad EPR spectrum. The signal intensities are related to the depth of the holes. The deeper the holes, the more intense the lines are in the EDNMR spectrum. This depth depends not only on I_f but also on the MW strength and the length of the HTA pulse, which determine the effective flip angle of the forbidden transition, β_f, (neglecting relaxation effects, see next) according to

$$\beta_f = \omega_1 t_{HTA}(I_f)^{1/2} = \beta_0(I_f)^{1/2} \tag{17.9}$$

where $\omega_1 = g_e \mu_e B_1/\hbar$ is the MW field strength and $\beta_0 = \omega_1 t_{HTA}$ is the nominal flip angle (corresponding to a transition probability of 1) and $\omega_{1f} = \omega_1 (I_f)^{1/2}$. Typically t_{HTA} is on the order of microseconds, as opposed to the observer pulses that are on the order of nanoseconds. The depth of the hole, h, which determines the intensity of the EDNMR spectrum, is[2]

$$h = 1 - I_a \cos\left(\beta_0 (I_f)^{\frac{1}{2}}\right) - I_f \cos\left(\beta_0 (I_a)^{\frac{1}{2}}\right) \tag{17.10}$$

To understand equation (17.10), one has to consider an inhomogeneous EPR linewidth, as shown in Figure 17.2(c), and take into account that at the frequencies of the HTA pulse and the observer pulses, both the allowed and forbidden transitions of different spin packets can overlap. Accordingly, the first term arises from a forbidden transition resonant with the HTA pulse for which the contribution to the hole is proportional to $\cos\left(\beta_0 (I_f)^{\frac{1}{2}}\right)$; the intensity of the observed allowed transition is taken into account by multiplying by I_a. The second term represents the case where an allowed transition is resonant with the HTA pulse and, therefore, its contribution to the hole is proportional to $\cos\left(\beta_0 (I_a)^{\frac{1}{2}}\right)$; in this case the observed transition is the forbidden one and, therefore, it is multiplied by I_f. In the case where $I_f/I_a \ll 1$, the second term can be neglected. Furthermore, for a very small I_f value and $\beta_0 (I_f)^{\frac{1}{2}} \ll 1$, namely, not a large HTA,

$$h \approx 1 - \cos\left(\beta_0 (I_f)^{\frac{1}{2}}\right) \approx 1 - 1 + \frac{1}{4}\beta_0^2 I_f \propto I_f \tag{17.11}$$

When the hyperfine splitting is large enough and resolved in the EPR spectrum, the observer pulses affect only one transition for all overlapping spin packets, and the EDNMR spectrum will be asymmetric with respect to $\Delta\nu = 0$, as shown in Figure 17.2(b). Such a situation is usually encountered in paramagnetic transition metal ions for the central-atom hyperfine interaction, such as $A_z(^{63,65}Cu)$ for Cu(II), and for organic radicals where there is a large spin density on magnetic nuclei like $A_z(^{14}N)$ for nitroxides. However, for ligand nuclei the hyperfine coupling is usually small and therefore the hyperfine splittings are not resolved in the EPR spectrum due to inhomogeneous broadening. Thus, the observed pulses affect both EPR transitions (though in different spin packets) and therefore, the EDNMR spectrum is symmetric with respect to $\Delta\nu = 0$ (Figure 17.2c and d). When the HTA pulse is on resonance with the allowed transition, it gives rise to the so-called central Lorentzian hole at $\Delta\nu = 0$.

An example of an experimental EDNMR spectrum of a frozen solution of Mn(II)ADP (adenosine diphosphate) is shown in Figure 17.4(b). Because the spectrum is symmetric with respect to $\Delta\nu = 0$, only the $\Delta\nu > 0$ region is displayed. Often the intense hole at $\Delta\nu = 0$ is removed by fitting it to a Lorentzian line and subtracting it from the original spectrum[4], as illustrated in Figure 17.4(b). Further multiplication of the resulting spectrum by −1 generates a spectrum with positive-looking signals, as shown for the expanded ^{31}P region displayed in Figure 17.4(c), bottom trace. Although Mn(II) has a spin of $S = 5/2$, at W-band its spectrum is often dominated by the central $m_S = -1/2 \rightarrow 1/2$ transitions, which give a sharp sextet due to the isotropic ^{55}Mn ($I = 5/2$) hyperfine coupling. Therefore, when the EDNMR spectrum is acquired by setting the magnetic field to one of these transitions, it behaves effectively like an $S = 1/2$ system. In Figure 17.4(c) the Davies ENDOR spectrum of the same sample is also depicted for comparison. The spectra differ in their lineshape; whereas the EDNMR spectrum shows no lineshape structure, the ENDOR spectrum does.

Because β_f is a function of orientation, θ, the relative intensities in the EDNMR spectrum in an orientationally disordered sample will deviate from the common Pake patterns observed in ENDOR. Figure 17.4(d) shows calculated ENDOR and EDNMR spectra, under conditions where the intensities are linearly proportional to I_f [equation (17.11)]. The loss of the $A_\perp$ and $A_\parallel$ singularities in the EDNMR spectra are clear in both the experimental and calculated spectra.

17.3 EDNMR FOR $I > 1/2$

For $S = 1/2$, $I = 1/2$, it is the anisotropic hyperfine interaction, with its non-secular terms [see equation (17.2)] that is responsible for the mixing of the nuclear states and consequently, the presence of forbidden transitions ($\beta_f > 0$). For systems with $I > 1/2$, the nuclear quadrupole interaction also contributes non-secular terms to the spin Hamiltonian, which can increase β_f and the nuclear-state mixing in the presence of the hyperfine interaction. The spin Hamiltonian then becomes

$$\hat{\mathcal{H}} = \omega_S \hat{S}_z + \omega_I \hat{I}_z + \hat{\mathbf{S}}\mathbf{A}\hat{\mathbf{I}} + \hat{\mathbf{I}}\mathbf{P}\hat{\mathbf{I}} \tag{17.12}$$

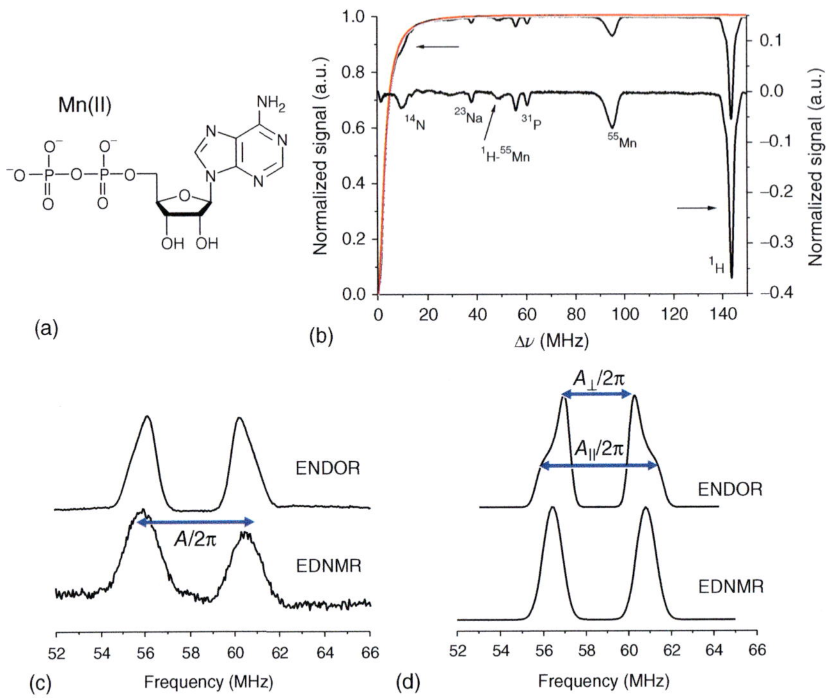

Figure 17.4. (a) Mn(II)ADP. (b) W-band EDNMR spectrum (only the $\Delta\nu > 0$ part) of a frozen solution of Mn(II)ADP in a water–glycerol solution. The top spectrum is the raw spectrum (left vertical axis), along with the function (in red) used for the background subtraction to yield the bottom spectrum (right vertical axis). The spectrum was recorded at 10 K with an HTA pulse of 1 μs and echo detection pulses of 100/200 ns. The spectrum shows various peaks assigned in the figure. (c) The ^{31}P region of the EDNMR spectrum, compared with the 10 K W-band ENDOR spectrum. The difference in the signal/noise ratio due to different number of accumulations, which were considerably larger for the ENDOR spectrum. (d) A comparison of the calculated W-band ENDOR and EDNMR powder patterns for ^{31}P with $A_{||}/2\pi = 6$ MHz and $A_{\perp}/2\pi = 4$ MHz

In equation (17.12) the last term describes the quadrupole interaction, where $\boldsymbol{P}$ is the quadrupole tensor in angular frequency units (see Chapter 6). We will illustrate the EDNMR characteristics for $I > 1/2$ through the ^{14}N ($I = 1$) of nitroxide spin probes (see Chapter 14) measured at W-band. Figure 17.5 shows a diagram of the associated energy levels for the strong coupling case, $A > 2\omega_I$.[5] The six energy levels are denoted by $|\beta, +1\rangle$, $|\beta, 0\rangle$, $|\beta, -1\rangle$ for the lower electron spin manifold and $|\alpha, +1\rangle$, $|\alpha, 0\rangle$, $|\alpha, -1\rangle$ for the upper one. The three allowed EPR transitions for $m_I = -1, 0, +1$ and the associated forbidden transitions are labeled in the figure. Each allowed transition is associated with four forbidden transitions, with which it shares a common energy level. The frequency difference between these allowed and forbidden transitions yields the nuclear frequencies, denoted by solid red arrows. The first-order expressions for the ^{14}N nuclear frequencies for $A > 2\omega_I$ are as follows:

$$\omega^{\alpha}{}_{\text{sq1}} = 2\pi\nu^{\alpha}{}_{\text{sq1}} = A/2 - \omega_I - 3P/2 \qquad (17.13a)$$
$$\omega^{\alpha}{}_{\text{sq2}} = 2\pi\nu^{\alpha}{}_{\text{sq2}} = A/2 - \omega_I + 3P/2 \qquad (17.13b)$$
$$\omega^{\alpha}{}_{\text{dq}} = 2\pi\nu^{\alpha}{}_{\text{dq}} = A - 2\omega_I \qquad (17.13c)$$
$$\omega^{\beta}{}_{\text{sq1}} = 2\pi\nu^{\beta}{}_{\text{sq1}} = A/2 + \omega_I - 3P/2 \qquad (17.13d)$$
$$\omega^{\beta}{}_{\text{sq2}} = 2\pi\nu^{\beta}{}_{\text{sq2}} = A/2 + \omega_I + 3P/2 \qquad (17.13e)$$
$$\omega^{\beta}{}_{\text{dq}} = 2\pi\nu^{\beta}{}_{\text{dq}} = A + 2\omega_I \qquad (17.13f)$$

where P is the orientation-dependent nuclear quadrupole coupling. For an axially symmetric quadrupole tensor $P = [(e^2qQ)/4\hbar](3\cos^2\theta' - 1)$, where eq is the largest magnitude component of the electric field gradient (EFG) at the nucleus and Q is the electric quadrupole moment of the nucleus. θ' is the angle between the external magnetic field and the principal direction of the tensor. The first-order expressions are usually valid when $|\omega_I \pm A| \gg |P|$ and $|\omega_I| \gg |A|$

Table 17.1. Summary of the allowed transitions and the associated forbidden transitions, the nuclear frequencies generated by their differences, and the side of the EDNMR spectrum where they appear ($\Delta\nu$ positive or negative)

Allowed EPR transition m_I	Forbidden transition $\Delta m_I = \pm 1$	Forbidden transition $\Delta m_I = \pm 2$	Nuclear frequency $\Delta\nu > 0$	Nuclear frequency $\Delta\nu < 0$
−1	$\vert\beta,-1\rangle \rightarrow \vert\alpha,0\rangle$ $\vert\alpha,-1\rangle \rightarrow \vert\beta,0\rangle$	$\vert\beta,-1\rangle \rightarrow \vert\alpha,+1\rangle$ $\vert\alpha,-1\rangle \rightarrow \vert\beta,+1\rangle$	ν^{α}_{sq1}, ν^{α}_{dq} ν^{β}_{sq2}, ν^{β}_{dq}	
0	$\vert\beta,0\rangle \rightarrow \vert\alpha,+1\rangle$ $\vert\beta,0\rangle \rightarrow \vert\alpha,-1\rangle$ $\vert\alpha,0\rangle \rightarrow \vert\beta,+1\rangle$ $\vert\alpha,0\rangle \rightarrow \vert\beta,-1\rangle$		ν^{α}_{sq2} ν^{β}_{sq1}	ν^{α}_{sq1} ν^{β}_{sq2}
+1	$\vert\beta,+1\rangle \rightarrow \vert\alpha,0\rangle$ $\vert\alpha,+1\rangle \rightarrow \vert\beta,0\rangle$	$\vert\beta,+1\rangle \rightarrow \vert\alpha,-1\rangle$ $\vert\alpha,+1\rangle \rightarrow \vert\beta,-1\rangle$		ν^{α}_{sq2}, ν^{α}_{dq} ν^{β}_{sq1}, ν^{β}_{dq}

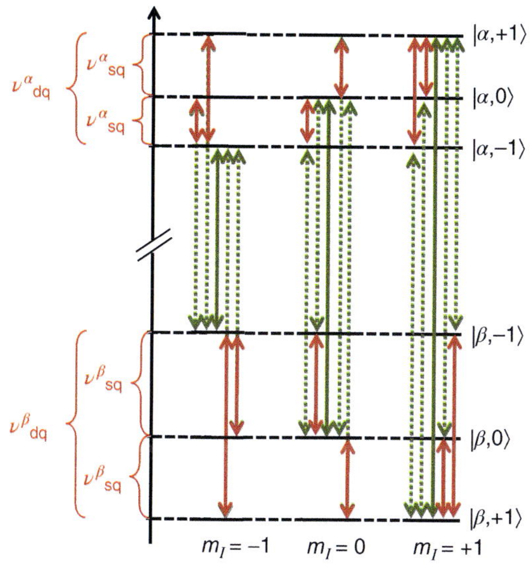

Figure 17.5. Energy level diagram for $S = 1/2$, $I = 1$ and $A > \omega_I$. The green lines correspond to allowed EPR transitions, the green dotted lines to forbidden ones, and red lines to NMR transitions; they are all labeled in red as well

or $|\omega_I| \ll |A|$. When the experiment is selective with respect to the ^{14}N hyperfine splitting, as encountered for the z direction for which A_z is large, the EDNMR spectrum is not symmetric, and negative and positive frequencies in the EDNMR spectrum are distinguishable. The nuclear frequencies, given in equation (17.13a–f), and their locations with respect to $\Delta\nu = 0$, are listed in Table 17.1 for observe pulses set to the $m_I = -1$, 0, or 1 spectral features. The EDNMR spectrum recorded at the $m_I = -1$ component of the g_z singularity (see Figure 17.6a) is depicted in Figure 17.6(b). In accordance with Table 17.1, the signals appear only on the $\Delta\nu > 0$ side with $\nu^{\beta}_{sq2} \approx 56.1$ MHz and $\nu^{\alpha}_{sq1} \approx 40.5$ MHz. The weak peaks at 75.3 and 116 MHz correspond to the double-quantum transitions, with frequencies ν^{α}_{dq} and ν^{β}_{dq}, respectively. From these frequencies, and using equation (17.13a–f), it is possible to determine $P_z = -1.82 \pm 0.7$ MHz. There is also a peak at 96 MHz, which does not correspond to any nuclear frequency; it is due to the off-resonance excitation of the nearby allowed EPR transitions ($m_I = 0$) by the HTA pulse. Indeed, the frequency difference between the two allowed EPR transitions is the hyperfine coupling, which equals 96 MHz.

Table 17.1 shows that by setting the magnetic field to the $m_I = 0$ component in the g_z direction, signals will appear at both $\Delta\nu > 0$ and $\Delta\nu < 0$, featuring all single-quantum transitions but not the double-quantum ones.[5,7] Figure 17.6(c) shows an example of a spectrum recorded at a field position where both $m_I = 0$ and $m_I = -1$ contribute. The spectrum also shows peaks at $\Delta\nu = \pm 100$ MHz, assigned to neighboring allowed transitions, as observed in Figure 17.6(b). At this field position we expect to see some contribution from the double-quantum transitions ν^{α}_{dq} and ν^{β}_{dq} on the $\Delta\nu > 0$ side arising from the $m_I = -1$ manifold (see Figure 17.6b). These are expected at the position of the peaks marked with *. However, such peaks appear also at the $\Delta\nu < 0$ side. These, in principle, could arise from the $m_I = 1$ manifold, but the $m_I = 1$ manifold is not within the observed pulse bandwidth. Therefore, these are tentatively assigned to ^{2}H satellites at $(A(m_I = \pm 1) \pm \omega_I(^2\text{H}))/2\pi$. The frequencies of these satellites accidentally overlap with those of the double-quantum transitions. These satellites arise

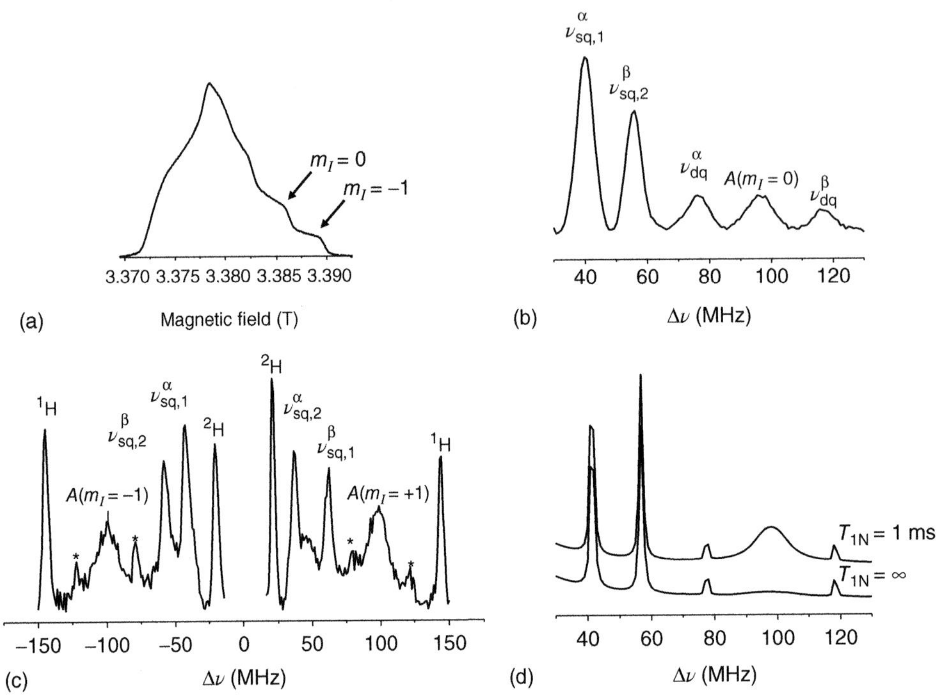

Figure 17.6. W-band EDNMR for $S = 1/2$ coupled to one ^{14}N in nitroxide-labeled polyethylene oxide, PEO–NO, in 3% F127 micelles in D_2O/glycerol-d_8 (7 : 3).[6] (a) Echo-detected EPR spectrum showing the positions at which the EDNMR spectra were recorded. (b) EDNMR spectrum for $m_I = -1$ position indicated in (a) and (c) for the position indicated by $m_I = 0$ in (a). The signal assignments are shown in the figure. On the $\Delta\nu > 0$ side, there are also contributions from overlapping $m_I = -1$ peaks that are not indicated. Experimental parameters: HTA pulse length 300 μs, echo detection pulses 100/200 ns. (d) Simulated EDNMR spectra for $m_I = -1$ spin packets at the outermost singularity of the g_z region calculated with and without taking T_{1N} into account as noted in the figure. Simulation parameters: HTA pulse length 100 μs, $\omega_1 = 3.5$ MHz, $T_{2e} = 20$ μs. The assignment of the peaks labeled with * is given in the text. (Reproduced with permission from Ref. 6. © Elsevier, 2014)

from forbidden transitions involving a ^{2}H spin flip associated with the neighboring allowed transitions. When EDNMR nitroxide spin probes are measured at the g_z region under conditions optimized for high resolution (see the next session), it is possible to distinguish signals arising from nitroxides having hydrogen bonds with the solvent from those without hydrogen bonds based on the differences in their A_z values.[7]

Examples of EDNMR spectra acquired under nonselective conditions are the ^{31}P spectrum of a frozen solution of Mn(II)ADP, shown in Figure 17.4(b) and (c), and the spectrum of a frozen solution of Cu(II)-histidine, shown in Figure 17.7. Here, the ^{14}N hyperfine signals are identified at 8 and 28 MHz ($\nu_I = 10$ MHz), yielding $A(^{14}N)/2\pi = (8+28)$ MHz = 36 MHz.[8] The ^{2}H signal at 21.3 MHz is due to solvent, D_2O, and it appears at the Larmor frequency of ^{2}H. The broad signal at 50–80 MHz is due to ^{63,65}Cu ($I = 3/2$).

When the hyperfine coupling is small and not resolved in the single-quantum transitions spectral region, owing to further inhomogeneous broadening induced by the quadrupole interaction, the hyperfine splitting can be resolved through the double-quantum transitions[9] [see equation (17.13)], as shown in the spectrum of a Mn(II) complex with ^{14}N ligands in Figure 17.8.

An important feature of EDNMR is that the intensity of a particular signal (measured under the same experimental condition) linearly depends on the number of nuclei that contributes to the signal, and it can therefore be used to determine the number of equivalent nuclei in the vicinity of one or more unpaired electrons. This

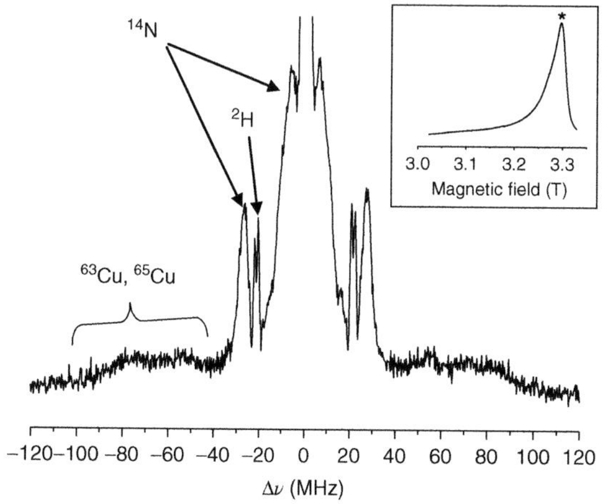

Figure 17.7. W-band EDNMR of frozen solution of Cu(II) histidine. The insert shows the corresponding echo-detected EPR spectrum and the * marks the field at which the measurement was carried out. In this spectrum the central peak at $\Delta\nu = 0$ was not removed[8]

is illustrated for the number of ^{14}N atoms coordinated to Mn(II), and for the percentage of D_2O in a Mn(II) aqueous solution of H_2O/D_2O in Figure 17.9a.[9]

EDNMR is also highly effective for other low-γ quadrupole nuclei such as ^{17}O, which has a nuclear spin $I = 5/2$.[11,12] Figure 17.10 presents the W-band EDNMR spectrum of $Mn(H_2{}^{17}O)_6{}^{2+}$.[4] The spectrum shows two types of ^{17}O nuclei: the directly coordinated ^{17}O with a resolved hyperfine splitting corresponding to $A/2\pi \approx 7.5$ MHz (highlighted in red) and the more remote ^{17}O nuclei with a small and unresolved hyperfine splitting appearing at the ^{17}O Larmor frequency (19.6 MHz). The double-quantum lines are also clearly visible, with a splitting of $2A/2\pi \approx 15$ MHz. The spectrum also shows a ^1H signal and two of the ^{55}Mn signals, which correspond to the $m_S = -1/2 \rightarrow 1/2$ appearing at $A(^{55}Mn)/4\pi \pm \nu_{Mn}$, indicated by purple lines. The ^{55}Mn signal is easy to observe and allows determining $A(^{55}Mn)$ more accurately than from the hyperfine splittings in the EPR spectrum. The EDNMR spectrum in Figure 17.10 resolves two additional doublets centered around the two ^{55}Mn single-quantum lines indicated by dashed purple lines. These are associated with transitions involving the nuclear flips of both the ^{55}Mn and ^{17}O nuclei, i.e., $\Delta m_I(^{55}Mn) = \pm 1$ and $\Delta m_I(^{17}O) = \pm 1$, and they are m_S specific.

17.4 EXPERIMENTAL CONSIDERATIONS

Adjusting the experimental parameters for the EDNMR experiment is not straightforward. The experimental settings are determined by a trade-off between resolution and sensitivity.[8] The depth of the hole burnt in an inhomogeneously broadened

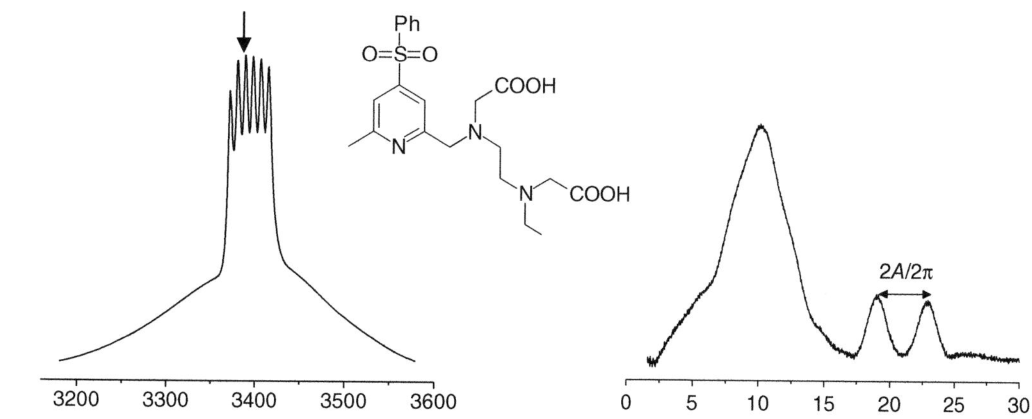

Figure 17.8. (a) W-band echo-detected EPR spectrum of a frozen solution of Mn(II) complex with *N*-(pyrid-2-yl-methyl)ethylenediamine-*N*,*N*′,*N*′-triacetic acid (PEDTA, shown on the right). (b) The EDNMR spectrum, measured at a field position indicated by the arrow in (a), showing the ^{14}N signals (single-quantum transitions at ≈10 MHz) and a doublet of the double-quantum transitions. (Reproduced with permission from Ref. 10. © Royal Society of Chemistry, 2015)

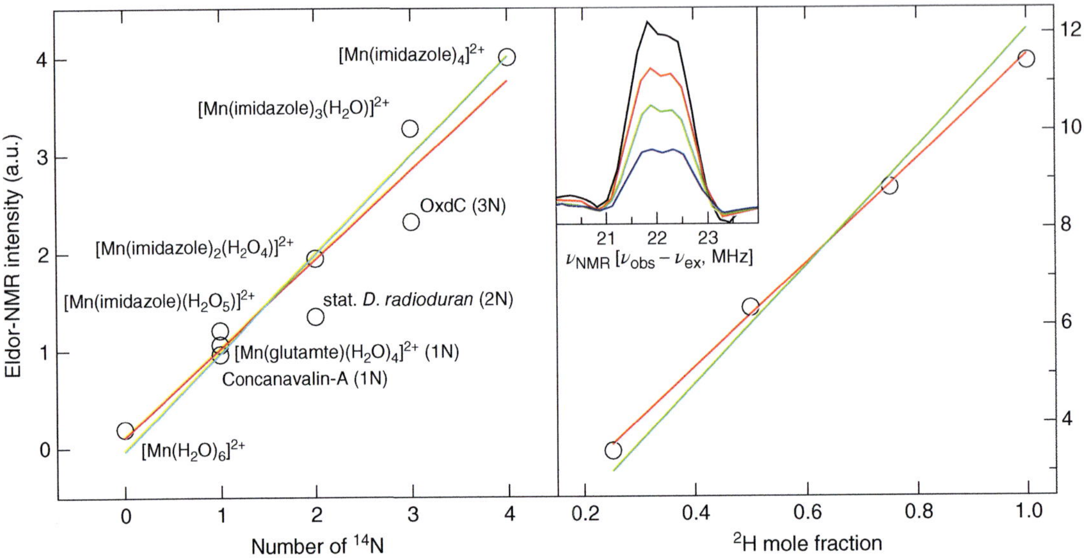

Figure 17.9. ^{14}N (left panel) and ^{2}H (right panel) EDNMR intensities as a function of the number of coupled nuclei (circles). The ^{14}N peaks are due to the N-coordinated ligands from the indicated complexes and proteins. The ^{2}H peaks come from water coordinated to Mn(II) in 50 µM Mn(II) and 250 mM sucrose in varying mixtures of $^{2}H_2O/^{1}H_2O$. The red lines represent the regression fit of the data, and the green, the ideal correlations. The ^{2}H EDNMR spectra are shown in the inset. (Reproduced with permission from Ref. 9. © American Chemical Society, 2015)

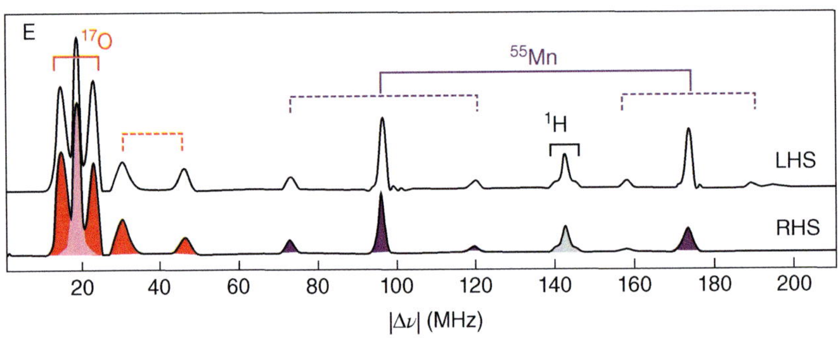

Figure 17.10. W-band EDNMR of $Mn^{II}(H_2{}^{17}O)_6$ at 20 K, inverted with the central blind spot subtracted from the raw data. The two halves of the spectrum, the left-hand side (LHS, $\Delta\nu < 0$) and right-hand side (RHS, $\Delta\nu > 0$), which are essentially identical, are stacked for easy comparison. Transitions from all three magnetic nuclei of the complex (^{17}O, ^{55}Mn, and ^{1}H) are denoted by red, magenta, and gray, respectively. (Reproduced with permission from Ref. 4. © Taylor and Francis, 2013)

spectrum by a pulse consists of contributions from both overlapping forbidden and allowed transitions, arising from different spin packets, as given in equation (17.10). Thus, intense and longer pulses produce deeper holes. On the other hand, the width of the hole depends on the MW irradiation amplitude ω_1; the HTA pulse length, t_{HTA}; and the phase memory time, T_m. A narrow central hole can be obtained by applying a weak HTA pulse such that $\omega_1 T_m \ll 1$, thereby achieving the best resolution. The lowest limit of the width of central hole is determined by the electron spin–spin relaxation rate, $1/T_{2e}$, (remember $T_m < T_{2e}$). The same arguments apply for the EDNMR signals, where one should consider ω_{1f}, which is ω_1 scaled by the transition probability. Therefore, the linewidth of EDNMR signals can be narrower

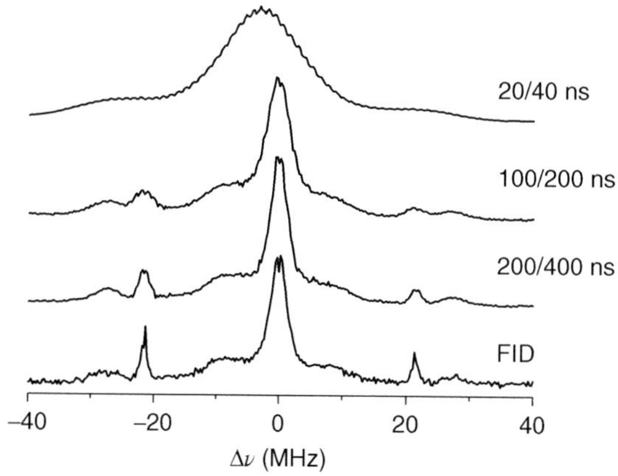

Figure 17.11. Comparison of W-band EDNMR spectra of a frozen solution of a Cu(II)-histidine complex acquired with a spin echo detection (top three) with different pulse lengths, as noted in the figure and FID detection following a 3 μs pulse (bottom). Other conditions: the HTA pulse had $\omega_1/2\pi = 1.6$ MHz and a length of 20 μs. (Reproduced with permission from Ref. 8. © Journal of Chemical Physics, 2008)

than the central hole. In addition, the $1/T_{2e}$ value, which provides the lowest limit, should be the one corresponding the forbidden transition, which may differ from $1/T_{2e}$ of the allowed transition. Another important parameter is the method used to detect the hole pattern burnt by the HTA pulse. Two possible ways of detection are the use of the FID intensity (integration of the FID) created by a single, long, and weak pulse, as proposed in the original version of EDNMR[2], or of the integrated intensity of a two-pulse echo (see Figure 17.1a). In both cases, pulses with the smallest possible bandwidth are desired, otherwise the acquired spectrum is broadened, as shown in Figure 17.11. Finally, one also has to consider the width of the echo integrating window. The wider the integration window, the better the resolution. Long detection pulses and broad integration windows come at the expense of signal-to-noise ratio (SNR); therefore, with relatively broad lines, sensitivity may be gained by applying short pulses, without significantly compromising the resolution. It is recommended to acquire full echo (or FID) transients and carry out the integration post-measurement such that the optimum SNR and resolution can be achieved.

FID detection provides improved resolution, again at the expense of SNR. One important aspect of FID detection is the width of the integration window. Setting the window in such a way that maximizes SNR may lead to artifacts because the shape of the FID may contain oscillatory signals appearing when the linewidth is smaller than the detection pulse bandwidth. This can be avoided by integrating over the whole FID in the case of a very broad EPR spectrum, which lasts for the same time as the detection pulse, but it degrades the SNR.

The delay time, τ_d, between the HTA pulse and the detection sequence (see Figure 17.1a) should be kept at the minimum possible (dictated by instrumental factors such as ringing after the pulse) to minimize T_{1e} relaxation effects and spectral diffusion (see Chapter 11) that will reduce the population differences induced by the HTA pulse.

17.5 EDNMR VERSUS PULSE ENDOR

Because of the similar spectroscopic information content of EDNMR and pulse ENDOR, it is appropriate to compare the two techniques and highlight their similarities and differences, along with the associated pros and cons. While the various ENDOR techniques are well established, EDNMR has been gaining popularity only recently, with the advent of high-field EPR, particularly W-band (95 GHz), and its full potential has not yet been realized. EDNMR and Davies ENDOR are based on the same generation of non-Boltzmann distributions of the population of energy levels within the nuclear sublevels of a particular electron spin manifold. The two methods do differ in the way this population transfer is generated. In the Davies ENDOR experiment, the population transfer is achieved by selectively inverting allowed EPR and allowed NMR transitions (see Chapter 16), employing MW and RF pulses. In EDNMR a shortcut is employed by applying the HTA pulse at a different MW frequency than the one used to detect the spins, employing pulses at two MW frequencies. Consequently, EDNMR and ENDOR differ considerably in terms of the hardware they require. Next, we compare the two techniques in terms of (i) instrumentation, (ii) resolution, and (iii) sensitivity.

17.5.1 Instrumentation

The ENDOR experiment requires an RF transmitter, which includes a rather expensive high power, broadband RF amplifier, and specially designed probe-head that houses an RF coil in the MW resonator. In terms of hardware, this is quite a demanding and costly setup. Detecting ENDOR frequencies at the low-frequency end of the spectrum (<5 MHz) is difficult because of hardware restrictions such as limited RF power. The ENDOR spectral bandwidth depends on the bandwidth of the RF source and the RF amplifier as well as the RF coil, but it is independent of the bandwidth of the EPR resonator. Therefore, the latter can be designed to have a relatively high Q factor, as long as it is not degraded by the need for an RF coil. This means increased sensitivity. Usually, it is difficult to record an ENDOR spectrum efficiently over a range of 200 MHz with a single RF excitation coil. Occasionally, the appearance of spurious signals owing to RF harmonic frequencies originating from high-power RF amplifier nonlinearities leads to artifacts in the ENDOR spectra.

EDNMR requires two MW channels with independently adjustable MW power and frequency. The spectral bandwidth is determined by the bandwidth of the transmitter part of the spectrometer and the bandwidth of the resonator (the narrower of the two); the broader they are, the larger $\Delta\nu$ can be. Accordingly, to acquire broadband spectra, one may need to reduce the Q value of the resonator, which reduces the sensitivity. Recording wide-range nuclear spectra is easier for EDNMR than for ENDOR, where designing an efficient broadband RF coil presents a challenge. The EDNMR does not require more power than what is usually required for pulse EPR, and the MW sources of the two channels are not coherent. Most spectrometers that have DEER (PELDOR) (see Chapter 19) capabilities are able to carry out effective EDNMR measurements.

Both ENDOR and EDNMR are better resolved at high magnetic fields because of the increased separation between Larmor frequencies of different types of nuclei. However, for EDNMR the need for a high field is more crucial because of the width of the central hole at $\Delta\nu = 0$, which often masks the frequency range from 0 to 5–10 MHz, depending on the length and power used for the HTA pulse. At high fields, the Larmor frequencies of most of the nuclei will be outside this range.

17.5.2 Resolution

Davies and Mims ENDOR suffer from blind spots at specific hyperfine couplings, as discussed in the ENDOR chapter (Chapter 16). The pulse ENDOR resolution is determined by the RF pulse; short RF pulses may introduce broadening for spectra with narrow lines, like that encountered in single-crystal work. In orientationally disordered samples, the RF pulse length is usually not an issue. The electron spin polarization decay, which is determined by the electron spin lattice relaxation time, T_{1e}, and spectral diffusion limit the length of the RF pulse that can be applied, as the RF pulse should be completed before the polarization returns to its equilibrium value. Accordingly, the ENDOR resolution depends on nuclear spin lattice relaxations T_{1N} and T_{1e}, depending on which one is shorter. A very nice feature of the ENDOR spectra of orientationally disordered samples is the lineshape (powder pattern) of the ENDOR spectrum, where spectral singularities at $\theta = 0$ and $\pi/2$ allow immediate determination of $A_{\perp}$ and $A_{\|}$ just by inspection.

Except for the central hole, EDNMR does not suffer from hyperfine-specific blind spots, although the intensity of the signals does depend on the specifics of the hyperfine and nuclear quadrupole interactions, as given by I_f [see equation (17.7)]. The resolution of the EDNMR spectrum is limited by the homogeneous EPR linewidth, T_{2e}. Accordingly, the resolution of EDNMR will often be lower than that of ENDOR. Here we note that the width of the echo integration gate plays a very important role and covering the full width of the echo will increase the resolution. In addition, the orientation dependence of EDNMR signals is less straightforward compared with ENDOR. This is particularly relevant in orientationally disordered samples, where the very clear powder singularities observed in the ENDOR spectra do not exist in EDNMR because the forbidden transition probabilities are zero for $\theta = 0$ and $\pi/2$ for the simple case of $S = 1/2$, $I = 1/2$ [equation (17.7)]. Double-quantum transitions are easily observed in EDNMR, and they can be used to enhance the resolution (see Figure 17.8b). Such transitions are usually not detected by ENDOR, unless special and rather complex sequences are applied.[13] Finally, EDNMR also shows combination peaks that may contain important information, which has not yet been addressed.

17.5.3 Sensitivity

In general, EDNMR has a significantly better sensitivity than ENDOR, particularly for low-γ quadrupolar nuclei, where I_f is large due to the contribution of the quadrupole interaction to the nuclear-state mixing. Short T_{1e} poses less of a sensitivity issue for

EDNMR than for ENDOR because there is no need to insert a long RF pulse (tens of microseconds) during the sequence. This is particularly important for low-γ nuclei (like ^{14}N or ^{17}O) that require a longer RF pulse than do high-γ nuclei (like ^{1}H and ^{19}F), where for an insufficiently long T_{1e}, the electron spin polarization achieved will vanish before it has been transferred to nuclear polarization. In addition, spectra with a broad spectral width are easily recorded with EDNMR. Finally, it is easier to relate the signal intensities to the number of nuclei in EDNMR (see Figure 17.9) than in ENDOR spectra because peak intensities in the former depend on fewer instrumental and relaxation parameters. In general, the setup for achieving the optimal conditions for an EDNMR spectrum requires mainly adjusting the HTA pulse (length and power) that balances sensitivity and resolution. Since the SNR in EDNMR is usually high, one can easily optimize it. This can be done by observing the reduction of the echo as a function of the HTA pulse length for a particular power setting.

To summarize, EDNMR at W-band has advantages in terms of ease of operation, instrument simplicity, and sensitivity, particularly for low-γ quadrupolar nuclei. It is yet to be seen how good it is at higher fields or at Q-band. Its major limitation lies in limited resolution. ^{1}H ENDOR is usually preferable over EDNMR because it is sufficiently sensitive and one usually strives for high resolution because several different types of ^{1}H nuclei often contribute to the spectrum. For all other nuclei, it is worth trying both methods and, according to the results and needs, deciding which one to use.

17.6 EDNMR VERSUS ESEEM

As mentioned in Section 17.2, there are some similarities between ESEEM-type experiments (see Chapter 18) and EDNMR, mostly in their dependence on the excitation of EPR-forbidden transitions. In this section we compare EDNMR with ESEEM. ESEEM is a single-frequency technique and therefore its hardware requirements are simpler. It relies on hard MW pulses with a sufficient bandwidth that can cover both the allowed and the associated forbidden transitions. Therefore, due to instrumental limitations on available power and pulse length, it cannot access large hyperfine couplings. This is in contrast to EDNMR, which does not require high power and can detect both small and large hyperfine couplings. As EDNMR is less stringent in terms of power, it is highly effective at high frequencies where MW power is limited but the nuclear Larmor frequencies are better resolved. In EDNMR it is possible to increase the length of the HTA pulse and thus drive more effectively forbidden transitions with very low transition probabilities. Another advantage of EDNMR is its higher tolerance to spectrometer dead time (see Chapter 11). One of the drawbacks of EDNMR is the central hole that masks signals of low frequency. This is not an issue in ESEEM, where low frequency nuclear modulations below 1 MHz are easily seen. Accordingly, while ESEEM is most effective at X band, EDNMR is not. ESEEM, like EDNMR, can detect double-quantum transitions, which are highly informative. Except for the central hole, EDNMR does not suffer from blind spots in the spectrum, while many of the ESEEM experiments do and require the acquisition of several experiments under different timing conditions to overcome this drawback.

ESEEM, as a time-domain technique where the sampling rate is dictated by the highest nuclear frequency through the Nyquist criterion, requires the full nuclear spectrum to be collected, including harmonics and/or sum frequencies, which affect the accumulation time. In contrast, as EDNMR is a frequency domain technique, it is possible to focus just on a single, narrow spectral region of interest. In terms of resolution, the ESEEM resolution is considerably higher, particularly in the HYSCORE version, when applicable, which is currently superior to the 2D EDNMR discussed below. In term of sensitivity, the comparison is more involved than the case of ENDOR, as this usually involves different spectrometer frequencies. To summarize, all the hyperfine spectroscopy techniques, ENDOR, ESEEM, and EDNMR should be used in concert, depending on the specific problem at hand, and often they are complementary.

17.7 SIMULATIONS

Often the analysis of the various kinds of spectra generated by the various EPR techniques requires spectral simulations to extract the spin-Hamiltonian parameters that are of interest, i.e., the hyperfine and nuclear quadrupole parameters in the case of hyperfine spectroscopy.

The simplest way to simulate EDNMR spectra is by just considering the allowed and forbidden transition frequencies and their transition probabilities, ignoring

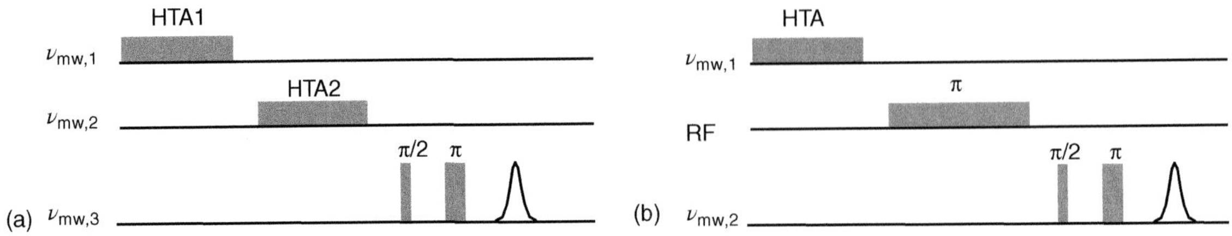

Figure 17.12. (a) The 2D EDNMR pulse sequence. The frequencies of the two HTA pulses are swept independently. (b) The THYCOS pulse sequence. The frequencies of the HTA and the RF pulse are swept independently

the specific effects of the pulses, spin dynamics during the pulse sequence, and associated relaxation times and pathways. This can be obtained by setting up the appropriate spin Hamiltonian in the rotating frame:

$$\hat{H}_0^{R} = \Delta\omega_S\,\hat{S}_z + \omega_I\,\hat{I}_z + A\,\hat{S}_z\hat{I}_z + B\,\hat{S}_z\hat{I}_x + \hat{\boldsymbol{I}}\boldsymbol{P}\hat{\boldsymbol{I}} \qquad (17.14)$$

where $\Delta\omega_S$ is the angular frequency offset (off-resonance frequency) given by

$$\Delta\omega_S = \frac{1}{\hbar}\mu_B B_0 g_{eff}(\theta, \phi) - \omega_0 \qquad (17.15)$$

where ω_0 is the applied MW frequency, and $g_{eff}(\theta, \phi)$ is the effective g value for a B_0 orientation of (θ,φ) (see Chapter 2) that corresponds to the MW frequency. The MW irradiation Hamiltonian in the rotating frame is

$$\hat{H}_1 = \omega_1\hat{S}_x \qquad (17.16)$$

The diagonalization of $\hat{H}_0^{R}$ yields the eigenvalues and eigenfunctions according to

$$\hat{H}_0^{D} = \boldsymbol{D}^{-1}\hat{H}_0\boldsymbol{D} \qquad (17.17)$$

The diagonalizing matrix, $\boldsymbol{D}$, comprises k eigenvectors $|\lambda_i\rangle$ corresponding to k eigenvalues λ_i. This is followed by calculating all transition frequencies given by $\omega_{ij} = \lambda_i - \lambda_j$ and the transition probabilities P_{ij}, with $P_{ij} = (\lambda_i|\hat{H}_1^{D}|\lambda_j)^2$ and $\hat{H}_1^{D} = \boldsymbol{D}^{-1}\hat{H}_1\boldsymbol{D}$.

The next step is to identify the allowed transitions through their high transition probabilities and then calculate the difference between the frequencies of the allowed and their associated forbidden transitions.[5] This yields the EDNMR frequencies (see, e.g., Figure 17.4d) and their intensities, taken as the corresponding forbidden transition probability [see equation (17.11)]. This can be done using programs like EasySpin.[14] When calculating EDNMR spectra under hyperfine-selective conditions, only specific allowed transitions should be selected. In the absence of hyperfine selectivity, all allowed transitions should be considered. For a more rigorous simulation of spectra measured under conditions of large HTA values, the signal intensity should be given by the hole intensity as given in equation (17.10), taking into account ω_1 and without the limitation of a very small I_f. However, under such conditions relaxation and off-resonance excitation of the HTA pulse are important and should be taken into account.

Another more fundamental method[6] of calculating EDNMR spectra recorded under saturation conditions (large HTA) is based on the density matrix formalism and the stochastic Liouville equation[6] (see Chapter 8). It also reproduces signals arising from off-resonance effects on nearby allowed transitions that are responsible, for example, for the signal labeled $A(m_I = 0)$ in Figure 17.6(b), and include relaxation processes. This approach was used to calculate EDNMR spectra of a nitroxide[6] and a Cu(II) complex.[15] The spectra shown in Figure 17.6(d) were calculated using this approach. They show the effect of T_{IN} on the intensity of the $A(m_I = 0)$ line, which reflects the difference between two allowed transitions; they appear at a frequency corresponding to the hyperfine coupling. The shorter is $T_{IN,}$ the stronger is the signal.

17.8 TWO-DIMENSIONAL EXPERIMENTS

When many different nuclei interact with an unpaired electron or when more than one paramagnetic center is present in the sample, the EDNMR spectra may consist of many overlapping signals. Such congested spectra present a challenge for assigning signals to their specific paramagnetic center and/or specific nuclear transition. 2D correlation spectroscopy alleviates this problem by enhancing resolution and aiding in signal assignment. One such highly popular experiment is HYSCORE, which is a 2D version of the three-pulse ESEEM experiment (see Chapter 18).

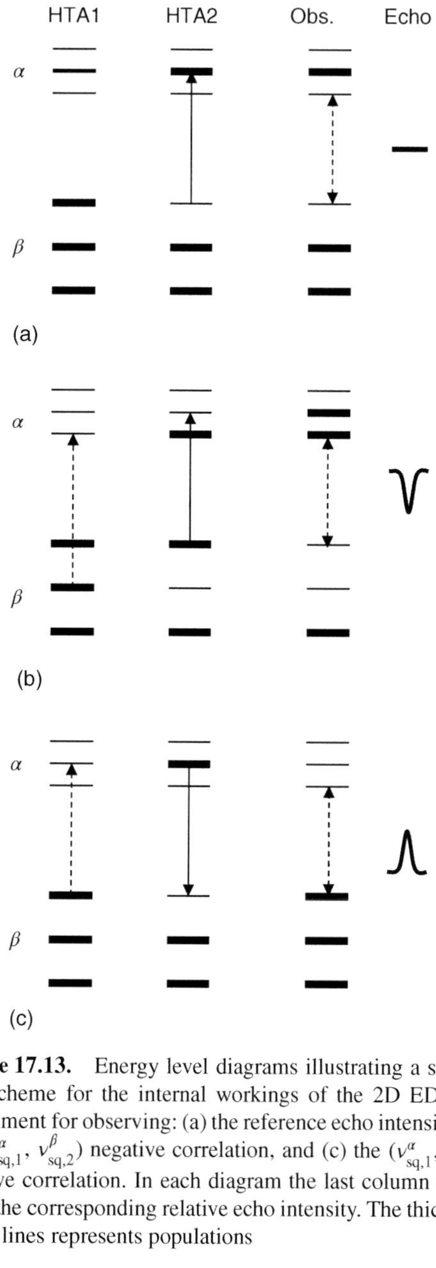

Figure 17.13. Energy level diagrams illustrating a simplified scheme for the internal workings of the 2D EDNMR experiment for observing: (a) the reference echo intensity, (b) the ($\nu^{\alpha}_{sq,1}$, $\nu^{\beta}_{sq,2}$) negative correlation, and (c) the ($\nu^{\alpha}_{sq,1}$, $\nu^{\alpha}_{sq,1}$) positive correlation. In each diagram the last column represents the corresponding relative echo intensity. The thickness of the lines represents populations

The EDNMR experiment can also be extended to a 2D experiment that correlates nuclear frequencies that belong to different m_s manifolds by the addition of a second HTA pulse. The frequencies of the two HTA pulses are swept, as shown in Figure 17.12(a).[6] In such an experiment, for each value of HTA1, an EDNMR spectrum is generated with peaks intensities that will differ from the case in which HTA1 has not been applied, namely, the normal EDNMR spectrum.

We use the $S = 1/2$, $I = 1$ spin system to illustrate the mechanism underlying the 2D EDNMR experiment and to better understand the origin of the observed correlations. Figure 17.13 shows the changes in populations observed during the 2D EDNMR pulse sequence, assuming that each HTA pulse inverts a forbidden transition and neglecting relaxation processes that can redistribute the population among the levels. In this example the detection is set to the $m_I = -1$ allowed EPR transition and three illustrative cases where HTA2 is resonant with the forbidden EPR transition corresponding to $\nu^{\alpha}_{sq,1}$ are presented: (i) HTA1 is not resonant with any EPR transition (≈48 MHz) as shown in Figure 17.13(a); thus, it acts as the reference signal (equivalent to 1D EDNMR). (ii) HTA1 is resonant with the forbidden transition corresponding to $\nu^{\beta}_{sq,2}$ (Figure 17.13b), affecting the population distribution across the detected allowed EPR transition. Consequently, the HTA2 pulse will decrease the echo intensity compared with the reference case shown in Figure 17.13(a). This will produce a negative cross-peak between $\nu^{\alpha}_{sq,1}$ and $\nu^{\beta}_{sq,2}$ in the 2D EDNMR spectrum. (iii) HTA1 is resonant with the forbidden transition corresponding to $\nu^{\alpha}_{sq,1}$ (Figure 17.13c), i.e., both HTA1 and HTA2 are on resonance with the same forbidden EPR transition. Here HTA2 reverses the effect of HTA1, leaving the populations almost unaltered from their equilibrium values, thus leading to a stronger echo than in the reference case in Figure 17.13(a); hence it generates a positive cross-peak on the diagonal in the full 2D EDNMR spectrum.

An example of a 2D EDNMR spectrum of a nitroxide spin probe recorded under the same conditions as the EDNMR spectrum (Figure 17.6b) is shown in Figure 17.13. The main diagonal is equivalent to the 1D EDNMR spectrum, showing only positive cross-peaks. Strong negative cross-peaks are observed at ($\nu^{\beta}_{sq,2}$, $\nu^{\alpha}_{sq,1}$), and weaker negative cross-peaks are observed between the ν_{sq} and ν_{dq} lines belonging to different electron manifolds. Weak positive cross-peaks are also observed for (ν_{sq}, ν_{dq}) within the same manifold. In addition, positive correlations between the $A(m_I = 0)$ peak (≈97 MHz) along $\Delta\nu_2$ and $\nu^{\alpha}_{sq,1}$, and $\nu^{\beta}_{sq,2}$ along the $\Delta\nu_1$ dimensions are evident. Interestingly, the latter do not have symmetric counterparts on the other side of the main diagonal. Although this simple example did not provide any new information

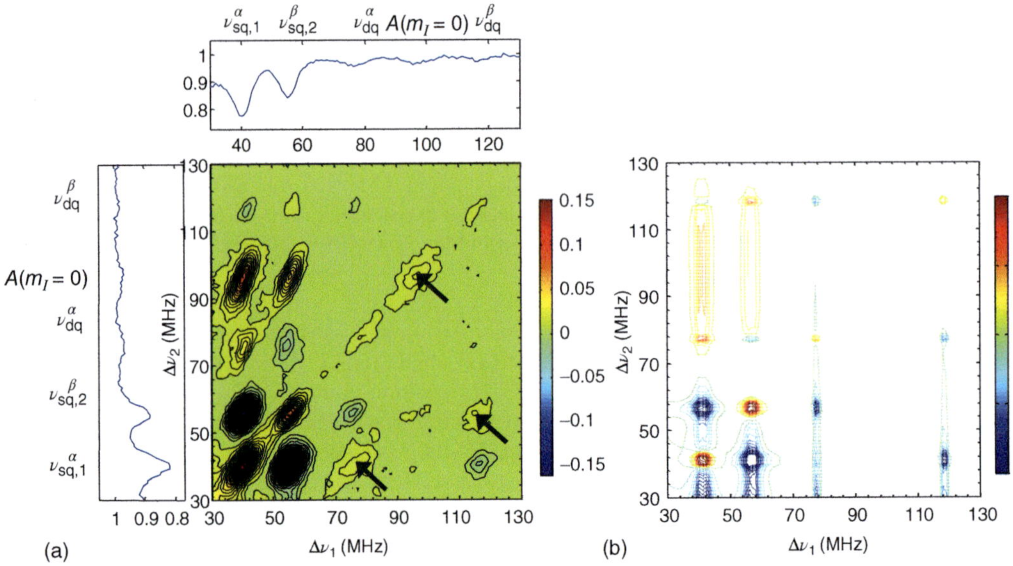

Figure 17.14. (a) Experimental 2D EDNMR spectrum (40 K) of PEO–NO in 3% F127 micelles acquired at the $m_I = -1$ singularity of the g_z field region. Experimental parameters: $t_{HTA1} = t_{HTA2} = 300\,\mu s$, echo detection pulses 100/200 ns. The spectra on the top and on the left are one-dimensional slices corresponding to the maximum offsets along the $\Delta\nu_1$ and $\Delta\nu_2$ dimensions, respectively. The signal assignments are denoted on the slices. (b) The corresponding simulated 2D EDNMR spectrum. Simulation parameters: ^{14}N hyperfine coupling principal values: [16, 16, 100] MHz, ^{14}N quadrupole tensor principal values: [1.65, 0.25, −1.90] MHz. The tensors were taken as collinear. $t_{HTA1} = t_{HTA2} = 300\,\mu s$; $\omega_1(HTA1)/2\pi = \omega_1(HTA2)/2\pi = 3.5$ MHz; T_{2e}=12 μs; $T_{1e} = T_{1N} = \infty$. The arrows in (a) point to peaks that were not reproduced in the simulations. (Reproduced with permission from Ref. 6. © Elsevier, 2014)

compared with the 1D EDNMR because of its simplicity, it is easy to envision more complex cases where the EDNMR spectrum is crowded and it is difficult to assign the lines so 2D EDNMR becomes useful. To date, only one such example was reported: resolving ^{33}S and ^{14}N signals in the Cu(II) site of azurin.[6] The information contained in this experiment is similar to the ENDOR analog, the 2D TRIPLE, except that the SNR is expected to be better but resolution may be lower.[16,17]

Although many of the cross-peaks observed in the 2D EDNMR spectrum (shown in Figure 17.14a) can be accounted for by the simplified model (Figure 17.13), there are some peaks that cannot be explained by it, in particular, positive off-diagonal cross-peaks. In addition, the assumption that the HTA pulses invert the forbidden transitions is not justified because long HTA pulses were employed to generate the 2D spectrum (shown in Figure 17.14a) – the situation is closer to saturation than inversion. These peaks can be accounted for if saturation, off-resonance excitation, and relaxation are taken into account.[6]

The behavior of the populations of the energy level under these conditions, for the three cases shown in Figure 17.13, is illustrated in Figure 17.15. Figure 17.15(a) corresponds to the observation of the $\nu^{\alpha}_{sq,1}$ line and the HTA1 pulse slightly affect the population difference across the $|\alpha,-1\rangle \rightarrow |\beta,-1\rangle$ and $|\alpha,0\rangle \rightarrow |\beta,0\rangle$ transitions through off-resonance irradiation. The HTA2 pulse, as expected, quickly saturates the $|\alpha,-1\rangle \rightarrow |\beta,0\rangle$ transition. The measured signal intensity is proportional to the population difference across the $|\alpha,-1\rangle \rightarrow |\beta,-1\rangle$ transition at the end of the pulse sequence, as marked by an arrow. This is the reference situation. This changes when the HTA1 pulse is on resonance with the forbidden transition corresponding to $\nu^{\beta}_{sq,2}$, as shown in Figure 17.15(b). Consequently, a more intense $\nu^{\alpha}_{sq,1}$ line (weaker echo) is observed. Here both forbidden EPR transitions corresponding to the observation of $\nu^{\alpha}_{sq,1}$ and $\nu^{\beta}_{sq,2}$ become saturated at the end of the pulse sequence with four levels, $|\alpha,-1\rangle$, $|\beta,0\rangle$, $|\alpha,0\rangle$, and $|\beta,-1\rangle$, being nearly equally populated. This practically eliminates the

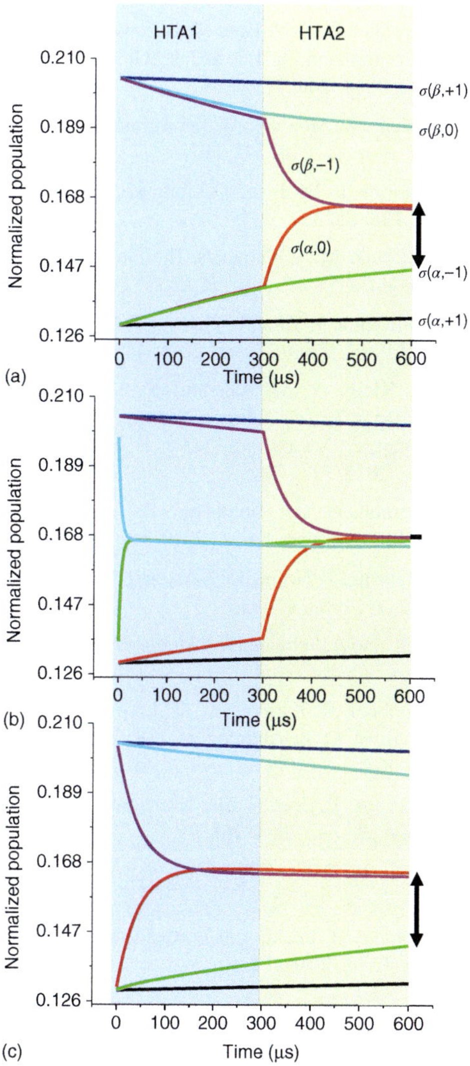

Figure 17.15. Evolution of individual population levels, given by $\sigma(m_S,m_I)$, where $\hat{\sigma}$ is the density matrix, during the 2D EDNMR pulse sequence applied to a nitroxide. (a) $\Delta\nu_1 = 48$ MHz (none), $\Delta\nu_2 = 40$ MHz ($\nu^{\alpha}_{\text{sq},1}$); (b) $\Delta\nu_1 = 55$ MHz ($\nu^{\beta}_{\text{sq},2}$), $\Delta\nu_2 = 40$ MHz ($\nu^{\alpha}_{\text{sq},1}$); (c) $\Delta\nu_1 = \Delta\nu_2 = 40$ MHz ($\nu^{\alpha}_{\text{sq},1}$). Other simulation parameters: $\omega_1(\text{HTA1})/2\pi = \omega_1(\text{HTA2})/2\pi = 3.5$ MHz; $t_{\text{HTA1}} = t_{\text{HTA2}} = 300$ μs; $T_{2e} = 12$ μs; $T_{1e} = T_{1N} = \infty$. The black arrow indicates the population difference between the levels corresponding to the detected EPR transition, $|\alpha,-1\rangle \rightarrow |\beta,-1\rangle$. (Reproduced with permission from Ref. 6. © Elsevier, 2014)

population difference across the $|\alpha,-1\rangle \rightarrow |\beta,-1\rangle$ transition and results in a much weaker echo across the $|\alpha,-1\rangle \rightarrow |\beta,-1\rangle$ transition, as compared with that in Figure 17.15(a), generating a negative cross-peak ($\nu^{\alpha}_{\text{sq},1}$, $\nu^{\beta}_{\text{sq},2}$) in the 2D spectrum. In the last case presented in Figure 17.15(c), the two HTA pulses are on resonance with the $|\alpha,-1\rangle \rightarrow |\beta,0\rangle$ transition. This corresponds to a weaker $\nu^{\alpha}_{\text{sq},1}$ line (stronger echo). As expected, the two levels, $|\alpha,-1\rangle$ | and $|\beta,0\rangle$, quickly become equally populated and remain so until the end of the pulse sequence. Note that the population difference across the allowed $|\alpha,-1\rangle \rightarrow |\beta,-1\rangle$ transition is higher in this case as compared with that in Figure 17.15(a), generating a positive peak on the diagonal. This is mainly due to the different extent of the off-resonance irradiation on the allowed $|\alpha,-1\rangle \rightarrow |\beta,-1\rangle$ and $|\alpha,0\rangle \rightarrow |\beta,0\rangle$ EPR transitions.

A simulated 2D EDNMR spectrum is shown in Figure 17.14(b). Most of the cross-peaks that appear in the experiment are reproduced in the simulations, as are their signs. The width of the cross-peaks involving $A(m_I = 0)$ is given by the degrees of the off-resonance excitation, which is determined by the values of ω_1 and T_{2e}.

Another correlation experiment is *t*riple resonance *hy*perfine *s*ublevel *co*rrelation *s*pectroscopy (THYCOS), shown in Figure 17.12(b), which combines EDNMR and ENDOR.[8,18] Its information content is similar to that of the 2D EDNMR and TRIPLE ENDOR experiment (see Chapter 16), where two RF pulses are employed and one obtains correlations between NMR frequencies belonging to the same electron spin manifold. In the THYCOS experiment, the non-equilibrium population level is achieved by a HTA pulse that selectively inverts the population of a forbidden transition, i.e., it moves populations from the β spin manifold to the α spin manifold (for a $S = 1/2$ case). Next, the RF pulse moves populations within the nuclear subsystem of each manifold. Two cases can be differentiated: (i) The RF matches nuclear transitions within the β electron spin submanifold such that the intensity of the detected echo (or FID) following a selective MW pulse on the corresponding allowed EPR transition will increase. (ii) The RF matches transitions within the α-electron spin manifold; here, the echo intensity is reduced. The THYCOS sequence can be implemented in several versions. The most general one is a 2D experiment where both the frequencies of the HTA and the RF are swept. Alternatively, 1D experiments can be performed with either the MW or RF frequencies fixed on one of the nuclear transitions while the other is swept. It was found that sweeping the RF is beneficial because this approach yields a resolved ENDOR sub-spectrum. Furthermore, it does not require subtracting the spectrum recorded without the HTA pulse from that obtained with the HTA pulse because without the HTA pulse, there are no signals. In this version, the spectrum shows all the NMR transitions in the manifold opposite the one excited by the HTA pulse. This experiment can also be used to resolve overlapping NMR frequencies belonging to different paramagnetic species with an overlapping EPR signal, as demonstrated for the two types of Cu(II) present in the enzyme ascorbate oxidase.[18]

ACKNOWLEDGMENTS

I wish to thank the members of my group: Thorsten Bahrenberg, Alberto Collauto, Aleksei Litvinov, and Marie Ramirez-Cohen for their critical reading of the manuscript and Shimon Vega and Alberto Collauto for many enlightening discussions. I thank Alberto Collauto and Aleksei Litvinov for providing some of the figures.

REFERENCES

1. N. Cox , A. Nalepa, M.-E. Pandelia, W. Lubitz, and A. Savitsky, *Methods Enzymol.*, 2015, **563**, 211.
2. P. Schosseler, T. Wacker, and A. Schweiger, *Chem. Phys. Lett.*, 1994, **224**, 319.
3. A. Schweiger and G. Jeschke, Principles of Pulse Electron Paramangetic Resonance, Oxford University Press: Oxford, UK, 2001.
4. N. Cox, W. Lubitz, and A. Savitsky, *Mol. Phys.*, 2013, **111**, 2788.
5. M. Florent, I. Kaminker, V. Nagarajan, and D. Goldfarb, *J. Magn. Reson.*, 2011, **210**, 192.
6. I. Kaminker, T. D. Wilson, M. G. Savelieff, Y. Hovav, H. Zimmermann, Y. Lu, and D. Goldfarb, *J. Magn. Reson.*, 2014, **240**, 77.
7. A. Nalepa, K. Moebius, W. Lubitz, and A. Savitsky, *J. Magn. Reson.*, 2014, **242**, 203.
8. A. Potapov, B. Epel, and D. Goldfarb, *J. Chem. Phys.*, 2008, **128**, 052320.
9. E. M. Bruch, M. T. Warner, S. Thomine, L. C. Tabares, and S. Un, *J. Phys. Chem. B.*, 2015, **119**, 13515.
10. A. Martorana, Y. Yang, Y. Zhao, Q. F. Li, X. C. Su, and D. Goldfarb, *Dalton Trans.*, 2015, **44**, 20812.
11. E. L. Klein, A. M. Raitsimring, A. V. Astashkin, A. Rajapakshe, K. Johnson-Winters, A. R. Arnold, A. Potapov, D. Goldfarb, and J. H. Enemark, *Inorg. Chem.*, 2012, **51**, 1408.
12. I. Kaminker, H. Goldberg, R. Neumann, and D. Goldfarb, *Chem. Eur. J.*, 2010, **16**, 10014.
13. M. Mehring, P. Hofer, H. Kass, and A. Grupp, *Europhys. Lett.*, 1988, **6**, 463.
14. S. Stoll and A. Schweiger, *J. Magn. Reson.*, 2006, **178**, 42.
15. A. Aliabadi, R. Zaripov, K. Salikhov, V. Voronkova, E. Vavilova, M. A. Abdulmalic, T. Ruffer, B. Buchner, and V. Kataev, *J. Phys. Chem. B.*, 2015, **119**, 13762.
16. D. Goldfarb, B. Epel, H. Zimmermann, and G. Jeschke, *J. Magn. Reson.*, 2004, **168**, 75.
17. B. Epel and D. Goldfarb, *J. Magn. Reson.*, 2000, **146**, 196.
18. A. Potapov, I. Pecht, and D. Goldfarb, *Phys. Chem. Chem. Phys.*, 2010, **12**, 62.

Chapter 18
Hyperfine Spectroscopy – ESEEM

Sabine Van Doorslaer
University of Antwerp, Antwerp, Belgium

18.1 INTRODUCTION

Similar to ENDOR-detected (see Chapter 16) and ELDOR-detected NMR (see Chapter 17), electron spin echo envelope modulation (ESEEM) spectroscopy is a so-called hyperfine spectroscopy and hence targeted at the detection of nuclear frequencies in paramagnetic systems. Although the first observation of the ESEEM effect and the development of the corresponding basic theory dates back to 1965 and 1972, respectively,[1,2] the more extended applications of ESEEM only came about in the late 1980s of the last century and were catalyzed by the commercialization of pulse EPR spectrometers and by the introduction of two-dimensional ESEEM experiments.[3] Since then, ESEEM techniques have been extensively employed in very diverse research fields.

EPR Spectroscopy: Fundamentals and Methods.
Edited by Daniella Goldfarb and Stefan Stoll

ISBN: 978-1-119-16299-5
Also published in eMagRes (online edition)
DOI: 10.1002/9780470034590.emrstm1517

All ESEEM techniques follow the same general scheme as outlined in Figure 18.1(a). They consist of a sequence of microwave (MW) pulses at a single MW frequency, which can be subdivided in different building blocks. The sequence starts with a *preparation period* consisting of one or more MW pulses in which the spin system is prepared in a coherence or polarization nonequilibrium state. This is followed by a first *evolution time* in which the spin system freely evolves under the influence of the spin Hamiltonian. The optional MW pulse sequence forming the *mixing period* induces transfer of magnetization of different types. This may be succeeded by a second *evolution period* and even repeated. Finally, the *detection period* serves to transfer coherences and polarizations to electron coherence (EC), i.e., a detectable electron spin echo (ESE). This echo is recorded as a function of the evolution time(s), and, as will be explained in later sections in detail, the sequence of pulses can induce a modulation of the echo amplitude. This results then in one-dimensional or two-dimensional modulated time-domain signals (Figure 18.1a). Subsequent Fourier transformation allows identification of the nuclear frequencies. Through simulation of these frequency-domain spectra, the hyperfine (***A***) and quadrupole (***P***) tensors of the nuclei surrounding the electron spin can be determined. By using intuitive models or advanced quantum-chemical computations (see Chapter 7), these parameters can then be translated into a structural model. In many cases, these ESEEM techniques give complementary information to the ENDOR experiments. For instance, low-frequency transitions (<5 MHz) are best detected by ESEEM. Therefore, ESEEM will be the method

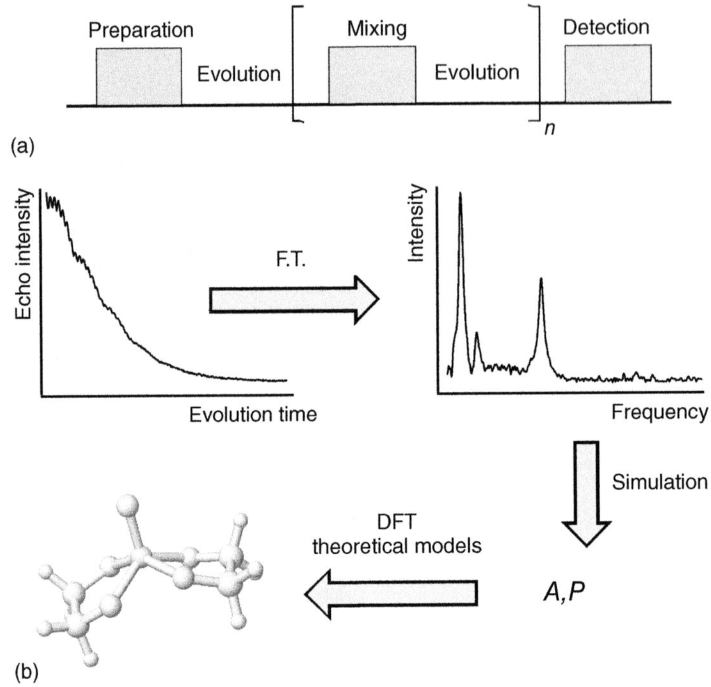

Figure 18.1. Basic concept of ESEEM. (a) General outline of an ESEEM experiment. The preparation, mixing, and detection building blocks consist of one or more MW pulses. (b) Cartoon showing the general process leading from the recorded ESEEM time trace to the final interpretation

of choice for the study of weakly coupled nuclei, such as many of the ^{14}N nuclei of protein-embedded radicals or paramagnetic transition-metal complexes. Moreover, in some cases, the electronic relaxation times may be too short to allow for enough time for an RF pulse with the appropriate length in the ENDOR sequences (see Chapter 16), but the relaxation conditions may still be favorable to detect an ESEEM spectrum.

Although the above paragraph reflects the essence of ESEEM spectroscopy, it will be of little use to a beginner in the field. The following paragraphs are therefore aimed at introducing some of the key points necessary to understand and use ESEEM spectroscopy and at providing a solid basis for further reading.

18.2 BASIC 1D ESEEM EXPERIMENTS

18.2.1 Two-pulse ESEEM

The simplest ESEEM experiment is based on the primary-echo or Hahn-echo sequence shown in Figure 18.2(a).[1,2] In this pulse sequence, we readily recognize the building blocks introduced in Figure 18.1(a). The preparation MW $\pi/2$ pulse generates EC that evolves during a time period of length τ. The π pulse mixes the ECs, which are detected via the ESE after a second evolution time τ. In two-pulse ESEEM, the ESE intensity is measured as a function of τ, whereby the echo signal will not only decay with τ due to relaxation effects (phase-memory time T_m of the electron spin) but also be modulated with the nuclear frequencies of magnetic nuclei hyperfine coupled to the unpaired electron(s).

In order to understand this, let us consider a simple $S = 1/2$, $I = 1/2$ system in the weak-coupling case, i.e., the hyperfine coupling is considerably smaller than the nuclear Larmor frequency.[4–6] In the laboratory frame, the spin Hamiltonian $\hat{\mathcal{H}}_0$ (in angular frequency units) for this system is given by

$$\hat{\mathcal{H}}_0 = \omega_S \hat{S}_z + \omega_I \hat{I}_z + A\hat{S}_z\hat{I}_z + B\hat{S}_z\hat{I}_x \qquad (18.1)$$

where the frame is chosen such that the electron spin lies at the origin and the nucleus lies in the xz plane. $A = A_{zz}$ and $B = \sqrt{A_{zx}^2 + A_{zy}^2}$ are the secular

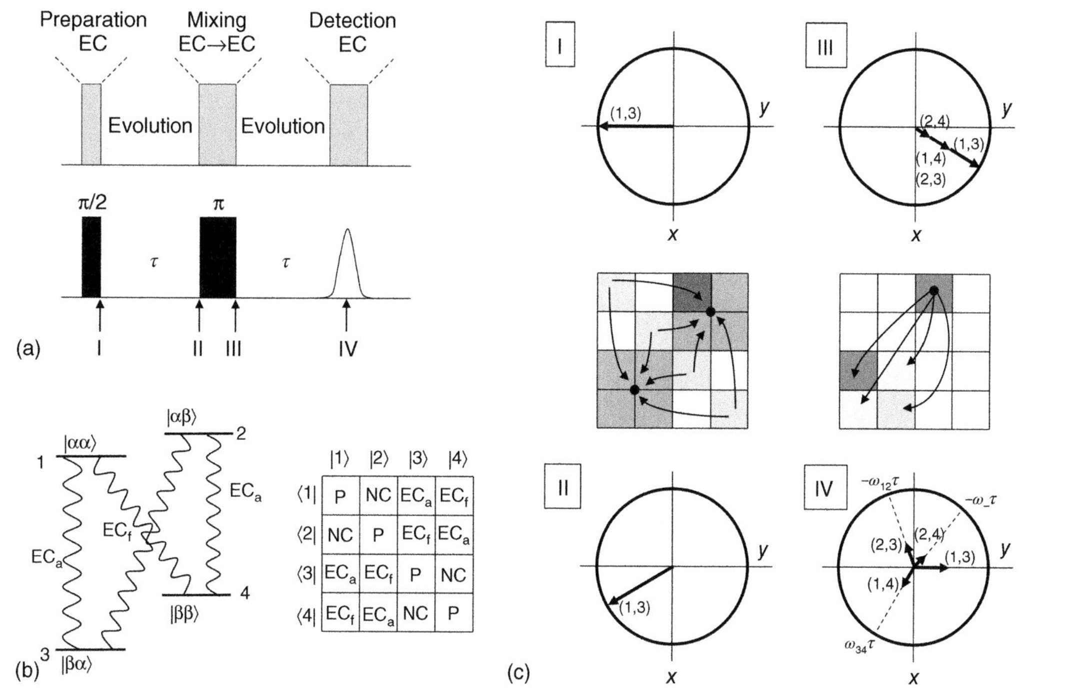

Figure 18.2. The basics of two-pulse ESEEM. (a) Two-pulse ESEEM sequence with essential building blocks highlighted. (b) (left) Energy-level scheme for an $S=1/2$, $I=1/2$ system (with its eigenstates $|m_S m_I\rangle$) for the weak-coupling case ($\omega_I<0$ ($g_n>0$) and A, $B>0$), (right) pictorial representation of the density matrix for this four-level system with P = polarization, NC = nuclear coherence, EC_a and EC_f = allowed and forbidden electron coherence, respectively. (c) Cartoon showing the evolution of the magnetization vector corresponding to EC on the (1,3) transition for time points I–IV indicated in the two-pulse echo sequence in Figure 18.2(a). For stages I and III, the transfer of EC is also illustrated in terms of the density matrix

and pseudo-secular hyperfine coupling constants (see Chapter 5), and ω_I is the Larmor or nuclear Zeeman frequency, all expressed in angular frequency units. $\hat{\mathcal{H}}_0$ can be transformed to its eigenbasis (or in other words diagonalized) using the unitary transformation operator

$$\hat{U}=\exp\left(-\mathrm{i}\left(\xi\hat{I}_y+\eta 2\hat{S}_z\hat{I}_y\right)\right) \tag{18.2}$$

with

$$\eta_{\alpha,\beta}=\tan^{-1}\left(-\frac{\pm B/2}{\pm A/2+\omega_I}\right)=\tan^{-1}\left(-\frac{m_S B}{m_S A+\omega_I}\right),$$
$$\xi=\frac{\eta_\alpha+\eta_\beta}{2},\qquad \eta=\frac{\eta_\alpha-\eta_\beta}{2} \tag{18.3}$$

leading to the diagonalized spin Hamiltonian

$$\begin{aligned}\hat{\mathcal{H}}_0^{\mathrm{d}}&=\omega_S\hat{S}_z+\omega_{12}\hat{S}^\alpha\hat{I}_z+\omega_{34}\hat{S}^\beta\hat{I}_z\\&=\omega_S\hat{S}_z+\frac{\omega_+}{2}\hat{I}_z+\frac{\omega_-}{2}2\hat{S}_z\hat{I}_z\end{aligned} \tag{18.4}$$

In equation (18.4), the signed nuclear frequencies are

$$\omega_{12}=\left(\omega_I+\frac{A}{2}\right)\cos\eta_\alpha-\frac{B}{2}\sin\eta_\beta \tag{18.5}$$

$$\omega_{34}=\left(\omega_I-\frac{A}{2}\right)\cos\eta_\alpha+\frac{B}{2}\sin\eta_\beta \tag{18.6}$$

and the combination frequencies are

$$\omega_\pm=\omega_{12}\pm\omega_{34} \tag{18.7}$$

The polarization operator matrices for an $S=1/2$ system are defined as

$$\hat{S}^\alpha=\begin{pmatrix}1&0\\0&0\end{pmatrix},\quad \hat{S}^\beta=\begin{pmatrix}0&0\\0&1\end{pmatrix} \tag{18.8}$$

The energy scheme for the system defined in equation (18.4) is depicted in Figure 18.2(b) for the weak-coupling case with $\omega_I<0$ ($g_n>0$) and $A,B>0$, alongside a pictorial representation of the corresponding density matrix. For simplicity, all relaxation processes will be disregarded, and ideal pulses (infinitely short with infinite excitation bandwidth) will be assumed in the following explanation. At the start of the experiment, the electron magnetization is in its equilibrium state, oriented along the external magnetic-field axis, z. This corresponds to a diagonal density matrix, i.e., no electron or nuclear coherences (NC) exist at this stage. The π/2 pulse along the x axis in the rotating frame then turns the magnetization to the $-y$ axis. In terms of the density-operator formalism (see Chapter 8), this can be written in the following way in the rotating frame[4]:

$$\hat{\sigma}_{\mathrm{eq}} \propto -\hat{S}_z \xrightarrow{\pi/2\hat{S}_x} \hat{S}_y \tag{18.9}$$

Transformation to the eigenbasis of the spin Hamiltonian, $\hat{\mathcal{H}}_0^{\mathrm{d}}$, gives

$$\hat{S}_y \xrightarrow{\xi\hat{I}_y+\eta 2\hat{S}_z\hat{I}_y} \cos\eta\,\hat{S}_y - \sin\eta\, 2\hat{S}_x\hat{I}_y \tag{18.10}$$

Since the matrix representations of $\hat{S}_y$ and $2\hat{S}_x\hat{I}_y$ have only nonzero matrix elements on the positions marked in Figure 18.2(b) with $\mathrm{EC_a}$ and $\mathrm{EC_f}$, respectively, the resulting density matrix in equation (18.10) will also have nonzero off-diagonal elements. This creation of EC is illustrated in Figure 18.2(c,I) in the pictorial representation of the density matrix.

We now focus on the evolution of the EC created on the (1,3) transition, keeping in mind that a similar derivation can be done for all other ECs. Transition (1,3) is an allowed EPR transition ($|\Delta m_S|=1$, $|\Delta m_I|=0$). After the first pulse, the allowed EC for this transition can be visualized as an electron magnetization vector (black vector in Figure 18.2(c,I) with its corresponding density matrix element highlighted in the pictorial representation of the density matrix). Note that the magnetization and electron spin are antiparallel. Evolution of the EC under $\hat{\mathcal{H}}_0^{\mathrm{d}}$ during time τ can be represented as a precession of the magnetization vector in the xy plane. After time τ, the vector has gained a phase $(\omega_{13}-\omega_{\mathrm{MW}})\tau$ with respect to the $-y$ axis of the rotating frame (Figure 18.2c,II). The mixing π pulse does not only rotate the magnetization vector 180° about the x axis but also redistributes the EC among all the electron spin transitions (so-called *branching*), since there is a finite probability that the nuclear spin is simultaneously inverted by the π pulse (Figure 18.2c,III). During the second evolution time, the magnetization vectors corresponding to the different ECs will now precess with their specific frequencies. As a consequence, at time τ after the π pulse, not all of the initial magnetization will be refocused along the y axis (Figure 18.2c,IV). As an example, the coherence on the (1,4) transition gained a phase $(\omega_{14}-\omega_{13})\tau=\omega_{34}\tau$ with respect to the coherence on the (1,3) transition. In other words, the echo intensity will differ from the one expected for an $S=1/2$, $I=0$ system, and the observed decrease in the echo intensity will depend on the phase that the different magnetization vectors gained during the experiment, which in turn depends on τ and on the difference between their precession frequency before $(\omega_{13}-\omega_{\mathrm{MW}})$ and after the application of the π pulse. These differences are the nuclear frequencies, and, hence, detection of the echo intensity as a function of τ will lead to a time-domain signal that is modulated with these frequencies.

For a simple $S=1/2$, $I=1/2$ system, the echo modulation is given by

$$V_{2\mathrm{p}}(\tau) = -\langle\hat{S}_y\rangle = 1-\frac{k}{4}[2-2\cos(\omega_{12}\tau)-2\cos(\omega_{34}\tau) + \cos(\omega_-\tau)+\cos(\omega_+\tau)] \tag{18.11}$$

with the so-called modulation depth k equal to

$$k = \sin^2 2\eta \tag{18.12}$$

with η defined in equation (18.3). Note that the basic frequencies $\omega_{12,34}$ in equation (18.11) can be replaced by $\omega_{\alpha,\beta}=|\omega_{12,34}|$, since the signs of the transition frequencies can be neglected.

Remember that equation (18.11) is derived assuming ideal pulses. This is still valid for hard pulses, i.e., short strong pulses with $\omega_1 \gg |\omega_I|, |A|, |B|$. Indeed, the two-pulse ESEEM experiment involves simultaneous excitation of a pair of EPR transitions. If $\omega_1 < \min(|\omega_{12}|, |\omega_{34}|)$, no modulation can be observed, whereas the condition $\min(|\omega_{12}|, |\omega_{34}|) < \omega_1 < \max(|\omega_{12}|, |\omega_{34}|)$ leads to observation of only the smaller of the two nuclear frequencies. This effect can be exploited to simplify and unravel the ESEEM spectra.

For more complex systems in which the electron spin is interacting with N different magnetic nuclei, the echo modulation can be obtained from the *product rule*

$$V_{2\mathrm{p,tot}}(\tau) = \prod_{i=1}^{N} V_{2\mathrm{p},i}(\tau) \tag{18.13}$$

with the specific echo-modulation expressions $V_{2\mathrm{p},i}(\tau)$ for the individual interaction with each of the nuclei.

Since the whole two-pulse ESEEM experiment is based on evolutions of ECs, the echo decay depends on the phase-memory time, T_m, of the electron spin (T_m being dominated by the electron spin–spin relaxation, T_{2e}), leading to an expression of the form

$$V'_{2p}(\tau) = \exp\left(-\frac{2\tau}{T_m}\right) V_{2p}(\tau) \qquad (18.14)$$

In practice, $V'_{2p}(\tau)$ is sampled at discrete, equidistant points in time τ, separated by the *dwell time* $\Delta\tau$. A practical example is shown in Figure 18.3(a). The time-domain signal can be converted into a frequency-domain spectrum by discrete Fourier transformation. This requires that first the relaxation-dependent exponential decay should be removed from the signal using a correct background correction. Next, the signal is apodized with a suitable window function (e.g., a Hamming window), and zeroes are added at the end of the time trace (so-called *zero filling*), which allows interpolation in the frequency domain. After FFT, the magnitude spectrum is calculated. Since the frequency range after FFT is $(-(\pi/\Delta\tau), (\pi/\Delta\tau))$, one should always choose the dwell time carefully, such that

$$\Delta\tau < \frac{\pi}{\max(|\omega_{nuc}|)} \qquad (18.15)$$

with $\max(|\omega_{nuc}|)$, the maximum nuclear frequency expected. This is the well-known *Nyquist criterion*. Failing to meet this criterion will lead to back folding (or aliasing) of peaks, e.g., for all nuclear frequencies ω_{nuc} for which $(2\pi/|\omega_{nuc}|) \geq \Delta\tau \geq (\pi/|\omega_{nuc}|)$, the peaks will appear at frequencies $(\pi/\Delta\tau) - |\omega_{nuc}|$. Incorrect background subtraction and/or insufficient apodization will lead to artifacts after Fourier transformation. Furthermore, in practice, the echo can only be detected starting from a nonzero τ value (due to instrument-dependent deadtime), causing serious spectral distortions after FFT, and solutions to circumvent this problem come with a price in spectral resolution.[7] Figure 18.3(b) shows the frequency-domain magnitude spectrum derived from the time signal in Figure 18.3(a).

Two-pulse ESEEM provides a very quick and easy method to obtain information about the nuclear frequencies, namely, the identity of coupled nuclei, but it has a number of specific drawbacks. Indeed, the (usually short) T_m of the electron spin limits the time for which the modulation can be observed [equation (18.14)], which inevitably gives rise to broadened frequency-domain signals after Fourier transformation. This considerably reduces the spectral resolution in two-pulse ESEEM. Furthermore, equation (18.11) and Figure 18.2(c,IV) illustrate that, besides the basic nuclear frequencies, also the sum and difference of these basic frequencies (ω_- and ω_+) will be detected, even for a simple $S = 1/2$, $I = 1/2$ system. The number of such combination frequencies will increase dramatically for systems with N nuclei due to the product rule given in equation (18.13). The wealth of detected frequencies together with the relatively broad linewidths will then often lead to overlapping and uninterpretable features in the one-dimensional two-pulse ESEEM spectra.

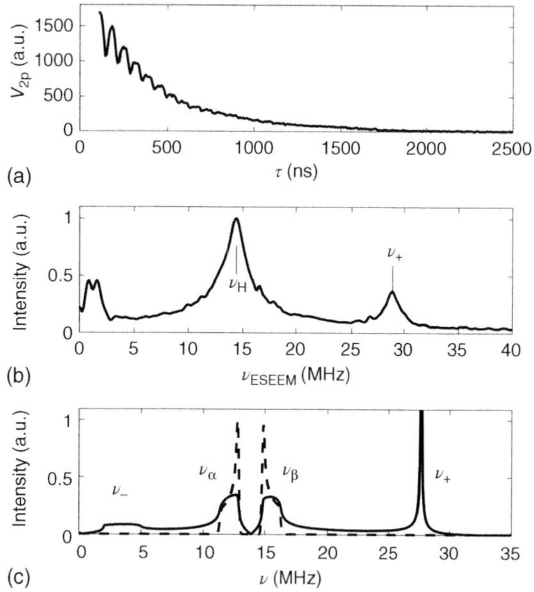

Figure 18.3. Examples of two-pulse ESEEM. The experimental time-domain (a) and frequency-domain (b) X-band two-pulse ESEEM trace of copper(II) complex of the (*R,R*)-enantiomer of *N,N*′-bis(3,5-di-*tert*-butylsalicylidene)-1,2-cyclohexanediamino with *R*-methylbenzylamine in toluene. The proton Larmor frequency (ν_H) and the sum peaks ν_+ ($\approx 2\nu_H$) are indicated. The low-frequency peaks cannot be assigned on the basis of the two-pulse ESEEM alone. (c) Comparison between a simulated X-band two-pulse ESEEM (solid line) and ENDOR (dashed line) spectrum for an $S = 1/2$, $I = 1/2$ (^{1}H) system with isotropic $g(=2)$ and the proton hyperfine coupling tensor [2, 2, 5] MHz. Magnetic-field setting is 324.9 mT (center of EPR line)

Finally, a number of issues valid for all ESEEM experiments can be derived considering the simple case of a two-pulse ESEEM experiment for an $S=1/2$, $I=1/2$ case. First of all, standard rectangular MW pulses used for ESEEM experiments have an excitation bandwidth that is often narrower than the spectral width of an EPR spectrum. Indeed, a 12 ns pulse has a bandwidth of about 100 MHz ($\approx$3.6 mT), while powder EPR spectra of transition-metal complexes, for instance, can be several hundreds of mT wide. Each magnetic-field value in a powder EPR spectrum can be linked to specific orientations in space (see Chapter 1). Hence, the ESEEM spectrum taken at a fixed magnetic field will only reflect the hyperfine (and nuclear quadrupole) values linked to the orientations excited at this magnetic field and not the full tensor information (so-called *orientation selection*; see Chapter 11). In order to obtain the full hyperfine and quadrupole tensors, ESEEM experiments should be performed at different magnetic-field settings throughout the whole EPR spectrum. Remember also that, as outlined earlier, the limited excitation width of the standard MW pulses also puts a limitation to the upper value of the nuclear frequencies that can be observed.

Second, from equations (18.11) and (18.12), it becomes immediately clear that no echo modulation can be obtained at room temperature for paramagnetic systems in low-viscosity solutions. Indeed, the modulated part of the expressions in equation (18.11) is multiplied with a modulation depth factor k that becomes zero for an isotropic hyperfine coupling arising from motional averaging ($B=0$ in equation (18.3) leads to $\eta=0$). We will later see that the k factor is common to all ESEEM experiments. In principle, one can observe two-pulse echo modulation at room temperature for solids, but in this case, T_m may be too short to allow detection of an echo.

Analogously, k also becomes zero for the principal orientations of the hyperfine tensor. Hence, the spectral patterns of powder samples will be less clear to interpret. This is illustrated in Figure 18.3(c) where the two-pulse ESEEM and the ENDOR spectrum for the theoretical system $S=1/2$, $I=1/2$ (^{1}H) with an axial hyperfine tensor [2, 2, 5] MHz are compared. The spectra are simulated assuming a powder EPR line at a magnetic-field setting where all spatial orientations are excited (isotropic $g=2$). For the ESEEM case, a deadtime-free experiment ($\tau_0=0$ ns) was taken to exclude deadtime-related distortions. The ENDOR spectrum has the typical features reflecting the fact that more spatial orientations agree with a hyperfine coupling of 2 MHz than with the unique principal value of 5 MHz (see Chapter 16). However, since no ESEEM effect can be detected for the principal orientations of the hyperfine tensor, the shape of the spectral pattern of the basic frequencies is clearly different in the two-pulse ESEEM case, making an identification of the parallel and perpendicular components of the hyperfine tensor less obvious. Note also that the two-pulse ESEEM experiment has extra signals at the sum and difference frequencies, which are not observed in the ENDOR spectrum.

Furthermore, although it is in general beneficial to perform ESEEM at different MW frequencies in order to disentangle more complicated spectra (the nuclear Larmor frequency varies with the external magnetic field, whereas the hyperfine and nuclear quadrupole interactions are independent of this field), one should be aware that some ESEEM peaks might have vanishing amplitude at some MW frequencies. Indeed, for an $S=1/2$, $I=1/2$ system, the modulation depth k in equation (18.12) can be rewritten as[4]

$$k=\left(\frac{B\omega_I}{\omega_\alpha\omega_\beta}\right)^2 \tag{18.16}$$

with B the pseudo-secular hyperfine interaction.

From this equation, it is clear that at the lower MW frequencies (i.e., low magnetic fields and small ω_I), ESEEM is not very suitable for the detection of hyperfine couplings that are large compared to ω_I as is illustrated in Figure 18.4(a) for the ^{15}N case. In contrast, k is very low or almost zero at high MW frequencies for nuclei with a large g_n value, such as protons (Figure 18.4b,c). The higher the anisotropy of the hyperfine coupling (i.e., higher B), the broader the MW frequency range is, for which a considerable k is found (Figure 18.4b,c). The modulation depth k becomes maximal in the *exact cancellation* case, where the hyperfine coupling is approximately twice the Larmor frequency, and both interactions cancel each other in one of the m_S manifolds. If the hyperfine coupling is considerably smaller, we have *weak hyperfine coupling*; on the other hand, in the *strong hyperfine coupling* regime, the hyperfine term dominates the nuclear Zeeman interaction. For both cases, k will be smaller than in the cancellation condition.

Finally, it has been long ignored that a cross-suppression effect may occur in ESEEM experiments when measuring a system where the unpaired electron is interacting with differently coupled magnetic nuclei. The contributions of weakly modulating nuclei (e.g., weakly coupled protons) can be completely suppressed by the contributions of strongly modulating nuclei (e.g., ^{14}N nuclei in the

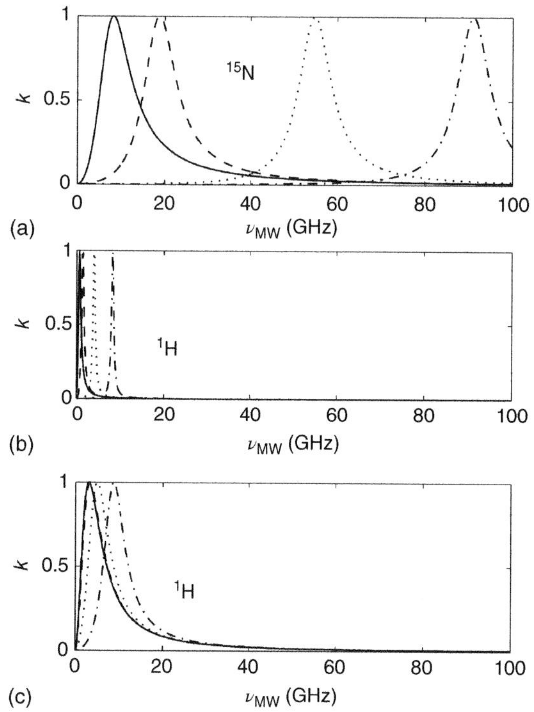

Figure 18.4. Illustration of the dependence of the modulation depth k on the MW frequency for an $S = 1/2$, $I = 1/2$ system [Equation (18.16)] with $g = 2$ for an interaction with (a) a ^{15}N nucleus and (b,c) a ^{1}H nucleus. The pseudo-secular part of the hyperfine coupling, B, is taken as 1 MHz (a,b) or 9 MHz (c). In all graphs, the secular part, A, is 1.5 MHz (solid line), 4 MHz (dashed line), 12 MHz (dotted line), and 25 MHz (dash-dotted line)

cancellation condition).[8] This is an important effect and anyone using ESEEM should be aware of it when interpreting ESEEM data.

18.2.2 Three-pulse ESEEM

The three-pulse ESEEM technique offers a better alternative to the two-pulse ESEEM technique. It is based on the stimulated-echo sequence, whereby the first two $\pi/2$ pulses create NC as depicted in Figure 18.5. This NC will evolve during a time t that is varied in the experiment. The last $\pi/2$ pulse turns the NC into detectable EC. The limitation in the time domain is now imposed by the phase-memory time of the nuclear spins ($T_{m,n}$), which is approximately of the order of the electron spin–lattice relaxation (T_{1e}). For many systems, $T_{m,n} \approx T_{1e} > T_{2e}$ and the three-pulse ESEEM spectral lines will thus be narrower in the frequency domain than for the two-pulse ESEEM case.

For the $S = 1/2$, $I = 1/2$ system introduced earlier, the echo-modulation formula is given by

$$\begin{aligned} V_{3p}(\tau,t) &= \frac{1}{2}\left\{\left[1-\frac{k}{2}\left[1-\cos\left(\omega_\beta\tau\right)\right]\left[1-\cos\left(\omega_\alpha(t+\tau)\right)\right]\right]\right. \\ &\quad \left. +\left[1-\frac{k}{2}\left[1-\cos\left(\omega_\alpha\tau\right)\right]\left[1-\cos\left(\omega_\beta(t+\tau)\right)\right]\right]\right\} \\ &= \frac{1}{2}\{V^{\alpha}_{3p}(\tau,t)+V^{\beta}_{3p}(\tau,t)\} \end{aligned} \quad (18.17)$$

with k defined as before [equation (18.12)]. Since the signs of the transition frequencies can be neglected, $\omega_{\alpha,\beta} = |\omega_{12,34}|$ is used.

In three-pulse ESEEM, fewer frequencies are observed than in a two-pulse ESEEM experiment. As is evident from equation (18.17) for a simple $S = 1/2$, $I = 1/2$ system, the three-pulse ESEEM experiment will only yield the basic nuclear frequencies; combination frequencies at $\omega_\pm$ are not observed. Furthermore, for systems in which the unpaired electron interacts with several magnetic nuclei, the three-pulse ESEEM experiment will, besides all basic nuclear frequencies, only allow detection of sum or difference combinations between the basic frequencies within one m_S manifold (*product rule*). For an $S = 1/2$ system with N nuclear spins, this is

$$V_{3p,tot}(\tau,t) = \frac{1}{2}\left[\prod_{i=1}^{N} V^{\alpha}_{3p,i}(\tau,t) + \prod_{i=1}^{N} V^{\beta}_{3p,i}(\tau,t)\right] \quad (18.18)$$

with $V^{\alpha,\beta}_{3p,i}(\tau,t)$ relating to the terms in the echo-modulation expression in which the echo is modulated in the t dimension with a specific frequency from the α or β manifold as defined in equation (18.17). This contrasts the two-pulse ESEEM case [equation (18.13)], in which combinations between all nuclear frequencies can occur, irrespective of the m_S manifold. The fact that three-pulse ESEEM gives fewer and narrower lines in the frequency domain makes it a preferred method over two-pulse ESEEM. This is illustrated in Figure 18.6(a) that shows a comparison between the two-pulse and three-pulse ESEEM spectra of a frozen solution of ferric cytoglobin recorded at the low-field position. In this heme-containing protein, the unpaired electron of the iron center is coupled to ^{1}H and ^{14}N nuclei. While the two-pulse ESEEM spectrum shows a multitude of broad peaks, also at the proton sum frequencies (ν^{H}_{+}) and at several ^{14}N combination frequencies (ν^{N}_{comb}),

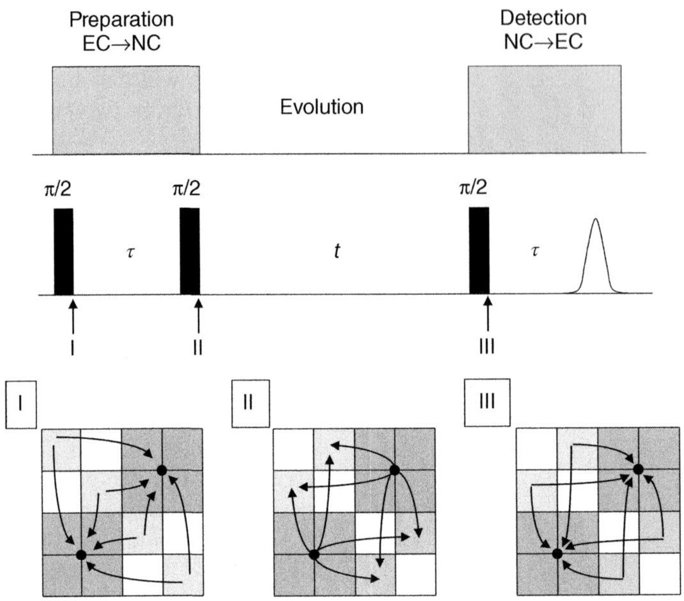

Figure 18.5. Three-pulse ESEEM sequence with its essential building blocks highlighted. The transfer of polarization and coherences at the different stages I–III is illustrated using the pictorial representation of the density matrix

the three-pulse ESEEM spectrum has fewer and narrower lines with the ^{14}N DQ ($|\Delta m_I| = 2$) transitions dominantly observed (vide infra, equation (18.27), for a more detailed explanation of the DQ frequencies).

Three-pulse ESEEM has, however, also drawbacks. Inspection of equation (18.17) reveals that the amplitude of the echo-modulation signal is strongly influenced by the first evolution time τ, and specific combinations of nuclear frequencies and τ will lead to τ-dependent blindspots. This is illustrated in Figure 18.6(b), where the τ-dependence of the three-pulse ESEEM spectrum is shown for the cytoglobin case introduced in Figure 18.6(a). Hence, one should always record three-pulse ESEEM experiments for a series of different values of τ, which makes it a more time-consuming experiment compared with the two-pulse experiment. The duration of the experiment is further prolonged by the fact that pulse phase cycling is necessary to remove the disturbing signals of the three primary echoes and one refocused primary echo generated by the three $\pi/2$ pulses.[4] Although in most cases, the loss in time by the phase cycle is compensated by the need of less signal accumulations, the phase cycle can form a drawback if the reprogramming overhead time for phase cycling is significant or if the signal-to-noise ratio is so good that there is no need for extra signal accumulations. Moreover, one loses signal intensity with every MW pulse that is added to the sequence due to pulse imperfections and different relaxation pathways, which lead to the ‘unwanted’ echoes mentioned above. This limits the sensitivity of the three-pulse ESEEM experiment, which is particularly crucial for very low spin concentrations. Furthermore, similar as in two-pulse ESEEM, the experimental deadtime (now with respect to $t+\tau$) will lead to spectral distortions after FFT, although this problem is partially circumvented by the summation over different τ values.[7]

The three-pulse ESEEM experiment is a one-dimensional experiment. Hence, in the case of many (partially) overlapping signals in complex spin systems, we will still be facing many difficulties to assign the peaks and extract the information needed. By variation of both τ and $t+\tau$ separately, one can easily turn the three-pulse ESEEM experiment into a two-dimensional experiment,[4] which will, after Fourier transformation along both dimensions, lead to similar spectral information as can be obtained by hyperfine sublevel correlation (HYSCORE) spectroscopy (vide infra, Section 18.3.1). However, as the τ domain concerns the evolution of EC and not NC, it results in broadened signals in one of the dimensions. This renders the two-dimensional three-pulse ESEEM experiment less useful in comparison to HYSCORE.

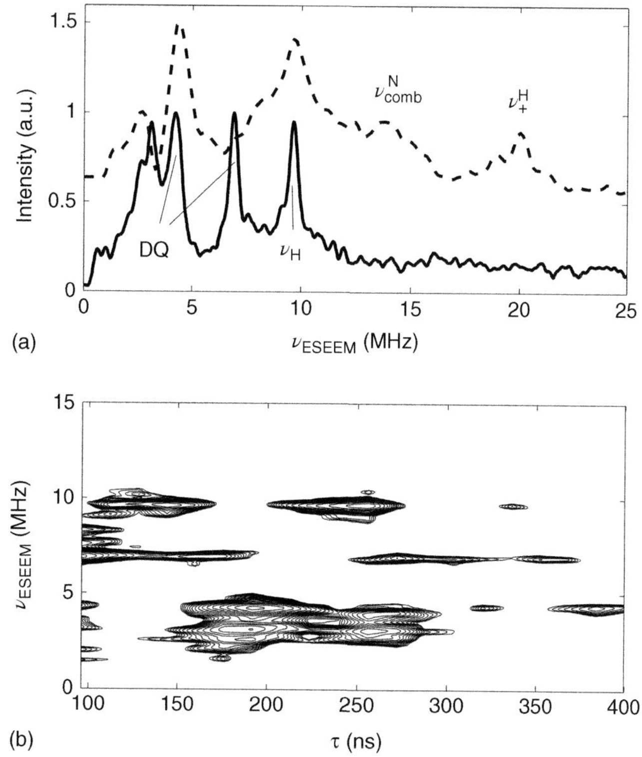

Figure 18.6. Advantages and disadvantages of three-pulse ESEEM. (a) Comparison of the X-band two-pulse (dashed line) and three-pulse (solid line) ESEEM spectrum of a frozen solution of ferric cytoglobin recorded at an observer position agreeing with $g = g_z = 3.2$. The three-pulse ESEEM spectrum is the sum of 20 spectra recorded for different τ-values. The three-pulse ESEEM spectrum has narrower and fewer lines as discussed in the text. (b) τ-Dependence of the three-pulse ESEEM spectra for the system in panel (a), illustrating the blindspot effect

18.3 FOUR-PULSE ESEEM

18.3.1 Basics of HYSCORE Spectroscopy

The drawback of the one-dimensionality of three-pulse ESEEM can be overcome by the four-pulse ESEEM experiment (Figure 18.7a).[3] In this experiment, the π pulse will transfer the NC created by the first two π/2 pulses from one m_S manifold to another. In the 2D version of this experiment, the so-called HYSCORE experiment, the two intervals t_1 and t_2 are independently varied. Two-dimensional Fourier transformation of the data results in a 2D spectrum featuring cross peaks linking the nuclear frequencies of the different m_S manifolds (Figure 18.7a). By introducing a second dimension, the spectral information hidden in many overlapping features observed in three-pulse ESEEM spectra can now be revealed. Note that only the positive-ν_2 half of the frequency-domain spectrum is generally depicted, since the negative-ν_2 half contains exactly the same information. Usually one uses the terms (+,+) and (−,+) quadrants of the HYSCORE spectrum, with the signs indicating the positive (+) or negative (−) half of the (ν_1, ν_2) axes (Figure 18.7b).

In more detail, the echo modulation of the four-pulse ESEEM experiment for $S = 1/2$, $I = 1/2$ can be written as[9]

$$V_{4p,tot}(\tau, t_1, t_2) = 1 - \frac{k}{4}(V_I + V_{IIa} + V_{IIb} + V_{IIIa} + V_{IIIb}) \quad (18.19)$$

with k as defined before [equation (18.12)], and the different modulation contributions stemming from the different coherence-transfer pathways are

$$V_I = 3 - \cos(\omega_\beta \tau) - \cos(\omega_\alpha \tau) - \sin^2\eta \cos(\omega_+ \tau) - \cos^2\eta \cos(\omega_- \tau) \quad (18.20)$$

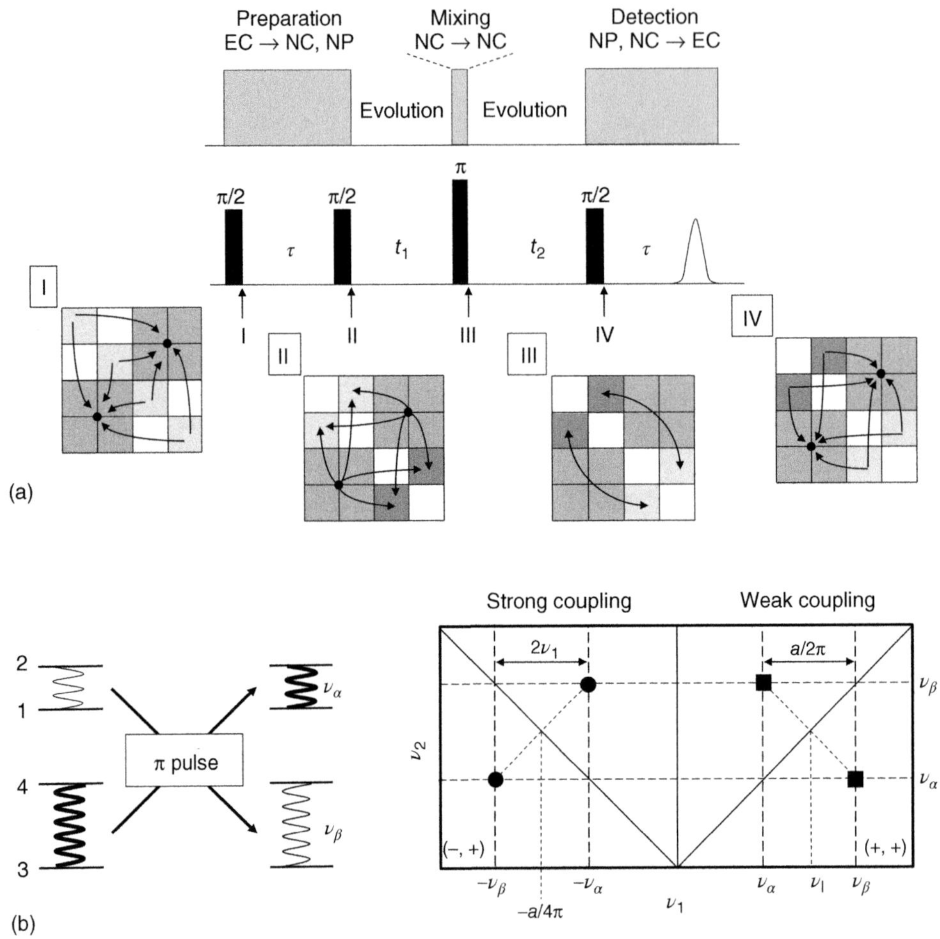

Figure 18.7. Basic concepts of four-pulse ESEEM spectroscopy. (a) MW pulse sequence with essential building blocks highlighted. The transfer of polarization and coherences at the different stages I–IV is illustrated using the pictorial representation of the density matrix. (b) (left) Cartoon visualizing the NC transfer induced by a π pulse for an $S = 1/2$, $I = 1/2$ system. (right) Schematic representation of a HYSCORE spectrum for an $S = 1/2$, $I = 1/2$ system. The appearance of the cross peaks in different quadrants in case of weak hyperfine coupling (■) and strong hyperfine coupling (•) is also indicated. $a/2\pi$ is the hyperfine coupling at the given observer position

$$V_{\text{IIa}} = C_\alpha(\tau)\cos\left(\omega_\alpha\left(t_2 + \frac{\tau}{2}\right)\right) + C_\beta(\tau)\cos\left(\omega_\beta\left(t_2 + \frac{\tau}{2}\right)\right) \quad (18.21)$$

$$V_{\text{IIb}} = C_\alpha(\tau)\cos\left(\omega_\alpha\left(t_1 + \frac{\tau}{2}\right)\right) + C_\beta(\tau)\cos\left(\omega_\beta\left(t_1 + \frac{\tau}{2}\right)\right) \quad (18.22)$$

$$V_{\text{IIIa}} = C_c(\tau)\cos^2\eta\left[\cos\left(\omega_\alpha t_1 + \omega_\beta t_2 + \omega_+\tau/2\right) + \cos\left(\omega_\beta t_1 + \omega_\alpha t_2 + \omega_+\tau/2\right)\right] \quad (18.23)$$

$$V_{\text{IIIb}} = -C_c(\tau)\sin^2\eta\left[\cos\left(\omega_\alpha t_1 - \omega_\beta t_2 + \omega_-\tau/2\right) + \cos\left(\omega_\beta t_1 - \omega_\alpha t_2 - \omega_-\tau/2\right)\right] \quad (18.24)$$

The coefficients C_α, C_β, and C_c are τ-dependent terms that induce blindspots similar to the three-pulse ESEEM case.

The different terms in equations (18.20)–(18.24) can be understood in the following way. The $\pi/2-\tau-\pi/2$

in the preparation block (Figure 18.7a) generates NC. In case of nonselective π/2 pulses applied to a system with *anisotropic* hyperfine couplings, also EC (which is not relevant for the current experiment) and nuclear polarization (NP) will be generated by this sequence.[4] The contribution V_I [equation (18.20)] then stems from the pathway in which NP is created by the preparation block, transferred to NP by the π pulse, and converted to EC by the detection π/2 pulse. Since polarization remains unchanged during the two evolution times t_1 and t_2, the term V_I only depends on τ and will not contribute to the HYSCORE modulation. Note that the expression in equation (18.20) is very similar to the one of the two-pulse ESEEM modulation formula [equation (18.11)]. This is not a coincidence, since contribution V_I describes in fact a case in which the $t_1-\pi-t_2$ block of the HYSCORE sequence does not involve evolution of coherences. When we neglect relaxation, we can ignore the existence of this part of the sequence for our description, leaving a $\pi/2-\tau-\pi/2-\pi/2-\tau$–echo scheme, which is the two-pulse scheme. The term V_{IIa} describes the case where NP during t_1 is transferred to NC by the π pulse. This NC then evolves during time t_2, giving a modulation term that depends only on τ and t_2. Term V_{IIb} is due to the reverse situation in which NC is created by the preparation block but converted to NP by the π pulse. Both terms are irrelevant for the HYSCORE experiment (after two-dimensional FFT they lead to peaks on the two frequency axes), but the terms play a role in other experiments based on the four-pulse ESEEM scheme (vide infra, Section 18.3.3). The terms of interest for HYSCORE are V_{IIIa} and V_{IIIb}, representing the case where NC is evolving during t_1, is transferred to NC on another m_S manifold by the π pulse (Figure 18.7b (left)), and then evolves during t_2 with a different frequency. Fourier transformation leads to cross peaks linking the basic frequencies ω_α and ω_β. V_{IIIa} gives rise to cross peaks in the (+,+) quadrant (Figure 18.7b (right)) and becomes dominant in the case that η is close to 0 (*weak (hyperfine) coupling* case). Alternatively, V_{IIIb} is dominant when η is close to π/2 (*strong coupling*), and the corresponding cross peaks then appear in the (−,+) quadrant. In the *exact cancellation* condition, the cross peaks will show up in both quadrants. In this case, the cross peaks will be very intense, due to the large k value [see equation (18.16) and its discussion].

Since the four-pulse ESEEM scheme is based on the three-pulse ESEEM scheme, a similar product rule for the echo amplitude of the electron spin interacting with N nuclei is valid:

$$V_{4p,tot}(\tau,t_1,t_2)=\frac{1}{2}\left[\prod_{i=1}^{N}V_{4p,i}^{\alpha\beta}\left(\tau,t_1,t_2\right)+\prod_{i=1}^{N}V_{4p,i}^{\beta\alpha}\left(\tau,t_1,t_2\right)\right] \quad (18.25)$$

where $V_{4p}^{\alpha\beta}$ indicates the contributions in equations (18.21)–(18.24) with NC in the α manifold during t_1 and/or NC in the β manifold during t_2 and $V_{4p}^{\beta\alpha}$ indicates the opposite case. The above implies that for cases in which the electron spin is interacting with several nuclear spins, the relative sign of the hyperfine coupling can potentially be revealed by the observation of specific combination frequencies. In order to understand this, let us consider a three-spin system, $S=1/2$, $I_1=1/2$, $I_2=1/2$ ($g_{n,1}$, $g_{n,2}>0$) with basic frequencies $\omega_{a,1}<\omega_{b,1}$ and $\omega_{a,2}<\omega_{b,2}$ belonging to the interaction of S with I_1 and I_2, respectively. If both hyperfine couplings have the same sign, $\omega_{a,1}$ and $\omega_{a,2}$ will belong to the same m_S manifold (see Figure 18.2b), and combination frequencies $\omega_{a,1}\pm\omega_{a,2}$, and hence also $\omega_{b,1}\pm\omega_{b,2}$ will be observable due to the product rule in equation (18.25). In contrast, if the hyperfine couplings have opposite sign, $\omega_{a,1}\pm\omega_{b,2}$ and also $\omega_{b,1}\pm\omega_{a,2}$ will be observed. The intensity of the cross peaks involving these combination frequencies is usually low, and a good signal-to-noise ratio is needed to observe them.

For the most common case of disordered systems, such as frozen solutions, the cross peaks are no longer sharp features but broad ridges stemming from the sum of the individual contributions of all molecular orientations excited at a certain observer position. Figure 18.8(a) illustrates schematically the powder HYSCORE pattern that can be obtained for an $S=1/2$, $I=1/2$ system with isotropic g value (i.e., excitation of all orientations in space) and an axial hyperfine tensor described by an isotropic part, $a_{iso}/2\pi$, and a dipolar hyperfine part, $T/2\pi$. Here, the interaction with a proton was considered, and the principal hyperfine values were chosen to be [−3,−3,10] MHz, and a MW frequency in the X-band (9.47 GHz) was assumed ($g=2$, $B_0=338.3$ mT). Since we are dealing with a weak hyperfine coupling, the HYSCORE signals appear in the (+,+) quadrant. The correlation pattern consists of two arcs centered around (ν_H, ν_H) (ν_H being the proton Larmor frequency). The horizontal differences between the end points of the two arcs are $|2T+a_{iso}|/2\pi=10$ MHz and $|-T+a_{iso}|/2\pi=3$ MHz

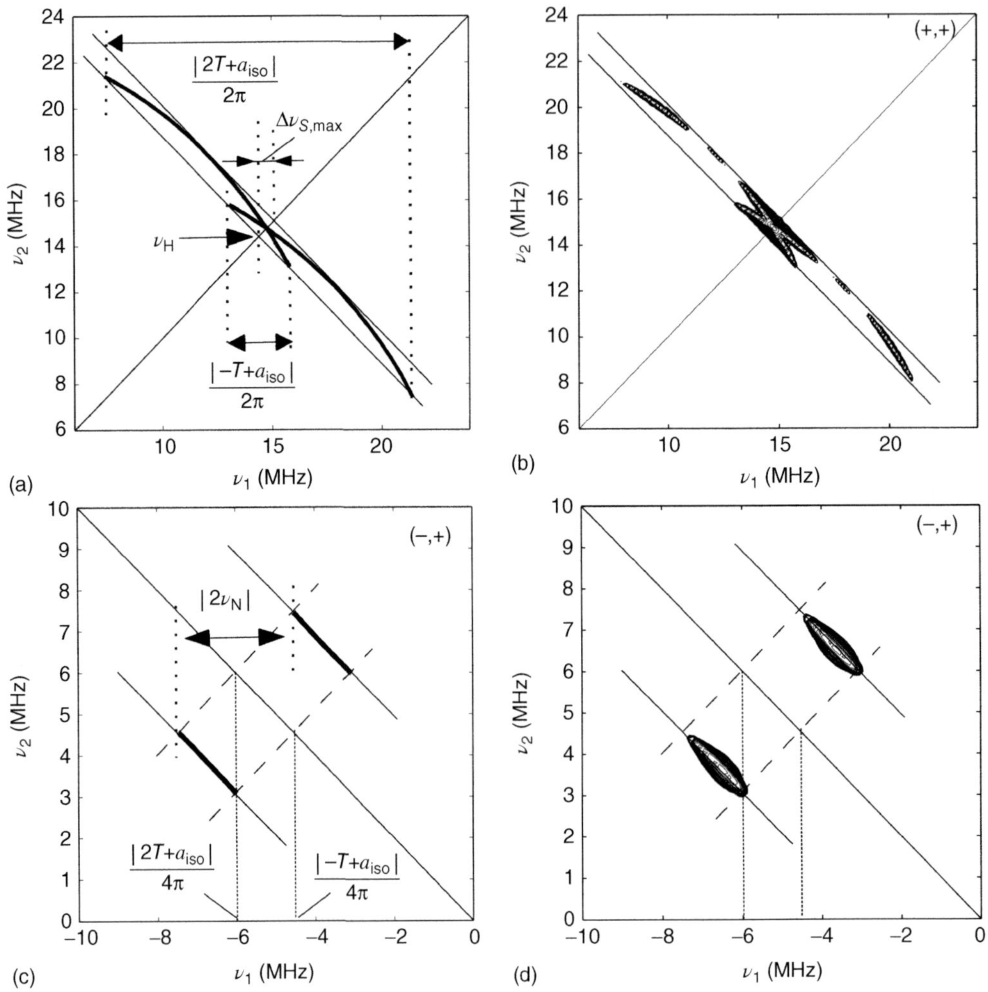

Figure 18.8. HYSCORE powder patterns for an $S = 1/2$, $I = 1/2$ spin system with $g = 2$ and a microwave frequency of 9.47 GHz (X-band). Sketch of appearance of HYSCORE signals in the case of (a) the weak hyperfine coupling case (^{1}H nucleus with principal hyperfine values [−3, −3, 10] MHz) and (c) the strong hyperfine coupling case (^{15}N nucleus with principal hyperfine values [9, 9, 12] MHz). (b,d) Simulation of the corresponding HYSCORE ridges for the cases shown schematically in (a,c) taking (b) $\tau = 176$ ns and (d) $\tau = 180$ ns. The effect of the τ-dependent blindspots is clearly visible in spectrum shown in panel (b)

and are in agreement with the hyperfine tensor values. The maximum horizontal frequency shift from the anti-diagonal at (ν_H, ν_H) is given by[4]

$$\Delta\nu_{S,\max} = \frac{9T^2}{32(2\pi)^2|\nu_I|} \tag{18.26}$$

Figure 18.8(b) shows the corresponding simulated HYSCORE spectrum of this case for a specific value of τ. The earlier-mentioned τ-dependent blindspots can be clearly seen. Hence, standard HYSCORE experiments should always be performed at several different τ values in order to avoid misinterpretations due to blindspot effects. Adversely, one can turn the blindspot effect into an advantage by deliberately suppressing dominating spectral features in order to analyze the weaker cross peaks. Figure 18.8(c) shows schematically the strong-coupling case here chosen for the ^{15}N nucleus at X-band with a hyperfine coupling of [9,9,12] MHz. We now see two ridges appear in the (−,+) quadrant that are separated by twice the Larmor frequency ν_N. The principal values of the hyperfine tensor can be derived as shown in Figure 18.8(c). Figure 18.8(d) shows the corresponding simulation for one specific τ value chosen now to minimize the blindspot effect. Figure 18.8(d) also illustrates an effect that we discussed earlier for the two-pulse ESEEM case, namely, the ESEEM effect disappears for the principal orientations of the hyperfine tensor. Because of this, the maximum signal intensity does not lie at the lower end of the HYSCORE ridges but more in the center of the cross peak. This is a similar effect as also found in Figure 18.3(c) for the two-pulse ESEEM spectrum.

So far, we considered the simplest spin system that can be encountered. However, in many cases, the electron spin is interacting also with high-spin nuclei ($I \geq 1$). The number of nuclear frequencies is then much larger than for the $S=1/2$, $I=1/2$ case, and, hence, the number of possible cross peaks in the HYSCORE spectra increases very rapidly. Let's consider the common example of an unpaired electron coupled to a ^{14}N ($I=1$) nucleus encountered in many metalloproteins and inorganic complexes. In each m_S manifold, two single-quantum (SQ) ($|\Delta m_I|=1$) nuclear transitions and one double-quantum (DQ) ($|\Delta m_I|=2$) nuclear transition will contribute to the ESEEM spectra. This then results in potentially $2\times 9=18$ cross peaks per quadrant in the HYSCORE spectra. However, as for the $S=1/2$, $I=1/2$ system, not all cross peaks will appear (equally intense) in the different quadrants. This is illustrated in Figure 18.9 for an ammonia-bound VO(II) ion in a mesoporous material. The most dominant cross peaks appear here in the (−,+) quadrant and stem from the DQ nuclear frequencies. In general, the quadrant in which the cross peaks between the highest multi-quantum frequencies are most intense identifies the hyperfine coupling case in a similar way as for the $S=1/2$, $I=1/2$ system. In the example at hand, we thus have a strong-coupling case (the hyperfine coupling will be larger than twice the Larmor frequency). The DQ nuclear frequencies can be used to estimate the spin Hamiltonian parameters using[10]

$$\nu_{\alpha,\beta}^{DQ} = 2\sqrt{\left(\nu_I \pm \frac{a/2\pi}{2}\right)^2 + \left(\frac{e^2qQ}{4h}\right)^2 (3+\eta_P^2)} \tag{18.27}$$

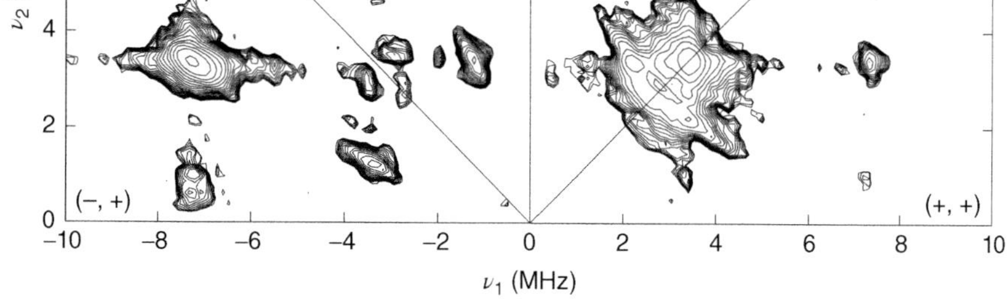

Figure 18.9. Example of an experimental X-band ^{14}N HYSCORE spectrum of reduced vanadium silicate-1 nanoparticles in SBA-15 treated with ammonia. The spectrum is recorded using a standard HYSCORE sequence with $t_\pi = 16$ ns, $t_{\pi/2} = 16$ ns, and $\tau = 176$ ns at an observer position where all orientations are excited (347.8 mT). The cross peaks stemming from the DQ nuclear frequencies are indicated

with η_P being the quadrupole asymmetry parameter and $a/2\pi$ the hyperfine coupling for the observer orientation. Equation (18.27) allows determining $a/2\pi$ and setting boundaries to the nuclear quadrupole coupling parameter $(e^2qQ)/h$, since $0 \leq \eta_P \leq 1$. These can then be used as first input for spectral simulations. From equation (18.27), it is clear that the nuclear quadrupole interaction is only contributing in second order to the DQ frequencies. The SQ ($|\Delta m_I| = 1$) frequencies depend in first order on the nuclear quadrupole tensor. This explains why the DQ cross peaks are less broad and thus stronger in intensity than the SQ ridges in the HYSCORE spectrum. In the example at hand, the SQ–SQ cross peaks are found in the [−4 − 4, 0−4] MHz region (Figure 18.9). These peaks are in this case less obvious to assign, but their interpretation can be facilitated using simulations with the starting values obtained from the analysis of the DQ–DQ cross peaks. More detailed information about HYSCORE of $S = 1/2$, $I \geq 1/2$ can be found in Refs 11,12.

18.3.2 Pros and Cons of Standard HYSCORE Spectroscopy

From the above, it is clear that the biggest advantage of HYSCORE spectroscopy is its two-dimensionality, allowing for a disentanglement of complex spectral features. However, this comes with a cost. Some of the drawbacks of standard HYSCORE spectroscopy include its long recording time due to the variation of two time intervals and the need to record spectra for different τ values due to the inherent blindspot effect. For similar reasons as outlined for the three-pulse ESEEM experiment, the necessity to use a phase cycle can sometimes also lead to an increase in recording time.

The π pulse plays an essential role in the HYSCORE experiment. The flip angle of this mixing pulse is in practice deviating from π due to off-resonance effects, leading to peaks on the diagonal in the HYSCORE spectrum because of incomplete transfer of NC. These peaks, in turn, can mask cross peaks due to small hyperfine couplings. As an example, one can see such diagonal peaks in the experimental spectrum shown in Figure 18.9 at positions $(\nu_\alpha^{DQ}, \nu_\alpha^{DQ})$ and $(\nu_\beta^{DQ}, \nu_\beta^{DQ})$. Because of this, it is advisable to use a π pulse with an as large as possible bandwidth, i.e., a short and strong pulse. The shortest length is spectrometer specific and depends on the maximum MW power. While the π pulse can still be easily made sufficiently strong at lower MW frequencies (S, X, Q band), this is more difficult to achieve at W-band frequency because of power limitations.[13,14]

Due to the larger number of MW pulses, the echo signal will be lower in the HYSCORE case than in the two-/three-pulse ESEEM case. Introduction of nonuniform sampling combined with maximum-entropy reconstruction may be helpful in case of low echo intensity.[15] Adapting this NMR approach to EPR allows shortening the experimental times of HYSCORE spectroscopy by an order of magnitude as compared to conventional linear sampling with no significant loss of information. This becomes important in cases of low-concentration samples (e.g., many metalloproteins) for which the spin echo is weak. A single-standard HYSCORE measurement then easily takes up to 24 h. This can be reduced to a fraction of this time by the nonuniform sampling.

Furthermore, the analysis of HYSCORE spectra of systems involving many nuclear spins and/or high-spin nuclei remains still very challenging. In fact, most of the published HYSCORE studies to date are related to the analysis of $^{1}H/^{2}H$, $^{14}N/^{15}N$, ^{13}C, ^{31}P couplings, whereby the provided references serve only as examples and are not exhaustive.[16–26] Examples of HYSCORE spectra involving $S = 1/2$, $I > 1$ or $S > 1/2$, $I \geq 1/2$ systems are in comparison quite limited.[12,16,19,20,27] Although sometimes valuable information can easily be obtained from a simple visual evaluation of a HYSCORE spectrum, in most cases a thorough HYSCORE analysis implies tedious simulations using appropriate simulation packages, like the very versatile EasySpin toolbox developed for Matlab users, which was used in this chapter for all simulations.[28] Often, HYSCORE analyses need to be complemented by experimental and theoretical methods, such as ENDOR, ELDOR-detected NMR, or DFT (see Chapters 7, 16 and 17), to extract the full spin Hamiltonian information.

18.3.3 Other Four-pulse ESEEM Schemes

The HYSCORE approach is not the only way how one can make use of the four-pulse ESEEM sequence in Figure 18.7(a). In a deadtime-free ESEEM by nuclear coherence-transfer echoes (DEFENCE) experiment, the time t_1 is kept constant and the time t_2 is varied.[9] The spectrum then represents the projection of the HYSCORE spectrum on the ν_2 axis, and, unlike three-pulse ESEEM, the technique allows measurement of nuclear-frequency spectrum without deadtime artifacts, because the coherence transfer described by $V_{\mathrm{IIIa,b}}$ [equations (18.23) and (18.24)] forms a coherence-transfer echo close to $t_1 = t_2$. As a consequence, with an appropriate choice of t_1, the modulation will not occur at the start of the time-domain signal, and deadtime problems can be largely circumvented.[4,9]

In combination-peak (CP) experiments, the times t_1 and t_2 are stepped simultaneously ($t_1 = t_2 = t$). The 1D-CP spectrum results in a projection of the HYSCORE spectrum on its diagonal.[29] It allows the detailed analysis of the combination frequencies $\omega_{\pm}$. This can, for instance, be easily seen from equations (18.23) and (18.24) when replacing t_1 and t_2 by t. For interactions with $I = 1/2$ nuclei, this allows facile evaluation of the hyperfine anisotropy B from the observation of the shift of ω_+ from $2\omega_I$ via[4]

$$\omega_+ = 2|\omega_I| + \frac{|\omega_I|B^2}{4\omega_I^2 - A^2} \qquad (18.28)$$

with the maximal shift being

$$(\omega_+)_{\max} = 2|\omega_I| + \frac{9T^2}{16|\omega_I|} \qquad (18.29)$$

Note that equation (18.29) can be immediately linked to equation (18.26) in the HYSCORE analysis. Each HYSCORE spectrum includes a 1D-CP spectrum (i.e., the Fourier transformation of the diagonal of the time-domain matrix) and many DEFENCE spectra (i.e., Fourier transform of individual columns of the time-domain matrix). It is, however, advisable to perform separate CP or DEFENCE experiments whereby the τ value is varied in a second dimension. Summation of the different 1D-CP or DEFENCE spectra (after Fourier transformation and computation of magnitude spectrum!) obtained for separate τ values is in any case needed to remove blindspots similar as in the three-pulse ESEEM experiment. Moreover, after 2D Fourier transformation (in t and τ) of the CP data, a blindspot-free 2D-CP spectrum can also obtained.[30] This contains in principle the same information as a HYSCORE spectrum and is recorded in a shorter time. However, the dependence on T_{m} ($T_{2\mathrm{e}}$) in the τ dimension reduces the resolution in the corresponding frequency domain, and this will be a limiting factor for systems with small phase-memory time.

18.4 IMPROVING THE PERFORMANCE OF ESEEM

18.4.1 Improving ESEEM Spectroscopy with Matched Pulses

Significant ESEEM sensitivity enhancement can be obtained by the use of matched MW pulses.[31–33] These pulses are high-turning-angle (HTA) pulses. The basic idea behind matched pulses stems from the finding that a semi-selective MW pulse, i.e., a pulse that simultaneously excites two transitions that have a common level, creates NC from electron polarization (EP), while this does not occur for an ideally selective or nonselective MW pulse.[4] There thus exists an optimal MW-field strength for the creation of NC. Keeping this in mind, in matched HYSCORE (Figure 18.10b), the second and fourth pulse are replaced by HTA pulses in order to optimize the creation of NC and then transfer back to EC. The optimal HTA conditions will differ depending on the system and interactions one wants to analyze. As a rule of thumb, the MW-field strength, ω_1, of the HTA pulse should be taken close to the Larmor frequency for weakly coupled nuclei to be enhanced. In case of strong coupling, ω_1 should be taken as strong as possible. Optimization of the pulse length is done via a matched three-pulse ESEEM experiment (Figure 18.10a). This is illustrated in Figure 18.10(d) for a $NOBF_4$-doped PPV-like oligomer. The value of ω_1 was chosen to match the proton Larmor frequency ($\omega_1/2\pi = 15.625$ MHz). The top graph shows the ESEEM intensity at the proton Larmor frequency, ν_{H}, as a function of the length of the HTA pulse. For the standard pulse lengths (16 ns), no proton signal could be resolved, while this changes when the HTA pulse length is increased. The bottom graph in Figure 18.10(d) shows the difference between the standard three-pulse ESEEM and the one recorded with an HTA pulse length of 40 ns. In the latter, a clear ESEEM signal centered around ν_{H} is observed. This illustrates that the use of matching pulses may not only enhance signals but may also be even required to observe the ESEEM effect. Matched HYSCORE presents also one of the options to overcome the earlier-mentioned cross-suppression effect.[8]

In single pulse matched resonance transfer (SMART) HYSCORE, the NC generator of the HYSCORE experiment ($\pi/2-\tau-\pi/2$) is replaced by a single HTA pulse (Figure 18.10c).[34] Since an HTA pulse can be seen as a series of $\pi/2$ pulses one

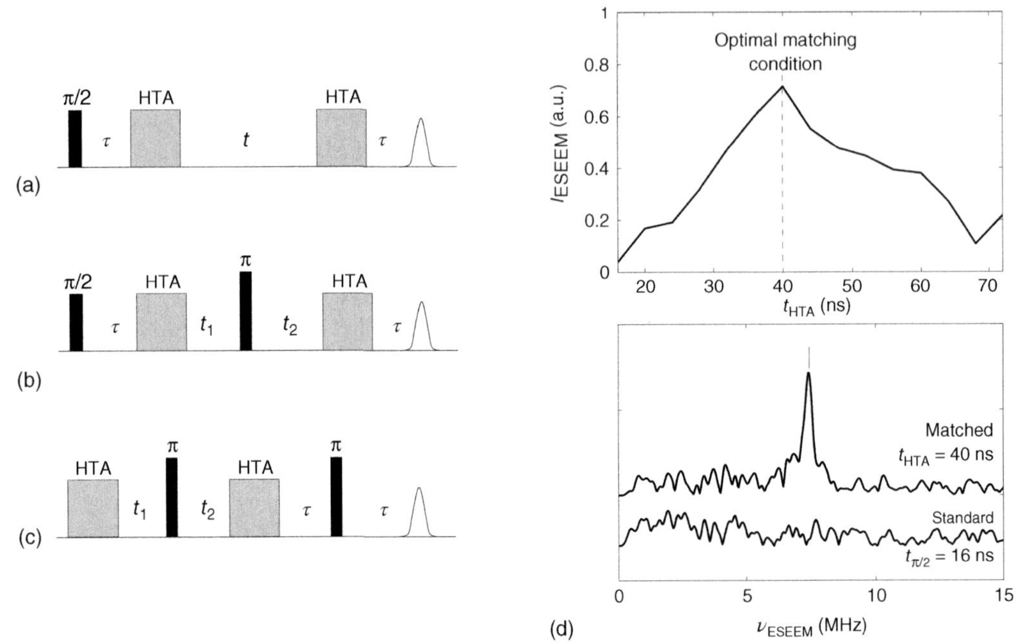

Figure 18.10. Improving ESEEM using matched pulses. Pulse sequences of (a) matched three-pulse ESEEM, (b) matched HYSCORE, and (c) SMART-HYSCORE spectroscopy. (d) Experimental example of the optimization of the HTA pulse in a matched three-pulse ESEEM experiment recorded of a $NOBF_4$-doped PPV-like oligomer at a magnetic-field setting of 348.2 mT. $\omega_1/2\pi$ was taken as 15.625 MHz ($\approx\nu_H$). The top graph shows the intensity of the ESEEM peak at the proton Larmor frequency, ν_H, as a function of the length of the HTA pulse. The optimal pulse length is here clearly 40 ns. The bottom figure shows the corresponding ESEEM spectra. While the standard three-pulse ESEEM shows no proton contributions, a clear signal appears in the matched three-pulse ESEEM

directly after the other, the first three pulses in the SMART-HYSCORE sequence can be considered as the sum of different standard HYSCORE sequences with different τ values and make the experiment virtually blindspot-free. However, this also means that the echoes of these different individual standard HYSCORE sequences will appear at different times during and after the second HTA pulse, hence the need for the last π pulse to induce a refocusing of the signal. Note that in many cases, SMART-HYSCORE spectra may need to be symmetrized after Fourier transformation, since quite often a strong asymmetry about the diagonal is found. In fact, all HYSCORE experiments are prone to signal asymmetry about the diagonal due to the fact that the NC generator does not generate NC equally well on all nuclear transitions. The weight of the coherence-transfer pathways depends on the type of the NC generator (HTA pulse in the SMART-HYSCORE, a $\pi/2-\tau-\pi/2$ sequence in the standard HYSCORE) and on the spin system determining the weight factors of the transfers. In general, symmetrization of HYSCORE spectra can be a way to circumvent this problem and further increase signal-to-noise ratio, but one should always be aware that artificial cross peaks may be induced by this process. Indeed, noise peaks will also be symmetrized and, hence, they will be hard to distinguish from genuine cross peaks in case of noisy spectra.

18.4.2 Six-pulse HYSCORE Spectroscopy

The earlier-mentioned cross-suppression effect in standard HYSCORE experiments can also be largely circumvented using six-pulse HYSCORE (Figure 18.11a).[35] A corresponding five-pulse ESEEM missing the second NC→NC transferring π pulse exists as alternative sequence for standard three-pulse ESEEM. In the five-pulse ESEEM and six-pulse HYSCORE, the reduction of the SQ peak intensities of weakly modulating nuclei in the presence of strongly modulating nuclei is less severe than in the corresponding standard ESEEM and HYSCORE experiments. Although five-pulse ESEEM and six-pulse HYSCORE also suffer from blindspots, it can be shown that taking $\tau_1 = \tau_2$ is usually sufficient to average them out.[35] The product rules given in equations (18.18) and (18.25) for three- and four-pulse ESEEM experiments result in different internuclear combination frequencies that will further complicate the analysis of these standard ESEEM spectra. In the case of five-pulse ESEEM and six-pulse HYSCORE, these combination frequencies are absent. Hence, these methods offer solutions for a number of problems encountered in the standard ESEEM experiments. However, as always, they also come with a cost. As mentioned earlier, due to the non-ideality of the pulses, every addition of an extra MW pulse to a sequence also implies loss of signal intensity. This may become an issue for very low spin concentrations.

18.4.3 DONUT-HYSCORE Spectroscopy

In order to facilitate the assignment of the different cross peaks in a HYSCORE experiment, it may be helpful to record a double nuclear coherence transfer (DONUT) HYSCORE spectrum (Figure 18.11b).[36] Compared to the HYSCORE experiment, the mixing block in DONUT HYSCORE is now $\pi - \tau_2 - \pi$, which means effectively a double transfer of NC between the different manifolds. This leads to correlations between nuclear frequencies *within* m_S manifolds. This sequence is completely useless for a simple $S = 1/2$, $I = 1/2$ system, since it will generate two peaks on the diagonal at (ν_α,ν_α) and (ν_β,ν_β) (two π pulses bounce NC from one manifold to the other and back again). However, the experiment becomes interesting for multinuclear systems or for systems where the electron spin is interacting with $I > 1/2$.

Assume the following case where we detect a system of two organic radicals with overlapping EPR signals. In this case, the HYSCORE experiment will contain contributions of both radicals. Assume that we detect the HYSCORE spectrum depicted schematically in Figure 18.11(c) (top left). We see two pairs of cross peaks (numbered 1 and 2 in the figure) stemming from the interaction with two nuclear spins I_1 and I_2. Performing a subsequent DONUT-HYSCORE experiment will help to discern the following cases. If

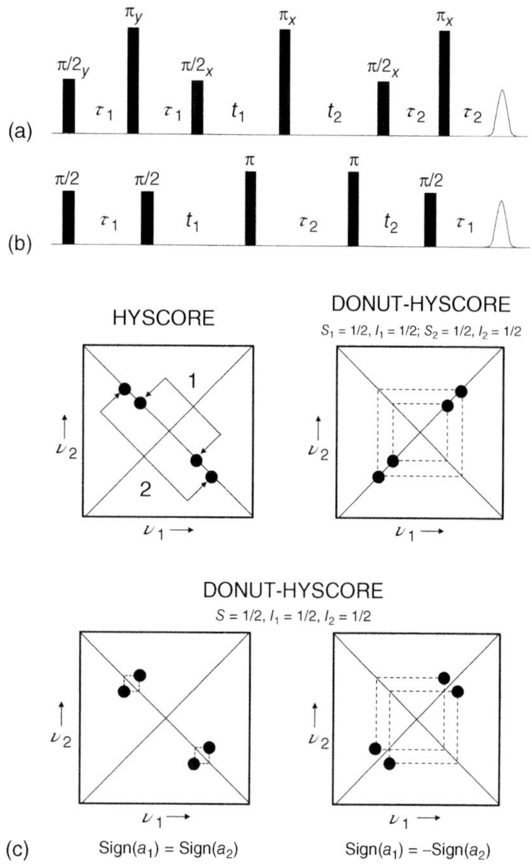

Figure 18.11. ESEEM experiments with more than four pulses. Pulse sequences of (a) six-pulse HYSCORE and (b) DONUT-HYSCORE spectroscopy. (c) Schematic representation of a HYSCORE spectrum that could stem both from a three-spin $S = 1/2$, $I_1 = 1/2$, $I_2 = 1/2$ system or from two two-spin systems ($S_1 = 1/2$, $I_1 = 1/2$; $S_2 = 1/2$, $I_2 = 1/2$) with overlapping EPR spectra (upper left). The corresponding DONUT-HYSCORE spectra are represented for the overlapping two-spin systems (upper right) and for two options of the three-spin system, i.e., with the same sign (bottom left) or opposite sign (bottom right) for the two hyperfine coupling constants a_1 and a_2 with the nuclei I_1 and I_2, respectively

the HYSCORE spectrum results from the sum of the HYSCORE contribution of two radicals that are individual $S = 1/2$, $I = 1/2$ systems, i.e., I_1 and I_2 belong to different radicals, the DONUT-HYSCORE spectrum will only show diagonal peaks (Figure 18.11c, upper right). If the two nuclear spins belong to one of the two radicals, off-diagonal cross peaks will also appear in the DONUT-HYSCORE spectrum (Figure 18.11c, bottom). These cross peaks result from different coherence-transfer pathways, such as the one in which NC of nucleus 1 is transferred from the α manifold to the β manifold by the first π pulse, while the second π pulse transfers this to NC of nucleus 2 on the α manifold. Depending on the relative sign of the corresponding hyperfine interactions $a_{1,2}/h$, the cross peaks will appear at different positions in the spectrum, since only cross peaks within one manifold are possible. This is illustrated in Figure 18.11(c) (bottom). We earlier saw in Section 18.3.1 that HYSCORE also provides a way of determining the relative sign of different hyperfine couplings. However, in the latter case, this is based on (often) weak cross peaks linking combination frequencies. The advantage of the DONUT-HYSCORE approach is that this determination is now based on cross peaks between basic frequencies.

Besides the off-diagonal cross peaks, the spectrum will also show peaks on the diagonal. These are, however, uninformative and left out of the scheme in Figure 18.11(c) (bottom).

DONUT HYSCORE can also be very informative when studying $S = 1/2$, $I > 1/2$ systems, since one m_S manifold then already harbors multiple nuclear transitions, and the mixing block can link these frequencies. Combination with the HYSCORE data allows assignment of the different nuclear frequencies.[36] This is illustrated in Figure 18.12. We here compare the HYSCORE (Figure 18.12a) and DONUT-HYSCORE (Figure 18.12b) spectrum of a concentrated frozen solution of a cobaltous tetraphenyl porphyrin (pyridine) complex (CoTPP(py)) in toluene taken at the same magnetic-field position. In the HYSCORE spectrum, we can identify cross peaks stemming from the porphyrin ^{14}N nuclei (e.g., the peaks labeled 1–3) of the Co(II)TPP(py) fragment and even from the porphyrin ligand of an oxidized $[Co(III)TPP(py)]^+$ unit that is π–π stacked with this complex (e.g., cross peak 4).[37,38] Some of the HYSCORE peaks are retained in the DONUT-HYSCORE spectrum (peaks 1–4). This is due to the imperfection of the π pulses. If one of the two π pulses is not affecting the spin system, a HYSCORE instead of a DONUT-HYSCORE peak will occur. It is therefore paramount to compare each DONUT-HYSCORE experiment carefully with its corresponding HYSCORE spectrum in order to identify residual HYSCORE peaks. Furthermore, we notice the appearance of strong cross peaks on the anti-diagonal of the DONUT-HYSCORE spectrum. They arise from the coherence pathways in which we generate NC of frequency $\nu^i_{\alpha,\beta}$ that is transferred by the first π pulse in NC of frequency $\nu^j_{\beta,\alpha}$ and by the second π pulse back into NC of the same initial frequency $\nu^i_{\alpha,\beta}$ (with *i,j* referring to a specific nuclear transition). These peaks will be always abundant in the DONUT-HYSCORE spectrum and are of little practical use. Inspection of Figure 18.12(b) shows, however, that also new cross peaks are found. We observe, for instance, cross peaks at (−1.7, 3.4) MHz (peak 5), (−0.6, 3.4) MHz (peak 6), (−0.6, 1.7) MHz (peak 7), and (0.6, 1.1) MHz (peak 8). DONUT-HYSCORE peaks only occur between frequencies within the same m_S manifold, whereby the DQ frequency will be the sum of the two SQ frequencies. We also know for the example at hand that there are four magnetically equivalent porphyrin nitrogens contributing to the spectrum, so we can potentially also observe combination frequencies. Combining this knowledge leads to the assignment that $\nu_\beta^{SQ,1} = 0.6$ MHz, $\nu_\beta^{SQ,2} = 1.1$ MHz, $\nu_\beta^{DQ} = \nu_\beta^{SQ,1} + \nu_\beta^{SQ,2} = 1.7$ MHz, and $2\nu_\beta^{DQ} = 3.4$ MHz. The latter frequency is a combination frequency. In the assignment of these frequencies to the β manifold, we assumed negative hyperfine couplings; the absolute sign of the hyperfine couplings is not obtainable from this experiment. HYSCORE peaks 1 (at (−3.4, 5.4) MHz) and 2 (at (−1.7, 5.4) MHz) are then identified as $\nu_\alpha^{DQ} = 5.4$ MHz. This illustrates the usefulness of DONUT HYSCORE.

Note, however, that similar to HYSCORE, DONUT-HYSCORE experiments suffer from blindspots and acquisition is time consuming. Furthermore, due to the fact that each additional pulse will cause signal loss, a DONUT-HYSCORE spectrum tends to have a worse signal-to-noise ratio than the corresponding HYSCORE spectrum for the same settings.

18.4.4 Decoupling ESEEM

Assignment of the nuclear frequencies to a specific type of nucleus is not always straightforward. Examples of such hampered analyses occur, for instance, if only one of the nuclear frequencies is identified in the spectrum or if we are dealing with $S > 1/2$ systems. To illustrate the latter, consider a high-spin system with a half-integer spin. If the zero-field splitting is significantly larger than the MW quantum at X-band, the X-band EPR spectrum will

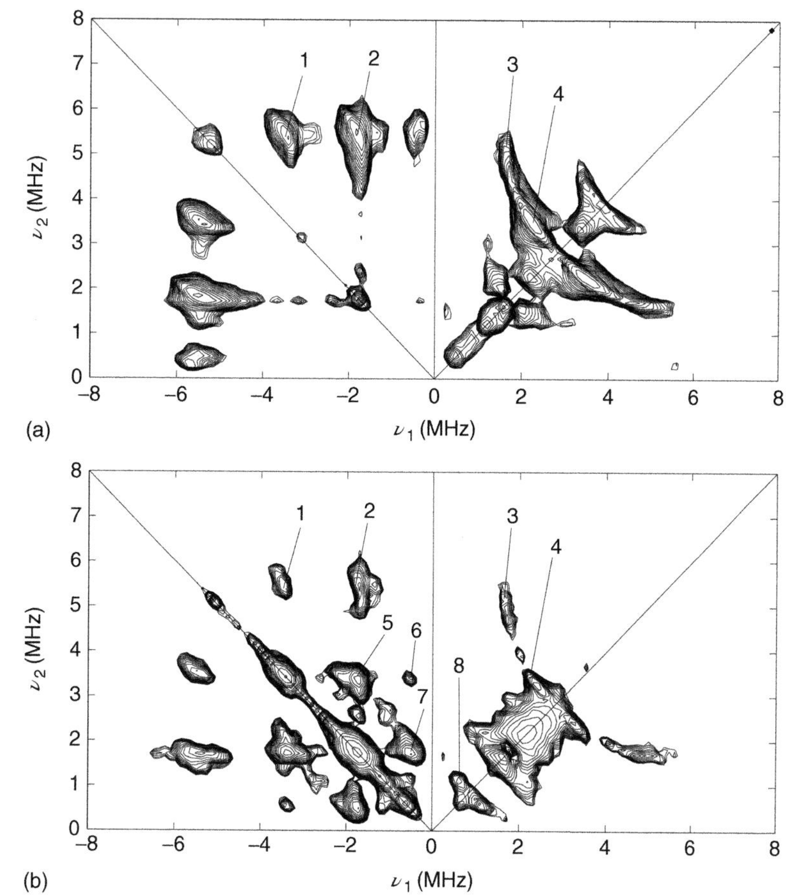

Figure 18.12. Comparison between the experimental X-band HYSCORE (a) and DONUT-HYSCORE (b) spectrum of a frozen toluene solution of 10 mM CoTPP(py) in air. Both spectra are recorded at 360 mT. The top spectrum is recorded using a standard HYSCORE sequence with $t_\pi = 16$ ns, $t_{\pi/2} = 24$ ns, and $\tau = 96$ ns. The bottom DONUT-HYSCORE spectrum is recorded with $t_\pi = 16$ ns, $t_{\pi/2} = 24$ ns, $\tau_1 = 96$ ns, and $\tau_2 = 304$ ns. The numbering of the peaks is explained in the main text

arise from the transitions of the lowest Kramers doublet only and can be described by an *effective* electron spin $^1/_2$ with a related *effective* spin Hamiltonian that is formally identical to the one of a standard $S = 1/2$ system but with effective spin Hamiltonian parameters that deviate from the real ones. This is also the case for the nuclear Larmor term; hence the effective g_n value (and thus corresponding ω_I) will deviate from the intrinsic g_n value, and ESEEM interpretation will become difficult (see Ref. 20 for an example). In these cases, a technique able to link nuclear frequencies to the corresponding ω_I is of high value. This can be achieved by so-called hyperfine-decoupling ESEEM experiments (Figure 18.13).[39,40] These experiments rely on the use of decoupling pulses (i.e., MW pulses with ω_1 as high as possible and with their length being varied during the experiment), which decouple the electron spin S from the nuclear spin I. Under this strong resonant MW irradiation, the quantization axis of S is rotating with resonance (Larmor) frequency ω_S in the xy plane of the laboratory frame. This implies that the nuclear spins will sense a time-dependent local electron spin-induced field. For times longer than $(\omega_S/2\pi)^{-1}$, this field will average out, decoupling the electron spin from the nuclear spin. Full decoupling will, however, only occur for on-resonant spin packets. In a 'real' case, the off-resonance component will result in a residual hyperfine coupling. For an

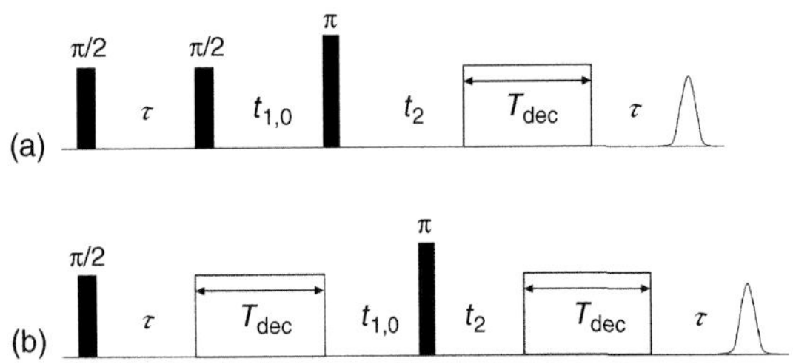

Figure 18.13. Hyperfine-decoupled DEFENCE. (a) Original pulse sequence introduced in Ref. 30 and (b) new pulse sequence allowing elimination of the residual hyperfine splitting as was introduced in Ref. 40

$S = 1/2$, $I = 1/2$ case, the decoupled frequency is given by[32]

$$\omega_{\alpha,\beta}^{\mathrm{dec}} = \omega_I \pm \frac{A\Delta\omega}{2\omega_1} - \frac{B((\Delta\omega)^2 - \omega_I^2)}{8\omega_1^2\omega_I} \qquad (18.30)$$

with the second term being the residual hyperfine splitting, $\Delta\omega = \omega_S - \omega_{MW}$ the resonance offset frequency from the electron Zeeman frequency ω_S, and A and B as defined earlier [see equation (18.1)].

Similarly, decoupling in case of $S = 1/2$, $I > 1/2$ systems leads to ω^{dec} frequencies that depend on the Larmor frequency and the nuclear quadrupole coupling only.[40]

Figure 18.13 shows two possible hyperfine-decoupling DEFENCE schemes.[39,40] Both are 2D experiments. Both schemes have the advantage of inducing a deadtime-free spectrum along the ESEEM dimension (DEFENCE dimension, variation of t_2). In the second dimension, the length of the decoupling pulse(s) is varied, leading to a time-domain signal that is modulated by the decoupled nuclear frequencies. After 2D Fourier transformation, the correlation peaks between the nuclear frequencies and the decoupled frequencies significantly simplify ESEEM peak assignments. The advantage of the scheme in Figure 18.13(b) over the one in Figure 18.13(a) lies in the simultaneous variation of the length of two decoupling pulses. Similar to the case of the CP experiments (vide supra, Section 18.3.3), this approach leads to spectra in which the basic nuclear frequencies are coupled not only to the basic decoupled frequencies, $\omega_{\alpha,\beta}^{\mathrm{dec}}$, but also to their combination frequencies, $\omega_{\pm}^{\mathrm{dec}} = \omega_{\alpha}^{\mathrm{dec}} \pm \omega_{\beta}^{\mathrm{dec}}$. The latter frequencies depend less on the residual hyperfine couplings, which leads to more defined cross peaks and hence better interpretations. This difference between the two-pulse sequences in Figure 18.13 is illustrated in Figure 18.14 for a polycrystalline bis(n-benzene)vanadium(0) sample. In this case, interactions with both ^{1}H and ^{13}C nuclei can be detected by ESEEM. The assignment of the ESEEM peaks to the two types of nuclei is immediately evident from the cross peaks at the Larmor frequencies, ν_H and ν_C, in the decoupling (ν_{dec}) dimension. These cross peaks are the only ones observed if the sequence in Figure 18.13(a) is used, and the large width of the peaks in the decoupling dimension is due to residual hyperfine couplings (Figure 18.14a). If the pulse sequence of Figure 18.13(b) is used, additional narrower cross peaks at twice the ^{1}H and ^{13}C Larmor frequency are observed in the decoupling dimension (Figure 18.14b).[40]

18.4.5 Relaxation-filtered ESEEM

When ESEEM experiments are performed for two paramagnetic compounds with strongly overlapping EPR spectra, it is very difficult, if not impossible, to separate the corresponding overlapping ESEEM spectra. As seen in Section 18.4.3, DONUT HYSCORE may help. Also, when the two paramagnetic compounds have different longitudinal relaxation times, T_{1e}, as is the case, for instance, for an organic radical in a protein with a paramagnetic metal center, spectral separation can also be achieved by making use of the REFINE filter.[41] This filter consists of an inversion π pulse or train of π pulses followed by a time T_F after which the ESEEM experiment of choice is performed (Figure 18.15a). The key idea behind the experiment is shown schematically in Figure 18.15(a) for the case of two paramagnetic compounds (1) and (2) with $T_{1e}^{(1)} < T_{1e}^{(2)}$. The initial π pulse(s) invert(s) the electron magnetization of both compounds. Due to the difference in T_{1e}, the magnetization of (1) and (2) will return to equilibrium with different rates. At time $T_F = t_I$, the magnetization of the compound (1) is in the xy plane ($M_z = 0$). If an ESEEM experiment is started at this time, the resulting ESEEM trace will not contain any contribution stemming from center (1). Indeed, all ESEEM experiments rely on a preparation block (mostly starting with a π/2 pulse) that turns EP into coherences. However, at $T_F = t_I$ this π/2 pulse will turn the magnetization of center (1) to the z axis instead of keeping it in a coherence state, and any ESEEM contribution will then stem from center (2). Similarly, the contributions of center (2) can be suppressed at time $T_F = t_{II}$. The REFINE filter can also be combined with other pulse sequences, such as the pulse

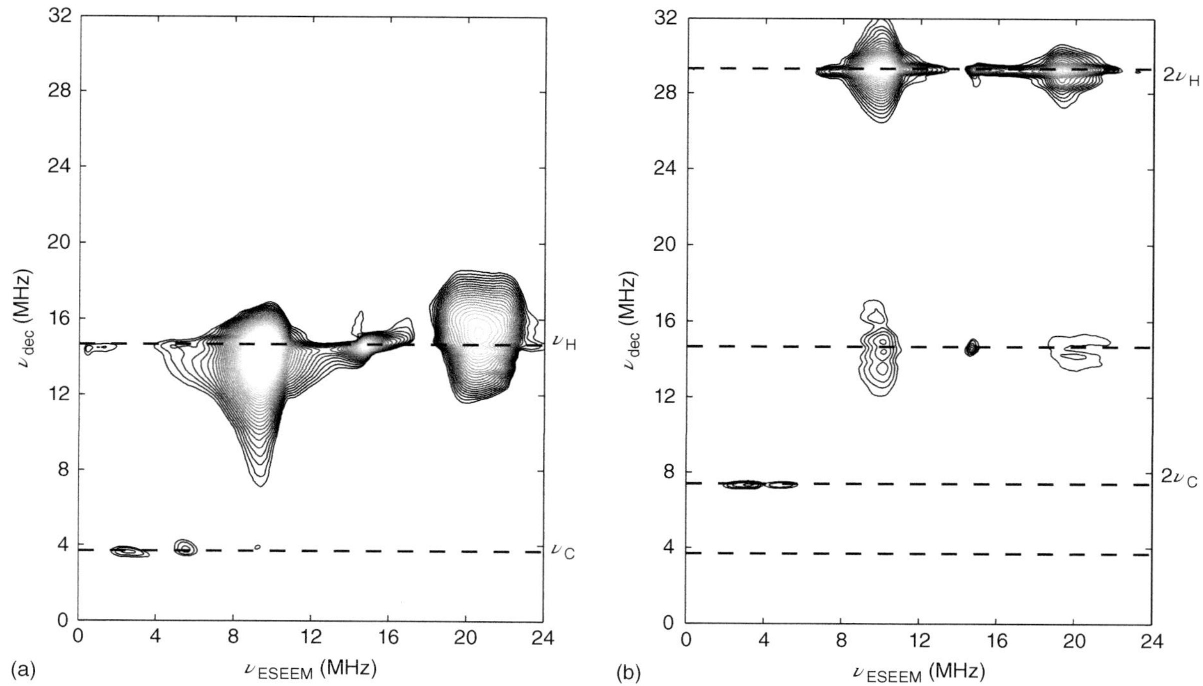

Figure 18.14. Experimental hyperfine-decoupling experiments on polycrystalline bis(n-benzene)vanadium(0) at X-band.[40] Experiment using (a) the hyperfine-decoupled DEFENCE of Figure 18.13(a) and (b) the hyperfine-decoupled DEFENCE of Figure 18.13(b). (Reprinted from J. Magn. Reson., 168, Hyperfine decoupling in electron paramagnetic resonance as a powerful tool for unraveling complicated ESEEM spectra of $S = 1/2$, $I \geq 1/2$ systems, 88. © 2004, with permission from Elsevier)

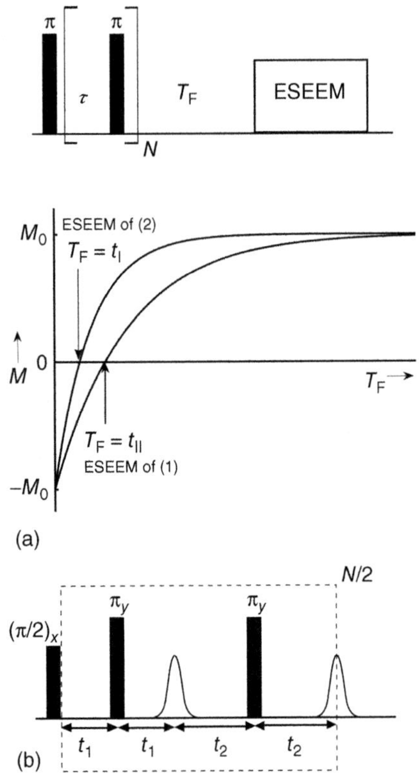

Figure 18.15. Optimizing ESEEM schemes. (a) Schematic representation outlining the basic idea behind REFINE-filtered ESEEM. (1) and (2) refer to the two paramagnetic compounds with overlapping EPR spectrum contributing to the ESE at the magnetic-field position of measurement. M_z is the z component of the magnetization, which has a value of M_0 at equilibrium. (b) Pulse-train-based 2D ESEEM sequence with enhanced ESEEM modulation depth. $N=2$, 4, 6, ... For $N=2$, the sequence is the refocused primary ESEEM method

ENDOR sequences (see Chapter 16). The coupled REFINE–ESEEM experiment can be performed in a 2D manner by varying independently the ESEEM time domain and the time T_F. Fourier transformation along the ESEEM dimension and inverse Laplace transform along the T_F dimension allows disentanglement of the ESEEM contributions of species with strongly differing T_{1e} times. This is illustrated in Figure 18.16 for a mixture of 2,2,6,6-tetra-methyl-piperidine-1-oxyl (TEMPO), α,β-bisphenylene-β-phenylallyl-benzolate (BDPA), and a perdeuterated copper–histidine complex (CuHis) that have clearly differing electronic spin–lattice relaxations.[42] Three sets of ESEEM patterns can be observed corresponding with the relaxation rates of the contributing species. These ESEEM spectra excellently agree with those of the pure model compounds.

18.4.6 ESEEM Modulation Depth Enhancement with Pulse Trains

As was discussed already in Section 18.2.1, one of the limits for the detection of the ESEEM effect is the size of the modulation depth k. A novel 2D scheme based on the refocused primary echo gives remarkable enhancement of k (Figure 18.15b).[43] In this experiment, t_1 and t_2 are varied independently. A detailed derivation of the echo-modulation formula for an $S=1/2, I=1/2$ system and $N=2$ shows that Fourier transformation leads not only to HYSCORE-like cross peaks between the basic frequencies, e.g., $(\omega_\alpha,\omega_\beta)$, but also to 2D-CP-like cross peaks, e.g., $(\omega_{\alpha,\beta},\omega_\pm)$.[43] With an optimal choice of N, this method increases intensity of the basic correlation peaks $(\omega_\alpha,\omega_\beta)$ and $(\omega_\beta,\omega_\alpha)$ and gives blindspot-free patterns. Furthermore, no phase cycling is required for this experiment. As mentioned in Section 18.3.2, non-ideality of the π pulse in HYSCORE experiment leads to diagonal artifact peaks. This 2D technique also solves this problem.[43]

The 1D version for $t_1=t_2$, the so-called CPMG sequence, selectively enhances the CP at ω_+ and its higher harmonics. This allows better determination of the anisotropic hyperfine coupling, T, through[43]

$$(\omega_+)_{\max} = 2|\omega_I| + \frac{9T^2}{16|\omega_I|} \tag{18.31}$$

Hence, the CPMG sequence gives an alternative to the 1D-CP experiment discussed in Section 18.3.3.

Note that the CPMG sequence, well known from NMR, has many other advantages, such as its use to increase sensitivity of various pulse EPR experiments.[44] This is treated also in other chapters (see Chapter 22).

18.4.7 The Next Generation of ESEEM

As mentioned earlier, all ESEEM experiments described in the above sections suffer from a limited excitation bandwidth of the standard MW pulses, which limits the magnitude of the nuclear

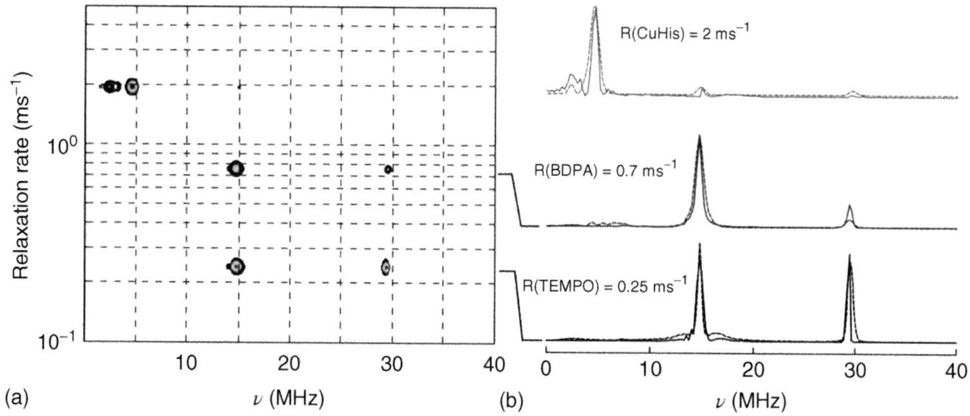

Figure 18.16. Example of a 2D-REFINE experiment (two-pulse ESEEM scheme was used) for a mixture of TEMPO, BDPA, and copper-histidine complex. (a) Contour plot of the spectrum obtained after Fourier transformation along the τ dimension and inverse Laplace transform along the T_F dimension. (b) Individual slices taken from the contour plot at the indicated relaxation times (solid lines) compared to the obtained two-pulse ESEEM spectra of the pure model compounds (dashed lines).[42] (Reprinted from J. Magn. Reson., 192, 2D-REFINE spectroscopy: Separation of overlapping hyperfine spectra, 78. © 2008, with permission from Elsevier)

frequencies that can be detected and, often, induces strong orientation-selection issues. Indeed, standard high-power rectangular pulses of typical lengths of 10 ns have an excitation width of ≈120 MHz, which is very small compared with the spectral width of many EPR spectra that can extend over several gigahertz. One of the consequences is that these pulses cannot excite all the hyperfine lines in the case of large hyperfine couplings. While standard ESEEM schemes allow detection of the hyperfine couplings of the weakly coupled nuclei, they will seldom provide information about the large hyperfine interactions of the central transition metal or the most strongly bound ligand nuclei. Recent development allowed overcoming this limitation by using frequency-swept chirp pulses with ultra-wideband (UWB) excitation provided by arbitrary-wave generators (AWG) (see Chapter 21).[45] For a bispicolinate ^{63}Cu(II) complex in a diamagnetic host, ESEEM frequencies far beyond the usual proton frequency area (<50 MHz) can be observed (frequencies up to 300 MHz) using UWB pulses, allowing detection and analysis of the ^{63}Cu hyperfine interaction. Furthermore, the full digitization of the echo transient provides an additional Fourier transformation EPR dimension, facilitating 3D EPR/HYSCORE experiments. While 3D pulse techniques are by now common in NMR, this was a first in pulsed EPR.[45] The outcome of the experiment gives the HYSCORE spectra versus the EPR spectrum.

ACKNOWLEDGMENTS

Bert Cuypers is thanked for the help with the figure preparations.

REFERENCES

1. L. G. Rowan, E. L. Hahn, and W. B. Mims, *Phys. Rev.*, 1965, **A137**, 61.
2. W. B. Mims, *Phys. Rev.*, 1972, **B5**, 2409.
3. P. Höfer, A. Grupp, H. Nebenführ, and M. Mehring, *Chem. Phys. Lett.*, 1986, **132**, 279.
4. A. Schweiger and G. Jeschke, Principles of Pulse Electron Paramagnetic Resonance, Oxford University Press: Oxford, 2001.
5. S. A. Dikanov and Y. D. Tsvetkov, Electron Spin Echo Envelope Modulation (ESEEM) Spectroscopy, CRC Press: Boca Raton, 1992.
6. A. Schweiger, in Modern Pulsed and Continuous-Wave Electron Spin Resonance, eds L. Kevan and M. K. Bowman, John Wiley & Sons, Inc: New York, 1990, Chap. 2.
7. S. Van Doorslaer, G. A. Sierra, and A. Schweiger, *J. Magn. Reson.*, 1999, **136**, 152.
8. S. Stoll, C. Calle, G. Mitrikas, and A. Schweiger, *J. Magn. Reson.*, 2005, **177**, 93.

9. A. Ponti and A. Schweiger, *J. Chem. Phys.*, 1995, **102**, 5207.
10. S. A. Dikanov, Y. D. Tsvetkov, M. K. Bowman, and A. V. Astashkin, *Chem. Phys. Lett.*, 1982, **90**, 149.
11. A. G. Maryasov and M. K. Bowman, *J. Magn. Reson.*, 2006, **179**, 120.
12. M. Gutjahr, R. Böttcher, and A. Pöppl, *Appl. Magn. Reson.*, 2002, **22**, 401.
13. P. A. S. Cruickshank, D. R. Bolton, D. A. Robertson, R. I. Hunter, R. J. Wylde, and G. M. Smith, *Rev. Sci. Instrum.*, 2009, **80**, 103102.
14. D. Goldfarb, Y. Lipkin, A. Potapov, Y. Gordetsky, B. Epel, A. M. Raitsimring, M. Radoul, and I. Kaminker, *J. Magn. Reson.*, 2008, **194**, 8.
15. K. K. Nakka, Y. A. Tesiram, I. M. Brereton, M. Nobli, and J. R. Harmer, *Phys. Chem. Chem. Phys.*, 2014, **16**, 16378.
16. J. Harmer, C. Finazzo, R. Piskorski, C. Bauer, B. Jaun, E. C. Duin, M. Goenrich, R. K. Thauer, S. Van Doorslaer, and A. Schweiger, *J. Am. Chem. Soc.*, 2005, **127**, 17744.
17. M. Brecht, M. Van Gastel, T. Buhrke, B. Friedrich, and W. Lubitz, *J. Am. Chem. Soc.*, 2003, **125**, 83.
18. D. Nakazawa, A. Ishii, J. Minagawa, and A. T. Ono, *Chem. Phys. Lett.*, 2005, **405**, 318.
19. J. McCracken, B. E. Eser, D. Mannikko, M. D. Krzyaniak, and P. F. Fitzpatrick, *Biochemistry*, 2015, **54**, 3759.
20. M. Fittipaldi, I. García-Rubio, F. Trandafir, I. Gromov, A. Schweiger, A. Bouwen, and S. Van Doorslaer, *J. Phys. Chem. B*, 2008, **112**, 3859.
21. B. K. S. Rao, A. M. Tyryshkin, A. G. Roberts, M. K. Bowman, and D. M. Kramer, *Biochemistry*, 2000, **39**, 3285.
22. S. Kababya, J. Nelson, C. Calle, F. Neese, and D. Goldfarb, *J. Am. Chem. Soc.*, 2006, **128**, 2017.
23. S. Grimaldi, F. MacMillan, T. Ostermann, B. Ludwig, H. Michel, and T. Prisner, *Biochemistry*, 2001, **40**, 1037.
24. M. Fittipaldi, H. J. Wijma, M. P. Verbeet, G. W. Canters, E. J. J. Groenen, and M. Huber, *Biochemistry*, 2005, **46**, 15193.
25. M. M. Roessler, R. M. Evans, R. A. Davies, J. Harmer, and F. A. Armstrong, *J. Am. Chem. Soc.*, 2012, **134**, 15581.
26. A. Adamska-Venkatesh, S. Roy, J. F. Siebel, T. R. Simmons, M. Fontecave, V. Artero, E. Reijerse, and W. Lubitz, *J. Am. Chem. Soc.*, 2015, **137**, 12744.
27. S. Maurelli, E. Morra, S. Van Doorslaer, V. Busico, and M. Chiesa, *Phys. Chem. Chem. Phys.*, 2014, **16**, 19625.
28. S. Stoll and A. Schweiger, *J. Magn. Reson.*, 2006, **178**, 42.
29. A. Schweiger, *Angew. Chem. Int. Ed.*, 1991, **30**, 265.
30. S. Van Doorslaer and A. Schweiger, *Chem. Phys. Lett.*, 1997, **281**, 297.
31. G. Jeschke and A. Schweiger, *J. Chem. Phys.*, 1996, **105**, 2199.
32. G. Jeschke and A. Schweiger, *Mol. Phys.*, 1996, **88**, 355.
33. G. Jeschke, R. Rakhmatullin, and A. Schweiger, *J. Magn. Reson.*, 1998, **131**, 261.
34. L. Liesum and A. Schweiger, *J. Chem. Phys.*, 2001, **114**, 9478.
35. B. Kasumaj and S. Stoll, *J. Magn. Reson.*, 2008, **190**, 233.
36. D. Goldfarb, V. Kofoman, L. Libman, A. Schanzer, R. Rahmatouline, S. Van Doorslaer, and A. Schweiger, *J. Am. Chem. Soc.*, 1998, **120**, 7020.
37. S. Van Doorslaer, R. Bachmann, and A. Schweiger, *J. Phys. Chem. A*, 1999, **103**, 5446.
38. S. Van Doorslaer, D. M. Murphy, and I. A. Fallis, *Res. Chem. Intermed.*, 2007, **33**, 807.
39. S. Van Doorslaer and A. Schweiger, *Chem. Phys. Lett.*, 1999, **308**, 187.
40. G. Mitrikas and A. Schweiger, *J. Magn. Reson.*, 2004, **168**, 88.
41. T. Maly and T. F. Prisner, *J. Magn. Reson.*, 2004, **170**, 88.
42. A. Cernescu, T. Maly, and T. F. Prisner, *J. Magn. Reson.*, 2008, **192**, 78.
43. G. Mitrikas and G. Prokopiou, *J. Magn. Reson.*, 2015, **254**, 75.
44. F. Mentink-Vigier, A. Collauto, A. Feintuch, I. Kaminker, V. Tarle, and D. Goldfarb, *J. Magn. Reson.*, 2013, **236**, 117.
45. T. F. Segawa, A. Doll, S. Pribitzer, and G. Jeschke, *J. Chem. Phys.*, 2015, **143**, 044201.

Chapter 19

Dipolar Spectroscopy – Double-resonance Methods

Gunnar Jeschke

ETH Zürich, Zürich, Switzerland

19.1 INTRODUCTION

The measurement of distance distributions in the nanometer range by electron paramagnetic resonance (EPR) techniques is valuable for structural characterization of macromolecules, macromolecular complexes, and supramolecular materials, in particular in cases where diffraction and NMR approaches fail due to structural disorder, size, or high complexity of the system. Information on the distance between two spins is encoded in their dipole–dipole coupling. In this chapter, we assume that the exchange coupling between two electron spins A and B is negligible and that both spins are aligned with the static magnetic field, which is a good approximation for g anisotropies of less than 10%. In this case, the coupling ω_{AB} between the two spins in angular frequency units is given by $\omega_{AB} = 2\pi(1-3\cos^2\theta)\nu_\perp$, where θ is the angle between the spin–spin vector and the external magnetic field and $\nu_\perp$ is the dipole–dipole interaction at orientations of the spin–spin vector perpendicular to the magnetic field ($\theta = 90°$). We further assume that ω_{AB} is much smaller than the difference $|\omega_A - \omega_B|$ between the two electron Zeeman angular frequencies. The energy level scheme and spectrum for an isolated spin pair in that situation are given in Figure 19.1. As many spin pairs with different electron Zeeman frequencies ω_A and ω_B contribute to the EPR spectrum of the sample, the dipolar coupling ω_{AB} is usually not resolved.

This coupling can be measured in an experiment where the B spin is inverted by a $(\pi)_B$ pulse that excites both transitions marked red in Figure 19.1(a) but does not excite the A spin transitions marked blue. Such a pulse exchanges coherence of the A spin between the two transitions, leading to a frequency change by $+\omega_{AB}$ for the left blue transition in Figure 19.1(b) and by $-\omega_{AB}$ for the right blue transition. With an appropriate pulse sequence (Section 19.3), this

EPR Spectroscopy: Fundamentals and Methods.
Edited by Daniella Goldfarb and Stefan Stoll

ISBN: 978-1-119-16299-5
Also published in eMagRes (online edition)
DOI: 10.1002/9780470034590.emrstm1518

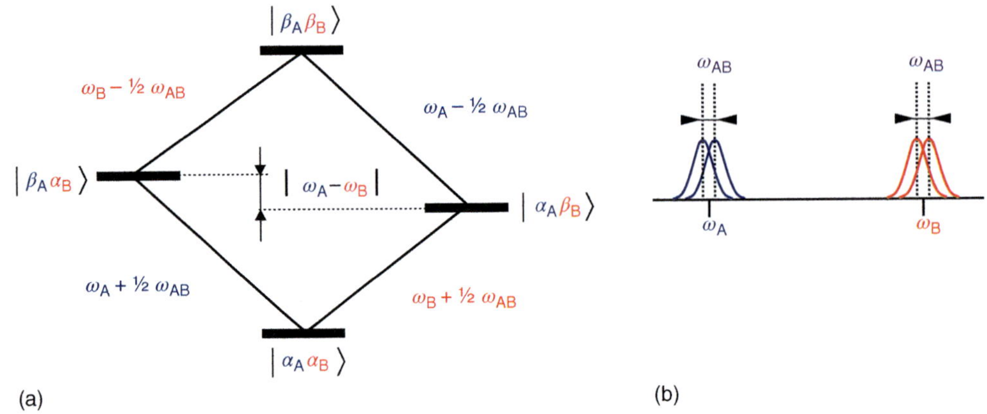

Figure 19.1. Energy level scheme (a) and schematic EPR spectrum (b) of an isolated pair of an observer spin A (blue) and a pumped spin B (red) with dipole–dipole coupling ω_{AB} and electron Zeeman frequencies ω_A and ω_B

frequency change can be measured as an echo amplitude oscillation of the form $\cos(\omega_{AB}t)$, where t is a pulse delay.

Taking into account the dependence of the splitting $\nu_{AB} = \omega_{AB}/2\pi$ on angle θ (Figure 19.2a) and that the probability of encountering an orientation with angle θ is proportional to $\sin\theta$, the dipolar spectrum of a spin pair with a uniform angular distribution of the spin–spin vector with respect to the magnetic field is a Pake pattern as shown in Figure 19.2(b). This Pake pattern has singularities at the frequencies $\pm\nu_\perp$. Note that the Pake pattern that would be directly observed in an EPR spectrum is only half as wide. Pulsed dipolar spectroscopy measures dipolar splitting rather than the frequency shift of an individual transition.

The distance between the two spins can be computed from $\nu_\perp$ as

$$r = \sqrt[3]{g_1 g_2 \mu_B^2 \frac{\mu_0}{4\pi h \nu_\perp}} \tag{19.1}$$

where g_1 and g_2 are the g values of the two spins, $\mu_B = 9.274\,009\,996 \times 10^{-24}$ J T^{-1} is the Bohr magneton, $\mu_0 = 4\pi \times 10^{-7}$ H m^{-1} is the magnetic constant, and $h = 6.626\,070\,04 \times 10^{-34}$ J s is the Planck constant. At a distance of 1 nm, one has $\nu_\perp \approx 52$ MHz and at a distance of 10 nm, $\nu_\perp \approx 52$ kHz. The upper limit of the distance range accessible by direct measurement of the dipole–dipole interaction is set by the phase memory time T_m of the observer A spins. Typical phase memory times of electron spins allow for observation of dipolar evolution for a few microseconds, in very favorable cases up to a few tens of microseconds. For a pair of electron spins with g values close to the g value of the free electron, this translates to an upper limit r_{max} of the distance range between about 5 and 15 nm.

The lower limit r_{min} depends on the excitation bandwidth of the pulses. With single-resonance methods (see Chapter 20), it is necessary to excite a large fraction p_A of all observer spins by all pulses in the sequence, as the wanted signal scales as $p_A{}^2$. With double-resonance methods, the system is partitioned into observer A spins and pumped B spins. The wanted signal now scales as $p_A\, p_B$, where p_B is the fraction of excited B spins. As the B spins can be spectroscopically distinct from the A spins and as the pump pulses can be incoherent, this tends to reduce requirements for large excitation bandwidth. As only the A spins are observed, the ratio between the wanted signal and unwanted signals scales with p_A rather than with $p_A{}^2$ as in single-resonance experiments. In practice, the lower distance limit for double-resonance techniques is set by the requirement to excite the two lines that are split by ω_{AB} for the B spin (red in Figure 19.1b) with the pump and for the A spin (blue in Figure 19.1b) with the observer pulses.[1] In other words, the excitation bandwidth must be at least comparable to the largest dipole–dipole coupling $\nu_\parallel = 2\ \nu_\perp$ that corresponds to the outer shoulders of the Pake pattern (Figure 19.2b). For typical pulse lengths of 10–32 ns, this sets a lower distance limit of 1.5 to 2 nm, which is virtually independent of the spectral widths of the A and B spins.[2]

Many molecular systems that are inaccessible or at least cannot fully be characterized by other structure determination techniques feature some extent of structural disorder. Even where this is not the case, the conformation of the spin label itself often features

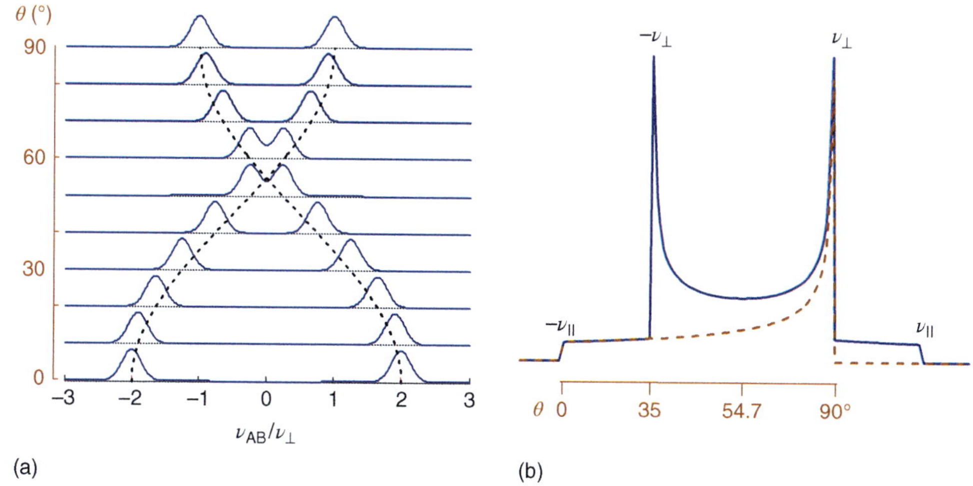

Figure 19.2. Dependence of the dipole–dipole coupling ν_{AB} on the angle θ between the spin–spin vector and the static magnetic field $\boldsymbol{B}_0$ (a) and the Pake pattern that is obtained in pulsed dipolar spectroscopy if the spin–spin vector with respect to $\boldsymbol{B}_0$ is uniformly distributed (b)

some distribution, as labels with flexible linkers are used that tend to be less perturbative to the system of interest than labels with rigid linkers. As a result of structural disorder and conformational distribution of the spin label, distances are distributed. In some cases, characterization of structural disorder or of populations of alternative structural states is the main point of a study of the system by EPR techniques. It is thus important to obtain not only mean distances but also distance distribution information from the primary data. This is complicated by the dependence of the dipole–dipole interaction on the polar angle θ between the external magnetic field $\boldsymbol{B}_0$ and the spin–spin vector $\boldsymbol{r}_{AB}$ (Figure 19.2). While the dipolar evolution function $D(t)$ for an isolated spin pair can be predicted from a known distance distribution $P(r_{AB})$ easily and with high fidelity (see Section 19.4.1), the inverse problem of computing $P(r_{AB})$ from $D(t)$ is ill-posed.[3] This ill-posedness implies that small deviations of $D(t)$ from the theoretical model, as they invariably arise from noise, can cause large deviations of the apparent distance distribution $P_{app}(r_{AB})$ from the true distribution $P(r_{AB})$. Among the different mathematical approaches for stabilizing the solution and thus reducing the deviation of $P_{app}(r_{AB})$ from $P(r_{AB})$, Tikhonov regularization[4] with the additional constraint $P_{app}(r_{AB}) \geq 0$[5] is most transparent and has thus become the most popular approach. Regularization does not fully eliminate the deviation of $P_{app}(r_{AB})$ from $P(r_{AB})$, which tends to increase strongly with an increase in deviation of $D(t)$ from the theoretical model. Therefore, the first and foremost requirement on a dipolar spectroscopy experiment consists in providing a high-fidelity dipolar evolution function $D(t)$. This requirement constrains experimental schemes, sample preparation, and all data processing steps that convert primary measured data $V(t)$ into a sum-of-pairs dipolar evolution function $D(t)$. These constraints need to be discussed in a common framework, as is attempted in the following. The complexity of the problem increases with the complexity of the spin system and of the excitation scheme. After a brief discussion of the general case, we shall thus start with a description of the experiment and of data processing for the simplest case of an isolated A–B spin pair with completely distinct spectra of the A and B spins and we shall later discuss additional complications step by step.

19.2 DILUTE CLUSTER DESCRIPTION OF THE SAMPLE

Distance distribution information can be obtained in the interval (r_{min}, r_{max}). In general, dipolar spectroscopy provides interpretable data for spin systems where all electron spins S_i ($i > 0$) are at distances $r \geq r_{min}$ from the observer spin S_0. The set of M spins, $S_1, S_2, \ldots S_{M-1}$ at distances $r \leq r_{max}$ together with

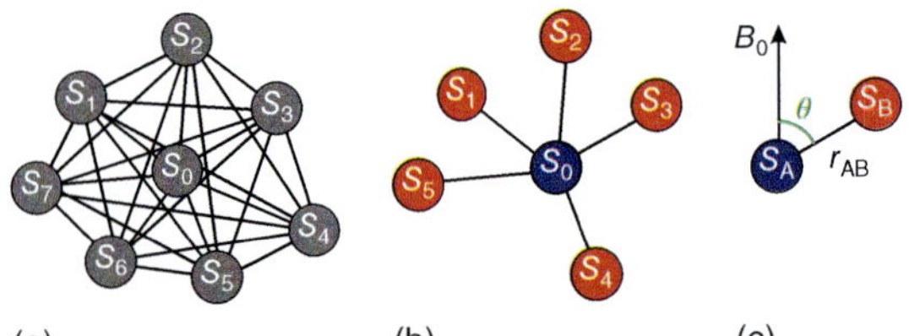

Figure 19.3. Topology of spin clusters. (a) All M electron spins in a cluster of spins with maximum dimension on the order of 10 nm have significant pairwise couplings. Visualization for $M = 8$. (b) Observer spin S_0 is excited by the observer subsequence at frequency ν_A and a subset of N of the M spins is excited by the pump pulse(s) at frequency ν_B. Only the pairwise couplings between spin S_0 and each individual spin S_i ($i = 1 \ldots N$) are relevant. Visualization for $N = 5$. (c) The signal factors into contributions from pairs of an observer spin A ($S_A = S_0$) and one of the pumped spins B_i ($S_B = S_i$, $i > 0$). The dipole–dipole coupling between the two spins, which encodes the distance, depends on angle θ between the magnetic field vector $\boldsymbol{B}_0$ and the spin–spin vector $\boldsymbol{r}_{AB}$

the observer spin S_0, is called a cluster henceforth (Figure 19.3a). In measurements on macromolecules or complexes thereof, it is often, but not always, possible to achieve an r_{max} that exceeds the largest dimension of the molecule or complex. In this common experimental situation, M is the number of native paramagnetic centers or labels within the same molecule or complex. We assume that the sample consists of a large number of clusters of the same type that may differ in the individual configurations of the M spins.

Spins S_i at distances $r > r_{max}$ ($i > M$), i.e., spins in clusters other than the one that harbors the observer spin S_0, do contribute to the decay of the primary signal $V(t)$, although their spatial distribution cannot be inferred with confidence from $V(t)$. Distance distributions $P(r)$ within the cluster can be obtained if and only if $V(t)$ can be separated into an intracluster contribution $F(t)$, from which $D(t)$ can be inferred, and an intercluster contribution $B(t)$ that corresponds to the background contribution from spins in other clusters.[6] The intracluster contribution $F(t) = 1 - \Delta + \Delta\ D(t)$ contains information on the number of spins in the cluster (modulation depth Δ) and on the distance distribution (dipolar evolution function $D(t)$). In analogy to scattering experiments, F(t) is also termed form factor. Pulse electron double-resonance (PELDOR) methods,[6] also termed double electron–electron resonance (DEER), have a particular advantage for this separation. As long as only one pump pulse is applied at frequency ν_B, the signal $V(t)$ strictly factorizes into contributions from individual spin pairs. Hence,

$$V(t) = F(t)B(t) \tag{19.2}$$

Equation (19.2) is often a good approximation even in cases where several pump pulses are applied, because the effect of pump pulses on A-spin evolution does not depend on B-spin polarization, whereas any correlations between B spins that are introduced by pump pulses scale with this polarization, which is usually small at the temperatures where the experiments are performed.

As we shall see later, $F(t)$ decays to a constant value within a time t_{dec} (Figure 19.4). This dipolar decay time depends on the longest distance within the cluster and on the width of the distribution of this distance over the whole ensemble of clusters. If the maximum dipolar evolution time t_{max}, corresponding to the last data point in the primary signal $V(t)$, is significantly larger than t_{dec}, $B(t)$ can usually be inferred to a good approximation by fitting $V(t)$ at $t > t_{dec}$ (red line in Figure 19.4b). Such fitting is aided by another favorable consequence of factorization into spin pair contributions. For a homogeneous distribution of the extracluster spins S_i ($i > M$) in three dimensions, $B(t)$ is an exponentially decaying function with a decay constant that is proportional to the concentration of spins excited by the pump pulse(s). In heterogeneous environments, such as lipid bilayers, and in cases where the homogeneous distribution is distorted by excluded volume at short distances, $B(t)$ can be modeled by a stretched exponential function[7]

$$B(t) = \exp[-(kt)^{d/3}] \tag{19.3}$$

where k is a decay rate constant and d is a fractal dimension that takes the value of 3 for a homogeneous distribution in three-dimensional space and is typically somewhat larger than 2 for cluster distributions in lipid bilayers. Values of $d > 3$ indicate excluded volume or longitudinal relaxation of B spins on the time scale of t_{max}. It is good practice to use $t_{max} > 2t_{dec}$ whenever possible and $t_{max} > 4t_{dec}$ if this still allows for measurements with a reasonable signal-to-noise ratio.

Background fits require that an estimate for t_{dec} can be obtained by visual inspection of the primary data. In general, this is possible only if the mean cluster–cluster distance, which can be estimated from cluster concentration and the expected fractal dimension d, is much longer than the maximum spin–spin

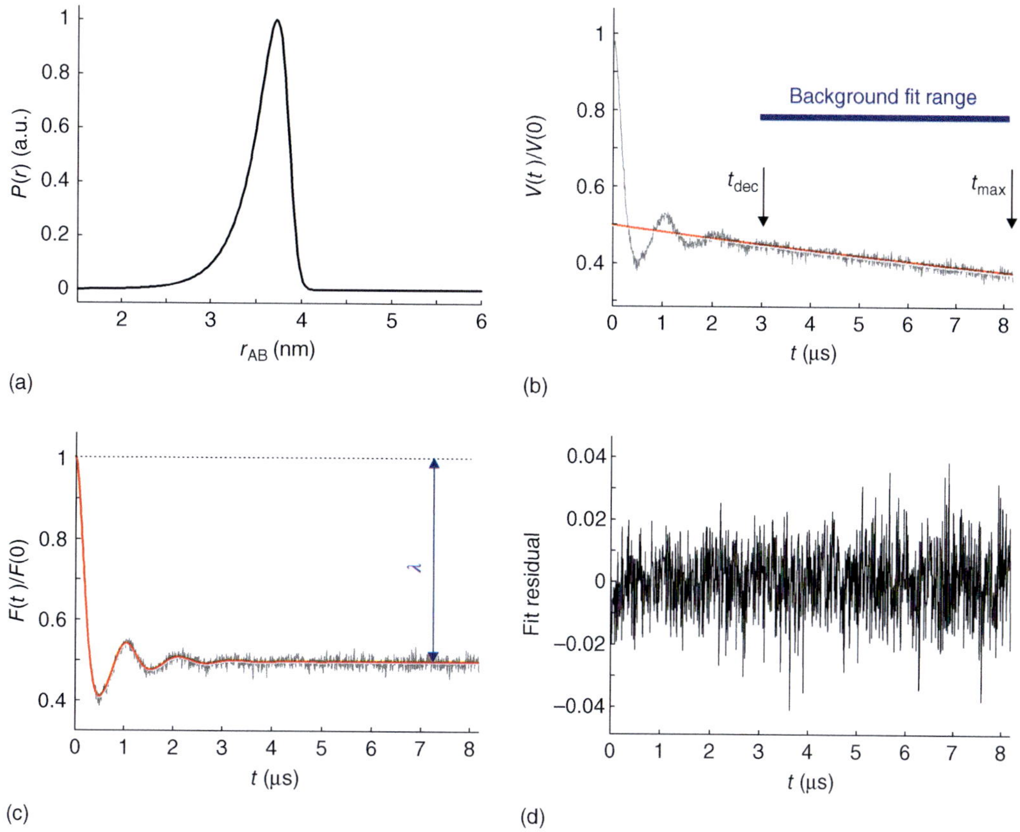

Figure 19.4. Signals at different data processing steps in dipolar spectroscopy (numerical simulation). Signal-to-noise ratio (signal amplitude to standard deviation of Gaussian white noise) is 100 in the primary data. (a) Asymmetric distance distribution corresponding to spins attached at the ends of a 4-nm-long worm-like chain with a persistence length of 6 nm. Additional Gaussian broadening with standard deviation of 0.1 nm was assumed. (b) Primary data $V(t)/V(0)$ assuming monoexponential background decay (see equation (19.3), $d = 3$, $k = 0.0352$) with $B(t_{max}) = 0.75$ and pair modulation depth $\lambda = 0.5$. The background function $b(t)$ is shown in red. (c) Form factor $F(t)/F(0)$ obtained by background fitting of primary data with variable background dimension d ($d = 2.96$ optimum), subsequent division by $b(t)$, and renormalization. (d) Residual of the fit obtained with Tikhonov regularization at optimal regularization parameter $\alpha = 75.1$

distance within the cluster. This requires sufficient dilution, typically to concentrations of 100–200 μM. As far as sensitivity permits, background fitting becomes more reliable by decreasing concentration even further.

Factorization into pair contributions also simplifies description of the intracluster contributions. With double-resonance techniques, the topology of the spin system reduces from the dense network topology shown in Figure 19.3(a) to the star topology shown in Figure 19.3(b). The network topology corresponds to an analytically intractable multibody problem whereas the star topology corresponds to a set of analytically solvable two-body problems. Each two-body problem is a problem of an isolated spin pair A–B as visualized in Figure 19.3(c).

Furthermore, double-resonance techniques permit simplification of the cluster spin system by spectroscopic selection. If the spins $S_0, S_1, S_2, \ldots S_{M-1}$ within the cluster are grouped into subsets of different chemical identity and thus with different EPR spectra, it is often possible to select only one of the subsets as pumped spins by a suitable choice of ν_B and the same or another subset as observer spins by a suitable choice of observer frequency ν_A.[8] The selection can be optimized by adjusting other experimental parameters, such as temperature or pulse parameters.[9] Because of such selection, the spin system taking part in a given double-resonance experiment can have $N \leq M$ spins (Figure 19.3b).

19.3 PULSE SEQUENCES

19.3.1 The SEDOR Experiment

In an ideal double-resonance experiment, A spins are excited only by pulses at frequency ν_A and B spins only by pulses at frequency ν_B. In this limit, a two-pulse echo experiment $(\pi/2)_A-\tau-(\pi)_A-\tau$-*echo* on the A spins completely refocuses dipole–dipole coupling in the A–B spin pair. When the interpulse delay τ is varied, the echo signal decays smoothly with phase memory time T_m. The dipole–dipole coupling can be reintroduced by applying a π pulse to the B spins. This spin echo double-resonance (SEDOR) experiment was first introduced in heteronuclear solid-state NMR in a version where the $(\pi)_B$ pulse is applied simultaneously with the $(\pi)_A$ pulse and dipolar evolution is observed by varying interpulse delay τ.[10] As a consequence, the dipolar oscillations are damped by echo decay, which is tolerable in heteronuclear NMR because T_m is typically much larger than the period of dipolar oscillations. For electron spin pairs, T_m is typically shorter than this period and, moreover, frequencies ν_A and ν_B are typically so close that overlap of the $(\pi)_A$ and $(\pi)_B$ pulses causes deviations of the signal amplitude from the one of an experiment with ideal flip angles π for both pulses. This applies in particular if the same amplifier or resonator mode is used for pulses at both frequencies. For these reasons, the original SEDOR experiment is not well suited for electron spin pairs.

19.3.2 PELDOR with Three Pulses

The influence of relaxation decay on the time-domain data can be removed and pulse overlap can be largely avoided in the original PELDOR experiment (Figure 19.5a).[6,11] Now the interpulse delay τ is fixed and the $(\pi)_B$ pulse is applied at a variable time $0 \leq t < \tau$ after the $(\pi/2)_A$ pulse. Phase memory time T_m still limits $t_{max} \leq \tau$, as the echo intensity is proportional to $\exp[-(2\tau/T_m)^{\xi}]$ with a stretch exponent ξ typically in the range between 1 and 2. However, except for this constant reduction in signal amplitude, $V(t)$ is now unaffected by relaxation. Pulse overlap now occurs at $t = 0$. This causes a dead time t_{dead}, which is at least as long as half of the sum of the two π pulse lengths. In practice, signal distortions are often observable at $t < 60–80$ ns. This dead time makes it impossible to characterize broad distance distributions unless $t_{dead}\, \nu_{\perp} \ll 1$, i.e., problems arise for distances shorter than about 3 nm. Furthermore, dead time introduces uncertainties in the determination of the zero time of the dipolar oscillation and in normalization of the amplitude of the time-domain trace. Because conversion of the time-domain data to a distance distribution is an ill-posed problem, it is important to reduce such uncertainties to a minimum.

The consequences of dead time can be reduced using separate amplifiers for pulses at ν_A and ν_B or by applying a signal correction.[12] Nevertheless, the signal amplitude at times when the two pulses overlap still deviates somewhat from the true amplitude and it is not obvious from the primary data which time corresponds to $t = 0$. As pulse lengths are finite and pulse shapes nonideal, this time cannot exactly be predicted from the interpulse delays.

19.3.3 Four-Pulse DEER

At the expense of moderate signal loss, pulse overlap can be completely avoided by extending the observer sequence to a refocused echo sequence $(\pi/2)_A-\tau_1-(\pi)_A-\tau_1-\tau_2-(\pi)_A-\tau_2$-*echo*.[13,14] A dipolar evolution time $t = 0$ now corresponds to applying the $(\pi)_B$ pulse at time τ_1 after the first $(\pi)_A$ pulse (Figure 19.5b), so that $t_{max} < \tau_2$. It suffices to choose a delay τ_1 that avoids any effects of pulse overlap in the amplifier and makes it possible to observe the signal maximum at $t = 0$. In most cases $\tau_1 \ll \tau_2$, so that relative signal loss due to relaxation is acceptable. Further moderate signal loss occurs as the additional $(\pi)_A$ pulse reduces the bandwidth of the observer echo sequence. The effect of a moderately reduced signal-to-noise ratio on data quality is usually less serious than the effect of a time-shift error in t and of errors in signal amplitude near $t = 0$. Four-pulse DEER has become the most widely applied double-resonance technique in electron spin dipolar spectroscopy, because it avoids such deviations of the data from the ideal model and obviates additional data correction steps in signal processing. To overcome the sensitivity loss compared to the three-pulse experiment, combination of four-pulse

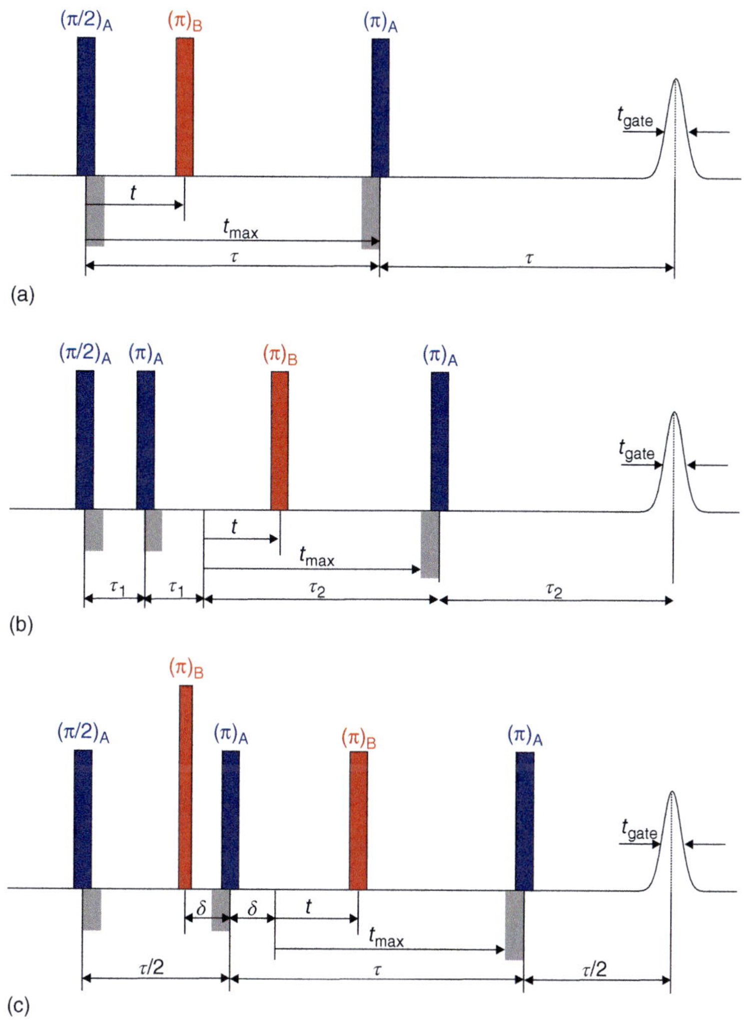

Figure 19.5. Pulse sequences for double-resonance dipolar spectroscopy. In each case, the interpulse delay t is varied while the other delays are constant. The optimal gate width for echo integration (echo width at half height) is denoted by t_{gate}. Observer pulses (blue) are applied at frequency ν_A and pump pulses at frequency ν_B. Gray boxes denote the minimum distance between pulses for obtaining an undistorted signal, even if separate amplifiers and a bimodal resonator are used or a correction is applied. (a) DEER with three pulses. (b) Four-pulse DEER. (c) Five-pulse DEER. To diminish the contributions from a four-pulse DEER artifact pathway, the first pump pulse at time δ before the first observer $(\pi)_A$ pulse should have a larger excitation bandwidth than the second pump pulse

DEER data near $t = 0$ with three-pulse data measured up to a longer t_{max} has been proposed.[15]

Background fitting according to equation (19.3) requires that any receiver offset is eliminated. Therefore, four-pulse DEER is performed with a phase cycle $[+(+x)-(-x)]$ on the $(\pi/2)_A$ pulse. The pump pulse may require a four-step cycle $[+(+x)+(+y)+(-x)+(-y)]$ if it is not incoherent and pump and observer excitation bands overlap.

19.3.4 Multipulse DEER

The asymmetric refocused echo sequence of four-pulse DEER is typically not optimal for reducing relaxation losses at a given total length of the refocused echo sequence $2(\tau_1+\tau_2)$. If phase coherence decays with a stretch exponent $\xi > 1$, as is usually the case at least for proton-containing environments of the electron spin, the optimum corresponds to $\tau_1 = \tau_2$. With a single $(\pi)_B$ pulse, this setting reduces t_{max} at given total sequence length, as dipolar refocusing occurs at the center of the second interpulse delay $(\tau_1 = \tau_2 = 2\tau)$. This can, however, be remedied by the five-pulse DEER experiment, where a second $(\pi)_B$ pulse inverts the dipole–dipole interaction and thus shifts the point of dipolar refocusing to a time close to the first $(\pi)_A$ refocusing pulse (Figure 19.5c).[16] In principle, relaxation losses can be further reduced by multiple refocusing.[17] Each additional refocusing $(\pi)_A$ pulse requires an additional $(\pi)_B$ pulse in order to keep t_{max} close to half the total sequence length.

A problem arises because each $(\pi)_B$ pulse inverts only a fraction $p_B < 1$ of all B spin packets. If each B spin packet would be either inverted by all $(\pi)_B$ pulses or not inverted by any $(\pi)_B$ pulse, DEER experiments with more than one $(\pi)_B$ pulse would provide the same primary signal $V(t)$ as four-pulse DEER. However, there exists a fraction of B spin packets that are inverted with probability $0 < p_B < 1$ by each pulse. This leads to artifact contributions from coherence transfer pathways where the B spin is inverted by some, but not all $(\pi)_B$ pulses. Signal expressions discussed in Section 19.4 apply to the three- and four-pulse schemes, but not directly to multipulse schemes. Procedures that correct for signal contributions from B spins that are flipped by at least one pump pulse, but not by all of them, have been proposed for five- and seven-pulse DEER.[16,17] It has not yet been established whether such signal correction generally leads to data with the same fidelity as can be achieved by four-pulse DEER. However, as the gain in t_{max} can be substantial, multipulse DEER can be of advantage in cases where separation of $V(t)$ into $F(t)$ and $B(t)$ is unreliable for four-pulse DEER data. Distance distributions obtained from multipulse DEER cannot currently be interpreted to the same level of detail as those from four-pulse DEER.

In the symmetric refocused echo sequence of the five-pulse DEER experiment, the stimulated echo that results from incomplete coherence transfer by the two $(\pi)_A$ pulses is exactly superimposed with the refocused observer echo. This stimulated echo must be eliminated by phase cycling to allow for background fitting by equation (19.3). A four-step phase cycle $[+(+x,+x)-(-x,+x)+(+x,-x)-(-x,-x)]$ on the first two observer pulses is required. Additional refocusing pulses on the observer spin generate additional echoes that exactly superimpose with the wanted observer echo. To cancel all overlapping echoes for n_A observer π pulses, the first n_A-2 $(\pi)_A$ pulses must be cycled $[+(+x)-(+y)+(-x)-(-y)]$, the second-to-last one $[+(+x)\ +(-x)]$, and the last one does not need to be cycled. If at least one $(\pi)_A$ pulse undergoes a four-step cycle, the receiver offset is canceled by this cycle and the two-step cycle on the $(\pi/2)_A$ pulse can be skipped. This leads to an eight-step phase cycle for the seven-pulse experiment with $n_A = 3$ and an extension of the phase cycle by a factor of four for each additional $(\pi)_A$ pulse beyond that.

19.4 EXPRESSIONS FOR DIPOLAR EVOLUTION

19.4.1 Dipolar Evolution Function of the Isolated Spin Pair

The simplest case is an isolated pair of a spin A with resonance offset $\Delta\omega_A$ from the observer frequency ω_A and a spin B with resonance offset $\Delta\omega_B$ from the pump frequency ω_B at a fixed distance $r = r_{AB}$. In this case, the dipole–dipole coupling is given by

$$\omega_{AB} = 2\pi\frac{1}{r^3}\frac{g_A g_B \mu_B^2 \mu_0}{4\pi h}(1 - 3\cos^2\theta) \qquad (19.4)$$

The factor including the g values g_A and g_B of the two electron spins, the Bohr magneton μ_B, the magnetic constant μ_0, and Planck's constant h can usually be considered as constant for given $\Delta\omega_A$ and $\Delta\omega_B$ and takes a value of about $C_{AB} = 52$ MHz·nm^3 if both g values are close to g_e. With the probability $p_B(\Delta\omega_B)$ that the B spin is inverted by the pump pulse, the form factor $F(t)$ of the isolated pair, normalized to its value at zero dipolar evolution time, is given as

$$\frac{F(t)}{F(0)} = 1 - p_B + p_B \cos(\omega_{AB} t) \tag{19.5}$$

The inversion probability p_B takes on values between 0 and 1 and is generally smaller than 1 for three reasons. First, the nominal flip angle of the pump pulse may differ from π. This allows for reducing p_B to reduce dipolar combination frequencies in systems with more than one pumped spin (see Section 19.4.3). Second, the resonance offset $\Delta\omega_B$ is usually of the order of the excitation bandwidth of the pump pulse, which is roughly the inverse pulse length $2\pi/t_{p,B}$, so that spin packets experience a reduced flip angle that depends on $\Delta\omega_B$. This effect can be described by an excitation profile $E_B(\Delta\omega_B)$, see Figure 19.6. Third, in the ensemble of spin-carrying macromolecules, some may carry the A spin, but a fraction f_C may miss the B spin due to incomplete spin labeling or heterogeneity of the redox state of native paramagnetic centers. Such molecules can be treated as A–B spin pairs in which the B spin is not inverted. In the following, the degree of labeling of B spins, $d_{label} = 1 - f_C$, is used to account for this.

In general, $\Delta\omega_A$, $\Delta\omega_B$, and θ are correlated through the relative orientation of the molecular frame of paramagnetic center B to the one of paramagnetic center A (Euler angles α, β, γ) and the orientation of the spin vector with respect to the molecular frames (Section 19.5.1). If this correlation is significant, p_B depends on θ via its dependence on $\Delta\omega_B$. In many cases, the correlation is insignificant, either because of a broad distribution of relative orientations of the two paramagnetic centers or because the anisotropy of the resonance offsets is unresolved on the scale of the excitation bandwidths of the observer and pump pulses. In this situation, equation (19.5) can be integrated over all spin pair packets. By substituting $\cos\theta$ by z and thus the orientation-weighted polar angle $\sin\theta\, d\theta$ by dz (with a corresponding change of the integration limits), one finds

$$\frac{F(t)}{F(0)} = 1 - \tilde{\lambda}\left[1 - \int_0^1 \cos\left(\frac{C_{AB}}{r^3}\left(1 - 3z^2\right)t\right)\mathrm{d}z\right] \tag{19.6}$$

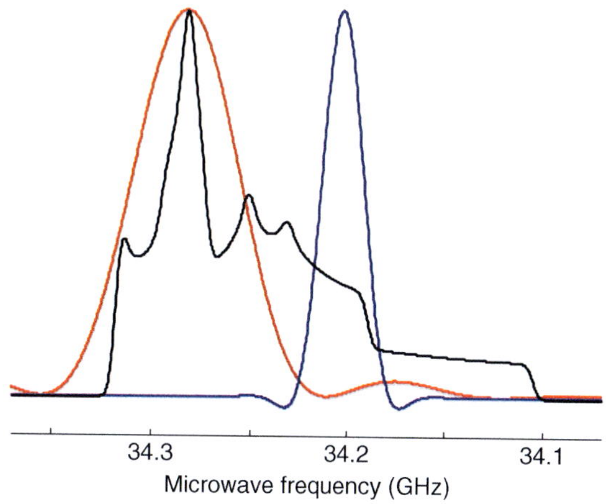

Figure 19.6. Observer (blue) and pump (red) excitation bands for all pulses having a length of 12 ns and their relation to the nitroxide spectrum at Q-band (numerical simulation). Amplitude in the observer excitation band corresponds to the weight by which a spin packet contributes to the echo signal. Amplitude in the pump excitation band corresponds to the probability of inversion of a B spin packet. A four-pulse DEER observer sequence with echo integration over a 24-ns gate and a 2-ns rise/fall time for the pulse flanks is considered. Even if pump and observer spin are of the same chemical species, A and B spins can be selected due to the dispersion of resonance frequencies by electron Zeeman anisotropy, hyperfine anisotropy, and nuclear spin magnetic quantum number m_I

where the net inversion efficiency $\tilde{\lambda}$ is given by

$$\tilde{\lambda} = d_{label}\int_{-\infty}^{\infty} E_B(\Delta\omega_B) I_B(\Delta\omega_B)\mathrm{d}\Delta\omega_B \Big/ \int_{-\infty}^{\infty} I_B(\Delta\omega_B)\mathrm{d}\Delta\omega_B \tag{19.7}$$

and $I_B(\Delta\omega_B)$ is the intensity in the absorption spectrum of the B spins.

Equation (19.6) together with equation (19.7) applies only if all spin pairs are A–B pairs. In general, a fraction $f_C < 1$ of all pairs may miss the B spin and, if the EPR spectra of A and B spins overlap or even coincide, a fraction $f_A < 1$ of all pairs may behave as A–A pairs. In A–A pairs, both coupled spins are excited by the observer pulses and thus give rise to dipolar modulation of the observer echo.[18,19] If the total pulse sequence length is sufficient for these dipolar oscillations to decay, the signal from fraction f_A is lost, whereas the signal from fraction f_C is constant. equation (19.6) then needs to be replaced by

$$\frac{F(t)}{F(0)} = 1 - f_A - \tilde{\lambda}(1 - f_C)\left[1 - \int_0^1 \cos\left(\frac{C_{AB}}{r^3}\left(1 - 3z^2\right)t\right)\mathrm{d}z\right] \Big/ (1 - f_A) \tag{19.8}$$

Equation (19.8) can be written in the same form as equation (19.6) by substituting the net inversion efficiency $\tilde{\lambda}$ by the *pair modulation depth*

$$\lambda = \frac{\tilde{\lambda}}{1 - f_A} \tag{19.9}$$

The normalized form factor is obtained by dividing the normalized primary data $V(t)/V(0)$ by $B(t)$. A background fit provides $b(t) = (1-\lambda)B(t)$ and, as $B(0) = 1$, λ can be inferred from $b(0)$. Hence, we have

$$\frac{F(t)}{F(0)} = \frac{b(0)V(t)}{b(t)V(0)} \tag{19.10}$$

Finally, we obtain the normalized dipolar evolution function

$$D(t) = \frac{\frac{F(t)}{F(0)} - (1 - \lambda)}{\lambda} = \int_0^1 \cos\left(\frac{C_{AB}}{r^3}\left(1 - 3z^2\right)t\right) dz \tag{19.11}$$

In an ensemble of spin pairs, the distance r may be distributed. Figure 19.4 illustrates the primary signal and form factor for a case with such a distance distribution $P(r)$.

19.4.2 Sensitivity Considerations

The distance distribution $P(r)$ is determined from $D(t)$. The signal-to-noise ratio of $D(t)$ is proportional to the signal-to-noise ratio of $V(0)$ and to the net inversion efficiency $\tilde{\lambda}$. Furthermore, the signal-to-noise ratio at $t > 0$ is reduced by a factor $1/B(t)$. Signal loss by the factor $(1-f_A)$ does not deteriorate signal-to-noise ratio, as it exclusively reduces the signal of A spins that are coupled to other A spins and thus leads to an increase in the depth of modulations due to B spins by the same factor.

It is important to realize that $V(0)$ is not a monotonously increasing function of concentration. This is because DEER data are usually measured to $t_{max} > T_m$. At high concentrations, a π refocusing pulse causes significant stochastic changes of the dipolar field at the observer spin due to flips of randomly distributed neighboring electron spins. This process, termed instantaneous diffusion, shortens T_m. At some concentration, exponential signal loss from this mechanism on further increase in concentration becomes stronger than linear signal gain. As $B(t_{max})$ is related to concentration in the same way, too high concentrations can often be recognized by a too low $B(t_{max})$, in particular if observer and pump spins are the same chemical species. As a rule of thumb, $B(t_{max})$ should not be smaller than 0.75. If it is smaller than 0.5 and data quality is not already impeccable, dilution of the sample by at least a factor of two is likely to improve it strongly. Note that for $t_{max} > t_{dec}$, $B(t_{max}) = V(t_{max})/[V(0)\cdot(1-\lambda)]$.

Furthermore, the signal-to-noise ratio of $V(0)$ can be improved by an increase of observer bandwidth, i.e., by shortening observer pulses. The scope for such improvement is limited by the requirement for nonoverlapping excitation bands for the A and B spins (see Section 19.6.3). Shaped excitation pulses (see Chapter 21) can potentially increase the fraction of excited A spins while at the same time reducing overlap of excitation bands.

Regarding detection, $V(0)$ is maximized if the observer frequency ν_A coincides with the center of the resonator mode. This choice may, however, conflict with maximization of the net inversion efficiency, $\tilde{\lambda}$, and optimum placement of ν_A and ν_B with respect to the resonator mode may depend on available microwave power and resonator characteristics.[20] Data quality is limited not only by noise but also by artifact contributions that are not described by the theoretical expressions for dipolar evolution, such as nuclear modulation contributions. Although signal-to-noise ratio scales with the product $\tilde{\lambda}V(0)$, signal-to-artifact ratio increases more strongly with $\tilde{\lambda}$ than with $V(0)$, as some artifact contributions are proportional to $V(0)$. Hence, it may be advantageous to increase net inversion efficiency even if this entails a stronger loss in $V(0)$.

The quality of $D(t)$, as opposed to its signal-to-noise ratio, is not a monotonously increasing function of $\tilde{\lambda}$. Usually, the by far leading cause of deviations of the experimentally determined $D(t)$ from the true $D(t)$ comes from uncertainties in background fitting. Background fitting becomes precarious if $(1 - \lambda) \ll 1$. As a rule of thumb, best data quality is obtained by measuring under conditions leading to $\lambda \approx 0.5$, which can be achieved for nitroxide labels at X-band frequencies around 9.6 GHz with a pump pulse length $t_{p,B} = 12$ ns and at Q-band frequencies around 34 GHz with $t_{p,B} = 10$ ns, if in both cases, ν_B is set to the maximum of the nitroxide absorption spectrum. For metal ions as B spins, usually much lower pair modulation depths have to be accepted and shaped pulses (see Chapter 21) may help in increasing them.

As mentioned above, samples should be diluted if $B(t_{max}) < 0.5$. For $B(t_{max}) \approx 0$, noise is so much enhanced toward the end of $D(t)$ that data quality becomes unacceptable.

Because of the exponential decay of $V(0)$ along τ_2 with time constant T_m, DEER experiments are usually performed at or near the low-temperature limit where T_m attains its maximum value set by temperature-independent nuclear spin-diffusion processes, i.e., at cryogenic temperatures. However, it has been demonstrated that such experiments can also be performed in a liquid matrix at or near ambient temperature if the spin-carrying macromolecule is immobilized.[21,22]

19.4.3 Form Factor of a Spin Cluster

For a cluster with observer spin S_0 and pumped spins $S_1, S_2, \ldots S_N$, the form factors of the spin pairs S_0, S_i $(i = 1 \ldots N)$ multiply. Originally, this had been shown disregarding the fraction $1 - d_{\text{label},i}$ of observer spins with missing B spins and the occurrence of A–A$_i$ pairs, but the product rule generally applies if $p_{\text{B,i}}$ is replaced by $d_{\text{label},i}\, p_{\text{B},i}/(1 - f_{\text{A},i})$. This is because each additional spin S_i leads to a reduction of all terms in the expanded product by a factor $1 - f_{\text{A},i}$, except for the terms containing a factor $\tilde{\lambda}_i$, as these terms are explicitly due to A–B$_i$ pairs. The latter terms are reduced by a factor $d_{\text{label},i}$. Hence, after renormalization by $\prod_{i=1}^{N} (1 - f_{\text{A},i})^{-1}$, all $p_{\text{B,i}}$ are replaced by the corresponding $d_{\text{label},i}\, p_{\text{B},i}/(1 - f_{\text{A},i})$. If orientation selection is negligible, this corresponds to substitution of net inversion efficiencies $\tilde{\lambda}_i$ by the corresponding pair modulation depths λ_i. For a single set of spin packets, one has[23]

$$\frac{F_N(t)}{F_N(0)} = \frac{1}{Q+1} \sum_{k=0}^{Q} \prod_{i \neq k}^{N} f_{ki}(t) \tag{19.12}$$

with the normalized pair form factors

$$f_{ki} = 1 - \lambda_i [1 - \cos(\omega_{ki} t)] \tag{19.13}$$

The sum in equation (19.12) runs over the $Q+1$ spins in the cluster that can take the role of an observer spin. If all spins are like spins, then $Q = N$. Assuming that all λ_i take the same value, as is reasonable if all pumped spins are the same chemical species, expansion of the products in equation (19.12) provides

$$\frac{F_N(t)}{F_N(0)} = \sum_{o=1}^{N-1} \lambda^o (1 - \lambda)^{N-o} C_o \tag{19.14}$$

where o is the order of the combination frequency term C_o with

$$C_o = \frac{1}{Q+1} \frac{1}{o!} \sum_{k=1}^{Q} \sum_{i_1 \neq k}^{N} \sum_{i_2 \neq k, i_1}^{N} \cdots \sum_{i_o \neq k, i_1, \cdots i_{o-1}}^{N} \prod_{l=i_1}^{i_o} \cos(\omega_{kl} t) \tag{19.15}$$

for $o > 0$ and $C_o = 1$.

All C_o with $o > 1$ depend on angles between spin vectors $\boldsymbol{r}_{ki}$ with different i. Thus, these terms contain information on cluster geometry beyond only the vertex lengths, but in general this information is hard to extract.[23,24] The contribution of each of the terms in equation (19.14) at $t = 0$ is given by

$$\Delta_o = \binom{N}{o} \lambda^o (1 - \lambda)^{N-o} \tag{19.16}$$

which leads to a total modulation depth

$$\Delta_N = 1 - (1 - \lambda)^N \tag{19.17}$$

Note that, unlike in some previous treatments, N is defined as the number of B spins coupled to the observer spin, not as the total number of spins in the cluster. Equation (19.17) can be used to estimate the mean number of spins in the cluster, provided that clusters with different numbers of active paramagnetic species have the same relaxation behavior.[25,26]

The distance distribution

$$P(r) = \sum_{i=1}^{N} P(r_i) \tag{19.18}$$

can be extracted if the pair contribution C_1 can be isolated. It has been shown that at sufficiently low λ,

$$\frac{F_n}{F_0} \approx \left(\frac{F_N(t)}{F_N(0)} \right)^{1/N} \tag{19.19}$$

is a reasonable approximation for the normalized sum of pair form factors F_n/F_0, where F_N is the form factor measured for the multispin system. As a rule of thumb, the total modulation depth Δ_N should not exceed 0.5. This can be achieved by deliberately reducing the flip angle of the pump pulse, which is a strategy incompatible with multipulse DEER, or by deliberately reducing the degree of labeling d_{label}.

19.5 ORIENTATION SELECTION

19.5.1 Orientation Selection as a Source of Information

At a given static magnetic field B_0 and a given observer frequency, the spins in each spin packet with resonance offset $\Delta\omega_A$ from the observer frequency have a two-dimensional probability density distribution $\rho_A(\theta_A,\phi_A, \Delta\omega_A)$ of orientations (θ_A,ϕ_A) of the magnetic field vector $\boldsymbol{B}_0$ in the molecular frame of spin A (for definition of the spin system geometry, refer to Figure 19.7).[27] If the observer pulses excite the whole A spin spectrum uniformly, $\rho_A(\theta_A,\phi_A)$ is uniform as a function of cos θ_A and ϕ_A, but in most cases, the probability density distribution is far from uniform. Analogously, the probability density of exciting a B spin coupled to an A spin with orientation (θ_A,ϕ_A) is described by a distribution $\rho_B(\theta_A,\phi_A,\Delta\omega_B)$. The contribution $\rho_0(\theta_A,\phi_A)$ of a certain orientation (θ_A,ϕ_A) to V_0 is given by

$$\rho_0(\theta_A, \phi_A, \Delta\omega_A) = E(\Delta\omega_A)\rho_A(\theta_A, \phi_A, \Delta\omega_A) \tag{19.20}$$

where the excitation profile $E(\Delta\omega_A)$ is the contribution of A spin packets at resonance offset $\Delta\omega_A$ to the signal (Figure 19.6), whereas the contribution to dipolar modulation is given by

$$\rho_\lambda(\theta_A, \phi_A, \Delta\omega_B) = E(\Delta\omega_B)\rho_B(\theta_A, \phi_A, \Delta\omega_B) \tag{19.21}$$

where $E(\Delta\omega_B)$ is the B spin excitation profile. Net inversion efficiency $\tilde{\lambda}$ is now given by

$$\tilde{\lambda}(\nu_A, \nu_B) = d_{\text{label}} \frac{\int_0^{\pi/2}\int_0^{2\pi} \rho_0(\theta_A, \phi_A, \Delta\omega_A)\rho_\lambda(\theta_A, \phi_A, \Delta\omega_B) \sin\theta d\phi_A d\theta_A}{\int_0^{\pi/2}\int_0^{2\pi} \rho_0(\theta_A, \phi_A, \Delta\omega_A) \sin\theta\, d\phi_A d\theta_A} \tag{19.22}$$

where the dependence on $\Delta\omega_A$ and $\Delta\omega_B$ was substituted by a dependence on ν_A and ν_B. Because ρ_A and ρ_B are related by the relative orientation of the A and B spin molecular frames, $\tilde{\lambda}$ and thus the pair modulation depth λ depend on Euler angles α, β, and γ. By studying modulation depth as a function of ν_A and ν_B, one can thus obtain information on these Euler angles. The task is complicated in many cases by a distribution of α, β, and γ. Furthermore, inversion symmetry of the spin Hamiltonian with respect to the magnetic field direction makes several sets of α, β, and γ equivalent,

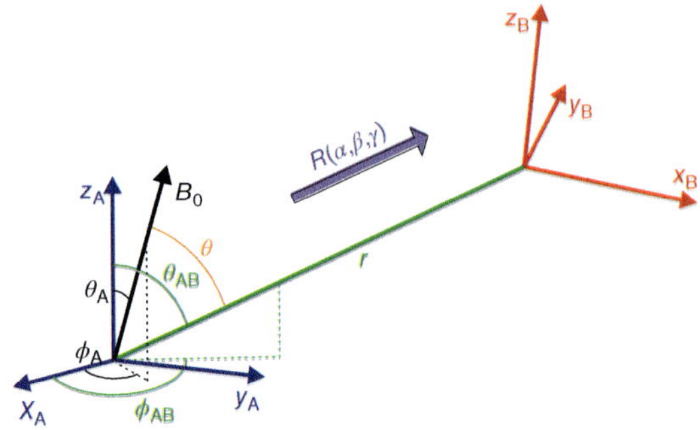

Figure 19.7. Geometry of a two-spin system. The molecular frame of spin A (blue) is the frame in which orientation selection is discussed. The molecular frame of spin B (red) is related to this frame by three Euler angles α, β, γ. Orientation of the magnetic field vector $\boldsymbol{B}_0$ (black) is described by polar angles θ_A and ϕ_A. Orientation of the spin–spin vector (green) is described by polar angles θ_{AB} and ϕ_{AB}. The angle θ between the magnetic field and spin–spin vector is a function of θ_A, ϕ_A, θ_{AB}, and ϕ_{AB}

which causes an ambiguity in determining the relative orientation of the A and B spin molecular frames.[28,29]

If deviations of the spin Hamiltonian of one of the two spins from axial symmetry are much smaller than the excitation bandwidth, the number of relevant Euler angles reduces to two. These two angles describe the orientation of the unique axis of the Hamiltonian with pseudoaxial symmetry in the molecular frame of the spin with lower symmetry of the Hamiltonian. If both spins have axially symmetric Hamiltonians, only the Euler angle β is relevant, which is then the angle between the two unique axes.[30] In the case where at least one of the spins has an isotropic Hamiltonian, orientation selection does not influence modulation depth.

Even if one of the spins has an isotropic Hamiltonian, orientation selection does influence the dipolar spectrum, if the other spin has an anisotropic Hamiltonian. This is because orientation selection on either of the two spins influences the probability density distribution $\rho(\theta)$ of the angle θ between the A–B spin–spin vector $\boldsymbol{r}_{AB}$ and the magnetic field vector $\boldsymbol{B}_0$, which is $\rho(\theta) = \sin\theta$ in the absence of orientation selection. The orientation of the spin–spin vector in the molecular frame of spin A is described by two polar angles θ_{AB} and ϕ_{AB} (Figure 19.7), which determine the angle θ for any set of angles (θ_A,ϕ_A). The contribution of this set to the dipolar modulation scales with $\rho_0\rho_\lambda$, which is not, in general, proportional to sinθ. As a consequence,

$D(t)$ depends on the distance r_{AB}, on the relevant Euler angles discussed above, and on θ_{AB} and ϕ_{AB}.

In the most general case, the modulation depth λ thus depends on three angles while the dipolar evolution function $D(t)$ depends on the distance and five angles. All these parameters may be distributed and their distributions can be mutually dependent. In practice, the distributions of r_{AB}, θ_{AB}, and ϕ_{AB} are often, but not always, narrow in cases where dipolar evolution functions observed for different choices of ν_A and ν_B differ significantly. Model-free determination of $\rho(\theta)$[31] and of the relevant angles has been attempted in cases of different complexities.[28–30,32] Less elaborate data analysis is required for testing consistency of a structural model with a set of form factor traces $F(t)/F(0)$.[33–36]

It appears advantageous to apply a divide and conquer approach. First, one determines the distance distribution by suppression of orientation selection (Section 19.5.2); second, the relevant Euler angles are determined by fitting the dependence of the modulation depth λ on observer and pump position; and third, angles θ_{AB} and ϕ_{AB} are determined by fitting the set of dipolar evolution traces. The data set for the second step is best acquired by keeping ν_A and ν_B constant and varying only static magnetic field B_0, as this strategy avoids uncertainties in the modulation depths stemming from different scaling of the excitation profile by the microwave mode. Note also that the second step is not required if one of the spins features an isotropic Hamiltonian. The resulting problem with only the distance r_{AB} and the angles θ_{AB} and ϕ_{AB} as free parameters is much easier to solve and provides more reliable parameter sets, as these parameters tend to have narrower distributions than the Euler angles.

There exist cases where it does not make much sense to determine angle parameters, even if orientation selection is observable, because such orientation selection may be too weak to obtain reliable estimates for the parameters. Even if such reliable estimates can be obtained, interpretation in terms of a structural model is hampered if the orientation of the spin A and spin B molecular frames with respect to the macromolecular backbone is unknown or too uncertain. In these cases, it is still necessary to suppress orientation selection in order to obtain reliable distance distributions.

19.5.2 Suppression of Orientation Selection

In the context of site-directed spin labeling (SDSL),[37] orientation selection can be remedied using spin labels that have an isotropic Hamiltonian in the sense defined in Section 19.5.1 both as the A and the B spins. Gd(III) complexes at any frequency or trityl labels up to Q-band frequencies are suitable candidates. However, often the choice of label is dominated by other considerations, such as steric hindrance and polarity in the environment of the labeling site or selectivity of the labeling reaction. With respect to these considerations, nitroxide labels with a flexible linker, in particular the methanethiosulfonato spin label (MTSL), are still most versatile. Averaging of the molecular frame orientation due to the flexible linker often, but not always, suppresses orientation selection.

Orientation selection can also be diminished by performing experiments at lower magnetic fields, where electron Zeeman anisotropy relative to excitation bandwidth is smaller. The scope of this strategy is limited by sensitivity considerations, as at a given sample volume and concentration, sensitivity with state-of-the-art spectrometers is optimum in the range between 17 and 94 GHz.[38–40] Furthermore, field-independent orientation selection due to hyperfine anisotropy may be significant, as is the case for nitroxides, where anisotropy of the ^{14}N hyperfine coupling dominates the dispersion of resonance offsets at X-band frequencies and below.

The simplest way of suppressing orientation selection is variation of the magnetic field B_0 and summation of the primary data traces $V(t)$ (Figure 19.8).[41] The measurement is best performed as a two-dimensional experiment with separate storage of the traces at different B_0, as this allows for background fitting of individual traces and summation of form factors $F(t)$ and for checking and potentially further analyzing orientation selection. Background fitting of individual traces is advantageous if the background is monoexponential (dimensionality d = 3), as in this case background fitting of the summed primary data requires an additional parameter, the stretch exponent $d/3$. Summing traces does not necessarily lead to complete suppression of orientation selection[42] and entails some loss in signal-to-noise ratio compared to an experiment at a single magnetic field setting with an optimal excitation scheme and the same total measurement time, as only one of the traces in the two-dimensional experiment corresponds to such optimized positions of the excitation bands

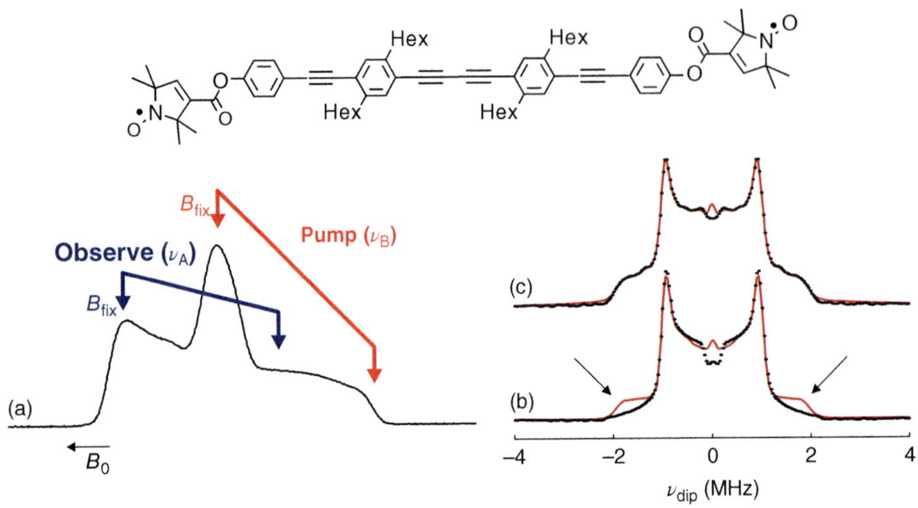

Figure 19.8. Averaging of orientation selection for a rigid nitroxide biradical at an X-band frequency of ≈9.3 GHz (data from Ref. 41). (a) At fixed observer frequency ν_A and pump frequency ν_B, the excitation positions in the spectrum are varied by a magnetic field sweep. (b) Experimental dipolar spectrum at the fixed observer and pump position B_{fix}, corresponding to optimal sensitivity (black), and theoretically expected spectrum without orientation selection (red). (c) Experimental dipolar spectrum averaged over the whole-frequency sweep (black), and theoretically expected spectrum without orientation selection (red)

for the A and B spins with respect to their spectral line shapes. Such loss is unavoidable if orientation selection is significant for the optimized settings. It is also possible to vary ν_A and ν_B at fixed B_0. However, this strategy requires manual readjustment of pulse power for measuring each individual trace and thus tends to result in a lower number of traces that are summed.

19.6 COMPLICATIONS AND REMEDIES

19.6.1 Insufficient Maximum Dipolar Evolution Time

The most serious limitation of DEER data quality is usually a too short t_{max} relative to that needed for reliably determining the distances encountered in $P(r)$. As a rule of thumb, upper distance limits for interpretation of $P(r)$ can be expressed in the form

$$r_{max} = L_{type} \cdot \left(\frac{t_{max}}{2\ \mu s}\right)^{1/3} \qquad (19.23)$$

where L_{type} is an upper distance limit for a given type of interpretation for a 2-μs-long dipolar evolution trace.[43] For recognizing whether long distances are at all present in the cluster, L_{type} takes the value of 6 nm, for determining the mean distance of a peak in $P(r)$ the value of 5 nm, for determining the width of such a peak the value of 4 nm, and for discussing the shape (asymmetry, shoulders) the value of 3 nm.

Owing to the stretched exponential decay of the observer echo roughly with $\exp[-(t_{max}/T_m)^{d/3}]$ and the scaling of the signal-to-noise ratio with the square root of total measurement time, the achievable t_{max} for a given sample is limited roughly to the value at which a signal-to-noise ratio of 5–20 in $D(t)$ can be obtained with a measurement time between 48 and 60 h. If this value of t_{max} is insufficient, either the sample needs to be measured with a more sensitive spectrometer or sample preparation needs to be changed. In most samples, phase relaxation of electron spins in the low-temperature limit (at temperatures of 80 K or below for trityl, at 50 K or below for MTSL, and at 10 K or below for transition and earth metal ions) is dominated by nuclear spin-diffusion-driven fluctuations of the hyperfine field. Hence, T_m can be lengthened by lowering the gyromagnetic ratio of the most abundant nuclear spin species. The simplest remedy is thus deuteration of the buffer, i.e., replacement of H_2O by D_2O and of the cryoprotectant, commonly glycerol, by a deuterated cryoprotectant. For membrane proteins, this strategy may complicate separation of liposomes by centrifugation unless lipids are also deuterated, which is expensive. For such samples, t_{max} can often be significantly lengthened by adding 10% deuterated glycerol as a cryoprotectant.[44] Even larger gains are achieved by deuterating not only the solvent but also the macromolecule itself. In such samples, measurement of mean distances up to 12 nm is realistic.[45,46]

In some samples, instantaneous diffusion[47,48] contributes significantly to phase relaxation. This happens if bulk concentration is too high (see Section 19.4.2) or if distribution of the clusters in the sample volume is strongly heterogeneous, as, for instance, for preparations of membrane proteins in liposomes.[49] In the latter case, local concentration may be too large for measuring at a sufficiently long t_{max} even at bulk concentrations that are close to the sensitivity limit. If an increase in the lipid-to-protein ratio does not solve the problem, it may be necessary to change to a different solubilization procedure, such as using bicelles,[50] nanodiscs,[51] or detergent micelles.[52]

19.6.2 Nuclear Modulation

The forbidden electron-nuclear transitions of the observer spin A are located roughly at $\nu_A \pm \nu_I$, where ν_I is the Zeeman frequency of nearby nuclei coupled to the observer spin. Hence, one of them is closer to the pump frequency than the allowed transitions by roughly $|\nu_I|$. Even if the excitation bands of observer and pump pulses do not significantly overlap for the allowed transitions, they may overlap for the forbidden transitions. Excitation of forbidden transitions of the observer spin A by the pump pulse causes nuclear modulation of the observer echo that lies in the same frequency range as dipolar modulation.[14]

Similarly to electron spin echo envelope modulation (ESEEM) experiments (see Chapter 18), the nuclear modulation exhibits blind spot behavior with respect to the fixed interpulse delays. Hence, it can be strongly suppressed by varying one of these delays in a manner that does not shift the zero time of the dipolar modulation.[53] In four-pulse DEER, usually τ_1 is varied at X-band in eight steps with an 8-ns increment in order to average proton modulation or in eight steps with a 56-ns increment in order to average deuterium modulation. At Q-band, eight steps with 8-ns increments are efficient for averaging deuterium modulation. At even higher frequencies, modulation from the abundant matrix nuclei is negligible. If nuclear modulation contributions are suspected, it is good practice to measure DEER data at two frequency bands. Dipolar modulation frequencies do not depend on the static magnetic field, while nuclear modulation frequencies do.

19.6.3 Overlap of A and B Spin Excitation Bands

Complete separation of A and B spin excitation can be achieved only using a bimodal resonator[6,11] or, potentially, using shaped pulses (see Chapter 21) on both observer and pumped spins. For reasons of versatility and ease of setup, four-pulse DEER is usually performed with strongly overcoupled single-mode resonators. In order not to lower detection sensitivity and conversion efficiency too much, overcoupling is limited to the narrowest bandwidth that still allows performing the experiment with optimum placement of observer and pump excitation bands in the spectra of A and B spins, respectively. In other words, the difference between observer frequency ν_A and pump frequency ν_B is kept as low as possible given the excitation bandwidths that can be achieved with the available microwave power. The rather broad wings of the excitation profiles of rectangular microwave pulses (Figure 19.6) slightly overlap under these conditions.

For spin pairs where the pump pulse also inverts A spins, spin dynamics is the same as in the '2+1' pulse train experiment.[54] Such spin dynamics causes a dipolar modulation contribution that overlaps with the one in the DEER pathway and another dipolar modulation contribution that has its zero time at the point where the pump pulse and the last observer π pulse overlap. The effect can thus be recognized by a telltale increase of $F(t)/F(0)$ near t_{max}. In many cases, the best remedy is a decrease of t_{max} at fixed τ_2, or a slight increase of τ_2 at fixed t_{max}. The decrease of t_{max} can also be applied after the measurement during signal processing. If both a shortening of t_{max} and a lengthening of τ_2 are unacceptable, the difference between ν_A and ν_B must be increased.

19.6.4 Heterogeneous Relaxation

Different contributions in the distance distribution can stem from different cluster configurations that can correspond, in turn, to large-scale conformational changes in the macromolecule. Phase memory time T_m of the observer spin A is not necessarily the same for all conformations. For instance, exposure of a spin label to the aqueous phase or to the lipid phase may change during a conformational transition. Differences in T_m are even more likely in clusters with more than two spins, where interaction between electron spins may significantly contribute to phase relaxation. As DEER data are usually measured at $t_{max} > T_m$, such differences can influence the relative contributions of the various conformations to the experimentally determined $P(r)$.[55]

Hence, if distance distributions are to be interpreted in terms of relative contributions of several distinct conformations, it is good practice to check whether the form factor depends on the choice of t_{max}. In cases where the dependence of T_m on conformational state is caused by a change in solvent accessibility, deuteration of the solvent may strongly reduce the effect.[55]

19.7 SPINS $S > 1/2$

Spins $S > 1/2$ may occur as native paramagnetic centers, such as Mn(II) ($S = 5/2$) in proteins, may act as good paramagnetic substitutes for native diamagnetic metal centers, such as Mn(II) for Mg(II) and Gd(III) ($S = 7/2$) for Ca(II), or may be advantageous as spin labels[56,57] because of their better redox stability compared to nitroxide radicals, line narrowing rather than broadening at high fields, and an isotropic Hamiltonian of the central $-1/2 \leftrightarrow 1/2$ transition even at high fields.[58] Polarization from noncentral transitions can be used to enhance the signal of the central transition.[59] Furthermore, high-spin species are particularly valuable in the context of spectroscopically orthogonal labeling of complex systems, as their higher transition moments and different relaxation properties as compared to organic radicals enhance selectivity.[9,60]

On the downside, even if the central transition is narrower than the lines of organic radicals at fields as high as 3.5 T, the whole spectrum is considerably broader implying low modulation depth, and in order to reach the low-temperature limit of T_m, the experiments may require cooling to 6–15 K at much higher helium consumption than required with organic radicals. This disadvantage is compensated by the much shorter T_1 even at these low temperatures, which allows for faster repetition of the experiment at much larger Boltzmann polarization than is available with organic radicals.

DEER involving high-spin half-integer spins can be analyzed similarly to a $S = 1/2$ pair as discussed in this chapter, taking into account potential caveats. The main complications with $S > 1/2$ species arise from the presence of noncentral transitions and from the possibility of mixing of electron spin levels with different magnetic quantum numbers m_S. In particular, the (m_{SA}, m_{SB}) = $(-1/2, 1/2)$ and $(1/2, -1/2)$ levels may have a small energy difference and may be mixed by the pseudosecular $S_A^+S_B^- + S_A^-S_B^+$ term of the dipole–dipole coupling, even if the frequency difference $|\nu_A - \nu_B|$ is much larger than this coupling, because either the pump or observer transition may involve an $m_S = +3/2$ or $m_S = -3/2$ level.[61,62] This effect leads to broadening of the distance distribution, which is particularly serious at short distances and small ZFS. Although the broadening still persists even if $|\nu_A - \nu_B| \gg \omega_{AB}/2\pi$, it should decrease with increasing $|\nu_A - \nu_B|$, as such an increase leads, on average, to excitation of spin packets with larger zero-field splitting and thus to a larger second-order splitting between the $(-1/2, 1/2)$ and $(1/2, -1/2)$ transitions. In agreement with this expectation, less broadening is observed in relaxation-induced dipolar modulation enhancement[63] (RIDME) experiments,[64] where relaxation-induced B spin flips occur for a distribution of ν_B that covers all satellite transitions up to the largest zero-field splittings.

Potential artifacts at integer multiples of the dipole–dipole coupling can be introduced by inversion of $|\Delta m_S| > 1$ transitions by the pump pulse.[65] Such excitation is, for instance, possible if the pump pulse is applied near the maximum of the $-1/2 \leftrightarrow 1/2$ transition, where some spin packets have small splittings between the different single-quantum transitions. For a well-defined distance r, the artifacts are expected at distances $r/|\Delta m_S|^{1/3}$, with an amplitude that decreases with increasing $|\Delta m_S|$. This complication, however, has not been observed in the many W-band Gd(III)–Gd(III) applications reported so far. At the expense of modulation depth, the artifacts can be suppressed by applying the pump pulse at a sufficiently large offset from the $-1/2 \leftrightarrow 1/2$ transition.

The pump pulse may also excite a transition that shares one level with the observer transition. This leads to partial transfer of coherence from the observer transition to a double-quantum transition and thus to a loss in observer echo intensity. This missing echo intensity can be partially recovered, at the expense of lower modulation depth, by decreasing the flip angle of the pump pulse. At Q-band frequencies, pump pulses with reduced flip angle have been found to provide better total sensitivity than pump pulses $(\pi)_B$ in both Gd(III)–Gd(III) and Gd(III)–nitroxide

DEER measurements.[65,66] The effect depends on second-order broadening of the central transition and thus on both the magnitude of the zero-field splitting and the magnetic field, and furthermore on relative positioning of the pump and observer excitation bands. Hence, an optimal pump pulse flip angle corresponds to given settings of these experimental parameters and to a given ligand.

19.8 DATA ANALYSIS

19.8.1 Tikhonov Regularization

The main advantage of dipolar spectroscopy of electron spin pairs over fluorescence resonance energy transfer (FRET) measurements on chromophore pairs in the same distance range lies in the fact that EPR techniques can provide much more reliable distance distributions with much less effort in data analysis. The least biased way of obtaining such distance distributions $P(r)$ from the dipolar evolution function $D(t)$ is Tikhonov regularization[4,67] under the constraint $P(r) \geq 0$.[5,68] Tikhonov regularization rather than ordinary least-squares fitting is required because the problem of determining $P(r)$ from $D(t)$ is ill-posed, as is explained in the following.

Experimental data for $D(t)$ are obtained as a vector $\boldsymbol{D}$ with n data points defined at times $t_i = i\Delta t$ ($i = 0 \ldots n-1$), where Δt is the increment for t during data acquisition. The physical model for such data is elegantly formulated by a matrix equation

$$\boldsymbol{D} = \boldsymbol{KP} \tag{19.24}$$

where $\boldsymbol{P}$ is the distance distribution as a vector with m points defined at distances $r_j = r_{\min} + j\Delta r$ ($j = 0 \ldots m-1$) and $\boldsymbol{K}$ the $n \times m$ kernel matrix with elements defined by

$$K_{ij} = \int_0^1 \cos\left(\frac{C_{AB}}{\left(r_{\min} + j\Delta r\right)^3}(1 - 3z^2)i\Delta t\right) \mathrm{d}z \tag{19.25}$$

If $\boldsymbol{D}$ perfectly corresponded to this model, $\boldsymbol{P}$ could be computed simply by choosing $m = n$ and inverting the kernel $\boldsymbol{K}$. However, it turns out that even with the best possible choice of $r_{\min}$ and Δr, the kernel matrix $\boldsymbol{K}$ has a condition number (ratio of the largest to the smallest singular value) significantly larger than 1.[3] This indicates an ill-posed problem: A small contribution to $\boldsymbol{D}$ that does not fit the model will lead to a large error in $\boldsymbol{P}$ (Figure 19.9a). In less abstract terms, the rows of $\boldsymbol{K}$ are not orthogonal, so that an increase in P_j can be largely compensated by a decrease in P_{j-1} and P_{j+1}. That way the solution partially fits deviations of the data from the model, such as noise. To quite some extent, the solution can be stabilized by requiring $P_j \geq 0$ for all j, which must apply for any probability distribution. However, even with this constraint, the calculated distance distribution is unrealistically ragged already at a low noise level.

The solution can be strongly stabilized by imposing that $P(r)$ is a smoothly varying function, which is reasonable if it stems from a large number of underlying conformations of the macromolecule and possibly of the linkers of the labels. Mathematically, smoothness can be defined as a small square norm of the first or second derivative of $P(r)$. Tikhonov regularization provides the vector $\boldsymbol{P}$ that minimizes the weighted sum G of the square deviation between the data and the model and the square norm of the first or second derivative,

$$G(\alpha) = \|\boldsymbol{D} - \boldsymbol{KP}\|^2 + \alpha\|\hat{\boldsymbol{L}}\boldsymbol{P}\|^2 = \rho + \alpha\eta \tag{19.26}$$

where α is the regularization parameter and $\hat{\boldsymbol{L}}$ a discrete differential operator matrix corresponding to the first or second derivative. For regularization of the time-distance problem, the second derivative was found to provide somewhat better results.[5]

The optimal value of the regularization parameter is usually not known beforehand. In most cases, a parametric plot of log η at minimum $G(\alpha)$ versus log ρ at minimum $G(\alpha)$ for a range of α values is L shaped (Figure 19.9b).[69] The regularization parameter corresponding to the mathematically best compromise between fitting experimental data and keeping the distance distribution smooth is located at the corner of the L curve. This is not necessarily the physically most plausible solution. If information on the expected widths of peaks in $P(r)$ is available, it can make sense to override the regularization parameter determined from the L curve and select it by adjusting the distance distribution to the expected peak width.

Regularization generally entails some artificial broadening of the distance distribution (Figure 19.9c). Too large regularization parameters α cause strong broadening (Figure 19.9d) and a clearly visible deviation of the simulated form factor from the experimental form factor (Figure 19.9e). In contrast, the fit residual does not deviate from white noise when using the regularization parameter corresponding to the corner of the L curve (Figure 19.4d). It is important to realize

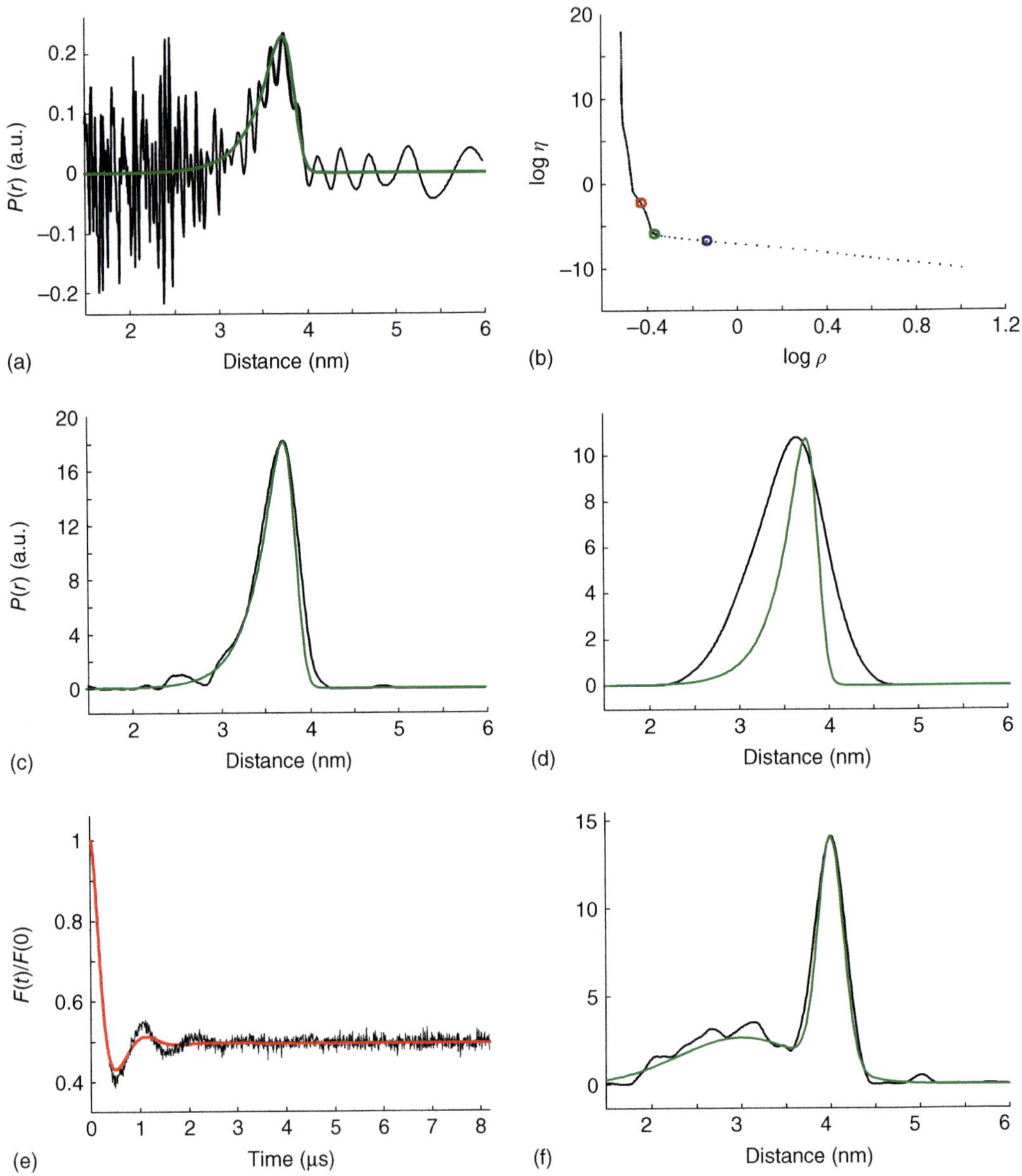

Figure 19.9. Tikhonov regularization. Panels (a–e) refer to the data displayed in Figure 19.4. (a) Distance distribution $P(r)$ obtained from $D(t)$ in Figure 19.4(d) without nonnegativity constraint and a too small regularization parameter $\alpha = 1$ (black) and original distance distribution (green). (b) L curve obtained from the $D(t)$ displayed in Figure 19.4(d). The corner is marked by a green circle, insufficient regularization corresponding to panel (a) by a red circle, and excessive regularization is marked by a blue circle. (c) Distance distribution $P(r)$ obtained by Tikhonov regularization with optimum regularization parameter (black) and the constraint $P(r) > 0$ and input distance distribution (green). (d) Distance distribution $P(r)$ obtained by excessive regularization (black) and input distance distribution (green). (e) Input form factor (black) and fit with excessive regularization (red). (f) Bimodal input distance distribution (simulation input) consisting of a narrow ($\langle r_2 \rangle = 4$ nm, $\sigma_{r1} = 0.2$ nm) and a broad ($\langle r_2 \rangle = 3$ nm, $\sigma_{r2} = 1$ nm) Gaussian peak (green) and distance distribution recovered by Tikhonov regularization with optimal regularization parameter $\alpha = 75$ (black). Other simulation parameters are as in Figure 19.4

that α cannot be optimal at the same time for narrow and broad peaks (Figure 19.9f). This situation is frequently encountered with protein preparations, where the narrow peak corresponds to the properly folded and solubilized molecule, whereas the broad peak corresponds to unfolded or aggregated material. It is good practice to resist interpretation of the artificially ragged broad peak in terms of minor well-defined conformations.

Tikhonov regularization without the nonnegativity constraint $P(r) \geq 0$ is computationally very efficient and provides L curves with a more distinct corner than regularization with this constraint. With the solution $\tilde{\boldsymbol{P}}$ obtained without the constraint and the regularization parameter α already known, the solution with the constraint can be obtained by solving the least-squares problem with nonnegativity constraint

$$\tilde{\boldsymbol{D}} = \boldsymbol{C}\boldsymbol{P} \tag{19.27}$$

where $\tilde{\boldsymbol{D}}$ is the vector that results from appending $\hat{\boldsymbol{D}}\tilde{\boldsymbol{P}}$ to $\boldsymbol{D}$ and $\boldsymbol{C}$ is a matrix that results from appending $\alpha\hat{\boldsymbol{D}}$ to $\boldsymbol{K}$.

19.8.2 Fitting of Model Distributions

Solution of the ill-posed problem can also be stabilized by fitting parameterized distribution functions with a small number of parameters.[70–72] Sometimes, it is overlooked that, by implying a less flexible physical model than kernel $\boldsymbol{K}$, this approach introduces a deviation between simulated and experimental data that adds to noise and experimental artifacts. As the relation between time and distance domain corresponds to an ill-posed problem, this additional deviation may cause errors in the parameters that are hard to estimate.

By far the most popular choice of parameterized distributions is a small number g of Gaussian peaks. The number of free fit parameters is $3g-1$, composed of g mean distances, g standard deviations, and $g-1$ relative populations. For the standard deviations (widths) to be stable, all mean distances must be shorter than the limit given by equation (19.23) with $L_{\text{type}} = 4$ nm. Without validation (Section 19.8.3), it is hard to predict errors in relative populations and standard deviations. If data quality is too low to recognize the g peaks in a distribution $P(r)$ obtained by Tikhonov regularization (at any regularization value α) and to estimate their widths and relative amplitudes at least roughly and no validation was performed, there is no basis for the assumption that the fit parameters are reliable.

Model-based fits may reasonably reproduce the experimental form factor and yet yield a distance distribution that is in much worse agreement with the true distribution than the one obtained by Tikhonov regularization, as is illustrated in Figure 19.10. The peak shape in the input distribution that is quite nicely reproduced by Tikhonov regularization (Figure 19.9c) is not matched very well by fitting two Gaussian peaks. Note also that the two peaks do not correspond to physical reality, as the asymmetry of the input

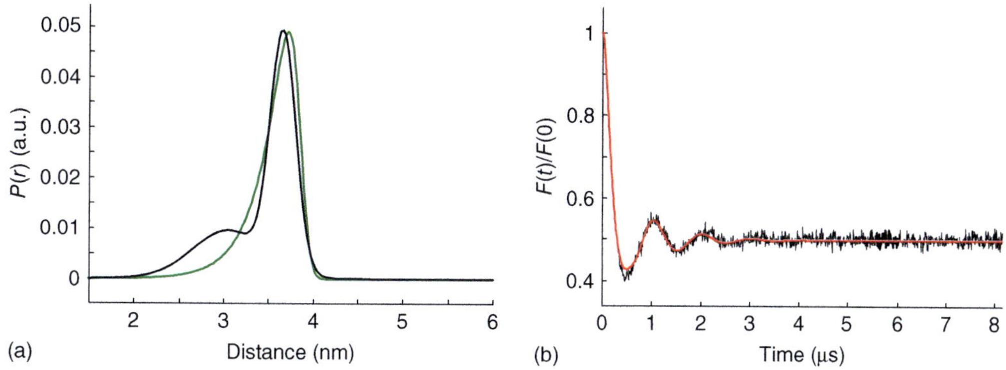

Figure 19.10. Fitting of data from an underlying asymmetric distance distribution by two Gaussian peaks. (a) Input distance distribution of the simulation (green) and distance distribution recovered by Gaussian fitting (black). Compare to Figure 19.9(c) for the result of Tikhonov regularization. (b) Input form factor (black) and fit by the Gaussian distance distribution (red)

distribution is explained by a continuous distribution of conformations of a single polymer chain.

It is good practice to use model-based fitting only in cases where good reasons exist to assume a particular model or where only the mean distance and width of a monomodal distribution are to be discussed. Even in such cases, the distance distribution obtained by Tikhonov regularization should be inspected and shown.

19.8.3 Validation

The relation between error in the dipolar evolution function $D(t)$ and error in the distance distribution $P(r)$ is strongly nonlinear. Hence, no analytical theory exists for predicting uncertainty of features in $P(r)$. This uncertainty can be estimated by numerical tests. In such tests, all uncertain parameters in data processing, such as background fit ranges and background dimensionality, are systematically varied. The influence of noise is considered by a Monte Carlo approach, in which pseudorandom numbers with a normal distribution with standard deviation corresponding to the experimental noise level are added to the data points in $V(t)/V(0)$. Each individual data set is then processed in the same way, preferably by Tikhonov regularization, and the mean distance distribution $P(r)$, standard deviations, and lower and upper limits for each data point are computed. This validation approach assumes that different error sources are uncorrelated. If the uncertainty estimates for input data and processing parameters are correct and complete, uncertainty in $P(r)$ is somewhat overestimated. Validation may thus turn out false negatives, i.e., some features that are significant may appear insignificant. Unless input uncertainties are underestimated, validation does not result in false positives.

Validation can thus tell from which features in $P(r)$ safe conclusions may be drawn. It is good practice to discuss features that vanish on validation only with cautionary remarks. Validation can be skipped if it is clearly recognized from the input data that $t_{max} > 2t_{dec}$, if any features that are discussed are at distances below the limit set by equation (19.23) with the appropriate L_{type}, and if signal-to-noise ratio in $D(t)$ exceeds 20 at all points. If only a single peak exists in $P(r)$, then its mean distance $\langle r \rangle$ is a very stable parameter that can be estimated quite well with any type of data analysis,[5] as long as $t_{max} > 2t_{dec}$ and equation (19.23) is fulfilled with $L_{type} = 5$ nm.

19.9 CONCLUSIONS

Dipolar spectroscopy of electron spin pairs is a valuable technique for characterization of structure and nanoscale morphology of soft matter, biomacromolecules, and macromolecular complexes in the range from about 1.5 to 12 nm. The upper distance limit varies strongly with sample class, with the longest distances accessible only with the most sensitive spectrometers at complete deuteration and for a homogeneous distribution of the molecules at concentrations of about 10 μM. For demanding samples, such as membrane proteins in liposomes, the upper limit is in the range between 5 and 6 nm.

Double-resonance methods of dipolar spectroscopy simplify the effective topology of the spin system from a fully coupled network to a system where only pairwise interactions between the observer spin and any one of the pumped spins are relevant. As a result, the signal factors into pair contributions and, in particular, into a factor from pumped spins in the same cluster (molecule) as the observer spin and a background factor from remote pumped spins. If background decay during observation of the dipolar evolution is significant, double-resonance methods can thus provide higher quality distance distributions than single-resonance methods, at the expense of somewhat lower sensitivity and a somewhat longer lower distance limit.

The distinction between observer spins A and pumped spins B in double-resonance methods allows for using spectroscopically orthogonal labels, a valuable addition to the arsenal of dipolar spectroscopy for complex systems. Some artifact contributions can be suppressed by an appropriate choice of the pump-observer frequency difference $|\nu_A - \nu_B|$. Orientation selection provides access to relative molecular orientation of the two coupled spins and to the orientation of the spin–spin vector with respect to the molecular frames. An estimate for the number of spins per cluster is available from the modulation depth.

Processing of primary data $V(t)$ to obtain the dipolar evolution function $D(t)$ is often fraught with uncertainties regarding the background fit. Computation of the distance distribution $P(r)$ from $D(t)$ is an ill-posed problem whose solution is best stabilized by Tikhonov regularization. As the problem is ill posed, discussion of any minor features in $P(r)$ requires careful validation.

ACKNOWLEDGMENTS

I am grateful to Andrin Doll, Yevhen Polyhach, and Maxim Yulikov for valuable discussions and to Frauke Breitgoff for careful proofreading of the manuscript.

FURTHER READING

P. P. Borbat and J. H. Freed, *Struct. Bond.*, 2014, **152**, 1.

A. Feintuch, G. Otting, and D. Goldfarb, *Methods Enzymol.*, 2015, **563**, 415.

G. Jeschke, *Struct. Bond.*, 2014, **152**, 83.

G. Jeschke and Y. Polyhach, *Phys. Chem. Chem. Phys.*, 2007, **9**, 1895.

G. Jeschke and H. W. Spiess, *Lect. Notes. Phys.*, 2006, **684**, 21.

T. F. Prisner, A. Marko, and S. T. Sigurdsson, *J. Magn. Reson.*, 2015, **252**, 187.

O. Schiemann and T. F. Prisner, *Q. Rev. Biophys.*, 2007, **40**, 1.

Y. D. Tsvetkov, A. D. Milov, and A. G. Maryasov, *Russ. Chem. Rev.*, 2008, **77**, 487.

M. Yulikov, *Electron Paramagn. Reson.*, 2015, **24**, 1.

REFERENCES

1. A. D. Milov, B. D. Naumov, and Y. D. Tsvetkov, *Appl. Magn. Reson.*, 2004, **26**, 587.
2. J. E. Banham, C. M. Baker, S. Ceola, I. J. Day, G. H. Grant, E. J. J. Groenen, C. T. Rodgers, G. Jeschke, and C. R. Timmel, *J. Magn. Reson.*, 2008, **191**, 202.
3. G. Jeschke, A. Koch, U. Jonas, and A. Godt, *J. Magn. Reson.*, 2002, **155**, 72.
4. Y. Chiang, P. P. Borbat, and J. H. Freed, *J. Magn. Reson.*, 2005, **172**, 279.
5. G. Jeschke, G. Panek, A. Godt, A. Bender, and H. Paulsen, *Appl. Magn. Reson.*, 2004, **26**, 223.
6. A. D. Milov, K. M. Salikhov, and M. D. Shchirov, *Fiz. Tverd. Tela (Leningrad)*, 1981, **23**, 975.
7. A. D. Milov and Y. D. Tsvetkov, *Appl. Magn. Reson.*, 1997, **12**, 495.
8. E. Narr, A. Godt, and G. Jeschke, *Angew. Chem,. Int. Ed.*, 2002, **41**, 3907.
9. P. Lueders, G. Jeschke, and M. Yulikov, *J. Phys. Chem. Lett.*, 2011, **2**, 604.
10. D. E. Kaplan and E. L. Hahn, *J. Phys. Rad.*, 1958, **19**, 821.
11. A. D. Milov, A. B. Ponomarev, and Y. D. Tsvetkov, *Chem. Phys. Lett.*, 1984, **110**, 67.
12. A. D. Milov, Y. A. Grishin, S. A. Dzuba, and Y. D. Tsvetkov, *Appl. Magn. Reson.*, 2011, **41**, 59.
13. R. E. Martin, M. Pannier, F. Diederich, V. Gramlich, M. Hubrich, and H. W. Spiess, *Angew. Chem. Int. Ed.*, 1998, **37**, 2834.
14. M. Pannier, S. Veit, A. Godt, G. Jeschke, and H. W. Spiess, *J. Magn. Reson.*, 2000, **142**, 331.
15. J. E. Lovett, B. W. Lovett, and J. Harmer, *J. Magn. Reson.*, 2012, **223**, 98.
16. P. P. Borbat, E. R. Georgieva, and J. H. Freed, *J. Phys. Chem. Lett.*, 2013, **4**, 170.
17. P. E. Spindler, I. Waclawska, B. Endeward, J. Plackmeyer, C. Ziegler, and T. F. Prisner, *J. Phys. Chem. Lett.*, 2015, **6**, 4331.
18. K. M. Salikhov, I. T. Khairuzhdinov, and R. B. Zaripov, *Appl. Magn. Reson.*, 2014, **45**, 573.
19. K. M. Salikhov and I. T. Khairuzhdinov, *Appl. Magn. Reson.*, 2015, **46**, 67.
20. E. Salvadori, M. W. Fung, M. Hoffmann, H. L. Anderson, and C. W. M. Kay, *Appl. Magn. Reson.*, 2015, **46**, 359.
21. Z. Yang, Y. Liu, P. Borbat, J. L. Zweier, J. H. Freed, and W. L. Hubbell, *J. Am. Chem. Soc.*, 2012, **134**, 9950.
22. G. Y. Shevelev, O. A. Krumkacheva, A. A. Lomzov, A. A. Kuzhelev, O. Y. Rogozhnikova, D. V. Trukhin, T. I. Troitskaya, V. M. Tormyshev, M. V. Fedin, D. V. Pyshnyi, and E. G. Bagryanskaya, *J. Am. Chem. Soc.*, 2014, **136**, 9874.
23. G. Jeschke, M. Sajid, M. Schulte, and A. Godt, *Phys. Chem. Chem. Phys.*, 2009, **11**, 6580.
24. T. von Hagens, Y. Polyhach, M. Sajid, A. Godt, and G. Jeschke, *Phys. Chem. Chem. Phys.*, 2013, **15**, 5854.
25. B. E. Bode, D. Margraf, J. Plackmeyer, G. Duerner, T. F. Prisner, and O. Schiemann, *J. Am. Chem. Soc.*, 2007, **129**, 6736.
26. D. Hilger, H. Jung, E. Padan, C. Wegener, K. Vogel, H. Steinhoff, and G. Jeschke, *Biophys. J.*, 2005, **89**, 1328.
27. R. G. Larsen and D. J. Singel, *J. Chem. Phys.*, 1993, **98**, 5134.

28. C. Abe, D. Klose, F. Dietrich, W. H. Ziegler, Y. Polyhach, G. Jeschke, and H. J. Steinhoff, *J. Magn. Reson.*, 2012, **216**, 53.
29. I. Tkach, S. Pornsuwan, C. Hobartner, F. Wachowius, S. T. Sigurdsson, T. Y. Baranova, U. Diederichsen, G. Sicoli, and M. Bennati, *Phys. Chem. Chem. Phys.*, 2013, **15**, 3433.
30. A. Marko, D. Margraf, P. Cekan, S. T. Sigurdsson, O. Schiemann, and T. F. Prisner, *Phys. Rev. E*, 2010, **81**, 021911.
31. A. Marko, D. Margraf, H. Yu, Y. Mu, G. Stock, and T. Prisner, *J. Chem. Phys.*, 2009, **130**, 064102.
32. A. Marko and T. F. Prisner, *Phys. Chem. Chem. Phys.*, 2013, **15**, 619.
33. V. P. Denysenkov, T. F. Prisner, J. Stubbe, and M. Bennati, *Proc. Natl. Acad. Sci U. S. A.*, 2006, **103**, 13386.
34. A. Savitsky, A. A. Dubinskii, M. Flores, W. Lubitz, and K. Mobius, *J. Phys. Chem. B*, 2007, **111**, 6245.
35. Y. Polyhach, A. Godt, C. Bauer, and G. Jeschke, *J. Magn. Reson.*, 2007, **185**, 118.
36. J. E. Lovett, A. M. Bowen, C. R. Timmel, M. W. Jones, J. R. Dilworth, D. Caprotti, S. G. Bell, L. L. Wong, and J. Harmer, *Phys. Chem. Chem. Phys.*, 2009, **11**, 6840.
37. W. L. Hubbell, D. S. Cafiso, and C. Altenbach, *Nat. Struct. Biol.*, 2000, **7**, 735.
38. P. P. Borbat, R. H. Crepeau, and J. H. Freed, *J. Magn. Reson.*, 1997, **127**, 155.
39. Y. Polyhach, E. Bordignon, R. Tschaggelar, S. Gandra, A. Godt, and G. Jeschke, *Phys. Chem. Chem. Phys.*, 2012, **14**, 10762.
40. P. A. S. Cruickshank, D. R. Bolton, D. A. Robertson, R. I. Hunter, R. J. Wylde, and G. M. Smith, *Rev. Sci. Instrum.*, 2009, **80**, 103102.
41. A. Godt, M. Schulte, H. Zimmermann, and G. Jeschke, *Angew. Chem. Int. Ed.*, 2006, **45**, 7560.
42. G. Jeschke, M. Sajid, M. Schulte, N. Ramezanian, A. Volkov, H. Zimmermann, and A. Godt, *J. Am. Chem. Soc.*, 2010, **132**, 10107.
43. G. Jeschke, *Annu. Rev. Phys. Chem.*, 2012, **63**, 419.
44. E. Bordignon and Y. Polyhach, *Methods Mol. Biol.*, 2013, **974**, 329.
45. R. Ward, A. Bowman, H. El-Mkami, T. Owen-Hughes, and D. G. Norman, *J. Am. Chem. Soc.*, 2009, **131**, 1348.
46. R. Ward, A. Bowman, E. Sozudogru, H. El-Mkami, T. Owen-Hughes, and D. G. Norman, *J. Magn. Reson.*, 2010, **207**, 164.
47. J. R. Klauder and P. W. Anderson, *Phys. Rev.*, 1962, **125**, 912.
48. A. Schweiger and G. Jeschke, Principles of Pulse Electron Paramagnetic Resonance, Oxford University Press: Oxford, 2001.
49. R. Dastvan, B. E. Bode, M. P. R. Karuppiah, A. Marko, S. Lyubenova, H. Schwalbe, and T. F. Prisner, *J. Phys. Chem. B*, 2010, **114**, 13507.
50. E. R. Georgieva, T. F. Ramlall, P. P. Borbat, J. H. Freed, and D. Eliezer, *J. Am. Chem. Soc.*, 2008, **130**, 12856.
51. P. Zou and H. S. Mchaourab, *Biophys. J.*, 2010, **98**, L18.
52. G. Jeschke, A. Bender, T. Schweikardt, G. Panek, H. Decker, and H. Paulsen, *J. Biol. Chem.*, 2005, **280**, 18623.
53. G. Jeschke, A. Bender, H. Paulsen, H. Zimmermann, and A. Godt, *J. Magn. Reson.*, 2004, **169**, 1.
54. V. V. Kurshev, A. M. Raitsimring, and Y. D. Tsvetkov, *J. Magn. Reson.*, 1989, **81**, 441.
55. J. L. Baber, J. M. Louis, and G. M. Clore, *Angew. Chem. Int. Ed. Engl.*, 2015, **54**, 5336.
56. A. M. Raitsimring, C. Gunanathan, A. Potapov, I. Efremenko, J. M. L. Martin, D. Milstein, and D. Goldfarb, *J. Am. Chem. Soc.*, 2007, **129**, 14138.
57. A. Potapov, H. Yagi, T. Huber, S. Jergic, N. E. Dixon, G. Otting, and D. Goldfarb, *J. Am. Chem. Soc.*, 2010, **132**, 9040.
58. D. Goldfarb, *Phys. Chem. Chem. Phys.*, 2014, **16**, 9685.
59. A. Doll, M. Qi, S. Pribitzer, N. Wili, M. Yulikov, A. Godt, and G. Jeschke, *Phys. Chem. Chem. Phys.*, 2015, **17**, 7334.
60. M. Yulikov, P. Lueders, M. F. Warsi, V. Chechik, and G. Jeschke, *Phys. Chem. Chem. Phys.*, 2012, **14**, 10732.
61. A. Dalaloyan, M. Qi, S. Ruthstein, S. Vega, A. Godt, A. Feintuch, and D. Goldfarb, *Phys. Chem. Chem. Phys.*, 2015, **17**, 18464.
62. A. G. Maryasov, M. K. Bowman, and Y. D. Tsvetkov, *Appl. Magn. Reson.*, 2006, **30**, 683.
63. L. V. Kulik, S. A. Dzuba, I. A. Grigoryev, and Y. D. Tsvetkov, *Chem. Phys. Lett.*, 2001, **343**, 315.
64. S. Razzaghi, M. Qi, A. I. Nalepa, A. Godt, G. Jeschke, A. Savitsky, and M. Yulikov, *J. Phys. Chem. Lett.*, 2015, **5**, 3970.

65. A. Doll, M. Qi, N. Wili, S. Pribitzer, A. Godt, and G. Jeschke, *J. Magn. Reson.*, 2015, **259**, 153.
66. L. Garbuio, E. Bordignon, E. K. Brooks, W. L. Hubbell, G. Jeschke, and M. Yulikov, *J. Phys. Chem. B*, 2013, **117**, 3145.
67. A. N. Tikhonov, *Dokl. Akad. Nauk SSSR*, 1963, **151**, 501.
68. J. Weese, *Comput. Phys. Commun.*, 1992, **69**, 99.
69. P. C. Hansen, *SIAM Rev.*, 1992, **34**, 561.
70. M. Pannier, V. Schadler, M. Schops, U. Wiesner, G. Jeschke, and H. W. Spiess, *Macromolecules*, 2000, **33**, 7812.
71. G. Jeschke, V. Chechik, P. Ionita, A. Godt, H. Zimmermann, J. Banham, C. R. Timmel, D. Hilger, and H. Jung, *Appl. Magn. Reson.*, 2006, **30**, 473.
72. K. I. Sen, T. M. Logan, and P. G. Fajer, *Biochemistry*, 2007, **46**, 11639.

Chapter 20

Dipolar Spectroscopy – Single-resonance Methods

Peter P. Borbat and Jack H. Freed

Cornell University, Ithaca, NY, USA

20.1 INTRODUCTION

Distance measurements on the nanometer scale by pulse EPR have developed into a widely used biophysical method, for which a number of applications and completed studies exist, benefiting particularly the field of structural biology. A key early application of pulse EPR was concerned with determining the spatial distributions of paramagnetic centers produced in solids by radiolysis. The main approaches to reporting on spatial distributions of radicals in those cases were based on measuring relaxation effects on the primary or stimulated echo, from which concentration-dependent contributions to relaxation could be estimated.[1,2] The local concentrations affected spectral diffusion in a complicated way due to the inherent complexity of the underlying relaxation processes, but in a more tractable way by the mechanism of 'instantaneous diffusion' (ID),[3–5] which is due to spin echo dephasing caused by the static dipole–dipole interactions among unpaired electron spins of the radicals generated in the radiation tracks. These studies yielded insights into the nature of the dominant solid-state relaxation mechanisms and underlying stochastic processes.[1,2] Notably, the distinct electron spin-echo envelope modulation (ESEEM) in the primary echo (PE) decay due to the dipolar interactions in spin pairs of $SO_4{}^-$ radicals produced in single crystal of $K_2S_2O_8$ was reported,[3,6] showing that the echo decay could be used to determine distances. However, in typical amorphous solids, the echo decays were dominated by phase relaxation and *nuclear* ESEEM,[7–9] making extracting distance information a very difficult task that was further complicated by long dead times ≈0.3–0.5 μs after the microwave (MW) pulses, leading to a loss of the key part of the signal emanating from the *electron spin dipolar coupling* related ESEEM.

EPR Spectroscopy: Fundamentals and Methods.
Edited by Daniella Goldfarb and Stefan Stoll

ISBN: 978-1-119-16299-5
Also published in eMagRes (online edition)
DOI: 10.1002/9780470034590.emrstm1519

Ideally, such experiments should be performed with very strong or 'hard' pulses that irradiate the whole spectrum, although the dead time could still be a serious impediment. These issues were overcome when a method originally developed in NMR[10] was implemented in EPR (in 1981)[11] as DEER ESE (*d*ouble *electron–electron* *r*esonance in *e*lectron *s*pin-*e*cho) or simply as DEER (*d*ouble *electron–electron* *r*esonance), and since 1998 another name PELDOR (*p*ulsed *el*ectron *do*uble *r*esonance) has been in use.[12] In this method, the 'pump' pulse from an MW power source is applied to affect the amplitude of the electron spin echo formed by the two-pulse 'detection' sequence derived from another MW pulse source operating at a different frequency (see Figure 20.1a and Chapter 19).[11] The spectral excitations at the two frequencies should not overlap, so the response to the pump pulse is then primarily determined by the electron–electron spin dipolar (as well as electron spin exchange) couplings, producing distinct changes in the amplitude of the spin-echo recorded vs time delay of the pump pulse. The instrumental setup used two independent high-power MW sources, a TWTA (traveling wave tube amplifier) and a magnetron, applied to an MW bimodal cavity resonator housing the sample.

Relaxation effects were small because a fixed delay was used between the detection MW pulses, which were relatively soft and applied at a substantial frequency separation from the pump pulse leading to much reduced nuclear ESEEM.[8] It was later shown that a single TWTA operating in the linear regime can be used for the task,[14] but before that the single-frequency technique of '2+1' (Figure 20.1a) was introduced and applied to biological systems[3,15] to overcome the need for two high-power MW sources and associated bimodal resonators and to enable distance measurements on narrow EPR spectra. Subsequently, in another development, the addition of a refocusing pulse to the three-pulse sequence yielded the four-pulse DEER sequence (Figure 20.1b) simplifying technical issues by eliminating the need for the second MW power source.[16–18] Earlier in an initial study,[19] the fully coherent single-resonance technique of DQC EPR was introduced. It was successfully implemented at K_u band (≈17 GHz) as a powerful highly sensitive method offering several distinct advantages based on the use of double-quantum filtering of the dipolar signal as well as the use of 'constant time' pulse sequences[14,20–22] minimizing relaxation effects. It yields clean dipolar signals, zero dead time, and a broad distance range. Since then, there have been two principal *single- and double-resonance techniques* in existence, which are known collectively as *p*ulse *d*ipolar EPR *s*pectroscopy (or PDS for short).

PDS has since been enriched with several single- and double-resonance type techniques,[21–25]

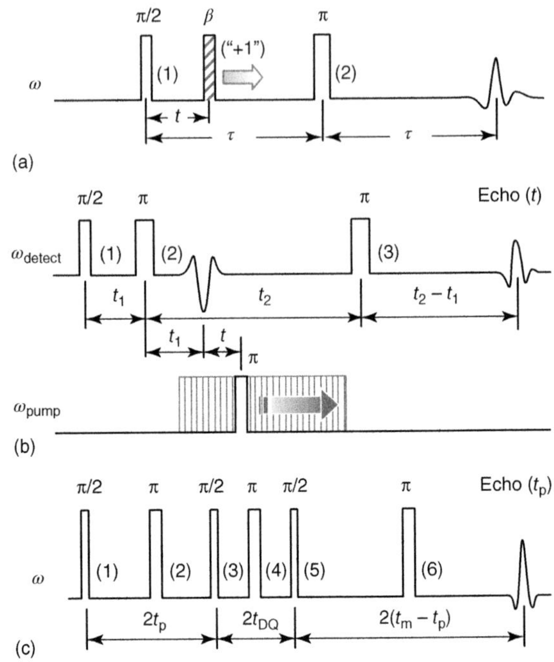

Figure 20.1. The DQC; '2+1', and DEER pulse sequences are shown. (a) The three-pulse DEER (PELDOR), and its analog, '2+1' rely on PE formed by the fixed-position pulses 1 and 2. The dipolar modulation of the echo is produced by letting the spin-flipping 'pump' pulse in the middle scan the whole range $(0,\tau)$ between the pulses 1 and 2. In PELDOR, the pump pulse labeled β is at a frequency shifted far enough so as not to affect the spins forming the echo, whereas in '2+1' all the pulses are at (about) the same frequency, so the pulses must not overlap. (b) The four-pulse DEER sequence is formed by adding a π pulse to three-pulse DEER indicated as pulse 3; this converts the PE sequence to a refocused echo. The pump pulse no longer needs to be brought close to pulse 2 as in (a). (c) The six-pulse DQC sequence is much different from the abovementioned sequences by having all six pulses derived from the same coherent source so that each spin participates in 'pumping' and detection. The distance between the first pulse and the echo is kept constant, while other pulse positions are varied in a special way to produce the dipolar modulation. (Reproduced with permission from Ref. 13. © Elsevier, 2014)

as well as not so distinct types among which RIDME (relaxation-induced dipolar modulation enhancement)[26,27] recently underwent extensive study and further development.[28–30] PDS has proved its potential as a sensitive and robust technology to study the structure and function of a broad range of biomolecules[14,31–35] and a variety of other systems. In the past two decades, there has been a rapid development of newer technical aspects of PDS[22,24,36–41] including the commercial implementation of four-pulse DEER at X-band (9.4 GHz) and, more recently, the more sensitive Q-band (34 GHz).

20.2 BASIC THEORETICAL ASPECTS OF PDS METHODS

20.2.1 Single and Double Resonance

As we outlined in the Introduction, different PDS methods have been developed to measure electron spin dipolar interactions from which distances can be obtained, to help solve a broad range of molecular structural issues. The whole family of PDS methods continues to expand. This article focuses on *single-resonance techniques* as a subset. As we note in the following section, the distinction between single- and double-resonance experiments is not always obvious. In this context, we cite the definition of double-resonance techniques as given by Slichter in p. 248 of his book[42] which includes: '… *The third category* [of double resonance] *depends in general on the existence of spin–spin couplings which in many cases must not be unduly obscured by either spin–lattice relaxation or cross-relaxation. We shall therefore call it spin coherence double resonance because it depends on the ability of spins to precess coherently for a sufficient time to reveal the spin–spin splitting. Typically, one here makes use of the fact that when two nuclei* [or electron spins] *are coupled, changing the spin orientation of one nucleus* [or electron spin] *changes the precession frequency of the nuclei* [or an electron spin] *to which it is coupled, so that the second nucleus* [spin] *can reveal in this way when the first nucleus* [spin] *is being subjected to a resonant alternating magnetic field. …. Moreover, 2D-FT NMR* [or EPR] *involving only one nuclear* [or electronic] *species is not a double-resonance experiment (only one oscillator is used), but it can be conceptually viewed as one in which the ability of a large* H_1 *… obviates the necessity of having a separate oscillator for each NMR line* [each separate portion of the EPR spectrum].' The reader may note that we introduced EPR parlance in parallel with Slichter's original quote, as it is equally applicable for EPR and NMR. We already sense a vagueness in trying to define precisely double resonance as just the existence of coherent precession effects due to coupling, which may encompass the whole PDS EPR. However, we do know that there are clearly two distinct approaches to PDS, and having a 'single oscillator' is not necessarily the main criterion that sets them apart. The issue of a 'single-frequency' experiment is somewhat complicated, particularly so with the advent of modern MW technology where using an NCO (numerically controlled oscillator), one can generate complex time-dependent gigahertz-wide spectrum pulses, which could be polychromatic[41]; so we do not necessarily equate single-frequency excitation with single-resonance PDS methods, and we center our discussion around the latter. Our focus here is on 'genuine' single-resonance EPR methods, which for the most part employ coherent pulses to control two coupled spins in a coordinated manner and are single-frequency only in that they use a 'single oscillator' and are typically best performed with strong B_1 MW fields. Specifically, we would like to mention in this context that such a single-resonance measures coupling that exists in a single group of (electron) spin centers (*A-spins*), which may be of different types but all contribute to the detected signal formed with a sequence of MW pulses, preferably intense ones. All these spins are manipulated by the pulses to transcribe the effect of the coupling onto the evolution of coherences, yielding amplitude modulation of the signal to the maximum effect. On the contrary, in pure double resonance, spins are separated into two groups; one is used to detect the signal, whereas the coupling is revealed typically by using the selective effect of pulse(s) at another frequency or other factors acting on another group of spins, *B-spins* that are dipolar coupled to the A-spins. However, double-resonance experiments could in principle be conducted as a single-frequency experiment, for example, 'light-induced magnetic dipolar spectroscopy'[43] and RIDME,[27,30,44] both are single-frequency experiments, but the evolution of couplings to B-spins is measured.

20.2.2 Spin Hamiltonian for Coupled Electron Spins

The theoretical background for PDS has been developed from the concepts utilized in NMR based on the elegant formalism of density matrices, coherences, and product operators (POs) (see Chapter 8).[45–47] We introduce a simplified form of this here.

For an isolated pair of electron spins, 1 and 2, connected by the vector $\boldsymbol{r}_{12}$ at angle θ relative to the static field $\boldsymbol{B}_0$, the spin Hamiltonian can be written as

$$\begin{aligned}\hat{H}_0 \equiv \hat{H}_0/\hbar = (\mu_B/\hbar)\boldsymbol{B}_0 \cdot \boldsymbol{g}_1 \cdot \hat{\boldsymbol{S}}_1 + \hat{\boldsymbol{I}}_1^{(k)} \cdot \boldsymbol{A}_1^{(k)} \cdot \hat{S}_{1z} \\ + (\mu_B/\hbar)\boldsymbol{B}_0 \cdot \boldsymbol{g}_2 \cdot \hat{\boldsymbol{S}}_2 + \hat{\boldsymbol{I}}_2^{(l)} \cdot \boldsymbol{A}_2^{(l)} \cdot \hat{S}_{2z} \\ - \gamma_{N1}^{(k)} B_0 \hat{I}_{1z}^{(k)} - \gamma_{N2}^{(l)} B_0 \hat{I}_{2z}^{(l)} + \hat{\boldsymbol{I}}_1^{(k)} \cdot \mathbf{P}_1^{(k)} \cdot \hat{\boldsymbol{I}}_1^{(k)} \\ + \hat{\boldsymbol{I}}_2^{(l)} \cdot \mathbf{P}_2^{(l)} \cdot \hat{\boldsymbol{I}}_2^{(l)} + d(3\hat{S}_{1z}\hat{S}_{2z} - \hat{\mathbf{S}}_1 \cdot \hat{\mathbf{S}}_2)/2 \\ + J(1/2 - 2\hat{\mathbf{S}}_1 \cdot \hat{\mathbf{S}}_2) \qquad (20.1)\end{aligned}$$

Here, the first two lines contain electron Zeeman terms for both electron spins and their couplings to their respective sets of nuclei, $\{k\}$ and $\{l\}$ with the summation over these indices implied; $\boldsymbol{A}_1^{(k)}$, $\boldsymbol{A}_2^{(l)}$ and $\boldsymbol{g}_1, \boldsymbol{g}_2$ are the hyperfine (hf) and $\boldsymbol{g}$-tensors of the coupled spins both assumed to be spin 1/2. These are followed by nuclear Zeeman and quadrupole terms, which generally cannot be discounted in single-resonance PDS, e.g., due to the development of nuclear ESEEM.[7–9] However, keeping these terms would be too cumbersome a matter or necessitate numerical treatment. The last two terms describes the electron spin dipolar interaction with the coupling constant $d = \omega_{dd}(1 - 3\cos^2\theta)$ with $\omega_{dd} = \gamma_e^2\hbar/r_{12}^3$ and the electron exchange with exchange integral $J(r_{12})$. In the dipolar coupling term, only the so-called secular terms A and B were retained, while the nonsecular terms referred to as C, D, E, and F were neglected as is appropriate for high fields.[4,48] We will simplify the discussion by ignoring other complications and dropping the nuclear spin terms (Zeeman and quadrupole) in line 3. This is the level of approximation used to analyze most PDS experiments, i.e., nitroxides, for which they are insignificant, and also the $\boldsymbol{g}$-tensor anisotropy is small. These simplifications will work well for the purpose of this article where we wish to avoid unnecessary complexity. We further simplify with

$$\begin{aligned}\hat{H}_0 = \hat{H}_{12} + \hat{H}_{dd}, \text{ where } \hat{H}_{12} = \Omega_{L1}\hat{S}_{1z} + \Omega_{L2}\hat{S}_{2z} \\ \text{and } \hat{H}_{dd} = a\hat{S}_{1z}\hat{S}_{2z} + b_{ff}(\hat{S}_1^+\hat{S}_2^- + \hat{S}_1^-\hat{S}_2^+)/2 \qquad (20.2)\end{aligned}$$

Here, $\hat{H}_{12}$ describes the Zeeman and hf terms in lines 1 and 2 of equation (20.1) very simply; this gives us the frequency spectrum, in the EPR spectral dimension. $\hat{H}_{dd}$ contains the couplings, with $a = d + 2J$ the secular (A) part and $b_{ff} = -d/2 + 2J$, the pseudosecular (or flip-flop, B) part. In the frame of reference rotating with the frequency ω_0 of the applied MW field (the rotating frame, for short), we replace the Larmor frequencies Ω_{Lk} with their offsets $\Delta\omega_k = \Omega_{Lk} - \omega_0$. The dipolar coupling term in equation (20.2) splits each of the two resonant lines at $\Delta\omega_1$ and $\Delta\omega_2$ into two lines separated by d for $\omega_{dd} \ll \Delta\omega_{12} = |\Delta\omega_1 - \Delta\omega_2|$; but by $3d/2$ in the opposite case of $\omega_{dd} \gg \Delta\omega_{12}$ (Figure 20.2). In samples with an isotropic distribution of orientations, the dipolar lineshape averaged over all orientations takes the form of a Pake doublet[8,49,50] in both cases. In intermediate cases, when $\omega_{dd} \approx \Delta\omega_{12}$, the lineshape is more complex and can be obtained numerically.

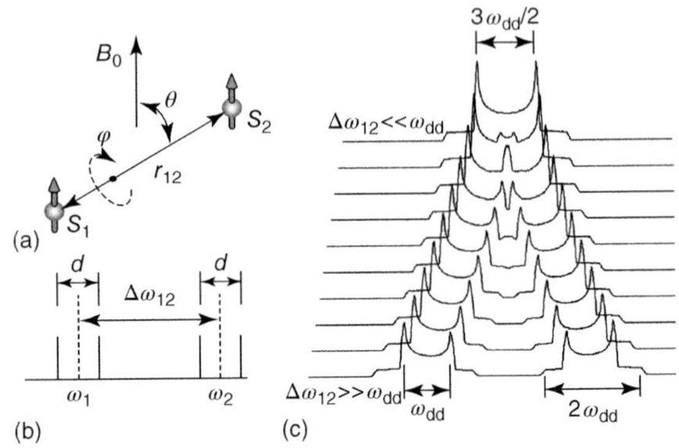

Figure 20.2. (a) Electron spins $\boldsymbol{S}_1$ and $\boldsymbol{S}_2$ coupled via the electron spin dipole–dipole interaction. Vector $\boldsymbol{r}_{12}$, connecting the spins, is aligned with the z-axis defined in the molecular frame of reference. This frame is given by Euler angles $\Omega = (0, \theta, \varphi)$ relative to the direction of the external magnetic field $\boldsymbol{B}_0$. (b) Dipolar coupling d splits the spectral lines of the electron spins into doublets at offsets $\Delta\omega_1$ and $\Delta\omega_2$. (c) The lineshape in disordered samples resulting from electron spin dipolar couplings for all Ω's. (Adapted from, Borbat & Freed 2014. (Reproduced with permission from Ref. 22. © Springer, 2014)

20.2.3 Density Operator and Coherences

The ensemble of electron spin pairs in a PDS sample is defined by the density operator (see Chapter 8). At equilibrium temperature T, it is given by[4]

$$\hat{\sigma}_{eq} = \exp(-\hat{H}_0/kT)/\text{tr}(\exp(-\hat{H}_0/kT)) \qquad (20.3)$$

where $\hat{H}_0 = \hbar\hat{H}_0$ is generally a many-body spin-Hamiltonian, which for a spin pair is given by equation (20.2). In a magnetically dilute sample, the magnetic interaction of all the spins pairs in the sample is described by the secular dipolar Hamiltonian.[48] The intramolecular couplings within each spin pair are dominant in PDS, so that we omit for now intermolecular interactions with the electron spins on surrounding molecules, but we introduce them later, as needed. The equilibrium density operator at high magnetic field and high temperature for a spin pair can be reduced to the form of just a series expansion of the exponential operator where in lowest order[48]

$$\hat{\sigma}_0 = -(\hat{S}_{1z} + \hat{S}_{2z}) \tag{20.4}$$

(where we ignore extra constants). The equivalent density matrix σ_0 at equilibrium contains only diagonal elements corresponding to energy levels with equilibrium populations. In a perturbed system, e.g., in the presence of resonant MW radiation (or by a light pulse), the populations are no longer at equilibrium, and there are off-diagonal elements (or coherences) representing the transitions.[45,51,52] They oscillate with their transition frequencies corresponding to the connected levels and coherence orders, $\overline{p}$, particularly important for larger spin systems. The order depends on the number of single-spin transitions involved in a (multiquantum) transition and is characterized by the change $\Delta\overline{p}$ in magnetic quantum number, e.g., for a single spin $\Delta\overline{p} = \pm 1$ for single-quantum transitions.

20.2.4 Evolution of Coupled Spins

PDS is a pulse EPR experiment, i.e., it is conducted by subjecting the equilibrium spin system to a sequence of MW pulses separated by periods of free evolution and after a certain time detecting the precessing magnetization $M_y(t)$. The initial transverse magnetization is usually created by a $(\pi/2)_x$ pulse from the equilibrium state of equation (20.4), yielding $\hat{\sigma}_{12}(0+) = \hat{S}_{1y} + \hat{S}_{2y}$. Then the free precession of spins proceeds with their (Larmor) frequency offsets $\Delta\omega_k$, $k = 1, 2$. With each precessing spin, there are associated coherences $\overline{p} = \pm 1$ evolving as $\exp(\mp i\Delta\omega_k t)$. The detected signal is given by $M_y(t) \propto -2\mathrm{Im}\{\mathrm{tr}(\hat{\sigma}_{12}(t)\hat{S}^+)/\mathrm{tr}(\hat{S}^-\hat{S}^+)\}$ where $\hat{S}^\pm = \hat{S}_1^\pm + \hat{S}_2^\pm$. (The denominator normalizes $M_y(0)$ to unity). The observable $\hat{S}_x$ and $\hat{S}_y$ components in $\hat{\sigma}_{12}(t)$ both contain the $\hat{S}^-$ coherence. In solids, nearly always the spin-echo is detected (see Chapter 11).

The spin dynamics of the ensemble of noninteracting spin pairs, leading to the echo, is described by the Liouville–von-Neumann equation

$$\frac{d\hat{\sigma}_{12}}{dt} = -i[\hat{H}, \hat{\sigma}_{12}] \tag{20.5}$$

which for the time-independent $\hat{H}$, e.g., $\hat{H}_0$ of equation (20.2) gives the free evolution of $\hat{\sigma}_{12}(t)$. The solution then is expressed as a unitary transformation of $\hat{\sigma}_{12}(0) = \hat{\sigma}_0$

$$\hat{\sigma}_{12}(t) = \exp(-i\hat{H}t)\hat{\sigma}_{12}(0)\exp(i\hat{H}t) \tag{20.6}$$

The pulse experiment gives the combined effect of a sequence of pulse propagators separated by the periods of free evolution described by the free evolution propagator $\hat{U}(t) = \exp(-i\hat{H}_0 t)$, during which coherence orders and polarizations are preserved in the absence of relaxation. The free evolution of each matrix element is determined by its transition frequency and coherence order. As the density operator is evolved by the pulse sequence, any element of the density matrix can be populated or changed by the pulses.

In the presence of the nth MW pulse, $\hat{H}$ becomes $\hat{H}_n = \hat{H}_0 + \hat{H}_{1n}$, where n numbers the pulses. In the rotating frame, the pulse spin-Hamiltonian is

$$\begin{aligned}\hat{H}_{1n} &= \omega_{1n}(\hat{S}^+ e^{-i\varphi_n} + \hat{S}^- e^{i\varphi_n})/2 \\ &= \omega_{1n}(\hat{S}_x \cos\varphi_n + \hat{S}_y \sin\varphi_n)\end{aligned} \tag{20.7}$$

The nth pulse is characterized by its nutation frequency $\omega_{1n} = -\gamma_e B_{1n}$, duration Δt_n, and phase φ_n. The action of a pulse is then described as a rotation. (If the nth pulse frequency is shifted by $\Delta\omega_{0n}$, equation (20.7) may need to add $\Delta\omega_0 t$ to φ_n or, as usually done, to change ω_0 by $\Delta\omega_0$ in our treatment). In the more complicated case of a time dependence in the rotating frame $\hat{H}_{1n}$ such as encountered with shaped pulses[53] (see Chapter 21), the pulse amplitude and phase temporal envelopes can be approximated by a sequence of discrete time-independent $\hat{H}_{1n}$ and the complex evolution is obtained by integrating the sequence of small-angle rotations. For constant ω_{1n} during the pulse and the spins at $\Delta\omega = 0$ equation (20.6) represents rotation by angle $\beta_n = \omega_{1n}\Delta t_n$, for $\Delta\omega \neq 0$ there are off-resonance effects[8] (see Chapter 11). We will avoid unnecessary complications by assuming hard (nearly infinite B_1) pulses, except as noted. The finite pulses in PDS can be treated using existing approaches.[21]

The evolution of coherences is often depicted as a coherence pathway transfer (CPT) chart where pulses change coherence orders. A perfect π-pulse changes signs of coherence orders as $\overline{p} \rightarrow -\overline{p}$, which can be viewed as time reversal that enables refocusing. It also inverts polarizations represented by polarization operators, $\hat{S}_k^{\alpha} \equiv \hat{E}_k/2 + \hat{S}_{kz}$ and $\hat{S}_k^{\beta} \equiv \hat{E}_k/2 - \hat{S}_{kz}$ as $\hat{S}_k^{\alpha} \leftrightarrow \hat{S}_k^{\beta}$, which is essential for PDS. (Here, $\hat{E}_k$ is the identity operator for spin k). An arbitrary pulse can change a given coherence order $\overline{p}$ to any and all available orders. For a single spin, it can change coherence order $\pm 1 \leftrightarrow 0$, refocus $\pm 1 \leftrightarrow \mp 1$, or invert population. A $\pi/2$ pulse, for example, can refocus (or flip) half of the spins, i.e., $\hat{S}^{+} \xrightarrow{(\pi/2)_x} (\hat{S}^{+} + \hat{S}^{-})/2 + \mathrm{i}\hat{S}_z$, producing both coherence orders and z-magnetization (which we will refer to as 'coherence order 0'). The picture becomes richer for coupled spins where the coupling can be refocused, higher orders with $|\overline{p}| > 1$ may occur, and coherence transfer (CT)[45] between spins is possible.

Relaxation effects are very important and could be included phenomenologically[19] using spin–lattice and spin–spin relaxation times T_1, T_2, and T_{m} denoting the phase memory time. If relaxation needs to be considered rigorously, one should resort to the superoperator form solving the stochastic Liouville equation (SLE) $\dot{\hat{\sigma}}(t) = -(\mathrm{i}/\hbar)\hat{\hat{L}}\hat{\sigma}(t) + \hat{\hat{\Gamma}}(\hat{\sigma}(t) - \hat{\sigma}_{\mathrm{eq}})$, which includes the Liouville, $\hat{\hat{L}}\hat{\sigma} = \hat{H}\hat{\sigma} - \hat{\sigma}\hat{H}$, and relaxation $\hat{\hat{\Gamma}}$ superoperators, and is generally a many-body problem.[45,54,55] We do not need to bring the SLE into this article, as relaxation will only be treated phenomenologically.

20.2.5 Product Operators

Here, we are mostly concerned about PDS based on *coherent single-resonance* methods. Our simplified treatment is conducted based on the spin Hamiltonian of equation (20.2). A formal description of PDS uses the density operator, whose time evolution is followed by solving equation (20.5) in the rotating frame using propagators based on respective spin-Hamiltonians $\hat{H}_0$ and $\hat{H}_n$ for free evolution and pulses, usually assumed to be hard (see Section 20.2.4). This issue is often simplified by employing the PO method (see Chapter 8),[45,46,52,56,57] (although a numerical treatment is hard to avoid in EPR). Throughout the text, we assume $J = 0$ and weak dipolar coupling by dropping the flip-flop terms (i.e., $b_{\mathrm{ff}} = 0$), but one may consider them where this is needed, for which a modified PO treatment exists.[21] That is, we use $\hat{H}_{\mathrm{dd}} = a\hat{S}_{1z}\hat{S}_{2z}$. In this case, $\hat{H}_{\mathrm{dd}}$ and the rest of $\hat{H}_0$ in equation (20.2) commute, which enables us to use the PO method in its basic form originally developed for NMR.

In the PO method, the density operator is constructed using a suitable operator basis.[45] For a spin pair, the density operator $\hat{\sigma}_{12}$ may be expressed in the operator basis taken as a direct product of the individual density operator bases for $\hat{\sigma}_1$ and $\hat{\sigma}_2$ for each spin, e.g., taken as the Cartesian $\hat{E}_k/2, \hat{S}_{kz}, \hat{S}_{kx}, \hat{S}_{ky}$ basis, where $k = 1, 2$.[45,47] To follow the evolution of coherences, we prefer a modified spherical basis $\hat{E}_k/2, \hat{S}_{kz}, \hat{S}_k^{+}, \hat{S}_k^{-}$,[21,52] where $\hat{S}_k^{\pm} = \hat{S}_{kx} + \mathrm{i}\hat{S}_{ky}$ are raising and lowering operators. The direct product basis contains a set of 16 operators making a complete set of 15 basis operators plus the identity operator. The polarization and coherence states are represented, respectively, by diagonal and off-diagonal elements in the matrix representation of the two-spin density operator. Off-diagonal elements correspond to $\overline{p} = 0$, ± 1, ± 2. The diagonal elements correspond to $\overline{p} = 0$. Polarization states are represented by the POs $\hat{S}_{1z}, \hat{S}_{2z}, 2\hat{S}_{1z}\hat{S}_{2z}$ with the first two related to populations, whereas the last represents 'dipolar order' $\hat{P}_{12} = 2\hat{S}_{1z}\hat{S}_{2z}$, which can be created by pulses or exists in the expansion of the density operator at high fields and low temperature, where the spin polarization is large. Or it could be produced in a reaction, such as initiated by a light pulse.[58,59]

We will need to follow coherences to fully describe the pulse sequences: single-quantum in-phase ($I^{\pm}$); single-quantum antiphase ($A^{\pm}$); and double-quantum ($DQ^{\pm}$). Their respective POs are as follows:

$$\hat{I}_1^{\pm} = \hat{S}_1^{\pm},\ \hat{I}_2^{\pm} = \hat{S}_2^{\pm};\ \hat{A}_1^{\pm} = 2\hat{S}_1^{\pm}\hat{S}_{2z},\ \hat{A}_2^{\pm} = 2\hat{S}_2^{\pm}\hat{S}_{1z};\ \hat{DQ}^{\pm} = \hat{S}_1^{\pm}\hat{S}_2^{\pm} \tag{20.8}$$

20.2.6 Evolution of SQC

As $\hat{H}_{12}$ and $\hat{H}_{\mathrm{dd}}$ commute at the level of approximation used here, the evolution of coherences due to offset and coupling can be treated independently. The POs defined in Cartesian form[45–47] are convenient for treatment of the pulses, while the spherical PO form is usually better for treatment of the free evolution and following the coherence orders. In both cases, the respective propagators for $\hat{H}_{12}$ and $\hat{H}_{\mathrm{dd}}$ expressed using a PO basis are independently applied to describe the evolution of the density operator $\hat{\sigma}_{12}$. The free evolution may thus be conveniently followed in the spherical basis. It is given as:

$$\hat{S}_k^{\pm} \xrightarrow{\hat{H}_{12}t} \hat{S}_k^{\pm} \exp(\mp \mathrm{i}\Delta\omega_k t) \tag{20.9}$$

$$\hat{S}_k^{\pm} \xrightarrow{\hat{H}_{\mathrm{dd}}t} \hat{S}_k^{\pm}(\cos\, at/2 \mp \mathrm{i}2\hat{S}_{jz} \sin\, at/2) \tag{20.10}$$

where the subscript $j \neq k$ is for the other spin (j, $k=1$, 2) and a is their coupling. The $\hat{S}_k^{\pm}$ evolves due to $\hat{H}_{12}$ as given by equation (20.9), describing two first-order coherences ($\overline{p}=\pm1$) associated with precession at offset $\Delta\omega_k$. We characterize this evolution by the phase Φ_{off} or 'offset phase', with time derivative $\dot{\Phi}_{\mathrm{off}} = \Delta\omega_k(t)\overline{p}$. The evolution due to $\hat{H}_{\mathrm{dd}}$ in equation (20.10) describes the evolution as interconversion between two terms of which the first is observable, whereas the second term, $2\hat{S}_k^{\pm}\hat{S}_{nz}$, describes the 'antiphase' order that develops between the two spins and is not directly observable, as its trace product with $\hat{S}_k^{-}$ vanishes. (Such coherences represent antiphase doublets in NMR,[45] but discussion of such details is not needed here.) We can describe this evolution by the more difficult to track 'dipolar phase' Φ_{dip} for which $\dot{\Phi}_{\mathrm{dip}} = \overline{p}\,\overline{z}(a/2)$, where $\overline{z}$ is $+1$ for $\hat{S}_n^{\alpha}$ and -1 for $\hat{S}_n^{\beta}$. (Note, it is not the sign of $\overline{z}$ but its change that matters, as the in-phase component in $\cos \Phi_{\mathrm{dip}}$ is detected. We can select $\hat{S}_n^{\alpha}$ as initial condition before the start of the pulse sequence.) Both phases describe evolution that can be reversed, so that the conditions of offset or coupling refocusing could be achieved.

A real (arbitrary) pulse can produce all possible coherence orders available for the observed spins, yielding a number of coherence pathways, from which the pathway of interest is then selected by performing phase cycling and by restricting some interpulse intervals.[8] Each pulse acting along this pathway may produce alternative (reversed/unreversed) routes for $\Phi_{\mathrm{dip}}(t)$ for a fraction of the spins, leading to a set of different $\Phi_{\mathrm{dip}}(t)$'s, which we will call 'dipolar pathways' or more distinctly 'trajectories'. Such dynamics is accounted for automatically by the PO for ideal hard pulses, as well as in a rigorous density matrix treatment. For arbitrary pulses, the standard PO method should be applied with consideration of such effects. In double resonance, in particular, the effect of pulses on the B-spins should account for what is the arbitrary nature of the pulse, which can be done first by calculating the probability p to flip this spin and then use it with equation (20.10) cast into a form conducive to the analysis of dipolar phase evolution.

$$\hat{S}_1^{\pm} \xrightarrow{H_{\mathrm{dd}}t} \hat{S}_1^{\pm}(c_t \mp \mathrm{i}s_t 2\hat{S}_{2z}) \equiv \hat{S}_1^{\pm}\hat{D}_{\pm t} \tag{20.11}$$

where $c_t = \cos(at/2)$ and $s_t = \sin(at/2)$. The propagators $\hat{D}_{\pm t}$ (or $\hat{D}(\pm t)$) can be used to derive trajectories Φ_{dip} for a coherence pathway of interest taking into account that $\hat{D}_t$ has the following convenient properties:

$$\hat{D}_t \xrightarrow{H_{nc}} q_{kn}^{(c)}\hat{D}_t + p_{kn}^{(c)}\hat{D}_t^*$$
$$\hat{D}_t^* = \hat{D}_{-t},\ \hat{D}_{t_1+t_2} = \hat{D}_{t_1}\hat{D}_{t_2} \tag{20.12}$$

where $p_{kn}^{(c)}$ is the probability for pulse n to change polarization state of spin k ('flip' it). In equation (20.12), $q_{kn}^{(c)}$ is the probability for the spin to be unaffected by the pulse. Superscript $c = (\mathrm{A, B})$ denotes the frequency at which the pulse is applied in cases of chirp pulses or DEER, i.e., to A or B spins. Then, it is straightforward to evolve the pulse sequence along the pathway, but many trajectories with different amplitudes and timing variables may be generated. For example, there are four distinct trajectories with time dependence in four-pulse DEER or refocused '2+1', three of which are unwanted. This is also the case with DEER when there is more than one pump pulse.[22,24] It is also a major problem with single-resonance and single-frequency methods in general.

20.2.7 Refocusing Offsets and Coupling

We showed above with POs the free evolution for a spin pair; now the effect of pulses should be elucidated. We consider two coupled spins A and B at Larmor frequencies ω_{A} and ω_{B} with offsets $\Delta\omega_{\mathrm{A}}$ and $\Delta\omega_{\mathrm{B}}$. (We will denote the A- and B-spin operators (and offsets) by subscript $k=1$, 2 for consistency throughout the chapter.) In single-resonance experiments ($\omega_{\mathrm{A}} \approx \omega_{\mathrm{B}}$), both spins are assumed to be at or near the detection frequency ω_{A}, and for double resonance, spin 1 is an A spin. As we detect an echo, refocusing by π-pulse(s) is necessary. The simplest refocusing sequence is an echo refocusing sequence τ–π–τ or 'sandwich' corresponding to a PE (see Chapter 11). With POs, the refocusing is expressed as:

$$S_k^{+} \xrightarrow{H_0\tau} S_k^{+}\exp(-\mathrm{i}\Delta\omega_k\tau)$$
$$\xrightarrow{\pi} S_k^{-}\exp(-\mathrm{i}\Delta\omega_k\tau) \xrightarrow{H_0 t} S_k^{-}\exp(-\mathrm{i}\Delta\omega_k(\tau - t)) \tag{20.13}$$

At $t = \tau$, the offset refocuses, and an echo is formed. Equation (20.13) describes the coherence pathway $+1 \rightarrow -1$. In Figure 20.3, we show three principal cases when π-pulses act on one of the two spins or both. The refocusing of an A spin is performed in panels (a) and (c); the pulse at ω_B in (b, c) affects spin 1 at ω_A indirectly through the evolution of the coupling. The effect of a pulse on dipolar evolution occurs through the antiphase $2\hat{S}_1^+\hat{S}_{2z}$ term in equation (20.10). In panel (a), the pulse at ω_A flips spin A, refocusing both the echo, according to equation (20.13), and the coupling by the refocusing $2\hat{S}_1^+\hat{S}_{2z}$ to $2\hat{S}_1^-\hat{S}_{2z}$. The coupling is refocused in panel (b) but, in this case, through population inversion $\hat{S}_{2z} \rightarrow -\hat{S}_{2z}$.

We can treat both cases in panels (a) and (c) simultaneously by introducing the probability p to flip the B-spin. Then, $q = 1 - p$ is the probability not to flip. We follow the evolution starting with $\hat{S}_1^+$,

$$\begin{aligned}\hat{S}_1^+ &\xrightarrow{H_{dd}\tau} \hat{S}_1^+(c_\tau - i2\hat{S}_{2z}s_\tau)\\ &\xrightarrow{(\pi)} \hat{S}_1^-[(q+p)c_\tau - i2(q-p)\hat{S}_{2z}s_\tau]\\ &\xrightarrow{H_{dd}t} \hat{S}_1^-[q(c_\tau - i2\hat{S}_{2z}s_\tau)\\ &\quad + p(c_\tau + i2\hat{S}_{2z}s_\tau)](c_t + i2\hat{S}_{2z}s_t)\\ &= \hat{S}_1^-[q(c_{\tau-t} - i2\hat{S}_{2z}s_{\tau-t}) + p(c_{\tau+t} + i2\hat{S}_{2z}s_{\tau+t})]\end{aligned} \tag{20.14}$$

For $t = \tau$, the result is $\hat{S}_1^-[q + p(c_{2\tau} + i2\hat{S}_{2z}s_{2\tau})]$, which for $p = 1$, corresponding to hard π-pulse at ω_B, is $\hat{S}_1^-(c_{2\tau} + i2\hat{S}_{2z}s_{2\tau})$, i.e., $\Phi_{dip}(t)$, unaffected by the π-pulse, is evolving as $at/2$. For $p = 0$, equivalent to the absence of any pulse resonant with B-spin, we have just $\hat{S}_1^-$ corresponding to complete refocusing.

Panel (b) has a π-pulse only at ω_B and it shows that Φ_{dip} is refocused (reverses direction) by the pulse. This sets the stage for evolving dipolar coupling by moving the pulse(s) at ω_B along the time axis while keeping the echo-forming 'observe' pulse sequence at A invariant. With the pulse assets of Figure 20.3, one can realize three-pulse DEER by combining panels (a) and (b). The sequence of panel (c), which actually best represents the single-resonance case, i.e., $\omega_A \approx \omega_B$ (with spin resonance offsets denoted $\Delta\omega_1$ and $\Delta\omega_2$ in this article), does not provide any means to evolve the coupling other than to expand the sandwich starting from its minimal width, limited by the dead time in practice. There are known issues with three-pulse DEER caused by pulse overlap, so it was modified by adding one more sandwich[16,18] to produce a refocused echo (RE) (see Chapter 19). A similar approach based on RE can be applied to the single-frequency case. This we show in Figure 20.4(a). The four-pulse DEER sequence is depicted for comparison in panel (b) together with its Φ_{off} and Φ_{dip} evolution graphs. The particular timing $\tau - 2\tau - \tau -$ echo[16] was not the most efficient to evolve couplings; therefore, the timing was changed

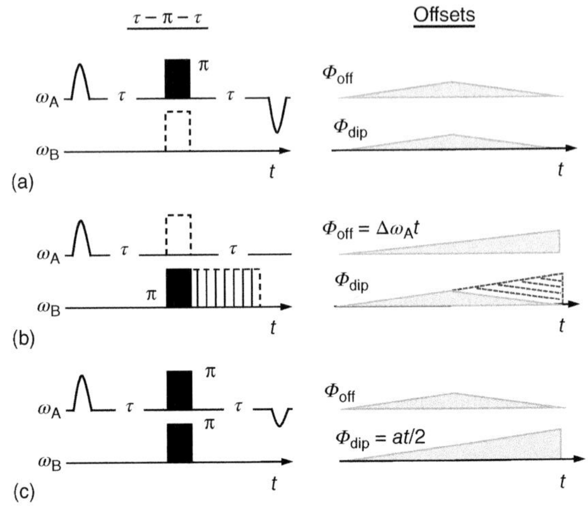

Figure 20.3. Coherence evolution of two coupled spins in the echo refocusing (or 'spin-echo') pulse sequence (or 'sandwich' for short) $\tau - \pi - \tau$. The pair of coupled spins, A (at ω_A) and B (at ω_B), is characterized by their Larmor frequency offsets $\Delta\omega_A$ and $\Delta\omega_B$ and their coupling a. Here, we are interested in refocusing the coherence for A-spins. In the absence of a pulse, first-order coherence phases evolve (from echo at $t = 0$) as $\Phi_{off} = \Delta\omega_A t$ due to offsets and $\Phi_{dip} = at/2$ due to coupling. (a) Refocusing π pulse is applied at the offset $\Delta\omega_A$ to flip an A spin. Both frequency offset and coupling are refocused, and both phases are zero at $t = 2\tau$, where the inverted echo is observed. This is complete reversal of the time evolution; that is, coherence amplitude at $t = 0$ fully recovers at $t = 2\tau$ for any a. (b) If the pulse is applied at $\Delta\omega_B$, there is no offset refocusing, but the coupling is refocused as shown. Refocusing the offset of spin A then requires applying additional π pulses outside of the sandwich. However, the pulse at ω_B is free to move along the time axis without any effect on offsets at $\Delta\omega_A$, thereby allowing one to control refocusing of Φ_{dip}, e.g., delaying it as indicated by dashed lines. This is what is used in double resonance. (c) The π-pulses flip both spins. (It can be just one pulse if $\omega_A \approx \omega_B$.) The offsets are refocused, but the coupling is not, leading to incomplete reversal of the evolution. The first-order coherence amplitude V_0 (e.g., spin echo) at time $t = 0$ will be $V_0 \cos a\tau$ at time $t = 2\tau$, so it does not refocus in full for all a's and will decay for a distribution in a. The evolution is characterized by linear 'dipolar phase' $\Phi_{dip} = at/2$. This is the case that single-resonance deals with

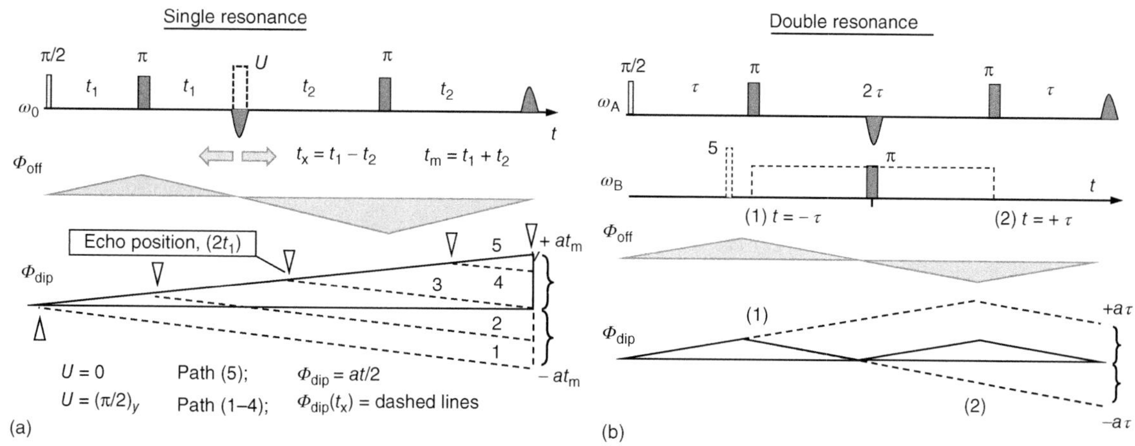

Figure 20.4. Two principal PDS schemes based on RE. (a) In a single-resonance sequence based on a sequence of two echo sandwiches, the evolution of the coupling is not refocused by any of the π pulses; thus, the progress of the dipolar phase Φ_{dip} (line 5 in a) is $at/2$, reaching at_m at the RE formed at $2t_m$. Here, Φ_{dip} does not depend on $t_x = t_1 - t_2$, so no evolution can be produced at constant t_m by varying t_1. A more complex spin manipulation, such as by a propagator $\hat{U}$ inserted between the refocusing sequences can change this picture by modifying one or both of the first-order coherences (i.e., of in- and antiphase). For example, just a single $(\pi/2)_y$ negates the antiphase term, hence refocusing the coupling. The envelopes of Φ_{dip} are shown for five positions of $(\pi/2)_y$ and echo at $2t_1$, as indicated by the wedges. Lines 1–4 are all different, leading for t_x in $[-t_m, t_m]$ to the maximum range of change at_m possible for $|\Phi_{dip}|$ over $2t_m$ evolution period. Other forms of propagator $\hat{U}$ are discussed in the following sections. (b) Double-resonance approach to the RE sequence. The in-phase coherence created by the initiating π/2 pulse is refocused in each of the two sandwiches, which, in this example, have equal durations. The π pulse at frequency ω_B can be applied at any time point, unlike in **a**, but does not need to be outside points 1 and 2. As it moves between the two π-pulses at $\Delta\omega_A$, Φ_{dip} changes from $-a\tau$ to $a\tau$, crossing the coupling refocusing point in the middle at $t = 0$. When π pulses at the two frequencies coincide, evolution does not reverse (dashed lines 1 and 2). The range of $|\Phi_{dip}|$ is $a\tau$, i.e., only half that shown in **a** for $t_m = 2\tau$. Making this pulse sequence asymmetric, as in **a** by making $t_1 \ll t_2$; or adding a π pulse (5) at $\Delta\omega_B$ and position near 1 or 2 recovers the full range of $|\Phi_{dip}|$ by shifting the range of Φ_{dip} (4τ) to be nearly completely above or below zero

to $t_1 - (t_1 + t_2) - t_2$ – echo, with $t_1 \ll t_2$.[60] In a different approach, another pulse was added for the B-spins, creating the DEER-5 method.[24] This demonstrates the flexibility of PDS based on double resonance.

We will use two notations for the timing in RE-based constant-time pulse sequences. We will use $2t_m$ for the length of a sequence (from the beginning to the echo) and t_1, t_2 for the two sandwiches in RE (i.e., $t_1 - \pi - t_1$, $t_2 - \pi - t_2$). Thus $t_1 + t_2 = t_m$. We will also use t_p for t_1, while $t_m - t_p$ will be used for t_2 to emphasize the constant-time sequence. Finally, we introduce the dipolar evolution variable $t_x = t_1 - t_2 = 2t_p - t_m$. (We also used t_ξ for $-t_x$ in the past.[21]) The range of change for $t_x = [-t_m, t_m]$. This is the preferred (constant-time) way to record evolution with single-resonance pulse sequences, which we detail in the following sections.

It is very clear from Figure 20.4 that there is no evolution of dipolar coupling in RE as a function of t_x.

The number of refocusing pulses can be any, but the result is the same, Φ_{dip} evolves as $at/2$ reaching at_m at the echo. To see this, the evolution can be expressed as follows. In the first sandwich, the evolution (for spin 1) proceeds according to:

$$\begin{aligned}\hat{S}_1^{\pm} &\xrightarrow{\hat{H}_{dd}\tau} \hat{S}_1^{\pm}(c_\tau \mp i2\hat{S}_{2z}s_\tau) \xrightarrow{(\pi)} \hat{S}_1^{\mp}(c_\tau \pm i2\hat{S}_{2z}s_\tau)\\ &\xrightarrow{\hat{H}_{dd}t} \hat{S}_1^{\mp}(c_\tau \pm i2\hat{S}_{2z}s_\tau)(c_t \pm i2\hat{S}_{2z}s_t)\\ &= \hat{S}_1^{\mp}(c_{\tau+t} \pm i2\hat{S}_{2z}s_{\tau+t}) \to \hat{S}_1^{\mp}(c_{2\tau} \pm i2\hat{S}_{2z}s_{2\tau})\end{aligned} \quad (20.15)$$

(For spin 2, the result $\hat{S}_2^{\pm} = \hat{S}_2^{\mp}(c_{2\tau} \pm i2\hat{S}_{1z}s_{2\tau})$ is obtained by swapping the subscripts.) Note that we omitted evolution due to offsets in equation (20.15) as they are refocused in equation (20.13) for $t = \tau$. (They should be retained, however, if one is considering 2D experiments that rely on the time variable $\delta t = t - \tau$.) Having obtained the evolution for $\hat{S}_1^{\pm}$ and $\hat{S}_2^{\pm}$ at 2τ, we then switch to Cartesian PO's

$$\hat{S}_{1y} = (\hat{S}_1^+ - \hat{S}_1^-)/2\mathrm{i} = [\hat{S}_1^-(c_{2\tau} + \mathrm{i}2\hat{S}_{2z}s_{2\tau}) - \hat{S}_1^+(c_{2\tau} - \mathrm{i}2\hat{S}_{2z}s_{2\tau})]/2\mathrm{i} = -\hat{S}_{1y}c_{2\tau} + \mathrm{i}2\hat{S}_{1x}\hat{S}_{2z}s_{2\tau} \quad (20.16)$$

Similar to equation (20.15), $\hat{S}_{2y} = -\hat{S}_{2y}c_{2\tau} + \mathrm{i}2\hat{S}_{2x}\hat{S}_{1z}s_{2\tau}$ and finally for $(\hat{S}_{1y} + \hat{S}_{2y})$,

$$-(\hat{S}_{1y} + \hat{S}_{2y})c_{2\tau} + \mathrm{i}(2\hat{S}_{1x}\hat{S}_{2z} + 2\hat{S}_{2x}\hat{S}_{1z})s_{2\tau} \quad (20.17)$$

or simply $\cos a\tau$ for evolution. The evolution of coupling in an RE can be described as follows. We will track spin 1, then after the first sandwich ($t_1 - \pi - t_1$) according to equation (20.16), there is in-phase coherence $\hat{I}_1(t_1) = -\hat{S}_{1y}\cos at_1$ and antiphase coherence $\hat{A}_1(t_1) = 2\hat{S}_{1x}\hat{S}_{2z}\sin at_1$. After the second sandwich, these terms evolve into detectable in-phase terms as

$$\hat{I}_1(t_1) + \hat{A}_1(t_1) \rightarrow \hat{I}_1(t_1 + t_2) = \hat{S}_{1y}(\cos at_1 \cos at_2 - \sin at_1 \sin at_2) \quad (20.18)$$

The term in cosines can be viewed as the evolution path $\hat{I}_1(0) \rightarrow \hat{I}_1(t_1) \rightarrow \hat{I}_1(t_1 + t_2)$, while that in sines as $\hat{I}_1(0) \rightarrow \hat{A}_1(t_1) \rightarrow \hat{I}_1(t_1 + t_2)$. The outcome is $\hat{S}_{1y}\cos at_{\mathrm{m}}$, the same as in the case with PE; the only dipolar evolution possible is that by varying t_{m}, starting from the end of the dead time, t_{min}. To overcome this trend, we need somehow to refocus Φ_{dip}, a task that unlike in double resonance just cannot be accomplished with π-pulses. However, this can be done, for example, by inserting between the sandwiches a propagator $\hat{U}$, which is able to refocus directly or indirectly, as we will soon show. $\hat{U}$ will be allowed to evolve coherence orders 0 and ±2. It may contain several pulses, may have a long duration, and can make use of phase relaxation or of any other spin manipulation. We describe its transfer properties in an arbitrary manner as $C_{\mathrm{I}}\hat{I}_k + C_{\mathrm{A}}\hat{A}_k$, where $k = 1, 2$ numbers the two spins and $|C_{\mathrm{I(A)}}| \leq 1$. As the evolution is for $\overline{p} = 0, \pm 2$ orders, no dipolar evolution takes place and the (stored) dipolar phase is carried through. Relaxation can render zero either C_{I} or C_{A}, as we will see. (Phase cycling can produce an equivalent result.) Then, equation (20.18) transforms to

$$\hat{I}_1(t_1) + \hat{A}_1(t_1) \xrightarrow{\hat{U}} C_{\mathrm{I}}\hat{I}_1(t_1) + C_{\mathrm{A}}\hat{A}_1(t_1) \xrightarrow{t_2} \hat{S}_{1y}(C_{\mathrm{I}}\cos at_1 \cos at_2 - C_{\mathrm{A}}\sin at_1 \sin at_2) = \hat{S}_{1y}[(C_{\mathrm{I}} + C_{\mathrm{A}})\cos at_{\mathrm{m}} + (C_{\mathrm{I}} - C_{\mathrm{A}})\sin at_{\mathrm{x}}] \quad (20.19)$$

If $C_{\mathrm{I}} = -C_{\mathrm{A}}$, only *cos* at_x remains; this is the case of coupling completely being refocused at $t_{\mathrm{x}} = 0$. This can be achieved by solid-echo refocusing,[61] i.e., applying $\hat{U} = (\pi/2)_y$ refocuses the coupling according to

$$-(\hat{S}_{1y} + \hat{S}_{2y})c_{2t_1} + \mathrm{i}(2\hat{S}_{1x}\hat{S}_{2z} + 2\hat{S}_{2x}\hat{S}_{1z})s_{2t_1} \xrightarrow{(\pi/2)_y} -(\hat{S}_{1y} + \hat{S}_{2y})c_{2t_1} - \mathrm{i}(2\hat{S}_{2x}\hat{S}_{1z} + 2\hat{S}_{1x}\hat{S}_{2z})s_{2t_1} \quad (20.20)$$

This is equivalent to $t_1 \rightarrow -t_1$ in $\sin at_{\mathrm{p}}$, i.e., time reversal, but it comes at the expense of swapping antiphase coherences of the spins. A $(\pi/2)_x$ pulse instead stores $\hat{S}_{ky}$ as $-\hat{S}_{kz}$, plus it generates pure $D\hat{Q}_y = (2\hat{S}_{1x}\hat{S}_{2y} + 2\hat{S}_{2x}\hat{S}_{1y})/2$.[47] This requires one to add more pulses to $\hat{U}$ in order not to lose the signal.

It is possible to selectively make either C_{I} or C_{A} zero, producing a sum or difference of the terms in cos at_{x} and cos at_{m}. In all cases, the dipolar signal can be acquired using fixed t_{m} by changing just t_{x}. This would alleviate (but not remove) issues with relaxation, and nuclear ESEEM should be suppressed to the extent possible. It usually decays as t_{m} increases. This is what single-resonance PDS sequences have to do in their work. We will show the effects of particular $\hat{U}$'s when discussing the respective pulse sequences. Briefly, $\hat{U}$'s in use are $[(\pi/2)_x - t_{\mathrm{d}} - \pi - t_{\mathrm{d}} - (\pi/2)]$; $(\pi/2)_y$; $[(\pi/2)_x - T - (\pi/2)]$; and $[(\pi/4)_y - T - (\pi/4)]$. They make up, respectively, six-, four-, and five-pulse single-resonance experiments that we discuss in the following section.

20.3 DOUBLE-QUANTUM COHERENCE EPR, SIX-PULSE SEQUENCE

20.3.1 Development of DQC EPR

The need to improve sensitivity inspired efforts to develop DQC EPR, which was expected to produce strong dipolar signals, but the initial attempts were not particularly convincing.[19] Renewed effort resolved all issues[20,21] with the demonstration of strong primary DQC signals, which are well in line with the basic theoretical concepts. Furthermore, the implementation of the method at $\mathrm{K_u}$ band at ≈17 GHz using an efficient dielectric resonator and intense nanosecond MW pulses[22,24,62] yielded a very sensitive technique. It is not unusual to obtain a DQC dipolar signal with an SNR in the thousands for a typical spin concentration in the range ~100–200 μM or as high as several hundreds on lower concentration samples (≲50 μM).

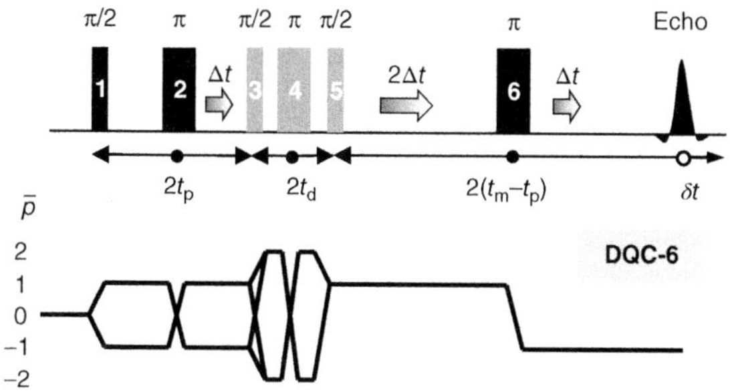

Figure 20.5. The six-pulse DQC sequence is based on RE formed by pulses 1, 2, and 6. The three remaining pulses 3, 4, and 5 constitute a composite propagator $\hat{U}$, which is a fixed refocusing sequence ($t_d - \pi - t_d$) flanked by π/2 pulses. The purpose of $\hat{U}$ here is to enable DQ filtering (DQF) by creating DQC from refocused antiphase coherence by $(\pi/2)_x$ pulse 3, refocusing it with the π pulse 4, and then converting it back to antiphase coherence by π/2 pulse 5, which then evolves to contribute into the detectable echo. This contribution from the spins that have passed through DQC paths is then isolated by performing phase cycling. The pulses are advanced in steps, as shown, as the timing variable is $t_x = t_m - 2t_p$. The recorded signal is symmetric with respect to $t_x = 0$; thus, it usually can be varied just from slightly less than 0 to t_m

20.3.2 Six-pulse DQC EPR Sequence: Basic Principles

The details of the *constant-time* six-pulse DQC sequence depicted in Figure 20.5 are explained elsewhere,[14,17,22] but a basic description useful for understanding single-resonance methods has been provided in Section 20.2. Here, we limit the discussion to key aspects involved in the function of this important DQC technique before discussing other single-resonance techniques. The pulse sequence in Figure 20.5 is composed of three refocusing sandwiches separated by π/2 pulses serving to change coherence levels in the desired way. This sequence can be viewed as a constant-time RE sequence of Section 20.2 with two sandwiches centered at π-pulses 2 and 6 separated by the three-pulse propagator, $\hat{U}$ (pulses 3–5), which is used to generate and refocus second-order coherence and then return it back to the $\bar{p} = +1$ pathway where it evolves into detectable coherence at time $2t_m$ after the first pulse labeled 1.

This sequence can also be viewed (as in MQC NMR) as a *preparation – evolution – detection* sequence,[45] where the first sandwich prepares antiphase coherence, which is converted to DQC, refocused in the evolution sandwich, and converted back to antiphase coherence evolving in the last sandwich into observable in-phase coherence for detection.

Starting from the equilibrium state $\hat{\sigma}_{12}(0)$, in-phase coherence $\hat{I}_{12} = (\hat{S}_{1y} + \hat{S}_{2y})$ is produced by the first $(\pi/2)_x$ pulse. In the first sandwich, $\hat{I}_{12}$ evolves into $\hat{I}_{12} + \hat{A}_{12}$, with $\hat{I}_{12} = -(\hat{S}_{1y} + \hat{S}_{2y}) \cdot \cos at_p$ and $\hat{A}_{12} = (\hat{S}_{1x}\hat{S}_{2z} + \hat{S}_{2x}\hat{S}_{1z}) \cdot \sin at_p$, and offsets are refocused. The $(\pi/2)_x$ pulse 3 stores the $\hat{I}_{12}$ as $-(\hat{S}_{1z} + \hat{S}_{2z}) \cdot \cos at_p$ and converts $\hat{A}_{12}$ to $DQ_y = -\sin at_p \cdot (\hat{S}_1^- \hat{S}_2^- - \hat{S}_1^+ \hat{S}_2^+)/2\mathrm{i}$, which evolves due to the frequency offsets of both spins as $DQ_y(t) = \sin at_p \cdot (\hat{S}_1^- \hat{S}_2^- e^{\mathrm{i}(\Delta\omega_1 + \Delta\omega_2)t} - \hat{S}_1^+ \hat{S}_2^+ e^{-\mathrm{i}(\Delta\omega_1 + \Delta\omega_2)t})/2\mathrm{i}$, with the time t referred to the third pulse. $DQ_y(t)$ is refocused in the ($t_d - \pi - t_d$) sandwich, and the third $(\pi/2)_x$ pulse 5 converts the resulting $DQ_y(2t_d) = (\hat{S}_1^- \hat{S}_2^- - \hat{S}_1^+ \hat{S}_2^+) \cdot \sin at_p$ back to $\hat{A}_{12} = (\hat{S}_{1x}\hat{S}_{2z} + \hat{S}_{2x}\hat{S}_{1z}) \cdot \sin at_p$ labeled with the phase of the dipolar evolution at $2t_p$. Finally, the last refocusing sandwich evolves $\hat{A}_{12}$ into in-phase $(\hat{S}_{1y} + \hat{S}_{2y}) \cdot \sin(at_m - at_p) \cdot \sin at_p$ and the antiphase $-(\hat{S}_{1x}\hat{S}_{2z} + \hat{S}_{2x}\hat{S}_{1z}) \cdot \cos(at_m - at_p) \cdot \sin at_p$ coherences, of which only the in-phase term is observable.

$\hat{I}_{12}$ at $2t_p$ passes through the $U(\pi/2 - t_d - \pi - t_d - \pi/2)$ propagator, which we will call for convenience a 'double-quantum filter' or DQF, for short. The DQF does not actually remove any signal component, rather it 'labels' the DQC signal leaving it to the phase cycling to achieve the final filtering. $\hat{I}_{12}$ passes through the DQF as $(\hat{S}_{1z} + \hat{S}_{2z})$ carrying with it the phase $\Phi_{\mathrm{dip}}(2t_p) = \cos at_p$. After the filter, it is again $\hat{I}_{12} = -(\hat{S}_{1y} + \hat{S}_{2y}) \cos at_p$, which evolves into the observable $(\hat{S}_{1y} + \hat{S}_{2y}) \cdot \cos at_p \cdot \cos(at_m - at_p)$. Note that the sum of the two contributions to observable $\hat{I}_{12}$ coherence is $(\hat{S}_{1y} + \hat{S}_{2y}) \cdot \cos at_m$, just as what one expects from the basic RE for pulses 1, 2, and 6 if the DQF is simply viewed as the equivalent of a 2π rotation, which could be a reasonably accurate picture in the absence of coupling and relaxation. It should be emphasized that there is no refocusing of dipolar coupling in the DQC-6 sequence. The refocusing is 'virtual' by separating two components constituting the RE into complementary contributions of the refocused type. It is the DQC path enabled by the coupling that makes it possible to separate them by deeply suppressing the unwanted in-phase component by constructing a phase cycle that selects only the pathways passing through DQC order.[21,24] The basic phase cycle would involve cycling the phases ϕ_k, ($k = 1 \ldots 3$) of the pulses 1–3 and the receiver phase ϕ_R in four steps as $(\phi_1, \phi_2, \phi_3, \phi_R) = \{(x, x, x, y); (y, y, y, -y); (-x, -x, -x, y); (-y, -y, -y, -y)\}$. The last $\pi/2$ pulse as well as all π pulses can be cycled independently to improve the performance, enabling an extended phase cycle sequence up to 256 steps.[21] This very robust filtering technique provides deep suppression of unwanted coherence pathways irrespective of the arbitrary nature of the pulses or their imperfections. This was also demonstrated at X band[31] where a standard TE_{012} cavity resonator and 16 ns π pulses were used.

The filtering properties of the DQF have other very important implications. First, we stress that adequate phase cycling[21,24,63,64] ensures very deep suppression of unwanted pathways. The second key property is that even in the case of strong but finite pulses, π pulses are not able to refocus all the spins, and consequently, they do not invert all the S_z spins in the antiphase coherence terms. This spawns a spurious dipolar trajectory in the evolution, but in this case, the dipolar coupling is refocused in the preparation sandwich, and the respective trajectory does not generate antiphase coherence; consequently, it does not contribute to the DQC pathway selected by the phase cycling and is suppressed. Similarly, the finite refocusing pulse in the detection sandwich after the DQF does not contribute another dipolar trajectory to the amplitude of the detected echo. The same holds for single spins: there is no DQC from them (!) and they can only contribute to the extent that DQC is formed with other spins on surrounding molecules. (This effect becomes insignificant in dilute samples.) This property means that the treatment based on hard pulses describes the performance of this pulse sequence very adequately.

The omitted flip-flop terms in equation (20.2), which exchange dipolar evolutions of coupled spins in an offset-dependent manner, are more difficult to consider in a closed-form analysis, but this has been done under minimal assumptions.[21] It has been shown that the closed-form expression for the DQC amplitude[21,65] is adequate for accurate representation of the signal for distances as short as ≈1.0 nm given large enough B_1 (>40 G) for such a case of strong coupling. It was also shown that for most practical cases their effects are relatively insignificant. Finally, in a rigorous numerical treatment, one can consider the dipolar coupling during the pulses. What one qualitatively expects from this case is a production of all five coherence pathways including $\overline{p} = \pm 2$ by a finite-length pulse. This case was considered as a central focus in the framework of the generation of 'forbidden' DQCs.[19] For the short intense pulses that we currently employ these effects can safely be neglected for distances greater than ≈1.2 nm.

20.3.3 DQC-6: Experimental Examples

We carried out PDS measurements based on the DQC-6 sequence on a variety of spin-labeled systems from synthetic molecules such as biradicals[21] and peptoids[66] to proteins.[31,32,63,67,68] In Figure 20.6, we show examples of DQC measurements conducted on very different systems. In top panel (a), the raw DEER and DQC data obtained on 47 μM of a 95-kDa soluble LOX-1 enzyme are shown before any background corrections were made. (MTSL-labeled LOX-1 was prepared by B.J. Gaffney and studied in the context of work published in Biophys. J.[63]; This open access work is notable by introducing to PDS a multidimensional scaling approach to perform trilateration for 3D localization by distance geometry of a spin-labeled substrate.) The dipolar modulations are very similar, but the background (see Chapter 19) is considerably smaller in DQC; this subsequently minimizes errors in making background corrections, thus leading to better reconstruction of distances for DQC. This is also helped by the fact that any residual orientational effects are also smaller in DQC due to much less selective pulses (see Section 20.6). In addition, the sensitivity was a factor of nearly 7 higher for the distances in the range 3.5–5.5 nm. The exact reason for the larger factor from that expected of ≈2–4[21] is likely due to a partial suppression of phase relaxation caused by nuclear spin diffusion of the protons,[24,69–71] yielding a longer phase relaxation time in DQC.

The second example Figure 20.6(b) compares K_u-band DEER and DQC data obtained in measuring the distance between two Cu^{2+} ions.[13] This case was challenging for both methods used. DEER using strong $B_1 = 30$ G pulses was able only to achieve ≈0.05 modulation depth given about 1 kG Cu^{2+} spectral width. It is clear that the large $B_1 = 45$ G we used for DQC could excite only a fraction of the spins. There is still good agreement between the two methods, and again, the sensitivity of DQC was clearly better notwithstanding the fact that π pulses of comparable intensity were used with DEER. In addition, this is an additional demonstration that the DQC technique is not much affected by the fact that there is incomplete spectral excitation. That is, the extent of spectral excitation is not a principal factor for DQC, as

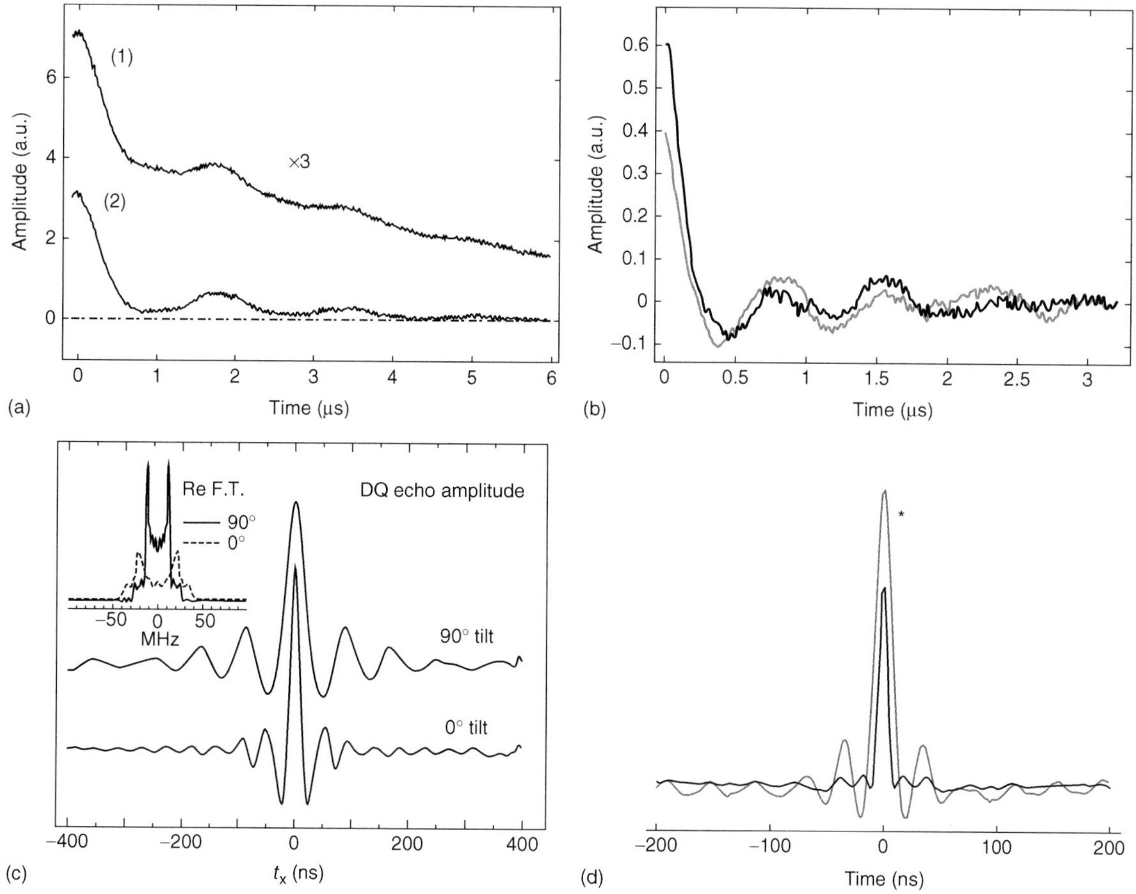

Figure 20.6. (a) Raw 17.3 GHz DQC (2) and DEER (1) signals from soybean lipoxygenase LOX-1 A569R1/A619R1 double mutant in 30% w/v glucose-d_6/D_2O, 0.1 M tricine buffer, pD 8.4. Protein concentrations were 47 and 160 μM, respectively. Data averaging time was 5.5 h for (2) and 13.5 h for (1). S/N for equal data averaging time and concentration is a factor of nearly 7 better for DQC. DEER used 16 ns pump pulse, while DQC used 6 ns π-pulses ($B_1 \approx 30$ G). (b) K_u-band DEER (gray) and DQC (black) for Cu^{2+}/Cu^{2+} pair of H48Q human superoxide dismutase SOD-1 fALS mutant. The data were recorded at 20 K at the same (g_y) field position and plotted after making background corrections and scaling to be close in amplitude of dipolar oscillations. DEER was recorded in 13.4 h using 6 ns pump pulse and 140 MHz frequency separation. DQC used 4 ns π-pulses ($B_1 \sim 45$ G) and was recorded in 1.8 h. The dipolar signal is about 6–7 times stronger for DQC, requiring an order of magnitude shorter averaging time than DEER for the same SNR. (Reproduced with permission from Ref. 13. © Elsevier, 2014.) (c) A rigid 1.63 nm nitroxide biradical aligned in nematic LC (phase V) at two director orientations with respect to $\boldsymbol{B}_0$ recorded at 200 K. (Reproduced with permission from Ref. 21. © Plenum Publishers – Books, 2000.) (d) Ku-band DQC measurements on BTurea and BTXA (starred), which are, respectively, 1.05 and 1.25 nm nitroxide biradicals developed for DNP NMR. (Derived from sample courtesy of R.G. Griffin.) All DQC data (other than (a)) were acquired at 17.3 GHz and 60 K using 2 ns π/2 and 4 ns π-pulses

the phase cycling provides much greater attenuation of unwanted pathways than other single-resonance methods we consider in the following section, ensuring relatively clean selection of just the desired signal.

One of the first successful DQC measurements included a rigid biradical oriented in a nematic liquid crystal shown in Figure 20.6(c).[21] The measurements were carried out at −70°C. In the parallel orientation to $\boldsymbol{B}_0$, the splitting was ≈20 MHz with spectral components at 32 MHz due to the pseudosecular term also visible.

In Figure 20.6(d), very short distances 1.05 and 1.25 nm were recorded on biradicals developed for DNP[22,72] without encountering any significant problem. The first distance is about the shortest DQC can measure as the dipolar coupling constant $\omega_{dd}/2\pi$ is 45 MHz ($\approx$16 G), which according to our numerical simulations[65] is close to maximum for a B_1 of 45 G. For greater dipolar coupling, its presence during the pulse should be considered and $J(r_{12})$ is likely to become large.

All the abovementioned measurements used a 64-step phase cycle for DQ filtering. This phase cycle was derived from the full 256-step phase cycle described in[21] as follows: the first 32 lines in the phase cycle table are copied to make lines 33–64. In this table for the 32 added lines, the phase of the sixth pulse is y and the receiver phases inverted. The CYCLOPS steps used for constructing the 128- and 256-step phase tables were omitted. Nuclear ESEEM from protons was suppressed by summing four data sets with t_m incremented for the subsequent set by half a period (19 ns) of the ESEEM. This describes one of the standard ESEEM suppression routines we developed to record DQC (and DEER) data[14]; other approaches used with low-resolution timing are also known.[63]

20.4 FOUR- AND FIVE-PULSE 'SINGLE-QUANTUM COHERENCE' PDS SEQUENCES

20.4.1 The Four-pulse SIFTER Experiment

The adoption of key NMR pulse methods added to PDS a group of four- and five-pulse constant-time single-resonance experiments.[8,21,23] Similar to DQC, they benefit from applying strong (almost) nonselective pulses, potentially covering nearly the entire spectral width for a nitroxide. The methods are based on simultaneous manipulation with the coherences of the two spins by switching the course of evolution of dipolar coupling in the RE sequence (see Section 20.2). The specifics are in how the refocused signal is modified in each case.

We now discuss single-resonance pulse sequences that can benefit from strong pulses by first assuming ideal infinite pulses to simplify the initial discussion. In Figure 20.4, we show an RE, which is at the

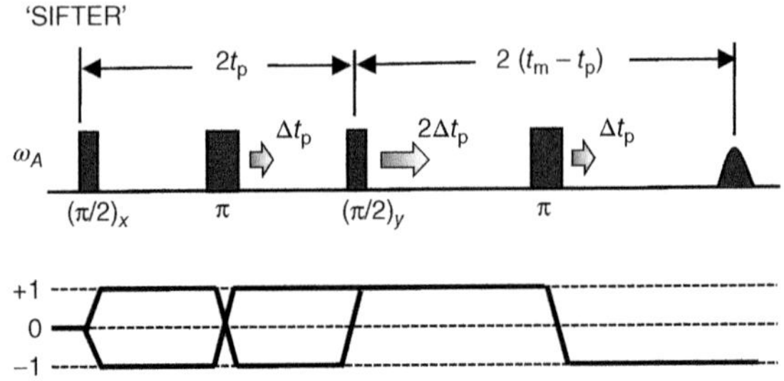

Figure 20.7. The four-pulse SIFTER SQC sequence. The CPT chart shows that the sequence is essentially an RE discussed in Section 20.2. The refocusing propagator U in this cases is just a $(\pi/2)_y$ pulse. It exchanges the antiphase coherences of the two spins, thus making coherence transfer. (The transfer also includes electron–nuclear but not nuclear coherences.) The sequence is 'constant time' with the position of the echo at $2t_m$ being constant. The pulses advance in steps shown, starting, e.g., at $t_p = 0$. The relevant time variable $t_x = t_p - (t_m - t_p) = 2t_p - t_m$ spans the range of $[-t_m, t_m]$

basis of DQC-6 and several other single-resonance techniques including the four-pulse single-frequency technique for refocusing (SIFTER) pulse sequence depicted in Figure 20.7 and discussed in this section. The pulse sequence described here was implemented in EPR in 2000[23] and was named SIFTER. According to Figure 20.4(a) and the subsequent discussion, it is not possible with nonselective π-pulses to refocus simultaneously frequency offsets and dipolar coupling. We have shown there and reiterate here the argument that a single nonselective $(\pi/2)_y$ pulse can, however, completely refocus the coupling.

The sequence based on RE has two refocusing $t_p - \pi - t_p$, and $(t_m - t_p) - \pi - (t_m - t_p)$, sandwiches. The evolution after the first sandwich produces

$$\hat{\sigma}_{12}(2t_p) = -(\hat{S}_{1y} + \hat{S}_{2y})\cos at_p + (2\hat{S}_{1x}\hat{S}_{2z} + 2\hat{S}_{2x}\hat{S}_{1z})\sin at_p \qquad (20.21)$$

In the absence of the $(\pi/2)_y$ pulse, this would then evolve, as we showed in Section 20.2.7 to produce $\hat{\sigma}_{12}(2t_m) = (\hat{S}_{1y}+\hat{S}_{2y})\cos at_m$ and is not very useful as the dipolar coupling is in the t_m variable, but this $(\pi/2)_y$ pulse changes the sign of the antiphase term by effecting CT according to $2\hat{S}_{1x}\hat{S}_{2z} \leftrightarrow -2\hat{S}_{2x}\hat{S}_{1z}$ so that the evolution through the second sandwich now gives $\hat{\sigma}_{12}(2t_m) = (\hat{S}_{1y}+\hat{S}_{2y})\cos at_x$, permitting one to record the evolution as a function of the new variable $t_x = 2t_p - t_m$. It also permits one to record the full echo, 100% modulated with the dipolar oscillations, which is a very attractive property of this pulse sequence. Unlike DQC-6, there is no constant term in $(\hat{S}_{1y} + \hat{S}_{2y})\cos at_{\mathrm{m}}$.

20.4.2 Effects of Finite Pulses in SIFTER

We have discussed SIFTER in the hard-pulse limit using POs to describe the spin dynamics. Performance with real finite pulses complicates the case to some extent. We will now look into the main implications of finite pulses with SIFTER. SIFTER does not have a mechanism for filtering (see Section 20.3.3). Therefore, with real finite pulses, all possible dipolar trajectories (see Section 20.2) will be produced and detected. Specifically, what happens during each refocusing sandwich in the case of finite pulses can be described as follows. We assume that the finite π-pulse can flip a fraction p of spins while the fraction $q = 1 - p$ remains unaffected (for a subensemble of single spins $q = 1$). Otherwise, the pulses are taken as ideal. The evolution proceeds as

$$\begin{aligned}\hat{S}_{1y} &\xrightarrow{t_1-\pi-t_1} -q\hat{S}_{1y} + p(\hat{S}_{1y}\cos at_1 + 2\hat{S}_{1x}\hat{S}_{2z}\sin at_1)\\ &\xrightarrow{(\pi/2)_y} -q\hat{S}_{1y} + p(\hat{S}_{1y}\cos at_1 - 2\hat{S}_{1x}\hat{S}_{2z}\sin at_1)\\ &\xrightarrow{t_2-\pi-t_2} \hat{S}_{1y}[(q + p\cos at_1)(q + p\cos at_2)\\ &\quad + p^2(-\sin at_2)\sin at_1] + 2\hat{S}_{1x}\hat{S}_{2z}(\dots) \qquad (20.22)\end{aligned}$$

The $(\pi/2)_y$ pulse changes the sign of the antiphase term as $(\hat{S}_{2x}\hat{S}_{1z} + \hat{S}_{1x}\hat{S}_{2z}) \rightarrow -(\hat{S}_{1x}\hat{S}_{2z} + \hat{S}_{2x}\hat{S}_{1z})$. The evolution for the second spin is obtained by swapping subscripts 1 and 2. One thus detects

$$\begin{aligned}(\hat{S}_{1y} + \hat{S}_{2y})\{q^2 &+ pq\cos[a(t_{\mathrm{m}} - t_{\mathrm{x}})/2]\\ &+ pq\cos[a(t_{\mathrm{m}} + t_{\mathrm{x}})/2] + p^2\cos at_{\mathrm{x}}\} \qquad (20.23)\end{aligned}$$

The last term in $\cos at_{\mathrm{x}}$ in equation (20.23) has its maximum at $t_{\mathrm{x}} = 0$ in the center of evolution interval $[-t_{\mathrm{m}}, t_{\mathrm{m}}]$ as we found from this sequence with hard pulses. The first term is a constant background, whereas the next two terms are dipolar signals oscillating at half the dipolar frequency, and they start at the opposite ends of the full evolution interval in t_{x}. For the pulse sequence to be useful, the condition $p \gg q$ for spins contributing to the signal should be met, so only the last term is significant. This would be difficult to accomplish even for nitroxides but is more feasible for much narrower spectra such as trityl or some organic radicals (e.g., tyrosyl or flavin radical cofactors). Single spins, if present, will add to the background term in q^2. The only way to deal with these issues is to achieve the conditions to ensure small q for spins contributing to the signal using very intense pulses or shaped pulses designed to minimize the product term $p(\Delta\omega)q(\Delta\omega)$ by achieving uniform profile of $q(\Delta\omega) \approx 0$ for flipped spins[73] (see Chapter 21).

The situation is markedly different in DQC-6. As discussed in Section 20.3.3, phase cycling selects only those coherences that have passed through the DQC channel. Consequently, single spins do not contribute to this signal. For finite pulses, as we discussed, this is also true for that fraction of coupled spins where only one offset-refocused spin was effected by the first π pulse; consequently, they do not contribute to any DQ-filtered pathway and are suppressed, so only the p^2 term is detected. Thus, with finite pulses, the performance of DQC-6 is not beset with the problems one has to address with SIFTER.

20.4.3 An Example of SIFTER and DQC-6 with Strong-B_1 Pulses

We demonstrate the abovementioned discussion with (not previously published) experimental SIFTER and DQC-6 data obtained using a typical case of a soluble MTSL double-labeled protein T4-lysozyme (T4L), for which extensive DEER[74,75] and DQC[14,22,31] data have been published. The measurements were carried out at K_u band (17.3 GHz).

We show in Figure 20.8 the data from SIFTER and DQC-6 applied to T4L double labeled at sites 80 and 128 and prepared in both H_2O and D_2O. Intense B_1 pulses of 45 G were used to produce π pulses of 4 ns in all four cases. Yet, even such intense pulses are still insufficient to attain complete inversion of the ^{14}N nitroxide spectrum; therefore, significant background appearing as double frequency modulations is still present in SIFTER, particularly visible for H_2O, because the spin labeling efficiency was ≈0.8 in this case. Single spins do not add any type of dipolar signal, except for intermolecular effects. In the case of protons, the background is strongly decaying with an approximately Gaussian profile, the decay being mostly due to nuclear spin diffusion[22,70,76,77] but also to ID. The decays are generally slower than for a two-pulse PE, as SIFTER uses two refocusing pulses that partially suppress nuclear spin diffusion, making the phase memory time T_{m} about a factor of $2^{1/2}$ longer than for the

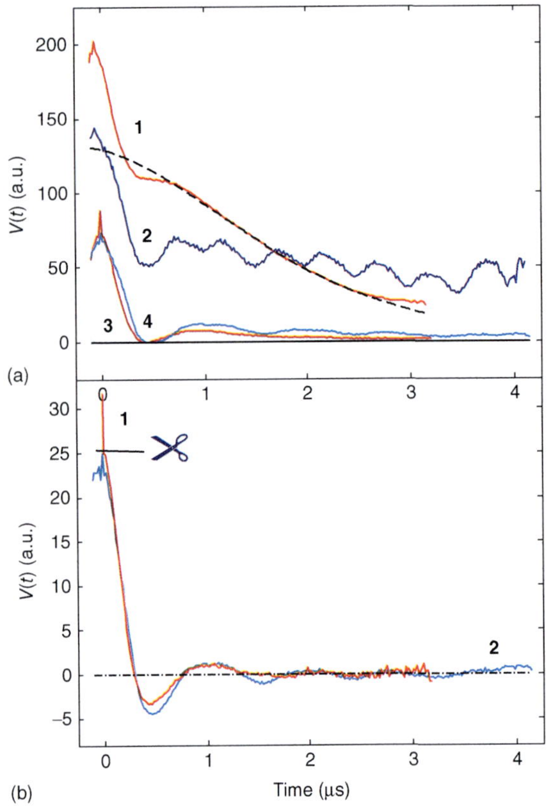

Figure 20.8. (a) 17.3 GHz SIFTER (**1**, **2**) and DQC (**3**, **4**) data taken on a 15-μl sample of T4 Lysozyme, double labeled with MTSL at positions 80 and 128. Samples were prepared in either H_2O or D_2O working buffers. Protein concentration was 48 μM in H_2O buffer (**1**, **3**) and 42 μM for D_2O (**2**, **4**). Spin-labeling efficiency was 0.8 for the H_2O sample and ≈1 in case of D_2O, as estimated using the DEER modulation depth. (The data for H_2O and D_2O are not plotted to the same scale.) The DQC data were scaled up by the factor of ≈2.5 to match the intensity of dipolar oscillation in SIFTER. The dashed curve for 1 is then fit to the baseline, whose temporal profile is dominated by nuclear spin diffusion (~$\exp[-(2t/T_{nsd})^{\kappa}]$, $\kappa \approx 2$, $T_{nsd} \approx 4$ μs for H_2O solutions) and ID (see Section 20.7). This was done by subtracting (**3**) from (**1**) and then fitting to a third-order polynomial the logarithm of the remainder without using the latter 1 μs of the record distorted by spurious contributions. All pulse sequences used π/2 pulses of 2 ns and π pulses of 4 ns, sample temperature was 60 K. A 64-step phase cycle was used for DQC, as well as for SIFTER. Data averaging times were 1 h for **1** and **3** and 4.6 h for **2** and **4**. (b) The DQC data sets shown in (a) are plotted after making corrections for the small background present in the raw data (**3** and **4**)

PE.[22,69] In cases of stretched exponential decay[24] as is the case here, the relaxation envelope is present in the data because the time intervals $t_1 = t_p$ and $t_2 = t_m - t_p$ are variable.[22,69] There are no such problems in DEER as the detection pulse sequence is fixed.

DQC-6 exhibits a very small baseline, so that even deep ESEEM by deuterium nuclei does not appear as a significant problem. The data for H_2O and D_2O are similar, with somewhat deeper dipolar oscillations in D_2O. Note that for H_2O the DQC data are also modified by relaxation caused by nuclear spin diffusion,[31] although the contribution to the decay due to ID is expected to be less than in SIFTER (see Section 20.7). The decay could result in a small broadening of the reconstructed distance distributions for distances <5 nm and can be partly corrected,[31] if so desired. As the background is very small, it is much easier to correct for it than in DEER and introduces smaller errors. We do find that Tiknonov regularization[22,78] can often be applied directly to DQC data to reconstruct distances without any baseline correction, leading to low-intensity broad humps at long distances, but with little if any effect on the main distance peaks at moderate and short distances.

20.4.4 Five-pulse SQC Sequences, Jeener–Broekaert Sequence

There are two five-pulse constant-time sequences (Figure 20.9) that operate with respect to the pulse position in the same manner as six-pulse DQC or SIFTER in that they have similar time variables, which in the case of DQC-6 are the sum (t_m) and difference (t_x) of two interpulse intervals. The two pulse sandwiches in RE are connected by a pulse propagator U, which performs the desired transfer of $\hat{I}_{12}$, $\hat{A}_{12}$ from the first (preparation) to the second (detection) sandwich to accomplish the refocusing of dipolar coupling evolution according to the discussion in Section 20.2.7. All these propagators start from a pulse, for which we use $(\pi/2)_x$ in DQC and $(\pi/2)_y$ in SIFTER. But now the pulse propagator U is much different, as it relies on phase relaxation to remove unwanted coherences. The sequences store selected coherences as polarizations for long enough time for unwanted transverse coherences to decay by phase relaxation. The storage period also allows one to manipulate other coupled spins, if such are present.[38] In general form, we expect the sequence to be described

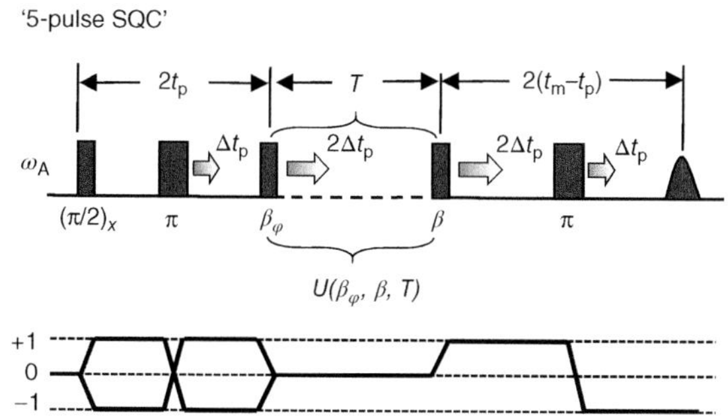

Figure 20.9. Five-pulse PDS 'single-quantum coherence' (SQC) sequences designed to store dipolar evolution for manipulation with coherences. The first $t_p - \pi - t_p$ sandwich prepares in-phase and antiphase coherences labeled by $\cos at_p$ and $\sin at_p$, respectively. When the β_φ pulse is $(\pi/2)_x$, it stores these coherences as polarizations, i.e., in longitudinal magnetizations and dipolar order with β, φ, B_1 (and spectral shape) determining their ratio. A hard $(\pi/2)_x$ pulse stores in-phase coherence as z-magnetization, whereas hard $(\pi/4)_y$ stores antiphase coherence in dipolar order $2S_{1z}S_{2z}$. In the latter case, the sequence functions as the Jeener–Broekaert sequence. After a storage period, $T \gg T_m$ all other coherences decay owing to phase relaxation with memory time T_m. A subsequent β-pulse ($\beta = \pi/2$, or $\pi/4$ for dipolar order) reads stored coherences back into SQCs, which then evolve into a detectable echo

by an arbitrary pulse sequence: $\alpha - t_p - \beta_1 - t_p - \beta - T - \beta - (t_m - t_p) - \beta_1 - (t_m - t_p)$-echo,[21] where we do not define the pulse phases but nominally let $\alpha = \pi/2$ and $\beta_1 = \pi$. Then, the first $\hat{S}_{1y} + \hat{S}_{2y}$ evolves as given by equation (20.21); the coherences $(\hat{S}_{1y} + \hat{S}_{2y})$ and $(\hat{S}_{1x}\hat{S}_{2z} + \hat{S}_{2x}\hat{S}_{1z})$ are transformed by the β-pulse and remain stored during the time $T \gg T_m$ as

$$\hat{S}_{1z} + \hat{S}_{2z},\quad 2\hat{S}_{1z}\hat{S}_{2z},\ + \text{(transverse terms)} \xrightarrow{T} \hat{S}_{1z} + \hat{S}_{2z},\quad 2\hat{S}_{1z}\hat{S}_{2z} \tag{20.24}$$

Then, $(\hat{S}_{1y} + \hat{S}_{2y})$ and $-2\hat{S}_{1x}\hat{S}_{2z} - 2\hat{S}_{2x}\hat{S}_{1z}$ are produced by the β-pulse and evolve to the final $\hat{S}_{1y} + \hat{S}_{2y}$ labeled with the respective dipolar evolution-dependent factors.[21]

If the DQC-6 refocusing pulse is omitted and a long pulse interval $2t_d = T$ is used, all transverse coherences will decay by phase relaxation except for the in-phase coherence that passes through the filter as S_z irrespectively of refocusing. Thus, in such a five-pulse single-quantum coherence sequence (in-phase SQC-5), the $(\pi/2)_x$ pulse stores coherence S_y as S_z, and the subsequent $\pi/2$ pulse generates the coherences contributing to the echo. These coherences are not always labeled by the dipolar interaction, e.g., for a single spin or if the π pulse in the preparation sandwich refocuses dipolar coupling. Therefore, significant background may be present. This background can be modulated by nuclear ESEEM, which is particularly detrimental in the case of deuterons (Figure 20.10a) or nitrogen ligands.[7,13,44,79]

As we showed (see Section 20.2.7) and recapitulate here, two refocusing propagators produce:

$$\hat{S}_{1y} + \hat{S}_{2y} \xrightarrow{t_1-\pi-t_1} \xrightarrow{t_2-\pi-t_2} (\hat{S}_{1y} + \hat{S}_{2y}) \times (\cos at_1 \cos at_2 - \sin at_1 \sin at_2) \tag{20.25}$$

which is proportional to $\cos at_m$. This is the same result as for RE. The similar echo signal is detected in DQC-6 before the second term in equation (20.25) is extracted by filtering out the rest using appropriate phase cycling to give $-\sin at_1 \sin at_1 = (\cos at_m - \cos at_x)/2$, which presents the refocused dipolar signal that we need for recording its dependence on a time variable, e.g., $t_x = t_1 - t_2$ in a constant-time PDS pulse sequence. If any of the two terms in equation (20.25) are removed 'physically', e.g., by relaxation, the outcome is similar, in that a large fraction of spin pairs may have the coupling refocused. As the terms in equation (20.25) depend on in-phase and antiphase coherences at time $2t_1$, modifying them separately solves the problem. This was implemented in SIFTER by complete refocusing of dipolar coupling accomplished by inverting the sign of just the antiphase coherence term. As in-phase and antiphase coherences refocus on different axes in the rotating frame, they can be stored selectively as polarization and then converted back to transverse coherence that will evolve into one of the terms in equation (20.25).

A better use of the five-pulse sequence was reported as a PDS pulse sequence[23] based on the Jeener–Broekaert (JB) technique[42,80] for creating two-spin dipolar order, $2\hat{S}_{1z}\hat{S}_{2z}$. In this technique, a $(\pi/4)_y$ pulse is employed for β_ϕ to convert the antiphase coherence $(\hat{S}_{1x}\hat{S}_{2z} + \hat{S}_{2x}\hat{S}_{1z})$ to $2\hat{S}_{1z}\hat{S}_{2z}$, which immediately follows from

$$\begin{aligned} 2(\hat{S}_{1x}\hat{S}_{2z} + \hat{S}_{2x}\hat{S}_{1z}) &\xrightarrow{(\pi/4)_y} 4(\hat{S}_{1x} + \hat{S}_{1z})2^{-1/2} \times (\hat{S}_{2x} + \hat{S}_{2z})2^{-1/2} \\ &= 2\hat{S}_{1z}\hat{S}_{2z} + \text{transverse terms} \end{aligned} \tag{20.26}$$

The transverse terms (*ZQ*, *DQ*, and antiphase *SQ*) vanish after a sufficient storage time $T \gg T_m$ so that

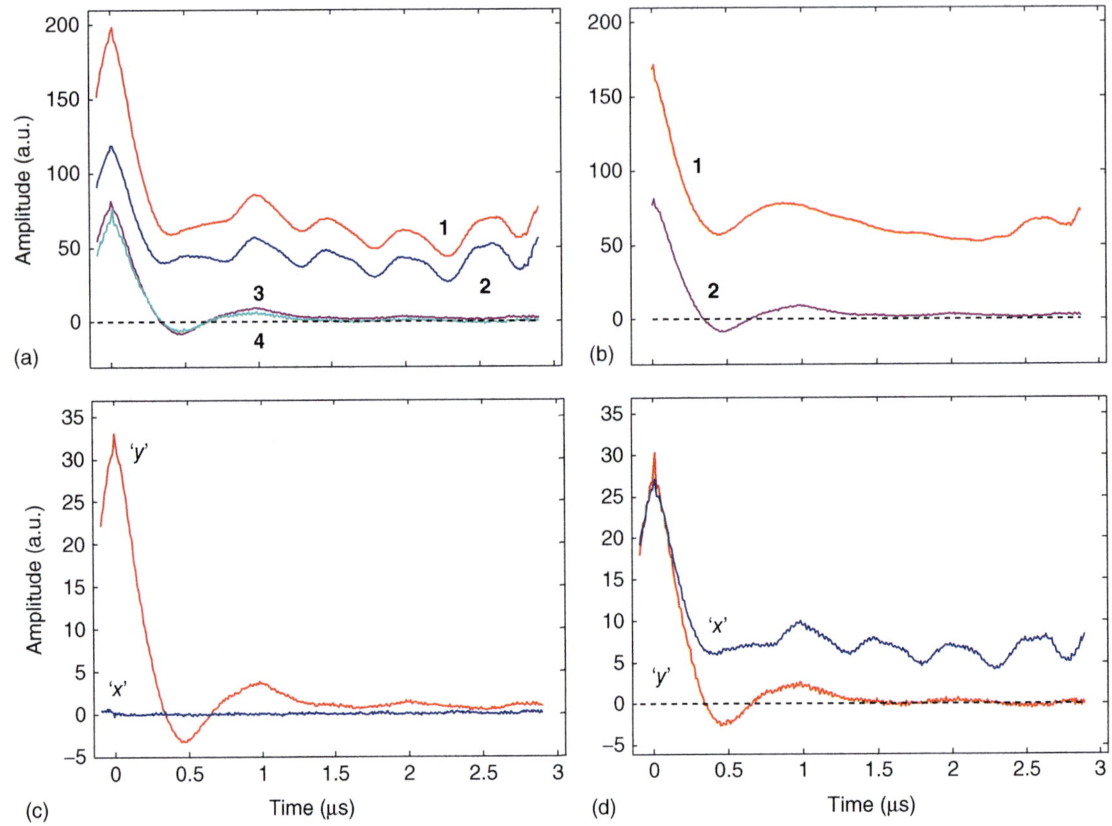

Figure 20.10. The six-pulse DQC and three SQC PDS sequences are compared. For the purpose of this article, comparative measurements were carried out. All data were taken on a 15-μl sample of 38 μM 80/135 MTSL-labeled mutant of T4-Lysozyme in 30% (w/v) glycerol-d_8/D_2O based buffer at 17.3 GHz and 60 K. (a) Shown are the data for SIFTER (**1**); five-pulse in-phase SQC (**2**); DQC-6 (**3**); and JB (**4**) sequences. The data were recorded in succession using the same sample and fixed receiver gain. In all cases, the π/2 and π pulses were 2 and 4 ns; for π/4 pulses in the JB sequence, the pulse duration was adjusted to about 1.5 ns to maximize the signal (b) DQC (**2**) is compared to SIFTER (**1**) with ^{2}H ESEEM filtered out using 2 and ≈4 MHz bandpass filters. The DQC signal was multiplied by 2.5. SIFTER exhibits a strong unwanted dipolar pathway visible at the end of the record, as the pulses are not yet ideal hard pulses. (c) Raw DQC data with *y* and *x* components shown. (d) JB data shows both *y* and *x* components. Plots (c) and (d) have a common vertical scale

only the dipolar order term $2\hat{S}_{1z}\hat{S}_{2z}$ labeled with $\sin at_\mathrm{p}$ remains. Then another $(\pi/4)_y$ converts half of dipolar order to first-order antiphase coherence according to

$$2\hat{S}_{1z}\hat{S}_{2z} \xrightarrow{(\pi/4)_y} 2(\hat{S}_{1z} - \hat{S}_{1x})2^{-1/2} \times (\hat{S}_{2z} - \hat{S}_{2x})2^{-1/2} = -(2\hat{S}_{1x}\hat{S}_{2z} + 2\hat{S}_{2x}\hat{S}_{1z})/2 + \text{other orders} \quad (20.27)$$

The antiphase term labeled with $\sin at_\mathrm{p}$ evolves into an observable signal, while the other orders (ZQ, DQ, dipolar order) do not lead to any detectable signal. The fact that antiphase coherence gets refocused by having its sign changed after a storage time T is of little interest, as the refocused in-phase coherence $-(\hat{S}_{1y} + \hat{S}_{2y})$ produced by the preparation sandwich simply does not survive the storage time, and in this case of ideal pulses, the lone antiphase term evolves into

$$\hat{S}_{y,\mathrm{detect}} = (1/2)(\hat{S}_{1y} + \hat{S}_{2y}) \sin at_\mathrm{p} \sin(at_\mathrm{m} - at_\mathrm{p}) = -(\hat{S}_{1y} + \hat{S}_{2y})(\cos at_x - \cos at_\mathrm{m})/4 \quad (20.28)$$

This is half the amplitude of the DQC-6. There is no in-phase component in S_x's if the pulses are nonselective. (The detected echo would be an out-of-phase echo (i.e., on the x-axis) if $(\pi/4)_x$ were used to read out dipolar order.) As in DQC-6, the dipolar coupling is only partly refocused. The JB sequence thus relies on 'physical' filtering out (by relaxation) of the first term in equation (20.25).

20.4.5 Comparison of DQC and SQC Single-resonance PDS Sequences

The literature data,[23] obtained at X-band on rigid biradicals using relatively soft pulses, demonstrated the feasibility of SQC methods but yielded obscure signals unlike the results on biological samples by DQC-6 conducted under similar conditions.[31] To compare all these methods on biological samples under better experimental conditions that we now routinely employ with DQC, we conducted several measurements on a spin-labeled protein using all four single-resonance methods described earlier. The results are compared in Figure 20.10(a), where the raw data for three SQC PDS sequences and DQC are shown. All experiments were conducted on the same sample and at an otherwise identical instrument setting. The SIFTER pulse sequence and the in-phase SQC-5 in Figure 20.10(a) and (d) are very similar in their outcome, except the SIFTER signal is about a factor of 2 stronger, as expected. We used deuterated solvent, making relaxation slow on the timescale used. Both signals have significant background, which is deeply modulated by deuterons. Filtering ESEEM out numerically (Figure 20.10b) reveals for the SIFTER at the end of the record a large spurious dipolar signal oscillating with half the dipolar frequency (equation (20.23)). However, the DQC-6 signal is clean, has low background, and shows very little presence, if at all, of nuclear ESEEM.

In Figure 20.10(c) and (d), the x and y components of the DQC-6 and JB pulse sequences are shown. Note that in the JB sequence the echo is formed on the axis orthogonal to that for in-phase SQC-5. That is, a so-called out-of-phase echo, similar to that observed in light-induced spin-correlated radical pairs (SCRP), is discussed later in text.

This property helps to separate the signal in the JB sequence from those that involve in-phase coherence stored as $\hat{S}_z$. It should be noted that, expectedly, there is a contribution of the SQC-5 type in the x projection of the signal in JB, as it is difficult to adjust short and intense but still finite pulses to completely avoid having an admixture of the two signals. However, these signals should be significantly out-of-phase as the longitudinal terms $\hat{S}_{1z} + \hat{S}_{2z}$ and $2\hat{S}_{1z}\hat{S}_{2z}$ after, e.g., the $(\pi/4)_y$ pulse will evolve to produce echoes on the x and y axes, respectively. A careful look at the JB signal recorded using different pulses shows that there is indeed an admixture of signals on each axis, so some additional empirical signal phasing and referencing is needed to ensure clean data. But then the relevant data for DQC and JB are close to each other. With ideal pulses, there should be a factor of 2 smaller signal for JB sequence, but DQC refocusing at the two spin frequencies with a third π-pulse reduces this amplitude by a factor of $\approx 2/3$, as the spectral excitation for N π-pulses can be shown to depend on N as $N^{-1/2}$ and leads to a factor of $(2/3)^{1/2}$ for each spin and 2/3 overall.[21]

The filtering properties of these two pulse sequences (DQC-6 and JB) reject the signal from single spins, so ideally there should be no background due to their presence. However, antiphase coherences are formed in the bath of single spins, so there will be some background increasing with concentration and evolution time. SIFTER and SQC-5 do allow the single-spin signal to go straight to the output, but its time dependence does not carry any dipolar modulation; however, it does contain relaxation, ESEEM, and intermolecular decay effects. Thus, this means that there is uncertainty with the nature of the background. Such uncertainty to a varying extent is inherent in all PDS methods. For DQC and JB, the background is minimal, so these issues are less important.

In both DQC and JB sequences, complete spin mixing occurs during DQ or dipolar order evolution periods. These orders are symmetric with respect to spin permutations; therefore, half of the spins will undergo CT as in SIFTER, i.e., the evolution starts on one spin and ends on its partner, whereas the other half remains unchanged. This should have consequences for orientational correlation experiments described in Section 20.6. Therefore, there is a mixture of signals that could depend differently on orientational correlations between the two spins. In the ideal case of four-pulse SIFTER, there is no such mixture, as only refocusing of dipolar coupling takes place, and the principal signal is of the CT type. In the real case, there are unwanted pathways, and the background, all of which do not have the CT property, and they need to be suppressed in order not to obscure the main signal.

At present, with rectangular pulses, DQC-6 and JB sequences present a better choice for distance measurements. In our limited experience with SQC sequences, we find it is more complex to implement JB filtering compared to DQ filtering. The latter is very forgiving for pulse and phase mistuning. A long storage time T is needed in the JB sequence creating some extra complications with the pulse amplifier, more so when it is a TWTA.

20.4.6 SIFTER and DQC on Trityl Biradical

Even though SIFTER yields the strongest dipolar signals, there are currently only a few examples of its applications in the literature. This is not entirely unexpected but is due to the extent of unwanted effects that are difficult to avoid in a multipulse EPR pulse sequence with finite pulses. This greatly complicates what could work perfectly otherwise. This has also been our experience with this group of methods. Over the course of 15 years, we have conducted a substantial number of DQC experiments, and in many of these cases, SIFTER has also been attempted, but even with the largest B_1's available to us (45 G), the outcome was as a rule too convoluted, with an admixture of unwanted dipolar signals, making the task of removal of unwanted signals and background correction an uncertain procedure leading to large residual errors. The background, as we see in Figures 20.8 and 20.10, in addition to relaxation decay, could be modulated with a strong nuclear ESEEM, for which a higher working frequency than K_u or even K_a band would be needed to make them smaller.

While proton ESEEM is easily suppressed by four-point averaging for protons,[21,31] often deep low-frequency ESEEM of the background by deuterium or nitrogen nuclei is very objectionable, easily overwhelming the dipolar signal, especially when spin labeling efficiency is not high enough, as is frequently the case. This was also true for in-phase SQC-5, which also requires (nearly) ideal pulses. Nevertheless, systems with narrow EPR spectra can be more amenable to SIFTER and SQC-5. A good target is the trityl radical, which could be studied even at ambient temperature in liquids.[68] One does not necessarily expect efficient spin labeling with trityl, so the single-label background could substantially hamper these two methods. Both, JB and DQC-6, however, have good filtering properties. DQC-6 yields slightly higher signal and (at least in our experience) is easier to perform. In part, this is caused by the need to set up three types of pulses, and the second issue is with the TWTA, which has to stay in transmission mode longer by $\approx T$, so that one has to break it into a sequence of shorter transmissions, leading to signal artifacts. DQC indeed was very successful for trityl-labeled immobilized T4L[68] and RNA[81] in solutions at ambient temperatures. None of these single-resonance sequences (DQC included) is expected to work in the case of high-spin Gd^{3+} due to its very large pseudosecular term and a potential to exhibit multiple higher coherences within each spin. (All our attempts to detect at 17 GHz and 10 K a DQC signal on the narrow central transition of Gd^{3+}/DOTA in 60 μM sample of doubly labeled T4L 8C/128C mutant thus proved unsuccessful. The DEER signal, however, was readily recorded yielding an SNR of about 30% of that for these sites labeled with MTSL, being recorded at 60 K. The lower SNR for Gd^{3+} compared to MTSL may be due to the insufficiently high working frequency of 17 GHz, as Gd^{3+} DEER becomes efficient at frequencies at Q-band (33–50 GHz) and above.)

SIFTER was successfully conducted on a trityl biradical in frozen solution with the results shown in Figure 20.11 where it is compared to those of DQC-6.[82] The time-domain data from both techniques are very similar but not identical, as modulation is more pronounced in SIFTER. The distance distributions reconstructed with DeerAnalysis 2013[83] are also different. The supplementary information for this work indicated large enough backgrounds so that even for this narrow spectrum there is still greater potential for uncertainty in extracting the dipolar signal in SIFTER compared to DQC, thus showing SIFTER requires still harder pulses than the 22 ns π-pulses used as well as tougher requirements for spin labeling.

20.4.7 New Approaches Based on Uniform Spectral Excitation by Adiabatic Pulses

As we have shown, even for sufficiently intense pulses, there still is a problem with the outcome of SQC (as distinct from DQC) PDS experiments. Even more intense pulses may be desirable, but this would require smaller resonators than we currently employ with some loss of sensitivity possible. It is also worth mentioning that relaxation, nuclear ESEEM, and flip-flop

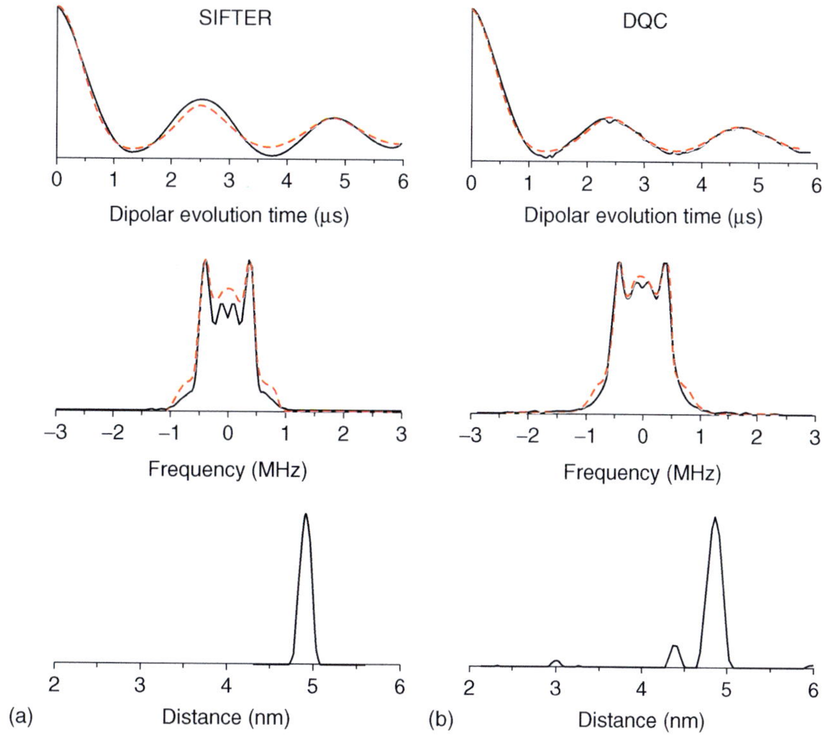

Figure 20.11. SIFTER (a) and DQC-6 (b) experiments performed on a trityl biradical at Q-band ≈34 GHz and 50 K. Widths of π-pulses were 22 ns. (Top) Shown here are background-corrected data traces (solid) and time traces (dashed) generated from the distance distribution as reconstructed using Tikhonov regularization included with DeerAnalysis software. (Middle) Fourier transforms of the background corrected SIFTER and DQC time traces plotted as normalized to the maxima of the Pake patterns. The transforms of the experimental time traces (solid) and the fitted traces (dashed) are shown. (Bottom) Distance distributions reconstructed from the data (top row) using Tikhonov regularization in DeerAnalysis. (Reproduced with permission from Ref. 82. © Royal Society of Chemistry, 2015)

terms in the secular dipolar Hamiltonian contribute a considerably more severe problem in single-resonance techniques compared to double resonance, where they can be avoided or minimized by having the detection pulse sequence fixed and pump pulses applied at large enough spectral separation from the detection sequence. On the other hand, these problems may in principle be overcome using adiabatic pulses that can provide more uniform spectral excitation to help to bring the performance closer to that expected from ideal hard pulses (see Chapter 21). Various types of shaped pulses present a possible approach permitting more uniform excitation, refocusing, and population inversion with spectral coverage that is wider than the typical nitroxide spectral width, or alternatively the shaped pulses could have a sufficiently rectangular excitation profile over a limited spectral extent to reduce the contribution of unwanted dipolar trajectories. Single spins would then remain the only source of the background but, unlike in DEER, may be more severely affected by nuclear ESEEM; thus, a higher working frequency would be of help with alleviating this issue. Such approaches, indeed, were recently tested with SIFTER.[73,84] In one approach tested on a biradical (Figure 20.12a),[73] the back-to-back $(\pi/2)_x - t_1 - \pi - t_1$ and $(\pi/2)_y - t_2 - \pi - t_2$ adiabatic PE sequences with 2 : 1 pulse width (or more precisely, inverse frequency sweep rates) ratio were employed for echo offset refocusing without time dispersion.[85] Using such adiabatic-pulse-based SIFTER at X-band with a 1-kW power source, a background of only 5% (Figure 20.12b) survived, but at Q band, with only 10

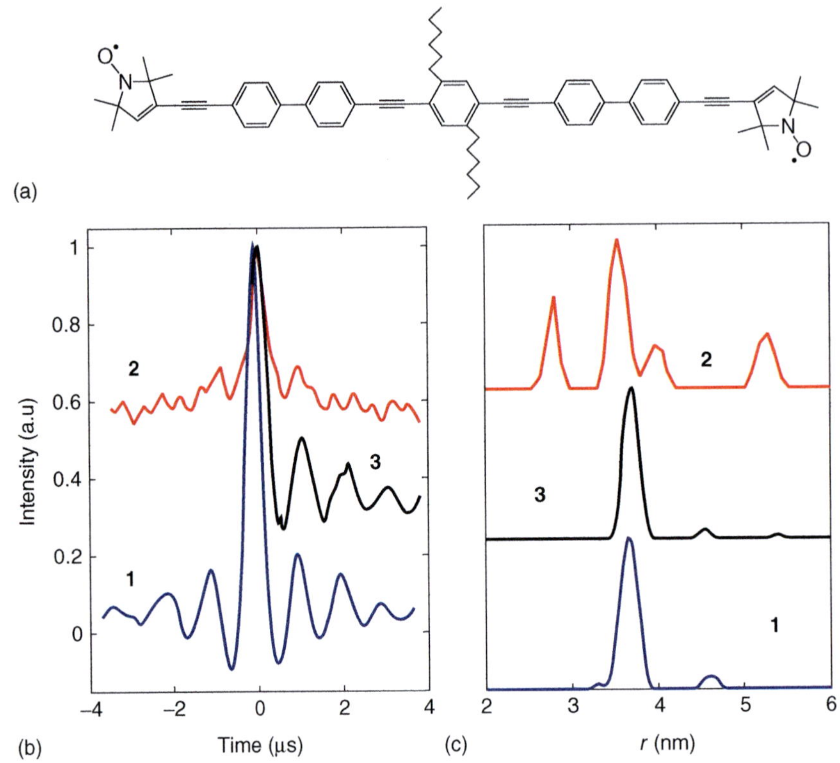

Figure 20.12. (b) X-band broadband SIFTER (**1**); rectangular pulsed SIFTER (**2**); and four-pulse DEER (**3**) – all measured at 50 K on the rigid nitroxide biradical in (a). The plotted data are background corrected and normalized to unity. For rectangular pulsed SIFTER, π/2 pulse was 8 ns; for four-pulse DEER, π/2 pulse was 16 ns, and pump pulse of 14 ns was applied at 70 MHz offset at the center maximum. (c) Distance distributions for corresponding time traces (**1–3**) produced by Tikhonov regularization included in DeerAnalysis 2013 software suite. (Reproduced with permission from Ref. 73. © Elsevier, 2015)

W pulses available, the outcome while encouraging was less convincing. Nevertheless, the improvement at Q-band was apparent as the time-domain data were much closer to DEER as compared to the case of low-power rectangular pulses shown in Figures 20.12(b,2) and (c,2).

Broadband SIFTER with adiabatic pulses has provided distance distributions of comparable quality to the four-pulse DEER (Figures 20.12b, 1 and 3; 20.12c, 1 and 3). The experimental results for SIFTER are thus generally encouraging. Systems with shorter distances, high local concentrations, and short relaxation times (e.g., membrane proteins) as well as short distances (<2.5 nm) may continue present a challenge. These cases, however, are handily addressed with DEER and DQC pulse sequences, which are far more robust techniques at present.

20.5 OTHER SINGLE-FREQUENCY PDS METHODS

20.5.1 Different Types of PDS Experiments

In a PDS experiment, first-order coherences are created for one or both spins, usually with a π/2 pulse followed by a pulse sequence that produces, after time $2t_m$, an echo that is used to sense the dipolar interactions. In Section 20.2, we outlined two types of PDS experiment, i.e., *single-resonance* and *double-resonance*, where we showed that the key difference between them is in how they evolve the dipolar coupling. Clearly, there is greater flexibility in manipulating pulse sequences in double resonance. In addition, it is possible to avoid certain artifacts that complicate single-resonance methods, viz. nuclear

ESEEM and sometimes relaxation effects, and also problematic are spurious dipolar trajectories, with only the DQC-6 and JB sequences able to successfully address them. Traditionally, double-resonance and single-frequency PDS methods were considered as distinct. However, our discussion and subsequent analysis in Section 20.2 indicated that double resonance is different from single-resonance more significantly in the manner in which it is conducted rather than in using pulses at a different frequency. For example, we have mentioned single-frequency methods of light-driven DEER and RIDME, neither of which use MW pulses for pumping, but they are otherwise conducted much like three- or four-pulse DEER acting (predominantly) on spins at a frequency different from that of the detected spins.

We now list several single-frequency methods, some are pure single resonance, while others function as double-resonance, but all are conducted at a single-frequency vis-à-vis MW pulses.

20.5.2 Single-frequency Single-resonance Methods

20.5.2.1 DQC Family

In addition to the DQC-6 pulse sequence described in Section 20.3, there are four- and five-pulse DQC sequences of the DQF COSY type, of which the latter is a constant-time sequence.[21] They are currently not in use as DQC-6 is a far better technique. Furthermore, in addition to DQC-6, as already described, there is a variant used in the so-called DQM (double-quantum modulation) mode.[86] This is not a constant-time sequence; therefore, it entails relaxation decay and stronger ESEEM, which is addressed by adjusting the pulses.[79] However, similar to variable-time DEER,[25] it may produce stronger signals at an earlier time. DQC-6 has also been used in a DQ-filtered refocused PE mode, DQF-RPE.[69] This is essentially the recording of $\cos(at_m)$ as one would have with just RE, but the background is better suppressed.

20.5.2.2 Spin-correlated Radical Pairs

There are several such examples when dipolar spectroscopy was performed on light-induced transients, especially the case of SCRP.[59,87,88] A laser pulse creates nonequilibrium spin populations that contain dipolar order. Subsequently, a two-pulse sequence is used with the first pulse being a ≈π/4 pulse that converts this dipolar order into antiphase coherence, and then it evolves in a refocusing sandwich to produce an out-of-phase echo (Figure 20.13).

The detection part of the JB sequence is similar to the sequence used to generate an out-of-phase echo in SCRP, which is 90° out-of-phase to a normal PE. The initial nonequilibrium population of radical pair spin states gives rise to dipolar order $2\hat{S}_{1z}\hat{S}_{2z}$ with some amount of $\hat{S}_{1z}+\hat{S}_{2z}$ also present. If we consider weak dipolar coupling, the evolution is the same as that described in Section 20.4.4; therefore, only the reconversion part of the evolution of $\sin at_2$ is available, leaving the researcher with an antiphase Pake doublet. This is a true single-resonance case, but the practical role is limited to systems where reasonably long-lived SCRP are generated and are amenable to PDS measurements. ZQC is also photo-produced in this case and can be refocused to evolve into observable single-quantum echo.

20.5.2.3 Out-of-phase Echo

An equilibrium spin state could itself entail dipolar order, e.g., at low temperatures and high fields due to the greater importance of higher-order terms in the density matrix expansion, leading to results similar to SCRP, but one can use any bilabeled molecule with acceptable relaxation properties. Nitroxides are certainly included. An experiment[89] has been conducted recently at G-band (180 GHz) at 6.4 T and 5 K, where the spin system is highly polarized. The experiment could have been an out-of-phase echo as with SCRPs, but (as the spectrum was not too narrow) was performed instead as out-of-phase PELDOR to benefit from the absence of relaxation decay and zero dead time. This certainly is not necessarily an optimum experiment in terms of spin sensitivity and is not suited for the MW frequency range but is an interesting approach, nevertheless.

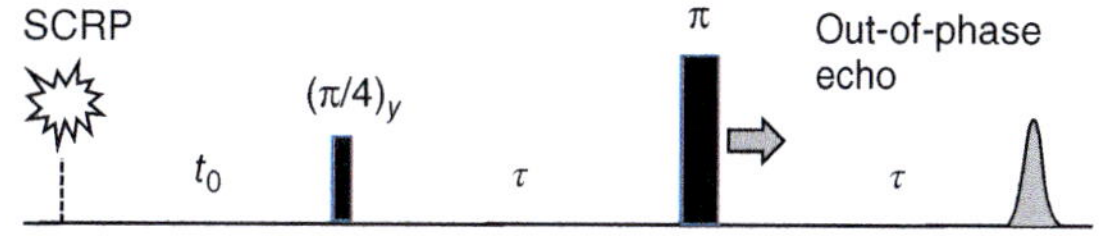

Figure 20.13. Out-of-phase echo can provide distances from $\sin(a\tau)$ evolution of the antiphase coherence produced with a π/4 pulse from the dipolar order formed in a spin-polarized SCRP

20.5.3 Single-frequency Double-resonance Type Methods

20.5.3.1 '2+1' and Refocused '2+1' Sequences

The first useful PDS EPR technique was, as previously noted, the three-pulse DEER (PELDOR), which is a *double-resonance method.* Early DEER experiments made use of two MW power sources tuned to different frequency modes of a bimodal MW cavity resonator, as low-Q cavities were inadequate.[11] This was later addressed by Bruker with more efficient single-mode broadband split ring and dielectric resonators conveniently housed in flow cryostats.[62,90–93] Before this happened, the '2+1' technique was introduced.[3,94] It used one or two MW power sources operating at the same or close frequencies utilizing a standard TE_{012} resonator. The pump pulse was generated at the same frequency (unlike DEER) but with an independent frequency source (like DEER). This was suitable for a typical pulse spectrometer and the technique was used in several studies[15] despite several significant shortcomings. Applications that have benefited from the '2+1' method were cases of relatively narrow EPR spectra, such as the tyrosyl radical encountered in the study of photosynthesis or other radical cofactors, and the method was extensively applied to map the location of radicals in photosystems[15] before the technique went out of use.

We note that '2+1' has a short but finite dead time. It is interesting whether this sequence with so much in common with DEER could be modified into a dead-time-free four-pulse '2+1' as was done to three-pulse DEER. Such a sequence was tested for this article. It turns out that the problem with the finite dead time in '2+1' cannot be as easily overcome by the addition of the refocusing pulse as in four-pulse DEER. With such a revision (Figures 20.1 and 20.14), one observes the same four dipolar trajectories as in four-pulse DEER,[22] i.e., two signals $V_1(t)$ and $V_2(t-\tau)$ of which V_2 has zero dead time and their counterparts appear reversed in time, $V_3(T-t)$ and $V_4(T-t+\tau)$. Here, $V_k(t) = V_0[1 - p(1 - \cos at)]$ is a standard signal in DEER. However, unfortunately, the intensities of these 'dipolar pathways' are different for the two cases. Both types, zero- and nonzero dead time, are comparable in the modified '2+1'. In both cases shown in Figure 20.14, the incoherent pump pulse from a DEER oscillator was used. The hard pulses had a B_1 of 45 G. The performance of the '2+1' sequence in this setup is quite good, but it does not reach the performance of DQC-6 conducted on the same sample for reference to provide a feeling for why such newer coherent methods are superior.

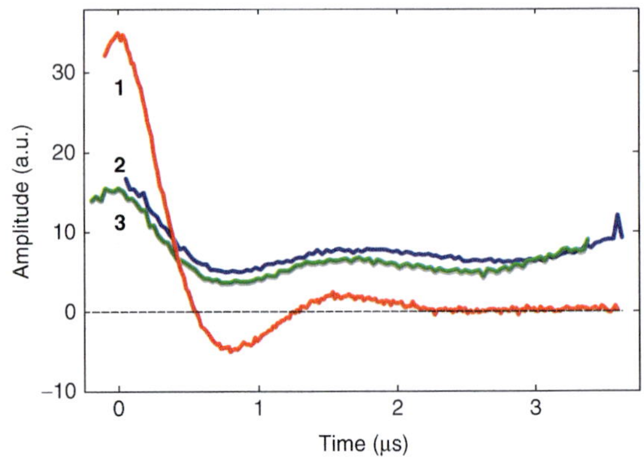

Figure 20.14. Measurements with '2+1' pulse sequence at K_u band based on intense pulses (2). Also plotted for comparison data (3) for '2+1' modified by adding a refocusing pulse in the same manner as in the standard four-pulse DEER or five-pulse RIDME to achieve zero dead time in dipolar evolution. The raw six-pulse DQC data (1) are shown for comparison on a common scale. All measurements were taken in succession with the same receiver gain and plotted as recorded using a reversed timescale to look as in standard DQC or DEER. The identical pump and detection frequency in all cases was 17.3 GHz, and π/2 and π pulses were 2 and 4 ns, respectively. The measurements were conducted at 60 K on a 15-μl sample of 35 μM of T4L doubly labeled with MTSL at 44 and 144 positions prepared in Tris-HCl buffer containing 30% w/v glycerol (this lab, previously unpublished)

20.5.3.2 The RIDME Experiment

The simplest single-resonance PDS experiment is based on a PE, but extracting dipolar coupling is rarely possible, as it is masked by larger effects. A successful solution was DEER, but alternatives were also developed, viz. DQC and SQC techniques described such as DQF-COSY and DQC-6,[20,21] but also three-pulse RIDME.[95] RIDME relies on spin flips caused by relaxation rather than by a pump pulse. Therefore, it may involve all the B-spins (i.e., those coupled spins not observed) independent of how wide the spectrum could be, thus, making it advantageous for, e.g., metal ions. The initial form of RIDME had the second refocusing pulse in a PE split into a $\pi/2 - T - \pi/2$ sequence, essentially producing a stimulated echo

(SE) sequence, $\pi/2 - \tau - \pi/2 - T - \pi/2 - \tau$–echo with a long enough T to ensure relaxation of the B-spin to complete. By repeating the experiment with two values of T (short and long), it was possible to factor out the dipolar signal by dividing two traces, in one of which with long T spin–lattice relaxation caused a spin flip, thus substituting for a pump pulse. The efficiency of a relaxation flip was sufficient to produce clear dipolar modulation well above the level of residual postrefferencing effects.[95] The performance could be further improved if the coupled spin has a broader spectrum and shorter spin–lattice relaxation time T_1 than the observing spin A, making the spin B contribution dominant. A more sensitive A-spin (e.g., nitroxide) can be used to form the echo. Further pulse-sequence developments soon followed, reducing dead time by adding a fourth refocusing pulse[27] and then eliminating dead time altogether with the addition of one more refocusing fifth pulse as was done in four-pulse DEER,[26] although three-pulse RIDME is still in use with high-field high-frequency (HFHF) EPR.[38] These elaborations yielded a '2+1'-like, and its refocused version type, single-frequency sequences, but the pump pulse was replaced with the storage period $\pi/2 - T - \pi/2$ providing a variety of opportunities to invert the coupled spin with a broad spectrum and reduce some of the unwanted effects. These effects include nuclear ESEEM, spectral diffusion, unwanted dipolar signals of the '2+1' type, and dipolar frequency harmonics for high spin. Several approaches have been tested to reduce unwanted effects, especially ESEEM from deuterium typically used in PDS for increasing T_m to improve the sensitivity and distance range. This includes conducting the experiment using two different T's or at two temperatures, using Q-band as well as using soft pulses at Q-band.[29]

RIDME as a pump pulse-free scheme is particularly valuable in relaxation-based single-frequency experiments, when most of the coupled spins are that of a metal ion, thus predominantly at a different frequency than the observed one, but it was also applied to nitroxides, where this is not the case.

Dead-time-free RIDME[44] (Figure 20.15) is an important development and was used in several studies on metalloproteins.[29,30,38,44] It would be too difficult to pump the broad spectral extent of an ion, but relaxation accomplishes this task very efficiently. The maximum pump efficiency of 0.5 is due to the fact that this is relaxation, not population inversion by a pulse that could ideally flip all spins. The use of relaxation has peculiarities for high spins, where relaxation involving all

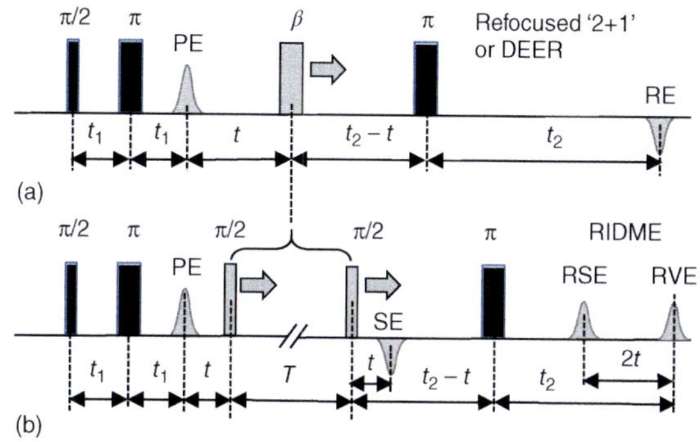

Figure 20.15. (a) Refocused '2+1' sequence (or four-pulse DEER) can be transformed to the RIDME sequence by replacing its β-pulse with a period T of longitudinal storage. Note that, unlike in DEER, the β-pulse in '2+1' is not a π-pulse in order not to suppress RE. (b) A single-frequency RIDME pulse sequence includes a pulse block – $\pi/2 - T - \pi/2$ – to store coherences as z-magnetization; operate on it; and then transform to transverse coherence for detection. While this block is used in the same manner as a B-spin-flipping pulse, this function and the nature of coherence pathways and detected echoes are different than in (b). A refocused virtual echo (RVE), whose position is fixed in time is typically sampled, but the use of a refocused stimulated echo (RSE) was also reported. All spin manipulations that occur during the storage period T are aimed to flip the coupled spin, giving rise to other methods than RIDME

levels can produce $\Delta m_B > 1$, leading to harmonics of the dipolar frequency, so this should be addressed in distance reconstruction.[28,96]

The sequence is used in five-pulse constant-time dead-time-free mode, i.e., using RE as a detection pulse sequence and a composite propagator $\pi/2 - T - \pi/2$ to make a storage period for relaxation-based spin pumping. However, there are principal differences here from detecting RE with no interfering pump pulse. As the sequence is of the double-resonance type, it cannot use pulse timing of a single-resonance five-pulse SQC sequence keeping the sum of refocusing sandwiches fixed. Instead, it uses four-pulse DEER timing, with the pump pulse block scanned between the last two pulses of RE sequence. This produces a refocused SE (RSE) and refocused virtual echo (RVE), separated by $2t$, where t is the evolution variable. Both can be used, the latter is usually preferred, but each is half of the RE intensity for the same t_m. Note that five-pulse SQC does not form SE, both RSE and RVE collapse into RE at $t = 0$.

Therefore, the storage block is just a delay in SQC-5 in the absence of phase relaxation whose existence is necessary for this propagator to attain its purpose. Moreover, pumping B-spin during storage is basically of no use for SQC-5, as it is truly different from double resonance in dipolar evolution.

As for using moving pulses at the detection frequency, there are several artifacts; particularly, nuclear ESEEM issues have to be addressed for deuterium. This was investigated[29] as we previously noted and the referencing was made based on using different temperatures, soft pulses, and different durations of storage period. In such methods, 'pumping' on detected spins (A-spins) produces artifacts, so this is undesirable.[3,22,40] And when they are intentionally applied to operate on detected spins (e.g., RIDME on nitroxides), their performance has not been yet as convincing as that of standard robust techniques such as DEER or DQC due to artifacts and significant loss of signal.

All in all, the use of a storage propagator is enabling approaches to improve sensitivity in the case of broad spectra of metal ions, for which there are few alternatives, and in addition, it offers a variety of methods to flip B-spins other than by relaxation in RIDME. That is, one can use other suitable means that can change the dipolar coupling – e.g., light pulse, electron beam, relaxation, chemical reaction, or some other process that can change polarization or create, modify, or remove spin density.

20.5.3.3 Field-jump PDS Experiment and Its Modern Analogs

Another example of single-frequency dipolar spectroscopy is the field-jump (FJ) DEER experiment based on a $(\pi/2)_A - \tau - (\pi/2)_A - T - (\pi/2)_A - \tau$–echo-stimulated echo (SE) pulse sequence with the pumping of a different spectral region performed during the $\pi/2 - T - \pi/2$ storage period.[27,38] It does not require a different frequency source to pump, but a field jump or sweep produced by a current pulse is utilized to make the equivalent of pumping B-spins at a different spectral region or creating an adiabatic population-inversion pulse[41] using a long MW pulse and a field sweep in a collective action during the storage period, thereby presenting a rather complex propagator, which flips spins at different spectral regions, as does DEER. The pulse sequence, however, was exercised as SEDOR,[38] i.e., by advancing pulse sequence length similar to PE or RE. The method entails phase relaxation decay and nuclear ESEEM issues. Signal referencing, resembling RIDME, was utilized to remove most of ESEEM and to account for relaxation decay. The dead-time issues were partly mitigated by adding a fourth refocusing pulse to the stimulated-echo sequence used. The four-pulse sequence resembled '2+1' with a pump pulse replaced with a $\pi/2 - T - \pi/2$ magnetization storage period and all the pumping being located there. However, the constant-time sequence mode was not used. Recent rethinking of the field sweep approach produced an experiment dubbed CIDME (chirp-induced dipolar modulation enhancement), which uses frequency-chirped adiabatic MW pulses[40] instead of magnetic field sweep for population inversion. However, it also adds a fifth pulse[26] enabling zero-dead-time implementation based on the scheme developed and now routinely employed with the latest RIDME technique discussed earlier. It was shown that for nitroxides it could be beneficial for low-power spectrometers as the adiabatic pulse could be long but without delay effects (see Section 20.6.5).

20.5.3.4 Laser-induced DEER

There are more single-frequency PDS experiments known; however, some (or most) of these single-frequency methods are actually double resonance. This includes laser-induced PDS,[43] which can use a high-Q single-mode resonator and provide pump action optically by effecting a change in spin polarization, e.g., by optical generation of a porphyrin triplet with a laser pulse. With a short enough optical pulse, it is possible to access a shorter range of distances than in standard DEER, especially if the detection pulses are made as intense as in DQC. A triplet-state EPR spectrum is much wider than that of a nitroxide, so there are relatively few molecules in the triplet state at the same frequency as a nitroxide, which will otherwise contribute '2+1'-like type spin dynamics as noted in Section 20.5.1. Therefore, it is still a double-resonance experiment, even though no MW pulses at the second frequency are needed. In addition, refocusing and pulse evolution (fixed detection sequence) are particularly in line with a double-resonance (DEER) approach. The only potential issue is the quantum yield, as it is desirable that at least 10% of the nitroxides have their partner with its spin flipped.

20.6 2D-FT ORIENTATION-CORRELATION PDS

20.6.1 Orientational Aspects in PDS

The primary goal of any PDS experiment is to determine distances between electron spins. The outcome could be affected by the fact that the time-domain signal recorded in a PDS experiment can be sensitive to orientations of the coupled magnetic dipoles in the molecular frame, which gives rise to effects known as orientation selection or correlation[97] (see also Chapters 11 and 19). In the fully isotropic case for a pair of coupled electron spins for each interspin orientation of the vector $\boldsymbol{r}_{12}$ connecting the spins, all orientations of $\boldsymbol{r}_{12}$ relative the static field $\boldsymbol{B}_0$ are described by the polar and azimuthal angles θ and φ and are equally probable, e.g., for two point dipoles. The dipolar coupling then depends only on θ as $(3\cos^2\theta - 1)$, and this gives rise to the characteristic dipolar line shape (Section 20.2.2) known as a Pake doublet.[49,50] However, for a spin label the orientation of the spin-bearing moiety may be fixed and tilted relative to $\boldsymbol{r}_{12}$ if it is not flexibly tethered. The EPR spectra in solids are determined by the hf and $\boldsymbol{g}$-tensors whose anisotropies are defined in their respective principal frames. (The hf and $\boldsymbol{g}$-tensor principal frames may not necessarily coincide, but in the case of nitroxides, the differences are typically small and can largely be ignored.) The orientations $\Omega_k = (\alpha_k, \beta_k, \gamma_k)$, $k = 1, 2$ of these frames are then defined relative to $\boldsymbol{r}_{12}$ taken as the polar axis giving five independent Euler angles with $\alpha_1 = 0$.[65] Nearly always, the finite pulses in PDS excite only a fraction, frequently a small fraction of the EPR spectrum; this particularly is true at high fields where EPR spectra are often dominated by the $\boldsymbol{g}$-factor anisotropy.[38] Therefore, in cases of pronounced anisotropy with well-defined orientations of the magnetic frames relative to $\boldsymbol{r}_{12}$, each $\boldsymbol{r}_{12}$ orientation selects a range of orientations in $\boldsymbol{B}_0$ and consequently a subset of possible dipolar couplings, so that the dipolar spectrum no longer has the shape of an ideal Pake doublet. Thus, PDS experiments at high fields, which are usually DEER or RIDME[28,38,40,44] type double-resonance experiments, could be very sensitive to molecular orientations and can be explicitly directed toward obtaining additional details of molecular structure.

Our focus will be on nitroxides as they have been used in most studies. One expects less pronounced orientation effects (correlations) as well as a greater relative contribution of hf tensor anisotropy as compared to the $\boldsymbol{g}$-tensor at fields up to Q-band. This is certainly the case for nitroxide spin labels, where such effects are not highly pronounced and would require magnetic frames of electron spins to be well fixed relative to the molecular frame. Typically, nitroxide spin labels exhibit low anisotropy because their flexible side chains give rise to several rotamers.[98,99] Nevertheless, orientation correlations were revealed by conducting DEER experiments from X to Q band using several pump and detection frequencies over the EPR spectrum. This could be a very time-consuming endeavor.

20.6.2 2D-DQC Correlation Maps

The standard DQC-6 experiment under nonselective spectral excitation is focused on accurate distances and suppresses the manifestation of orientational effects in the dipolar evolution signal, as the spins of molecules at all orientations relative to $\boldsymbol{B}_0$ contribute equally to the signal. This ideal picture of eliminating the orientation effects in the DQC-6 experiment is limited by the finite nature of the pulses. (The contribution of the flip-flop terms in the secular part of the dipolar spin-Hamiltonian depends on the orientationally dependent frequency separation between spectral components underlying the spectrum. However, these effects are usually relatively small unless the coupling becomes strong, i.e., at distances as short as 1.0–1.5 nm.) The dipolar evolution is acquired as the amplitude of the spin echo, i.e., each at refocusing point; thus, there is no dependence on frequency offset. Nevertheless, the spin-echo envelope, determined by the EPR spectrum shape, does encode the information on orientations of the magnetic frames of coupled electron spins, enabling an orientation selection 2D-DQC technique. This hidden orientational information can be revealed if one elicits the contributions to the full dipolar signal made by each spectral component contributing to the spin-echo signal. This is revealed by taking the Fourier transformation (FT) of the spin-echo shape, providing the EPR spectrum, and enabling a 2D-FT plot of the dipolar spectrum vs the rigid-limit nitroxide spectrum, each point of which corresponds to a set of molecular orientations. This yields a 2D-FT frequency correlation plot, which is a fingerprint of the orientation-dependent effects caused by the anisotropy of magnetic tensors. In the actual

DQC experiment, the full echo envelope is captured by a fast ADC, sampling at GHz rates and can be stored for 2D processing. (For further FT techniques, see Chapter 15.)

Assuming ideal pulses, the dipolar signal relevant for 2D-DQC experiment can be written out. For spin 1, antiphase coherence $2\hat{S}_{1x}\hat{S}_{2z}$ produced after the DQF by the $\pi/2$ pulse 5 evolves into observable in-phase coherence in the last $t_2 - \pi - t_2$ sandwich ($t_2 = t_m - t_p$) as

$$2\hat{S}_{1x}\hat{S}_{2z} \rightarrow (\hat{S}_{1y} \cos \Delta\omega_1 \delta t - \hat{S}_{1x} \sin \Delta\omega_1 \delta t)$$
$$(s_{t_2} c_{t_2+\delta t} + c_{t_2} s_{t_2+\delta t}) \qquad (20.29)$$

Here, δt is the time variable taken from the refocusing time $2t_2$. After adding to equation (20.29), the dipolar evolution phase $\sin at_p$ acquired through the evolution in the preparation period before the filter, the detected in-phase component in equation (20.29) is

$$\hat{S}_{1y} \cos(\Delta\omega_1 \delta t) \sin(at_1) \sin(at_2 + a\delta t/2)$$
$$= \hat{S}_{1y} \cos(\Delta\omega_1 \delta t) \sin(at_p) \sin(at_m - at_p + a\delta t/2)$$
$$= \hat{S}_{1y} \cos(\Delta\omega_1 \delta t) \cdot [\cos(at_x - a\delta t/2)$$
$$- \cos(at_m + a\delta t/2)] \qquad (20.30)$$

For the second spin, $\Delta\omega_2$ is replaced by $\Delta\omega_1$, but the dipolar part is the same. We now define the spin-echo time $t_{echo} \equiv \delta t$ and the dipolar evolution time $t_{dip} \equiv t_x$ variables, and also $t_{dip}' = t_{dip} + t_{echo}/2$. The 2D signal $F(t_{echo}, t_{dip}')$ is in terms of these variables. After a 2D FT, followed by a shear transformation of the frequency variables,[65] a 2D-FT plot $V(\nu_{echo}, \nu_{dip})$ is produced. The resulting 2D plot (or map) gives the dipolar spectrum for each frequency in the spectrum. Figure 20.16(a) and (b) shows examples of such distinctive 2D maps simulated for uncorrelated and strongly orientationally correlated cases. On the side of the plots are the spectral sums calculated along ν_{dip} and ν_{echo}. They give, respectively, the 1D EPR spectrum (with no dipolar effects in it) and the pure the 1D dipolar spectrum (Pake doublet without the orientation effects), as for any spin all its partners contribute to the sum, so all orientations contribute equally.

In Figure 20.16(c), we show the 2D DQC data obtained at 17.25 GHz and 60 K on 15 μl of 15 μM solution of a 3-nm rigid rod biradical[21] in *o*-terphenyl-d_{24} glass. The shortest π pulse available for a sample of this size was 4 ns, yielding a B_1 of 45 G. Yet, this B_1 is insufficient to provide *completely* uniform spectral coverage for the whole ^{14}N nitroxide spectrum. However, this is not a very serious problem. As most of the spectrum is covered, the outcome can be simulated for the finite B_1 used.[65] In addition, one could increase the B_1 to ≈60 G using a somewhat smaller resonator. Nevertheless, two separate data collections were made to widen the coverage. The pulse sequences were applied 30 G apart at the low- and high-field parts of the spectrum, respectively. The individual 2D spectra were then truncated on opposite sides at a selected frequency and combined into the single spectrum shown in Figure 20.16(c,**2**). This manipulation is not absolutely necessary, as fitting each of the 2D spectra suffices to reveal the key orientational properties. This approach based on recording two spectra does not incur a significant penalty in acquisition time, as most of it is used to produce good SNR for the high-field side of the spectrum, which has a factor of ≈3 lower intensity than the more intense central and low-field regions.

For this nearly linear biradical, the nitroxide moieties have their beta angles, $\beta_{1(2)}$ at about 90° relative to the interspin vector, $\boldsymbol{r}_{12}$, connecting the points located at three-fourth of the NO bond length toward the nitrogens. As confirmed by simulations, the remaining angles are relatively unimportant. Molecular modeling[21] has determined $|\boldsymbol{r}_{12}| = 2.89$ nm and the following set of Euler angles: $\Omega_1 = (0°, 75°, 0°)$, $\Omega_2 = (90°, 111°, 0°)$.[21,22] We show in Figure 20.16(c) the correlation maps for experimental and calculated data, rendered as normalized filled contour plots. The plots are quite similar, i.e., nitroxide orientations are of the expected type as simulated in Figure 20.16(b), but there are some visible differences that could be addressed by modeling. One may also need to account for the effects of the resonator bandwidth, which would require more substantial computational efforts than used in this work. In summary, we note that acquiring orientational information by DEER necessitates recording several traces,[100] whereas 2D DQC can do it in a single pass (or just two). The outcome is not the full equivalent of ideal hard pulses but is deemed to be close enough in information content as we have shown above. However, DEER, to its advantage, is not limited to just nitroxides and can be applied to infer orientational information from very broad spectra of metal ions and at high fields from orientationally better resolved spectra.[101] It is, however, worth mentioning that the 2D-DQC technique, as for the 1D-DQC on which it is based, presents a robust sensitive and low-artifact technique covering three octaves in distance (i.e., a factor of ≈500 in dipolar frequency). The upper range of distances is limited by the phase memory time T_m

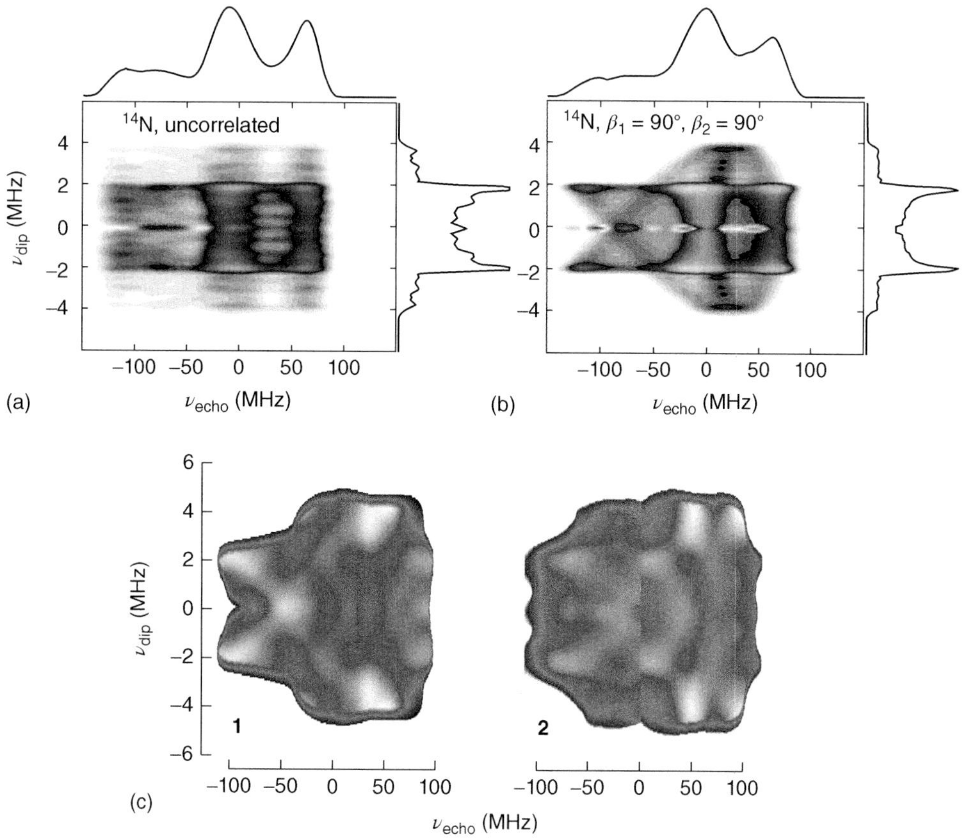

Figure 20.16. Illustration of the 2D-DQC experiment: (a, b) Simulated 2D DQC magnitude filled contour plots obtained by 2D FT with respect to t_{dip} and t_{echo}. Shown are uncorrelated (a) and correlated (b) cases for $B_0 = 6200$ G and $\nu_{dip} = \omega_{dd}/2\pi \approx 2$ MHz ($r_{12} = 2.96$ nm). B_1 was set to infinity (i.e., assuming perfect delta-function pulses), pseudosecular terms were neglected. In (b) Euler angles beta (polar to $\boldsymbol{r}_{12}$) were (90°, 90°) corresponding to strong correlations. The remaining Euler angles were set to zero. The side plots are a Pake doublet (on the right) and 1D EPR absorption spectrum decoupled from dipolar broadening (at the top). Both were produced by summing the 2D signal along the orthogonal frequency dimension. (c) Experimental (**2**) and calculated (**1**) 2D correlation maps from 17.25 GHz DQC data. Map (**2**) is composed of two correlation plots obtained from 2D data acquired for MW pulses applied at 28 and −56 MHz frequency offsets from the central maximum of the ^{14}N nitroxide EPR spectrum. The cutoff line for each map is at the offset of −42 MHz. Map (**1**) was computed using B_1 of 80 G, $r = 2.9$ nm, and the set of Euler angles $\Omega_1 = (0°, 75°, 0°)$, $\Omega_2 = (90°, 111°, 0°)$ from molecular modeling. (Reproduced with permission from Ref. 22. © Springer, 2014)

and can reach ≈8 nm or likely more in deuterated systems. DEER was shown to reach beyond 10 nm in fully deuterated systems or even more in special cases.[102]

20.6.3 Perspectives for 2D-SQC

The same type of experiment can be conducted using SQC sequences of Section 20.4, but it would take a much larger effort to isolate the dipolar part of the signal for most of them. The least effort is expected for the JB sequence due to good background suppression, although we do not expect significant differences from DQC as the nature of the signal is similar. The performance of the rest of the SQC sequences described in Section 20.4 needs to be improved by using B_1's even greater than 45 G. Such high B_1's can be combined with more uniform broadband excitation from amplitude-modulated pulses, such as

Hermite pulses.[103] Alternatively, the use of moderate B_1 frequency-chirping quasi-adiabatic pulses could help to suppress unwanted dipolar signals[73,84] and reduce backgrounds, although such techniques (see next section) in 2D format are yet to be developed to the extent comparable to routinely performed 2D-DQC, as coupled spin dynamics is more complex with adiabatic pulses and the time dispersion in dipolar evolution limits the strength of the dipolar coupling that can be faithfully reconstructed by this approach.

20.6.4 2D-SIFTER

With the advent of frequency-swept (chirp) quasi-adiabatic pulses in EPR, this approach has been applied to improve SIFTER (see Section 20.4). Recent work has applied chirped-pulse SIFTER approach to provide the 2D mode in the manner of 2D-DQC. We comment on specifics of this new method. For each contributing spin, the time of pulse excitation depends on its frequency offset and the (inverse) rate of sweep, $k = (\mathrm{d}f/\mathrm{d}t)^{-1} \approx \Delta t_{\mathrm{pulse}}/\Delta f$, where Δf is the sweep range and $\Delta t_{\mathrm{pulse}}$ the pulse width. This results in a time dispersion of Φ_{off}, spreading the echo. The problem was solved for PE by Böhlen *et al.* using a 2 : 1 chirping rate scheme for pulse k's ratio.[85] For RE, a 2 : 2 : 1 scheme is efficient[73] (see also the following section). For coupled spins, situated at different offsets, excitation is applied at different times. While spin dynamics in this case is quite complex, for weak coupling, we can employ a model based on the simplified picture of two hard pulses, each acting on its spin when the frequency is equal to the spin resonance frequency. This allows a qualitative treatment to be made, illustrating the specifics of this approach. We can employ the methods for tracking dipolar pathways outlined in Section 20.2, as was developed for DEER with multiple pulses (excitations) at the pump frequency.[22] We show in the following section an example sketched for chirp-RE to illustrate the evolution of dipolar coupling.

In Figure 20.17(a), the chirp pulses are depicted in the second row as a 2 : 2 : 2 : 1 inverse sweep rate scheme.[84] As in the Böhlen–Bodenhausen 2 : 1 scheme,[85] only the last π-pulse needs to have a faster rate to refocus offsets without phase dispersion. The evolution of dipolar phases for spins S_2 and S_1 experience CT by the $(\pi/2)_y$ pulse. The switching occurs in two steps. One of the two spins, S_1 that started evolution later than the other, passes through

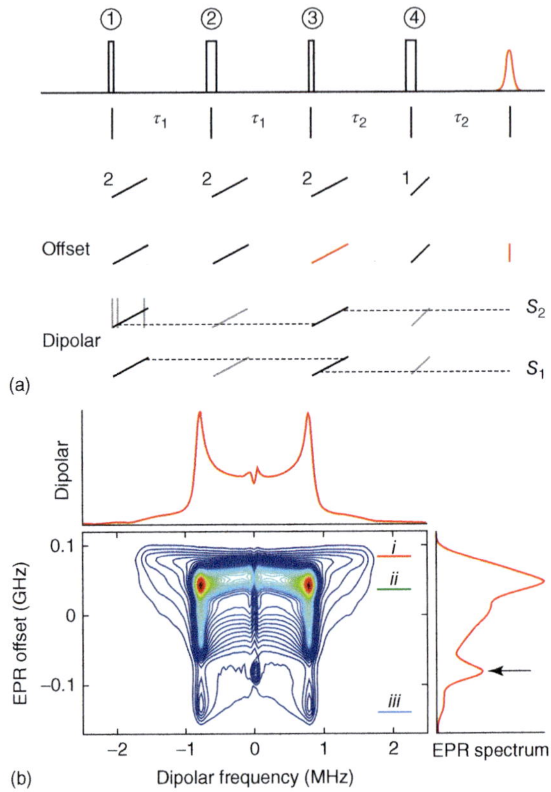

Figure 20.17. 2D-SIFTER with adiabatic pulses. (a) Under the standard rectangular-pulse sequence, the pulse sequence with frequency chirping pulses is shown as each pulse frequency vs time. The 2 : 2 : 2 : 1 inverse chirp rate ratio scheme is employed. The chart under this graph shows timing of pulse action on offset progression (creation, refocusing), essentially replicating the pulse frequency ramps (red lines indicate PE from pulses 1, 2 and RPE, $(\pi/2)_y$ pulse has no effect on refocusing offsets). For two coupled spins 1 and 2, their dipolar evolutions are charted with spin 2 shown starting first at the low-frequency path. At the third $(\pi/2)_y$ pulse, there is coherence transfer (CT) between the spins. For spin 2, the gap in dipolar evolution is produced in CT to spin 1 as the state of the dipolar order existed between the spins, when they waited for the transfer to complete. An overhanging trajectory is observed in the delayed path starting at the higher frequency on spin 1; in this case, it corresponds to a period of evolution as DQC/ZQC. These gaps are equal and do not contribute dipolar evolution; the offsets are exchanged but refocusing is unaffected. This pulse is skewless as the dipolar variable, $t_x = \tau_1 - \tau_2$ does not depend on frequency offsets. (b) The correlation map of the 2D-DQC type, as shown in Figure 20.16. The cross-sections at spectral points i–iii are available in Ref. 84. (Reproduced with permission from Ref. 84. © Royal Society of Chemistry, 2016)

DQ and ZQ coherence orders, whereas its partner passes through dipolar order. There is no effect on evolution of offsets, as they simply continue as if CT had not happened, so the offsets behave as in RPE (see the following section). Out-of-phase components caused by pulse imperfections and those that can be produced, e.g., by Bloch–Siegert phase shifts could evolve differently, and pseudosecular terms can contribute more issues. These effects could be studied numerically. Overall, there is enthusiasm that this four-pulse single-resonance 2D-method could work well for sufficiently weak couplings, starting from about 3 nm as was demonstrated. There is indeed good coverage achieved of the Q-band nitroxide spectrum, and the Pake doublet for the ≈3.5 nm distance in panel (**b**) does not show any problem.

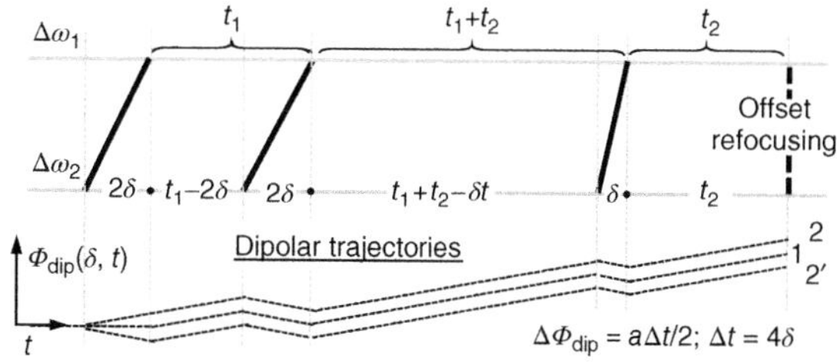

Figure 20.18. The RE pulse sequence based on adiabatic (frequency-chirping) pulses and the phase-invariant refocusing scheme with the (inverse) frequency sweep rate ratio 2 : 2 : 1. Each pulse has its frequency equal to spin offsets $\Delta\omega_1$ and $\Delta\omega_2$ at times shifted by 2δ (for the first two pulses) or δ (for the last π pulse). The pulse action is assumed to be that of two hard pulses positioned at the times when the pulse frequency equals the offsets. The first π/2 pulse produces two dipolar pathways with equal amplitudes, i.e., the two dipolar phase trajectories denoted 2 and 2′. Each refocusing π pulse introduces to the evolution of Φ_{dip}, a time lag equal to twice the time shift by reversing dipolar evolution twice during the pulse. The spin at $\Delta\omega_1$ has just one phase trajectory intermediate to 2 and 2′ for $\Delta\omega_2$

20.6.5 Chirp-pulse RE: Evolution and Refocusing

For a standard monochromatic pulse, the effect of the pulse on the dipolar evolution given by the set of trajectories $\{\Phi_{dip}(t)\}$ in the detected signal can be calculated using the simple means of Section 20.2 based on dipolar evolution operators $\hat{D}(t)$. Chirp pulses (see Chapter 21) act on both spins at different times, so each pulse can be viewed as a double pulse acting first on $\hat{S}_1^{\pm}$ of the first spin, changing $\bar{p}$ to stay on the selected pathway and after a delay acting on $\hat{D}(t)$ of the second spin to modify the trajectory set. Assuming nearly adiabatic pulses, we can approximate their action by hard pulses separated by a delay. This does not multiply the pathways for π pulses but introduces $\Phi_{dip}(t)$ dispersion into echo-forming spin ensemble. The evolution of $\Phi_{dip}(t)$ will be reversed twice for a π-pulse introducing delay to $\Phi_{dip}(t)$'s. A π/2 pulse flips half the spins, thereby doubling the number of trajectories and delaying those flipped. The delay (phase shift) depends on the pulse frequency sweep rate and the offsets of the spins. We focus on dipolar evolution in RE and will consider this case here using as an example the sequence $\pi/2 - t_1 - \pi - (t_1 + t_2) - \pi - t_2 - echo$. The dispersion effects for adiabatic PDS sequences, such as shaped-pulse SIFTER, have been analyzed in the literature.[84]

The sequence (Figure 20.18) is a 2 : 2 : 1 chirp RE described in the work on broadband SIFTER,[73] where the ratio is that of inverse frequency sweep rates, $(df/dt)^{-1}$. The PE in the Böhlen–Bodenhausen scheme[85] has inverse chirping rate ratio 2 : 1 to refocus the offsets without phase dispersion; and so does the 2 : 2 : 1 scheme for RE, as for any value of δ the refocusing occurs at time $2(t_1 + t_2) + 2\delta$ after the initial time lead of 2δ, so the refocusing point is at $2(t_1 + t_2)$ for all δ. The π/2 pulse first engages spin at $\Delta\omega_2$ and after a delay of 2δ another at $\Delta\omega_1$, flipping it with probability $p \approx 0.5$. This produces two dipolar trajectories 2 and 2′, of which one whose evolution was reversed is delayed by 4δ, hence introduces $-2a\delta$ phase shift to Φ_{dip}. This picture repeats with each π pulse and is similar for the $\Delta\omega_1$ and $\Delta\omega_2$ pathways. We illustrate this point by calculating part of the evolution for trajectory 1 of spin 1 at $\Delta\omega_1$ to the point in time t after the end of the first π pulse. For the RE in Figure 20.18, the pathway to echo is (−1, +1, −1). Referring time to the end of the π/2 pulse, we write the dipolar evolution as follows:

$$\begin{aligned} &\hat{S}_1^- \xrightarrow{t_1-2\delta} \hat{S}_1^- \hat{D}^*_{t_1-2\delta} \xrightarrow{\pi(\Delta\omega_2)} \hat{S}_1^+ \hat{D}^*_{t_1-2\delta} \xrightarrow{2\delta} \hat{S}_1^+ \hat{D}^*_{t_1-2\delta} \hat{D}_{2\delta} \\ &\xrightarrow{\pi(\Delta\omega_1)} \hat{S}_1^+ \hat{D}_{t_1-2\delta} \hat{D}^*_{2\delta} \xrightarrow{t} \hat{S}_1^+ \hat{D}_{t_1-2\delta} \hat{D}^*_{2\delta} \hat{D}_t \\ &= \hat{S}_1^+ \hat{D}(t_1 + t - 4\delta) \end{aligned} \tag{20.31}$$

We see that the π pulse introduced a delay 4δ to the evolution of the dipolar phase, as one expects for a 'dipolar refocusing sandwich' (2δ − π − 2δ). Full dipolar trajectories are shown in Figure 20.18; there are two denoted as 2 and 2′ for the 'lower' $\Delta\omega_2$ path, 2 but only one trajectory exists for the spin at $\Delta\omega_1$ (cf. Figure 20.18). Collectively, they produce a skew of $4a\delta$ which depends on the excited spectral shape and extent. For a 200 ns pulse, the skew could be as large as 400 ns. Assuming it to be ≈ 200 ns on average, and permitting the error in dipolar phase of a quarter period of dipolar oscillation, we estimate that distances less than 3.5 nm may be too short for such a pulse sequence. Nevertheless, for moderate and long distances this pulse sequence may be useful.

20.7 RELAXATION AND INSTANTANEOUS DIFFUSION

20.7.1 Instantaneous Diffusion

As we noted in Section 20.2.2, the intramolecular dipolar coupling is the main term in $\hat{H}_{dd}$. However, surrounding spins usually cannot be discounted. Considering intermolecular couplings leads to a secular dipolar Hamiltonian, where each spin k is in principle coupled to all N spins in the sample, so that $\hat{H}_{dd}$ becomes

$$\hat{H}_{dd} = \sum_{n \neq k} a(\boldsymbol{r}_{kn})(3\hat{S}_{kz}\hat{S}_{nz} - \hat{\boldsymbol{S}}_k\hat{\boldsymbol{S}}_n) \tag{20.32}$$

where $k, n \leq N$. In PDS, the flip-flop terms in equation (20.32), which could lead to spin diffusion, should be neglected to give just the secular part of dipolar coupling $\hat{H}_{dd} = \sum_{n \neq k} a(\boldsymbol{r}_{kn})\hat{S}_{kz}\hat{S}_{nz}$. This certainly holds for most cases, including nitroxides, unless the concentration is approaching the molar range, but other concentration effects impair the methods well before it is reached.[98] The upper level of concentration may become even less for cases such as trityl radicals or H-atoms, whose lines are narrow. This case may require the SLE treatment to account for relaxation caused by these terms.[99]

As was shown by Klauder and Anderson[5] and confirmed in a many-body analysis by Nevzorov and Freed,[99] the dipolar coupling in equation (20.32) leads to an inhomogeneously broadened line having a Lorentzian dipolar line shape. For a spin system at concentration C having hypothetically zero-width dipolar spectrum, applying a hard π pulse 'instantaneously' produces dephasing whose FT yields an inhomogeneous line width k_0C, where $k_0 = 2\pi\mu_0\gamma_e^2\hbar/9\sqrt{3} \approx 10^{-3}\mu M^{-1}\mu s^{-1}$. This explains the term of 'ID' applied to this case.[5] In the primary spin-echo $\pi/2 - \tau - \pi - \tau - echo$, if only a fraction p_A of spins is excited (flipped), i.e., the A spins, ID contributes to the echo decay as

$$V(\tau) = V(0)\exp(-p_A k_0 C\tau) \tag{20.33}$$

The signal decay appears as a relaxation decay but is based on coherent spin dynamics and therefore can be at least partially refocused.[48] ID affects single- and double-resonance pulse sequences differently. In DEER, pump pulses are applied to the B-spins, refocusing the dipolar coupling of A-spins with the B-spins; but the dipolar phase at the point of detection (i.e., the echo) depends on the varying position of the pump pulse, leading to time dependence in the echo. The (pumped) B-spins include the partner to an A-spin and all the surrounding spins at the frequency of the pump pulse. All these dipolar interactions of the A-spin with unaffected B-spins are refocused in the signal, which becomes a product

$$\begin{aligned} V(t) &= V_{intra}V_{inter} \\ &= V_{intra}\left\langle \prod_{n \neq 1,2} \left(q_B + p_B \cos a\left(\boldsymbol{r}_{1n}\right)t\right)\right\rangle_{\{\boldsymbol{r}_{1n}\}} \end{aligned} \tag{20.34}$$

where for partner spins 1 and 2, $V_{intra} = V_0(q_B + p_B \cos a_{12}t)$ is a standard DEER signal with p_B the probability to flip a B-spin and $q_B = 1 - p_B$. Markov averaging of V_{inter} in equation (20.34) produces,

$$V(t) = V_{intra}(t)\exp(-k_0Cp_Bt) \tag{20.35}$$

where C is spin concentration in the (uniform) sample. Thus, equation (20.35) is an estimate of ID for the DEER experiment. For observing the RE sequence with a length of $2t_m$, ID from A-spins, whose excitation probability is p_A as in equation (20.33), contributes to the signal a constant factor of $V(t_m) = V(0)\exp(-p_Ak_0Ct_m)$ with t_m in place of τ to give the total ID contribution to DEER: $V(t) = V_{intra}(t)\exp[-k_0C(p_Bt + p_At_m)]$ (see Chapter 19).

Because single-resonance pulse sequences consist of refocusing segments, the static dipolar coupling to B-spins refocus in each of them. However, when relaxation cannot be neglected, B-spins contribute spectral diffusion to the dynamics of A-spins, leading to decay.[1,98] Note that for ideal hard pulses there are no B-spins, but they exist for real nonideal pulses. The single-frequency techniques such as PE and RE do contain an ID contribution from A-spins as

$$V(t_m) = V(0)\exp(-p_A k_0 C t_m) \quad (20.36)$$

However, modification of the RE sequence to form a PDS pulse sequence by inserting a propagator $\hat{U}$ (see Section 20.2.7) contributes sequence-specific time dependence to the ID, as refocusing of coupling may be introduced by the nature of $\hat{U}$ and may partially refocus ID.[99] Thus, equation (20.36) will now depend on t_x, such as contributed by the term in $\cos at_x$ in equation (20.23) and could introduce a slope to the background. The task of writing out this term rigorously is quite complicated for any PDS method,[99] as the accurate result also depends on generation of higher order coherences, flip-flop effects, and spectral diffusion – all constituting a many-body problem. It has been shown[99] that for a solid-echo there is a decay of the signal due to these MQCs. We estimate that for the *solid-echo-based* SIFTER sequence that refocuses ID at $t_x = 0$ (assuming an ideal $(\pi/2)_y$ pulse and not including higher order coherences) introduces a time dependence

$$V(t_x) \cong V(0)\exp(-pk_0 C|t_x|) \quad (20.37)$$

independent of t_m, which adds to the relaxation decay when it is significant, leading to strong attenuation of the weak signal at large t_x's. Interestingly, the DQC-6 pulse sequence appears to have no such decay in t_x, as the ID term is constant and the same as in RE; that is,

$$V(t_x, t_m) = V_0(t_x)\exp(-pk_0 C t_m) \quad (20.38)$$

where V_0 is the signal in isolated pair. Generally, proper phase cycling does not modify ID, which otherwise would degrade the performance; this is the case with DQC-6. However, $(\pi/2)_x$ pulses do generate even higher order coherences with other spins, and as they are real pulses they also act to an extent as $(\pi/2)_y$ to generate odd coherence orders and causing refocusing, so one could also expect some dependence on the t_x variable.

20.7.2 Local Concentrations; Freezing Effects

If the sample is not uniform, $V_{\text{inter}} \approx \exp(-pk_0\overline{C}_{\text{loc}}t)$ in DEER, where $\overline{C}_{\text{loc}}$ is an effective local concentration, so from averaging over the sample volume the expression $\langle C_{\text{loc}}(\boldsymbol{r})\exp(-k_0 p C_{\text{loc}}(\boldsymbol{r})t)\rangle_r$, will possibly produce a more complex dependence than a simple exponential factor. A recent study[75] was conducted to determine the effect of adding glycerol to an aqueous sample on the intermolecular decay (or 'background slope') in DEER for a spin-labeled protein. It showed an increase in $\overline{C}_{\text{loc}}$ when the glycerol concentration was below ≈50 wt% and was typically decreasing with increasing freezing rate. Thus, caution needs to be exercised when estimating concentrations from slopes in DEER. It should not be forgotten that similar concentration effects are expected for single-resonance techniques, but they are of least concern for the methods that suppress background, such as DQC-6.

20.7.3 Fractional Dimension

When spatial distributions of spins are not locally isotropic, this can lead to time dependence in the ID term that is described by a stretched exponential with fractional dimensionality, α. Thus $V_{\text{inter}} \approx \exp(-p\overline{C}_{\text{loc}}^{(\alpha)}k_0^{(\alpha)}t^{\alpha/3})$.[12] This can correspond, e.g., to an amyloid fibril; a self-avoiding polymer chain containing randomly spin-labeled sites ($\alpha = 1$); or lipid bilayers, particularly liposomes ($\alpha = 2$). Usually, α is in the range 2–3 with a lower value observed for lipid membranes; while $k_0^{(\alpha)}$ varies from case to case, being known explicitly for uniform fractal spaces.[12] This type of relaxation behavior may vastly complicate its removal from PDS data; therefore, low values of $\overline{C}_{\text{loc}}^{(\alpha)}$ are desirable.

20.7.4 Relaxation Effects in PDS

Unlike ID, which in principle is a coherent process, there are relaxation mechanisms that impose a limit to what PDS can accomplish in terms of distance measurements. The distance range and resolution is limited by the achievable duration t_m of dipolar evolution that one can record for a given sample. Typically, for biomolecules, a simple-exponential phase-relaxation law is observed with T_m's of the order 1–2 μs. When relaxation is relatively slower, e.g., for solvent-exposed spin labels, the nuclear spin-diffusion relaxation mechanism dominates for a longer time, leading to a signal decay, which in aqueous solutions is described by a stretched exponential such as $\exp[-(2t_m/T_m)^{\kappa}]$, with κ in the range 1.5–3 and a phase memory time $T_m \approx 4$ μs. This sets a limit of about 5 nm to the upper range of distances that can be measured accurately. Resolution for longer distances require solvent deuteration or in critical cases could include deuteration of the biomolecule as well,[100,102,104] virtually eliminating the effects of nuclear spin diffusion. It is well known that a CPMG pulse sequence is efficient in suppression of such spin diffusion processes; hence, this is relevant to PDS. As single-resonance methods described in this article are based on RE, there is a partial suppression of nuclear spin diffusion in them. This also was realized in the form of the DQ-filtered RE pulse sequence.[69] A more efficient approach, however, was undertaken for DEER, where multipulse CPMG-based PDS pulse sequences were applied for deeper suppression of spin diffusion.[24]

20.7.5 Relaxation in Single- and Double-resonance

As in double resonance the detection pulse sequence has its pulse fixed, the only time-dependent signal decay is caused by ID due to spins B, thus yielding the background slope. (The methods that use long coherence storage time T, experience effects of spectral diffusion.[44]) The echo signal, however, decays due to phase relaxation as $\exp[-(2t_m/T_m)^\kappa]$. The behavior is different for single resonance, where the need to vary the duration of refocusing sequences (see Section 20.2.6) for $\kappa \neq 1$ produces relaxation decay that has an effect on the dipolar signal shape given by

$$R(t_x) = \exp[-((t_m - t_x)^\kappa + (t_m + t_x)^\kappa)/T_m^\kappa] \quad (20.39)$$

Such behavior is evident in Figure 20.8(a). For $t_x = 0$, equation (20.39) yields $\exp(-2t_m^\kappa/T_m^\kappa)$, which corresponds to a longer $T_m' = 2^{1-1/\kappa}T_m$. In single-resonance pulse sequences, the t_x dependence in equation (20.39) favors the signal amplitude at small t_x, which de facto introduces signal windowing vs t_x. Signal acquisition as a function of t_m at condition $t_x = 0$ was used to extend the evolution in the DQ-filtered RE pulse sequence.[69]

20.8 CONCLUSIONS

A variety of modern PDS methods implemented at several frequencies are available today, allowing one to find the best conditions for accurate distance measurement as well as improved distance resolution and orientation selection. Although many new PDS methods are being developed, double- and single-resonance methods constitute the two major groups, which are not always clearly distinguishable. The pure, coherent, single-resonance techniques described in this article are making an increasing contribution as pulse technology rapidly progresses. For nearly the past two decades, DQC with hard pulses was a 'gold standard' for this class of techniques, as it provides high-quality distance data at high spin sensitivity, while at the same time allowing robust implementation at several convenient working frequencies. Recent improvements in SIFTER based on broadband adiabatic pulses are helping to develop this potentially useful and sensitive technique. This includes adding 2D-SIFTER experiments to the methods for the study of orientational correlations represented by 2D-DQC. This extends the methods to look at orientations, which also include HFHF DEER[38,101,105,106] and RIDME.[38,101]

So far, the standard four-pulse DEER method and its recent extensions such as multiple-pulse dipolar refocusing CPMG-type DEER[24,107,108] are the mainstay techniques as a matter of application flexibility, modest requirements of peak power, and reach to long distances. However, there are an increasing number of cases where DQC, SIFTER, or RIDME are better suited or may have no alternatives. For DQC and SIFTER, this particularly includes cases such as trityl label, especially in liquid solutions,[68] radical cofactors, low-concentration samples where sensitivity is a key issue, whereas broad spectra such as those of transition metal ions may benefit from RIDME, which otherwise is not an alternative to DQC or DEER. For nearly two decades, we have used DEER and DQC techniques, implemented at K_u band for high sensitivity, which has enabled many studies on important biological systems, and we have a positive outlook into the future use of these PDS techniques. We have emphasized DQC of the single-resonance methods, as with current technology, it is the most effective in avoiding unwanted signals. While there was a large effort to increase the sensitivity of PDS methods, a minimal level of signal artifacts may be even more important than modest SNR improvements paid for by the loss of signal fidelity. Relatively small signal artifacts may contribute a more serious impediment to extraction of distance information from the dipolar evolution traces than a moderate loss of SNR, as evidenced by the reconstruction of simulated vs experimental data.[78,109]

Nevertheless, spin sensitivity is of utmost importance in PDS and it usually is studied for nitroxides, in particular using MTSL as the main spin label.[24,110]

Furthermore, we mention new methods of signal processing the PDS data, which are an additional approach to significantly improving the SNR of the experiment, somewhat relaxing the need of high SNR.[109,111] However, today, there is a large and increasing variety of spin labels based on nitroxides, trityls, photoexited states, cofactors, or metal ions, both endo- and exogenous, as well as a plurality of vastly different PDS methods.[82,97,106,112] The particular experiment or spin label often benefits from using the most suitable technique, and single-resonance methods such as DQC and SIFTER and others described here will play increasingly valuable roles.

ACKNOWLEDGMENTS

The authors acknowledge NIH NIGMS Grant P41 GM103521 to ACERT National Biomedical Center for Advanced EPR Technology. The authors are grateful to Dr. Elka R. Georgieva for providing spin-labeled T4-lysozyme protein samples that enabled us to obtain model DQC and SQC data shown in this chapter. The authors are thankful to our multiple collaborators and group members that over the course of 15 years made a great contribution to the development of biomolecular PDS spectroscopy at ACERT. In particular the efforts of Brian R. Crane, Elka R. Georgieva, David Eliezer, Betty J. Gaffney, Hassane S. Mchaourab, Olga Boudker, Sang-Youn Park, Jaya Bhatnagar, Yun-Wej Chiang, Wayne L. Hubbell, Samuel E. Butcher, Dale Edmondson, Anup Upadhyay, Shaogeng Tang, Scott D. Emr, Charles P. Scholes, Ria Sircar, Dipanjan Samanta, Gregory E. Merz, Kent Kirshenbaum, Boris D. Dzikovski, Yeon-Kyun Shin, Jiansong Tong, Vladimir M. Grigoryants, Trudy F. Ramlall, Zhongyu Yang, Alexandrine M. Crane, Songi Han, Neil A. Eschmann, Sushil K. Misra, Aaron T. Fafarman, Michael V. Airola, David E. Budil, Michael G. Malkowski, Benjamin J. Orlando, Elizabeth D. Getzoff, Nattakan Sukomon are highly appreciated.

REFERENCES

1. A. M. Raitsimring and K. M. Salikhov, *Bull. Magn. Reson.*, 1985, **7**, 184.
2. L. Kevan and R. N. Schwartz, in Time Domain Electron Spin Resonance, eds L. Kevan and R. N. Schwartz, John Wiley & Sons, Inc.: New York, 1979.
3. A. Raitsimring, in Biological Magnetic Resonance, eds L. J. Berliner, G. R. Eaton and S. S. Eaton, Kluwer Academic: New York, 2000, Vol. 19, p 385.
4. A. Abragam, The Principles of Nuclear Magnetism, Clarendon Press: Oxford, 1961.
5. J. Klauder and P. Anderson, *Phys. Rev.*, 1962, **125**, 912.
6. V. F. Yudanov, K. M. Salikhov, G. M. Zhidormirov, and Y. D. Tsvetkov, *Theor. Exp. Chem.*, 1972, **5**, 451.
7. S. A. Dikanov and Y. Tsvetkov, Electron Spin Echo Envelope Modulation (ESEEM) Spectroscopy, Taylor & Francis: London, 1992.
8. A. Schweiger and G. Jeschke, Principles of Pulse Electron Paramagnetic Resonance, Oxford University Press: Oxford, 2001.
9. W. B. Mims, *Phys. Rev. B Solid State*, 1972, **5**, 2409.
10. M. Emshwiller, E. L. Hahn, and D. Kaplan, *Phys. Rev.*, 1960, **118**, 414.
11. A. D. Milov, K. M. Salikhov, and M. D. Shirov, *Sov. Phys. Solid State*, 1981, **23**, 565.
12. A. D. Milov and Y. D. Tsvetkov, *Appl. Magn. Reson.*, 1997, **12**, 495.
13. G. E. Merz, P. P. Borbat, A. J. Pratt, E. D. Getzoff, J. H. Freed, and B. R. Crane, *Biophys. J.*, 2014, **107**, 1669.
14. P. P. Borbat and J. H. Freed, in Two-Component Signaling Systems, Pt B, eds M. I. Simon, B. R. Crane and A. Crane, Elsevier Academic Press Inc: San Diego, 2007, Vol. 423, p 52.
15. A. V. Astashkin and A. Kawamori, in Biophysical Techniques in Photosynthesis, eds T. Aartsma and J. Matysik, Springer: Dordrecht, Netherlands, 2008, Vol. 2, p 325.
16. M. Pannier, S. Veit, A. Godt, G. Jeschke, and H. W. Spiess, *J. Magn. Reson.*, 2000, **142**, 331.
17. L. J. Berliner, G. R. Eaton, and S. S. Eaton, Distance Measurements in Biological Systems by EPR, Kluwer Academic: New York, 2000.
18. R. E. Martin, M. Pannier, F. Diederich, V. Gramlich, M. Hubrich, and H. W. Spiess, *Angew. Chem. Int. Ed.*, 1998, **37**, 2834.
19. S. Saxena and J. H. Freed, *J. Chem. Phys.*, 1997, **107**, 1317.
20. P. P. Borbat and J. H. Freed, *Chem. Phys. Lett.*, 1999, **313**, 145.
21. P. P. Borbat and J. H. Freed, in Biological Magnetic Resonance, eds L. J. Berliner, G. R. Eaton and S.

S. Eaton, Academic/Plenum Publishers: New York, 2000, Vol. 19, p 385.

22. P. P. Borbat and J. H. Freed, in Structural Information from Spin-Labels and Intrinsic Paramagnetic Centres in the Biosciences, eds C. R. Timmel and J. R. Harmer, Springer-Verlag: Berlin, 2014, Vol. 152, p 1.
23. G. Jeschke, M. Pannier, A. Godt, and H. W. Spiess, *Chem. Phys. Lett.*, 2000, **331**, 243.
24. P. P. Borbat, E. R. Georgieva, and J. H. Freed, *J. Phys. Chem. Lett.*, 2013, **4**, 170.
25. G. Jeschke, A. Bender, H. Paulsen, H. Zimmermann, and A. Godt, *J. Magn. Reson.*, 2004, **169**, 1.
26. S. Milikisyants, F. Scarpelli, M. G. Finiguerra, M. Ubbink, and M. Huber, *J. Magn. Reson.*, 2009, **201**, 48.
27. L. V. Kulik, Y. A. Grishin, S. A. Dzuba, I. A. Grigoryev, S. V. Klyatskaya, S. F. Vasilevsky, and Y. D. Tsvetkov, *J. Magn. Reson.*, 2002, **157**, 61.
28. S. Razzaghi, M. Qi, A. I. Nalepa, A. Godt, G. Jeschke, A. Savitsky, and M. Yulikov, *J. Phys. Chem. Lett.*, 2014, **5**, 3970.
29. D. Abdullin, F. Duthie, A. Meyer, E. S. Muller, G. Hagelueken, and O. Schiemann, *J. Phys. Chem. B*, 2015, **119**, 13534.
30. A. Collauto, V. Frydman, M. D. Lee, E. H. Abdelkader, A. Feintuch, J. D. Swarbrick, B. Graham, G. Otting, and D. Goldfarb, *Phys. Chem. Chem. Phys.*, 2016, **18**, 19037.
31. P. P. Borbat, H. S. Mchaourab, and J. H. Freed, *J. Am. Chem. Soc.*, 2002, **124**, 5304.
32. S.-Y. Park, P. P. Borbat, G. Gonzalez-Bonet, J. Bhatnagar, A. M. Pollard, J. H. Freed, A. M. Bilwes, and B. R. Crane, *Nat. Struct. Mol. Biol.*, 2006, **13**, 400.
33. G. Jeschke, C. Wegener, M. Nietschke, H. Jung, and H.-J. Steinhoff, *Biophys. J.*, 2004, **86**, 2551.
34. B. G. Dzikovski, P. P. Borbat, and J. H. Freed, *Biophys. J.*, 2004, **87**, 3504.
35. O. Schiemann, N. Piton, Y. G. Mu, G. Stock, J. W. Engels, and T. F. Prisner, *J. Am. Chem. Soc.*, 2004, **126**, 5722.
36. P. A. S. Cruickshank, D. R. Bolton, D. A. Robertson, R. I. Hunter, R. J. Wylde, and G. M. Smith, *Rev. Sci. Instrum.*, 2009, **80**, 103102.
37. G. W. Reginsson, R. I. Hunter, P. A. S. Cruickshank, D. R. Bolton, S. T. Sigurdsson, G. M. Smith, and O. Schiemann, *J. Magn. Reson.*, 2012, **216**, 175.
38. K. Möbius and A. Savitsky, High-Field EPR Spectroscopy on Proteins and Their Model Systems : Characterization of Transient Paramagnetic States, Royal Society of Chemistry: Cambridge, 2009.
39. P. E. Spindler, P. Schops, A. M. Bowen, B. Endeward, and T. F. Prisner, *eMagRes*, 2016, **5**, 1477.
40. A. Doll, M. Qi, A. Godt, and G. Jeschke, *J. Magn. Reson.*, 2016, **273**, 73.
41. A. Doll, M. A. Qi, S. Pribitzer, N. Wili, M. Yulikov, A. Godt, and G. Jeschke, *Phys. Chem. Chem. Phys.*, 2015, **17**, 7334.
42. C. P. Slichter, Principles of Magnetic Resonance, Springer-Verlag: Berlin, 1990, p 248.
43. C. Hintze, D. Bucker, S. D. Kohler, G. Jeschke, and M. Drescher, *J. Phys. Chem. Lett.*, 2016, **7**, 2204.
44. A. V. Astashkin, in Electron Paramagnetic Resonance Investigations of Biological Systems by Using Spin Labels, Spin Probes, and Intrinsic Metal Ions, Pt A, eds P. Z. Qin and K. Warncke, 2015, Vol. 563, p 251.
45. R. R. Ernst et al., Principles of Nuclear Magnetic Resonance in One and Two Dimensions, Clarendon Press: Oxford [Oxfordshire], 1987.
46. O. W. Sorensen, G. W. Eich, M. H. Levitt, G. Bodenhausen, and R. R. Ernst, *Prog. Nucl. Magn. Reson. Spectrosc.*, 1983, **16**, 163.
47. J. Cavanagh, Protein NMR Spectroscopy: Principles and Practice, Academic Press, 1996.
48. C. P. Slichter, Principles of Magnetic Resonance, 3rd edn, Springer-Verlag: Berlin-Heidelberg-New-York, 1990.
49. G. E. Pake, *J. Chem. Phys.*, 1948, **16**, 327.
50. H. W. Spiess and K. Schmidt-Rohr, Multidimensional Solid-State NMR and Polymers, Academic Press: London, 1994.
51. A. D. Bain, D. A. Fletcher, and P. Hazendonk, *Concepts Magn. Reson.*, 1998, **10**, 85.
52. J. Shriver, *Concepts Magn. Reson.*, 1992, **4**, 1.
53. M. H. Levitt, eMagRes, John Wiley & Sons, Ltd: Chichester, 2007.
54. J. H. Freed, *Annu. Rev. Phys. Chem.*, 2000, **51**, 655.
55. J. H. Freed, *Annu. Rev. Phys. Chem.*, 1972, **23**, 265.
56. M. H. Levitt, Pulse Methods in 1D & 2D Liquid-Phase NMR, Academic Press: San Diego, 1988, p 111.
57. M. H. Levitt, Spin Dynamics: Basics of Nuclear Magnetic Resonance, John Wiley & Sons, Ltd: Chichester, 2008.

58. M. Di Valentin, M. Albertini, E. Zurlo, M. Gobbo, and D. Carbonera, *J. Am. Chem. Soc.*, 2014, **136**, 6582.
59. S. A. Dzuba and A. Hoff, in Biological Magnetic Resonance, eds L. J. Berliner, G. R. Eaton and S. S. Eaton, Academic/Plenum Publishers: New York, 2000, Vol. 19, p 569.
60. G. Jeschke, M. Pannier, and H. W. Spiess, in Distance Measurements in Biological Systems by EPR, eds L. J. Berliner, G. R. Eaton and S. S. Eaton, Springer: Boston, MA, 2002, p 493.
61. J. G. Powles and P. Mansfield, *Phys. Lett.*, 1962, **2**, 58.
62. P. P. Borbat, R. H. Crepeau, and J. H. Freed, *J. Magn. Reson.*, 1997, **127**, 155.
63. B. J. Gaffney, M. D. Bradshaw, S. D. Frausto, F. Y. Wu, J. H. Freed, and P. Borbat, *Biophys. J.*, 2012, **103**, 2134.
64. C. E. Tait and S. Stoll, *Phys. Chem. Chem. Phys.*, 2016, **18**, 18470.
65. S. K. Misra, P. P. Borbat, and J. H. Freed, *Appl. Magn. Reson.*, 2009, **36**, 237.
66. A. T. Fafarman, P. P. Borbat, J. H. Freed, and K. Kirshenbaum, *Chem. Commun.*, 2007, 377.
67. M. Bennati, J. H. Robblee, V. Mugnaini, J. Stubbe, J. H. Freed, and P. Borbat, *J. Am. Chem. Soc.*, 2005, **127**, 15014.
68. Z. Y. Yang, Y. P. Liu, P. Borbat, J. L. Zweier, J. H. Freed, and W. L. Hubbell, *J. Am. Chem. Soc.*, 2012, **134**, 9950.
69. P. P. Borbat, J. H. Davis, S. E. Butcher, and J. H. Freed, *J. Am. Chem. Soc.*, 2004, **126**, 7746.
70. M. Huber, M. Lindgren, P. Hammarstrom, L.-G. Martensson, U. Carlsson, G. R. Eaton, and S. S. Eaton, *Biophys. Chem.*, 2001, **94**, 245.
71. A. D. Milov, K. M. Salikhov, and Y. D. Tsvetkov, *Fiz. Tverd. Tela*, 1973, **15**, 1187.
72. K. N. Hu, C. Song, H. H. Yu, T. M. Swager, and R. G. Griffin, *J. Chem. Phys.*, 2008, **128**.
73. P. Schoeps, P. E. Spindler, A. Marko, and T. F. Prisner, *J. Magn. Reson.*, 2015, **250**, 55.
74. S. M. Islam, R. A. Stein, H. S. Mchaourab, and B. Roux, *J. Phys. Chem. B*, 2013, **117**, 4740.
75. E. R. Georgieva, A. S. Roy, V. M. Grigoryants, P. P. Borbat, K. A. Earle, C. P. Scholes, and J. H. Freed, *J. Magn. Reson.*, 2012, **216**, 69.
76. M. Lindgren, G. R. Eaton, S. S. Eaton, B.-H. Jonsson, P. Hammarstrom, M. Svensson, and U. Carlsson, *J. Chem. Soc. Perkin Trans.*, 1997, **2**, 2549.
77. A. D. Milov, K. M. Salikhov, and Y. D. Tsvetkov, *Sov. Phys. Solid State*, 1973, **15**, 802.
78. Y.-W. Chiang, P. P. Borbat, and J. H. Freed, *J. Magn. Reson.*, 2005, **172**, 279.
79. S. Ruthstein, M. Ji, B. K. Shin, and S. Saxena, *J. Magn. Reson.*, 2015, **257**, 45.
80. J. Jeener and P. Broekaert, *Phys. Rev.*, 1967, **157**, 232.
81. G. Y. Shevelev, O. A. Krumkacheva, A. A. Lomzov, A. A. Kuzhelev, O. Y. Rogozhnikova, D. V. Trukhin, T. I. Troitskaya, V. M. Tormyshev, M. V. Fedin, D. V. Pyshnyi, and E. G. Bagryanskaya, *J. Am. Chem. Soc.*, 2014, **136**, 9874.
82. D. Akhmetzyanov, P. Schops, A. Marko, N. C. Kunjir, S. T. Sigurdsson, and T. F. Prisner, *Phys. Chem. Chem. Phys.*, 2015, **17**, 24446.
83. G. Jeschke, V. Chechik, P. Ionita, A. Godt, H. Zimmermann, J. Banham, C. R. Timmel, D. Hilger, and H. Jung, *Appl. Magn. Reson.*, 2006, **30**, 473.
84. A. Doll and G. Jeschke, *Phys. Chem. Chem. Phys.*, 2016, **18**, 23111.
85. J. M. Bohlen and G. Bodenhausen, *J. Magn. Reson. Ser. A*, 1993, **102**, 293.
86. M. Bonora, J. Becker, and S. Saxena, *J. Magn. Reson.*, 2004, **170**, 278.
87. A. J. Hoff, P. Gast, S. A. Dzuba, C. R. Timmel, C. E. Fursman, and P. J. Hore, *Spectrochim. Acta Mol. Biomol. Spectrosc.*, 1998, **54**, 2283.
88. J. Tang, M. C. Thurnauer, and J. R. Norris, *Chem. Phys. Lett.*, 1994, **219**, 283.
89. A. Marko, V. Denysenkov, and T. F. Prisner, *Mol. Phys.*, 2013, **111**, 2834.
90. R. R. Mett, J. W. Sidabras, I. S. Golovina, and J. S. Hyde, *Rev. Sci. Instrum.*, 2008, **79**, Article number: 094702.
91. R. R. Mett, J. W. Sidabras, and J. S. Hyde, *Appl. Magn. Reson.*, 2007, **31**, 573.
92. J. Forrer, I. Garcia-Rubio, R. Schuhmam, R. Tschaggelar, and J. Harmer, *J. Magn. Reson.*, 2008, **190**, 280.
93. S. Pfenninger, J. Forrer, A. Schweiger, and T. Weiland, *Rev. Sci. Instrum.*, 1988, **59**, 752.
94. V. V. Kurshev, A. M. Raitsimring, and Y. D. Tsvetkov, *J. Magn. Reson.*, 1989, **81**, 441.

95. L. V. Kulik, S. A. Dzuba, I. A. Grigoryev, and Y. D. Tsvetkov, *Chem. Phys. Lett.*, 2001, **343**, 315.

96. A. Meyer and O. Schiemann, *J. Phys. Chem. A*, 2016, **120**, 3463.

97. A. M. Bowen, C. E. Tait, C. R. Timmel, and J. R. Harmer, in Structural Information from Spin-Labels and Intrinsic Paramagnetic Centres in the Biosciences, eds C. R. Timmel and J. R. Harmer, Springer: Berlin, 2013, p 283.

98. V. V. Kurshev and T. Ichikawa, *J. Magn. Reson.*, 1992, **96**, 563.

99. A. A. Nevzorov and J. H. Freed, *J. Chem. Phys.*, 2001, **115**, 2401.

100. E. R. Georgieva, T. F. Ramlall, P. P. Borbat, J. H. Freed, and D. Eliezer, *J. Biol. Chem.*, 2010, **285**, 28261.

101. A. Savitsky, J. Niklas, J. H. Golbeck, K. Mobius, and W. Lubitz, *J. Phys. Chem. B*, 2013, **117**, 11184.

102. T. Schmidt, M. A. Walti, J. L. Baber, E. J. Hustedt, and G. M. Clore, *Angew. Chem. Int. Ed.*, 2016, **55**, 15905.

103. W. S. Warren, *J. Chem. Phys.*, 1984, **81**, 5437.

104. W. Richard, A. Bowman, E. Sozudogru, H. El-Mkami, T. Owen-Hughes, and D. G. Norman, *J. Magn. Reson.*, 2010, **207**, 164.

105. C. L. Motion, J. E. Lovett, S. Bell, S. L. Cassidy, P. A. S. Cruickshank, D. R. Bolton, R. I. Hunter, H. El Mkami, S. Van Doorslaer, and G. M. Smith, *J. Phys. Chem. Lett.*, 2016, **7**, 1411.

106. D. Goldfarb, in Structural Information from Spin-Labels and Intrinsic Paramagnetic Centres in the Biosciences, eds C. R. Timmel and J. R. Harmer, Springer: Heidelberg, 2013, p 163.

107. P. E. Spindler, I. Waclawska, B. Endeward, J. Plackmeyer, C. Ziegler, and T. F. Prisner, *J. Phys. Chem. Lett.*, 2015, **6**, 4331.

108. J. M. Soetbeer, Dynamical Decoupling in Distance Measurements by Double Electron–electron Resonance, Springer Spektrum: Wiesbaden, 2016.

109. M. Srivastava, E. R. Georgieva, and J. H. Freed, *J. Phys. Chem. A*, 2017, **121**, 2452.

110. R. Tschaggelar, B. Kasumaj, M. G. Santangelo, J. Forrer, P. Leger, H. Dube, F. Diederich, J. Harmer, R. Schuhmann, I. Garcia-Rubio, and G. Jeschke, *J. Magn. Reson.*, 2009, **200**, 81.

111. M. Srivastava, C. L. Anderson, and J. H. Freed, *IEEE Access*, 2016, **4**, 3862.

112. Z. Y. Yang, M. Ji, T. F. Cunningham, and S. Saxena, in Electron Paramagnetic Resonance Investigations of Biological Systems by Using Spin Labels, Spin Probes, and Intrinsic Metal Ions, Pt A, eds P. Z. Qin and K. Warncke, 2015, Vol. 563, p 459.

Chapter 21
Shaped Pulses in EPR

Philipp E. Spindler, Philipp Schöps, Alice M. Bowen, Burkhard Endeward, and Thomas F. Prisner

Goethe University Frankfurt, Frankfurt am Main, Germany

21.1 INTRODUCTION

Although the birth of pulsed electron paramagnetic resonance (EPR) and nuclear magnetic resonance (NMR) happened within an interval of 8 years in 1958 and 1950, respectively,[1,2] the advent of shaped pulse methodology, beginning in the 1980s for NMR, only emerged in the field of EPR recently. The primary reason why the application of shaped pulses in EPR lagged behind NMR is the fact that the dynamic modulation of the pulse shape has to occur in the low nanosecond (ns) time regime for coherent manipulation of electron spins compared to the microseconds (μs) necessary for nuclear spin manipulation. In addition, the excitation frequency for the electron spin is usually in the microwave (MW) regime, while nuclear spins are excited in the radio frequency (RF) range. Both factors make the technical realization of shaped pulses much more demanding in the field of EPR spectroscopy (this will be described in more detail in Section 21.3). In the past few years, arbitrary waveform generators (AWGs), which operate on the nanosecond timescale, have become commercially available, allowing the implementation of such pulses in EPR spectroscopy. The opportunities afforded by commercially available AWGs that can produce and modulate waveforms at frequencies suitable for pulsed EPR measurements open up the possibilities for scientists studying a diverse range of systems to explore this technology and render EPR a technique on the brink of a suite of new methodologies. In this chapter, we aim to discuss the types of shaped pulses that are currently applied (Section 21.2) and exemplify the instrumental setup required to generate these pulses (Section 21.3).

Shaped pulses have already been used in NMR for several decades; naturally, some of the NMR shaped pulse methodology[3–5] has been translated and used in EPR. This includes, for example, pseudostochastic phase-modulated pulses,[6,7] Gaussian amplitude-modulated pulses,[8] composite pulses,[9] adiabatic and chirp pulses,[10,11] and optimum control theory (OCT)-derived pulses.[12] Here, we take a broad definition of shaped pulses to cover any pulse shape that is not rectangular (or intended to be rectangular, as generating a truly rectangular pulse is difficult

EPR Spectroscopy: Fundamentals and Methods.
Edited by Daniella Goldfarb and Stefan Stoll

ISBN: 978-1-119-16299-5
Also published in eMagRes (online edition)
DOI: 10.1002/9780470034590.emrstm1520

and will most likely show finite rise and fall times resulting from instrumental limitations, which lead to an overall shape more akin to a 'shark fin' than a rectangle). In our definition, shaped pulses are pulses that alter their amplitude, phase, or frequency within the pulse length. As excellent reviews on this topic already exist for NMR applications, only a short theoretical description of the pulses mostly used for EPR spectroscopy will be given in Section 21.2.

In NMR, shaped pulses are usually utilized either to control the bandwidth of the excitation with a high degree of precision or to suppress signals (e.g., those resulting from the solvent). In EPR, where the signal is specific to the unpaired electron spin center and the solvent is EPR silent, the second aspect is less important. On the other hand, in EPR, the inhomogeneous linewidth often exceeds by far the excitation bandwidth of classical rectangular pulses. Therefore, the ability to enlarge and control the excitation bandwidth of pulses and to tailor this bandwidth to compensate for experimental and instrumental imperfections, such as resonator bandwidth or field inhomogeneity, presents exciting opportunities to improve current pulsed EPR methodology based on rectangular pulses.

Early examples of the application of shaped pulses in EPR include the use of pseudostochastic phase-modulated pulses,[13] composite pulses[14] for 1D and 2D EPR spectroscopy, and the use of Gaussian amplitude shaped MW pulses for ELDOR (electron-electron double resonance)-detected NMR (EDNMR)[15] and Fourier transform (FT) EPR-detected NMR.[16]

More recently, fast AWGs have been used in EPR spectroscopy to generate MW pulses with extended excitation bandwidths, both for generation of transverse magnetization and spin inversion. Implementation of such pulses has shown improved performance with respect to classical rectangular pulses. This is not only due to the increased bandwidth but also due to the increased uniformity of excitation of spins within the given bandwidth. This is especially important for pulse sequences with an extended number of pulses where an offset independent flip angle achieved with shaped pulses strongly enhances the weight of the desired spin trajectory. In Section 21.4, this aspect will be illustrated by some current applications of shaped pulses in the field of dipolar and hyperfine spectroscopies.

Until recently, the great technical hurdle to shaped pulses in the nanosecond regime was the lack in availability of AWGs capable of generating these pulse shapes in the time domain. Until recently, pulse shaping in EPR has been accomplished by home-built modifications of pulsed EPR spectrometers. Prisner and coworkers modified a commercial Bruker E580 spectrometer using an AWG from Innovative Technical Systems, Inc. (AWG 8200) and have applied this at both X- and Q-band frequencies by positioning the pulse shaping unit either between the bridge and the TWT for X-band measurements[17] or between the X-band bridge and the 24 GHz local oscillator (LO) for upconversion to Q-band frequencies.[18] Jeschke and coworkers added an ultra-wide bandwidth channel to commercial Bruker E580/680 X-band and Q-band spectrometers by mixing signals from an Agilent M8190A AWG (12 GS/s) with an 8 GHz LO and using the output from this setup to replace one of the channels in the spectrometer, and detection is performed using a 1.5 or 2 GS/s digitizer, synchronized with the AWG.[19–21] Furthermore, Doll *et al.* used the advanced pulse programming capabilities of the Agilent M8190A, in combination with MATLAB scripts to generate the necessary trigger pulses, making the Bruker console obsolete.[22] Freed and coworkers modified a home-built Ku-band spectrometer with an AWG from Anlogic, Inc. (DBS2050A) in order to produce phase- and amplitude-modulated pulses over a 300 MHz bandwidth.[23] Han and coworkers have built a digital-to-analog converter (DAC) board-based X-band EPR spectrometer.[24] Recently, spectrometers with built-in AWG units have become commercially available from Bruker Biospin for EPR spectrometers ranging from 9 (X-band) to 263 GHz (J-band).

AWGs capable of generating shaped MW pulses have also found applications in more specialized forms of EPR. These include rapid-scan EPR, described in Chapter 23, electrically detected magnetic resonance (EDMR), covered in Chapter 26, and quantum information processing (QIP), as described in Chapter 22. Within QIP applications, Gaussian pulses have been used to suppress off-resonance excitation.[25] Composite pulses, such as the BB1 pulse,[26] and adiabatic pulses have been used to increase the fidelity of gate operations, acting on P donors in a high-purity ^{28}Si crystal, compared to rectangular pulses.[27] The pulses were generated using an AWG to modulate a fixed-frequency X-band MW source. At lower frequencies (300 MHz), an AWG was used to generate shaped pulses for time-domain EPR imaging.[28] An AWG operating to produce chirped MW pulses in the frequency range 7.5–18.5 GHz (after frequency upconversion by a factor of 8) has been incorporated into a home-built spectrometer used to study the rotational spectra of gas molecules.[29] An AWG has also been implemented to control the MW signals used in a dynamic nuclear polarization (DNP) experiment.[30] In this DNP setup, an AWG is used to control the input voltage to a varactor-tuned Gunn oscillator, with a frequency range of about 400 MHz around a fixed frequency, which allows the frequency output to be modulated.

Of course, alternative methods to shaped pulses for improving the excitation bandwidth also exist. These include rapid magnetic field variation[31] and the application of additional frequency sources to extend the frequency bandwidth of rectangular pulses, as demonstrated on a DEER (double electron–electron resonance) or PELDOR (pulsed electron–electron double resonance), experiment on manganese.[32]

21.2 SPECIFIC TYPES OF PULSES

Although shaped pulses potentially have many possible benefits over rectangular pulses, they are also more complex to understand. In particular, while the effect of a rectangular pulse on a two-level system can be modeled analytically at any offset ($\Delta\omega$) of the spin Larmor frequency (ω_0) from the MW frequency (ω_{MW}), only specific shaped pulses have analytical solutions.[3,4] Nevertheless, the effect of a shaped pulse on an isolated spin can be approximately calculated by dividing the pulse into short segments of length dt with constant amplitude and phase and integrating the rotating-frame Bloch equations (see Chapter 8)

$$\frac{\mathrm{d}M_x}{\mathrm{d}t} = \gamma(M_yB_z - M_zB_y) - \frac{M_x}{T_2}$$
$$\frac{\mathrm{d}M_y}{\mathrm{d}t} = \gamma(M_zB_x - M_xB_z) - \frac{M_y}{T_2}$$
$$\frac{\mathrm{d}M_z}{\mathrm{d}t} = \gamma(M_xB_y - M_yB_x) - \frac{(M_z - M_0)}{T_1} \quad (21.1)$$

successively for each segment and for every frequency offset $\Delta\omega$. M_i and B_i correspond to the projections of the magnetization and magnetic field on the i axis ($i = x, y, z$) of the rotating frame. Solution of the Bloch equations shows that a spin packet with a frequency offset $\Delta\omega$ will precess in the rotating frame around an effective field $\boldsymbol{B}_{\mathrm{eff}}$, which is composed of the off-resonance component $\Delta\omega/\gamma$ in the z-direction and the MW field $\boldsymbol{B}_1$ in the transversal plane (Figure 21.1).

The amplitude of the effective field is given by

$$B_{\mathrm{eff}} = |\boldsymbol{B}_{\mathrm{eff}}| = \sqrt{B_1^2 + \left(\frac{\Delta\omega}{\gamma}\right)^2} \quad (21.2)$$

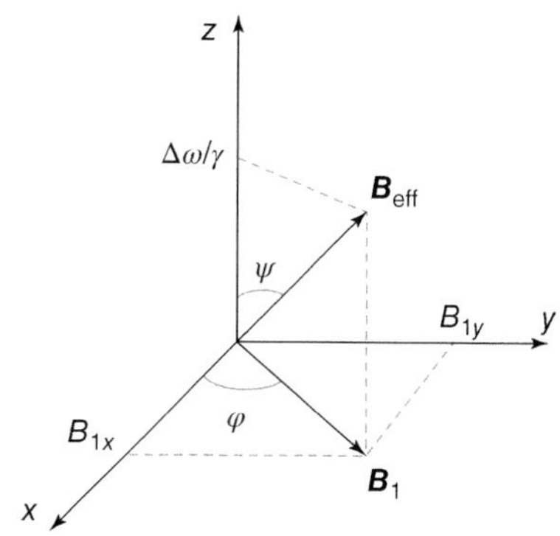

Figure 21.1. Magnetic field vectors in the rotating frame. The microwave pulse $\boldsymbol{B}_1$ is applied in the x–y-plane and described by the phase angle φ and the amplitudes B_{1x} and B_{1y}. The resulting effective field vector $\boldsymbol{B}_{\mathrm{eff}}$ is a combination of the off-resonance component $\Delta\omega/\gamma$ and $\boldsymbol{B}_1$

Considering the angle φ to be defined according to Figure 21.1 as the angle between B_1 field and the x-axis and calculating the angle ψ with respect to the z-axis according to

$$\psi = a\tan\left(\frac{\gamma B_1}{\Delta\omega}\right) \quad (21.3)$$

the Bloch equation can be solved numerically via

$$\tilde{M}(t_{n+1}) = \boldsymbol{R}_z(\varphi)\,\boldsymbol{R}_y(\psi)\boldsymbol{R}_z(\gamma B_{\mathrm{eff}}\cdot \mathrm{d}t)\,\boldsymbol{R}_y(-\psi)\boldsymbol{R}_z(-\varphi)\boldsymbol{M}(t_n)$$
$$M_x(t_{n+1}) = \tilde{M}_x(t_{n+1})\mathrm{e}^{-\mathrm{d}t/T_2}$$
$$M_y(t_{n+1}) = \tilde{M}_y(t_{n+1})\mathrm{e}^{-\mathrm{d}t/T_2}$$
$$M_z(t_{n+1}) = M_0 - (M_0 - \tilde{M}_z(t_{n+1}))\mathrm{e}^{-\mathrm{d}t/T_1} \quad (21.4)$$

with $\boldsymbol{R}_i(\theta)$ being the rotation matrix for rotation by an angle of θ around the ith axis. Solving the Bloch equations allows us to understand the effect of shaped pulses on a single transition. For simulations of coupled spin systems or spins $S > 1/2$, the concept of short time-independent sections can be extended to other equations of motions, such as the Liouville–von Neumann equation or the stochastic Liouville equation (see Chapter 8).[33,34]

Two important categories of shaped pulses can be distinguished. The first one is universal rotation (UR) pulses, which rotate any initial magnetization vector around a given rotation axis by a defined angle. Such pulses have been used in NMR for refocusing and excitation of higher coherence orders. Rectangular, Gaussian, and sinc (*sinus cardinalis*) pulses fall into this category. The other important pulse category are point-to-point (PP) rotation pulses, which rotate a given initial magnetization to a fixed target orientation, e.g., from the z-axis to the $-z$-axis or from the z-axis to the x-axis. The first type represents spin inversion and the second type excitation of transverse magnetization from Boltzmann equilibrium. Adiabatic pulses and some composite pulses are examples of PP rotation pulses. It has been shown that a UR pulse can be constructed from two PP rotation pulses of half the flip angle desired for the UR pulse and where the first one is time and phase inversed with respect to the second one.[35]

21.2.1 Rectangular Pulses

In the case of a hard rectangular pulse, $|\gamma B_1| \gg |\Delta\omega_{max}|$, the off-resonance effects can be neglected. In this case, the nutation angle θ is directly proportional to the pulse duration (t_p) and to the MW pulse strength (γB_1):

$$\theta = \gamma B_1 t_p \tag{21.5}$$

However, this condition is rarely fulfilled in EPR owing to the relative large spectral width of most EPR spectra. The inversion profile of a rectangular π pulse as a function of the frequency offset is illustrated in Figure 21.2. This function can be approximated as a sinc function.

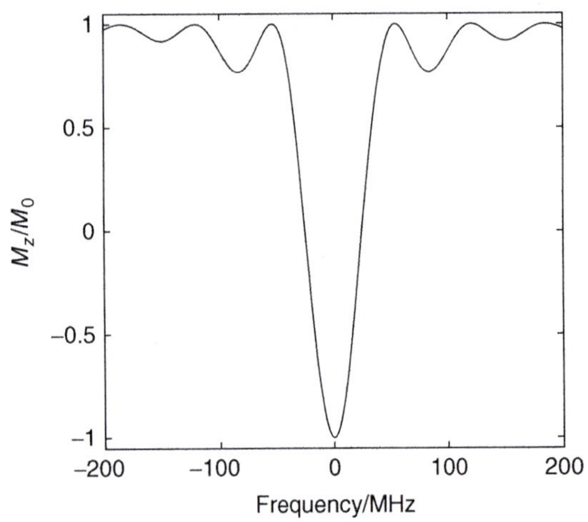

Figure 21.2. Frequency domain ($\Delta\omega$) inversion profile of an ideal rectangular π pulse with a pulse length of 16 ns. For comparison, an X-band nitroxide powder spectrum has a width of approximately 200 MHz

The FWHM (full width at half maximum) of the longitudinal (z) magnetization excitation profile of such an inversion pulse can be approximated by

$$\Delta\nu_\pi \approx \frac{0.8}{t_p} \tag{21.6}$$

Taking the same considerations for a $\pi/2$ pulse, which generates transverse magnetization, leads to an excitation bandwidth of[36]

$$\Delta\nu_{\pi/2} \approx \frac{1.2}{t_p} \tag{21.7}$$

In pulsed EPR, MW resonant structures are used to increase the MW field strength and to achieve more homogeneous B_1 amplitudes over the sample. The enhancement of the MW field strength counteracts the requirement of fast switching MW, leading to distortions of the rectangular pulse shape. The effect of the resonator bandwidth on pulses will be described in more detail in Section 21.3.

There are three shortcomings of rectangular pulses. The first one is the limited bandwidth, because very short pulses cannot be generated, the second one is the existence of sidebands, and the third one is the offset-dependent orientation of the magnetization at the end of the pulse, illustrated in Figure 21.3. In conclusion, in typical EPR cases, where the inhomogeneous linewidth exceeds the MW excitation field strength, rectangular pulses are not the optimum tool for manipulating electron spins.

These problems can be overcome by using shaped pulses. There is a vast number of shaped pulses and

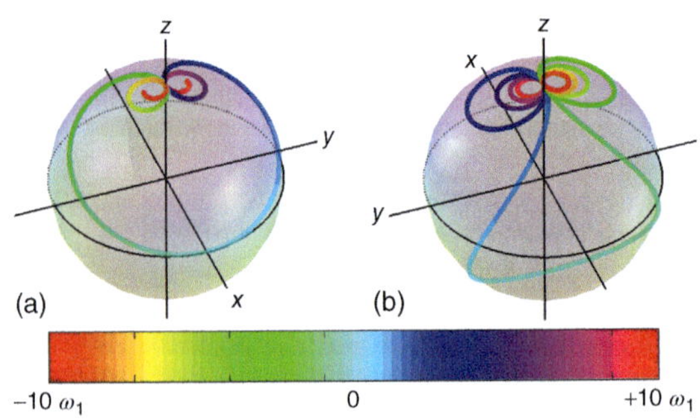

Figure 21.3. Distribution of spin packets with resonance offset frequencies as multiples of the on-resonance nutation frequency ω_1 from $-10\omega_1$ to $+10\omega_1$ on the Bloch sphere after the application of a rectangular $\pi/2$ pulse (a) and π pulse (b) applied in the y-direction of the rotating frame. Before the pulse, the spin packets are all aligned with the z-axis

different techniques to determine and optimize the pulse for a certain property; for example, OCT[12] or composite pulses.[9] Here, we will describe three groups of pulses often used in the field of EPR: Gaussian, sinc, and adiabatic pulses/fast-passage.

21.2.2 Gaussian Pulses

Gaussian pulses can be used for selective excitation/inversion without creating sidebands. Examples of their application in EPR include ELDOR-detected NMR[15] and quantum control experiments.[37] A Gaussian MW pulse can be described by

$$B_1(t) = B_{1\,\max} e^{-\rho(t/t_p)^2}\ t\varepsilon \left[\frac{-t_p}{2}, \frac{t_p}{2}\right] \qquad (21.8)$$

with ρ being the adjustable parameter. In order to achieve spin inversion with a Gaussian pulse using the same $B_{1\max}$ field strength as for a rectangular pulse, the length of the pulse has to be increased. The resulting pulse is characterized by a slightly smaller bandwidth compared to the corresponding rectangular pulse (Figure 21.4a).

Gaussian pulses are UR pulses and can easily replace any rectangular pulse in classical EPR pulse sequences without leading to loss of phase coherence.

21.2.3 Sinc Pulses

Sinc inversion pulses also provide practically sideband-free M_z profiles as shown in Figure 21.4(b). This property was used to improve RF time-domain EPR imaging.[28] The amplitude profile of these pulses is

$$B_1(t) = B_{1\,\max}\frac{\sin(n2\pi t/t_p)}{2\pi t/t_p}\ t\varepsilon \left[\frac{-t_p}{2}, \frac{t_p}{2}\right] \qquad (21.9)$$

Owing to the horizontal asymptote, a perfect sinc pulse would require infinite pulse duration. For practical reasons, the pulse is truncated at the nth zero crossing, where n is an integer.

The FWHM for the sinc inversion pulse profile is approximately the same as that of a comparable rectangular inversion pulse with the same $B_{1\max}$. A more rectangular shape of the sinc inversion pulse profile in the frequency domain can be obtained by combining the sinc function with a Hamming amplitude function. This combination reduces truncation effects of pulses with finite length.[38] As with Gaussian pulses, the phase coherence is conserved using sinc pulses, as both are UR pulses.

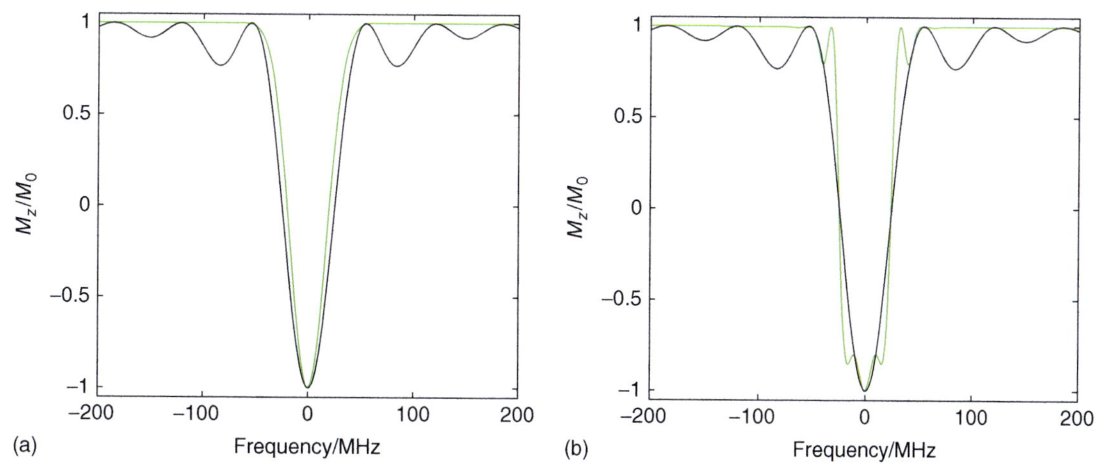

Figure 21.4. (a) Frequency domain ($\Delta\omega$) excitation profiles of rectangular (black) and Gaussian (green, $\rho = 5$) inversion pulses with the same maximum MW field strength of $\frac{\gamma B_{1\max}}{2\pi} = 31.25$ MHz. This corresponds to a pulse length of 16 ns for the rectangular inversion pulse and 40 ns for the Gaussian pulse. (b) Comparison of the frequency domain ($\Delta\omega$) excitation profiles of a rectangular inversion pulse (black) and a sinc inversion pulse (green) truncated at $n = 3$. The maximum MW field strength is $\frac{\gamma B_{1\max}}{2\pi} = 31.25$ MHz for both pulses. This leads to a pulse length of 16 ns for the rectangular pulse and 90 ns for the sinc pulse

21.2.4 Adiabatic and Fast-passage Pulses

Adiabatic pulses were introduced in NMR to compensate for B_1 inhomogeneity.[39] These pulses have a 180° flip angle over a certain bandwidth independent of B_l field strengths down to a certain threshold. This property is equivalent to a high bandwidth for inversion pulses, which is the main reason why adiabatic pulses are of interest for EPR. Adiabatic pulses are defined by analytical expressions for the instantaneous frequency offset $\Delta\omega_{MW}(t)$ and amplitude $B_1(t)$ during the pulse. The instantaneous MW frequency offset $\Delta\omega_{MW}(t)$ is defined as the difference between the instantaneous MW frequency $\omega_{MW}(t)$ and the center frequency ω_c of the pulse; $\Delta\omega_{MW}(t) = \omega_{MW}(t) - \omega_c$. In order to explain the concept of adiabatic pulses, it is convenient to introduce a new accelerated coordinate system, called the frequency frame (x^*, y^*, z^*).[10] The alignment of the z^*-axis is, as for the conventional rotating frame, collinear with the external magnetic field $\boldsymbol{B}_0$. However, the frequency at which the frame rotates is not constant but defined by the instantaneous MW excitation frequency $\omega_{MW}(t)$ of the adiabatic pulse. Hence, the orientation of $\boldsymbol{B}_1(t)$ is fixed in this frequency frame (here arbitrarily chosen to lie along the x^*-axis, as shown in Figure 21.5).

The z^* component of $\boldsymbol{B}_{\mathrm{eff}}(t)$, given by $(\omega_{MW}(t) - \omega_0)/\gamma = \Delta\omega(t)/\gamma$, changes with time, because the frequency offset $\Delta\omega$ is now time dependent. If the Larmor frequency ω_0 coincides with the center frequency ω_c, the z component is given by $\Delta\omega_{MW}/\gamma$. The angle between the magnetization vector $\boldsymbol{M}$ and the effective field is small at the beginning of the adiabatic pulse and $\boldsymbol{M}$ starts to precess around $\boldsymbol{B}_{\mathrm{eff}}(t)$. During the pulse, $\boldsymbol{B}_{\mathrm{eff}}(t)$ is rotated from its initial direction almost collinear to z^* toward $-z^*$. If the precession of $\boldsymbol{M}$ around $\boldsymbol{B}_{\mathrm{eff}}(t)$ is always faster than the rotation of $\boldsymbol{B}_{\mathrm{eff}}(t)$ in the frequency frame, adiabatic tracking is accomplished and $\boldsymbol{M}$ follows $\boldsymbol{B}_{\mathrm{eff}}(t)$ and ends up almost collinear to $-z^*$ at the end of the pulse. This so-called adiabatic condition is quantified by the adiabaticity factor Q:[10]

$$\left|\frac{\mathrm{d}\alpha}{\mathrm{d}t}\right| \ll |\gamma B_{\mathrm{eff}}(t)| \qquad Q = \left|\frac{\gamma B_{\mathrm{eff}}(t)}{\mathrm{d}\alpha/\mathrm{d}t}\right| \gg 1 \qquad (21.10)$$

If the adiabatic condition is satisfied, the magnetization will stay essentially parallel to the effective field vector $\boldsymbol{B}_{\mathrm{eff}}(t)$. If the MW frequency of the adiabatic pulse starts from a frequency far below the resonance frequencies of the inhomogeneous EPR line and ends at frequency far above it, the angle α changes from 0° to 180°. The adiabatic behaviour is not obvious in the rotating frame (see Figure 21.6a). In the frequency frame (Figure 21.6b) the effective field, $\boldsymbol{B}_{\mathrm{eff}}(t)$, along with the magnetization $\boldsymbol{M}$, will move from the z^*- to the $-z^*$-direction.

Pulses derived from adiabatic pulse forms with reduced MW power, below the adiabatic region, are known as fast-passage pulses. These pulses can be used for uniform flip angles below 180°, exciting transverse magnetization.[18,22]

Numerous adiabatic pulse forms exist. Here, we describe only a small selection that has been used in EPR so far. The simplest pulse showing adiabatic behavior consists of a linear frequency sweep function and a constant amplitude, $B_{1\max}$, and is called a chirp pulse.[40] The instantaneous frequency function is given by

$$\omega_{MW}(t) = \omega_c + kt \qquad (21.11)$$

k is called the sweep parameter and ω_c the center frequency of the chirp pulse, with $t \in [-t_p/2, t_p/2]$. If the center frequency corresponds to the Larmor frequency of the spin, then the time-dependent frequency offset is given by

$$\Delta\omega(t) = \Delta\omega_{MW}(t) = kt \qquad (21.12)$$

and the phase angle of the MW excitation field $\boldsymbol{B}_1$ in the rotating frame with constant rotation frequency ω_0,

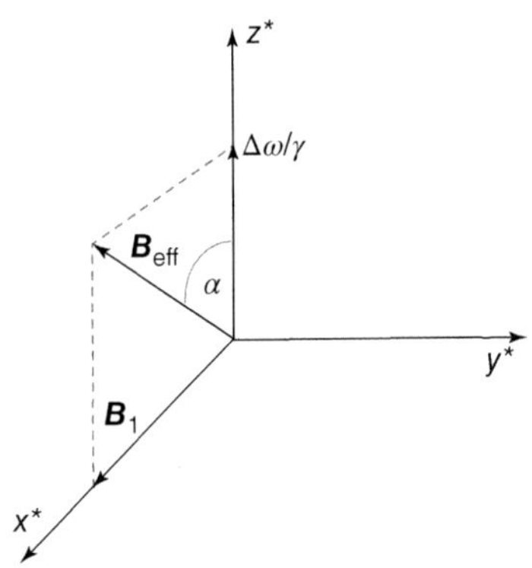

Figure 21.5. Magnetic field vectors in the frequency frame. The frequency frame rotates with the instantaneous MW frequency $\omega_{MW}(t)$ of the applied adiabatic pulse $\boldsymbol{B}_1(t)$. The z^* component of $\boldsymbol{B}_{\mathrm{eff}}(t)$ is given by $\Delta\omega/\gamma$, where $\Delta\omega = \omega_{MW}(t) - \omega_0$ is the instantaneous frequency offset of the MW pulse with respect to the Larmor frequency ω_0. The orientation of $\boldsymbol{B}_1(t)$ is constant and $\boldsymbol{B}_{\mathrm{eff}}(t)$ is tilted in the x^*–z^*-plane with the angle α

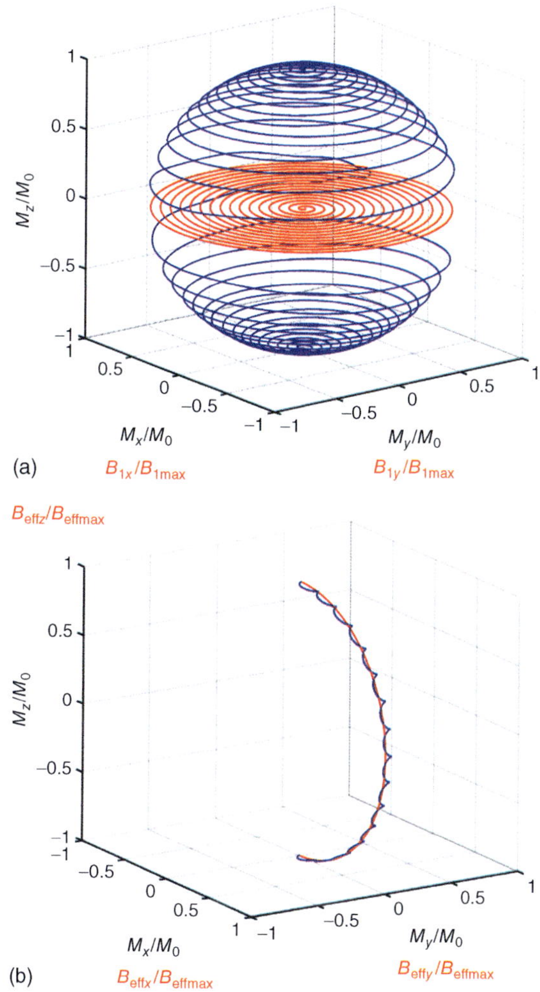

Figure 21.6. Simulation of the time-dependent trajectory of the magnetization ($\boldsymbol{M}$) (blue) and the effective field (red) in the conventional rotating (a) and frequency (b) frames during an adiabatic pulse with a constant adiabaticity Q [as defined by equations (21.10) and (21.17)] and a Larmor frequency ω_0 equal to the center frequency of the adiabatic pulse

the Larmor frequency, is given by

$$\varphi(t) = \int_{-t_{p/2}}^{t} \Delta\omega_{MW}(t')\mathrm{d}t' \qquad (21.13)$$

Therefore, the MW field oscillates in the x–y-plane of the rotating frame and can be described as

$$\boldsymbol{B}_1(t) = B_1(t)\begin{pmatrix}\cos(\varphi)\\ \sin(\varphi)\\ 0\end{pmatrix} \qquad (21.14)$$

Linear chirped pulses have a varying adiabaticity factor throughout the pulse. The effective magnetic field is lowest when the pulse is on resonance with the spin packet under consideration and its value at this instance is $B_{1\,max}$. The adiabaticity factor, at its minimum, is therefore given by[10]

$$Q_{min} = \frac{\omega_1^2}{k} = \frac{\omega_1^2 t_p}{\Delta\omega} \qquad (21.15)$$

Equation (21.15) is useful for estimating the MW power needed for spin inversion within a given bandwidth $\Delta\omega$ and pulse duration t_p assuming that $Q_{min} \geq 5$ should be achieved for quantitative inversion.[10] In order to smooth the inversion profile of the chirp pulse, additional amplitude modulation functions can be implemented. The combination of a linear frequency chirp pulse with the amplitude modulation given by equation (21.16) is called the WURST (wide-band, uniform rate, smooth truncation) pulse.[39]

$$|B_1(t)| = B_{1max}\left(1 - \left|\sin\left(\frac{\pi t}{t_p}\right)\right|^n\right) \qquad (21.16)$$

with n being an adjustable parameter used to round-off the edges of the pulse shape. This causes $\boldsymbol{B}_{eff}(t)$ to initially lie more collinear with z, which enhances the adiabaticity for the isochromats within the pulse bandwidth. The inversion profile of a WURST pulse compared to a rectangular pulse with the same B_{1max} is shown in Figure 21.7(c).

There are two clear advantages of WURST and other adiabatic pulses over rectangular pulses: the approximately rectangular inversion profile in the frequency domain and the larger bandwidth. However, this comes at a price, namely adiabatic pulses require longer pulse lengths. Owing to the short relaxation times of electron spins in paramagnetic samples, these pulses often have to be shortened in EPR experiments, leading to a lower Q factor and losing their ideal adiabatic behavior. Nevertheless, it has been shown that improved inversion performance over rectangular pulses can still be retained.[19,41]

The CAP (constant adiabaticity pulse) pulse is mentioned here to illustrate the principle of adiabaticity as shown in Figure 21.6. The adiabaticity (Q) is constant throughout the pulse, which can be seen in Figure 21.6. The CAP pulse is defined by the amplitude ($B_1(t)$) and frequency ($\Delta\omega_{MW}(t)$) modulation functions:[10]

$$B_1(t) = \frac{\pi Q_0}{\gamma t_p}\cos\left(\frac{\pi t}{t_p}\right), \quad \Delta\omega_{MW}(t) = \frac{\pi Q_0}{t_p}\sin\left(\frac{\pi t}{t_p}\right) \qquad (21.17)$$

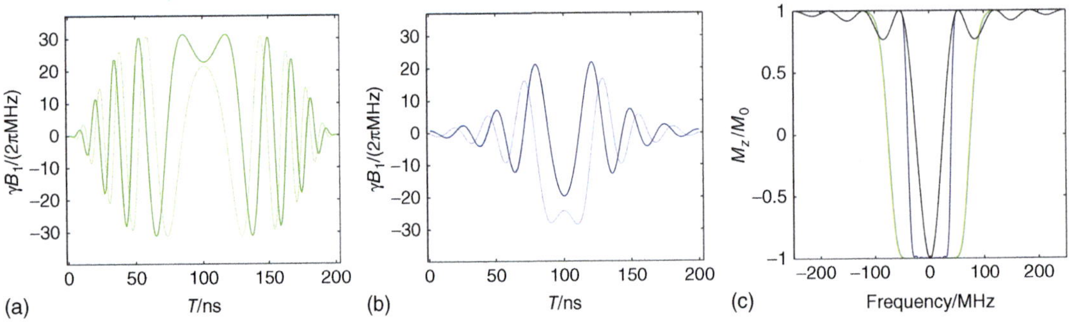

Figure 21.7. Comparison of a WURST, sech/tanh, and a rectangular pulses. The time-domain shape of the WURST and sech/tanh pulses are shown in (a) and (b), respectively (x and y components as full and dotted lines). The frequency domain ($\Delta\omega$) inversion profiles of the rectangular pulse (black), the WURST pulse (green), and the sech/tanh pulse (blue) are compared in (c). The pulse length is 16 ns for the rectangular pulse (black) and 200 ns for the WURST and sech/tanh pulses. The same maximum microwave field strength $\frac{\gamma B_{1\max}}{2\pi} = 31.25$ MHz was used for all three pulses The parameters for the adiabatic WURST pulse were set to $n = 10$ and $k = 0.006$ GHz ns^{-1} whereas the sech/tanh pulse was calculated according to equation (21.18) with $\beta = 0.043$ ns^{-1} and BW = 0.04 rad ns^{-1}

The pulse time axis is defined by $t \in \left[\frac{-t_p}{2}, \frac{t_p}{2}\right]$ and the adiabaticity factor Q_0 determines the pulse bandwidth $\frac{2\pi Q_0}{t_p}$ as well as $B_{1\max} = \frac{\pi Q_0}{\gamma t_p}$ for a given pulse duration t_p. An adiabatic pulse widely used in the field of EPR is the sech/tanh pulse,[42] which is defined by the amplitude and frequency functions:

$$B_1(t) = B_{1\max}\,\text{sech}(\beta t),\quad \Delta\omega_{MW}(t) = 0.5 \cdot BW\,\tanh(\beta t) \tag{21.18}$$

β is an adjustable parameter used to truncate the waveform and BW the bandwidth for infinite pulse duration. Sech/tanh pulses have the highest selectivity among the family of adiabatic pulses, whereas the bandwidth of a WURST pulse is larger for a given pulse length and MW power. A typical value used for β is $10.6/t_p$, which causes a truncation of the amplitude at 1% of the maximum peak amplitude. The minimum pulse duration t_p and the maximum bandwidth BW depend on each other for a given inversion efficiency and are best evaluated by simulating the inversion profile. It has been shown that a sech/tanh pulse is not adiabatic in the frequency frame; it is, however, adiabatic in the SORF (second-order rotating frame).[43]

21.2.5 Optimum Control Theory Pulses

The application of OCT to calculate pulse shapes, which drive a system from an initial state to a specific target state results in pulse shapes with strong improvements over analytical pulses shapes, especially for excitation or manipulation of transverse magnetization.[12] A distinct feature of this technique is that multiple experimental restrictions such as T_1 and T_2 relaxation,[44] B_1 inhomogeneity,[45] or pulse distortions by bandwidth limitations[17] can be incorporated into the optimization algorithm.

21.3 INSTRUMENTATION

21.3.1 Generation of Amplitude- and Phase-modulated Microwave Pulses

The bandwidths of shaped pulses for NMR are typically in the range 10–100 kHz, which is at least a factor of 10^3 smaller than the bandwidth needed for EPR applications. This is mainly due to the larger gyromagnetic ratio of the electron spin compared to that of a proton nuclear spin ($|\gamma_e/\gamma_H| = 660$). Thus, the pulse shaping in EPR has to take place on a nanosecond timescale. Commercial AWGs producing arbitrary signals with fast enough switching time (sampling rate) and vertical resolution (in bits) have become available in recent years. They make use of DACs, which directly synthesize the output voltage from binary code loaded into the AWG memory. The DAC converts the digital number into an output voltage proportional to this value, which can be changed at a specified sampling rate. Amplitude- and phase-modulated MW pulses for EPR applications can be generated using three different types of setups. They depend on the sampling rate of the AWG and the excitation bandwidth required.

An AWG with a sampling rate in the range of the targeted frequency modulation bandwidth can be used in a setup with two independent AWG channels to generate the shaped pulses in conjunction with an IQ (in-phase quadrature) modulator as shown in Figure 21.8(a).[17] For example, an AWG sampling rate of 1 GS/s enables a baseband modulation of up to 500 MHz according to Nyquist's theorem. A two-channel setup with an IQ modulator can create sideband frequencies from $\nu_c - 500$ MHz to $\nu_c + 500$ MHz, thus achieving the full modulation bandwidth of 1 GHz.

An IQ modulator consists of two modulators operating with an LO phase difference of 90° with respect to each other. The two components of the carrier MW signal with a 90° phase difference are generated by a 90° hybrid and separately mixed with the waveforms U_x and U_y synthesized by the two AWG channels before finally being recombined. IQ modulators have amplitude and phase imbalances, which need to be compensated by calibration before use. Another potential issue is that IQ modulators let a fraction of the LO (or carrier frequency, ω_{LO}) pass even though the control voltages are zero and this could have a large influence on the pulse performance. This LO leakage has to be minimized by applying DC (direct current) offsets to the I and Q channels of the AWG. Typically, a LO-RF isolation < -23 dB can easily be achieved (without applying a DC offset) but may affect experiments involving long pulses and saturated MW amplifiers. Furthermore, the I and Q channels might have an amplitude imbalance, which can be corrected by a scaling factor. Despite that, the I and Q channels of commercial available IQ modulators are not exactly 90° out of phase. The deviation ranges from 3° to 10°, which must be considered upon waveform calculation.

If the sampling rate of the AWG is significantly larger than the target modulation bandwidth, an approach with only a single AWG channel and a mixer is sufficient as shown in Figure 21.8(b). The AWG creates the pulse shape $U(t) = A(t) \cdot \cos(\omega_\Delta t + \varphi(t))$ at an arbitrary carrier frequency ω_Δ. This signal created by a single AWG channel generates a time-dependent phase of the pulse in addition to the amplitude modulations. This waveform is mixed with a LO source with frequency ω_{LO} to generate the final shaped MW pulse at the frequency $\omega_{LO} + \omega_\Delta$.

The mixer output is

$$B_1(t) = 0.5 \cdot A(t) \cdot (\cos((\omega_{LO} + \omega_\Delta)t + \varphi(t)) + \cos((\omega_{LO} - \omega_\Delta)t + \varphi(t))) \qquad (21.19)$$

The sum of the frequencies $\omega_{LO} + \omega_\Delta = \omega_{MW}$ is the operating frequency of the spectrometer; the other signal is removed, if necessary, by a filter. This design synthesizes the spectrometer frequency from the heterodynes ω_{LO} and ω_Δ. The heterodyne ω_Δ is amplitude and phase modulated; hence, this approach only requires a single channel AWG to generate pulses with arbitrary phase and amplitude. A calibration of the mixer is not necessary if the two heterodynes are well separated in frequency.

Common practice for detection in pulsed EPR is a down-conversion of the induction signal from the electron spins with the spectrometer carrier frequency to remove the fast oscillation close to the Larmor precession frequency. The final signal recorded corresponds to the expected time-domain signal in the rotating frame. It is typically in the megahertz frequency range and can be digitized easily. In a setup where the amplitude- and phase-modulated pulses are generated at an intermediate frequency, there is no

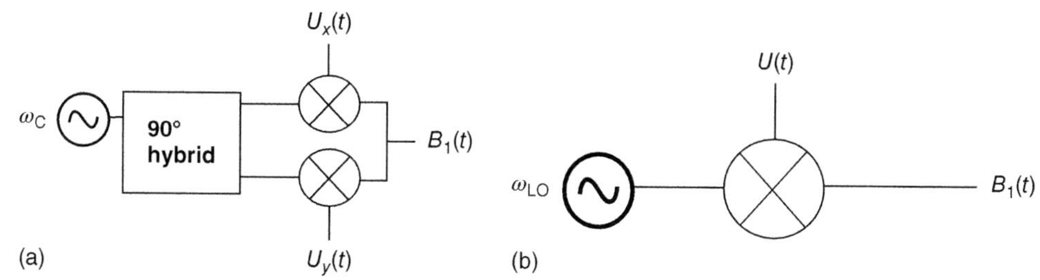

Figure 21.8. (a) IQ modulator setup for shaped pulses with two AWG channels, which independently modulate the carrier frequency ν_c of two 90° out of phase components of the MW. (b) Direct sideband generation of shaped pulses by mixing the AWG output $U(t)$ with a frequency ω_{LO}

LO with the spectrometer frequency. The lack of a coherent LO does not matter when the pulses generated are used solely for pure inversion, such as the pump pulse in DEER or PELDOR.[19,41] However, when shaped pulses are required for coherent spin manipulations, such as excitation or refocusing, the phase of the AWG with respect to the phase of the LO must be known and constant. Jeschke and coworkers solved this issue by down-mixing the spin signal oscillating at about $\omega_{LO} + \omega_{\Delta} = 9.5$ GHz with the continuous wave source ω_{LO} at 8 GHz and recording the resulting transients oscillating around ω_{Δ} with a fast digitizer.[22] The remaining ω_{Δ} frequency component is removed by a digital down-conversion. Another possibility to solve this problem would be a second MW signal with constant amplitude and frequency, ω_{Δ}, which is phase coherent with the AWG. This signal can be mixed with the LO at frequency ω_{LO} used for the pulse generation to achieve a coherent LO for the detection IQ modulator/demodulator.

In cases where the sampling rate is more than double the operating frequency, an arbitrary shaped MW pulse can be generated by the AWG directly without any further mixing processes. Currently, AWGs with sampling rates of up to 65 GS/s have become commercially available, which could easily produce X-band EPR waveforms directly. However, digitizers with comparable sampling rates, which are needed to record the spin signal directly, do not exist. Hence, a second channel of the AWG would be needed to produce the phase-coherent LO to down-convert the signal for detection.

21.3.2 Pulse Distortion by Active and Passive Microwave Components

Every component in the MW excitation and signal detection pathway has a nonflat frequency response and therefore gives rise to pulse and signal distortions. This causes deviations of the pulse shape seen by the electrons spins from the calculated shape and will lead to deviations of the final state of the spin system from the desired target state. Therefore, such effects should ideally be quantified and taken into account for pulse optimization.

The distortions can be split into a nonlinear part, such as the saturation of an amplifier, and a linear part, due to a nonflat frequency response. It was shown that these contributions can be dependent on each other.[22] The whole spin excitation pathway contributes to the frequency response; this includes the AWG, modulators, MW cables, passive components, and amplifiers as well as, to a large extent, the MW resonator. These power-independent pulse distortions can be compensated by calibration. All time-invariant linear distortions, meaning they have no memory regarding their input, can be treated by linear response theory. This theory states that once the answer of the system to a delta-shaped input function is known, the output $y(t)$ for any input function $x(t)$ can be calculated.[46] The response to the delta function is called the impulse response function $h(t)$ and contains all linear phase and frequency information of the system:

$$y(t) = \int_{-\infty}^{\infty} h(t')x(t - t')\mathrm{d}t' \tag{21.20}$$

In principle, every component has its own impulse response function, causing the total impulse response function of a chain of linear time-invariant components to be the convolution of all individual response functions. As a delta function cannot be created as an input function and the numerical deconvolution of the impulse response function from a measured signal is mathematically unstable, more practically feasible ways have to be used to determine the impulse response function $h(t)$ of the spectrometer. Two different possibilities will be discussed here.

The FT $H(\nu)$ of the impulse response function $h(t)$ is the frequency response function or transfer function of the system. It can be determined by measuring the amplitude and phase response during a frequency sweep of the excitation over the whole bandwidth. To do this, a MW B_1 detector has to be placed close to the resonator, where the actual spin excitation takes place, which may disturb the field distribution and frequency response in the probehead (Figure 21.9).

Alternatively, as proposed by Jeschke and coworkers, the magnitude response function $|H(\nu)|$ can be measured by means of an excitation frequency-dependent nutation experiment.[19] This method uses the actual spin system under investigation thus avoiding distortions of the setup by the implementation of an additional MW detector. The real and imaginary parts of the transfer function of such a system depend on each other, as described by the Kramers–Kronig relation. The amplitude and phase responses are connected by the Bode relation, which was used to obtain the frequency-dependent phase delay of the spin excitation path.[19] The transfer function obtained can be used to derive corrections for phase and sweep

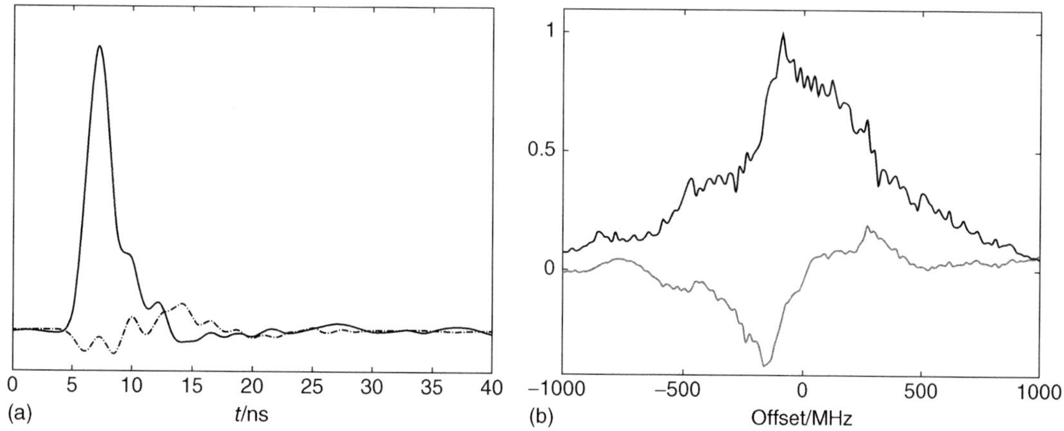

Figure 21.9. (a) Measured impulse response function $h(t'')$ and (b) its Fourier transform, which is the frequency response $H(\nu)$ for a fully overcoupled Bruker MS3 Resonator.[17] (Reprinted from J. Magn. Reson., 218, Spindler, P. E.; Zhang, Y.; Endeward, B.; Gershernzon, N.; Skinner, T. E.; Glaser, S. J.; Prisner, T. F. Shaped optimal control pulses for increased excitation bandwidth in EPR, 49. © 2012, with permission from Elsevier)

rates and to design bandwidth-compensated adiabatic chirp pulses. The amplitude-transfer function gives the values used to adjust the sweep rate of linear chirped adiabatic pulses to compensate for the frequency-dependent B_1 field strength in order to achieve the same adiabaticity factor Q for all spin packets. The phase response can be used to alter the phase during the pulse in such a way that the effective field stays in the x–z-plane and does not build up a y-component during the pulse, which would reduce the adiabaticity.[22]

A different approach was followed by Prisner and coworkers.[17] A pseudostochastic maximum length sequence $m(t')$ is created by a linear feedback register and has the property that the autocorrelation function approximates the delta function.

$$\int_{-\infty}^{\infty} m(t')m(t'+t)\mathrm{d}t' = \delta(t) \qquad (21.21)$$

The measured output generated by the excitation $m(t')$ is $y(t)$. Thus, $y(t)$ is the convolution of $m(t')$ with the impulse response $h(t')$:

$$y(t) = \int_{-\infty}^{\infty} h(t')m(t-t')\mathrm{d}t' \qquad (21.22)$$

The cross correlation of the excitation function $m(t')$ with the measured response function $y(t)$ yields the impulse response $h(t'')$:

$$\begin{aligned}&\int_{-\infty}^{+\infty} y(t)m(t+t'')\mathrm{d}t\\ &= \int_{-\infty}^{+\infty}\int_{-\infty}^{+\infty} h(t')m(t-t')m(t+t'')\mathrm{d}t\,\mathrm{d}t'\\ &= \int_{-\infty}^{+\infty} h(t')\delta(t''-t')\mathrm{d}t' = h(t'') \qquad (21.23)\end{aligned}$$

The parameters t, t', and t'' are time axes that are linearly shifted with respect to each other by experimental delays. In this case, the response function $y(t)$ to the pseudostochastic excitation $m(t')$ was measured using a pick-up coil brought into the stray field of the MW resonator.[17] The impulse response function $h(t'')$ obtained (see Figure 21.9) was incorporated into the OCT algorithm to optimize a prefocused pulse sequence.[17]

The first procedure has the advantage that the optimization can be performed under the momentary experimental conditions and directly before performing the actual experiment; however, only indirect phase information can be obtained from this measurement. Using a B_1 probe such as the pick-up coil relies on the assumption that the stray field is proportional to the field created inside the resonator where the spins are located. Further developments may lead to resonant structures enabling online B_1 monitoring allowing the determination of the spectrometer response function under experimental conditions.

21.4 APPLICATIONS OF SHAPED PULSES IN EPR

The experiments described here can be divided into two different categories: the first one uses shaped broadband inversion pulses, which move the magnetization from the z- to the $-z$-direction; the second one uses shaped pulses to create transverse magnetization in the x–y-plane. Broadband inversion is easier to realize, because phase coherence of the spins after the pulse is not required. Adiabatic pulses are very well suited for this purpose even if they are partially truncated to reduce the length of the pulse, such that the adiabaticity, Q, is not strictly conserved (low Q factor).

21.4.1 Shaped Pulses for Magnetization Inversion

In many applications of pulsed EPR in disordered frozen solution, the excitation bandwidth of the MW pulses is not large enough to cover the full inhomogeneous linewidth of the paramagnetic species. A typical pulse length for a π pulse is 8 ns for the common Bruker MS3 resonator in overcoupled mode in conjunction with a 1 kW TWT at X-band. This corresponds to a usable bandwidth of roughly 60 MHz. This is especially true for transition metal ions, where the linewidth easily exceeds 10 mT, corresponding to 280 MHz. In such cases, standard rectangular pulses only invert narrow spectral holes, strongly reducing the efficiency of pulsed dipolar (see Chapters 11 and 19) and hyperfine (see Chapters 16–18) spectroscopy experiments. Broadband shaped pulses have been used to increase the inversion efficiency for both types of experiments.

The four-pulse PELDOR pulse sequence is well established for the measurement of distances between two paramagnetic centers in the nanometer range.[47–49] It is described in Chapter 19. If the inversion pump pulse acting on the coupled, but not observed, second spin is replaced by a broadband shaped inversion pulse, a larger modulation depth of the dipolar evolution function can be achieved. This has been demonstrated on a copper-nitroxide biradical with fast-passage chirp pulse[19] and on a cobalt-nitroxide model compound with a sech/tanh pump pulse.[41] Figure 21.10 shows that the modulation depth is more than doubled by replacing an 8 ns rectangular pump pulse acting on the copper spin by a fast-passage pulse with a length of 64 ns and a bandwidth of 600 MHz.[19] It is worthwhile to note that the enhanced off-resonance irradiation by the prolonged pump pulses may induce a pump–probe offset-dependent additional loss in signal intensity of the detected spins compared to rectangular pump pulses.[50]

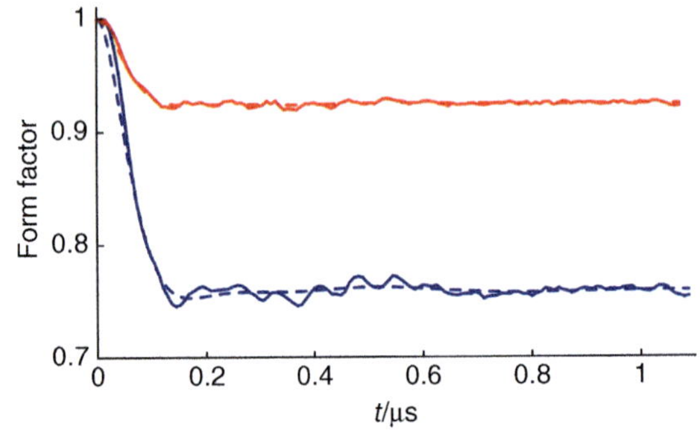

Figure 21.10. Four-pulse PELDOR time traces (solid lines) obtained with a rectangular MW inversion pulse (red) and with a fast-passage chirp inversion pulse (blue) on a copper-nitroxide biradical. In both cases, the detection is performed at the center of the nitroxide line. The pump frequency of the rectangular pulse is 200 MHz above the detection frequency and 100–700 MHz above for the fast-passage pulse.[19] (Reprinted from J. Magn. Reson., 230, Doll, A.; Pribitzer, S.; Tschaggelar, R.; Jeschke, G., Adiabatic and fast passage ultra-wideband inversion in pulsed EPR, 27. © 2013, with permission from Elsevier)

Chirp inversion pump pulses at Q-band frequencies have also been applied to enhance the modulation depth for dipolar distance measurements of a Gd(III)–Gd(III) biradical.[20] Here, owing to the large electron spin ($S=7/2$), the situation is more complex. The chirp pump pulse on the one hand increases the modulation depth (up to 10% of the original echo intensity) depending on the adiabaticity factor Q (up to $Q=10$ for the central $m_S=-1/2 \leftrightarrow +1/2$ transition). On the other hand, pulses with larger adiabaticity factors also induce double-quantum coherence and therefore lead to loss in echo signal intensity. Optimum performance of the chirp pump pulse was achieved with Q values between 1 and 2, meaning that the pulses can no longer be considered adiabatic and leading to a reduction in inversion efficiency. For such half-integer high-spin systems, the signal intensity of the observable central transition $m_S=-1/2 \leftrightarrow +1/2$ can also be increased by polarization transfer from other transitions (for example the $m_S=3/2 \leftrightarrow +1/2$ transition) beforehand (Figure 21.11).[21] The signal

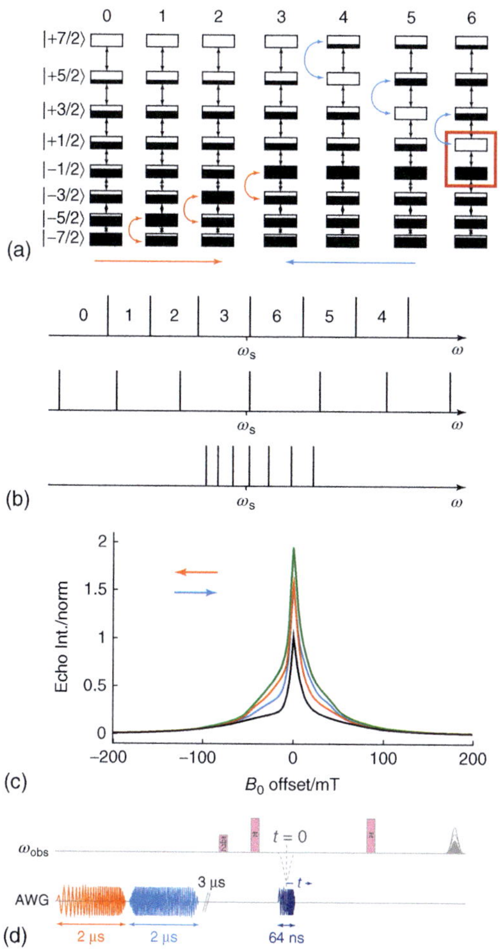

Figure 21.11. Enhanced sensitivity of spectroscopy with Gd(III) spin labels achieved by chirp polarization transfer pulses. (a) The population of the different m_S sublevels of the Gd(III) spin system is shown. Application of up-sweep chirp pulses on the negative m_s states (orange) and down-sweep on the positive m_s states (blue) optimizes the population difference of the finally observed central transition $m_S = -1/2 \leftrightarrow m_s = +1/2$, as illustrated in the rightmost level diagram. (b) The difference in Larmor frequency of the different $m_s \leftrightarrow m_S+1$ transitions depends on the orientation of the Gd(III) center in the external magnetic field. (c) The experimental echo-detected field-sweep spectrum prepolarized with up-chirp (orange), down-chirp (blue), and sequential combination of both (green) as well as equilibrium reference obtained without prepolarization (black). (d) The actual pulse sequence used for bulldozer-enhanced four-pulse PELDOR on Gd(III). (Reproduced from Ref. 21 with permission of the PCCP Owner Societies)

intensity could be improved by a factor of two with one or two frequency-swept adiabatic pulses with 1 GHz bandwidth and 1 μs length compared to the four-pulse PELDOR experiment.

CP (Carr–Purcell) PELDOR is a new pulse sequence for dipolar spectroscopy,[23,51] utilizing adiabatic sech/tanh inversion pulses not because of their potentially enhanced bandwidth but rather for their superior performance when it comes to quantitative inversion of spins within a certain bandwidth (Figure 21.7). In CP-PELDOR, a CP train of inversion pulses[52] is applied within the evolution time of the observer spin instead of the refocused echo sequence in four-pulse PELDOR. This allows the observation time window of the dipolar evolution function to be prolonged. The maximum possible dipolar defocusing for such a sequence is achieved if a pump pulse inverts the coupled second spin whenever the observed spin is refocused. Rectangular pulses can only accomplish such multiple inversions efficiently if the inhomogeneous broadening is much smaller than the excitation field strength of the pump pulses due to the rapidly decaying inversion efficiency toward larger offset frequencies (Figure 21.2). The spectral width of nitroxide radicals is substantially larger than the typical excitation field strength achievable at X- or Q-band frequencies, giving rise to artifacts in the dipolar evolution function when several rectangular pump pulses are used.[20] A CP-PELDOR experiment with three rectangular refocusing pulses for the probe spin and three sech/tanh inversion pulses for the pump spin was performed at Q-band frequencies on a betaine transporter membrane protein trimer with a significantly prolonged time window and improved signal-to-noise ratio, SNR (Figure 21.12).

21.4.2 Shaped Pulses for Generating Transverse Magnetization

Designing broadband shaped pulses for generating transverse magnetization is more demanding, especially if phase coherence for spins with different Larmor resonance frequencies is also required at a specific time after the pulse. A BEBOP (broadband excitation by optimized pulses) shaped pulse developed by OCT (see Section 21.2.5) for NMR applications[53] was adapted and used at X-band frequencies for FT-EPR.[17] This shaped pulse has superior bandwidth compared to classical rectangular pulses with the same maximum B_1 field strength, allowing

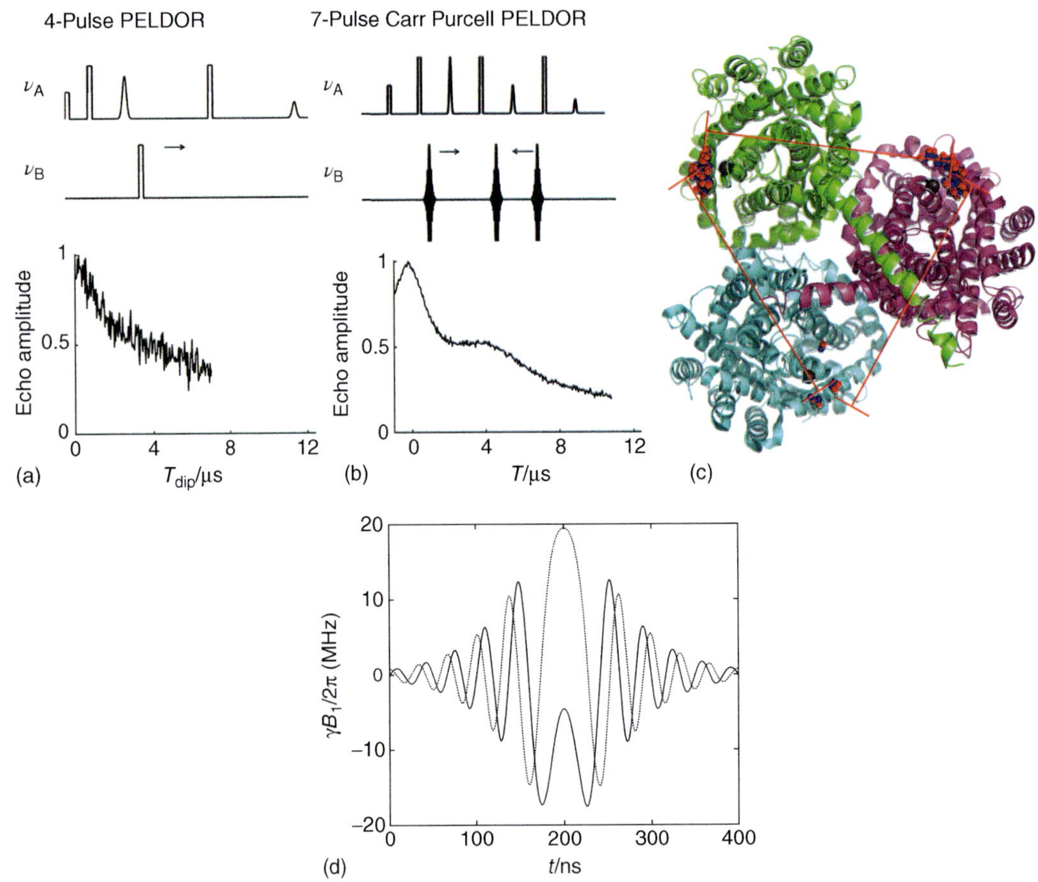

Figure 21.12. Comparison of dipolar evolution time traces recorded with a classical four-pulse PELDOR (a) and a seven-pulse CP-PELDOR (b) sequence at Q-band frequencies on a trimeric betaine membrane transporter complex *BetP* (c).[51] Time-domain shape of the sech/tanh pulse used as pump pulse in the seven-pulse CP-PELDOR experiment (d). (Reprinted with permission from Spindler, P. E.; Waclawska, I.; Endeward, B.; Plackmeyer, J.; Ziegler, C.; Prisner, T. F. J. Phys. Chem. Lett. 2015, 6, 4331. © 2015, American Chemical Society)

the EPR spectra of hyperfine resolved radicals in solution to be recorded with much better amplitude accuracy (Figure 21.13).

The BEBOP pulse has nonlinear phase dispersion for different offset frequencies. Therefore, a FT of the magnitude has to be taken to collect the FT-EPR spectra. Adiabatic and passage pulses (as discussed above) can also be utilized to create and refocus transversal magnetization. As they sweep through the resonance frequencies rather than turning the magnetization of spins at all resonance offsets at once, each spin packet is actually flipped at a different point in time, causing linear phase dispersion for a linearly swept pulse. Refocusing this phase dispersion can be achieved with an echo sequence where the refocusing pulse has half the length of the first excitation pulse.[11] In this Böhlen–Bodenhausen scheme, the offset-dependent linear phase dispersion produced by the first broadband shaped $\pi/2$ excitation pulse is compensated by a second broadband π pulse. The second pulse has to be swept at twice the rate of the first pulse to accomplish refocusing. It is common to make the refocusing pulse half as long as the excitation pulse and giving both pulses equal bandwidth. It was shown that such pulses allow excitation of the full inhomogeneous linewidth of nitroxide radicals at X-band frequencies[18,22] and phase-coherent echo signal detection. The same principle can also be applied to refocused and stimulated echoes and thus to dipolar and hyperfine spectroscopy.

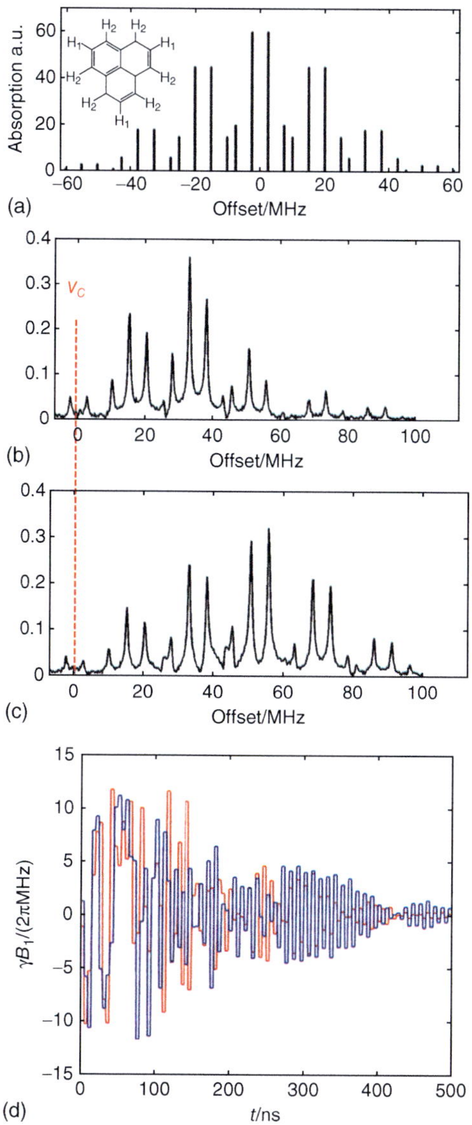

Figure 21.13. FT-EPR spectra of the PNT radical. (a) Stick spectra of the hyperfine pattern of the PNT radical. (b) FT-EPR spectra recorded after a 20 ns rectangular $\pi/2$ pulse. (c) FT-EPR spectra recorded after a 500 ns BEBOP pulse (the real (blue) and imaginary (red) parts of the BEBOP pulse are shown in (d)). The MW excitation frequency ν_c (red line in b and c) is 51 MHz off resonance with respect to the center of the spectrum.[17] (Reprinted from J. Magn. Reson., 218, Spindler, P. E.; Zhang, Y.; Endeward, B.; Gershernzon, N.; Skinner, T. E.; Glaser, S. J.; Prisner, T. F., Shaped optimal control pulses for increased excitation bandwidth in EPR, 49. © 2012, with permission from Elsevier)

Application of such pulses to 1D ESEEM (electron spin-echo envelope modulation) and 2D HYSCORE (hyperfine sublevel correlation spectroscopy) has been demonstrated at X-band frequencies.[54] ESEEMs from copper hyperfine couplings could be observed due to the large bandwidth of 500 MHz of the chirp pulses (Figure 21.14).

FT of the echo signal gives an additional dimension to this spectrum as the echo signal contains all the spectral information, which usually would have to be collected by a step-by-step field sweep. Thus, a 3D HYSCORE has been recorded directly in the time domain for the first time.

Shaped pulses consisting of a combination of a sech/tanh pulse[42] with a WURST amplitude profile[39] have been used in the SIFTER (single frequency technique for refocusing dipolar couplings) experiment,[55] where a solid-echo-based pulse sequence is used to detect dipolar couplings between two electron spins. Despite the fact that full refocusing for this echo

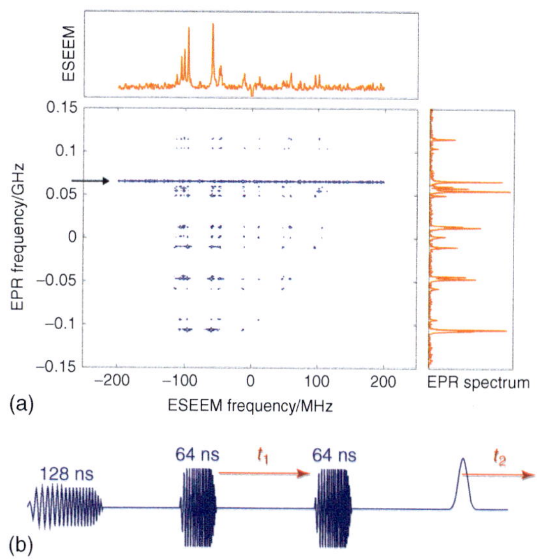

Figure 21.14. (a) Experimental 2D EPR three-pulse ESEEM spectra of Cu(II) in a rutile crystal with pulse lengths of 128, 64, and 64 ns for the stimulated echo sequence. The sweep range of the adiabatic pulse was 500 MHz. The line at 0.06 GHz (arrow) is an artifact from a dielectric resonance of the rutile crystal. The pulse sequence is shown in (b).[54] (Reproduced from Segawa, T. F.; Doll, A.; Pribitzer, S.; Jeschke, G., J. Chem. Phys., 2015, 143, 044201. http://dx.doi.org/10.1063/1.4927088, with the permission of AIP Publishing)

sequence is not obtained with the adapted broadband frequency-swept pulses, considerable improvement of the time traces containing the dipolar oscillations was observed.[18] A comparison between the experiment with classical rectangular pulses and broadband pulses performed at X-band frequencies on a bis-nitroxide model compound is shown in Figure 21.15. The artifacts visible in the SIFTER experiment with rectangular pulses, arising from the incomplete excitation and inversion of the nitroxide spin system by the rectangular pulses, are absent in the broadband pulse experiment. The modulation depth and the signal-to-noise ratio of the SIFTER experiment are superior to the four-pulse PELDOR experiment under these experimental conditions.

Another advantage of shaped pulses is the opportunity to compensate for B_1 inhomogeneity over the sample. This was the main purpose of the BIR4 (B_1 insensitive rotation) composite pulses,[56] consisting of four adiabatic half-passage pulses with tanh amplitude

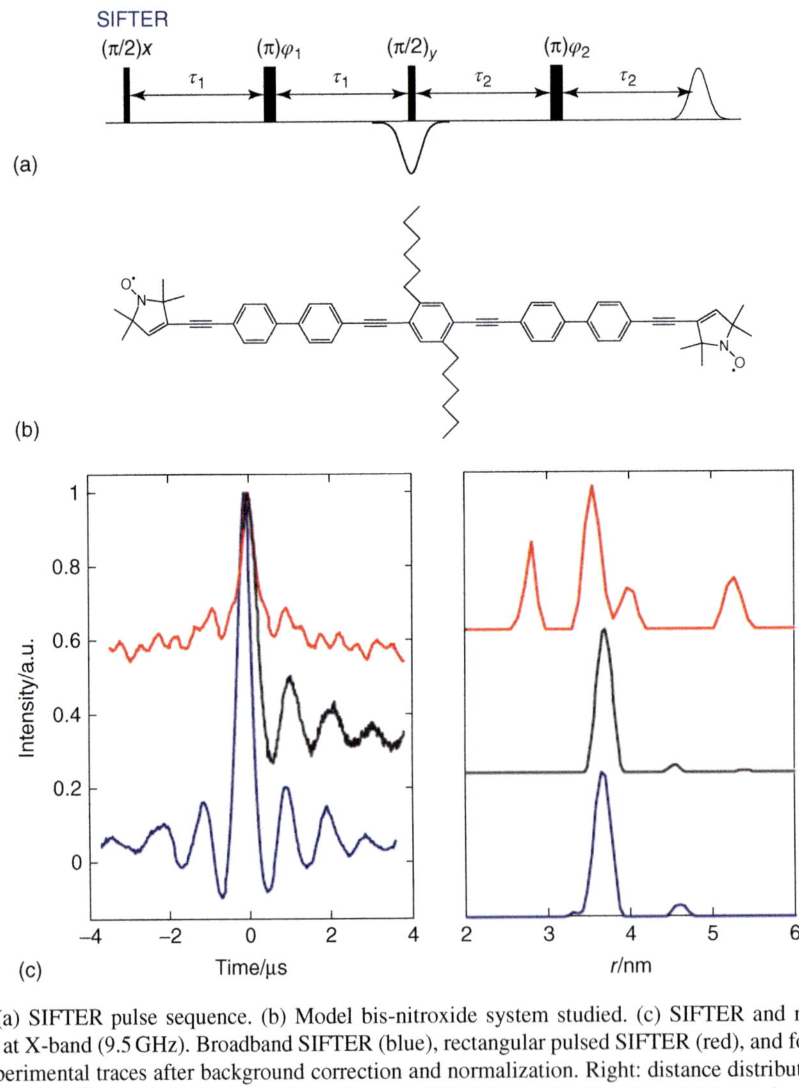

Figure 21.15. (a) SIFTER pulse sequence. (b) Model bis-nitroxide system studied. (c) SIFTER and reference PELDOR results measured at X-band (9.5 GHz). Broadband SIFTER (blue), rectangular pulsed SIFTER (red), and four-pulse PELDOR (black). Left: experimental traces after background correction and normalization. Right: distance distributions obtained from Tikhonov regularization.[18] (Reprinted from J. Magn. Reson., 250, Schöps, P.; Spindler, P. E.; Marko, A.; Prisner, T. F., Broadband spin echoes and broadband SIFTER in EPR, 55. © 2015, with permission from Elsevier)

and tan frequency modulation, used in an EDMR (see Chapter 26) setup.[57] As the stripline MW excitation is very broadband in this design, no distortions from MW resonant structures have to be taken into account. Rabi oscillations show the compensation for the B_1 inhomogeneity of the antenna (Figure 21.16). A modified electrically detected Davies ENDOR (electron-nuclear double resonance) sequence employing such pulses also on the RF channel was applied to ionized phosphorus donors in silicon.

21.5 FUTURE OUTLOOK

The recent applications selected and illustrated in the previous section demonstrate the potential to improve the performance of EPR experiments with shaped pulses. The experimental setups required for such experiments have only become commercially available and been installed in the past few years, and so many more applications and new methodologies can be expected in the future. One of the limiting aspects for shaped pulses is the experimental bandwidth, which is mainly restricted by the MW resonator. Lowering the quality factor of the resonator increases this bandwidth, but also lowers the MW field strength and the sensitivity. In this respect, applications at high magnetic fields in the subterahertz frequency range might be very attractive. MW resonators at frequencies above 100 GHz have bandwidths much larger than the excitation bandwidth of classical rectangular pulses. This is due to the low MW power available at subterahertz frequencies from pulsed and phase-coherent MW sources. In such cases, the use of shaped broadband pulses could considerably improve the performance of the spectrometer. The same might be true for applications with microresonators, which also have low intrinsic quality factors. In this case, it is not the MW field strength but the B_1 inhomogeneity that is of concern; but this can be compensated using OCT pulses. Finally, shaped pulses might allow new types of EPR experiments to be performed which were, up to now, out of reach or only accessible under specific conditions. This category includes experiments that make use of the direct Fourier domain or multiquantum experiments with more than two electron spins. The transfer of pulse sequences and pulse shapes from NMR is nevertheless not straightforward in these cases. Anisotropic spin interactions, such as hyperfine, dipolar, zero-field splitting, and g-tensor, have to be included in the optimization procedure,

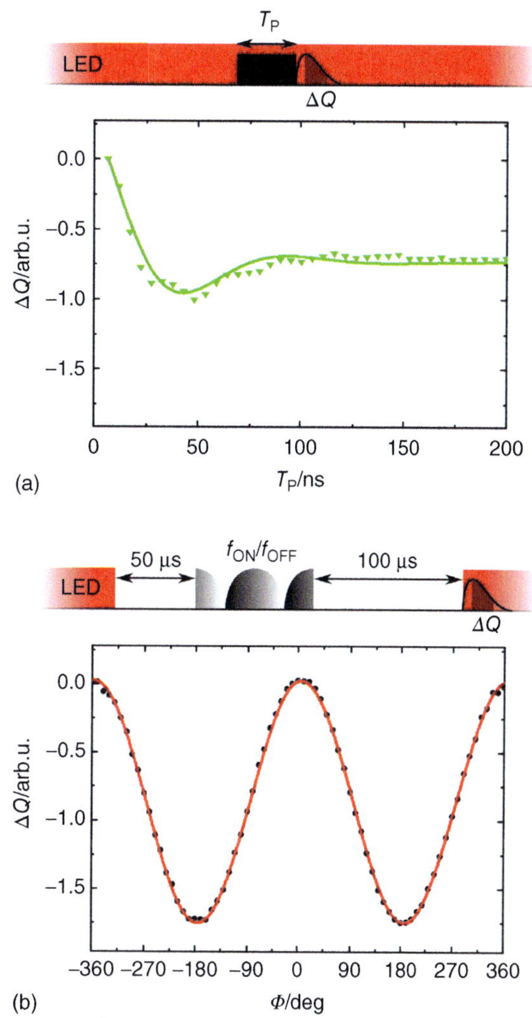

Figure 21.16. Rabi oscillation on the high-field phosphorus resonance of Si : P detected with rectangular pulses (a) and with the BIR4 pulse (b). In the second experiment, the microwave is applied in the dark and the EDMR signal is measured during the transient induced by the optical refill pulse (indicated as LED in the pulse sequences). The resulting EDMR intensity (ΔQ) is plotted versus the rotation angle ϕ of the BIR4 pulse on a scale corresponding to the time evolution shown in (a).[57] (Reprinted from J. Magn. Reson., 254, Hrubesch, F. M.; Braunbeck, G.; Voss, A.; Stutzmann, M.; Brandt, M. S., Broadband electrically detected magnetic resonance using adiabatic pulses, 62. © 2015, with permission from Elsevier)

making it more demanding compared to liquid-state NMR. It should be noted that shaped pulses already (and will further) contribute to many other application areas that have not been addressed explicitly in this chapter, including EPR imaging and microscopy, quantum computing, and optical or AFM (atomic force microscopy)-detected EPR of single molecules.

ACKNOWLEDGMENTS

The authors like to thank Steffen Glaser (TU Munich) and Thomas Skinner (Wright State University) and their coworkers for the cooperation in setting up the OCT pulses for experiments performed in Frankfurt. Financial support of the project by the German Research Society (DFG, SPP 1601: *New Frontiers in Sensitivity in EPR Spectroscopy*) is gratefully acknowledged.

REFERENCES

1. R. J. Blume, *Phys. Rev.*, 1867, **1958**, 109.
2. J. T. Arnold, S. S. Dharmatti, and M. E. Packard, *J. Chem. Phys.*, 1951, **19**, 507.
3. W. Warren, *Science*, 1988, **242**, 878.
4. W. S. Warren, in, eds M. S. Silver and S. W. John, Adv. Magn. Reson.: 1988, Vol. **12**, 1.
5. S. McDonald and W. S. Warren, *Concept Magn. Reson.*, 1991, **3**, 55.
6. R. R. Ernst, *J. Magn. Reson.*, 1970, **3**, 10.
7. B. Blümich, *Prog. Nucl. Magn. Reson. Spectrosc.*, 1987, **19**, 331.
8. C. Bauer, R. Freeman, T. Frenkiel, J. Keeler, and A. J. Shaka, *J. Magn. Reson.*, 1984, **58**, 442.
9. M. H. Levitt, *Prog. Nucl. Magn. Reson. Spectrosc.*, 1986, **18**, 61.
10. J. Baum, R. Tycko, and A. Pines, *Phys. Rev. A*, 1985, **32**, 3435.
11. J.-M. Bohlen, M. Rey, and G. Bodenhausen, *J. Magn. Reson.*, 1989, **84**, 191.
12. H. Corns and S. K. Branch, *NMR Spectroscopy in the European and US Pharmacopeias*, John Wiley & Sons, Ltd., 2015.
13. T. Prisner and K.-P. Dinse, *J. Magn. Reson.*, 1989, **84**, 296.
14. R. H. Crepeau, A. Dulcic, J. Gorcester, T. R. Saarinen, and J. H. Freed, *J. Magn. Reson.*, 1989, **84**, 184.
15. P. Schosseler, T. Wacker, and A. Schweiger, *Chem. Phys. Lett.*, 1994, **224**, 319.
16. T. Wacker and A. Schweiger, *Chem. Phys. Lett.*, 1991, **186**, 27.
17. P. E. Spindler, Y. Zhang, B. Endeward, N. Gershernzon, T. E. Skinner, S. J. Glaser, and T. F. Prisner, *J. Magn. Reson.*, 2012, **218**, 49.
18. P. Schöps, P. E. Spindler, A. Marko, and T. F. Prisner, *J. Magn. Reson.*, 2015, **250**, 55.
19. A. Doll, S. Pribitzer, R. Tschaggelar, and G. Jeschke, *J. Magn. Reson.*, 2013, **230**, 27.
20. A. Doll, M. Qi, N. Wili, S. Pribitzer, A. Godt, and G. Jeschke, *J. Magn. Reson.*, 2015, **259**, 153.
21. A. Doll, M. Qi, S. Pribitzer, N. Wili, M. Yulikov, A. Godt, and G. Jeschke, *Phys. Chem. Chem. Phys.*, 2015, **17**, 7334.
22. A. Doll and G. Jeschke, *J. Magn. Reson.*, 2014, **246**, 18.
23. P. P. Borbat, E. R. Georgieva, and J. H. Freed, *J. Phys. Chem. Lett.*, 2013, **4**, 170.
24. T. Kaufmann, T. J. Keller, J. M. Franck, R. P. Barnes, S. J. Glaser, J. M. Martinis, and S. Han, *J. Magn. Reson.*, 2013, **235**, 95.
25. G. D. Fuchs, V. V. Dobrovitski, D. M. Toyli, F. J. Heremans, and D. D. Awschalom, *Science*, 2009, **326**, 1520.
26. S. Wimperis, *J. Magn. Reson. A*, 1994, **109**, 221.
27. H. Wu, E. M. Gauger, R. E. George, M. Möttönen, H. Riemann, N. V. Abrosimov, P. Becker, H.-J. Pohl, K. M. Itoh, M. L. W. Thewalt, and J. J. L. Morton, *Phys. Rev. A*, 2013, **87**, 032326.
28. N. Devasahayam, R. Murugesan, K. Matsumoto, J. B. Mitchell, J. A. Cook, S. Subramanian, and M. C. Krishna, *J. Magn. Reson.*, 2004, **168**, 110.
29. G. G. Brown, B. C. Dian, K. O. Douglass, S. M. Geyer, S. T. Shipman, and B. H. Pate, *Rev. Sci. Instrum.*, 2008, **79**, 053103.
30. Y. Hovav, A. Feintuch, S. Vega, and D. Goldfarb, *J. Magn. Reson.*, 2014, **238**, 94.
31. I. Kaminker, A. Potapov, A. Feintuch, S. Vega, and D. Goldfarb, *Phys. Chem. Chem. Phys.*, 2009, **11**, 6799.
32. Y. Asada, R. Mutoh, M. Ishiura, and H. Mino, *J. Magn. Reson.*, 2011, **213**, 200.

33. H. J. Hogben, M. Krzystyniak, G. T. P. Charnock, P. J. Hore, and I. Kuprov, *J. Magn. Reson.*, 2011, **208**, 179.
34. S. Pribitzer, A. Doll, and G. Jeschke, *J. Magn. Reson.*, 2016, **263**, 45.
35. B. Luy, K. Kobzar, T. E. Skinner, N. Khaneja, and S. J. Glaser, *J. Magn. Reson.*, 2005, **176**, 179.
36. A. Schweiger and G. Jeschke, *Principles of Pulse Electron Paramagnetic Resonance*, Oxford University Press: 2001.
37. Y. Tabuchi, M. Negoro, K. Takeda, and M. Kitagawa, *J. Magn. Reson.*, 2010, **204**, 327.
38. M. A. Bernstein, K. F. King, and X. J. Zhou, *Handbook of MRI pulse sequences*, Academic Press: Burlington, 2004, 1.
39. E. Kupče and R. Freeman, *J. Magn. Reson.*, 1995, **115**, 273.
40. J.-M. Böhlen, I. Burghardt, M. Rey, and G. Bodenhausen, *J. Magn. Reson.*, 1990, **90**, 183.
41. P. E. Spindler, S. J. Glaser, T. E. Skinner, and T. F. Prisner, *Angew. Chem., Int. Ed.*, 2013, **52**, 3425.
42. M. S. Silver, R. I. Joseph, and D. I. Hoult, *Phys. Rev. A*, 1985, **31**, 2753.
43. D. Rosenfeld and Y. Zur, *J. Magn. Reson.*, 1998, **132**, 102.
44. N. I. Gershenzon, K. Kobzar, B. Luy, S. J. Glaser, and T. E. Skinner, *J. Magn. Reson.*, 2007, **188**, 330.
45. T. E. Skinner, M. Braun, K. Woelk, N. I. Gershenzon, and S. J. Glaser, *J. Magn. Reson.*, 2011, **209**, 282.
46. D. D. Rife and J. Vanderkooy, *J. Audio Eng. Soc.*, 1989, **37**, 419.
47. O. Schiemann and T. F. Prisner, *Q. Rev. Biophys.*, 2007, **40**, 1.
48. R. E. Martin, M. Pannier, F. Diederich, V. Gramlich, M. Hubrich, and H. W. Spiess, *Angew. Chem., Int. Ed.*, 1998, **37**, 2833.
49. M. Pannier, S. Veit, A. Godt, G. Jeschke, and H. W. Spiess, *J. Magn. Reson.*, 2000, **142**, 331.
50. M. K. Bowman and A. G. Maryasov, *J. Magn. Reson.*, 2007, **185**, 270.
51. P. E. Spindler, I. Waclawska, B. Endeward, J. Plackmeyer, C. Ziegler, and T. F. Prisner, *J. Phys. Chem. Lett.*, 2015, **6**, 4331.
52. H. Y. Carr and E. M. Purcell, *Phys. Rev.*, 1954, **94**, 630.
53. T. E. Skinner, T. O. Reiss, B. Luy, N. Khaneja, and S. J. Glaser, *J. Magn. Reson.*, 2003, **163**, 8.
54. T. F. Segawa, A. Doll, S. Pribitzer, and G. Jeschke, *J. Chem. Phys.*, 2015, **143**, 044201.
55. G. Jeschke, M. Pannier, A. Godt, and H. W. Spiess, *Chem. Phys. Lett.*, 2000, **331**, 243.
56. R. S. Staewen, A. J. Johnson, B. D. Ross, T. Parrish, H. Merkle, and M. Garwood, *Invest. Radiol.*, 1990, **25**, 559.
57. F. M. Hrubesch, G. Braunbeck, A. Voss, M. Stutzmann, and M. S. Brandt, *J. Magn. Reson.*, 2015, **254**, 62.

PART D
Special Techniques

Chapter 22

Pulse Techniques for Quantum Information Processing

Gary Wolfowicz[1] and John J.L. Morton[2]

[1] *University of Chicago, Chicago, IL, USA*
[2] *University College London, London, UK*

Since the early 1980s, it has been recognized[1,2] that representing information within quantum systems offers the potential for profound enhancements in information processing.[3,4] By initializing, controlling, and measuring arrays of coupled quantum systems, it is possible, in principle, to build efficient tools for the simulation of quantum systems, ranging from purpose-built analog models through to 'digital' quantum simulators, which follow algorithmic approaches to solve the evolution of a Hamiltonian. So-called universal quantum computers possess a sufficient set of logical operations, or gates, to perform arbitrary quantum algorithms such as those for factoring large numbers and searching datasets, and represent a generalization of the concept of a Turing machine. Motivated by these opportunities, the past 20 years has seen a great deal of experimental research into realizing quantum information processors, based on a wide range of physical systems, including spins, atoms, ions, and superconducting circuits. More recently, applications of individual highly coherent quantum systems as sensors have been identified, potentially yielding nearer-term technologies. In this chapter, we will review pulse techniques in electron paramagnetic resonance (EPR) related to research in quantum information processing and sensing using electron spins.

EPR Spectroscopy: Fundamentals and Methods.
Edited by Daniella Goldfarb and Stefan Stoll

ISBN: 978-1-119-16299-5
Also published in eMagRes (online edition)
DOI: 10.1002/9780470034590.emrstm1521

22.1 SPIN QUBITS

In analogy to digital binary logic, the fundamental unit of quantum information is the quantum bit, or qubit, consisting of a two-level quantum system labeled with the basis states $|0\rangle$ and $|1\rangle$. The ability of a quantum system to exist in a superposition of states, with some phase relationship, provides the qubit with a much richer state space to occupy than its classical counterpart. We can write the state of the qubit in general as

$$|\psi\rangle = \cos\frac{\theta}{2}|0\rangle + e^{i\phi}\sin\frac{\theta}{2}|1\rangle \qquad (22.1)$$

which maps out the surface of the Bloch sphere (Figure 22.1a).

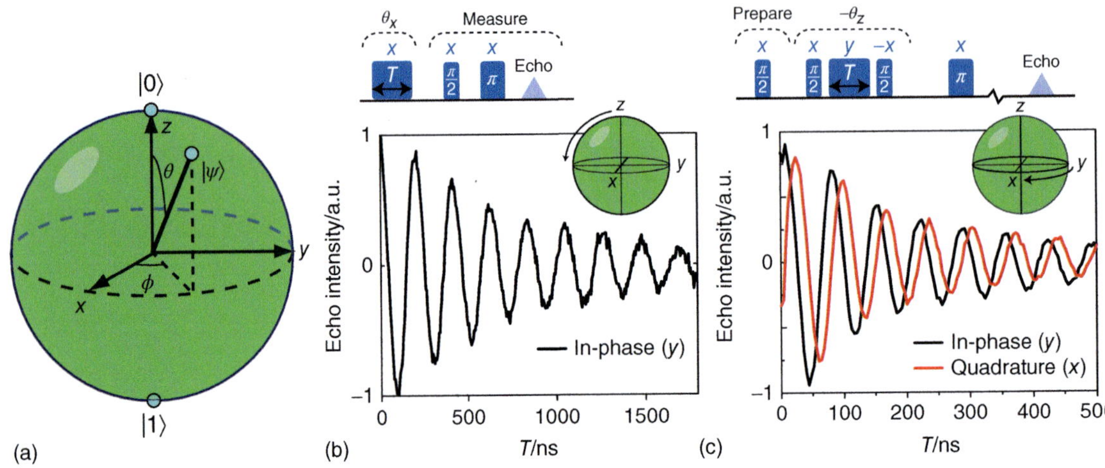

Figure 22.1. Single-qubit representation and control. (a) The Bloch sphere representation of a qubit state $|\psi\rangle$, defined by angles θ and ϕ. Spin-down and spin-up eigenstates can be mapped onto the qubit states $|0\rangle$ and $|1\rangle$. (b) Rotations of an electron spin qubit about the x-axis can be performed using a simple resonant microwave pulse, observable (using a subsequent 2-pulse echo sequence) as Rabi oscillations where the duration of the pulse is swept. (c) Rotations about the z-axis can be performed by decomposing into x- and y-rotations. Experimental data taken using P donors in Si, at 8 K

A spin-1/2 in a magnetic field is a natural realization of a qubit, with the eigenstates $\left|m_S = -\frac{1}{2}\right\rangle \equiv |\downarrow\rangle \equiv |0\rangle$ and $\left|m_S = +\frac{1}{2}\right\rangle \equiv |\uparrow\rangle \equiv |1\rangle$ (representing, respectively, the spin projection, the spin orientation, and the qubit state – (see Chapter 8). Much early progress was made in exploring quantum algorithms and control using liquid-state NMR, including the application of Shor's factoring algorithm to the number 15 in Ref. 5. However, thermal initialization of NMR systems becomes exponentially harder with increasing number of qubits, limiting the scalability of the approach. Focus on spin qubits has therefore shifted to a large extent to electron spins, though nuclear spin qubits remain attractive for memory applications due to their long quantum state lifetimes of up to 6 h.[6] Electron spins being explored for quantum information applications include the nitrogen vacancy (NV) and other color centers in diamond[7,8] and SiC,[9] donors in silicon,[10] and quantum dots in silicon,[11] GaAs,[12] and other semiconductors.[13] Molecular electron spins[14–16] have also been explored as useful systems to test concepts in quantum information, owing to their reproducible nature and the ability, in principle, to engineer coupled systems.

22.1.1 Single-qubit Gates

In classical binary logic, the only nontrivial operation, or 'gate', which can be applied to a single bit is the NOT gate (where $0 \rightarrow 1$ and $1 \rightarrow 0$). Thanks to the much richer state space of the qubit (Figure 22.1a), there is a much greater set of single-qubit gates, which (neglecting a global phase) can be represented as rotations of the Bloch sphere around some axis and by some angle. In pulse EPR, and in the rotating frame, rotations about axes in the x–y-plane can be trivially performed through simple microwave pulses, where the pulse phase determines the rotation axis, while the pulse power and duration set the angle (Figure 22.1b). Rotations about other axes (and thus any unitary operation) can be performed by decomposition into a series of x and y rotations ($\hat{\theta}_{x,y}$), or through more elaborate methods such as pulse shaping, as discussed in Chapter 21. For example, a z-rotation (Figure 22.1c) can be implemented as

$$\hat{\theta}_z = (\pi/2)_x(\theta)_y(\pi/2)_{-x} \tag{22.2}$$

while the Hadamard gate, a common single-qubit gate used to generate superposition states (and representing a rotation by π about an axis in the direction $x + z$), can

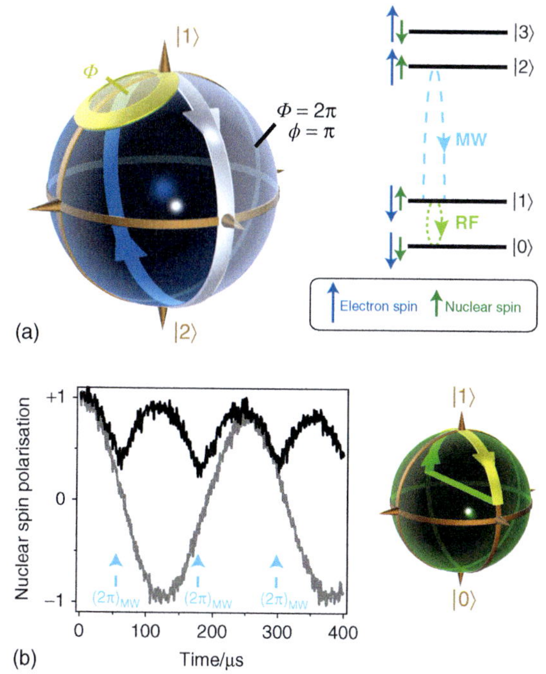

Figure 22.2. Aharonov–Anandan geometric phase gates using coupled electron and nuclear spins. (a) A microwave pulse may be applied resonant with the $|1\rangle \leftrightarrow |2\rangle$ transition to drive states $|1\rangle$ and $|2\rangle$ around a closed trajectory on the corresponding Bloch sphere. The solid angle, Φ, enclosed by this trajectory defines the geometric phase $\phi = \Phi/2$ acquired by both the states. Example closed trajectories are shown for the cases where the transition is driven resonantly (white) or nonresonantly (yellow). (b) The effect of this phase shift on a qubit defined by states $|0\rangle$ and $|1\rangle$ can be observed by driving nuclear spin Rabi oscillations without (gray) and with (black) fast phase gates. After each 2π microwave pulse applied on the electron spin, the nuclear spin acquires a π phase and is prevented from making a complete rotation around the Bloch sphere. Experiment performed using N@C_{60}. (Reprinted by permission from Macmillan Publishers Ltd: Nature Physics, Ref. 14 © 2006)

be decomposed as follows:

$$\hat{U}_{\text{Hadamard}} = \frac{1}{\sqrt{2}}\begin{pmatrix} 1 & 1 \\ 1 & -1 \end{pmatrix} = \mathrm{i}(\pi)_x \left(\frac{\pi}{2}\right)_y \qquad (22.3)$$

An additional approach to single-qubit operations employs the geometric phase acquired by spins driven through some closed trajectory. In particular, when a spin qubit is driven through a cyclic evolution (i.e., the initial and final states coincide on the Bloch sphere), it acquires a geometric, or 'Berry', phase that is equal to half the solid angle enclosed by its evolution on the Bloch sphere (Figure 22.2).[17] Because this phase is global (the same phase is acquired by both eigenstates), it serves no purpose for an isolated spin-1/2 qubit. However, given the availability of additional levels (e.g., thanks to hyperfine coupling, or for $S > 1/2$), it can be exploited to create the so-called Aharonov–Anandan (AA) phase gate, which is an effective z-rotation of the qubit.[18] AA phase gates have been performed on coupled electron–nuclear spin systems by applying a selective 2π microwave pulse to an electron spin transition, resulting in a nuclear spin phase gate that is orders of magnitude faster than that which could be achieved using conventional ENDOR/NMR pulse sequences.[14]

22.1.2 Quantum State Tomography

The complete state of a qubit, taking into account effects such as decoherence, must in general be described by the density matrix (see Chapter 8), $\hat{\boldsymbol{\sigma}}$, which can be decomposed as follows:

$$\hat{\boldsymbol{\sigma}} = \frac{\mathbf{1}}{2} + r_x\hat{S}_x + r_y\hat{S}_y + r_z\hat{S}_z \qquad (22.4)$$

where $\mathbf{1}$ is the identity matrix, $\hat{S}_{x,y,z}$ are the spin operators, and $r_{x,y,z}$ are the real coefficients that determine the state. In order to characterize the practical implementation of the gates described above, it is necessary to measure the resulting state of the qubit, determining its density matrix through a process known as state tomography. Owing to the collapsing effect of measurement on quantum systems, state tomography is achieved only by combining many repetitions of an experiment on a single qubit, or using an ensemble of many such qubits. In conventional EPR, values proportional to r_x and r_y can be determined upon formation of a spin echo (Figure 22.3). A corresponding value for r_z cannot be directly measured simultaneously (at least not without some additional measurement technique, such as a SQUID), but its value can be inferred if the echo is followed by an ideal $\pi/2$-pulse, effectively mapping the desired $\hat{S}_z$ component onto the $\hat{S}_x$ observable.

The measured values $r'_{x,y,z}(= \alpha r_{x,y,z})$ must then be normalized to produce a physical density matrix–this can be done in various ways, but a common approach in ensemble EPR is to make the 'pseudo-pure' approximation,[19] which works as follows: At typical

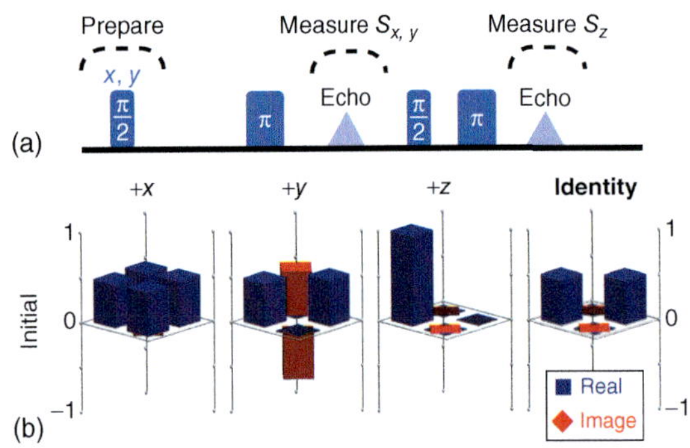

Figure 22.3. Quantum state tomography using EPR pulses to obtain the density matrix of a qubit. (a) Two consecutive Hahn echo sequences provide separately the x, y, and z components of the qubit state, necessary to reconstruct the density matrix. (b) Extracted spin-1/2 density matrices (matrix elements plotted on a vertical scale), given different state preparation pulses

temperatures and magnetic fields, electron spins are in a weakly polarized state such that $r_{x,y,z} \ll 1$. However, it is possible to neglect the large **1** component of the density matrix, as it is invariant under unitary transformations, and treat the remaining part of the state as if it were pure. In this way, we can assume a given state to be 'pseudo-pure' and impose the normalization condition for a pure state, $||\boldsymbol{r}|| = 1$, in order to extract a value of the proportionality constant, α. This assumption is usually applied to the initial state in the pulse sequence, such that subsequent states produced (which in general will not be pure) are normalized using the same value of α.

As multiple qubits are coupled together, the state space exponentially expands. This gives an illustration of the increased power of quantum information processors, and also brings new challenges in performing state tomography due to the number of operations required (see Section 22.6). Having discussed methods to extract the full state of a spin qubit, we next turn to ways in which this state can be corrupted–first through 'natural' processes in the system and its environment (decoherence) and second through the nonideality of the applied operations (pulse errors).

22.2 DECOHERENCE

One of the primary criteria in choosing a particular physical representation of a qubit or quantum sensor is maximizing the coherence lifetime: decoherence, defined as undesired phase shifts in the qubit state, are errors in quantum computation, while a short coherence time prevents long-phase accumulation necessary for sensing applications.

Although electron spin (2-pulse echo) coherence times (T_2) as long as several hundred microseconds have been observed in certain molecular systems,[20] the longest coherence lifetimes have emerged from defect spins in solid-state materials, in particular those with a low natural abundance of nuclear spins such as carbon- and silicon-based materials. The ability to grow pure crystals with low charge defect concentration and the possibility of isotopically enriching samples to minimize the nuclear spin concentration have led to measured electron spin coherence times of up to 1.8 ms for NV centers in diamond at room temperature[21] and up to 20 ms for P donors in silicon at 6 K.[22]

Notably, when measuring such long times in spin ensembles by conventional EPR, instantaneous diffusion often becomes the limiting decoherence mechanism[23] (see Chapter 11). This can be confirmed by measuring the decoherence time as a function of the refocusing pulse angle in the 2-pulse echo sequence (Figure 22.4). Instantaneous diffusion[23] can be avoided using alternative spin measurement techniques with greater sensitivity such as optically detected magnetic resonance, where more dilute ensembles,[24] or indeed single spins,[7,25] can be measured.

The T_2 values measured in bulk crystals can become orders of magnitude shorter when the electron spins are brought near interfaces, as they often must for practical applications. There is good motivation therefore to develop methods for extending coherence times and protecting the spin qubit from noise sources. In the following section, we describe two approaches: first, using a technique known as dynamical decoupling (DD), which is a natural extension of the 2-pulse (Hahn) echo to a large number of refocusing pulses and second, by identifying spin transitions that are inherently insensitive to external perturbations such as magnetic field noise. These methods can be used to achieve gains in coherence time between one and two orders of magnitude.

22.2.1 Dynamical Decoupling (DD)

The single refocusing π-pulse in the Hahn echo is the most basic example of DD, able to protect the

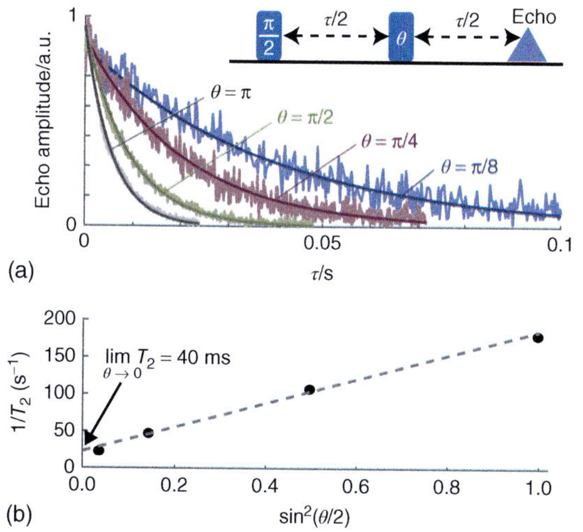

Figure 22.4. Instantaneous diffusion. (a) Measuring echo decay curves for different refocusing pulse angles (θ), as shown in the pulse sequence (inset), can be used extract the instantaneous diffusion contribution to T_2. (Note: Curves are normalized to unit intensity (for $\tau \approx 0$) and the actual echo intensity falls as θ decreases). (b) Extrapolating the measured decay constants (T_2) for the case of $\theta \to 0$ provides an estimate for T_2 in the absence of instantaneous diffusion (i.e. in the dilute spin limit).[22] Experiment performed at 5 K using Bi donors in silicon, at a concentration of 4×10^{15} cm^{-3}

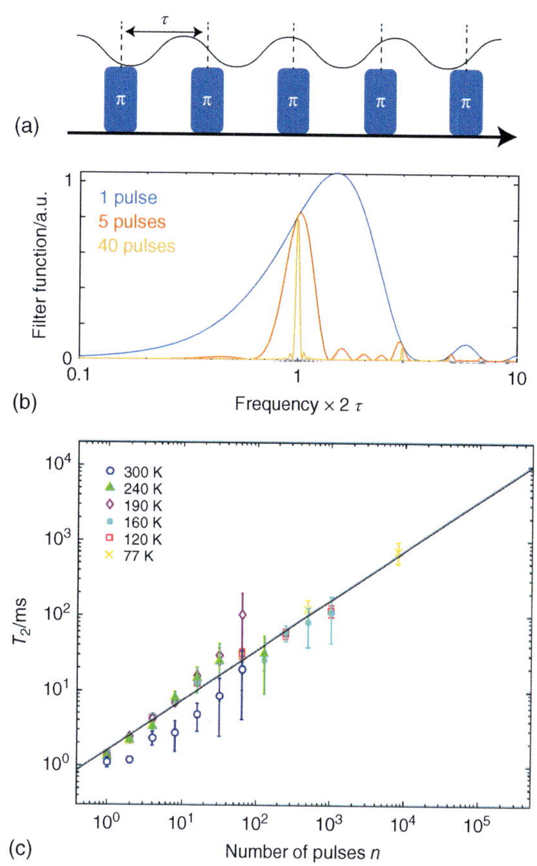

Figure 22.5. Dynamical decoupling (DD). (a) A typical DD sequence consists of repeated π refocusing pulses whose inverse period $1/\tau$ must be smaller than the frequency of the noise (black curve). (b) DD acts as a filter function, where larger pulse numbers lead to narrower bandpass filters. (c) Electron spin coherence time T_2 as a function of pulse number in an NV center showing an increase of up to three orders of magnitude using 10 000 DD pulses. (Reprinted by permission from Macmillan Publishers Ltd: Nature Communications, Ref. 29 © 2013)

spin qubit from dephasing caused by variations in the spin transition frequency that are static (at least, on the timescale of the experiment). More general DD sequences can be used to protect the spin from more elaborate noise sources and are adaptable to various noise spectra and spin interactions.[26]

Decoherence from dynamic noise can be suppressed by repeated application of refocusing π-pulses, with an inverse period $1/\tau$ chosen to be less than the high-frequency cut-off of the noise spectrum.[26] Such a sequence corresponds to a band-pass filtering effect centered around harmonics of $f = \frac{1}{2\tau}$, and with a bandwidth narrowing with large pulse number,[27,28] as shown in Figure 22.5. The higher harmonics of f have a lesser contribution to the decoherence as the noise frequency becomes too fast compared to the spin evolution.

DD sequences have been applied with millions of pulses in NMR[30] and up to 10 000 pulses in EPR[29] (Figure 22.5), and in such cases, careful consideration must be given to the effect of pulse errors. In order for the DD pulses not to significantly introduce errors themselves, either pulses with very high fidelity are required or the DD sequence should be robust to pulse error. For example, when the basic DD sequence $\left(-\frac{\tau}{2} - \pi_x - \frac{\tau}{2}-\right)^n$ is applied to an input state oriented along x, the pulse errors are additive leading to rapid corruption of the state. On the other hand, if the same sequence is applied to a $\pm y$ input state, it is robust to pulse errors as they cancel out to first order – indeed, this is why Meiboom and Gill[31] adapted the CP sequence[32] to produce what we now term CPMG: $\left(\frac{\pi}{2}\right)_x\left(-\frac{\tau}{2}-\pi_y-\frac{\tau}{2}-\right)^n$.

DD sequences of most interest in quantum information processing possess robustness to pulse errors for all input states, and examples include the XYXY sequence and higher orders:[33]

$$XY\text{-}4 \equiv \pi_y\pi_x\pi_y\pi_x \tag{22.5}$$

$$XY\text{-}8 \equiv \pi_y\pi_x\pi_y\pi_x\pi_x\pi_y\pi_x\pi_y \tag{22.6}$$

where the delay times τ between the pulses are omitted for clarity.

The above sequences are suitable for decoupling a 'central' spin from some noise source, such as a bath of different spins, but they are not able to refocus interactions between identical spins, which also evolve under the DD pulses. Sequences to decouple such interactions involve $\pi/2$ rotations, and examples to decouple dipolar interactions include the Waugh–Huber–Haeberlen sequence (WAHUHA) and its higher order extension MREV.[34]

DD sequences have applications beyond simply preserving spin-qubit states–their filtering properties can also be used in spectroscopic applications as a 'lock-in amplifier' to pick out certain interactions[35] (see Section 22.5), while they can also be used to indirectly drive rotations of coupled spins. For example, given an electron spin (with operator $\hat{\boldsymbol{S}}$) and a nuclear spin (with operator $\hat{\boldsymbol{I}}$) coupled by an anisotropic interaction, the repeated π-pulses on the electron spin under a DD sequence can, as a result of the $\hat{S}_z\hat{I}_x$ or $\hat{S}_z\hat{I}_y$ terms in the spin Hamiltonian, be seen by the nuclear spin as an oscillating field along the x- or y-axis. If the decoupling period matches the Larmor precession of the nuclear spin, there is a resonant condition, and the nuclear spin rotates while the electron spin coherence is simultaneously preserved.[36]

22.2.2 Parameter-insensitive Transitions

Decoherence can be directly related to the loss of phase caused by uncontrolled fluctuations in the qubit transition frequency, altering the Larmor precession rate. Such fluctuations further originate from variations in free (i.e., externally controllable) parameters present in the spin Hamiltonian, such as the external magnetic or electric field. An important characteristic is the frequency sensitivity of the relevant spin transition to these parameters. A simple and obvious example here is the difference in coherence times typically seen for electron and nuclear spins: the greater decoherence rate of the electron spin is attributed to its much higher gyromagnetic ratio, which (for a system with only the Zeeman interaction) is equal to $d\omega/dB_0$, the first-order dependence of the transition frequency ω with respect to the external magnetic field B_0. In systems where various interactions compete, this dependence can be less than the free-spin gyromagnetic ratio, and even reach zero, to first order, for certain parameter values (e.g., orientation and magnetic field strength). In the case of magnetic field insensitivity, such transitions are called 'zero first-order Zeeman shift (ZEFOZ)' or 'clock transitions (CTs)' and can be used to suppress decoherence (Figure 22.6).

CTs at, or close to, EPR frequencies have been studied in two separate electron–nuclear spin systems, first theoretically in rare-earth-doped crystals[37] and later demonstrated for Bi donors in silicon.[38,39] In the case of Bi donors, the hyperfine interaction between the electron spin ($S = 1/2$) and nuclear spin ($I = 9/2$) is close to 1.5 GHz, resulting in the presence of four CTs at frequencies between 5.2 and 7.3 GHz (Figure 22.6a). For rare-earth ions (Er^{3+}), the weaker hyperfine interaction ($|A| < 700$ MHz, anisotropic) provides CTs below 1 GHz. In both these systems, the limiting decoherence mechanism is spectral diffusion (see Chapters 9 and 11) from either nuclear or electron spins in the environment, arising from dipolar or hyperfine interactions, which can be treated as magnetic field fluctuations. At the clock transition, decoherence due to spectral diffusion is suppressed linearly with decreasing $d\omega/dB_0$, leading to coherence times in Bi donors as long as 2.7 s at 4.3 K, measured using a 2-pulse Hahn echo (Figure 22.6b).[38] In addition to being a strategy to mitigate decoherence, the ability to tune the sensitivity of a spin transition to various parameters can also be used as a way to distinguish various decoherence mechanisms.

22.3 MEASURING GATE FIDELITIES

As described above, the decoherence times in many candidate electron spin-qubit systems have reached such a level that natural decoherence can cease to be a major error source in a quantum algorithm. In such cases, imperfections of the qubit control place a limit on the achievable error rate–for example, the decay in the Rabi oscillations of Figure 22.1(b) and (c) is due to the accumulation of pulse error, and not intrinsic decoherence.

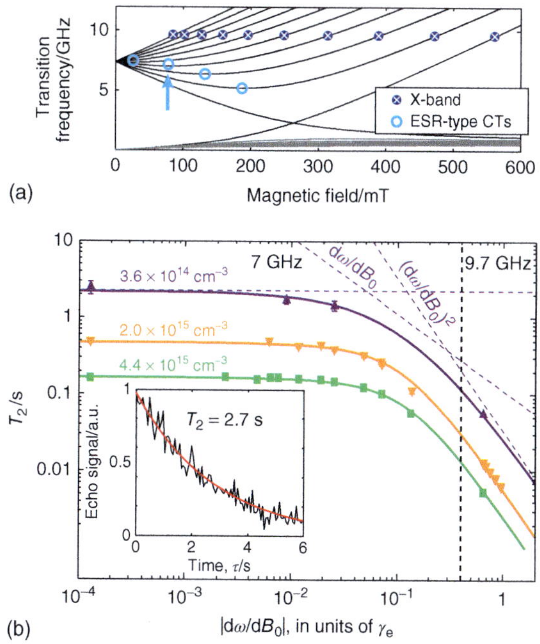

Figure 22.6. EPR clock transitions. (a) The allowed spin transition frequencies for Bi donor spins in silicon ($S = 1/2$, $I = 9/2$, $A \approx 1.5$ GHz) show four 'clock transitions' (CTs) at microwave frequencies (open circles) at which the first-order dependence on magnetic field goes to zero. (b) Electron spin coherence time, T_2, shown as a function of the first-order sensitivity of the transition frequency to magnetic field ($\mathrm{d}\omega/\mathrm{d}B_0$) in Bi donors in silicon, for 3 different Bi concentrations. Measurements for $\mathrm{d}\omega/\mathrm{d}B_0 \approx \gamma_e$ were obtained from the 10 X-band (9.7 GHz) transitions (100–600 mT), while the remaining points were taken in the vicinity of the clock transition highlighted in panel (a) ($B_0 \approx 80$ mT, $\omega_0/2\pi = 7.03$ GHz). Lines are fit to a model with a quadratic dependence in $\mathrm{d}\omega/\mathrm{d}B_0$ due to instantaneous diffusion and a linear dependence due to spectral diffusion. As $\mathrm{d}\omega/\mathrm{d}B_0 \to 0$, the coherence time becomes limited by direct flip-flop processes that are not suppressed at the clock transition and are spin concentration dependent. The inset shows a Hahn echo decay measured at 4.3 K using [Bi] = 3.6×10^{14} cm^{-3}

Practical gate errors in EPR are typically systematic in nature, arising, for example, from nonidealities in the microwave components (switches, IQ mixers, etc.) used to form the pulses, or the finite bandwidth of the cavity. In the usual case in EPR where spin ensembles are addressed, the dominant source of error often arises from the inhomogeneity of the microwave field intensity across the sample, leading to a variation in the achieved rotation angle across the ensemble. In principle, errors arising from field inhomogeneity are not relevant at the level of single-spin control.[25] However, owing to the challenges of confining microwave magnetic fields to small regions of space, many quantum computing schemes based on spins employ microwaves applied to large arrays of spins, each of which can be individually tuned into resonance with the field.[40]

Assuming the experimentally applied operator $\hat{B}$ is unitary, its fidelity $\mathcal{F}$ with respect to an ideal operator $\hat{A}$ can be expressed as (for a spin-1/2)[3]

$$\mathcal{F} = \frac{1}{2}\mathrm{Tr}(\hat{A}\hat{B}^{-1}) \tag{22.7}$$

In the following sections, first, we describe the methods that have been used to characterize gate fidelities in EPR and then we discuss the strategies for maximizing the fidelity.

22.3.1 CP vs CPMG

Minimum gate fidelities for fault-tolerant quantum computation are around 99% (based on the latest, most forgiving error-correction schemes identified so far[41]) but are ideally at the level of 99.99% and above. When the error per gate is so low, accurately measuring the gate fidelity requires repeated application of the gate so the errors can build up and be more readily measured. For example, we have already seen how different DD sequences, such as CP and CPMG, have different tolerances to rotation angle errors and this difference can be exploited to measure pulse errors associated with rotation angle.[42,43] Measuring the decay of echo intensity in a train of spin echoes can provide a basic estimate for rotation angle errors (from which gate fidelity can be derived). However, for a more complete characterization of the possible gate errors, more elaborate techniques are required.

22.3.2 Quantum Process Tomography

An experimentally applied operation can in general be described by a process matrix, χ, which describes how a given input state described by the density matrix, $\hat{\boldsymbol{\sigma}}_{\mathbf{in}}$, is transformed into an output state:

$$\hat{\boldsymbol{\sigma}}_{\mathbf{out}} = \sum_{m,n=0}^{3} \chi_{mn}\hat{A}_m\hat{\boldsymbol{\sigma}}_{\mathbf{in}}\hat{A}_n^{\dagger} \tag{22.8}$$

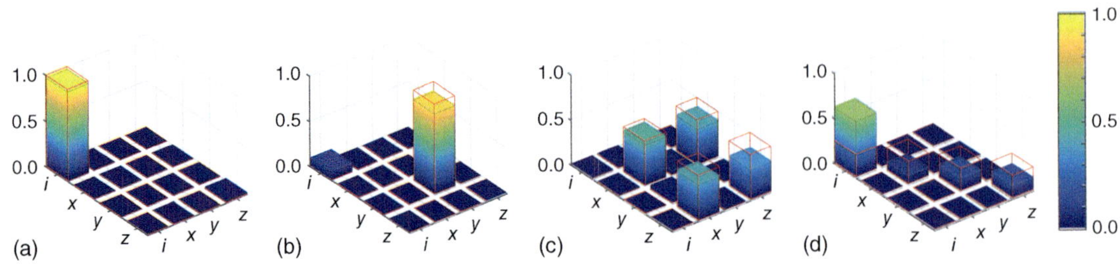

Figure 22.7. Quantum process tomography. Experimentally obtained processes matrices, χ, for (a) the identity operator, (b) π_y rotation, (c) a Hadamard gate, and (d) a combination of processes, such as T_1, T_2 and pulse errors, leading to a partially mixed state. Red outlines correspond to the ideal processes. Experiments performed using (a, b) Sb : Si,[44] (c) P : Si; and (d) Nd^{3+} : YSO[46]

where $\hat{A}_i$ is the Pauli basis $\{\hat{\mathbf{1}}, \hat{\sigma}_x, \hat{\sigma}_y, \hat{\sigma}_z\}$. Quantum process tomography (QPT) aims to obtain this matrix, χ, by performing state tomography of the output state, given a suitably representative set of input states.[3] Owing to state preparation and measurement (SPAM) errors, or variations in the applied operation, the process matrix directly constructed by linear inversion may not be physical (Hermitian positive). In such cases, the closest physical operation can be found by least squared fitting.[44,45] Some example process matrixes are shown in Figure 22.7.

22.3.3 Randomized Benchmarking

The above methods to characterize gates have a number of limitations: QPT is sensitive to SPAM errors (although in principle they can be separated from the applied process with appropriate calibration steps) and requires an exponentially increasing number of operations as the number of qubits grows. On the other hand, the method of comparing CP and CPMG assesses only one kind of error and gives an unrealistic picture of how errors may combine in a practical computation.

Randomized benchmarking is a technique that solves these issues by applying a sequence with increasing number of randomly selected pulses that eventually depolarize the qubit. The pulses are chosen from the Clifford group, a set of rotations that span uniformly the Hilbert space, thus providing an average pulse error.[47] For a single qubit, the Clifford gates are the identity, π, $\pi/2$, and $2\pi/3$ rotations.[48] After each sequence, the qubit state is measured along some axis (e.g., x or y). The decay in fidelity as a function of pulse number gives the average pulse error, with randomization ensuring the fidelity is not correlated to any particular pulse. In order to test a specific gate, the fidelity decay due to an increasing number of Clifford gates is compared to the decay with this gate interleaved between the Clifford gates.

Randomized benchmarking was first performed using liquid-state NMR,[49] followed by EPR experiments on single-electron spins of NV centers in diamond[50] and donors in silicon,[11] and in ensemble EPR using a loop-gap resonator.[51]

22.4 HIGH-FIDELITY OPERATIONS

As mentioned above, and discussed in more detail in Section 22.6.5, required gate fidelities for quantum computation are typically 99.9–99.99%, much higher than is typically required for spectroscopic applications. In conventional dielectric and loop-gap cavities in EPR, inhomogeneities of 10–20% in the microwave field strength are common, leading to fidelities of less than 95–99% for a simple π-pulse. In stripline resonators and microresonators, the fidelities can be substantially less.[52]

The fidelity of a gate achieved by a single rectangular pulse can be increased using composite pulses (comprising several pulses) or by various pulse shaping methods (see Chapter 21) including adiabatic pulses and the use of geometric phases. In each case, common considerations when designing a pulse shape or sequence to yield a high-fidelity operation include (i) minimizing the pulse duration, given the available power and bandwidth of cavity and microwave components; (ii) achieving the desired frequency selectivity (e.g., to perform a conditional operation); (iii) maximizing the robustness to the dominant imperfections in the system (such as microwave field amplitude, or resonant frequency in the case of inhomogeneous broadening in the spin system); and (iv) independence on the initial spin state–in other words, the sequence should be correcting the operator, not the final state.

Many studies, often originating in NMR, have attempted to fulfill these criteria, offering robustness to errors in frequency/detuning and pulse amplitude, at the cost of longer pulses. In the following section, we discuss various approaches to performing high-fidelity gates in EPR.

22.4.1 Composite Pulses

Composite pulses attempt to minimize errors by concatenating multiple (rectangular) pulses with various phases and rotation angles,[53] based on the assumption that the error is systematic, at least over the duration of the composite pulse. The overall evolution is equivalent to a single rotation whose dependence on variations in parameters such as frequency detuning or pulse amplitude cancels up to a desired order (in average Hamiltonian theory).[54] Composite pulses are an appealing solution as they are relatively simple to apply, often universal, and for high concatenation levels can theoretically achieve arbitrary robustness. In practice, however, higher levels are rarely applied as they require exponentially more pulses.

BB1 (short for 'broadband 1') is a common composite pulse to tackle pulse amplitude error,[54] where the sensitivity of fidelity to rotation angle errors in each pulse is reduced to sixth order. It is defined with four pulses:

$$(\theta)_0^{\mathrm{ideal}} = (\pi)_\phi(2\pi)_{3\phi}(\pi)_\phi(\theta)_0 \tag{22.9}$$

$$\text{with} \quad \phi = \cos^{-1}(-\theta/4\pi) \tag{22.10}$$

where θ is the desired rotation angle, while the phases ϕ and 3ϕ are defined with respect to the phase of the desired ideal rotation. Using BB1, fidelities above 99.9% have been measured in conventional dielectric EPR resonators, despite amplitude errors up to 10%.[43] Other variations exist such as NB1 or PB1 (short for 'narrowband 1' and 'passband 1') that provide high fidelity but over a restricted deviation of the amplitude,[54] which can be useful to spatially select spin ensembles in a sample using microwave field gradients.

For robustness against frequency detuning (e.g., given inhomogeneous broadening), the CORPSE (compensation for off resonance with a pulse sequence) pulse can be used,[55] defined, for an ideal rotation $(\theta)_0$, as

$$\left(\frac{\theta}{2} - \psi\right)_0 (2\pi - 2\psi)_\pi \left(2\pi + \frac{\theta}{2} - \psi\right)_0 \tag{22.11}$$

$$\text{with} \quad \psi = \sin^{-1}(\sin(\theta/2)/2) \tag{22.12}$$

If the desired pulse is a π-pulse (as is commonly used in many DD sequences), the so-called Knill pulse[56,57] can be very effective, with robustness to both detuning and pulse amplitude errors. The Knill pulse, which results in a $\pi/3$ phase shift in additional to the desired rotation, consists of a symmetric sequence of five π-pulses:

$$\pi_0^{\mathrm{ideal}}\left(\frac{\pi}{3}\right)_z = (\pi)_{\frac{\pi}{6}}(\pi)_0(\pi)_{\frac{\pi}{2}}(\pi)_0(\pi)_{\frac{\pi}{6}} \tag{22.13}$$

The performance of BB1, CORPSE, and Knill sequences is compared in Figure 22.8. Composite pulses can themselves be used as the base pulses of

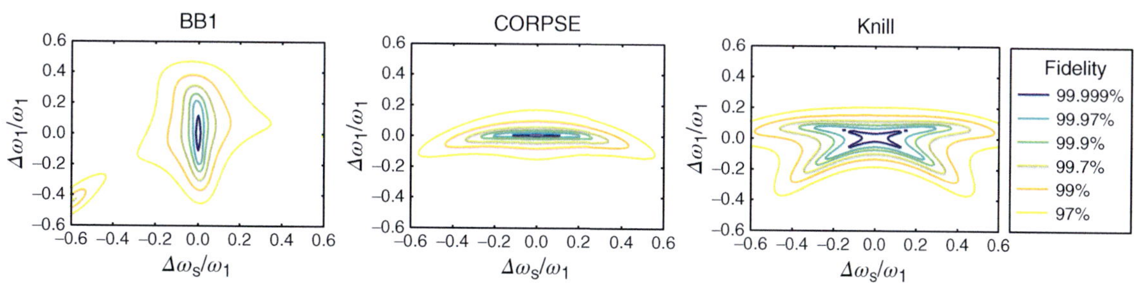

Figure 22.8. Robustness of composite pulses for frequency detuning and pulse amplitude errors. The fidelity of BB1, CORPSE, and Knill sequences are compared as a function of frequency detuning ($\Delta\omega_S$) and pulse amplitude detuning ($\Delta\omega_1$) expressed as a fraction of $\omega_1(=\gamma_e B_1)$, the microwave field strength (in angular frequency units). Both the BB1 and Knill sequences are five times the duration of a simple π-pulse ($5t_{180}$), while the CORPSE sequence is $\frac{13}{3}t_{180}$ in duration

other composite pulses, in order to combine different type of robustness.[58]

22.4.2 Adiabatic Pulses

As discussed in Chapters 8 and 11, an applied microwave field resonant with a spin appears, in the rotating frame, as a static magnetic field in the x–y-plane about which the spin then precesses. If the microwave field is off-resonant, the static field in the rotating frame acquires a residual z-component. As an alternative to abruptly turning on and off such fields to drive spin rotations, the microwave field amplitude and frequency can be gradually adjusted to change the direction of the effective magnetic field in the rotating frame. In this way, thanks to the adiabatic theorem in quantum mechanics,[59] spin eigenstates can be moved along determined trajectories around the Bloch sphere, defining some unitary operation. (The adiabatic theorem states that a physical system remains in its instantaneous eigenstate if a given perturbation is acting on it slowly enough and if there is a gap between the eigenvalue and the rest of the spectrum of the Hamiltonian.)

A prototypical adiabatic pulse is the adiabatic fast passage (AFP), introduced by Bloch[60]: a chirp pulse where the microwave frequency is swept across the spin resonance frequency. In the rotating frame, all spins whose resonant frequency lies within the range swept by the chirp pulse experience an effective magnetic field that smoothly rotates from up to down, creating an effective π-pulse (see Chapter 21).

AFPs are robust to variations in the microwave irradiation amplitude as well as detuning; however, care must be taken when applying them to coherent states–for example, the AFP includes a inhomogeneous phase shift due to frequency broadening. This phase shift is canceled when pairs of AFPs are applied, making them suitable for multipulse DD sequences.[38,61]

While an AFP can be used to achieve a π-pulse, other rotations can be performed with more complex adiabatic pulses such as the BIR-1 or BIR-4 pulse, which are composite AFP pulses.[62] These have the disadvantage (in samples with short T_2) of being even longer than AFPs, which are already an order of magnitude or more longer than a standard rectangular pulse, and are only robust to pulse amplitude errors and not frequency. BIR pulses have been used, for example, for EPR using coplanar waveguides in Ref. 52 to perform 2-pulse echo experiments in the presence of over one order of magnitude variation in pulse amplitude.

22.5 SENSING

Research toward developing spin qubits has helped motivate the identification and control of spins with long coherence lifetimes, as well as their measurement at the single-spin level (e.g., by optical[63] or electrical[25] means). These achievements have opened up new possibilities for single spins to act as sensitive local probes for parameters such as magnetic field,[64] electric field,[65] or even temperature.[66] The use of an electron spin as a probe of local environment is well known in EPR spectroscopy, for example, through techniques such as ENDOR (see Chapter 16) and DEER (double electron-electron resonance, see Chapter 19). However, the applicability of such methods at the single-spin level holds the potential for high-spatial resolution studies, for example, using scanning probe techniques to create spatial magnetic field images.

The development of electron spin-based 'sensors' has primarily arisen out of research on NV centers in diamond, where the electron spin of a single center can be optically measured (see Chapter 25) at room temperature with coherence times around a millisecond (Figure 22.9).[21] Figure 22.9 (a, inset) shows the energy level diagram of the ($S = 1$) NV center, in the low-field limit–the difference of the fluorescence intensities of the $m_S = \pm 1$ and $m_S = 0$ states form the basis for the optical measurement. Such systems couple to magnetic fields via the Zeeman interaction and to other spins via dipolar and hyperfine interactions, while, thanks to the crystal field splitting term, the NV center spin Hamiltonian also depends on electric field, temperature, and strain. Energy shifts arising from variations in these parameters impart phase shifts to the electron spin, and thus the longer the T_2^* and T_2 times of the system is, the weaker will be the perturbation that can be measured. The NV centers are brought in close proximity to the sample that is to be imaged, either by implanting them near the surface of a diamond substrate (Figure 22.9b)[67] or within diamond nanocrystals on the tip of an AFM (Figure 22.9a).[61] Recent achievements and proposals include nanoscale imaging of living cells (Figure 22.9c),[66] measurement of action potentials in neurons,[68] and NMR-like spectroscopy distinguishing multiple nuclear species.[69] Sensing techniques can

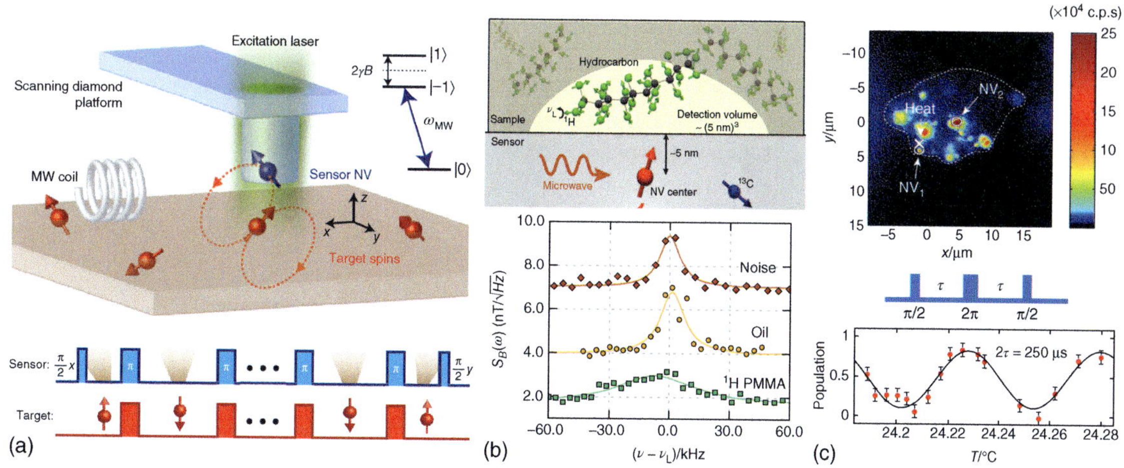

Figure 22.9. Quantum sensing using single-electron spins of nitrogen-vacancy (NV) centers in diamond. (a) Magnetic sensing with an NV center embedded in a scanning diamond nanopillar, allowing for 2D imaging with nanometer resolution (Reprinted by permission from Macmillan Publishers Ltd: Nature Physics, Ref. 61 © 2013). Here, a single-electron spin within the sample, also diamond, is targeted using DD (bottom) on both sensor and target spins (lock-in method). (b) (Top) Schematic of a shallow NV center as a magnetic sensor for molecules at the surface of the diamond[67] (From T Staudacher, F Shi, S Pezzagna, J Meijer, J Du, C A Meriles, F Reinhard, and J Wrachtrup. Nuclear Magnetic Resonance Spectroscopy on a (5-Nanometer) 3 Sample Volume. Science, 339(6119):561–563, Feb 2013. Reprinted with permission from AAAS.). The detection sensitivity can reach volumes down to 5 nm^3, corresponding to about 10^4 protons. (Bottom) Frequency spectra obtained using DD on the NV electron spin and shown as sensitivity against detuning from the expected Larmor frequency (ν_L) of spins in the test molecules: protons in oil (yellow) and ^{1}H polymethyl methacrylate (PMMA) (green). The spectrum from an external magnetic field noise source at a fixed frequency, simulating a precessing spin, is also shown (red) for reference. (c) (Top) Confocal microscopy measurement of a single human embryonic fibroblast cell, outlined by the dotted line, with injected nanodiamonds (NV) as local thermometers (Reprinted by permission from Macmillan Publishers Ltd: Nature, Ref. 66 © 2013). (Bottom) Spin-state projection as a function of temperature, using the given modified Ramsey sequence for the spin-1 NV system, where the 2π-pulse here refocuses phase shifts from static magnetic fields but not from temperatures. The projection amplitude oscillates with temperature, hence this technique is used only for relative temperature sensing. In addition, because the delay τ determines the oscillation period, it can be varied to adjust the balance between greatest dynamic range (small τ) and maximum sensitivity (large τ)

be categorized into two approaches: DC and AC sensings.

22.5.1 DC Sensing

DC sensing aims at measuring parameters that are static, at least on the timescale of the measurement. Measurements are realized either by detecting changes in the electron spin resonance frequency in the CW EPR spectrum or in the time domain via the free induction decay ($\pi/2 - \tau$). In both cases, the sensitivity is determined by the inhomogeneous spin linewidth or T_2^* – this includes the effective broadening caused by randomly fluctuating fields (aside from that to be detected). Depending on the source of inhomogeneous broadening and the parameter to be measured, this limit can be extended to T_2 when the refocusing pulse in a Hahn echo sequence does not reverse the interaction between the spin and the measured quantity. For example, using NV centers in diamond, DC sensing of the temperature is possible with measurement times limited only by T_2 (Figure 22.9c and Ref. 66), because while phase shifts caused by magnetic field inhomogeneities are reversed following the refocusing π-pulse, those due to temperature-dependent shift in the zero field splitting are not.

22.5.2 AC Sensing

In AC sensing, the sensor spin is tuned to sense the local environmental noise at some defined frequency. This can be achieved, for example, by applying DD to the sensor spin and exploiting the bandpass filtering effect of repeated π-pulses in DD, as was described in Section 22.2.1 and Figure 22.5(a, b). As the pulse periodicity is swept, so is the filter frequency (centered around $1/2\tau$ in Figure 22.5b); and when the filter coincides with a dominant component in the noise spectrum, a reduction is observed in the echo intensity. Mapping out the noise spectrum experienced by a spin can be useful to investigate sources of decoherence. For sensing, the technique has been described as the quantum analog of a lock-in amplifier.[35] The detection frequency of the sensor spin could be set, for example, to match the natural Larmor precession of a nuclear spin.[36] In Figure 22.9(b), ^{1}H spins in hydrocarbon molecules were detected using this method with NV electron spins. Alternatively, DD can be applied simultaneously to some spin that is to be probed, as well as to the sensor spin, in order to single out that particular interaction[61] (pulse sequence shown in Figure 22.9a, bottom), as an extension of multipulse DEER methods.

While measurements in the frequency domain are the most common, time-resolved sensing is also possible. In this case, DD sequences are used to obtain the different components in a time series expansion (e.g. the polynomial or Walsh series) from which the signal can be reconstructed.[68] This has been proposed to monitor the magnetic field produced by action potentials in neurons.

For magnetic field sensing (or magnetometry), the sensitivity (in $\mathrm{T}/\sqrt{\mathrm{Hz}}$) of an ensemble of N independent electron spin-1/2 is approximately given by the relation:[64]

$$\eta \approx \frac{\pi}{2|\gamma_e| C\sqrt{NT_2}} \tag{22.14}$$

where C is the measurement efficiency, which is the fraction of spins that can be measured in a single experiment with a signal-to-noise ratio of 1. In EPR, C encompasses factors such as the efficiency of the microwave detection, the resonator quality and filling factors, and inhomogeneous broadening. The relation above assumes the sensed field is fully synchronized with the applied DD sequence (i.e., same frequency and phase, see bottom of Figure 22.9b).

22.6 MULTIPLE QUBITS

Up to this point, we have been concerned primarily with single qubits, the measurement and control of their states, and their application as quantum sensors. Applications in quantum information processing rely on the ability to engineer interactions between multiple spin qubits in order to perform useful computations.

22.6.1 Multiple Qubit Gates

In classical logic, the so-called universal gates such as NAND and NOR operations can be used to construct any combinatorial logic function, and thus to perform any classical algorithm. Analogously, although there exist a vast set of possible two-qubit operations that can be performed, we can focus on a subset that possesses this 'universal' property and thus provides the necessary toolbox to perform universal quantum computation. One such universal set of gates is combination of the single-qubit Hadamard gate (see Section 22.1.1) and the two-qubit controlled-NOT, or C-NOT gate, expressed as

$$U_{\text{C-NOT}} = \begin{pmatrix} 1 & 0 & 0 & 0 \\ 0 & 1 & 0 & 0 \\ 0 & 0 & 0 & 1 \\ 0 & 0 & 1 & 0 \end{pmatrix} \tag{22.15}$$

in the basis $\{|00\rangle, |01\rangle, |10\rangle, |11\rangle\}$. In other words, the C-NOT corresponds to an inversion (or $\pi_{x,y}$ rotation) of the second ('target') qubit if the first ('control') qubit is in state $|1\rangle$, and no operation (the $\hat{\mathbf{1}}$ process) if the control qubit is in state $|0\rangle$. A C-NOT gate can be performed in EPR (modulo some phase factor) as a selective π_x-pulse, which can be readily implemented between pairs of spins if their coupling is large enough to be resolved. Another universal gate (when combined with the H-gate) is the controlled-phase (C-PHASE) gate:

$$U_{\text{C-PHASE}} = \begin{pmatrix} 1 & 0 & 0 & 0 \\ 0 & 1 & 0 & 0 \\ 0 & 0 & 1 & 0 \\ 0 & 0 & 0 & -1 \end{pmatrix} \tag{22.16}$$

which has the effect of applying a π phase shift to $|11\rangle$ with respect to the other three basis states. A C-PHASE can be converted to a C-NOT gate by applying a Hadamard operation to the 'target' qubit before and after the C-PHASE, and both the C-NOT and C-PHASE

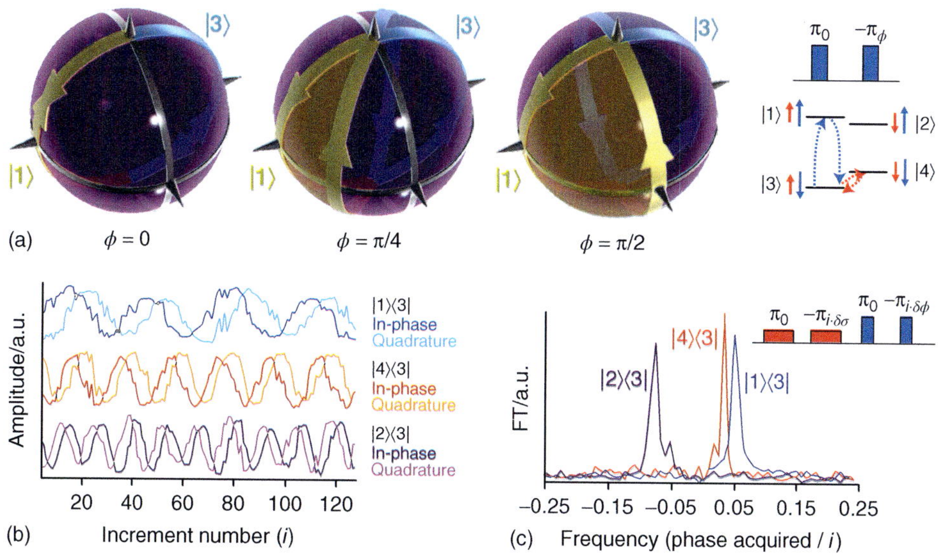

Figure 22.10. State tomography of a coupled electron and nuclear spin. (a) A geometric phase ϕ can be imparted onto two states driven by a pair of pulses: $(\pi)_0(-\pi)_\phi$. (b) Such phase gates are applied to both the $|1\rangle \leftrightarrow |3\rangle$ and $|3\rangle \leftrightarrow |4\rangle$ transitions, where the phase applied is incremented from shot to shot by $\delta\phi$ and $\delta\sigma$, respectively (thus, the phases applied on the ith shot are $i\delta\phi$ and $i\delta\sigma$). The effect of these phase gates is shown for three different initial states: $|1\rangle\langle 3|$ (an electron spin coherence); $|3\rangle\langle 4|$ (a nuclear spin coherence); and $|2\rangle\langle 3|$ (a zero quantum coherence)–in each case, the state is first prepared, then the phase gate is applied, and then the state is transformed to be observable in the electron spin echo. (c) The Fourier transform of the resulting phase oscillations yields distinct frequencies for the different quantum coherences, equal to $\delta\phi$ for the $|1\rangle\langle 3|$ coherence, $\delta\sigma$ for $|3\rangle\langle 4|$, and $-\delta\phi - \delta\sigma$ for $|2\rangle\langle 3|$. (Reprinted by permission from Macmillan Publishers Ltd: Nature, Ref. 71 © 2011)

gates can be used to generate entangled states.[70,71] A fast C-PHASE gate between two nuclear spin qubits coupled to an electron spin can be performed using the principle of the AA phase gate described in Section 22.1.1 and has been demonstrated using optically excited triplet states[72] and NV centers.[73]

22.6.2 State Tomography of Multiple Qubits

Extracting the density matrix for a set of coupled qubits requires going beyond the basic approach outlined in Section 22.1.2 and carries some fundamental challenges. First, the number of independent elements increases exponentially with the number of qubits, making full density matrix tomography practical for only small numbers of qubits. Second, these elements must be mapped to observable quantities in the system (in the same way that the $\hat{S}_z$ component was mapped onto $\hat{S}_x$ in the single-qubit case). Given imperfect pulses, this mapping process can become increasingly sensitive to pulse errors as the number of qubits increases.

While these challenges cannot be readily overcome, it is at least possible to reduce the sensitivity to pulse error, to some extent, by 'labeling' elements of the density matrix using phase shifts, before mapping them onto observables.[70] In this method, illustrated in Figure 22.10, a phase shift ϕ_i is applied to state $|i\rangle$, such that coherences between such states $|i\rangle$ and $|j\rangle$ acquire phases $\phi_i \pm \phi_j$. The phases ϕ_i are each incremented by some fixed amount $\delta\phi_i$ from measurement to measurement, leading to phase oscillations, and the phase increments $\delta\phi_i$ are selected so that each type of coherence acquires phase at a distinct effective frequency. Various approaches can be used to apply the phase shift, for example, using geometric phases and a $(\pi)_0(-\pi)_\phi$ sequence.[71]

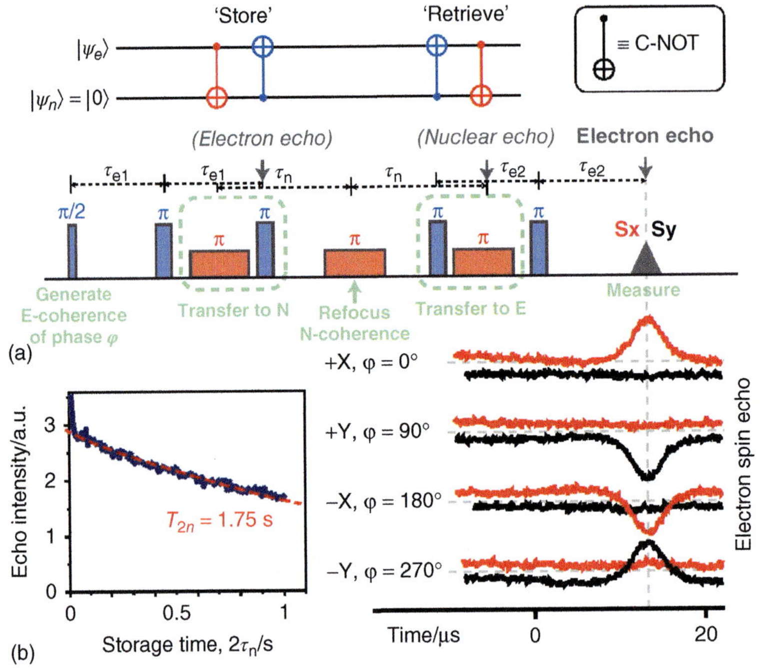

Figure 22.11. SWAP operations between electron and nuclear spin degrees of freedom. (a) (top) Two C-NOTs approximate a SWAP and can be used as the basis for a quantum memory where a coherent state of the electron spin $|\psi_e\rangle$ is transferred to a nuclear spin degree of freedom and subsequently retrieved. In the implementation of this scheme (bottom), C-NOT gates are achieved using selective π rotations, while additional refocusing pulses are needed to account for inhomogeneous broadening. The final electron spin echo is shown to vary its phase in accordance with the original $\pi/2$-pulse applied at the beginning of the sequence (50 ms earlier) (Reprinted by permission from Macmillan Publishers Ltd: Nature, Ref. 76 © 2008). (b) Measuring the recovered echo intensity as a function of the total storage time $2\tau_n$ yields a nuclear spin T_2 of almost 2 s, compared to the electron spin T_2 of 5 ms under the same conditions. In later experiments, nuclear spin coherence times as long as 3 min were measured for neutral P donors in silicon at 1.7 K.[24]

22.6.3 Quantum Memory

The ability to transfer a qubit state between different degrees of freedom (such as between electron spin and nuclear spin transitions) has numerous applications, from enhanced fidelity spin measurement,[74] initialization and sensing, quantum repeaters for quantum communication,[75] and to preserve qubit states for longer periods of time.[76] This state transfer can be achieved by the SWAP logical operation, which exchanges the states of two qubits:

$$U_{\mathrm{SWAP}} = \begin{pmatrix} 1 & 0 & 0 & 0 \\ 0 & 0 & 1 & 0 \\ 0 & 1 & 0 & 0 \\ 0 & 0 & 0 & 1 \end{pmatrix} \qquad (22.17)$$

and can be performed in general via three C-NOT operations (although typically in practice only two are used), as illustrated in Figure 22.11.

In NV centers in diamond, nuclear spin quantum memory operations have also been achieved using fast passages across Landau-Zener transitions,[77] or exploiting anisotropic hyperfine interactions in the spin system.[8]

22.6.4 Quantum Error Correction

Identifying and correcting errors is essential for any practical realization of an information processor–this is especially important in quantum information processing where the expected error rates due to decoherence and gate fidelity are expected to be high, and also particularly demanding, as we are prohibited from directly measuring the state of any logical qubit during the calculation (to avoid collapse of the qubit state). Fortunately, strategies have been put forward since the mid-1990s to identify and correct errors in qubits without learning about their state.[78,79] These strategies have the common feature of redundancy, representing one *logical* qubit using several *physical* qubits, and have different fault-tolerant thresholds, corresponding to the maximum intrinsic error rates per gate that can be effectively corrected (see Section 22.4).

One approach to quantum error correction is based on 'majority voting' and requires at least three physical qubits per logical qubit. This scheme has been implemented using NV centers in diamond, employing the NV center electron spin and two nearby nuclear spins to represent the logical qubit.[73,80] A qubit state is first mapped from the electron spin to all three spins via two C-NOT gates, at which point it becomes robust to spin-flip errors (Figure 22.12). After some period, further C-NOT gates are applied to distill any single spin-flip error onto the nuclear spins, leaving the electron spin qubit in its original state. The nuclear spins must then be refreshed into an initial state before being reused.

Notably, among the assumptions in such error-correction schemes is that of uncorrelated errors, and in the case of several nuclear spins with significant coupling to an electron spin, this assumption may not be justified. Nevertheless, such demonstrations highlight the potential benefit of small 'registers' of nuclear spins coupled to an electron spin.

More recent approaches to quantum error correction based on topological schemes such as the surface code[41] have high threshold error rates, although at the expense of even larger resource overheads (e.g., in terms of the number of physical qubits used to represented one logical qubit).

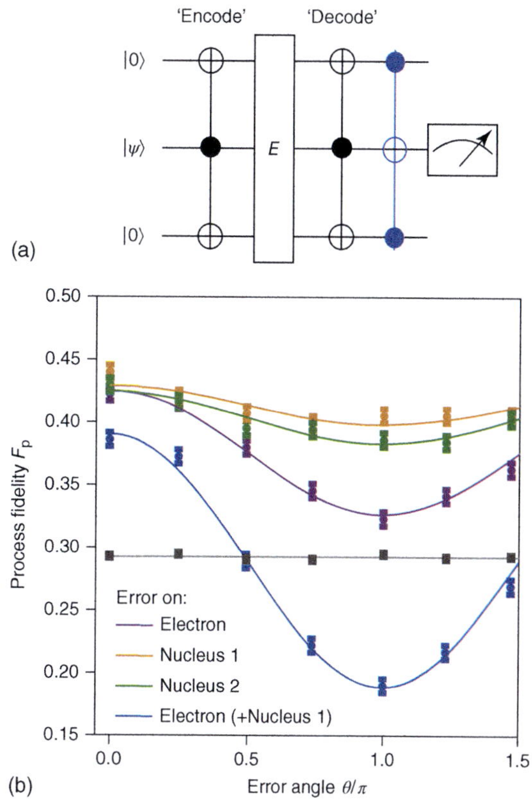

Figure 22.12. Quantum error correction using an electron spin coupled to two nuclear spins.[80] (a) Theoretical error-correction protocol for bit-flip error (box labeled 'E'). The initial 'encoding' step consists of entangling the qubit (in state $|\psi\rangle$) with two nuclear spin ancillae, while the decoding stage leaves the qubit in its original state, $|\psi\rangle$, regardless of whether a single bit-flip error occurred. (b) Process fidelity of the protocol considering a single or two flip errors among the three spins. The gray line gives the average fidelity without correction

22.6.5 Spin-qubit Networks

Controlling scalable arrays of interacting spin qubits remains one of the greatest challenges in realizing spin-based quantum information processors. Approaches such as those based on electrically tuneable exchange couplings in nanoelectronic devices[12,81] and dipolar couplings between spins[82,83] are being explored; however, the short-range nature of the spin–spin interactions, combined with the need to address individual spins for measurement and control, provide considerable technical challenges. For these reasons, quantum 'interconnects' to couple distant spins are an attractive ingredient to create scalable networks of spin qubits.

Candidates for such interconnects include optical photons, which can be entangled with the state of an electron spin and then interfered to entangle spins separated by as much as 1.3 km.[75] Such schemes require spins systems with strong coupling to light, for example, quantum dots or certain defect centers in solids. Another possibility, albeit over shorter length scales, is to use single microwave photons strongly coupled to spins in microwave cavities. This represents a new regime for EPR, but one in which there is increasingly progress and many opportunities.[84–86]

22.7 CONCLUSIONS

Pulse techniques from NMR and EPR have played a major role in the development of spin qubits, from the characterization of candidate systems and measurement of coherence lifetimes, to the high-fidelity control of spins to achieve logical gates. We are now at the stage that innovations arising from research into quantum information processing may feed back into EPR methodology, whether in the form of instrumentation such as quantum-limited microwave amplifiers,[86] magnetic field sensing techniques to complement structural characterization by DEER, and the use of DD methods as a spectroscopic technique, and, ultimately, it may provide advanced simulators to model the dynamics of coupled spin systems at scales not possible today.

ACKNOWLEDGMENTS

We thank Richard Brown and Stephanie Simmons for the data shown in Figure 22.1(c), as well as Sofia Qvarfort, Padraic Calpin, David Wise, and Philipp Ross for the data shown in Figure 22.7.

REFERENCES

1. R. P. Feynman, *Int. J. Theor. Phys.*, 1982, **21**, 467.
2. D. Deutsch, *Proc. R. Soc.A: Math. Phys. Eng. Sci.*, 1985, **400**, 97.
3. I. L. Chuang and M. A. Nielsen, '*Quantum Computation and Quantum Information*', Cambridge University Press: Cambridge, UK, 2000.
4. V. Vedral, '*Introduction to Quantum Information Science*', Oxford University Press: Oxford, 2013.
5. L. M. Vandersypen, M. Steffen, G. Breyta, C. S. Yannoni, M. H. Sherwood, and I. L. Chuang, *Nature*, 2001, **414**, 883.
6. M. Zhong, M. P. Hedges, R. L. Ahlefeldt, J. G. Bartholomew, S. E. Beavan, S. M. Wittig, J. J. Longdell, and M. J. Sellars, *Nature*, 2014, **517**, 177.
7. F. Jelezko, T. Gaebel, I. Popa, A. Gruber, and J. Wrachtrup, *Phys. Rev. Lett.*, 2004, **92**, 076401.
8. M. V. G. Dutt, L. Childress, L. Jiang, E. Togan, J. Maze, F. Jelezko, A. S. Zibrov, P. R. Hemmer, and M. D. Lukin, *Science*, 2007, **316**, 1312.
9. D. J. Christle, A. L. Falk, P. Andrich, P. V. Klimov, J. Ul Hassan, N. T. Son, E. Janzén, T. Ohshima, and D. D. Awschalom, *Nat. Mater.*, 2014, **14**, 160.
10. J. J. L. Morton, D. R. McCamey, M. A. Eriksson, and S. A. Lyon, *Nature*, 2011, **479**, 345.
11. M. Veldhorst, J. C. C. Hwang, C. H. Yang, A. W. Leenstra, B. de Ronde, J. P. Dehollain, J. T. Muhonen, F. E. Hudson, K. M. Itoh, A. Morello, and A. S. Dzurak, *Nat. Nanotechnol.*, 2014, **9**, 981.
12. J. R. Petta, A. C. Johnson, J. M. Taylor, E. A. Laird, A. Yacoby, M. D. Lukin, C. M. Marcus, M. P. Hanson, and A. C. Gossard, *Science*, 2005, **309**, 2180.
13. J. Tribollet, J. Behrends, and K. Lips, *EPL (Europhysics Letters)*, 2008, **84**, 20009.
14. J. J. L. Morton, A. M. Tyryshkin, A. Ardavan, S. C. Benjamin, K. Porfyrakis, S. A. Lyon, G. Andrew, and D. Briggs, *Nat. Phys.*, 2006, **2**, 40.
15. G. A. Timco, S. Carretta, F. Troiani, F. Tuna, R. J. Pritchard, C. A. Muryn, E. J. L. McInnes, A. Ghirri, A. Candini, P. Santini, G. Amoretti, M. Affronte, and R. E. P. Winpenny, *Nat. Nanotechnol.*, 2009, **4**, 173.
16. C. J. Wedge, G. A. Timco, E. T. Spielberg, R. E. George, F. Tuna, S. Rigby, E. J. L. McInnes, R. E. P. Winpenny, S. J. Blundell, and A. Ardavan, *Phys. Rev. Lett.*, 2012, **108**, 107204.
17. M. V. Berry, *Proc. R. Soc. A: Math. Phys. Eng. Sci.*, 1984, **392**, 45.
18. Y. Aharonov and J. Anandan, *Phys. Rev. Lett.*, 1987, **58**, 1593.
19. E. Knill, I. Chuang, and R. Laflamme, *Phys. Rev. A*, 1998, **57**, 3348.
20. J. J. L. Morton, A. M. Tyryshkin, A. Ardavan, K. Porfyrakis, S. A. Lyon, and G. Andrew, and D. Briggs, *J. Chem. Phys.*, 2006, **124**, 014508.
21. G. Balasubramanian, P. Neumann, D. J. Twitchen, M. L. Markham, R. Kolesov, N. Mizuochi, J. Isoya, J. Achard, J. Beck, J. Tissler, V. Jacques, P. R. Hemmer, F. Jelezko, and J. Wrachtrup, *Nat. Mater.*, 2009, **8**, 383.
22. A. M. Tyryshkin, S. Tojo, J. J. L. Morton, H. Riemann, N. V. Abrosimov, P. Becker, H.-J. Pohl, T. Schenkel, M. L. W. Thewalt, K. M. Itoh, and S. A. Lyon, *Nat. Mater.*, 2011, **11**, 143.

23. A. Schweiger and G. Jeschke, '*Principles of Pulse Electron Paramagnetic Resonance*', Oxford University Press: Oxford, 2001, p 216.
24. M. Steger, K. Saeedi, M. L. W. Thewalt, J. J. L. Morton, H. Riemann, N. V. Abrosimov, P. Becker, and H.-J. Pohl, *Science*, 2012, **336**, 1280.
25. J. J. Pla, K. Y. Tan, J. P. Dehollain, W. H. Lim, J. J. L. Morton, D. N. Jamieson, A. S. Dzurak, and A. Morello, *Nature*, 2012, **489**, 541.
26. L. Viola, E. Knill, and S. Lloyd, *Phys. Rev. Lett.*, 1999, **82**, 2417.
27. Ł. Cywiński, R. M. Lutchyn, C. P. Nave, and S. D. Sarma, *Phys. Rev. B*, 2008, **77**, 174509.
28. J. Bylander, S. Gustavsson, F. Yan, F. Yoshihara, K. Harrabi, G. Fitch, D. G. Cory, Y. Nakamura, J.-S. Tsai, and W. D. Oliver, *Nat. Phys.*, 2011, **7**, 565.
29. N. Bar-Gill, L. M. Pham, A. Jarmola, D. Budker, and R. L. Walsworth, *Nat. Commun.*, 2013, **4**, 1743.
30. K. Saeedi, S. Simmons, J. Z. Salvail, P. Dluhy, H. Riemann, N. V. Abrosimov, P. Becker, H.-J. Pohl, J. J. L. Morton, and M. L. W. Thewalt, *Science*, 2013, **342**, 830.
31. S. Meiboom and D. Gill, *Rev. Sci. Instrum.*, 1958, **29**, 688.
32. H. Y. Carr and E. M. Purcell, *Phys. Rev.*, 1954, **94**, 630.
33. T. Gullion, D. B. Baker, and M. S. Conradi, *J. Magn. Reson. (1969)*, 1990, **89**, 479.
34. J. S. Waugh, L. M. Huber, and U. Haeberlen, *Phys. Rev. Lett.*, 1968, **20**, 180.
35. S. Kotler, N. Akerman, Y. Glickman, A. Keselman, and R. Ozeri, *Nature*, 2011, **473**, 61.
36. T. H. Taminiau, J. J. T. Wagenaar, T. van der Sar, F. Jelezko, V. V. Dobrovitski, and R. Hanson, *Phys. Rev. Lett.*, 2012, **109**, 137602.
37. D. L. McAuslan, J. G. Bartholomew, M. J. Sellars, and J. J. Longdell, *Phys. Rev. A*, 2012, **85**, 032339.
38. W.-L. Ma, G. Wolfowicz, S.-S. Li, J. J. L. Morton, and R.-B. Liu, *Phys. Rev. B*, 2015, **92**, 161403.
39. G. Wolfowicz, A. M. Tyryshkin, R. E. George, H. Riemann, N. V. Abrosimov, P. Becker, H.-J. Pohl, M. L. W. Thewalt, S. A. Lyon, and J. J. L. Morton, *Nat. Nanotechnol.*, 2013, **8**, 561.
40. A. Laucht, J. T. Muhonen, F. A. Mohiyaddin, R. Kalra, J. P. Dehollain, S. Freer, F. E. Hudson, M. Veldhorst, R. Rahman, G. Klimeck, K. M. Itoh, D. N. Jamieson, J. C. McCallum, A. S. Dzurak, and A. Morello, *Sci. Adv.*, 2015, **1**, e1500022.
41. A. G. Fowler, A. M. Stephens, and P. Groszkowski, *Phys. Rev. A*, 2009, **80**, 052312.
42. J. J. L. Morton, A. M. Tyryshkin, A. Ardavan, K. Porfyrakis, S. A. Lyon, G. Andrew, and D. Briggs, *Phys. Rev. A*, 2005, **71**, 012332.
43. J. J. L. Morton, A. M. Tyryshkin, A. Ardavan, K. Porfyrakis, S. A. Lyon, G. Andrew, and D. Briggs, *Phys. Rev. Lett.*, 2005, **95**, 200501.
44. G. Wolfowicz, M. Urdampilleta, M. L. W. Thewalt, H. Riemann, N. V. Abrosimov, P. Becker, H.-J. Pohl, and J. J. L. Morton, *Phys. Rev. Lett.*, 2014, **113**, 157601.
45. J. L. O'Brien, G. J. Pryde, A. Gilchrist, D. F. V. James, N. K. Langford, T. C. Ralph, and A. G. White, *Phys. Rev. Lett.*, 2004, **93**, 080502.
46. G. Wolfowicz, H. Maier-Flaig, R. Marino, A. Ferrier, H. Vezin, J. J. L. Morton, and P. Goldner, *Phys. Rev. Lett.*, 2015, **114**, 170503.
47. E. Knill, D. Leibfried, R. Reichle, J. Britton, R. B. Blakestad, J. D. Jost, C. Langer, R. Ozeri, S. Seidelin, and D. J. Wineland, *Phys. Rev. A*, 2008, **77**, 012307.
48. R. Barends, J. Kelly, A. Megrant, A. Veitia, D. Sank, E. Jeffrey, T. C. White, J. Mutus, A. G. Fowler, B. Campbell, Y. Chen, Z. Chen, B. Chiaro, A. Dunsworth, C. Neill, P. O'Malley, P. Roushan, A. Vainsencher, J. Wenner, A. N. Korotkov, A. N. Cleland, and J. M. Martinis, *Nature*, 2014, **508**, 500.
49. C. A. Ryan, M. Laforest, and R. Laflamme, *New J. Phys.*, 2009, **11**, 13034.
50. F. Dolde, V. Bergholm, Ya. Wang, I. Jakobi, B. Naydenov, S. Pezzagna, J. Meijer, F. Jelezko, P. Neumann, T. Schulte-Herbrüggen, J. Biamonte, and J. Wrachtrup, *Nat. Commun.*, 2014, **5**, Article number:3371.
51. D. K. Park, G. Feng, R. Rahimi, J. Baugh, and R. Laflamme, 'Randomized Benchmarking of Quantum Gates for Optimizing Microwave Control in ESR', 2015. *ArXiv e-prints*.
52. A. J. Sigillito, H. Malissa, A. M. Tyryshkin, H. Riemann, N. V. Abrosimov, P. Becker, H.-J. Pohl, M. L. W. Thewalt, K. M. Itoh, J. J. L. Morton, A. A. Houck, D. I. Schuster, and S. A. Lyon, *Appl. Phys. Lett.*, 2014, **104**, 222407.
53. M. H. Levitt and R. Freeman, *J. Magn. Reson. 1969*, 1979, **33**, 473.
54. S. Wimperis, *J. Magn. Reson., Ser. A*, 1994, **109**, 221.
55. H. Cummins, G. Llewellyn, and J. Jones, *Phys. Rev. A*, 2003, **67**, 042308.

56. C. A. Ryan, J. S. Hodges, and D. G. Cory, *Phys. Rev. Lett.*, 2010, **105**, 200402.
57. A. M. Souza, G. A. Álvarez, and D. Suter, *Phys. Rev. Lett.*, 2011, **106**, 240501.
58. M. Bando, T. Ichikawa, Y. Kondo, and M. Nakahara, *J. Phys. Soc. Jpn.*, 2013, **82**, 014004.
59. M. Born and V. Fock, *Z. Phys.*, 1928, **51**, 165.
60. F. Bloch, *Phys. Rev.*, 1946, **70**, 460.
61. M. S. Grinolds, S. Hong, P. Maletinsky, L. Luan, M. D. Lukin, R. L. Walsworth, and A. Yacoby, *Nat. Phys.*, 2013, **9**, 215.
62. A. Tannús and M. Garwood, *NMR Biomed.*, 1997, **10**, 423.
63. A. Gruber, A. Drabenstedt, C. Tietz, L. Fleury, J. Wrachtrup, and C. Borczyskowski, *Science*, 1997, **276**, 2012.
64. J. M. Taylor, P. Cappellaro, L. Childress, L. Jiang, D. Budker, P. R. Hemmer, A. Yacoby, R. Walsworth, and M. D. Lukin, *Nat. Phys.*, 2008, **4**, 810.
65. F. Dolde, H. Fedder, M. W. Doherty, T. Nöbauer, F. Rempp, G. Balasubramanian, T. Wolf, F. Reinhard, L. C. L. Hollenberg, F. Jelezko, and J. Wrachtrup, *Nat. Phys.*, 2011, **7**, 459.
66. G. Kucsko, P. C. Maurer, N. Y. Yao, M. Kubo, H. J. Noh, P. K. Lo, H. Park, and M. D. Lukin, *Nature*, 2013, **500**, 54.
67. T. Staudacher, F. Shi, S. Pezzagna, J. Meijer, J. Du, C. A. Meriles, F. Reinhard, and J. Wrachtrup, *Science*, 2013, **339**, 561.
68. A. Cooper, E. Magesan, H. N. Yum, and P. Cappellaro, *Nat. Commun.*, 2014, **5**, 3141.
69. S. J. DeVience, L. M. Pham, I. Lovchinsky, A. O. Sushkov, N. Bar-Gill, C. Belthangady, F. Casola, M. Corbett, H. Zhang, M. Lukin, H. Park, A. Yacoby, and R. L. Walsworth, *Nat. Nanotechnol.*, 2015, **10**, 129.
70. M. Mehring, J. Mende, and W. Scherer, *Phys. Rev. Lett.*, 2003, **90**, 153001.
71. S. Simmons, R. M. Brown, H. Riemann, N. V. Abrosimov, P. Becker, H.-J. Pohl, M. L. W. Thewalt, K. M. Itoh, and J. J. L. Morton, *Nature*, 2011, **470**, 69.
72. V. Filidou, S. Simmons, S. D. Karlen, F. Giustino, H. L. Anderson, and J. J. L. Morton, *Nat. Phys.*, 2012, **8**, 596.
73. G. Waldherr, Y. Wang, S. Zaiser, M. Jamali, T. Schulte-Herbrüggen, H. Abe, T. Ohshima, J. Isoya, J. F. Du, P. Neumann, and J. Wrachtrup, *Nature*, 2014, **506**, 204.
74. P. Neumann, J. Beck, M. Steiner, F. Rempp, H. Fedder, P. R. Hemmer, J. Wrachtrup, and F. Jelezko, *Science (New York)*, 2010, **329**, 542.
75. B. Hensen, H. Bernien, A. E. Dréau, A. Reiserer, N. Kalb, M. S. Blok, J. Ruitenberg, R. F. L. Vermeulen, R. N. Schouten, C. Abellán, W. Amaya, V. Pruneri, M. W. Mitchell, M. Markham, D. J. Twitchen, D. Elkouss, S. Wehner, T. H. Taminiau, and R. Hanson, *Nature*, 2015, **526**, 682.
76. J. J. L. Morton, A. M. Tyryshkin, R. M. Brown, S. Shankar, B. W. Lovett, A. Ardavan, T. Schenkel, E. E. Haller, J. W. Ager, and S. A. Lyon, *Nature*, 2008, **455**, 1085.
77. G. D. Fuchs, G. Burkard, P. V. Klimov, and D. D. Awschalom, *Nat. Phys.*, 2011, **7**, 789.
78. P. W. Shor, *Phys. Rev. A*, 1995, **52**, R2493.
79. A. M. Steane, *Phys. Rev. Lett.*, 1996, **77**, 793.
80. T. H. Taminiau, J. Cramer, T. van der Sar, V. V. Dobrovitski, and R. Hanson, *Nat. Nanotechnol.*, 2014, **9**, 171.
81. B. E. Kane, *Nature*, 1998, **393**, 133.
82. N. Y. Yao, L. Jiang, A. V. Gorshkov, P. C. Maurer, G. Giedke, J. I. Cirac, and M. D. Lukin, *Nat. Commun.*, 2012, **3**, 800.
83. J. O'Gorman, N. H. Nickerson, P. Ross, J. J. L. Morton, and S. C. Benjamin, *NPJ Quantum Inf.*, 2016, **2**, 15019.
84. D. I. Schuster, A. P. Sears, E. Ginossar, L. DiCarlo, L. Frunzio, J. J. L. Morton, H. Wu, G. A. D. Briggs, B. B. Buckley, D. D. Awschalom, and R. J. Schoelkopf, *Phys. Rev. Lett.*, 2010, **105**, 140501.
85. Y. Kubo, C. Grezes, A. Dewes, T. Umeda, J. Isoya, H. Sumiya, N. Morishita, H. Abe, S. Onoda, T. Ohshima, V. Jacques, A. Dréau, J.-F. Roch, I. Diniz, A. Auffeves, D. Vion, D. Esteve, and P. Bertet, *Phys. Rev. Lett.*, 2011, **107**, 220501.
86. A. Bienfait, J. J. Pla, Y. Kubo, X. Zhou, M. Stern, C. C. Lo, C. D. Weis, T. Schenkel, D. Vion, D. Esteve, J. J. L. Morton, and P. Bertet, *Nature*, 2016, **531**, 74.

Chapter 23
Rapid-scan EPR

Gareth R. Eaton and Sandra S. Eaton

University of Denver, Denver, CO, USA

23.1 INTRODUCTION

In conventional continuous-wave (CW) EPR spectroscopy, magnetic field modulation is used to encode the signal, which is recorded by phase-sensitive detection at the modulation frequency (see Chapter 1). To acquire spectra with lineshapes that are not broadened by the modulation, the modulation amplitude should be less than about one tenth of the peak-to-peak linewidth of the first-derivative spectrum. In addition, the modulation frequency should be much less than the linewidth in hertz and less than the electron spin–lattice relaxation rate.[1] These constraints mean that the experimentalist needs to be mindful of the characteristics of the sample when selecting instrumental parameters, which may be difficult when dealing with unknown samples. These constraints also may limit the signal-to-noise ratio (*S/N*) of the resulting spectrum, which is especially problematic for samples with low concentrations of paramagnetic species.

An alternative detection method designated as 'rapid-scan' EPR is compared with conventional CW spectroscopy in Figure 23.1. In a CW experiment (Figure 23.1a), the magnetic field is stepped slowly through the spectrum. At each point in the spectrum, phase-sensitive detection at the modulation frequency produces a signal with an amplitude that is proportional to a small segment of the absorption spectrum. If the modulation amplitude is sufficiently small, the output is proportional to the slope. This signal is one point in the CW spectrum (see Section 23.3.1.1).

In a rapid-scan experiment (Figure 23.1b), the main magnetic field is held constant. An auxiliary set of coils is used to generate a rapidly varying magnetic field that sweeps through the spectrum hundreds or thousands of times per second (Figure 23.1b). The range of magnetic field scan frequencies used for rapid scan overlap with the modulation frequencies that are commonly used for conventional CW. If the CW modulation amplitude is less than the linewidth,[1] the magnetic field scan rate typically will be less than the electron spin relaxation rate, and the rapid-scan effects discussed in this chapter will not be present. However, for spin

EPR Spectroscopy: Fundamentals and Methods.
Edited by Daniella Goldfarb and Stefan Stoll

ISBN: 978-1-119-16299-5
Also published in eMagRes (online edition)
DOI: 10.1002/9780470034590.emrstm1522

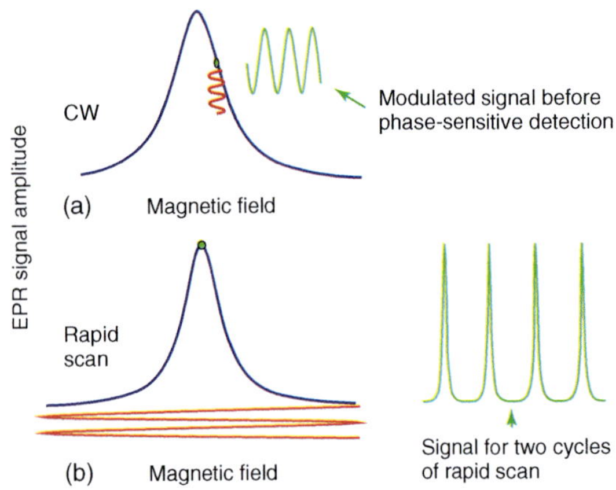

Figure 23.1. Schematic comparison of CW and rapid scans for a single-line EPR spectrum. (a) At each point in a CW spectrum, the signal is encoded by phase-sensitive detection at the modulation frequency. (b) In a rapid-scan experiment, the center field is held constant and an additional magnetic field is rapidly varied across the full spectrum. The full amplitude of the spectrum is detected in each sweep. The example shows two full cycles of the rapid scan, which is composed of two upfield scans and two downfield scans through the single peak

systems with long relaxation times, normal CW modulation frequencies may cause signal distortion, which is why lower modulation frequencies are important to achieve accurate CW lineshapes for those species. These distortions are examples of passage effects (see Section 23.2.2). For example, T_1 for nitroxide radicals in glassy solutions at 80 K is about 1 ms, which means that a modulation frequency of 100 kHz is rapid relative to $1/T_1$ (1 kHz) and the use of 100 kHz modulation results in CW spectra at 80 K that are distorted by passage effects. Unlike CW spectra, analysis of rapid-scan signals takes into account the passage effects and produces accurate lineshapes as discussed in this chapter.

The fundamentals of rapid-scan EPR can be traced back to the observation of transient effects by Bloembergen, Purcell, and Pound (BPP).[2] The famous 'wiggles' that are observed on the trailing edges of each peak were interpreted qualitatively by BPP and quantitatively by Jacobsohn and Wangsness[3] as a relaxation phenomenon. The magnetic moment does not follow the magnetic field as the field passes through resonance, and hence ends up pointing in a nonequilibrium direction. After the scanning field has passed through resonance, the spins precess about B_0 at the instantaneous Larmor frequency determined by the B_0 field. This frequency beats with the microwave/RF field, resulting in a series of damped oscillations. The spacing between the oscillations decreases as the field scans away from the Larmor frequency, in contrast to the case for an FID, in which the oscillation frequencies are constant and define the static offset from resonance.

For many years, CW NMR simply avoided scanning at a rate that resulted in transient effects. Dadok, Gupta, and coworkers showed that one could deconvolve the transient effects to recover undistorted NMR spectra.[4,5] The rapid-scan experimental conditions can be viewed as a driving function $a(t)$ that acts upon the spin system for which the slow-scan spectrum is $s(t)$, to produce a signal $o(t)$. The output power and frequency, ω, of the RF source are constant. The magnetic field, B_0, is swept, so the frequency offset, $\omega_0 - \omega$, between the electron spin resonance frequency, $\omega_0 = \left(\frac{g\mu_B}{\hbar}\right) B_0$, and the source frequency varies with time. For a linear (triangular) scan, the driving function is given by equation (23.1).[6]

$$a(t) = \exp\left(\frac{\mathrm{i}bt^2}{2}\right) \tag{23.1}$$

where the EPR scan rate b is given by $b = 2(g\mu_B/h)B_m(2\pi f_s)$ (in rad s^{-2}), $g\mu_B/h$ is the conversion from millitesla to hertz, B_m the scan width in millitesla, $2\pi f_s$ the angular scan frequency, and the additional factor of 2 takes account of the fact that the field passes through resonance twice in each cycle of the triangular scan.

The detected signal is given by equation (23.2)

$$o(t) = a(t) * s(t) \tag{23.2}$$

where * is the linear convolution operator.

Deconvolution is facilitated by the fact that in Fourier space, deconvolution becomes division, giving

$$O(\omega) = A(\omega)S(\omega) \tag{23.3}$$

where $O(\omega)$, $A(\omega)$, and $S(\omega)$ are the inverse Fourier transforms of $o(t)$, $a(t)$, and $s(t)$, respectively.

Equation (23.3) can be rearranged to give

$$S(\omega) = \frac{O(\omega)}{A(\omega)} \tag{23.4}$$

Fourier transformation of $S(\omega)$ gives $s(t)$, which is the slow-scan EPR spectrum.[6] The time dependence of the driving function for a sinusoidal scan requires different mathematics for the deconvolution than for linear scans, but the general principles remain the same.[7]

Provided that the incident microwave power is chosen such that the EPR signal increases linearly with B_1, which is called the linear response regime, this mathematical treatment of deconvolution is valid. If the relaxation times are short enough that the rapid scan of the field does not produce transient effects, the postacquisition deconvolution has no effect on the recovered slow-scan EPR signal. Thus, the rapid-scan method can be used for mixed samples in which some transitions have long relaxation times and some short. The technology for rapid scan can be applied generally to any spin system, whether passage effects are important or not.

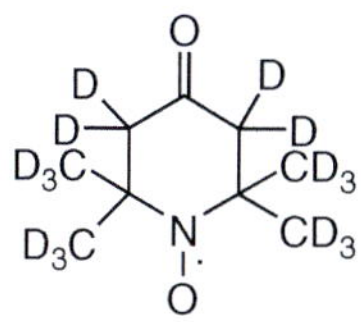

PDT (I)

As the fluid-solution EPR spectra of nitroxide radicals may be familiar to many readers, the three-line spectrum of tempone-d_{16}, 4-oxo-2,2,6,6-tetramethyl-piperidin-1-oxyl-d_{16} (CDN Isotopes, Quebec, Canada, PDT) is used as the introductory example of rapid-scan EPR (Figure 23.2). The perdeuterated radical **I** was used because the peak-to-peak linewidths are about 12 μT,[8] which is substantially narrower than for the same radical with natural isotope abundances. Because of the potential use of rapid scan for in vivo imaging, the spectrum was acquired at a Zeeman frequency of 256 MHz, although rapid scan can be performed at any resonance frequency. In the rapid-scan experiment, the full spectrum is recorded in each up- or downfield scan (Figure 23.1). The downfield half-cycle for a sinusoidal rapid scan of aqueous PDT (**I**) is shown in Figure 23.2(a). The rapid scans cause oscillations on the trailing edge of each hyperfine line. For a sinusoidal scan, the rate of scan is faster at the center than toward the wings of the spectrum so the

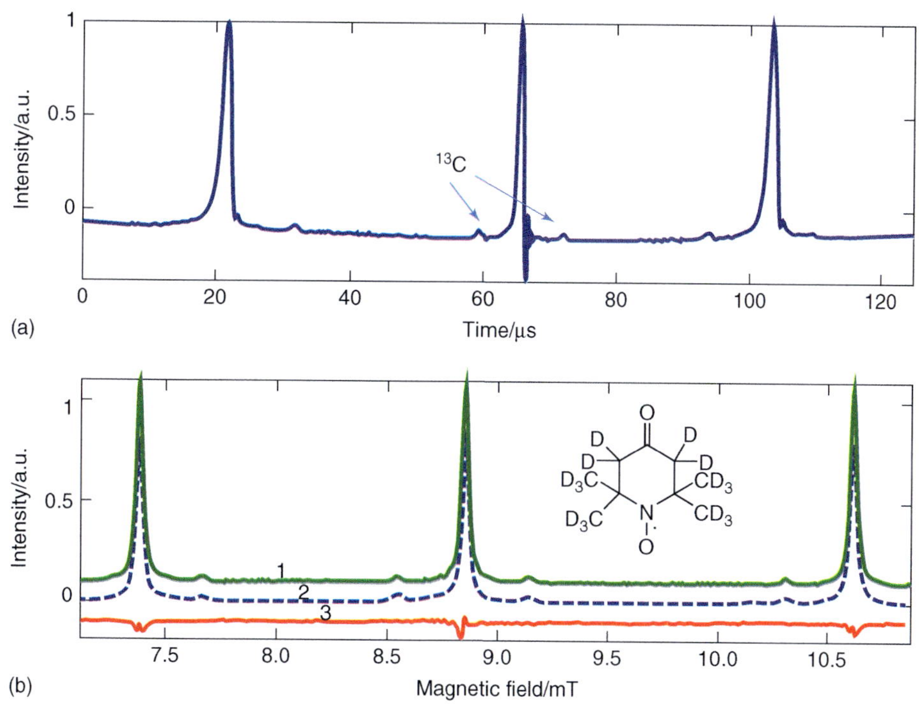

Figure 23.2. Rapid-scan and CW spectra of aqueous tempone-d_{16} (PDT, **I**) obtained at 256 MHz. (a) Downfield half-cycle of the sinusoidal rapid scan recorded with a scan frequency of 4.0 kHz. (b) Comparison of the sum of spectra obtained by deconvolution of up- and downfield signals (trace 1) with first integral of CW spectrum (trace 2), and the difference between traces 1 and 2 (trace 3). The arrows highlight the positions of the ^{13}C hyperfine lines on the nitrogen $m_I = 0$ line

rapid-scan oscillations are more pronounced for the line near the center of the spectrum than for the lines toward the edges of the scan (Figure 23.2a). The up- and downfield scans were deconvolved independently and combined to give the absorption spectrum (Figure 23.2c, trace 1), which is in good agreement with the first integral of the conventional CW spectrum (Figure 23.2b, trace 2). Note that at this low magnetic field, the magnitude of the nitrogen hyperfine coupling is not negligible relative to the magnitude of the external magnetic field, so the three nitrogen hyperfine lines are not equally spaced. Simulation of the hyperfine-split spectrum requires either second-order perturbation theory or full matrix diagonalization (see Chapter 5). The weak ^{13}C hyperfine lines are well defined in both the CW and rapid-scan spectra.

The spectra in Figure 23.2 demonstrate several features of rapid-scan EPR.[9] (i) Unlike conventional CW, the rapid-scan signal is displayed as the absorption spectrum because phase-sensitive detection at the modulation frequency is not used. The same spectral information is available from the absorption spectrum as from the first derivative. (ii) For this example, the magnetic field was scanned sinusoidally at 4 kHz with a scan width of about 3.8 mT. As the magnetic field goes through resonance twice per period of the sinewave (up- and downfield scans), the magnetic field passed through resonance 8×10^3 times per second. (iii) The time on resonance (about 250 ns for the center line) was short relative to the electron spin relaxation times (T_1 about 1 μs), which resulted in oscillations on the trailing edge of the signal that are analogous to the FID following a pulse (see Section 23.3.2). These oscillations were removed by the deconvolution procedure discussed above.[9] (iv) The rapid-scan lineshape, including the oscillations on the trailing edge of the signal, can be simulated using the Bloch equations (see Section 23.4.4).[10] (v) The time constant for the decay of the amplitude of the oscillations is T_2^*, which is a function of the T_2 of the sample and of the inhomogeneous broadening of the line. As the inhomogeneous broadening can be calculated from slow-scan spectra,[11] simulation of the rapid-scan oscillations provides a method to determine T_2.[12]

In the following sections of this chapter, we provide examples of the advantages of rapid scan relative to conventional CW spectroscopy, explanations of the principles behind the improved *S/N* that can be obtained by rapid scan, and an overview of the instrumentation and software that is used in performing the rapid-scan experiments.

23.2 ADVANTAGES OF RAPID SCAN RELATIVE TO CONVENTIONAL CW SPECTROSCOPY

Rapid scan is a CW experiment in the sense that the microwave frequency and power are set to constant values. However, the signal detection methods are different, as discussed below. To distinguish between the two techniques, we use CW to refer only to the conventional detection method.

23.2.1 Linewidths Are Not Broadened by Overmodulation

In conventional CW spectroscopy, the use of a modulation amplitude that is greater than about one tenth of the peak-to-peak linewidth broadens the line.[1] Although small modulation amplitudes are required to avoid distortion of lineshapes, *S/N* increases with increasing modulation amplitude up to the point that modulation amplitude is comparable to the smallest peak-to-peak linewidth. Modulation larger than the narrowest line severely distorts the narrowest line, while broader lines continue to increase in amplitude, resulting in incorrect ratios of line heights in the spectra. Thus, for low-concentration samples of radicals with narrow linewidths that are characteristic of many organic radicals, the experimentalist is faced with a dilemma in selecting parameters. The problem is exacerbated if the sample is an unknown and the user needs to explore a range of modulation amplitudes to find a value that does not distort the spectrum. For example, triarylmethyl radicals such as trityl-CD_3 (**II**), OX63 (**IV**), and OX31 (**V**) are used in EPR imaging because the linewidths are about 5–15 μT.[13] Very low modulation amplitude and low modulation frequency are required to obtain accurate CW lineshapes for these radicals. In rapid-scan EPR, scan widths that are much larger than linewidths can be used with subsequent deconvolution of the oscillations to accurately determine the lineshapes.[6] As discussed in Section 23.5, there are, however, constraints on the selection of scan rates for rapid scans of narrow-line spectra because the signal bandwidth increases with scan rate and must be small enough that the bandwidth of the resonator does not attenuate high-frequency components.

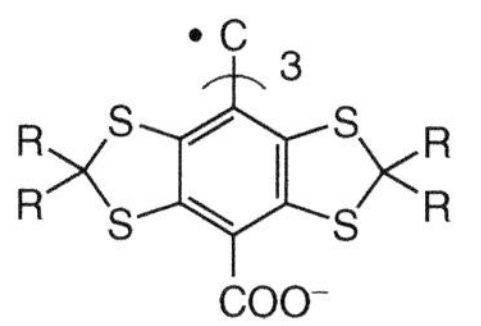

R = CD_3	trityl-CD_3	**II**
R = CH_3	trityl-CH_3	**III**
R = CH_2CH_2OH	OX63	**IV**
R = $CH_2OCH_2CH_2OH$	OX31	**V**

23.2.2 Passage Effects Are Taken into Account

Relaxation-dependent spectral phenomena that occur when the rate of scan through an EPR signal is rapid relative to relaxation times are collectively called 'passage effects'.[14,15] The term relates to the rate of passage through resonance. Some 'strange' spectra can be observed. For example, at low temperatures, under CW EPR conditions that would normally yield a derivative signal, one can observe a signal that looks like an absorption signal. In addition, a signal may invert when the direction of the magnetic field scan is reversed. These are but two of many situations that can be created by various combinations of T_1, T_2, B_1, modulation frequency, modulation amplitude, and scan rate of the main magnetic field. Weger[14] described many cases of passage effects in EPR. 'Wiggles' analogous to those observed in NMR by BPP are one example.

An example of a radical with a long spin–lattice relaxation time is the E′ center in irradiated fused quartz, which has T_1 about 200 μs at room temperature.[16] In conventional CW spectroscopy of the E′ center, it is difficult to record a room-temperature spectrum that is free of passage effects – low modulation frequency, low modulation amplitude, low microwave power, and slow scans are needed.[17,18] Room-temperature X-band rapid-scan spectra for the E′ center are compared with a CW spectrum in Figure 23.3. The CW spectrum (Figure 23.3d) was recorded in 1 min with 10 kHz modulation frequency to avoid passage effect distortions that were observed if 100 kHz modulation was used. The rapid-scan signal was recorded in 5 s (Figure 23.3a). The 'wiggles' in the rapid-scan signal are a manifestation of passage effects. The slow-scan absorption spectrum (Figure 23.3b) was obtained by deconvolution of the rapid-scan signal, which takes

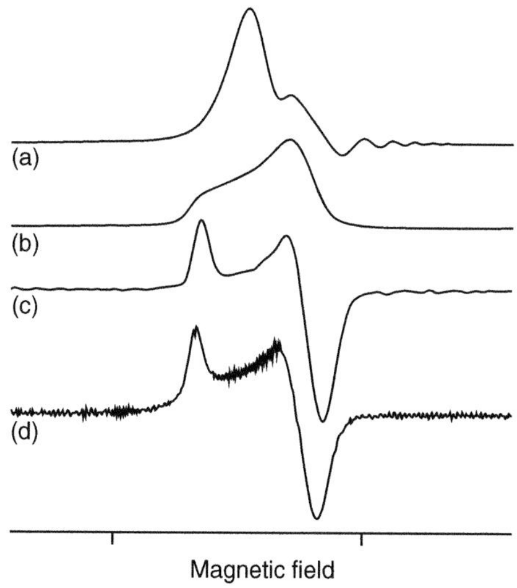

Figure 23.3. Comparison of room-temperature rapid-scan and conventional CW EPR spectra of the E′ center in irradiated fused quartz. (a) As-recorded sinusoidal rapid-scan signal obtained with a scan frequency of 29 kHz and a scan width of 5.9 mT. 1024 averages were recorded in about 5 s. The incident microwave power was about 3.3 mW. A 1.0 mT segment of the scan is displayed. (b) Slow-scan absorption spectrum obtained by deconvolution of signal in (a). (c) First-derivative spectrum obtained by pseudomodulation[51] of the signal in (b). (d) Single scan of a conventional field-modulated first-derivative CW EPR spectrum of the same sample, obtained in 1 min using about 0.02 mW incident microwave power, 10 kHz modulation frequency, and 0.005 mT modulation amplitude.[18] (Reprinted from Radiat. Meas., 46, Mitchell, D. G.; Quine, R. W.; Tseitlin, M.; Meyer, V.; Eaton, S. S.; Eaton, G. R, Comparison of continuous wave, spin echo, and rapid scan EPR of irradiated fused quartz, 993. © 2011, with permission from Elsevier)

account of the passage effects. The first derivative (Figure 23.3c) of the rapid-scan absorption signal has substantially higher *S/N* than the CW spectrum taken with 12 times as long data acquisition time and using the restricted conditions that were required to avoid passage effects.

23.2.3 For the Same Data Acquisition Time, *S/N* is Higher for Rapid Scan than for CW

The spectra in Figure 23.3 show that even when the data acquisition time for the CW spectrum is 12 times longer than for rapid scan, the *S/N* is higher for rapid scan. For a wide range of samples, it has been shown that for the same data acquisition time, the *S/N* for rapid-scan spectra are substantially improved relative to that for CW spectra recorded under conditions chosen to preserve lineshape fidelity.[9] Two examples for distinctly different types of samples are selected for discussion here. Additional examples are in reference 9. The reasons why *S/N* is improved are discussed in Section 23.3.

Nitroxide spin labels are widely used as probes of motion[19] and of local polarity in proteins. Obtaining the best possible *S/N* for small quantities of spin-labeled samples is a high priority. The *S/N* advantage for rapid scan relative to CW for PDT (**I**) immobilized in sucrose octaacetate is shown in Figure 23.4.[20] CW and rapid-scan data were recorded with total signal acquisition time of 10 s. The *S/N* for the rapid-scan absorption signal is 13 times better than for the CW first-derivative spectrum and the *S/N* for the first derivative of the rapid-scan spectrum is about 7 times better than for CW.

Improving *S/N* also improves the accuracy with which signal intensity can be quantitated. One of the techniques that is being developed to monitor accidental radiation exposure is the measurement of the stable radiation-induced signal in human teeth.[21] This task presents additional challenges because the radiation-induced signal is superimposed on a naturally occurring EPR signal that has similar amplitude at the low radiation doses that are clinically significant. To test the rapid-scan methodology for this application, experiments were performed on irradiated samples of tooth enamel (Figure 23.5). The radiation-induced signal in 60 mg of a sample irradiated with a dose of 0.5 Gy was readily characterized in rapid-scan spectra recorded with 34 min data acquisition times, but was not recognizable in CW spectra recorded with the same data acquisition time. Comparison of the rapid-scan and CW spectra was based on samples with 1 Gy doses. After subtraction of the native signal, the difference spectrum for the 1 Gy sample has the lineshape expected for the dose-dependent signal, which is seen more clearly in the first derivative (Figure 23.5b) than in the absorption spectrum (Figure 23.5a). The *S/N* is good enough to estimate the dose. The coefficient of variation (the ratio of the standard deviation to the mean value) of the calculated dose for a sample with 1 Gy radiation based on simulation of the first-derivative spectra for three replicates as the sum of native and radiation-induced signals was 3.9% for CW and 0.4% for rapid scan. This result demonstrates that the improved *S/N* that can be achieved by rapid scan substantially improves signal quantitation.

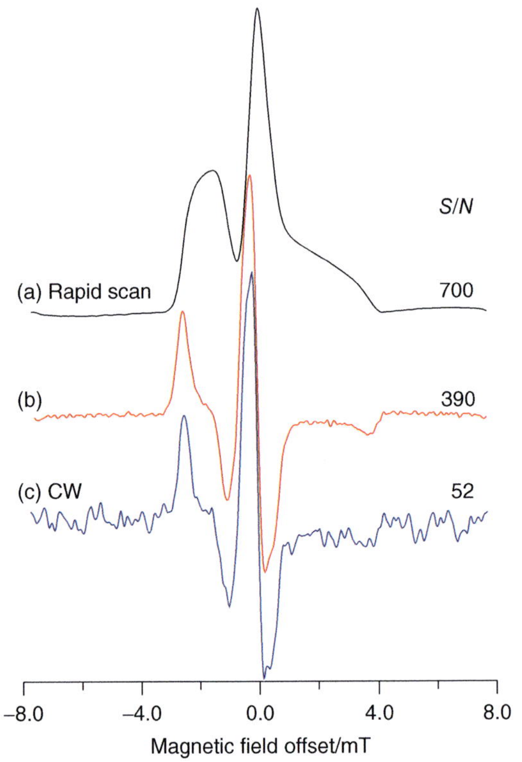

Figure 23.4. CW and rapid-scan spectra of 0.15 mM ^{14}N-PDT (**I**) in sucrose octaacetate at 293 K obtained with 10 s acquisition time. (a) Absorption spectrum obtained by rapid scan with 13.4 kHz scan frequency and 15.5 mT scan width; (b) first-derivative spectrum obtained from (a) by numerical differentiation; and (c) field-modulated CW spectrum obtained with 100 kHz and 0.063 mT modulation amplitude, which is 20% of $\Delta B_{pp} = 0.32$ mT.[20] (Reprinted from J. Magn. Reson, 247, Yu, Z.; Quine, R. W.; Rinard, G. A.; Tseitlin, M.; Elajaili, H.; Kathirvelu, V.; Clouston, L. J.; Boratyński, P. J.; Rajca, A.; Stein, R.; Mchaourab, H.; Eaton, S. S.; Eaton, G. R, Rapid-scan EPR of immobilised nitroxides, 67. © 2014, with permission from Elsevier)

23.2.4 Advantages for EPR Imaging Experiments

In EPR imaging (see Chapter 13), magnetic field gradients are used to encode spatial information. The

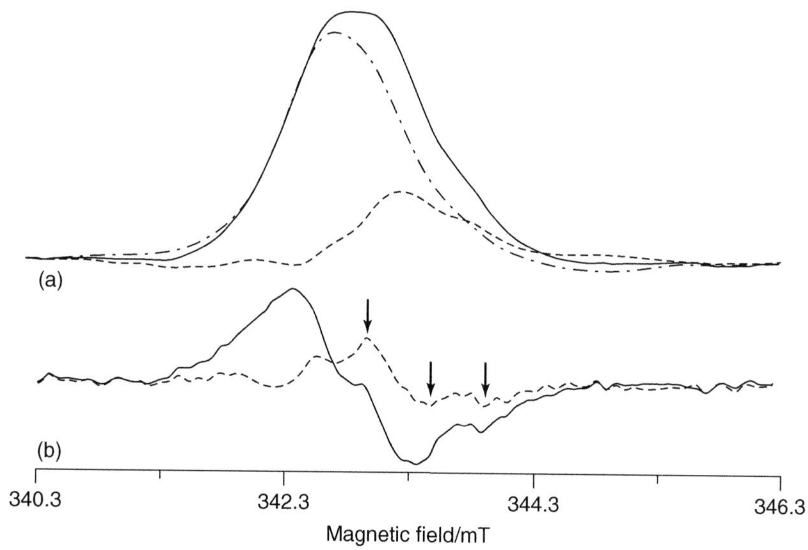

Figure 23.5. Rapid-scan EPR spectra of 60 mg of irradiated tooth enamel obtained with 34 min acquisition time. (a) Absorption spectra of sample irradiated to 1 Gy (—), native signal (— — • — — •), and difference between spectra irradiated to 1 Gy and native signal (— —). (b) First derivatives of spectra in part a for sample irradiated to 1 Gy (—) and the difference spectrum (— —). The arrows highlight the features of the radiation-induced signal

signal is broadened as the gradient increases, which degrades *S/N* especially at higher gradients.[22] The spatial resolution of an image is strongly dependent on the *S/N* for the data at higher gradient. If the relative weightings of Gaussian and Lorentzian contributions to a first-derivative lineshape are constant, then the area is proportional to $(\text{width})^2 \times (\text{amplitude})$. Thus, the amplitude decreases quadratically as the signal is broadened. For the absorption signal, the area is proportional to $(\text{width}) \times (\text{amplitude})$ so the signal amplitude decreases linearly as the signal is broadened. If the weightings of Gaussian and Lorentzian components change as the signal is broadened, the relationship between width and amplitude is more complicated. However, it remains true that broadening of the signal by a gradient has substantially less impact on the amplitude and on the *S/N* of the absorption spectrum than of the first-derivative spectrum. Although the absorption spectrum can be obtained by integration of the first derivative, integration introduces low-frequency 'noise' that is especially problematic when *S/N* is low.[23] The improved *S/N* for rapid scan and the direct detection of the absorption lineshape are major advantages for EPR imaging relative to CW.[24,25]

23.3 WHY DOES RAPID SCAN GIVE IMPROVED *S/N*?

23.3.1 Comparison with Conventional CW Spectroscopy

23.3.1.1 Fraction of Signal Amplitude Detected

In a CW experiment, the magnetic field modulation adds a rapidly varying component to the slowly scanned external magnetic field (Figure 23.1a). The microwave signal is detected, usually with a microwave crystal diode detector, which rectifies the signal to a rapidly changing DC signal. The component of the signal that is in phase with the magnetic field modulation is detected with a lock-in amplifier.[1] This signal is the 'average' slope over the region of the signal that is encompassed by the field modulation, which is why CW spectra are displayed as the first-derivative signals. If the modulation amplitude is sufficiently small relative to the linewidth, the phase-sensitive-detected signal is an excellent approximation for the true slope of the line. However, as the modulation amplitude increases, the detected signal averages the slope over a field range in which the slope is changing, and the signal is no longer a good approximation for the slope of the true lineshape. The requirement that modulation amplitude be kept low relative to the linewidth means that only a fraction of the possible signal is detected in a scan.

In a rapid-scan experiment, the EPR signal is detected with a quadrature mixer at the resonance frequency. The signal is converted to a DC component that is in phase (I) and out of phase (Q) with the reference arm (LO input) of the mixer. The use of quadrature detection, as is common in pulse EPR but not in CW, has the benefit that the microwave phase does not need to be adjusted before spectra are acquired, and the two channels (sometimes called real and imaginary, or in-phase (I) and quadrature (Q)) can be combined in postacquisition processing.

In the rapid-scan experiment, the magnetic field scan width can be much larger than the linewidth, permitting detection of the full signal amplitude during each pass of the magnetic field through resonance (Figure 23.1b). The detection of the full signal amplitude on each scan rather than the small segment detected by the limited modulation amplitude of the CW experiment is a major advantage in *S/N* for rapid scan.

The phase-sensitive detection at the modulation frequency that is used in CW experiments (see Chapter 1) acquires only the signal that is in phase with the modulation, which is called the first harmonic. As the ratio of modulation amplitude to linewidth is increased, the amplitudes of signals at higher harmonics of the modulation frequency increase.[26,27] These higher harmonics are rejected by the phase-sensitive detection. The *S/N* of a CW spectrum can be increased substantially using higher modulation amplitudes and reconstruction of the spectrum using digital phase-sensitive detection and inclusion of the higher harmonics.[26,27] Rapid-scan EPR can be viewed as the extreme case of overmodulation. All harmonics that are within the bandwidth of the resonator are acquired and are included automatically in the deconvolved signal, thereby improving *S/N* relative to acquisition of only the first harmonic.

The quadrature detection in rapid-scan EPR also provides an advantage because the deconvolution procedures discussed in the data analysis section (see Section 23.4.4) use both the I and Q signals from the quadrature mixer. Full advantage from the quadrature detection is enhanced using a cross-loop resonator (CLR), discussed in the hardware section (see Section 23.4.2), which provides similar *S/N* in both channels. The noise in the two channels is not correlated, except for the special case where the noise is dominated by the source. The dispersion signal can be converted to an equivalent absorption signal by means of the Kramers–Kronig relations, which is also known as a Hilbert transform.[24] The converted signal is added to the directly measured absorption signal to improve the *S/N* by a factor of $\sqrt{2}$. To combine the data from the two channels, it is important to calibrate the gains of the two channels and to correct for deviations from precise orthogonality.[28]

23.3.1.2 *Differences in Power Saturation of Signals*

In a conventional CW experiment, the magnetic field is scanned slowly enough that the field is on field/frequency resonance for each line for a long time relative to that for the rapid-scan method. To avoid power saturation of the signal, the microwave magnetic field B_1 must be small enough that net magnetization along the direction of the external field is perturbed very little. This means that relaxation processes should return spins to equilibrium approximately as fast as the B_1 field perturbs it. The extent to which the system is perturbed by increasing B_1 can be displayed in a power saturation curve in which signal amplitude is plotted as a function of B_1 (Figure 23.6). As B_1 is proportional to the square root of power incident on the resonator, the *x*-axis of a power saturation curve sometimes is shown as $\sqrt{P}$. To obtain spectra in which the integrated area under the spectrum is proportional to the number of spins in the sample, spectra must be recorded in the regime where signal intensity increases linearly with B_1 (see Chapter 1), which is designated as the unsaturated regime.[1] This criterion is the same for CW and rapid-scan EPR. In both methods, lineshapes in data recorded at higher B_1 than the unsaturated regime are broadened.

Power saturation curves for an aqueous solution of the nitroxide PDT (**I**) obtained by CW and rapid scans are shown in Figure 23.6. For CW spectra, the power saturation curve is independent of scan rate. The regions of the plots for which signal intensity increases linearly with B_1 extend to higher power for rapid scans than for CW scans, and the difference increases with increasing scan rate (Figure 23.6). The scan rate, a_t, for a triangular scan is constant through the scan and can be calculated using equation (23.5).

$$a_t = 2f_s B_m \tag{23.5}$$

where f_s is the scan frequency and B_m the scan width.

For a sinusoidal scan, the rate varies through the scan. The rate at the center of the scan, a_s, is largest

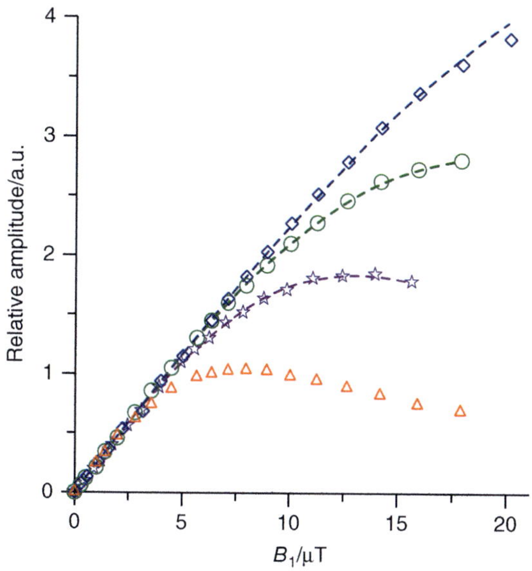

Figure 23.6. Amplitude of CW and rapid-scan spectra of the low-field nitrogen hyperfine line for 0.2 mM ^{15}N-PDT (**I**) solution as a function of microwave B_1 for 3 scan rates: (◇) 460 mT ms^{-1}, (○) 126 mT ms^{-1}, (☆) 42 mT ms^{-1}, and (△) CW. The rapid-scan frequency was ≈30 kHz and the scan width was varied to change the scan rate. Rapid-scan signals were 1024 averages, collected in less than 1 s. To achieve adequate *S/N*, CW spectra were single scans, collected in ~82 s. The *y*-axis scale is the same for all of the rapid scans. The dashed lines represent the calculated power saturation curves, which were simulated by solving the Bloch equations. The amplitude of the CW spectra is scaled to match that obtained for the rapid scans at low B_1.[33] (Reprinted from J. Magn. Reson, 214, Mitchell, D. G.; Quine, R. W.; Tseitlin, M.; Eaton, S. S.; Eaton, G. R., X-band rapid-scan EPR of nitroxyl radicals, 221. © 2012, with permission from Elsevier)

and is given by equation (23.6).

$$a_s = \pi f_s B_m \tag{23.6}$$

The ability in rapid-scan EPR to use higher B_1 compared to CW EPR without saturating the signal increases the relative signal amplitude and contributes to an improved *S/N* for rapid scan. For example, the data in Figure 23.6 show that for this sample, the linear region of the power saturation curve extends to $B_1 \approx 10\,\mu$T at a scan rate of 460 mT ms^{-1}. At this B_1 and scan rate, the signal amplitude is about 2.5 times larger than could be obtained without saturation of the CW spectra for which the linear region extends only to about 4 μT.

23.3.1.3 *Filtering via Coherent Averaging of Multiple Cycles of the Scan*

There are almost limitless sources of noise in EPR data, from low-frequency building vibrations and power line frequencies to thermal noise, which has uniform power spectral density across the spectrum. Each component of the spectrometer, from the source to the detector, contributes noise. Microwave frequency sources usually exhibit noise with a power spectral density that is inversely proportional to frequency. This is called *1/f* noise. Historically, most environmental noise has been of low frequency, but modern communication systems increasingly cause interfering signals at specific frequencies. In a CW experiment, the use of field modulation and phase-sensitive detection at the modulation frequencies moves the EPR signal into a frequency range where there is less source and environmental noise and thereby reduces noise in the spectra.[1] In rapid scan, data are acquired continuously for multiple cycles of the field scans. Noise components, including source and environmental noise, that are not coherent with the periodic signal are efficiently attenuated.[29] This continuous averaging of multiple cycles of a rapid-scan signal is inherently different from the signal averaging of CW spectra in which there is a delay of undefined length between successive scans that are averaged.

In summary, the *S/N* advantages of rapid scan relative to CW EPR arise from the detection of the full signal amplitude on every scan through the signal, the ability to use higher microwave power, and the coherent averaging of multiple scans. These factors are quantitatively analyzed in Ref. 9. In addition to improved *S/N*, rapid scan has the advantage that once the appropriate scan rate is selected to fit the spectral bandwidth and resonator bandwidth as described in Section 23.5.4, passage effects are taken into account by the deconvolution procedure. The selection of the amplitude of the rapid magnetic field scan does not need to be determined based on the width of the spectral lines in the way the CW modulation amplitude is determined.

A disadvantage of rapid scan is that for narrow-line spectra, faster scans require lower resonator Q in order to have enough bandwidth, as discussed in Section 23.5.3. The higher signal bandwidth of rapid-scan signals also requires higher filter bandwidth in the signal detection pathway to accurately preserve the high-frequency components of the rapid-scan signal. These high bandwidths include more high-frequency noise in the data. However, after deconvolution, the bandwidth of the resulting signal is based on the slow-scan lineshape. Thus, after deconvolution, the noise-reduction filter that can be applied in post-processing is independent of the method by which the data were acquired.[9] The noise-reduction filter removes much of the high-frequency noise that was present in the original rapid-scan data.

23.3.2 Comparison with Pulse EPR Spectroscopy

Rapid-scan EPR can be viewed as a type of pulse EPR experiment (see Chapters 11 and 15). Consider the following example for a semiquinone sample.[30] A sinusoidal scan with 7.7 kHz scan frequency and 1 mT scan width has, at the center of the scan, a scan rate of 24.2 mT ms^{-1} according to equation (23.6). The spin packet width ($1/T_2$) for the semiquinone sample is about 3.6 μT. Thus, the magnetic field traverses each spin packet in less than 150 ns. The spin response to the short time on resonance is equivalent to an FID after a microwave pulse at constant field. As the magnetic field continues to scan, the frequency of the oscillations after passing through resonance increases continually, unlike the constant frequency of the FID at constant field. In addition, unlike an FID, there continues to be off-resonance excitation because the B_1 is still on.

If the spectrum can be fully excited by the microwave pulse, pulse EPR detects the full signal. Standard X-band pulse EPR spectrometers use amplifiers that can deliver tens or hundreds of watts of microwave power to the resonator. This power is sufficient to generate 20 ns π/2 pulses that can excite about 50 MHz of spectral width, provided that resonator Q is sufficiently low and that the resonator bandwidth is larger than the bandwidth of the excitation pulse.[31] Some spectrometers can achieve shorter pulses and correspondingly larger spectral coverage. However, the relatively few spectra that are narrow enough to be fully excited with pulse EPR can be studied with rapid-scan EPR with much lower power and hence less expensive instrumentation. Rapid-scan signals can also be obtained for spectra that are too wide to be fully excited by a 20 ns pulse.[32] Pulse EPR also is limited to signals with T_2 that is substantially longer than the dead time of the resonator, which is not a limitation for rapid scan.[9]

23.4 HARDWARE AND SOFTWARE USED IN RAPID SCAN

A block diagram for a rapid-scan spectrometer is shown in Figure 23.7. The principal differences relative to a conventional CW EPR spectrometer are the absence of the magnetic field modulation and lock-in detection system. The new features are the rapid-scan drive and coil system and the use of quadrature (I and Q) detection. X-band rapid-scan experiments can be performed with commercial instrumentation, such as the transient mode of the Bruker E500T. Performance is improved by addition of custom-designed scan coils and coil driver as discussed in the following paragraphs.

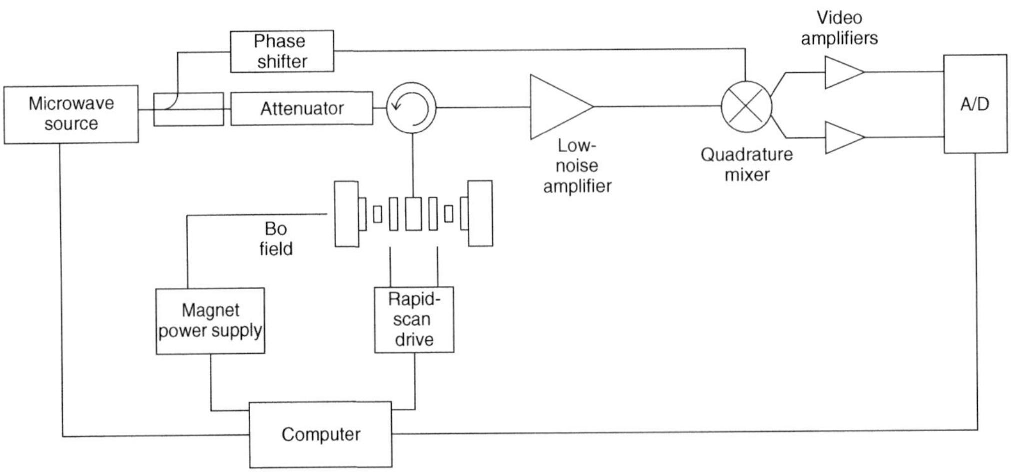

Figure 23.7. Block diagram for a rapid-scan spectrometer. The modules are discussed in Section 23.4

23.4.1 Scan Coils and Scan Drivers

Commercial EPR spectrometers have resonator assemblies with built-in modulation coils that can be calibrated to provide as much as 4 mT peak-to-peak sinusoidal scans at 100 kHz. This is a maximum scan rate of 1.25×10^3 mT ms^{-1}. If the EPR signal for a narrow-line sample acquired at this scan rate with a standard spectrometer were re-routed to go to a mixer instead of going to the phase-sensitive detector at 100 kHz, the passage effects described above could be seen.[16] However, the sample would have to be very small because the standard modulation coils are not designed to give a uniform magnetic field over the full height of the resonator. Spectra showing the degradation of the lineshape of the deconvolved rapid-scan signal when the scanned field is not uniform over the length of the sample have been recorded.[9,33]

The deconvolution algorithms assume that the rapidly swept magnetic field is uniform over the sample so the rapid-scan coils need to be larger than the dimensions of the sample. For the rapid-scan experiments described in Refs 33 and 34, the coils were designed to provide better than 0.1% homogeneity over the length of a 10-mm-long sample. For the Bruker FlexLine resonators, sample lengths were selected to be shorter than the resonator to achieve uniform B_1 and scan field. To maximize field uniformity, the spacing of the coils should have a Helmholtz geometry (spacing between coils equal to coil radius), which is constrained by the diameter of the resonator assembly. The smaller the coils are, the lower will be the power that is required to drive them, which is advantageous. For X-band resonators that are less than 25 mm in diameter, coils with a diameter of 76 mm, spaced 38 mm apart, have been found to be convenient. For large resonators at lower frequencies, 9.5-cm-diameter coils separated by 4.5 cm have been used.[9]

In addition to the size of the coils, the number of turns and the type of wire can be selected based on the goals of the experiment. For higher scan frequencies, lower resistance can be achieved using Litz wire, which has many strands of very fine wire and reduces the demands on the rapid-scan driver. When the scan coils are used in an iron-core magnet, the scanning field induces eddy currents in the poles of the magnet, which dissipates power that otherwise would go into the scanned field. This power dissipation can be decreased by placing thin, highly conducting aluminum plates between the scan coils and the poles of the magnet.[20]

To create accurate scan waveforms, a carefully designed scan driver is needed. In principle, magnetic scans could have any shape designed for a particular experiment. For example, trapezoids with wait periods might be the shape of choice to exploit or study a system with a particular relaxation time. In practice, linear (triangular) and sinusoidal scans have been implemented most commonly. The feedback system in the linear coil driver produces a voltage pattern that is designed to produce a current with the desired triangular or sinusoidal time dependence. The time-dependent voltage pattern is very different from the desired current pattern.[35,36] Sinusoidal scans have the advantage that the driver and coil system can be resonated, which substantially decreases the power required to create the scans and facilitates generation of wider scans.[36]

23.4.2 Resonators

The EPR resonator is a crucial component that is selected or designed for the experiment. Rapid-scan resonators should be as transparent as feasible to the rapidly scanned magnetic fields. Metal components are minimized to decrease eddy currents induced by the rapidly changing fields. At X-band, the Bruker flexline dielectric resonator has been used, which has few metallic parts near the sample. Rapid-scan resonator designs have been published for experiments at 250 MHz[25] or L-band (about 1 GHz).[37] Open-style resonators can minimize the amount of metal near the sample.[38]

Each resonator assembly has some mechanical vibration frequencies that are within the bandwidth of the EPR experiment. These mechanical resonances are problematic for rapid scan because they can be major contributors to the rapid-scan background that is discussed in Section 23.4.4.2. Minimizing mechanical resonances (microphonics) is a major part of the effort involved in refining a rapid-scan resonator. After characterizing the resonator, the operator deliberately selects scan frequencies that are well separated from the known mechanical resonances.

CLRs are a style of bimodal resonator that is highly advantageous for rapid scan.[9] In these resonators, the signal excitation is via one resonator and signal detection is via a second, orthogonal resonator, and there is high isolation between the two resonators.[39] Typically much of the source noise is variation in frequency rather than amplitude of the output. With the usual reflection resonators, this frequency variation contributes to higher noise in the out-of-phase (dispersion) channel of the quadrature mixer. Because of the high isolation of the CLR, the source noise is efficiently isolated from the detection resonator. Noise levels are then similar in the detected I and Q channels, as the noise in the detection pathway dominates. This similarity in *S/N* for the two channels makes it advantageous to use the signals from both the I and Q channels to improve the *S/N* of the final spectrum.

23.4.3 Microwave Bridge

Several features distinguish a rapid-scan bridge from a standard CW bridge. These features are typically also found in transient or pulse bridges. (i) The narrow-band detection and amplification system that directs the EPR signal to the phase-sensitive detector and filter system in a CW bridge is not used. Instead, the signal from the resonator goes to a high-gain low-noise first-stage amplifier. The noise figure for this stage in the amplification is crucial to the performance of the spectrometer because it sets the noise floor for the system. (ii) A quadrature mixer is used in the rapid-scan system in place of the diode detector that is standard for CW. (iii) The bandwidth of the bridge is comparable to that of a pulse EPR bridge, and the output of the bridge goes to a fast digitizer. If a CLR is used, tuning paths are needed for both resonators.

23.4.4 Data Analysis

23.4.4.1 *Simulation of Rapid-scan Effects*

It has been known since the NMR work of BPP[2,4–6] and Jacobsohn and Wangsness[3] that the spin response to a rapid scan can be described mathematically. Simulations of the rapid-scan experiment can be performed by numerical integration of the rotating-frame representation of the Bloch equations including the magnetic field scans as shown in equation (23.7)[10]

$$\frac{dM_u}{dt} = \frac{-M_u}{T_2} - (\Delta\omega + \Omega_m \cos(\omega_m t))M_v$$
$$\frac{dM_v}{dt} = (\Delta\omega + \Omega_m \cos(\omega_m t))M_u - \frac{M_v}{T_2} - \gamma B_1 M_z$$
$$\frac{dM_z}{dt} = \frac{M_0}{T_1} + \gamma B_1 M_v - \frac{M_z}{T_1} \tag{23.7}$$

where γ is the electron magnetogyric ratio; $\Delta\omega = \omega_0 - \omega$ is the offset of a spin packet from the center of the scan in angular frequency; Ω_m is the amplitude of the field scan in angular units and equals $0.5\ \gamma\ B_m$, where B_m is the peak-to-peak scan amplitude; ν_m is the scan frequency in hertz; $\omega_m = 2\pi f_s$ is the angular scan frequency; and B_1 is the RF magnetic field amplitude (peak-to-peak). The two axes of the rotating frame u and v are often denoted as x and y (see Chapters 8 and 11). The time evolution of the magnetization for a spin packet can be evaluated using fourth-order Runge–Kutta numerical integration[40] performed at 20 000 to 40 000 points per field scan cycle.

Rapid-scan spectra at 252 MHz of lithium phthalocyanine (LiPc) are shown in Figure 23.8(a) for a range of more than 100 in field scan rate. Spectra are labeled with the scan rate at the center of the scan, which was set at the resonant field for LiPc. The scan rate slows toward the extremes of the scans. At a scan rate of 0.13 mT ms^{-1}, the rapid-scan signal is superimposable on the first integral of the conventional phase-sensitive-detected CW signal (Figure 23.8a). Simulation of the integral of the CW spectrum (Figure 23.8a) or of the rapid-scan signals required inclusion of a small inhomogeneous broadening. Inhomogeneous broadening was included in the rapid-scan simulation by summing the contributions from spin packets that differed in offset from the center of the signal. As the scan rate is increased, an oscillation appears on the trailing edge of the signal. When the scan rate is rapid relative to $1/T_2$, the net magnetization does not relax during the time that the spin is on resonance. As the signal is traversed, a significant component of magnetization is created in the x–y plane. This component precesses in the x–y plane and a signal analogous to an FID is detected. As the field moves off resonance, the offset between the magnetic field and the resonant field increases and the interval between the peaks of the oscillation decreases (Figure 23.8a). The amplitude of the oscillations increases as the scan rate increases. For a magnetic field scan rate of 0.63 mT ms, the value of $a = \gamma\ dB_0/dt$ [equation (23.5)] is 1.16×10^{11} and $a^{1/2}\ T_2 = 0.85$. Thus, the onset of oscillations at this scan rate is consistent with the expectation that they will be observed when $a^{1/2}T_2 \geq 1$.[3] The oscillations on the left side of the scan at 34 mT ms^{-1} (Figure 23.8a) are due to the previous traverse of the signal.

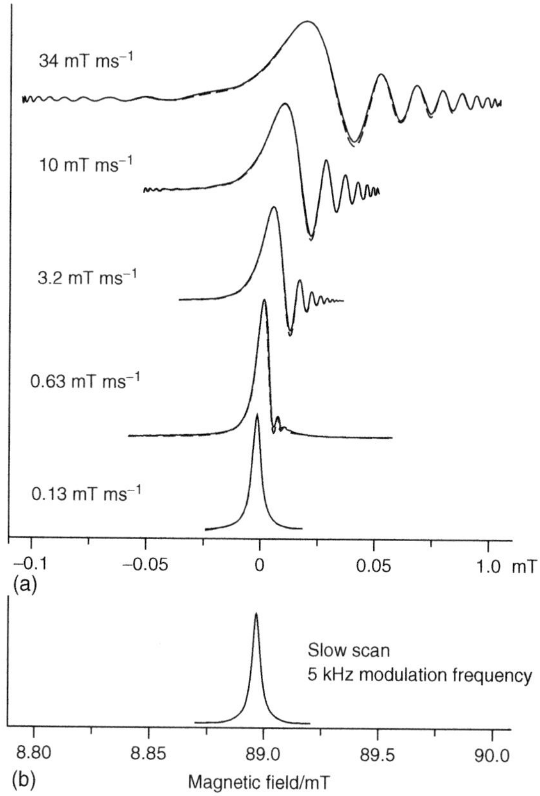

Figure 23.8. EPR spectra of LiPc at about 252 MHz. (a) Rapid-scan spectra obtained at the scan rates shown. The x-axis is the offset from resonance, and the widths of the traces are the rapid-scan widths. The direction of the field scan was left to right in each trace. The number of scans for each scan rate was selected to give a total signal acquisition time of 84 s. The y-axis scales are arbitrary. Simulated spectra calculated using numerical integration of the Bloch equations [equation (23.8)] are overlaid on the experimental data. (b) First integral of conventional slow-scan spectrum obtained with 5 kHz modulation frequency.[10] (Reprinted from J. Magn. Reson, 170, Stoner, J. W.; Szymanski, D.; Eaton, S. S.; Quine, R. W.; Rinard, G. A.; Eaton, G. R., Direct-detected rapid-scan EPR at 250 MHz, 127. © 2004, with permission from Elsevier)

At the faster scan rates, the magnetic field has moved further from resonance during the time that the oscillation persists, so when the signal is displayed on a magnetic field axis (as in Figure 23.8a), the signal is broadened. The broadening of the signal increases with increasing scan rate. Although the broadening is modest at relatively low scan rates, it becomes very large at higher scan rates. The changes in lineshape as a function of scan rate agree well with the simulations (Figure 23.8a).

For linear (triangular) and sinusoidal[7] scans, the effect of the scan on the spin response can be deconvolved (see introduction) to calculate the equivalent slow-scan spectrum. If the time on resonance is long relative to electron spin relaxation, rapid-scan oscillations are not observed and there is no need to deconvolve the spectrum. However, no information is lost by applying the deconvolution procedure to a slow-scan spectrum, so the postacquisition data processing does not need to distinguish between overlapping signals from species with different relaxation times. In real-world 'linear scans', there is rounding of the scan waveform at the extremes of the scans, so deconvolution is applied only to the central region of the scan that is linear.[6] The details of the algorithms are described in Ref. 9.

23.4.4.2 Background Correction

The rapidly scanned magnetic fields generate a background signal, the shape of which depends on many factors including resonator design. To first approximation, the background is a sinusoid with a frequency that is equal to the scan frequency, but may be offset in phase (timing) relative to the timing of the scan.[28,41] The amplitude of the background increases at higher scan frequencies and higher scan widths. At higher scan frequencies and scan widths, higher harmonics of the scan frequency also contribute to the background, especially for linear scans. The background signal also increases with increasing B_1. Work is on-going, but there appear to be multiple contributions to the background signal. Optimizing resonator rigidity and avoiding mechanical resonances are important steps to decreasing the background. Several approaches to separating the background signal from the EPR signal have been successfully implemented.[9,28,41] For example, data for linear scans have been acquired at two different center fields that are offset by an amount that is small relative to the scan width.[41] The signals shift in a predictable way, but the background signal is the same for the two data sets, which provides a means to distinguish background from signal. For sinusoidal scans, Fourier transformation of the combined I and Q data permits separation of the signals from up- and downfield scans.[28] Comparison of these signals can then be used to calculate the background signal.

23.4.4.3 Noise Filtering

Because of the large bandwidth for rapid-scan signals, data are acquired with wide detection bandwidths. After deconvolution, noise can be filtered in postprocessing with parameters that are the same as for the corresponding CW signal, based, for example, on maximum acceptable broadening of the linewidth. For the same lineshape, the signal bandwidth for the absorption is smaller than for the first derivative.[23] Thus, if the intent is to calculate the first-derivative spectrum, the derivative should be calculated from the original rapid-scan data with final filtering done afterward.

23.5 PARAMETER SELECTION

In addition to the usual overall consideration of microwave frequency/magnetic field and temperature needed to obtain the desired information about the paramagnetic species of interest, there are considerations of scan rate, resonator Q, and signal bandwidth that are important for obtaining rapid-scan spectra. As is the case for CW and pulse EPR, one can record a rapid-scan signal with parameters that are far removed from optimum values. In addition, similarly, the optimization depends on the information desired with the usual tradeoffs between maximizing *S/N* versus accuracy in lineshapes. The focus of the following paragraphs is on preserving lineshape fidelity, with the recognition that constraints can be relaxed if lineshape broadening is of lesser concern than extracting a weak signal from the noise.

23.5.1 Linear versus Sinusoidal Scans and Scan Widths

Unless acquisition of only a segment of a spectrum is desired, the scan width is determined by the spectrum of interest. Linear scans have the advantage that the scan rate is uniform throughout the spectrum, which can be helpful if changes in scan rate would result in differential saturation of components of a spectrum. However, the high voltage required to drive the scan coils for wide linear scans becomes problematic as the scan width increases. In addition even with very carefully designed coils and coil drivers, the turning points for a triangular scan are rounded, which limits the fraction of the scan width that is useable.[35] Linear (triangular) scans have been used to combine multiple rapid-scan segments into a wider spectrum as discussion in Section 23.6, taking advantage of the constant scan rate through the scan.

Sinusoidal scans have the advantage that the scan driver and scan coil assembly can be resonated by selecting capacitors appropriate for the scan frequency.[36] Resonating the coils reduces the power required to create wider scans. The widest sinusoidal rapid scans that have been recorded in a single segment in our lab were 15.5 mT for an immobilized nitroxide at X-band[20] and 18.2 mT for radicals in irradiated solids at L-band.[37] The wider scans were possible at L-band because there was space in the magnet for coils with a larger number of turns.

23.5.2 Scan Frequency and Scan Rate

After the scan width has been selected, a scan frequency needs to be chosen taking into consideration several constraints. The scan frequency should avoid mechanical resonances of the resonator (see Section 23.4.2). Another constraint is the capabilities of the scan driver. In general, the wider the scan, the lower the scan frequency that can be used.[36] In addition, there are constraints based on the signal and resonator bandwidths as discussed in the following sections. Those constraints are based on the scan rates calculated using the equations given in Section 23.3.1.2.

23.5.3 Resonator *Q* and Bandwidth

The resonator Q has a dominant role in selecting rapid-scan rates because the resonator acts as a filter – if the signal bandwidth exceeds the resonator bandwidth, high-frequency components are attenuated, which broadens the signal.[42] A convenient expression relating the (loaded) resonator Q and bandwidth $\Delta\nu$ is given in equation (23.8).

$$Q = \frac{\nu}{\Delta\nu} \tag{23.8}$$

where ν is the resonant frequency and $\Delta\nu$ the bandwidth for the resonator, which is the full width at half-height of the resonator dip (see Chapters 1 and 12). As each half-cycle of a rapid-scan experiment is recorded with either increasing or decreasing magnetic field, the rapid-scan signal has frequencies that are only above or below the center frequency of the resonator and the available bandwidth for the rapid-scan signal, BW_{rs}, is half of the resonator bandwidth $\Delta\nu$, equation (23.9).

$$BW_{rs} = \frac{\nu}{2Q} \tag{23.9}$$

For an X-band resonator with $\nu = 9.5$ GHz and $Q = 1000$, $\Delta\nu = 9.5$ MHz and the bandwidth for rapid scan (BW_{rs}) would be 4.7 MHz. To take advantage of this resonator bandwidth, the spectrometer detection and amplification bandwidths should be significantly larger than 4.7 MHz.

The Q of a critically coupled X-band Bruker dielectric resonator is typically about 5000 when a relatively low-loss sample is in the resonator, which gives BW_{rs} about 1 MHz. Thus, this resonator does not have sufficient bandwidth for some rapid scans as discussed in the following section. The Q of the dielectric resonator can be lowered by various means, including using lossy solvents.[9] As microwave power is on continuously in a rapid-scan experiment, lowering Q by overcoupling, as is done in pulse EPR, is not an option for a reflection resonator because of the resulting increase in reflected power. However, overcoupling could be used to lower the Q of a CLR as high power is incident on one resonator and signal detection is from the second resonator.

23.5.4 Signal Bandwidth

The signal bandwidth, BW_{sig}, is given by equation (23.10)

$$BW_{sig} = \left|\frac{N\gamma}{2\pi}\right| aT_2^* \tag{23.10}$$

where $\gamma = -g_e\mu_B/\hbar$ is the magnetogyric ratio, a the scan rate, either a_t (triangular) or a_s(sinusoidal), and N an adjustable parameter that depends on the acceptable extent of lineshape broadening. T_2^* is the constant for the decay of the oscillations on the trailing edge of the signal and is approximately equal to the time constant that would be observed for an FID in a pulsed experiment for the same sample. For a Lorentzian lineshape, T_2^* can be calculated from the first-derivative peak-to-peak linewidth ΔB_{pp} using equation (23.11).

$$T_2^* = \frac{2}{\sqrt{3}\gamma\Delta B_{pp}} \tag{23.11}$$

Although equation (23.11) is not applicable to lines that are not Lorentzian, it provides an approximate starting point for estimating T_2^* and signal bandwidth. The larger the value of N, the less the signal is broadened. A value of $N = 5$ is a conservative starting point for selection of scan parameters. To avoid broadening of the line, BW_{sig} should be less than BW_{rs} for the resonator. Substitution of equation (23.11) into equation (23.10) gives

$$BW_{sig} = \frac{Na}{\sqrt{3}\pi\Delta B_{pp}} \tag{23.12}$$

Two practical examples may help to clarify the concepts. The rapidly tumbling nitroxide PDT (**I**) in a deoxygenated solution has a very narrow linewidth with $\Delta B_{pp} = 15\ \mu T$.[43] A 6.0 mT scan of the full hyperfine-split spectrum with a sinusoidal scan frequency of 20 kHz has $a_s = 3.8 \times 10^2$ mT ms^{-1}, which corresponds to $BW_{sig} = 23$ MHz for $N = 5$ or $BW_{sig} = 9$ MHz for $N = 2$. These bandwidths are substantially higher than BW_{rs} for the X-band dielectric resonator ($Q \approx 5000$ with nonlossy samples, $BW_{rs} \approx 1$ MHz), which would, therefore, require lower scan frequency, narrower scans, or lowering Q to avoid broadening this very narrow line. By contrast, the narrowest features in the immobilized nitroxide signal shown in Figure 23.3 has $\Delta B_{pp} = 0.32$ mT. For a sinusoidal scan frequency of 13.4 kHz and scan width of 15.5 mT, $a_s = 6.5 \times 10^2$ mT ms^{-1}. For $N = 2$, $BW_{sig} = 0.8$ MHz, which is less than the BW_{rs} of the X-band dielectric resonator. This calculation is consistent with the observation that the lineshape for the rapid-scan spectrum of immobilized PDT (**I**) obtained under these scan conditions is in good agreement with the CW spectrum recorded at low modulation amplitude (Figure 23.3).

23.6 EXTENDING RAPID SCAN TO WIDER SPECTRA

Many paramagnetic species have EPR spectra that extend over much wider ranges than the widest rapid scan discussed in this chapter.[20,37] In this case, one could apply a method called nonadiabatic rapid scan (NARS) in which spectra are acquired in segments.[44–47] A field-stepped direct-detection EPR method that is based on rapid-scan technology has been developed with scan widths up to 620 mT and applied to spectra of Cu^{2+}, VO^{2+}, and Mn^{2+} complexes.[48] This method has the potential to replace conventional CW spectroscopy for rigid-lattice disordered samples in powders, crystals, and frozen solution.

In the field-stepped rapid-scan method, triangular rapid-scan signals are acquired at a series of magnetic fields with extensive overlap of the spectral regions. Data from the triangular scans are combined by matching DC offsets for overlapping regions of successive scans. The amplitude of the DC component of the EPR signal, caused by RF reflected from the resonator, is typically quite large compared to the detected EPR signal. The spectrometer has a high-pass filter to eliminate the high DC level of the recorded signal without impacting the EPR signal. Therefore, it is necessary to restore the relative DC level in each segment before combining them to obtain the complete spectrum.[48] The field-stepped rapid-scan and CW spectra of Mn^{2+} doped into $Mg(acac)_2(H_2O)_2$ (1 : 200 Mn^{2+} to Mg^{2+}, acac is acetylacetonate) are compared in Figure 23.9. If the wide CW scan is integrated, there are several broad regions of the spectrum in which it is unclear whether there are real spectral components or if the baseline is poor. For example, the broad feature that is evident (and reproducible) in the rapid-scan spectrum (Figure 23.9a) was also observed in the integral of the first-derivative CW spectrum, but could easily have been mistaken for background (Figure 23.9b).

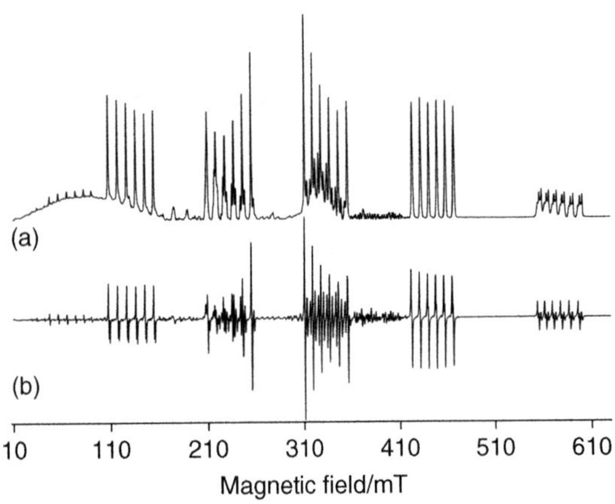

Figure 23.9. (a) Spectrum of an oriented crystal of Mn^{2+}-doped 1 : 200 into $Mg(acac)_2(H_2O)_2$ obtained by field-stepped direct detection in 6201 steps over 620 mT. (b) CW spectrum obtained with a modulation amplitude of 0.11 mT and field resolution of 0.10 mT.[48] (Reprinted from J. Magn. Reson, 258, Yu, Z.; Liu, T.; Elajaili, H.; Rinard, G. A.; Eaton, S. S.; Eaton, G. R., Field-stepped direct detection electron paramagnetic resonance, 58. © 2015, with permission from Elsevier)

23.7 RAPID FREQUENCY SCANS

The principles of rapid scan apply equally well to rapid scans of frequency or magnetic field. As indicated in equation (23.8), the frequency bandwidths of typical EPR resonators are relatively narrow, which limits the accessible scan ranges. Another major limitation on frequency scans is that the reflected power and phase follow the resonator Q curve. The cavity reflection coefficient increases significantly as the frequency is swept further from the center frequency of the resonator, which results in a frequency-dependent baseline. For the same Q, the resonator bandwidth increases with frequency [equation (23.12)], so frequency sweep is more practical at higher frequencies. The 1 GHz bandwidth of a 95-GHz resonator has been exploited to perform rapid frequency-sweep EPR.[49] The microwave frequency can be swept very quickly, as in 'chirp' experiments, as has been demonstrated with frequencies generated by an arbitrary waveform generator.[50]

23.8 FUTURE

For rapid-scan EPR to become widely available, it will need to be made available as an accessory to a commercial spectrometer. Work is in progress toward this goal. Substantial engineering effort will be needed to permit rapid-scan experiments at cryogenic temperatures. The dewars for most current cryogenic systems are largely metallic, and will need to be replaced with nonmetallic regions that permit penetration of the rapidly scanned fields.

The ability of rapid-scan EPR to acquire data rapidly permits higher temporal resolution for kinetics than can be achieved with CW EPR spectroscopy. The combination of rapid scan with improvements in digital electronics provides opportunities to revolutionize the way that EPR will be done in the future.

The *S/N* performance of CW has been the standard for comparison of other methods and often is the method of choice for finding a signal before performing pulse measurement. Magnetic field modulation has been an undesirable necessity. Rapid scan can replace CW and provide improved *S/N* for all samples for which comparisons have been made so far. This is a new regime for EPR, and new aspects of methodology are being developed. Accessories to commercial EPR spectrometers are envisioned.

ACKNOWLEDGMENTS

Partial support of the development of rapid-scan technology and its application has been provided by the National Science Foundation grant IDBR 0753018, National Institutes of Health grants EB000557 and CA177744, and an NSF graduate fellowship to Deborah G. Mitchell. The development of rapid-scan EPR is based on the talents of Engineers Richard W. Quine and George A. Rinard and our students and postdoctoral associates Mark Tseitlin (now at West Virginia University), Joshua Biller (now at National Institute of Standards and Technology), Laura Buchanan, Deborah G. Mitchell, Joseph McPeak, Yilin Shi, and Zhelin Yu. The other coauthors of the papers cited have made invaluable contributions. Experience with prototype rapid-scan systems in the laboratories of Professor Howard J. Halpern and Boris Epel (University of Chicago) and Dr. Ralph T. Weber, Arthur H. Heiss, and Peter Höfer (Bruker BioSpin) has contributed to refinement of our ideas.

REFERENCES

1. G. R. Eaton, S. S. Eaton, D. P. Barr, and R. T. Weber, *Quantitative EPR*, Springer-Verlag/Wein: New York, 2010.
2. N. Bloembergen, E. M. Purcell, and R. V. Pound, *Phys. Rev.*, 1948, **73**, 679.
3. B. A. Jacobsohn and R. K. Wangsness, *Phys. Rev.*, 1948, **73**, 942.
4. J. Dadok and R. F. Sprecher, *J. Magn. Reson.*, 1974, **13**, 243.
5. R. K. Gupta, J. A. Ferretti, and E. D. Becker, *J. Magn. Reson.*, 1974, **13**, 275.
6. J. P. Joshi, J. R. Ballard, G. A. Rinard, R. W. Quine, S. S. Eaton, and G. R. Eaton, *J. Magn. Reson.*, 2005, **175**, 44.
7. M. Tseitlin, G. A. Rinard, R. W. Quine, S. S. Eaton, and G. R. Eaton, *J. Magn. Reson*, 2011, **208**, 279.
8. J. R. Biller, V. Meyer, H. Elajaili, G. M. Rosen, J. P. Y. Kao, S. S. Eaton, and G. R. Eaton, *J. Magn. Reson.*, 2011, **212**, 370.
9. S. S. Eaton, R. W. Quine, M. Tseitlin, D. G. Mitchell, G. A. Rinard, and G. Eaton, in *Multifrequency Electron Paramagnetic Resonance: Data and Techniques*, ed S. K. Misra, John Wiley & Sons, Inc: Hoboken, NJ, 2014, Chap. 2.
10. J. W. Stoner, D. Szymanski, S. S. Eaton, R. W. Quine, G. A. Rinard, and G. R. Eaton, *J. Magn. Res.*, 2004, **170**, 127.
11. C. Mailer, B. H. Robinson, B. B. Williams, and H. J. Halpern, *Magn. Reson. Med.*, 2003, **49**, 1175.
12. M. Tseitlin, A. Dhami, R. W. Quine, G. A. Rinard, S. S. Eaton, and G. R. Eaton, *Appl. Magn. Reson.*, 2006, **30**, 651.
13. L. Yong, J. Harbridge, R. W. Quine, G. A. Rinard, S. S. Eaton, G. R. Eaton, C. Mailer, E. Barth, and H. J. Halpern, *J. Magn. Reson.*, 2001, **152**, 156.
14. M. Weger, *Bell System Technical J.*, 1960, **39**, 1013.
15. A. M. Portis, *Phys. Rev.*, 1955, **100**, 1219.
16. J. R. Harbridge, G. A. Rinard, R. W. Quine, S. S. Eaton, and G. R. Eaton, *J. Magn. Reson.*, 2002, **156**, 41.
17. P. Ludowise, S. S. Eaton, and G. R. Eaton, *J. Magn. Reson.*, 1991, **93**, 410.
18. D. G. Mitchell, R. W. Quine, M. Tseitlin, V. Meyer, S. S. Eaton, and G. R. Eaton, *Radiat. Meas.*, 2011, **46**, 993.
19. J. H. Freed, *Biol. Magn. Reson.*, 2005, **24**, 239.
20. Z. Yu, R. W. Quine, G. A. Rinard, M. Tseitlin, H. Elajaili, V. Kathirvelu, L. J. Clouston, P. J. Boratyński, A. Rajca, R. Stein, H. Mchaourab, S. S. Eaton, and G. R. Eaton, *J. Magn. Reson.*, 2014, **247**, 67.
21. G. A. Alexander, H. M. Swartz, S. A. Amundson, W. F. Blakely, B. Buddemeier, B. Gallez, N. Dainiak, R. E. Goans, R. B. Hayes, and P. C. Lowry, *Radiat. Meas.*, 2007, **42**, 972.
22. G. R. Eaton, S. S. Eaton, and K. Ohno (eds), *EPR Imaging and in Vivo EPR*, CRC Press: Boca Raton, FL, 1991.
23. M. Tseitlin, S. S. Eaton, and G. R. Eaton, *Conc. Magn. Reson.*, 2012, **40A**, 295.
24. M. Tseitlin, R. W. Quine, G. A. Rinard, S. S. Eaton, and G. R. Eaton, *J. Magn. Reson.*, 2010, **203**, 305.
25. J. R. Biller, M. Tseitlin, R. W. Quine, G. A. Rinard, H. A. Weismiller, H. Elajaili, G. M. Rosen, J. P. Kao, S. S. Eaton, and G. R. Eaton, *J. Magn. Reson.*, 2014, **242**, 162.
26. M. Tseitlin, S. S. Eaton, and G. R. Eaton, *J. Magn. Reson.*, 2011, **209**, 277.
27. Z. Yu, M. Tseitlin, S. S. Eaton, and G. R. Eaton, *J. Magn. Reson.*, 2015, **254**, 86.
28. M. Tseitlin, D. G. Mitchell, S. S. Eaton, and G. R. Eaton, *J. Magn. Reson.*, 2012, **223**, 80.

29. M. P. Klein and B. W. Barton, *Rev. Sci. Instrum.*, 1963, **34**, 754.
30. H. Elajaili, G. A. Rinard, Z. Yu, D. Mitchell, R. W. Quine, S. S. Eaton, and G. R. Eaton, *J. Magn. Reson.*, 2015, **259**, 20.
31. A. Schweiger, *Angew. Chem. Int. Ed. Engl.*, 1991, **30**, 265.
32. D. G. Mitchell, M. Tseitlin, R. W. Quine, V. Meyer, M. E. Newton, A. Schnegg, B. George, S. S. Eaton, and G. R. Eaton, *Mol. Phys.*, 2013, **111**, 2664.
33. D. G. Mitchell, R. W. Quine, M. Tseitlin, S. S. Eaton, and G. R. Eaton, *J. Magn. Reson.*, 2012, **214**, 221.
34. J. R. Biller, M. Tseitlin, D. G. Mitchell, Z. Yu, L. A. Buchanan, H. Elajaili, G. M. Rosen, J. P. Y. Kao, S. S. Eaton, and G. R. Eaton, *ChemPhysChem.*, 2015, **16**, 528.
35. R. W. Quine, T. Czechowski, and G. R. Eaton, *Conc. Magn. Reson., Magn. Reson. Engineer*, 2009, **35B**, 44.
36. R. W. Quine, D. G. Mitchell, S. S. Eaton, and G. R. Eaton, *Conc. Magn. Reson., Magn. Reson. Engineer*, 2012, **41B**, 95.
37. Y. Shi, G. A. Rinard, R. W. Quine, S. S. Eaton, and G. R. Eaton, *Radiation Measurements*, 2016, **85**, 57.
38. G. A. Rinard, R. W. Quine, J. R. Biller, and G. R. Eaton, *Concepts Magn. Reson. B, Magn. Reson. Engineer*, 2010, **37B**, 86.
39. G. A. Rinard, R. W. Quine, B. T. Ghim, S. S. Eaton, and G. R. Eaton, *J. Magn. Reson. A*, 1996, **122**, 50.
40. W. H. Press, S. A. Teukolsky, W. T. Verrling, and B. P. Flannery, *Numerical Recipes in Fortran: The Art of Scientific Computing*, Cambridge University Press: Cambridge, 1992, 704.
41. M. Tseitlin, T. Czechowski, R. W. Quine, S. S. Eaton, and G. R. Eaton, *J. Magn. Reson.*, 2009, **196**, 48.
42. J. P. Joshi, G. R. Eaton, and S. S. Eaton, *Appl. Magn. Reson.*, 2005, **28**, 239.
43. J. R. Biller, V. M. Meyer, H. Elajaili, G. M. Rosen, S. S. Eaton, and G. R. Eaton, *J. Magn. Reson.*, 2012, **225**, 52.
44. A. W. Kittell, T. G. Camenisch, J. J. Ratke, J. W. Sidabras, and J. S. Hyde, *J. Magn. Reson.*, 2011, **211**, 228.
45. A. W. Kittell, E. J. Hustedt, and J. S. Hyde, *J. Magn. Reson.*, 2012, **221**, 51.
46. J. S. Hyde, B. Bennett, A. W. Kittell, J. M. Kowalski, and J. W. Sidabras, *J. Magn. Reson.*, 2013, **236**, 15.
47. A. W. Kittell and J. S. Hyde, *J. Magn. Reson.*, 2015, **255**, 68.
48. Z. Yu, T. Liu, H. Elajaili, G. A. Rinard, S. S. Eaton, and G. R. Eaton, *J. Magn. Reson.*, 2015, **258**, 58.
49. J. S. Hyde, R. A. Strangeway, T. G. Camenisch, J. J. Ratke, and W. Froncisz, *J. Magn. Reson.*, 2010, **205**, 93.
50. M. Tseitlin, G. A. Rinard, R. W. Quine, S. S. Eaton, and G. R. Eaton, *J. Magn. Reson.*, 2011, **211**, 156.
51. J. S. Hyde, A. Jesmanowicz, J. J. Ratke, and W. E. Antholine, *J. Magn. Reson.*, 1992, **96**, 1.

Chapter 24
EPR Microscopy

Aharon Blank
Schulich Faculty of Chemistry, Technion, Haifa, Israel

24.1 PREFACE

This section presents some theoretical and experimental aspects of EPR microscopy (EPRM). While an article with similar title has appeared quite recently[1] and described the state of the art in the field, as of 2009, the fast developments, triggered mainly by new applications in semiconductor and quantum devices, stimulated us to provide a significantly revised and updated version portraying the field. Furthermore, in contrast to the previous article, which focused on both CW and pulse induction–detection-based EPRM, this work will examine only the field of pulse EPRM. In adition to that, it will also be relating to generalized EPR signal detection approaches employing either the conventional induction-type or any alternative coherent spin detection methods (such as electrical or optical).

Section 24.2 will provide a general formal definition of the field of EPRM and a brief historical overview. In the interest of space limitations, the discussion will not expand on concepts that are common to all MRI approaches, such as magnetic field gradients and projections (which were already explained in details in Chapter 13). Following Section 24.2, Sections 24.3 and 24.4 will outline some general theoretical and experimental aspects that are unique to pulse EPRM (in contrast to conventional mm-scale EPR imaging methods that are described in Chapter 13). Section 24.3 will provide a theoretical analysis of signal-to-noise ratio (SNR) and spatial resolution in the context of EPRM, while Section 24.4 will describe in detail the experimental requirements for pulse EPRM. Section 24.5 will provide some state-of-the-art experimental results that demonstrate the current capabilities of EPRM. Finally, some general conclusions and future prospects for this field of research will be provided.

24.2 INTRODUCTION

24.2.1 EPR Microscopy: Definition and Short Historical Overview

In the broad context of EPR imaging, it is common to define EPRM as related to images that provide any EPR-related spatially resolved data (e.g., spin concentration, T_1, T_2, EPR spectrum) that are acquired with

EPR Spectroscopy: Fundamentals and Methods.
Edited by Daniella Goldfarb and Stefan Stoll

ISBN: 978-1-119-16299-5
Also published in eMagRes (online edition)
DOI: 10.1002/9780470034590.emrstm1523

resolution that is better than ≈100 μm. This definition is compatible with what is customary in the field of NMR microimaging.[2]

The field of EPRM closely follows the developments of EPR imaging in general. The first published account of EPR imaging, dated from 1978, already showed EPRM capability.[3] In that early experiment, Karthe and Wehrsdorfer employed a simple static magnetic field gradient to resolve in one-dimensional (1D) two grains of DPPH, separated by 200 μm. The 1D resolution in their CW-based method, which employed a conventional TE_{102} rectangular resonator, was calculated to be ≈12 μm. Other researchers that closely followed this early work expanded CW EPRM to 2D by acquiring multiple EPR spectra with various magnetic field gradient directions (called the projection reconstruction (PR) method; see Chapter 13). These measurements were carried out with a DPPH sample and a diamond sample with paramagnetic defects.[4,5] The 2D resolution achieved in that work was ≈100 μm × 100 μm. Most CW EPR images are still obtained today with the PR method. However, it should be noted that early CW EPR microimaging work also employed the modulated field gradient (MFG) method, which acquires EPR spectra with field modulation having a steep gradient spatial dependence.[6] The oscillating magnetic fields from these gradients are added to the regular CW field modulation and cause a distortion of the EPR signal apart from a single point, which can be scanned over the sample, where all the gradients are canceled out. This approach is much less efficient time-wise and thus rarely used today.

These early developments were closely followed by some more advanced experimental setups, based on CW and pulse systems, which further improved the resolution. Significant past achievements in the field, resolution-wise, include the following works: In CW EPRM, the MFG method achieved a resolution better than ≈10 μm in one dimension,[6,7] whereas 2D images exhibit resolution of ≈100 μm × 100 μm. The CW technique with static magnetic field gradients, achieved a resolution of ≈10 μm for 1D experiments[3] and the 2D resolution, employing CW with PR, is on the order of ≈100 μm × 100 μm.[8] Pulse EPR imaging requires fast gradient switching and spin probes with sufficiently long T_2. A key work in pulse high-resolution EPRM was conducted with a $(FA)_2X$ crystal, a unique organic conductor with T_2 of ≈6 μs at room temperature. Using this material, 1D imaging with ≈10 μm resolution was reported for pulse X-band.[9] Low-frequency pulse RF EPR, employing standard NMR microscopy gradients, achieved 2D and 3D images with resolution of ≈20–30 μm (in all three dimensions) [10,11] after ≈10 h of data acquisition. The unique crystals used in these experiments and the long acquisition time for 2D/3D images prevented it from being attractive for biophysical, botanical, and many other potential applications. An early 1D imaging work at high field (5 T, 140 GHz) achieved a resolution of ≈200 μm, limited by the gradient system (with maximum gradient of ≈30 T m^{-1}).[12] One should also note the early efforts in pulse X-band spin echo imaging.[13]

24.2.2 EPRM: From Test Samples to Real-life Applications

The above mentioned experimental work clearly showed the potential of EPRM; however, all these experiments were carried out on simple test samples, showing the methodology but not expanding it toward real-life applications. This is in contrast to conventional EPR imaging that is different from EPRM since it is carried out on cm-scale samples with mm-scale resolution, which soon after the introduction of the methodology in the 1980s found many applications that provided unique information, unavailable by any other technique. For example, conventional EPR imaging can show the spatially dependent oxygen partial pressure in tissues of small animals in vivo and could even be used for clinical applications (see Chapter 13).[14] On the other hand, EPRM seemed for many years to be just a nice experiment but lacking in terms of the required resolution to provide real unique advantage over other imaging techniques operating at the micron and sub-micron scale that are already well established (e.g., fluorescence and electron microscopy). Therefore, recent efforts in the field of EPRM focused both on significantly improving the available imaging resolution as well as seeking out unique scientific applications of EPRM that may be of broad interest. While at this point there are still many remaining challenges to address, the current state of EPRM is such that it can already provide unique scientific information for quite a long list of scientific applications, as will be exemplified in Section 24.5. This is especially true in view of the recent developments that make use of EPRM not only with the conventional induction–detection method but also with inherently more sensitive electric and optical detection methods (see Chapters 25 and 26), as will also be described in Section 24.5.3.

24.2.3 EPRM vs Conventional EPR Imaging

Chapter 13 already provided many explanations and details regarding the methodology of EPR imaging, and therefore here the focus will be only in the unique aspects of EPRM. These include not only unique applications but also some significant methodological differences. First, EPRM is often pursued in pulse mode, with *pulse field gradients* (as opposed to *quasi-static gradients* with conventional mm-scale CW or pulse EPR imaging). This unique imaging technique will be described in detail in Section 24.3.2. While pulse field gradients, which provide much higher versatility and better imaging resolution, are standard practice for NMR imaging, it is far less common for EPR. This is mainly because it is very difficult to generate short enough gradient pulses (much less than $\approx$1 μs, to be much shorter than T_2 of the measured spins; see Section 24.3.2), over volumes larger than a few millimeters, thus making this imaging approach unique and relevant only for EPRM.

A second unique methodological feature in EPRM is the drive to improve the absolute spin sensitivity rather than the concentration sensitivity. Absolute spin sensitivity relates to the minimal number of spins that can be measured by the system for 1 s of acquisition time and is reported in units of spins/$\sqrt{\text{Hz}}$. (This is because sensitivity improves with the square root of the accumulation time, assuming normal random thermal noise source, and the longer the acquisition time, the smaller the detection bandwidth, in Hz; see also Chapter 12.) On the other hand, concentration sensitivity relates to the minimal measured spin concentration by the system, denoted here in units of mol·L^{-1}/$\sqrt{\text{Hz}}$ (for the same square root time dependence reason mentioned above). In general, different applications would dictate different sensitivity requirements and thus steer toward the preferable spectrometer configuration. For example, if sample quantity is not a limitation, but one cannot increase its concentration (for various reasons: aggregation, mimicking biological conditions, or measurements on intact cells), then a configuration with optimal concentration sensitivity should be sought, as it is often the case in conventional EPR imaging. However, if there are severe limitations on the size of the sample available for measurements, then usually the goal is to measure it with imagers having high spin sensitivity. This is the more common case in EPRM since it often strives to improve more and more the image resolution. Such improvements are ultimately limited by the number of spins in each voxel (a 3D pixel), which becomes smaller and smaller as the image voxel size is decreased.

A third methodological uniqueness for EPRM is related to sample preparation and handling. Unlike optical microscopy, where samples can be placed on relatively large glass slides and imaged from the bottom or top, in EPRM the sample must fit into very small resonators, making it a mechanical challenge by itself. A few examples will be given in this chapter to show some available methodologies for sample preparation and handling, which solve some of these issues.

24.3 PULSE EPR MICROSCOPY: THEORY

This section provides the theoretical background about the factors governing the performances and capabilities of pulse EPRM systems. Following that, the pulse sequences and algorithms used for acquiring pulse EPR images are described in detail.

24.3.1 Pulse EPR Microscopy: Sensitivity Issues

As noted above, the ultimate limiting factor for EPRM capabilities is the absolute spin sensitivity, because as one strives to improve the spatial resolution, fewer and fewer spins are left in any given image voxel. The literature includes many expressions for the calculated absolute spin sensitivity of a given setup. Most of them are based on classical electrodynamics reasoning,[15–20] while more recently expressions were developed in the frameworks of cavity quantum electrodynamics.[21,22] Here the classical expression is briefly presented, which is valid under all relevant EPR and EPRM experimental conditions.

The general expression for pulse EPR spin sensitivity is given by:[23]

$$\text{Sensitivity (spins}/\sqrt{\text{Hz}})\approx \frac{8\sqrt{V_c}\sqrt{k_B T(1/\pi T_2^*)}}{\mu_B\omega_0\sqrt{2\mu_0}}\sqrt{\frac{\omega_0}{Q_L}}\sqrt{T_1}B_F \quad (24.1)$$

where V_c represents the resonator's effective volume,[16,24] which is equal to the volume of a small hypothetical sample V_v (e.g., [1 μm]^3, usually located at the point where the resonator's microwave magnetic field is maximal), divided by the filling factor[25] of this small sample (which is approximately the ratio between the volume of the sample and V_c). k_B is the Boltzmann constant, T is the temperature at which the experiment is carried out (assumed to be the same for the spins and the resonator), and $(1/\pi T_2^*) = \Delta f$ is the bandwidth of signal acquisition, chosen to match the spin–spin relaxation time constant (including static inhomogeneities), $T_2{}^*$. If T_2 is long enough to enable multi-echo acquisition sequence (such as CPMG), the sensitivity is accordingly improved and one can replace $T_2{}^*$ by T_2. The variable ω_0 is the Larmor angular frequency, Q_L is the loaded quality factor of the resonator, T_1 is the spin-lattice relaxation time, B_F is the Boltzmann population factor $B_F = \frac{1+e^{-\frac{\hbar\omega_0}{k_B T}}}{1-e^{-\frac{\hbar\omega_0}{k_B T}}}$, μ_B is the Bohr magneton, and μ_0 is the magnetic constant.

Analogous expressions for pulse EPR sensitivity appear in the literature.[15–17] Equation (24.1) (developed in Refs 19, 23, 24, 26) closely follows the original derivation by Mims[16] but without getting into the details of exciting just part of the EPR spectrum (i.e., full spectral excitation is assumed). Moreover, a variant of the principle of reciprocity, which characterizes the sensitivity of a receiver coil based on the strength of the magnetic field generated by a unit of current in it (common in pulse NMR literature),[27] is employed. This results in more simplified expressions than the ones of Mims.

Based on this equation it is evident that, for a given sample, spin concentration, and static magnetic field, sensitivity is determined mainly by the properties of the microwave resonator being used. *Thus, the optimum resonator should be tailored to a given application*. For example, if sample size is not limited, then it is evident via equation (24.1) that it is best to use the largest possible resonator with large V_c, filled by the sample, to obtain the highest SNR (thus maximizing the concentration sensitivity of the setup). In the case of size-limited samples, where the sample size is much smaller than the resonator size (such as micron- or nanoscale samples, or voxel sizes in imaging applications), the recommended devices are microwave resonators with the highest reasonable Q_L (although not too high, if the spectrometer dead time is of concern; see Chapter 12), and the smallest possible physical size (thereby maximizing the filling factor of the small sample). The size of conventional resonators is usually of the order of the wavelength, so at a typical frequency (e.g., 30 GHz with a wavelength $\lambda \approx 10$ mm, corresponding to $B_0 \approx 1$ T), the typical resonator dimensions would be $\approx\lambda/2 \approx 5$ mm. This is far too large for high-sensitivity/high-spatial resolution EPR. So the challenge of high absolute spin sensitivity EPR boils down to making resonant structures that 'focus' the microwave magnetic field component on to a size that is much smaller than the wavelength but still have good Q_L. Such miniature resonators have recently been developed by several groups, either in conjunction with EPRM or just in the general framework of enhancing absolute spin sensitivities,[28–30] as will be described in Section 24.4.

The results of equation (24.1) are summarized in Figure 24.1, which presents universal plots describing the ultimate limits to enhancing the spin sensitivity and the corresponding available image resolution for various static fields, temperatures, resonator volumes, and sample concentrations. Some experimental data (for spin sensitivity only) are also provided for reference. In general, equation (24.1) was verified in several experiments at a variety of static fields from $\approx$0.1 to $\approx$1.2 T. Based on Figure 24.1, it is clear that in order to obtain high sensitivity and thus good image resolution, the resonator dimensions must be decreased. Thus, for example, for a typical biological application with spin concentration of $\approx$1 mM measured at room temperature, one can hope to achieve 1 μm voxel resolution only with rather small resonator of $\approx$100 nL. Alternatively, for typical semiconductor application, with sample concentration of 10^{16} spins cm^{-3} at $\approx$10 K, one may reach resolution in the 10 s of nm range but only with very small resonator volume of $\approx$0.01 nL, whose typical dimensions are not larger than a few tens of microns.

24.3.2 Pulse EPR Microscopy: Resolution Issues

In addition to the sensitivity issues, the second limiting factor for EPRM resolution is the strength of the gradients employed to acquire the image. In order to understand the nature of this limitation, the image acquisition approach in pulse EPRM must first be briefly described. This will be carried out in conjunction with a typical imaging pulse sequence, shown in Figure 24.2. Let us examine the effect of this sequence on a simple system of 3 spins, located

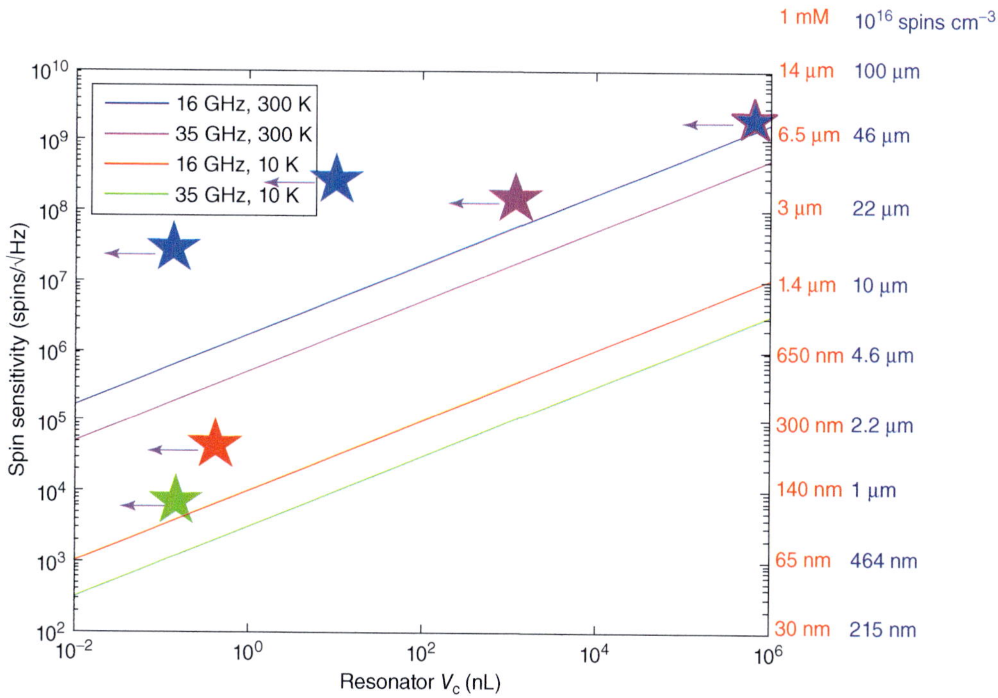

Figure 24.1. Calculated and measured absolute spin sensitivities (in spins/$\sqrt{\text{Hz}}$ – on the left scale) and the corresponding theoretical optimal image resolution for various resonator volumes, static fields, temperatures, and sample spin concentration. All calculations (solid lines) assume a resonator with $Q_L = 1000$ and an optimal sample with a single narrow line having $T_1 = T_2 = T_2^*$. The four different lines represent four different conditions of resonance frequency and temperature, as described in the figure legend. The calculated resolution scale on the right are representative for two types of samples, one with a spin concentration of 1 mM (units common in biological applications of EPR) and one with 10^{16} spins cm^{-3} (units common is EPR of semiconductors and solids). The value for the pixel resolution was calculated assuming a 3D image voxel with equal size in all axes and assumed an SNR of 10 is required for the voxel after 1 h of averaging time. A few experimental results of achievable spin sensitivity only (and NOT image resolution) are also placed on the plot for comparison and reference. The green star indicates measurements of P-doped ^{28}Si at Q-band and 10 K,[31] red star is for P-doped ^{28}Si at Ku-band and 10 K.[32] The pink star describes room-temperature experiments at Q-band with dielectric resonator and lithium phthalocyanine (LiPc).[26] The two blue stars are for E′ centers in irradiated SiO_2 at room temperature.[28] The pink/blue star represents typical spin sensitivity of a commercial Bruker cavity at room temperature

at three different positions along the laboratory x-axis (Figure 24.3). When the static field is homogenous, following a nonselective $\pi/2$ pulse, all the spins in the sample start to precess in-phase and generate an FID signal (Figure 24.3b). Upon application of a magnetic field gradient along the x-axis $\left(\frac{\partial B_z}{\partial x}\right)$, the spins in the left part of the sample are subjected to smaller local field, precess more slowly and thus lag in phase with respect to the spins on the right side (Figure 24.3c). The magnitude of these phase differences depends upon the strength of the gradient and its duration (more precisely, the time integral of the gradient). These phase variations affect the overall FID signal of all the spins that can be collected immediately after the end of the gradient pulses. By repeating this sequence for varying magnitudes of the gradients, the phase differences between the spins can be varied accordingly. These phase oscillations (shown in Figure 24.3d for one of the spin species) can be used to obtain the spatial distribution of the spins via simple Fourier analysis (see below). The same approach can also be employed in Hahn echo sequences, where the gradient

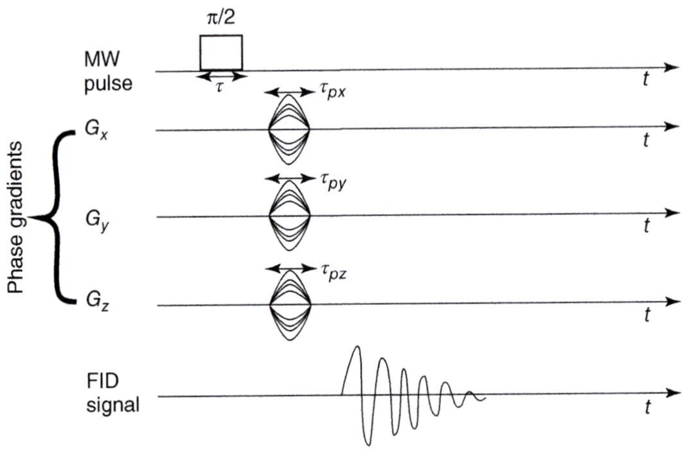

Figure 24.2. A typical simple pulse EPR microscopy imaging sequence including a $\pi/2$ pulse followed by phase gradient pulses that can be applied in all three axes and generate gradients with components of $G_x = \frac{\partial B_z}{\partial x}$; $G_y = \frac{\partial B_z}{\partial y}$; $G_z = \frac{\partial B_z}{\partial z}$. The FID of course starts immediately after the $\pi/2$ pulse, but for imaging purposes its data is collected only after the phase gradient pulses, so that it can be fully subjected to their effects (Figure 24.3)

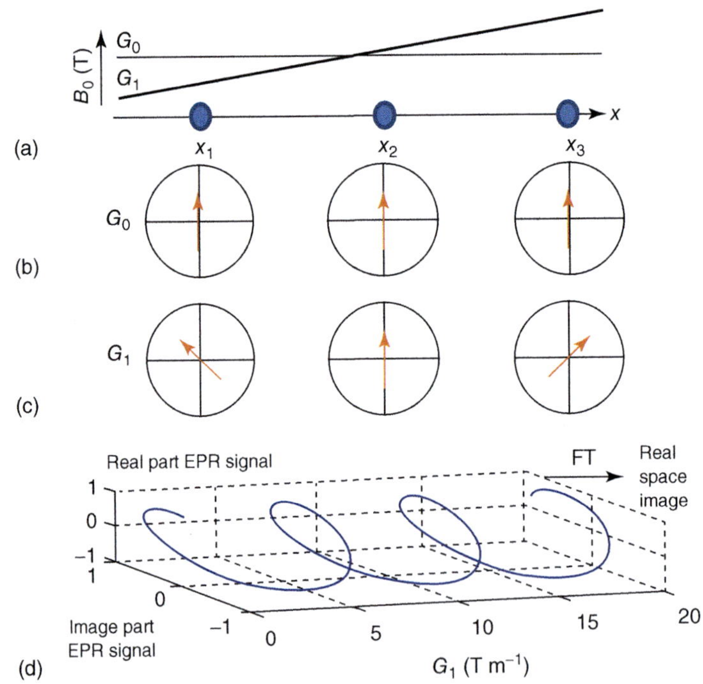

Figure 24.3. The effect of phase gradient pulses on spins located on the x-axis. (a) The static field vs the position along the x-axis without (G_0) and with (G_1) gradient applied. The blue dots represent three specific different point locations of spins in the sample. (b) The phase of the EPR FID signal of each individual spin, when no gradient is applied. (c) The phase of the EPR FID signal of each individual spin, when gradient G_1 is applied. Spins that were subjected to weaker local static field due to the gradient lag in phase vs spins that were subjected to stronger local static field. (d) The signal of a specific spin located at position x_1 along the x-axis, as a function of the strength of the applied gradient G_1, which shows phase oscillations between the real and imaginary parts

pulse can be applied in between the π/2 and π pulses, which is more suitable for species where $T_2^* \ll T_2$.[26]

The above qualitative discussion can be cast into more quantitative terms. The EPR signal for a given gradient strength and time duration (e.g., half-sine as shown in Figure 24.2) can be seen as the sum of signals from all the locations x_j ($j = 1 \cdots N$) in the sample with spin density $\rho(x_j)$:

$$S(G_x) = \sum_{j=1}^{N} \rho(x_j) e^{ix \cdot \gamma_e \cdot \int_{t=0}^{\tau_{px}} G_x(t) dt} \tag{24.2}$$

where γ_e is the electron gyromagnetic ratio and G_x is the gradient strength along the x-axis. The argument in the exponent, $x \cdot \gamma_e \cdot \int_{t=0}^{\tau_{px}} G_x(t) dt$, is the added phase term due to the gradient pulse, which is position-dependent. Inspection of equation (24.2) reveals that the parameters G_x and x can be seen as reciprocal Fourier variables (through the relation $k_x \equiv \frac{\gamma_e}{2\pi} \cdot \int G_x(t) dt$) and that the EPR signal as a function of the gradient pulse strength, $S(k_x)$, and the spin concentration as a function of x, $\rho(x)$, are their corresponding Fourier pairs. This means that indeed if one collects information in several values of G_x and then performs Fourier transform on the corresponding collected EPR signal $S(k_x)$, the value of $\rho(x)$ is obtained. The same notion can be generalized to three-dimensions by applying differing gradients along all 3 axes and then obtaining $\rho(\boldsymbol{r})$ via 3D Fourier analysis. This form of so-called phase gradient or Fourier imaging also preserves the spectral information because at the time of signal acquisition no gradient is applied. Thus, the FID time-domain data can be used via Fourier analysis to obtain the EPR spectrum as is done in simple 1D pulse EPR spectral analysis (see Chapter 15), and this spectrum can serve as a fourth dimension of the EPR image. Having formulated the imaging algorithm as a simple Fourier analysis enables us to provide a compact expression for the corresponding imaging resolution, which relies on analogous Fourier analysis of, e.g., time-frequency signals common in signal processing (providing the spectral resolution for a given time window). The resulting expression for the available imaging resolution, Δx, for a system that can produce maximum gradient strength G_x^{max} is:[2]

$$\Delta x = \frac{1}{(\gamma_e/2\pi) \cdot \int_{t=0}^{\tau_{px}} G_x^{max}(t) dt} \tag{24.3}$$

It is interesting to note that the relaxation time of the spin system does not explicitly appear in equation (24.3). However, since the gradient duration, τ_p cannot be much longer than T_2^* (or T_2, in case of an echo rather than FID imaging sequence),[26] it is clear that the relaxation time *does* affect the possible resolution of a given setup, since it limits the time integral of the gradient.

As a numerical example, one can consider a half-sine-shaped gradient pulse, with duration of 1 μs and peak amplitude of $50\,T\,m^{-1}$. In this case the resolution will be ≈1.13 μm. In Section 24.4.2 a typical EPRM experiment is described, with gradient efficiency of $\approx 4.6\,T\,m^{-1}\,A^{-1}$, meaning that a peak current of ≈12 A is required to reach a resolution of 1 μm. Such currents are readily available from 'homemade' drivers (see Section 24.4.2) that charge a capacitor and then discharge it into the gradient coil.[33,34]

24.3.3 Pulse EPR Microscopy: Available Image Contrasts

The imaging sequence shown in Figure 24.2 can provide either a 3D map of the spin concentration or, as noted above, can make use of the Fourier transform of the time-domain FID signal to extract the 3D spatially resolved EPR spectrum of the paramagnetic species in the measured sample as a fourth dimension in the image. This information can be used, for example, to extract the 3D oxygen concentration map in the measured sample, as shall be described in Section 24.5. The EPR spectrum, however, is not the only sample-related parameter that can be imaged. Namely, just as is common in the field of medical MRI, one can use other imaging pulse sequences to obtain other parameters of interest, such as the spatially resolved T_1, T_2, and potentially even the spin diffusion coefficient, in specific special cases. A detailed account of such imaging sequences can be found in the book of Callaghan[2] in the context of NMR imaging, but they are all readily applicable to pulse EPRM.

24.4 PULSE EPR MICROSCOPY: PRACTICE

Following the theoretical details of how the EPR image can be acquired and what are the primary limiting factors for this approach, the description will move into the world of practice. Here an outline is provided for the main parts of a typical experimental pulse EPRM system, namely, the sensitive resonator, the gradient coils and gradient drivers, and the supporting EPR imaging spectrometer. This 'homemade' system operates at Q-band and supports EPRM experiments at a wide range of temperatures and also enables to combine EPRM experiments with electrical and optical detection. Commercial pulse EPRM systems are available from Bruker, based on their pulse X-band spectrometer configuration, with additional static field gradients that can vary from acquisition to acquisition (in the ms time scale) but not during the sequence (see Chapter 13).

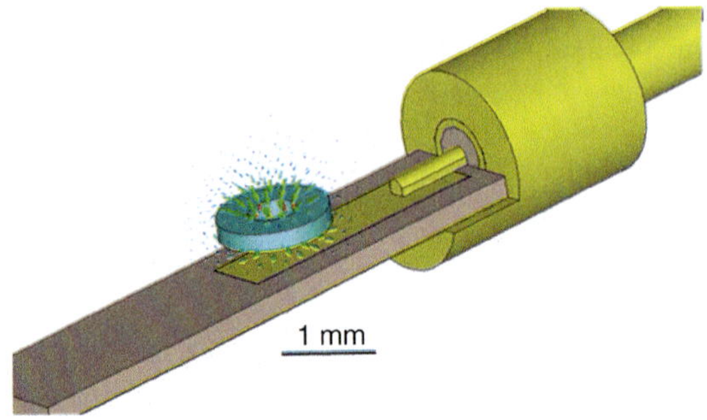

Figure 24.4. Miniature Q-band dielectric resonator. The microwave energy comes via a thin ≈1 mm coaxial line (on the right) that is converted to a microstrip line, which efficiently excites the dielectric resonator.[26] The microwave magnetic field (shown in colored arrows) is focused mainly in the center of the resonator

24.4.1 Sensitive Resonators

As noted above, EPRM often requires the use of miniature resonators with high absolute spin sensitivity. Based on Figure 24.1 and equation (24.1), it is clear that such a goal can be achieved using small structures with a highly confined resonance mode for the B_1 field. Two classes of such miniatures resonators used for EPRM are presented here. The first one is based on forming a high-permittivity dielectric material in the shape of a ring. Such structures, denoted as dielectric resonators (see Chapter 12),[35] can become very compact, provided that the permittivity of the material is very high. For example, at Q-band, a resonator made of a single crystal of TiO_2, which has permittivity of ≈130 (depending on crystal orientation), will be just ≈1.1 mm in outer diameter and ≈0.2 mm in height, with effective volume, V_c, of ≈500 nL. Figure 24.4 depicts the details of such a resonator. This type of resonator achieved at RT an experimental sensitivity of $\approx 10^8$ spins/$\sqrt{\text{Hz}}$, with a sample of LiPc,[26] which is worse by about a factor of ≈5 compared with the expected theoretical sensitivity for an optimal sample. At cryogenic temperatures, with a cryogenic low-noise amplifier (see below), the experimentally achievable spin sensitivity with a sample of phosphorus-doped ^{28}Si with a total number of $\sim 10^{10}$ spins was $\approx 10^6$ spins/$\sqrt{\text{Hz}}$.

A second type of resonator, which is much more compact in terms of its mode volume, is depicted in Figure 24.5. This type of resonator, denoted as 'surface loop-gap micro-resonator,' is made of a 'Packman-like-shaped' metallic layer deposited on a high-permittivity substrate.[28] Such a structure enables efficient focusing of the microwave magnetic field,

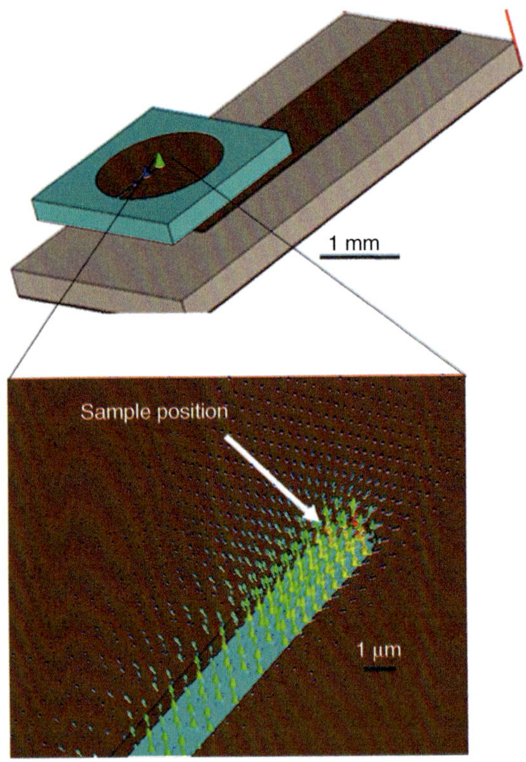

Figure 24.5. A surface loop-gap micro-resonator for Q-band with inner dimension of ≈2 µm. The upper plot shows the resonator placed on a high-permittivity substrate with the magnetic field, B_1, maximal in its center. The lower plot is a zoom-in showing the details of the inner part of the resonator, including the B_1 field distribution

B_1, to the center of the resonator, which is just a few microns in size. The corresponding effective volume of such resonator is ≈0.1 nL and its spin sensitivity at cryogenic temperature was experimentally found to be ≈6700 spins/$\sqrt{\text{Hz}}$ (for a sample containing ~2×10^6 spins within the volume of the resonator).[31]

24.4.2 Gradient Coils and Gradient Drivers

As noted above, the second ingredient of importance to EPRM is the capability to apply fast and strong magnetic field gradients. This is achieved by a combination of two unique features of the system. First is the use of miniature gradient coils (Figure 24.6) that have small inductance, L (≈few μH at most), and are very close to the resonator. The small inductance makes them very agile and supportive of the use of short strong current pulses (since the induced voltage drop on a coil with current i is $L\frac{di}{dt}$), as required to image radicals with short T_2. Furthermore, the close proximity to the resonator results in a high efficiency to produce high gradients of up to $\approx 4.6\,\text{T}\,\text{m}^{-1}\,\text{A}^{-1}$. The second feature is the use of uniquely fast pulse gradient drivers that are based on charging a capacitor and then discharging it into the gradient coil.[34]

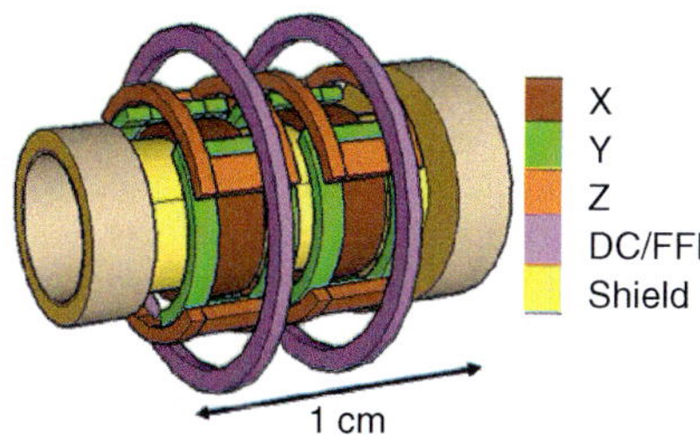

Figure 24.6. A set of gradient coils used for 3D EPRM at Q-band. The drawing shows the X, Y, and Z gradient coils that are placed in layers around a 2.8 mm o.d. cylindrical coil former that is covered by a thin (≈1 μm) gold shield to minimize radiation losses from the resonator inside the cylinder. The X-coil is the most inner one and is based on a Maxwell coil pair,[36] while the Y and Z coils are on outer shells and are based on Golay geometry.[36] The most outer coil (DC/FFL) is configured as Helmholtz pair (i.e., generates homogenous field rather than a gradient) and can be used either for field modulation in CW applications or to change the local field on the sample. Each coil typically has 20–40 turns, inductance of a few μH and resistance of ≈1 Ω

This type of circuit produces a half-sine current pulse, such as the one shown in Figure 24.2. While the basic architecture of this driver is fairly simple, namely, a capacitor–coil discharge, the actual implementation of it requires fairly sophisticated auxiliary electronics to quickly dump the energy in the circuit and avoid additional ring-down currents well beyond the nominal duration of the half-sine pulse (see Ref. 34 for full details). Furthermore, in order to minimize eddy-currents-based distortions of the applied gradient pulses, one must eliminate from the vicinity of the sample conductive elements with thickness larger than ≈2–3 μm (keeping in mind the upper gradient pulse frequency is ≈5 MHz, which corresponds to a skin depth of ≈30 μm in copper).

24.4.3 Pulse EPR Spectrometer for Microimaging

While the most important aspects of pulse EPRM instrumentation are related to the sensitive resonators and to the pulse gradients, which ultimately determine the performance of the system, there are still many complementary aspects necessary to facilitate the image acquisition, essential for the overall pulse EPRM spectrometer system. Figure 24.7 provides a simplified block diagram of a typical system. Some parts of it, like the pulse MW bridge, the magnet, and the gauss meter, can be found in any pulse EPR system, as described in Chapter 12. Other parts, like the control computer, need to have additional functionalities to what is found in 'simple' pulse EPR systems, while some items, such as the gradient drivers and the imaging probehead, are quite unique for pulse EPRM. Some details of the individual parts of the system will be briefly provided here, while the interested reader can find more details in Refs 26, 31, 37.

Control computer: The control computer includes a timing system to synchronize the multiple events in any given pulse imaging process. Events such as charge and discharge of capacitors for the gradients drivers, and opening and closing of fast MW switches must all be synchronized with time resolution of a few nanoseconds. In addition, the computer has an arbitrary waveform generator (AWG) card with time resolution of 0.5 ns that can be used to generate the required MW pulses and control their envelope shape and phase. These pulses are then sent to the MW bridge for up-conversion and further amplification. The control computer also determines the

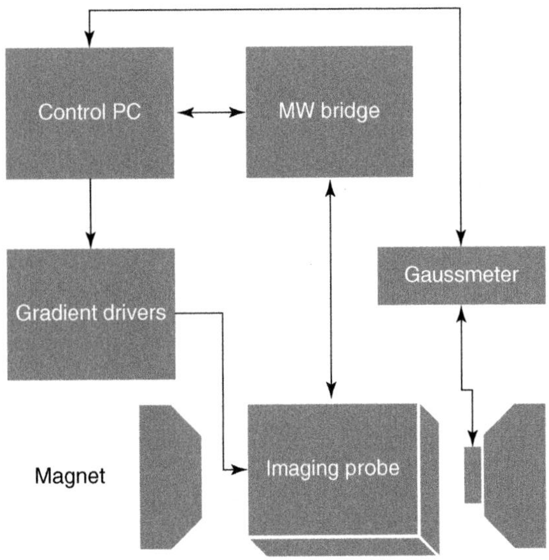

Figure 24.7. Block diagram of a typical pulse EPRM system. The imaging probe includes both the resonator and the gradient coils structure

amplitude of the gradients pulses, for any given axes, as they are being varied during the imaging procedure. It also controls and gathers information about the static magnetic field. Finally, the control computer collects the acquired EPR signal from the MW bridge and processes it via the Fourier analysis described above to generate the EPR image.

MW bridge: The bridge receives baseband signals from an AWG at ≈700 MHz and up-converts them to the required MW band. It then amplifies the MW signal and sends it out to the resonator at the imaging probehead. Following spin excitation, the acquired EPR signal is amplified back in the MW bridge and then down-converted to be digitized at the control computer.

Gradient drivers: The drivers generate intense and fast current pulses at the gradient coils, as was described above. The amplitude of the gradients is set by the computer and can change its value within a few microseconds. The gradient pulse events are controlled by the computer.

Imaging probehead: The main components of the probehead (resonator and gradient coils) were already described above. In addition, many times the probehead includes a circulator and a sensitive cryogenic amplifier to maximize the sensitivity of the setup. It also includes a mechanism to control the coupling of the resonator to the MW bridge. Details of the various imaging probeheads used in EPRM can be found in Refs 31, 37, 38.

The above mentioned instrumental setup employs the conventional induction–detection to acquire the EPR signal and construct the EPR image. EPRM can also be pursued in conjunction with electrical[39] or optical detections,[40] which can have improved sensitivity. Experimental examples for images obtained with these alternative detection methods are given in Section 24.5. In terms of the spectrometer architecture, they require additional parts such as sensitive current amplifier for electrical detection and a laser source and detector for optical detection. The full details of these alternative setups are given in Refs 39, 40 (see Chapters 25 and 26).

24.5 EXPERIMENTAL EXAMPLES FOR EPRM APPLICATIONS

24.5.1 Materials Science

The new capabilities described above can be used in a variety of scientific and technological applications. For example, in the field of semiconductor devices, high-resolution EPR imaging enables one to observe diffusion and migration phenomena of point defects in amorphous oxides.[41] Amorphous oxides (serving as isolators or superconductors) are key ingredients in electronic and optical devices. Such oxides include a variety of point defects that greatly affect their electrical and optical properties. Many of these defects are paramagnetic and, as such, the best tool to identify them and characterize their structure is EPR. However, due to its limited sensitivity and spatial resolution, traditional EPR could not provide information about the defects' migration properties, which are of crucial importance for device fabrication. Ultra-high resolution imaging modalities such as transmission electron microscope (TEM), as well as theoretical calculations, are severely limited in amorphous media, resulting in a wide experimental and theoretical knowledge gap in this field. EPRM was applied to examine unique samples that are prepared using e-beam irradiation and have well-defined point defect patterns. This provides a capability to unambiguously identify the defects and at the same time track their migration with high spatial resolution,

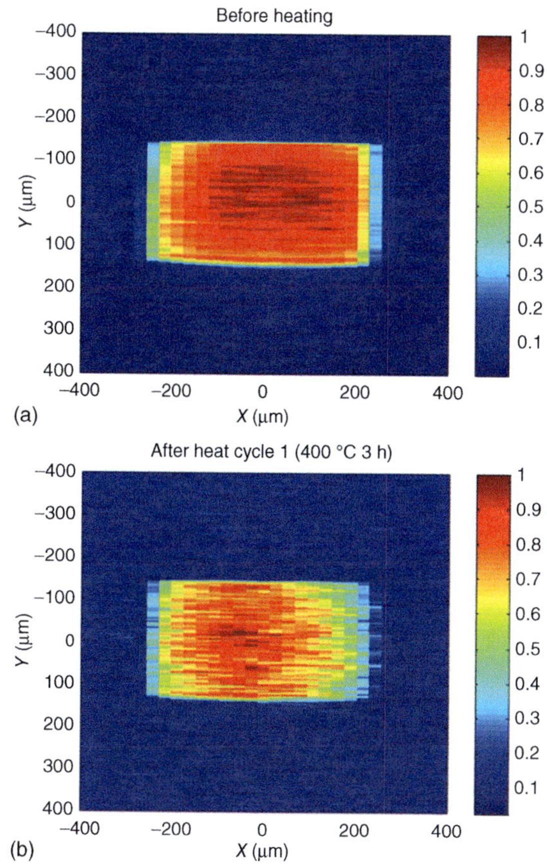

Figure 24.8. Pulse EPR 2D images of the irradiated SiO_2 sample before (a) and after (b) heating cycle at 400 °C for 3 h. The heat cycle causes some changes in the spatial pattern that can be analyzed to provide information about forces acting on the defects in the SiO_2 and processes they undergo.[41] Image resolution is ≈1.6 × 27 µm and acquisition time is 17 h

revealing new information about their properties, such as the elastic force acting on the defects, their binding energy to the lattice, and the activation barrier for their diffusion. Figure 24.8 shows a typical example of EPR imaging results of an amorphous SiO_2 piece on which a rectangular pattern of point defects was created by e-beam irradiation. The piece was imaged immediately after preparation and then following a 3 h heat cycle at 400 °C. Changes in the image were analyzed to obtain valuable information about atomic-level potentials and forces between point defects in SiO_2.[41]

24.5.2 Life Science

Another example, taken from a completely different field, is focused on the use of EPRM for mapping oxygen in sub-mm-sized tissues.[42] Oxygen (O_2) plays a central role in most living organisms. The concentration of O_2 is important in physiology and pathology. Despite the importance of accurate knowledge of the O_2 levels, there is very limited capability to measure with micron-scale spatial resolution its distribution in millimeter-scale live biological samples. Many of the current oximetric methods, such as oxygen microelectrodes and fluorescence lifetime imaging, are compromised by O_2 consumption, sample destruction, invasiveness, and calibration difficulty. In the case of biological samples, EPR imaging requires the incorporation of a suitable stable and inert paramagnetic spin probe into the desired object. Microcrystals of a paramagnetic spin probe such as tg-LiNc-BuO (lithium naphthalocyanine in a new crystallographic packing form) can be used for this application. These paramagnetic species interact with the paramagnetic oxygen molecules, causing spectral line broadening to the tg-LiNc-BuO that is linearly proportional to the oxygen concentration. This new oximetry microimaging method addresses all the problems mentioned above. It is noninvasive, sensitive to physiological oxygen levels, and easy to calibrate. Furthermore, in principle, it can be used for repetitive measurements without causing cell damage. The tissue model used in this research was grown in a shape of a sphere (known as cancer spheroids) of human colorectal carcinoma cell line (HCT-116) with a typical diameter of ≈600 µm. Most studies of the microenvironmental O_2 conditions inside such viable spheroids carried out in the past used microelectrodes, which require invasive puncturing of the spheroid and are also not applicable to 3D O_2 imaging. High-resolution 3D oxygen maps made it possible to evaluate the relationship between morphological and physiological alterations in the spheroids, which would help in understanding the oxygen metabolism in solid tumors and its correlation with the susceptibility of tumors to various oncologic treatments. Figure 24.9 shows a typical example of an oxygen map measured for one of the spheroids with the imaging sequence shown in Figure 24.2. The spectral line broadening was extracted from the Fourier transform of the FID time-domain data as the fourth dimension of the image (3D spatial–1D spectral). The oxygen concentration was evaluated based on this spectral broadening via

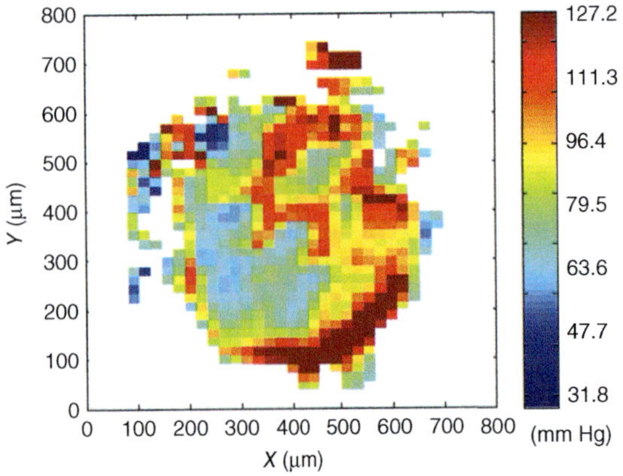

Figure 24.9. Oxygen map of a colorectal carcinoma spheroid containing tg-LiNc-BuO. The map shows the oxygen concentration in various parts of the spheroid, as derived from their measured spatially resolved EPR spectrum.[42] Image resolution $\approx 11 \times 17.9 \times 80$ µm (only one of the z-slices is shown), acquisition time 20 min

a calibration curve. It indicates that there is more oxygen on the exterior parts than in the inner parts, as one would expect for such tissues. It also shows the significant heterogeneity such spheroids may possess with respect to their oxygen concentration.

24.5.3 EPR Microscopy with Alternative Detection Methods

EPRM-based approaches can be pursued also with alternative detection methods, such as electrically detected magnetic resonance (EDMR; see Chapter 26) or optically detected magnetic resonance (ODMR; see Chapter 25).

EDMR microimaging can provide images of paramagnetic states in an operating real-world electronic device. Figure 24.10 shows how this method can be applied to the imaging of a device-grade 1-µm-thick amorphous silicon (a-Si : H) solar cell and an identical device that was degraded locally by an electron beam. The degraded areas with increased concentrations of paramagnetic defects lead to a local increase in recombination that is mapped by EDMR imaging with ≈20-µm-scale pixel resolution.

ODMR provides an ultrasensitive means to detect and image a small number of electron and nuclear spins, down to the single-spin level with nanoscale resolution. It can be employed only with specific samples whose fluorescence signal depends on their electron spin's state. Figure 24.11 shows an example of ODMR image of a diamond sample having specific fluorescent paramagnetic defects called nitrogen vacancies (NVs). The image shows how ODMR can be combined with pulse EPRM to spatially encode the sample. This results in what is denoted as an 'optically detected magnetic resonance imaging' (ODMRI) technique. It offers the advantage that the image is acquired in parallel from all parts of the sample, with well-defined three-dimensional point-spread function, and without any loss of spectroscopic information. In addition, this approach may be used in the future for parallel but yet spatially selective efficient addressing and manipulation of the spins in the sample. Such capabilities are of fundamental importance in the field of quantum spin-based devices and sensors.

24.6 CONCLUSIONS AND FUTURE PROSPECTS

The field of pulse EPRM has made significant progress over the last few years, due to the development of ultrasensitive resonators, imaging capability at cryogenic temperatures, fast and powerful gradient drivers, miniature gradient coils, and the introduction of novel stable spin probes. It is now possible to routinely provide images of samples of interest with a spatial resolution of 1 µm or less, combined with spin sensitivity that is better than $\approx 10^4$ spins/$\sqrt{\text{Hz}}$. This opens the door to a wealth of applications, ranging from life to materials science, as was demonstrated here. Furthermore, EPRM is useful in basic physics applications related to quantum computing.[43] Namely, EPRM could be used to selectively control and manipulate a small ensemble and even a single spin, out of a given sample, which is of significant importance in quantum information processing. The combination of EPRM-based approaches with electrical and optical detection techniques further increase its power and can lead, for example, to a variety of new applications in the field of metrology of electronic devices. Finally, EPRM has been the catalyst for the development of methodologies for ultrasensitive induction–detection EPR, which by itself can find uses in applications such as measurements of micro-single crystals of biological materials, measurements of spin monolayers, and in general the measurements of small spin-limited samples. Future progress, going to W-band frequencies and using even smaller

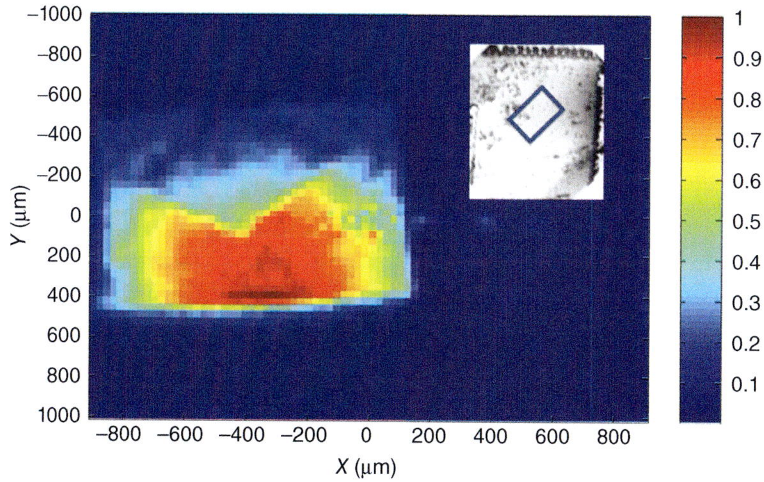

Figure 24.10. Pulse EDMR images of an e-beam-degraded solar cell, measured at 10 K. The degraded area is shown in the SEM image in the inset (upper right), framed by the blue diamond shape. The amplitude of the signal in each pixel is proportional to the concentration of the paramagnetic defects, weighted by the light irradiation profile, which was not homogenous on all the cell surface. The color code is linearly scaled and normalized to the largest pixel signal in the image.[39] Image resolution ≈20 × 20 μm, acquisition time 4 h

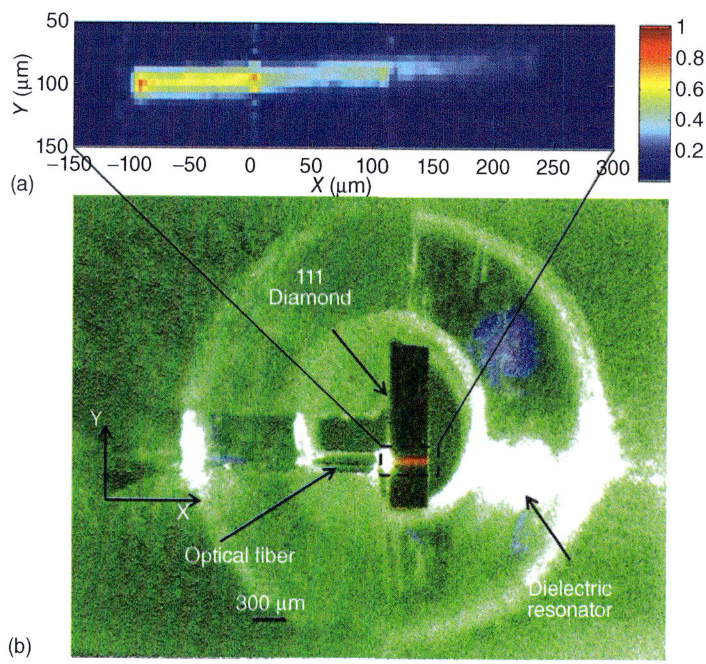

Figure 24.11. (a) ODMR image of the fluorescence pattern generated by fiber illumination of the diamond crystal. (b) Optical image of the diamond in the resonator during laser illumination, showing the fluorescence red signal at the center of the crystal.[40] Image resolution ≈3.5 × 5 μm, acquisition time 15 min

micro-resonators, should enable EPRM to enter the regime of nanoscale resolution and single-spin EPR.

REFERENCES

1. S. K. Misra, *Multifrequency Electron Paramagnetic Resonance, Theory and Applications*, Berlin: Wiley-VCH, 2011.
2. P. T. Callaghan, *Principles of Nuclear Magnetic Resonance Microscopy*, Oxford University Press: Oxford, England, 1991.
3. W. Karthe and E. Wehrsdorfer, *J. Magn. Reson.*, 1979, **33**, 107.
4. M. J. R. Hoch, *J. Phys. C-Solid State Phys.*, 1981, **14**, 5659.
5. M. J. R. Hoch and A. R. Day, *Solid State Commun.*, 1979, **30**, 211.
6. T. Herrling, N. Klimes, W. Karthe, U. Ewert, and B. Ebert, *J. Magn. Reson.*, 1982, **49**, 203.
7. T. Herrling, J. Fuchs, and N. Groth, *J. Magn. Reson.*, 2002, **154**, 6.
8. G. R. Eaton, S. S. Eaton, and K. Ohno, *EPR Imaging and In Vivo EPR*, CRC Press: Boca Raton, 1991, 321.
9. G. G. Maresch, M. Mehring, and S. Emid, *Physica B & C*, 1986, **138**, 261.
10. A. Coy, N. Kaplan, and P. T. Callaghan, *J. Magn. Reson. Ser. A*, 1996, **121**, 201.
11. A. Feintuch, G. Alexandrowicz, T. Tashma, Y. Boasson, A. Grayevsky, and N. Kaplan, *J. Magn. Reson.*, 2000, **142**, 382.
12. A. I. Smirnov, O. G. Poluectov, and Y. S. Lebedev, *J. Magn. Reson.*, 1992, **97**, 1.
13. A. D. Milov, A. Y. Pusep, S. A. Dzuba, and Y. D. Tsvetkov, *Chem. Phys. Lett.*, 1985, **119**, 421.
14. G. L. He, A. Samouilov, P. Kuppusamy, and J. L. Zweier, *Mol. Cell. Biochem.*, 2002, **234**, 359.
15. P. P. Borbat, R. H. Crepeau, and J. H. Freed, *J. Magn. Reson.*, 1997, **127**, 155.
16. W. B. Mims, in *Electron Paramagnetic Resonance*, ed S. Geschwind, Plenum Press: New York, 1972, Chap. Electron spin echoes, 263.
17. G. A. Rinard, R. W. Quine, J. R. Harbridge, R. T. Song, G. R. Eaton, and S. S. Eaton, *J. Magn. Reson.*, 1999, **140**, 218.
18. T. Yalcin and G. Boero, *Rev. Sci. Instrum.*, 2008, **79**, 094105.
19. A. Blank and J. H. Freed, *Isr. J. Chem.*, 2006, **46**, 423.
20. G. A. Rinard, R. W. Quine, R. T. Song, G. R. Eaton, and S. S. Eaton, *J. Magn. Reson.*, 1999, **140**, 69.
21. D. I. Schuster, A. P. Sears, E. Ginossar, L. DiCarlo, L. Frunzio, J. J. L. Morton, H. Wu, G. A. D. Briggs, B. B. Buckley, D. D. Awschalom, and R. J. Schoelkopf, *Phys. Rev. Lett.*, 2010, **105**, 140501.
22. A. Bienfait, J. J. Pla, Y. Kubo, M. Stern, X. Zhou, C. C. Lo, C. D. Weis, T. Schenkel, M. L. W. Thewalt, D. Vion, D. Esteve, B. Julsgaard, K. Mølmer, J. J. L. Morton, and P. Bertet, *Nat. Nanotechnol.*, 2015, **11**, 253.
23. Y. Twig, E. Suhovoy, and A. Blank, *Rev. Sci. Instrum.*, 2010, **81**, 104703.
24. A. Blank, C. R. Dunnam, P. P. Borbat, and J. H. Freed, *J. Magn. Reson.*, 2003, **165**, 116.
25. C. P. Poole, Electron Spin Resonance: A Comprehensive Treatise on Experimental Techniques, 2nd edn, John Wiley & Sons, Inc: New York, 1983.
26. L. Shtirberg, Y. Twig, E. Dikarov, R. Halevy, M. Levit, and A. Blank, *Rev. Sci. Instrum.*, 2011, **82**, 043708.
27. D. I. Hoult, *Concept Magn. Reson.*, 2000, **12**, 173.
28. Y. Twig, E. Dikarov, and A. Blank, *Mol. Phys.*, 2013, **111**, 2674.
29. R. Narkowicz, D. Suter, and I. Niemeyer, *Rev. Sci. Instrum.*, 2008, **79**, 084702.
30. H. Malissa, D. I. Schuster, A. M. Tyryshkin, A. A. Houck, and S. A. Lyon*arXiv:1202.6305*, 2012.
31. Y. Artzi, Y. Twig, and A. Blank, *Appl. Phys. Lett.*, 2015, **106**, 084104.
32. A. Blank, E. Dikarov, R. Shklyar, and Y. Twig, *Phys. Lett. A*, 2013, **377**, 1937.
33. M. S. Conradi, A. N. Garroway, D. G. Cory, and J. B. Miller, *J. Magn. Reson.*, 1991, **94**, 370.
34. L. Shtirberg and A. Blank, *Concept Magn. Reson. B*, 2011, **39B**, 119.
35. D. Kajfez and P. Guillon, Dielectric Resonators, Artech House: Dedham, MA, 1986, 539.
36. J.-M. Jin, Electromagnetic Analysis and Design in Magnetic Resonance Imaging, Boca Raton: CRC Press, 1999.
37. Y. Twig, E. Dikarov, and A. Blank, *J. Magn. Reson.*, 2012, **218**, 22.
38. A. Blank, E. Suhovoy, R. Halevy, L. Shtirberg, and W. Harneit, *Phys. Chem. Chem. Phys.*, 2009, **11**, 6689.

39. I. Katz, M. Fehr, A. Schnegg, K. Lips, and A. Blank, *J. Magn. Reson.*, 2015, **25**, 26.

40. A. Blank, G. Shapiro, R. Fischer, P. London, and D. Gershoni, *Appl. Phys. Lett.*, 2015, **106**, 034102.

41. E. Dikarov, R. Shklyar, and A. Blank, *Phys. Status Solidi Appl. Mater. Sci.*, 2014, **211**, 2177.

42. M. Hashem, M. Weiler-Sagie, P. Kuppusamy, G. Neufeld, M. Neeman, and A. Blank, *J. Magn. Reson.*, 2015, **256**, 77.

43. A. Blank, *Quantum Inf. Process.*, 2013, **12**, 2993.

Chapter 25

Optically Detected Magnetic Resonance (ODMR)

Etienne Goovaerts

University of Antwerp, Antwerp, Belgium

25.1 INTRODUCTION

Detection of the resonant interaction between microwaves (MWs) and electron spins is in most EPR techniques performed by measuring changes in the transmitted or reflected MW field (CW EPR) or by measuring emitted MW (pulse EPR). However, it has been soon realized that changes in the electronic spin states can influence other observables, allowing for alternative detection schemes.[1,2] The most important examples are optical detection of magnetic resonance (ODMR) and electrical detection of magnetic resonance (EDMR), which are discussed in this and the following chapter, respectively, and are very often complementary. Indeed, many processes in atomic, molecular, and solid-state systems depend on the electron spin, and this is reflected in the optical and electronic properties of materials, leading to well-studied effects such as magnetic circular dichroism (MCD) and magnetoresistance, respectively. In EPR experiments, we are able to manipulate the electronic spin state, which provides us with a golden opportunity to alter readily accessible observables, e.g., intensity and polarization of light in absorption or emission spectra and electrical conductivity. In this chapter, I will focus on the influence of MW resonances on optical properties important in ODMR, but in quite a number of cases, the same processes are involved in electronic transport properties, which will be discussed in the chapter on EDMR (see Chapter 26).

The range of systems that can be studied by ODMR techniques is limited compared to EPR in general. The optical detection scheme restricts its application to systems with characteristic optical absorption or emission that are influenced by the spin state of the unpaired electron(s). This can result directly from spin dependence of the transition probabilities or indirectly by the changes in the population of the states involved as a result of spin-dependent generation and decay processes. When applicable, the optical detection methods offer important advantages:

I. They give access to a number of systems, such as excited states of molecules and charge carriers in various organic and inorganic semiconductors

EPR Spectroscopy: Fundamentals and Methods.
Edited by Daniella Goldfarb and Stefan Stoll
 ISBN: 978-1-119-16299-5
Also published in eMagRes (online edition)
DOI: 10.1002/9780470034590.emrstm1524

(SCs) that are hard or impossible to detect otherwise.

II. As thoroughly discussed elsewhere,[3,4] they hold the potential for very high sensitivity of spin detection. This essentially rests on the transformation of quanta from the MW to the optical range. For the latter, detectors with high quantum efficiency (QE) are readily available, and, contrary to MW quanta, the optical photons are practically free of thermal noise. As a result, the measurements approach the shot-noise limit and their sensitivity is ultimately determined by the available optical intensity on the detector. The high sensitivity that can be reached is illustrated by single-molecule ODMR measurements on triplets states of organic molecules, achieved already two decades ago.[5,6]

III. As a further bonus of ODMR, there is selectivity for detection of only those spins that are involved in the probed optical transition against a background of unrelated paramagnetic species, helping to clarify the origin of the measured signals. As discussed later on, the so-called EPR-tagged optical spectroscopy also provides for attribution of features in absorption or emission spectra to specific paramagnetic species observed in ODMR.

IV. As in the case of transient EPR of excited states (Chapter 10), optical excitation in many cases leads to preferential population of spin states. One then achieves much larger population differences within the spin multiplet, i.e., larger spin polarization, than expected from Boltzmann statistics at a given temperature, significantly boosting the signals.

A large number of variants of ODMR spectroscopy have been developed over the past decennia, based on different mechanisms by which the spin state influences the optical observable, each demanding a specific approach for the excitation and detection. These ODMR variants all rely on the spin dependence of optical transition probabilities in absorption and emission and eventually of nonradiative relaxation between states. The latter includes as an important class the so-called intersystem crossing (ISC) between states of different total spin. Finally, a basic feature is the occurrence of population differences between the spin states, i.e., the spin polarization, either thermally induced or as a result of optical pumping.

Table 25.1 presents a general overview of the different types of ODMR, their basic characteristics and specific denomination, many of which will be explained in more detail in the following sections. The following section consists of a basic description of the instrumentation for ODMR measurements in zero magnetic field (ZF) and high magnetic fields (HFs: the energy from the electron Zeeman interaction is large compared to the zero-field splitting of the spin multiplet), ending with a short survey of combined techniques providing additional information on kinetics, spin relaxation,

Table 25.1. Overview of possible approaches to ODMR spectroscopy: Changes in an optical observable are induced by microwave excitation at an EPR resonance of a paramagnetic system, either in a ground or an excited electronic state

Microwave-induced changes in:	Detected in:	
	Absorption	Emission
Light intensity (both in ZF and HF ODMR)	Absorption: **ADMR**	Fluorescence (or photoluminescence): **FDMR** (or **PLDMR**) Delayed fluorescence: **DFDMR** Phosphorescence: **PDMR**
Degree of circular polarization (in HF ODMR)	Magnetic circular dichroism in absorption: **MCDA**	Magnetic circular polarization in emission: **MCPE**

The observable can be the intensity of light transmitted by the sample (absorption-detected magnetic resonance: ADMR) or emitted by the sample (either fluorescence-, delayed-fluorescence- or phosphorescence-detected magnetic resonance: FDMR, DFDMR, or PDMR, respectively). Besides these IUPAC-backed terms and abbreviations, the term photoluminescence-detected magnetic resonance, PLDMR, is also quite widely in use for FDMR. When the ODMR is based on the circular polarization of light, this is denoted by MCDA (in absorption) and MCPE (in emission). In specific systems, optical excitation, eventually with circularly polarized light, can induce a nonthermal population of the states. This so-called optical pumping process can enhance the ODMR signal. Applied to recombination emission in a semiconductor, this special case of MCPE is called conduction-electron-spin resonance (CESR).

or electron–nuclear interactions. Then, I will discuss ODMR spectroscopy of triplet states involving ISC to or from singlet states.[7–9] In this chapter, the spin resonance can be detected through changes in total emitted optical intensity and is often observable in zero field as well as under an applied magnetic field. In half-integer spin systems, straightforward intensity detection is only possible when the Kramers' degeneracy can be sufficiently lifted by a magnetic field, selecting individual spectral features within a Zeeman split multiplet. This approach has been followed in particular systems with very narrow optical spectra. In condensed matter however, the spectral line widths are in the majority of cases much larger than the attainable Zeeman splitting, and spectral separation has to be replaced by polarization-selective schemes, using magnetic circular dichroism in optical absorption (MCDA) or magnetic circular polarization in emission (MCPE).[3,4] I will further pay specific attention to the case of weakly coupled pairs of $S = 1/2$ spins, which has been studied for donor–acceptor pairs in inorganic SCs.[4,10] Finally, another class of spin pairs will be considered in organic SCs involving different charge excitations, positive and negative polarons (P^+ and P^-), and triplet excitons (TEs). Interactions between pairs of these excitations are important for the understanding of the optoelectronic properties of such materials.[11–15]

25.2 INSTRUMENTATION

25.2.1 Zero-field ODMR

As described already in Chapter 3, spin triplet ($S = 1$) systems with lower-than-cubic local symmetry, in molecules or in crystals, display splitting of the energy levels even in ZF. Magnetic-dipole transitions induced by MWs are then possible between any pair of states within the triplet. In the most common scheme (see Figure 25.1a), the ODMR spectrum is therefore measured as a function of the MW frequency ω_{MW} by monitoring the MW-induced change in the intensity of the emitted light, either fluorescence detection of magnetic resonance (FDMR), or delayed fluorescence detection of magnetic resonance (DFDMR), or phosphorescence detection of magnetic resonance (PDMR).[8,16] Alternatively, the intensity of a transmitted light beam can be detected to probe the absorption of the sample (see Figure 25.1b).[17] In absorption-detected magnetic resonance (ADMR), the wavelength of the probe beam can be chosen to match an absorption band of the studied system. This can be either from the ground state or from a long-lived excited state, the so-called excited-state absorption (see also Section 25.3.1). In the latter case, simultaneous photoexcitation is applied to populate this excited state, as shown in Figure 25.1(b).

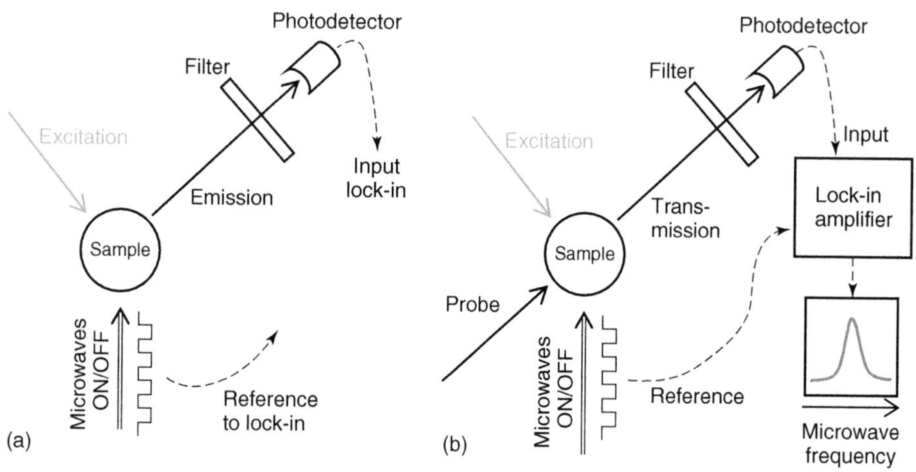

Figure 25.1. Detection scheme for ZF ODMR based on intensity changes of (a) emission (FMDR, DFDMR, and PDMR) and (b) the transmission of a light beam through a sample probing its optical absorption (ADMR). The microwave amplitude is typically ON/OFF modulated by means of a microwave switch. Phase-sensitive detection using a lock-in amplifier allows for sensitive detection of the intensity changes induced by the microwaves. The spectrum is obtained by scanning the microwave frequency

As tuning of the MW frequency is required, one needs the source and all components to be broadband, and thus, contrary to most EPR setups and also to those for HF ODMR (see Section 25.2.2), the use of high-Q resonators is excluded. In a classical experimental configuration, the source consists of a tunable generator in the gigahertz region followed by an MW amplifier and a device to irradiate the sample, which can be a broadband antenna such as a simple loop or a helix, or a more sophisticated narrowband MW device with resonance frequency adjusted to that of the generator. Optical access is required for light excitation from a lamp or a laser and for collection of the emitted light onto a detector of high sensitivity in the wavelength region of interest. Wavelength selectivity and/or tuning will often be implemented in excitation and/or in collection, by optical filtering or using monochromators, as required by the optical properties of the system under investigation and by the specific design of the experiment.

In most ODMR experiments, modulation techniques are applied to detect the fractional changes in intensity induced by the resonant MWs, usually small and often in the range 10^{-3} to 10^{-4}. The field modulation technique generally used in CW EPR is not applicable in ZF and replaced by amplitude modulation of the MWs, often by ON/OFF modulation. The modulation frequency needs to be adapted to the response time of the ODMR signal, determined by the spin population relaxation time, T_1, eventually combined with other bottlenecks in the optical cycle. Indeed, intermediate states with long lifetimes can occur especially at low temperatures, which will often determine this response time. This can be exploited to determine characteristic times of such processes or discriminate between signals with different response times. In ZF ODMR, the optimal modulation frequency will often be lower than 100 kHz, which is the near-standard value in CW EPR. Low-noise detectors and amplifiers are combined with phase-sensitive detection to discriminate the ODMR signal from the noise, fundamental shot noise on the detected optical intensity, and thermal, $1/f$, and other sources of noise.

Finally, even in ZF ODMR spectrometers, the researchers may want to apply and vary a static magnetic field as an additional condition allowing for a more detailed evaluation of the properties of the spin system. Indeed, frequency shifts of the resonances and their dependence on field orientation can be exploited. Fortunately, only small fields (in the tens of millitesla range) are usually required because of the typical limited frequency range, and it is therefore not hard to reach acceptable field homogeneity. Such a static field can be obtained from simple coils or even a permanent magnet. It is also possible to control the orientation of such small fields without moving parts by varying the currents through a set of coils surrounding the sample area, such as applied in imaging setups.

25.2.2 High-field ODMR: Zeeman Energy Larger than Zero-field Splitting

Different from the ZF case, ODMR spectroscopy at higher fields has often, but not exclusively, been performed in modified EPR spectrometers, taking advantage of the extensive technological developments and the commercial availability of these instruments.[3,4,9,10,12,18,19] In a typical configuration (Figure 25.2), the sample is placed in a high-Q/high filling factor MW resonator in which optical access is implemented by various approaches. One can obtain single-mode MW cavities from commercial sources, featuring semitransparent metalized windows, slits in the side walls, or openings adjusted in size and geometry to keep the MWs sufficiently confined. More dedicated designs include the well-known (multimode) Fabry–Pérot resonator,[20–22] open resonator structures,[23–25] or light guides or fibers accommodated in the sample holder.[19]

Magnetic fields induce distinct optical properties for light with right circular polarization (RCP) or left circular polarization (LCP), propagating along the field (Faraday configuration, see Figure 25.3). Absorption of light with RCP and LCP becomes different, which is known as MCD, and the effect is proportional to the strength of the magnetic field. Besides the (at least for our purpose) uninteresting diamagnetic effect, there is a paramagnetic contribution to the difference in absorption coefficients $\alpha_L - \alpha_R$ for the two circular polarizations. This contribution depends on the populations of the spin states quantized along the field direction. Similarly, emission along the field from paramagnetic centers shows different intensities for LCP and RCP, which is called magnetic circular polarization (MCP). This effect can be exploited in the MCDA and MCPE detection schemes measuring the MW-induced changes in circular polarization of the transmitted or emitted light, using elements to manipulate and select optical polarization (see Figure 25.3), which can be linear polarizers, half- and quarter-wave plates, and optical phase modulators

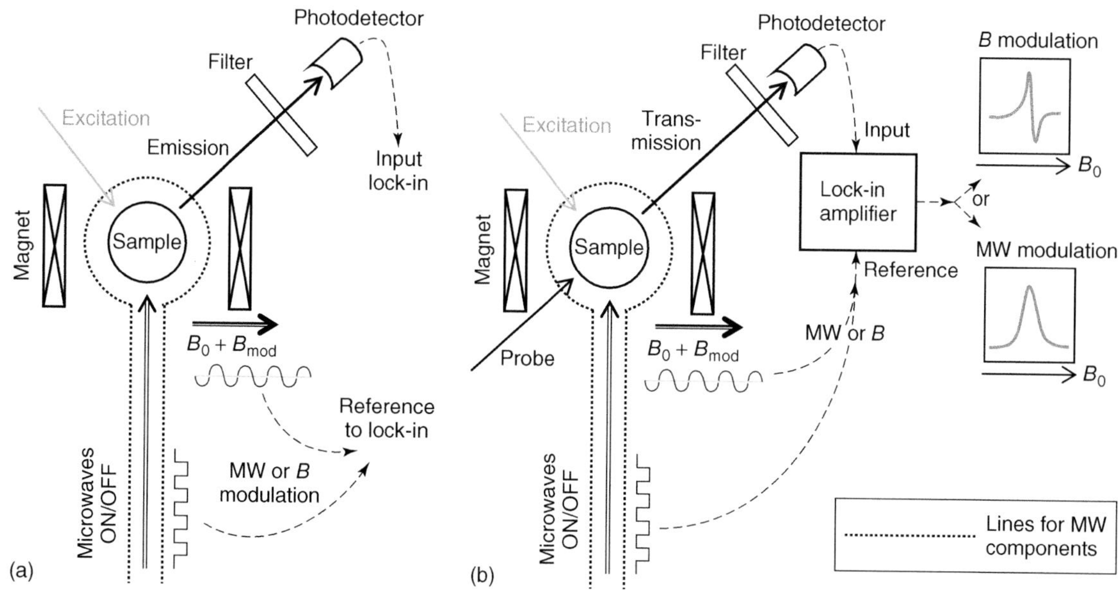

Figure 25.2. Detection scheme for HF ODMR based on intensity changes of (a) emission (FMDR, DFDMR, and PDMR) and (b) the transmission of a light beam through a sample probing its optical absorption (ADMR). Two modulation modes can be envisaged, either microwave amplitude as described for ZF ODMR (see Figure 25.1) or modulation of the static magnetic field as is applied in continuous-wave EPR. In the latter case, the first derivative of the ODMR signal is registered. In HF ODMR, the microwave frequency is typically kept constant and matched to the mode of a high-Q resonant cavity. The spectrum is registered as a function of the strength of the static magnetic field, as in continuous-wave EPR

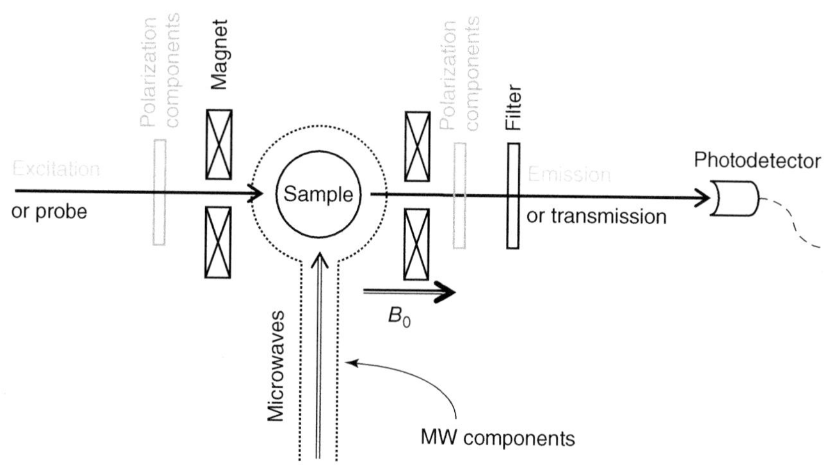

Figure 25.3. Scheme for the detection of HF ODMR via magnetic circular dichroism in absorption (MCDA) or magnetic circular polarization in emission (MCPE). The light beams propagate along the magnetic field (Faraday geometry). A combination of a linear polarizer and a phase modulator is employed to periodically vary the selected polarization between LCP and RCP. In MCDA, this polarization modulation is performed on the incident light, combined with linear polarization analyzer in the transmitted beam. In MCPE, the excitation is linearly polarized, and polarization modulation is applied in the path of the emitted light

(acousto- or electro-optical).[3,4,9,10,12] A necessary condition for the application of these schemes is the existence of population differences between the spin states with opposite magnetic quantum number, i.e., $+m_S$ and $-m_S$ in spin multiplets or $+m_J$ and $-m_J$ in J manifolds, namely spin polarization. This already exists in thermal equilibrium when the Zeeman splitting is comparable or larger than k_BT. Besides the thermal spin polarizations, population differences can be induced by optical pumping in specific systems possessing spin preferential transitions. The sensitivity of MCDA/MCPE measurements increases with applied field owing to the linear field dependence of the circular polarization effects, which follows from the field-induced spin polarization in thermal equilibrium.

Contrary to ZF ODMR, the MW frequency range is limited by the use of a high-Q resonator. Therefore, the static magnetic field is scanned to match the resonance conditions and spectra are typically taken in field-swept mode as in standard CW EPR. It is also possible to apply the conventional field modulation combined with phase-sensitive detection for effective noise rejection, which leads to first-derivative spectra (Figure 25.2). However, MW amplitude modulation, also depicted in Figure 25.2, can be profitably applied in HF ODMR particularly for spectra broader than the achievable modulation amplitudes, providing direct signal spectra rather than the first derivative. As mentioned for ZF ODMR earlier, the modulation frequency has to be adapted to the response time of the system, determined by the electronic transition rates and spin relaxation time T_1. Quite interestingly, field modulation at higher frequency (such as the 100 kHz that is standard in EPR instruments) can be used in HF ODMR to increase the signal in systems with inhomogeneously broadened resonances. Affording sufficient MW power, a larger number of centers then contributes to the ODMR signal as the field is rapidly scanned over spin packets with different resonance fields.

Moreover, in ODMR, hand in hand with conventional EPR, there has been a development toward increasing MW frequency from Q-band upward, in correspondingly HFs. This was realized sometimes in ODMR-dedicated instruments or else as an extension to EPR spectrometers.[19–22,26–32] The reasons valid for high-field EPR (spectral resolution, separation of contributions in the spin Hamiltonian, and absolute sensitivity) apply here as well, but in addition, the MCD effects linearly increase with applied field, which increases sensitivity in MCDA and MCPE detection schemes.

There may be various reasons for a dedicated design of HF ODMR instruments. One motivation is the realization of the instrumentation at lower cost, taking into account that the sophisticated detector and circuits in the MW bridge of EPR instruments are not needed. On the contrary, there can be demanding requirements, such as versatility in the mode of operation, or high specifications (stability and low noise) of optical excitation and detection, or finally demanding combinations with other functions such as classical or scanning-probe microscopy. In this case, the conventional EPR spectrometer may not be suitable as starting point for the design of the ODMR instrument.

25.2.3 Temperature in ODMR Measurements

ODMR spectroscopy has most often been performed at low temperature, at liquid He temperatures and down to the λ-point, although experiments in the range up to 100 K are not exceptional. The ODMR signal is usually rapidly lost at higher temperatures because of the increasingly fast spin relaxation. Population (T_1) relaxation tends to spoil the optically induced spin polarization or alignment, and dephasing processes may lead to excessive broadening of the transitions. While for many of the triplet excited states in organic molecules the best conditions are found at the very lowest temperatures, experiments can be performed at higher temperatures in systems with slower relaxation. A quite exceptional and intensively studied system, further discussed in the following, is the negatively charged nitrogen-vacancy (NV^-) defect in diamond,[33] which possesses a triplet ground state with extremely long room-temperature T_1 values in the millisecond range and still appreciably long spin–spin relaxation time, T_2, in the microsecond range, and as a result is easily detectable even at room temperature. Taking aside such special cases, the use of cryostats with optical access is required in ODMR and is an important aspect of the design of the spectrometers.

25.2.4 Related and Combined Methods

Finally, a range of techniques can be applied to obtain additional information on the spin systems under scrutiny, adding time-resolved and pulsed techniques to the different ODMR approaches.

Transient measurements of the optical signal after sudden modification of conditions, such as optical or MW excitation, or magnitude of the static magnetic field (between on- and off-resonance), give access to kinetic properties of the photophysical system. It was understood early on that the relevant rates of population relaxation within spin multiplets as well as rates of ISC and radiative and nonradiative decay could be extracted from well-devised experiments of this type.[3,16,34]

Complementary to ODMR in many of the studied systems, but not strictly a magnetic resonance technique as no MWs are applied, is level anticrossing (LAC) spectroscopy.[18] In systems with higher spin multiplicity, the energy levels as a function of magnetic field may cross or, as in most cases, undergo avoided crossings under the influence of the Zeeman interaction. Even small matrix elements of the Hamiltonian are efficiently mixing the states close to the field of near-crossing. This modifies the rates of spin relaxation and of transitions to other electronic states, leading to changes of optical observables in a similar way as in ODMR. The so-called LAC is detected by scanning the magnetic field while monitoring the optical observable (emission intensity, MCDA or MCPE). In addition, transient detection after a sudden change in static field can be used to determine kinetic parameters of the system.

While first CW MW excitation was used, pulse techniques have also soon been implemented in ODMR[9,35–38] to study spin dynamics with various pulse sequences (see Chapter 11). They have been applied to investigate spin relaxation in the excited or ground state and also the interaction with other electronic spins or with nearby nuclear spins (hyperfine spectroscopy) became accessible. Pulsed laser excitation can be part of the pulse sequence eventually creating spin polarization far more pronounced than in thermal equilibrium, as is also exploited in transient EPR (see Chapter 10). In addition, the optical read-out provides information on spin populations. Thus, to study the coherences, one needs to apply dedicated MW pulses transferring spin polarization into coherence and vice versa. Besides investigation of the spin dynamics, pulse ODMR techniques have in recent years been intensively applied for the manipulation of spin states in their read-out for quantum information applications (see Chapter 22). Various paramagnetic centers have been proposed for this purpose, among which the NV^- center in diamond is the most successful one. Its basic properties from the ODMR point of view will be discussed in Section 25.3.2.

ENDOR is a powerful double-resonance method providing a high-resolution tool for hyperfine spectroscopy (see Chapter 16). When RF excitation of the sample is added to the possibilities of the ODMR instrument, one can implement optically detected ENDOR (ODENDOR).[4,21,22,39–41] The approaches for implementation are very similar to those described for ENDOR, but further developments in combination with open resonators are being reported.[42] As shown in Chapter 16, the resonant excitation of nuclear spins by radiowaves is able, under double resonance conditions, to influence the spin polarization of electrons that are coupled to the nuclear spins by hyperfine interactions. This will in turn show up in the ODMR effect as the latter is depending directly on the spin polarization. ODENDOR combines all the useful features discussed here for ODMR, with the opportunity to determine the interactions between the involved electron spin and nuclear spins in its neighborhood, giving detailed clues to unravel the microscopic structure of the paramagnetic center.

25.3 ODMR OF TRIPLET STATES IN MOLECULES AND CRYSTALS

25.3.1 Excited-state Triplet

Excited triplet states can be produced in systems with a singlet ground state by optical excitation to higher singlet states and subsequent ISC crossing (see Figure 25.4). This can be molecules in solution or in the solid state (molecular crystals or macromolecules). In fact, conjugated organic molecules, and the crystals thereof, have been a playground of ODMR spectroscopy since the start of the field. In Chemistry, these triplet states are often important intermediate states in reactions, leading to intensive research.[7,9,43] Conjugated ligands are found in many relevant biomolecules such as proteins, DNA, and RNA, and excited triplet states take part in biochemical reactions. Therefore, triplet ODMR was developed as an important method for biochemistry and biomedicine.[8,9,44,45] Triplet excited states are essential in the photophysics of ionic crystals as well, e.g., the formation of self-trapped excitons (STEs) in ionic crystals such as alkali halides[46,47] and silver chloride[20,48] and also in the related nanostructures.[48,49] Excited triplet states in crystals of conjugated organic

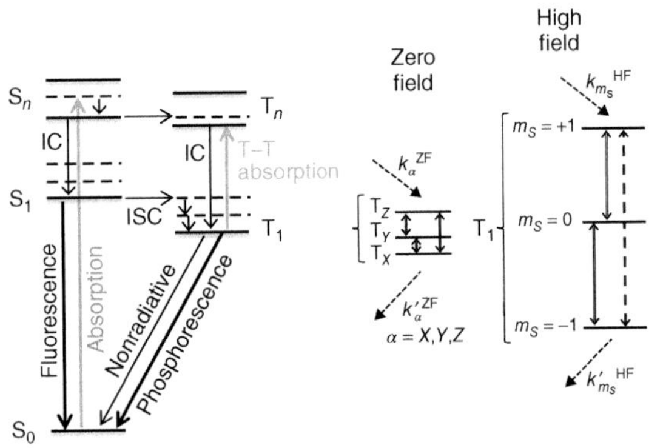

Figure 25.4. Level scheme of a system with singlet ground state S_0 and triplet excited state T_1 ($D < 0$, arbitrary choice). Optical transitions in absorption and emission are indicated in gray and black, respectively, and thin arrows stand for nonradiative transitions. Optical excitation can lead to the lowest excited singlet S_1 or at higher incident photon energies to vibrationally excited (hot) S_1 states or higher singlet states S_n. The latter decay rapidly to S_1 by internal conversion (IC). From either of the singlet states, ISC can occur to a vibrational sublevel of a triplet state, subsequently decaying to the lowest triplet T_1 by vibrational relaxation and (from the higher-lying T_n) by IC. Decay to the ground state can proceed radiatively (phosphorescent) or nonradiatively. The magnetic sublevels of T_1 in the ZF and HF cases are shown on expanded scales. Sublevel-dependent rate constants k and k' are indicated, respectively, from the excited singlet to the triplet and from the triplet to the ground state, as discussed in the text. All EPR transitions (double arrows) within the triplet are allowed in zero field, but only two allowed transitions remain after quantization in a sufficiently high field. The so-called half-field transition (dashed double arrow) is often observed as a weak but characteristic triplet feature

molecules or in conjugated polymers, often called TEs, will be discussed separately because different processes are involved in ODMR (see Section 25.5.2).

As shown in Figure 25.4, in systems with a diamagnetic spin-singlet ground state, S_0, optical transitions are spin conserved and lead to excited electronic singlet states S_n with total spin $S = 0$. Next to these singlet states, there is a manifold of excited triplet levels T_n, in which the spins of the two unpaired electrons sum up to a total $S = 1$. Transitions between the singlet and triplet levels become possible as a result of state mixing by the spin–orbit coupling (see Section 2.4 of Ref. 8). The triplet levels can be populated under optical excitation through ISC (Figure 25.4). Owing to fast relaxation among states of the same total spin, the system commonly ends up in the lowest triplet level T_1 (Kasha's rule), which possesses a long lifetime because transitions to the singlet ground state are (nearly) spin forbidden.

In the absence of a magnetic field and in sufficiently low symmetry, the triplet is split into three sublevels with corresponding states denoted as T_X, T_Y, and T_Z. The rates of ISC, indicated in Figure 25.4 (zero-field scheme), depend on the specific sublevel, both in the singlet-to-triplet transition and in the decay toward the ground state (the latter defining the lifetime of these levels). In each of the steps between the states of singlet and triplet manifolds, the transition probabilities will be governed by matrix elements of spin-dependent interactions (e.g., the spin–orbit and spin–spin interactions), which will be different for different sublevels. In HF, i.e., when the Zeeman energy becomes much larger than the zero-field splitting (ZFS) of the triplet, the field determines the quantization leading to eigenstates $|S = 1, m_S\rangle$ of the spin system (see Figure 25.4, high-field scheme). Moreover, in this case, the ISC rates will in general depend on the spin quantum number m_S and the orientation of the field. To a good approximation, the ISC rates in HF can be found from those in ZF by projection of the field-quantized states on the ZF eigenstates. For example, for the transitions from S_1 to the $|S = 1, m_S\rangle$ states of T_1, this yields the rate constants: $k_{m_S}^{\mathrm{HF}} = \sum_{\alpha=X,Y,Z} |\langle T_1, \alpha | T_1, m_S\rangle|^2 k_\alpha^{\mathrm{ZF}}$, in which the k_α^{ZF} stands for the corresponding ZF probabilities from S_1 to the T_X, T_Y, and T_Z states. Similar relations hold for the outgoing transitions between the HF and ZF rates, $k'^{\mathrm{HF}}_{m_S}$ and k'^{ZF}_α, respectively. In the high-field limit, the $m_S = \pm 1$ states are related by time-reversal symmetry, and as a result, the probabilities of ISC transition between each of these states and a singlet state should be the same. A particular preferential population ($n_0 \neq n_{-1} = n_{+1}$, for the $m_S = 0$, -1 and $+1$ states, respectively) is obtained. This does not involve a net magnetic magnetization and is sometimes called spin alignment rather than spin polarization. However, it is possible to obtain selective population of either the $m_S = -1$ or $+1$ states by optical pumping with circularly polarized light in the Faraday geometry, as discussed for the case of Kramers systems in Section 25.4.

Continuous optical excitation produces a steady state of populations that is determined by the excitation intensity, the direct decay rate from the excited singlet S_1 back to S_0, the ISC rates to and from the triplet, and the population (T_1) relaxation between the triplet states. Assuming different ISC rates for the three states of the triplet, optical excitation induces population differences between them. If the intra-triplet relaxation is slow compared to the decay to the ground state, these population differences can be much larger than usual in thermal equilibrium, which is in general favorable for ODMR detection. Faster relaxation within the triplet will lead to partial or complete thermalization, tending to a Boltzmann distribution of the populations with weaker ODMR signals.

In an ODMR experiment, MW excitation is brought in resonance with a transition between two states within the triplet, which will affect their steady-state populations, tending to decrease the difference between them. In zero field, all transitions between pairs of states are allowed as described in Chapter 3. In HF ODMR field, one finds two allowed $\Delta m_S = \pm 1$ transitions, although a third transition between the outer states, with $\Delta m_S = 2$, is often appearing with lower intensity and is used as a characteristic feature of triplet states. This becomes weakly allowed by perturbation of the dominant Zeeman interaction by the ZFS term in the spin Hamiltonian (see Chapter 3).

The most obvious observable for optical detection is the change in intensity of the phosphorescence, the radiative decay to S_0 starting from each of states in the lowest triplet, T_X, T_Y, and T_Z in zero field or the $|S=1, m_S\rangle$ states in high field. This directly probes redistribution of their populations by the MWs, each having their own probability for radiative decay. Either increase or decrease in the emission intensity is possible depending on the redistribution of the populations and on the relative radiative transition probabilities from the different states of the triplet.

MW transitions can also be detected as changes in the fluorescence intensity corresponding to the $S_1 \rightarrow S_0$ transition. Indeed, the decay probabilities from the three triplet states correspond to different lifetimes and therefore also the total triplet population is affected, as is the corresponding amount of depletion of the singlet ground state population. This ground state depletion can be detected in ADMR, as the rate of S_0 to S_1 excitations is proportional to the ground state population. Indirectly, the latter can be probed by FDMR via the induced change in the fluorescence intensity, which is also proportional to the rate of excitations and thus to the S_0 population.

Another scheme is detection via the total triplet population by excited-state absorption spectroscopy, which is a standard technique to characterize triplet states in molecules.[50] The population of the T_1 level is often significant owing to its long lifetime. If the transition probability of the spin-allowed optical transition $T_1 \rightarrow T_n$ is independent of the initial spin state within T_1, the photoinduced absorption yields a signal proportional to the total triplet population. This scheme clearly requires additional features in the ODMR instrument: In addition to the optical excitation creating the triplet state, the sample is now irradiated with light at the appropriate $T_1 \rightarrow T_n$ wavelength, while monitoring the MW-induced change of the transmitted intensity at this wavelength. Detecting fluorescence or triplet excited-state absorption can be applied to systems that do not or weakly emit phosphorescence.

As a ZF example, fluorescence-detected ODMR spectra are shown in Figure 25.5 of chlorophyll *a* molecules in binding proteins involved in the photosynthesis of algae organisms.[51] Two transitions

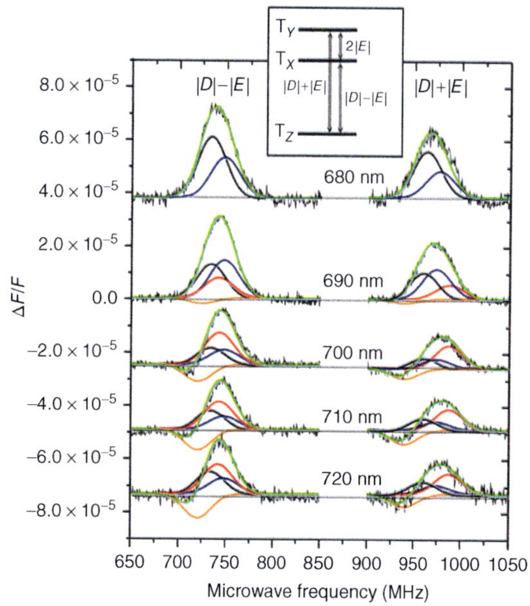

Figure 25.5. ZF ODMR spectra of the excited triplet state (denoted ^{3}Chl) of chlorophyll *a* molecules in a binding protein that acts as light harvesting center of an organism of the algae family. The fluorescence-detected spectra show the $|D|-|E|$ and $|D|+|E|$ transitions (see inset) within the ^{3}Chl state. The experimental spectra are compared to simulation results (colored curves) assuming a sum (green curve) of spectra of four subpopulations of chlorophyll sites. (Reproduced with permission from D. Carbonera, Biochem. Biophys. Acta – Bioenerg., 2014, 1837, 1235–1246. © Elsevier, 2014)

between the spin states in the triplet are visible (see inset), which consist of several components that change in relative intensity upon variation of the excitation wavelength. Together with transient and pulse EPR measurements, the ODMR results on systematically varied samples of the binding protein were providing information on protective mechanisms based on triplet–triplet energy transfer (TTET) processes.

To illustrate HF ODMR of excited triplets, Figure 25.6 shows the W-band (95 GHz) spectrum of an AgCl single crystal under UV-light inducing band-to-band transitions. Detection is performed via the fluorescence emitted throughout the visible region, with the magnetic field along a cubic axis, ***B*** || <100>. Transitions of strongly localized self-trapped excitons (STEs) are marked for the excitonic states oriented parallel and perpendicular to the field. In addition to the STE spectrum, one observes ODMR signals corresponding to free electrons and self-trapped holes (STHs), which are also strongly localized. In these W-band measurements, the different species yield well-separated spectra, which would not be the case at lower MW frequencies. The appearance of the free electron and STH signals is attributed to recombination processes (see Section 25.5.1).

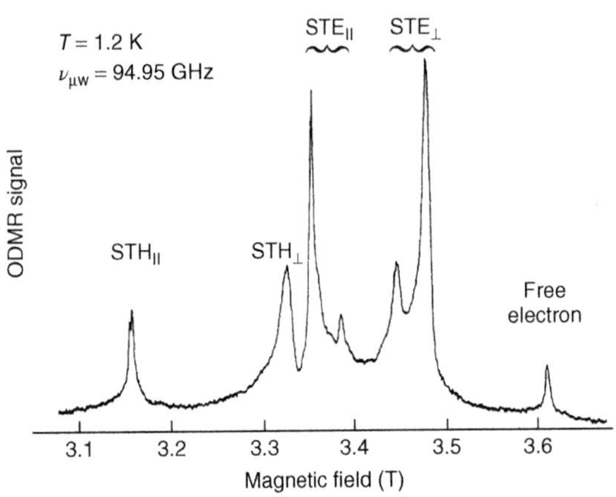

Figure 25.6. HF ODMR spectrum detected via the fluorescence ($\lambda_{em} \approx 510$ nm) of a single crystal of AgCl (***B*** || <100>) measured at W-band (95 GHz), showing contributions from the triplet STE together with doublet resonances of the STH and free electron. Parallel and perpendicular resonances result from *g*-anisotropy (for both STE and STH) and ZFS (for STE). (Reprinted figure with permission from M. C. J. M. Donckers, O. G. Poluektov, J. Schmidt, P. G. Baranov, Phys. Rev. B, 1992, 45, 13061–13063. Copyright © 1992, American Physical Society)

25.3.2 Ground State Triplet – The Case of the NV$^-$ Center in Diamond

A system with a triplet ground state presents an analogous, in a sense inverted, scheme of ISC between singlet and triplet states (see level scheme in Figure 25.7a), but a number of differences appear. For the eigenstates T_X, T_Y, T_Z and T'_X, T'_Y, T'_Z of the ground and excited triplets, one considers again either the ZF case (T_X, T_Y, T_Z and T'_X, T'_Y, T'_Z, respectively) or the case of quantization by a large magnetic field ($|S=1, m_S\rangle$ states). Optical excitation is essentially spin conserving and now leads to higher states T_n of the triplet manifold eventually relaxing into the lowest excited triplet state T_1. From here, ISC can now occur toward the states S_n of the singlet manifold at a rate that will depend on the state from which it originates, ending up in the lowest and long-living singlet state S_1. From either T_1 or S_1, decay can occur to the ground triplet T_0, radiatively (fluorescence or phosphorescence, respectively) or nonradiatively at rates that differ for the states of the initial/final triplets. Very much like in the previous case of an excited-state triplet system, unequal populations of the states of T_0 and T_1 will be produced by optical excitation. The population differences are then altered by resonant MW excitation, which can be probed via the optical emission intensity. An important difference with the previous system is the comparatively short lifetime of the excited triplet, owing to the allowed transition to the ground state. This makes it usually tougher to populate the excited triplet and reach detectable ODMR signals. However, in the ground state, the population differences can be appreciable if the ISC is selective and the optical pumping rate is sufficient to compete with the T_1 relaxation time.

The most prominent system in the literature with such an optical cycle is the NV$^-$ center in diamond, consisting of an electron trapped on a carbon vacancy and a substitutional nitrogen impurity in nearest-neighbor positions (see model in Figure 25.7a). It has been the subject of extensive research over the last decennia, with perspectives for a wide range of applications.[33] The NV$^-$ center, having two unpaired electrons with $S=1$ in the ground state, possesses a trigonal symmetry along the <111> crystal

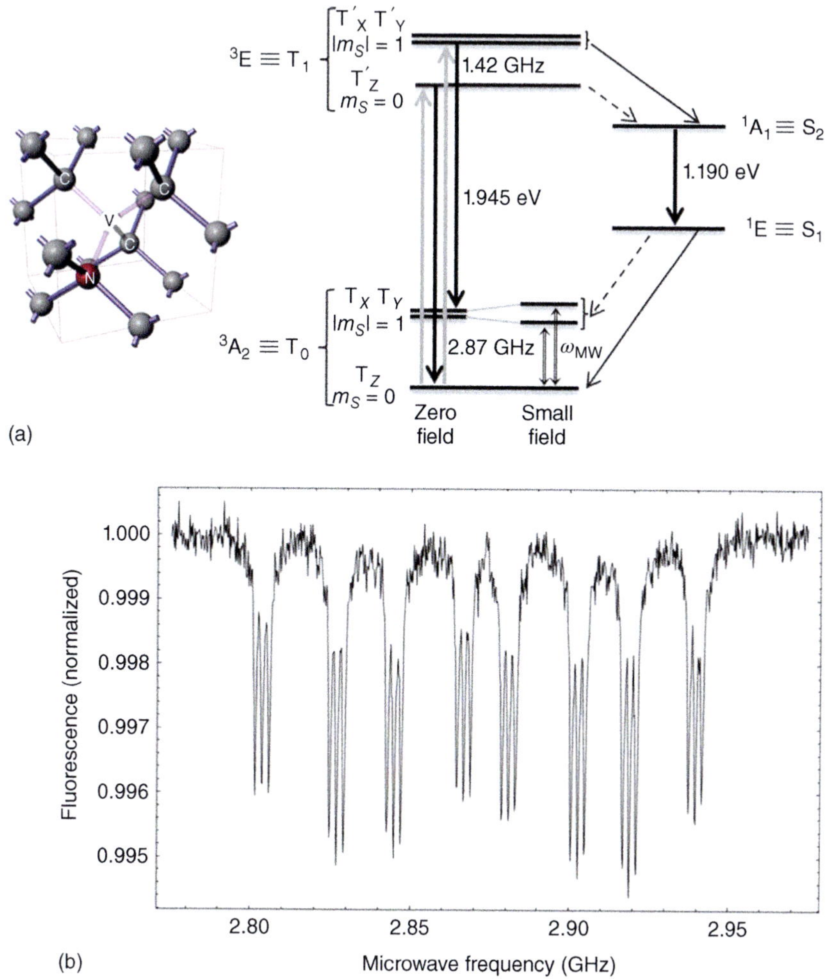

Figure 25.7. (a) Model and energy level scheme of the NV^- center in diamond showing the ZFS in trigonal symmetry of the ground and excited triplet states ($D = 2.87$ GHz and 1.42 GHz, respectively), the transitions for optical excitation (gray) and fluorescence (black), and the ISC transitions (thin arrows) to and from the singlet states. The ISC rates depend on the spin state as indicated by full and broken lines. This induces spin polarization in the ground triplet level. Microwave-induced (ω_{MW}) transitions are indicated in the ground triplet. (b) ODMR spectrum of NV^- centers in a single-crystal diamond under a weak (≅5 mT) magnetic field lifting the degeneracy of the T_X, T_Y doublet. The field is applied in a low-symmetry direction at different angles with each of the four <111> defect orientations, leading to a pair of transitions for each defect orientation, each showing a triplet substructure due to nitrogen hyperfine interactions. (Reprinted figure with permission from V. M. Acosta, E. Bauch, M. P. Ledbetter, C. Santori, K.-M. C. Fu, P. E. Barclay, R. G. Beausoleil, H. Linget, J. F. Roch, F. Treussart, S. Chemerisov, W. Gawlik, D. Budker, Phys. Rev. B, 2009, 80, 115202. Copyright © 2009, American Physical Society)

direction and hence an axial ZFS matrix leading in ZF to a nondegenerate lowest T_Z state (corresponding to $m_S = 0$ in quantization along the defect axis) separated by 2.87 GHz from the twofold degenerate level of the T_X and T_Y states (consisting of the $m_S = \pm 1$ states in quantization along the defect axis). An energy level scheme for ZF ODMR is shown in Figure 25.7(a). A weak magnetic field with component along the defect axis lifts the twofold degeneracy as illustrated in this figure for the ground triplet, which results in a pair of transitions with splitting proportional to this magnetic field component. In Figure 25.7(b), the

ODMR spectrum of NV^- centers in a single crystal is shown.[52] A symmetrical pattern of four pairs of ODMR lines is observed, corresponding to the four <111> defect orientations, which are at different angles to the applied weak magnetic field. Each of the lines is split by ^{14}N ($I = 1$) hyperfine interactions. Very strong spin alignment can be reached near zero field in this system, even at room temperature, leading to high contrast between the fluorescence intensities without and with MWs at the resonance frequency and thus high ODMR sensitivities.

25.4 ODMR OF HALF-INTEGER SPIN SYSTEMS

25.4.1 Ground-state EPR Detected by Magnetic Circular Dichroism

Detecting ground-state spectra is in many cases possible by more straightforward EPR methods and one may wonder why ODMR should be applied. In addition to the higher sensitivity that can be achieved, there is an important added value in the selectivity of ODMR as a function of photon energy, which allows a more efficient separation of overlapping EPR spectra and very often a direct correlation with bands in optical absorption or MCD spectra. In the opposite way, one has been able to assign defect structures to previously known but not yet attributed optical bands.

In paramagnetic systems with half-integer spin, stemming from an odd number of unpaired electrons, the eigenstates in ZF are pairwise related by time-reversal symmetry, leading to at least twofold degeneracy, forming Kramers doublets. In ZF, there is no selectivity between the $+m_S$ and $-m_S$ spin states, neither for optical excitation nor for probing. The Zeeman interaction splits these doublets energetically, as the two states are carrying magnetic moments of opposite direction, but in condensed matter, it is rather exceptional that transitions from these states can be spectrally resolved. In absorption as well as emission spectra, various broadening mechanisms result in optical linewidths larger than the Zeeman splitting, often by orders of magnitude. Usually, the broadening is dominated by electron–vibrational coupling (in solids: electron–phonon), strongly limiting the applicability of ODMR on the basis of spectrally selective excitation or detection of the spin states. However, in the presence of the magnetic field, an alternative is available: spin-specific transitions can be selected on the basis of RCP or LCP of the light propagating along the direction of the magnetic field (the so-called Faraday configuration), using the effects of MCD and MCP in absorption and emission, MCDA and MCPE, respectively.

An example of this selectivity is shown in the level scheme in Figure 25.8, approximately describing the situation in F-centers in alkali halides; an unpaired electron in an s-type orbital that can be promoted to a p-type orbital. These correspond to a ground ^2S term and an excited ^2P term, which, in turn, is split by spin–orbit coupling (Δ) in a doublet $^2P_{1/2}$ and a quartet $^2P_{3/2}$ level. The linewidths of the optical transitions are much larger than the Zeeman energies. The allowed electric-dipole transitions for circularly polarized light are indicated together with their relative probabilities. The absorption of light in each transition is proportional to the population of the initial state (neglecting the small excited state populations). Taking the doublet-to-doublet transition (^2S to $^2P_{1/2}$) in the case of thermal equilibrium, absorption of RCP light is stronger than LCP, i.e., the MCDA is negative. On the contrary, in the transition to the quartet, the LCP light will be absorbed more strongly, leading to the inverse sign for the MCDA. As described in Section 25.2, the MCDA signal is sensitively recorded using phase-sensitive (lock-in) detection following periodic modulation of the polarization between RCP and LCP. In this chapter, we are discussing the paramagnetic contribution to the MCDA (next to this, there is a temperature independent diamagnetic contribution which is not influenced by resonant microwave excitation), in which the difference between the absorption coefficients, α, for right and left polarized lights, α_R and α_L, respectively, is proportional to the spin polarization P in the ground state,

$$\alpha_R - \alpha_L \propto P = \frac{n_- - n_+}{n_- + n_+} = \tanh(g\mu_B B_0/2k_B T)$$

in thermal equilibrium ($n_\pm$ stands for the populations in the ground doublet and $g \approx g_e$ may be assumed). The MCDA signal is therefore increasing with magnitude of the field. Low temperatures with respect to the Zeeman energy are required to obtain significant spin polarization. In this approach of ODMR, the change in MCDA is measured under the influence of MWs resonant with the spin transition in the ground state, tending to equalize the spin populations, i.e., to decrease the spin polarization P.

This specific example is only one illustration of the analysis needed to identify the mechanism for

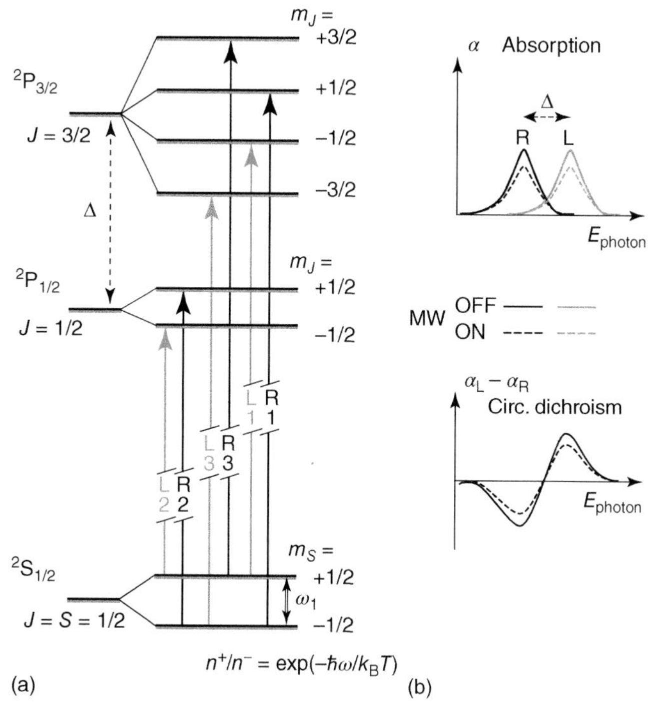

Figure 25.8. (a) Energy level diagram for a simple atomic system with a single unpaired electron ($S = 1/2$). The atom can be in an ^{2}S ($L = 0$) ground state or in ^{2}P ($L = 1$) excited states split into $^2P_{1/2}$ and $^2P_{3/2}$ by spin–orbit coupling (Δ). The allowed optical transitions are indicated with their R/L handedness and underneath numbers indicating their relative transition probabilities. (b) The top graph schematically shows the two absorption bands for the transitions to the $^2P_{1/2}$ and $^2P_{3/2}$ states. They are broadened by strong electron phonon coupling, blurring the magnetic substructure. In the absence of microwaves, thermal equilibrium for the population n^+ and n^- of the ground doublet results in opposite net circular polarization for the two bands and the resulting MCD spectrum (bottom, solid line). The change in circular dichroism induced by the microwave excitation of the ground EPR transition (bottom, broken line) yields the MCDA-detected ODMR signal. (Reproduced with permission from J.-M. Spaeth, J. R. Niklas, R. H. Bartram, in Structural Analysis of Point Defects in Solids, Spring Series in Solid State Physics, 1992. © Springer, 1992)

ODMR detection in a given system. In this example, opposite signs are expected for the ODMR signals detected by MCDA in the two absorption bands. In addition, a direct link can be made experimentally between magnetic resonance spectra and optical bands by measurements of the ODMR signal of a specific spin transition as a function of absorption wavelength, the so-called EPR-tagged MCD spectroscopy. This powerful double-resonance (optical and MW) technique is able to identify the optical bands with paramagnetic species possessing specific EPR transitions. Figure 25.9(a) shows the MCDA-detected spectra of the $Tl^0(1)$ defect in KCl:TlCl in which a single unpaired electron is essentially in a 6p orbital on a substitutional Tl-ion with a chlorine vacancy in nearest-neighbor position.[53] While not fully discussed here, anisotropic properties of the paramagnetic center can have important effects on the ODMR spectrum, as shown here for the features in the $Tl^0(1)$ spectra for defect orientations parallel and perpendicular to the static field, showing opposite signs for the ODMR signal. Note the linearly varying background from diamagnetic nonresonant contributions to MCDA. In Figure 25.9(b), the ODMR signal of given EPR transitions is measured as a function of photon energy of the transmitted light, tagging the electronic transitions belonging to the $Tl^0(1)$ defect.

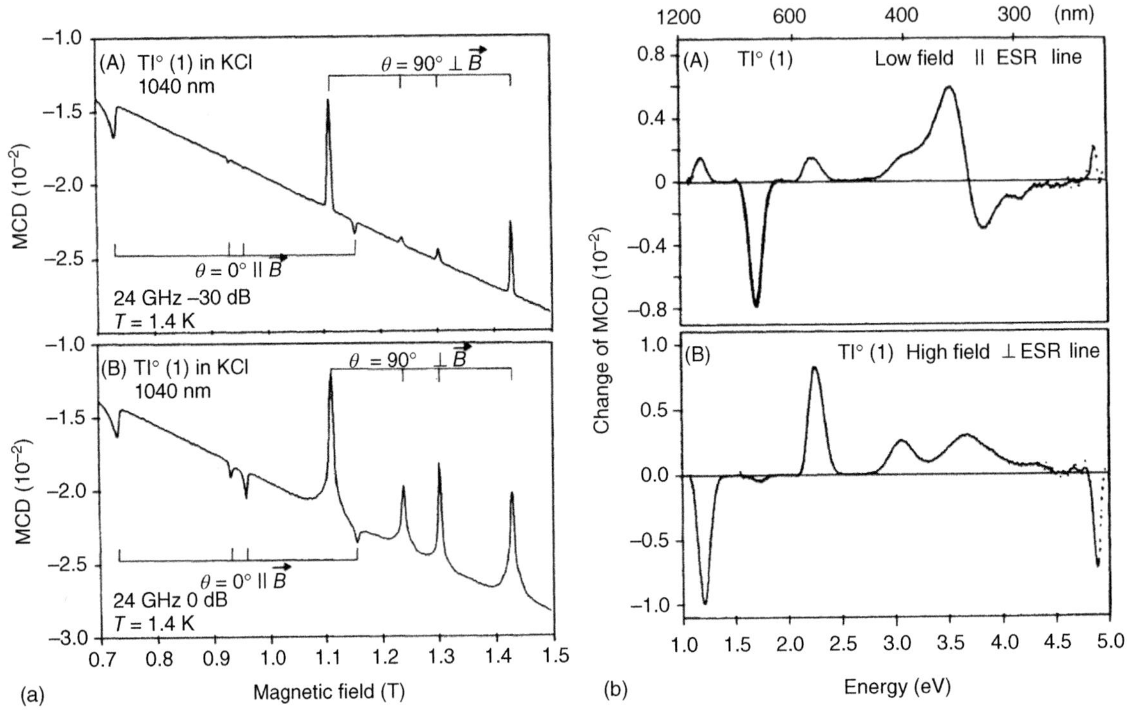

Figure 25.9. (a) ODMR spectrum of an $S = 1/2$ electron trapped center in KCl:TlCl consisting of a Tl atom substituting a lattice cation and flanked by a nearest-neighbor halogen vacancy. The spectrum is detected by MCDA (at 1040 nm, $T = 1.5$ K) in the K-band (24 GHz): (A) at low and (B) high microwave power. For each of the two defect orientations in the crystal, parallel and perpendicular to the applied field, four lines can be discerned. The outer ones form a hyperfine pair ($I = 1/2$, ^{203}Tl and ^{205}Tl) of two allowed transitions ($\Delta m_I = 0$) and the inner ones are 'forbidden' transitions ($|\Delta m_I| = 1$), centered around different fields due to the pronounced g-anisotropy. (b) MCD spectra tagged on one of the allowed transitions of the defects oriented (A) parallel with and (B) perpendicular to the magnetic field. (Reprinted figure with permission from F. J. Ahlers, F. Lohse, J.-M. Spaeth, Phys. Rev. B, 1983, 28, 1249. Copyright © 1983, American Physical Society)

25.4.2 Excited-state EPR Detection Based on Magnetic Circular Polarization

Spin polarization in Kramers systems was assumed up to this point to follow from thermal population differences between the states, and the transmitted light to act only as a probe of the spin polarization. However, as in the case of triplet states, one can envision optical pumping schemes leading to enhanced spin polarization in either excited or ground state. Under excitation with circularly polarized light, a net population in one of the states of an excited doublet can be reached if the excited-state lifetime is sufficiently long and the spin–lattice relaxation not too fast. As most direct result, one expects different circularly polarized emission, giving the opportunity for MCPE detection of the EPR transitions between the spin states in the excited level.[3,4,40,54]

However, even excitation with unpolarized light can lead to spin polarization by optical pumping, taking into account that other transitions between the states, e.g., nonradiative decay, or cross relaxation to other excited states, can be spin dependent under the applied magnetic field and can contribute to different populations in two states of the Kramers doublet. This can affect the populations (and spin polarization) in the ground levels as well, with several consequences[3,4,40] First, spin pumping allows for an increased sensitivity in the ODMR detection of spin transitions in the ground state.[3] Moreover, the excited-state spin transitions can under some conditions be indirectly observed together with those of the ground state in MCDA.[4,40] Similarly, spin transitions in the ground level may appear in the MCPE-detected spectra as they can also indirectly affect the excited-state spin polarization.[4,40]

25.5 ODMR OF INTERACTING SPIN PAIRS

In the examples treated till now, a single center, e.g., a molecule or a crystal defect that is paramagnetic in the ground or excited state, was considered and the mechanisms allowing for ODMR detection were innate to the center's electronic properties. In this section, I will treat a class of systems in which a pair of spin-carrying entities are interacting weakly or during a very short time. They have in common that states of the coupled system have to be taken into account to describe the spin dependence of the optical transitions. I will limit the discussion here to two types of interacting pairs of spin-carrying entities: (i) donor–acceptor (D–A) pairs as studied mainly in inorganic SCs and (ii) pairs of charge excitations in organic SCs, such as polarons and TEs.

25.5.1 Donor–Acceptor Pairs in Inorganic Semiconductors

Donor and acceptor levels in SCs, either intentionally introduced by doping with selected elements or from impurities and defects in the crystal structure, are technologically important and as such have been the subject of intensive scrutiny. EPR methods are powerful in the determination of the composition and local structure of these centers, but the low concentrations appearing in high-grade SC materials call for sensitive detection, which can be expected from ODMR. When electrons and holes are created in an SC by optical excitation or electrical injection, they can recombine either in band-to-band transitions or after relaxation of one or both of the carriers in donor and acceptor levels, respectively, eventually emitting photons at or below the bandgap energy. Shallow donor and acceptor states are defect levels very close in energy to the conduction and valence band, respectively. Shallow donors easily release electrons in the conduction band, leading to n-type doping. In the presence of shallow acceptors, electrons are easily promoted from the valence band to the acceptor states, resulting in p-type conductivity. In contrast with shallow centers, the trapping energy for deep donors and acceptors is larger than the thermal energy, making it very hard to release the carriers. Owing to their low binding energies for the electrons or holes, shallow traps typically have electronic states extending over a large region around the defect, while these are localized for deep traps.

As described by Cavenett,[10] two approaches were initially developed for EPR detection via recombination emission. In conduction electron-spin resonance (CESR), also called optical pumping, one applies circularly polarized light to induce population differences between Zeeman-split levels of the conduction electrons, and the MW resonances are detected by MCPE as described above for half-integer spin systems.

Figure 25.10(a) shows a simple scheme for electron–hole recombination, each of the carriers in doublet levels, one deriving from the s-type conduction band orbitals ($S=1/2$) and the other from $m_J=\pm 3/2$ states of the $J=3/2$ multiplet of the p-type valence band orbitals in axial symmetry. Allowed recombination transitions with circularly polarized emission along the magnetic field direction are indicated, which can be used in the CESR-type approach. In this scheme, resonant MW excitation of the donor spin will decrease the emission in the same circular polarization as used for excitation, while increasing it in the other circular polarization.

The second approach exploits steady-state population differences, eventually in thermal equilibrium, requiring low temperatures, or more often induced as described in previous sections by sufficient optical excitation combined with spin-dependent transition rates to the ground state (radiative or radiationless) or between excited states (compare to ISC in the triplet/singlet systems). However, large population differences can also occur under linear or unpolarized illumination. The level scheme in Figure 25.10(b) represents the spin states of a shallow donor weakly interacting with a deep acceptor with $S=1/2$, in a magnetic field. Two eigenstates are pure triplet states with $m_S=\pm 1$, from which decay to the singlet ground state is forbidden. The remaining two product states are quasi-degenerate and can be considered as mixtures of total singlet and triplet states possessing $m_S=0$, with partially spin-allowed recombination to the singlet ground state. For sufficiently slow spin relaxation within the doublets, this eventually leads to spin polarization. The induced EPR transition in either the donor or the acceptor doublet will then increase the intensity of recombination fluorescence. This scheme is not universal, and other combinations of donor and acceptor states have been considered for specific cases.[4,10]

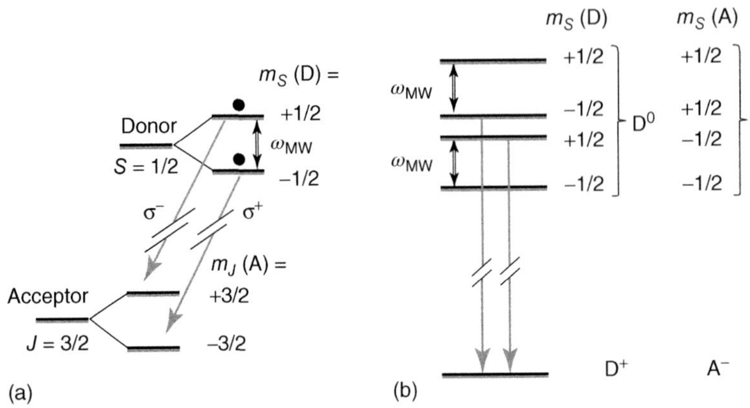

Figure 25.10. High-field ODMR detected via donor–acceptor recombination emission. (a) Orbital energy scheme of a close pair of a shallow donor with an unpaired electron in an s-type orbital and a shallow acceptor with a hole in a p-type orbital deriving from the conduction and valence band states, respectively. In a uniaxial crystal such as CdS, the $^4P_{3/2}$ (with total angular momentum $J = 3/2$) splits and the $m_J = \pm 3/2$ doublet emerges. Conservation of angular momentum only allows for two transitions for electron–hole recombination with circularly polarized emission (gray arrows). Starting from thermal population of the donor level, the microwave resonance on the donor spin enhances σ^- and decreases σ^+ emission. (b) Energy level diagram of the combined system of a weakly interacting electron on a shallow donor and hole on a deep acceptor, both carrying $S = 1/2$. Electron–hole recombination is most efficient from the singlet-containing states with total $m_S = m_S(\mathrm{D}) + m_S(\mathrm{A}) = 0$, which therefore are depleted relative to the upper and lower states. The microwave transitions toward these depleted states lead to enhanced recombination and emission intensities. (Reproduced with permission from Ref. 10. © Taylor and Francis, 1981)

The magnetic resonance spectra of shallow acceptors (or donors) are strongly determined by the properties of the corresponding edge conduction and valence orbitals, and it is not straightforward to obtain more detailed information about the local structure of these centers by ODMR spectroscopy.[55] In most cases, these have resulted in the determination of *g*-values for electrons in states at the band edge or in orbitals of shallow trap orbitals. This is different for deep levels in which the wavefunctions are strongly localized around the impurity or defect, and the local structure can be often derived in more detail on the basis of hyperfine interactions with involved nuclei and anisotropic spectral properties. One should remember that, contrary to the well-defined unique configurations assumed in previous sections, one will find here a distribution of configurations with different D–A distances and thus variable properties of the centers and interactions between the spins. Consequently, linewidths in recombination-detected ODMR spectra are usually quite broad, which can hamper the discrimination between centers and the interpretation in terms of composition and local structure. The ODMR spectra in Figure 25.11(a), detected through the recombination of a shallow donor with a deep Mg-acceptor (both with effective spin $S = 1/2$) in a hexagonal GaN crystal, illustrate the observation of axial anisotropy along the crystal axis.[29] The variation of line positions with field orientation is shown in Figure 25.11(b) together with possible fits from a spin Hamiltonian analysis, and the proposed level scheme comparable to Figure 25.10(b). Here, unlike most D–A pairs studied in SCs, a relatively large spin–spin coupling was derived, pointing to a close and probably quite specific configuration of the pair. The linewidths were ascribed mainly to hyperfine interaction with a Ga interstitial involved in the donor center.

An interesting variant on D–A recombination studies exploits the afterglow intensity in crystals previously submitted to ionizing radiation (e.g., UV-light and X-rays) at low temperature. Afterglow in ionic crystals related to a tunneling recombination of trapped electrons and holes can last for many hours at low temperature. In such experiments, there is no optical excitation during the ODMR measurement. The spin dependence of the recombination process makes it possible to detect EPR transitions of the donor and acceptor states by changes in the afterglow intensity.[31,56,57]

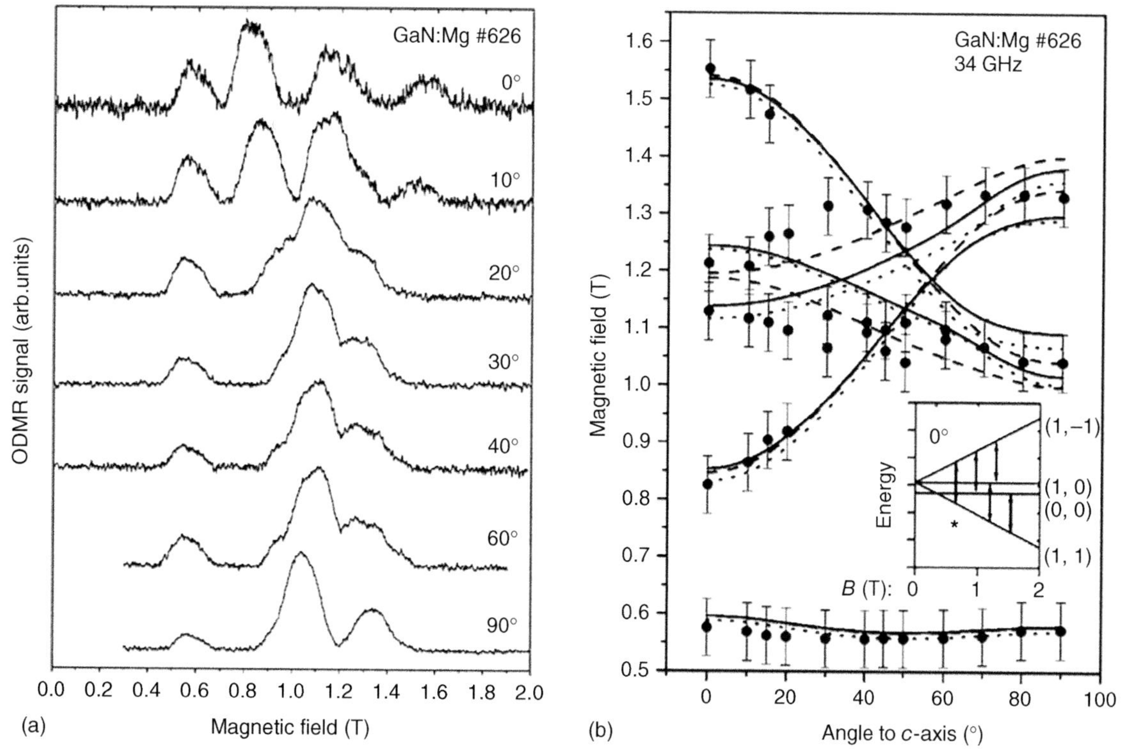

Figure 25.11. (a) ODMR spectra measured in GaN via emission from D–A recombination between a Mg acceptor and an interstitial Ga deep-donor center as a function of magnetic field orientation. (b) Angular variation of the experimental resonance line positions (dots) of the observed transitions (see attribution in inset level scheme), both allowed (in the 0.8–1.6 T region) and half-field (around 0.58 T; starred in inset). The solid, dashed, and dotted curves correspond to three possible sets of spin Hamiltonian parameters (g-values and spin–spin interaction parameters) providing similarly close fits of the angular variation. (Reprinted figure with permission from G. N. Aliev, S. Zeng, S. J. Bingham, D. Wolverson, J. J. Davies, T. Wang, P., J. Parbrook, Phys. Rev. B, 2006, 74, 235205. Copyright © 2006, American Physical Society)

25.5.2 Pairs of Charge Excitation in Organic Semiconductors

ODMR is also used for the investigation of interacting pairs of charge excitations in conjugated organic molecules in the solid state, initially in molecular crystals and later also in conjugated polymers. In these organic SCs, the singlet exciton (SE) created after photon absorption can dissociate to independent positive and negative charges, holes and electrons usually called positive and negative polarons (P^+ and P^-), respectively, because of the strong coupling to the molecular vibrations. The polarons are the primary $S=1/2$ entities involved here. However, ISC from the SE can produce TE states with $S=1$ and, furthermore, combined entities have been recognized such as bipolarons (polaron pairs (PPs) of equal charge) and trions (pairs of an exciton and a polaron).

It is worth noting the relationship to another kind of interacting spin pairs that will not be treated here but has given rise to extensive chemistry-related research in a subfield called reaction-yield detected magnetic resonance (RYDMR).[11] Spin-dependent reactions were investigated during encounters of molecules in the triplet-excited state with others of the same species or with radicals. Magnetic resonance was also in this case monitored by optical means (phosphorescence or fluorescence) to obtain information on the type of interactions and on relevant intermediate states. The concepts and methods developed in this field are transferable to the study of charge excitations in organic SCs, including molecular crystals and polymers.

The simplest case that can be considered here is the recombination of distant positive and negative polarons, occurring because of weak overlap of the wavefunctions without formation of the excitonic bound state. The spin dependence of the recombination can be treated in the same scheme as in Figure 25.10(b), where both D and A have $S=1/2$. EPR transitions within any of the doublets will yield an increase in the recombination rate that gives rise to a narrow positive ODMR peak at the corresponding g-values, close to the free-electron value g_e. In many cases, P^+ and P^- transitions cannot be distinguished because of the close relationship between highest occupied molecular orbital (HOMO) and lowest unoccupied molecular orbital (LUMO) states forming the valence and conduction bands, which in the basic treatment of conjugated molecules are linear combination of the same atomic states. This relationship is perturbed in thiophene-based polymers in which the two resonances could be disentangled.[15] Pulse ODMR or EDMR methods allow demonstrating the correlation between the two interacting spins[58] and distinguish this situation from that of a polaron interacting with other charge excitations, for example, interactions with TEs.

A level scheme for interaction between two TEs is schematically shown in Figure 25.12(a), leading to possible singlet, triplet, and quintet intermediate states

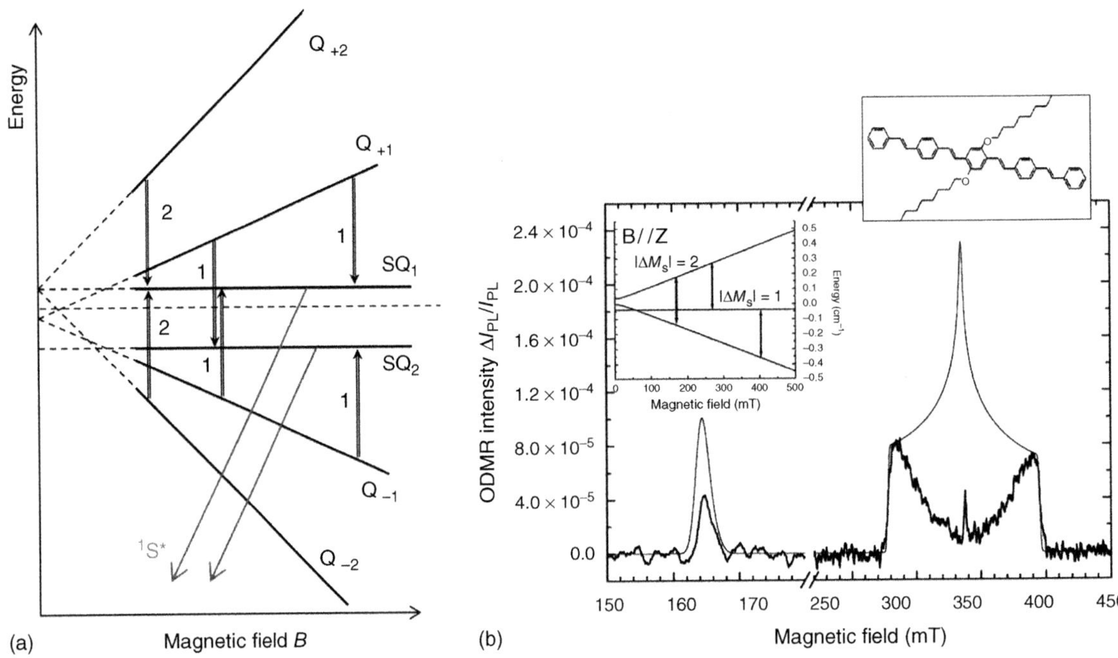

Figure 25.12. (a) Energy level scheme for a weakly interacting pair of identical triplets, with combined states of total angular momentum singlet S, triplet T, and quintet Q states as intermediate states in triplet–triplet annihilation (TTA) process. At sufficiently high field compared to the ZFS within the triplets, the only mixed states are SQ1 and SQ2, derived from the S and the Q states with $m_S=0$. Transitions with $\Delta m_S=1$ and 2 between SQ and Q states are marked by double arrows and give rise to the full- and half-field spectral features. Gray arrows indicate decay from the emissive SQ states, which are depleted in population relative to the others. (Reproduced with permission from Ref. 11. © John Wiley and Sons, 1982). (b) ODMR spectrum of triplet excitons on PPV oligomers in a polystyrene film, attributed to TTA of pairs of triplet excitons (TEs) giving rise to emissive singlet excitons ($\lambda_{exc}=476$ nm, $\lambda_{fluo}\approx 530$ nm). The sharp peak at $g\approx 2$ derives from a polaron spin transition observed through distant polaron pair interactions. The TE spectrum is compared to a spectral simulation (gray lines) based on EPR transition probabilities. The spectral features at full and half field are well reproduced, but significant differences are seen in the region around 338 mT, which are attributed to the different angular dependence of ODMR and EPR intensities. (Reprinted figure with permission from J. De Ceuster, E. Goovaerts, A. Bouwen, V. Dyakonov, Phys. Rev. B, 2003, 68, 125202. Copyright © 2003, American Physical Society)

for the total spin of the pair, which may be expected to be formed statistically according to their multiplicities, 1 : 3 : 5. A well-studied reaction for these conjugated molecules is triplet–triplet annihilation (TTA), in which the interaction of two TEs, each with $S = 1$, leads to possible intermediate states of total $S = 0$, 1, and 2, allowing for a transition to a final singlet state with one SE. This process is also considered for upconversion (UC) because it converts two low-energy TEs to one SE. Finally, this SE can decay and contribute to delayed fluorescence intensity. This transition is spin forbidden for all but the total $m_S = 0$ states (with singlet character) of the pair, which leads to relative depletion of these states. MW-induced transitions to these states then lead to an increase in the delayed singlet emission and allows for DFDMR-type ODMR detection of the TE spectra. The high-field spectra ascribed to this TTE-UC process in a para-phenylene vinylene oligomer[59] hosted by a nonconjugated polymer are shown in Figure 25.12(b). Because of the quasi-random orientation of the molecules, a powder triplet spectrum is found in the $g \approx 2$ region together with the characteristic half-field transition (see inset). The spectrum, attributed to a rhombic spin triplet with ZFS parameters $D/hc = 0.056$ cm^{-1} and $E/hc = 0.018$ cm^{-1}, is not well described by the corresponding EPR simulation, which may be due to the TTA-UC mechanism for ODMR with intensities varying with the orientation of the center main axes in the field, and also could result from a partial preferential orientation of the molecules in the spin-coated polymer film. Both effects have been further studied by Cambré *et al.*[60] measuring the dependence on field orientation for different materials and preparation conditions.

A discussion of the triplet ODMR spectra in conjugated molecular materials and their origin can be found in review papers.[11,12,15] A large variety of reaction mechanisms, including the already mentioned TTA and also PP interactions and the so-called TE-P quenching reaction, can be involved. Even within one of these classes, careful distinction between cases should be made. Indeed, the TTA can occur between nongeminate pairs of TEs obeying the abovementioned production statistics. Alternatively, geminate pairs of TEs can be present as a result of the so-called singlet fission process (SE → TE + TE), in a sense the reverse of the TTA reaction. Singlet fission necessarily produces the two TEs in a total spin-singlet state, and for this case, a negative ODMR signal is expected. As a final remark, it is worth noting that the different mechanisms that can be involved may show very distinct response times that can be separated from each other by smart modulation schemes.[15]

ACKNOWLEDGMENTS

I wish to first thank Dirk Schoemaker to whom I am indebted for introducing me to the topic of paramagnetic centers in solids and gave me the initial and solid training in the techniques of EPR spectroscopy. Acknowledgments should also go to a multitude of colleagues and collaborators over the years, too many to list here, but who will know. Finally, the continuing and patient support of my wife and life companion has been invaluable over the years, and this applies in particular for the period dedicated to this work.

REFERENCES

1. S. Geschwind, R. J. Collins, and A. L. Schawlow, *Phys. Rev. Lett.*, 1959, **3**, 545.
2. J. Brossel, G. Geschwind, and A. L. Schawlow, *Phys. Rev. Lett.*, 1959, **3**, 548.
3. S. Geschwind, in Electron Paramagnetic Resonance, ed S. Geschwind, Plenum Press: New York-London, 1972, Chap. 5, pp. 353–425, 'Optical Techniques in EPR in Solids'.
4. J.-M. Spaeth, J. R. Niklas, and R. H. Bartram, in Springer Series in Solid State Sciences, Springer: Heidelberg & New York, 1992, Vol. 43. Structural analysis of point defects in solids, Chap. 4, pp. 77–280. 'Optical Detection of Electron Paramagnetic Resonance', Chap. 9, pp. 309–328, 'Experimental Aspects of Optically Detected EPR and ENDOR'.
5. J. Köhler, J. A. J. M. Disselhorst, M. C. J. M. Donckers, E. J. J. Groenen, J. Schmidt, and W. E. Moerner, *Nature*, 1993, **363**, 242.
6. J. Wrachtrup, C. von Borczyskowski, J. Bernard, M. Orrit, and R. Brown, *Nature*, 1993, **363**, 244.
7. R. H. Clarke (ed), Triplet State ODMR Spectroscopy, John Wiley & Sons: New York, NY, 1982.
8. A. H. Maki, *Biol. Magn. Reson.*, 1984, **6**, 187.
9. A. J. Hoff, in Advanced EPR: Applications in Biology and Biochemistry, ed A. J. Hoff, Elsevier: Amsterdam, The Netherlands, 1989, Chap. 5, pp. 633–684, 'Optically-detected Magnetic Resonance of Triplet States'.

10. B. C. Cavenett, *Adv. Phys.*, 1981, **30**, 475.
11. E. L. Frankevich and S. I. Kubarev, Spectroscopy of Reaction-Yield Detected Magnetic Resonance, John Wiley & Sons: New York, NY, 1982, in Ref. 7, Chap. 5, pp. 137–183.
12. J. Shinar, in Handbook of Organic Molecules and Polymers: Vol. 3 Conductive Polymers: Spectroscopy and Physical Properties, ed H. S. Nalwa, John Wiley & Sons: New York, NY, 1997, Chap. 7, pp. 319–366, 'Optically Detected Magnetic Resonance (ODMR) Studies of π-conjugated Polymer Films, Solutions, Blends, and Devices'.
13. V. Dyakonov and E. L. Frankevich, *Chem. Phys.*, 1998, **227**, 203.
14. R. Osterbacka, M. Wohlgenannt, M. Shkunov, D. Chinn, and Z. V. Vardeny, *J. Chem. Phys.*, 2003, **118**, 8905.
15. J. Shinar, *Laser Photonics Rev.*, 2012, **6**, 767.
16. I. Y. Chan, Zero-Field ODMR Techniques- Phosphorescence Detection, John Wiley & Sons: New York, NY, 1982, in Ref. 7, Chap. 1, pp. 1–24.
17. R. H. Clarke, Zero-Field ODMR Techniques- Nonphosphorescence Detection, John Wiley & Sons: New York, NY, 1982, in Ref. 7, Chap. 2, pp. 25–42.
18. D. W. Pratt, ODMR Studies of Excited Triplet States in High Field, John Wiley & Sons: New York, NY, 1982, in Ref. 7, Chap. 3, pp. 43–82.
19. G. Janssen, A. Bouwen, P. Casteels, and E. Goovaerts, *Rev. Sci. Instrum.*, 2001, **72**, 4295.
20. M. C. J. M. Donckers, O. G. Poluektov, J. Schmidt, and P. G. Baranov, *Phys. Rev. B*, 1992, **45**, 13061.
21. I. Tkach, U. Rogulis, S. Greulich-Weber, and J.-M. Spaeth, *Rev. Sci. Instrum.*, 2004, **75**, 4781.
22. J.-M. Spaeth, I. Tkach, S. Greulich-Weber, and H. Overhof, *Magn. Reson. Chem.*, 2005, **43**, S153.
23. M. Chamel, R. Chicault, and Y. M. d'Aubigné, *J. Phys. E*, 1976, **9**, 87.
24. K. M. Lee, *Rev. Sci. Instrum.*, 1982, **53**, 702.
25. G. Annino, M. Fittipaldi, M. Martinelli, H. Moons, S. Van Doorslaer, and E. Goovaerts, *J. Magn. Reson.*, 2009, **200**, 29, and references therein.
26. H. J. Reyher, U. Hellwig, and O. Thiemann, *Phys. Rev. B*, 1993, **47**, 5638.
27. B. Briat, H. J. Reyher, A. Hamri, N. G. Romanov, J. C. Launay, and F. Ramaz, *J. Phys. Condens. Matter*, 1995, **7**, 6951.
28. B. Briat, A. Watterich, F. Ramaz, L. Kovacs, B. C. Forget, and N. Romanov, *Opt. Mater.*, 2002, **20**, 253.
29. G. N. Aliev, S. Zeng, S. J. Bingham, D. Wolverson, J. J. Davies, T. Wang, and P. J. Parbrook, *Phys. Rev. B*, 2006, **74**, 235205.
30. V. Y. Ivanov, M. Godlewski, D. R. Yakovlev, M. K. Kneip, M. Bayer, S. M. Ryabchenko, and A. Waag, *Phys. Rev. B*, 2008, **78**, 085322.
31. N. G. Romanov, D. O. Tolmachev, A. G. Badalyan, R. A. Babunts, P. G. Baranov, and V. V. Dyakonov, *Physica B*, 2010, **404**, 4783.
32. V. Stepanov, F. H. Cho, C. Abeywardana, and S. Takahashi, *Appl. Phys. Lett.*, 2015, **106**, 063111.
33. M. W. Doherty, N. B. Manson, P. Delaney, F. Jelezko, J. Wrachtrup, and L. C. L. Hollenberg, *Phys. Rep.*, 2013, **528**, 1.
34. M. C. J. M. Donckers, J. Schmidt, W. M. Chen, and B. Monemar, *Phys. Rev. B*, 1990, **42**, 1684.
35. M. Glasbeek, *Radiat. Eff. Def. Sol.*, 1983, **72**, 13.
36. M. Glasbeek, *Top. Curr. Chem.*, 2001, **213**, 95.
37. C. Boehme and K. Lips, *J. Mater. Sci. Electron.*, 2007, **87**, S285.
38. K. J. van Schooten, J. Huang, D. V. Talapin, C. Boehme, and J. M. Lupton, *Phys. Rev. B*, 2013, **87**, 125412.
39. K. P. Dinse and C. J. Winscom, Optically Detected ENDOR Spectroscopy, John Wiley & Sons: New York, NY, 1982, in Ref. 7, Chap. 4, pp. 83–136.
40. J.-M. Spaeth and F. Lohse, *J. Phys. Chem. Solids*, 1990, **51**, 861.
41. M. C. J. M. Donckers, A. M. Schwencke, E. J. J. Groenen, and J. Schmidt, *J. Chem. Phys.*, 1992, **97**, 110.
42. J. S. Colton and L. R. Wienkes, *Rev. Sci. Instrum.*, 2009, **80**, 035106.
43. J. Schmidt, *Semicond. Insul.*, 1978, **4**, 193.
44. G. Giacometti, G. Agostini, S. Santabarbara, and D. Carbonera, *Appl. Magn. Reson.*, 2007, **31**, 179.
45. D. Carbonera, *Photosynth. Res.*, 2009, **102**, 403.
46. A. Wasiela, G. Ascarelli, and Y. Merle d'Aubigné, *Phys. Rev. Lett.*, 1973, **31**, 993.
47. M. Shirai and K. Kan'no, *J. Phys. Soc. Jpn.*, 1998, **6**, 2112.
48. P. G. Baranov, G. Romanov, O. G. Poluektov, and J. Schmidt, *Appl. Magn. Reson.*, 2010, **39**, 453.
49. G. Janssen, A. Bouwen, and E. Goovaerts, *Phys. Rev. B*, 2005, **71**, 03415.

50. I. Carmichael and L. Hug, *J. Phys. Chem. Ref. Data*, 1986, **15**, 1.

51. D. Carbonera, A. Agostini, M. Di Valentin, C. Gerotto, S. Basso, G M. Giacometti, and T. Morosinotto, *Biochem. Biophys. Acta Bioenerg.*, 2014, **1837**, 1235.

52. V. M. Acosta, E. Bauch, M. P. Ledbetter, C. Santori, K.-M. C. Fu, P. E. Barclay, R. G. Beausoleil, H. Linget, J. F. Roch, F. Treussart, S. Chemerisov, W. Gawlik, and D. Budker, *Phys. Rev. B*, 2009, **80**, 115202.

53. F. J. Ahlers, F. Lohse, and J.-M. Spaeth, *Phys. Rev. B*, 1983, **28**, 1249.

54. K. Imanaka, T. Wada, M. Tanaka, and H. Ohkura, *J. Phys. Soc. Jpn.*, 1978, **45**, 2041.

55. It was shown for several shallow donors centers that detailed mapping of the wavefunction can be achieved by W-band ENDOR measurements, also showing the confinement effects in SiC nanoparticles, see e.g.: S. B. Orlinskii, J. Schmidt, P. G. Baranov, D. M. Hofmann, C. D. Donega, and A. Meijerink, *Phys. Rev. Lett.*, 2004, **92**, 047603.

56. N. G. Romanov, Y. P. Veshchunov, V. A. Vetrov, and P. G. Baranov, *Phys. Status Solidi B*, 1981, **107**, K119.

57. D. O. Tolmachev, A. G. Badalyan, R. A. Babunts, V. A. Khramtsov, N. G. Romanov, P. G. Baranov, and V. V. Dyakonov, *J. Phys. Condens. Matter*, 2010, **22**, 295306.

58. M. E. Limes, J. Wang, W. J. Baker, B. Saam, and C. Boehme, *Phys. Rev. B*, 2013, **87**, 165204.

59. J. De Ceuster, E. Goovaerts, A. Bouwen, and V. Dyakonov, *Phys. Rev. B*, 2003, **68**, 125202; In this paper an unconventional set of axes was is used, leading to ZFS parameters, $D = 0$ and $E/hc = 0.037$ cm^{-1}, equivalent after permutation of axis labels to the values cited in this chapter, $D/hc = 0.056$ cm^{-1} and $E/hc = 0.018$ cm^{-1} (obeying $|E/D| \leq 1/3$).

60. S. Cambré, J. De Ceuster, E. Goovaerts, A. Bouwen, and H. Detert, *Appl. Magn. Reson.*, 2007, **31**, 343.

Chapter 26

Electrically Detected Magnetic Resonance (EDMR) Spectroscopy

Christoph Boehme and Hans Malissa

University of Utah, Salt Lake City, UT, USA

26.1 INTRODUCTION

Electrically detected magnetic resonance (EDMR) spectroscopy can be loosely defined as an electron magnetic resonance experiment [that means either an EPR or a ferromagnetic resonance (FMR) experiment] where magnetic resonance is detected through the change in a sample's conductivity σ and thus, resistivity. This is in contrast to the traditional inductively detected CW EPR or FMR spectroscopy, which are both based on absorption (dispersion) or reflection (transmission) of microwave (MW) and radio frequency (RF) intensities. For EDMR, the change in conductivity $\Delta\sigma/\sigma$ induced by magnetic resonance is for many examples found in the literature the result of spin-dependent recombination or transport processes; however, the exact nature of any spin-dependent process observed with EDMR and thus, the observation itself depends largely on the studied sample. While most EDMR signals are governed by spin-selection rules such as Pauli blockade, i.e., the suppression of charge transitions into doubly occupied electron states due to the Pauli principle, there is a variety of other spin-dependent processes that are also known under various materials and environmental sample conditions. Thus, EDMR gives direct insight into how material properties are influenced by paramagnetic centers. EDMR spectroscopy is highly sensitive because in many cases, the magnitude of EDMR signals is not determined by spin polarization as it is the case for inductively detected EPR signals.[1,2] Both continuous wave (CW) and pulse (p) electrically detected magnetic resonance (pEDMR) have been implemented in a wide range of material systems and devices, including organic semiconductors; mono-, poly-, microcrystalline, amorphous silicon; and silicon carbide.[3–13]

An EDMR experiment consists of electric circuits for spin excitation and for detection that are electrically independent, as is illustrated in Figure 26.1(a). The setup for spin excitation can be very similar to what is used in conventional (inductively detected) EPR. Quite frequently, commercial spectrometer

EPR Spectroscopy: Fundamentals and Methods.
Edited by Daniella Goldfarb and Stefan Stoll

ISBN: 978-1-119-16299-5
Also published in eMagRes (online edition)
DOI: 10.1002/9780470034590.emrstm1525

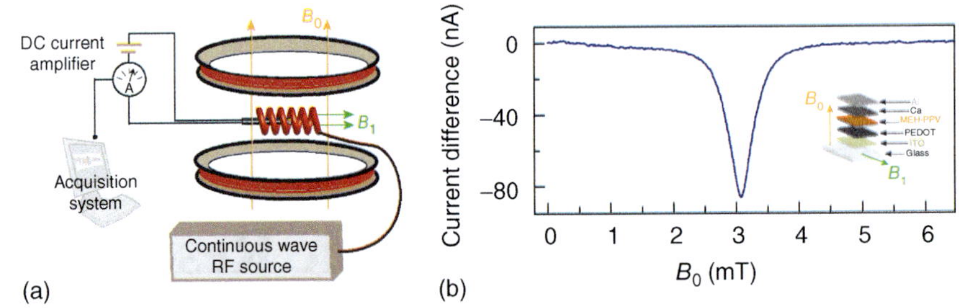

Figure 26.1. (a) Illustration of the most basic setup for electrically detected magnetic resonance, which consists of a circuit for MW excitation, a separate circuit for the measurement of sample resistivity, and an adjustable magnet to induce a Zeeman splitting. The experiment resembles an EPR setup where the electric current in a material is monitored rather than the absorption of electromagnetic waves. Additional components (not shown) that may be used include a lock-in amplifier with field modulation coils or radiation modulators (e.g., PIN diodes for MW modulation) for continuous-wave measurements. (b) Plot of the change in the room-temperature steady-state forward current of an MEH-PPV-based OLED due to the application of an oscillating magnetic RF field as a function of the applied static magnetic field B_0. (Reproduced with permission from Ref. 9. © Nature Publishing Group, 2015.)

setups are used for this purpose. In the simplest case, only an RF or MW source with an attenuator, a waveguide, and a resonant or nonresonant structure to apply an MW magnetic field B_1 to the sample are needed.[9] Other components such as circulators, isolators, or phase shifters are not required. Therefore, EDMR is inherently suitable for multifrequency applications because the required components are not necessarily narrowband.

Magnetic resonance is detected electrically through a change in sample conductivity when the magnetic resonance condition is fulfilled. In the simplest case, this is achieved through a sensitive measurement of sample resistivity, for example, by applying a constant bias voltage and recording the device current with a sufficiently fast current amplifier. In most cases this requires an electrical circuit that is independent from the MW excitation circuit, and care must be taken in order to ensure that crosstalk between the two circuits is minimized.

Figure 26.1(b) displays an example for a CW EDMR spectrum recorded on charge carrier states (so-called polarons, which are localized but mobile electronic states) in an organic bipolar injection device [i.e., a diode, also called an organic light-emitting diode (OLED)] based on the polymer poly[2-methoxy-5-(2-ethylhexyloxy)-1,4-phenylenevinylene] (MEH-PPV) that is sandwiched between a Ca layer for electron injection and the polystyrene sulfonate-doped poly(3,4-ethylenedioxythiophene) (PEDOT : PSS), which acts as a hole injection layer.[5] The plot displays the change of an $I = 50\,\mu A$ forward current as a function of the applied magnetic field in the presence of a continuously applied oscillating RF field with amplitude $B_1 = 0.071$ mT and a frequency of 85 MHz. On resonance, the device current decreases. The experiment was conducted for a B_0 field range between 0 and 6 mT.[9]

In contrast to the example shown in Figure 26.1, most reports of CW EDMR experiments found in the literature are conducted as quasi-adiabatic experiments, where lock-in detection is used for noise reduction with modulation of the B_0-field or the B_1-amplitude, while the modulation–frequency component of the resistivity is measured (see Chapter 1). In the case of pEDMR, the measurement circuit must be fast enough to reproduce the dynamics of the system on the relevant time scales (spin–lattice relaxation and spin–spin relaxation times, as well as time constants related to electronic recombination and transport). Delays and distortions due to the transfer characteristic of the measurement setup must be taken into account as well.

The details of the experimental setup, in particular related to the signal detection, depend to a large degree on the microscopic details of the spin-dependent recombination and transport processes in the sample under investigation. Besides the spin–lattice and spin–spin relaxation times, the electronic relaxation times and the time response of the detection and amplification system must be considered for a quantitative interpretation of CW and pEDMR measurements.

26.2 A BRIEF HISTORY OF EDMR SPECTROSCOPY

EDMR spectroscopy was first conducted in mid-1960s, within a few years after first optically detected magnetic resonance (ODMR) experiments had demonstrated in 1959 that electronic transitions in condensed matter systems can be controlled by spin-selection rules.[14,15] While Guéron and Solomon had reported a first EDMR experiment due to a bolometric effect, it was Hirose *et al.* as well as Maxwell and Honig in 1966 who first demonstrated EDMR based on spin-selection rule controlled electronic transitions.[16–18] In 1972, Lepine and Prejean realized that EDMR signals can depend on the square of the polarization P^2, indicating that pairs of spin states (in this case, electrons and holes) can control electric currents in semiconductors.[19,20]

Following Lepine's work, various EDMR signals were reported, whose magnitudes exceeded the upper bounds predicted by his model,[21,22] and various other attempts to theoretically account for EDMR signals were made. These included the reconsideration of bolometric effects caused by magnetic resonant heating, clusters of paramagnetic centers, strong ferromagnetic exchange, exchange-coupled extended defects causing spin systems with $S > 1/2$, or combinations of several of these models.[23–28] Most of these models still failed to explain the weak or nonexistent temperature dependence exhibited by many of the observed EDMR signals. Eventually, in 1978, Kaplan, Solomon, and Mott (KSM) revisited the Pauli blockade-based permutation–symmetry model by Lepine.[29] They realized that charge carriers that recombine through localized electron states (e.g., point defects) always undergo a 'readjustment time' in which they form an 'exclusive pair', before the actual recombination transition takes place.[30] This refinement to the pair model made by introduction of intermediate-pair states accounted for both large EDMR signals and weak temperature dependencies, and it has since then been the most widely applied mechanisms for the description of EDMR signals.

Most of the EDMR studies reported in the literature are CW adiabatic field sweep experiments. Because some limitations of CW EDMR are similar to limitations to CW EPR, efforts were taken to develop pEDMR spectroscopy, which allows for electrical access to coherent spin-motion effects, analogously to pulse EPR (pEPR) spectroscopy. In contrast to pEPR spectroscopy, the introduction of pEDMR appeared only decades after the development of CW EDMR[31,32] owing to the difficulty that small changes to the conductivity of a conductor are detectable only with a time resolution that is fundamentally limited by the material's dielectric relaxation time. First time-domain EDMR experiments were not reported until 1999, and only since 2002 has pEDMR-detected coherent spin motion been performed.[31,33]

26.3 ELECTRONIC MECHANISMS THAT CAUSE EDMR SIGNALS

While a plethora of fundamentally different physical mechanisms are known, which can cause EDMR signals, an overwhelming number of EDMR studies are based on spin-dependent transitions between two paramagnetic single-electron states that are controlled by the 'relative spin orientation' of the two partners within an electron–hole pair; in other words, spin-permutation symmetry is assumed to govern the transition probability between singlet (S) and triplet (T_0, T_+, T_-) states when weak spin–orbit coupling dictates spin conservation. This situation is illustrated for the example of two electrons localized in two adjacent singly occupied localized states in Figure 26.2(a). When the two spins form a spin-pair state with high S content, the transition probability becomes maximal. This particular spin-selection rule is known today as Pauli blockade for which spin-pair states that form with higher S content are more likely to undergo a transition than pairs with higher T content.

For Lepine's recombination model, the Pauli blockade implies that when the likelihood of any arbitrary electron within a material to encounter any arbitrary hole is approximately equal, then a change of the recombination rate ΔR will take place solely by the change of the charge carrier ensemble's overall permutation symmetry of the pair (S, T_0, T_+, and T_-). Thus, when the equilibrium charge carrier ensemble state (see Chapter 8) changes under magnetic resonance to a reduced-polarization state, a change of the net transition rate within this ensemble occurs (see a detailed review of this in Ref. 34). Lepine's model provided a first quantitative prediction for the polarization dependency of EDMR signals, and thus, predictions for temperature T and magnetic field B_0 dependencies including an upper bound

$$\frac{\Delta\sigma}{\sigma} = \frac{g^2\mu_B^2B_0^2}{16k_B^2T^2} \tag{26.1}$$

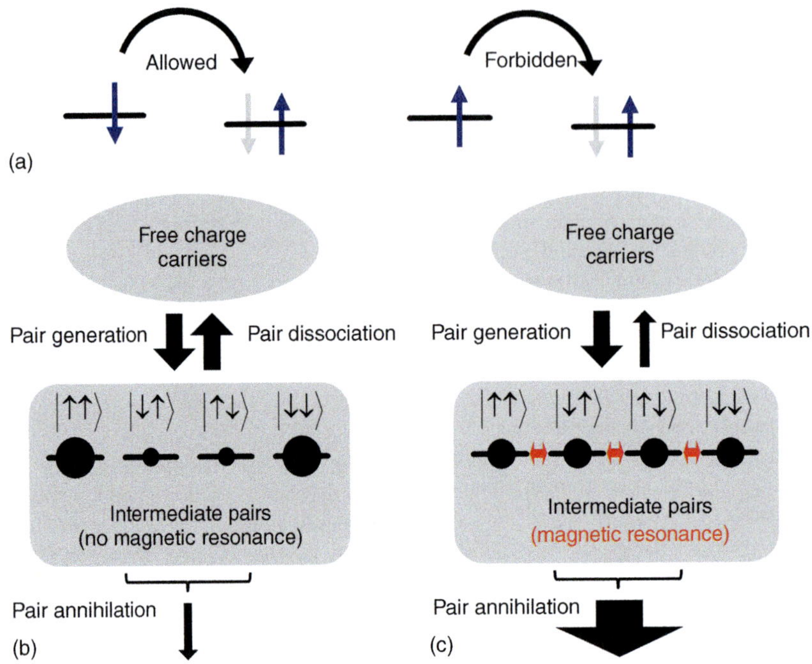

Figure 26.2. (a) Illustration of a spin-dependent electronic transition between two paramagnetic electron states with $S = \frac{1}{2}$. When spin-exchange and spin-dipolar interaction within the pair of adjacent charge carriers is negligible and spin–orbit interaction is small, too, the transition into a diamagnetic singlet state strongly depends on the singlet content of the electron pair spin state before the transition. This is called a Pauli blockade. (b,c): The transition rate diagrams as described by Kaplan, Solomon, and Mott (KSM) involving the ensemble of uncorrelated free charge carriers, the ensemble of intermediate pairs, and the ensemble of singlet states in absence (b) and presence (c) of magnetic resonance.[29] The change of the singlet formation rates between (b) and (c) is significant, and it causes a large change of the singlet generation rate and thus, a change of the dissociation rate and along with this a change of the material conductivity (the EDMR signal). The respective populations of the intermediate pairs are indicated by black spheres

for the magnitude of $\Delta\sigma$ of an EDMR signal relative to the baseline conductivity σ, when g, μ_B, and k_B represent the g-factor of the involved paramagnetic states, the Bohr magneton, and the Boltzmann factor, respectively.[34]

Compared with the Lepine pair mechanism, the KSM pair process only differs by the introduction of the intermediate-pair states, which were first described for recombination in the context of Shockley–Read–Hall recombination. This seemingly minor difference turns out to have dramatic consequences for the statistics of spin-selection rules. Characteristic for this mechanism that is illustrated in Figure 26.2(b) and (c) is that it typically consists of two spatially adjacent localized electronic states that, due to the Pauli blockade, can only undergo a transition into an S state or dissociate into uncorrelated charge carriers. The intermediate pair constitutes a weakly spin–spin coupled system of two spin-1/2 particles. In contrast, the S state that forms during a spin-dependent transition consists of two strongly exchange-coupled electrons.

According to KSM, before two uncorrelated charge carriers can form an S state, they have to form the intermediate pair first. While the spin-permutation symmetry of intermediate pairs does not (or not directly) influence their dissociation probability, it strongly governs their likelihood to undergo transitions into the strongly bound S states.

The distinction of uncorrelated charge carrier pairs from intermediate pairs may appear to be purely semantic. However, in the presence of intermediate pairs, the execution of EDMR experiments becomes a nonequilibrium experiment whose observables (the spin-dependent recombination or transport rates), are – as long as thermal polarization is weak – not governed by the thermal equilibrium in either the presence or the absence of magnetic resonance. This is illustrated by Figure 26.2(b): When spin-dependent transition rates are governed by the KSM mechanism, intermediate pairs with high S state character have a large likelihood to annihilate into strongly coupled S states (for instance, due to electron–hole recombination) and thus, they will have shorter lifetimes compared with intermediate pairs with high T character. Therefore, in the absence of magnetic resonance, the steady state of an intermediate-pair ensemble will assume a very high T density, while the magnetic polarization of such an ensemble may still be negligible. An intermediate pair in a T state will increase its S content under magnetic resonance of one or both pair constituents (note that, in order for the magnetic resonance-induced S–T transition strength to be significant, sufficiently weak spin-exchange and spin-dipolar coupling is required) and thus, it will enhance the electronic transition probability into the pure S state.[35] When magnetic resonance is applied to an ensemble of intermediate pairs (Figure 26.2c), the high steady-state T content of the ensemble in the absence of magnetic resonance will be quenched, while the low S content and therefore, the pair annihilation rate will be enhanced. Because of this, EDMR signals caused by the KSM mechanism can become significantly larger compared to EDMR signals that are governed by thermal polarization effects. Since the T content of intermediate-pair ensembles does not directly depend on the magnetic field or the temperature, EDMR signals caused by the intermediate-pair mechanism display only weak or no temperature and magnetic field dependencies.

The KSM mechanism can account for most EDMR signals reported in the literature, and it has been applied to the investigation of charge carrier transport and recombination in many semiconductor materials and devices including crystalline, microcrystalline, and amorphous silicon as well as silicon nitride, silicon/silicon dioxide interfaces, various organic semiconductors, fullerenes, and semiconductor heterostructures.[5,6,31,32,36–46] Since spin-selection rules require spin–orbit coupling to be weak, most applications of EDMR are found for studies of electronic processes in materials with low atomic order number Z such as silicon-based and organic semiconductor materials. Many studies invoking intermediate-pair-based spin-dependent electronic transition mechanisms have introduced adjustments of this model taking into account the specific nature of particular material and device conditions as well as the particular electronic states that are involved in these mechanisms.[47–51] These adjustments typically pertain to the transition probabilities of the intermediate-pair model, specifically, spin–lattice and spin–spin relaxation times, g-factors, pair dissociation and annihilation times, the strength of spin-dipolar and spin-exchange coupling parameters (which, albeit low, will never vanish entirely), and the distributions of all these parameters throughout given intermediate-pair ensembles.[35,50,52–59] The resulting wealth of physical parameters that characterizes spin-dependent intermediate-pair processes can lead to ambiguous interpretations of EDMR spectra and, in fact, it imposes the strongest limitation for EDMR spectroscopy. These ambiguities exist in particular for EDMR spectra obtained for the still predominantly used CW EDMR spectroscopy, and they have been among the strongest motivators for the development of pEDMR techniques.[60]

In spite of the KSM pair mechanism being the most reported underlying physical process that is found with EDMR experiments, there are plenty of examples for EDMR that are not governed by a KSM mechanism. Spin-pair systems that do not form intermediate pairs were observed among the first reported EDMR experiments in the 1960s.[18,61] In contrast to intermediate-pair processes, these mechanisms exhibit a strong dependency on the Boltzmann factor (i.e., on temperature and external field) as described by Lepine. Furthermore, the original Lepine model is still of importance even for intermediate pairs, when EDMR is conducted under non-negligible spin-polarization conditions where the polarization-related Lepine effect exceeds the magnitude of the KSM effect.[62]

Next to EDMR-pair mechanisms based on intermediate pairs of charge carriers with $S = {}^1\!/_2$, there are also reports of EDMR experiments that involve either intermediate pairs of spin manifolds with $S > {}^1\!/_2$, such as spin-dependent T-exciton/polaron pair recombination or pairs consisting of T states with $S = 1$.[63]

Finally, there are also examples of EDMR signals not caused by spin-pair processes[64–68], which therefore cannot be described by theoretical descriptions of intermediate-pair-based EDMR spectroscopy and that, instead, require a separate theoretical treatment.[65,67] Examples for these include EDMR based on hyperfine interactions between two-dimensional electron gases and nuclear spins in AlGaAs–GaAs heterojunctions where edge state transport in quantum

Hall experiments as observed through changes in differential Hall resistance reveals nuclear magnetic resonance under RF excitation[69,70] and EDMR based on the electrical detection of FMR through the inverse spin Hall effect (ISHE).[71–73] Here, FMR is induced in a ferromagnetic layer via resonant MW excitation. The precession of the magnetization during the FMR excitation induces a charge-free spin current in the adjacent nonmagnetic semiconductor layer via spin pumping. This spin-polarized current gives rise to an electrically detectable transverse charge current via the ISHE, which is a consequence of the spin–orbit coupling in the host material. This inverse spin Hall current is therefore directly related to the FMR signal in the ferromagnetic layer. This recently demonstrated technique can be used to investigate the strength of spin–orbit coupling in the nonmagnetic semiconductor layer.

26.4 CONTINUOUS-WAVE EDMR

Most of the CW EDMR studies found in the literature are based on adiabatic or quasi-adiabatic (lock-in detected) magnetic field sweep experiments where either the amplitude of the oscillating magnetic field or the magnitude of the external magnetic field is modulated.[37,39,74–79] Because the typical transition times of electronic processes are often slower than spin relaxation times associated with the electronic mechanisms that cause strong CW EDMR signals, the modulation frequencies for CW EDMR are usually lower than those typical for CW EPR spectroscopy (Hz to low kHz range).[60] Few CW EDMR studies are based on the measurement of true steady-state currents in the presence of CW radiation without any lock-in detection as they are prone to low-frequency noise, especially $1/f$ noise.[9] Another way to conduct CW EDMR spectroscopy is through an adiabatic frequency sweep experiment in the presence of a static magnetic field B_0. This type of experiment is possible due to the wide frequency bandwidth of EDMR, in contrast to EPR spectroscopy, albeit still not technically straightforward and therefore rarely found in the literature.[6]

In addition to the components shown in the illustration of an EDMR experiment in Figure 26.1, most laboratory setups for CW EDMR are also equipped with a lock-in amplifier and an additional set of Helmholtz coils for the modulation of the magnetic field B_0, cryostat facilities, and optical instrumentation allowing for the photoinjection of charge carriers into the probed materials and for the simultaneous measurement of CW EDMR and CW ODMR. While pEDMR has increasingly been utilized in the past 15 years, CW EDMR spectroscopy is still preferable when highest sensitivities are needed owing to the combination of lock-in detection with the already high spin sensitivity that spin detection via spin-selection rules provides. CW EDMR, like CW ODMR, allows for far superior spin sensitivity compared with EPR experiments, for selected materials and device systems even at the level of single-spin sensitivity.[80,81]

The data shown in Figure 26.1(b) is an example for the high spin sensitivity that EDMR spectroscopy exhibits, even without the use of lock-in detection. It was obtained on a $2\,mm \times 3\,mm$ large device area with an active MEH-PPV layer of less than 100 nm thickness, yielding a sample volume $<6 \times 10^{-7}\,cm^3$. Assuming a charge density of $\rho = I/(vA) = I/(\mu EA) = Id/(\mu VA)$, typical MEH-PPV mobility $\mu = 10^{-4}\,cm^2\,V^{-1}\,s^{-1}$, and a bias voltage above 2 V, we obtain an upper estimate for an ensemble size of $N = \rho Ad/e = Id^2/(e\mu V) < 1.56 \times 10^8$ electron spins (where ρ is the charge density, I the device current, v the charge carrier drift velocity, A the active area, μ the charge carrier mobility, E the electric field, d the device thickness, V the bias voltage, N the number of charge carriers, and e the elementary charge). The plot in Figure 26.1(b) displays an average of 20 sweeps. It shows a pronounced magnetic resonance peak with a signal-to-noise ratio of >30. This peak is well fitted by two overlapping Gaussian resonance lines and is due to electron and hole states that have identical positions (very close to the vacuum g-factor of $g \approx 2.0023$ due to the weak spin–orbit coupling in MEH-PPV) but different widths due to different hydrogen-induced hyperfine fields.[9] Given that this experiment was conducted at very low magnetic fields and room temperature with a nearly vanishing ensemble polarization and no lock-in detection, the data in Figure 26.1(b) show that spin sensitivity of EDMR spectroscopy can be much higher compared with EPR spectroscopy.[82]

The much higher sensitivity of EDMR compared with EPR techniques and its independence on the applied radiation frequency that exists for many spin-dependent electronic processes limits the applicability range of EDMR only to the range of magnetic field strengths and RF frequencies that the equipment provides.[9,83,84] Most of these experiments have been carried out as CW EDMR experiments with or without lock-in detection. Due to the broadband character of EDMR spectroscopy, the variation of a magnetic resonance spectrum as a function of the applied frequency is much easier to observe compared with EPR spectroscopy, which requires separate resonators with narrow bandwidth for sensitive detection. Often this is only possible for a distinct set of frequency ranges (e.g., X-, Q-, W-bands) and only under significant use of resources. In fact, using EDMR, magnetic resonance spectroscopy conducted in a similar manner as ODMR becomes possible where the photon flux rather than an electric current is observed as a function of the applied oscillating magnetic field frequency (see Chapter 25).[81,83]

Figure 26.3(a) displays EDMR spectra of non-ionized phosphorus (P) donor atoms at a temperature of 5 K in proximity of a silicon–silicon dioxide interface.[85] The P-donor state is also coupled via isotropic hyperfine interaction to the ^{31}P nucleus with $I = 1/2$. In the presence of photoinduced excess charge carriers, the paramagnetic ($S = 1/2$) P-donor state allows for spin-dependent intermediate-pair recombination into an adjacent silicon dangling bond (a so-called P_b center), which is also a single-electron state with $S = 1/2$. The measurement of EDMR spectra and the subsequent determination of the observed resonance peak centers at various RF frequencies allows for the determination of the transition frequencies between each of the four eigenstates of the ^{31}P–P_b pairs as a function of the applied magnetic field, as displayed in Figure 26.3(b). This in turn allows for a very clear identification of transition and coupling parameters within the pair.[86]

26.4.1 Limitations of CW EDMR

The advantages of CW EDMR spectroscopy compared with CW EPR in terms of the magnetic field and frequency dependence and sensitivity are contrasted by a number of limitations. Unlike inductive spin

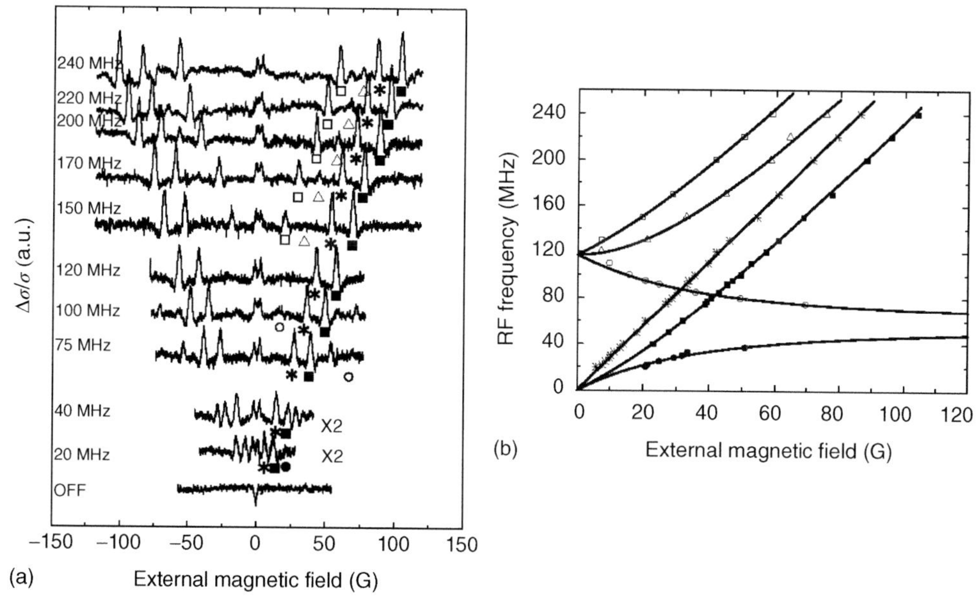

Figure 26.3. (a) Low temperature ($T = 5$ K) CW EDMR spectra obtained for various RF excitations from the magnetic field modulated and lock-in detected photocurrent of spin-dependent recombination through P–P_b pairs at the crystalline silicon to silicon dioxide interface. The individual peaks have been attributed to the transitions $|1\rangle \leftrightarrow |3\rangle$ (□), $|2 \leftrightarrow |4\rangle$ (■), $|2\rangle \leftrightarrow |3\rangle$ (Δ), $|1\rangle \leftrightarrow |2\rangle$ (•), and $|3\rangle \leftrightarrow |4\rangle$ (o), with $|1\rangle = |\uparrow\uparrow\rangle$, $|2\rangle = \alpha|\uparrow\downarrow\rangle + \beta|\downarrow\uparrow\rangle$, $|3\rangle = -\beta|\uparrow\downarrow\rangle + \alpha|\downarrow\uparrow\rangle$, and $|4\rangle = |\downarrow\downarrow\rangle$ being the eigenstates of the Si : P spin Hamiltonian. (b) Plot of the resonance peak frequencies as a function of the applied magnetic fields for the six different transitions within the P–P_b pair that is hyperfine coupled to the ^{31}P nucleus ($I = 1/2$). (Reprinted with permission from H. Morishita, L. S. Vlasenko, H. Tanaka, K. Semba, K. Sawano, Y. Shiraki, M. Eto, and K. M. Itoh, Phys. Rev. B, 2009, 80, 205206. Copyright 2009 by the American Physical Society)

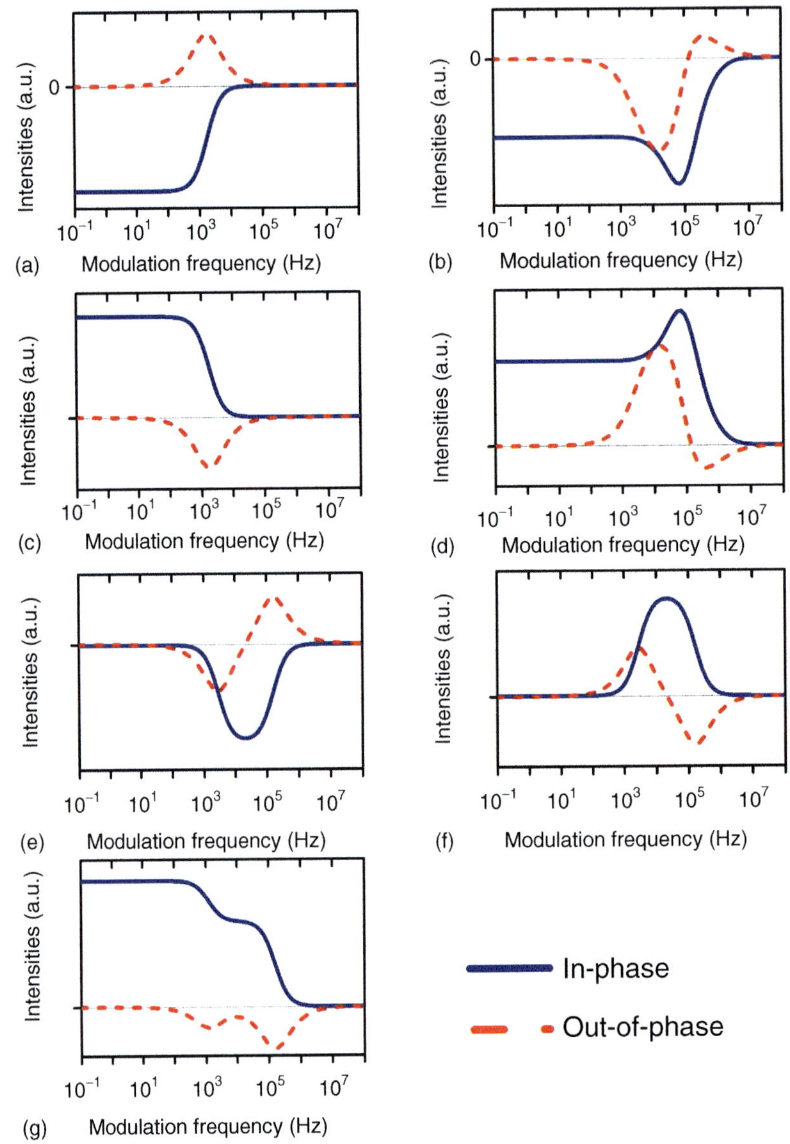

Figure 26.4. Plots of the qualitatively different modulation–frequency dependencies of analytical simulations of in-phase and out-of-phase components of lock-in detected CW EDMR spectra for spin-dependent transition rates governed by an intermediate-pair process as obtained from hundreds of quantitatively different rate coefficient scenarios. The data show that all the different scenarios can be reduced to a set of seven qualitatively different behaviors for which the rate change (referred to as intensity) can exhibit quenching and enhancement for either of the two components. (Reprinted with permission from S. Y. Lee, S. Y. Paik, D. R. McCamey, and C. Boehme, Phys. Rev. B, 2012, 86, 115204. Copyright 2012 by the American Physical Society)

detection used for EPR spectroscopy and like ODMR, EDMR cannot be applied to any paramagnetic center as it requires a paramagnetic state to be involved in a spin-dependent process that allows the spin manifold to control electronic transitions. The way by which an electrically detectable spin resonance translates into a measured EDMR spectrum is very much dependent on the nature of a given spin-dependent process, which, as discussed in the last section, varies greatly. For instance, the superior sensitivity of EDMR compared with EPR that was mentioned above is given for EDMR signals due to intermediate-pair mechanisms under nonequilibrium conditions. In contrast, EDMR signals due to polarization-controlled mechanisms suffer from the same limitations as inductively detected EPR experiments.[20,29] Moreover, in contrast to EPR spectroscopy, EDMR spectra are solely determined not only by the spin Hamiltonian (e.g., spin-coupling parameters) and the dynamics (e.g., spin-relaxation times) of the observed paramagnetic centers, but also by the dynamics of the electronic transitions. Thus, compared with EPR spectra, EDMR spectra are usually determined by a much larger number of parameters, and for convoluted spectra this can cause significant ambiguities for both the qualitative interpretation and the quantitative analysis of EDMR signals.

Examples for this ambiguity are room-temperature spin-dependent processes in polymer materials, including the MEH-PPV signal that is responsible for the data shown in Figure 26.1.[87] For the underlying physical nature of these processes, mutually contradicting microscopic models have evolved.[45,50,85,88–91] While the data in Figure 26.1 unambiguously shows that weakly spin–orbit coupled paramagnetic centers with $g \approx 2$ can govern conductivity via spin-selection rules, it does not narrow down the plethora of qualitatively different models that can account for this CW EDMR data. Attempts to resolve this problem by studying the dynamics of EDMR signals and their temperature dependency or by comparing with ODMR data have brought little resolve. Similarly, an interpretation of the sign of a spin-dependent signal, arguing that a charge transport process causes a conductivity increase under magnetic resonance, while a conductivity decrease must be observed when charge carrier recombination is present is inaccurate because of the complexity of the dynamics of spin-dependent processes. Based on the intermediate-pair model as outlined in Figure 26.2, Lee *et al.*[60] carried out generalized numerical simulations of modulated, i.e., lock-in detected CW EDMR or ODMR experiments assuming an intermediate-pair process in the presence of spin-selection rules. Figure 26.4 displays seven qualitatively distinct patterns for the modulation–frequency dependence of in-phase and out-of-phase CW EDMR and CW ODMR on-resonance signals that were found in this study. More than 1000 simulations were conducted with the rate coefficients defined by the rate model in Figure 26.2 being varied such that most relative quantitative scenarios were covered. Remarkably, across all the quantitatively different models, only the seven qualitatively distinguishable patterns displayed in Figure 26.4 were found. This alone is indicative of the ambiguity that exists when it comes to the interpretation of CW EDMR spectra.

Figure 26.4 also shows that the sign for any intermediate-pair mechanism (transport or recombination) can be positive or negative, depending on the actual values of the various transition rate coefficients. For certain rate constellations, the sign of these rate changes may even be positive at one modulation frequency and negative at another.[60]

The work by Lee is based on the assumption that an intermediate-pair process is the origin of an observed CW EDMR signal; however, for many experimentally observed signals, it is not known whether this is a correct assumption. Thus, any interpretation of the sign of a rate change observed with CW EDMR as well as an assignment of transition rate coefficients based on CW EDMR spectra is impossible without proper analysis of the dynamics of this signal. A dynamical analysis of spin-dependent processes oftentimes requires access to the sub-microsecond time domain that is hardly accessible by lock-in schemes where modulation frequencies of most EDMR spectrometers have upper limits below 1 MHz.

26.5 PULSE EDMR

Many materials that show pronounced spin-selection rules have low-charge carrier mobilities and consequently, their dielectric relaxation limited rise times are longer than the time scales on which coherent charge carrier spin motion takes place. This circumstance, which was likely the most important reason for the late development of pEDMR experiments, can be overcome experimentally by the separation of the preparation and the detection of coherent spin states within a pEDMR experiment. Figure 26.5 illustrates

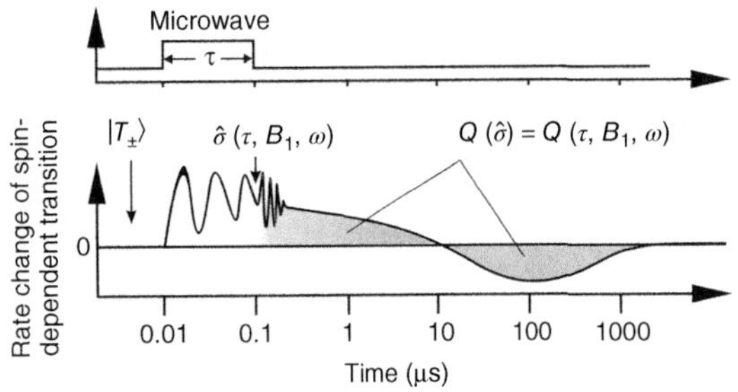

Figure 26.5. Timing illustration of a pEDMR experiment using a logarithmic time scale. A magnetic resonant excitation pulse or pulse sequence is applied to the ensemble of paramagnetic charge carriers that lead to the preparation of a pulse parameter (length, strength B_1, or frequency ω)-dependent final state σ. This coherent ensemble state is then probed by integration of the current transient as a function of time on time scales orders of magnitudes longer than the pulse excitation. The charge Q collected over these long time scales, on which the gradual decay of the prepared coherent spin ensemble takes place, reflects the projection of the spin-pair ensemble state onto the singlet operator. (Reprinted from C. Boehme, and K. Lips, Physica B, 2006, 376–377, 930–935, with permission from Elsevier)

the timeline of a pEDMR experiment on a logarithmic time scale: The pEDMR excitation requires coherent spin manipulation that leads to the preparation of a charge carrier ensemble spin state (characterized by a density operator $\hat{\sigma}$) that depends on the length τ, the amplitude B_1 as well as the frequency ω of the excitation pulse. This excitation takes place on timescales shorter than spin relaxation, while pEDMR detection takes place by long-time integration of the current, corresponding to a charge Q of resonantly induced conductivity changes during which the charge carriers are counted that undergo spin-dependent electronic transitions due to the magnetic resonant excitation. Since the detected charge Q depends on the spin ensemble state σ right after the pulse excitation, it also depends on the excitation pulse parameters. The detection of Q has to be done on time scales longer than the dielectric relaxation time of the probed material in order to prevent that the material filters its own spin-induced electric signals. pEDMR became possible after the realization was made that, for a spin-dependent current controlled by a particular spin ensemble, the integrated current transient caused by a spin-resonantly excited charge carrier ensemble that amounts to a charge ΔQ that is directly proportional to the S content of the probed spin manifold at the end of the spin-excitation sequence.[35] Thus, the observation of coherent spin motion becomes possible by repetition of a 'pulsed spin-resonant excitation pump/charge integration-probe' experiment where the spin excitation sequence is carried out repetitively, while one or more the experimental parameters (e.g., the pulse length) is or are slightly changed. Such measurements drastically increase the measurement duration of pEDMR experiments compared with inductively detected pEPR experiments. However, at the same time, they allow to overcome the fundamental time resolution limits posed by dielectric relaxation and thus, they have opened the door for the same range of spectroscopy tools that is known from traditional pEPR spectroscopy.

The advent of pEDMR has given access to a broad series of quantitative parameters that define the nature of spin-dependent electronic transitions. Since 2002, a series of theoretical studies has matured this technique to a degree where many pulse sequences known from pEPR can be applied in order to obtain characteristic parameters of intermediate-pair process and other investigated spin-dependent electronic processes. These experiments include rotary echoes and Hahn echo sequences, echo decay measurements, inversion recovery, electron–electron double resonance (ELDOR), electron-nuclear double resonance (ENDOR), electron spin echo envelope modulation spectroscopy, and other pulse techniques, including a recent demonstration of adiabatic pulsing.[7,11,32,40,92–94] Furthermore, since EDMR does not require the detection of radiation and since most observed EDMR signals are based on intermediate spin pairs, EDMR has been applied to a broad range of frequencies, ranging from the lower RF domain all the way into the sub-terahertz domain, from 40 MHz to 336 GHz.[7,9,62,83]

The exact nature of how coherent spin motion is observed in a pEDMR experiment depends on the microscopic spin-dependent processes that produce a pEDMR signal. There are many systems to which pEDMR has been applied, including point defects in crystalline, microcrystalline and amorphous silicon, and a variety of organic materials and devices (e.g., OLEDs and solar cells).[5,13,45,46,62,95,96] In its most basic form, a pEDMR experiment consists of a single coherent MW pulse that is on resonance with a transition of charge carrier spins after which the device current is recorded as a function of time. Before the MW pulse is applied, the device is in a steady state and the current is constant. During the coherent manipulation, charge

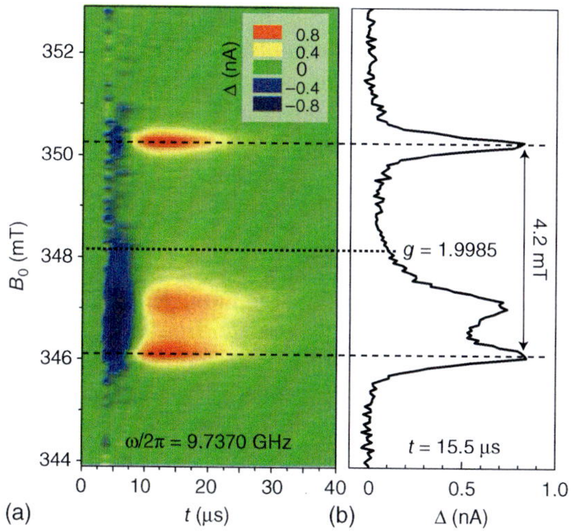

Figure 26.6. Coherent manipulation of spin-dependent charge carrier recombination between a ^{31}P donor and paramagnetic localized states at the Si/SiO_2 interface. (a) Contour plot of the transient current change after a 480 ns long MW pulse with a power of 1 W as a function of the magnetic field and after baseline correction in order to remove MW artifacts. (b) A slice through the contour plot at $t = 15.5\,\mu s$. Every time slice has the shape of an EPR spectrum, corresponding to a superposition of the hyperfine split resonance of the ^{31}P donors and the resonance of the P_{b0} centers. (Reproduced with permission from Ref. 13. © Nature Publishing Group, 2006)

carrier spin states are changed and the steady state is disturbed. As a result, an enhancement or quenching of the current is observed. At longer time scales, the spin states and electronic states undergo relaxation transitions, and the system returns to its steady state. The time scales involved in these processes are not only the spin–lattice and spin–spin relaxation times but also the charge carrier lifetimes as well as time constants related to the detection system (i.e., amplifier response times).[44,56] An example, measured on P-doped silicon is shown in Figure 26.6.[13] Here, the ^{31}P-donor spins are weakly coupled to P_{b0} centers near the SiO_2 interface, and magnetic resonant manipulation changes the spin-dependent recombination rates, which is detected as a change in device current. Figure 26.6(a) shows a contour plot of the current transients as a function of magnetic field across the resonances. Each horizontal slice corresponds to the experiment illustrated in Figure 26.5 (although no time integration is carried out here). Figure 26.6(b) shows a single-field slice that corresponds to the magnetic resonance spectrum, which exhibits the P_{b0} resonance as well as the hyperfine-split ^{31}P resonances. The time dynamics exhibits an initial current decrease due to the enhanced S content and a slower increase due to the quenching of the number of T states. A free induction decay, as in pEPR, is not visible in pEDMR experiments.

The simplest way to observe coherent spin motion using a single-pEDMR experiment is to measure the current response after the MW pulse as a function of the pulse length. In this case, Rabi oscillations are observed as the nutation angle is increased.[5,6] Figure 26.7 shows an example of this, measured at low temperature ($T = 5$ K) on diodes made of PEDOT:PSS, which contains only a low fraction of hydrogen atoms and thus exhibits small local hyperfine fields and, consequently, narrow charge carrier line widths compared with other organic semiconductor materials.[6] The plots in Figure 26.7 display the Fourier transforms of the charge $Q(\tau)$ (time-integrated change in device current) after the pulse excitation as a function of the applied pulse length τ, measured at various static magnetic fields B_0. Panel (a) and (b) display these Fourier-transformed functions $Q(\tau)$ for the on-resonance field $B_0 = 347.1$ mT and for a detuned field $B_0 = 346.04$ mT. The fundamental frequency component corresponds to a single spin with $S = 1/2$ undergoing Rabi nutation. Under strong B_1 conditions a higher-frequency harmonic is visible that corresponds to a spin-pair precessing together, i.e., the double quantum transition between T states T^+ and T^-.[9,91,97] Figure 26.7(c) shows a contour plot of the Fourier-transformed charge signals for continuous detuning.[57] Through a careful analysis, it was possible to extract estimates for exchange and dipolar coupling strengths of $|J| < 7.3$ MHz and $|D| = 5.7 \pm 0.4$ MHz.[6]

The interplay between electronic and nuclear spins can be investigated using pENDOR experiments (see Chapter 16).[7,11,98,99] Figure 26.8 summarizes an electrically detected pENDOR measurement on P-doped silicon at 240 MHz.[7] In this system, the degeneracy of the spin resonance conditions is lifted by the presence of contact hyperfine interaction and non-negligible nuclear Zeeman energy, and consequently the transitions can be independently addressed.[7] The nuclear resonance is observed by comparing the current signals with and without manipulation of the nuclear spin ensemble by an RF pulse. Figure 26.8(a) illustrates the level diagram, pulse sequence, and current transient for the case where no nuclear resonant pulse is applied; the pulse sequence consists of two MW π-pulses only. Figure 26.8(b) shows the experiment where an RF π-pulse that is resonant with the nuclear spins of

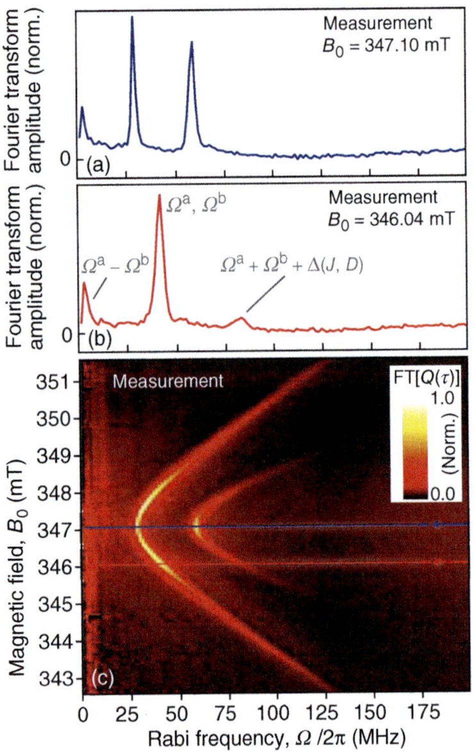

Figure 26.7. Rabi oscillations as a function of B_0 in the frequency domain, measured on a PEDOT : PSS injection device. (a and b) Fourier transform of the time-integrated device current following a coherent excitation with constant MW power and progressively increasing pulse length for (a) B_0 on resonance and (b) B_0 detuned from resonance by 1.06 mT. Next to the fundamental frequencies $\Omega_{a,b}$, there is a harmonic with about twice the fundamental frequency at $\Omega_a + \Omega_b + \Delta(J,D)$ due to the strongly driven pair process in which both spins are resonant with B_1. The frequency peak close to the origin is due to difference-beat oscillations. Fundamental and harmonic shift to higher frequencies under detuning. (c) A contour plot of Fourier-transformed time-integrated device current for continuous detuning of B_0. The horizontal blue and red lines correspond to the field positions at which the data in (a and b) were taken [(a) blue, (b) red]. (Reproduced with permission from Ref. 6. © Nature Publishing Group, 2015)

^{31}P is applied between the two MW pulses. For both cases, the corresponding NMR signal is obtained from the difference of the detected signals (cf. Figure 26.8c; Figure 26.8d shows the dependence of the signal on the applied RF). Using this approach, it was possible to establish a nuclear spin–lattice relaxation time $T_{1N} = 112$ s by varying T_{Store}, the delay between the RF pulse and the MW readout pulse.

EDMR lends itself to broadband multifrequency EPR applications because the resolution depends on a lesser degree on the spin polarization compared with conventional EPR. MW circuitry or RF excitation schemes tend to be much simpler for EDMR, and narrow-banded MW components can be mostly avoided. When a resonator is replaced by a nonresonant structure for MW excitation, EDMR can be used over a wide frequency range, from the lower RF domain up to several GHz.[83,94] Inhomogeneities in the MW field B_1 can be mitigated by using adiabatic pulses that are specifically designed to compensate for that (see Chapter 21).[94,100]

26.5.1 Electrically Detected Electron-spin Echoes

The ability to detect spin echoes is crucial to many established high-fidelity pulse magnetic resonance techniques, such as measurements of spin–lattice and spin–spin relaxation times, electron spin-echo envelope modulation (ESEEM), ENDOR, ELDOR, and many others (see Chapters 11, 16, 18, and 19).[101] Spin-echo experiments are feasible in EDMR measurements, however, with a modified pulse technique.

Initial attempts to observe coherent spin effects in the presence of inhomogeneities in the MW field B_1 were based on so-called rotary (Rabi) spin echoes.[32,102] In such an experiment, spins in thermal equilibrium (along $+z$-direction) are subjected to an MW pulse of a certain duration t_p, which (i) rotates the spins by an angle that depends on t_p and B_1 and (ii) leads to a dephasing of the spins depending on the inhomogeneity of B_1. When subsequently a pulse of opposite phase is applied, the spins will refocus at a time $2t_p$ independent of the size of the field inhomogeneities. These rotary echoes form along the $+z$-direction, which can be directly observed as a change in device resistivity. Since these initial pEDMR experiments, much progress has been made on the design of sample device and contact structures for pEDMR experiments that do not distort the B_1 field distribution.[5,96,103] Most notably, the use of lithographically defined thin-film wiring for device operation and current detection rather than macroscopic bond wires has to a large degree eliminated distortions of the resonator modes due to the skin effect. These developments have enabled pEDMR

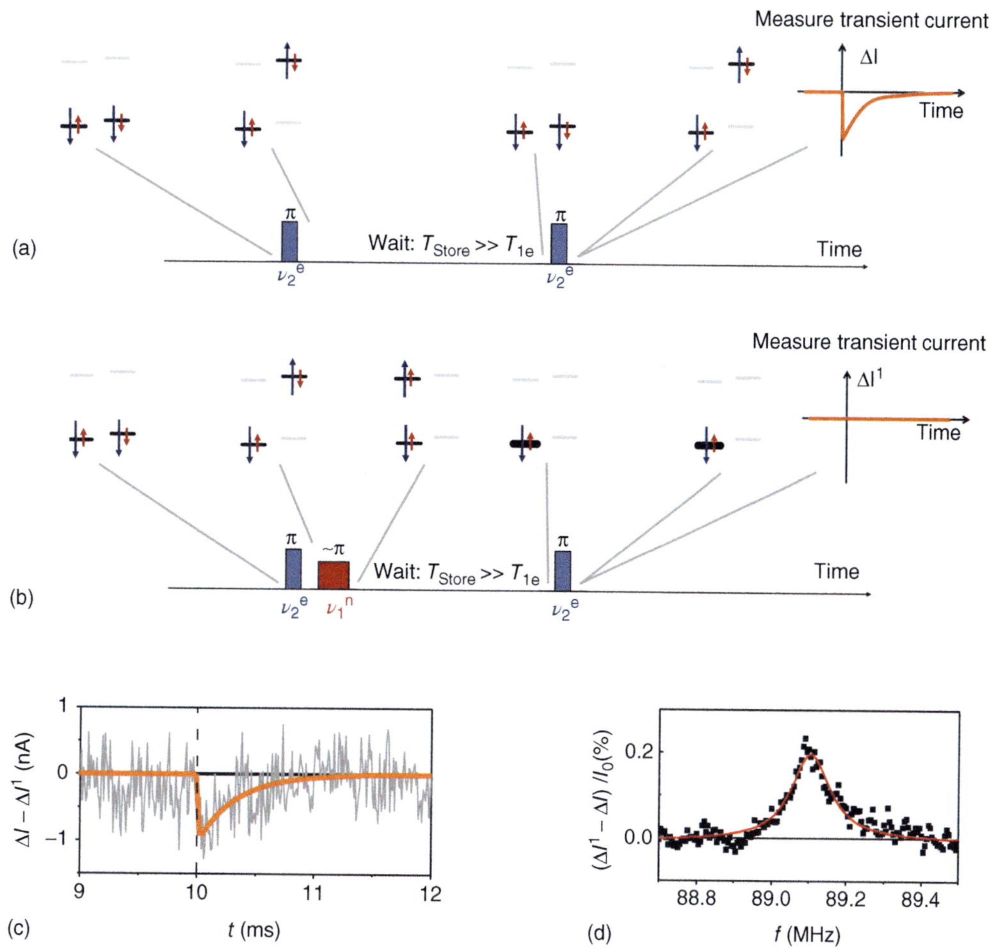

Figure 26.8. Electrically detected electron–nuclear double resonance experiments on P-doped silicon at a MW frequency of 240 GHz. (a and b) The pulse sequence [consisting of MW ($\nu_2{}^e$) and RF ($\nu_1{}^n$) pulses] and corresponding level populations at each step (a) without and (b) with radio frequency nuclear storage pulse. At the end of the sequence, the current transient ΔI is recorded as a function of time. If no nuclear spin storage is undertaken, the initial spin state has no impact on the final spin state obtained. However, if a nuclear storage pulse is applied, the magnitude of the current change after the readout pulse is reduced (ΔI^1). T_{Store} is the delay between the RF pulse and the MW readout pulse, and T_{1e} is the electronic spin–lattice relaxation time. (c) The ENDOR signal corresponds to the difference between a and b ($\Delta I - \Delta I^1$). (d) The relative current change as a function of radio frequency. Note that an additional long radiofrequency pulse is applied at the end of the pulse sequence in order to thermalize the nuclear spins faster. (Reproduced with permission from Ref. 7. © The American Association for the Advancement of Science, 2010)

experiments with almost the same B_1 homogeneity as pEPR experiments and the rotary echo experiments have therefore become increasingly obsolete.

Hahn echoes occur in conventional magnetic resonance experiments when the spins that precess in the *xy*-plane, perpendicular to the external magnetic field along the +*z*-direction in the lab frame, are refocused. At the point in time when the spins rephase, macroscopic magnetization builds up and a brief spike in the detected radiation signal can be observed (see Chapter 11). For EDMR or ODMR detection, such collective magnetization effects do not give rise to a signal, as it does not influence spin-pair permutation symmetry. In order for this to happen,

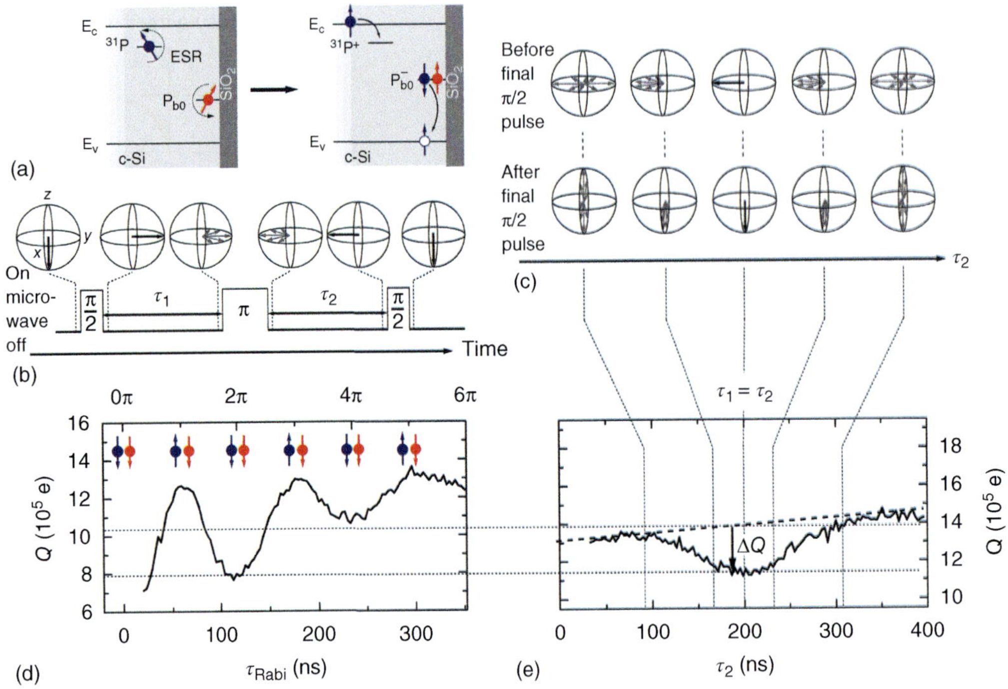

Figure 26.9. (a) Illustration of spin-dependent recombination between a ^{31}P donor and a P_{b0} center at the crystalline Si–SiO_2 interface. (b) Pulse sequence and corresponding evolution of an ensemble of identical spins plotted on Bloch spheres, with $-z$ being the thermal equilibrium orientation of the spins. (Note that the notation used in Ref. 40 – from where this figure originates – is different from the rest of this chapter: in subsequent figures and in the text, τ is used instead of τ_1, and τ' is used instead of τ_2.) At a time τ_2 after the π-pulse a so-called tomography $\pi/2$ pulse is applied, which rotates the spins from the x–y plane into z-direction. (c) Bloch spheres before and after the tomography pulse for various values of τ_2 corresponding to the pulse sequence shown in (b). The echo maximum occurs for $\tau_2 = \tau_1$. (d) Time-integrated device current as a function of pulse length τ_{Rabi}. The modulation depth corresponds to the echo amplitude observed with the modified Hahn echo sequence. (e) Time-integrated device current after the pulse sequence shown in (b) as a function of τ_2. The echo maximum occurs for $\tau_2 = \tau_1$. The dashed lines indicate the corresponding Bloch sphere illustrations in (c). (Reprinted with permission from H. Huebl, F. Hoehne, B. Grolik, A. R. Stegner, M. Stutzmann, and M. S. Brandt, Phys. Rev. Lett., 2008, 100, 177602. Copyright 2008 by the American Physical Society)

the spins need to be rotated by a projection pulse into their eigenstates along $+z$ or $-z$-direction. This can be achieved with an additional $\pi/2$-pulse right at the time when the echo occurs.[11,40,92] The resulting current changes are then time-integrated (over several microseconds) in order to obtain a charge signal. This improves the signal-to-noise ratio and eliminates time-transient signal contributions. This scheme is related to the so-called longitudinally detected spin echoes, where the longitudinal component of the magnetization along z-direction was detected using an additional pickup coil in order to reduce spectrometer dead time in a pEPR experiment.[104]

Figures 26.9 and 26.10 show early implementations of this measurement scheme for pEDMR. Figure 26.9 shows experiments on P-donor electrons in silicon at the vicinity of a SiO_2 interface.[40] The spin-dependent transitions involving the P donor and the P_{b0} defect at the crystalline silicon to silicon dioxide interface are illustrated in Figure 26.9(a). Figure 26.9(b) and (c) shows the pulse sequence and the Bloch spheres indicating the corresponding magnetization orientations. The first part of the pulse sequence, $\pi/2-\tau_1-\pi$, is analogous to the conventional Hahn echo sequence and leads to a refocusing of the spins in the xy-plane at a time τ_1 after the π-pulse.[101] Here, the so-called echo

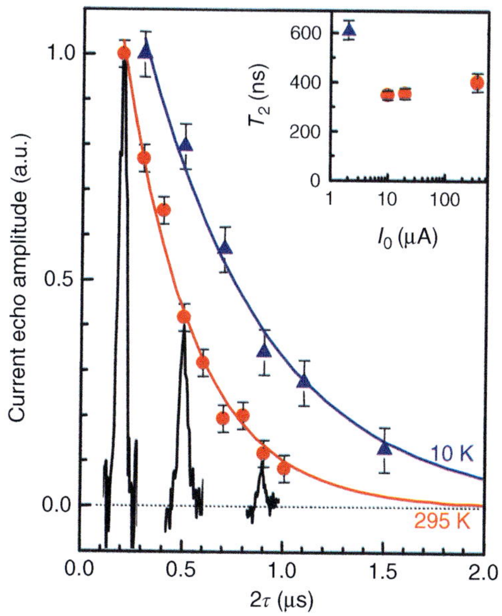

Figure 26.10. Modified Hahn echoes obtained from forward currents of MEH-PPV-based diodes similar to those described in Figure 26.1. Echo measurements were performed for various values of τ (as defined in Figure 26.9). The black curves show the echo signal scaled to the time axis, with the echo maximum occurring at 2τ. From the decay of the echo amplitude, the intrinsic spin–spin relaxation time T_2 is determined. The echo intensity follows an exponential decay with time and depends only weakly on temperature. All measurements were performed at 295 K unless otherwise marked. The inset shows T_2 values at different device currents to exclude the possibility of current-induced dephasing. (Reprinted with permission from W. J. Baker, T. L. Keevers, J. M. Lupton, D. R. McCamey, and C. Boehme, Phys. Rev. Lett., 2012, 108, 267601. Copyright 2012 by the American Physical Society)

tomography pulse with length $\pi/2$ is applied at a time τ_2 after the π-pulse, which rotates the spins onto the z-direction, as necessary for electrical detection. The strongest echo signal is obtained at $\tau_1 = \tau_2$, which is shown in Figure 26.9(e). For comparison, single-pulse Rabi oscillations, i.e., the charge (time-integrated current signal) after a pulse of duration τ_{Rabi} are shown in Figure 26.9(d). The depth of the nutation signal is thus the upper limit of the echo signal in Figure 26.9(e). When this experiment is repeated for various values of τ_1 and τ_2, the spin–spin relaxation time T_2 can be determined. (Note that the notation used in Figure 26.9 is different from the rest of the chapter: subsequently, τ is used instead of τ_1, and τ' is used instead of τ_2.)

Figure 26.10 shows the application of this technique to organic semiconductors (MEH-PPV devices similar to the illustration in Figure 26.1), where a value of $T_2 = 320$ ns was obtained.[92]

The signal-to-noise ratio of the basic EDMR-based echo experiment can be improved in several ways. On one hand, the use of analog boxcar integration to obtain the charge signal rather than digital integration can help to mitigate influences of the frequency-dependent transfer characteristic of the current amplifier.[11] On the other hand, varying spin-independent echo backgrounds can be eliminated by using proper phase-cycling schemes.[11,105,106] Typically, in electrically detected spin-echo experiments, the echo appears on top of a large background that arises from RF and MW artifacts, as well as from the time constants of the EDMR detection circuit. These additional contributions do not depend on spin orientation and are not affected by a change in the MW phase of the pulses. The projection pulse that is used to rotate the spins from the xy-plane in which the echo occurs to the z-direction can be applied in such a way that the spins are oriented along $+z$ or $-z$, resulting in a change of sign of the echo shape, but without influencing the background signal. Experimentally, this is achieved by using MW phases that differ by 180°. A subtraction of both echo measurements cancels the spin-independent background signal out while adding the measured echo signals. This phase-cycling scheme adapted for electrical detection is summarized in Figure 26.11.[11] In conventional pEPR so-called phase cycling is employed in order to eliminate unwanted signals, and established phase-cycling schemes (see Chapter 11) are extended with the two different orientations of the projection pulse.[101,107] Figure 26.11(b) shows the full phase-cycling sequence for electrically detected Hahn echoes, and the resulting echo shape free from background signals is shown in Figure 26.11(c).

Another commonly used pulse technique is the so-called stimulated echo, consisting of three $\pi/2$-pulses (Figures 26.11d and 26.12a) (see Chapter 11). The first two pulses, $\pi/2 - \tau - \pi/2$, generate electronic spin polarization. After a mixing time T, another $\pi/2$-pulse is applied, and after a time τ the stimulated echo forms in the xy-plane. In order to utilize this sequence for EDMR, a projection $\pi/2$ pulse is applied at the echo maximum, analogous to the case of the Hahn echo sequence discussed above. The corresponding phase-cycling sequence consists of eight steps as shown in Figure 26.11(d) and the resulting background-corrected echo in Figure 26.11(e).

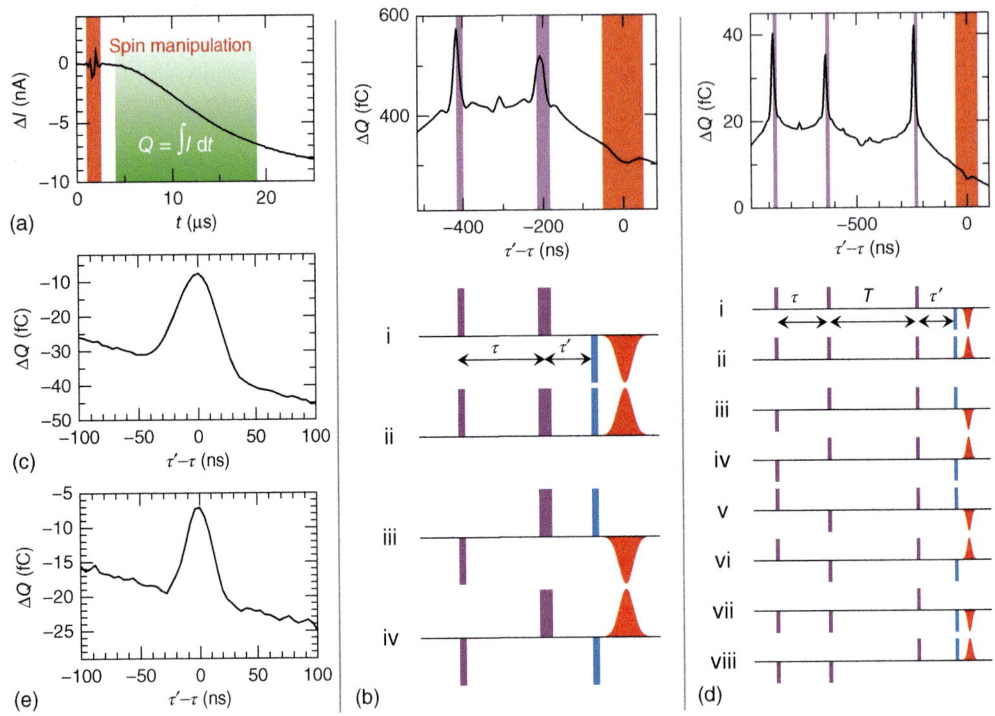

Figure 26.11. Electrical detection of electron–hole spin echoes in the current of an organic polymer diode as illustrated in Figure 26.1. (a) Sketch of the measurement sequence. After manipulation with coherent MW pulses, the resulting transient current response is recorded by boxcar integration. (b) Schematic representation of the two-pulse (Hahn) echo sequence showing all four steps (i–iv) of the phase cycle that is used for background subtraction. The fixed $\pi/2$- and π-pulses are shown in purple, and the $\pi/2$ pulse-projection pulse, which is swept across the echo (red), is shown in blue. A pulse pointing up/down represents a pulse in the $+x/-x$ pulse channel. When the echo points up (or down), the contribution of the phase cycle step is added to (or subtracted from) the sum. The actual temporal evolution of the echo as a function of τ' is also shown, corresponding to the initial phase cycle step (i). Here, the projection pulse is swept across the entire time range and also covers the temporally fixed pulses. The echo forming at $\tau' = \tau$ is visible as the dip in the black curve, marked by the red bar. (c) The background-corrected Hahn echo transient resulting from the full four-stage phase cycle described in b. (d) The three-pulse (stimulated) echo sequence consisting of three fixed $\pi/2$-pulses (purple) and one $\pi/2$-projection–pulse (blue) that is swept across the echo (red). All eight steps (i–viii) of the phase cycle are shown. The temporal evolution of the echo (non-background corrected) as a function of τ' is shown, corresponding to phase cycle step (i), with the echo marked at time zero. (e) The shape of a measured stimulated echo corrected by the eight-pulse sequence described in (d) as a function of τ'. (Reproduced with permission from Ref. 11. © The American Association for the Advancement of Science, 2014)

Using the projection–pulse electrically detected echo technique, it is straightforward to implement many of the high-resolution pulse sequences that have been established for pEPR spectroscopy with electrical detection. The spin–spin relaxation time T_2 is directly established from the Hahn echo experiment described in Figures 26.9 and 26.10 by varying the delay between the pulses. As τ increases, the echo amplitude decays exponentially with a time constant T_2. In pEPR, it is common to sweep τ over a wide range and directly record echo amplitudes or areas rather than the exact line shape. Using the background correction scheme outlined above, this method can be applied to electrically detected echoes as well. However, while in conventional pEPR one preferably records the echo area for an improved signal-to-noise ratio, this is not practical for electrical detection, because with the additional projection pulse, a separate measurement is required for each point along the echo shape.

Therefore, only the echo amplitude is detected in these experiments.

For measurements of the spin–lattice relaxation time T_1, a so-called inversion-recovery pulse sequence $\pi - T - \pi/2 - \tau - \pi - \tau -$ echo is used, and the echo amplitude is measured as a function of time T. The second part of the pulse sequence is a Hahn echo with fixed delay τ, which means that the whole sequence can be easily adopted for electrically detected experiments by appending a projection $\pi/2$ pulse at the position of the echo maximum (see Chapter 11).[36]

ESEEM experiments are used to investigate hyperfine coupling between charge carrier spins and nuclear spins.[10,11,108] For this experiment, nuclear spins are excited indirectly through flipping of the electron spins, which induces a non-adiabatic change of the magnetic field at the position of the nucleus and leads to precession. This is detected through a beat oscillation that is imposed on the echo amplitude as a function of pulse delay. Experimentally, this is observed in pulse sequences where the delay between pulses is varied. For example, the beat oscillations can be observed in T_2 measurements as an envelope modulation that is superimposed on the exponential decay of echo amplitude. In a similar way, the oscillations also appear in stimulated echo experiments, which has the advantage that the echo amplitude decays at a rate slower than T_2, and the modulations can be observed over a longer pulse delay range. The nuclear resonance frequencies will then appear as frequency components of the beat oscillation and are resolved by Fourier transformation of the echo envelope.[10,11] Figure 26.12 shows electrically detected ESEEM experiments on MEH-PPV devices similar to the structure displayed in Figure 26.1. In this material, the charge carrier spins couple to the nuclear spin of hydrogen. The coupling is observed through stimulated echo experiments and appears as an envelope modulation with a frequency close to 15 MHz, which is the nuclear Larmor frequency at the magnetic fields displayed in Figure 26.12(b) and (c). When using a polymer where some fraction of the hydrogen atoms is replaced by deuterium, its chemical and electronic properties remain unchanged except for the influence of the nuclear spin-induced hyperfine fields on the electronic spin states. This results in a markedly different oscillation imposed on the spin-echo envelope, corresponding to the nuclear Larmor frequency of deuterium close to 2 MHz (Figure 26.12d and e) (see Chapter 18).

Electrical spin-echo detection can also be applied to the ENDOR experiments. For example, for a Davies-ENDOR experiment (see Chapter 16), an RF pulse is applied during the mixing period T of an inversion–recovery experiment (see above). This is analogous to the ENDOR experiment discussed above, as the projection Hahn echo detection sequence effectively corresponds to a π-pulse. This scheme differs from the EDMR–ENDOR approach described above only because it relies on electrical Hahn echo detection and thus features superior baseline correction and sensitivity. Adaptation of established pENDOR techniques for EDMR is therefore straightforward and has been applied to P-doped silicon and for polymer materials.[11,98,99]

Another example for the application of well-established pEPR techniques for EDMR is double electron–electron resonance (DEER) spectroscopy that is used for the investigation of coupling (exchange and dipolar) between coupled electron spins (see Chapter 18). DEER relies on the simultaneous selective excitation of well-separated individual spin resonances using different MW frequencies. Since this experiment is also related to the Hahn echo technique, it can be utilized for pEDMR in a similar way as pENDOR and ESEEM.[109] Figure 26.13 displays electrically detected double electron–electron resonance (EDDEER) revealing the coupling between P_{b0} (A spins) and ^{31}P (B spins) in silicon near the SiO_2 interface, where both A and B spins are selectively excited using two MW frequencies. Figure 26.13(a) and (b) shows the pulse sequence used along with the measured echo signal. The pulse sequence for the MW frequency that is resonant with the A spins (red) induces a projection Hahn echo signal that is described above. In addition, a pulse with a different MW frequency (blue) that excites only B spins is inserted during the mixing time before (Figure 26.13a) and after (Figure 26.13b) the refocusing pulse at different times t_p. Any change in the echo amplitude detected by signals from the A spins can only arise from the coupling between the A and the B spins. From the observed changes of echo amplitude, an exchange coupling between P_{b0} and ^{31}P spins is extracted from a numerical simulation of exchange coupling that takes the distribution of $^{31}P–P_{b0}$ in the sample geometry into account. The resulting exchange coupling strength ranges from 25 kHz to 3 MHz, which corresponds to distances between 14 and 20 nm.

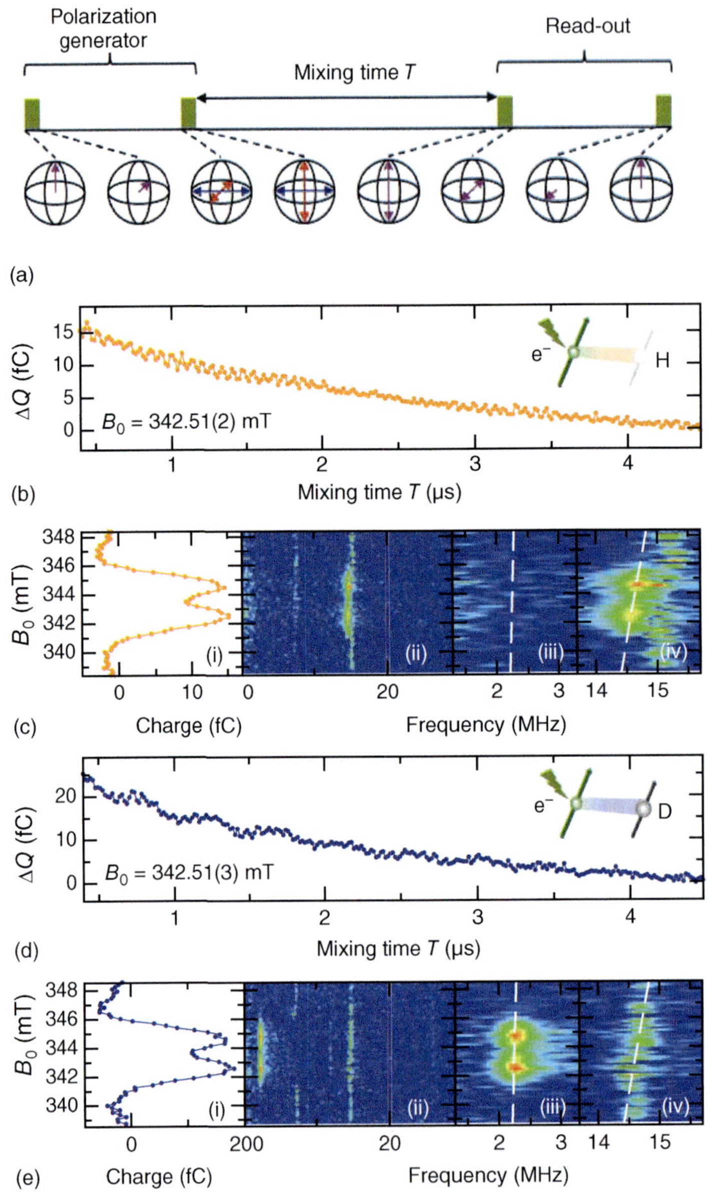

Figure 26.12. Electrically detected ESEEM measurements performed at a MW frequency of 9.63 GHz for an organic polymer diode based on a device stack as shown in Figure 26.1. (a) Pulse scheme for the electrically detected ESEEM experiments. (b and d) Electrically detected ESEEM signal as a function of the mixing time, *T*. The data reveals the nuclear spin precession-induced oscillations that modulate the exponential decay of the spin echo signal. (c.i and e.i) A slice of the data shown in (b) and (d) as a function of the static magnetic field B_0 after background correction. (ii) Fourier transform of ESEEM signal. (iii and iv) Magnified views of the data in (i) around approximately 2 and 15 MHz, where the deuteron and proton resonances are expected, respectively. The dotted lines represent the frequencies of the proton and deuteron resonances as a function of B_0 as calculated from the respective nuclear gyromagnetic ratios. Note that the additional nonresonant features at 15 and 7.5 MHz correspond to higher harmonics of the Nyquist critical frequency used for data acquisition. (Reproduced with permission from Ref. 11. © The American Association for the Advancement of Science, 2014)

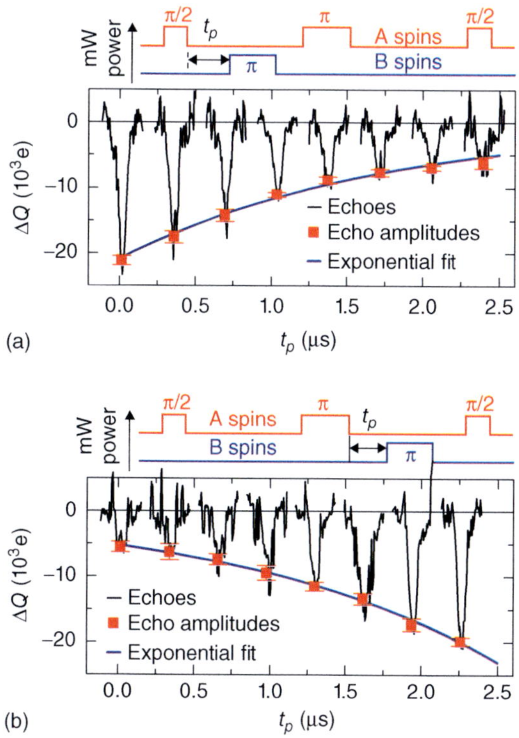

Figure 26.13. Electrically detected double electron–electron resonance (EDDEER) experiments on the electron spin of a ^{31}P donor coupled to a P_{b0} center at the interface between crystalline Si and SiO_2. A modified Hahn echo sequence is applied on resonance with the P_{b0} centers (A spins), with an additional inversion pulse that is resonant with ^{31}P (B spins) at various positions t_p in the free evolution interval $t = 2.5\,\mu s$ (a) before and (b) after the Hahn-echo *p*-pulse [the actual pulse sequences are shown in (a) and (b)]. The echo amplitudes (red) decreases/increases with a time constant (a) $T = 1750 \pm 100$ ns and (b) $T = 1700 \pm 50$ ns. (Reproduced with permission from Ref. 109. © Taylor and Francis, 2013)

26.6 RADIO FREQUENCY EDMR

The detection of resonant changes in sample resistivity via RF reflectometry has been demonstrated on the ^{31}P–P_{b0} recombination in silicon in both CW and pulse experiments.[110] This method allows for the direct observation of coherent spin oscillations during MW excitation, which is not detectable in conventional resistivity measurements due to the limited bandwidth caused by the detection setup and dielectric relaxation times. In this particular approach the sample, together with its stray capacitance and an on-chip inductor, forms a resonant LCR oscillator circuit, and a change in sample resistivity is detected as a change in reflected MW excitation power on resonance. Based on the experimental setup, RF-detected pEDMR and pEPR experiments appear to be very similar since both are based on radiation-detected magnetic resonance spectroscopy. However, RF detection is, in the context of these experiments, still electrical detection since conductivity is the monitored physical observable. As for any EDMR experiment that is governed by spin-selection rules of intermediate pairs, spin polarization does not play a role for signal strength.

26.7 CONCLUSIONS AND OUTLOOK

Fifty years after the first EDMR experiments were recorded and more than 15 years after first pEDMR experiments showed the electrical detectability of coherent spin motion, a broad variety of EDMR studies has been conducted on many spin-dependent processes in different materials and electronic device systems for the purpose of defect spectroscopy, the pursuit of spin-readout concepts for spin-based information and quantum information technologies, and the exploration of efficiency-limiting processes in optoelectronic materials and devices as well as for the utilization of spin-dependent electronic mechanisms for new sensor applications.[13,32,61,83] With the advent of pEDMR, further development of EDMR spectroscopy has become particularly diverse, as a continuously growing set of inductively detected pEPR experiments is gradually applied to pEDMR as well. In the near future, various broadband detection techniques, almost continuously ranging over more than five orders of magnitude (lower MHz range to near-THz range), pulse shaping, and pEDMR-based imaging techniques are expected to dominate the development of EDMR spectroscopy. In the long range, almost any methods development taking place for EPR spectroscopy can be expected to affect EDMR methods development in the same way, with little to no delay.

ACKNOWLEDGMENTS

The compilation of this chapter took place with support of the US Department of Energy, Office of Basic

Energy Sciences, Division of Material Sciences and Engineering under Award DESC0000909. For the contributions to the work reviewed in this chapter, C. B. acknowledges in particular Klaus Lips, who is the coinventor of pEDMR.

REFERENCES

1. D. R. McCamey, H. Huebl, M. S. Brandt, W. D. Hutchison, J. C. McCallum, R. G. Clark, and A. R. Hamilton, *Appl. Phys. Lett.*, 2006, **89**, 182115.
2. J. T. Muhonen, J. P. Dehollain, A. Laucht, F. E. Hudson, R. Kalra, T. Sekiguchi, K. M. Itoh, D. N. Jamieson, J. C. McCallum, A. S. Dzurak, and A. Morello, *Nat. Nanotechnol.*, 2014, **9**, 986.
3. W. Akhtar, A. Schnegg, S. Veber, C. Meier, M. Fehr, and K. Lips, *J. Magn. Reson.*, 2015, **257**, 94.
4. C. C. Lo, M. Urdampilleta, P. Ross, M. F. Gonzalez-Zalba, J. Mansir, S. A. Lyon, M. L. W. Thewalt, and J. J. L. Morton, *Nat. Mater.*, 2015, **14**, 490.
5. D. R. McCamey, H. A. Seipel, S. Y. Paik, M. J. Walter, N. J. Borys, J. M. Lupton, and C. Boehme, *Nat. Mater.*, 2008, **7**, 723.
6. K. J. van Schooten, D. L. Baird, M. E. Limes, J. M. Lupton, and C. Boehme, *Nat. Commun.*, 2015, **6**, 6688.
7. D. R. McCamey, J. van Tol, G. W. Morley, and C. Boehme, *Science*, 2010, **330**, 1652.
8. C. J. Cochrane, P. M. Lenahan, and A. J. Lelis, *J. Appl. Phys.*, 2011, **109**, 014506.
9. D. P. Waters, G. Joshi, M. Kavand, M. E. Limes, H. Malissa, P. L. Burn, J. M. Lupton, and C. Boehme, *Nat. Phys.*, 2015, **11**, 910914.
10. M. Fehr, J. Behrends, S. Haas, B. Rech, K. Lips, and A. Schnegg, *Phys. Rev. B*, 2011, **84**, 193202.
11. H. Malissa, M. Kavand, D. P. Waters, K. J. van Schooten, P. L. Burn, Z. V. Vardeny, B. Saam, J. M. Lupton, and C. Boehme, *Science*, 2014, **345**, 1487.
12. P. Kanschat, K. Lips, and W. Fuhs, *J. Non-Cryst. Solids*, 2000, **266–269**, 524.
13. A. R. Stegner, C. Boehme, H. Huebl, M. Stutzmann, K. Lips, and M. S. Brandt, *Nat. Phys.*, 2006, **2**, 835.
14. S. Geschwind, R. J. Collins, and A. L. Schawlow, *Phys. Rev. Lett.*, 1959, **3**, 545.
15. J. Brossel, S. Geschwind, and A. L. Schawlow, *Phys. Rev. Lett.*, 1959, **3**, 548.
16. M. Guéron and I. Solomon, *Phys. Rev. Lett.*, 1965, **15**, 667.
17. T. Hirose, H. Yoshioka, and K. Horai, *J. Phys. Soc. Jpn.*, 1966, **21**, 559.
18. R. Maxwell and A. Honig, *Phys. Rev. Lett.*, 1966, **17**, 188.
19. S. P. Keller, J. C. Hensel, and F. Stern, eds, in *Proceedings of the 10th International Conference on the Physics of Semiconductors*, U.S. Atomic Energy Commission, Division, 1970.
20. D. J. Lepine, *Phys. Rev. B*, 1972, **6**, 436.
21. I. Solomon, *Solid State Commun.*, 1976, **20**, 215.
22. I. Solomon, D. Biegelsen, and J. C. Knights, *Solid State Commun.*, 1977, **22**, 505.
23. T. Wosiński and T. Figielski, *Phys. Status Solidi B*, 1977, **83**, 93.
24. G. Mendz, D. J. Miller, and D. Haneman, *Phys. Rev. B*, 1979, **20**, 5246.
25. G. Mendz and D. Haneman, *J. Phys. C*, 2000, **13**, 6737.
26. R. M. White and J. F. Gouyet, *Phys. Rev. B*, 1977, **16**, 3596.
27. V. S. L'vov, O. V. Tretyak, and I. A. Kolomiets, *Sov. Phys. Semicond.*, 1977, **11**, 661.
28. F. G. Fumi, in *Proceedings of the 13th International Conference on the Physics of Semiconductors*, North-Holland Pub. Co., 1976.
29. D. Kaplan, I. Solomon, and N. F. Mott, *J. Phys. Lett.-Paris*, 1978, **39**, 51.
30. W. Shockley and W. T. Read, *Phys. Rev.*, 1952, **87**, 835.
31. C. Boehme and K. Lips, *Phys. Status Solidi B*, 2002, **233**, 427.
32. C. Boehme and K. Lips, *Phys. Rev. Lett.*, 2003, **91**, 246603.
33. I. Hiromitsu, Y. Kaimori, M. Kitano, and T. Ito, *Phys. Review B*, 1999, **59**, 2151.
34. C. Boehme, *Dynamics of Spin-Dependent Charge Carrier Recombination*, Ph.D. Dissertation, University of Marburg, 2003.
35. C. Boehme and K. Lips, *Phys. Rev. B*, 2003, **68**, 245105.
36. S. Y. Paik, S. Y. Lee, W. J. Baker, D. R. McCamey, and C. Boehme, *Phys. Rev. B*, 2010, **81**, 075214.
37. J. Shinar, *Laser & Photonics Rev.*, 2012, **6**, 767.

38. P. M. Lenahan and J. F.Conley Jr, *J. Vac. Sci. Technol. B*, 1998, **16**, 2134.
39. G. B. Silva, L. F. Santos, R. M. Faria, and C. F. O. Graeff, *Physica B*, 2001, **308–310**, 1078.
40. H. Huebl, F. Hoehne, B. Grolik, A. R. Stegner, M. Stutzmann, and M. S. Brandt, *Phys. Rev. Lett.*, 2008, **100**, 177602.
41. H. Morishita, W. J. Baker, D. P. Waters, R. Baarda, J. M. Lupton, and C. Boehme, *Phys. Rev. B*, 2014, **89**, 125311.
42. T. W. Herring, S. Y. Lee, D. R. McCamey, P. C. Taylor, K. Lips, J. Hu, F. Zhu, A. Madan, and C. Boehme, *Phys. Rev. B*, 2009, **79**, 195205.
43. S. Y. Paik, S. Y. Lee, D. R. McCamey, and C. Boehme, *Phys. Rev. B*, 2011, **84**, 235305.
44. W. Harneit, C. Boehme, S. Schaefer, K. Huebener, K. Fostiropoulos, and K. Lips, *Phys. Rev. Lett.*, 2007, **98**, 216601.
45. J. Behrends, A. Schnegg, K. Lips, E. A. Thomsen, A. K. Pandey, I. D. W. Samuel, and D. J. Keeble, *Phys. Rev. Lett.*, 2010, **105**, 176601.
46. S. Y. Lee, S. Y. Paik, D. R. McCamey, J. Hu, F. Zhu, A. Madan, and C. Boehme, *Appl. Phys. Lett.*, 2010, **97**, 192104.
47. A. V. Barabanov, V. A. L'vov, and O. V. Tretiak, *Preprints of the Bogolyubov Institute for Theoretical Physics*, 1994, **ITP-93-68E**, 1–13.
48. A. V. Barabanov, V. A. L'vov, and O. V. Tretyak, *Phys. Status Solidi B*, 1999, **207**, 419.
49. A. V. Barabanov, O. V. Tretiak, and V. A. L'vov, *Phys. Rev. B*, 1996, **54**, 2571.
50. E. L. Frankevich, A. A. Lymarev, I. Sokolik, F. E. Karasz, S. Blumstengel, R. H. Baughman, and H. H. Hörhold, *Phys. Rev. B*, 1992, **46**, 9320.
51. R. Haberkorn and W. Dietz, *Solid State Commun.*, 1980, **35**, 505.
52. R. Glenn, W. J. Baker, C. Boehme, and M. E. Raikh, *Phys. Rev. B*, 2013, **87**, 155208.
53. R. C. Roundy and M. E. Raikh, *Phys. Rev. B*, 2013, **88**, 125206.
54. C. Michel, A. Gliesche, S. D. Baranovskii, K. Lips, F. Gebhard, and C. Boehme, *Phys. Rev. B*, 2009, **79**, 052201.
55. R. Glenn, M. E. Limes, B. Saam, C. Boehme, and M. E. Raikh, *Phys. Rev. B*, 2013, **87**, 165205.
56. D. R. McCamey, S. Y. Lee, S. Y. Paik, J. M. Lupton, and C. Boehme, *Phys. Rev. B*, 2010, **82**, 125206.
57. M. E. Limes, J. Wang, W. J. Baker, S. Y. Lee, B. Saam, and C. Boehme, *Phys. Rev. B*, 2013, **87**, 165204.
58. V. Rajevac, C. Boehme, C. Michel, A. Gliesche, K. Lips, S. D. Baranovskii, and P. Thomas, *Phys. Rev. B*, 2006, **74**, 245206.
59. A. Gliesche, C. Michel, V. Rajevac, K. Lips, S. D. Baranovskii, F. Gebhard, and C. Boehme, *Phys. Rev. B*, 2008, **77**, 245206.
60. S. Y. Lee, S. Y. Paik, D. R. McCamey, and C. Boehme, *Phys. Rev. B*, 2012, **86**, 115204.
61. A. Honig, *Phys. Rev. Lett.*, 1966, **17**, 186.
62. D. R. McCamey, G. W. Morley, H. A. Seipel, L. C. Brunel, J. van Tol, and C. Boehme, *Phys. Rev. B*, 2008, **78**, 045303.
63. W. J. Baker, T. L. Keevers, C. Boehme, and D. R. McCamey, *Phys. Rev. B*, 2015, **92**, 041201.
64. K. Ando, S. Watanabe, S. Mooser, E. Saitoh, and H. Sirringhaus, *Nat. Mate.*, 2013, **12**, 622.
65. T. L. Keevers and D. R. McCamey, *Phys. Rev. B*, 2015, **91**, 245208.
66. W. J. Baker, D. R. McCamey, K. J. van Schooten, J. M. Lupton, and C. Boehme, *Phys. Rev. B*, 2011, **84**, 165205.
67. T. L. Keevers, W. J. Baker, and D. R. McCamey, *Phys. Rev. B*, 2015, **91**, 205206.
68. S. Watanabe, K. Ando, K. Kang, S. Mooser, Y. Vaynzof, H. Kurebayashi, E. Saitoh, and H. Sirringhaus, *Nat. Phys.*, 2014, **10**, 308.
69. M. Dobers, K. von Klitzing, J. Schneider, G. Weimann, and K. Ploog, *Phys. Rev. Lett.*, 1988, **61**, 1650.
70. T. Machida, T. Yamazaki, S. Ishizuka, S. Komiyama, K. Muraki, and Y. Hirayama, *J. Phys. Soc. Jpn.*, 2013, **72**, 44.
71. D. Sun, K. J. van Schooten, M. Kavand, H. Malissa, C. Zhang, M. Groesbeck, C. Boehme, and Z. V. Vardeny, *Nat. Mater.*, 2016, **15**, 863.
72. K. Ando, S. Takahashi, J. Ieda, H. Kurebayashi, T. Trypiniotis, C. H. W. Barnes, S. Maekawa, and E. Saitoh, *Nat. Mater.*, 2011, **10**, 655.
73. E. Saitoh, M. Ueda, H. Miyajima, and G. Tatara, *Appl. Phys. Lett.*, 2006, **88**, 182509.
74. H. Dersch, L. Schweitzer, and J. Stuke, *Phys. Rev. B*, 1983, **28**, 4678.
75. V. Dyakonov, N. Gauss, G. Rösler, S. Karg, W. Rieß, and M. Schwoerer, *Chem. Phys.*, 1994, **189**, 687.

76. C. J. Cochrane and P. M. Lenahan, *Appl. Phys. Lett.*, 2014, **104**, 093503.
77. B. Stich, S. Greulich-Weber, and J. M. Spaeth, *Appl. Phys. Lett.*, 1996, **68**, 1102.
78. M. A. Anders, P. M. Lenahan, C. Cochrane, and A. J. Lelis, *IEEE T. Electron Dev.*, 2015, **62**, 301.
79. F. A. Castro, G. B. Silva, L. F. Santos, R. M. Faria, F. Nüesch, L. Zuppiroli, and C. F. O. Graeff, *J. Non-Cryst. Solids*, 2004, **338–340**, 622.
80. M. Xiao, I. Martin, E. Yablonovitch, and H. W. Jiang, *Nature*, 2004, **430**, 435.
81. G. Joshi, R. Miller, L. Ogden, M. Kavand, S. Jamali, K. Ambal, S. Venkatesh, D. Schurig, H. Malissa, J. M. Lupton, and C. Boehme, *Appl. Phys. Lett.*, 2016, **109**, 103303.
82. C. P. Poole, Electron Spin Resonance, 2nd edn, John Wiley & Sons, Inc: New York, 1983.
83. W. J. Baker, K. Ambal, D. P. Waters, R. Baarda, H. Morishita, K. J. van Schooten, D. R. McCamey, J. M. Lupton, and C. Boehme, *Nat. Commun.*, 2012, **3**, 898.
84. A. Honig and M. Moroz, *Rev. Sci. Instrum.*, 1978, **49**, 183.
85. P. A. Bobbert, T. D. Nguyen, F. W. A. van Oost, B. Koopmans, and M. Wohlgenannt, *Phys. Rev. Lett.*, 2007, **99**, 216801.
86. H. Morishita, L. S. Vlasenko, H. Tanaka, K. Semba, K. Sawano, Y. Shiraki, M. Eto, and K. M. Itoh, *Phys. Rev. B*, 2009, **80**, 205206.
87. U. E. Steiner and T. Ulrich, *Chem. Rev.*, 1989, **89**, 51.
88. V. Ern and R. E. Merrifield, *Phys. Rev. Lett.*, 1968, **21**, 609.
89. J. M. Lupton and C. Boehme, *Nat. Mater.*, 2008, **7**, 598.
90. C. Boehme and J. M. Lupton, *Nat. Nanotechnol.*, 2013, **8**, 612.
91. D. R. McCamey, K. J. van Schooten, W. J. Baker, S. Y. Lee, S. Y. Paik, J. M. Lupton, and C. Boehme, *Phys. Rev. Lett.*, 2010, **104**, 017601.
92. W. J. Baker, T. L. Keevers, J. M. Lupton, D. R.break McCamey, and C. Boehme, *Phys. Rev. Lett.*, 2012, **108**, 267601.
93. F. Hoehne, H. Huebl, B. Galler, M. Stutzmann, and M. S. Brandt, *Phys. Rev. Lett.*, 2010, **104**, 046402.
94. F. M. Hrubesch, G. Braunbeck, A. Voss, M. Stutzmann, and M. S. Brandt, *J. Magn. Reson.*, 2015, **254**, 62.
95. A. J. Kupijai, K. M. Behringer, F. G. Schaeble, N. E. Galfe, M. Corazza, S. A. Gevorgyan, F. C. Krebs, M. Stutzmann, and M. S. Brandt, *Phys. Rev. B*, 2015, **92**, 245203.
96. C. Boehme, D. R. McCamey, K. J. van Schooten, W. J. Baker, S. Y. Lee, S. Y. Paik, and J. M. Lupton, *Phys. Status Solidi B*, 2009, **246**, 2750.
97. S. Y. Lee, S. Y. Paik, D. R. McCamey, J. Yu, P. L. Burn, J. M. Lupton, and C. Boehme, *J. Am. Chem. Soc.*, 2011, **133**, 2019.
98. D. R. McCamey, C. Boehme, G. W. Morley, and J. van Tol, *Phys. Rev. B*, 2012, **85**, 073201.
99. F. Hoehne, L. Dreher, H. Huebl, M. Stutzmann, and M. S. Brandt, *Phys. Rev. Lett.*, 2011, **106**, 187601.
100. A. J. Sigillito, H. Malissa, A. M. Tyryshkin, H. Riemann, N. V. Abrosimov, P. Becker, H. J. Pohl, M. L. W. Thewalt, K. M. Itoh, J. J. L. Morton, A. A. Houck, D. I. Schuster, and S. A. Lyon, *Appl. Phys. Lett.*, 2014, **104**, 222407.
101. A. Schweiger and G. Jeschke, Principles of Pulse Electron Paramagnetic Resonance, 1st edn, Oxford University Press: New York, 2001.
102. J. Behrends, K. Lips, and C. Boehme, *Phys. Rev. B*, 2009, **80**, 045207.
103. C. Boehme and K. Lips, *Physica B*, 2006, **376–377**, 930.
104. A. Schweiger and R. R. Ernst, *J. Magn. Reson. 1969*, 1988, **77**, 512.
105. F. Hoehne, L. Dreher, J. Behrends, M. Fehr, H. Huebl, K. Lips, A. Schnegg, M. Suckert, M. Stutzmann, and M. S. Brandt, *Rev. Sci. Instrum.*, 2012, **83**, 043907.
106. F. Hoehne, L. Dreher, M. Suckert, D. P. Franke, M. Stutzmann, and M. S. Brandt, *Phys. Rev. B*, 2013, **88**, 155301.
107. S. Stoll and B. Kasumaj, *Appl. Magn. Reson.*, 2008, **35**, 15.
108. F. Hoehne, J. Lu, A. R. Stegner, M. Stutzmann, M. S. Brandt, M. Rohrmüller, W. G. Schmidt, and U. Gerstmann, *Phys. Rev. Lett.*, 2011, **106**, 196101.
109. M. Suckert, F. Hoehne, L. Dreher, M. Kuenzl, H. Huebl, M. Stutzmann, and M. S. Brandt, *Mol. Phys.*, 2013, **111**, 2690.
110. F. Hoehne, C. Huck, M. S. Brandt, and H. Huebl, *Phys. Rev. B*, 2014, **89**, 161305.

Chapter 27
Very-high-frequency EPR

Alexander Schnegg
Helmholtz-Zentrum Berlin für Materialien und Energie, Berlin, Germany

27.1 INTRODUCTION

Recent decades have witnessed an ever-growing number of very-high-frequency EPR (VHF-EPR) applications in such diverse fields as biology, organic and inorganic chemistry, catalysis, as well as functional and magnetic materials research. Important achievements demonstrating the unique capabilities of VHF-EPR include the structural and functional elucidation of the primary processes in photosynthesis,[1] or the explanation and characterization of single-molecule-magnet behavior in exchange-coupled transition-metal-ion (TMI) clusters.[2]

Among the different fields of application, two main targets of VHF-EPR stand out. Firstly, the determination and exploitation of small $\boldsymbol{g}$-tensor anisotropies in $S = 1/2$ states of organic radicals and paramagnetic defect states and, secondly, studies on the spin Hamiltonian (SH) parameters – in particular zero-field splittings (ZFS) – of high-spin (HS) ($S > 1/2$) states.

The limit at which EPR frequencies and corresponding magnetic fields are regarded as high can be defined from either a technical perspective or with respect to the properties of the spin system. Historically, employing superconducting magnets – above the limit of normal electromagnets – and microwave (MW) frequencies around 90–150 GHz was one of the first major steps in the development of VHF-EPR spectrometers. Therefore, W-band conditions (94 GHz/3.4 T) are often chosen to distinguish between conventional and high-frequency or VHF-EPR experiments. In this chapter, any EPR at resonance frequencies around 90 GHz or higher is subsumed under the term VHF-EPR, although other sources distinguish between high-frequency EPR, ranging from W-band to D-band frequencies (90–150 GHz), and VHF-EPR above this range.

In addition, different definitions related to the SH parameters of a particular sample are in use. For $S = 1/2$ systems with small $\boldsymbol{g}$-tensor anisotropies, the high-field condition is defined as the resonance condition where the $\boldsymbol{g}$-tensor resolution exceeds the field-independent inhomogeneous line broadening. In such cases, the resonance field scales linearly with the external magnetic field, and high-resonance frequencies correspondingly require high external magnetic fields. As a result, EPR studies that target the resolution of small $\boldsymbol{g}$-tensor anisotropies are ideally performed on optimized high-field/single-high-frequency EPR (SHF-EPR) spectrometers. This spectrometer type provides optimal resolution and sensitivity for continuous-wave

EPR Spectroscopy: Fundamentals and Methods.
Edited by Daniella Goldfarb and Stefan Stoll
 ISBN: 978-1-119-16299-5
Also published in eMagRes (online edition)
DOI: 10.1002/9780470034590.emrstm1526

(CW) and pulse EPR studies in spin systems with g-values close to 2, at one (or a few) resonance frequencies.

Alternatively, the term high-field regime is used when the Zeeman interaction dominates all other interactions. For organic radicals, the limit where $\boldsymbol{g}$-tensor anisotropies dominate hyperfine couplings is typically reached between Q-band and 360 GHz. However, for HS states, ZFS can span from a few hundred MHz to several THz. In some cases, magnetic-field strengths beyond the currently available magnet technology are thus required to reach the high-field regime. In the intermediate region between zero-field and the high-field regime, the EPR of HS states can be very complicated and distributed over a very broad magnetic-field/transition-energy range. EPR mapping of these states requires the availability of EPR spectrometers covering a broad frequency-(hundreds of GHz or even THz) and magnetic-field range (up to tens of tesla). These conditions are met both in true multi-high-frequency EPR (MHF-EPR) spectrometers, which provide field-domain spectra at many different resonance frequencies, or frequency-domain EPR (FD-EPR) spectrometers, sweeping the resonance frequency at a fixed external magnetic field.

Throughout this chapter, SHF-EPR, MHF-EPR, and FD-EPR are used to classify the respective spectrometer types, whereas VHF-EPR denotes any EPR beyond 90 GHz. The following sections introduce important science drivers of VHF-EPR, together with dedicated EPR detection strategies. Section 27.2 summarizes the major benefits of VHF-EPR experiments for studies in $S = 1/2$ and HS states, respectively. Section 27.3 outlines instrumental requirements to harvest the benefits of different VHF-EPR approaches, while 27.4 provides suggestions for further reading along with a brief outlook concerning the prospects of VHF-EPR in view of recent instrumental breakthroughs.

27.2 BENEFITS OF VHF-EPR

27.2.1 $S = {}^1\!/_2$ States

27.2.1.1 *Spin Hamiltonian*

For an unpaired electron spin ($S = 1/2$) coupled to an external magnetic field, $\boldsymbol{B}_0$, and i surrounding nuclei, the static SH is

$$\begin{aligned}\widehat{H} &= \widehat{H}_{\text{ZeeEl}} + \widehat{H}_{\text{ZeeNuc}} + \widehat{H}_{\text{HFI}} \\ &= \mu_{\text{B}}\widehat{\boldsymbol{S}}\boldsymbol{g}\boldsymbol{B}_0 - \mu_{\text{n}}\sum_i \widehat{\boldsymbol{I}}_i g_{\text{n},i}\boldsymbol{B}_0 + \sum_i \widehat{\boldsymbol{S}}\boldsymbol{A}_i\widehat{\boldsymbol{I}}_i \end{aligned} \tag{27.1}$$

In this equation, the terms $\widehat{H}_{\text{ZeeEl}}$ and $\widehat{H}_{\text{ZeeNuc}}$ are field dependent and represent the electron and nuclear Zeeman interaction of electron and nuclear spins with $\boldsymbol{B}_0$, respectively. μ_{B} is the Bohr magneton and μ_{n} the nuclear magneton. $\boldsymbol{g}$ is the electron $\boldsymbol{g}$-tensor and $g_{\text{n},i}$ the g-factor of the ith nucleus. The term $\widehat{H}_{\text{HFI}}$ is the field-independent hyperfine interaction (HFI) that couples the electron and nuclear spin operators, $\widehat{\boldsymbol{S}}$ and $\widehat{\boldsymbol{I}}$, via the anisotropic HFI tensor, $\boldsymbol{A}$ (see Chapter 5).

27.2.1.2 *g-Tensor Resolution*

$\boldsymbol{g}$-tensors of organic radicals, paramagnetic defect states, or TMIs are sensitive probes of the electronic structure and its modulation by the environment (see Chapter 2 for further information). Furthermore, overlapping or interacting paramagnetic states can be assigned by their characteristic $\boldsymbol{g}$-tensors. However, precise $\boldsymbol{g}$-tensor measurements require external magnetic fields that fulfill the high-field condition:[3]

$$\frac{\Delta g}{g_{\text{iso}}} B_0 > \Delta B_{\frac{1}{2}} \tag{27.2}$$

where Δg is the $\boldsymbol{g}$-tensor anisotropy (difference between g_x and g_z) or orthorhombicity (difference between g_x and g_y) while g_{iso} and $\Delta B_{1/2}$ are the isotropic g-value and the inhomogeneous line broadening, respectively. The latter can be field independent, e.g., if unresolved HFI dominate the line width of a delocalized spin system, or field dependent if a variation of g-values is the dominant distribution. A corresponding definition is valid for the limit at which two spin centers with different g-values are separated.

Since the parameters in equation (27.2) are sample dependent (aside from B_0), the same applies to the high-field condition. Consequently, the unambiguous assignment of g-values requires SHF-EPR spectrometers providing sufficient resolving power. For example, g-anisotropies of semiquinone or tyrosine radicals are well resolved in frozen solution W-band EPR spectra. Other molecules like chlorophylls, tryptophans, or flavins (Figure 27.1) exhibit much smaller g-anisotropies, which can only be resolved at resonance frequencies above 300 GHz.[4]

Figure 27.1 demonstrates the increase of $\boldsymbol{g}$-tensor resolution achievable at 360 GHz/12 T as compared to

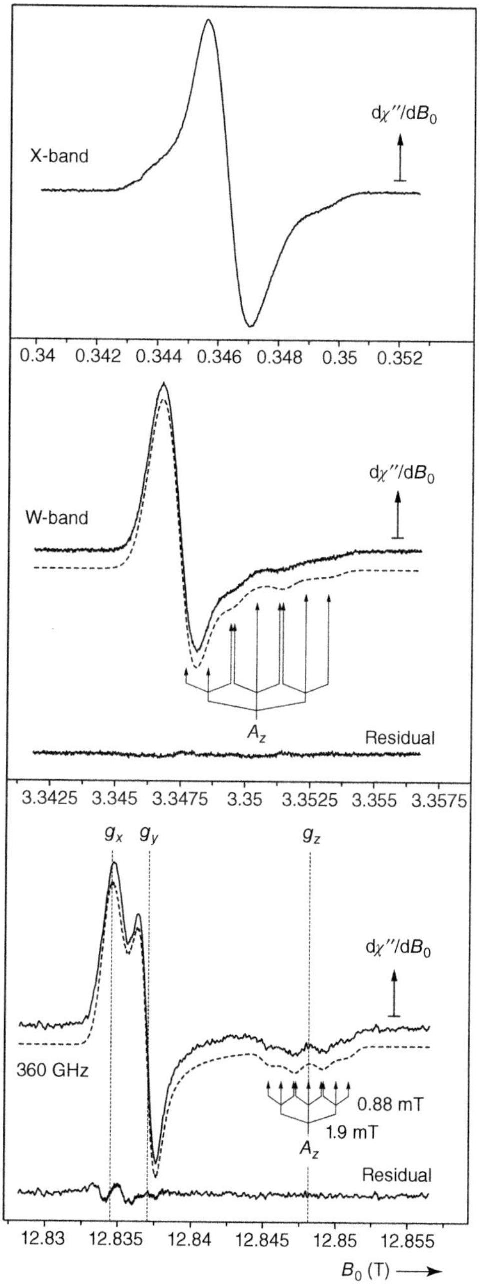

Figure 27.1. X-band (upper), W-band (middle), and 360 GHz (lower) experimental (solid lines) and simulated (dashed lines) CW-EPR spectra of the FAD cofactor of glucose oxidase in its stable anionic radical form, $FAD^{\bullet -}$, at pH 10. The 360 GHz EPR spectrum was detected in induction mode without a resonator. Due to the superior resolving power of 360 GHz EPR, the **g**-tensor anisotropy ($g_x = 2.00429(3)$, $g_y = 2.00389(3)$, and $g_z = 2.00216(3)$) could be fully resolved. Orientation selection at the g_z-component of the W-band and 360 GHz spectrum also provided the A_z-values of two ^{14}N atoms, as shown by the stick spectrum. (Reprinted with permission from A. Okafuji, A. Schnegg, E. Schleicher, K. Moebius, S. Weber, J. Phys. Chem. B, 2008, 112, 3568. Copyright 2008 American Chemical Society)

X-band and W-band EPR for the FAD cofactor anion in glucose oxidase.[5] In the present case, not only the **g**-tensor resolution increases upon increasing the resonance frequency but also the hyperfine resolution due to orientation selection on the g_z-component. A similar case is shown in Figure 27.2, which depicts EPR spectra of the nitroxide radical TEMPOL at different resonance conditions.

The excellent resolution of g-anisotropy achievable in protein-based organic radicals and spin labels was one of the major motivations behind developing VHF-EPR methods for applications in structural biology[4] and material sciences.[6] Subsequently, g-values of many stable and transient protein-based amino-acid radicals and cofactor radicals have been determined by VHF-EPR. An elaborate overview of these studies is provided in a review.[4] In combination with site-directed spin labeling (SDSL) techniques (see Chapter 14), studies of the **g**-tensors of nitroxide labels attached to defined protein sites can be employed to study function-induced dynamic changes of the local protein polarity or the formation and release of hydrogen bonds.[1] Furthermore, VHF-EPR measurements on paramagnetic states with resolved g-anisotropy are a powerful tool to study tumbling motions[7] (see also Section 2.4).

However, according to equation (27.2), increasing the resonance field does not necessarily enhance the g-resolution for cases where the inhomogeneous line broadening increases linearly with B_0. This is the case for paramagnetic species featuring site-to-site disorder of the microenvironment. Disorder can result in a distribution of g-values – often referred to as 'g-strain' – which leads to field-dependent inhomogeneous line

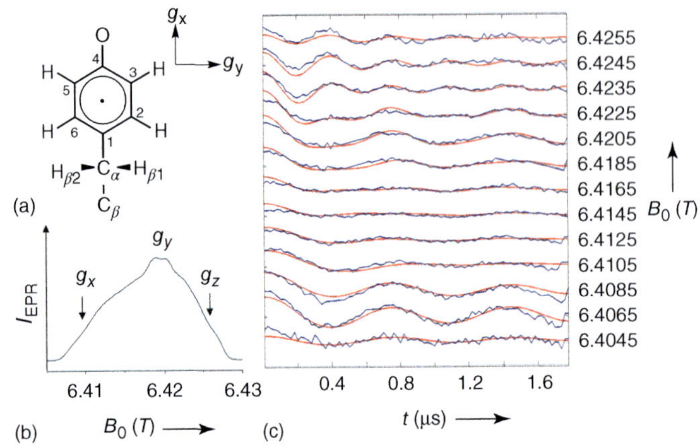

Figure 27.2. (a) Tyrosyl structure alongside with arrows indicating the $\boldsymbol{g}$-tensor principal axes. (b) 180 GHz echo-detected EPR spectrum of the tyrosyl radical in mouse ribonucleotide reductase (RNR) R2 dimer. (c) Experimental (blue lines) and simulated (red lines) 180 GHz orientation-dependent PELDOR traces of the tyrosyl biradical in mouse RNR recorded across the tyrosyl EPR spectrum from $B \| g_z$ (top) to $B \| g_x$ (bottom). (Reproduced with permission from V. P. Denysenkov, D. Biglino, W. Lubitz, T. F. Prisner, and M. Bennati, *Angew. Chem. Int. Ed.*, 2008, **47**, 1224. © John Wiley & Sons, 2008)

widths in the EPR spectrum. g-strain is most evident in TMI complexes,[8] although it has also been observed in frozen solution spectra of organic radicals, e.g., in chlorophyll $a^{\bullet +}$ in frozen methylene chloride.[9] In cases where the inhomogeneous line width is fully dominated by g-strain, a further increase of B_0 does not improve the resolution. However, for many protein-based radicals like tryptophan[10] or the primary bacteriochlorophyll donor in bacterial photosynthetic reaction centers,[11] it has been shown that the $\boldsymbol{g}$-tensor resolution still increases even up to 600 GHz/21.4 T.

Finally, the $\boldsymbol{g}$-tensor resolution in VHF-EPR spectra can be used to perform orientation-selective measurements even in powders or frozen solutions. This benefit is particularly relevant for high-field electron nuclear double resonance (ENDOR). Orientation-selective ENDOR (see Chapter 16) measurements on defined positions within the EPR powder pattern yield single crystal-like information and allow for the assignment of overlapping ENDOR signals. Equally importantly, they are sensitive to the orientation of the HFI tensors with respect to the electron $\boldsymbol{g}$-tensor and thereby the molecular frame, thus providing additional important structural information. Orientation selection in combination with the increased spectral separation of the ENDOR spectra of different nuclei, with increasing magnetic field, renders high-field ENDOR a powerful method for protein[12] and material research.[13] It has been shown that orientation selection at high field can also increase the information content of high-field pulse electron–electron double resonance (PELDOR)[14] (Figure 27.2) and relaxation time measurements.[16,17]

27.2.2 High-spin States

27.2.2.1 Mononuclear High-spin States

For an HS state with quenched orbital angular momentum, the SH in equation (27.1) (neglecting the interaction with nuclear spins) extends to[18]

$$\widehat{H} = \widehat{H}_{\mathrm{ZeeEl}} + \widehat{H}_{\mathrm{ZFS}} = \mu_{\mathrm{B}}\widehat{\boldsymbol{S}}\boldsymbol{g}\boldsymbol{B}_0 + \widehat{\boldsymbol{S}}\boldsymbol{D}\widehat{\boldsymbol{S}} \qquad (27.3)$$

Here, $\widehat{H}_{\mathrm{ZeeEl}}$ again denotes the electron-Zeeman interaction and $\widehat{H}_{\mathrm{ZFS}}$ the ZFS, with $\boldsymbol{D}$ being the ZFS tensor. This contribution lifts the degeneracy of the $2S+1$ sublevels even in the absence of an external magnetic field. ZFS results from two contributions: dipole–dipole interactions of open-shell electron spins and spin–orbit coupling, mixing high-lying excited states into the electronic ground state (see Chapters 3 and 7).[19–21] It is convenient to define $\boldsymbol{D}$ in a local frame, in which it is diagonal and the z-axis is the main anisotropy axis:[18]

$$\hat{H}_{ZFS} = \hat{S}\boldsymbol{D}\hat{S} = D\left(\hat{S}_z^2 - \frac{1}{3}S(S+1)\right) + E(\hat{S}_x^2 - \hat{S}_y^2) \tag{27.4}$$

In this equation, $D = 3/2D_z$ and $E = (D_x - D_y)/2$ are the axial and the rhombic ZFS parameters, respectively, and D_x, D_y, and D_z are the principal values of the ZFS tensor. $\hat{H}_{ZFS}$ can also include higher order terms expressed by the Stevens operators.[18] For a more detailed insight into the phenomenon of ZFS and the relationship between D, E and the components of the ZFS tensor, see Chapter 3.

Due to their differences in the spin-energy-level splittings, spin states in EPR spectroscopy are often divided into half-integer ($S = 1/2$, $3/2$, $5/2$, ...) and integer ($S = 1$, 2, 3, ...) states. According to the Kramers theorem, spin levels of half-integer spin states (Kramers ions) are at least doubly degenerate in the absence of an external magnetic field. The double degeneracy of these characteristic Kramers doublets is removed by an external magnetic field, but not by ZFS. By contrast, integer spin states (non-Kramers ions) can be split by ZFS to nondegenerate spin levels (Figure 27.3).[18]

27.2.2.2 Coupled Multinuclear High-spin Clusters

Major targets of VHF-EPR are multinuclear HS clusters with coupled electron spins[24] (see Chapter 4), due to their importance in (bio-)catalysis[25] and molecular magnetism.[2]

For a complex containing j exchange-coupled electron spins with quenched orbital momentum, the SH writes.[24]

$$\begin{aligned}\hat{H}_{cluster} &= \hat{H}_{ZeeEl} + \hat{H}_{ZFS} + \hat{H}_{Ex} \\ &= \sum_j \mu_B \boldsymbol{B}_0 \boldsymbol{g}_j \hat{\boldsymbol{S}}_j + \sum_j \hat{\boldsymbol{S}}_j \boldsymbol{D}_j \hat{\boldsymbol{S}}_j + \sum_{j<k} J_{jk} \hat{\boldsymbol{S}}_j \hat{\boldsymbol{S}}_k\end{aligned} \tag{27.5}$$

Here, the terms $\hat{H}_{ZeeEl}$ and $\hat{H}_{ZFS}$ account for the electron–Zeeman interactions and the ZFS of the j individual electron spins, respectively. $\hat{H}_{Ex}$ represents the Heisenberg–Dirac–van Vleck Hamiltonian, where J_{jk} is the isotropic exchange interaction coupling between the jth and kth electron spins. Chapter 4 contains a detailed discussion on how the spin eigenenergies and eigenfunctions for different coupling situations and strengths may be obtained. For HS clusters containing several coupled spins, e.g., in TMI-based single-molecule magnets, SH matrices can become very large and their diagonalization computationally challenging. However, the description of the ground-state spin levels and the resulting EPR transitions may be strongly simplified if the exchange term dominates the SH. In this case, the resulting spin levels form multiplets, each with a well-defined value of S. Splittings induced by smaller terms in the SH can subsequently be described by an effective SH acting within each multiplet. Such a giant SH takes the form of equation (27.3), with effective S, ZFS, and $\boldsymbol{g}$-tensors replacing the single-ion values. These effective SH parameters may be calculated from the values of the uncoupled spins according to relationships outlined in Chapter 4.[2,24]

27.2.2.3 EPR of High-spin States with Very Large Zero-field Splittings

EPR generally requires the application of radiation, matching the energy of an allowed spin transition. However, since ZFS may range from $\approx 10^{-2}$ cm^{-1} (hundreds of MHz, e.g., in highly symmetric Mn^{II} complexes) to $\approx 10^3$ cm^{-1} (several THz, e.g., in HS Co^{II} complexes), it can induce spin-energy-level splittings that are so large that the high-field regime cannot be reached with current VHF-spectrometer technology. In the case of integer-spin states (e.g., Mn^{III} or Fe^{II}), the splittings may exceed the excitation-energy range of many SHF-EPR spectrometers (Figure 27.3). Due to this fact, the latter cases were classified as 'EPR silent' in former days. In the case of half-integer HS states with large ZFS, some transitions are accessible by conventional EPR, e.g., between the $m_S = \pm 1/2$ levels (m_S being the spin projection quantum number). Note that despite the doublets being labeled with the quantum number m_S, in the case of non-vanishing E, they are linear combinations of the different levels. However, transitions between these levels are usually insufficient for a precise determination of the desired ZFS parameters. To increase the information content extractable from EPR spectroscopy for both half-integer and integer states, wide resonance-frequency and/or field scans are required to map out the complicated field/frequency landscape of HS complexes with large ZFS. In the following, this situation is discussed for an integer ($S = 1$) and a half-integer ($S = 3/2$) state, respectively.

Figure 27.3 depicts field-dependent $S = 1$ and $S = 3/2$ energy levels simulated for the same SH parameters ($D/h = 535$ GHz, $E/h = 12.5$ GHz, and $g = 2$).

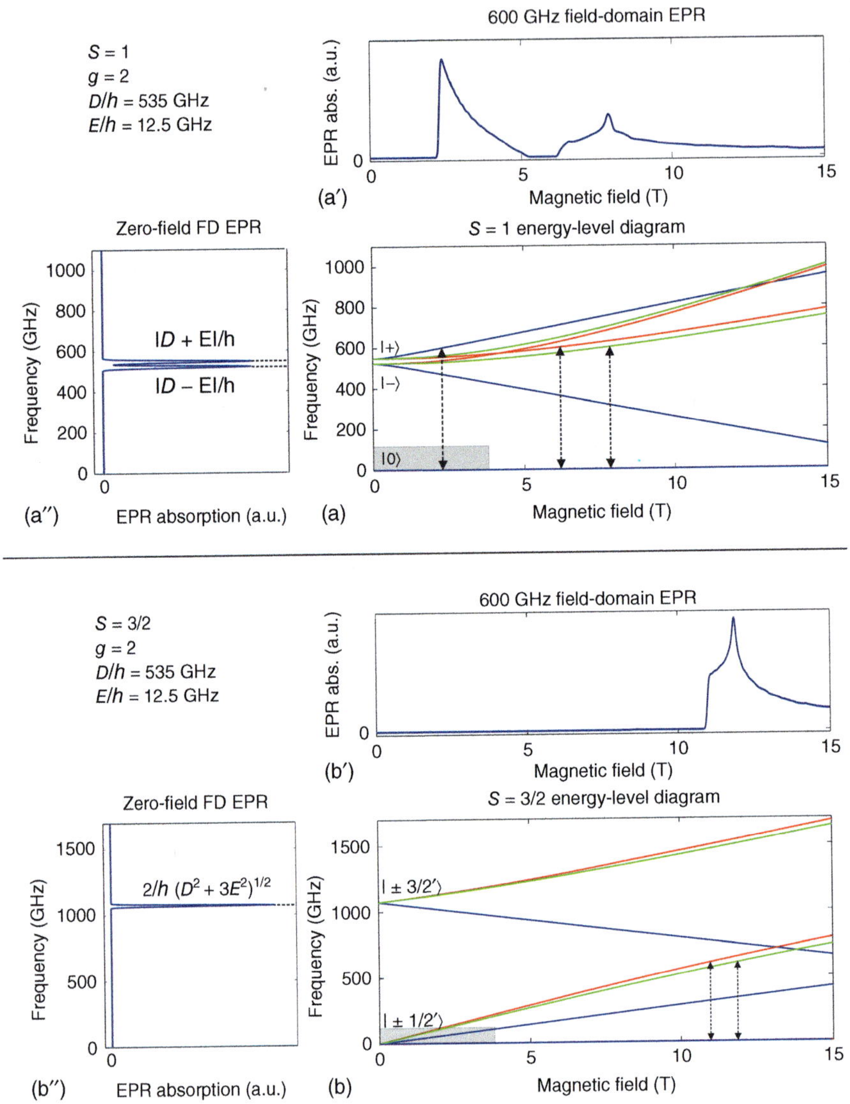

Figure 27.3. Spin-energy levels for (a) an integer ($S=1$) and (b) a half-integer ($S=3/2$) state, both simulated with the same SH parameters.[22,23] Energy levels are plotted for the three canonical orientations of the ZFS tensor relative to $\boldsymbol{B}_0$ (green: $\boldsymbol{B}_0 \| x$, red: $\boldsymbol{B}_0 \| y$, and blue: $\boldsymbol{B}_0 \| z$) and normalized to the corresponding ground state, i.e., the difference in energy compared to the ground state is plotted. Vertical dashed arrows indicate allowed ground-state transitions for resonance frequencies of 600 GHz. Gray boxes indicate the spectral range accessible by X-, Q-, and W-band EPR. (a′) and (b′) depict calculated 600 GHz field-domain powder EPR absorption spectra at $T=5$ K. Note that the field-domain EPR spectra would extend to even much higher fields, which are not depicted here. (a″) and (b″) exhibit their zero-field FD-EPR counterparts

These values correspond to $[Tp_2Mn]SbF_6$,[26] a Mn^{III} ($S=1$) complex, for which experimental MHF-EPR and FD-EPR results are depicted in Figure 27.4. To compare the peculiarities of VHF-EPR on integer and half-integer HS states, the same values were chosen for the simulation of the $S=3/2$ example. For both states, the energy levels for field alignments parallel to the molecular x-, y-, and z-axes are plotted relative

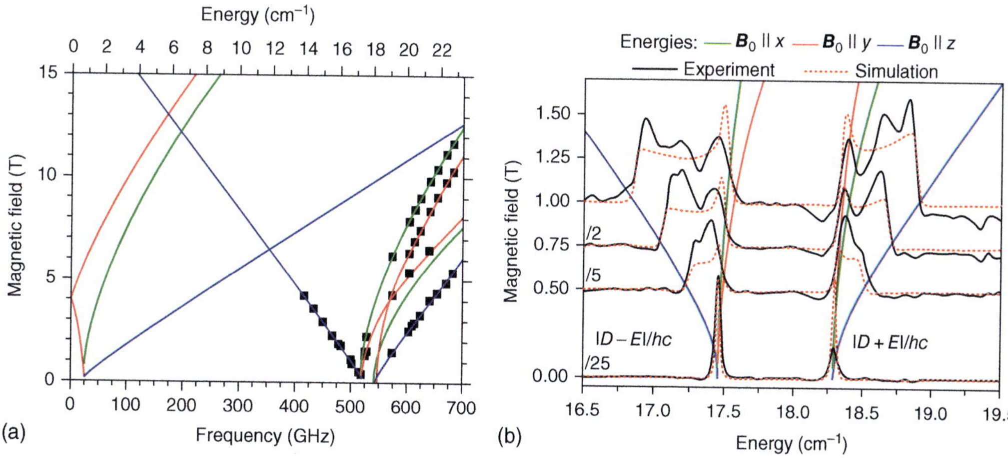

Figure 27.4. (a) Calculated EPR-transition energies (green: $\boldsymbol{B}_0 \| x$, red: $\boldsymbol{B}_0 \| y$, and blue: $\boldsymbol{B}_0 \| z$) and experimentally obtained turning points of MHF-EPR spectra (black squares, $T = 4.5$ K) for $[Tp_2Mn]SbF_6$. (Reprinted with permission from A. P. Forshaw, J. M. Smith, A. Ozarowski, J. Krzystek, D. Smirnov, S. A. Zvyagin, T. D. Harris, H. I. Karunadasa, J. M. Zadrozny, A. Schnegg, K. Holldack, T. A. Jackson, A. Alamiri, D. M. Barnes, J. Telser, Inorg. Chem., 2013, 52, 144. Copyright 2013 American Chemical Society.) (b) Experimental (black solid lines) and calculated (red dotted lines) FD-FT THz-EPR absorbance spectra of $[Tp_2Mn]SbF_6$ taken at the indicated magnetic fields along with calculated EPR-transition energies. Experimental spectra were obtained by dividing raw spectra at 40 and 2 K for each field point. This procedure is necessary to discriminate EPR signals from nonmagnetic absorptions in the beam path. (Reprinted with permission from J. Nehrkorn, J. Telser, K. Holldack, S. Stoll, A. Schnegg, J. Phys. Chem. B, 2015, 119, 13816. Copyright 2015 American Chemical Society.) Transition energies in both plots were calculated with SH parameters ($D/h = +535(3)$ GHz, $E/h = 12(1)$ GHz, $g_x = 2.065(6)$, $g_y = 2.073(8)$, and $g_z = 1.978(6)$) and plotted for the three canonical orientations of the ZFS tensor relative to $\boldsymbol{B}_0$. For better visibility, FD-FT THz-EPR spectra were scaled with division factors given together with the respective spectra

to their respective ground-state level. Along with the spin-energy levels, the simulated field-domain[23] (a′ and b′) and frequency-domain[22] (a″ and b″) powder EPR spectra are plotted.

Spin Triplet State

For $S = 1$, the three spin sublevels are split by D into a ground-state singlet ($|S, m_S\rangle = |1, 0\rangle$) and a doublet ($|1, +1\rangle$ and $|1, -1\rangle$). Rhombic ZFS further splits the degenerate doublet levels and mixes the eigenfunctions. In the presence of an external magnetic field aligned along the ZFS z-axis and under the assumption of isotropic g, the eigenfunctions and eigenvalues E_i are[18]

$$
\begin{aligned}
|-\rangle &= \sin\alpha|1,+1\rangle - \cos\alpha|1,-1\rangle,\\
E_{|-\rangle} &= \frac{D}{3h} - \sqrt{\left(\frac{\mu_B B_0 g}{h}\right)^2 + \left(\frac{E}{h}\right)^2}\\
|+\rangle &= \cos\alpha|1,+1\rangle + \sin\alpha|1,-1\rangle,\\
E_{|+\rangle} &= \frac{D}{3h} + \sqrt{\left(\frac{\mu_B B_0 g}{h}\right)^2 + \left(\frac{E}{h}\right)^2}\\
|0\rangle &= |1,0\rangle\\
E_{|0\rangle} &= -\frac{2D}{3h}
\end{aligned}
\tag{27.6}
$$

with $2\alpha = \frac{E}{(\mu_B B_0 g)}$.[18] At zero magnetic field, the three eigenfunctions in equation (27.6) are $\{1/\sqrt{2}\{|1,+1\rangle - |1,-1\rangle\}; 1/\sqrt{2}\{|1,+1\rangle + |1,-1\rangle; |1,0\rangle\}$,[18] with zero-field transition energies $\Delta E_1 = |D+E|/h$ and $\Delta E_2 = |D-E|/h$. This significantly simplifies the EPR spectrum and makes it possible to determine ZFS with high accuracy (Figure 27.3a″). The corresponding expressions for the eigenfunctions and eigenvalues for an alignment of $\boldsymbol{B}_0$ parallel to the ZFS x- and y-axes are given in Chapter 3.

For the case depicted in Figure 27.3(a), the splitting between singlet and doublet states is so large that X-, Q-, and even W-band spectrometers would fail to detect any ground-state EPR transitions. This restriction may be lifted by VHF-EPR experiments in the frequency domain (Figure 27.3a′) or field domain (Figure 27.3a″), provided that the applied MW frequency is sufficiently high. Both modalities yield characteristic EPR spectra, from which the ZFS values can be extracted (Figure 27.4). Field-domain EPR requires spectral simulations for the assignment of SH parameters, whereas ZFS can be directly determined from the resonance positions of the zero-field FD-EPR spectrum (Figure 27.3b″). Field-domain EPR simulations of HS systems are well established and routinely applied. However, in some cases, no unique set of SH parameters can be extracted, even from MHF-EPR data sets obtained over a broad field/frequency range. In addition, g-strain may further complicate the assignment of ZFS in the presence of an external magnetic field. Zero-field FD-EPR can simplify the situation,[26] though, for both modalities, the unambiguous simultaneous determination of ZFS and g-values requires EPR measurements at several frequencies (MHF-EPR) or fields (FD-EPR) along with spectral simulations.[22] Corresponding MHF-EPR and frequency-domain Fourier-transform THz-EPR (FD-FT THz-EPR) data sets for the Mn^{III}, $S=1$ state of $[Tp_2Mn]SbF_6$ are depicted in Figure 27.4.

Spin Quartet State

The corresponding situation for a half-integer $S=3/2$ state is shown in Figure 27.3(b). For $S=3/2$ and $\boldsymbol{B}_0$ parallel to z, ZFS splits the four sublevels into two doublets with energies of[18]

$$\left|\pm\frac{3'}{2}\right\rangle = \cos\alpha\left|\pm\frac{3}{2}\right\rangle + \sin\alpha\left|\mp\frac{1}{2}\right\rangle,$$

$$E_{\left|\pm\frac{3'}{2}\right\rangle} = \pm\frac{\mu_B B_0 g}{2h} + \sqrt{\left(\frac{D}{h}\pm\frac{\mu_B B_0 g}{h}\right)^2 + 3\left(\frac{E}{h}\right)^2}$$

$$\left|\mp\frac{1'}{2}\right\rangle = -\sin\alpha\left|\pm\frac{3}{2}\right\rangle + \cos\alpha\left|\mp\frac{1}{2}\right\rangle,$$

$$E_{\left|\mp\frac{1'}{2}\right\rangle} = \pm\frac{\mu_B B_0 g}{2h} - \sqrt{\left(\frac{D}{h}\pm\frac{\mu_B B_0 g}{h}\right)^2 + 3\left(\frac{E}{h}\right)^2} \tag{27.7}$$

Rhombic ZFS mixes the m_S levels with $2\alpha = \sqrt{3}E/(D \pm 2\mu_B B_0 g)$.[18] For $D>0$, the $|\mp 1/2'\rangle$ doublet is lowest in energy. $D<0$ inverts this order. In both cases the two doublets are separated by the zero-field transition energy $\Delta E_1 = 2/h\sqrt{D^2+3E^2}$. Analytic expressions for the field-dependent eigenvalues for field alignments along the remaining principle axes can be found in Ref. 27. In contrast to the $S=1$ case, zero-field EPR cannot distinguish between axial and rhombic ZFS. However, the field-induced splittings of the two zero-field doublets are sensitive to the rhombicity of the ZFS. A ground-state EPR transition within the $|\mp 1/2'\rangle$ doublet can be observed at any frequency that matches the Zeeman splitting. However, for direct determination of the ZFS, a transition between the doublets has to be observed. Considerations for triplet and quartet states can be extended to higher spin states (see, e.g., Chapter 3 and Refs 18, 28, 29).

27.2.2.4 EPR of HS States in the High-field Regime

The spectra of HS states may be significantly simplified when the ZFS is much smaller than the Zeeman interaction. VHF-EPR in the high-field regime proved to be particularly useful for studies in the SH parameters of half-integer HS states with small or intermediate ZFS (e.g., in Mn^{II} ($S=5/2$) or Gd^{III} ($S=7/2$)). These states usually already exhibit pronounced EPR at conventional frequencies. However, in the intermediate regime between zero-field regime, where the electron spin is quantized in the ZFS frame, and the high-field regime, where $\boldsymbol{B}_0$ defines the quantization axis, the interpretation of the spectrum can be particularly difficult, as has been outlined in the previous subsection. Since the two bases are not eigenstates of the spin system, so-called 'forbidden' transitions contribute to the EPR spectrum, which frequently prohibit an unambiguous extraction of SH parameters. The situation may be simplified in the high-field regime where the central $m_S=-1/2$ to $m_S=+1/2$ transition for a Kramers system is only affected in second order by ZFS and substantially narrows with increasing magnetic field ($\propto D^2/B_0$). This renders VHF-EPR very sensitive to these states and allows for an improved assignment of hyperfine and g-values as well as super-HFIs to nuclei in the ligand sphere. Due to their spectroscopic properties, the six narrow lines of the central EPR transition of $^{55}Mn^{II}$ ($S=5/2$, $I=5/2$) are frequently used as field standards in VHF-EPR.[1] In addition HS ions with small ZFS are currently being introduced as spin labels for pulsed SHF-EPR distance measurements in structural biology.[30]

For integer and half-integer HS states, the extractable information content can be further increased by temperature-dependent measurements in the high-field regime. If $kT \ll h\nu_0$ (where ν_0 is the MW frequency), only the $m_S=-S$ level is populated

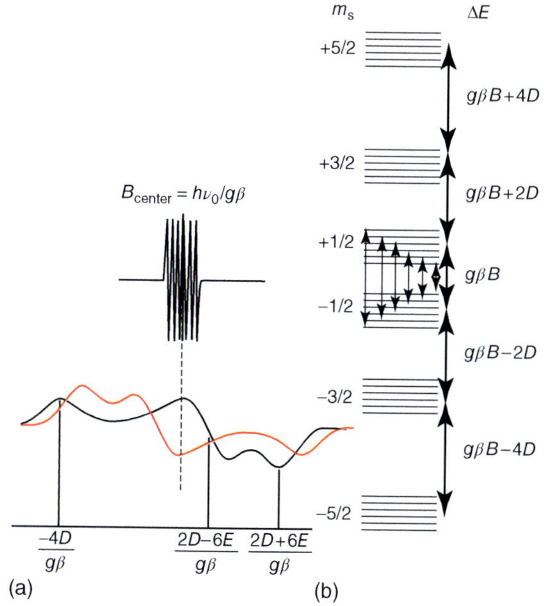

Figure 27.5. (a) Simulated Mn^{II} ($S=5/2, I=5/2$) CW-EPR spectra for $h\nu_0 \ll kT$ (upper trace) and $h\nu_0 \gg kT$ (lower traces). (b) Energy levels of a Mn^{II} ion when the Zeeman interaction is the dominant interaction. The energy diagram is for the case where $E=0$. The bold arrows denote the five formally allowed electronic spin transitions. On the left, the powder spectrum of the $m_S=-5/2\rightarrow-3/2$ transition is shown for $D<0$ (red) and $D>0$ (black). The $m_S=-1/2\rightarrow1/2$ spectrum and the associated electron–nuclear hyperfine transitions (fine arrows) are also shown. Both the $m_S=-5/2\rightarrow-3/2$ and $m_S=-1/2\rightarrow1/2$ spectra are centered at $B_{center}=h\nu_0/g\beta$ (with $\beta=\mu_B$ being the Bohr magneton). The labeled field positions are relative to B_{center}. (Reprinted with permission from L. C. Tabares, J. Gätjens, C. Hureau, M. R. Burrell, L. Bowater, V. L. Pecoraro, S. Bornemann, S. Un, J. Phys. Chem. B, 2009, 113, 9016. Copyright 2009 American Chemical Society)

and the $m_S=-S$ to $m_S=-S+1$ transition is the only significant contribution to the EPR spectrum. Approximate values of the resulting frequency-dependent Zeeman temperature are $T_{Zee}=h\nu_0/k_B$ $=0.5$ K (X-band), 5 K (W-band), and 13 K (263 GHz). Below this temperature, the shape of the resulting EPR spectrum is defined by the values of D and E and the absolute sign of D,[1,31] which allows for a clear-cut assignment of the latter parameters. Furthermore, spin polarization under VHF-EPR conditions and at cryogenic temperatures allows for the determination of the sign of HFI by ENDOR.[13] Temperature dependence of the EPR spectrum of an HS ion is demonstrated for Mn^{II} in Figure 27.5.

27.3 VHF-EPR SPECTROMETERS

First reports about VHF-EPR spectrometers date back to the 1960s.[32,33] However, the improvement of these instruments toward higher detection sensitivity and experimental versatility proved a very challenging task that required the redesign of all major spectrometer components. The necessity to generate very high fields as well as intense yet low-noise MWs at mm and sub-mm wavelengths, the construction of dedicated sample probes and low-loss MW transmission lines, and, finally, sensitive MW detection were challenges that could only be mastered based upon significant progress in the fields of MW and magnet technology. Major instrumental breakthroughs in this direction were achieved by Lebedev and his coworkers, who constructed the first sensitive 140 GHz/4.5 T SHF-EPR spectrometers.[34] Subsequently, EPR groups worldwide met the challenge and constructed VHF-EPR spectrometers for CW and pulse operation at ever-higher resonance fields and frequencies. A detailed review of the history of VHF-EPR instrument development and their construction principles can be found, e.g., in Refs 1 and 35. In addition to these laboratory-built spectrometers, a growing number of commercial W-band[36] and 263 GHz/12 T[37] SHF-EPR spectrometers – both introduced by Bruker BioSpin in 1996 and 2009, respectively – entered EPR labs.

To harvest the benefits of VHF-EPR for studies in $S=1/2$ and HS states, researchers constructing VHF spectrometers followed two main strategies. First, SHF-EPR mainly targets the resolution of small **g**-tensor anisotropies in $S=1/2$ systems and the simplification of EPR spectra of HS states in the high-field regime. CW and particularly advanced pulse SHF-EPR are employed to extract spectroscopic information with ultimate resolution and sensitivity. The second approach aims to detect HS states with EPR transitions distributed over a very broad transition-energy/magnetic-field landscape. Respective spectrometers are realized as either true MHF-EPR spectrometers, in which broad field sweeps can be performed at many high MW frequencies, or

FD-EPR spectrometers – where the resonance frequency is swept at constant or even zero-magnetic field.

In the following subsections, we introduce the design concepts for ultrasensitive SHF-EPR (see Section 27.3.1) and variable-frequency MHF-EPR and FD-EPR (see Section 27.3.2) spectrometers, before we finally compare different spectrometer concepts in Section 27.3.3. Since the general principles of EPR instrumentation are outlined in Chapter 12, only aspects relevant for VHF instruments are treated in the following.

27.3.1 Single-High-frequency EPR

27.3.1.1 Probes

To boost the detection sensitivity of SHF-EPR spectrometers, resonant cavities like single-mode TE_{011} resonators, Fabry–Pérot resonators, whispering-gallery resonators, and loop-gap resonators are in use. Characteristics and application ranges of different EPR-resonator types – including the above mentioned – are presented in Chapter 12. Overviews of the features of different SHF-EPR resonators can be found, e.g., in Refs 1 and 35.

In both TE_{011} and Fabry–Pérot resonators, RF circuits for ENDOR excitation have been integrated. Due to their high fill factor, large achievable B_1 fields, and compatibility with robust tuning and matching strategies, TE_{011} resonators have proven a very attractive solution for SHF-EPR and ENDOR. In addition, nonresonant probes are an effective solution to achieve very good concentration sensitivity for CW SHF-EPR applications[10,35,38] (Figure 27.1) and can even be employed for pulse experiments if sufficient MW power is available.[39] Since, for mm and sub-mm wavelengths, the waveguide sizes are compatible with typical sample volumes, the resonator can be replaced by a nonresonant structure completely filled with the sample. In such configurations, restrictions with respect to sample handling (e.g., in the dark or under cryogenic temperatures) and sample dimensions are less strict than for sample tubes with an inner diameter of a few hundred μm (for SHF-EPR single-mode resonators) or films with the thickness of a few hundred micrometers (for SHF-EPR Fabry–Pérot resonators). In addition, nonresonant, open structures can accommodate comparably large samples (e.g., an X-band tube with 4 mm diameter) and facilitate the implementation of modulation coils, optical fibers, facilities for sample rotation, electrical wiring, or additional RF coils.[40]

27.3.1.2 Magnets

State-of-the-art SHF-EPR spectrometers are based upon sweepable superconducting magnets with high homogeneity (typically 10^{-4} relative homogeneity in a sphere of 1 cm^3 volume). To realize broad field sweeps over many tesla as well as very accurate sweeps over a small range (typically 0.2–0.15 T), a combination of main and sweep coils is successfully used in many SHF-EPR spectrometers. Sweepable superconducting magnets equipped with additional narrow sweep coils have been constructed for magnetic fields of up to 14 T.[1] For higher fields, one has to resort to either resistive Bitter magnets or hybrid combinations of resistive and superconducting magnets, which are available for EPR experiments at user facilities, e.g., for magnetic fields of up to 45 T (continuous) at the National High Magnetic Field Lab (NHMFL) (Tallahassee, USA), up to 33 T (continuous) at the Laboratoire National des Champs Magnétiques Intenses (Grenoble, France), up to 37.4 T (continuous) at the High Field Magnet Laboratory (HFML) (Nijmegen, The Netherlands), and up to 70 T (pulsed) at the High Magnetic Field Lab (HZR, Rossendorf, Germany) (see Ref. 41 for further details).

27.3.1.3 Microwave Bridges and Transmission Lines

A major challenge in the construction of SHF-EPR spectrometers are losses occurring in MW transmission lines. These losses can be significantly reduced by employing quasi-optical MW components and transmission lines (Figure 27.6).[42–46] The propagation of electromagnetic radiation under quasi-optical conditions may be conveniently described by Gaussian beams. These are plane waves whose amplitudes decrease with increasing distance from the optical axis.[47] In a quasi-optical EPR spectrometer, Gaussian beams are usually inserted by corrugated feed horns.[47] This beam propagates in open space and can be directed and focused by off-axis mirrors (Figures 27.6 and 27.9) or lenses. For MW-transmission over longer straight sections with a fixed diameter, which is usually required to couple the sample situated in the magnet to the MW bridge, thin-walled corrugated waveguides are the optimum solution.[42] However,

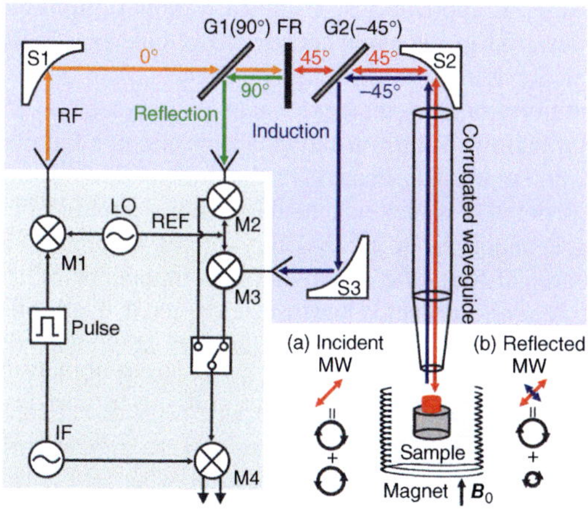

Figure 27.6. Quasi-optical SHF heterodyne spectrometer scheme.[37] Gray box: basic MW unit with IF and LO sources, pulse-forming unit, and mixers for up-conversion (M1) and detection (M2–M4). Light blue box: Quasi-optical front end, in which the RF transmits as a Gaussian beam focused by off-axis parabolic mirrors (S1–S3). Corrugated horn antennas guarantee low-loss insertion and extraction of the MW beam from the basic unit to the front end and back. A combination of Faraday rotator (FR) and wire grids (G1 and G2) is used to discriminate between different MW polarizations. Orange, red, green, and blue arrows indicate beams with MW polarization angles given in the same color. MW polarization is indicated as the angle between *E*-field vector and propagation normal. At the sample probe, the polarization of incident (a) and reflected (b) MWs and their circular components are depicted as arrows and circles, respectively[42]

quasi-optical transmission lines are not compatible with waveguide-based passive MW components like isolators, attenuators, phase and polarization shifters, and circulators. These components are crucial for the function of low-frequency EPR spectrometers (see Figure 12.10). This challenge has been mastered by the development of quasi-optical analogues.[42] Key components of such schemes are free-standing wire grids. These devices reflect MWs when the electric component (*E*-field) of a linear polarized beam is oriented parallel to the wires. For perpendicular orientation they are transparent. Based on this simple principle, beam splitters, combiners, and attenuators can be replaced by wire grids properly oriented to the propagation direction and *E*-field of the MW beam. Furthermore, movable rooftop mirrors have proved a suitable replacement for phase shifters, and combinations of wire grids with a Faraday rotation plate can be configured as a circulator or isolator (Figure 27.6). Alternatively, rooftop mirrors rotatable around the propagation axis of the MW beam can be used as variable polarization rotators[48] (Figure 27.9). Nonetheless, it should be noted that the proper function of these components requires the conservation of the MW polarization throughout the entire MW transmission pathway.

Significant advantages of quasi-optical bridges are that they can be configured to operate over a very wide frequency range and that they support detection schemes based upon polarization manipulation (see Figure 27.6 for further details).

27.3.1.4 Sources and Detectors

To extend the superior noise characteristics of MW sources at lower MW frequencies (up to 100 GHz) to ever-higher frequencies, most modern SHF-EPR spectrometers make use of stable low-noise semiconductor sources operating at an intermediate frequency (IF) and further up-conversion to the final MW frequency by multiplication or mixing.[35] These semiconductor-based heterodyne SHF-EPR bridges (see Figure 12.10) provide output powers ranging from up to 2 W at W-band to 100 mW at around 250 GHz, with rapid progress toward higher power levels. Even significantly higher powers for ns-pulse operation without a resonant cavity are accessible through vacuum-tube amplifiers or accelerators. In particular, extended interaction klystrons (EIKs) with kW output powers and a few ns long $\pi/2$ MW pulses have been successfully integrated in W-band SHF-EPR spectrometers.[39,49] This technology could be potentially employed at even higher frequencies. A free-electron laser (FEL)-based SHF-EPR spectrometer recently demonstrated 6 ns $\pi/2$ pulse lengths at 240 GHz.[50] In this setup, fast MW switching is achieved by high-purity Si wafers, in which the MW transmittance is modulated over many orders of magnitude via illumination by 532 nm laser pulses.[50] For the detection of VHF MWs, the highest sensitivity is achieved by bolometers cooled to liquid-He or even superfluid-He temperatures (see Section 27.3.2.3 for further details). However, since the maximum detection bandwidth (BW) of a bolometer is typically limited to a few MHz, they are less suitable for pulse-EPR detection. To realize both CW and pulse operation in the same spectrometer, mixer detectors, with BWs up to the GHz range, provide the optimum combination of instantaneous dynamic range and sensitivity.[35]

27.3.1.5 *Heterodyne Quasi-optical SHF-spectrometer Configuration*

Aiming at different sample classes and resonance conditions, different CW/pulse SHF spectrometers have been realized.[1,35] Despite differences in their design concepts, latest-generation quasi-optical heterodyne CW/pulse SHF-EPR spectrometers operating above 150 GHz share several important commonalities. These common features can be outlined in three main parts, depicted in Figure 27.6 for a prototype spectrometer. The depicted scheme shares basic features with the Bruker ELEXSYS E780 spectrometer[37] and other instruments of this class.[1,35,37,51–53]

The first main part is a basic MW unit (gray box), in which MW generation, pulse forming, and detection are accomplished by active solid-state MW components. The second main component is a quasi-optical front end, through which MWs are further transmitted as a polarized Gaussian beam (light blue box in Figure 27.6). Low-loss transmission of the MWs toward the sample inside the magnet is achieved by a corrugated waveguide. The third main part is the superconducting magnet equipped with a cryostat, in which a resonant or nonresonant sample probe may be immersed.

A crucial feature of quasi-optical spectrometers is that they enable two powerful strategies for the decoupling of the incident excitation and the reflected detection beams. This can be achieved either in conventional reflection mode, or alternatively in induction mode. The latter makes use of the fact that a linear polarized incident beam can be decomposed into two beams with counteracting circular polarization. At resonance, an EPR sample absorbs only one circular component. The attenuation of one of the two counteracting components in the reflected beams leads to a cross-polarized component. This component is perpendicular to the linear polarization of the incident beam.

In an induction-mode EPR setup, the quasi-optical components are arranged such that they allow for a discrimination of both polarization components. Induction-mode detection can drastically reduce the source noise even without a critically coupled cavity. However, an additional reflection channel is required if fundamental-mode cavities are employed for, e.g., pulse-EPR applications. Technically, both options can be realized in the same spectrometer. One possibility to achieve this is depicted in Figure 27.6. In this configuration, an incident beam experiences a 45° polarization shift toward the floor normal in a Faraday rotation plate. This beam is further transmitted through a corrugated waveguide to the sample, situated in a nonresonant probe or a Fabry–Pérot resonator. In Figure 27.6(a), the polarization components of the incident and reflected beams are depicted. Under the depicted conditions, EPR absorption attenuates the MW component with the incident polarization and leads to a cross-polarized component in the reflected beam (Figure 27.6b). Note that this scheme is only applicable in Faraday configuration when $\boldsymbol{B}_0$ is oriented parallel to the propagation direction of the MW beam. After being back-reflected through the corrugated waveguide, a wire grid (G2) separates these two components. The cross-polarized component – the induction-mode signal – is reflected, whereas the reflection signal with perpendicular polarization to the cross-polarized signal and the wires of grid G2 is transmitted. After the grid, the reflection signal experiences a further 45° polarization shift before it is reflected by wire grid G1. Both components are then further transmitted by the quasi-optical bridge and focused on the feed horns of the detector mixers for down-conversion and further processing of the EPR signals.

27.3.2 Broadband MHF-EPR and FD-EPR

Despite their unsurpassed detection sensitivity and spectral resolution, SHF-EPR spectrometers may fail to map out EPR transitions distributed over a very wide transition-energy/field range. This limitation results from the fact that the optimization of SHF-EPR spectrometers to a particular frequency reduces the tunability of the resonance frequency to a few percent. Therefore, SHF-EPR can only provide a one-dimensional scan through the two-dimensional energy/field landscape of an HS sample. In the worst case, the EPR transitions of a given HS compound may be completely missed by such an approach (Figure 27.3). Since the transition energies are not known in advance, an ideal spectrometer would allow free choice of the resonance conditions with tuning capabilities in both field and frequency domains. To increase the variability with respect to the resonance condition, MHF- or FD-EPR spectrometers are employed.

In a true MHF-EPR spectrometer,[54–60] arrays of sources operating at many different frequencies or tunable frequency sources allow for broad field sweeps at many high MW frequencies. As an example, MHF-EPR data for the compound $Tp^{Np,H}Co(NCO)$ is depicted in Figure 27.7. Similar and often complementary information can be obtained from FD-EPR approaches, like frequency-domain magnetic resonance (FDMR)[61–63] or FD-FT THz-EPR (Figure 27.8).[33,64–67]

In FDMR, a series of tunable MW sources are used to excite the sample immersed in a magnet (see Sections 27.3.2.1 and 27.3.2.4 for further details on the employed sources and magnets, respectively). Absolute transmission coefficients are determined as the ratio of frequency-dependent detector signals between the case of the sample being in the beam path and the case of an empty cryostat.[61,63]

Alternatively, FD-FT THz-EPR employs a single broadband excitation source in combination with a Michelson or Martin–Puplett interferometer. In FD-FT THz-EPR, time-dependent changes of the detector signal as a function of the mirror position in the interferometer are recorded. Fast Fourier transformation (FFT) converts this signal into a frequency-domain spectrum. Further details on the FD-FT THz-EPR working principle are given in Section 27.3.2.5. Like in FDMR, FD-FT THz-EPR monitors absolute EPR-induced transmission or reflection changes as the ratio of the sample spectrum to a reference. The latter can be either the empty-cryostat spectrum or a spectrum taken on the same sample, albeit at a different temperature or external magnetic field.[68] By dividing spectra taken at low temperature by spectra taken at elevated temperature, changes in the population of the spin-energy levels may be recorded as EPR-induced transmission changes. This method can be used even without an external magnetic field. However, many nonmagnetic absorption processes also depend on temperature. Sometimes they are difficult to distinguish from spin transitions. Alternatively, reference spectra may be obtained by taking raw spectra at different magnetic fields. Since the application of an external magnetic field shifts the EPR resonance,

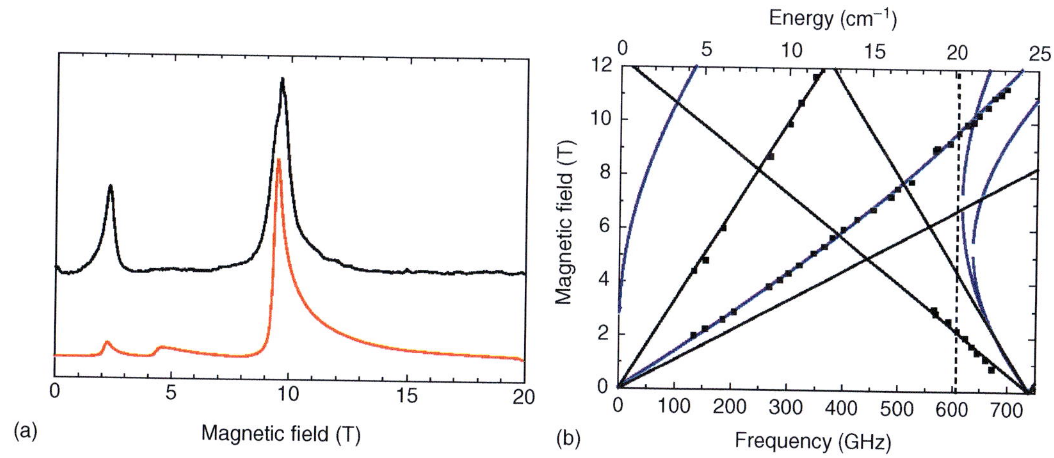

Figure 27.7. (a) Experimental (black trace) and simulated (red trace) 608 GHz powder EPR spectrum of polycrystalline $Tp^{Np,H}Co(NCO)$ ($Tp^{NP,H}$ = hydro(3-Np,5-H-trispyrazolyl)borate anion; Np = neopentyl) at 4.2 K. SH parameters for the simulations ($S = 3/2$, $D/hc = +12.32(5)$ cm^{-1}, $E/hc = 0.07(9)$ cm^{-1}, $g_\perp = 2.54(3)$, $g_\parallel = 2.14(2)$) were obtained from the least-square fit to the multifrequency data set as shown on the right. Chopping the sub-THz radiation causes the absorptive line shapes. (b) Field vs frequency map of MHF-EPR turning points in polycrystalline $Tp^{Np,H}Co(NCO)$ at 4.2 K. The squares are experimental points while the curves are simulated using the best-fitted SH parameters as for the spectra depicted on the left. Turning points with the magnetic field oriented parallel to the x- and z-axes are depicted as blue and black curves, respectively. The dashed vertical line represents the frequency at which the spectrum shown on the left was recorded. (J. Telser, A. Ozarowski, J. Krzystek, in 'Electron Paramagnetic Resonance: Volume 23', ed. B. C. Gilbert, M. J. Davies, D. M. Murphy, The Royal Society of Chemistry, 2013, Vol. 23, 209 – Reproduced by permission of The Royal Society of Chemistry)

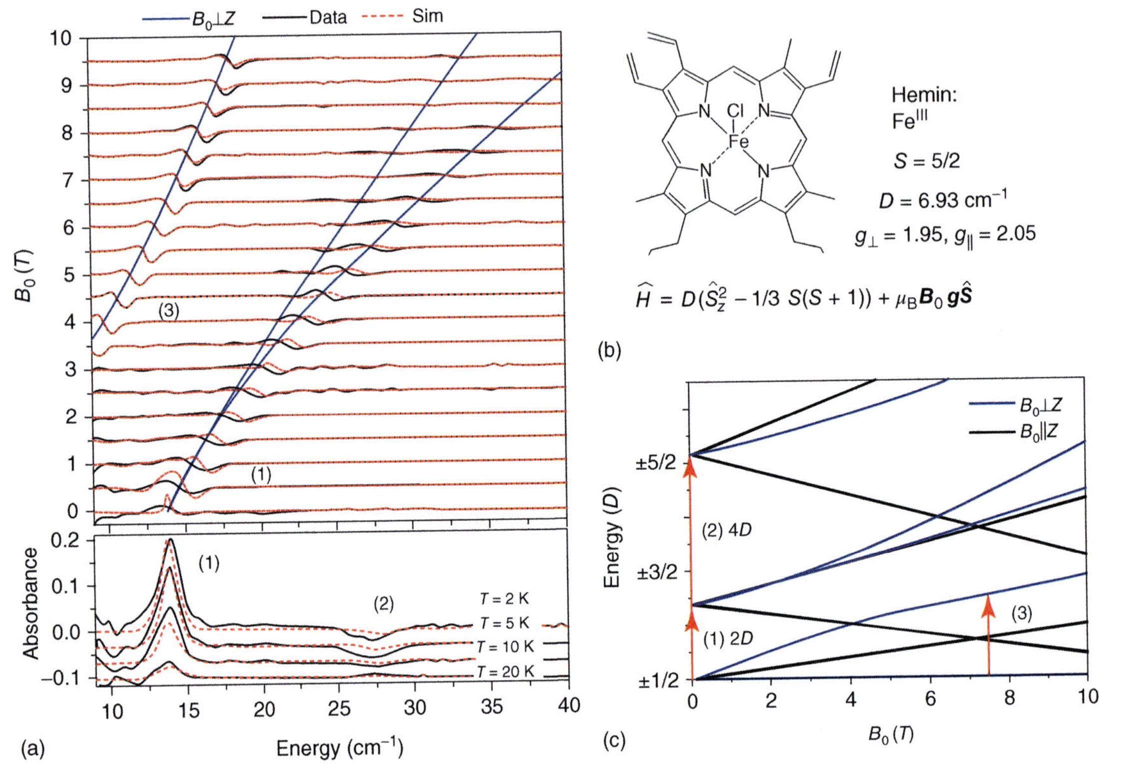

Figure 27.8. (a) Experimental (black solid traces) and simulated (red dotted traces) FD-FT THz-EPR spectra of Fe^{III} ($S = 5/2$) in hemin. Upper left panel: FD-FT THz-EPR spectra, obtained by dividing two raw spectra, both taken at 5 K but at two different external magnetic fields with a field offset of 0.5 T. Spectra are indexed by the lower field value given on the *y*-axis. Blue lines indicate EPR-transition energies with respect to the ground state for $\boldsymbol{B}_0$ alignment perpendicular to the *z*-axis. Lower left panel: FD-FT THz-EPR absorbance spectra where the absorbance is $A = \log_{10}(I_{36\,\mathrm{K}}/I_{\mathrm{low}})$, with $I_{36\,\mathrm{K}}$ and I_{low} the THz intensities measured at 36 K and the indicated low temperatures, respectively. Data is shown with an offset for clarity. FD-EPR simulations have been obtained with EasySpin[22] using the given ZFS and *g*-values. (b) Hemin structure, (c) $S = 5/2$ spin-energy levels for the given axial ZFS vs $\boldsymbol{B}_0$, for parallel (black lines) and perpendicular (blue lines) alignment with respect to *z*. m_S quantum numbers are plotted on the *y*-axis. (1), (2), and (3) indicate EPR transitions visible in the spectra on the left

the division of raw spectra taken at different external magnetic fields results in negative and positive parts in the spectrum. These two strategies are shown in Figure 27.8, which depicts FD-FT THz-EPR spectra of hemin obtained by varying the external magnetic field (Figure 27.8a upper box) and the temperature (Figure 27.8a lower box).

MHF- and FD-EPR experiments target optimum frequency tunability and are thus typically performed with nonresonant sample probes. Further instrumental requirements with respect to magnets, transmission lines, detectors, and sources employed in MHF- and FD-EPR spectrometers are summarized in the following.

27.3.2.1 Magnets

MHF- and FD-EPR have proven particularly powerful in combination with magnets that can be swept over a wide field range in a reasonable time (~0.1–1 T min^{-1}). FD-EPR is possible even without an externally applied magnetic field. However, the possibility to apply an external magnetic field significantly increases the extractable spectroscopic information. Furthermore, in transmission-type FD-EPR experiments, the application of magnetic fields is often required to discriminate between magnetic and nonmagnetic absorption changes. For MHF- and FD-EPR studies of HS systems – which frequently exhibit broad inhomogeneous line widths – the requirements for field homogeneities are less stringent than for high-resolution SHF-EPR. For example, for TMI-HS complexes, excellent results could be obtained with field homogeneities of 0.6% in a sphere of 1 cm^3 volume.[22,61] Appropriate magnets are available for lab applications up to 20 T. Even higher fields are available at dedicated EPR user facilities. An updated list of labs running MHF- and FD-EPR facilities has recently been compiled by Telser *et al.*[41]

27.3.2.2 Transmission Lines

Earlier MHF-EPR spectrometers employed oversized waveguides and sometimes simply copper pipes as broadband MW transmission lines.[69,70] However, this configuration suffers from standing waves. The latter reduce the MW power incident on the sample as well as the control of MW phase and polarization. Phase changes can lead to admixtures of absorption and dispersion in the detected MHF-EPR signals, while loss of the MW polarization may complicate the EPR data interpretation, when both 'allowed' ($\boldsymbol{B}_1 \perp \boldsymbol{B}_0$) and 'forbidden' ($\boldsymbol{B}_1 \parallel \boldsymbol{B}_0$) transitions appear in the same spectrum with comparable intensity.

Standing waves are also a serious challenge in FD-EPR since they induce baseline modulations, reducing the effective spectrometer sensitivity. Sources of standing waves can be reflections in the beam path, at the surfaces separating different compartments of the spectrometer (Figure 27.9) or even on the surface of the sample,[71] which is often pressed in flat pellets for FD-FT and FDMR applications. These modulations can be partly corrected by dividing raw data spectra by reference spectra taken at the same optical configuration but at a different temperature or external magnetic field.[61,68] Nevertheless, these corrections can only partly remove standing waves, since minute changes of the spectrometer-path length – induced by changes in temperature, magnetic field, or other unknown influences – can shift the standing-wave pattern between two scans. To correct for standing waves in FDMR spectra, a simulation program was introduced that accounts for the contribution of dielectric and magnetic dipoles to the detected signals.[71] Nevertheless, there is a growing consensus that the optimum way to improve the situation involves a reduction of standing waves already in the spectrometer. Therefore, quasi-optical bridges – which have been successfully used in SHF-EPR – are integrated in an increasing number of MHF- and FD-EPR spectrometers.[41,68,72]

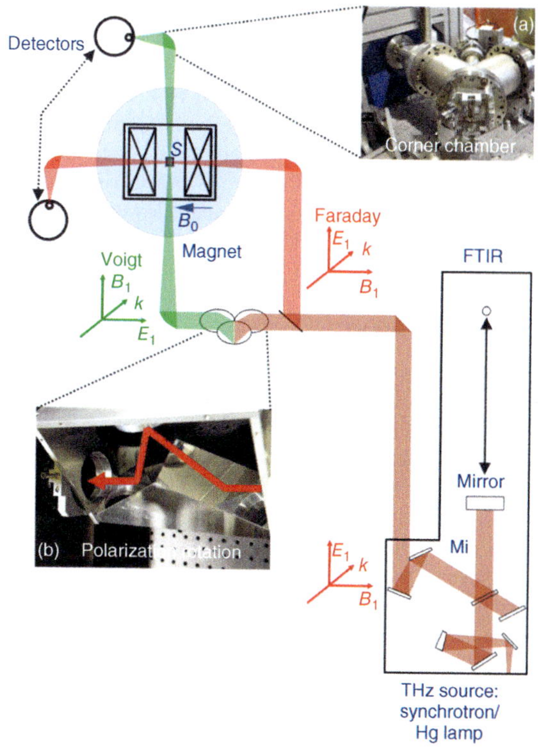

Figure 27.9. FD-FT THz-EPR setup, with FTIR spectrometer, magnet, and detectors as main components. Light from the synchrotron or an Hg arc lamp is passing through a Michelson interferometer (Mi) inside the FTIR spectrometer and transmitted through the sample (*s*) inside the magnet (light blue) toward the detector, either in Faraday (red beam) or Voigt geometry (green beam). Green and red coordinate systems indicate the orientation of the MW E_1- and B_1-field components and the propagation direction (*k*) of the THz radiation, respectively. The upright coordinate (either E_1 or B_1 for the respective beams) is oriented perpendicular to the floor plane. All quasi-optical components are immersed in a fully evacuated beam line. Photos (a) and (b) depict one of the vacuum corner chambers and the polarization rotator with the THz beam pictured in red, respectively. Focusing of the THz beam is achieved by off-axis quasi-optical mirrors inside the corner chambers. These mirrors can be adjusted from outside with the mechanical mechanism shown in photo (a)

27.3.2.3 Detectors

For MHF- and FD-EPR detection, liquid He cooled Si (noise equivalent power (NEP) 1.2×10^{-13} W/Hz$^{1/2}$(4.2 K) and 3.6×10^{-15} W/Hz$^{1/2}$ (1.3 K)) and InSb bolometers (NEP 8×10^{-13} W/Hz$^{1/2}$) are extensively used due to their unrivaled sensitivity and reasonable dynamic range. Superfluid-He cooled bolometers show the highest absolute sensitivity. However, due to their faster response time, InSb bolometers (≈1 MHz as compared to a few hundred Hz for their Si counterparts) are better suited for fast lock-in detection employing amplitude[67] or field modulation. Alternatively, Golay cells are in use, which are also broadband but exhibit much lower detection sensitivities (NEP 0.8×10^{-10} W/Hz$^{1/2}$).

27.3.2.4 Sources

Differences between MHF and FD approaches mainly concern the radiation sources employed and the detection schemes. While FD-EPR requires continuously tunable (in the case of FDMR) or broadband (for FD-FT EPR) sources, MHF-EPR can be carried out with either tunable sources or sources operating at many different discrete frequencies.

A variety of different sources have been tested in MHF-EPR applications, including YIG oscillators,[54] voltage-controlled Gunn oscillators with frequency multipliers,[54,55,73] tunable far-infrared lasers,[55,73] mm-wave network vector analyzers (MVNAs),[54] FELs,[74] and backward wave oscillators (BWOs).[57,73,75–77]

With a wide tunability of ≈30% around their central frequency, availability from 100 GHz up to 1.3 THz, relatively high power levels of ≈80 mW in the low-frequency range and ≈1 mW in the high-frequency range as well as frequency resolution of less than 0.01 GHz, BWOs are very well suited for both FDMR[61,62] and MHF-EPR. BWOs are employed in MHF- and FD-EPR spectroscopy along with semiconductor sources (e.g., Gunn diodes or completed VHF MVNAs), with both often in combination with multiple amplification/multiplication stages.[29] Such combinations allow even for MHF and FD-EPR experiments in the same spectrometer, as has been demonstrated recently in the van Slageren group at Stuttgart University.[72]

By contrast, FD-FT THz-EPR requires broadband sources. The spectral width of the excitation source directly determines the maximum accessible frequency range. Standard FD-FT measurements in the VHF-range utilize Hg arc lamps, which emit in the far infrared. The disadvantage of these sources for FD-FT THz-EPR measurements is that their emitted intensity drastically drops below ≈500 GHz, which limits their application range to EPR experiments at higher frequencies. This restriction can be lifted by employing intense broadband THz radiation emitted by synchrotrons.[65,67] Optimum conditions for FD-FT THz-EPR in the 120 GHz–1.5 THz range are reached in synchrotron-operation modes with drastically reduced electron-bunch lengths, like the specialized low α coherent synchrotron radiation (CSR) mode,[67,78] which can provide up to 10^3 times higher average power as compared to Hg arc lamps.[67]

27.3.2.5 FD-FT THz-EPR Setup

Figure 27.9 depicts a simplified scheme of the main components of the FD-FT THz-EPR facility installed at the synchrotron BESSY II.[79] In the spectrometer, linearly polarized broadband THz CSR extracted from the synchrotron is employed as the radiation source. This radiation covers the excitation-energy range from 4 to 50 cm^{-1} (corresponding to 120 GHz–1.5 THz). For higher energies (1–6 THz), globar or Hg arc lamps provide sufficient power for FD-FT THz-EPR experiments.[80] In spite of its higher power levels, THz CSR provides the additional benefit of linear polarization, while globars or Hg arc lamps emit unpolarized light. THz radiation emitted from one of these sources is coupled into the high-resolution FTIR spectrometer (Bruker IFS 125, max. resolution: 0.0063 cm^{-1} or 200 MHz) as a freely propagating Gaussian beam (depicted in red). In a Michelson interferometer (Mi), the incoming beam is split in two. These beams are reflected by a stationary and a movable mirror, respectively, and interfere with each other. The interfering radiation is directed toward the sample inside the superconducting split-coil magnet by a quasi-optical transmission line. The magnet housing is equipped with a variable temperature insert (VTI) and four THz-transparent z-cut quartz windows. This configuration allows for FD-FT THz-EPR measurements from $T = 1.5$ to 300 K, at external magnetic field variable between −10 and +10 T (±11 T with pumped He reservoir). Before passing through the sample compartment, the radiation can be switched between two different pathways. These directions correspond to Faraday configuration (red beam), in which the external magnetic field is aligned parallel to the propagation direction of the THz radiation $\boldsymbol{k}(\boldsymbol{B}_0 \| \boldsymbol{k})$, and Voigt configuration (green beam), where $\boldsymbol{B}_0 \perp \boldsymbol{k}$. In Voigt geometry, the polarization of the linearly polarized THz light can be rotated around the axis of the $\boldsymbol{k}$-vector by a broadband rooftop mirror polarization shifter (see photo (b) in Figure 27.9).[48] Thereby, EPR detection with $\boldsymbol{B}_0$ oriented parallel or

Table 27.1. Pros and cons of different VHF-EPR spectrometer concepts

Concept	Pros	Cons
SHF-EPR (with resonator)	High absolute and concentration sensitivity High spectral and time resolution pulse and double resonance operation	Fixed to a limited excitation-energy range Small active sample volume
MHF-EPR	Broad excitation-energy range High spectral resolution	Low absolute sensitivity Low excitation power above 1 THz Requires series of MW sources Mainly CW
FDMR	Broad excitation-energy range High spectral resolution possible Zero-field and high-field EPR	Low absolute sensitivity Low excitation power above 1 THz Requires series of MW sources Only CW
FD-FT THz-EPR	Broadest excitation range (MW to IR) Single excitation source Zero-field and high-field EPR	Low absolute sensitivity Low resolution (>200 MHz) Only CW

perpendicular to the magnetic-field component of the beam, $\boldsymbol{B}_1$, or any orientation in between can be realized.

Highly sensitive detection can be achieved with a Si bolometer (IR labs, lHe-cooled to 4.2 K or superfluid He cooled to 1.6 K) or an InSb bolometer (QMC, lHe-cooled to 4.2 K). These detectors monitor the THz radiation, which is modulated in time. FFT of the interferogram yields the raw FD spectrum. The latter is dominated by the emission spectrum of the source as well as absorptions occurring in the THz beam path. To discriminate the EPR resonances from other contributions like standing waves or nonmagnetic excitations in the sample, the raw spectrum is divided by a reference (Figure 27.8).

27.3.3 Comparison of Different Spectrometer Concepts

The major advantages of SHF-EPR compared to MHF- and FD-EPR are superior absolute sensitivity and versatility with respect to pulse and double resonance methods. This particular option of combining the application range of advanced pulse-EPR, ENDOR, and PELDOR methods with the increased resolution at very high frequencies/field dramatically boosted the application range of SHF-EPR. Furthermore, if the sample amount is limited and sensitivity is an issue – like in most applications to proteins – SHF-EPR spectrometers are often the only option to achieve satisfying results. However, for applications to concentrated HS samples with large ZFS, broadband methods may be the better choice to map out the spin transition energy/field landscape. Pros and cons of the different spectrometer types discussed in this chapter are summarized in Table 27.1.

When comparing the sensitivity of different spectrometer types, both the spectrometer specifications and the sample properties have to be considered. Since these parameters may decide whether a projected experiment is feasible, a proper estimation is required in each individual case.

Under the assumption that the spectrometer specifications are frequency independent in their noise characteristics and the available MW power levels (which is not the case in reality), simple dependencies can be derived for the minimum number of detectable spins ($N_{\min}$) and the minimum spin concentration ($N_{\min}/V_S$ with V_S being the sample volume). For non-lossy, non-saturating samples and constant incident MW power, $N_{\min}$ is proportional to $\omega_0^{-9/2}$ and $N_{\min}/V_S$ to $\omega_0^{-3/2}$.[81] However, for a realistic sensitivity estimate, the frequency dependence of the spectrometer components (e.g., available MW power, resonator quality factor (Q), and filling factor (η)) have to be considered. Nevertheless, the given relations confirm a general trend in VHF-EPR. While $N_{\min}$ can be dramatically decreased by going to higher frequencies, the gain in concentration sensitivity is much smaller.

Under realistic conditions the N_{min}/V_S in CW-VHF experiments is often comparable to their low-frequency counterparts. For experiments on small samples that cannot be increased in size (e.g., sub-mm-sized protein crystals), resonator-based SHF-EPR is the combination of choice to achieve optimum sensitivity. However, for powder samples or frozen solutions, nonresonant probes may yield comparable or even higher CW VHF-EPR concentration sensitivities. Good concentration sensitivity in the range of 100 μM in a 10 μl volume and 0.1 T spectral width in combination with better long-term stability render nonresonant probes a very competitive option for CW VHF-EPR experiments. More detailed discussions of the parameters determining sensitivity in CW[42,81] and pulse[82] SHF-EPR experiments have been detailed in the past.

For MHF-EPR and FD-EPR applications, one usually omits a sample resonator to achieve broad MW tunability. A further decline in the detection sensitivity results from the fact that the employed transmission lines cannot be optimized for all frequencies covered by the spectrometer and that the performance of the employed sources may be strongly frequency dependent. As a result, MHF-EPR spectrometers can provide concentration sensitivities down to 0.1 mM, while broadband FD-EPR experiments covering several hundred GHz or more require sample concentrations in the mM range. The higher sample concentration required for broadband techniques results from instrumental limitations, as well as sample-specific properties of HS systems. In particular, the spread of powder MHF-EPR spectra over a very broad field and frequency range in combination with pronounced broadening due to g- and ZFS strain frequently lower the achievable concentration sensitivity of high-field MHF-EPR.[29] In certain cases this situation may be significantly improved by zero-field FD-EPR experiments, where EPR transitions coincide in few characteristic transitions and g-strain does not play a role.

Among the three broadband methods discussed here, MHF-EPR and FDMR potentially exhibit the highest spectral resolution, which is determined by the homogeneity of the magnet, the resolution of the magnet power supply, and the excitation source, respectively. The resolution in FD-FT THz-EPR is determined by the range of the movable mirror in the FTIR spectrometer (typically >200 MHz). In most cases, this resolution is sufficient for studies in the ZFS of HS states, which frequently exhibit broad EPR resonances. However, it is clearly not adequate to resolve small **g**-tensor anisotropies or hyperfine splittings. Major limitations of BWO- or semiconductor-based MHF-EPR and FDMR spectrometers compared to FD-FT THz-EPR approaches include the necessity to use a multitude of sources and the maximum accessible frequency/energy of 1.2 THz (40 cm^{-1}) for a BWO source, or even 900 GHz, which is a more realistic limit based upon power availability. By contrast, the presented FD-FT THz-EPR setup allows accessing spectra from ≈200 GHz to 5 THz with a single scan. The upper limit originates from the choice of the detectors and the window materials. However, for measurements below 200 GHz, both MHF-EPR and FDMR are superior to FD-FT THz-EPR in most cases. A clear advantage of FD-FT THz-EPR and FDMR over MHF-EPR is that they directly yield absolute transmission values. This allows comparing absolute EPR intensities over a broad field/frequency range, which is difficult to obtain with field sweep EPR approaches. A comparison of FDMR, FD-FT THz-EPR, and THz time-domain spectroscopy (THz-TDS) EPR[83] for studies in the magnetic properties of molecular nanomagnets was published by van Slageren.[62]

27.4 FURTHER READING AND OUTLOOK

More detailed information on general and specific aspects of VHF-EPR can be found in the following books and review articles that have appeared during the last 15 years.

A special volume of *Biological Magnetic Resonance* themed *Very High Frequency (VHF) ESR/EPR* contains reviews on the instrumentation and application of VHF-EPR, contributed by the leading groups in the field.[84] The book *Principles of Pulse Electron Paramagnetic Resonance* by Schweiger and Jeschke treats VHF-EPR in a special chapter.[85] *High-Field EPR Spectroscopy on Proteins and their Model Systems* – a monograph by Möbius and Savitsky – provides a detailed description of the history and working principles of SHF-EPR instrumentation, as well as its application to protein research.[1]

Multifrequency Electron Paramagnetic Resonance: Theory and Application[86] – a book edited by Misra in 2011 – contains chapters on the design principles of VHF spectrometers, their application to different material classes, and the interpretation of VHF-EPR data. A special issue of *Magnetic Resonance in Chemistry* themed *High field EPR in Biology, Chemistry and Physics*[87] covers a broad range of applications.

The following reviews provide further insights into general aspects of VHF-EPR instrumentation,[35,58,88–93] quasi-optical EPR bridges,[44,45] pulse SHF-EPR,[82,94,95] SHF-EPR resonators,[45] and FD-EPR.[61,62] Reviews of VHF-EPR applications to single-molecule magnets,[2,62,96,97] TMI complexes,[29,41,98,99] metal proteins,[70,100] *g*-tensors of bioorganic radicals,[4,73] photosynthesis,[3,38,101] structural biology,[1,12,16,102,103] spin labeling,[7,104,105] catalysis,[106] and inorganic materials[89,90] have been published.

These exciting applications were only possible based upon experimental breakthroughs achieved in VHF-EPR method development. Nevertheless, the full application range of VHF-EPR remains far from fully exploited, and VHF methods are still rapidly progressing. Emerging trends in spectrometer development are the employment of arbitrary wave form generators (AWGs) to generate tailor-made pulse shapes (see Chapter 21) as well as the application of tube-[39] and accelerator-based sources[50] for powerful broadband VHF pulses. VHF-EPR experiences further important stimuli from the development of combined EPR/NMR spectrometers for dynamic nuclear polarization (DNP) experiments.[107] These advances are currently pushing the application range and detection sensitivity of pulse SHF-EPR methods, particularly for applications to biomolecules.

Broadband MHF and FD techniques benefit from the availability of higher and higher magnetic fields and the employment of novel excitation sources – both on the lab scale[72] and at large-scale facilities[67,74] – which bring an increasing number of previously classified 'EPR silent' paramagnetic centers within the reach of EPR spectroscopy.

Furthermore, the increased application range of VHF-EPR results from the design of dedicated sample probes for VHF-EPR experiments, e.g., under high pressure,[60] ultrahigh vacuum,[108] externally applied voltage bias, or light. In addition, indirect detection methods like ODMR (see Chapter 25) and EDMR (see Chapter 26) under VHF conditions are becoming more widespread. The latter can further boost the detection sensitivity for certain sample classes by orders of magnitude, thereby extending the application range even to miniaturized, fully processed electronic devices.[109]

ACKNOWLEDGMENTS

Over the last years a number of colleagues, MSc, Diploma, and PhD students, postdocs at Free University Berlin and HZB as well as colleagues from all over the world whose names appear in the respective references contributed to the work presented in this chapter. To all of them I would like to express my deep gratitude. I'm further grateful to Igor Gromov (Bruker BioSpin, Germany), Stefan Weber (Freiburg University, Germany), Marina Bennati (Max Planck Institute for Biophysical Chemistry, Germany), Sun Un (CEA, France), and Joshua Telser (Roosevelt University, USA) for the permission to reproduce figures from their published work. I am furthermore thankful to Joscha Nehrkorn, Karsten Holldack, and Jannik Möser for critically reading this manuscript. This work was funded by the German Federal Ministry of Education and Research (BMBF, network project EPR-Solar), Deutsche Forschungsgemeinschaft (DFG, SPP 1601) and HZB's user program.

REFERENCES

1. K. Möbius and A. Savitsky, High-Field EPR Spectroscopy on Proteins and their Model Systems: Characterization of Transient Paramagnetic States, Royal Society of Chemistry: Cambridge, UK, 2009.
2. D. Gatteschi, A. L. Barra, A. Caneschi, A. Cornia, R. Sessoli, and L. Sorace, *Coord. Chem. Rev.*, 2006, **250**, 1514.
3. K. Möbius, *Chem. Soc. Rev.*, 2000, **29**, 129.
4. S. Stoll, in Electron Paramagnetic Resonance, eds B. C. Gilbert, M. J. Davies and D. M. Murphy, The Royal Society of Chemistry: Cambridge, 2011, Vol. 22, 107.
5. A. Okafuji, A. Schnegg, E. Schleicher, K. Moebius, and S. Weber, *J. Phys. Chem. B*, 2008, **112**, 3568.
6. M. Fehr, A. Schnegg, B. Rech, K. Lips, O. Astakhov, F. Finger, G. Pfanner, C. Freysoldt, J. Neugebauer, and R. Bittl, *Phys. Rev. B*, 2011, **84**, 245203.
7. K. A. Earle and A. I. Smirnov, in Very High Frequency (VHF) ESR/EPR, eds O. Grinberg and L. J. Berliner, Kluwer/Plenum Publishers: New York, 2004, Vol. 22, 96.
8. W. R. Hagen, in Advanced EPR. Applications in Biology and Biochemistry, ed. A. Hoff, Elsevier: Amsterdam, 1989, 785.

9. P. J. Bratt, O. G. Poluektov, M. C. Thurnauer, J. Krzystek, L. C. Brunel, J. Schrier, Y. W. Hsiao, M. Zerner, and A. Angerhofer, *J. Phys. Chem. B*, 2000, **104**, 6973.
10. S. Stoll, H. S. Shafaat, J. Krzystek, A. Ozarowski, M. J. Tauber, J. E. Kim, and R. D. Britt, *J. Am. Chem. Soc.*, 2011, **133**, 18098.
11. P. J. Bratt, E. Ringus, A. Hassan, H. v. Tol, A.-L. Maniero, L.-C. Brunel, M. Rohrer, C. Bubenzer-Hange, H. Scheer, and A. Angerhofer, *J. Phys. Chem. B*, 1999, **103**, 10973.
12. M. Bennati and T. F. Prisner, *Rep. Prog. Phys.*, 2005, **68**, 411.
13. D. Goldfarb, *Phys. Chem. Chem. Phys.*, 2006, **8**, 2325.
14. I. Tkach, K. Halbmair, C. Höbartner, and M. Bennati, *Appl. Magn. Reson.*, 2014, **45**, 969.
15. V. P. Denysenkov, D. Biglino, W. Lubitz, T. F. Prisner, and M. Bennati, *Angew. Chem. Int. Ed.*, 2008, **47**, 1224.
16. K. Möbius, A. Savitsky, A. Schnegg, M. Plato, and M. Fuchs, *Phys. Chem. Chem. Phys.*, 2005, **7**, 19.
17. T. Prisner, M. Rohrer, and F. MacMillan, *Annu. Rev. Phys. Chem.*, 2001, **52**, 279.
18. A. Abragam and B. Bleaney, Electron Paramagnetic Resonance of Transition Ions, Dover Publications: New York, 1986.
19. R. Boča, *Coord. Chem. Rev.*, 2004, **248**, 757.
20. F. Neese, *Coord. Chem. Rev.*, 2009, **253**, 526.
21. F. Neese, in Calculation of NMR and EPR Parameters. Theory and Applications, eds M. Kaupp, M. Bühl and V. G. Malkin, Wiley-VCH: Weinheim, 2004, 541.
22. J. Nehrkorn, J. Telser, K. Holldack, S. Stoll, and A. Schnegg, *J. Phys. Chem. B*, 2015, **119**, 13816.
23. S. Stoll and A. Schweiger, *J. Magn. Reson.*, 2006, **178**, 42.
24. A. Bencini and D. Gatteschi, EPR of Exchange Coupled Systems, Springer Verlag: Berlin, 1990.
25. N. Cox, M. Retegan, F. Neese, D. A. Pantazis, A. Boussac, and W. Lubitz, *Science*, 2014, **345**, 804.
26. A. P. Forshaw, J. M. Smith, A. Ozarowski, J. Krzystek, D. Smirnov, S. A. Zvyagin, T. D. Harris, H. I. Karunadasa, J. M. Zadrozny, A. Schnegg, K. Holldack, T. A. Jackson, A. Alamiri, D. M. Barnes, and J. Telser, *Inorg. Chem.*, 2013, **52**, 144.
27. J. R. Pilbrow, *J. Magn. Reson.*, 1978, **31**, 479.
28. J. R. Pilbrow, Transition Ion Electron Paramagnetic Resonance, Clarendon Press; Oxford University Press: Oxford, 1990.
29. J. Telser, J. Krzystek, and A. Ozarowski, *J. Biol. Inorg. Chem.*, 2014, **19**, 297.
30. D. Goldfarb, *Phys. Chem. Chem. Phys.*, 2014, **16**, 9685.
31. L. C. Tabares, J. Gätjens, C. Hureau, M. R. Burrell, L. Bowater, V. L. Pecoraro, S. Bornemann, and S. Un, *J. Phys. Chem. B*, 2009, **113**, 9016.
32. J. B. Mock, *Rev. Sci. Instr.*, 1960, **31**, 551.
33. P. L. Richards, W. S. Caughey, H. Eberspaecher, G. Feher, and M. Malley, *J. Chem. Phys.*, 1967, **47**, 1187.
34. O. J. Grinberg, A. A. Dubinskii, V. F. Shuvalov, L. G. Oranskii, V. I. Kurochkin, and J. S. Lebedev, *Dokl. Akad. Nauk SSSR*, 1976, **230**, 884.
35. E. Reijerse, *Appl. Magn. Reson.*, 2010, **37**, 795.
36. P. Höfer, A. Kamlowski, G. G. Maresch, D. Schmalbein, and R. T. Weber, in Very High Frequency (VHF) ESR/EPR, eds O. Grinberg and L. Berliner, Springer: New York, 2004, Vol. 22, 401.
37. Bruker Biospin, ELEXSYS E780 product webpage, 2016. https://www.bruker.com/products/mr/epr/elexsys/e780/overview.html
38. S. Un, P. Dorlet, and A. W. Rutherford, *Appl. Magn. Reson.*, 2001, **21**, 341.
39. P. A. S. Cruickshank, D. R. Bolton, D. A. Robertson, R. I. Hunter, R. J. Wylde, and G. M. Smith, *Rev. Sci. Instr.*, 2009, **80**, 103102.
40. J. van Tol, L. C. Brunel, and R. J. Wylde, *Rev. Sci. Instr.*, 2005, **76**, 074101.
41. J. Telser, A. Ozarowski, and J. Krzystek, in Electron Paramagnetic Resonance, eds B. C. Gilbert, M. J. Davies and D. M. Murphy, The Royal Society of Chemistry: Cambridge, 2013, Vol. 23, 209.
42. G. M. Smith, J. C. G. Lesurf, R. H. Mitchell, and P. C. Riedi, *Rev. Sci. Instr.*, 1998, **69**, 3924.
43. W. B. Lynch, K. A. Earle, and J. H. Freed, *Rev. Sci. Instr.*, 1988, **59**, 1345.
44. A. F. Gullá and D. E. Budil, *Concepts Magn. Reson. B*, 2004, **22B**, 15.
45. D. Budil and K. Earle, in Very High Frequency (VHF) ESR/EPR, eds O. Grinberg and L. Berliner, Springer: New York, 2004, Vol. 22, 353.

46. K. A. Earle, D. E. Budil, and J. H. Freed, in Advances in Magnetic and Optical Resonance, ed W. S. Warren, Academic: San Diego, 1996, Vol. 19, 253.
47. P. F. Goldsmith, Quasioptical Systems: Gaussian Beam Quasioptical Propagation and Applications, Wiley-IEEE Press: New York, 1997.
48. C. E. Greninger, *Appl. Opt.*, 1988, **27**, 774.
49. W. Hofbauer, K. A. Earle, C. R. Dunnam, J. K. Moscicki, and J. H. Freed, *Rev. Sci. Instr.*, 2004, **75**, 1194.
50. S. Takahashi, L. C. Brunel, D. T. Edwards, J. van Tol, G. Ramian, S. Han, and M. S. Sherwin, *Nature*, 2012, **489**, 409.
51. H. Blok, J. Disselhorst, S. B. Orlinskii, and J. Schmidt, *J. Magn. Reson.*, 2004, **166**, 92.
52. G. W. Morley, L. C. Brunel, and J. van Tol, *Rev. Sci. Instr.*, 2008, **79**, 064703.
53. F. H. Cho, V. Stepanov, and S. Takahashi, *Rev. Sci. Instr.*, 2014, **85**, 075110.
54. A. K. Hassan, L. A. Pardi, J. Krzystek, A. Sienkiewicz, P. Goy, M. Rohrer, and L. C. Brunel, *J. Magn. Reson.*, 2000, **142**, 300.
55. A. L. Barra, L. C. Brunel, and J. B. Robert, *Chem. Phys. Lett.*, 1990, **165**, 107.
56. M. Mola, S. Hill, P. Goy, and M. Gross, *Rev. Sci. Instr.*, 2000, **71**, 186.
57. Y. Alpert, Y. Couder, J. Tuchendler, and H. Thome, *Biochim. Biophys. Acta*, 1973, **322**, 34.
58. L.-C. Brunel, J. van Tol, A. Angerhofer, S. Hill, J. Krzystek, and A. L. Maniero, in Very High Frequency (VHF) ESR/EPR, eds O. Grinberg and L. Berliner, Springer: New York, 2004, Vol. 22, 465.
59. NHMFL EPR webpage; 31.05.2016. https://nationalmaglab.org/user-facilities/emr
60. H. Ohta, S. Okubo, E. Ohmichi, T. Sakurai, W. M. Zhang, and T. Shimokawa, *J. Low Temp. Phys.*, 2013, **170**, 511.
61. J. van Slageren, S. Vongtragool, B. Gorshunov, A. A. Mukhin, N. Karl, J. Krzystek, J. Telser, A. Müller, C. Sangregorio, D. Gatteschi, and M. Dressel, *Phys. Chem. Chem. Phys.*, 2003, **5**, 3837.
62. J. van Slageren, in EPR Spectroscopy, eds M. Drescher and G. Jeschke, Springer: Berlin Heidelberg, 2012, Vol. 321, 199.
63. J. van Slageren, S. Vongtragool, B. Gorshunov, A. A. Mukhin, and M. Dressel, *J. Magn. Magn. Mater.*, 2004, **272**, E765.
64. R. R. Joyce and P. L. Richards, *Phys. Rev.*, 1969, **179**, 375.
65. L. Mihály, D. Talbayev, L. F. Kiss, J. Zhou, T. Fehér, and A. Jánossy, *Phys. Rev. B*, 2004, **69**, 024414.
66. D. Talbayev, L. Mihály, and J. Zhou, *Phys. Rev. Lett.*, 2004, **93**, 017202.
67. A. Schnegg, J. Behrends, K. Lips, R. Bittl, and K. Holldack, *Phys. Chem. Chem. Phys.*, 2009, **11**, 6820.
68. J. Nehrkorn, B. M. Martins, K. Holldack, S. Stoll, H. Dobbek, R. Bittl, and A. Schnegg, *Mol. Phys.*, 2013, **111**, 2696.
69. F. Muller, M. A. Hopkins, N. Coron, M. Grynberg, L. C. Brunel, and G. Martinez, *Rev. Sci. Instr.*, 1989, **60**, 3681.
70. W. R. Hagen, *Coord. Chem. Rev.*, 1999, **190–192**, 209.
71. N. Kirchner, J. van Slageren, and M. Dressel, *Inorg. Chim. Acta*, 2007, **360**, 3813.
72. P. V. Neugebauer, J. van Slageren, AG van Slageren group webpage, 2015. http://www.ipc.uni-stuttgart.de/AGVanSlageren/research/methods/
73. A.-L. Barra, A. Gräslund, and K. K. Andersson, in Very High Frequency (VHF) ESR/EPR, eds O. Grinberg and L. Berliner, Kluwer/Plenum Publishers: New York, 2004, Vol. 22, 145.
74. S. A. Zvyagin, M. Ozerov, E. Čižmár, D. Kamenskyi, S. Zherlitsyn, T. Herrmannsdörfer, J. Wosnitza, R. Wünsch, and W. Seidel, *Rev. Sci. Instr.*, 2009, **80**, 073102.
75. G. S. Shakurov and V. F. Tarasov, *Appl. Magn. Reson.*, 2001, **21**, 597.
76. J. Krzystek, S. A. Zvyagin, A. Ozarowski, S. Trofimenko, and J. Telser, *J. Magn. Reson.*, 2006, **178**, 174.
77. S. A. Zvyagin, J. Krzystek, P. H. M. van Loosdrecht, G. Dhalenne, and A. Revcolevschi, *Physica B: Condens. Matter*, 2004, **346–347**, 1.
78. M. Abo-Bakr, J. Feikes, K. Holldack, P. Kuske, W. B. Peatman, U. Schade, G. Wüstefeld, and H. W. Hübers, *Phys. Rev. Lett.*, 2003, **90**, 094801.
79. K. Holldack and A. Schnegg, *JLSRF*, 2016, **2**, 51.
80. Y. Rechkemmer, F. D. Breitgoff, M. van der Meer, M. Atanasov, M. Hakl, M. Orlita, P. Neugebauer, F. Neese, B. Sarkar, and J. van Slageren, *Nat. Commun.*, 2016, **7**, 10467.

81. C. P. Poole, Electron Spin Resonance: A Comprehensive Treatise on Experimental Techniques, 2nd edn, John Wiley & Sons, Inc: New York, 1983.

82. G. M. Smith, P. A. S. Cruickshank, D. R. Bolton, and D. A. Robertson, in Electron Paramagnetic Resonance, eds B. C. Gilbert, M. J. Davies and D. M. Murphy, The Royal Society of Chemistry: Cambridge, 2008, Vol. 21, 216.

83. B. Parks, J. Loomis, E. Rumberger, D. N. Hendrickson, and G. Christou, *Phys. Rev. B*, 2001, **64**, 184426.

84. O. Grinberg and L. J. Berliner (eds), Very High Frequency (VHF) ESR/EPR, Kluwer/Plenum Publishers: New York, 2004.

85. A. Schweiger and G. Jeschke, Principles of Pulse Electron Paramagnetic Resonance, Oxford University Press: Oxford, UK, 2001.

86. S. Misra (ed.), Multifrequency Electron Paramagnetic Resonance: Theory and Application, Wiley-VCH Verlag GmbH & Co. KGaA: Weinheim, 2011.

87. W. Lubitz, K. Möbius, and K.-P. Dinse, *Magn. Reson. Chem.*, 2005, **43**, S2.

88. S. K. Misra, H. A. Buckmaster, E. Reijerse, S. Subramanian, and M. C. Krishna, in Multifrequency Electron Paramagnetic Resonance, ed S. K. Misra, Wiley-VCH Verlag GmbH & Co. KGaA: Weinheim, 2011, 115.

89. P. C. Riedi, in Electron Paramagnetic Resonance, eds B. C. Gilbert, M. J. Davies and D. M. Murphy, The Royal Society of Chemistry: Cambridge, 2007, Vol. 20, 245.

90. P. C. Riedi and G. Smith, in Electron Paramagnetic Resonance, eds B. C. Gilbert, M. J. Davies and D. M. Murphy, The Royal Society of Chemistry: Cambridge, 2004, Vol. 19, 338.

91. P. C. Riedi and G. M. Smith, in Electron Paramagnetic Resonance, A Specialist Periodical Report, eds N. M. Atherton, M. J. Davies, B. C. Gilbert and K. A. McLauchlan, The Royal Society of Chemistry: London, 2002, Vol. 18, 254.

92. G. M. Smith and P. C. Riedi, in Electron Paramagnetic Resonance, A Specialist Periodical Report, eds N. M. Atherton, M. J. Davies, B. C. Gilbert and K. A. McLauchlan, The Royal Society of Chemistry: London, 2000, Vol. 17, 164.

93. L. C. Brunel, A. Caneschi, A. Dei, D. Friselli, D. Gatteschi, A. K. Hassan, L. Lenci, M. Martinelli, C. A. Massa, L. A. Pardi, F. Popescu, I. Ricci, and L. Sorace, *Res. Chem. Intermed.*, 2002, **28**, 215.

94. T. F. Prisner, in Very High Frequency (VHF) ESR/EPR, eds O. Grinberg and L. J. Berliner, Kluwer/Plenum Publishers: New York, 2004, Vol. 22, 249.

95. S. S. Eaton and G. R. Eaton, *Concepts Magn. Reson. A*, 2009, **34A**, 315.

96. M. Baker, S. Blundell, N. Domingo, and S. Hill, in Molecular Nanomagnets and Related Phenomena, ed S. Gao, Springer: Berlin Heidelberg, 2015, Vol. 164, 231.

97. D. Gatteschi and R. Sessoli, *J. Magn. Magn. Mater.*, 2004, **272–276**, 1030.

98. J. Krzystek, A. Ozarowski, and J. Telser, *Coord. Chem. Rev.*, 2006, **250**, 2308.

99. J. Krzystek, A. Ozarowski, J. Telser, and D. C. Crans, *Coord. Chem. Rev.*, 2015, **301–302**, 123.

100. S. Lyubenova, T. Maly, K. Zwicker, U. Brandt, B. Ludwig, and T. Prisner, *Acc. Chem. Res.*, 2010, **43**, 181.

101. A. Schnegg, A. A. Dubinskii, M. R. Fuchs, Y. A. Grishin, E. P. Kirilina, W. Lubitz, M. Plato, A. Savitsky, and K. Mobius, *Appl. Magn. Reson.*, 2007, **31**, 59.

102. K. Moebius, W. Lubitz, and A. Savitsky, *Prog. Nucl. Magn. Reson. Spectrosc.*, 2013, **75**, 1.

103. I. D. Sahu, R. M. McCarrick, and G. A. Lorigan, *Biochemistry*, 2013, **52**, 5967.

104. A. I. Smirnov, in Electron Paramagnetic Resonance, eds B. C. Gilbert, M. J. Davies and D. M. Murphy, The Royal Society of Chemistry: Cambridge, 2002, Vol. 18, 109.

105. J. H. Freed, in Very High Frequency (VHF) ESR/EPR, eds O. Grinberg and L. J. Berliner, Kluwer/Plenum Publishers: New York, 2004, Vol. 22, 19.

106. D. M. Murphy and C. C. Rowlands, *Curr. Opin. Solid State Mater. Sci.*, 2001, **5**, 97.

107. R. G. Griffin and T. F. Prisner, *Phys. Chem. Chem. Phys.*, 2010, **12**, 5737.

108. J. Rocker, D. Cornu, E. Kieseritzky, A. Seiler, O. Bondarchuk, W. Hänsel-Ziegler, T. Risse, and H.-J. Freund, *Rev. Sci. Instrum.*, 2014, **85**, 083903.

109. W. Akhtar, A. Schnegg, S. Veber, C. Meier, M. Fehr, and K. Lips, *J. Magn. Reson.*, 2015, **257**, 94.

Index